McGRAW-HILL
ELECTRONICS
DICTIONARY

About the Authors

NEIL SCLATER began his career as a microwave components engineer before turning to writing and editing. He first became an editor for *Electronic Design* magazine and then *Product Engineering* magazine before becoming a regular contributor to *Electronic Engineering Times, Electronic Buyers News,* and many other publications.

As a consultant in marketing communications for more than 25 years, he has served a varied list of clients that includes many electronics manufacturers, publishers, and public relations agencies.

He is the author of the fifth and sixth editions of this dictionary, as well as the author or coauthor of six other books. Of these, *Gallium Arsenide IC Technology, Electrostatic Discharge Protection for Electronics, Wire and Cable for Electronics,* and *The Encyclopedia of Electronics* (Second Edition, coauthor) were published by McGraw-Hill.

The late JOHN MARKUS was a professional writer who had a long association with McGraw-Hill and was the author of many of its best-selling technical books. He was a feature editor of *Electronics* magazine before serving as a technical director for the McGraw-Hill Book Company, responsible for developing new methods for publishing and information retrieval.

He was the author or coauthor of the first four editions of this dictionary, as well as the author, coauthor, or editor of books including *Television and Radio Repairing, How To Make More Money in Your TV Servicing Business, Sourcebook of Electronic Circuits, Electronics Circuits Manual,* and *Guidebook of Electronic Circuits,* all published by McGraw-Hill. Mr. Markus died in 1982.

McGRAW-HILL ELECTRONICS DICTIONARY

SIXTH EDITION

NEIL SCLATER

JOHN MARKUS

McGraw-Hill

New York   San Francisco   Washington, D.C.   Auckland   Bogotá
Caracas   Lisbon   London   Madrid   Mexico City   Milan
Montreal   New Delhi   San Juan   Singapore
Sydney   Tokyo   Toronto

Library of Congress Cataloging-in-Publication Data

Sclater, Neil.
    McGraw-Hill electronics dictionary / Neil Sclater, John Markus. —
6th ed.
        p.   cm.
    Markus' name appears first on the earlier edition.
    Includes index.
    ISBN 0-07-057837-0
    1. Electronics—Dictionaries.   I. Markus, John, 1911–   .
II. Title.
TK7804.M354   1997
621.38′03—dc21   97-16168
                                                                CIP

1 2 3 4 5 6 7 8 9 0   DOC/DOC   9 0 2 1 0 9 8 7

ISBN 0-07-057837-0

*The sponsoring editor for this book was Steve Chapman, the editing
supervisor was M. R. Carey, and the production supervisor was Clare
Stanley. It was set in Times Roman by North Market Street Graphics.*

Printed and bound by R. R. Donnelley & Sons Company.

Contents

Preface

The *Electronics Dictionary* is an easy-to-understand, up-to-date compilation of electronics terminology. Published as a handy single volume, it serves either as a personal desktop reference or as a library resource. Extensively illustrated and essentially nonmathematical, the dictionary defines even the most complex electronics hardware and technical concepts in comprehensible English. Moreover, it does not include the often confusing etymological derivations found in standard dictionaries.

Most of the new words, phrases, abbreviations, and acronyms added to this sixth edition cover the most active sectors of electronics technology that have grown significantly over the past five years. These sectors include personal computers and their peripherals, computer networks, satellite communications, cellular mobile telephones, and consumer electronics, particularly digital television. Many of these new terms are already in the public domain because of their appearance in magazine articles and books devoted to electronics and computers. They have also been topics for discussion at conventions and are found in business and technology articles in daily newspapers and news magazines. Some also show up in print and TV ads for computers and consumer electronic products these days, so pervasive is this technology. Unlike many technical words and phrases introduced in the past, their use is not confined to the writing and speech of engineering specialists.

The first edition of this dictionary was published in 1945, just after the end of World War II. At that time, electronics was just emerging as a discipline distinct from electrical engineering. Before the war, electronics—if the term was used at all—meant radio broadcasting, amateur or professional, and some experimental work in TV. For the average person, however, it might have meant a vacuum-tube radio and, perhaps, a phonograph player.

But the war changed all that. Thousands of civilian men and women had been trained in electronics for work in defense plants, and thousands of members of the armed forces had gone to military technical schools. The object of this training was to learn how to build, operate, and maintain radio, radar, sonar, and other electronic equipment, as well as how to use the instruments for testing them. In the course of their training, all of these people had to learn a strange new language.

Consequently, by war's end there was a pool of trained personnel who had some inkling of the vast growth potential of the field. Those who had mastered the skills needed to build or operate and maintain electronic equipment were anxious to take advantage of that knowledge in ongoing military contract work or consumer industries. Thus, the stage was set for the unprecedented expansion of the newly emerging electronics industry by the early 1950s. As their first postwar commercial product, many companies began to manufacture low-cost TV receivers to meet a growing consumer demand. This had become feasible because so many had acquired technical know-how building military electronic equipment, and a manufacturing infrastructure had been established.

We need only look around us to see how this burgeoning industry has changed the lives of people everywhere in the world over the past half century. Some believe that the developments in electronics have been more significant than the industrial revolution. No other industry in history has ever grown so fast. Nearly everyone living in a developed country is now in close contact with some form of electronics every day. It is obvious that at least a minimal understanding of electronics and a knowledge of its terminology is a requisite for success in all walks of life today. The term *computer literacy* expresses that concept, and it could reasonably be expanded to mean *electronics and computer literacy.*

Today, more American homes have TV sets than telephones, and millions of those homes now have stereos, VCRs, cordless phones, phone answering machines, microwave ovens, electronic clocks, and smoke detectors. Add to that many personal computers, tape recorders, and camcorders in both homes and offices. The latest appliances, from blenders to air conditioners, now include "smart" electronic controls. And, of course, there are the millions of persons of all ages walking around with quartz watches, pocket calculators, cell phones, pagers, portable radio/CD players, and electronic games.

Do not overlook the electronics in automotive engine controls, antilock brakes, and security systems. Some cars now have satellite receivers that can pinpoint their precise location for emergency help, and others have video map displays for finding the best way to get around in large cities. Boating and sports enthusiasts have their depth finders, fishfinders, UHF transceivers, GPS receivers, and night-vision scopes. ATMs dispense cash and permit banking transactions 24 hours a day, and grocery bills are compiled by supermarket holographic code readers. In addition, point-of-sale terminals record purchases and debit credit cards.

Medical diagnosis and the study of physiology have been revolutionized by new computerized X-ray and MRI scanners capable of showing the human body as colorful slices or three-dimensional projections. Vast computer networks now link universities, banks, financial institutions, airlines, travel agencies, and industries. They have made the Internet possible. Moreover, data from weather satellites and radar are computer-processed to create animated weather displays for TV weathercasts. Less conspicuous, however, are many different specialized instruments designed and built to advance the frontiers of science and support research in industry and academia. Most of the recent dramatic discoveries in astronomy, meteorology, oceanography, particle physics, and the earth sciences have been made possible with computerized instrumentation.

The impact of electronics and computers has been felt in industry with the rise of robots and computer-aided design, manufacturing, and management. Even the publishing industry has been revolutionized with computerized typesetting, printing, and desktop publishing. The marriage of computers and TV has resulted in a new concept for simulation called *virtual reality.* The viewer, wearing a head-mounted display, can gain a sense of being present in a computer-generated scene. He or she can then "move" in the scene to view objects from different perspectives or, perhaps, perform some task or training exercise.

World instability has resulted in ongoing development and production of military electronic equipment and electronics-based weaponry, although the pace has slowed since the end of the cold war. New systems are being developed and older systems are being upgraded to meet the ever-changing requirements for warfare in the air, on the land, and on and under the sea in all parts of the world. Included are electronics for missile and weapons fire control, navigation and guidance, communications, control, and surveillance.

Public attention seems to be fixated on personal computers. Estimates are that only about 30 percent of American homes now have them, but that number is rising. New generations of PCs are being introduced in about 18-month cycles, making it difficult and expensive for most people to keep up. Paradoxically, the latest and most powerful PCs turn out to be easier to use than their predecessors, thanks to more memory and improved software.

Faster and more powerful microprocessors with even larger transistor populations are being introduced, and there is intense competition in the development of memory media and drives that offer higher density and faster data access at lower unit prices. Here, optical-storage methods are vying with the magnetic-storage approach.

There are indications that the present PC market could fracture into high-end and low-end segments. The most powerful machines will attract business and professional users who need capability while a more specialized home-entertainment and multimedia computer evolves. It could become a combination computer and TV receiver, now a possibility because of newly adopted compatibility standards. There has also been a lot of discussion lately about network computers without hard drives, which will depend on the Internet for data storage. These systems might attract businesses seeking to avoid the high cost of hundreds of complete computers for general employee use.

Yet another visible PC trend is the downward migration of notebook PCs to even smaller, less powerful hand-held PCs (HPCs) or personal digital assistants (PDAs). These might contain the fax and E-mail functions of larger PCs but act principally as satellite or auxiliary PCs. Some are intended as organizers to store schedules, messages, and phone numbers while others are intended for information gathering in the field. Their data can be up-loaded to larger PCs for further processing, in a so-called "docking maneuver."

The words, phrases, abbreviations, and acronyms in this dictionary tie together the many diverse specialty areas in electronics today. A knowledge of these definitions is crucial to the understanding of what is happening in the industry today. This dictionary will save readers the task of trying to decipher definitions by themselves from the context of articles and Internet displays. Even those readers with current knowledge and recent formal education or training in electronics will often find it necessary to look up an unusual word or phrase, or perhaps a newly coined acronym or abbreviation.

Electronics was spun off from physics and electrical engineering, so its lingua franca is saturated with classical physics and electrical engineering terms. However, words and phrases have entered the electronics lexicon from other sciences and engineering disciplines—chemistry, mathematics, and mechanical engineering, for example. Most recently, terms from the biological sciences have entered because of the perceived analogies between computers and living organisms, particularly brain cells.

Learning the language of a discipline, like learning the language of any country or ethnic group, is an important step toward understanding that culture. Those persons likely to gain the most from this dictionary are probably now, or hope to be, employed in the electronics industry or a closely allied field. They could be engineers or engineering managers, or they might have duties in marketing, sales, purchasing, publications, training, or customer service. But they might also be technicians or hobbyists who want a handy reference to help them keep up with changes in the language. Others for whom this dictionary will doubtless have value are teachers, students, secretaries, librarians, journalists, technical writers, economists, stock analysts, ad copy writers, or just about anyone with a need to know or write about electronics.

This sixth edition contains more than 12,000 entries and more than 800 illustrations. More than 1000 new or revised entries and over 285 new illustrations have been added. The newest additions reflect the latest technology or document evolutionary changes in the meanings of words introduced years ago. While many entries and illustrations have been deleted from this edition to make room for the flood of new entries, a substantial number of the classical terms and illustrations have been retained because of their importance in the history of electronics and because of frequent references to them in books and articles.

This sixth edition follows a practice established in the first edition of keeping all entries, words, and phrases, as well as acronyms and abbreviations, in alphabetical order to help the reader find the desired term. In cases where a device, circuit, product, or system is known by more than one name, the commonly accepted alternatives or synonyms are included with appropriate cross-referencing. When a word or phrase has multiple meanings, commonly accepted alternative definitions are given.

The first *Electronics Dictionary* was prepared as a collaborative effort between the late Nelson M. Cook and the late John Markus, both well-known authors of electronics books and publications. Their first edition defined only 6400 words. However, in subsequent editions, the title was changed to *Electronics and Nucleonics Dictionary* to reflect what was anticipated in the 1950s to be parallel growth in two separate but equal fields.

However, after the passing of Mr. Cook, the name of the fourth edition was changed back to *Electronics Dictionary*.

Mr. Markus realized that progress in nuclear technology had stalled, and the explosive growth in electronics terminology demanded more pages. Consequently, the only nuclear technology terms carried over to this edition relate to radiation measurements made with electronic instrumentation.

The most significant milestones since the publication of the first edition are arguably the invention of the transistor, the integrated circuit, and the microprocessor (MPU). The invention of the transistor and, subsequently, the IC drove advancements in both analog and digital circuitry by eliminating the power, heat, size, and weight constraints imposed by vacuum-tube circuitry. During this period the digital computer evolved from a room full of tube circuits to transistorized circuits in desktop cases.

The original MPU had a clock speed of 100 kHz. The latest versions have speeds of 200 MHz or more, and they contain millions of transistors. In addition, transistor density in semiconductor memory devices has doubled every 18 months, and this trend is expected to continue. It has been predicted that by 2000 PCs will have billion-bit memories and their processors will run at gigahertz speeds. Moreover, it is likely that flat-panel displays will replace the time-honored CRT monitors.

Microcontrollers (MCUs), those low cost derivatives of the MPU, made possible whole new generations of "intelligent" products. These self-contained but limited "computers-on-a-chip" do not need additional peripherals. Factory-programmed, they are, for the most part, invisible to the user. They can be found in cameras, camcorders, VCRs, cell phones, test instruments, appliances, and automotive controls. Keypads on some of these products permit storing data or changing product functions.

The extraordinary changes in electronics since 1945 have resulted in a revolution in the way electronic components and products are made and packaged. This, in turn, has resulted in an upheaval of the industry as it existed 40 years ago, changes more drastic than have occurred in any other industry over the same period of time. Not only have the methods changed, but the locations of the principal factories have changed. Labor-intensive tasks have been sent offshore where wage scales are lower, and automation has eliminated much of the work that was once done in the manufacturer's home country.

Intensive global competition has driven many once-prominent manufacturers out of the industry, and their places have been taken by newer, more aggressive firms. The rapid pace of development has accelerated the life cycle of most electronic products and made many obsolete before they paid for themselves or were effectively mastered by their users. This has been most prevalent in computers, telecommunications equipment, and consumer products.

Global competition, plus the high cost of new product development and manufacturing facilities, has led to increased interdependence between multinational corporations. Close examination of the latest electronic products today will reveal the many different countries of origin of their parts. Moreover, the product probably was not even assembled in the home country of the manufacturer whose name is on the label.

Preface

Like some giant living, growing organism, the electronics industry will continue to divide and multiply. New innovative companies will spring up or be spun off from older enterprises, and unsuccessful competitors will continue to fall by the wayside. Only time will tell if the newly emerging products are truly cost-effective and productive, or simply passing fads that were overpromoted and oversold. Nevertheless, words, phrases, and abbreviations will be coined for those new developments, and future editions of this dictionary will define them.

Neil Sclater

A

a Abbreviation for *atto-*.

A 1. Abbreviation for *ampere*. 2. Symbol for *anode*. 3. Symbol for *argon*.

Å Abbreviation for *angstrom*.

A.A. Abbreviation for *angular aperture*.

ABC 1. Abbreviation for *automatic bass compensation*. 2. Abbreviation for *automatic brightness control*.

aberration A deviation from the expected state of an optical field.

ABM Abbreviation for *antiballistic missile*.

abnormal glow discharge A glow discharge characterized by an increase in the voltage drop as the current increases. It occurs when the current is increased beyond the point at which the cathode of the gas tube is completely covered with glow.

abnormal propagation Radio-wave propagation in which unstable atmospheric and/or ionospheric conditions interfere with communication.

abnormal reflection A sharply defined reflection of radio waves from an ionized layer of the ionosphere, occurring at frequencies higher than the critical or penetration frequency of the layer. It is also called *sporadic reflection*.

abrasion resistance The ability of a material to withstand mechanical wear, such as that produced by movement of a contact, brush, or wiper arm.

abrupt junction A junction in which the transition from P- to N-type material is effectively discontinuous in a single-crystal semiconductor.

ABS 1. Abbreviation for *antilock brake system*. 2. Abbreviation for *acrylonitrile-butadiene-styrene*.

abscissa The horizontal distance from a point on a graph to the zero reference line. The units of this distance are indicated on a scale at the bottom or top of the graph.

A/B signaling A special case of 8-bit (LSB) signaling in a μ-law system that allows four logic states to be multiplexed with voice on PCM channels.

absolute maximum ratings Electrical limits specified for a circuit, product, or system that, if exceeded, could cause damage or destruction. They are not continuous ratings, and they are independent of proper operation.

absolute error The magnitude of an error, disregarding its algebraic sign or direction.

absolute pressure Pressure with respect to a vacuum.

absolute shaft-angle optical encoder An electromechanical encoder whose digital word output uniquely defines each shaft-angle. It contains a light-emitting diode (LED) and an array of photodetectors separated by an *absolute encoder* disk mounted on the shaft. When the shaft rotates, a unique digital word is generated as the multiple radial transparent and opaque rings on the disk "chop" the optical paths to the photosensor array. The output code word is read in a radial line. It will retain the last angular position of the encoder shaft if it stops because of power interruption. See *absolute shaft-angle optical encoder disk*.

absolute shaft-angle optical encoder disk A thin glass or plastic disk divided into concentric radial rings with varying percentages of opaque and transparent area. When rotating, the disk "chops" the optical paths from the encoder's light-emitting diode (LED) to the photodetectors, which generate a digital word uniquely defining shaft position. There are disks that produce natural binary or Gray code. Shaft position accuracy is proportional to the number of rings or channels on the disk. See also *absolute shaft-angle optical encoder, binary code,* and *Gray code*.

absolute temperature scale A temperature scale in which zero is the absolute zero of temperature, −273.16°C or −459.69°F. The most commonly used scale is the Kelvin scale, based on Celsius (centigrade) degrees; absolute zero is 0 K, water freezes at 273.16 K and boils at 373.16 K. The Rankine scale is based on Fahrenheit degrees; water freezes at 491.69°R and boils at 671.69°R.

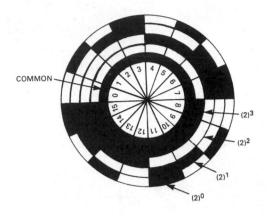

Absolute optical shaft-angle encoder disk can generate a digital word uniquely defining shaft position.

absolute unit   A unit defined in terms of fundamental units of mass, length, time, and charge, such as the centimeter-gram-second electromagnetic and electrostatic units and the meter-kilogram-second-ampere electromagnetic units.

absolute value   The numerical value of a number without regard to sign.

absolute zero   The lowest temperature that can exist, corresponding to a complete absence of molecular motion. Absolute zero is approximately –273.16°C or –459.69°F.

absorbed dose   *Dose.*

absorbed dose rate   The dose per unit of time, measured in rads per unit time.

absorber   A material or device that takes up and dissipates radiated energy. It can be used to shield an object from that energy, prevent reflection of the energy, determine the nature of the radiation, or selectively transmit one or more components of the radiation. Examples are acoustic absorbers and microwave absorbers.

absorptance   The ratio of the radiant energy absorbed in a body of material to the incident radiant energy.

absorptiometer   An instrument that determines the concentration of substances by their absorption of nearly monochromatic radiation at a wavelength selected by filters or by a simple radiation-dispersing system.

absorption   The dissipation of energy by radiation passing through a medium. Electromagnetic energy is lost when radio waves travel through the atmosphere and acoustic energy is lost when sound waves pass through an object. The kinetic energy of a nuclear particle is reduced when it passes through a body of matter.

absorption band   A region of the absorption spectrum of a material in which the amount of absorption passes through a maximum.

absorption circuit   A series resonant circuit that absorbs power at an unwanted signal frequency. The circuit provides a low impedance to ground at this frequency.

absorption coefficient   The fraction of the intensity of a radiation that is absorbed by a unit thickness of a particular substance.

absorption current   The component of dielectric current that is proportional to the rate of accumulation of electric charges within the dielectric.

absorption discontinuity   A discontinuity in the absorption coefficient of a substance for a particular type of radiation.

absorption edge   The wavelength that corresponds to an absorption discontinuity.

absorption fading   Gradual changes in the strength of a received radio signal, caused primarily by slow changes in absorption by the atmosphere along the signal path.

absorption frequency meter   *Absorption wavemeter.*

absorption line   A dark line corresponding to a peak in the absorption spectrum of a gas or a vapor.

absorption loss   1. That part of the transmission loss that is converted into heat when radiated energy is transmitted or reflected by a material.   2. Power loss in a transmission circuit caused by coupling to an adjacent circuit.

absorption modulation   A system of amplitude modulation in which a variable-impedance device is inserted in or coupled to the output circuit of the transmitter to absorb carrier power in accordance with the information to be transmitted. In one system the modulator tubes or transistors control the absorption of the transmission line directly by means of stub connections, to achieve the same result. It is also called loss modulation.

absorption peak   Abnormally high attenuation at a particular frequency as a result of absorption loss.

absorption spectroscopy   *Spectroscopy* that includes measurement of the energies and wavelengths of radiation absorbed by atoms and molecules of matter under various conditions.

absorption spectrum   The spectrum obtained when continuous radiation is passed through an absorbing medium before it enters a spectroscope. The resulting recorded spectrum shows dark lines at wavelengths corresponding to maximum absorption.

absorption trap   A parallel-tuned circuit that absorbs and attenuates interfering signals.

absorption wavemeter   A wavemeter that consists of a calibrated tuned circuit and a resonance indicator. When the wavemeter is lightly coupled to a signal source and tuned to resonance, maximum energy is absorbed from the source. The unknown wavelength or frequency can be read on the calibrated tuning dial. With waveguides, a cavity-type resonant circuit is used. Also called an absorption frequency meter.

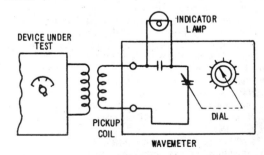

Absorption wavemeter. A lamp indicates a resonance condition.

absorptive attenuator   A waveguide section that contains dissipative material which gives a desired transmission loss.

absorptivity   A measure of that part of incident radiation or sound energy absorbed by a material.

AB test A method of comparing two sound systems by switching inputs so the same recording is heard in rapid succession over one system and then the other.

AC Abbreviation for *alternating current.*

AC adapter A small power supply that plugs into an AC power outlet and delivers the low DC voltage to power the electronics in a portable calculator, tape recorder, or other portable battery-operated device.

AC bias An AC signal that is applied to a magnetic tape recording head, along with the signal being recorded, to improve frequency response and minimize distortion and noise. The bias frequency must be several times the highest frequency value being recorded.

accelerated life test Operation of a device, circuit, or system above maximum ratings to produce premature failure. It is used to estimate normal operating life.

accelerating electrode An electrode in cathode-ray and other electron tubes that increases the velocity of the electrons which constitute the space current or form a beam. This electrode is operated at a high positive potential with respect to the cathode.

acceleration The time rate of change of the *velocity* of a body. It represents motion in which the velocity changes from point to point. When the velocity of a mass moving in a straight line changes by equal amounts in equal intervals of time, the acceleration is said to be *constant,* and the motion is said to be accelerated *uniformly.*

acceleration space The region in an electron tube just outside the output aperture of the electron gun in which electrons are accelerated to a desired higher velocity.

accelerometer A device that measures the acceleration of a moving body and translates it into a corresponding electrical quantity.

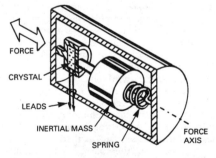

Accelerometer, based on the piezoelectric effect, measures acceleration as the inertial mass opposes the axial force and compresses the crystal to generate an electric signal.

accentuation *Preemphasis.*

accentuator A circuit that provides preemphasis of certain audio frequencies.

acceptable quality level [AQL] The percentage of defects that will be accepted over a predetermined time period by a sampling plan during the inspection or test of a product or system.

acceptable reliability level [ARL] The percentage of failures allowed per thousand operating hours for acceptance of production parts or equipment. It is a measure of the reliability that will be accepted over a predetermined period of time by a reliability sampling plan.

acceptance angle The solid angle within which all received light reaches the light-sensitive area of a phototube, photodiode, optical fiber or other light-sensitive device. It is also called the *acceptance cone.*

acceptance cone *Acceptance angle.*

acceptance sampling plan A plan that specifies the sample sizes for incoming inspection and the test criteria for acceptance, rejection, or taking of another sample.

acceptance test A test that determines conformance of a product to design specifications, as a basis for customer acceptance.

acceptor An impurity element that increases the number of holes in a semiconductor crystal such as germanium and silicon. Current flow is then essentially limited to the transfer of holes. Because these holes are equivalent to positive charges, the resulting alloy is called a P-type semiconductor. Aluminum, gallium, and indium are examples of acceptors. See *donor.*

acceptor circuit A series resonant circuit that has a low impedance at the frequency to which it is tuned and a higher impedance at all other frequencies. Used in series with a signal path to pass the desired frequency.

acceptor level An intermediate level close to the normal band in the energy-level diagram of an extrinsic semiconductor. It is empty at absolute zero. At other temperatures some electrons corresponding to the normal band can acquire energies corresponding to this intermediate level.

access control An interactive technology that enables a cable operator, satellite programmer, or other program provider to authorize a customer to receive one or more television programs or other services.

access time 1. The time interval between the instant that information in a computer is called for from memory and the instant when it is delivered (*read time*). 2. The time interval between the instant information in a computer is made ready for storage and the instant when storage is complete (*write time*). In a disk drive it includes the amount of time required to position the read/write head at the correct location on the disk and carry out a read or write operation.

AC coupling A coupling arrangement that will not pass direct current or a DC component of a signal.

accumulator 1. A computer register that stores a number and, on receipt of another number, adds it to the number already stored and stores the sum. In another version, stored integers can be increased by unity or an arbitrary integer. An accumulator can be reset to either zero or an arbitrary integer. Also called counter. 2. British term for *storage battery.*

accuracy 1. The quality of being free from errors. 2. The extent to which the indications of an instrument approach the true values of the quantities measured.

AC/DC An abbreviation indicating that a receiver, instrument, or appliance will operate from either an AC or DC power line.

AC/DC receiver A radio receiver that operates from either an AC or DC power line. It is also called a universal receiver.

AC erasing head A magnetic head that uses alternating current to produce the gradually decreasing magnetic field necessary for erasing recorded signals.

acetal A plastic resin offering a high modulus of elasticity, low coefficient of friction, excellent resistance to abrasion and impact, low moisture absorption, and ease of machining. It is used to make insulating bodies for electronic components, washers, and seals.

acetate An odorless, tasteless, nontoxic plastic that offers high grease resistance and impact strength. It is used to make insulating parts of electronic components. It is also called *cellulose* and *cellulose acetate*.

acetate base A transparent backing film for magnetic recording tape and motion-picture film, made from cellulose acetate. Also called *safety base*.

acetate tape A magnetic recording tape that has an acetate base.

AC fan-out The fan-out limit of a logic circuit under high-speed conditions. Parasitic capacitances can reduce the permissible number of fan-outs to almost half that for DC conditions.

AC generator A rotating electric machine that converts mechanical power into AC electric power.

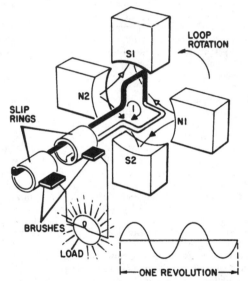

AC generator with four poles, simplified to show the operating principle.

achromatic 1. Without color. 2. Capable of transmitting light without breaking it up into constituent colors.

achromatic antenna An antenna whose characteristics are uniform in a specified frequency band.

achromatic color A shade of gray.

achromatic lens A lens combination that corrects chromatic aberration. It is usually a convex lens of crown glass and a concave lens of flint glass: one lens corrects for the errors of the other. The combination brings all colors of light rays nearer to the same focus point.

achromatic locus An area on a chromaticity diagram that contains all points that represent acceptable reference white standards. It is also called the *achromatic region*.

achromatic point A point on a chromaticity diagram that represents an acceptable reference white standard.

achromatic region *Achromatic locus.*

achromatic stimulus A visual stimulus that gives the sensation of white light and thus has no hue.

ACIA Abbreviation for *asynchronous communications interface adapter.*

ACL Abbreviation for *advanced CMOS logic.*

ACM Abbreviation for *Association for Computing Machinery.*

acoustic Containing, producing, arising from, actuated by, related to, or associated with sound. The adjective acoustic is used (rather than acoustical) when the term being qualified designates something that has the properties, dimensions, or physical characteristics associated with sound waves.

acoustic absorption coefficient *Sound absorption coefficient.*

acoustic absorption loss Energy lost by conversion into heat or other forms when sound passes through or is reflected by a medium.

acoustic absorptivity *Sound absorption coefficient.*

acoustical Containing, producing, arising from, actuated by, related to, or associated with sound. The adjective acoustical (rather than acoustic) is used when the term being qualified does not explicitly designate something that has the properties, dimensions, or physical characteristics associated with sound waves.

acoustical attenuation constant The real part of the acoustical propagation constant. The commonly used unit is the neper per section or per unit distance.

acoustical ohm A unit of acoustic resistance, acoustic reactance, or acoustic impedance. The magnitude is 1 acoustical ohm when a sound pressure of 1 dyn/cm² (1 μbar) produces a volume velocity of 1 cm³/s. It is also called an *acoustic ohm*.

acoustical phase constant The imaginary part of the acoustical propagation constant. The commonly used unit is the radian per section or per unit distance.

acoustical propagation constant A rating for a sound medium. It is the natural logarithm of the complex ratio of particle velocities, volume velocities, or pressures at two points in the path of a sound wave. The ratio is determined by dividing the value at the point nearer the sound source by the value at the more remote point. The real part of this constant is the acoustical attenuation constant, and the imaginary part is the acoustical phase constant.

acoustical reciprocity theorem A theorem that applies to an acoustic system. The theorem states that a simple sound source at point *A* in a region will produce the same sound pressure at another point *B* as would have been produced at *A* had the source been located at *B*.

acoustic amplifier An amplifier that increases the strength of a bulk or surface acoustic wave by an interaction involving energy transfer from traveling electric fields generated by acoustic waves in or on a piezoelectric semiconductor. The resulting charge carriers lose velocity during bunching if the drift field is optimized for maximum amplification, and the excess kinetic energy is transferred to the acoustic wave. It is also called an *acoustoelectric amplifier*.

acoustic compensator A device that matches acoustical path lengths in binaural or stereophonic audio equipment.

acoustic compliance The reciprocal of acoustic stiffness.

acoustic coupler A device used between the modem of a computer terminal and a standard telephone line to permit

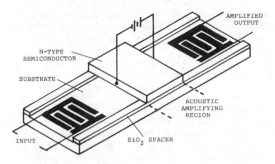

Acoustic amplifier has interdigital transducers at its input and output terminals.

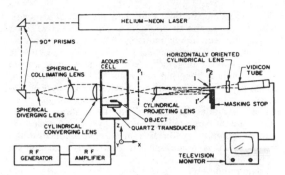

Acoustic imaging for object immersed in acoustic cell.

transmission of digital data in either direction without making direct connections. When the handset is placed in the coupler, a loudspeaker converts modem output pulses to sounds for the handset microphone. Similarly, a microphone in the coupler converts computer return tone data to audio signals for amplification to the correct level for the modem.

acoustic delay line A device capable of transmitting and delaying sound pulses by recirculating them in a liquid or solid medium.

acoustic dispersion The separation of a complex sound wave into its frequency components. It is usually caused by variation of the wave velocity of the medium with frequency. The rate of change of the velocity with frequency is a measure of the dispersion.

acoustic feedback The feedback of sound waves from a loudspeaker to a preceding part of an audio system, such as to the microphone, to aid or reinforce the input. When feedback is excessive, a howling sound is heard from the loudspeaker. It is also called howling.

acoustic filter A sound-absorbing device that selectively suppresses certain audio frequencies.

acoustic generator A transducer that converts electric, mechanical, or other forms of energy into sound. Buzzers, headphones, and loudspeakers are examples.

acoustic homing The tracking and interception of a target or object, generally underwater, by the use of sound energy. An acoustic transmitter and receiver can be mounted on the target-seeking platform, or the tracking of the target by the target-seeking platform can be done solely by listening. Target bearing can be determined with a trainable (passive) hydrophone, but an active sound transmitter is needed to determine the range from the transmitter to the target.

acoustic imaging The production of real-time images of the internal structure of a metallic or nonmetallic object that is opaque to light. In a Bragg-diffraction version, the object is immersed in water and irradiated by plane waves of ultrasound. The resulting scattered waves generate a Bragg-diffracted laser beam that produces an optical image. It is also called ultrasonic imaging.

acoustic impedance The sound pressure on a unit area of surface divided by the sound flux through that surface, expressed in acoustical ohms. The real component of acoustic impedance is acoustic resistance, and the imaginary component is acoustic reactance. The two types of acoustic reactance are acoustic compliance and acoustic mass.

acoustic inertance *Acoustic mass.*

acoustic interferometer An instrument that measures the velocity of sound waves in a liquid or gas. Variations of sound pressure are observed in the medium between a sound source and a reflector as the reflector is moved or the frequency is varied. Interference between direct and reflected waves produces standing waves that are related to the velocity of sound in the medium.

acoustic jamming Generation of sound waves that interfere with enemy ground or underwater listening or acoustic homing devices.

acoustic labyrinth A loudspeaker baffle that consists of a long absorbent-walled duct folded into the volume of a cabinet, with a loudspeaker mounted at one end. The other end is open to the air in front of or underneath the cabinet. Used to reinforce bass response and prevent cavity resonance.

acoustic lens An array of obstacles that refracts sound waves in the same way that an optical lens refracts light waves. The dimensions of the obstacles are small compared to the wavelengths of the sounds being focused.

acoustic mass The quantity that when multiplied by 2π times the frequency gives the acoustic reactance associated with the kinetic energy of a medium. The unit is the gram per centimeter to the fourth power. Also called acoustic inertance.

acoustic memory A computer memory that is an acoustic delay line in which a train of pulses travels through a medium such as mercury or quartz.

acoustic mirage The distortion of a sound wavefront by a large temperature gradient in air or water, creating the illusion of two sound sources.

acoustic mode A type of thermal vibration of a crystal lattice in which neighboring points in the lattice move almost in unison.

acoustic ocean-current meter An instrument that measures current flow in rivers and oceans by transmitting acoustic pulses in opposite directions parallel to the flow and then measures the difference in pulse travel times between transmitter-receiver pairs.

acoustic ohm *Acoustical ohm.*

acoustic pickup A pickup that transforms phonograph-record groove modulations directly into sound. The phonograph needle is mechanically linked to a flexible diaphragm.

acoustic position reference system An acoustic system used in offshore oil drilling, oceanographic exploration, or salvage to help the base ship maintain itself at a specific location above the ocean floor. A shipboard active sonar transducer sequentially interrogates three or more acoustic transponders positioned at known locations around the periphery of the target area. With the aid of these known slant distances the ship can make slight maneuvers to maintain its position with respect to the desired ocean-floor coordinates.

acoustic radiation pressure The steady-state unidirectional pressure exerted on a surface by a sound wave.

acoustic radiator A vibrating surface that produces sound waves, such as a loudspeaker cone and a headphone diaphragm.

acoustic radiometer An instrument that measures sound intensity by determining the unidirectional steady-state pressure caused by the reflection or absorption of a sound wave at a boundary.

acoustic reactance The imaginary component of acoustic impedance. The unit is the acoustical ohm.

acoustic reflection coefficient *Sound reflection coefficient.*

acoustic reflectivity *Sound reflection coefficient.*

acoustic refraction A bending of sound waves when they pass obliquely from one medium to another in which the velocity of sound differs from warm water to cool water in the ocean or from warm air to cool air.

acoustic regeneration *Acoustic feedback.*

acoustic resistance The real component of acoustic impedance. The unit is the *acoustical ohm.*

acoustic resonator A resonator in the form of an enclosure that exhibits resonance at a particular frequency of acoustic energy.

acoustics 1. The science concerned with the production, transmission, and effects of sound, including its absorption, reflection, refraction, diffraction, and interference. 2. The properties of a room or location that affect reflections of sound waves and therefore determine the character of sounds heard in that location.

acoustic scattering The irregular and diffuse reflection, refraction, or diffraction of sound in many directions.

acoustic sounding 1. The use of sound waves to determine water depth by measuring the time required for a sound pulse to travel from the surface to the bottom and back. 2. The use of sound waves to study the lower atmosphere, as with acoustic water sounding. See *depth sounder.*

acoustic spectrograph A spectrograph used with sound waves of various frequencies to study the transmission and reflection properties of thermal layers and marine life in the ocean.

acoustic stiffness The quantity that when divided by 2π times the frequency gives the acoustic reactance associated with the potential energy of a sound medium. The unit is the *dyne* per centimeter to the fifth power. The reciprocal of acoustic stiffness is *acoustic compliance.*

acoustic storage *Acoustic delay line.*

acoustic surface wave *Surface acoustic wave.*

acoustic surveillance The use of sound pickup, amplifying, recording, and/or transmitting instruments to obtain intelligence from enemy sound sources.

acoustic transmission coefficient *Sound transmission coefficient.*

acoustic transmission system An assembly of elements adapted for the transmission of sound.

acoustic transmittivity *Sound transmission coefficient.*

acoustic treatment The use of sound-absorbing materials to eliminate or reduce echoes and reverberations from a closed room.

acoustic velocity *Velocity of sound.*

acoustic wave An elastic nonelectromagnetic wave that can have a frequency in the gigahertz range. One type is a surface acoustic wave, which travels on a surface that is an interface between two media (such as between a piezoelectric crystal and air). The other type is a bulk or volume acoustic wave, which travels through the material (as in a quartz delay line). It is also called an *elastic wave.* See also *transverse wave.*

acoustic wave amplifier An amplifier that contains a semiconductor device whose charge carriers are coupled to an acoustic wave which is propagated in a piezoelectric material to produce amplification.

acoustic-wave filter A filter that separates sound waves of different frequencies.

acoustic well logging Use of sound waves to determine depth and other properties of a bore-hole or liquid level in a well.

acoustoelectric amplifier *Acoustic amplifier.*

acoustoelectric effect The development of a DC voltage in a semiconductor or metal by an acoustic wave traveling parallel to the surface of the material. It is also called an *electroacoustic effect.*

acoustoelectronics *Pretersonics.*

acoustooptical cell An electric-to-optical transducer in which an acoustic or ultrasonic electric input signal can modulate or otherwise act on a beam of light.

acoustooptical filter An optical filter that is tuned across the visible spectrum by acoustic waves in the frequency range of 40 to 68 MHz.

acoustooptical material A material whose refractive index or some other optical property can be changed by an acoustic wave.

acoustooptical modulator An arrangement for modulating a beam of light by passing it through an acoustic wave in a light-transparent solid or gaseous medium. The resulting amplitude modulation of the light beam is generally produced by diffraction, with the frequency of the acoustic wave determining the amount of diffraction or bending of the light beam.

acoustooptics The science that deals with interactions between acoustic waves and light.

AC power supply A power supply that provides one or more AC output voltages, such as an AC generator, dynamotor, inverter, or transformer.

acquisition 1. The process of locating and tracking an aircraft or missile by radar or laser to obtain fire-control data necessary for taking offensive or defensive action and perhaps for the setting of a fuse on a shell or missile. 2. The process of a surface ship locating a submarine or a submarine locating and determining the position of a surface ship with active or passive sonar. 3. The process of locating and tracking a satellite or space debris by radar to calculate its position and condition.

acquisition and tracking radar A radar set that locks onto a strong signal, tracks the object emitting the signal, and feeds position data directly and continuously to gun or missile control systems.

acquisition laser In an optical guidance system, a laser that radiates over a relatively large solid angle, such as 10°, for picking up the target during search or chase. When the target comes within range, a narrow-beam tracking laser takes over.

acquisition radar A radar set that detects an approaching target and feeds approximate position data to a fire-control or missile-guidance radar, which takes over the function of tracking the target.

AC receiver A radio receiver that operates only from an AC power line.

AC resistance *High-frequency resistance.*

acrylic resin A thermoplastic resin made by polymerizing esters of acrylic or methacrylic acid. It exhibits excellent optical and dielectric properties and high tensile strength. It is used to make lenses and electrical and electronic components. It is the generic term for the tradenames of Plexiglas (Rohm & Haas) and Lucite (General Electric).

acrylonitrile-butadiene-styrene [ABS] A plastic that offers high-rigidity, high-impact strength, excellent abrasion resistance, excellent insulating properties, and is resistant to many inorganic salts, alkalies, and acids. It is used to make tough, electrically insulated cases, containers, and parts of electronic components.

activation 1. The process of treating the cathode or target of an electron tube to create or increase its emission. It is also called *sensitization.* 2. The process of inducing radioactivity by bombardment with neutrons or other types of radiation. 3. The process of adding acid to a cell or battery to make it operative.

activation energy The excess energy required to initiate a particular molecular process. An example is the energy needed by an electron to reach the conduction band in a semiconductor.

activation time The time interval from the moment activation is initiated to the moment the desired operating voltage is obtained in a cell or battery.

activator 1. An impurity atom that increases the luminescence of a solid material, such as copper in zinc sulfide and thallium in potassium chloride. 2. An impurity atom used to activate the target of a camera tube. It is also called a *sensitizer.*

active 1. Contributing to signal energy, as in transistors, electron tubes, repeaters, and other amplifying devices and systems. 2. *Radioactive.*

active analog filter A filter that includes an operational amplifier as well as external resistors and capacitors which permit the filter to operate without any inductor. It can amplify an input signal or provide gain, but unlike a *passive filter,* it requires a power source. Incomplete analog active filters have been integrated on a single silicon chip, but the required resistors and capacitors are external to obtain the required precision. See also *switched capacitor filter* [SFC].

active area That part of the rectifying junction of a metallic rectifier which carries forward current.

active component A component capable of controlling voltages or currents to produce gain or switching action in a circuit. Examples include diodes, electron tubes, transistors, and integrated circuits. It is also called an *active device.*

active device *Active component.*

active display A display capable of rapid responses in the visual subject matter being displayed—graphics or alphanumeric characters, usually under electronic control. Examples include cathode-ray tubes (CRTs) and active-matrix liquid-crystal, electroluminescent, and plasma panels for the display of graphics and characters for television receivers, computers, and radar systems.

active electronic countermeasures Procedures to neutralize or destroy enemy search or fire-control radar. Radar signals can be jammed with powerful broadband radio transmitters, or radio-signal-seeking missiles can be launched to home on enemy transmitters or their antennas.

active homing The procedure for locating and making contact with a target or object to bring about its damage or destruction by means of an active radar, sonar, or ladar mounted within the missile or other weapon. It is the opposite of *passive homing,* which depends on the detection of emission of energy by an appropriate receiver to direct a responsive weapon.

active infrared detection The location and tracking of a target or object with an active infrared [IR] transmitter (e.g., IR laser or other illuminator) and a receiver. A technique used when emitted IR is insufficient to provide a useful source for the detection and ranging of the target. The active section can determine both range and bearing, and the passive section can detect, recognize, and identify the target. See also *FLIR* and *passive infrared system.*

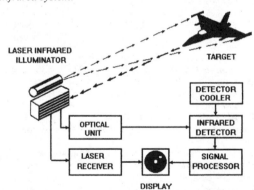

Active infrared detection system includes a cryogenically cooled detector for accurate ranging.

active jamming The deliberate interference with the transmission and reception of radar or radio frequencies by means of broadband radio noise transmission that prevents the enemy from transmitting and receiving intelligent signals or tracking and homing on friendly missiles, aircraft, or ships. The jamming transmitter is usually located on a ship or an aircraft in the local airspace.

active length The distance traveled by an electron in a transistor moving from emitter or source to collector or drain.

active line A horizontal line that carries picture information in television, as opposed to retrace lines which are blanked out during horizontal and vertical retrace.

active logic Logic that incorporates active components which provide such functions as level restoration, pulse shaping, pulse inversion, and power gain.

active material 1. A fluorescent material used in screens for cathode-ray tubes. Examples include calcium tungstate, zinc phosphate, and zinc silicate. 2. The lead oxide or other energy-storing material used in the plates of a battery.

active-matrix display A liquid-crystal video display developed for *notebook computers* that can display text and images in color. It was developed to replace the *cathode-ray-tube* monitor and is a higher-priced alternative to the *passive matrix screen*. It offers higher brightness and contrast than the passive screen because there is a transistor at each pixel location which stores the on or off state of each pixel. Pixels in active matrix panels are turned on and off by addressing them from computer memory. An example is the *thin-film transistor liquid-crystal display* [AMLCD]. Screen size is from about 10 to 12 in, measured diagonally.

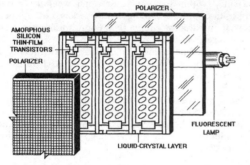

Active-matrix liquid crystal display includes thousands of thin-film transistors.

active-matrix screen *Active-matrix display.*

active-matrix thin-film transistor [AMTFT] A display matrix that includes thin-film, typically amorphous silicon transistors for switching display pixels on and off. See also *thin-film liquid-crystal display, thin-film transistor.*

active network A network whose output is dependent on a source of power other than that associated with the input signal.

active region The region in which amplifying, rectifying, light-emitting, or other dynamic action occurs in a semiconductor device.

active sonar Underwater sonar equipment that generates bursts of ultrasonic sound and listens for echoes reflected from submarines, fish, and other submerged objects within range. It is capable of determining both range and bearing of objects it detects. Targets can be displayed on *plan-position indicator* (PPI) displays. See also *passive sonar* and *sonar.*

active-switch modulator A modulator for pulsing the cathode of a radar magnetron that can provide varying pulse widths within the limitations of the energy stored in a high-voltage power supply. If it includes a gasless vacuum tube capable of passing high current and holding off high voltage, it is called a *hard-tube modulator.* See also *line-type modulator* and *magnetic modulator.*

active transducer A transducer that has a power source.

activity 1. The intensity of a radioactive source. It can be expressed as the number of atoms disintegrating in unit time or the number of scintillations or other effects observed per unit time. One unit of activity is the curie, equal to 3.7×10^{10} disintegrations per second. 2. A measure of the amplitude of vibration of a crystal unit, generally expressed as the rectified base or gate current of the oscillator circuit in which the crystal is used. 3. A computer transaction that results in the use or modification of information in the master file. 4. Short form of *radioactivity.*

activity dip A decrease in the value of crystal activity, other than a band break, that occurs over a small temperature interval. It is usually the result of loose coupling to other modes of vibration or variations in the mounting system.

activity ratio The fraction of the total records in a computer master file that is updated or otherwise processed in a given period.

AC transducer A transducer that needs an AC voltage source for proper operation.

AC transmission A mode of television transmission in which a fixed setting of the controls makes any instantaneous value of signal correspond to the same value of brightness for only a short time.

ACTS Abbreviation for *Advanced Communications Satellite,* a NASA-sponsored communications satellite.

actual frequency The measured frequency of a crystal-controlled oscillator, as distinguished from the nominal frequency value that is marked on the crystal unit.

actuating signal The reference input minus the primary feedback in a control system.

actuating transfer function The transfer function that relates a feedback control-loop actuating signal to the corresponding loop input signal.

actuator A motor, solenoid, other active device, or circuit that converts electrical, hydraulic, or pneumatic energy to a form that will cause motion or initiate some desired response. A TV remote control is a form of actuator.

acute angle An angle numerically smaller than a right angle, hence less than 90°.

AC voltage *Alternating voltage.*

acyclic Following no regularly repeated cycles of variations.

acyclic machine *Homopolar generator.*

A/D Abbreviation for analog-to-digital, usually used in connection with converters.

Ada A standardized computer language used primarily by the U.S. Department of Defense.

adaptable A circuit or system capable of making self-directed corrections. In a robot this can be accomplished with visual, force, or tactile sensors.

adapter A device that makes electric or mechanical connections between items not originally intended for use together.

adaptive array An antenna array in which each element is independently phased according to information received from incoming signals. It is used in automatic beam steering and focusing, retrodirective steering, and adaptive radar arrays. It is also called self-phased array.

adaptive control A control method by which system control parameters are continuously and automatically adjusted in response to measured process variables to improve the performance of the host equipment. An example is an aircraft or ship automatic pilot.

adaptive control system A control system based on concepts of *adaptive control.*

adaptive differential pulse-code modulation [ADPCM] A method for transcoding from a 64-kb/s pulse-code modulation channel to either a 16-, 24-, 32-, or 64-kb/s channel, or the reverse.

adaptive filter A filter circuit that automatically adjusts itself to respond to fixed-waveform signals which occur in a random manner and are completely buried in noise. The filter does this without prior knowledge of the existence or shape of the waveform. It can extend the detection range of sonar, radar, and electromagnetic reconnaissance systems. It is also called *adaptive waveform recognition.*

adaptive quantizer A quantizer in which step size is matched to signal variance by comparing each new signal segment with the previous step size as stored in memory.

adaptive radar A radar system capable of adjusting its own performance parameters automatically to meet specific situations.

adaptive system A system that can change itself to meet new requirements. This requires identification of a change, comparison with previous conditions, and initiation of the corrective action needed to restore optimum performance.

adaptive waveform recognition *Adaptive filter.*

ADC Abbreviation for *analog-to-digital converter.*

ADCCP Abbreviation for *advanced digital communications control protocol (or procedure).*

Adcock antenna A directional antenna that consists of two vertical wires spaced one-half wavelength apart or less, connected in phase opposition to give a figure-eight radiation pattern.

Adcock direction finder A radio direction finder that includes one or more pairs of Adcock antennas.

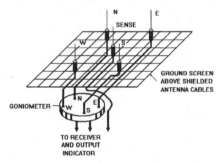

Adcock direction-finding antenna with a narrow aperture has two ports and four elements.

Adcock radio range An A-N radio range that includes Adcock antennas arranged at the four corners of a square on the ground. The vertical antennas at one set of opposite corners transmit the letter A in international Morse code, and the other two antennas transmit the letter N.

A/D converter Abbreviation for *analog-to-digital converter.*

addend A number or quantity to be added to another (the augend) to produce a result or sum.

adder 1. A circuit in which two or more signals are combined to give an output signal amplitude that is proportional to the sum of the input signal amplitudes. In a color television receiver, the adder combines the chrominance and luminance signals. 2. A computer circuit that can form the sum of two or more numbers or quantities. It is also called a *full adder.*

adder-subtracter A circuit whose output is equal or proportional to the sum or difference of quantities represented by its two inputs, as determined by a control signal.

additive color system A system that adds two colors to form a third.

additive primaries Sources of color or light that, by additive mixture in varying proportions, can be made to match a large range of colors. The three additive primaries used in television are red, green, and blue.

additive printed circuit-board process A process for forming the conductive traces and pads on a chemically prepared circuit board substrate by plating or depositing copper only where it is required. This eliminates the need for chemical etching or mechanical removal of excess copper to leave the desired traces and pads as in the conventional subtractive printed circuit board fabrication process. See also *plated-through holes.*

address A specific location in memory where a block of data is stored. A disk drive address generally specifies cylinder, track, and sector. It is given by a unique sequence of letters or numbers that designates the location of the data in a computer or the identification of a specific peripheral device.

address computation A computation that produces or modifies the address part of a computer instruction.

address format The arrangement of the address parts of a computer instruction.

address register A register that stores an address in a computer.

ADF Abbreviation for *automatic direction finder.*

adhesion *Bond strength.*

adiabatic Occurring without change in heat content.

adiabatic demagnetization A technique used to obtain temperatures within thousandths of a degree of absolute zero. A strong magnetic field is applied to a precooled paramagnetic salt to align its atomic magnets. The material is then thermally insulated from its surroundings, and the magnetic field is removed. The resulting disorientation of the atomic magnets absorbs heat energy, thereby reducing the temperature.

adion An ion that has been adsorbed on a surface and cannot move out of it.

A display A radar display in which targets appear as vertical deflections from a horizontal line that represents a

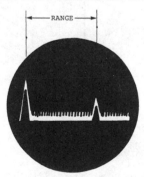

A display for radar shows the target as a vertical spike on a horizontal time base to the right of the transmission spike, with the separation proportional to range.

time base. Target distance is proportional to the horizontal position of the deflection from one end of the time base, and target echo signal intensity is proportional to the amplitude of the vertical deflection.

adjacent channel The channel immediately above or below the channel under consideration.

adjacent-channel attenuation *Selectance*.

adjacent-channel interference Interference caused by a transmitter operating in an adjacent channel. It is recognized as a peculiar garbled sound heard along with the desired program when the sidebands of the adjacent-channel transmitter beat with the carrier signal of the desired station. It is also called monkey chatter, sideband interference, and sideband splash.

adjacent-channel selectivity The ability of a receiver to reject signals on channels adjacent to that of the desired station.

adjacent sound carrier The RF carrier that carries the sound modulation for the television channel immediately below that to which the receiver is tuned.

adjacent video carrier The RF carrier that carries the picture modulation for the television channel immediately above the channel to which the receiver is tuned.

adjustable resistor *Variable resistor*.

adjustable short A waveguide section in which a movable wiper acts as a variable short circuit that changes the reactance for tuning or other purposes.

adjustable voltage divider A variable resistor that has one or more movable terminals which can be slid along the length of the exposed resistance wire until the desired voltage values are obtained.

admittance [Y] A measure of how readily alternating current will flow in a circuit. Admittance is the reciprocal of impedance and is expressed in siemens. The real part of admittance is conductance, and the imaginary part is susceptance.

admittance meter A null-type instrument for measuring complex impedance and admittance in coaxial systems. It can also be used for measuring standing-wave ratios and reflection coefficients.

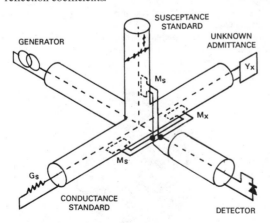

Admittance meter for frequency range of 20 MHz to 1.5 GHz. Three identical loops (*M*) in parallel, magnetically coupled to three coaxial lines, drive the null detector.

ADP Abbreviation for *automatic data processing*.

ADPCM Abbreviation for *adaptive differential pulse-code modulation*.

ADPCM transcoder A full-duplex, single-channel transcoder circuit (typically integrated) that performs ADPCM transcoding at rates from 64 kb/s to 5.12 Mb/s in accordance with ITU-TS (formerly) CCITT G.721 and American National Standard T1.301-1987.

ADP crystal Abbreviation for *ammonium dihydrogen phosphate crystal*.

ADP microphone A crystal microphone that includes an ammonium dihydrogen phosphate crystal which has piezoelectric properties.

advanced integrated landing system [AILS] An all-weather instrument landing system that provides precise three-dimensional guidance for aircraft during approach, flareout, landing, and rollout, as well as ground-based radar monitoring.

advanced intelligent network [AIN] In cellular telephony, an advanced version of the intelligent network. It has an independent architecture that allows telecommunications operators to create and modify services for both network performance and the customer's needs.

advanced low-power Schottky A bipolar digital transistor-transistor logic (TTL) family whose transistors include a Schottky diode to increase logic switching speed above that of standard gold-doped TTL. See also *advanced Schottky, Schottky,* and *transistor-transistor logic*.

Advanced Research Projects Agency [ARPA] A research and development arm of the U.S. Department of Defense, formerly called DARPA (for Defense Research Projects Agency).

Advanced Self-Protection Jammer [ASPJ] An airborne electronic-countermeasures [ECM] jamming system developed by the U.S. Navy as the ALQ-65. It was canceled in 1992 before it reached production.

AEGIS A fully automatic and integrated air defense system for U.S. Navy combat ships. It includes a surface-to-air missile system that supplies fire-control tracking data on more than 250 airborne and surface targets while maintaining surveillance of the hemisphere about the ship. It also keeps track of all targets for other weapons, evaluates threats, assigns weapons, and issues commands. It is based on the AN/SPY-1A/B multifunctional phased-array radar system that includes four fixed, flat-antenna arrays, providing hemispherical coverage. Each 10-ft-diameter array contains over 4400 radiating elements. Computers rapidly adjust the phase relationships between groups of elements to shift each array's beam direction electronically to track targets in about a 100° sector.

aelotropic *Anisotropic*.

aeronautical mile A unit of length equal to 6076.11549 ft (1852 m) or about 1.15 mi, the same length as a nautical mile. It is also called an air mile.

aeronomy The study of the upper atmosphere, where physical and chemical reactions due to solar radiation take place.

aeropause A region of indeterminate limits in the upper atmosphere, considered the boundary between the atmosphere and outer space.

aerophysics Physics related to the design, construction, and operation of vehicles that move rapidly through the atmosphere of the earth.

aerosol monitor An instrument capable of detecting particles smaller than 0.3 μm, to monitor air contamination in clean rooms. Particles in the monitored air cause scattering of light from a laser beam, with scattered light being converted to a proportional pulse by a multiplier phototube.

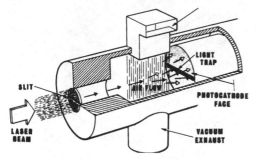

Aerosol monitor. A laser serves as a light source for a photoelectric pickup.

aerospace The earth's atmosphere and the space beyond.

AES Abbreviation for *auger electron spectroscopy*.

AEW radar Abbreviation for *airborne early-warning radar*.

AF Abbreviation for *audio frequency*.

AFC Abbreviation for *automatic frequency control*.

AF noise Any electric disturbance, in the audio-frequency range, that is introduced from a source extraneous to the signal.

AFSK Abbreviation for *audio-frequency-shift keying*.

afterglow 1. *Phosphorescence*. 2. *Persistence*.

afterpulse A spurious pulse induced in a multiplier phototube by a previous pulse.

AGC Abbreviation for *automatic gain control*.

aging Operating a circuit or device under controlled conditions for a preset time to screen out failures.

AGP Abbreviation for *aperture grille pitch*.

agravic Pertaining to zero gravity.

Ah Abbreviation for *ampere-hour*.

AI Abbreviation for *artificial intelligence*.

aided tracking A radar antenna control system in which a constant rate of motion of the tracking mechanism is maintained by a DC motor and selsyn system so an equivalent constant rate of movement of a target in bearing, elevation, distance, or any combination of these variables can be followed. An operator can adjust the rate of motion with a potentiometer in the DC motor circuit, as required to compensate for target speed and course changes.

AILS Abbreviation for *advanced integrated landing system*.

AIN Abbreviation for *advanced intelligent network*.

AIP Abbreviation for *American Institute of Physics*.

AI radar Abbreviation for *airborne intercept radar*.

airborne beacon *Radar safety beacon*.

airborne early-warning radar [AEW radar] An early-warning radar carried by aircraft. The radar signals are relayed from the aircraft to surface stations, or their signif-

icance is reported by radio to a ground control center. See also *airborne warning and control system* (AWACS).

airborne intercept radar [AI radar] Airborne radar for detecting and tracking other aircraft at night or in clouds. It might also include computers that provide fire-control data.

airborne magnetometer A magnetometer designed and built for detecting large metal objects (anomalies) underwater from low-flying aircraft. The U.S. Navy version for detecting submarines is called the *magnetic anomaly detector* [MAD].

airborne moving-target indicator [AMTI] A moving-target indicator system for airborne radar operating close to the ground, where moving targets are obscured by ground clutter and both the ground and the target are moving with respect to the radar in the airplane.

airborne radar A self-contained radar installed in aircraft. It can provide information about ground landmarks, ships at sea, shoreline contours, nearby aircraft, storm clouds, or weather fronts.

airborne warning and control system [AWACS] A U.S. Air Force system that includes airborne radar to detect low-flying and high-altitude enemy aircraft and missiles. It also provides command and control functions for directing intercepting aircraft.

airbridge A microminiature gold bridge formed by deposition and etching during the manufacture of gallium arscnide ICs that permits electrical contact between two conductors by spanning a third conductor. Clearance is sufficient for air to act as the insulator.

air capacitor A capacitor that has only air as the dielectric material between its plates.

air cell A power cell in which depolarization at the positive electrode is accomplished chemically by reduction of the oxygen in the air. See *zinc-air cell*.

air cleaner *Precipitator*.

air column The air space within a horn or acoustic chamber for a loudspeaker.

air-core coil A coil wound on a fiber, plastic, or other nonmagnetic tube or bobbin.

air-core transformer A transformer with two or more coils wound on a fiber, plastic, or other nonmagnetic bobbin with no iron or ferrite in its assembly. It is a typical construction for radio frequency [RF] and intermediate frequency [IF] transformers and oscillator coils.

air gap 1. A short gap or equivalent filler of nonmagnetic material across the core of a choke, transformer, or other magnetic device. The gap prevents the core from being saturated by direct current or permits required mechanical movement of coils or an armature. 2. A spark gap that consists of two conducting electrodes separated by air.

air-gap crystal unit A crystal unit whose electrodes are separate metallic plates rigidly spaced apart by an amount slightly greater than the thickness of the quartz plate.

air ionizer A machine that emits positive and negative air ions to neutralize static-charge buildup on nonconductive surfaces such as paper, plastic, or glass. The ions are typically distributed by a fan that circulates the air over the ion source. See also *electrostatic discharge* [ESD] *protection*.

air-launched cruise missile See *cruise missile.*

airport surveillance radar [ASR] A radar located on or near an airport to provide an indication of the bearing and distance of each aircraft within the terminal area. It is used by itself for air traffic control and with precision approach radar to form a ground-controlled approach system.

air-position indicator [API] An airborne computing system that presents a continuous indication of aircraft position based on aircraft heading, airspeed, and elapsed time. Position is indicated in latitude and longitude values or other coordinates. True heading and air mileage flown are also shown.

air route surveillance radar [ARSR] A long-range (approximately 150-mi or 240-km) radar used by the Federal Aviation Agency to control air traffic between terminals.

air search radar Radar that is assigned to detect aircraft targets more than 50 mi (80 km) away and give range and bearing of each while maintaining a complete 360° azimuth search.

air sounding Measuring atmospheric pressure, humidity, and other characteristics of the atmosphere with instruments carried aloft.

air-spaced coax Coaxial cable whose coaxial conductor is centered by beads, spirally wound plastic, or other dielectric material electrically equivalent to air.

air-to-air guided missile A guided missile that is fired at an airborne target from an airborne aircraft. Examples include Amraam, Sidewinder, and Sparrow.

air-to-ground missile [AGM] A guided missile that is fired from an aircraft to a target on the ground or a ship at sea.

air traffic control [ATC] A service that monitors and controls the flow of air traffic.

air traffic control radar beacon system [ATCRBS] A modern version of IFF, used worldwide to track aircraft for air traffic control purposes. All interrogation is at 1.03 GHz and all replies at 1.09 GHz, as established by the International Civil Aviation Organization (a United Nations agency). The reply pulses from an aircraft identify it and give its altitude. Also called secondary surveillance radar.

air-transportable radar A radar system that can be disassembled and have its antenna stowed within its control console shelter for transport by a cargo aircraft or heavy-lifting helicopter to a war zone or site of a natural disaster. An example is the AN/TPN-24, an S-band radar (2.7 to 2.9 GHz). It was designed for surveillance of the airspace in the vicinity of an advanced or temporary airfield as part of an aircraft ground-controlled approach system. The 2-D information is displayed on a plan-position indicator [PPI]. The shelter contains all the radar electronics as well as the display, a radio, and microwave-relay equipment.

AJ Abbreviation for *antijamming.*

A law A European companding-encoding law commonly used in PCM systems.

albedo The ratio of the amount of light or other radiation reflected from an object to the total radiation falling on it.

albedometer An instrument that measures the albedo of an object or surface.

Alford loop antenna A multielement antenna that has approximately equal in-phase currents uniformly dis-

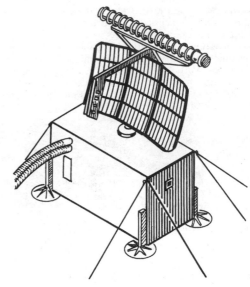

Air-transportable radar. The AN/TPN-24 S-band radar for a ground-controlled approach system can be disassembled, and its antenna stowed in the shelter, for transport by cargo aircraft.

tributed along each of its peripheral elements. Its radiation pattern is very nearly circular in the plane of polarization.

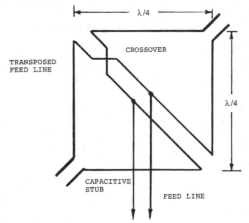

Alford loop antenna is suitable for transmitting and receiving high frequencies.

Alford slotted tubular antenna A horizontally polarized antenna that consists of a metal cylinder which has a full-length slot. Currents flow in horizontal circles, simulating the operation of a vertical stack of in-phase loop antennas.

Alfven wave An electromagnetic wave that propagates along lines of force in magnetized plasma.

AlGaAs Abbreviation for *aluminum gallium arsenide.*

algebraic language The conventional method of writing the symbols, parentheses, and other signs of formulas and mathematical expressions. Many scientific calculators accept algebraic language directly for keying in problems.

ALGOL [ALGOrithmic Language]   An internationally accepted arithmetic language that permits numerical procedures to be precisely presented to a computer in a standard form.

algorithm   A set of well-defined rules for solving a problem in a finite number of steps.

alias   An alternate designation for a given data element or a given point in a computer program.

alias frequency   An erroneous lower frequency obtained when a periodic signal is sampled at a rate less than twice per cycle.

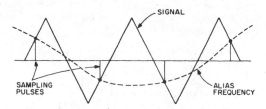

Alias frequency caused by low sampling rate.

aliasing   Sampling of a time function at too low a rate, so high-frequency components can impersonate low frequencies.

aliasing noise   A distortion component that is created when frequencies present in a sampled signal are greater than one-half the sample rate.

align   To adjust two or more sections of a circuit or system so their functions are properly synchronized. Trimmers, padders, or variable inductances in tuned circuits are adjusted to give a desired response for fixed-tuned equipment or to provide tracking for tunable equipment.

aligner   A tool used in semiconductor-device manufacturing for transferring *lithographic* patterns from a photomask to a semiconductor wafer. Four types of aligners are in use: contact, proximity, projection, and stepper. Contact aligners place the photomask in direct contact with the wafer, introducing possible particle contamination. The others avoid direct mask contact and bring increasing line-width control and resolution. See *lithography, mask,* and *stepper.*

aligning plug   The plug in the center of the base of an octal, loktal, or other tube; it has a single vertical projecting rib that prevents the tube from being inserted incorrectly into its socket.

alignment   The process of aligning.

alignment chart   *Nomograph.*

alignment pin   A pin that ensures the correct mating of a tube in its socket or the correct mating of connectors.

alkali metal   Alkali-producing metal such as lithium, cesium, or sodium that have photoelectric characteristics for use in phototubes and camera tubes.

alkaline battery   A battery made by packaging two or more *alkaline cells* in series in a rigid case and providing positive and negative terminals to obtain no-load voltages that are higher than the 1.5-V output of a single alkaline cell. An example is the 9-V battery suitable for powering electronics that contains six cells. It is also called a *transistor battery.*

alkaline cell   A *primary cell* that contains a zinc–potassium hydroxide–manganese dioxide chemical system. It

is now recognized as the premium primary zinc anode, and is especially recommended for powering electronic circuitry. Its major advantages compared with the zinc-carbon system are high energy density, the ability to operate continuously at relatively high discharge rates over a wide temperature range, and a shelf life in excess of 3 years. The open-circuit voltage ranges from 1.5 to 1.6 V, and the operating voltage ranges from 1.3 to 1.1 V. Energy density is 60 h/lb (130 Wh/kg); 5.8 Wh/in^3 (350 Wh/L). It is also called the *alkaline–manganese dioxide cell.*

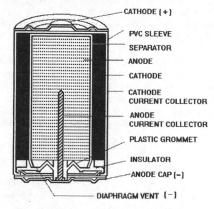

Alkaline cell is a zinc–potassium hydroxide–manganese dioxide primary-battery system.

alkaline–manganese dioxide battery   *Alkaline battery.*

alkaline–manganese dioxide cell   *Alkaline cell.*

all-channel tuning   The ability of a television set to receive UHF channels 14 to 83 as well as VHF channels 2 to 13.

Allen screw   A screw that has a hexagonal hole in its head.

Allen wrench   A wrench made from a straight or bent hexagonal rod, used to turn an Allen screw.

alligator clip   A long, narrow, spring-loaded metal clip with toothed jaws which, when used for terminating a test lead, permits temporary but secure connections to be made as well as rapid disconnection for placement at other test points.

Alligator clip provides a secure connection with test points for meter leads.

allochromatic   The fluorescent property of some crystals caused by the presence of internal microscopic particles that occur naturally or can result from the crystal's exposure to certain forms of radiation.

allochromy   Fluorescence in which the wavelength of the emitted light differs from that of the absorbed light.

allowed band   A band in certain materials that contains a group of energy levels which electrons can occupy, for example, a conduction or valence band.

allowed transition   The most probable transition between two states of a quantum-mechanical system.

alloy   A mixture of two or more materials, at least one of which is a metal. It has the properties of a metal or a semiconductor.

alloy film   A thin film of an alloy such as nickel chromium that is used for forming resistors in thin-film circuits.

alloying   The process of making semiconductor junctions by melting an acceptor or donor on the surface of a semiconductor and letting it recrystallize.

alloy junction   A junction produced by alloying one or more impurity metals to a semiconductor. A small button of impurity metal is placed at each desired location on the semiconductor wafer, heated to its melting point, and cooled rapidly. The impurity metal alloys with the semiconductor material to form a P or N region, depending on the impurity used. Also called fused junction.

all-pass network   A network that introduces phase shift or delay without introducing appreciable attenuation at any frequency.

all-wave antenna   A radio receiving antenna that responds reasonably well to a wide range of frequencies, including the shortwave and broadcast bands.

all-wave receiver   A radio receiver that can be tuned to all of the communication and entertainment broadcasting bands, including FM stations.

all-weather landing system   An aircraft instrument landing system that has optimum operational capability in low and zero visibility.

alnico [ALuminum NIckel CObalt]   An alloy consisting chiefly of iron, aluminum, nickel, and cobalt suitable for making high-retentivity magnets required for microwave tubes and loudspeakers. The alloys are classified by a number indicating relative magnetic strength. Alnico V is an alloy with the strongest (fifth-level) magnetic flux.

alpha [Greek letter α]   1. The symbol for the current amplification factor of a transistor in a grounded-base circuit. It is the ratio of an incremental change in collector current to an incremental change in emitter current, when collector voltage is held constant.   2. The symbol for *attenuation constant*.

alpha chamber   *Alpha-counter tube.*

alpha counter   An instrument that counts alpha particles; it comprises an alpha counter tube, amplifier, pulse-height discriminator, scaler, and recorder.

alpha-counter tube   An electron tube consisting chiefly of a chamber for detecting alpha particles. It is usually operated in the nonmultiplying or proportional region. Pulse-height selection is used to discriminate against pulses due to beta or gamma rays. It is also called an *alpha chamber*.

alpha cutoff   The high frequency at which the alpha of a bipolar transistor drops 3 dB from its low-frequency value. The current amplification at alpha cutoff is thus 0.7 of the alpha rating for the transistor.

alpha laser   An orbiting space-based chemical laser operating at a wavelength of 2.7 microns in the infrared region intended for antimissile defense. Its lasing action is produced by the combustion of hydrogen and fluorine. The fluorine is obtained on demand by a reaction between nitrogen trifluoride, deuterium, and helium. The laser is designed to produce 2.2 million watts, and it can melt metal at long range in space. The energy intensity at the laser's core exceeds that of the surface of the sun. Made largely of aluminum, it contains a system of mirrors known as *LAMP* [Large Advanced Mirror Program].

alphameric   *Alphanumeric.*

alphanumeric (also called *alphameric*)   A designation or display readout that combines letters with numbers.

alphanumeric readout   A display of both letters and numbers that can be obtained with 7- or 14-segment "union jack" and dot-matrix LCD, LED, vacuum fluorescent, and neon gas-discharge display devices.

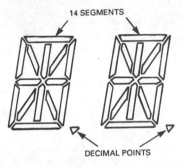

Alphanumeric display: 14 controllable illuminated or reflective segments can form all capital letters, the numbers 0 to 9, and other ASCII symbols.

alpha particle   A positively charged particle consisting of two protons and two neutrons emitted from the nucleus of an atom during the decay process. It has properties that are identical to those of the helium atom nucleus. Emitted from certain radioactive elements or isotopes, its ionizing power is high enough to ionize air molecules, but not to penetrate human skin. However, alpha particles can damage living cells. They can originate in alumina ceramic, enter IC transistor gates or memory cells, and change their charge states to cause *soft errors*.

alpha radiation   Alpha particles emerging from radioactive atoms.

alpha radiator   A radioactive substance that emits alpha particles.

alpha ray   A stream of alpha particles. It is only slightly deflected by a magnetic field.

alpha wave   A brain-wave current that has a frequency of 9 to 14 Hz. Also called alpha rhythm.

ALS   Abbreviation for *advanced low-power Schottky,* a digital transistor-transistor logic (TTL) logic family.

AlSb   Symbol for *aluminum antimonide.*

alternating current [AC]   An electric current that is continually varying in value and reversing its direction of flow at regular intervals, usually in a sinusoidal manner. Each repetition, from zero to a maximum in one direction and then to a maximum in the other direction and back to zero, is called a cycle. The number of cycles occurring in 1 s is called the frequency in hertz. The average value of an alternating current is zero. For alternating current terms, see AC entries.

alternating gradient A magnetic field in which successive magnets have gradients of opposite sign, so the field increases with radius in one magnet and decreases with radius in the next. Used in synchrotrons and cyclotrons.

alternating quantity A periodic quantity whose average value is zero over a complete cycle.

alternating voltage The voltage generated by an alternator or developed across a resistance or impedance through which alternating current is flowing. This voltage is continually varying in value and reversing its direction at regular intervals. Also called AC voltage.

alternation Half of an AC cycle, consisting of a complete rise and fall of voltage or current in one direction. There are 120 alternations per second in 60-Hz AC power.

alternator A machine that generates an alternating voltage when its armature or field is rotated by a motor, an engine, or other means. The output frequency is directly proportional to the speed at which the generator is driven. It is also called a synchronous generator.

alternator transmitter A radio transmitter that utilizes power generated by an RF alternator.

altimeter An instrument that indicates altitude above sea level, above ground level at the point of measurement, or above ground level at some other point for which the altimeter was calibrated. A conventional altimeter is an aneroid barometer that measures changes in barometric pressure with altitude. An absolute altimeter determines altitude above ground or water by measuring the time it takes for a radio or radar wave to travel straight down and be reflected back to the aircraft.

altitude chamber A chamber in which air pressure, humidity, and temperature can be controlled to simulate conditions at various altitudes for test purposes.

ALU Abbreviation for *arithmetic-logic unit.*

alumina An aluminum-oxide ceramic with excellent insulation properties used to make insulators for power electron tubes and substrates for hybrid circuits (particularly those operating at microwave frequencies) (HMIC). It is easily outgassed, and exhibits low dielectric losses over a wide frequency range. It can withstand continuous temperature of up to 1400°C.

aluminization Evaporation of a thin film of aluminum onto a substrate for a semiconductor device or integrated circuit, to provide conductive patterns corresponding to openings in a mask over the substrate.

aluminized screen A television picture tube screen that has a thin coating of aluminum on the back of its phosphor layer. Electrons in the beam readily penetrate the coating and activate the phosphors to produce an image. The aluminum reflects outward the light that would otherwise go back inside the tube, thereby improving the brilliance and contrast of the image. Also called metal-backed screen, metallized screen, and mirror-backed screen.

aluminum [Al] A lightweight, silver-white metal element with an atomic number of 13 and an atomic weight of 27. It is an *acceptor* element for the doping of silicon to be a *P-type* material. It is also used to make the plates and cases of aluminum electrolytic capacitors and the plates of air-dielectric tuning capacitors. It is deposited and etched on integrated circuits as an electrical conductor, and it is drawn as fine wire for semiconductor-device lead bonding.

Aluminum oxide or *alumina* ceramic is a common substrate material, insulator, and window for microwave power tubes.

aluminum antimonide [AlSb] A semiconductor that has a forbidden bandgap of 2.2 eV and a maximum operating temperature of 500°C when used in a semiconductor device.

aluminum electrolytic capacitor An electrolytic capacitor whose dielectric is formed from aluminum oxide by electrochemical reaction. It is made from high-purity aluminum foil that has been electrochemically etched to increase surface area.

aluminum gallium arsenide [AlGaAs] A ternary compound formed by alloying the elements aluminum and gallium from Group III of the periodic table with arsenic from Group V used in the fabrication of integrated circuits and optoelectronic devices. It is also referred to as gallium aluminum arsenide and abbreviated GaAlAs.

aluminum oxide *Alumina.*

A/m Abbreviation for *ampere per meter.*

AM Abbreviation for *amplitude modulation.*

amateur A person holding a license, issued by the Federal Communications Commission or by corresponding authorities in other countries, which authorizes that person to operate a licensed amateur radio station for pleasure and service only, not for profit. Also called ham (slang).

amateur band A band of frequencies assigned exclusively to amateur operators.

amateur teleprinting over radio [AMTOR] A technique that provides error-correction capabilities. See *automatic repeat request* and *forward error correction.*

ambient The surrounding environmental conditions such as air pressure, humidity, or temperature.

ambient condition The condition of the surrounding medium.

ambient light Normal room light.

ambient-light filter A filter placed in front of a computer monitor screen to reduce the amount of ambient light reaching the screen and to minimize reflections of light from the glass face of the tube. The filter, generally a dull color, can be incorporated into the glass faceplate of the tube or the safety-glass window, or it can be a separate sheet of plastic.

ambient noise Noise associated with a given environment, made up of more or less continuous sounds from near and far sources. It is generally undesirable.

ambient temperature The temperature of still air immediately surrounding a circuit or device taken about ½ in from the case of the object.

ambipolar photoconductivity Photoconductivity in which the probability of capture of photoionized carriers in a semiconductor is about the same for either N- or P-type material.

ambisonics The reproduction of sound as it would be heard at a given point in the concert hall, including reverberations and other ambience characteristics of the hall.

American Morse code A dot-dash code used in wire telegraphy. It has a different spacing method from the international Morse code used in radio and entirely different letter codes.

American National Standards Institute [ANSI] The organization that represents the United States in the International Standards Organization [ISO]. It was formerly known as United States of America Standards Institute and as the American Standards Association.

American Standard Code for Information Interchange [ASCII]   A standard code that has seven channels and an eighth channel for parity. Developed to simplify the interconnection of computers, communication circuits, and digital output equipment.

American Standards Association [ASA]   Former name of American National Standards Institute, Inc. [ANSI].

American wire gage [AWG]   A gage used chiefly for specifying nonferrous wire and sheet metal. Sizes range from 0000 for the largest (0.46 in or 1.17 cm) to 0 (0.325 in or 0.8255 cm) and 1 (0.289 in or 0.654 cm) to 50 (0.001 in or 0.00254 cm). Also called Brown and Sharpe gage.

AM/FM   An abbreviation that indicates a radio receiver can receive either AM or FM signals.

AML   Abbreviation for *automatic modulation limiting*.

AMM   Abbreviation for *antimissile missile*.

ammeter   An instrument that measures current flow. Its scale can be calibrated in amperes or smaller units. An ammeter that indicates milliampere values is called a *milliammeter*. An ammeter that indicates microampere values is called a *microammeter*.

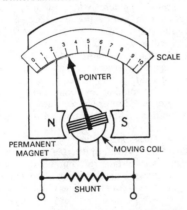

Ammeter: analog instrument based on moving-coil meter can measure current. The coil with pointer is deflected in proportion to the current passing in the shunted meter circuit.

ammonia-beam maser   See *maser*.

ammonium dihydrogen phosphate crystal [ADP crystal]   A piezoelectric crystal used in sonar transducers and crystal microphones.

amorphous film   A magnetically ordered metallic film that can be deposited on a semiconductor wafer or other materials.

amorphous laser   *Glass laser*.

amorphous silicon   A noncrystalline form of silicon used to fabricate transistors on large-area, flat, active displays. Its properties are inferior to those of crystalline silicon, but it is easier to deposit on a glass substrate.

amorphous silicon cell   A photovoltaic cell made from amorphous silicon with hydrogen atoms deposited as an irregular atomic structure on a substrate.

ampacity   Current-carrying capacity in amperes. Used as a rating for power cables.

amperage   The amount of current in amperes.

ampere [A]   1. A practical unit of electrical current. One ampere of current is 6.24×10^{18} electrons passing one point in 1 s. It equals 1 C/s. The result of 1 V across a resistance of 1Ω.   2. (SI): that constant current which, if maintained in two straight parallel conductors of infinite length, of negligible cross section, and placed 1 m apart in vacuum, would produce between these conductors a force equal to 2×10^{-7} N per meter of length.

ampere-hour [Ah]   A unit of quantity of electricity. Multiplying current in amperes by time of flow in hours gives ampere-hours. It indicates the amount of energy that a secondary battery can deliver before it needs recharging or that a primary battery can deliver before it must be replaced.

ampere-hour efficiency   The efficiency of a battery, equal to the ratio of ampere-hour output to ampere-hour input required for recharging.

ampere-hour meter   A meter that measures current drawn per unit of time, integrated for reading in ampere-hours.

ampere per meter [A/m]   The SI unit of magnetic field strength.

Ampere's law   The magnetic intensity at any point near a current-carrying conductor can be computed on the assumption that each infinitesimal length of the conductor contributes an amount which is directly proportional to that length and to the current it carries, inversely proportional to distance, and directly proportional to the sine of the angle.

Ampere's rule   The magnetic field surrounding a conductor will have a counterclockwise direction when the electron flow is away from the observer.

ampere-turn   The SI unit of magnetomotive force, equal to coil turns multiplied by coil current in amperes.

ampere-turn amplification   The ratio of the change in output ampere-turns of a magnetic amplifier to the change in control ampere-turns.

amplidyne [AMPLIfier DYNE]   A rotating magnetic amplifier that consists of a combination DC motor and generator which has special windings and brush connections to give power amplification, so small changes in power input to the field coils produce large changes in power output.

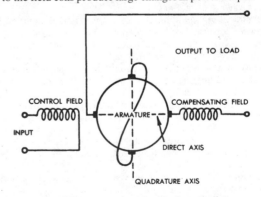

Amplidyne connections shown schematically.

amplification   The process of increasing the strength (current, voltage, or power) of a signal.

amplification factor [symbol μ]   The ratio of the change in anode voltage of an electron tube to a change in control-electrode voltage that produces the same change in anode current when other tube voltages and currents are held constant.

amplified AGC An automatic gain-control circuit in which the control voltage is amplified before being applied to the tube whose gain is to be controlled in accordance with the strength of the incoming signal.

amplifier A circuit based on an electron tube, transistor, magnetic unit, or other amplification-producing component that increases the strength of a signal without appreciably altering its characteristic waveform. An amplifier transfers power to the signal from an external source.

amplifier noise Unwanted signals existing in a completely isolated amplifier when there is no input signal.

amplify To increase in magnitude.

amplitron A Raytheon company tradename for a backward-wave, crossed-field, microwave-frequency amplifier. An anode cavity and pins form the resonator circuits. A pair of pins and the cavity are excited in opposite phase by the strap line. The electron beam and electromagnetic waves interact in the resonant circuits. The amplitron can deliver up to 3-MW pulses of 10-μs duration in S-band.

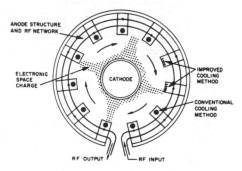

Amplitron interaction region, showing space-charge spokes (shaded areas) that rotate about center cathode. The magnetic flux is directed into the page.

amplitude The value of a varying quantity at a specified instant.

amplitude balance control *Sensitivity-time control.*

amplitude discriminator *Pulse-height discriminator.*

amplitude distortion *Frequency distortion.*

amplitude fading Fading in which all frequency components of a modulated carrier signal are uniformly attenuated in amplitude.

amplitude-frequency distortion *Frequency distortion.*

amplitude-frequency response A graph that shows how the gain or loss of a circuit or system varies with frequency. It is also called the frequency characteristic, frequency response, response, response characteristic, and sinewave response.

amplitude limiter *Limiter.*

amplitude-modulated transmitter A transmitter that transmits an amplitude-modulated wave.

amplitude-modulated wave A sinusoidal wave whose envelope contains a component similar to the waveform of the signal to be transmitted.

amplitude modulation [AM] A broadcast method for combining an information signal and a radio frequency

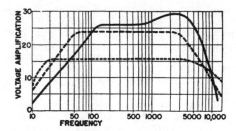

Amplitude-frequency response curves for three different AF amplifiers.

[RF] carrier wave to vary the amplitude of the RF envelope (carrier and sidebands) so that it is proportional to information signal strength.

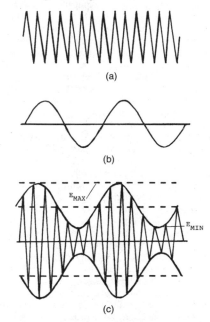

Amplitude modulation. The carrier signal (*a*) is modulated by the audio waveform (*b*), resulting in a modulated envelope (*c*).

amplitude-modulation noise level The noise level produced by undesired amplitude variations of an RF signal in the absence of any intended modulation.

amplitude-modulation rejection The ability of an FM radio receiver to reject amplitude-modulated RF interference from manmade sources or electric storms.

amplitude noise Noiselike variations in the amplitude of the signal reflected from a radar target, caused by changes in target aspect.

amplitude range The ratio between the upper and lower limits of audio signal amplitudes that contain all significant energy contributions.

amplitude-shift keying [ASK] Data transmission in which the signals from computers or data terminals produce many different amplitude levels of a sinewave carrier.

Amraam Acronym for *advanced medium-range air-to-air missile*. It contains an onboard inertial reference unit, using target coordinates provided in prelaunch by the launching aircraft's radar. For the final target interception the missile switches to its own active radar for guidance and tracking.

AM/SSB *SSB/AM.*

AM-TFT Abbreviation for *active-matrix, thin-film transistor.*

AMTI *Airborne moving-target indicator.*

AMTOR Abbreviation for *amateur teleprinting over radio.*

AMVER [Automated Merchant VEssel Report] A U.S. Coast Guard service that stores in a computer the positions of merchant vessels and aircraft, for locating quickly the vessels nearest the location from which an SOS is transmitted by a craft in distress.

anacoustic zone The zone of silence in space, starting at about 100 mi (160 km) altitude, where the distance between air molecules is greater than the wavelength of sound, and sound waves can no longer be propagated.

analog A continuous representation of physical phenomena that can be plotted as points of amplitude versus time with each point merging imperceptibly into the next.

analog channel A channel on which the information transmitted can have any value between the channel limits. A voice channel is an analog channel.

analog comparator A comparator that produces a high digital output signal (binary 1) when the sum of two analog voltages is positive, or a low output signal (binary 0) when their sum is negative.

analog computer A computer based upon analog circuits that contain precision potentiometers which can be varied to represent physical values and operational amplifiers to perform calculations. Input signals can be obtained from sensors or transducers or can be entered directly on the potentiometers. Mathematical operations such as addition, subtraction, multiplication, division, integration, and differentiation can be performed by the operational amplifiers when combined with active and passive feedback components and appropriate input impedances. The solutions to problems are analog values which can be displayed on an oscilloscope or conditioned to drive a graphic plotter. It has been replaced by digital computers except for certain kinds of laboratory experimentation such as wind tunnel and water tank testing or simulation. See also *hybrid computer.*

analog control The control of a product or system by analog signals that are derived from physical variables such as pressure, temperature, flow, frequency, voltage, current, power, and audio level.

analog data Data represented in a continuous form, as contrasted with digital data that have discrete values.

analog device A control device that operates with variables represented by continuously measured voltages or other quantities.

analog multiplier A circuit with three input terminals and one output terminal that can accept two inputs in analog form and, when the proper external feedback loops are completed, produce an output proportional to the product of the input variables (typically voltages). Some are organized as two-quadrant, and others as four-quadrant, circuits. Integrated-circuit versions based on transistor transconductance can also divide, square, and determine the square root of inputs in analog form when the prescribed changes are made in external connections.

analog panel meter The conventional meter in which the value being measured is indicated by a pointer moving over a calibrated scale, as contrasted to digital panel meters that show the exact value on a numerical display.

analog signal A control signal whose magnitude represents information content.

analog-to-digital converter [A/D converter or ADC] A circuit that converts analog input signals into digital signals. Examples of analog-to-digital converters include the *successive-approximation* ADC, *voltage-to-frequency converter* [V/F], *dual-slope* ADC, and *high-speed flash* ADC. It is also called a *digitizer.* See also *analog-to-frequency converter.*

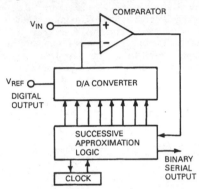

Analog-to-digital converter (ADC): a successive approximation ADC produces a digital output that is equivalent to the unknown input voltage value.

analog-to-frequency converter [AFC or A/FC converter] A circuit that accepts an analog input in some form other than a frequency and converts it into a frequency. It can be made as a hybrid or monolithic IC or as a discrete-component module.

analytical model A mathematical model of a semiconductor device based on an explicit function that yields approximate results in relatively short computer runs.

AND A logic operator whose output is a logic 1 if all the inputs are a logic 1.

Anderson bridge A six-branch modification of the Maxwell-Wien inductance bridge that can be used to measure a wide range of inductances with reasonable values of fixed capacitance. It can also be used to measure the residuals of resistors with a substitution method to eliminate the effects of residuals in the bridge elements. It can be balanced independent of frequency.

AND gate A multiple-input gate circuit whose output is energized only when every input is energized simultaneously in a specified manner.

AND NOT A logic operator that is equivalent to the EXCEPT operator.

AND/OR gate A gate that produces an output for one of several possible combinations of inputs. It combines the characteristics of AND and OR gates.

anechoic room 1. A room whose floor, ceiling, and all walls are lined with a sound-absorbing material to minimize reflections of sound. 2. A room completely lined with a material that absorbs radio waves at a particular frequency or over a range of frequencies.

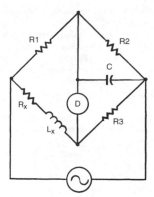

Anderson bridge can measure inductance with a fixed capacitance.

anemometer   An instrument that measures wind velocity.

angle diversity   Diversity reception in which beyond-the-horizon tropospheric scatter signals are received at slightly different angles, equivalent to paths through different scatter volumes in the troposphere.

angle modulation   Modulation in which the angle of a sinewave carrier is the characteristic varied from its normal value. Phase modulation and frequency modulation are forms of angle modulation.

angle noise   In radar, tracking error caused by variations in echo arrival area as the target changes its aspect.

angle of elevation   The angle between the horizontal plane and the line ascending to the object.

angle of incidence   The angle between a wave or beam arriving at a surface and the perpendicular to the surface at the point of arrival.

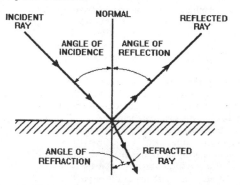

Angles of incidence, reflection, and refraction for electromagnetic waves.

angle of reflection   The angle between a wave or beam leaving a surface and the perpendicular to the surface.

angle of refraction   The angle between a refracted wave or beam and the perpendicular to the refracting surface.

angstrom [Å]   A unit of length; 10,000 Å = 1 cm; 1 Å = 10^{-8} cm. Also, 1 Å = 10^{-10} m = 10^{-4} μm = 3.937×10^{-9} in. The micrometer (μm) is now preferred to the angstrom as the unit of measure. Visible-light wavelength is between 4000 and 7800 Å (0.4 and 0.78 μm) = 400 to 780 nm; ultraviolet (UV) radiation is shorter than 0.4 μm, and infrared (IR) radiation is longer than 0.78 μm.

angular acceleration   The rate of change of angular velocity about a rotational axis.

angular accelerometer   An accelerometer that measures the rate of change of angle between two objects under observation.

angular deviation loss   The ratio of the response of a microphone or loudspeaker on its principal axis to the response at a specified angle from the principal axis, expressed in decibels.

angular deviation sensitivity   The ratio of the change in a course-indicator reading to the actual angular change in the course of an aircraft or ship.

angular distance   Distance expressed in radians or degrees. It is equal to the distance in wavelengths multiplied by 2π rad or 360°.

angular frequency   The frequency expressed in radians per second. It is equal to the frequency in hertz multiplied by 2π.

angular momentum   The angular velocity of a body multiplied by its moment of inertia.

angular-momentum quantum number   A quantum number that determines the total angular momentum of a molecule exclusive of nuclear spin.

angular resolution   The ability of a radar to distinguish between two targets solely by the measurement of angles. It is generally expressed as the minimum angle by which targets must be spaced to be separately distinguishable.

angular velocity [ω]   1. The speed of a rotating object measured in radians per second. It is equal to revolutions per second multiplied by 2π.   2. The rate of change of phase of an alternating quantity. It is equal to the frequency in hertz multiplied by 2π.

angular width   *Course width.*

anhysteresis   Magnetization with a unidirectional field upon which is superposed an alternating field of gradually decreasing amplitude.

anion   A negative ion.

anisotropic   The characteristic of having different properties in different directions. It is also called *aelotropic.*

anisotropic etching   A process of preferential directional etching of a material in the fabrication of silicon integrated circuits and bulk micromachining. It is done with liquid etchants such as potassium hydroxide [KOH] and water for etching silicon.

ANN   Abbreviation for *artificial neural networks.*

annular transistor   A transistor whose characteristic semiconductor channels are in concentric circles around the emitter.

annunciator   An electric-powered remote signaling device, such as a buzzer, signal lamp, or lighted push-button switch.

anode   1. The positive terminal of a cell or battery. Electron flow is toward the anode through the connected load; current flow is away from the anode to the load.   2. [Symbol A] The positive terminal of an electron tube. Electrons flow through the tube to its anode, and from there to the positive terminal of the connected voltage source. Also called plate.   3. In a semiconductor diode, the terminal toward which forward current flows from the external circuit. This anode terminal is normally marked negative. Current flow is from the anode to the cathode inside the diode, so electron flow is from the cathode to the anode inside the diode, and electrons flow away from the anode to the external circuit.

anode breakdown voltage   The anode voltage necessary to cause conduction across the main gap of a gas tube when the starter gap is not conducting and all other tube elements are at cathode potential.

anode characteristic   A graph plotted to show how the anode current of an electron tube is affected by changes in anode voltage.

anode circuit   A circuit that includes the anode voltage source and all other parts connected between the cathode and anode of an electron tube. Also called plate circuit.

anode current   The electron current flowing through an electron tube from the cathode to the anode. Also called plate current.

anode dark space   A narrow dark zone next to the surface of the anode in a gas tube.

anode detection   Detection in which rectification of RF signals takes place in the anode circuit of an electron tube. The grid bias is made sufficiently negative to bring anode current nearly to cutoff for no signal, so average anode current follows changes in signal amplitude. Also called plate detection.

anode dissipation   Power dissipated as heat in the anode of an electron tube because of bombardment by electrons and ions.

anode efficiency   The ratio of the AC load circuit power to the DC anode power input for an electron tube. Also called plate efficiency.

anode follower   A tube circuit with heavy feedback from anode to grid, such that the output voltage is nearly equal and opposite to the input voltage. The input impedance is then very high.

anode glow   A narrow bright zone on the anode side of the positive column in a gas tube.

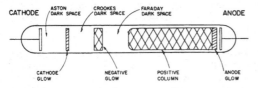

Anode glow in gas discharge tube.

anode input power   The product of the direct anode voltage applied to the tubes in the last radio stage of a transmitter and the total direct current flowing to the anodes of these tubes, measured without modulation. Also called plate input power.

anode load impedance   The total impedance between the anode and cathode of an electron tube, exclusive of the electron stream. Also called plate load impedance.

anode modulation   Modulation produced by introducing the modulating signal into the anode circuit of any tube in which the carrier is present. Also called plate modulation.

anode neutralization   Neutralization in which a portion of the anode-cathode AC voltage is shifted 180° and applied to the grid-cathode circuit through a neutralizing capacitor. Also called plate neutralization.

anode power input   The DC power delivered to the anode of an electron tube by the power supply. It is the product of average anode voltage and average anode current. Also called plate power input.

anode pulsing   An RF oscillator circuit whose anode voltage is normally reduced to such a low value that no anode current flows and no oscillations occur. A pulse equal to the full anode voltage is then introduced in series with the anode. Oscillations begin and last for the duration of the pulse. This circuit requires a modulator capable of supplying full anode power.

anode rays   Positive ions that come from the anode of an electron tube. They are generally due to impurities in the metal of the anode.

anode region   The positive column, anode glow, and anode dark space in a gas tube.

anode resistance   The resistance value obtained when a small change in the anode voltage of an electron tube is divided by the resulting small change in anode current.

anode saturation   The condition in which the anode current of an electron tube cannot be further increased by increasing the anode voltage. The electrons are then being drawn to the anode at the same rate as they are emitted from the cathode. Also called current saturation, plate saturation, saturation, and voltage saturation.

anode sheath   A layer of electrons that surrounds the anode of a gas tube when the anode current is high.

anode shield   A shield that partially surrounds the anode in a mercury-arc rectifier. It protects the anode from excessive ionization or radiation.

anode sputtering   The emission of fine particles from the anode of an electron tube as a result of electron bombardment.

anode strap   *Strapping.*

anode supply   The direct voltage source used in an electron-tube circuit to place the anode at a high positive potential with respect to the cathode. Also called plate supply.

anode terminal   The semiconductor-diode terminal that is positive with respect to the other terminal when the diode is biased in the forward direction.

anode voltage   The direct voltage that exists between anode and cathode of an electron tube. Also called plate voltage and high tension (British).

anode voltage drop   The voltage that exists between anode and cathode in a cold-cathode gas tube after conduction has been established in the main gap.

anodizing   An electrolytic process for producing a protective or decorative film on certain metals, chiefly aluminum and magnesium.

anomalous propagation   Freak propagation of VHF radio waves beyond the horizon, apparently caused by temperature inversion in the lower atmosphere.

A-N radio range   A radio range that provides four radial lines of position for aircraft guidance, each line identified aurally as a continuous tone resulting from the interlocking of equal-amplitude A and N international Morse code letters. The sense of deviation from one of these lines is indicated by deterioration of the steady tone into audible A or N code signals. The two types of A-N radio ranges are the Adcock and the loop.

ANSI   Abbreviation for *American National Standards Institute.*

antenna   A device that radiates or receives radio waves. British term is aerial.

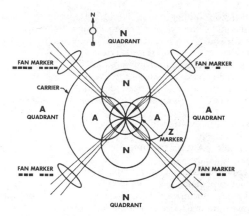

A-N radio range courses, radiation patterns, and fan marker codes.

antenna array An arrangement of two to several thousand individual radiating elements, appropriately spaced and energized to give desired directional characteristics. Also called array and beam antenna.

antenna beamwidth The angle in degrees between two opposite half-power points of an antenna beam.

antenna coil The first coil in a receiver, through which antenna current flows. When this coil is inductively coupled to a secondary coil, the combination of two coils becomes an antenna transformer or RF transformer.

antenna counterpoise *Counterpoise.*

antenna coupler An RF transformer, tuned line, or other circuit that transfers energy efficiently from a transmitter to a transmission line or from a transmission line to a receiver.

antenna cross section A microwave receiving-antenna rating, expressed as the area, perpendicular to incident radiation, that intercepts an amount of energy equal to that delivered to a receiver by the antenna.

antenna crosstalk A measure of undesired power transfer through space from one antenna to another. It is the ratio of the power received by one antenna to the power transmitted by the other, usually expressed in decibels.

antenna current The RF current that flows in a transmitting antenna. It is generally measured when there is no modulation.

antenna drive A motor or other device that rotates or positions an antenna, such as for tracking a radar target.

antenna duplexer A circuit that permits two transmitters to transmit simultaneously from the same antenna without interaction.

antenna feed The device that supplies energy to a transmitting antenna. It can be a transmission line for direct feed, or a dipole or horn for indirect feed to a reflector.

antenna field gain A Federal Communications Commission rating for a transmitting antenna. It is the effective free-space field intensity in millivolts per meter that is produced in the horizontal plane at a distance of 1 mi (1.609 km) by an antenna input power of 1 kW, divided by 137.6 mV/m (the value for a half-wave dipole).

antenna gain The effectiveness of a directional antenna as compared to a standard nondirectional antenna. It is usually expressed as the ratio in decibels of standard antenna input power to directional antenna input power that will produce the same field strength in the desired direction. For a receiving antenna, the ratio of signal-power values produced at the receiver input terminals is used. The more directional an antenna, the higher its gain.

antenna height above average terrain A Federal Communications Commission rating for transmitting antennas. It is the average of the antenna heights above the terrain from 2 to 10 mi (3.2 to 16 km) from the antenna for the eight directions, spaced evenly for each 45° of azimuth, starting with true north. The averages for each direction are averaged to get the final value.

antenna lens An arrangement of shaped metal vanes or dielectric material in front of a microwave antenna to concentrate the beam of transmitted or received radio waves. See also *Luneberg lens.*

antenna loading The use of lumped reactances to tune an antenna.

antenna power A transmitter rating equal to the square of the antenna current multiplied by the antenna resistance at the point where the current is measured.

antenna power gain A transmitting antenna rating equal to the square of the antenna gain, expressed in decibels.

antenna resistance A transmitting antenna rating that expresses the total resistance of the antenna system at the operating frequency. The antenna resistance in ohms is equal to the power in watts supplied to the entire antenna circuit divided by the square of the effective antenna current in amperes measured at the point where power is supplied to the antenna. Factors that affect antenna resistance include radiation resistance, ground resistance, RF resistance of conductors in the antenna circuit, and the equivalent resistance due to corona, eddy currents, insulator leakage, and dielectric power loss.

antenna resonant frequency The frequency or frequencies at which an antenna appears to be a pure resistance.

antenna series capacitor A capacitor used in series with an antenna to shorten the electrical length of the antenna.

antialiasing filter A low-pass filter that removes high frequencies from an input signal *before* sampling to prevent aliasing noise at lower frequencies.

anticapacitance switch A switch that has low capacitance between its terminals when open.

anticlutter circuit A radar circuit that attenuates undesired reflections, to permit detection of targets otherwise obscured by such reflections.

anticlutter gain control *Sensitivity-time control.*

anticoincidence Occurrence of a count in a specified detector unaccompanied, simultaneously or within an assignable time interval, by a count in one or more other specified detectors.

anticoincidence circuit A circuit that produces a specified output pulse when one of two inputs receives a pulse and the other receives no pulse within an assigned time interval.

anticoincidence counting The recording, from one or more counters of a coincidence circuit, of all counts except coincidences.

antielectron *Positron.*

antifading antenna An antenna that confines radiation mainly to small angles of elevation, to minimize radiation of the sky waves which cause fading.

antiferroelectricity A dielectric phenomenon in which neighboring lines of spontaneously polarized ions are aligned in antiparallel directions.

antiferromagnetic material A material in which spontaneous magnetic polarization occurs in equivalent sublattices. The polarization in one sublattice is aligned antiparallel to the other.

antiferromagnetism A form of magnetism, occurring primarily at low temperatures, in which interaction between elementary atomic magnets causes adjacent magnets to try to have their magnetic directions oppose each other.

antihunt circuit A stabilizing circuit in a closed-loop feedback system that prevents self-oscillations.

antihunt transformer A transformer used as a stabilizing network in a DC feedback system. Its primary winding is in series with the load. Its secondary winding provides a voltage, proportional to the derivative of the primary current, that is fed back into some other part of the loop to prevent self-oscillation.

antijamming [AJ] Active and passive techniques taken to overcome or minimize deliberate jamming or interference with radio or radar bands by the enemy. See *electronic countermeasures* [ECM].

antilog Abbreviation for *antilogarithm.*

antilock brake system [ABS] A vehicular braking system that prevents skidding on slippery road surfaces by automatically pumping the brakes on two or four wheels at high speed to prevent the brake lock-up that can be expected to occur in a car without ABS if the driver applies brakes prior to or during a skid. The ABS computer accepts speed signals from the wheels and compares them to detect skidding conditions. When detected, signals are sent to pump the brakes as many as five times per second to stabilize braking action until the vehicle slows to ABS control-release speed.

antilogarithm [antilog] The number corresponding to a given logarithm. Example: If the logarithm of 563.2 is 2.75066, then 563.2 is the antilogarithm of 2.75066.

antimony [Sb] A metallic element with an atomic number of 51 and an atomic weight of 121.75. It is used as a *donor* dopant of silicon to form an *N-type* material. It is also used to form indium antimonide, an important semiconductor for magnetoresistors and infrared detectors.

antinode A point, line, or surface in a standing-wave system at which some characteristic of the wave has maximum amplitude. Also called loop.

antinoise microphone A microphone that has characteristics which discriminate against undesired noise, such as a close-talking microphone, lip microphone, or throat microphone.

antiradar coating A coating applied to an aircraft to reduce reflection of radio waves and thereby minimize detection by radar.

antireflection coating A thin film applied to an optical surface to reduce reflectance and increase transmittance.

antiresonance *Parallel resonance.*

antiresonant circuit *Parallel resonant circuit.*

antisidetone circuit A circuit that has a balancing network for reducing sidetones in a telephone set.

antistatic coating A metallic or other conductive coating applied to nonconductive materials to reduce or prevent buildup of static electricity.

antistickoff voltage A small voltage applied to the rotor winding of the coarse synchro control transformer in a two-speed control system to eliminate the possibility of ambiguous behavior. It is also called an *antistat.*

antivoice-operated transmission [ANTIVOX] The use of a voice-actuated circuit to prevent the operation of a transmitter when an associated receiver is in use.

ANTIVOX Abbreviation for *antivoice-operated transmission.*

APC Abbreviation for *automatic phase control.*

APD See *avalanche photodiode.*

aperiodic A term meaning not responsive to any particular frequency.

aperiodic antenna An antenna that has essentially constant impedance over a wide range of frequencies. Examples are terminated rhombic antennas and terminated wave antennas. Also called untuned antenna.

aperiodic damping Damping so great that a disturbed system or instrument comes to a position of rest without passing through this position. The point of change between aperiodic and periodic damping is called critical damping.

aperture An opening through which electrons, light, radio waves, or other radiation can pass. The aperture in the electron gun of a cathode-ray tube determines the size of the electron beam. The aperture in a television camera is the effective diameter of the lens that controls the amount of light entering the camera tube. The dimensions of the horn mouth or parabolic reflector determine the aperture of a microwave antenna.

aperture antenna An antenna whose beamwidth is determined by the dimensions of a horn, lens, or reflector.

aperture distortion Attenuation of the high-frequency components of a television picture signal caused by the finite cross-sectional area of the scanning beam in the camera. The beam then covers several mosaic depositions in the camera simultaneously, causing loss of picture detail.

aperture grille *Slot mask.*

aperture illumination The strength distribution of an electromagnetic wave in an aperture.

aperture mask *Shadow mask.*

aperture time The time required for a sampling circuit to change to its hold mode after receipt of a control command. Aperture time should generally be less than 50 ns, to prevent error in hold level if the signal changes during this time.

apex step A flexible step or corrugation at the center of a loudspeaker cone to modify high-frequency response.

aphelion The point most distant from the sun on the orbit of a planet or spacecraft.

API Abbreviation for *air-position indicator.*

apochromatic lens A lens that has been corrected for chromatic aberration for three colors.

apogee 1. The point in an elliptical orbit of an earth satellite farthest from the earth. 2. The point in the trajectory of a ballistic missile farthest from the earth.

apostilb [asb] A unit of luminance equal to 1×10^{-4} lambert. The SI unit of luminance, the candela per square meter, is preferred.

apparent horizon The visible line of demarcation between land or sea and sky.

apparent power The power value obtained in an AC circuit by multiplying the effective values of voltage and current. The result is expressed in voltamperes and must be multiplied by the power factor to secure the average or true power in watts.

apparent precession The relative angular movement of the spinning axis of a gyroscope in relation to a line on the earth, resulting from the rotation of the earth.

Applegate diagram A diagram that illustrates the behavior of the electrons in a velocity-modulation tube such as a reflex klystron. The distances of electrons from the buncher in the drift space are plotted as vertical coordinates against time on the horizontal axis. Close spacing of these vertical lines indicates bunching of electrons.

Appleton layer *F layer.*

application-specific integrated circuit [ASIC] A class of dedicated semicustom ICs that includes gate arrays, standard cells, and programmable logic devices (PLDs) for specific applications, in contrast with standard ICs that have more universal applications.

APT 1. [Automatically Programmed Tools] A computer language developed primarily for programming numerically controlled machine tools. It is written in convenient Englishlike computer language, using such practical instructions as line, sphere, and tangent. 2. Abbreviation for *automatic picture transmission.*

APT satellite A low earth-orbiting [LEO] satellite with automatic picture-taking capabilities.

AQL Abbreviation for *acceptable quality level.*

A quadrant One of the two quadrants in which the A signal of an A-N radio range is heard.

ARAM An acronym for an *audio dynamic random-access memory.* (It is also called an *audio DRAM*).

arc *Electric arc.*

arcback The flow of a principal electron stream in the reverse direction in a mercury-vapor rectifier tube because of formation of a cathode spot on an anode. This results in failure of the rectifying action. Also called backfire. The action corresponds to reverse emission in electron tubes.

arc cathode A cathode in which electron emission is self-sustaining at a low voltage drop, approximately equal to the ionization potential of the gas.

arc-discharge tube A discharge tube in which a high-current arc discharge passes through the gas between the electrodes, generally for the purpose of producing an intense flash of light.

arc drop The voltage drop between the anode and cathode of a gas rectifier tube during conduction.

arc-drop loss The product of the instantaneous values of tube voltage drop and current averaged over a complete cycle of operation of a gas tube.

Archie A network communications tool for locating files on the *Internet* available on *file transfer protocol* [FTP] serving computers.

arcing The production of an arc, as at the brushes of a motor or the contacts of a switch.

arc spectrum The spectrum of light produced by vaporizing an element in an electric arc.

arc spraying Spraying with metal that has been melted by an electric arc.

arcthrough In multielectrode gas tubes, the loss of control that results from the flow of a principal electron stream in the normal direction during a scheduled nonconducting period.

arc welding A fusion welding process in which welding heat is obtained from an electric arc struck between an electrode and the metal being welded or between two separate electrodes, as in atomic hydrogen welding.

area control radar A radar set or system for air traffic control over a relatively large area, to provide a smooth flow of air traffic to the approach control radar.

areal density On a computer memory disk, the density (bits per inch) multiplied by track density (tracks per inch), or bits per square inch of disk surface.

argon [A or Ar] The most abundant inert (rare) gas element that has an atomic number of 18 and an atomic weight of 39.948. It makes up 0.9% of the air by volume. It is used as a fill in ionized gas signs, in lasers capable of penetrating seawater, and as a fill with nitrogen in electric lightbulbs, where it retards the evaporation of the lamp filament.

argon glow lamp A lamp containing argon that glows with a pale blue-violet light. Typical power ratings are from ¼ to 2 W.

argon laser A gas laser that contains ionized argon to produce strong radiation at 0.488 μm and infrared radiation.

arithmetic circuit A computer circuit that performs an arithmetic operation.

arithmetic-logic unit [ALU] A section of a microprocessor or microcontroller that performs mathematical operations such as addition, subtraction, multiplication, division, and logic on numbers (usually binary) presented to its inputs. It provides an output that is an appropriate function of its inputs.

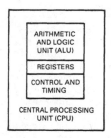

Arithmetic-logic unit (ALU) is that part of a computer's central processing unit (CPU) that performs all arithmetic calculations.

arithmetic mean *Mean.*

arithmetic operation A digital computer operation in which numerical quantities are added, subtracted, multiplied, divided, or compared.

arithmetic shift A computer operation in which a quantity is multiplied or divided by a power of the number base. Thus, binary 1011 represents decimal 11, and therefore two arithmetic shifts to the left is binary 101100, which represents decimal 44.

ARL Abbreviation for *acceptable reliability level.*

arm 1. An interconnected set of links and powered joints that supports and moves a mechanical wrist, hand, or end-effector tool in the performance of some task. It can be manually controlled, as in a *telercheric,* or under programmable control, as in a robot. It is also called a *manipulator.* 2. Branch as in a bridge circuit.

armature The rotating section of a generator or motor, the movable iron part of a relay, or the spring-mounted iron part of a vibrator or buzzer.

armature chatter Undesired vibration of the armature of a relay in an AC circuit.

armature contact *Movable contact.*

armature reaction Interaction between the magnetic flux produced by armature current and that of the main magnetic field in an electric motor or generator. The resulting distortion of the main magnetic field affects the speed of motors and the voltage regulation of generators.

armature relay *Electromechanical relay.*

Armstrong oscillator A tuned-grid, tuned-anode, vacuum-tube oscillator circuit developed by E. H. Armstrong. A parallel resonant circuit provides the inductive anode load when tuned slightly above the resonant frequency of the grid tank circuit or crystal. The interelectrode capacitance of the oscillator tube serves as the feedback path.

ARPA Abbreviation for 1. *Advanced Research Project Agency* and 2. *automatic radar plot aid.*

ARQ Abbreviation for *automatic repeat request.*

array *Antenna array.*

arrester *Lightning arrester.*

ARRL American Radio Relay League, an organization of licensed amateur radio operators.

arsenic [As] An element with an atomic number of 33 and an atomic weight of 74.92. It is a common *donor* doping element for silicon semiconductor devices. When alloyed with gallium, it forms the binary crystal *gallium arsenide,* and with gallium and aluminum it forms the trinary crystal *gallium aluminum arsenide.*

ARSR Abbreviation for *air route surveillance radar.*

artificial antenna *Dummy antenna.*

artificial delay line *Delay line.*

artificial dielectric A three-dimensional arrangement of metallic conductors, which are usually small compared to a wavelength. The resulting medium acts as a dielectric for electromagnetic waves.

artificial ear A device that presents an acoustic impedance to an earphone equivalent to the impedance presented by the average human ear. It is equipped with a microphone for measuring the sound pressures developed by the earphone.

artificial echo A radar echo signal generated by artificial means for test purposes, as by an *echo box.*

artificial intelligence [AI] The capability of a computer or other circuit to perform functions normally associated with human intelligence, such as reasoning, learning, and self-improvement.

artificial ionization The introduction of an artificial reflecting or scattering layer into the upper atmosphere to improve beyond-the-horizon radio communication.

artificial line A network that simulates the electrical characteristic of a transmission line.

artificial load A dissipative but essentially nonradiating device that has the impedance characteristics of an antenna, transmission line, or other practical load.

artificial mouth A small loudspeaker that simulates voice sounds coming from the center of the lips of a person, for such applications as testing microphones and telephone receivers.

artificial neural network [ANN] An integrated-circuit microprocessor with an architecture that attempts to model the actions and capabilities of the neurons of the human brain electronically. It can perform pattern recognition, provide associative memory, perform calculations, and control machines such as robots or industrial processes. It must be programmed with software to carry out its functions.

artificial neuron [AN] An electronic device that receives inputs on one or more lines and converts them to network outputs in what is called a *transfer function.* Groups of them can be connected in different patterns to form a network. See also *artificial neural network.*

artwork Accurately scaled and precisely outlined drawing of a pattern that is used to produce the master masks for printed circuit boards or integrated circuits.

ASA Abbreviation for *American Standards Association.*

ASCII Abbreviation for *American Standard Code for Information Interchange.*

A scope A radarscope that produces an A display.

ASDE Abbreviation for *airport surface detection equipment.*

asdic [Anti-Submarine Detection Investigation Committee] British term for *sonar and underwater listening devices.*

ASDL Abbreviation for *asymmetric digital subscriber link.*

ASIC Abbreviation for *application-specific integrated circuit.*

A signal A dot-dash signal heard in either a bisignal zone or an A quadrant of a radio range.

ASJP Abbreviation for *Advanced Self-Protection Jammer.*

ASK Abbreviation for *amplitude-shift keying.*

aspect ratio The ratio of frame width to frame height of a face of a cathode-ray tube for television or a computer monitor. An aspect ratio of 4:3 is typical for a television picture tube or computer monitor.

ASR Abbreviation for *airport surveillance radar.*

ASRM Abbreviation for *asynchronous send/receive monitor.*

ASROC Abbreviation for *antisubmarine rocket.*

assemble To prepare a machine-language program from a symbolic-language program by substituting absolute operation codes and addresses for symbolic operation codes and addresses.

assembly Parts or subassemblies joined together to perform a specific function.

assembly language A computer language that has a one-to-one relationship between machine instructions and the program supplied by the user.

assembly program *Assembler.*

assembly routine *Assembler.*

assigned frequency The center of the frequency band assigned to a radio station.

assigned frequency band The frequency band whose center coincides with the frequency assigned to a radio station. The width of the band is the necessary bandwidth plus twice the absolute value of the frequency tolerance.

AST Abbreviation for *asynchronous transfer mode.*

astable circuit A circuit that alternates automatically and continuously between two unstable states at a frequency dependent on circuit constants. It can be readily synchronized at the frequency of any repetitive input signal. Blocking oscillators and certain multivibrators are examples.

astable multivibrator A multivibrator in which each tube or transistor alternately conducts and is cut off for intervals of time determined by circuit constants, without use of external triggers. Also called free-running multivibrator.

astatic Without orientation or directional characteristics; having no tendency to change position.

astatic galvanometer A sensitive galvanometer that consists of two small magnetized needles mounted parallel to each other inside the galvanometer coil. The needles reduce errors caused by the earth's magnetic field.

astatic microphone *Omnidirectional microphone.*

astatic wattmeter An electrodynamic wattmeter that is insensitive to uniform external magnetic fields.

A station The loran station whose signal is always transmitted more than half a repetition period before the signal from the slave or B station of the pair. Also called master station.

astigmatism 1. A type of spherical aberration in which light rays from a single point of an object do not converge at the corresponding point in the image. 2. An electron-beam tube defect in which electrons in the beam come to a focus in different axial planes as the beam is deflected, so the spot on the screen is distorted in shape and the image is blurred.

ASTM Abbreviation for the *American Society for Testing and Materials.*

Aston dark space The dark space next to the cathode of a glow discharge tube, in which the emitted electrons do not have enough velocity to excite the gas.

Aston mass spectrograph A spectrograph that depends on successive electric and magnetic fields to focus rays of a constant charge-mass ratio at a focal line.

ASTOR Abbreviation for *antisubmarine torpedo.*

astrionics The science of electronics as applied to space travel.

astrocompass A compass that gives directions with respect to certain stars.

astrodynamics Dynamics as applied to the motions of bodies in space, including artificial satellites and deep-space probes, and the forces acting on these moving bodies.

astronomical unit The average distance from the earth to the sun—approximately 93 million miles or 149 million kilometers.

astrophysics Physics as applied to stellar astronomy.

ASW Abbreviation for *antisubmarine warfare.*

asymmetrical Not symmetrical.

asymmetrical distortion Distortion that affects two-condition modulation in which the intervals corresponding to one of the two significant conditions have longer or shorter durations than the original signal.

asymmetrical-sideband transmission *Vestigial-sideband transmission.*

asymmetric digital subscriber link [ASDL] A communications link that permits downloading computer data at 1.5 Mb/s, but uploads data at a slower 16 kb/s to 500 kb/s.

asynchronous *Nonsynchronous.*

asynchronous control A control method in which the time allotted for performing an operation depends on the time

required for the operation rather than on a predetermined fraction of a fixed machine cycle.

asynchronous data transmission A form of data transmission in which each character contains its own start and stop pulses, and there is no control over the time between characters.

asynchronous device A device whose speed of operation is not related to any frequency in the host system.

asynchronous logic Logic in which computer operations occur independently of time, in a sequential manner.

asynchronous machine An AC machine whose speed is not proportional to the frequency of the power line.

asynchronous multiplex A multiplex transmission system in which two or more transmitters occupy a common channel without provision for preventing simultaneous operation.

asynchronous operation Operation of a switching network by a free-running signal that triggers successive instruction. The completion of one instruction triggers the next one.

asynchronous transfer mode [ATM] In telecommunications, a packet multiplexing scheme that allows services operating at different bit rates to be carried efficiently on an optical fiber.

ATC Abbreviation for *air traffic control.*

ATCRBS Abbreviation for *air traffic control radar beacon system.*

AT-cut crystal A quartz crystal slab cut at a 35° angle to the Z axis of the mother crystal. Temperature variations have little effect on its natural vibration frequency.

ATE Abbreviation for *automatic test equipment.*

ATG Abbreviation for *automatic test generation.*

athermanous Opaque to infrared radiation.

Atlantic Missile Range A 6000-mi (9600-km) range for testing missiles and space vehicles. Instrument stations are at Cape Canaveral, Jupiter Inlet (Florida), Grand Bahama, Eleuthera, San Salvador, Mayaguana, Grand Turk, Dominican Republic, Mayaguez, Antigua, St. Lucia, Fernando de Noronha, and Ascension.

Atlas Centaur A U.S. ballistic missile designed to launch nuclear warheads that has been converted for launching low earth-orbiting (LEO) and geostationary satellites.

atm Abbreviation for *atmosphere.*

ATM 1. Abbreviation for *asynchronous transfer mode.* 2. *Automated teller machine.*

atmosphere [atm] The mixture of gases, chiefly oxygen and nitrogen, that surrounds the earth for a distance of about 100 mi (160 km) from the surface. Atmospheric pressure at sea level is about 14.7 lb/in^2 (a pressure of 760 mmHg at 0°C); this value is used as a unit of pressure called 1 atmosphere.

atmospheric absorption The attenuating effect of gases and moisture in the earth's atmosphere on the propagation of microwaves.

atmospheric duct An atmospheric layer that conducts radio waves in the same manner as a waveguide under certain conditions of temperature and humidity, to give signal transmission far outside the usual reception area.

atmospheric infrared window Those parts of the *infrared spectrum* that are relatively transparent, permitting the transmission of infrared energy beyond the earth's atmosphere or for use in terrestrial thermal imaging. One window occurs in the 3- to 5-μm band for thermal imaging at ranges up to about 900 m, and a second exists between 8 and 12 μm for ranges beyond 900 m.

atmospheric interference Static caused by natural electric disturbances such as lightning and northern lights. Heard as crackling and hissing noises in radios. Also called atmospherics and sferics.

atmospheric noise Noise heard during radio reception because of atmosphere interference.

atmospheric radio wave A radio wave that reaches its destination after reflection from the upper ionized layers of the atmosphere.

atmospheric radio window The band of the electromagnetic spectrum in which radio waves pass readily through the atmosphere of the earth. The major window extends from the longest radio waves (approaching zero frequency) up to about 19 GHz. Small windows for millimeter waves occur at 35 and 94 GHz.

atmospheric refraction The bending of the path of electromagnetic radiation from a distant point as the radiation passes obliquely through varying air densities.

atmospherics *Atmospheric interference.*

atom The smallest particle into which an element can be divided and still retain the chemical properties of that element.

atomic-beam frequency standard A frequency standard that provides one or more precise frequencies derived from an element such as cesium. Frequency is determined by measuring the nuclear magnetic moment of the element using resonance absorption techniques. When the element is acted on by an applied frequency at its atomic spectrum line frequency, the element absorbs a quantum of energy, indicating coincidence between the two frequencies. The applied frequency thus becomes a standardized frequency and can be used at its fundamental or harmonics for calibration of other instruments.

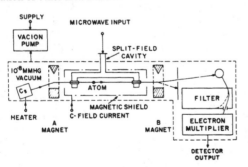

Atomic-beam frequency standard.

atomic charge The electric charge of an ion equal to the number of electrons the atom has gained or lost in its ionization multiplied by the charge on one electron.

atomic clock A highly accurate reference for frequency or time that depends on the unchanging nuclear resonance of atoms of elements such as cesium when subjected to an RF electromagnetic field. Accuracy of 1 part in 10 million is achievable.

atomic frequency The natural vibration frequency of an atom.

atomic frequency standard A frequency standard that consists of a quantum-mechanical resonator which is frequency-locked to an electronic frequency converter if the resonator is passive or phase-locked. The three types are atomic-beam, molecular-beam, and gas-cell.

atomic mass *Atomic weight.*

atomic nucleus *Nucleus.*

atomic number [Z] The number of elementary positive charges (protons) in the nucleus of an atom. It is a different number for each element, starting with 1 for hydrogen and going beyond 103. For a neutral atom, the atomic number is also the number of electrons outside the nucleus of the atom.

atomic photoelectric effect *Photoionization.*

atomic time Time based on atomic resonance.

atomic weight The relative weight of a neutral atom of an element, based on an atomic weight of 16 for the neutral oxygen atom. On this basis, hydrogen has an atomic weight of 1.0078. It is also called *atomic mass.*

ATR tube [Anti-Transmit-Receive tube] A gas-filled RF switching tube used in radar transmitters to isolate the transmitter from the antenna during the time needed for pulse reception.

attenuation A reduction in energy. Attenuation occurs naturally during wave travel through wires, cables, optical fibers, waveguides, spaces, or a medium such as water. It can be produced intentionally by inserting an attenuator in a circuit or placing an absorbing device in the path of the wave travel. The amount of attenuation is generally expressed in decibels or decibels per unit length.

attenuation coefficient The fraction of a beam of light that is removed by scatter or absorption in passing through unit thickness of a material.

attenuation constant [α] A rating for a line or medium through which a plane wave of a given frequency is being transmitted. It is the real part of the propagation constant and is the relative rate of decrease of amplitude of a field component (or of voltage or current) in the direction of propagation, in nepers per unit length. The imaginary part of the propagation constant is the phase constant.

attenuation distortion A deviation from uniform amplification or attenuation over a required frequency range.

attenuation equalizer An equalizer that makes the total transmission loss of a line or circuit essentially the same for all frequencies in the range being transmitted.

attenuation factor 1. The ratio of incident intensity to transmitted intensity for radiation passing through a layer of material. 2. The ratio of input current to output current for a transmission line or network.

attenuation ratio The magnitude of the propagation ratio.

attenuator A device that reduces the strength of an electromagnetic or audio signal without introducing appreciable distortion. It is designed so that its impedance matches that of the host circuit regardless of the amount of attenuation. For examples see *flap attenuator, rotary-vane attenuator,* and *pad.*

attenuator tube A gas-filled RF switching tube in which a gas discharge, initiated and regulated independently of RF power, controls this power by reflection or absorption.

attitude The position of an aircraft as determined by the inclination of its axes to some frame of reference, such as the earth.

attitude control A control system or mechanism, such as an automatic pilot, that puts or keeps an aircraft or missile in a desired attitude.

attitude gyro A gyro for the attitude control of aircraft.

attitude indicator An indicator that shows the roll and pitch angles of an aircraft in relation to the earth.

attitude sensor A sensor that provides a reference signal for keeping a spacecraft in a desired attitude with respect to a given direction, such as toward the sun or earth.

atto- [abbreviated a] A prefix representing 10^{-18}, which is 0.000 000 000 000 000 001, or one-millionth of a millionth of a millionth.

audibility A measure of the strength of a specified sound as compared with the strength of a sound that can just be heard, expressed in decibels.

audibility limit A threshold of hearing. The lower limit is the minimum effective sound pressure that can be heard at a specified frequency. The upper limit is the minimum effective sound pressure that causes pain in the ear.

audibility threshold The lower limit of audibility. For a specified signal it is the minimum effective sound pressure capable of producing an auditory sensation in a specified fraction of the trials. The characteristics of the signal, the manner in which it is presented to the listener, the point at which the sound pressure is measured, and the ambient noise all affect the value. Usually expressed in decibels above 0.0002 μbar or above 1 μbar.

audible Doppler enhancer A Doppler circuit that serves as a third detector for the received pulses. It separates the envelope from the pulses, thus giving an audio signal that can be amplified and fed to a loudspeaker.

audio [Latin for "I hear"] 1. A reference to signals, equipment, or phenomena that involve frequencies in the range of human hearing. 2. Slang for *sound*.

audio amplifier *Audio-frequency amplifier.*

audio frequency [AF] A frequency that can be detected as a sound by the human ear. The range of audio frequencies extends approximately from 15 to 20,000 Hz.

audio-frequency amplifier An amplifier that has one or more amplifier stages for amplifying an AF signal. In a superheterodyne receiver it follows the second detector and amplifies the AF signal after demodulation. It can separately amplify the AF output of a microphone, phonograph, tape recorder, or other AF signal source. It is also called an *audio amplifier*.

audio-frequency harmonic distortion Distortion in which integral multiples of a single AF input signal are generated by the amplifier.

audio-frequency-shift keying [AFSK] Radioteletype keying in which the RF carrier is transmitted continuously and pulses are transmitted by frequency-shifted tone modulation. Commonly used audio tones are 2.125 kHz for mark and 2.975 kHz for space.

audio-frequency-shift modulation A facsimile system whose picture tones are represented by audio frequencies. In one example a 1.5-kHz tone represents black, a 2.3-kHz tone represents white, and frequencies in between represent shades of gray.

audio-frequency signal generator A signal generator that can be set to generate a sinusoidal AF signal voltage at any desired frequency in the audio spectrum. Also called audio signal generator.

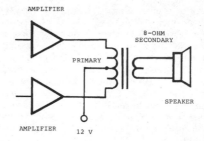

Audio-frequency transformer. A schematic of a center-tapped version.

audio-frequency transformer An iron-core transformer used for coupling between AF circuits. Also called audio transformer.

audiogram A graph that shows hearing loss, percent hearing loss, or percent hearing as a function of frequency.

audio jack One of a class of female connectors adapted for use with audio systems. See *DIN jack, phone jack, phono (RCA) jack,* and *XLR jack.*

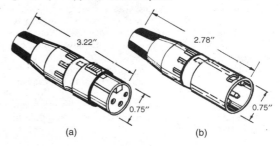

Audio XLR jack (*a*) and plug (*b*) are for microphones and public address systems.

audio-level meter An instrument that measures AF power with reference to a predetermined level. Its scale is usually calibrated in decibels.

audiology 1. The science of hearing. 2. The branch of medicine dealing with causes and treatment of defective hearing.

audio masking *Masking.*

audiometer An instrument that measures hearing ability. One form consists of an audio oscillator that has variable calibrated output and is capable of generating a wide range of audio tone frequencies. Recorded speech sounds can also be sound sources.

audiometry The study of hearing ability determined by audiometers.

audion The original three-element vacuum tube invented by Dr. Lee de Forest.

audio oscillator *Audio-frequency oscillator.*

audio peak limiter *Audio-frequency peak limiter.*

audio plug One of a class of male connectors adapted for use with audio systems. See *DIN plug, phone plug, phono (RCA) plug,* and *XLR plug.*

audio plug adapter One of a class of audio connector components that adapts stereo phone plugs to monaural jacks or stereo or monaural plugs to phone jacks with the same-size or smaller diameters. An XLR-plug-to-phone-jack transformer is an example.

audio rectification   Interference to an electronic circuit or device caused by a strong nearby radio-frequency field that is rectified and amplified by the device or circuit itself.

audio response unit   A magnetic or digital recording system that provides voice response to an inquiry made from a terminal connected to a computer by a data-transmission line. The appropriate audio response is selected by the computer from spoken words previously recorded on a magnetic disk or other storage device. Applications include automatic stock-price-quotation service from any telephone when the query code for a particular stock is punched or dialed.

audio signal   An electric signal that has an audio frequency.

audio signal generator   *Audio-frequency signal generator.*

audio spectrum   The continuous range of frequencies extending from the lowest to the highest audio frequency (from about 15 to 20,000 Hz).

audio subcarrier   A subcarrier whose frequency lies within the audio range.

audio taper   A nonlinearity in the resistance value of volume and tone controls. The resistance increases slowly at the beginning of shaft rotation and increases at a faster rate as the shaft or knob is rotated toward the limit of its clockwise rotation. It compensates for the lower frequency range of the human ear at low volume levels.

audio transformer   *Audio-frequency transformer.*

audiovisual   A reference to the combination of both sight and sound.

auditory perspective   Three-dimensional realism of sound, as produced by an actual orchestra or a stereophonic sound system.

auditory sensation area   The region enclosed by curves defining the thresholds of feeling and audibility as functions of frequency.

augend   The number to which a new quantity, called the addend, is to be added.

Auger coefficient   The ratio of Auger yield to fluorescence yield, or the ratio of the number of Auger electrons to the number of X-ray photons emitted from a large number of similarly excited atoms.

Auger effect   A nonradiative transition of an atom from an excited energy state to a lower energy state, accompanied by the emission of an electron.

Auger electron   An electron ejected from an atom by a photon in the Auger effect. It has a kinetic energy equal to the difference between the energy of the X-ray photon of the corresponding radiative transition and the binding energy of the ejected electron.

Auger electron microscopy   A method for examining materials and surfaces in very fine detail with electron beams within a vacuum chamber.

Auger yield   The ratio of the number of Auger electrons emitted to the number of events that result in an electron vacancy in the inner shell of an atom.

aural center frequency   The average frequency of a carrier modulated by an AF signal or the carrier frequency without modulation.

aural harmonic   A harmonic generated in the human ear.

aural masking   *Masking.*

aural null   The condition of weakest sound when tuning or otherwise adjusting a circuit that has an audio output.

aural-null direction finder   A radio direction finder that consists of a radio receiver and rotatable loop antenna. When the loop is rotated to give an aural null, the plane of the loop is at right angles to the direction of the transmitted signal.

aural radio range   A radio range station that provides lines of position by aural identification or comparison of signals at the output of a receiver, as in the A-N radio range.

aural signal   1. A signal that can be heard. 2. The sound portion of a television signal; the picture portion is called the visual signal.

aurora   The sporadic visible emission from the upper atmosphere over middle and high latitudes.

auroral absorption   Absorption of radio waves by abnormal particle radiation from the sun.

auroral electrojet   Currents of about 1 million amperes that occur naturally above the stratosphere and follow circular paths around the earth's geomagnetic field.

auroral reflection   Reflection of radio waves back to earth by a rapidly fluctuating ionized layer in the upper atmosphere of the polar regions, usually during magnetic storms accompanied by auroral displays.

auroral storm   *Ionospheric storm.*

autoalarm   *Automatic alarm receiver.*

autocall   A signal system that generates preset combinations of gong or tone signals for paging people in buildings or other locations.

autocode   *Automatic code.*

autocoder   The computer section that converts a symbolic input code language into the machine language used in that particular computer.

autocorrelation   A mathematical technique that detects cyclic activity in a complex signal.

autocorrelation function   A mathematical quantity defined as the time average of the product of a function of time and a delayed version of that function of time.

autocorrelator   A correlator whose input signal is delayed, then multiplied by the undelayed signal. The product is then smoothed in a low-pass filter to give an approximate computation of the autocorrelation function. It can detect a weak periodic signal hidden in noise if the chosen time delay is equal to the period of the signal. The autocorrelator can also be applied to the detection of nonperiodic signals.

autodyne circuit   A circuit in which the same transistor or tube elements serve as the oscillator and detector simultaneously. The output frequency is equal to the difference between the frequencies of the received signal and the oscillator signal, as in a conventional superheterodyne circuit.

autoelectric emission   The emission of electrons from a cold cathode under the influence of an intense electric field.

automated teller machine [ATM]   A computer banking terminal that will carry out banking transactions automatically and even dispense paper currency to anyone who inserts the proper identification card, usually a magnetically coded card, in a slot. Permitted transactions include the depositing of checks and the withdrawal of cash up to a specified amount. The terminal includes a cathode-ray-tube display, keypad, and slots for incoming checks and outgoing cash. The user follows instructions which appear on the display screen after selecting the desired transaction.

automated test generation [ATG]   A computer-aided process that automatically produces fault-test sequences when given a description of a circuit to be tested.

automatic aiming   *Automatic tracking.*

automatic back bias   One or more automatic gain control loops in a radar receiver to prevent overloading by strong radar echoes or jamming signals.

automatic background control   *Automatic brightness control.*

automatic bass compensation [ABC]   A circuit used in some radio receivers and audio amplifiers to make bass notes sound more natural at low volume-control settings. The circuit usually consists of a resistor and capacitor in series, connected between ground and a tap on the volume control. This circuit automatically compensates for the poor response of the human ear to weak low-frequency sounds.

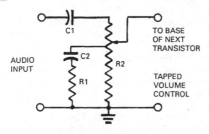

Automatic bass compensation.

automatic brightness control [ABC]   A television receiver circuit to keep the average brightness of the reproduced image essentially constant. Its action is like that of an automatic volume control circuit in a sound receiver. Also called automatic background control.

automatic chroma control   *Automatic color control.*

automatic chrominance control   *Automatic color control.*

automatic color control   A circuit in a color television receiver that keeps color intensity levels essentially constant despite variations in the strength of the received color signal. Control is usually achieved by varying the gain of the chrominance bandpass amplifier. It is also called automatic chroma control and automatic chrominance control.

automatic compact-disk changer   An electromechanical machine equipped with a mechanism that selectively moves compact disks into position, usually by rotation, for playing by a laser playback system. Commercial versions can pick out disks selected from a stack by the customer and place them in the playback position.

automatic contrast control   A circuit that varies the gain of the RF and video IF amplifiers so that the contrast of the television picture is maintained at a constant average level. Control is achieved by varying the bias on one or more variable transistors. The manual contrast control determines the average level and the automatic contrast control maintains this average, despite variations in signal strength as different stations are tuned in.

automatic control   Control in which regulating and switching operations are performed automatically in response to predetermined conditions. It is also called automatic regulation.

automatic controller   An instrument that continuously measures the value of a variable quantity or condition, then automatically acts on the controlled equipment to correct any deviation from a desired preset value.

automatic control system   A control system that has one or more automatic controllers connected in closed loops with one or more processes. Also called regulating system.

automatic degausser   An arrangement of degaussing coils mounted around a color television picture tube, combined with a special circuit that energizes these coils only while the set is warming up after being turned on. The coils demagnetize any parts of the receiver that have been affected by the earth's magnetic field or by the field of any nearby home appliance. Automatic degaussing permits a color television receiver to be moved around a home without readjusting purity controls. Also called automatic color purifier.

automatic degaussing control system   A system that automatically changes the degaussing current for a ship to compensate for changes in the ship's heading. A synchro transmitter installed in the gyrocompass equipment of the ship provides a signal that is rectified and amplified for use in controlling the degaussing current source.

automatic dialer   A circuit that enables a person to dial a telephone number automatically with up to 14 digits by pressing a single button.

automatic frequency control [AFC]   1. A circuit that maintains the frequency of an oscillator within specified limits, as in a transmitter.   2. A circuit that keeps a superheterodyne receiver tuned accurately to a given frequency by controlling its local oscillator, as in an FM receiver.   3. A circuit in a radar superheterodyne receiver that varies the local oscillator frequency to compensate for changes in the frequency of the received echo signal.   4. A circuit in television receivers that makes the frequency of a sweep oscillator correspond to the frequency of the synchronizing pulses in the received signal.

automatic gain control [AGC]   A control circuit that automatically changes the gain (amplification) of a receiver or other equipment so that the desired output signal remains essentially constant despite variations in input signal strength.

automatic light control   Automatic adjustment of illumination reaching a film, television camera, or other imaging device, as a function of scene brightness.

automatic loop radio compass   *Automatic direction finder.*

automatic message-switching center   A location at which an incoming message is automatically directed to one or more outgoing circuits according to intelligence contained in the message.

automatic modulation control   A transmitter circuit that reduces the gain for excessively strong audio input signals without affecting the strength of normal signals. This permits higher average modulation without overmodulation, equivalent to an increase in carrier-frequency power output.

automatic modulation limiting [AML]   A circuit that prevents overmodulation in some radio transmitters. This circuit reduces the gain of one or more audio amplifier stages when the voice signal becomes stronger, to keep the modulation level below 100%.

automatic noise limiter   *Noise limiter.*

automatic pattern recognition   *Pattern recognition.*

automatic peak limiter   *Limiter.*

automatic pedestal control   A process that automatically adjusts the pedestal height in a received television signal as a function of input signal strength or some other specified parameter.

automatic phase control [APC]   1. A circuit in color television receivers that reinserts a 3.58-MHz carrier signal with exactly the correct phase and frequency by synchronizing it with the transmitted color-burst signal.   2. An automatic frequency-control circuit that feeds the difference between two frequency sources to a phase detector which produces the required control signal.

automatic picture control   A multiple-contact switch in some color television receivers that disconnects one or more of the regular controls and makes connections to corresponding preset controls. Pushing one button corrects for accidental misadjustment of controls.

automatic picture transmission [APT]   A slow-scan television system in weather satellites. It is capable of transmitting conventional television pictures of clouds in the daytime and infrared pictures of clouds at night. Each image is stored for about 200 s in a vidicon while being scanned for transmission to earth.

automatic pilot   *Autopilot.*

automatic programming   A computer-aided method for translating computer programs from a form that is easily understood by humans to one that is more efficient for computer data processing.

automatic radar plot aid [ARPA]   A shipboard radar system that digitizes navigational return signals and permits mariners to designate and track the speed and course of other ships in the vicinity.

automatic radio compass   *Automatic direction finder.*

automatic radio direction finder   *Automatic direction finder.*

automatic ranging   *Autoranging.*

automatic record changer   An electromechanical machine equipped with the mechanism that selectively lifts and drops vinyl or shellac records from a stack and places them on a turntable for playing. Home versions generally drop the records only in the order in which they were stacked, but commercial versions (called *jukeboxes*) can select records from a stack according to customer preference. This is an obsolete technology generally replaced by *automatic compact-disk* [CD] *changers.*

automatic regulation   *Automatic control.*

automatic repeat request [ARQ]   One of two AMTOR communications modes. In ARQ, also called Mode A, the two stations are constantly confirming each other's transmissions. If information is lost, it is repeated until the receiving station confirms reception.

automatic request for information [ARQ]   A system employing an error-detecting code that causes any false signals to initiate a repetition of the transmission of the character incorrectly received

automatic-scanning receiver   A receiver that automatically sweeps back and forth through a preselected frequency range. It can be set to plot signal occupancy in the range or stop when a signal is found.

automatic selectivity control   A circuit that makes a receiver less selective when the received signal is strong and more selective when the signal is weak. The reduction in the emitter-collector resistance of a transistor in the presence of a strong signal damps a tuned circuit and makes the receiver less selective.

automatic sensitivity control   A circuit that maintains receiver sensitivity at a predetermined level.

automatic test equipment [ATE]   Test equipment, typically with an embedded computer, that automates the testing of integrated circuits, circuit boards, and complete products or systems yielding "pass-fail" results. It normally produces a printed readout.

automatic tracking   Tracking in which a servomechanism keeps the radar beam trained on the target. The servomechanism is actuated by circuits that respond to some characteristic of the echo signal from the target. It is also called automatic aiming and autotrack.

automatic tuning system   An electric, mechanical, or electromechanical system that tunes a radio receiver or transmitter automatically to a predetermined frequency when a button or lever is pressed, a knob is turned, or a telephone-type dial is operated.

automatic voltage regulator   *Voltage regulator.*

automatic volume compressor   *Volume compressor.*

automatic volume control [AVC]   An automatic gain control that keeps the output volume of a radio receiver essentially constant despite variations in input signal strength during fading or when tuning from station to station. A DC voltage proportional to audio output signal strength is obtained from the second detector and used to change the bias of one or more preceding RF and IF amplifier stages.

automatic volume expander [AVE]   *Volume expander.*

automation   Continuous automatic operation in which control functions are performed by mechanisms instead of people.

autonavigator   A navigation system that includes means for coupling the outputs of inertial and other navigation sensors to the control system of a vehicle.

autopilot   An arrangement of gyroscopes combined with amplifiers and servomotors to detect deviations in the flight of an aircraft and apply the required corrections directly to the controls. Also called automatic pilot.

autopilot coupler   A coupling system that links the output of the navigation system receiver to the automatic pilot in an aircraft.

autopolarity   Automatic interchanging of connections to a digital meter when polarity is wrong. A minus sign appears ahead of the value on the digital display if the reading is negative.

autoranging   Automatic switching of a multirange meter from its lowest range to the next higher range, with the switching process repeated until a range is reached for which the full-scale value is not exceeded. Automatic downranging is also provided in some multimeters, such as for choosing the resistance range that gives highest accuracy for a particular measurement. Also called automatic ranging.

autotrack   *Automatic tracking.*

autotracking   Single-control operation of two or more DC power supplies. When the master supply, which generally has the highest positive output voltage, is turned on or off or changed in voltage, the interconnected slave supplies are similarly switched or changed so that their output voltages are changed proportionally.

autotransformer   A power transformer that has one continuous tapped winding. Part of the winding serves as the primary, and all of it serves as the secondary, or its inverse.

auxiliary memory   Computer memory other than main memory. The term generally refers to a mass-storage subsystem containing disk drives and backup magnetic-tape drives, controllers, and buffer memory. It is also called *peripheral memory.*

avalanche breakdown   Nondestructive breakdown in a semiconductor diode when the electric field across the barrier region is so strong that current carriers collide with valence electrons to produce ionization and cumulative multiplication of carriers. Often confused with Zener breakdown, in which the electric field across the barrier region becomes high enough to produce field emission that suddenly increases the number of carriers in this region.

avalanche diode   A semiconductor breakdown diode, usually made of silicon, whose avalanche breakdown occurs across the entire PN junction. Voltage drop is then essentially constant and independent of current. The two most important examples are IMPATT and TRAPATT diodes. Their breakdown characteristics are considerably sharper than those of Zener diodes.

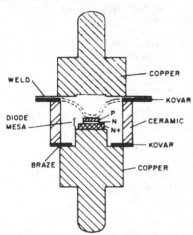

Avalanche diode with silicon mesa capable of handling peaks of up to 400 W as a pulsed microwave source.

avalanche effect   1. The cumulative process in which an electron or other charged particle accelerated by a strong electric field collides with and ionizes gas molecules, thereby releasing new electrons that in turn have more collisions. The discharge is thus self-maintained. Also called cascade, cumulative ionization, Townsend avalanche, and Townsend ionization.   2. Cumulative multiplication of carriers in a semiconductor because of avalanche breakdown.

avalanche-induced migration   A technique for forming interconnections in a field-programmable logic array by applying appropriate voltages for shorting selected base-emitter junctions.

avalanche ionization   Ionization in which a single ion creates a large number of additional ions through successive ionizing collisions in a gas.

avalanche noise   Noise produced when a PN junction diode is operated at the onset of avalanche breakdown.

avalanche oscillator   An oscillator that uses an avalanche diode as a negative resistance to achieve one-step conversion from DC to microwave outputs in the gigahertz range. The diode is mounted in a coaxial cavity or a waveguide structure.

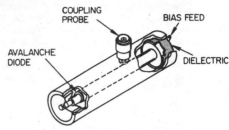

Avalanche oscillator with a coaxial cavity.

avalanche photodiode [APD]   A photodiode operated in the avalanche breakdown region to achieve internal photocurrent multiplication, thereby providing rapid light-controlled switching operation. It can have good infrared response, as required for detection of modulated light from lasers and light-emitting diodes.

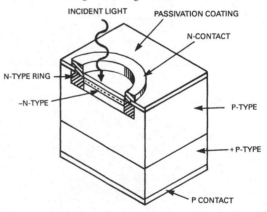

Avalanche photodiode absorbs photons in the P-type region near its PN junction and generates electrons that traverse the lighter doped P-type region below to cause avalanche electron multiplication.

AVC   Abbreviation for *automatic volume control.*

AVE   Abbreviation for *automatic volume expander.*

average access time   The average time required to make all possible length accesses (or seeks) in a memory device or system. A typical measure of performance.

average life   *Mean life.*

average noise factor   The ratio of the total delivered noise power of a linear system to the portion of the noise produced by the input termination. Also called average noise figure.

average noise figure   *Average noise factor.*

average pulse amplitude   The average of the instantaneous amplitudes taken over the duration of a pulse.

average speech power   The average of instantaneous speech power values during a given time interval.

average value   The average of many instantaneous amplitude values taken at equal intervals of time during half a cycle of alternating current. For a sine wave, the average value is 0.637 times the peak value.

averaging   A means for improving the precision of measurement of a given quantity by averaging a number of measured values. The chief drawback is the danger of smoothing out real perturbations of the variable, along with the fluctuations due to noise and other factors.

averaging multiplier   A multiplier whose output is a discrete function of the input variables sampled at specific intervals and averaged. Also called sampling multiplier.

avionics [AVIation electrONICS]   The field of airborne electronics.

AWACS   Abbreviation for *airborne warning and control system.*

AWG   Abbreviation for *American wire gage.*

axes   The directions of movements in a robot or other mechanical system, generally designated by *X, Y,* and *Z.*

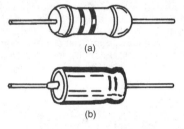

(a)

(b)

Axial-leaded component examples are (*a*) a resistor and (*b*) an electrolytic capacitor.

axial-leaded component   A component that has leads projecting axially from its body that must be bent at right angles for insertion in a conventional circuit board. Examples include discrete resistors, film capacitors, and some aluminum electrolytic capacitors. See also *radial-leaded components.*

axial ratio   A measure of polarization circularity in antennas, expressed in decibels or as the ratio of the larger of the vertical and horizontal field strengths to the smaller of the two.

Ayrton-Perry winding   The winding of two wires in parallel but opposite directions around a form or mandrel to cancel induced magnetic fields. The resistive wire in some wirewound resistors is wound in this way to reduce or cancel inductive reactance when it is in a high-frequency circuit. See also *bifilar winding.*

Ayrton shunt   A shunt that increases the range of a galvanometer without changing the damping. Also called universal shunt.

azimuth   *Bearing.*

azimuthal equidistant projection map   A map drawn showing its center at one geographic location on a continent with the other continents projected from that point. It is also called a *great-circle map,* and it is especially useful for determining where to aim a directional antenna to communicate with a specific receiver location.

azimuth blanking   1. The automatic blanking of a radar transmitter beam as the antenna scans a predetermined horizontal sector of its scanning region.   2. The automatic blanking of a radar PPI display for a selected sector of the horizontal region scanned by the antenna.

azimuth gating   Brightening of a selected sector of a radar PPI display, usually by applying a step waveform to the automatic-gain-control circuit.

azimuth marker   A radar receiver circuit that produces a bright radial line on a PPI display at an angle which can be adjusted by a control dial so the line passes through a target indication on the screen.

azimuth rate computer   A computer that calculates the rate of change of horizontal angular measurements from a base line.

azimuth resolution   The minimum azimuth angle that will permit two targets that have the same range to be distinguished by a given radar.

azimuth-stabilized PPI   A plan-position indicator that is stabilized by a gyrocompass so either true or magnetic north is at the top of the screen regardless of equipment orientation. In a north-stabilized PPI, the reference direction is magnetic north.

b Abbreviation for *bit*.

B 1. Symbol for the *base* on bipolar transistor circuit diagrams. 2. Abbreviation for *bel*.

B Symbol for *magnetic flux density*.

babble The aggregate crosstalk from a large number of channels.

backbone The principle infrastructure of the *Internet*.

back echo An echo signal produced on a radar screen by one of the minor back lobes of a radar beam.

back electromotive force *Counterelectromotive force*.

back emission *Reverse emission*.

back end In semiconductor manufacturing, the production stages that include packaging, burn-in, and environmental testing. See also *front end*.

backgating The percentage reduction in the drain current when a negative bias is applied to an ohmic contact at a specified distance from a conducting field-effect transistor (FET).

background count A count caused by ionizing radiation coming from sources other than that being measured.

background monitor An ionization chamber or other radiation counter used to measure prevailing background counts.

background noise Undesired noise heard along with desired signals or sound.

background radiation Natural radiation including that from the sun and cosmic ray elements of the earth.

background response The response caused by ionizing radiation from sources other than those to be measured by a radiation detector.

background return *Clutter*.

backheating The heating of a magnetron cathode by the high-velocity electrons that return to the cathode surface. Once the tube is in operation, backheating can be enough to keep the cathode at emitting temperature without heater current.

backlash The difference between the values obtained for a parameter when a control dial is set to an indicated value from opposite directions.

back lobe A radiation pattern lobe that is directed away from the intended direction.

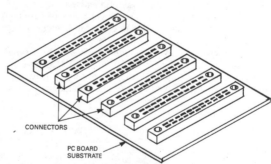

Backplane is a master circuit board with edge connectors that accept multiple plug-in "daughter" boards.

backplane In electronic packaging, a motherboard that forms a communications spine for a product or system into which complete circuit boards are plugged, generally at right angles. It can be made as a conventional printed circuit board, or it might be made from aluminum laminated on glass-fiber epoxy.

backplate The electrode to which the stored charge image of a camera tube is capacitively coupled.

back porch The portion of a composite picture signal that follows the horizontal sync pulse and extends to the trailing edge of the corresponding blanking pulse. The color burst, if present, is not considered part of the back porch.

back-porch effect The continuation of collector current in a transistor for a short time after the input signal has dropped to zero. The effect is due to storage of minority carriers in the base region. The effect also occurs in junction diodes.

back resistance　The contact resistance that opposes the inverse current of a contact rectifier.

back ripple current　*Reflected ripple current.*

backscatter　Signals that are reflected from the earth's surface after traveling through the ionosphere. The signals can be reflected back into the ionosphere along several paths and be refracted to earth again. It can provide communications into a radio station's skip zone, far beyond its normal range.

backscattering coefficient　The ratio of reflected power to incident power for a plane wave.

back-shunt keying　A method of keying a transmitter in which the RF energy is fed to the antenna when the telegraph key is closed and to an artificial load when the key is open.

back-to-back connection　A method of connecting a pair of diodes so that each operates on half of an AC cycle, thereby permitting control of alternating current. The anode of one diode is connected to the cathode of the other, and vice versa. Transistors are similarly connected in parallel in opposite directions to control current in either direction without causing rectification.

backup　A computer disk or magnetic tape that is frequently updated so that if the computer power is lost, erasing data from a volatile memory, the data can be recovered.

backup drive　A disk or tape drive that accepts data offloaded from a hard-disk drive to permit copying files, distributing data, and for redundancy in the event of computer system failure.

backup file　File copies made on removable media (either disk or tape) and kept to ensure the recovery of data lost due to equipment failure, human errors, updates, or fire or flood.

backward wave　A wave traveling opposite to the normal direction, such as a wave whose group velocity is opposite to the direction of electron-stream motion in a traveling-wave tube, or the reflected wave in a mismatched transmission line.

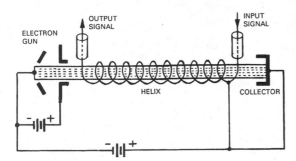

Backward-wave amplifier (BWA): the traveling radio-frequency signal is amplified by its interaction with the electron stream as it moves backward toward the BWA's electron gun.

backward-wave amplifier [BWA]　A microwave electron tube consisting essentially of an electron gun at one end directing a beam of electrons through a helix or slow-wave structure to a collector at the other end. An axial magnetic field confines the electron beam in the center of the helix. An RF signal is impressed on the helix near the collector, and the amplified signal is removed at the other end near the electron gun. The electron beam is velocity-modulated as in a normal traveling-wave tube. As the signal travels toward the gun, electrons are bunched near the collector. The bunches are decelerated by the signal, transferring their energy to the signal to amplify it. The BWA has separate power supplies for its anode and helix. See also *Amplitron.*

backward-wave magnetron　A magnetron oscillator whose electron beam travels in a direction opposite to the flow of RF energy. It can have high power output, and it can be voltage-tuned over a wide frequency band.

backward-wave oscillator [BWO]　A special vacuum-tube oscillator in which electrons are bunched by an RF magnetic field as they flow from cathode to anode. This bunching action produces a backward wave that becomes larger as it progresses toward the electron-gun end of the tube. The magnetic field is produced by a cylindrical magnet centered on the axis of the electron beam. The output signal is taken from the gun end of this folded line.

bakeout　The heating of an object to drive out absorbed gases.

balance control　A control for a stereo sound system that varies the volume of one loudspeaker system relative to the other while maintaining their combined volume essentially constant.

balanced amplifier　An amplifier circuit in which there are two identical signal branches connected to operate in phase opposition, with input and output connections each balanced to ground. A push-pull amplifier is an example.

balanced bridge　A bridge adjusted for zero output, occurring when bridge branch currents are equal.

balanced circuit　A circuit whose two sides are electrically alike and symmetrical with respect to a common reference point, usually ground.

balanced converter　*Balun.*

balanced currents　Currents flowing in the two conductors of a balanced line so that, at every point along the line, they are equal in magnitude and opposite in direction. Also called push-pull currents.

balanced detector　An FM receiver detector. In one form the audio output is the rectified difference between the voltages produced across two resonant circuits. One circuit is tuned slightly above the carrier frequency and the other slightly below.

balanced input　A two-terminal input circuit that has the same impedance between each terminal and ground.

balanced line　1. A transmission line that consists of two conductors which are capable of being operated so that the voltages of the two conductors at any transverse plane are equal in magnitude and opposite in polarity with respect to ground. The currents in the two conductors are then equal in magnitude and opposite in direction. It is also called a *balanced transmission line.*

balanced-line system　A system that consists of generator, balanced line, and load adjusted so the voltages of the two conductors at each transverse plane are equal in magnitude and opposite in polarity with respect to ground.

balanced low-pass filter　A low-pass filter designed to be used with a balanced line.

balanced method   A measuring method in which the reading is taken at zero or in the absence of a signal, as with a balanced bridge.

balanced mixer   A mixer circuit in superheterodyne receivers that reduces leakthrough and suppresses even-order harmonics of one input, usually the local oscillator. Balance is achieved with a bridge or ring configuration of diodes, a balanced pair of transistors, or tube circuits.

balanced modulator   A mixer circuit that combines an audio signal with a carrier oscillator signal. The output signal contains the two sidebands produced by this mixing, but does not include the original carrier-oscillator signal or the pure audio signal. A modulated RF signal contains some information to be transmitted. A circuit in a single-sideband suppressed-carrier transmitter combines a voice signal with an RF signal. The balanced modulator isolates the input signals from each other and the output, so that only the difference between the two input signals reaches the output. In color television transmitters it applies the I and Q signals to the subcarriers.

balanced network   A network that has equal impedances in opposite branches.

balanced oscillator   An oscillator in which the impedance centers of the tank circuits are at ground potential and the voltages between either end and their centers are equal in magnitude and opposite in phase. A push-pull oscillator is an example.

balanced ring modulator   A modulator with tubes or diodes to suppress the carrier signal while providing double-sideband output.

balanced termination   A load with two terminals that both present the same impedance to ground.

balanced transmission line   *Balanced line.*

balanced voltages   Voltages that are equal in magnitude and opposite in polarity with respect to ground. Also called push-pull voltages.

ballast   An iron-core inductance connected in series with a fluorescent lamp or other arc-discharge lamp to provide the required high starting voltage and limit the operating current.

ballast lamp   A lamp that increases in resistance when current increases, to maintain a nearly constant current.

ballast resistor   A resistor that increases in resistance when current increases, thereby maintaining essentially constant current despite variations in line voltage.

ball bonding   A method for bonding fine 0.7- to 1.0-mil gold wire between pads on a semiconductor chip and package leads. Wire fed from a *capillary* is cut by a gas flame or electric arc to form a molten ball on the end which is then *thermocompression* bonded to the pad. A loop is formed, and the second end is *wedge bonded* to a lead. It is also called *ball-wedge bonding*. See also *wire bonding*.

ball-grid array [BGA]   A flat package with a profile less than 0.1 in high for surface mounting large-scale ICs on circuit boards. Similar to the pin-grid array (PGA), it has solder bumps (balls) in place of pins, which are melted to bond the BGA to the circuit board.

ballistic electrons   Electrons that travel at high velocity through a semiconductor crystal lattice without being scattered.

ballistic galvanometer   An instrument that measures the total quantity of electricity in a transient current, such as the discharge current of a capacitor.

ballistic missile   A missile that is guided during powered flight in the upward part of its trajectory but becomes a free-falling body in the latter stages of its flight toward its target. The Russian Scud missile is an example.

ballistic-missile early-warning system [BMEWS]   An electronic system that provides detection and early warning of attack by enemy intercontinental ballistic missiles. One example is a network of three long-range-radar bases at Thule (Greenland), Clear (Alaska), and in the British Isles, to provide warning of missile attack across the polar region.

ballistic trajectory   The trajectory followed by a body acted upon only by gravitational forces and the resistance of the medium through which it passes.

ball-wedge bonding   *Ball bonding.*

balun [BALanced to UNbalanced]   A device that matches an unbalanced coaxial transmission line or system to a balanced two-wire line or system. It is a quarter-wavelength cylindrical sleeve placed over the end of a coaxial cable feed to an antenna, and isolates the outer conductor of the cable from ground. Also called balanced converter, bazooka (slang), and line-balance converter.

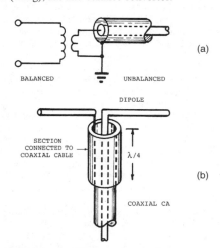

Balun. Examples of balanced-to-unbalanced line-coupling devices are (*a*) coupled coils and (*b*) "bazooka."

balun filter   An input-line filter that includes a differential wound transformer to present a low impedance to differential-mode signals and high impedance to common-mode signals.

banana jack   A jack that accepts a banana plug, typically for panel mounting.

banana plug   A low-voltage, low-frequency connector with a spring-metal tip formed so that it has a banana shape. It terminates test leads or is a terminal for plug-in components.

band   1. A range of frequencies between two definite limits.   2. One track, or a group of tracks, on a magnetic disk in a computer disk drive.

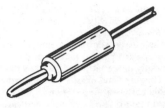

Banana plug provides a secure connection for meter test leads.

band-edge energy The energy of the edge of the conduction band or valence band in a solid. It is the minimum energy needed by an electron to move freely in a semiconductor or the maximum energy it can have as a valence electron.

band-elimination filter *Band-rejection filter.*

bandgap 1. The minimum energy that must be added to a valence electron held within a semiconductor crystal lattice to permit it to become a conduction electron able to move freely throughout the crystal. 2. The energy difference between the conduction band and the valence band in a material. The bandgap represents the wavelength or "color" of light that a compound will absorb or emit.

band-limited function A function whose Fourier transform is very small or vanishes outside some finite interval.

bandpass amplifier An amplifier that passes a definite band of frequencies with essentially uniform response.

bandpass filter A passive resistive-capacitive (RC) or inductive-capacitive (IC) circuit that allows only those signals to pass which are within a limited range of frequencies. It attenuates signals above and below that range of frequencies. See *filter characteristics.*

bandpass response A response characteristic in which a definite band of frequencies is transmitted with essentially uniform response. In IF transformers, this response is obtained by tuning the primary and secondary resonant circuits to slightly different frequencies. The response curve then usually has two humps. Also called double-hump response and flat-top response. See *filter characteristics.*

band plan An agreement for operating amateur radio within a certain band of the radio spectrum. It sets aside certain frequencies for the different modes of amateur operation: CW, SSB, FM, repeaters, and simplex.

band pressure level The effective sound pressure level of the sound energy contained within a specified frequency band.

band-rejection filter A filter that attenuates alternating currents whose frequencies are between given upper and lower cutoff values while transmitting frequencies above and below this band. It is the opposite of a bandpass filter. The band rejected is generally much wider than that suppressed by a trap. It is also called a *band-elimination filter, band-stop filter,* and a *rejector circuit.* See also *Filter characteristics.*

band selector A switch that selects any one of the bands in which a receiver, signal generator, or transmitter is designed to operate. It usually has two or more sections, to make the required changes in all tuning circuits simultaneously. Also called band switch.

B and S gage Abbreviation for *Brown and Sharpe gage.*

bandstop filter *Band-elimination filter.*

band switch *Band selector.*

bandwidth [BW] 1. The width of a band of frequencies assigned to a specific purpose, such as the bandwidth of a television station (6 MHz). 2. The range of frequencies that has been specified as performance limits for a filter, amplifier, or attenuator, defined as between the 3 dB (0.707) (half-power) points at the high-pass and low-pass ends of the frequency response curve. It is in the frequency domain for analog signals, expressed in hertz, and in the time domain for digital signals, expressed as bits per second.

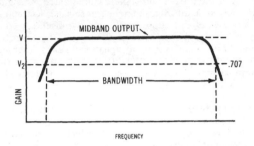

Bandwidth is determined at the 3 dB (0.707) half-power points.

bang-bang control A servomechanism control method in which the corrective actuator, motor, or drive is turned either full on or full off to correct an error and return the system to its desired or preset condition. The speed or intensity of response is not governed by the magnitude of the error. A home heating plant is a simple example.

bank An assembly or matrix of similar components wired or controlled together to perform a specific function. They might be connected either in series or parallel. Examples include banks of resistors, incandescent lamps, capacitors, or batteries.

banked winding A method of winding RF coils: single turns are wound one over the other in a flat outward spiral. The entire coil consists of many such spirals side by side, giving a multilayer coil without going back to the starting point. This construction reduces the distributed capacitance of the coil.

bar A unit of pressure equal to 1×10^6 dyn/cm² or 1×10^5 N/m² (slightly less than 1 atmosphere). The microbar, equal to 1×10^{-6} bar, was the unit of pressure formerly used in acoustics. The newton per square meter (N/m²) is now the SI unit of pressure.

bar-code reader An optical reader that reads combinations of printed bars which represent numerical characters.

bar generator A signal generator that delivers pulses uniformly spaced in time and synchronized to produce a stationary bar pattern on a television screen. A color-bar generator produces these bars in different colors on the screen of a color television set.

BARITT diode [BARrier Injection Transit-Time diode] A microwave diode whose carriers that traverse the drift region are generated by minority carrier injection from a forward-biased junction instead of being extracted from the plasma of an avalanche region. It can function as an

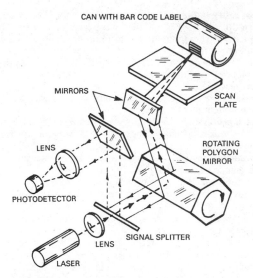

CAN WITH BAR CODE LABEL

MIRRORS

SCAN PLATE

LENS

ROTATING POLYGON MIRROR

PHOTODETECTOR

SIGNAL SPLITTER

LENS

LASER

Bar-code reader projects a laser beam through lenses and mirrors to "read" the bar code printed on the package and then translates the reflected signal to identify the product and give its price.

oscillator in the gigahertz range. One version has a thin slice of N-type silicon between platinum silicide Schottky barrier contacts.

barium [Ba]   A silver-white metal element with an atomic number of 56 and an atomic weight of 137.4. It is acted upon energetically by moist air and by carbon dioxide. It acts as a degassing agent in electron tubes, and when alloyed with nickel emits electrons in heated cathodes.

barium titanate   A ceramic that has piezoelectric properties and is capable of withstanding much higher temperatures than Rochelle salt crystals. It is used in crystal pickups, microphones, and sonar transducers.

Barkhausen criteria   The requirements for the oscillation of a device or circuit: The circuit or device must contain an amplifier with a gain of more than unity, and the output must be fed back to the input exactly 180° out of phase.

Barkhausen effect   The succession of abrupt changes in magnetization that occur when the magnetizing force acting on a piece of iron or other magnetic material is varied.

Barkhausen-Kurz oscillator   A retarding-field oscillator whose frequency of oscillation depends solely on the electron transit time within the tube. Electrons oscillate about a highly positive grid before reaching the less positive anode. It is also called a *Barkhausen oscillator.*

Barkhausen magnet   A permanent magnet mounted on the horizontal output tube of a television receiver to reduce Barkhausen oscillations.

Barkhausen oscillation   An undesired oscillation in the horizontal output tube of a television receiver; it causes one or more ragged dark vertical lines on the left side of the picture.

Barkhausen oscillator   *Barkhausen-Kurz oscillator.*

bar magnet   A bar of hard steel that has been strongly magnetized and holds its magnetism, thereby serving as a permanent magnet.

Barnett effect   The very slight magnetization produced in an iron rod when it is rotated at high speed about an axis perpendicular to its length.

barograph   A barometer that produces a continuous record of atmospheric pressure on a graph.

barometer   An instrument that measures the pressure of the atmosphere. Two common types are aneroid and mercury barometers.

barometric pressure   Atmospheric pressure as measured by a barometer.

barometric switch   *Baroswitch.*

bar pattern   The pattern of repeating color bars produced by a bar generator, for adjusting color television receivers.

barrel distortion   Distortion in which all four sides of a received television picture bulge outward like a barrel.

barrel shifter   Computer circuitry that allows arbitrary shifting of data.

barrel-stave reflector   A parabolic radar antenna reflector that has its horizontal top third and bottom third cut away, to give a high vertical "fan" beam so the roll of a ship will not cause the beam to miss the target. It is also called a *cut paraboloid reflector.*

barreter   A *bolometer* with a fine wire or metal film as its temperature-sensing element. The wire or film has a *positive temperature coefficient* of resistivity [PTC] (resistance increases with temperature). It is suitable for measuring power in microwave systems. It has a characteristic that is the inverse of a *thermistor,* which has a *negative temperature coefficient* [NTC] of resistivity (resistance decreases with temperature).

barrier   *Potential barrier.*

barrier diode   A diode formed by depositing a metal film on a high-resistivity semiconductor material. It can be integrated into a bipolar transistor die.

barrier voltage   The minimum voltage required for conduction through a PN junction.

base   1. One of the three regions of a bipolar junction transistor. It separates the emitter and collector regions. Minority carriers are injected from the emitter into the base, where they later either recombine or diffuse into the collector. It is analogous to the grid of a triode vacuum tube and the gate of a field-effect transistor.   2. The part of an electron tube that has the pins, leads, or other terminals to which external connections are made either directly or through a socket.   3. The plastic, ceramic, or other insulating board that supports a printed-wiring pattern.   4. An integer to which all digits are related in a positional notation system for computers. The successive digits are coefficients of successive powers of the base. The most common base values are 2, 8, and 10. It is also called the radix.   5. The number on which a system of logarithms is based, such as 10 or 2.718.

base address   A number that appears as an address in a computer instruction but serves as the starting point for subsequent addresses to be modified.

baseband   The frequency band occupied by all the transmitted signals that modulate a particular carrier. The band that transmits picture and synchronizing signals in television is one example; another is the band containing all the modulated subcarriers in a carrier system.

base bias   The direct voltage that is applied to the base of a transistor.

base charge The charge produced by excess minority carriers in the base region of a transistor or diode.

base film The plastic substrate that supports the magnetic coating of magnetic tape. The base film of most instrumentation and computer tapes is a polyester, such as Mylar. For less critical applications, cellulose acetate and polyvinyl chloride are used.

basegroup A group of carrier channels in a particular frequency range that forms a basic unit for further modulation to a final frequency band in a carrier communication system.

base insulator A heavy-duty insulator that supports the weight of an antenna mast and insulates the mast from the ground or some other surface.

baseline 1. A line that joins the two stations between which electric phase or time is compared in determining navigation coordinates, such as a line joining a master and a slave station in a loran system. 2. The line produced on the screen in the absence of an echo in certain radar displays.

baseline dwell time The time during which a pulse waveform coincides with its baseline.

base-loaded antenna A vertical antenna whose electrical height is increased by adding inductance in series at the base.

base modulation Amplitude modulation produced by applying the modulating voltage to the base of an amplifier transistor.

base region The interelectrode region of a bipolar junction transistor into which minority carriers are injected.

base station A fixed-location land station that provides radio communication service with mobile stations and other fixed radio stations.

BASIC Acronym for Beginner's All-purpose Symbolic Instruction Code; a high-level computer language commonly used in personal computers.

basic frequency The frequency of the sinusoidal component considered the most important, such as the fundamental frequency of an oscillator or the driving frequency of an acoustic transducer.

basic input-output system [BIOS] Software that translates operating system instructions into commands to and from computer components.

basic repetition rate The lowest pulse repetition rate of each of the several sets of closely spaced repetition rates in a loran system.

basic research Fundamental, theoretical, or experimental investigation to advance scientific knowledge. Immediate practical application is not a direct objective.

basic television service A charge for delivery of television broadcast by cable; typically a monthly fee for the lowest level of service.

basket winding A crisscross coil-winding pattern in which adjacent turns are far apart except at points of crossing, giving low distributed capacitance.

bass Sounds corresponding to frequencies at the lower end of the audio range, below about 250 Hz.

bass boost A circuit that emphasizes the lower audio frequencies, generally by attenuating higher audio frequencies.

bass compensation A circuit that offsets the lowered sensitivity of the human ear to weak low frequencies by making the bass frequencies relatively stronger than the high audio frequencies as volume is lowered.

bass control A manual tone control that changes the level of bass frequencies in an audio amplifier.

bass reflex baffle A loudspeaker baffle that has an opening dimensioned below the loudspeaker so that bass frequencies from the rear emerge to reinforce those radiated directly forward.

bass response The extent to which an audio-frequency amplifier or loudspeaker responds to low audio frequencies.

bassy A reference to overemphasis of bass notes in sound reproduction.

batch processing In computers, a technique in which data to be processed is coded and collected into groups prior to processing.

bat-handle switch A toggle switch whose actuating lever is shaped like a baseball bat.

bathtub capacitor A paper capacitor enclosed in a metal housing that has broadly rounded corners like those on a bathtub.

bathythermograph A sensitive recording thermometer that is lowered into the water to determine temperatures at different levels, as required in predicting sound conditions. The data can either be recorded or transmitted up the support cable as modulation on a carrier signal.

battery An electrochemical power source made by connecting two or more electrochemical power cells in series. Both primary and secondary batteries are used to power electronic circuits, products, and systems. Disposable alkaline primary batteries are the most popular power sources for portable consumer electronic entertainment products. However, rechargeable nickel-cadmium (Ni-Cd) and nickel metal-hydride (Ni-MH) batteries are widely used to power notebook and subnotebook computers, portable mobile radio and marine transceivers, and cellular telephones. Today the terms *cell* and *battery* are widely used interchangeably despite technical differences in their definitions. The term *storage battery* typically refers to an automotive lead-acid battery and is considered an obsolete term. See also *sealed lead-acid battery*.

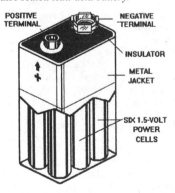

Battery. A 9-V secondary battery suitable for powering electronic circuitry consists of six 1.5-V alkaline cells. Terminal dimensions accept a standard battery clip.

battery analyzer A tester that automatically analyzes charge-discharge capacities of a secondary battery with appropriate logic, timing, and switching circuits.

battery backup 1. A power subsystem for electronic equipment that sustains it if there is a line power loss. 2.

A battery or power cells on a circuit board that prevents loss of data from volatile memories (DRAMs) during a power shutdown or failure.

battery charger A rectifier unit that changes AC power to DC power for charging a storage battery. Also called charger.

batwing antenna *Superturnstile antenna.*

baud [Bd] A unit of signaling speed in telecommunications equal to one element per second. Signaling speed in bauds is equal to the reciprocal of signal element length in seconds. A pulse and a space are separate elements; thus a teleprinter handling 22.5 pulses per second is operating at 45 Bd. The baud rate measurement includes data bits, parity bits, and stop bits. Baud rate is less than bits per second (b/s).

Baudot code A teleprinter code based on combinations of five or six mark and space intervals of equal duration. The five-unit code gives 32 possible characters, and the six-unit version gives 64. Used in radio and wire teleprinter operation.

BAW Abbreviation for *bulk-acoustic wave.*

Bayard-Alpert vacuum transducer A hot-cathode vacuum gage for measuring an ultra-high vacuum. It has a fine nickel wire collector at its center, and its filament and cathode are located outside the grid, but its operation is similar to that of the standard *hot-cathode vacuum gage.* It has an operating range of 4×10^{-10} to 5×10^{-2} torr. See also *cold-cathode vacuum gage* and *thermocouple vacuum gage.*

bayonet base A tube or lamp base that has two projecting pins on opposite sides of a smooth cylindrical base. The pins engage in corresponding slots in a bayonet socket and clamp the base firmly in the socket.

bayonet coupling A connector that consists of a cylindrical plug with projecting pin stubs that is inserted into a cylindrical socket or receptacle which contains a J-shaped slot. The coupling is secured by rotating the plug until the pins reach the end of the slot. See also *coaxial cable connector.*

bayonet socket A socket for bayonet-base tubes or lamps; it has J-shaped slots on opposite sides and one or more contact buttons at the bottom.

bazooka Slang term for *balun.*

BBD Abbreviation for *bucket brigade device.*

BCD Abbreviation for *binary-coded decimal.*

BCI Abbreviation for *broadcast interference.*

Bd Abbreviation for *baud.*

BDI Abbreviation for *bearing deviation indicator.*

B display A rectangular radarscope display in which targets appear as bright spots, with target bearing indicated by the horizontal coordinate and target distance by the vertical coordinate.

beacon A navigational aid that provides a radio, radar, light, acoustic, or other characteristic signal which gives bearing, course, or location guidance for ships or aircraft.

beacon presentation The radarscope presentation that results from RF waves transmitted by a radar beacon.

beacon stealing Interference between radar beacons, resulting in loss of beacon tracking by one tracking radar.

beacon system A system consisting of an active transmitter/receiver, called a *transponder,* located on the target or object to be located (or avoided) and a transmitter, *interrogator,* for eliciting replies from the transponder. Beacon systems can be based on radio frequency, microwave radar,

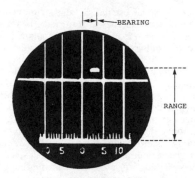

B display for radar presents the target as a pip, with the range displayed on the vertical axis and the bearing on the horizontal axis.

visible light, infrared, or acoustic energy and can provide range, bearing and homing guidance. See *Air Traffic Control Radar Beacon System* [ATCRBS], *Identification Friend or Foe* [IFF], *microwave beacon system, transponder,* and *interrogator.*

bead A glass, ceramic, or plastic insulator with a hole through its center that supports the inner conductor of a coaxial line in the exact center of the line. Ferrite beads are placed on wire connections of high-frequency circuits to suppress parasitic oscillation by providing inductance.

bead thermistor A thermistor that consists of a small bead of semiconducting material with axial leads. It is used for microwave power measurement, temperature measurement, and as a protective device.

beam 1. A concentrated unidirectional stream of particles, such as electrons or protons. 2. A concentrated unidirectional flow of electromagnetic waves, as from a radar antenna, a microwave relay antenna, or an A-N radio range antenna array. The beam here is a major lobe of the antenna radiation pattern and is restricted to a small solid angle in space. 3. A concentrated unidirectional flow of acoustic waves. 4. A parallel arrangement of light rays. Also called ray.

beam antenna *Antenna array.*

beam bender *Ion-trap magnet.*

beam blanking *Blanking.*

beam capture Entry of a missile into the beam of a radar beam-rider guidance system so it can receive coded guidance signals.

beam convergence The adjustment that makes the three electron beams of a three-gun color picture tube meet or cross at a shadow-mask hole.

beam coupling The production of an alternating current in a circuit connected between two electrodes that are close to, or in the path of, a density-modulated electron beam.

beam current The electric current determined by the number and velocity of electrons in an electron beam.

beam deflection tube An electron beam tube whose beam controls the current to an output electrode by means of a transverse movement.

beam finder A switch within a cathode ray tube (CRT) oscilloscope that puts the trace on the screen regardless of the settings of the horizontal, vertical, and intensity controls.

beam jitter A small oscillatory, angular movement of a radar antenna array and hence of the radar beam, required to develop accurate error signals for automatic tracking.

beam-lead technology An integrated circuit processing technology in which gold beams are formed as an inherent part of the chip during the processing of the silicon wafer. The beams are exposed and extend over the edge of the chip after it is separated from the wafer. The chip is then inverted (face down), and several beam leads are attached simultaneously to the substrate. The chip remains slightly elevated above the substrate. See also *flip-chip*.

beam loading Absorption of power by the beam of particles in an accelerator as they gain energy.

beam-lobe switching A radio direction-finding technique that is based on electronic switching of antenna elements to give reception with lobes of different beam angles. Signals are compared to determine the direction to their source. A similar lobe-switching technique is used in radar.

beam magnet *Convergence magnet.*

BEAMOS Abbreviation for *beam-accessed metal-oxide semiconductor.*

beam pattern *Directivity pattern.*

beam-plasma amplifier An amplifier based on penetration of a plasma by an electron beam. It is capable of operating at millimeter and shorter wavelengths.

beam-power tube An electron-beam tube built so that directed electron beams contribute substantially to its power-handling capability. The control and screen grids are essentially aligned, and special deflecting electrodes are used to concentrate the electrons into beams. Also called beam tetrode.

beam rider A missile that is directed to its target by remaining within a laser or microwave beam. Feedback circuitry determines if the missile has strayed from the beam and sets controls to return it to the beam.

beam-rider guidance A missile guidance system that directs the missile to its target by aiming a laser or microwave beam at the target to create an invisible "path" or "track" for the missile to follow. Feedback circuitry in the missile detects any deviation of the missile from the path because it senses a decrease in received energy and sends control signals to the missile's rudder and/or elevators, which change position to direct the missile back onto the centerline of the beam.

beam splitting A process for increasing radar target-locating accuracy. The bearings at which a target first reflects data and then stops reflecting data during a scan define a beam that is split mathematically to calculate mean bearing for the target.

beam steering Changing the direction of the major lobe of a radiation pattern, usually by switching antenna elements.

beam width The angular width of a radar, radio, or other beam, measured in bearing between points of half-power intensity.

beam-width error A radar error that occurs because the width of the scanning beam makes the target appear wider than it actually is. By covering two targets in a single sweep, beam width can also make two targets look like one on a radarscope.

bearing Angular position in a horizontal plane, expressed as the angle in degrees from true north in a clockwise direction. Also called azimuth. In navigation, azimuth and bearing have the same meaning; however, bearing is preferred for terrestrial navigation and azimuth for celestial navigation.

bearing cursor A radial line that extends from the center of a radar *plan-position indicator* [PPI] display to an outer circular scale, graduated in degrees, that can be positioned by manual controls over the spot on the screen representing a target or other object so that the bearing of the object can be read directly on the outer scale or from a readout on the PPI screen. This information is of value as a navigational aid to help the pilot set a course for interception or avoidance of the object designated.

bearing resolution The smallest angular difference in bearing at which radar is able to distinguish between close targets that have the same range.

bearing sensitivity The minimum signal strength required by a radio direction finder to give repeatable bearings.

beat frequency The sum or difference between two frequencies that are combined in a nonlinear circuit.

beat-frequency oscillator [BFO] An oscillator that produces a desired audio difference frequency by combining two different RF signals. Used in audio signal generators for test purposes and in communication receivers to produce an audible signal when tuned to continuous-wave signals.

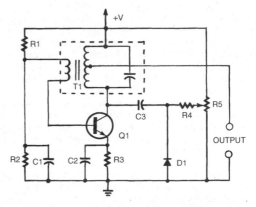

Beat-frequency oscillator includes a 465-Hz transformer and a tuning potentiometer.

beating A phenomenon in which two or more periodic quantities of different frequencies produce a resultant that has pulsations of amplitude.

beating-in Adjusting the frequency of one of two interconnected oscillators until no beat frequency is heard in a connected receiver. The oscillators are then operating at very nearly the same frequency.

beat note The difference frequency obtained when two sinusoidal waves of different frequencies are fed to a nonlinear circuit.

beat-note detector A detector that includes an oscillator or is fed by an external oscillator which has a frequency sufficiently close to the unmodulated incoming carrier frequency so that an audible signal frequency is produced.

beat reception *Heterodyne reception.*

beats Beat notes, generally at a sufficiently low audio frequency to be counted.

beavertail beam A fan-shaped radar beam, wide in the horizontal plane and narrow in the vertical plane. The beavertail beam is swept vertically for aircraft or missile height finding.

beeper *Pager.*

bel [B] The fundamental unit of sound level, equal to the logarithm to the base 10 of the ratio of two amounts of power. One power value is a reference value. The decibel, a smaller unit equal to $\frac{1}{10}$ B, is more commonly used.

benchmark A standardized test for comparing computer performance on a specified task. It simplifies the task by presenting a single piece of data that tells which product is better for a given application.

bend A smooth change in the direction of the longitudinal axis of a waveguide.

bender element A combination of two different thin strips of piezoelectric material bonded together in such a way that when voltage is applied, one strip increases in length and the other becomes shorter. This causes the combination to bend.

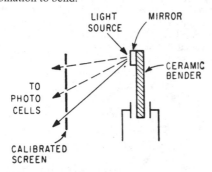

Bender element with mirror reflects light beam to different photocells, depending on voltage applied to bender strips.

bender transducer A transducer in which a voltage is generated when the bender element is bent. Conversely, application of voltage will cause bending.

bent antenna An open *dipole* antenna, except that one or both of its ends are bent to form a vertical section.

Bernoulli box A data storage drive for small computers with a removable cartridge, a hybrid of a floppy diskette and a hard disk, the size of four conventional 5.25-in floppy diskettes in a stack. Typical storage capability is 90 megabytes on a single cartridge. It is used primarily for backup and archives.

beryllium [Be] A gray, hard, toxic metal element with an atomic number of 4 and an atomic weight of 9. When alloyed with copper it forms beryllium-copper, a resilient, tough metal for making springs, washers, connectors, and other fittings. Integrated-circuit leadframes and potentiometer wipers are stamped from sheet beryllium-copper alloy.

Bessel filter An active filter that has linear phase characteristics and approximately constant time delay over a limited frequency range. Transient waveforms are passed with minimum distortion.

Bessel response A linear phase response (constant time delay) in the passband of a tunable active filter, with a shallower attenuation rate than for a Butterworth response and no overshoot.

beta [Greek letter β] The current gain of a transistor when it is connected as a grounded-emitter amplifier. It is the ratio of a small change in collector current to the resulting change in base current, with collector voltage remaining constant.

beta cutoff frequency The frequency at which the beta of a transistor is down 3 dB from its low-frequency value.

beta particle An electron or positron emitted from a nucleus during beta decay.

beta ray A stream of beta particles.

beta-ray spectrometer An instrument that determines the energy distribution of beta particles and secondary electrons.

beta-ray spectrum The distribution in energy or momentum of the beta particles emitted in a beta-decay process.

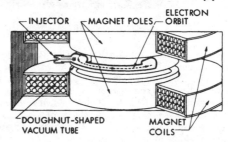

Betatron. The important parts of the accelerator are shown.

betatron An accelerator that consists of a horizontal doughnut-shaped vacuum enclosure, an electron gun as a source of electrons inside, and an external AC electromagnet which produces magnetic lines of force passing vertically through the enclosure. As emitted electrons travel in a circular path around the doughnut, they are continuously accelerated by the rapidly changing magnetic field. The resulting high-energy electron beam is then deflected out through the window for direct use as beta rays or directed against a target to produce high-energy X-rays.

beta wave A brain wave that has a frequency above 14 Hz.

Bethe hole directional coupler A directional coupler for microwave systems that permits electric and magnetic field coupling between two waveguides to be balanced. One waveguide is rotated with respect to the other about a common hole in both wide surfaces of opposing waveguide sections.

BeV Abbreviation for *billion electronvolt.*

bevatron *Proton synchrotron.*

Beverage antenna A nonresonant directional long-wire antenna, two or more wavelengths long, that is usually suspended 10 to 20 ft (3 to 6 m) above the ground. It is also called a *wave antenna.*

beyond-the-horizon communication *Scatter propagation.*

bezel A grooved rim that holds a transparent glass or plastic window or lens for a meter, tuning dial, or other indicating device.

BFO Abbreviation for *beat-frequency oscillator.*

BGA Abbreviation for *ball-grid array.*

B-H curve A characteristic curve that shows the relation between magnetic induction B and magnetizing force H for a magnetic material. It shows the manner in which the permeability of a material varies with flux density. Also called magnetization curve.

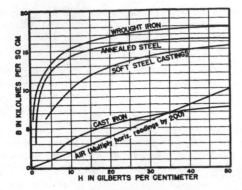

B-H curves for four different ferrous materials.

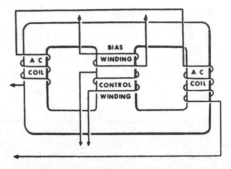

Bias winding on magnetic amplifier.

B-H meter A meter that measures the intrinsic hysteresis loop of a sample of magnetic material.

bias 1. The DC voltage applied to the control electrode of a transistor or electron tube to establish the desired operating point. 2. The direct current sent through the bias winding of a magnetic amplifier to establish the desired operating conditions. 3. An electric, mechanical, or magnetic force applied to a device to establish a desired electric or mechanical reference level for its operation. 4. The alternating current, usually 3 to 10 times the highest audio frequency, sent through the recording head of a magnetic tape recorder to linearize the recording process.

bias current An alternating electric current above about 40 kHz, added to the audio current being recorded on magnetic tape to reduce distortion.

biased automatic gain control *Delayed automatic gain control.*

bias modulation Amplitude modulation in which the modulating voltage is superimposed on the bias voltage of an RF amplifier tube.

bias oscillator An oscillator used in a magnetic recorder to generate an AC signal that has a frequency in the range of 40 to 80 kHz, as required for magnetic biasing to give a linear recording characteristic. The bias oscillator usually also serves as the erase oscillator.

bias resistor A resistor in the emitter or base circuit of a bipolar transistor, in the source or gate circuit of a field-effect transistor, or the cathode or grid circuit of an electron tube that provides a voltage drop which serves as a bias voltage.

bias stabilizer A device or circuit that maintains a stable bias current in various discrete bipolar junction and field-effect transistors, typically discrete RF stages operating from a low-voltage regulated supply. It provides a reference voltage and acts as a DC feedback element around the RF stage transistor. One example is an integrated circuit containing two PNP transistors.

bias winding A control winding that carries a steady direct current which establishes desired operating conditions in a magnetic amplifier or other magnetic device.

BiCMOS Acronym for *bipolar-CMOS*.

BiCMOS process An IC technology combining the linearity and speed of bipolar devices with the low power drain, low heat dissipation, and higher density of CMOS. It can operate at either ECL (emitter-coupled logic) or TTL (transistor-transistor logic) levels, and is widely used for mixed-signal devices.

biconical antenna An antenna that consists of two metal cones which have a common axis, with their vertices coinciding or adjacent and coaxial cable or waveguide feed to the vertices. The radiation pattern is circular in a plane perpendicular to the axis. It is considered to be a dipole if the vertex angle is less than 90°. It is similar to the *biconical horn.*

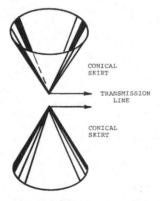

Biconical antenna is a form of dipole antenna if the included angle is less than 90°.

biconical horn A radiator similar to the biconical antenna but considered to be a horn if its vertex angle exceeds 90°.

biconjugate network A linear network that has four resistances, conjugate in pairs, so a voltage in series with one resistance causes no current to flow in the other of a pair. Examples include directional couplers, hybrid junctions, and hybrid rings.

bidirectional antenna An antenna that radiates or receives most of its energy in only two directions.

bidirectional breakdown diode A breakdown diode that has similar reverse characteristics for both polarities of the applied voltage.

bidirectional clamping circuit A clamping circuit that operates whenever the control pulse is applied, irrespective of the polarity of the input signal source at that time.

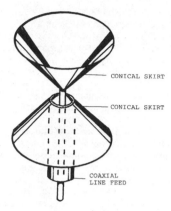

Biconical horn has an omnidirectional radiation pattern in the horizontal plane with the cone axes vertical.

bidirectional counter   A counter that has two or more operating modes, determined by gate control signals, to provide various combinations of addition and subtraction for pulses applied to either or both inputs.

bidirectional coupler   A device in a microwave circuit that measures direct and reflected power. It consists of a straight section of waveguide with an enclosed section attached to each side along its narrow dimension. Each section contains an RF pickup probe at one end and an impedance termination at the other end. Microwave energy from the main waveguide passes through three apertures, each spaced one-quarter wavelength apart on centers. The RF probe farthest away from the transmitter can measure direct power, and the nearest to the probe can measure reflected power. See also *directional coupler.*

bidirectional diode   A semiconductor junction diode that has essentially the same forward, reverse, and/or breakdown characteristics for both polarities of the applied voltage.

bidirectional microphone   A microphone that responds equally well to sounds coming from its front and rear, corresponding to sound incidences of 0 and 180°.

bidirectional pulse   A pulse whose variation from the normally constant value occurs in both directions.

bidirectional pulse train   A pulse train that consists of pulses which rise in two directions.

bidirectional thyristor   A three-terminal semiconductor device that acts as a thyristor for either polarity of the voltage applied to its anode and cathode. It can function as an AC switch. The *triac* is an example.

bidirectional transducer   A transducer capable of measuring in both positive and negative directions from a reference position.

BiFET   Abbreviation for *bipolar FET.*

bifilar resistor   A resistor wound with a wire doubled back on itself to reduce its inductance.

bifilar suspension   A meter movement that has two supporting conductors at each end of the moving element.

bifilar transformer   A transformer whose two windings are wound side by side to give extremely tight coupling. When used as television IF transformers to couple stagger-

tuned IF stages, the high coupling eliminates the need for a DC blocking capacitor.

bifilar winding   A winding or coil formed by doubling a single length of wire back on itself in hairpin fashion before winding it on a form or mandrel. The inductive effects of current passing in close proximity will be canceled, thus reducing the overall inductance of the coil. This winding technique is applied to wire wound resistors used in high-frequency circuits to reduce or eliminate inductive reactance. See *Ayrton-Perry winding.*

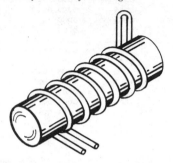

Bifilar winding reduces inductive reactance in wire-wound resistors at high frequencies.

bilateral   1. Having two sides.   2. *Bidirectional.*

bilateral network   A network in which current flow in either direction causes the same voltage drop.

bilateral transducer   A transducer capable of transmission simultaneously in both directions between at least two terminations.

billboard antenna   An antenna that consists of an array of *dipole* radiators mounted in front of a metal (generally mesh) reflecting screen. The larger the number of active dipoles, the greater the gain in the desired forward direction. It is not usually used at frequencies below 150 MHz because of its size. It is also called a *bedspring array.*

billion electronvolt [BeV]   Former name for *gigaelectronvolt.*

bimetallic strip   A length of two strips of sheet metal having different coefficients of temperature that has been welded together along the edges of its wide dimensions. The strip will curl or bend when exposed to increasing temperature due to the different expansion rates, permitting it to act as a temperature-sensitive armature and contact for thermostats or thermal-delay switches. It returns to its initial flat position when the source of heat decreases or is eliminated.

bimorph cell   Two piezoelectric plates cemented together so that an applied voltage causes one to expand and the other to contract; thus the cell bends in proportion to the applied voltage. Conversely, applied pressure will generate double the voltage of a single cell. It is a transducer in pickups and microphones.

BiMOS   Acronym for *bipolar metal-oxide semiconductor.*

binary   A number system with 2 as its base that uses only the digits 0 and 1. Binary logic is based on one of two states, "off" or "on," or 0 and 1, respectively. The binary system is the numerical coding used in most digital computers.

binary chain A series of binary circuits, each of which can affect the next circuit.

binary code A code in which each allowable position has one of two possible states. The common notation for binary states is 0 and 1. The binary number system is one of many binary codes.

Decimal	Binary	Binary Coded Decimal	Gray Code
0	0000	0000 0000	0000
1	0001	0000 0001	0001
2	0010	0000 0010	0011
3	0011	0000 0011	0010
4	0100	0000 0100	0110
5	0101	0000 0101	0111
6	0110	0000 0110	0101
7	0111	0000 0111	0100
8	1000	0000 1000	1100
9	1001	0000 1001	1101
10	1010	0001 0000	1111
11	1011	0001 0001	1110
12	1100	0001 0010	1010
13	1101	0001 0011	1011
14	1110	0001 0100	1001
15	1111	0001 0101	1000

Binary-coded decimal codes for 0 through 15, with binary and Gray codes shown for comparison.

binary-coded character An alphameric character represented by a predetermined configuration of consecutive binary digits.

binary-coded decimal [BCD] A system of number representation in which each decimal digit is represented by a group of four binary digits. Thus in the 8-4-2-1 coded decimal notation, the number 17 is 0001 0111 for 1 and 7, respectively.

binary code disk A disk that has patterns of concentric clear and opaque bars; the bars convert shaft angle directly to a nonambiguous natural binary code.

Binary code disk.

binary-coded octal An octal notation system in which each octal digit (0 through 7) is represented by a three-place binary-coded character.

binary compound semiconductor A semiconductor made of a compound of two elements, such as gallium arsenide (GaAs) or gallium phosphide (GaP). Binary compounds are the easiest semiconductor compounds to make, and they can be grown in bulk.

binary counter *Binary scaler.*

binary digit *Bit.*

binary encoder An encoder that changes angular, linear, or other forms of input data into binary-coded output characters.

binary file A file containing "raw" information not expressed in text form, as contrasted with a *text file.*

binary magnetic core A ferromagnetic core that can be switched to take either of two stable magnetic states.

binary notation *Binary number system.*

binary number A numerical value expressed in binary digits as a sequence of 0s and 1s representing 1, 2, 4, 8, 16, 32, 64, 128, and other powers of 2 according to position from right to left in a group. These positional values are added to get the equivalent decimal number. Thus 010 is 2; 101 is 5; 1010 is 10; 10000 is 16; 11110 is 30.

binary number system A system of positional notation in which the successive digits are interpreted as coefficients of the successive powers of the base 2, as in binary numbers. Also called binary notation.

binary phase-shift keying [BPSK] Keying of binary data or Morse-code dots and dashes by ±90° phase deviation of the carrier.

binary scaler A scaler that produces one output pulse for every two input pulses. Two binary scaler stages in sequence give an output pulse for every 4 input pulses; three in sequence give one for 8; four in sequence give one output pulse for every 16 input pulses. Also called binary counter and scale-of-two circuit.

binary search A search in which a set of items is divided into two parts, one of which is rejected. The process is repeated on the accepted part until the items that have the desired property are found.

binary signaling A communication system in which information is conveyed by the presence and absence, or positive and negative variations, of only one parameter of the signaling medium.

binary synchronous communication [BSC] A byte- or character-oriented communications protocol that uses a defined set of control characters for synchronized transmission of binary-coded data between stations in a data communications system. Examples include the control of the exchange of digital data between computers and/or terminals over telephone lines. It is also referred to as *bisynchronous transmission.*

binary-to-decimal conversion The mathematical process of converting a number written in binary notation to the equivalent number written in ordinary decimal notation.

binaural A reference to sound that reaches the listener over two paths, to give the effect of auditory perspective.

binaural effect The ability to determine the direction from which a sound is coming by sensing the difference in arrival times of a sound wave at each ear.

binder A resin or other adhesive material that holds particles together and provides mechanical strength.

binding energy 1. The net energy required to remove a particle from a system. 2. The net energy required to decompose a system into its constituent particles.

binding post A manually turned screw terminal for making electric connections.

binomial antenna array A broadside array that has major lobes in opposite directions and no side lobes. This is achieved by spacing the antennas of the array at half-wavelength intervals and feeding them all in phase, with the relative current amplitudes of the various elements being proportional to the coefficients of successive terms in a binomial series.

bioelectrogenesis The generation of electricity by living organisms.

bioelectronics The application of electronic theories and techniques to the problems of biology.

bioengineering The application of engineering knowledge to the fields of medicine and biology.

bioinstrument An instrument attached to the body for sensing and transmitting one or more forms of physiological data.

bionics A field combining the disciplines of electrical and electronics engineering with biology that seeks to simulate biological phenomena with electronic devices or circuits such as neural networks. See *neural networks*.

BIOS Abbreviation for *basic input/output system*.

biphase suppressed carrier A digital radio modulation scheme that uses one phase of a carrier to indicate a 0 and the opposite phase to indicate a 1. If the two phases are less than 180° apart, some residual carrier will be transmitted.

biosensor A device that senses a biological function like blood pressure or heart rate, generally for recording or telemetry.

biosphere The part of the earth and its atmosphere in which animals and plants live.

biotelemetry Telemetry of biological data, as from inside the human body, without interfering with the function or process being monitored.

biotron A chamber in which pressure, temperature, and other environmental factors can be accurately controlled for biological research.

Biot-Savart's law A law that predicts the intensity of the magnetic field produced by a current-carrying conductor.

bipolar Having two poles, polarities, or directions.

bipolar amplifier An amplifier capable of supplying a pair of output pulse signals corresponding to the positive or negative polarity of the input signal.

bipolar metal-oxide semiconductor [BiMOS] A combination of bipolar and MOS technology on a single device. See also *BiCMOS*.

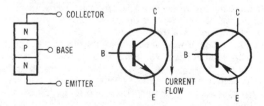

Bipolar junction transistor: (*a*) diagram of an NPN version, (*b*) schematic of an NPN version, and (*c*) schematic for a PNP version.

bipolar junction transistor [BJT] An active semiconductor signal amplifier with two PN junctions. There are two types, NPN and PNP, named for the way in which the junctions are combined. Its operation depends on the migration of both negative (electron) and positive (hole) carriers in contrast to the field-effect transistor that depends on the migration of only one carrier.

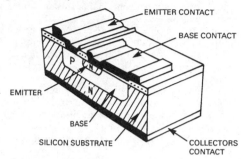

Bipolar junction transistor: this vertically constructed device has emitter, base, and collector terminals. There are NPN and PNP versions.

bipolar power supply A high-precision, regulated DC power supply that can be set to provide any desired voltage between positive and negative design limits, with a smooth transition from one polarity to the other. Some types can be remotely programmed with an external resistance or control voltage.

bipolar transistor See *bipolar junction transistor* [BJT].

bipotential cathode A cathode that has two different surface potentials to eliminate electron emission from portions of the cathode which are under the grid conductors while maintaining control of electron flow through the grid openings.

biquartic filter An active filter that includes operational amplifiers in combination with resistors and capacitors to provide infinite values of Q and simple adjustments for bandpass and center frequency.

biquinary notation A mixed-base notation system in which the first of each pair of digits counts 0 or 1 unit of five, and the second counts 0, 1, 2, 3, or 4 units. Thus decimal number 7 is biquinary 12; 43 is 04 03; 901 is 14 00 01; 4719 is 04 12 01 14. This is the code of the Japanese abacus, and it is used in binary digit form in some computers.

birdie A high-pitched whistle sometimes heard while tuning a radio receiver. It is due to beating between two carrier frequencies differing by about 10 kHz.

bis A French word for "more" or "encore" that in computer science means an extension of the standard. For example, 32 bis includes the V.32 standard plus additional refinements.

B-ISDN Abbreviation for *broadband integrated services digital network*.

bismuth [Bi] A brittle, heavy, metal element with an atomic number of 83 and an atomic weight of 208.98. It has a low melting point that makes it useful in cathodes, and when alloyed with other metals, it is a superconducting material.

bismuth telluride [Bi_2Te_3] An intermetallic compound that has high thermoelectric power.

bistable A property of a circuit that permits two stable output states for a given input. An example is a flip-flop circuit.

bistable multivibrator *Flip-flop circuit.*

bistable switching circuit An internally triggered switching circuit that has two operating levels or states.

bistatic radar A radar system whose receiver is separated by a long distance from the transmitter, with separate antennas for each. In conventional monostatic radar, the transmitter and receiver are at the same site.

bit An abbreviation for *binary digit*. There are two: 0 and 1. A bit is the basic data unit of most digital computers. A bit is usually part of a data byte or word, but bits can be used singly to control or read logic "on-off" functions. See also *binary*.

bit density The number of bits stored per unit length (generally bits per inch or bpi). It is also called *linear density*.

bit error rate The ratio of erroneous bits to total received bits in a data-transmission system.

bit interface unit [BIU] A functional circuit in a *microprocessor*.

bit map The bit pattern stored in a computer's memory that corresponds to the pixel pattern to be displayed on the computer's monitor where each pixel is being represented by one bit.

bit parallel Transmission of character-forming bits simultaneously over parallel paths, as contrasted to bit serial, where the bits for a character are transmitted in sequence over a single path.

bit rate The number of binary digits or equivalent pulses passing a given point in a data-transmission system per unit of time.

bit reversal 1. An addressing technique in signal processing in which the order of bits in an address is reversed during a computation (such as a fast Fourier transform). 2. A computer addressing technique in which the order of bits in an address is reversed during a computation.

bit serial Transmission of character-forming bits in sequence, as contrasted to bit parallel, where all the bits for a character are transmitted simultaneously.

bit-slice architecture A design concept in partitioning microprocessors that permits customized devices to be formed from two or more partial microprocessor chips.

bits per second [b/s] A unit of data transmission rate or speed in data transmission. The higher the b/s value, the more information that can be sent or received in a given amount of time. It is also abbreviated as *bps*.

BIU Abbreviation for *bit interface unit*.

BJT Abbreviation for *bipolar junction transistor*.

black after white A television receiver defect in which an unnatural black line follows the right-hand contour of any white object on the picture screen. The same defect also causes a white line to follow a sudden change from black to a lighter background. It is caused by receiver misalignment.

black-and-white television *Monochrome television.*

blackbody A perfect absorber of all incident radiant energy. It radiates energy solely as a function of its temperature.

black box A reference to any assembly or subassembly, usually a packaged electronic circuit module dedicated to a specific function, that can be inserted in or removed from a system by a user without a detailed knowledge of its internal structure.

black compression A reduction in television picture-signal gain at levels corresponding to dark areas in a picture. The effect reduces contrast in the dark areas of the picture as seen on monitors and receivers. Also called black saturation.

blacker-than-black region That part of the standard television signal in which the electron beam of the picture tube is cut off and synchronizing signals are transmitted. These synchronizing signals have greater peak power than those for the blackest portions of the picture.

black level The level of the television picture signal corresponding to the maximum limit of black peaks. This level is generally set at 75% of the maximum signal amplitude of the synchronizing pulses.

black light Obsolete term for invisible *ultraviolet radiation.*

black negative A television picture signal in which the voltage corresponding to black is negative with respect to the voltage corresponding to the white areas of the picture.

blackout A condition of zero voltage on the 120/240 VAC power line due to a failure or deliberate shutdown of the utility's generation and/or transmission facilities.

blackout effect A temporary loss of sensitivity of an electron tube after it handles a strong, short pulse.

black peak A peak excursion of the television picture signal in the black direction.

black positive A television picture signal in which the voltage corresponding to black is positive with respect to the voltage corresponding to the white areas of the picture.

black recording Facsimile recording in which the maximum received power corresponds to the maximum density of the record medium for amplitude modulation or to the lowest received frequency for frequency modulation.

black saturation *Black compression.*

black signal The signal at any point in a facsimile system produced by the scanning of a maximum-density area of the subject copy.

black transmission Facsimile transmission based on black recording.

blade A flat, moving conductor in a switch.

blade antenna An antenna shaped like the tail fin of an airplane for mounting on the fuselage of an airplane.

blank 1. The result of the final cutting operation on a natural crystal. 2. To cut off the electron beam of a cathode-ray tube. 3. A machine character that represents a space in the printout of a printer.

blanked picture signal The signal resulting from blanking a television picture signal. Adding the sync signal to the blanked picture signal gives the composite picture signal.

blanket area The area in the immediate vicinity of a broadcast station, where the signal of that station is so strong (above 1 V/m) that it interferes with reception of other stations.

blanketing Interference caused by a nearby transmitter whose signals are so strong that they override other signals over a wide band of frequencies.

blanking The process of cutting off the electron beam of a television picture tube, camera tube, or cathode-ray oscilloscope tube during retrace by applying a rectangular pulse voltage to the grid or cathode during each retrace interval. Also called beam blanking. The opposite action is called gating.

blanking level The level that separates picture information from synchronizing information in a composite television

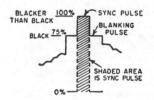

Blanking pulse at end of line, with horizontal sync pulse in its center.

picture signal. It coincides with the level of the base of the synchronizing pulses. Also called pedestal and pedestal level.

blanking pulse   One of the pulses that make up the blanking signal in television.

blanking signal   A wave of recurrent pulses, related in time to the scanning process, to effect blanking in television. The pulses occur at both the line and field frequencies, and cut off the electron beam during retrace at both transmitter and receiver.

bleeder current   Current drawn continuously from a voltage source to lessen the effect of load changes or provide a voltage drop across a resistor.

bleeder resistor   A large-value resistor connected across a power supply or other voltage source that improves voltage regulation by drawing a fixed current value continuously. It will also dissipate the charge remaining in filter capacitors when the equipment is turned off.

blip   1. To remove a portion of the recorded sound from a videotape of a television program, such as deleting an expletive or other undesired words.   2. *Pip.*

Bloch band   *Energy band.*

Bloch wall   The transition layer that separates adjacent ferromagnetic domains which are magnetized in different directions.

block   A group of bytes that is handled, stored, and accessed as a logical data unit, such as an individual file record.

block diagram   A diagram in which the principal divisions of an electronic system are indicated by rectangles or other geometric figures, and the signal paths are represented by lines.

blocked resistance   The real part of blocked impedance.

blocking   1. Applying a high negative bias to the grid of an electron tube to block its anode current, or producing an equivalent current-blocking effect in a transistor or other active solid-state device.   2. Overloading of a receiver by an unwanted signal so the automatic gain control reduces the response to a desired signal.   3. Combining two or more computer records into one block.   4. Preventing forward current flow in a semiconductor device.

blocking capacitor   *Coupling capacitor.*

blocking oscillator   An oscillator in which the negative base bias increases gradually during oscillation as a capacitor is charged, until a point is reached where collector current is cut off and oscillations stop. The capacitor then discharges until the base is unblocked and oscillation is resumed. This process produces a sawtooth voltage waveform that can be used as the sweep voltage for a cathode-ray tube. Also called squegging oscillator.

blocking-oscillator driver   A blocking oscillator that develops and shapes an essentially square pulse for driving radar modulator tubes.

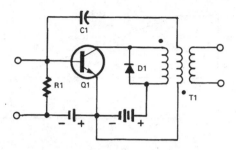

Blocking oscillator: transistorized version produces sawtooth waveforms that can sweep the electron stream repetitively across the inner face of a cathode-ray tube.

blooper   A radio receiver that is radiating an excessively strong oscillator signal.

blower   An electric fan that supplies air for cooling purposes. Air is delivered in the same plane as the impeller.

blown-fuse indicator   A warning light or lamp connected across a fuse so it lights when the fuse is blown.

blowout magnet   An electromagnet or permanent magnet that deflects and extinguishes the arc formed when a high-current circuit breaker or switch is opened.

blue-beam magnet   A small permanent magnet used as a convergence adjustment to change the direction of the electron beam for blue phosphor dots in a three-gun color television picture tube.

blue gain control   A variable resistor in the matrix of a three-gun color television receiver, used to adjust the intensity of the blue primary signal.

blue glow   A glow normally seen in electron tubes that contain mercury vapor; the glow is due to ionization of the molecules of mercury vapor. A blue glow near the electrodes of a vacuum tube means that the tube is gassy and hence defective. A soft blue fluorescent glow is normal on the glass envelopes of some vacuum tubes.

blue gun   The electron gun whose beam strikes phosphor dots emitting the blue primary color in a three-gun color television picture tube.

blue restorer   The DC restorer for the blue channel of a three-gun color television picture tube circuit.

blue video voltage   The signal voltage output from the blue section of a color television camera, or the signal voltage between the receiver matrix and the blue gun grid of a three-gun color television picture tube.

BMEWS   Abbreviation for *ballistic-missile early-warning system.*

BNC   Abbreviation for a miniature standard bayonet-coupling coaxial-cable connector.

bobbin   An insulated spool that supports a coil winding.

Bode diagram   A plot of the phase shift or gain of an amplifier vs. frequency to show the frequency response of a device or circuit. It is also called a *Bode plot.*

body capacitance   The capacitance that exists between the human body and the earth, generally between 100 and 300 pF. This value is used to estimate the charge on a person that can produce electrostatic discharge under favorable conditions of temperature and humidity.

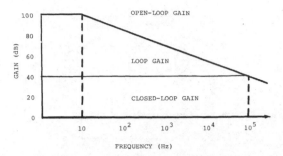

Bode diagram for a typical operational amplifier.

body-capacitance alarm   An alarm system that is triggered by the capacitance between the body of an intruder and a sensing wire or metal plate.

body resistance   The resistance of the human body, typically measured between the hands. Its value depends on variables, including the subject's perspiration and oils. For simulation of electrostatic discharge, its value is set at 1500 Ω.

body-section radiography   *Laminography.*

bogie   An indication of an enemy or unidentified aircraft on a radar screen.

bolometer   A device that measures microwave and infrared energy. It contains a resistance element that changes in resistance when heated by the radiant energy.

bolometer bridge   A bridge circuit that has a bolometer in one arm, to measure RF power.

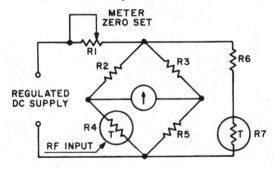

Bolometer bridge. Microwave energy to be measured is directed at bolometer R4. Thermistor R7 provides compensation for changes in ambient temperature.

Boltzmann's constant   A physical constant equal to 1.380662×10^{-23} J/K or 1.380662×10^{-16} erg/K.

Boltzmann's equation   The equation for particle conservation, based on the description of individual collisions.

Boltzmann's factor   A number, dependent on temperature and energy difference, that gives the ratio of the number of particles with one energy to the number of particles with another energy in an atomic system.

bombard   To direct a stream of high-energy particles or photons against a target.

bombardment   1. The process of directing high-speed electrons at an object, causing secondary emission of electrons, heating, fluorescence, disintegration, or production of X-rays.   2. The process of directing electrons or other high-speed particles at atoms or smaller particles.

bombardment-induced conductivity   An increase in the number of charge carriers in semiconductors or insulators, caused by bombardment with ionizing particles.

bond   A low-resistance junction of two conducting members. Semiconductor devices can have ball, die, face, stitch, thermocompression, ultrasonic, wedge, and wire bonds.

bonding   1. The process of connecting wires from the semiconductor chip (or die) bonding pads to the leadframe or package leads as in *wire bonding.*   2. The joining of metallic or nonmetallic materials by soldering, cementing or adhering, such as securing a semiconductor chip to a lead frame or substrate.

bond pad   A metallized area (typically 100×100 μm) on the periphery of a semiconductor die or chip for making a connection to one of the package pins. A small-diameter gold or aluminum wire is bonded to the pad area by application of heat and ultrasonic energy.

bond strength   1. A measure of the stress required to separate a layer of material from the base to which it is bonded, measured in kilograms per centimeter of width.   2. A measure of the stress required to separate a conducting wire bonded to a metallized semiconductor surface or metallized surface on an insulating substrate or lead frame, measured in kilograms.

bone conduction   The process by which sound is conducted to the inner ear through the cranial bones.

Boolean algebra   Algebra that deals with classes, propositions, ON/OFF circuit elements, and other nonnumerical elements associated with such operators as AND, NAND, NOR, NOT, and OR.

Boolean calculus   Boolean algebra that has been modified to include time, to permit the calculation of (a) states and events; (b) operators like "after," "while," "happen," "delay," and "before"; (c) classes whose members change with time; (d) circuit elements whose ON/OFF state changes from time to time, like delay lines, flip-flops, and sequential circuits; (e) step functions and their combinations.

Boolean function   A mathematical function in Boolean algebra.

boom   A movable mechanical arm for the support or suspension of a camera, microphone, or other instrument to permit it to function in a location where direct positioning is not practical or desirable. Examples are booms that position microphones overhead so they are unseen in telecasts or stage plays and booms for television or film cameras that permit operators to rapidly change overhead camera angles.

boost   To increase or amplify.

boost charge   A fast partial charge of a battery at a high current rate.

booster   1. A separate RF amplifier connected between an antenna and a television receiver to amplify weak signals.   2. An RF amplifier that amplifies and rebroadcasts a received television or communication radio signal at higher power without change in carrier frequency, for reception by the general public.

booster amplifier   An audio amplifier located between the mixer controls and the master volume control of a studio audio console. It compensates for mixing-circuit losses. A booster amplifier can increase the output voltage or current capability of an operational amplifier without polarity inversion or appreciable loss of accuracy.

booster voltage The additional voltage supplied by the damper transistor to the horizontal output, horizontal oscillator, and vertical output transistors of a television receiver to give greater sawtooth sweep output.

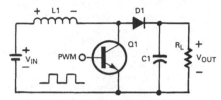

Boost converter is a form of switching power supply that has a pulse-width-modulated switching transistor in parallel with its load.

boost regulator A switching power supply with an input inductor that stores energy for transfer to the output when the shunt switching transistor is turned off. It accepts unregulated input voltage and produces higher regulated voltage. See also *buck regulator.*

boot 1. A protective flexible rubber or plastic sleeve that covers any part of a wire, cable, or connector. 2. To start up a computer by loading a program into memory from an external storage medium such as a disk or tape. This can be accomplished by first loading a small program whose purpose is to read the larger program into memory. The program is said to "pull itself in by its own bootstraps," and is the origin of the terms bootstrapping and booting.

boot disk *Startup disk.*

bootstrap A special coded instruction at the beginning of a computer routine to make the routine assemble itself in the computer.

bootstrap circuit A single-stage amplifier with its output load connected between the negative end of the drain or collector supply and the emitter or source. Signal voltage is applied between the gate and source or base and emitter. It refers to the "bootstrapping" action in gate bias that changes the voltage of the input source with respect to ground by an amount equal to the output signal.

bootstrap driver A circuit that produces a square pulse which drives a radar modulator tube. The duration of the square pulse is determined by a pulse-forming line. The circuit is called a bootstrap driver because voltages on both sides of the pulse-forming line are raised simultaneously with voltages in the output pulse, but their relative difference (on both sides of the pulse-forming line) is not affected by the considerable voltage rise in the output pulse.

bootstrapped sawtooth generator A circuit capable of generating a highly linear, positive, sawtooth waveform by the use of bootstrapping.

bootstrapping A technique for lifting a generator circuit above ground by a voltage value derived from its own output signal.

boresighting Initial alignment of a directional microwave or radar antenna system, using an optical procedure or a fixed target at a known location.

boresight tower A tower on which a visual target and an antenna fed from a signal generator are mounted. These targets are used for parallel alignment of the electrical axis of a receiving antenna and the optical axis of a telescope mounted on that antenna.

boron [B] An element with an atomic number of 5 and an atomic weight of 10.8 that is used as an *acceptor* dopant to form P-type silicon. It is also an ingredient in superconducting materials, and its oxides protect the melted crystal from oxidation during crystal growth.

BORS(C)HT An acronym for battery, overvoltage, ringing, supervision, (codec) hybrid, test; the functions performed by a subscriber line in a telephone exchange.

boule A pure crystal, as of silicon or other semiconductor, formed synthetically by rotating a small seed crystal while pulling it slowly out of molten material in a special furnace. The resulting atomic structure is that of a single crystal. For semiconductor devices, the boule is sawed into circular slices that in turn are cut into chips or dies.

bounce A sudden variation in television picture brightness or size, independent of illumination of the original scene.

boundary An interface between P- and N-type semiconductor materials, at which donor and acceptor concentrations are equal.

boundary element method A mathematical procedure for solving electromagnetic field problems by breaking the boundaries into smaller segments and using equations to calculate solution variables for those areas.

bound charge The residual charge held on a conductor by the inductive action of a neighboring charge.

bound circuit A circuit that limits the excursion of an output signal to an approximate maximum value for protection purposes or to a precise maximum value required for operational amplifier applications.

bound electron An electron bound to the nucleus of an atom by electrostatic attraction.

bow-tie antenna A dipole antenna that is made of stiff wire or flat sheet metal in the shape of a pair of triangles positioned vertically with their vertices close to each other. Each vertex is the connecting point to the transmission line. It is suitable for UHF television reception. Signal pickup is improved with a metal wire mesh screen behind the antenna.

boxcar circuit A radar circuit for sampling voltage waveforms and storing the latest value sampled. The term is derived from the flat, steplike segments of the output voltage waveform.

boxcar lengthener A pulse-lengthening circuit that lengthens a series of pulses without changing their height.

boxcars Long pulses separated by very short intervals.

bpi Abbreviation for *bits per inch.*

bps [b/s] Abbreviation for *bits per second.*

BPSK Abbreviation for *binary phase-shift keying.*

Bragg angle The angle determined by the propagation path of a light beam incident on a Bragg cell and a line perpendicular to the direction of sound propagation. It is used in X-ray orientation of quartz crystals for electronics applications.

Bragg cell

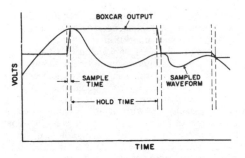

Boxcar-circuit output waveform.

Bragg cell   A thin piece of transparent crystal (such as tellurium oxide, lithium niobate, or gallium phosphide) with one or more piezoelectric transducers attached. A microwave signal excites a transducer to create a sound wave in the crystal at the microwave's frequency. The sound wave creates compressions and rarefactions in the crystal that change its optical index of refraction. Those changes in optical index diffract a portion of the incident light beam so that two beams emerge from the crystal: an undiffracted beam exiting at the angle of entry (the Bragg angle), and the other, diffracted, beam exiting at an angle proportional to the frequency of the sound wave. A Bragg cell can be a surface acoustic wave (SAW) device or a monolithic block.

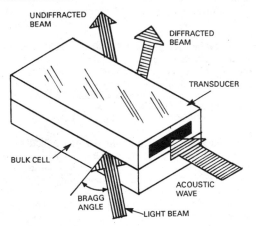

Bragg cell: the input acoustic wave diffracts light passing through the cell, modulating its amplitude and shifting its frequency to match those of the acoustic wave.

Bragg curve   1. A curve that shows the average number of ions per unit distance along a beam of initially monoenergetic ionizing particles, usually alpha particles, passing through a gas.   2. A curve that shows the average specific ionization of an ionizing particle of a particular kind as a function of its kinetic energy, velocity, or residual range.

Bragg scattering   Scattering of X-rays and neutrons by the regularly spaced atoms in a crystal, for which constructive interference occurs only at definite angles called Bragg angles.

Bragg's law   A statement of the conditions under which a crystal will reflect a beam of X-rays with maximum intensity.

Bragg spectrometer   An instrument for X-ray analysis of crystal structure, in which a homogeneous beam of X-rays is directed on the known face of a crystal, and the reflected beam is detected in a suitably placed ionization chamber. As the crystal is rotated, the angles at which Bragg's law is satisfied are identified as sharp peaks in the ionization current. Also called crystal spectrometer and ionization spectrometer.

braided wire   A tube of fine wires woven around a conductor or cable for shielding, or used alone in flattened form as a grounding strap.

brain wave   A rhythmic fluctuation of voltage between parts of the brain, ranging from about 1 to 60 Hz and 10 to 100 μV. It is called a delta wave when the frequency is below 9 Hz, an alpha wave when the frequency is 9 to 14 Hz, and a beta wave when the frequency is above 14 Hz.

branch   1. A portion of a network that consists of one or more two-terminal elements in series.   2. A product that results from one mode of decay of a radioactive nuclide which has two or more modes of decay.   3. A line segment that joins two nodes, or joins one node to itself.   4. A set of computer instructions executed between two successive decision instructions.   5. *Conditional jump.*

branch instruction   An instruction that makes the computer choose between alternative subprograms, depending on the conditions determined by the computer during the execution of the program.

branch point   1. A terminal common to two or more branches of a network, or a terminal on a branch of a network. Also called junction point.   2. A location in a computer routine at which one of two or more choices is made.

branch transmittance   The ratio of branch output signal to branch input signal.

brassboard circuit   See *breadboard circuit.*

breadboard circuit   A prototype or sample circuit intended to prove the feasibility of a principle, design, circuit, or system without regard to optimum or economic layout or packaging. This term is synonymous with *brassboard circuit.*

break   1. Interruption of a radio transmission, as for sending in the opposite direction.   2. A fault in a circuit.

break-before-make contacts   Contacts that interrupt one circuit before establishing another.

break contact   *Back contact.*

breakdown   1. A disruptive discharge through insulation, involving a sudden and large increase in current through the insulation because of complete failure under electrostatic stress.   2. Initiation of a desired discharge between two electrodes in a gas, occurring at a voltage dependent on gas density, electrode shape, electrode spacing, and polarity.   3. An undesired runaway increase in an electrode current in a gas tube.   4. Loss of blocking action in a reverse-biased semiconductor PN junction, causing a sudden current increase that is not normally destructive.

breakdown diode   A semiconductor diode in which the reverse-voltage breakdown mechanism is based either on the Zener effect or the avalanche effect.

breakdown region   The entire region of the semiconductor-diode voltampere characteristic beyond the initiation of breakdown for increasing magnitude of reverse current.

breakdown voltage 1. The voltage measured at a specified current in the breakdown region of a semiconductor diode. It is also called Zener voltage. 2. The voltage at which breakdown occurs in a dielectric or in a gas tube. 3. The maximum AC or DC voltage that can be applied from the input to output (or chassis) of a circuit without causing damage.

breakout A joint at which one or more conductors are brought out from a multiconductor cable.

breakout box [BOB] A device for transmission line testing.

breakover In a thyristor, the transition from the forward-blocking to the forward-conducting state.

breakpoint A point in a computer program at which conditional interruption can occur to permit visual check, printing out, or other special action.

breakpoint instruction An instruction that will cause a computer to stop or transfer control to a supervisory routine for monitoring the progress of the interrupted program.

breakpoint switch A manually operated switch that controls conditional operation at breakpoints.

breezeway The time interval between the trailing edge of the horizontal synchronizing pulse and the start of the color burst in the standard NTSC color television signal.

Brewster angle The angle of incidence for which a wave polarized parallel to the plane of incidence is wholly transmitted, with no reflection.

Brewster window A special glass window at opposite ends of some gas lasers to transmit the laser output beam while reflecting other light.

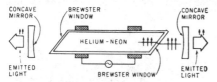

Brewster window in a gas laser.

[BRI] ISDN Abbreviation for *Basic Rate Interface for ISDN*. See *Integrated Services Digital Network*.

bridge An instrument or circuit that has four or more arms.

bridge amplifier An amplifier across the output of a bridge, in place of a meter. It can be an operational amplifier to which precision external resistors are added to set gain or an instrumentation amplifier that has its own gain-setting resistor network.

bridge circuit A circuit that consists basically of four sections connected in series to form a diamond. An AC voltage source is connected between one pair of opposite junctions, and an indicating instrument or output circuit is connected between the other pair of junctions. When the bridge is balanced, the output is zero. See also *Wheatstone bridge*.

bridge converter A switching converter topology for off-line power supplies with four switching transistors (full-bridge) or two transistors (half-bridge). It offers high output power and low ripple, but is complex and costly and can have low reliability.

bridged-T network A T network with a fourth branch connected across the two series arms of the T, between an input terminal and an output terminal.

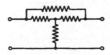

Bridged-T network.

bridge hybrid *Hybrid junction.*

bridge rectifier A *full-wave bridge* rectifier with four elements connected in series, as in a bridge circuit. Alternating voltage is applied to one pair of opposite junctions, and direct voltage is obtained from the other pair of junctions.

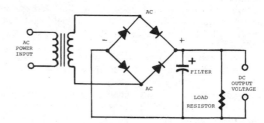

Bridge rectifier includes four silicon diodes to provide a smooth DC output.

bridge voltage doubler A voltage doubler that combines the conventional *voltage doubler* and the *bridge rectifier circuit*. If diodes D3 and D4 are removed, the circuit is a conventional voltage doubler. See also *cascade voltage doubler* and *voltage multiplier*.

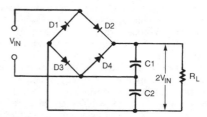

Bridge voltage doubler combines a conventional voltage doubler and a bridge rectifier circuit.

bridging 1. Connecting one electric circuit in parallel with another. 2. Selector-switch action in which the movable contact is wide enough to touch two adjacent contacts so that the circuit is not broken during contact transfer.

bridging amplifier An amplifier with an input impedance sufficiently high so that its input can be bridged across a circuit without substantially affecting its signal level.

bridging gain The ratio of the power a transducer delivers to a specified load impedance under specified operating conditions to the power dissipated in the reference impedance across which the input of the transducer is bridged, expressed in decibels.

bridging loss The reciprocal of the bridging gain ratio, typically expressed in decibels.

brightening pulse   A pulse applied either to the grid or cathode of a radar cathode-ray tube at the beginning of the sweep, to intensify the beam during the sweep.

brightness   1. The characteristic of light that gives a visual sensation of more or less light.   2. The former name for *luminance.*

brightness control   A control that varies the brightness of the fluorescent screen of a cathode-ray tube by changing the grid bias of the tube, thereby changing the beam current. It is also called a *brilliance control* and an *intensity control.*

brilliance   The degree to which higher audio frequencies are present when a sound recording is played back.

brilliance control   *Brightness control.*

Brillouin function   A mathematical function that relates the magnetic moment, applied magnetic field, and temperature of a paramagnetic material to its magnetic susceptibility.

Brillouin scattering   Interaction of sound waves at microwave frequencies with light waves from laser-generated coherent sources.

broadband   A transmission facility whose bandwidth is greater than that available on voice-grade facilities. It is also called *wideband.*

broadband amplifier   An amplifier that has essentially flat response over a wide range of frequencies. Also called wideband amplifier.

broadband antenna   An antenna that will function satisfactorily over a wide range of frequencies, such as for all 12 VHF television channels.

broadband channel   A data-transmission channel that can transmit frequencies higher than the normal voice-grade line limit of 3 to 4 kHz. A broadband channel can carry many voice or data channels simultaneously or it can carry high-speed single-channel data.

broadband interference   Interference distributed over a wider spectrum of frequencies than the tuning range of the affected receiver.

broadband klystron   A klystron that has three or more resonant cavities which are externally loaded and stagger-tuned to broaden the bandwidth.

broadband noise   Thermal noise that is uniformly distributed across the frequency spectrum at a wide range of energy levels.

broadcast   A television or radio transmission intended for public reception.

broadcast band   The band of frequencies extending from 535 to 1605 kHz, corresponding to assigned carrier frequencies that increase in multiples of 10 kHz between 540 and 1600 kHz for the United States. Also called standard broadcast band.

broadcasting   Transmission of television and radio programs by radio waves for public reception.

broadcasting-satellite service [BSS]   A radio and television service that provides signals that are relayed by satellite for direct reception by the general public.

broadcasting service   A radio communications service in which the transmissions, including sound and television, are intended for direct reception by the general public.

broadcast station   A television or radio station that transmits programs to the general public.

broadcast transmitter   A transmitter for a commercial AM, FM, or television broadcast channel.

broadside   Perpendicular to an axis or plane.

broadside array   An antenna array whose maximum radiation direction is perpendicular to the line or plane of the array. The spacing of the elements in the array is typically kept at less than the wavelength. If the array is horizontal, the dipole axes are vertical and the radiation is vertically polarized. If the array line is vertical and the dipole axes are horizontal, the radiation will be horizontally polarized.

broad tuning   Poor selectivity in a radio receiver, causing reception of two or more stations at a single setting of the tuning dial.

bromine [Br]   A nonmetallic liquid with an atomic number of 35 and an atomic weight of 79.9.

Brown and Sharpe gage [B and S gage]   *American wire gage.*

brownout   Normally a voltage reduction initiated by the utility to counter excessive demand on its electric power generation and distribution system.

Bruce antenna   *Rhombic antenna.*

brush   A conductive metal or carbon block that makes sliding electric contact with a moving armature.

brush encoder   An encoder with brushes that make contact with conductive segments on a rotating or linearly moving surface to convert positional information to digitally encoded data.

brushless DC motor   A permanent magnet motor for light loads that is commutated electronically, typically by switching transistors under *Hall-effect* sensor control.

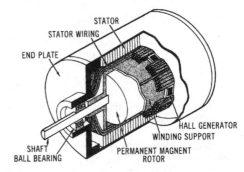

Brushless DC motor with a permanent-magnet rotor and Hall-effect generator for commutation.

b/s   Abbreviation for *bit per second.*

BSC   Abbreviation for *binary synchronous communication.*

B scope   A radarscope that produces a B display.

BSS   Abbreviation for 1. *base station subsystem* and 2. *broadcasting-satellite service.*

BST   Abbreviation for *binary synchronous transmission.*

B station   The loran station whose signal is always transmitted more than half a repetition period after the signal from the master or A station of the pair. It is also called a *slave station.*

B supply   A power source that provides a positive voltage for the anode and other electrodes of an electron tube.

BT-cut crystal   A crystal plate cut from a plane that is rotated about an X axis so the angle made with the Z axis is approximately −49°. This cut has an essentially zero temperature coefficient.

BTO   Abbreviation for *bombing through overcast.*

bubble memory   A computer memory in which the presence or absence of a magnetic bubble in a localized region of a thin magnetic film designates a 1 or 0. Bubbles representing stored data can be moved by selectively exciting thin-film conductive loops placed on the surface to produce localized magnetic fields. Storage capacity can be well over 1 Mb/in^3. Also called magnetic-bubble memory (MBM).

buck-boost converter   *Flyback converter.*

buck-derived converter   *Forward converter.*

buck regulator   A basic switching power supply with a series switching transistor that "chops" the input voltage and applies the pulses to an averaging inductive-capacitive (LC) filter. Its output voltage is lower than its input voltage. See also *boost regulator.*

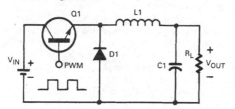

Buck converter is a switching power supply that has its pulse-width-modulated switching transistor in series with the load.

bucket brigade device [BBD]   A semiconductor device whose majority carriers store charges that represent information, and whose minority carriers transfer charges from point to point in sequence, much as buckets of water were once passed along a line of volunteer firemen. Applications include shift registers and delay lines.

bucket counter   A counter that replaces a binary counter in a ramp-type analog-to-digital converter to eliminate the need for a stable clock frequency.

bucking coil   A coil connected and positioned so that its magnetic field opposes the magnetic field of another coil. The hum-bucking coil of an excited-field loudspeaker is an example.

bucking voltage   A voltage that has exactly opposite polarity to that of another voltage against which it acts.

buffer   1. An area of a computer's memory reserved for such purposes as holding graphical information to be displayed on the screen or text characters being read from a peripheral device. It can also be a holding area for transferring data between devices operating at different speeds such as the CPU and printer or disk drive. 2. A storage circuit that compensates for differences in rates of data flow when it is transmitting information from one circuit to another. 3. *Buffer amplifier.*

buffer amplifier   An amplifier located after an oscillator or other critical stage to isolate it from the effects of load impedance variations in subsequent stages. It is also called a *buffer* and a *buffer stage.*

buffer stage   *Buffer amplifier.*

buffer memory   A synchronizing element between two different forms of memory in a computer. Computation

continues while transfers take place between buffer memory and the secondary or internal memory.

bug   1. A semiautomatic code-sending key designed so that movement of a lever to one side produces a series of correctly spaced dots, and movement to the other side produces a single dash.   2. Slang term for an error in computer software.   3. An electronic listening device, generally concealed, for gaining information surreptitiously.

bugging   The use of electronic eavesdropping devices. These include high-gain directional microphones, hidden microphones wired directly to listening or recording devices or miniature radio transmitters, and inductive pickups or direct wire taps on telephones or telephone lines to monitor both sides of conversations surreptitiously.

building-out section   A short section of transmission line, either open or short-circuited at the far end, shunted across another transmission line for tuning or matching purposes.

built-in antenna   An antenna located inside the cabinet of a radio or television receiver.

bulb   *Envelope.*

bulk acoustic wave   An acoustic wave that travels through a piezoelectric material, as in a quartz delay line. (With a surface acoustic wave, propagation is only on the surface.) Also called volume acoustic wave.

bulk-acoustic-wave delay line   A delay line whose delay is determined by the distance traveled by a bulk acoustic wave between input and output transducers mounted on a piezoelectric block.

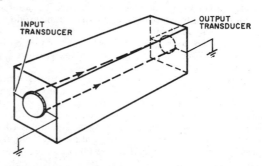

Bulk-acoustic-wave delay line.

bulk diode   A semiconductor microwave diode based on the bulk effect, such as Gunn diodes and diodes operating in limited space-charge-accumulation modes.

bulk effect   An effect that occurs within the entire bulk of a semiconductor material rather than in a localized region or junction.

bulk-effect device   A semiconductor device that depends on a bulk effect, as in Gunn and avalanche devices.

bulk micromachining   A micromachining technology based on single-crystal silicon etching. Micromechanical structures developed with this technology are made from either silicon crystal or from deposited or grown layers of silicon. See also *micromachining.*

bulk photoconductor   A photoconductor that has high power-handling capability and other unique properties which depend on the semiconductor and doping materials used. Examples include cadmium selenide, germanium, indium antimonide, indium arsenide, lead selenide, and silicon.

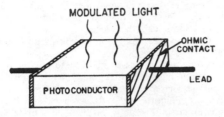

Bulk-photoconductor geometry for use as a microwave demodulator.

bulk resistor An integrated-circuit resistor whose N-type epitaxial layer of a semiconducting substrate functions as a noncritical high-value resistor. The spacing between the attached terminals and the sheet resistivity of the material together determine the resistance value.

bump Metal pads added to discrete device dies or integrated-circuit chips so they can be bonded to a substrate while elevating them above the upper surface of the hybrid microcircuit or printed circuit board substrate.

buncher resonator The first or input cavity resonator in a velocity-modulated tube, next to the cathode. Here the faster electrons catch up with the slower ones to produce bunches of electrons. Also called input resonator.

bunching The flow of electrons from cathode to anode of a velocity-modulated tube as a succession of electron groups rather than as a continuous stream. It is a direct result of the differences of electron transit time produced by the velocity modulation.

buried diffused layer A low-resistance layer formed by impurity diffusion in semiconductor material before formation of a surface epitaxial layer.

burned-in image An image that persists in a fixed position in the output signal of a television camera tube after the camera has been turned to a different scene.

burn-in The operation of a new circuit or product for a specified time under accelerated aging conditions (that could include both elevated temperature and humidity) prior to the product's shipment to stabilize its circuitry and sort out premature failures.

burnout Failure of a device because of excessive heat produced by excessive current.

burst amplifier In a color television receiver, an amplifier stage keyed into conduction and amplification by a horizontal pulse at the exact instant of each arrival of the 3.58-MHz color-burst signal. Also called chroma bandpass amplifier.

burst generator *Tone-burst generator.*

burst pedestal *Color-burst pedestal.*

burst rate The rate of transmission of the burst of bits during the transmission interval from a time-division multiaccess (TDMA) terminal.

burst separator The circuit in a color television receiver that separates the color burst from the composite video signal.

burst transmission Transmission in which messages are stored for a given time, then released at from 10 to 100

or more times the normal speed. The received signals are recorded and then slowed down to the normal rate for the user.

bus 1. Parallel conductor that forms a major interconnection route between the computer system CPU and its peripherals such as input/output (I/O) devices, monitors, or printers. It can carry data, addresses, and power. 2. A heavy, conductive wire or bar, typically made of copper, for power or ground connections.

bus architecture A physical layout or design guide for a bus system designed to operate according to established protocol rules governing the transfer of data. Examples of bus architectures are VMEbus, IBM bus, and S-bus. Plug-in circuit boards are designed and built to conform to the protocol standards that govern circuit board dimensions and termination configurations.

busbar A heavy, rigid, metallic conductor, usually uninsulated, that carries a large current or makes a common connection between several circuits.

bus driver An integrated circuit added to the data bus in a computer to provide sufficient drive to the CPU when several peripheral devices are connected to the bus. Drivers are necessary because of capacitive loading which slows down the data rate and prevents proper time sequencing in computer system operation.

busing Connecting a large number of data sources to a common bus.

bus sizer A semiconductor device that permits the user to choose between synchronous and asynchronous timing control when connecting microprocessors to access peripherals or buses. It replaces PALs, latches, and transceivers.

Butler oscillator A crystal-controlled oscillator that has the crystal in a feedback loop between the emitter and collector. The transistor's emitter current is directly proportional to the strength of the radio-frequency signal. The crystal behaves like a narrow-bandpass filter. It feeds some of the RF energy back into the emitter, and the emitter current forms clean sine waves that are low in harmonics.

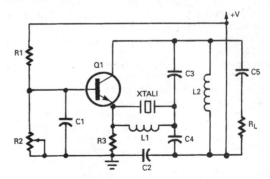

Butler oscillator: its frequency is controlled by a crystal in its emitter-to-collector circuit.

butt contact A hemispherically shaped contact that mates against a similarly shaped contact. The contacts are usually held together by spring pressure.

butterfly capacitor A variable capacitor that has stator and rotor plates shaped like butterfly wings. Each stator plate has an outer ring that forms an inductance which varies with rotor position. Both inductance and capacitance are at a minimum when the stator and rotor plates form the four quarters of a circle and increase simultaneously to a maximum when the plates are rotated to the fully meshed position. This greatly increases the tuning range when the capacitor functions as a tuned circuit in VHF and UHF circuits.

Butterworth filter A filter that has essentially flat amplitude response in the passband and an attenuation rate beyond cutoff of 6 dB per octave for a single-pole filter. Its transient response is much better than for a comparable Chebyshev filter.

Butterworth response A maximally flat amplitude response in the passband of an active filter, combined with moderate settling time and moderate overshoot.

button cell A small, low-voltage power cell made as a metal disk with a diameter of about 0.4 in (11 mm) and a height of about 0.2 in (5 mm). One face is the *positive terminal* [*anode*], and the opposite face is the *negative terminal* [*cathode*]. Examples include the *lithium cell, silver-oxide cell,* and the *zinc/air cell.*

buzzer An electromagnetic device that has an armature which vibrates rapidly, producing a buzzing sound.

BW Abbreviation for *bandwidth.*

BWA Abbreviation for *backward-wave amplifier.*

BWO Abbreviation for *backward-wave oscillator.*

bypass A low-impedance path provided around part or all of a circuit.

bypass capacitor A capacitor connected to provide a low-impedance path for RF or AF currents around a circuit element.

bypassed mixed highs The mixed-highs signal, containing frequencies between 2 and 4 MHz, that is shunted around the chrominance-subcarrier modulator or demodulator in a color television system.

B-Y signal A blue-minus-luminance color-difference signal used in color television. It is combined with the luminance signal in a receiver to give the blue color-primary signal.

byte From the expression "by eights." A group of eight contiguous bits (binary digits) treated as a unit in computer processing. A byte can store one alphanumeric character. A kilobyte (KB or Kbyte) is 1024 bytes or 8192 bits. A megabyte (MB or Mbyte) is 1024 kilobytes or 1,048,576 bytes or 8,388,608 bits. Eight bits can represent any number up to 255.

bytes per second [bytes/s] A rating that specifies the speed of a digital transmission system.

bytes/s Abbreviation for *bytes per second.*

C

c 1. Abbreviation for *centi-*. 2. Abbreviation for *curie*. 3. Abbreviation for *character*.

C 1. Abbreviation for *capacitor*. 2. Abbreviation for *capacitance*. 3. Abbreviation for *Celsius* (preferred) or *centigrade*. 4. Abbreviation for *coulomb*. 5. Symbol for transistor *collector* on schematic diagrams. 6. A general-purpose programming language descended from BCPL (Basic Combined Programming Language). The *Unix* operating system is written in C. There is also a later version called C++.

°C Abbreviation for degree *Celsius*.

CaAs Symbol for *cadmium arsenide*.

cabinet The housing for a radio receiver, television receiver, or other electronic equipment.

cable A transmission line, group of transmission lines, or group of insulated conductors mechanically assembled in compact flexible form.

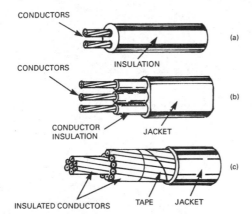

Cables for electronics include an insulated dual-conductor (*a*), triple-conductor (*b*), and multiconductor (*c*).

cable clamp A clamp that gives mechanical support to a cable at the rear of its plug or outlet.

cable television [CATV] A television program distribution system in which signals from all local stations and usually a number of distant stations are picked up by one or more high-gain antennas at elevated locations, amplified on individual channels, then fed directly to individual receivers of subscribers by overhead or underground coaxial cable. Used to improve reception and to make more stations available in a given area. The system sometimes includes facilities for originating local programs, time and weather reports, news bulletins, and other services. Also called community antenna television.

cache memory A small, high-speed semiconductor buffer memory located close to the central processing unit (CPU) and main memory of a computer to give the CPU faster access to blocks of data than could be taken directly from the larger, slower memory. The cache is automatically loaded with data or instructions from adjacent locations that are logically related to the most recently accessed data or instructions because it is the information most likely to be requested next. The cache is typically made from ECL logic to attain the highest, fastest access times.

cache miss rate The number of times that an operand needed by the CPU is not found in cache memory, and therefore a measure of cache algorithm performance.

CAD Abbreviation for *computer-aided design*.

cadmium [Cd] A metallic element widely used as a plated coating on steel hardware for electronic equipment because it improves solderability and surface conductivity and prevents corrosion. It has an atomic number of 48 and an atomic weight of 112.4.

cadmium cell A standard cell used as a voltage reference. At 20°C its voltage is 1.0186 V.

cadmium-selenide photoconductive cell A photoconductive cell that uses cadmium selenide as the semiconductor material. It has fast response time and high sensitivity to longer light wavelengths, such as those of incandescent lamps and some infrared light sources.

cadmium sulfide [CdS]   A semiconductor that has a forbidden bandgap of 2.4 eV and a maximum operating temperature of 870°C.

cadmium-sulfide photoconductive cell   A photoconductive cell in which a small wafer of cadmium sulfide provides an extremely high dark-light resistance ratio. Some models can function directly as a light-controlled switch that operates directly from a 120-VAC power line.

cadmium-telluride detector   A photoconductive cell capable of operating continuously at ambient temperatures up to 750°F (400°C) for solar cells and infrared, nuclear-radiation, and gamma-ray detectors.

CAE   Abbreviation for *computer-aided engineering*.

cage   To lock the gyroscope of a gyro-controlled instrument in a fixed position with reference to its case.

cage antenna   A variation of the open *dipole* except that it has many active elements arranged horizontally in a circular format attached to each dipole element. The active elements are connected in parallel by circular rings to form a cylindrical cagelike structure. The antenna is fed at its center by a two-wire transmission line or coaxial cable.

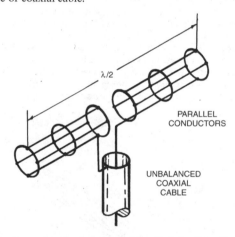

Cage antenna is a dipole antenna with added parallel active elements.

CAI   Abbreviation for *computer-aided instruction*.
cal   Abbreviation for *calorie*.
calcium [Ca]   A silver-white soft metallic element used in cathode coatings for some types of phototubes. It has an atomic number of 20 and an atomic weight of 40.08.

calculator   A circuit that performs arithmetic operations based on numerical data which is entered by pressing numerical and control keys. Pocket-size electronic versions operate from batteries or solar cells. Larger desk versions operate from AC power and might have a paper-tape printer.

calibrate   1. To determine, by measurement or comparison with a standard, the correct value of each scale reading on a meter or other device.   2. To determine the settings of a control that correspond to particular values of voltage, current, frequency, or some other characteristic.

calibration curve   A plot of calibration data that gives the correct value for each indicated reading of a meter or control dial.

calibration marker   A marker line or circle that divides a radar screen into accurately known intervals for determination of range, bearing, height, or time.

call   1. A radio transmission that identifies the transmitting station and designates the station for whom the transmission is intended.   2. To transfer control to a specified closed computer subroutine.

call letters   Identifying letters, and numerals, assigned to radio and television stations by the Federal Communications Commission and other regulatory authorities throughout the world.

calorescence   The production of visible light by infrared radiation. The transformation is indirect because the light is produced by heat, not by any direct change of wavelength.

calorie [cal]   The metric unit of quantity of heat. It is approximately the amount of heat required to raise the temperature of 1 g of water 1°C.

calorimeter   An instrument that measures quantity of heat. It can measure microwave power in terms of its heating effect.

CAM   1. Abbreviation for *computer-aided manufacturing*.   2. Abbreviation for *content-addressable memory*.

camcorder   A portable video camera that incorporates its own video tape and an audio tape recorder. It is capable of recording full-color video tapes with sound. Many are equipped with telephoto lenses to permit panning from near to distant objects or scenes.

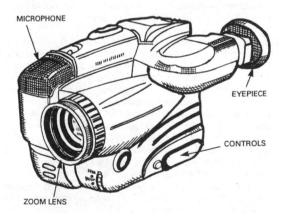

Camcorder is a handheld color video camera combined with an audio recorder that can also have a zoom lens to magnify distant objects.

Campbell bridge   An AC bridge for measuring mutual inductances. It permits a comparison of unknown and standard mutual inductances with different values.

Campbell-Colpitts bridge   An AC bridge that measures capacitance by the substitution method.

Canadian Standards Association [CSA]   An independent organization that establishes safety standards and tests materials, components, and products to determine their compliance for the Canadian market. Similar in to the Underwriters Laboratories (UL) in the United States.

candela [cd]   The SI unit of luminous intensity. One candela is defined as the luminous intensity of 1/60 cm² of a black-body radiator operating at the temperature of solidification of platinum. Formerly called candle.

candela per square meter [cd/m²]   The SI unit of luminance. The term nit is sometimes used for this unit.

candle   *Candela.*

candlepower   Luminous intensity expressed in candelas.

cannibalize   To remove serviceable parts from one piece of equipment to repair another piece of equipment.

capacitance [C; mathematical symbol *C*]   1. The electrical size of a capacitor. The basic unit is the farad, but the smaller microfarad, nanofarad, and picofarad units are commonly used.   2. The property that exists whenever two conductors are separated by an insulating material, permitting the storage of electricity.

capacitance bridge   A bridge for comparing two capacitances, such as a Schering bridge or a universal bridge.

capacitance level indicator   A level indicator that uses the material being monitored as the dielectric of a capacitor formed by a metal tank and an insulated electrode mounted vertically in the tank. The increase in capacitance with level can be measured accurately for depth ranges up to 200 ft (60 m).

capacitance-loop directional coupler   A directional coupler that contains a coupling link which is shorter than a quarter wavelength when positioned lengthwise in the waveguide.

capacitance meter   An instrument that measures the capacitance values of capacitors or circuits which contain capacitance.

capacitance multiplier   A circuit that uses operational amplifiers to multiply the value of an input capacitor by a fixed or adjustable factor.

capacitance-operated intrusion detector   A perimeter alarm system triggered by a signal generated by a change in antenna-to-ground capacitance caused by the presence of a moving object, presumed to be an intruder. An alarm is sent when the intruder approaches an antenna strung horizontally around a protected area.

capacitance relay   An electronic relay that responds to a small change in capacitance, such as that created by bringing a hand near a pickup wire or plate.

capacitance standard   *Standard capacitor.*

capacitive coupling   The transfer of energy from one circuit to another by a capacitor.

capacitive diaphragm   A resonant window in a waveguide that provides the equivalent of capacitive reactance at the frequency being transmitted.

capacitive-discharge ignition   An automotive ignition system that has a silicon controlled rectifier (SCR) which triggers a spark-plug discharge. Energy stored in a capacitor is discharged across the spark-plug gap through a step-up pulse transformer.

capacitive feedback   Feedback through a capacitor connected between the output and input of a circuit.

capacitive load   A load whose capacitive reactance exceeds its inductive reactance. This load draws a leading current.

capacitive post   A metal post or screw that extends across a waveguide at right angles to the E field, to provide capacitive susceptance in parallel with the waveguide for tuning or matching purposes.

capacitive reactance [X_c]   Reactance due to the capacitance of a capacitor or circuit. Capacitive reactance is measured in ohms and is equal to 1 divided by 6.28 *fC,* where *f* is in hertz and *C* is in farads.

capacitive transducer   A sensor that depends on changes in capacitance due to such variables as acceleration, audio sound field, displacement, flow, fluid level, force, vacuum, pressure, and velocity.

capacitive tuning   Tuning by means of a variable capacitor.

capacitive tuning screw   See *waveguide tuning screw.*

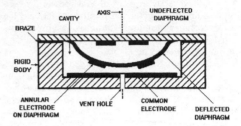

Capacitive (pressure) transducer made from quartz is rugged and has excellent thermal and elastic stability.

capacitive window   A conductive diaphragm that extends into a waveguide from one or both sidewalls, producing the effect of a capacitive susceptance in parallel with the waveguide.

capacitor [C]   A component that consists essentially of two conducting surfaces separated by a dielectric material such as air, paper, mica, ceramic, glass, or polyester film. A capacitor stores electric energy, blocks the flow of direct current, and permits the flow of alternating current to a

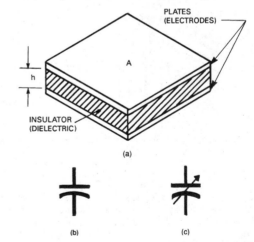

Capacitor is formed (*a*) when two parallel-spaced metal plates or electrodes are separated by a dielectric, (*b*). Symbol for a fixed capacitor (*c*) and variable capacitor.

degree dependent on its capacitance value and the frequency. It was formerly called a *condenser.*

capacitor color code   A method of marking the value on a capacitor by dots or bands of colors as specified in the EIA color code.

capacitor-input filter A power-supply filter whose shunt capacitor is the first element after the rectifier.

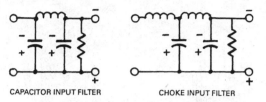

CAPACITOR INPUT FILTER   CHOKE INPUT FILTER

Capacitor-input filter and choke-input filter.

capacitor integrator A circuit containing a capacitor that is charged by a current which is proportional to a function to be integrated. The capacitor voltage after a specified integration period equals the result. The concept is applied in analog-to-digital circuitry.

capacitor ionization chamber *Capacitor R-meter.*

capacitor loudspeaker *Electrostatic loudspeaker.*

capacitor microphone A microphone consisting of a flexible metal diaphragm and a rigid metal plate that together form a two-plate air capacitor. Sound waves set the diaphragm in vibration, producing capacitance variations that are converted into audio signals by a suitable amplifier circuit. It is also called an *electrostatic microphone.*

capacitor-start motor A capacitor motor that has a capacitor in its circuit only during the starting period. The capacitor and its auxiliary winding are disconnected automatically by a centrifugal switch or other device when the motor reaches a predetermined speed. The motor then runs as an induction motor.

capacitor start-run motor *Permanent-split capacitor motor.*

capacity 1. The rated or maximum load or capability of a machine defined by such terms as revolutions per minute, miles per gallon, or load in tons. 2. The charge a battery (or cell) can hold (ampere-hours) or its energy (watt-hours). 2. In capacitors (also called capacitance), a measure of electrical energy storage in farads, typically microfarads, μF, or picofarads (pF).

capillary In *wire bonding,* a tungsten sleeve that functions as a feed tube for fine wire and as a compression tool for *wire bonding.*

capstan The drum that rotates against the tape in a magnetic-tape recorder, pulling the tape past the head at a constant speed during recording and playback.

capstan idler A rubber-tired roller that holds the magnetic tape against the capstan by spring pressure.

captive fastener A screw-type fastener that does not drop out after it has been unscrewed.

capture effect The suppression of a weak frequency modulation (FM) signal by a stronger FM signal when both are at or near the same frequency.

capture ratio A measure of the ability of an FM tuner to reject the weaker of two stations that are on the same frequency. The lower the ratio in decibels of desired and undesired signals, the better the performance of the tuner.

carbon [C] A nonmetallic element with an atomic number of 6 and an atomic weight of 12 that is used to make brushes for DC motors and electrodes in primary power cells and batteries. Powdered carbon mixed with clay

forms the bulk resistance material of discrete molded carbon resistors. Carbon-based ink or film is screened on ceramic mandrels or substrates to form carbon-film resistors. Granular carbon is the variable resistance element in carbon microphones. Graphite and diamonds are forms of carbon.

carbon-composition resistor A bulk resistor that is formed by mixing powdered carbon with a clay binder. The composition is typically molded into a cylinder with an axial lead inserted at each end before it is furnace-fired. Its resistance value decreases with increasing temperature and is affected by humidity.

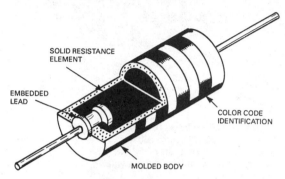

Carbon-composition resistor has a molded monolithic carbon-composition element and an insulating jacket.

carbon-dioxide laser [CO_2 laser] A gas laser whose principal active gas is carbon dioxide (CO_2). The addition of other gases such as helium and nitrogen increase its output power. Because it can operate continuously at kilowatt output levels, it can perform such applications as cutting, drilling, heating, and welding. It is also used for long-range communication and as an infrared illuminator for directing laser-guided bombs and missiles. This laser can be pumped by chemical, electric, or optical energy. The highest average output power is achieved by pulsing the laser.

carbon-film resistor A fixed or variable resistor made by screening a thin film of carbon mixed with a binder on a ceramic mandrel or substrate. Precise resistive values can be obtained by trimming away excess carbon film. A leaded

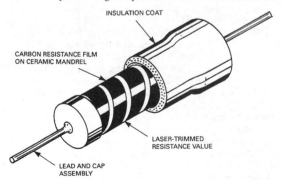

Carbon-film resistor: a thin film of carbon deposited on a ceramic mandrel is helically trimmed to a precise resistive value.

end cap is put on each end of the resistor. Carbon-film resistive elements are included in trimmer and control potentiometers. The most common fixed resistor value is ¼ W.

carbon microphone A microphone with a flexible diaphragm that moves in response to sound waves and applies a varying pressure to a container filled with carbon granules. This causes the resistance of the microphone to vary correspondingly.

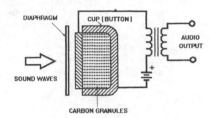

Carbon microphone contains carbon granules whose resistance value is altered by scattering by sound pressure.

carbon-monoxide laser [CO laser] A gas laser whose active gas is carbon monoxide (CO). Its highest output is achieved at wavelengths of 4.9 to 5.7 μm.

carbon resistor See *carbon-composition resistor* and *carbon-film resistor.*

carbon-zinc cell See *Leclanche cell.*

carborundum A compound of carbon and silicon that is an abrasive material. In a crystal form it will detect and rectify radio waves.

carborundum thermistor A thermistor with a sensing element made from a mixture of carborundum and ceramic formed into a disk or rod that offers a wide range of negative temperature coefficient (NTC) characteristics and resistance values.

carborundum varistor A varistor containing a voltage-dependent resistor that is a mixture of carborundum and ceramic materials fired at high temperature. The nonlinear voltampere characteristic is expressed as $I = KE^n$, where n is typically a value between 1 and 6 and K is a constant.

carcinotron oscillator A microwave M-type backward-wave oscillator. Electrons interact with the slow-wave structure in a crossed-field space. The slow-wave structure is in parallel with the *sole* electrode. A DC electric field exists between the grounded slow-wave structure and the negative sole. A DC magnetic field interacts with it, and electrons from the cathode are bent through a 90° angle by the magnetic field. The electrons interact with a backward-wave space harmonic, and the energy flows in a direction that opposes electron motion. The slow-wave structure is terminated at the collector end, and the RF signal is taken from the electron gun end. Efficiency ranges from 30 to 60% because it is a crossed-field device.

card A small circuit board.

card drive A peripheral device for microcomputers that allows memory cards to be read and written. These drive interfaces can emulate computer-disk-drive interfaces.

card-edge connector A rectangular connector that mates with pads or "fingers" which extend to the edge of a printed-circuit board on one or both sides to complete the connection with circuit board components. The connector has internal rows of opposing, spring-loaded contacts that

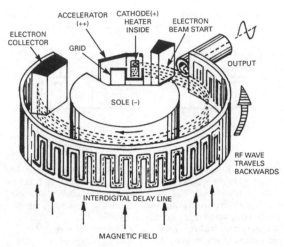

Carcinotron: a crossed-field magnetron (M)-type backward wave oscillator.

clamp securely to the edge of the circuit board. Some versions are fastened to the board with screws to assure reliable contact during shock and vibration.

card information structure [CIS] The header at the beginning of a PCMCIA/JEIDA-format memory card that supplies the basic format and card technology information to the host system for the card. It is also called the *metaformat.*

cardiac pacemaker *Pacemaker.*

cardiogram *Electrocardiogram.*

cardioid diagram A polar diagram in the shape of a heart. An example is the radiation pattern of a dipole antenna with a reflector.

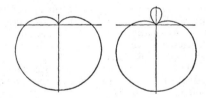

Cardioid a diagram representing a pickup pattern of directional microphone, and a modified cardioid diagram (right) having still greater suppression of sounds arriving from sides and rear. The microphone is at the intersection of the horizontal and vertical lines.

cardioid microphone A microphone that has a heart-shaped or cardioid response pattern. It has nearly uniform response for a range of about 180° in one direction and minimum response in the opposite direction. In one form it is a combination of dynamic-microphone and ribbon-microphone elements.

cardiometer A medical electronic instrument that measures the force of the action of the heart.

cardiotachometer An electronic amplifier that times and records pulse rates of the heart.

Carey-Foster bridge A bridge for measuring the mutual inductance of inductors in terms of capacitance and capacitance in terms of mutual inductance. It is also called the *Heydweiller bridge.*

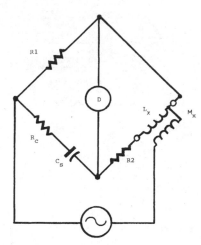

Carey-Foster bridge can measure the mutual inductance of inductors.

carnauba wax   A natural wax used as insulation with a melting point of 85°C.

car phone   *Mobile phone.*

carriage   A device that moves in a predetermined path and carries some other part, such as a recorder head.

carrier   1. The basic frequency of a transmitter when no modulating signal is present.   2. A mobile electron or a hole in a semiconductor.

carrier amplifier   A direct-current amplifier whose DC input signal is filtered by a low-pass filter, then used to modulate a carrier so it can be amplified conventionally as an AC signal. The amplified DC output is obtained by rectifying and filtering the rectified carrier signal. The chopper amplifier is a more common version of the carrier amplifier. It uses either one or two choppers to convert the DC input signal into a square-wave AC signal and synchronously rectify the amplified square-wave signal.

carrier-amplitude regulation   The change in amplitude of the carrier wave in an amplitude-modulated transmitter when modulation is applied symmetrically.

carrier channel   The equipment and lines that make up a complete carrier-current circuit between two or more points.

carrier chrominance signal   *Chrominance signal.*

carrier current   A higher-frequency alternating current superimposed on ordinary telephone, telegraph, and power-line frequencies for communication and control purposes. The carrier current is modulated with voice signals to provide telephone communication between points on the power system or tone-modulated to actuate switching relays or convey data.

carrier detect   An RS-232C modem signal that indicates to a connected terminal that the modem is receiving a signal from a remote modem.

carrier frequency   The frequency generated by an unmodulated radio, radar, carrier communication, or other transmitter, or the average frequency of the emitted wave when modulated by a symmetrical signal. Also called center frequency and resting frequency.

carrier-frequency pulse   A carrier that is amplitude-modulated by a pulse. The amplitude of the modulated carrier is zero before and after the pulse.

carrier-interference ratio   [C/I ratio]   An interference-specifying ratio used in microwave relays and other communication systems. It is based on measuring the desired signal, turning it off, and then measuring the undesired signal. The ratio of the two measurements is expressed in decibels.

carrier leak   The carrier frequency remaining in a suppressed-carrier system.

carrier level   The strength or level of an unmodulated carrier signal at a particular point in a radio system, expressed in decibels in relation to some reference level.

carrier line   Any transmission line used for multiple-channel carrier communication.

carrier mobility   The average drift velocity of carriers per unit electric field in a homogeneous semiconductor. The mobility of electrons usually differs from that of holes.

carrier modulation   The process of varying some characteristic of a carrier in accordance with a modulating wave.

carrier noise level   The noise level produced by undesired variations of an RF signal in the absence of any intended modulation. Also called residual modulation.

carrier-power transformer   A transformer that supplies AC carrier power to a magnetic amplifier.

carrier repeater   A one- or two-way repeater for a carrier channel.

carrier shift   Radioteletypewriter transmission in which the carrier frequency is shifted in one direction for a mark signal and in the opposite direction for a space signal.

carrier signaling   Use of tone signals for ringing and other signaling functions in a carrier communication system.

carrier suppression   1. Suppression of the carrier frequency after conventional modulation at the transmitter, with reinsertion of the carrier at the receiving end before demodulation.   2. Suppression of the carrier when there is no modulation signal to be transmitted.

carrier swing   The total deviation of a frequency- or phase-modulated wave from the lowest to the highest instantaneous frequency.

carrier system   A system that permits many independent communications over the same circuit.

carrier telegraphy   Telegraphy in which a single-frequency carrier wave is modulated by the transmitting apparatus for transmission over wire lines.

carrier telephony   Telephony in which a single-frequency carrier wave is modulated by a voice-frequency signal, for transmission over wire lines.

carrier-to-noise ratio   The ratio of the magnitude of the carrier to that of the noise after selection and before any nonlinear process such as amplitude limiting and detection.

carrier transmission   Transmission in which a single-frequency carrier wave is modulated by the signal to be transmitted.

carrier wave   *Carrier.*

carry   A signal produced in a computer when the sum of two digits in the same column equals or exceeds the base of the number system in use or when the difference between two digits is less than zero.

cartridge   1. *Phonograph pickup.*   2. *Tape cartridge.*

CAS   Abbreviation for *collision avoidance system.*

cascade   *Avalanche effect.*

cascade amplifier An amplifier that contains two or more stages arranged in the conventional series manner, in which the output of one stage is amplified by the succeeding stage.

cascade-amplifier klystron A klystron that has three resonant cavities to provide increased power amplification and output. The extra resonator, located between the input and output resonators, is excited by the bunched beam emerging from the first resonator gap and produces further bunching of the beam.

cascade connection A series connection of amplifier stages, networks, or tuning circuits, in which the output of one feeds the input of the next.

cascade control An automatic control system in which various control units are linked in sequence, each control unit regulating the operation of the next control unit in line.

cascade voltage doubler An AC voltage-doubler circuit. Capacitor C1 is charged to the peak value of the AC voltage through diode D2 during one half cycle. During the other half cycle it discharges in series with the AC source through diode D1 to charge C2 to twice the AC peak voltage. See also *conventional doubler, bridge voltage doubler,* and *voltage multiplier.*

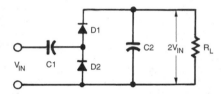

Cascade voltage doubler has a DC-voltage output that is twice the AC-voltage input.

cascode amplifier A common-emitter transistor amplifier in series with a common-base stage analogous to a grounded-cathode, grounded-grid amplifier. Its input resistance and current gain are nominally equal to the corresponding values for a single common-emitter stage, and the output resistance is approximately equal to the high output resistance of the common-base stage. Found in TV receiver-tuned amplifiers where the collector load is replaced with a tuned circuit, it is most effective in amplifying signals that are 25 MHz and higher.

case temperature The temperature of the outside of a device case, usually taken on a rectifier or power transistor in a metal case to determine its heat effect.

Cassegrain antenna A microwave antenna that has a feed radiator mounted at or near the surface of the main reflector aimed at a mirror at the focus. Energy from the feed first illuminates the mirror, then spreads outward to illuminate the main reflector. This technique, adapted from optical telescope technology, eliminates the need for mounting a heavy feed radiator far in front of the main reflector.

cassette A two-reel magnetic-tape cartridge designed for easy insertion into a tape recorder or player, without threading of tape. Also called tape cassette.

cassette player A magnetic-tape player designed for playback of prerecorded cassettes.

cassette recorder A magnetic-tape recorder designed for recording and playback of cassettes.

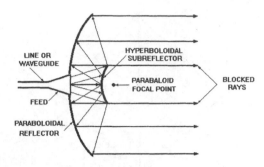

Cassegrain antenna feed for a parabolic reflector capable of transmitting and receiving microwave frequencies.

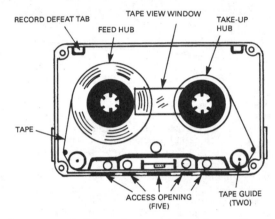

Cassette: a standard-size package containing audio recording tape that moves between supply and takeup reels to record and reproduce sound.

Cassini spacecraft A spacecraft to be launched in 1997 by NASA for the exploration of the planet Saturn and its giant moon Titan.

catalyst A material or condition that starts or speeds up a chemical or other reaction, such as the hardening of an epoxy cement.

cathode 1. [K] The primary source of electrons in an electron tube. In directly heated tubes the filament is the cathode. In indirectly heated tubes a coated metal cathode surrounds a heater. Other types of cathodes emit electrons under the influence of light or high voltage. 2. The negative electrode of a battery or other electrochemical device. 3. The terminal of a semiconductor diode that is negative with respect to the other terminal when the diode is biased in the forward direction.

cathode bias Bias obtained by placing a resistor in the common cathode return circuit, between cathode and ground. Flow of electrode currents through this resistor produces a voltage drop that makes the control grid negative with respect to the cathode of the vacuum tube.

cathode-coupled amplifier A cascade amplifier whose coupling between two stages is provided by a common cathode resistor.

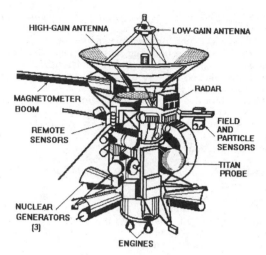

Cassini spacecraft is intended to explore the planet Saturn and its rings.

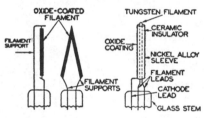

Cathodes for electron tubes. At left are two directly heated types, and at right is heater-type cathode.

cathode dark space   The relatively nonluminous region between the cathode and negative glows in a glow-discharge cold-cathode tube. Also called Crookes dark space.

cathode disintegration   The destruction of the active area of a cathode by positive-ion bombardment.

cathode follower   A vacuum-tube circuit in which the input signal is applied between the control grid and ground, and the load is connected between the cathode and ground. A cathode follower has a low output impedance, high input impedance, and a gain of less than unity. The anode is at ground potential at the operating frequency.

cathode glow   The luminous glow that covers all or part of the cathode in a glow-discharge cold-cathode tube.

cathode keying   Transmitter keying by a key in the cathode lead of the keyed vacuum-tube stage, opening the DC circuits for the grid and anode simultaneously.

cathode luminous sensitivity   The photoelectric emission current divided by the luminous flux on a photocathode under specified conditions of illumination.

cathode modulation   Amplitude modulation accomplished by applying the modulating voltage to the cathode circuit of an electron tube in which the carrier is present.

cathode poisoning   The chemical effect of residual gases on the emissivity of the cathode of an electron tube.

cathode preheating time   The minimum period of time during which heater voltage should be applied before electrode voltages are applied in an electron tube.

cathode pulse modulation   Modulation produced in an amplifier or oscillator by applying externally generated pulses to the cathode circuit.

cathode radiant sensitivity   The photoelectric emission current divided by the radiant flux on a photocathode at a given wavelength, under specified conditions of irradiation.

cathode ray   A stream of electrons, such as that emitted by a heated filament in a tube or emitted by the cathode of a gas-discharge tube when the cathode is bombarded by positive ions.

cathode-ray charge-storage tube   A charge-storage tube in which the information is written by a cathode-ray beam.

cathode-ray oscilloscope [CRO]   A test instrument that uses a cathode-ray tube to make visible on a fluorescent screen the instantaneous values and waveforms of electrical quantities which are rapidly varying as a function of time or another quantity. Also called oscilloscope and scope.

cathode-ray storage tube   A storage tube in which the information is written by a cathode-ray beam.

cathode-ray terminal   *Cathode-ray-tube terminal.*

cathode-ray tube [CRT]   An electron-beam tube in which the electrons emitted by a hot cathode are formed by an electron gun into a narrow beam that can be focused to a small cross section on a fluorescent screen. The beam can be varied in position and intensity by internal electrostatic deflection plates or external electromagnetic deflection coils to produce a visible trace, pattern, or picture on the screen.

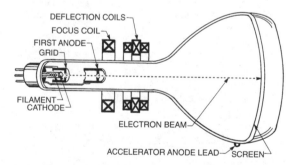

Cathode-ray tube (CRT): this version with magnetic deflection and focus coils around its neck is widely used in computer monitors, TV receivers, and radars.

cathode-ray-tube display   The presentation of a received signal on the screen of a cathode-ray tube.

cathode-ray-tube terminal [CRT terminal]   See *video display terminal* (VDS).

cathode resistor   A resistor in the cathode circuit of a vacuum tube; it has a resistance value selected so that the voltage drop across it due to tube current provides the correct negative grid bias for the tube.

cathode sputtering   *Sputtering.*

cathodoluminescence   Luminescence produced by high-velocity electrons. When these electrons bombard a metal in a vacuum, small amounts of the metal are vaporized in an excited state and emit radiation characteristic of the metal.

cathodophosphorescence A phosphorescence produced when high-velocity electrons bombard a metal in a vacuum.

CATV Abbreviation for *cable television* or *community antenna television.*

catwhisker A sharply pointed flexible wire that makes contact with the surface of a semiconductor crystal at a point which provides rectification.

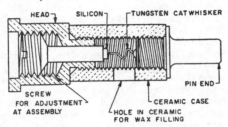

Catwhisker in microwave crystal diode.

cavity *Cavity resonator.*

cavity coupling A method of introducing or removing energy from a resonant cavity. Wire probes and loops are commonly used with coaxial lines; aperture or slot coupling is used with waveguides.

cavity frequency meter *Cavity-resonator frequency meter.*

cavity resonance 1. The natural resonant frequency of a loudspeaker baffle. If in the audio range, it is evident as unpleasant emphasis of sounds at that frequency. 2. The resonant frequency of a cavity resonator.

cavity resonator A space totally enclosed by a metallic conductor and excited so that it becomes a source of electromagnetic oscillations. The size and shape of the enclosure determine the resonant frequency.

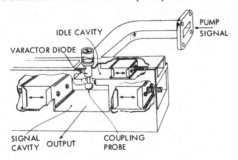

Cavity resonators as used in parametric amplifier.

CB Abbreviation for *citizens band.*

C band A band of frequencies extending from 4.0 to 8.0 GHz, corresponding to wavelengths of 7.5 and 3.75 GHz, respectively, in accordance with IEEE Standard 521-1976. It corresponds to both the G band (4.0 to 6.0 GHz) and H band (6.0 to 8.0 GHz) in the U.S. Military Joint Chiefs of Staff (JCS) triservice frequency designations (1970).

C-band waveguide A rectangular waveguide, 3.48 by 1.58 cm, used in the dominant mode for 3.7- to 5.1-cm wavelengths.

CCD Abbreviation for *charge-coupled device.*

CCIR Abbreviation for *Consultative Committee in International Radio.*

CCITT Abbreviation for *Consultative Committee in International Telegraphy and Telephony.*

CCM Abbreviation for *counter-countermeasure.*

C core A spirally wound magnetic core that is formed to a desired rectangular shape before it is cut into two C-shaped pieces and placed around a transformer or magnetic amplifier coil.

CCSL Abbreviation for *compatible current-sinking logic.*

CCTV Abbreviation for *closed-circuit television.*

C³L Abbreviation for *complementary constant-current logic.*

cd Abbreviation for *candela.*

CD-I System abbreviation for *compact-disk interactive.*

CD-4 sound Abbreviation for *compatible discrete four-channel sound.*

CDF Abbreviation for *cumulative distribution function.*

C display A rectangular radarscope display in which targets appear as bright spots, with target bearing indicated by the horizontal coordinate and target angle of elevation by the vertical coordinate.

CDM Abbreviation for *code-division multiplex.*

cd/m² Abbreviation for *candela per square meter.*

CDMA Abbreviation for *code-division multiaccess.*

CDPD Abbreviation for *cellular digital packet data.*

CD-ROM Abbreviation for *compact disk-read-only memory.*

CD-ROM XA Abbreviation for *compact-disk-ROM extended architecture.*

CdS Symbol for *cadmium sulfide.*

CDTV Abbreviation for *Commodore dynamic total vision,* a multimedia system developed by Commodore.

CDVCC Abbreviation for *coded digital verification color code.*

CE Abbreviation for *chip enable.*

ceiling-height indicator A photoelectric instrument for measuring the height of a cloud ceiling with the aid of a vertical beam of light. Also called ceilometer and cloud-height detector.

ceilometer *Ceiling-height indicator.*

celestial guidance Guidance of a long-range missile by reference to celestial bodies. The missile is equipped with gyroscopes, optical or radio star trackers, servos, computers, and other devices that together sight stars, calculate positions, and direct the missile. Also called stellar guidance.

cell 1. A single unit of a primary or secondary battery that converts chemical energy into electric energy. It is also called a power cell. 2. A single unit of a device that converts radiant energy into electric energy, such as a solar cell, or photovoltaic cell. 3. A single unit of a device whose resistance varies with radiant energy, such as a selenium cell. 4. An elementary unit of storage in a computer, such as a binary cell or decimal cell. 5. A generally hexagonal geographical area assigned to specific frequencies in a *cellular mobile telephone system.*

cell phone *Cellular mobile telephone.*

cell site The antennas and transmission and receiving equipment for communication between *cellular mobile telephones* and the *mobile telephone switching office* [MTSO] in a *cellular mobile telephone system.* It is also called a *land station.*

cellular mobile telephone A small, lightweight (typically less than 10 oz) portable transceiver that functions as a

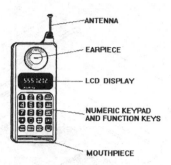

ANTENNA

EARPIECE

LCD DISPLAY

NUMERIC KEYPAD
AND FUNCTION KEYS

MOUTHPIECE

Cellular telephone is powered by a rechargeable battery.

telephone in a *cellular mobile telephone system*. Battery-powered, it includes a keypad and numerical display. It is also called a *cell phone*, a *mobile unit* and a *mobile station*.

cellular mobile telephone system A terrestrial mobile telephone system that uses RF links to transmit and receive voice and data from a *cellular mobile telephone* (*mobile unit* or *station*) and a *cell site* (*base* or *land station*) for relay back to a *mobile telecommunications switching office* [MTSO], which is connected to the land telephone network. The number of available radio channels is increased by dividing a geographical area into *cells*. Frequencies are allocated so that adjacent cells do not broadcast or receive on the same frequencies. In North America base stations transmit at 869–894 MHz, and mobile stations transmit at 824–849 MHz.

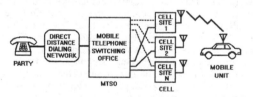

Cellular mobile telephone system includes a mobile telephone switching office that coordinates all switch functions and performs all internal cell switching.

Cellular Telecommunications Industry Association [CTIA] An industry organization that sets cellular telephone standards in North America.

cellulose acetate A thermoplastic material that is widely used as the base for magnetic tape and movie film. It can be made transparent or opaque in various colors. It is tough, flexible, slow-burning, and long-lasting.

CELP coding Abbreviation for *code-excited linear predictive coding.*

Celsius temperature scale [C] The international metric system standard for measuring temperature. The interval between the freezing point, set at 0.01°C, and boiling point, set at 100°C, is 99.9 degrees. Absolute zero is −237.16°C. The international standard was formerly the *centigrade temperature scale* and the change was made in 1948. The two scales differ slightly because the freezing point in the centigrade scale is 0°C but the Celsius cold-reference point of 0.01° is more convenient for making

calculations. To change degrees Celsius (or centigrade) to degrees Fahrenheit, multiply by 1.8 and add 32° to the result. The Fahrenheit and Celsius scales agree at −40°F = −40°C.

center-fed Hertz antenna A center-fed, half-wavelength dipole that can have spaced feeders for its transmission line. It is current-fed on its fundamental and voltage-fed on all even harmonics. It can transmit and receive in the 1.6- to 30-MHz range. It is also called a *tuned doublet* and a *center-fed zepp.*

center frequency *Carrier frequency.*

center-frequency stability The ability of a transmitter to maintain an assigned center frequency in the absence of modulation.

centering The process of adjusting the position of the trace or image on a cathode-ray-tube screen so it is centered on the screen.

centering control One of the two controls for positioning the image on the screen of a cathode-ray tube. The horizontal centering control moves the image horizontally, and the vertical centering control moves the image vertically. Centering is achieved by adjusting the DC voltage applied to deflection plates or by adjusting the direct current flowing through deflection coils.

center of gravity [CG] The point on or in a body through which the resultant of weight forces will always pass, regardless of the orientation of the body. Thus, the weight of an object can be represented by a single force acting downward at the CG, although in practice it is actually a system of parallel forces acting upon all of its component parts.

center tap [CT] A terminal at the electrical midpoint of a resistor, coil, or other device.

centi- A prefix representing 10^{-2}, that is 0.01 or one-hundredth.

centigrade temperature scale [C] See *Celsius temperature scale* [C].

centimeter [cm] A unit of length in the metric system, equal to 0.01 m or 0.394 in.

centimeter-gram-second unit [CGS unit] An absolute unit based on the centimeter, gram, and second as fundamental units.

centimetric wave A radio wave between the wavelength limits of 1 and 10 cm, corresponding to the superhigh-frequency (SHF) range of 3 to 30 GHz.

central office [CO] A main telephone office, usually within a few miles of a subscriber, that houses switching gear; most are capable of handling about 10,000 subscribers.

central processor unit [CPU] The heart of a computer system that executes programmed instructions. It includes the *arithmetic logic unit* [ALU] for performing all mathematical and logic operations, a control section for interpreting and executing instructions, and internal memory for temporary storage of program variables and other functions. See also *microprocessor.*

centronics parallel port An industry-accepted standard interface for the interconnection of personal computers to printers.

ceramic A mixture of inorganic, nonmetallic compounds that are prepared as a malleable substance (such as clay) and then fired to become a vitrified into solid form.

ceramic capacitor

Ceramics such as aluminum oxide (alumina) are formed as dielectric substrates for ICs or hybrid circuits; others such as barium titanate are formed into piezoelectric transducers.

ceramic capacitor A capacitor whose dielectric is a ceramic material such as steatite or barium titanate. Its composition can be varied to give a wide range of temperature coefficients. The electrodes are usually silver alloy coatings fired on opposite sides of the ceramic disk or slab or on the inside and outside of a ceramic tube. After connecting leads are soldered to the electrodes, the unit is usually given a protective insulating coating.

ceramic cartridge A device that contains a piezoelectric ceramic element, used in phonograph pickups and microphones. Ceramic cartridges deliver somewhat lower output voltage than crystal cartridges but are less affected by heat and humidity.

ceramic chip capacitor *Ceramic monolithic multilayer capacitor* [MLC].

ceramic dual-in-line package [CERDIP] A component package assembled with the leadframe sandwiched between two ceramic layers and sealed by firing powdered glass between them.

ceramic ladder filter A ladder filter that consists of many piezoelectric ceramic elements that are coupled electrically in a ladder network. The number of elements determines the width of the bandpass response.

ceramic leaded-chip carrier [CLCC] A square, flatpack for packaging integrated circuits made of ceramic with leads on all four sides and up to 84 leads.

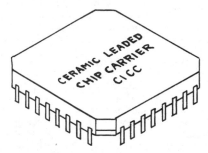

Ceramic-leaded chip carrier (CLCC) package has pins on all four sides for mounting a multipin VLSI integrated circuit on a conventional circuit board.

ceramic magnet A permanent magnet made from pressed and sintered mixtures of ceramic and magnetic powders.

ceramic microphone A microphone whose operation depends on the *piezoelectric* properties of a ceramic transducer such as barium titanate. It has characteristics that are similar to those of a *crystal microphone*.

ceramic monolithic multilayer capacitor [MLC] *Monolithic multilayer capacitor* [MLC].

ceramic package [CERPACK] A CERDIP-like ceramic package with pins extending from two or four sides of the package, typically for circuit board surface mounting.

ceramic tube An electron tube that has a ceramic envelope capable of withstanding operating temperatures of over 500°C, as required to withstand reentry temperatures of guided missiles.

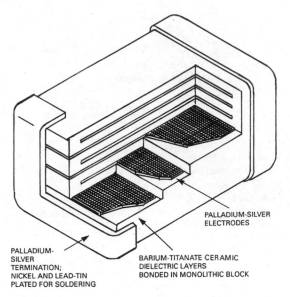

PALLADIUM-SILVER ELECTRODES

PALLADIUM-SILVER TERMINATION; NICKEL AND LEAD-TIN PLATED FOR SOLDERING

BARIUM-TITANATE CERAMIC DIELECTRIC LAYERS BONDED IN MONOLITHIC BLOCK

Ceramic monolithic multilayer capacitor is formed from layers of metalized "green" ceramic that are pressed and fired to form a small monolithic block.

CERDIP An acronym for *ceramic dual-in-line package*.

cermet An acronym for ceramic-metal. Cermet is a mixture of powdered precious metal such as palladium, powdered glass (frit), and a volatile binder mixed to form an ink which is painted or screened on a ceramic substrate or mandrel to form conductive thick films. The film is hardened by furnace firing. Cermet resistive elements are in axial-leaded resistors, resistor networks, chip resistors, potentiometers, and hybrid microcircuit substrates.

cermet resistor A metal-glaze resistor that consists of a mixture of finely powdered precious metals and insulating materials fired onto a ceramic substrate.

CERPACK An acronym for *ceramic package*.

cesium [Cs] A silvery, alkali, metallic element with an atomic number of 55 and an atomic weight of 132.9. When used as a cathode material, it can emit profuse amounts of electrons. Cesium atoms, when oscillating in a resonator, provide a standard for frequency and time measurement.

cesium atomic-beam resonator A resonator that forms atoms evaporated from liquid cesium into a beam that is acted on by a magnetic field, then passed into a microwave cavity where further magnetic interaction occurs. One application is in the frequency control of a microwave oscillator.

cesium-beam frequency standard A frequency standard that includes a precision quartz oscillator in combination with a frequency synthesizer and multiplier stages to generate standard frequencies such as 1 and 5 MHz with an accuracy of the order of 1 part in 10^{11}, by continuous comparison of the output with the 9192.631770-MHz output of a cesium atomic-beam resonator.

cesium clock An atomic clock regulated by the natural vibration frequency of atoms in a cesium atomic-beam resonator.

cesium phototube   A phototube that has a cesium-coated cathode. It has maximum sensitivity in the infrared frequency band.

cesium-vapor lamp   A lamp whose light is produced by the passage of current between two electrodes in ionized cesium vapor.

cesium-vapor rectifier   A gas tube in which cesium vapor serves as the conducting gas, and a condensed monatomic layer of cesium serves as the cathode coating. The tube is heated to about 180°C to give the desired vapor pressure.

CFA   Abbreviation for *CompactFlash Association.*

CGA   Abbreviation for *color graphics monitor.*

CGS unit   Abbreviation for *centimeter-gram-second unit.*

chaff   A passive electronics countermeasure (ECM) in the form of a "cloud" of thousands of fine wire clippings or paper-backed metal foil cut to lengths that act as dipoles that resonate over a wide range of frequencies. It is dispersed as a defensive measure to screen a target from the attempts of enemy radar to track that target to determine its position, course, and speed. It can also mislead or confuse radar-guided missiles seeking a target. Dispersed by a rocket launcher from ships or aircraft, a chaff "cloud" causes intense echoes over a wide arc on a radarscope to mask the intended target that launched the chaff. It can cause the radar to "bloom" and defeat tracking. When fired away from the target, it can misdirect a missile which detonates in the chaff cloud, overshoots, or bypasses the target. Chaff misleads radar-guided missiles in the same way as flares deceive infrared-guided missiles.

chaff rocket   A rocket filled with *chaff* that is fired from a ship or aircraft and explodes in the air to disperse thousands of the tiny dipoles to produce a large radar return and mask the position of the ship or aircraft that fired the rocket.

chain   A network of radio, television, radar, loran, or other stations connected by special telephone lines, coaxial cables, or radio relay links so they can operate as a group.

chain code   A cyclic sequence of *n*-bit words in which each word is derived from its neighbor by displacing the bits one digit position to the left or right, dropping the leading bit, and inserting a bit at the end. There is no repetition within the cyclic sequence. For 3-bit words, the example is 000 001 010 101 011 111 110 100 000.

chained list   A randomly arranged computer list in which each item contains an identifier that locates the next item to be considered.

channel   1. A band of radio frequencies allocated for specific purposes. The term typically applies to television broadcasting that has a 6-MHz-wide channel compared with the 10-kHz channel for an amplitude-modulation (AM) station.   2. An audio amplifier can have several input channels; a stereo amplifier has at least two complete channels.   3. A path for data flow in a computer.   4. The region separating the source from the drain of a field-effect transistor (FET). The channel in a *depletion-mode* FET is "normally on" (conducting) and in an *enhancement-mode* FET is "normally off" (insulating). The application of a voltage to the gate electrode alters the conduction of the channel, controlling the current through the channel. Chan-

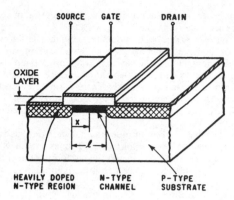

Channel in an insulated-gate MOS field-effect transistor.

nel length is an important parameter in determining FET current and speed. See also *drain, FET, gate,* and *source.*

channel bank   Communications equipment for multiplexing voice-grade channels into a digital transmission signal (typically 24 channels in the United States and 30 channels in Europe).

channel breakdown   Avalanche breakdown of a MOSFET transistor channel.

channel capacity   The maximum number of bits or other information elements that can be handled in a particular channel per unit time.

channel effect   A leakage current that flows over a surface path between the collector and emitter in some types of transistors.

channeled array   A semiconductor gate array base die with basic cells arranged in rows or columns. This arrangement permits routing in new spaces (channels) between rows and gates. Routing efficiency is typically near 90%. Routing is generally achieved by placing macros along single rows or columns.

channelized receiver   A radio receiver that performs a frequency-to-time transformation on received time-coincident signals and it produces a parallel format output. It is also called a channelizer. See *compressive intercept receiver.*

channelizing   The process of subdividing a wideband transmission facility to handle many different circuits requiring comparatively narrow bandwidths.

channel selector   A switch or other control that tunes in the desired channel in a television receiver.

channel separation   The electric or acoustic difference between the left and right channels in a stereo system.

channel shifter   A radiotelephone carrier circuit that shifts one or two voice-frequency channels from normal channels to higher voice-frequency channels to reduce crosstalk between channels. The channels are shifted back by a similar circuit at the receiving end.

channel stopper   A ring of opposite-polarity semiconductor material diffused around each transistor in a multiple-transistor integrated circuit to provide the electrical isolation required to prevent formation of parasitic devices in the field between transistors.

channel strip   An amplifier that has sufficient bandpass for one television channel. In cable television systems and fringe-area home locations it improves reception of a single desired station.

character [c]   A symbol that denotes a number, letter, symbol, or punctuation mark. When assembled in various combinations, characters can express information or lines in a program. In the binary-coded decimal (BCD) system, a group of characters can represent a single character. Characters can be stored in computer memory as one byte or eight bits of data and then be displayed or printed out. See also *ASCII.*

character density   The number of characters recorded per unit of length or area. Character densities for magnetic tape range from 200 to over 1000 characters per linear inch (80 to over 400 characters per linear centimeter). Also called record density.

character generator   A generator that creates letters, numerals, and symbols on a cathode-ray screen or other viewing surface in a desired sequence. It might also include facilities for recording the information on photographic negatives or paper.

characteristic   1. A measurable property of a device.   2. The integral part of a logarithmic value, at the left of the decimal point.

characteristic curve   A curve plotted on graph paper to show the relation between two changing values.

characteristic impedance   The impedance that, when connected to the output terminals of a transmission line of any length, makes the line appear to be infinitely long. There are then no standing waves on the line, and the ratio of voltage to current is the same for each point on the line. For a waveguide, the characteristic impedance is the ratio of RMS voltage to total RMS longitudinal current at specified points on a diameter when the guide is match-terminated. For an acoustic device it is the ratio of the effective sound pressure at a point to the effective particle velocity at that point. Also called surge impedance.

characteristic radiation   Radiation that originates in an atom following removal of an electron. The wavelength of the emitted radiation depends only on the element concerned and the energy levels involved.

characteristic X-rays   Electromagnetic radiation emitted as a result of rearrangements of the electrons in the inner shells of atoms. The spectrum of the radiation consists of lines that are characteristic of the element in which the X-rays are produced. The target of an X-ray tube will in general emit both continuous X-rays and characteristic X-rays.

character per inch [cpi]   A unit that specifies the number of characters that are printed or otherwise produced in 1 in.

character printer   *Serial printer.*

character reader   A device that scans printed or handwritten characters and delivers corresponding machine-readable code characters that can be fed to a computer or other data-processing circuit. The two most common commercial versions are optical and magnetic character readers.

character recognition   The technology permitting a machine to sense and encode into a machine language characters that are written or printed to be read by human beings.

character set   A list of characters acceptable for coding to a specific computer or input/output device.

character-writing storage tube   A character-writing tube that retains its display as long as the necessary operating voltages are supplied. It can be erased by lowering one electrode voltage momentarily.

character-writing tube   A cathode-ray tube that forms alphameric and symbolic characters on its screen for viewing or recording purposes.

Charactron   Trademark for a cathode-ray tube that produces a display in the form of letters or numbers.

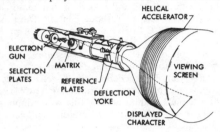

Charactron shaped-beam cathode-ray tube.

charge   1. The quantity of electric energy stored in a capacitor, battery, elementary particle, or insulated object. Also called electric charge.   2. The material or part to be heated by induction or dielectric heating.   3. The conversion of electric energy to chemical energy in a storage battery by sending direct current through the battery in the opposite direction to that of discharge current.

charge amplifier   An amplifier that converts capacitance changes of a transducer to corresponding changes in output voltage, as required for capacitor microphones and other capacitive transducers. Operational amplifiers, frequently used for this purpose, do this by converting changes in capacitor charge to changes in output voltage.

charge carrier   A mobile conduction electron or mobile hole in a semiconductor.

charge-coupled device [CCD]   A semiconductor charge-transfer device that consists basically of a bottom semiconductor layer, a metal semiconductor-oxide insulation layer, and a top layer of metal electrodes. A negative voltage on an electrode creates a depletion region under the electrode in the bottom layer for storing minority carriers that represent information. When the negative voltage is shifted to an adjacent electrode, the stored information moves correspondingly.

charge-coupled image sensor   A charge-coupled device whose charges are introduced when light from a scene is focused on the surface of the device. The image points are accessed sequentially to produce a television-type output signal. It is also called a *solid-state image sensor.*

charge density   The charge per area on a surface or per unit volume in space.

charged particles   Charged radiation products: alpha particles and protons have positive charges; beta particles and electrons have negative charges. Ions can be either negatively or positively charged.

charge-exchange phenomenon   The phenomenon in which a positive ion that possesses sufficient kinetic energy is neutralized by colliding with a molecule and capturing an electron from it. The molecule is transformed into a positive ion.

charge-mass ratio   The ratio of the electric charge of a particle to its mass.

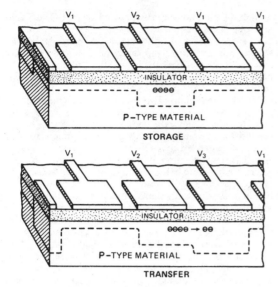

Charge-coupled device, showing how minority-carrier charge stored initially under electrode V_2 is transferred to electrode at right when voltage on V_3 is made greater than that on V_2.

charger   *Battery charger.*

charge storage   The buildup of carriers at the base of a semiconductor diode or bipolar transistor by charge-carrier flow, to give the concentration gradient required for an output current at the collector or back face.

charge-storage diode   A semiconductor diode whose turnoff time is substantially increased by charge storage.

charge-storage transistor   A transistor whose collector-base junction will charge when forward bias is applied with the base at a high level and the collector at a low level.

charge-transfer device   A semiconductor device that depends upon movements of stored charges between predetermined locations, as in charge-coupled and charge-injection devices.

charge trapping   A parasitic phenomenon that occurs in a semiconductor device to prevent electrons or holes from moving freely.

charging   1. The process of converting electric energy to chemical energy in a secondary battery.   2. The process of feeding electric energy to a capacitor or other device that can store electric energy.

charging current   The current that flows into a capacitor when a voltage is first applied.

chart   The paper or other material on which a graphic record is made by a recording instrument.

chart recorder   A recorder that plots a dependent variable against an independent variable with an ink-filled pen moving on plain paper, a heated stylus on heat-sensitive paper, a light beam or electron beam on photosensitive paper, an electrode on electrosensitive paper, or other means. The plot can be linear or curvilinear on a strip-chart recorder, or polar on a circular chart recorder.

chassis   A metal frame or boxlike structure for mounting electronic components to form a functional circuit such as a radio or television receiver, stereo tuner, or power sup-

ply. It was developed primarily for electron-tube circuitry, and has largely been replaced by the circuit board in modern electronics packaging. The plural form of this word is the same as the singular form.

chassis ground   A connection to the metal chassis on which the components of a circuit are mounted serve as a common return path to the power source. The chassis might or might not be connected to an earth ground.

chatter   Prolonged undesirable opening and closing of electric contacts, as on a relay. It is also called *contact chatter.*

cheater cord   A special extension cord that applies AC power to a television or radio receiver when the back cover with its protective power interlock is removed for servicing.

Chebyshev array   An antenna array whose elements are fed to produce an array factor that can be expressed in terms of a Chebyshev polynomial. For a given side lobe level, the width of the main beam is minimized.

Chebyshev filter   A *constant-k filter* that achieves sharp frequency cutoff in a tradeoff for amplitude ripple in the passband.

cheesebox antenna   *Pillbox antenna.*

chelate laser   A liquid laser based on the use of a rare-earth chelate (a metallo-organic compound). Initial excitation takes place within the organic part of the liquid molecule and then it is transferred to the metallic ions to give lasing action. The normal wavelength range is 0.3 to 1.2 µm. It is also called a *rare-earth chelate laser.*

chemical etching   The removal of material from a semiconductor surface by applying a suitable acid etching solution.

chemical gas sensor   A chemical gas sensor made with thin-film, metal-oxide silicon technology to provide warnings of dangerous levels of gas in homes, vehicles, and factories. A heater is embedded in the silicon dioxide (SiO_2) layer between the silicon substrate and the metal-oxide sensing film to raise the film temperature to its sensitivity level for the target gas. Conductivity in the external circuit across the sensor contacts provides the warning signal. A *micromachined* silicon diaphragm formed in the silicon substrate under the heater reduces power consumption. Depending on its composition, the film can sense the presence of carbon monoxide (CO) or methane (CH_4). A matching control chip provides signal conditioning, output driving, and interfacing.

chemical laser   A gas laser that produces the required population inversion for lasing directly from a chemical reaction between such gases as hydrogen and chlorine, deuterium and fluorine, or between either hydrogen fluoride or deuterium fluoride and carbon dioxide. The chemical reaction can be pure, with no external energy source, or it can be initiated by external energy inputs such as

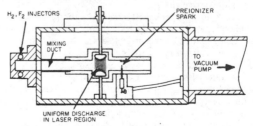

Chemical laser produces pulsed output. A spark-initiated discharge in the injected mixture of hydrogen and fluorine gases starts it.

electric discharges or light flashes. Its normal wavelength range is 2 to 100 μm.

chemically pumped laser A laser that depends on chemical reactions rather than electric energy to produce the pulses of light required for pumping the laser. In one method, the light is emitted by a chemical reaction. In another, shock waves from an explosion generate light in a flash tube that triggers the laser. The wave from a pyrotechnic squib compresses argon gas that emits intense radiation over a wide band in the ultraviolet region.

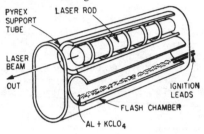

Chemically pumped laser. Chemical powders in a flash chamber are ignited electrically to provide the flash for pumping the neodymium glass laser rod.

chemical polishing The removal of several micrometers of semiconductor surface material with a suitable acid solution, usually after the semiconductor wafer has been lapped and mechanically polished.

chemical tracer A tracer that has chemical properties similar to those of the substance with which it is mixed.

chemical-vapor deposition [CVD] A process for coating an object by heating it close to its melting point in a vacuum chamber and injecting other materials, usually as gases into the chamber where they are attracted to the object and bond to it. The process is used to dope semiconductors with acceptors or donors to change their electrical characteristics. It is also used to apply a metal film to base metals or insulators or form oxide films.

chemiluminescence Luminescence produced by chemical action.

chemosphere That layer of the earth's atmosphere between altitudes of about 20 and 50 mi (30 and 80 km) where photochemical activity is present. Here nitric oxide forms an ionized cloud that reflects radio signals.

Child's law An equation stating that the current in a thermionic diode varies directly with the three-halves power of anode voltage and inversely with the square of the distance between the electrodes. Thus the current is limited only by the space charge.

chip 1. A bare, unpackaged semiconductor device cut from a wafer, typically an integrated circuit. The term is interchangeable with *die* for a discrete device—diode, transistor, cr power transistor. 2. It is also used (inaccurately) to refer to a packaged IC. 3. A miniature active or passive component, with a low profile, generally rectangular, and less than ¼ in long, for example, a surface-mount resistor made as resistive metal film on a ceramic chip, a multilayer capacitor (MLC), or tantalum solid-slug capacitor.

chip capacitor A miniature leadless capacitor packaged for placement and bonding to a hybrid circuit substrate or surface-mount circuit board. Examples are the leadless

monolithic multilayer ceramic capacitor (MLC) or leadless solid-slug tantalum capacitor.

chip carrier A low-profile component package, usually square, with an active chip cavity or mounting area and external leads that typically extend from four sides of the package. See *leaded-chip carrier* [LCC], *leadless-chip carrier* [LLCC].

chip-level integration Two or more integrated-circuit functions and/or technologies combined on a single IC for miniaturization, and reduction in assembly cost.

chip resistor A miniature leadless surface-mount resistor for bonding to a hybrid circuit substrate or circuit boards. It is made by deposition of thick-film resistive material on a ceramic substrate before being furnace-fired and diced.

chip-scale packaging A term to describe the processes and materials applied to the bare semiconductor chip to package and protect it. The outlines of these packages are smaller than those of dual-in-line (DIP) or surface-mount (SOT) packages, and are only slightly larger than the semiconductor dies or chips they protect. In addition, this technique provides alternative interconnects that are solderable by conventional surface-mount assembly techniques.

chirp 1. An undesirable variation in the frequency of a continuous-wave carrier when it is keyed. 2. The sound heard in a code receiver when the transmitted carrier frequency is increased linearly for the duration of a pulse code.

chirp filter A filter used for pulse compression in chirp radar, such as a surface-wave chirp filter.

chirp modulation Modulation in which the carrier signal is swept through the available band of frequencies repeatedly at a fixed rate. In the receiver, this wide pulse with high average power is compressed into a narrow pulse by delay-line techniques that enhance the desired signal while discriminating against unwanted signals. Used in radar and sonar to minimize the effects of multipath echoes, noise crosstalk, and frequency shifts. Also called swept-frequency modulation.

chirp radar Radar in which a swept-frequency signal is transmitted, received from a target, then compressed in time to give a final narrow pulse called the *chirp signal*. Compression is achieved with a network that introduces a delay proportional to frequency. Advantages include high immunity to jamming and inherent rejection of random noise signals.

chisel bond A thermocression bonding technique for attaching gold or aluminum contact wires from the pad of a semiconductor chip or die to a leadframe or package pin. Heat and pressure are applied to the wire end with a chisel-shaped bonding tool. It is normally used between terminals rather than from a semiconductor chip or die to a terminal pin.

chlorine [Cl] A gaseous element that has an atomic number of 17 and an atomic weight of 35.5. It is an important oxidizing agent in electronics manufacturing.

chlorofluorocarbons [CFCs] Nonflammable perfluorinated aliphatic compounds with different ranges of viscosity and boiling point from below room temperature to more than 200°C. They have low permittivities (near 2.0) and very low conductivity. They are inert chemically and have low solubilities for most other materials. The vapors of these liquids have high dielectric strengths. They are used as refrigerants in air conditioners and refrigerators, as cleaning agents for wave-soldered circuit board assemblies, and as heat-transfer agents in vapor-phase soldering processes.

choke 1. An inductance in a circuit that presents a high impedance to frequencies above a specified frequency range without appreciably limiting the flow of direct current. It is also called a *choke coil.* 2. A groove or other discontinuity in a waveguide surface shaped and dimensioned to impede the passage of guided waves within a limited frequency range.

choke coil *Choke.*

choke coupling Coupling between two parts of a waveguide system that are not in direct mechanical contact with each other.

choke flange A waveguide flange that has a round slot in its mating surface dimensioned to restrict leakage of microwave energy within a limited frequency range.

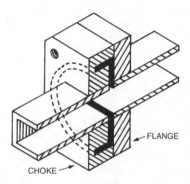

Choke flange. The choke has a circular groove that forms a quarter-wavelength choke when mated with a flat-faced flange.

choke-input filter A power-supply filter whose first filter element is a series choke.

choke joint A connection between two waveguides with two mating choke flanges that provides effective electric continuity without metallic continuity at the inner walls of the waveguide.

chopper A device that interrupts a current or beam of light or infrared radiation at regular intervals to permit amplification of the associated electrical quantity or signal by an AC amplifier. A chopper can also be a single-ended inverter with silicon controlled rectifiers for transforming DC to DC of a different voltage, or DC to AC.

chopper amplifier A carrier amplifier whose DC input is filtered by a low-pass filter, then converted into a squarewave AC signal by either one or two choppers. After amplification, the squarewave signal is synchronously rectified by chopper action to obtain the desired amplified DC output.

chopper-stabilized amplifier A direct-current amplifier that has a direct-coupled amplifier in parallel with a chopper amplifier. The chopper amplifier provides increased stability against drift in the direct-coupled amplifier, particularly when negative feedback is used. Conversely, the direct-coupled amplifier extends the frequency range beyond that of the chopper amplifier.

chopping Removing peaks of a waveform at a predetermined amplitude level.

chroma The dimension of the Munsell system of color that corresponds most closely to saturation, the degree of vividness of a hue. It is also called *Munsell chroma.*

chroma bandpass amplifier *Burst amplifier.*

chroma control The control that adjusts the amplitude of the carrier chrominance signal fed to the chrominance demodulators in a color television receiver, to change the saturation or vividness of the hues in the color picture. When in its zero position, the received picture becomes black and white. Also called color control and color-saturation control.

chroma oscillator A crystal oscillator used in color television receivers to generate a 3.579545-MHz signal for comparison with the incoming 3.579545-MHz chrominance-subcarrier signal being transmitted. Also called chrominance-subcarrier oscillator, color oscillator, and color-subcarrier oscillator.

chromatic Relating to color.

chromatic aberration 1. An optical lens defect that causes color fringes because the lens shape brings different colors of light to a focus at different points. An achromatic lens corrects for this error. 2. An electron-gun defect that causes enlargement and blurring of the spot on the screen of a cathode-ray tube because electrons leave the cathode with different initial velocities so they are deflected differently by the electron lenses and deflection coils.

chromatic dispersion The characteristic of glass or plastic light-conducting fiber that alters the wavelength of incident light traveling through the fiber at different speeds. It arises from both the fiber's properties and the difference in refractive index between the fiber's core and cladding.

chromaticity The color quality of light that can be defined by its chromaticity coordinates. Chromaticity depends only on hue and saturation of a color, not on its luminance (brightness). Chromaticity applies to all colors, including shades of gray, whereas chrominance applies only to colors other than grays.

chromaticity coordinate One of the two coordinates (x or y) that precisely specify the exact identity or chromaticity of a color on the CIE chromaticity diagram. It is also called a *color coordinate* and a *trichromatic coefficient.*

chromaticity diagram A diagram in which one of the three chromaticity coordinates is plotted against another. The most common version is the CIE chromaticity diagram for color television.

chromaticity flicker Flicker in a color television receiver caused by fluctuation of chromaticity only.

Chromel The trademark for a nickel-chromium alloy used in thermocouples.

chrominance The difference between any color and a specified reference color of equal brightness. In color television, this reference color is white, having coordinates $x = 0.310$ and $y = 0.316$ on the chromaticity diagram.

chrominance bandwidth *Chrominance-channel bandwidth.*

chrominance carrier *Chrominance subcarrier.*

chrominance-carrier reference A continuous signal that has the same frequency as the chrominance subcarrier in a color television system and fixed phase with respect to the color burst. This signal is the reference to which the phase of a chrominance signal is compared for modulation or demodulation. In a color receiver it is generated by a crystal-controlled oscillator. Also called chrominance-

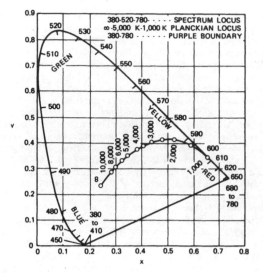

Chromaticity diagram as prepared by CIE. Temperature values on the curve are in kelvins, and wavelength values are in nanometers.

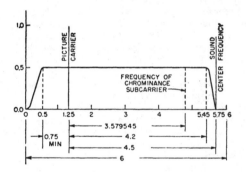

Chrominance subcarrier in standard 6-MHz television channel. The vertical scale gives relative maximum radiated field strength with respect to the picture carrier. Frequency values are in megahertz.

subcarrier reference, color-carrier reference, and color-subcarrier reference.

chrominance channel Any path that is intended to carry the chrominance signal in a color television system.

chrominance-channel bandwidth The bandwidth of the path intended to carry the chrominance signal in a color television system. Also called chrominance bandwidth.

chrominance demodulator A demodulator in a color television receiver for deriving the I and Q components of the chrominance signal from the chrominance signal and the chrominance-subcarrier frequency. Also called chrominance-subcarrier demodulator.

chrominance modulator A modulator used in a color television transmitter to generate the I or Q components of the chrominance signal from the video-frequency chrominance components and the chrominance subcarrier. Also called chrominance-subcarrier modulator.

chrominance primary The nonphysical color represented by either the I or Q chrominance signal component in a

color television system. These chrominance signals are chosen to be electrically convenient components remaining after the luminance signal is removed from a full-color signal at the transmitter.

chrominance signal That part of a TV signal containing the information about the color of the image.

chrominance subcarrier The 3.579545-MHz carrier whose modulation sidebands are added to the monochrome signal to convey color information in a color television receiver. The chrominance subcarrier is transmitted unmodulated in the form of color bursts that are used for synchronizing purposes in the receiver. Also called chrominance carrier, and subcarrier.

chrominance-subcarrier demodulator *Chrominance demodulator.*

chrominance-subcarrier modulator *Chrominance modulator.*

chrominance-subcarrier oscillator *Chroma oscillator.*

chrominance-subcarrier reference *Chrominance-carrier reference.*

chronistor A subminiature elapsed-time indicator that depends on electroplating principles to totalize operating time of equipment up to several thousand hours.

chronograph 1. An instrument that records intervals of time with a high degree of accuracy. 2. An instrument that measures and records projectile velocity by measuring the time required for the projectile to travel a known distance.

chronotron A device that measures millimicrosecond time intervals between pulses. In one type, interval-determining pulses are fed to a transmission line, and the spacing between the pulses on the line is measured.

Ci Abbreviation for *curie.*

CID Abbreviation for *charge-injection device.*

CIE Abbreviation for *Commission Internationale de l'Eclairage.*

CIE photoptic curve *Standard Observer Curve.*

CIE chromaticity diagram A chromaticity diagram established as an international standard by the Commission Internationale de l'Eclairage. In this diagram, that applies to color television, the color wavelengths are plotted as coordinates of x and y.

CIM Abbreviation for *computer-integrated manufacturing.*

c/in Abbreviation for *characters per inch.*

cipher A transposition or substitution code for transmitting messages secretly.

ciphony equipment [CIPHer + telephONY] Any equipment attached to a radio transmitter, radio receiver, or telephone for scrambling or unscrambling voice messages. In one form speech is converted into a series of ON/OFF pulses that are mixed with pulses from a key generator. To recover the speech at the receiving terminal, an identical key generator must be set to the daily key setting and its output subtracted from the received signal. The ON/OFF pulses can then be converted to the original speech patterns.

C/I ratio Abbreviation for *carrier-interference ratio.*

circuit An interconnected group of active and passive electrical and electronic components that accomplishes a desired function such as switching, amplification, filtering, or data conversion.

circuit board A sheet of insulating plastic, insulated metal, or ceramic material with or without printed wiring, on which circuit components are mounted.

circuit breaker   A device whose electrical contacts open automatically under *overload* current to protect a circuit. See *magnetic circuit breaker* and *thermal circuit breaker.*

circuit diagram   *Schematic circuit diagram.*

circuit noise   Undesired electric noise generated within a circuit or arriving from external sources.

circuit-noise level   The ratio of the circuit noise at a point in a transmission circuit to some arbitrary amount of circuit noise chosen as a reference. Usually expressed in decibels above reference noise (dBrn) or in adjusted decibels (dBa).

circuit-noise meter   An instrument that measures the electric noise level in a communication circuit, in decibels above reference noise. Also called noise-measuring set.

circuit parasitics   Unwanted effective resistance, capacitance, and inductance that results from the close proximity of conductors in circuit components operating at radio frequencies.

circuit simulation   A technique for simulating circuit behavior before a prototype is made. Accurate models of components such as transistors, resistors, and capacitors are stored in computer memory. The computer can integrate them into a circuit before it solves the fundamental circuit analysis equations with numerical analysis algorithms in the simulation process.

circuitry   The complete combination of circuits in a unit of electronic equipment.

circular antenna   A folded dipole that is bent into a circle so the transmission line and the abutting folded ends are at opposite ends of a diameter. When mounted horizontally, it radiates uniformly in all directions and has very little vertical radiation.

circular-beam multiplier   A multiplier that projects a circular beam of electrons on four isolated metallic quadrants. The currents collected from the quadrants are combined to give an output current proportional to the product of the input variables that deflect the beam.

circular-chart recorder   A recorder with one or more writing pens or other recording devices that are actuated by signals as a circular chart makes one complete revolution slowly under the control of a time-clock motor. After each revolution, a new chart must be placed on the recorder.

circular electric wave   A transverse electric wave whose lines of electric force form concentric circles.

circular horn   A circular-waveguide section that flares outward into the shape of a horn, to serve as a feed for a microwave reflector or lens. See *horn antenna.*

circularly polarized wave   An electromagnetic wave whose electric and/or magnetic field vectors at a point describe a circle. This term is usually applied to transverse waves.

circular magnetic wave   A transverse magnetic wave whose lines of magnetic force form concentric circles.

circular mil [abbreviated cmil]   A unit of area equal to the area of a circle whose diameter is 1 mil (0.001 in or 0.00254 cm). It is useful for specifying cross-sectional areas of round conductors.

circular orbit   An orbit that makes a complete constant-altitude revolution around the earth.

circular polarization   Polarization in which the vector representing the wave has a constant magnitude and rotates continuously about a point.

circular-polarization duplexer   A duplexer that achieves simultaneous transmission and reception at millimeter

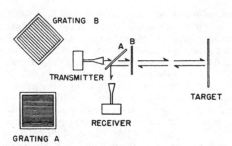

Circular-polarization duplexer for a pulsed or continuous-wave radar system.

wavelengths with two selective reflector gratings which consist of closely spaced metal slats. A vertically polarized wave leaving the transmitter is converted to a circularly polarized wave that has two components perpendicular to each other and 90° out of time phase by the combined action of the two gratings. A circularly polarized wave returning through the first grating is converted to a linear horizontally polarized wave that is completely reflected by the second grating before entering the receiving horn.

circular scanning   Scanning in which the radar antenna rotates in a complete circle so that the beam generates a plane or a cone whose vertex angle is close to 180°.

circular trace   A time base produced by applying sine waves of the same frequency and amplitude, but 90° out of phase, to the horizontal and vertical deflection plates of a cathode-ray tube. The trace is then a circle, and signals give inward or outward radial deflections from the circle.

circular waveguide   A waveguide whose cross-sectional area is circular.

circulator   A waveguide component that has many terminals arranged so that energy entering one terminal is transmitted to the next adjacent terminal in a specific direction. It is also called a *microwave circulator.*

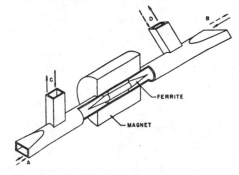

Circulator with a ferrite rod positioned in the longitudinal magnetic field in a circular section of waveguide to rotate the plane of polarization in a traveling plane wave.

CIS   Abbreviation for *card information structure.*

CISC   Abbreviation for *complex-instruction-set computer.*

citizens band [CB]   A band of frequencies assigned by the FCC for use by private citizens. It consists of 40 channels ranging in frequency from 26.965 MHz for channel 1 to 27.405 MHz for channel 40.

citizens radio service   A radio communication service for private or personal radio communication, including radio signaling, control of objects by radio, and other purposes. The frequency bands include 460 to 470 MHz for general use and the band around 27 MHz for ON/OFF unmodulated or tone-modulated carrier for remote control.

cladding   1. A process for covering one metal with another by clamping the metals together and rolling, extruding, drawing, or swaging until a bond is produced. An example is the copper coating on steel wire used as a conductor.   2. The bonding of two different kinds of materials to form a seamless bond, as in the coating of pure glass fibers with impurity-doped glass in the manufacture of graded- and step-index optical fibers.

clamp-and-hold digital voltmeter   *Sample-and-hold digital voltmeter.*

clamper   *DC restorer.*

clamping   The introduction of a reference level that has some desired relation to a pulsed waveform, as at the negative peaks or the positive peaks. Also called DC reinsertion and DC restoration.

clamping circuit   A circuit that reestablishes the DC level of a waveform. Used in the DC restorer stage of a television receiver to restore the DC component to the video signal after its loss in capacitance-coupled AC amplifiers, to reestablish the average light value of the reproduced image.

clamping diode   A diode that clamps a voltage at some level in a circuit.

clamping voltage   The sustained voltage held by a *clamp circuit* at some desired level.

clamp-on ammeter   *Snap-on ammeter.*

clamp-tube modulation   Amplitude modulation in which the AF signal is applied to the screen grid of a tetrode or pentode operated as a modulator.

clapper   A relay armature that is hinged or pivoted.

Clapp oscillator   A modified version of the Colpitts inductive-capacitive oscillator. Its resonant circuit contains a small value capacitor added in series with the resonant tank coil. The values of that capacitor and the tank coil determine the oscillator's frequency. This design minimizes the effects of amplifier capacitance on the output frequency. It can function with either bipolar junction or field-effect transistor amplifiers. It can generate sinewave frequencies up to UHF. See *Colpitts oscillator.*

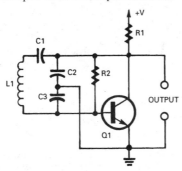

Clapp oscillator with an NPN transistor.

class A amplifier   An amplifier whose base bias and alternating base voltages are selected so that collector current

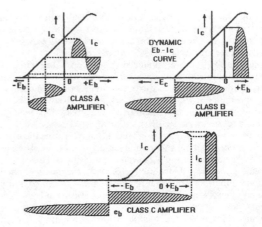

Class A, B, and C amplifier characteristics.

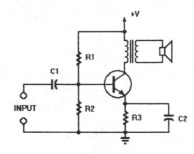

Class A amplifier as a receiver's output stage with a loudspeaker load.

in a specific transistor flows at all times. The suffix 1 is added to the letter or letters of the class identification to denote that base current does not flow during any part of the input cycle. The suffix 2 denotes that base current flows during some part of the cycle. In a class A transistor amplifier, each transistor is in its active region for the entire signal cycle.

class AB amplifier   An amplifier whose base bias and alternating base voltages are selected so that collector current in a specific transistor flows for appreciably more than half but less than the entire electric cycle. The suffix 1 denotes that base current does not flow. The suffix 2 denotes that base current flows during some part of the cycle. In a class AB transistor amplifier, operation is class A for small signals and class B for large signals.

class A insulation   Insulation that consists of cotton, silk, paper, and other similar organic materials which are impregnated with or immersed in a liquid dielectric and can withstand temperatures up to 105°C.

class A modulator   A class A amplifier that supplies the necessary signal power to modulate a carrier.

class A push-pull sound track   Two single photographic sound tracks side by side; the transmission of one is 180° out of phase with the transmission of the other. Both positive and negative halves of the sound wave are linearly recorded on each track.

class B   A screening process for electronic circuits that are intended for ground-based military electronic systems. A class B circuit must conform to screening standards per MIL-STD-883C and MIL-M-38510.

class B amplifier   An amplifier whose base bias is approximately equal to the cutoff value, so that collector current is approximately zero when no exciting base voltage is applied, and flows for approximately half of each cycle when an alternating base voltage is applied. The suffix 1 denotes that base current does not flow. The suffix 2 denotes that base current flows during some part of the cycle. In a class B transistor amplifier, each transistor is in its active region for approximately half the signal cycle.

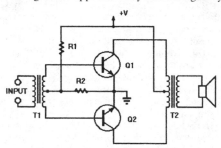

Class B amplifier as a push-pull amplifier driving a loudspeaker.

class B insulation   Insulation that consists of asbestos, mica, glass fiber, and other similar inorganic materials combined with organic binders and capable of withstanding temperatures up to 130°C.

class B modulator   A class B amplifier that supplies the necessary signal power to modulate a carrier. It is usually connected in push-pull.

class B push-pull sound track   Two photographic sound tracks side by side; one track carries the positive half of the signal only, and the other carries the negative half. During the inoperative half-cycle, each track transmits little or no light.

class C amplifier   An amplifier whose base bias is appreciably greater than the cutoff value so that collector current in each transistor is zero when no alternating base voltage is applied, and flows for appreciably less than half of each cycle when an alternating grid voltage is applied. The suffix 1 denotes that base current does not flow. The suffix 2 denotes that base current flows during some part of the

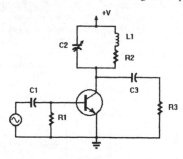

Class C amplifier as an RF power amplifier with an efficiency as high as 80%.

cycle. In a class C transistor amplifier, each transistor is in its active region for significantly less than half the signal cycle.

class C insulation   Insulation that consists of glass, mica, porcelain, quartz, and similar organic materials capable of withstanding temperatures over 220°C.

class D channel   A data-transmission circuit that can transmit punched-tape data at up to 240 words per minute, or 80-column punched-card data at about 10 cards per minute.

class D stage   A bipolar transistor stage operating as an ON/OFF switch in a circuit that changes suddenly between saturation and cutoff conditions.

class E channel   A data-transmission circuit that can transmit at up to 1200 b/s.

class H insulation   Insulation that consists of asbestos, glass fiber, mica, and similar inorganic materials combined with silicone or equivalent binders and capable of withstanding up to 180°C.

class S   A screening process for electronic circuits that are intended for military satellite electronic systems. A class S circuit must conform to screening standards per MIL-STD-883C and MIL-M-38510.

classical scattering cross section   *Scattering cross section.*

classical system   *Nonquantized system.*

classified   Containing information whose disclosure to a prospective enemy is not in the best interests of the nation.

classify   To sort into groups that have common properties. Thus intercepted electromagnetic radiations might be classified into radar, navigational, jamming, and missile-control signal groups.

CLCC   Abbreviation for *ceramic leaded-chip carrier.*

clean room   A sealed, pressurized room designed to prevent minute contaminants, such as airborne dust particles, from causing irregularities in film coatings, optical anomalies, and other problems. All entering air is filtered, and humidity and temperature are controlled within specified limits. Separate air-conditioning, heating, and ventilation systems prevent cross contamination. Room air pressure is kept higher than atmospheric to keep out dust and chemical vapors. The primary standard for clean-room design is Federal Standard 209B, which classifies clean rooms in terms of control of airborne particulates. Classification is based on the number of airborne particles ≥0.3 μm per cubic foot of air. A class 1 clean room can have only one particle of any kind in one cubic foot of space. The latest clean rooms are classes 1–10 for manufacturing ICs with a feature size close to a micrometer. Clean rooms are equipped with air locks and special foot-wiping mats. Workstations are individually filtered, and employees must wear special lint-free, hooded "bunny" clothes and slippers.

clear   1. To restore a memory device or binary circuit to a prescribed state, usually that representing logic 0. Also called reset.   2. A function key on calculators, to delete an entire problem or just the last keyboard entry. Usually labeled C. On some calculators, the first depression of the clear key deletes the previous entry, and the next depression clears everything. Other calculators have a separate key, labeled CE, for clearing the previous entry.

clearance   The minimum distance between two conducting objects or a conductor and ground.

clear channel   A standard broadcast channel whose dominant station or stations render service over wide areas. Stations on a clear channel are cleared of objectionable interference within their primary service areas and over all or a substantial portion of their secondary service areas.

clear text A message that is not coded and has no hidden meanings.

clear to send [CTS] A modem interface signal that commands the data terminal equipment to begin transmission.

clear-voice override The ability of a speech scrambler to receive a clear message even when the scrambler is set for scrambler operation.

click A short-duration electric disturbance, such as that sometimes produced by a code-sending key or a switch.

click tuner A mechanical television tuner that clicks into position for each of the 70 available UHF channels and the 12 VHF channels.

clipper *Limiter.*

clipper amplifier An amplifier that limits the instantaneous value of its output to a predetermined maximum.

clipper diode A bidirectional breakdown diode that clips signal voltage peaks of either polarity when they exceed a predetermined amplitude.

clipper-limiter A device whose output is a function of the instantaneous input amplitude for a range of values between two predetermined limits but is approximately constant, at another level, for input values above the range.

clipping 1. Cutoff of initial or final speech sounds in voice transmission, due to operation of voice-operated or other switching devices. 2. Voice distortion caused by severe overloading of amplifier circuits so that peaks of audio waveforms are cut off. 3. Any action that cuts off the peaks of a television signal. This can affect either the positive (white) or negative (black) picture-signal peaks or the synchronizing signal peaks. 4. *Limiting.*

clipping level The amplitude level at which a waveform is clipped.

clipping time The time constant of a limiter.

clock A source of accurately timed pulses for synchronization in a digital computer or as a time base in a transmission system.

clocked flip-flop Two flip-flops connected in a master-slave combination to eliminate the need for capacitors or other circuit delay elements. The master flip-flop stores the input information when the clock voltage is high, and transfers it to the slave when the clock voltage is low.

clocked gate A gate circuit that is actuated by a clock pulse.

clocked logic A logic circuit whose switching action is controlled by repetitive pulses from a clock.

clocked RS flip-flop A flip-flop that gives a 1 output if the set (S) input is enabled when the clock pulse arrives and a 0 output if the reset (R) input is enabled when clocked.

clock frequency The master frequency of the periodic pulses that schedule the operation of a digital computer.

clock rate The rate at which bits or words are transferred from one internal element of a computer to another.

clock skew A phase shift caused by different delays in various parts of a clock-signal distribution system. Excessive skew can cause data-processing errors.

clock track A recorded track that contains a signal pattern which acts as a time reference.

clockwise capacitor A variable capacitor whose capacitance increases with clockwise rotation of its rotor, as viewed from the end of the control shaft.

clockwise polarized wave *Right-hand polarized wave.*

clone A unit of electronic equipment, usually a personal computer, that emulates another model (typically made by IBM) because it has the same operating system (OS) and, therefore, can run the same software.

close-control radar Ground radar used with radio to position an aircraft over a target that is normally difficult to locate or invisible to the pilot.

close coupling The coupling obtained when the primary and secondary windings of an RF or IF transformer are close together.

closed architecture A computer with proprietary hardware, interface devices, or software that limit its compatibility with other computers and peripherals, making it essentially compatible only with other computers made by the same manufacturer.

closed circuit A complete path for current.

closed-circuit communication system A self-contained communication system that has no provision for interconnection with other systems. An intercom is an example.

closed-circuit signaling Signaling in which current flows in the idle condition, and a signal is initiated by increasing or decreasing the current.

closed-circuit television [CCTV] Any application of television that does not involve broadcasting for public viewing. Theater television and industrial television are examples. The programs can be seen only on specified receivers connected to the television camera by circuits that might or might not include microwave relays and coaxial cables.

closed-circuit voltage The voltage at the terminals of a source when a specified load current is being drawn.

closed-cycle control system A control system in which changes in the quantity being controlled are utilized to actuate the controller directly.

closed-cycle cryogenic cooling Cryogenic cooling in which a gas such as nitrogen is liquefied by a refrigerator-type compressor, for cooling infrared detectors and other cryogenic components. The gas recovered after cooling returns to the compressor for liquefaction.

closed loop 1. A computer program whose cycle of instructions is executed repeatedly. 2. A signal path for feeding the output of a control system back to the input for comparison with reference values, to achieve a desired form of regulation.

closed-loop control system A control system, consisting of one or more *feedback control loops,* that continuously compares system response with its input command, a desired speed or position of the load, or other response. A sensor in the feedback loop such as as an *encoder, tachometer,* or *thermostat* senses any difference between the input command and the system response and generates an *error signal,* which is sent to a controller/ampli-

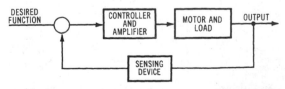

Closed-loop control system includes at least one feedback loop.

fier. The signal from the amplifier changes the input to the motor or other actuator so its output will cancel the error signal. The sensor's performance is directly proportional to the activity being controlled, such as the speed or position of the load or the ambient temperature. It is also called a *feedback control system*. See also *servosystem*.

closed magnetic circuit   A complete circulating path for magnetic flux around a core of ferromagnetic material.

closed subroutine   A computer subroutine that is stored away from the routine which refers to it. Such a subroutine is entered by a jump, and provision is made to jump back to the proper point in the main routine at the end of the closed subroutine.

closest point of approach [CPA]   A term used in shipboard radar navigation and station-keeping for estimating the closest distance that two moving ships on converging courses will be separated if they hold their present courses and speeds over time. If the two ships are on parallel courses or just begin to diverge, they are at the CPA. A zero estimated CPA is a *collision course*. The CPA can be determined graphically by plotting relative courses on a radar PPI display. The CPA does not take into account the effects of wind and current on either ship.

cloud   The mass of electrons circulating about a heated cathode in a vacuum tube in the absence of positive voltages on screens or plates.

cloud absorption   Attenuation of electromagnetic radiation by absorption within the water particles of a cloud.

cloud attenuation   Attenuation of electromagnetic radiation by scattering action in clouds.

cloud echo   The echo pattern produced on a radar screen by reflection of radar signals from a cloud.

cloud-height detector   *Ceiling-height indicator.*

cloud pulse   The output that results from space charge effects produced by turning the electron beam on or off in a charge-storage tube.

clover-leaf antenna   A nondirectional VHF transmitting antenna that consists of a number of horizontal four-element radiators stacked vertically a half-wave apart. Each horizontal unit has four loops arranged like a four-leaf clover. The units are energized to give maximum radiation in the horizontal plane.

clutter   Unwanted echoes on a radar screen, such as those caused by ground, sea, and rain clutter; stationary objects; chaff; enemy jamming transmissions; and grass. Also called background return and radar clutter.

clutter gating   A gating technique that gives a normal radar video display in regions which have no clutter, with moving-target-indicator video being switched in only for clutter areas.

cm   Abbreviation for *centimeter*.

CMAC   Abbreviation for *control mobile attenuation code*.

C message   A frequency weighting that evaluates the effects of noise based on its annoyance to the "typical" subscriber of standard telephone service or the effects of noise (background and impulse) on voice-grade data service.

cmil   Abbreviation for *circular mil*.

CML   Abbreviation for *current-mode logic*.

CMOS   Abbreviation for *complementary metal-oxide semiconductor*.

CMOS process   A digital logic IC fabrication technology that combines enhancement-mode N-channel (NMOS) and P-channel (PMOS) FET transistors on the same substrate to form logic gates or memory cells. There are standard metal gate processes and advanced polysilicon gate processes for higher switching speed and higher density.

CMOS RAM   A random-access memory that uses complementary metal-oxide semiconductor technology.

CMOS SOS   Silicon-on-sapphire technology combined with complementary metal-oxide semiconductor technology.

CMOS SOS RAM   A random-access memory that uses a combination of silicon-on-sapphire and complementary metal-oxide semiconductor technologies.

CMR   Abbreviation for *common-mode rejection*.

CMRR   Abbreviation for *common-mode rejection ratio*.

C network   A network composed of three impedance branches in series; the free ends of the network are connected to one pair of terminals, and the junction points are connected to another pair of terminals.

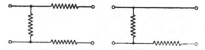

C network has three impedance branches, whereas L network at right has only two branches.

coarse chrominance primary   The least important of the two chrominance primaries in a color television signal, called the Q signal. Because its bandwidth is limited to 0.5 MHz, this signal affects only the larger, coarser variations in the color picture. The other primary is the fine chrominance primary or I signal, going up to 1.5 MHz.

coarse control   A control that makes a rough adjustment of some characteristic or quantity.

coastal refraction   The bending of the path of a direct radio wave as it crosses the coast at or near the ground due to changes in electrostatic conditions between earth and water.

coated cathode   A cathode that has been coated with compounds to increase electron emission. An oxide-coated cathode is an example.

coated filament   A vacuum-tube filament that has been coated with metal oxides to provide increased electron emission.

coated lens   A lens whose air-glass surfaces have been coated with a thin transparent film that has an index of refraction which minimizes light loss by reflection. Coatings used include magnesium fluoride, silicon oxide, sodium fluoride, and titanium oxide.

coax   *Coaxial cable.*

coaxial antenna   An antenna that consists of a quarter-wave extension of the inner conductor of a coaxial line and a radiating sleeve which is formed by folding back the outer conductor of the coaxial line for a length of approximately one quarter-wavelength. It is also called a *sleeve antenna*.

coaxial attenuator   An attenuator that has a coaxial construction and terminations suitable for use with coaxial cable.

coaxial cable   A transmission line in which one conductor is centered inside and insulated from a metal tube that

coaxial cable connector

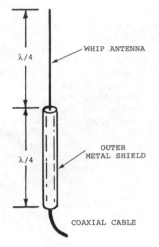

Coaxial antenna is a vertical radiator one-half wavelength long with a quarter-wave whip element.

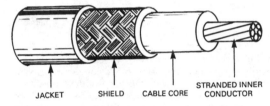

Coaxial cable has an inner conductor that is solid or stranded copper wire. Flexible metal braid acts as a return path, and the cable is protected by an insulating jacket.

serves as the second conductor. The insulation can be a continuous solid dielectric, or there can be dielectric spacers and an insulating gas. It is also called coax, coaxial line, and coaxial transmission line.

coaxial cable connector A connector that consists of a mating plug and receptacle for making a temporary or permanent connection between two lengths of coaxial cable without affecting impedance. The three most common types use bayonet, threaded, and push-pull quick-disconnect coupling methods.

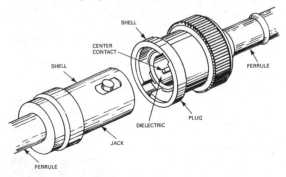

Coaxial cable connector consists of the jack (left) and the plug (right).

coaxial cavity A cylindrical resonating cavity that has a central conductor in contact with its pistons or other reflecting devices. The conductor picks up a desired wave.

coaxial diode A diode that has the same outer diameter and terminations as a coaxial cable or that is otherwise designed to be inserted into a coaxial cable.

coaxial-dipole antenna A dipole antenna that has lengths of metal tubing as the radiating elements. The twin-lead transmission line connects conventionally to the inner ends of the tubing. The outer ends of the tubing are connected by a metal rod centered in the radiating elements.

coaxial dry load A sand load for a coaxial cable.

coaxial filter A section of coaxial line that has reentrant elements which provide the inductance and capacitance of a filter section.

coaxial isolator An isolator in a coaxial cable that provides a higher loss for energy flow in one direction than in the opposite direction. All types use a permanent magnetic field in combination with ferrite and dielectric materials.

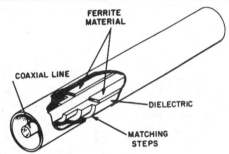

Coaxial isolator. A magnetic field can be provided by either rod-type permanent magnet inside the central conductor or an external horseshoe magnet with its poles near ferrite strips.

coaxial launcher A transducer that couples a coaxial cable to a waveguide, cavity, or other microwave device.

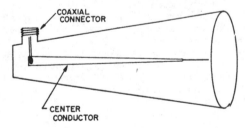

Coaxial launcher.

coaxial line *Coaxial cable.*

coaxial-line frequency meter A shorted section of coaxial line that acts as a resonant circuit and is calibrated in terms of frequency or wavelength. It is also called a *coaxial wavemeter*.

coaxial-line resonator A resonator that consists of a length of coaxial line short-circuited at one or both ends.

coaxial loudspeaker A loudspeaker whose tweeter is mounted in the center of the woofer.

coaxial magnetron A magnetron made with a coaxial cathode surrounded by multiple cavities, typical of mod-

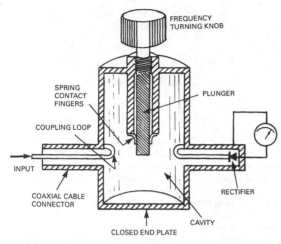

FREQUENCY TURNING KNOB

SPRING CONTACT FINGERS

PLUNGER

COUPLING LOOP

INPUT

COAXIAL CABLE CONNECTOR

RECTIFIER

CAVITY

CLOSED END PLATE

Coaxial-line frequency meter is a variable high-Q inductance-capacitance (LC) resonant cavity for measuring microwave frequencies.

ern construction. See *magnetron* and *crossed-field electron tubes*.

coaxial relay A relay that opens or closes a coaxial cable circuit without introducing a mismatch that would cause wave reflections.

coaxial sheet grating A sheet grating that consists of concentric metal cylinders each about one wavelength long, centered in a coaxial waveguide to suppress undesired modes of propagation.

coaxial stop filter A movable tuned filter placed around a conductor to limit its electrical radiating length at a given frequency.

coaxial stub A length of nondissipative cylindrical waveguide or coaxial cable branched from the side of a waveguide to produce some desired change in its characteristics.

coaxial switch A switch that changes connections between coaxial cables going to antennas, transmitters, receivers, or other high-frequency devices without introducing impedance mismatch.

coaxial transmission line *Coaxial cable.*

coaxial wavemeter *Coaxial-line frequency meter.*

cochannel interference Interference between two signals of the same type in the same radio channel.

code A system of symbols and rules for expressing information, such as the Morse and EIA color codes and the binary and other machine languages used in digital computers.

codec Acronym for *coder-decoder.*

code character A combination of code elements, such as dots and dashes or 0s and 1s, that represent a character.

code converter A converter that changes coded information to a different code system. Also called decoder.

coded decimal digit A decimal digit that is expressed by a pattern of four 1s and 0s.

coded decimal notation A form of notation in which each decimal digit is converted separately into a pattern of binary 1s and 0s. For example, in the 8-4-2-1 coded decimal notation, the number 13 is represented as 0001 0011, whereas in pure binary notation it is represented as 1101.

code division multiaccess [CDMA] A U.S. Standard for digital wireless telephone service and satellite communication. Segmented and coded data and/or speech travel to their destinations by different paths so that available radio frequencies can be shared. Data and/or speech are reassembled at the receiver with correlation-detection techniques.

code-division multiplex [CDM] Multiplex in which two or more communication links each occupy the entire transmission channel simultaneously, with code signal structures designed so a given receiver responds only to its own signals and sees the other signals as noise.

code-excited linear predictive [CELP] **coding** A digital speech coding method also called *vector-sum excited linear predictive* [VSELP] coding.

coded interrogator An interrogator whose output signal forms the code required to trigger a specific radio or radar beacon. The IFF interrogator is an example.

coded passive reflector A radar reflector whose reflecting properties can be varied according to a predetermined code, to produce a recognizable indication on a radarscope.

coder 1. A device that generates a code by generating pulses which have varying lengths and/or spacings, as required for radio beacons and interrogators. Also called moder, pulse coder, and pulse-duration coder. 2. A person who translates a sequence of computer instructions into codes that are acceptable to the machine.

coder-decoder [codec] The analog-to-digital and digital-to-analog function on a telephone subscriber line card in a telephone exchange.

code reader A reader that responds to printed, painted, magnetic, radioactive, or other types of coded marks placed on products, shipping cartons, freight cars, or other moving objects.

code recorder An instrument that makes a permanent record of code messages received by radio or wire, as by punching holes in a tape or by making dot-and-dash marks on a tape.

coder-filter-decoder [cofidec] The combination of a codec, the associated filtering, and voltage references required to code and decode voice in a subscriber line card.

code ringing Party-line telephone ringing in which the number and/or duration of rings identifies the station being called.

coding A list, in computer code, of the successive operations required to carry out a given routine or solve a given problem.

coefficient of coupling A numerical rating between 0 and 1 that specifies the degree of coupling between two circuits or the corresponding percentage value. Maximum coupling is 1, and no coupling is 0. Also called coupling coefficient.

coercimeter An instrument that measures the magnetic intensity of a natural magnet or electromagnet.

coercive force The magnetizing force required to reduce the flux density to zero in a magnetic material that has been magnetized alternately by equal and opposite magnetizing forces. It is the reverse magnetizing force needed to remove the residual magnetism.

coercivity The property of a magnetic material that is measured by the coercive force which corresponds to the saturation induction for the material.

cofired   A term for the furnace firing of two or more different objects together in accordance with the same processing schedule. An example is the formation of multilayer ceramic capacitors by firing the metallic ink on stacked layers of the "green" ceramic while simultaneously firing the ceramic into a monolithic chip.

cognitive intelligence   The ability to plan, establish goals, and model the environment based on received sensory input.

cognitive machine   A machine that has the ability to learn.

cohered video   The video detector output signal in a coherent moving-target-indicator radar system.

coherence   The existence of a statistical or time correlation between the phases of two or more waves.

coherence function   A measured spectrum value that determines if two measurements which produce a spectral line are correlated. The function is normalized at all frequencies, so it can have values only between 0 and 1. A value of 1 means that the spectral line at the monitored point is completely coherent with the measured source, and values of 0.5 and 0 mean that 50 and 0% of the power at a given frequency at the monitored point is coherent with the measured source.

coherent   Moving in unison or having some other fixed relationship, as between particles in a synchrotron or photons in a coherent laser beam.

coherent-carrier system   A transponder system whose interrogating carrier is retransmitted at a definite multiple frequency for comparison.

coherent decade frequency synthesizer   A frequency synthesizer that provides a wide range of output frequencies, such as from direct current to 100 kHz, in decimal steps.

coherent detection   Detection in which the received pulse-modulated signal is cross-correlated with locally generated signals that represent admissible signaling elements.

coherent detector   A detector in moving-target-indicator radar that gives an output signal amplitude which depends on the phase of the echo signal instead of on its strength, as required for a display that shows only moving targets.

coherent echo   A radar echo that has relatively constant phase and amplitude at a given range.

coherent frequency synthesizer   A frequency synthesizer that derives its frequency from a single source, for example, from an atomic resonance device.

coherent interrupted wave   An interrupted continuous wave in which the phase of the waves is maintained through successive wave trains.

coherent light   Light that has essentially a single wavelength, with definite phase relationships between different points in a beam. A laser beam is an example.

coherent-pulse radar   A radar whose RF oscillations of recurrent pulses bear a constant phase relation to those of a continuous oscillation.

coherent radiation   Radiation that exhibits definite phase relationships between different points in a cross section of the beam. In noncoherent radiation these relationships are random. Interference bands are observed only between coherent beams.

coherent reference   A reference signal, usually with a stable frequency, to which other signals are phase-locked to establish coherence throughout a system.

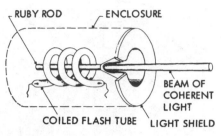

Coherent light produced by ruby laser.

coherent scattering   Scattering that exhibits a definite phase relationship between incoming and scattered particles or photons.

coherent-scattering cross section   The cross section for coherent scattering.

coherent system   A navigation system whose signal output is obtained by demodulating the received signal after the received signal is mixed with a local signal which has a fixed-phase relation to that of the transmitted signal. This permits the information carried by the phase of the received signal to be used.

coherent transponder   A transponder that maintains a fixed relation between frequency and phase of input and output signals.

coherent video   The video signal produced in a moving-target-indicator system by combining a radar echo signal with the output of a continuous-wave oscillator. After delay, this signal is detected, amplified, and subtracted from the next pulse train to give a signal that represents only moving targets.

coho   Abbreviation for *coherent oscillator.*

coil [L]   Turns of wire that introduce inductance into an electric circuit to produce magnetic flux or to react mechanically to a changing magnetic flux. In high-frequency circuits a coil might be only a fraction of a turn. The electrical size of a coil is called inductance and is expressed in henrys. The opposition that a coil offers to alternating current is called impedance and is expressed in ohms. The impedance of a coil increases with frequency. It is also called an *inductor.*

coil form   A rectangular or cylindrical hollow form made of plastic, waxed paper, or other suitable insulating material, on which an air-core coil can be wound. It is also called a *bobbin.*

coil loading   The insertion of loading coils at regular intervals in a transmission line to improve its transmission characteristics over the required frequency band.

coil neutralization   *Inductive neutralization.*

coil serving   *Serving.*

coil winder   A manual or motor-driven mechanism for winding coils individually or in groups.

coincidence circuit   A circuit that produces a specified output pulse when and only when a specified number or combination of two or more input terminals receive pulses within an assigned time interval. Also called coincidence counter and coincidence gate.

coincidence correction   *Dead-time correction.*

coincidence counter   *Coincidence circuit.*

coincidence gate   *Coincidence circuit.*

coincidence loss   Loss of counts due to the occurrence of ionizing events at intervals less than the resolution time of the counting system.

coincidence multiplier   A multiplier whose operation is based on the probability of the simultaneous occurrence of two independent events that are the product of the probabilities of the separate events. Each event signal is converted to a pulse whose width is proportional to the amplitude of the signal. The duration of time in which the two trains of pulses are in coincidence is then proportional to the product of the probabilities for the events. Also called probability multiplier.

coincident-current selection   The selection of a magnetic cell, for reading or writing in computer memory, by simultaneously applying two or more currents.

CO laser   *Carbon-monoxide laser.*

cold cathode   A cathode whose operation does not depend on its temperature being above the ambient temperature.

cold-cathode counter tube   A counter tube that has one anode and three sets of 10 cathodes. Two sets of cathodes act as guides that direct the glow discharge to each of the 10 output cathodes in correct sequence in response to driving pulses. It is used for data storage, preset counting, tuning, and gating.

cold-cathode magnetron ionization gage   *Cold-cathode vacuum gage.*

cold-cathode rectifier   A cold-cathode gas tube whose electrodes differ greatly in size, so electron flow is much greater in one direction than in the other. An example is the 0Z4 full-wave gas rectifier tube. Also called gas-filled rectifier.

cold-cathode tube   An electron tube that contains a cold cathode, such as a cold-cathode rectifier, neon tube, phototube, and voltage regulator.

cold-cathode vacuum gage   A vacuum gage that generates electrons to measure ionization current with a strong electric field rather than a hot cathode. A ring-shaped anode in the gage envelope has a zirconium- or thorium-plated cathode on each side. A high voltage applied between the anode and cathodes drives electrons from the cathodes, and a magnetic field from a permanent magnet produces a magnetic field that causes the electrons to spiral toward the positively biased ring anode. Most electrons make several passes through the ring anode and, while making these transits, they ionize the residual gas in the envelope. An ion current produced by the positive ions collected by

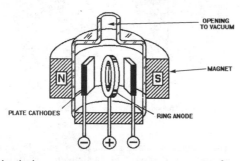

Cold-cathode vacuum gage can measure pressure of 10^{-8} to 10^{-10} torr.

the negatively biased cathode is proportional to the gas pressure (vacuum). The gage has an operating range of 1×10^{-8} to 2×10^{-2} torr. See also *hot-cathode vacuum gage* and *thermocouple vacuum gage.*

cold junction   The junction of thermocouple wires with conductors leading to the measuring instrument. This junction is normally at room temperature. See *Thermocouple.*

cold-plate heatsink   A flat, thermally conductive plate with a hollow core for circulating liquid or air to cool electronic circuitry, particularly a printed-circuit board (PCB). Typically, a flat aluminum plate is machined to accept copper tubing, which is flattened to keep the plate surface flat. Other versions are machined or cast from conductive alloys as two halves with channels or cavities and inlet and outlet pipes included. They are brazed together to form sealed units. *Heat pipes* and finned structures can be added to increase conduction and convection efficiency.

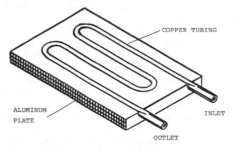

Cold-plate heatsink has a liquid pumped through its serpentine copper tubing to transfer heat.

collector   1. One of the three regions of a bipolar junction transistor. The base–collector PN junction is usually reverse-biased so that minority carriers injected into the base from the emitter are efficiently drawn into the collector. The pin or terminal connected to this region is also called the collector and corresponds to the anode of the electron tube. The other two regions are *base* and *emitter.* 2. In a traveling-wave tube, it is the electrode that collects electrons or ions that have completed their functions of transferring energy to the helix within the tube.

collector capacitance   The depletion-layer capacitance associated with the collector junction of a transistor.

collector characteristic curves   A set of characteristic curves of collector voltage versus collector current, for a fixed value of transistor base current.

collector-current runaway   The continuing increase in collector current as collector junction temperature is increased by collector-current flow.

collector efficiency   The ratio of useful power output to DC power input for a transistor, usually expressed as a percentage.

collector-follower effect   An effect used in constructing a transformerless single-transistor flip-flop with a conventional bipolar junction transistor. As the base of the saturated common-emitter transistor is moved rapidly toward cutoff, the collector voltage follows and remains at the extreme value for a time period before rising to the collector supply voltage.

collector junction   A semiconductor junction between the base and collector electrodes of a transistor. It is normally biased in the high-resistance direction, and the current through it is controlled by introducing minority carriers.

collector transition capacitance   The capacitance across the collector-to-base transition region, in which the impurity concentration changes.

collimate   To modify the paths of electrons in a flooding beam, or various rays of a scanning beam, to cause them to become more nearly parallel.

collimation   1. The precise alignment of the mechanical system of a radar antenna by comparison with an optical device aligned on known points in azimuth and elevation. 2. The process of focusing light rays or the paths of electrons or other particles into a parallel beam.

collimator   A device that confines the elements of a beam within an assigned solid angle.

collimator lens   1. In optics, a glass or plastic lens that produces parallel rays of light from a diverging light source. 2. In an electron tube, electrodes that form electrons into a parallel beam.

collinear array   An antenna array that consists of many half-wave dipoles mounted end to end and connected to operate in phase. It is also called a *linear array*.

collinear heterodyning   An optical processing system whose correlation function is developed from an ultrasonic light modulator. The output signal is derived from a reference beam so that the two beams are collinear until they enter the detection aperture. Variations in optical path length then modulate the phase of both signal and reference beams simultaneously, and phase differences cancel out in the heterodyning process.

collision avoidance system [CAS]   An airborne electronic system that provides automatic warning of a potential midair collision between planes.

collision-course homing   A technique for aiming a defensive missile on a course calculated to intercept an incoming enemy missile based on the projection of the course, speed, and trajectory of the incoming missile, as determined by a computer. The input data, obtainable from radar or laser tracking, permits programming the defensive missile before its launch. The launch program can be supplemented by mid-course corrections sent by RF or laser signals.

colloidal graphite   Powdered graphite, a form of carbon, suspended in water, glycerine, or other solvent for use as a conductive shield on the inside or outside surfaces of cathode-ray and other electron tubes for shielding and grounding purposes.

color   A characteristic of light that can be specified in terms of luminance, dominant wavelength, and purity. Luminance is the magnitude of brilliance. Wavelength determines hue and ranges from about 4000 Å for violet to 7000 Å for red (400 to 700 nm). Purity corresponds to chroma or saturation and specifies vividness of a hue.

color balance   Adjustment of the circuits that feed the three electron guns of a color picture tube to compensate for differences in light-emitting efficiencies of the three color phosphors on the screen of the tube.

color-bar code   A bar code based on one or more different bar colors in combination with black bars and white spaces, to increase the density of binary coding of data printed on merchandise tags or directly on products for inventory control and other purposes.

color-bar generator   A signal generator that delivers to the input of a color television receiver the signal needed to produce a color-bar test pattern on one or more channels.

color-bar test pattern   A test pattern of different colored vertical bars used to check the performance of a color television receiver.

color burst   A short series of oscillations at the chrominance subcarrier frequency of 3.579545 MHz, following each transmitted horizontal sync pulse in a color television system. This burst is a frequency reference for generating a continuous wave that is at the burst frequency and locked to it in phase. Also called burst.

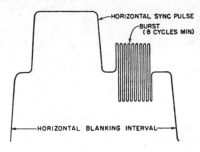

Color burst.

color-burst pedestal   The rectangular pulselike component that is part of the color burst when the axis of the color-burst oscillations does not coincide with the back porch. Also called burst pedestal.

color-carrier reference   *Chrominance-carrier reference.*

color code   A system of colors that indicates the electrical value of a component or identifies terminals and leads. The EIA color codes are the standards for the electronics industry.

color coder   *Matrix.*

color comparator   A photoelectric instrument that compares an unknown color with a standard color sample for matching purposes. The sample and unknown can be placed alternately in the measuring position, or two identical measuring systems can be used to feed a null indicator. Also called photoelectric color comparator.

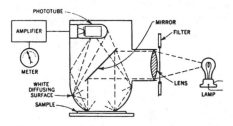

Color comparator: The sample and unknown are alternately placed in measuring position.

color contamination   Poor rendition of color in a color television receiver, caused by incomplete separation of color component paths.

color control   *Chroma control.*

color coordinate   *Chromaticity coordinate.*

color decoder   *Matrix.*

color-difference signal   A signal that is added to the monochrome signal in a color television receiver to obtain a signal representative of one of the three tristimulus values needed by the color picture tube. There are three color-difference signals: the G-Y signal feeds the green channel, the R-Y signal feeds the red channel, and the B-Y signal feeds the blue channel.

color edging   Spurious color at the boundaries of differently colored areas in a color television picture. Color edging includes color fringing and misregistration.

color encoder   *Matrix.*

color fidelity   The ability of a color television system to reproduce faithfully the colors in an original scene.

color field corrector   A device positioned outside a color picture tube to produce an electric or magnetic field that acts on the electron beam after deflection to produce more uniform color fields.

color filter   A sheet of material that absorbs certain wavelengths of light while transmitting others.

color flicker   Flicker due to fluctuations of both chromaticity and luminance in a color television receiver.

color fringing   Spurious chromaticity at boundaries of objects in a color television picture. Small objects might appear separated into different colors. It can be caused by a change in position of the televised object from field to field or by misregistration.

color graphics adapter [CGA]   A circuit board for the color monitors of personal computers that supplies red, green, blue, and intensified (RGBI) video and a composite video signal. Up to 16 colors can be formed on an RGBI TTL-compatible monitor, and up to 16 shades of gray can be formed on a composite video monitor.

colorimeter   An instrument that measures color by determining the intensities of the three primary colors which will give that color.

colorimetry   The science of color measurement.

color killer circuit   The circuit in a color television receiver that biases chrominance amplifier tubes or transistors to cutoff during reception of monochrome programs.

color oscillator   *Chroma oscillator.*

color phase   The difference in phase between a chrominance signal (I or Q) and the chrominance-carrier reference in a color television receiver.

color phase alternation [CPA]   The periodic changing of the color phase of one or more components of the chrominance subcarrier between two sets of assigned values after every field in a color television system.

color phase detector   The color television receiver circuit that compares the frequency and phase of the incoming burst signal with those of the locally generated 3.579545-MHz chroma oscillator, and delivers a correction voltage to a reactance tube to ensure that the color portions of the picture will be in exact register with the black and white portions on the screen.

color picture signal   The electric signal that represents complete color picture information, excluding all synchronizing signals. In one form it consists of a monochrome component plus a sub-carrier modulated with chrominance information.

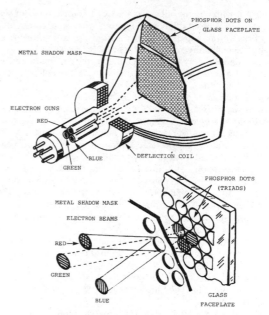

Color picture tube with three electron guns and a conventional shadow mask.

color picture tube   A cathode-ray tube that has three different phosphor colors. When these are appropriately scanned and excited in a color television receiver, a color picture is obtained.

colorplexer   The section of a color television transmitter in which red, green, and blue signals are combined by matrixing and multiplexing to produce a single compatible color television signal.

color primaries   The red, green, and blue primary colors that are mixed in various proportions to form all the other colors on the screen of a color television receiver.

color purity   Absence of undesired colors in the spot produced on the screen by each beam of a television color picture tube.

color-purity magnet   A magnet on the neck of a color picture tube, to improve color purity by changing the path of the electron beam.

color registration   The accurate superimposing of the red, green, and blue images used to form a complete color picture in a color television receiver.

color response   The sensitivity of a device to different wavelengths of light.

color sampling rate   The number of times per second that each primary color is sampled in a color television system.

color saturation   The degree to which a color is mixed with white. High saturation means little or no white, as in a deep red color. Low saturation means a lot of white, as in light pink. The amplitudes of the I and Q chrominance signals determine color saturation in a color television receiver. It is also called *saturation.*

color-saturation control   *Chroma control.*

color sidebands   The signals that extend for about 0.4 MHz above and below the 3.579545-MHz color subcarrier signal which is broadcast as part of a color television signal. The color sidebands contain picture chrominance information, which is removed from the color subcarrier by a synchronous detection process in receivers.

color signal   Any signal that controls the chromaticity values of a color television picture, such as the color picture signal and the chrominance signal.

color-subcarrier oscillator   *Chroma oscillator.*

color-subcarrier reference   *Chrominance-carrier reference.*

color-sync signal   A sequence of color bursts that is continuous except for a specified time interval during the vertical blanking period. Each burst occurs at a fixed time with respect to horizontal sync in a color television system.

color television   A television system that reproduces an image in its original colors. In the United States a compatible dot-sequential color television system is used. The bandwidth is 6 MHz, just as for monochrome television. Color synchronizing information is transmitted on a 3.579545-MHz subcarrier.

color temperature   The temperature of a black-body radiator that produces the same chromaticity as the light under consideration.

color transmission   The transmission of a signal wave for controlling both the luminance and chromaticity values in a picture.

color triangle   A triangle drawn on a chromaticity diagram, to represent the entire range of chromaticities obtainable as additive mixtures of three prescribed primaries.

Colpitts oscillator   An inductive-capacitive (LC) oscillator whose frequency-determining resonant tank circuit includes a capacitive voltage divider. A fraction of the current flowing in the tank circuit is regeneratively fed back to the base through a coupling capacitor. Developed as a vacuum tube circuit, it will function with bipolar junction or field-effect transistor amplifiers. It can generate sine-wave frequencies up to UHF. See *Clapp oscillator.*

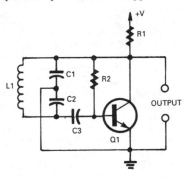

Colpitts oscillator is identified by the capacitor voltage divider in its feedback path.

columbium   Former name for *niobium.*

COM   Abbreviation for *computer-output microfilm.*

coma   A cathode-ray-tube image defect that makes the spot on the screen appear comet-shaped when the spot is away from the center of the screen.

coma lobe   A side lobe that occurs in the radiation pattern of a microwave antenna when the reflector alone is tilted back and forth to sweep the beam through space. The coma

lobe is produced under these conditions because the feed is no longer always at the center of the reflector. Used to eliminate the need for a rotary joint in the feed waveguide.

comb antenna   A broadband antenna for vertically polarized signals. Half of a fishbone antenna is erected vertically and fed against ground by a coaxial line.

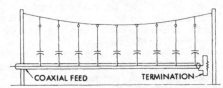

Comb antenna, supported from wire between two poles.

comb filter   A filter whose insertion loss causes its spectrum to form a sequence of equispaced narrow passbands or stop bands centered at multiples of some specified frequency. An oscilloscope display of the spectrum resembles the teeth of a comb. A comb filter separates luminance Y (black-and-white) and chrominance C (color) signals from the composite television video signal.

comb generator   A signal generator that converts a single-frequency RF input signal into an RF output signal which has a large number of spectral lines, each line harmonically related to the input frequency.

combinatorial logic   Logic whose outputs are dependent only on input states and the delays encountered in the logic path. In contrast, sequential logic produces outputs dependent on the previous state of the logic array, on the presence of a discrete timing interval, and on the input states and delays.

combined double-T   A combination of two T-shaped waveguide junctions, designed so the paths of power flow are determined by the loading of the arm and the matching between the arms.

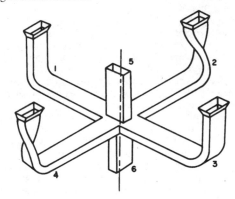

Combined double-T waveguide array for controlling power flow.

combiner circuit   The circuit that combines the luminance and chrominance signals with the synchronizing signals in a color television camera chain.

COMFET   *Insulated-gate bipolar transistor* (IGBT).

COMIT   A user-oriented, general-purpose, symbol-manipulation programming language for computers.

COMLOGNET [COMbat LOGistic NETwork] A long-line data-transmission network for rapid processing and distribution of logistic information at computer-equipped automatic switching centers. The network translates input data and transmits the information to the addressee, with messages routed on a priority basis.

command 1. An instruction sent by the *central processor unit* (CPU) to a *controller* for execution. It can be executed by pressing a key designated to perform a specific function, such as put a word in boldface or delete it, initiate spelling corrections, or print out a line or file. Commands can be given by clicking a *mouse* or other controller when a marker is moved on the screen to designate a command written in a language format or displayed as an icon on the computer screen. 2. A RF or infrared order or correction for a remote mobile platform, such as a spacecraft, drone, missile, or robot. It can also be an order to open a garage door or switch a TV channel with a remote controller. Commands can be given by infrared, sonic, ultrasonic, or radio signals, or they can be given as electrical signals over wires.

command guidance A missile guidance system that transmits flight direction information to the missile from a point external to the missile. The information needed to control missile performance can be obtained by telemetering, ground-based radar, optical tracking, or other means. The commands can be transmitted to the missile by radio, radar, or optical means.

command module The spacecraft module that carries the crew, the main communication and telemetry equipment, and the reentry capsule during cruising flight.

command resolution The maximum change in command that can be made in a feedback control system without causing a change in the ultimately controlled variable.

command set A radio set that receives or gives commands, as between one aircraft and another or between an aircraft and the ground.

Commission Internationale de l'Eclairage [CIE] An international group that has set most of the basic standards of light and color now used in color television.

common-base amplifier A basic bipolar amplifier whose base terminal is common to both the input and output ports. It is also called a *grounded-base amplifier*.

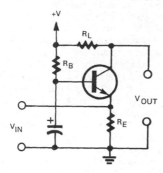

Common-base amplifier with an NPN transistor.

common cathode A cathode for two or more sections of an electrode tube, as in a duodiode pentode.

common-cathode amplifier An electron tube amplifier whose cathode is at ground potential at the operating frequency. The input signal is applied between the control grid and ground, and the output load is connected between anode and ground.

common-collector amplifier A bipolar junction transistor (BJT) amplifier whose collector terminal is common to both the input and output ports. Its input signal is applied between the base and ground, and its output signal is applied between its emitter and ground. The collector is normally connected to the power supply. Voltage gain from base to emitter is less than one, but current gain is high. It offers high input impedance and low output impedance. It is also called an *emitter follower* and a *grounded-collector amplifier*. See also *Darlington pair*.

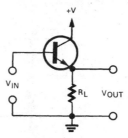

Common-collector amplifier with an NPN transistor.

common-drain amplifier A basic field-effect transistor (FET) amplifier whose drain terminal is common to both the input and output ports. The input signal is introduced between gate and ground, and the output is taken between source and ground. It is also called a source-follower amplifier. The common-drain amplifier is used in applications that require low input capacitance or the ability to handle large input signals. It is analogous to the bipolar transistor emitter-follower amplifier.

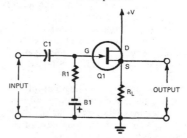

Common-drain amplifier that includes an N-channel JFET is also called a *source follower*.

common-emitter amplifier A basic bipolar amplifier whose emitter terminal is common to both the input and output ports. It offers medium-input impedance, high-output impedance, and both high voltage and current gain. It is also called a grounded-emitter amplifier.

common-gate amplifier A basic field-effect transistor (FET) amplifier whose gate terminal is common to both the input and output ports. The input signal is introduced between source and ground, and the output signal is taken

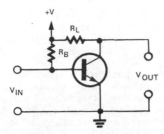

Common-emitter amplifier with an NPN transistor.

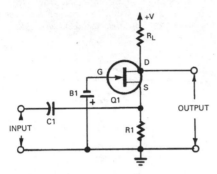

Common-gate N-channel JFET amplifier is analogous to a bipolar common-base amplifier.

between drain and ground. The common-gate amplifier is used for transforming a low input impedance to a high output impedance and for high-frequency amplification. It is analogous to a bipolar junction transistor-base amplifier.

common-grid amplifier An electron tube amplifier whose control grid is at ground potential at the operating frequency. The input signal is applied between the cathode and ground, and the output load is connected between anode and ground. It offers low-input impedance and freedom from oscillation because of feedback.

common language A machine-readable language that is common to a group of computers and associated equipment.

common mode Signals that are identical in amplitude and phase at both inputs, as in a differential operational amplifier.

common-mode coupling Coupling that introduces similar signals with respect to ground on different leads.

common-mode error The error voltage that exists at the output terminals of an operational amplifier due to the common-mode voltage at the input.

common-mode electromagnetic interference Interference in which the potentials of two signal or power leads change simultaneously and equally with respect to ground.

common-mode interference Interference signals that appear between both inputs of a differential circuit and a common reference like ground.

common-mode noise The noise component that is common to both the converter output and return lines with respect to the input common.

common-mode rejection [CMR] The ability of an amplifier to cancel a common-mode signal while responding to an out-of-phase signal. Usually expressed in decibels.

$CMR = 20 \log_{10} (CMRR)$; thus if the CMRR is 10^6, the corresponding CMR value is 120 dB. Also called in-phase rejection.

common-mode rejection ratio [CMRR] The ratio of the RMS value of the common-mode interference voltage at the input terminals of an operational amplifier to the effect or error produced by that common-mode interference at the output when referred to the input (when divided by amplifier gain). A high ratio is desirable. This ratio expresses the ability of the device to reject the effect of a voltage that is applied simultaneously to both input terminals.

common-mode signal A signal applied equally to both ungrounded inputs of a balanced amplifier stage or other differential device. Also called in-phase signal.

common-mode voltage A voltage that appears in common at both input terminals of a device with respect to the output reference (usually ground).

common return A return conductor that serves two or more circuits.

common-source amplifier A basic field-effect transistor (FET) amplifier whose source terminal is common to both the input and output ports. The input signal is introduced between the gate and the source, and the output signal is taken between the drain and the source. It offers high input impedance, medium to high output impedance, and voltage gain greater than unity. It is analogous to a bipolar transistor common-emitter amplifier.

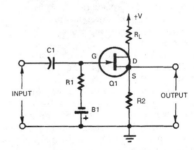

Common-source amplifier includes an N-channel JFET.

communication The transmission of intelligence between two or more points over wires or by radio. The terms telecommunication and communication are often used interchangeably, but telecommunication is usually the preferred term when long distances are involved.

communication channel The wire or radio channel that conveys intelligence between two or more terminals.

communication protocol The rules governing the exchange of information between devices on a data line.

communication receiver A receiver for reception of voice or code messages transmitted by radio communication systems.

communication satellite An orbiting satellite that relays radio, television, and other signals between ground terminal stations thousands of miles apart. A satellite is placed in orbit 22,300 mi (35,880 km) above the earth to give a 24-h orbital period that makes it appear stationary. A radio wave traveling at the speed of light then takes about 0.12 s to travel from an earth station to the satellite and another 0.12 s to return to another earth station.

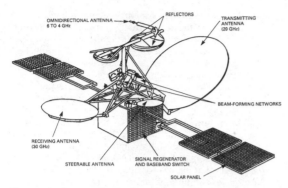

Communications satellite is equipped with receiving antennas, signal regeneration circuitry, and transmitting antennas.

Communications Satellite Corp. [COMSAT]   A common carrier created under the provisions of the Communications Satellite Act of 1962, to provide communication satellite service.

communications security [COMSEC]   Protection of transmitted information by converting it to a form that is unintelligible to an unauthorized listener yet can be reconverted to its original form at the intended receiving station, generally by using a crypto system.

communications security [COMSEC] **device**   A semiconductor integrated circuit that is designed to encode incoming communications. It is included in the transmission and receiving ends of a communications link (cable, fiberoptic, or radio frequency) to prevent unauthorized access to the information being transmitted.

community antenna television   *Cable television.*

commutated-antenna direction-finding   A radio receiver that is switched in sequence to one antenna after another in a circular array, to determine the direction of arrival of radio waves by sensing the commutation phase shifts.

commutating diode   A diode in a switching regulator circuit that provides current for the filter choke when the switching transistor is not conducting.

commutation   1. The continuous transfer of current from the brushes of a DC motor to its rotating commutator segments to supply current to the armature windings.   2. The sampling and sequential transmission of blocks of data in a computer at regular intervals, or the time sharing of a single-channel transmitter in telemetry or other communications system. See also *multiplexing.*

commutator   1. A circular arrangement of copper segments insulated from each other and the rotor of a DC motor or generator on which they are mounted. Current-carrying brushes bear on the exposed surfaces to provide commutation of direct current to armature coils in sequence.   2. A circuit for time-division multiplexing that provides repetitive sequential switching of signals from a multiplicity of channels.

commutator switch   A switch, electronic or mechanical, that performs a set of switching operations in repeated sequential order, for example, as required for telemetering many quantities. It is also called a *sampling switch* and a *scanning switch.*

compact disk [CD]   A plastic disk with a 4⅔-in diameter that stores voice, music, and other sound in the form of microscopic pits on spiral tracks produced by a laser and pressed into the CD. The recordings are played back by directing laser light on the moving pits, which modulate the reflected light to reproduce the original sound.

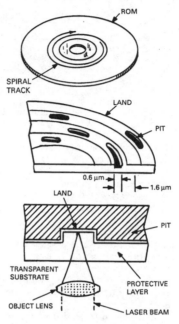

Compact disk has sound recorded as laser-generated microscopic pits on a spiral track. Reflected laser light, modulated by the pits, is converted back to a reproduction of the original sound. It is a read-only memory (ROM).

compact-disk-interactive [CD-I] **system**   A development based on consumer entertainment compact-disk (CD) technology whose output can be selected by user interaction with video and TV receivers, and sound can be played through home audio systems. Users can control video and sound with devices such as trackballs and joysticks. CD-I also has commercial and industrial applications.

compact disk [CD] **player**   An electromechanical apparatus that includes both a motor-driven turntable and an electrooptical playback system which converts reflected laser light from moving pits on the compact disk to an audio signal for amplification. It can be a battery-powered portable unit for headphone listening, or it can be included in a stereo receiver so that the output can be played back through the stereo speakers.

compact-disk read-only memory [CD-ROM] **player**   An electromechanical personal computer peripheral that includes both a motor-driven turntable and an electrooptical playback system for converting reflected laser light from the moving pits on the CD-ROM to audio and video signals for amplification. It can be played back through the color monitor of the personal computer supplemented by stereo speakers. It permits the interactive playback of recorded text, graphics, stories, or complete video games and movies.

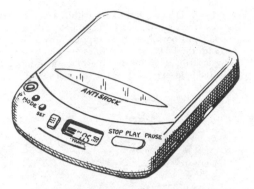

Compact disk player plays back compact disks and the signal is amplified for headphones or speaker.

compact-disk, read-only memory extended architecture [CD-ROM XA] A compact-disk ROM that specifies an encoding format (adaptive differential pulse-code modulation) for storing audio information in a digital format.

CompactFlash Association [CFA] An industry organization that sets standards for computer memory card manufacturers.

companding The process of compressing or reducing the dynamic range of a voice or music signal (typically at the transmitter) and then expanding or restoring that signal at the receiver in accordance with a specified transfer characteristic (usually a logarithmic companding law). It is performed primarily to improve signal-to-noise (S/N) ratio. See *compandor*.

compandor An acronym formed from the words *COMPressor* and *expanDOR*. A combination of compressor circuit for compressing the amplitude range of signals, followed by an expandor for restoring the original amplitude range. It can improve the signal-to-noise (S/N) ratio. Complete compandors are available on a single integrated circuit with access to both compressor and expansion sections. The compressor section of the IC by itself can perform automatic audio-level control (ALC) in radio receivers.

comparator 1. A device that compares two transcriptions of the same information to verify the accuracy of transcription, storage, arithmetic operation, or some other process in a computer and delivers an output signal of some form to indicate whether the two sources are equal or in agreement. 2. An electronic instrument that measures a quantity and compares it with a precision standard. 3. An operational amplifier without feedback to detect changes in voltage level, as required in analog-to-digital, digital-to-analog, and other types of converters.

comparison bridge 1. A bridge circuit that generates an error which corresponds with any change in output voltage with respect to a reference voltage. The error signal corrects the output voltage by *negative feedback,* restoring bridge balance. 2. A bridge that compares the values of two impedances and gives the result as a ratio.

compass bearing A bearing measured relative to compass north.

compatibility The ability of a new system to serve users of an old system.

compatible color television system A color television system that permits the substantially normal monochrome reception of the transmitted color picture signal on a typical unaltered monochrome receiver. This is accomplished in the U.S. color television system by dividing the color video information into a luminance signal and two chrominance signals. The luminance signal is the equivalent of a monochrome television picture signal and is used alone by a monochrome receiver.

compatible current-sinking logic [CCSL] A semiconductor integrated-circuit logic family that is interchangeable with diode-transistor logic for comparable functions.

compatible discrete four-channel sound [CD-4 sound] A sound system that maintains separate channels from each of the four sets of microphones at the recording studio or other input location to the four sets of loudspeakers that serve as the output of the system. Consequently four channels are therefore required on records and magnetic tape.

compatible integrated circuit An integrated-circuit family that has input/output logic levels and operating characteristics which make it compatible with one or more other families of integrated circuits.

compatible single-sideband system A single-sideband system that can be received by an ordinary amplitude-modulation radio receiver without distortion.

compatible stereo system A stereo system that gives satisfactory single-channel sound for single-channel audio tape or radio receivers.

compensated amplifier A broadband amplifier whose frequency range is extended by choice of circuit constants.

compensated impurity A donor or acceptor impurity that is neutralized by impurities of the opposite polarity in a semiconductor.

compensated-loop direction finder A radio direction finder that has a loop antenna whose polarization error is compensated by an additional antenna system.

compensation 1. Modification of the amplitude-frequency response of an amplifier to broaden the bandwidth or make the response more nearly uniform over the existing bandwidth. Also called frequency compensation. 2. The introduction of donors into a P-type semiconductor or acceptors into an N-type semiconductor.

complement A number whose representation for a computer is derived from the finite positional notation of another number by subtracting each digit from one less than the base, adding 1 to the least significant digit, and executing all carries required. This is the true complement. Thus the two's complement of binary 10010 is 01110; the ten's complement of decimal 2546 is 7454. In many machines a negative number is represented as a complement of the corresponding positive number.

complementary A term describing integrated circuits that employ components of opposing polarities connected so that the operation of either device is *complemented.* A complementary bipolar circuit includes both NPN and PNP transistors, and a complementary CMOS circuit includes both N-channel and P-channel transistors. Complementary devices operate with voltage and currents of opposite polarity.

complementary color A color that, when added to another given color in proper proportion, produces white.

complementary direct-coupled amplifier A direct-coupled amplifier whose PNP and NPN transistor stages are in alternating sequence, simplifying both direct coupling and negative feedback.

complementary functions Two driving-point functions whose sum is a positive constant.

complementary logic switch A complementary transistor pair with a common input and interconnections so that one transistor is on when the other is off, and vice versa. Load current flows only briefly during switching, when the voltage at the junction between the transistors changes from the level of logic 1 to the level of logic 0.

complementary metal-oxide semiconductor [CMOS] A combination of N- and P-channel enhancement-mode MOSFETs on a single silicon chip, connected as push-pull complementary digital circuits. Their advantages include low quiescent power dissipation and high operating speed.

complementary-output circuit A logic circuit that has two outputs, one of which is logic 0 when the other is logic 1.

complementary symmetry A circuit that has both PNP and NPN transistors or N-channel and P-channel MOS-FETs in a symmetrical arrangement which permits push-pull operation without an input transformer or other form of phase inverter.

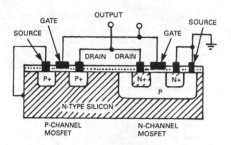

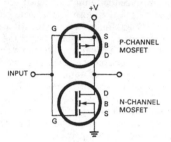

Complementary symmetry MOS (CMOS) contains both N- and P-channel enhancement-mode MOSFETs to minimize power loss. The simplest CMOS circuit is the NOT gate or inverter shown.

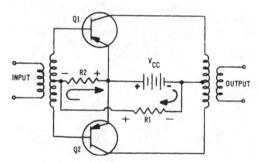

Complementary symmetry in class B push-pull amplifier with a small bias voltage developed across R2.

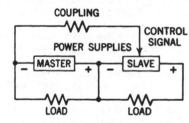

Complementary tracking of master and slave regulated power supplies.

complementary-symmetry metal-oxide semiconductor *Complementary metal-oxide semi-conductor.*

complementary tracking The interconnection of two regulated power supplies so that one is acting as a master to control the other. The output voltage of the slave supply is equal or proportional to that of the master but of opposite polarity with respect to a common point.

complementary-transistor amplifier An amplifier that uses the complementary symmetry of NPN and PNP transistors.

complementary-transistor logic [CTL] Logic that uses complementary transistors, generally in medium- and small-scale integration.

complementary transistors Two transistors whose characteristics and ratings are similar but opposite in sense, such as PNP and NPN bipolar transistors or P- and N-channel field-effect transistors. They will give push-pull output from a single input.

complementary unijunction transistor [CUJT] A semiconductor device with characteristics like those of a standard unijunction transistor except that the currents and voltages applied to it are of opposite polarity.

complementary wavelength The wavelength of light that, when combined with a sample color in suitable proportions, matches a reference standard light. The purples that have no dominant wavelengths, including nonspectral violet, purple, magenta, and nonspectral red colors, are specified by use of their complementary wavelengths.

complete carry The propagation in a computer of a carry that results from the addition of carry.

complex reflector A structure or group of structures having many radar-reflecting surfaces facing in different directions.

complex tone A sound wave produced by the combination of simple sinusoidal components of different frequencies.

complex value An impedance value that consists of magnitude and phase angle presented as a function of frequency.

compliance The acoustical and mechanical equivalent of capacitance. It is the opposite of stiffness.

compliance voltage The output voltage range needed in a regulated DC power supply to maintain a specified constant value of current for a specified range of load resistances.

component Any electronic device such as a coil, resistor, capacitor, transistor, integrated circuit, or diode that has distinct electrical characteristics and terminals for connecting it to other components to form a circuit.

component density The number of components per unit area or unit volume.

component-level bus A conductor for sending and receiving microprocessor signals whose input and output pins have defined functions and timing.

composite color signal The color picture signal plus all blanking and synchronizing signals. The composite color signal thus includes the luminance signal, the two chrominance signals, vertical and horizontal sync pulses, vertical and horizontal blanking pulses, and the color-burst signal.

composite color sync The signal comprising all the sync signals necessary for proper operation of a color receiver. This includes the horizontal and vertical sync and blanking pulses, and the color-burst signal.

composite line A line or circuit connecting a pair of *multiplexers* or *concentrators*.

composite modulation voltage The combined output voltage of the subcarrier oscillators in a telemetering system, applied as modulation to the transmitter.

composite picture signal The complete picture signal as it leaves the television transmitter. The picture consists of picture data, blanking pulses, synchronizing pulses for monochrome, and the color subcarrier, color burst, and other information needed for transmission of color pictures. Also called composite signal and composite video signal.

composite pulse A pulse composed of a series of overlapping pulses received from the same source over several paths in a pulse navigation system.

composite signal *Composite picture signal.*

composite video signal *Composite picture signal.*

composite wave filter A combination of two or more low-pass, high-pass, bandpass, or band-elimination filters.

composition resistor *Carbon composition resistor.*

compound modulation Modulation in which one or more signals modulate their respective subcarriers, and these subcarriers in turn modulate the carrier.

compound semiconductor A semiconductor made as a compound of two or more elements rather than a single element such as silicon or germanium. Group III–V semiconductor materials are made from Group III elements of the periodic table, which have three valence electrons (e.g., aluminum, gallium, and indium) and Group V elements which have five valence electrons (e.g., phosphorous, arsenic, and antimony). Binary compounds are made with two elements, ternary compounds have three elements, and quaternary compounds have four.

compression The reduction of the volume range of an audio signal. Weak signal components are made stronger so that they will not be lost in background noise, and loud passages are reduced in strength so they will not overload any part of the system. This result is achieved by making the effective gain vary automatically as a function of signal magnitude.

compressional wave A wave in an elastic medium that causes an element of the medium to change its volume without undergoing rotation. A compressional plane wave is a longitudinal wave.

compression ratio The ratio of the amplification at a reference signal level to that at a higher stated signal level.

compressive intercept receiver An electromagnetic surveillance receiver that instantaneously analyzes and sorts all signals within a broad RF spectrum with pulse compression techniques which perform a complete analysis up to 10,000 times faster than a superheterodyne receiver or spectrum analyzer. One version has three identical, highly dispersive, surface acoustic-wave filters. Two filters act as pulse expansion lines in a passive sweeping local-oscillator configuration, and the third filter acts as a pulse compression line in the signal-processing portion of the receiver.

compressive receiver See *compressive intercept receiver.*

compressor The section of a compandor circuit that compresses the intensity range of signals at the transmitting or recording end of the circuit. The compressor amplifies weak signals and attenuates strong signals to produce a smaller amplitude range.

compromise net A network used with a hybrid junction to balance a connected communication circuit such as a subscriber's loop. It gives a compromise between the extremes of impedance balance.

computed radiography A medical diagnostic imaging technique that is based on X-ray radiation. An X-ray image is scanned by a laser, and the output signal is digitized for immediate presentation on a cathode-ray tube. It permits rapid diagnosis by eliminating the time lost in developing an X-ray film. A permanent record can be obtained by storing the digital data or reproducing it directly on paper by xerographic copying methods.

computed tomography [CT] A diagnostic imaging technique in which a rotating projector directs X-rays axially around a patient, and a computer accepts the X-ray data and computes two-dimensional images of a "slice" through the patient's body for cathode-ray-tube display. Three-dimensional images can also be formed with this technique.

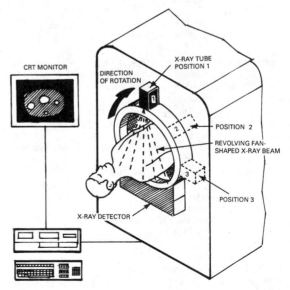

Computed tomography (CT) scanner and computer can produce X-ray section views of a patient's internal organs.

computer A machine capable of accepting information, processing the information, and supplying results in a desired form. Examples include analog, digital, and hybrid computers.

computer-aided design [CAD] A design technique that depends on interactive computer graphics to create, manipulate, and store a design. Large-scale integrated circuits can be designed on computer workstations by taking schematic symbol data from memory and arranging them on the monitor screen to form a schematic. Wafer maps can be generated to aid in storing and tracking test data, and final designs can be connected and located within an image of a package that is also generated on the screen.

computer-aided engineering [CAE] The application of computers and software to all aspects of engineering and in all engineering disciplines. Computers, typically workstations, with appropriate software can be used in the preparation of complete functional and block diagrams, circuit board layouts, integrated circuit masks, electronic schematics, electrical wiring, and mechanical piping diagrams. Workstations can also prepare three-view mechanical drawings of parts, subassemblies and assemblies, architectural elevations, and floor and plot plans. They can also translate data into isometric, perspective, exploded-view, and cutaway drawings or sections. The three-dimensional drawings can be rotated in space with contrasting colors to highlight features. Computers can also prepare specifications, parts lists, contracts, operating and maintenance instructions, proposals, and reports. See *computer-aided graphics* [CAG].

computer-aided graphics [CAG] The use of a computer, typically a workstation or advanced personal computer, to display objects or concepts in color and three dimensions. Examples include products such as automobiles or airplanes for design analysis and styling and animated weather maps. Scientific concepts such as the dynamics of thunderstorms, chemical bonding, atomic particle interaction, and volcanic and earthquake activity can be animated. CAG is also applied to simulation, cartooning, and special effects for moving pictures and television. See *computer-aided design* and *computer-aided engineering*.

computer-aided instruction [CAI] Use of a computer with a large number of on-line student terminals to supplement or replace classroom instruction. Also called computer-assisted instruction.

computer-aided manufacturing [CAM] The use of the computer and software to aid in the manufacture of a product or the processing of chemicals or materials. Integrated-circuit masks, printed circuit masks, and scaled drawings made during computer-aided design (CAD) can assist in the manufacture of integrated circuits, circuit boards and mechanical parts, and assemblies. Digital data can be sent directly to cutting, forming, turning, or milling machines and used in testing and checking quality on the finished product. A computer-controlled robot that can cut, weld, assemble, apply adhesive, paint, or sort is another example of CAM. Computer-based process control is also a form of CAM.

computer-aided prototyping The formation of three-dimensional models of products or casting molds for products by the successive buildup of three-dimensional objects in layers formed by extruding plastic, melting plas-

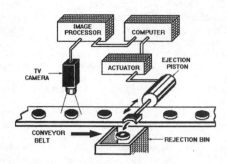

Computer-aided vision. A TV camera views parts passing on a conveyor belt and compares their images with a master profile stored in computer memory. Parts that do not conform are rejected.

tic beads, or hardening photoresistive emulsions under computer control. In some processes the heating or solidifying of the media is done by a scanning laser beam. Data obtained from 3-D drawings is converted electronically to control signals for depositing material in successive layers to form the prototype. After the removal of excess materials, the prototype can be finished by manual or automatic smoothing processes.

computer-assisted instruction *Computer-aided instruction.*

computer code The system of binary digits or other machine codes used by computers.

computer connector A multipin connector that permits the simultaneous connection or disconnection of 25 or more pins. They are typically configured in a D-format to assure the proper mating of the pins. The socket is mounted on the back panel of the computer, and the plug is attached to the end of the cable.

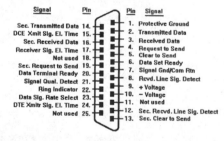

Computer connector pin diagram. A 25-pin RS-232C interface for the connection of data communications equipment to data terminal equipment.

computer control A general term applied to the control of industrial processes, machine tools, aircraft flight, automotive engines, instruments, home entertainment products, and other activities by some form of software-programmed computer central processor (CPU). Today, this term can mean an activity under the control of a complete computer or plug-in circuit board with an embedded *microprocessor* or *microcontroller.*

computer graphics *Computer-aided graphics.*

computerize The process of converting an activity formerly performed by mechanical means, such as clockwork, or electromechanical assembly, such as motors and relays, to computer control. Examples of computerized products include electronic test instruments, scales, calculators, panel meters, and cameras.

computer program A routine for solving a problem or carrying out an information-processing operation on a computer.

COMSAT Abbreviation for *Communications Satellite Corp.*

COMSEC Abbreviation for *communications security.*

concentrator A switching system that lets a large number of telephone or data-processing subscribers use a lesser number of transmission lines or a narrower bandwidth, such as by storing messages until lines become available and then selecting messages for transmission on a first-come or other priority basis.

concurrent engineering An engineering management concept for reducing time-to-market and improving new product quality by having teams work concurrently on different parts of the same project. The team consisting of engineering, manufacturing, marketing, and quality control personnel tries to involve the ultimate customer during the product definition phase.

condenser 1. A system of lenses that concentrates or focuses light rays to a point. It is also called a *condensing lens.* 2. Obsolete term for *capacitor.*

condensing lens *Condenser.*

conditioning Modifying data-transmission lines and equipment as required to match transmission levels and impedances or provide equalization between facilities.

conductance [G] A measure of the ability of a material to conduct electric current. It is the reciprocal of the resistance of the material and is expressed in siemens. Conductance is the resistive or real part of admittance.

conducted interference Interfering signals arriving on power or telephone lines.

conducting polymer A plastic whose high conductivity approaches that of metals.

conduction The transmission of energy by a medium without movement of the medium itself.

conduction band An energy band in which electrons can move freely in a solid. The conduction band is normally empty or partly filled with electrons.

conduction current A current due to a flow of conduction electrons through a body or charge carriers through a semiconductor.

conductive coating A coating that reduces surface resistance and thus prevents the accumulation of static electric charges. Its application prevents *electrostatic discharge.*

conductive elastomer A rubberlike silicone material whose suspended metal particles conduct electricity.

conductive gasket A flexible metallic gasket that reduces RF leakage at joints in shielding.

conductive pattern A design formed of any conductive or resistive material.

conductive silver paste Silver powder in a suitable vehicle for applying to ceramic or other insulating materials by silk-screening or other methods. Firing at appropriate temperatures fixes the silver to provide a hard conductive surface or joint.

conductivity The ability of a material to conduct electric current, as measured by the current per unit of applied voltage. It is the reciprocal of resistivity. In a semiconduc-

tor, N-type conductivity is associated with conduction electrons, and P-type conductivity is associated with holes.

conductivity modulation Variation of the conductivity of a semiconductor by variation of the charge-carrier density.

conductivity-modulation transistor A transistor whose active properties are derived from minority carrier modulation of the bulk resistivity of a semiconductor.

conductor 1. A material that conducts electric current easily because it offers little electrical resistance. Examples include such metals as aluminum, copper, gold, lead, nickel, platinum, silver, and tin, but carbon, carbon-impregnated plastic, and salt water are also conductors. 2. A material that conducts thermal energy or heat, in general the same as those materials that are good electrical conductors. 3. A material such as transparent glass or plastic, or a clear liquid that will transmit visible light.

conduit Solid or flexible metal or other tubing through which insulated electric wires are run.

cone The cone-shaped paper or fiber diaphragm of a loudspeaker.

cone antenna *Conical antenna.*

cone loudspeaker A loudspeaker that includes a magnetic driving unit which is mechanically coupled to a paper or fiber cone.

cone of nulls A conical surface formed by directions of negligible radiation from an antenna.

cone of radio silence A cone-shaped region, directly over the antenna of a radio-beacon transmitter, in which no signal is heard by the pilot of an aircraft.

cone resonance The frequency at which the diaphragm or cone of a loudspeaker vibrates most easily. Cone resonance must be minimized by careful design and by placing the loudspeaker in a proper enclosure to prevent abnormally high acoustic output at the resonant frequency.

conference call A telephone or radio call that permits simultaneous conversation between three or more locations.

confidence The degree of assurance that a specified failure rate is not exceeded.

confidence curve A curve from which the maximum failure rate of a component in the first 1000 h of operation can be determined with a given percentage of confidence.

confidence interval The probability, generally expressed as a percentage, that a characteristic or performance specification will be within a specified range of values.

configuration 1. The layout of machines interconnected and programmed to operate as a system. 2. The layout of components interconnected to perform a desired circuit function.

confocal resonator A wavemeter for millimeter wavelengths that consists of two spherical mirrors which face each other. Changing the spacing between the mirrors affects propagation of electromagnetic energy between them, permitting direct measurement of free-space wavelength. Confocal operation occurs when the center of curvature of one mirror is on the surface of the other.

conformal array A circular, cylindrical, hemispherical, or other shaped array of electronically switched antennas that provide the special radiation patterns required for tacan, IFF, and other air navigation, radar, and missile control applications.

conformal coating A thin layer of epoxy, silicone, or other insulating material applied by spraying or dipping that

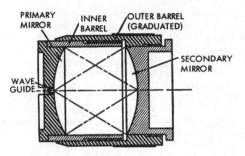

Confocal resonator mounted on end of waveguide for use as wavemeter in millimeter range.

gives a protective coating conforming to the irregular surfaces of an integrated circuit or other electronic device. It is generally cured by heating.

conical antenna   A wideband antenna whose driven element is conical in shape. It is also called a *cone antenna*.

conical helix antenna   A frequency-independent circularly polarized antenna that resembles an inverted cone and provides a unidirectional radiation pattern.

conical horn   A horn that has a circular cross section and straight sides.

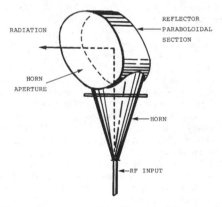

Conical horn reflector antenna is used in terrestrial communications relays.

conical horn reflector antenna   A horn antenna that has a circular cross section and straight sides, as in a cone. It is energized by a circular waveguide that feeds the smaller end of the horn.

conical scanning   Radar scanning in which the radar beam describes a cone whose axis coincides with the axis of the reflector.

conjugate branches   Two network branches arranged so that a voltage applied to either branch produces no response in the other.

conjugate bridge   A bridge whose detector circuit and the supply circuits are interchanged, as compared with a normal bridge topology.

conjugate impedances   Impedances that have resistance components which are equal and reactance components which are equal in magnitude but opposite in sign.

conjunction   1. The logic operation that uses the AND operator or logic product.   2. The condition wherein a planet or spacecraft has the same celestial longitude as the sun, when the center of the earth is the reference.

connected network   A network containing at least one path, composed of branches of the network, between every pair of nodes of the network.

connection   A direct wire path for current between two points in a circuit.

connection diagram   A diagram that shows the connections needed for the operation of an electronic system consisting of one or more assemblies, power supplies, and circuits being controlled.

connector   A general term that applies to any mechanical means for joining two conductors together to form a seamless, low-resistance conductive path. These components, which can be in many different sizes and shapes, are found in electrical and electronic products and systems. Connectors generally have two parts: one terminates a cable or conductor and the other is a mating element on another cable, receptacle, or power outlet. The male part is called the *plug* and the female part is called the *receptacle* or *jack*. Electronic connectors are designed for terminating coaxial cable, ribbon cable, multiwire cable, and interconnecting racks and panels, circuit boards, telephones, and audio and video components. There are also fiberoptic, cable television, and power line connectors. See also *card-edge connector, computer connector, phone plug, phone jack, phono plug, phono jack, RCA plug,* and *RCA jack.*

connect time   1. In a computer-based data communication system, the switching time required to set up a connection between two terminal points.   2. In a telephone system, the time duration of a connection between two points.

conoscope   An optical instrument that locates the optical or Z axis of a quartz crystal.

console   1. A large cabinet for a radio or television receiver that stands on the floor rather than on a table.   2. A main control desk for electronic equipment, as at a radar station, radio or television station, or airport control tower.   3. The section of a mainframe or supercomputer that controls the machine manually, corrects errors, manually revises the contents of memory and provides communication in other ways between the operator or service engineer and the central processing unit.

consortium (plural: **consortia**)   A combination or group of organizations formed to undertake a common objective that is beyond the resources of any single organization. An example is SEMATECH.

constellation   1. Any collection of similar satellites assigned to provide multiple coverage or multiple redundancy, such as the 21 *GPS satellites.*   2. A group of stars that are grouped together in a pattern that is recognizable by people on earth. The ancient peoples named constellations after mythical gods, heros, and animals. Examples are Orion the hunter and Ursa Major, the big bear.

constant   A value that does not change during a process.

constant-amplitude recording   A sound-recording method in which all frequencies that have the same intensity are

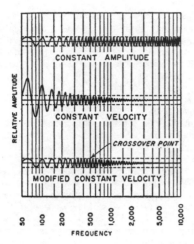

Constant-amplitude, constant-velocity, and modified constant-velocity recording having constant amplitude below the crossover frequency of 500 Hz.

recorded at the same amplitude. The resulting recorded amplitude is independent of frequency.

constantan An alloy that contains 60% copper and 40% nickel for making precision wirewound resistors because of its low temperature coefficient of resistance. It is paired with iron or copper in thermocouples.

constant-current generator A generator whose output current remains essentially constant despite variations in load resistance.

constant-current modulation A system of amplitude modulation in which the output circuits of the signal amplifier and the carrier-wave generator or amplifier are connected through a common coil to a constant-current source. Changes in anode current of the signal amplifier thus produce equal and opposite changes in anode current of the RF carrier stage, thereby giving the desired modulation of the carrier. Also called Heising modulation.

constant-current transformer A transformer that automatically maintains a constant current in its secondary circuit under varying loads when supplied from a constant-voltage source.

constant-*k* filter A filter whose series and shunt impedance product is a constant that is independent of frequency.

constant-*k* lens A microwave lens constructed as a solid dielectric sphere. A plane electromagnetic wave focused at one point on the sphere emerges from the opposite side as a parallel beam. Its focusing properties are similar to those of a *Luneberg lens.*

constant-*k* network A ladder network whose series and shunt impedance product is independent of frequency within the operating frequency range.

constant-potential accelerator An accelerator whose constant DC voltage is applied to an accelerating tube to produce high-energy ions or electrons.

constant-velocity recording A sound recording method in which input signals are recorded at an amplitude that is inversely proportional to the frequency.

constant-voltage power supply A regulated power supply that maintains a predetermined DC voltage across a load for a specified range of load resistance values, line voltages, temperatures, and other variables by automatically varying load current.

Consultative Committee in International Telegraphy and Telephony [CCITT] The former name of the International Standards Union, Telecommunications Standardization Sector (ITU-TS), an advisory committee established by the United Nations to recommend worldwide standards for data transmission. It sets worldwide communications standards such as V.21, V.22, and V.22 bis.

contact 1. A conducting part of a relay, connector, or switch that coacts with another part to make or break a circuit. 2. Initial detection of an enemy aircraft, ship, submarine, or vehicle on a radarscope or other detecting equipment.

contact bounce The uncontrolled making and breaking of contact one or more times, but not continuously, when relay contacts are moved to the closed position.

contact chatter *Chatter.*

contact electromotive force *Contact potential.*

contact force The force exerted by the moving contact of a switch or relay on a stationary contact.

contact gettering The absorption of gas by contact with a dispersed getter film in an electron tube.

contact material A metal that has high electric and thermal conductivity, low contact resistance, minimum sticking or welding tendencies, and high corrosion resistance. Commonly used contact materials include copper, silver, and gold and their alloys, platinum and palladium alloys, tungsten, and molybdenum.

contact microphone A microphone that picks up mechanical vibrations directly and converts them into corresponding electric currents or voltages. When used with wind, string, and percussion musical instruments, it is attached to the housing of the instrument. When used for vibration analysis of machinery, it is held against various parts of the machinery. When used as a throat microphone, it is strapped against the throat of the speaker. When used as a lip microphone, it is held against the lip of the speaker.

contact-modulated amplifier An amplifier that has a chopper at its input to change DC and very low-frequency AC signals to a higher frequency such as 60 or 400 Hz. The resulting modulated wave is amplified in an AC amplifier to a suitable level, then demodulated, sometimes by the same contact system that performed the original modulation.

contact noise The fluctuating electric resistance observed at the junction of two metals or at the junction of a metal and a semiconductor.

contactor A heavy-duty relay that controls electric power circuits. Its ratings typically exceed 20 W.

contact overtravel *Contact follow.*

contact plunger *Contact piston.*

contact potential 1. The voltage due to contact between two different metals, bodies in different physical states, or materials that have different chemical compositions. Also called contact electromotive force and Volta effect. 2. The voltage that exists between the control grid and cathode of an electron tube when there is no external grid bias, due to the difference in work functions of the electrode surfaces.

contact-potential barrier The potential hill at the contact surfaces of two bodies, due to formation of a barrier layer.

contact-potential difference The difference between the work functions of two materials in contact, divided by the electronic charge.

contact pressure The amount of pressure that holds a set of contacts together.

contact rectifier *Metallic rectifier.*

contact resistance The resistance in ohms between the contacts of a relay, switch, or other device when the contacts are touching each other. The value is generally a small fraction of an ohm.

contact wipe The distance that two mating contact surfaces slide with respect to each other while making or breaking contact.

contention A method of operating a multiterminal communication channel such as a local area network (LAN) in which any station can transmit if the channel is free. If the channel is in use, the queue of contention requests are maintained by a computer in chronological or other predetermined sequence.

continuity The presence of a complete path for current flow.

continuity test An electrical test that determines the presence and location of an open connection.

continuous carrier A carrier over which information is transmitted without interrupting the carrier.

continuous control Automatic control in which the controlled quantity is measured continuously and corrections are a continuous function of the deviation.

continuous-duty rating The rating that defines the load which can be carried for an indefinite time without exceeding a specified temperature rise.

continuous linear antenna array An antenna array that consists of an infinite number of infinitesimally spaced sources, as in some dielectric antennas.

continuous loading Loading in which the added inductance is distributed uniformly along a line by wrapping magnetic material around each conductor.

continuously variable slope delta [CVSD] **modulation** A technique for converting an analog signal (such as audio or video) into a serial bit stream. Modulator/demodulator circuits that encode and decode functions on the same chip with a digital input for selection.

continuous power The power-handling rating of an audio or other amplifier, expressed in watts RMS for a sinewave signal.

continuous power spectrum A power spectrum that can be represented by the indefinite integral of a suitable spectral density function. All power spectra of physical systems are continuous.

continuous recorder A recorder whose record sheet is a continuous strip or web rather than individual sheets.

continuous spectrum The spectrum of a wave whose components are continuously distributed over a frequency region without being broken up into lines or bands.

continuous-tone squelch Squelch in which a continuous subaudible tone, generally below 200 Hz, is transmitted by FM equipment along with a desired voice signal. The tone activates a frequency-sensitive circuit that unblocks the squelch circuit of the receiver to allow reception of the desired message. Signals without the correct tone frequency or with no tone are not heard.

continuous-transmission frequency-modulated sonar [CTFM sonar] A sonar system whose transmitted frequency is varied continuously in linear sawtooth fashion. The frequency received by reflection from an object is then proportional to the range to that object. The difference between the transmitted and received frequencies is measured with a multichannel frequency analyzer and the results are fed to a PPI cathode-ray display.

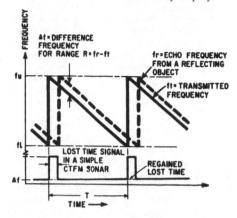

Continuous-transmission frequency-modulated sonar principles.

continuous wave [CW] A radio or radar wave that maintains a constant amplitude and a constant frequency.

continuous-wave Doppler radar *Continuous-wave radar.*

continuous-wave gas laser A laser that has a quartz envelope filled with a mixture of helium and neon at a low pressure, Brewster-angle mirrors at opposite ends, and an external optical system. An applied RF field excites the atoms in the tube, causing spontaneous emission of photons. These photons are reflected back into the gas to stimulate neon atoms, with the process repeating and building up to a self-sustained oscillation that becomes the desired coherent laser radiation. The useful portion of this radiation passes through the 1% transmissive mirrors in an extremely narrow beam. See also *gas laser.*

continuous-wave jamming The transmission of constant-amplitude, constant-frequency unmodulated jamming signals as a radar countermeasure to change the gain characteristics of enemy radar receivers.

continuous-wave laser [CW laser] A laser that generates a beam of coherent light continuously, as required for communication and other applications. The maximum average power is generally less than can be obtained with pulsed operation.

continuous-wave radar [CW radar] A radar system whose transmitter sends out a continuous flow of radio energy. The target reradiates a small fraction of this energy to a separate receiving antenna located and oriented to minimize the amount of transmitted power that can enter the receiver. The reflected wave is distinguished from the transmitted signal by a slight change in radio frequency called the *Doppler shift.* Continuous-wave radar can distinguish moving targets against a stationary reflecting background and it needs less bandwidth than pulsed radar. It is also called *continuous-wave Doppler radar.*

contrast The degree of difference in tone between the lightest and darkest areas in a television or facsimile picture. Contrast is measured in terms of *gamma,* a numerical indication of the degree of contrast. Pictures with high contrast have deep blacks and brilliant whites, and pictures with low contrast have an overall gray appearance.

contrast control A manual control that adjusts the range of brightness between highlights and shadows on the reproduced image in a television receiver. Usually the contrast control varies the gain of a video amplifier tube or transistor. In a color television receiver a dual control can have one section controlling the luminance signal and the other section controlling the chrominance signals; this permits adjustment of contrast without changing color.

contrast ratio The ratio of the maximum to minimum luminance values on a cathode-ray tube, liquid-crystal display or other active display for a television receiver, a computer monitor, or a video terminal. (A contrast ratio of at least 10:1 is needed for the best readability.)

contrast resolution The number of gray levels at each pixel in a digital image, determined by raising 2 to the power of the number of bits at each pixel.

control 1. A component that starts, stops, or adjusts a piece of equipment. 2. The section of a digital computer that carries out instructions in proper sequence, interprets each coded instruction, and applies the proper signals to the arithmetic unit and other parts in accordance with this interpretation. 3. A mathematical check used in some computer operations. 4. A test that determines the extent of error in experimental observations or measurements.

control accuracy The degree of correspondence between the ultimately controlled variable and the ideal value in a feedback control system.

control channel A channel for transmitting digital control information from a cellular mobile telephone *land station* to a cellular telephone *mobile station,* or vice versa.

control character A character whose occurrence in a particular context initiates, modifies, or stops a control operation in a computer or associated equipment.

control circuit 1. The circuit that feeds the control winding of a magnetic amplifier. 2. One of the circuits that responds to the instructions in the program for a digital computer. 3. A circuit that controls some function of a machine, device, or instrument.

control counter A computer circuit that records the storage location of the instruction word to be operated upon following the instruction word in current use.

control desk *Console.*

control diagram *Flowchart.*

control electrode An electrode that initiates or varies the current between two or more electrodes in an electron tube.

control element The section of a feedback control system that acts on the process or machine being controlled.

control-flow machine A computer with parallel processing architecture and a single central sequence of instruction, carried out by many processors.

control grid A grid, ordinarily placed between the cathode and an anode, that controls the anode current of an electron tube.

controlled avalanche device A semiconductor device that has rigidly specified maximum and minimum avalanche voltage characteristics and is able to operate and absorb momentary power surges in this avalanche region indefinitely without damage.

controlled-carrier modulation A method of modulation that holds the percentage modulation constant at all times by varying the amplitude of the carrier wave automatically to offset the variations produced by conventional amplitude modulation of the carrier wave. It is also called *floating-carrier modulation* and *variable-carrier modulation.*

controller 1. An electromechanical or electrical circuit or system that can control some function by transmitting signals through a conductor, the air, or water. Control can be carried out by audio, radio, acoustic or infrared transmission. Examples of controllers include infrared TV/stereo/VCR remotes, and radio controllers for opening doors or controlling model airplanes, cars, or boats. It can also be a cable-connected switch box. 2. A synonym for *microcontroller.*

control point The value of controlled variable that is maintained by an automatic control system.

control register A register that holds the identification of the instruction word to be executed next in time sequence, following the current operation in a computer.

control signal The signal applied to the circuit that makes corrective changes in a controlled process or machine.

control synchro *Control transformer.*

control system An arrangement of a sensing element, amplifier, and control device acting together to control some condition of a process or machine.

control transformer A synchro whose rotor output signal depends on both the shaft position and electric input to the stator. It is also called a *control synchro.*

control winding A winding on a magnetic amplifier or saturable reactor that applies magnetomotive forces to control the core.

convection cooling Heat transfer by natural upward flow of hot air from the device being cooled.

convection current The time rate at which the electric charges of an electron stream are transported through a given surface.

convection-current modulation The time variation in the magnitude of the convection current passing through a surface, or the process of directly producing such a variation in a microwave tube.

convective discharge The movement of a visible or invisible stream of charged particles away from a body that has been charged to a sufficiently high voltage.

convenience receptacle *Outlet.*

conventional voltage doubler A voltage doubler circuit in which capacitors C1 and C2 are each charged, during

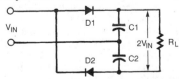

Conventional voltage doubler has a DC-voltage output that is twice the AC-voltage input.

alternate half cycles, to the peak value of the AC input voltage. The capacitors are discharged in series into the load resistor, producing an output across the load of approximately twice the AC peak voltage. See also *cascade voltage doubler, bridge voltage doubler* and *voltage multiplier*.

convergence The intersection of electron beams of a multibeam cathode-ray tube at a specified point, such as at an opening in the shadow mask of a three-gun color television picture tube. Both static and dynamic convergence are required.

convergence coil One of the coils that causes convergence of electron beams in a three-gun color television picture tube.

convergence control A color television receiver control that adjusts the potential on the convergence electrode of the three-gun color picture tube to achieve convergence.

convergence electrode An electrode whose electric field causes two or more electron beams to converge.

convergence magnet A magnet assembly whose magnetic field causes two or more electron beams to converge in three-gun television picture tubes. It is also called a *beam magnet*.

convergence plane A plane that contains the points at which the electron beams of a multibeam cathode-ray tube appear to deflect for convergence.

convergence surface The surface generated by the point of intersection of two or more electron beams in a multibeam cathode-ray tube during the scanning process.

convergence zone A sound transmission channel produced in sea water by a combination of pressure and temperature changes in the depth range between 2500 and 15,000 ft (760 and 4570 m). In this channel a downward sonar signal is refracted back toward the surface, to reach the surface about 30 nautical miles (55 km) away from the sonar transmitter. If the signal encounters a reflecting object along this path, the signal returns along the same route to the sonar set.

conversational mode A computer operating mode that permits interactive responses between the computer and human operators at keyboard terminals.

conversion efficiency 1. The ratio of AC output power to the DC power input to the electrodes of an electron tube or semiconductor device. 2. The ratio of the output voltage of a converter at one frequency to the input voltage at some other frequency.

conversion fraction The ratio of the number of internal conversion electrons to the total number of quanta plus the number of conversion electrons emitted in a given mode of deexcitation of a nucleus.

conversion gain ratio The ratio of signal power output to signal power input for a frequency converter or mixer.

conversion time The time of one complete measurement by an analog-to-digital converter.

convert To change the representation of data from one form to another, as from binary to decimal or from disks to tape.

converter 1. A circuit for converting AC power from one frequency to another. 2. A circuit for converting DC to AC. 3. A circuit for converting AC to DC. 4. A circuit for converting an analog function to a digital code. 5. A

circuit for converting a digital (binary) code to an analog value. See *analog-to-digital converter, digital-to-analog converter, AC-to-DC converter, DC-to-AC converter, DC-to-DC converter,* and *AC-to-AC converter*.

converter tube An electron tube that combines the mixer and local-oscillator functions of a heterodyne conversion transducer.

convolver A surface-acoustic-wave device that processes signals by a nonlinear interaction between two waves traveling in opposite directions. In one version, the two different input frequencies are applied at opposite ends of the structure, and the sum-frequency signal is detected by an output transducer structure on the surface between the two inputs. It is also called an *acoustic convolver*.

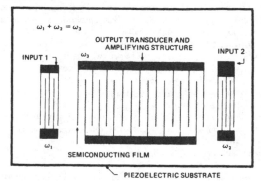

Convolver in which the output transducer also provides amplification at the sum frequency.

Coolidge tube An X-ray tube that produces electrons with a hot cathode.

Cooperative Engagement Capability [CEC] A triservice U.S. military concept for linking all radar systems in a given theater of operations together in a unified, computer-controlled command network. Theatre commanders will then have an overall view of all operations in their areas of responsibility and will be able to deploy any or all Army, Navy, Air Force, or Marine Corps units and weapons to counter a threat in that region.

coordinate Any one of two or more magnitudes that determine position relative to the reference axes of a coordinate system.

coordinate system A system for specifying the location of a point with two coordinates if they are on a surface and three coordinates if in space.

coplanar electrodes Electrodes mounted in the same plane.

copper [Cu] A ductile, malleable, red-orange metal with an atomic number of 29 and an atomic weight of 63.5. Next to silver, it is the best conductor of heat and electricity. Readily rolled into sheets and foil and drawn into wire, it is the most widely used metal in wire, cable, and bus bars. When plated on steel wire it forms a high-strength conductor. As foil on circuit boards, it can be etched to form conductive traces in the *subtractive process*. It can also be directly plated on chemically treated circuit boards to form conductors in the *additive process*. When alloyed with gold, it makes gold more malleable.

copper loss Power loss in a winding due to current flow through the resistance of the copper conductors. It is also called I^2R *loss*.

copper-oxide photovoltaic cell A photovoltaic cell whose sensitive element consists of layers of copper and cuprous oxide. Incident light falling on the cell produces a voltage.

cord A thin, very flexible insulated conductor. The term is applied to microphone cable, household telephone cable, two-conductor consumer appliance power cable, and cable that connects a *mouse* to a computer.

cordless telephone A telephone set whose handset is connected to a base unit by a short-range radio link. The handset is a radio transceiver that includes a keypad for dialing. It has a range of about 200 feet. The base unit, also a transceiver, is directly connected to the public dial-up telephone system. The *cellular mobile telephone* is also cordless, but it is normally called a cellular telephone. It transmits and receives all messages to and from a local radio relay station in a network which, in turn, is connected to the public dial-up telephone system.

core *Magnetic core* or *ferrite core*.

core iron A grade of soft iron suitable for cores of chokes, transformers, and relays.

core loss The power loss in an iron-core transformer or inductor caused by eddy currents and hysteresis effects in the iron core.

core memory *Ferrite core memory*.

core plane *Ferrite core memory*.

Coriolis force A natural force that can deflect a projectile during its flight over the surface of the earth, caused by the rotation of the earth.

corner-reflecting antenna An antenna that has a reflector with two conducting planes which intersect at right angles ("square corner"). It is usually fed by a dipole or colinear array of dipoles located on the bisector of the angle. The planes can be made of sheet metal, or wire mesh for reduced wind loading. Maximum pickup is obtained along the bisector. It is suitable for the transmission of UHF television. It is also called a *corner antenna*.

corner reflector A radar reflector that consists of three conducting surfaces mutually intersecting at right angles. This reflector returns electromagnetic radiation to its source. It makes a ship's position more conspicuous for radar observation.

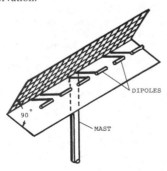

Corner reflecting antenna with a three-dipole collinear array feed.

corona A discharge of electricity that appears as a bluish purple glow on the surface of and adjacent to a conductor when the voltage gradient exceeds a certain critical value. It is due to ionization of the surrounding air by the high voltage.

corona shield A shield placed about a point of high potential to redistribute electrostatic lines of force.

corrected compass course *Magnetic course*.

corrected compass heading *Magnetic heading*.

correction A quantity added to a calculated or observed value to obtain the true value.

correction time The time required for the controlled variable to reach and stay within a predetermined band about the control point following any change of the independent variable or operating condition in a control system. Also called settling time.

corrective network An electric network inserted in a circuit to improve its transmission properties, its impedance properties, or both. Also called shaping network.

correlation In telecommunications, the degree of agreement between pairs of signals.

correlation direction finder A radar direction-finding receiver that receives jamming signals. By correlating signals received at several such stations, range and azimuth of an enemy jammer can be obtained.

correlation-type receiver *Correlator*.

correlator A circuit that detects weak signals in noise by performing an electronic operation which approximates the computation of a correlation function. Examples include the autocorrelator and crosscorrelator. It is also called a *correlation-type receiver*.

corrugated-surface antenna A microwave antenna that consists of a waveguide feed to a mode transformer or horn launcher and a transversely corrugated metal surface that guides surface waves.

cosecant antenna An antenna that forms a beam whose amplitude varies as the cosecant of the angle of depression below the horizontal. It can be a cheesebox antenna with a line source or a distorted parabolic reflector with a point source.

cosecant-squared antenna An antenna that forms a cosecant-squared radiation pattern.

cosecant-squared pattern A ground radar antenna radiation pattern that sends less power to nearby objects than to those farther away in the same sector. The field intensity varies as the square of the cosecant of the elevation angle. The pattern is achieved by either bending the top part of the parabolic reflector forward or placing a spoiler on the reflector. With this pattern, approximately equal echo signals are received from objects at the same altitude but at varying distances. In airborne radar antennas it can produce a uniform electric field along a line on the earth's surface.

cosine effect An effect that occurs with laser and microwave speed-measuring devices in which the measured speed and the actual speed are related by a factor equal to the cosine of the angle between the centerline of the beam and the line of travel of the target vehicle.

cosine emission law The energy emitted by a radiating surface in any direction is proportional to the cosine of the angle which that direction makes with the normal.

cosine winding   A winding used in the deflection yoke of a cathode-ray tube to prevent changes in focus as the beam is deflected over the entire area of the screen.

cosmic radio wave   A radio wave that originates in an extraterrestrial source. Examples include galactic radio noise and solar radio noise coming from the sun.

CO_2 laser   *Carbon dioxide laser.*

Coulmer antenna array   A high-grain planar antenna array that consists of nonresonant elements stacked vertically and horizontally to produce both vertically and horizontally polarized waves.

coulomb   [C]   The SI unit of electric charge. It is the amount of electric charge that passes through a given cross section of a conductor when a steady current of 1 A is flowing.

coulomb collision   The collision of two charged particles. The collision cross section is considerably larger than when one of the particles is neutral because the electric fields of the two particles can interact at much larger distances.

coulomb force   The electrostatic force of attraction or repulsion exerted by one charged particle on another.

coulomb friction   Friction that occurs between dry surfaces.

coulombmeter   A measuring instrument that measures quantity of electricity in coulombs by integrating a stored charge in a circuit which has very high input impedance. When incorporated into an electrometer, switches provide a choice of ranges.

coulomb potential   A scalar point function equal to the work per unit charge done against or by the coulomb force in transferring a particle that bears an infinitesimal positive charge from infinity to the field of a charged particle in a vacuum.

coulomb scattering   Scattering that occurs when charged particles passing through matter are acted on by the electrostatic forces of other charged particles.

Coulomb's law   The attraction or repulsion between two electric charges is proportional to the product of their magnitudes and inversely proportional to the square of the distance between them. It is also called the *law of electrostatic attraction.*

coulometer   An electrolytic cell that measures quantity of electricity in coulombs by the chemical action produced.

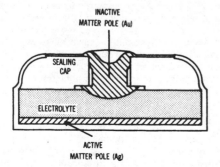

Coulometer with a dry solid electrolyte to integrate electric charge over periods of from 1.5 min to 8000 h.

coulometric titration   Measurement of the integrated current passing through an electrode during the chemical process of titration.

count   1. A single response of the counting system in a radiation counter.   2. The total number of events indicated by a counter.

countdown   1. The ratio of the number of interrogation pulses not answered by a transponder to the total number received.   2. The step-by-step process of preparing a missile for launching.

counter   1. A complete instrument for detecting, totalizing, and indicating a sequence of events. When used to measure frequency by counting the periods of a waveform for 1 s and displaying the count as a numerical value in hertz, it is usually known as an electronic counter or digital frequency meter. 2. *Accumulator.* 3. *Radiation counter.*

counter circuit   A circuit that receives pulses which represent units to be counted and produces a voltage proportional to the total count.

counterclockwise polarized wave   *Left-hand polarized wave.*

counter-countermeasure   [CCM]   Any technique, circuit, or system used against enemy electronic countermeasures, such as methods of extracting radar target signals from a jamming environment.

counterelectromotive force   The voltage developed in an inductive circuit by a changing current. The polarity of the voltage is at each instant opposite that of the generated or applied voltage. Also called back electromotive force.

countermeasures   Devices and/or techniques intended to impair the operational effectiveness of enemy activity.

counterpoise   A system of wires or other conductors that is elevated above and insulated from the ground to form a lower system of conductors for an antenna. Used as a substitute for a ground connection. Also called antenna counterpoise.

counter tube   1. An electron tube that converts an incident particle or burst of incident radiation into a discrete electric pulse, generally by measuring the current flow through a gas that is ionized by the radiation. It is also called a *radiation-counter tube.* 2. An electron tube that has one signal input electrode and 10 or more output electrodes, with each input pulse transferring conduction sequentially to the next output electrode. Beam-switching tubes and cold-cathode counter tubes are examples.

counting circuit   A circuit that counts pulses by frequency-dividing techniques by charging a capacitor to produce a voltage proportional to the pulse count.

counting dial   A dial that fits on the end of a control shaft of a precision potentiometer to give an accurate indication of shaft position. Some dials provide digital indications of up to 1000 turns and fractions of a turn.

counting efficiency   The ratio of the average number of photons or ionizing particles that produce counts to the average number incident on the sensitive area of a radiation counter.

counting ionization chamber   *Pulse ionization chamber.*

counting rate The average rate of occurrence of events as observed by means of a counting system.

counting-rate curve A curve that shows how counting rate varies with applied voltage in a radiation counter. It generally starts with a sharp rise at the threshold voltage, has a flat region known as the plateau, then ends with a sudden sharp rise. The counter is usually operated in the plateau region where the counting rate is not appreciably affected by changes in applied voltage.

counting-rate meter An instrument that indicates the time rate of occurrence of input pulses to a radiation counter, averaged over a time interval.

couple 1. Two metals placed in contact, as in a thermocouple. See *bimetallic strip.* 2. To connect two circuits so signals are transferred from one to the other.

coupler A component that transfers energy from one circuit to another.

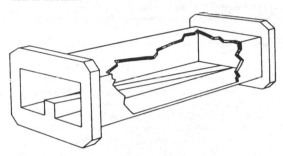

Coupler for insertion between rectangular and ridge waveguides.

coupling 1. A mutual relation between two circuits that permits energy transfer from one to the other. Coupling can be direct through a wire, resistive through a resistor, inductive through a transformer or choke, or capacitive through a capacitor. 2. A flexible or rigid device that fastens together two shafts end to end.

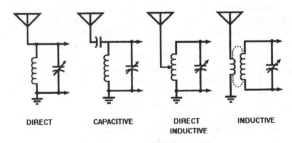

Coupling methods for an antenna shown schematically.

coupling capacitor A capacitor that blocks the flow of direct current while allowing alternating or signal currents to pass. Widely used for joining two circuits or stages. Also called blocking capacitor.

coupling coefficient 1. The ratio of the maximum change in energy of an electron traversing an interaction space to the product of the peak alternating gap voltage and the electronic charge. 2. *Coefficient of coupling.*

coupling hole *Coupling aperture.*

coupling loop A conducting loop that projects into a waveguide or cavity resonator for the transfer of energy to or from an external circuit.

coupling probe A probe that projects into a waveguide or cavity resonator for the transfer of energy to or from an external circuit.

coupling slot *Coupling aperture.*

covalent bond A pair of electrons shared by two neighboring atoms.

covalent crystal A crystal held together by covalent bonds.

covalent radius The effective radius of an atom in a covalent bond.

coverage *Service area.*

CPA Abbreviation for 1. *color-phase alteration* and 2. *closest point of approach.*

cpm Abbreviation for *cycle per minute.*

cps Abbreviation for *cycle per second,* now called hertz and abbreviated Hz.

CPU Abbreviation for *central processing unit.*

CRC Abbreviation for *cycling redundancy check.*

creep Any slow change in a dimension or characteristic with time or usage.

creepage The conduction of electricity across the surface of a dielectric.

creep recovery The slow return to an original dimension or characteristic with time, after removal of the load or other condition that caused the original creep.

crest *Peak.*

crest factor The ratio of the peak value to the effective value of any periodic quantity such as a sinusoidal alternating current.

crest value *Peak value.*

crest voltmeter A voltmeter that can read the peak value of the voltage applied to its terminals.

crimp contact A contact whose back portion is a hollow cylinder that will accept a wire. After a bared wire is inserted, a swedging tool is applied to crimp the contact metal firmly against the wire. Also called solderless contact.

crippled leapfrog test A variation of the leapfrog test, modified so the computer tests are repeated from a single set of memory locations rather than a changing set of locations.

critical absorption wavelength The wavelength characteristic of a given electron energy level in an atom of a specified element at which an absorption discontinuity occurs.

critical angle 1. The smallest angle measured from the vertical at which a radiated radio wave will be reflected by the ionosphere. At a smaller angle, the radio waves will pass through the *ionosphere* and not be returned to earth. 2. The smallest angle measured from the perpendicular to the axis of an *optical fiber* at which light incident to the core will be propagated along the fiber. At a smaller angle, the light will be absorbed by the side-wall *cladding* and not be returned to the fiber conductor.

critical coupling The degree of coupling between tuned circuits at which the resonant current in the secondary circuit reaches its maximum value. If the coupling is less than critical (*undercoupling*), the secondary current has a lower value, and the circuit has a narrower bandwidth.

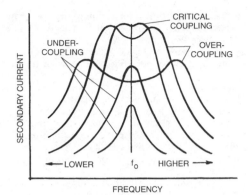

Critical coupling provides the maximum transfer of signal energy from one resonant circuit to another.

When the coupling is greater than critical (*overcoupling*), the response curve widens (greater bandwidth), but displays two peaks. It is also called *optimum coupling*.

critical current   The current in a superconductive material above which the material is normal and below which the material is superconducting, at a specified temperature and in the absence of external magnetic fields.

critical damping   The degree of damping required to give the most rapid transient response without overshooting or oscillation. Thus, a critically damped meter moves its pointer to a new value without going past it.

critical dimension   *Broad dimension.*

critical field   The smallest theoretical value of steady magnetic flux density that would prevent an electron emitted from the cathode of a magnetron at zero velocity from reaching the anode. Also called cutoff field.

critical flicker frequency   The frequency at which a flickering light source is perceived by the eye to be changing from pulsating to continuous. The lowest frequency at which the TV picture does not flicker is about 60 Hz.

critical frequency   The highest frequency at which a vertically projected radio wave will be reflected and returned from the ionosphere. Radio signals with a frequency higher than the critical frequency will pass through the ionosphere and will not return to earth. It is also called the *cutoff frequency.*

critical inductance   The minimum input choke inductance required to prevent the input choke current from going to zero during any part of the cycle in a choke-input filter for a full-wave rectifier.

critical magnetic field   The field below which a superconductive material is superconducting and above which the material is normal, at a specified temperature and in the absence of current.

critical path   In computer circuits, the conductive path that determines the circuit's overall clock rate, typically the longest path in the circuit.

critical temperature   The temperature below which a superconductive material is superconducting and above which the material is normal, in the absence of current and external magnetic fields.

critical wavelength   The free-space wavelength that corresponds to the critical frequency.

Crookes dark space   *Cathode dark space.*

Crookes radiometer   A radiometer that demonstrates that radiant energy from the sun can produce motion. A miniature four-vane windmill is mounted in a glass-envelope vacuum tube. Each vane is polished on one side and black on the other side. Absorption of radiant energy by the black sides warms these sides and makes adjacent residual molecules of gas rebound more rapidly than from the polished sides. The black sides then rotate away from the source of radiation.

Crookes tube   An early cathode-ray tube made by the English physicist W. Crookes. In 1879 he found that by passing current through the tube that contained traces of gas, cathode rays cast a shadow on fluorescent material at the end of the tube. The electron flow was initiated by an induction coil. By varying the voltage, the state of the vacuum, and the amount of gas, both streamers and pulses of colored light were produced. X-rays were also produced, but Roentgen, not Crookes, discovered them with a similar tube.

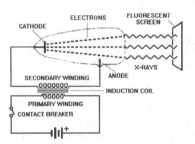

Crookes tube, a primitive cathode-ray tube, was the first discovered source of X-rays.

cross antenna   An antenna formed from a pair of horizontal elements, each with a right-angle center bend. The transmission line is connected at the bends. If it is mounted in a horizontal plane, it produces a horizontally polarized signal.

crossband   Two-way communication in which one radio frequency is transmitted in one direction and a frequency that has different propagation characteristics is transmitted in the opposite direction.

crossbanding   Use of one interrogation frequency with several reply frequencies or one reply frequency with several interrogation frequencies.

crossband transponder   A transponder that replies in a different frequency band from that of the received interrogation.

crossbar switch   A telephone switch that has a three-dimensional arrangement of contacts and a magnet system which selects individual contacts according to their coordinates in the matrix.

cross-channel communication   Two-way communication in which one radio frequency is propagated in one direction and a different frequency that has similar characteristics is propagated in the opposite direction.

cross-color interference   Interference produced in the chrominance channel of a color television receiver by crosstalk from the monochrome signal.

cross-control circuit A compandor circuit whose input signals to the compressor also control the operation of the expandor at the same end of the circuit.

cross-correlation function A mathematical quantity defined as the product of two functions of time.

cross-correlator A correlator whose locally generated reference signal is multiplied by the incoming signal, and the result is smoothed in a low-pass filter to give an approximate computation of the crosscorrelation function. It can detect weak signals in noise in cases where the important signal characteristics are known prior to detection. It is also called a synchronous detector.

cross-coupling A measure of the undesired power transferred from one channel to another in a transmission medium.

crossed-field amplifier A forward-wave, beam-type microwave amplifier whose crossed-field interaction achieves good phase stability, high efficiency, high gain, and wide bandwidth for most of the microwave spectrum.

crossed-field backward-wave oscillator One of several types of backward-wave oscillators whose operation depends on a crossed field. Examples are the amplitron and carcinotron.

crossed-field multiplier phototube A multiplier phototube in which repeated secondary emission is obtained from a single active electrode by the combined effects of a strong RF electric field and a perpendicular DC magnetic field. Laser beams modulated in the gigahertz range have been detected and amplified by this tube.

crossed-field tube [M-type] An electron tube whose operation depends on a DC electrical field and a magnetic field that are perpendicular to each other. It is called M-type from the French term for tubes that depend on a magnetic field for propagation. The common-crossed-field tubes are *magnetrons, forward-wave crossed-field amplifiers* [FWCFA], *backward-wave crossed-field amplifiers* [BWCFA or *amplitrons*], and *backward-wave crossed-field oscillators* [BWCFO or *carcinotrons*].

crossed pinning A system configuration that allows two data-terminal equipment devices or two data communications devices to communicate.

crossed-pointer indicator A two-pointer indicator in an instrument-landing system that indicates the position of an airplane with respect to the glide path.

crossed stripline cavity A cavity in which two striplines intersect at right angles, with a sphere of yttrium-iron gar-

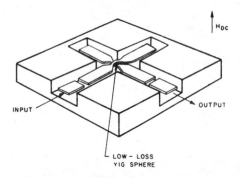

Crossed stripline cavity showing input and output terminals.

net between them at the intersection to provide coupling that is maximum at low power levels and negligible at high power levels.

crosshatch generator A signal generator that generates a crosshatch pattern for adjusting color television receiver circuits.

cross-modulation A type of interference in which the carrier of a desired signal becomes modulated by the program of an undesired signal on a different carrier frequency. The program of the undesired station is then heard in the background of the desired program. Cross-modulation occurs if the first circuit in the receiver is non-linear and acts as a detector for the strong undesired signal.

cross-neutralization A method of neutralization for push-pull amplifiers in which a portion of the AC anode-cathode voltage of each tube is applied to the grid-cathode circuit of the other tube through a neutralizing capacitor.

crossover 1. A location in a circuit diagram where two conductors cross, designated either by a curved line where one conductor crosses the other or no dot at the junction where the lines representing conductors cross. 2. A conductor that runs the length of a cable and connects to different connector-pin numbers at each end of the cable.

crossover frequency 1. The frequency at which a dividing network delivers equal power to the upper and lower frequency channels when both are terminated in specified loads. 2. *Transition frequency.*

crossover network A selective network that divides the audio-frequency output of an amplifier into two or more bands of frequencies. The band below the crossover frequency is fed to the woofer loudspeaker, and the high-frequency band is fed to the tweeter. Also called dividing network and loudspeaker dividing network.

crossover region A zone in space, close to the localizer on-course line or guide slope of an instrument approach system, in which the pointer of the indicator is in a position between the full-scale indications.

crosspoint In telephony, the operating contacts or other low-impedance path connection for the routing of telephone messages.

crosspoint reed relay A reed relay that has from two to five reed switches, one holding coil, and two coincident-count coils. Simultaneous energization of the two coincident-count coils closes one reed switch that is permanently wired in series with the holding coil. The holding coil then takes over and keeps all contacts in all reed switches closed until the holding coil is externally interrupted or until a reverse-polarity pulse is applied to one of the coincident-count coils.

cross-polarization The component of the electric field vector normal to the desired polarization component.

cross-polarized operation Operation of two independent microwave digital data transmitters on the same carrier frequency and the same antenna by using cross-polarized feeds.

crosstalk 1. The sound heard in a receiver along with a desired program because of cross-modulation or other undesired coupling to another communication channel. 2. Interaction of audio and video signals in a television system, causing video modulation of the audio carrier or audio modulation of the video signal at some point. 3.

Interaction of chrominance and luminance signals in a color television receiver.

crosstalk unit [CU]   A measure of the coupling between two circuits. The number of crosstalk units is 1 million times the ratio of the current or voltage at the observing point to the current or voltage at the origin of the disturbing signal, the impedances at these points being equal.

crowbar   A device or circuit that monitors the output of a power supply and rapidly places a low-resistance shunt (crowbar) across the output terminals whenever a preset voltage limit is exceeded, to provide protection until slower fuses or circuit breakers can act.

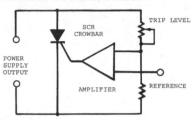

Crowbar circuit. A silicon controlled rectifier protects the power supply from damage by overload, and the load from power-supply malfunction and line-voltage surges.

CRT   Abbreviation for *cathode-ray tube*.

CRT terminal   Abbreviation for *cathode-ray-tube terminal*.

cruciform core   A transformer core that has all of its windings on one center leg, and four additional legs arranged in the form of a cross serve as return paths for magnetic flux.

cruise missile   A pilotless air-breathing missile that can be launched from a submarine, surface ship, or bomber-type aircraft. There are two versions: the Navy *Tomahawk* that can be launched from a submarine or surface ship and the Air Force air-launched version that can be launched from a high-altitude bomber. The Tomahawk has a range of 1150 mi (1850 km) and can carry a 1000-lb (425-kg) warhead. The longer *air-launched* version has a range of 750 mi (1200 km) and can carry a 2000-lb (850-kg) warhead. Once programmed and launched, both missiles can fly circuitous courses at different altitudes, some as low as 60 m, to avoid detection by enemy radar. Originally designed to navigate by matching internally stored computer maps with terrain features recognized during the flight, they have been modified for more precise navigation with GPS satellite receivers. Accuracy is stated as within plus or minus 24 ft (7 m). Both versions can carry either conventional or nuclear warheads.

cryoelectronics   A branch of electronics concerned with the study and application of superconductivity and other low-temperature phenomena to electronic devices and systems.

cryogenic device   A device whose operation depends on superconductivity as produced by temperatures near absolute zero.

cryogenic laser   A laser designed for operation at cryogenic temperatures, usually below 80 K, to increase the average output.

cryogenic parametric amplifier   A parametric amplifier that is cooled to about 17 K in a refrigerator to increase operating life.

cryogenics   The science of physical phenomena at very low temperatures, approaching absolute zero. At such temperatures, small changes in magnetic field strength can produce large current changes in superconducting materials.

cryogenic temperature   A temperature within a few degrees of absolute zero, which is −273.16°C.

cryogenic thermistor   A thermistor designed for operation in a cryogenic liquid. Applications include liquid-level detection, for which the thermistor is heated slightly by passing a small current through it. This heat is dissipated more rapidly in the liquid than when the thermistor is above the liquid, producing a change in resistance that can be used to detect or control the level of the cryogenic liquid. Germanium is a common thermistor element because of its excellent stability at cryogenic temperatures.

cryogenic transformer   A transformer that operates in digital cryogenic circuits. A controlled-coupling transformer is an example.

cryophysics   Physics that is restricted to phenomena occurring at very low temperatures, approaching absolute zero.

cryopump   A vacuum pump dependent on a cold surface (77 K or less) within the chamber being pumped. The surface condenses gas molecules from the chamber, thus lowering internal pressure. Periodic warmup removes those condensed gases.

cryosistor   A cryogenic semiconductor device with a reverse-biased PN junction that controls the ionization between two ohmic contacts. It can serve as a three-terminal switch, pulse amplifier, or oscillator after ionization.

cryostat   Equipment that maintains cryogenic temperatures.

cryotron   A current-controlled switching device, based on superconductivity, for possible use in computer circuits. The earlier version has been replaced by a *tunneling cryotron*, which consists basically of a *Josephson junction*.

cryotronics   The branch of electronics that deals with the design, construction, and use of cryogenic devices.

cryptanalysis   The process of converting intercepted encrypted text into plain text without initial knowledge of the key used for encryption.

cryptochannel   A complete system of communication that uses electronic encryption and decryption equipment and has two or more radio or wire terminals.

cryptogear   Electronic equipment that includes cryptographic circuits or logic for encryption and decryption of messages so that they are meaningless when intercepted by enemy agents.

cryptography   The encryption and decryption of messages for secret transmission. This term is usually shortened to crypto for both the noun and adjective (cryptographic) forms.

crystal   A natural or synthetic piezoelectric or semiconductor material whose atoms are arranged with some degree of geometric regularity.

crystal activity   A measure of the amplitude of vibration of a piezoelectric crystal plate under specified conditions.

crystal blank   The result of the final cutting operation on a piezoelectric or semiconductor crystal.

crystal cartridge   *Crystal pickup.*

crystal control   Control of the frequency of an oscillator by a quartz crystal unit.

crystal-controlled transmitter

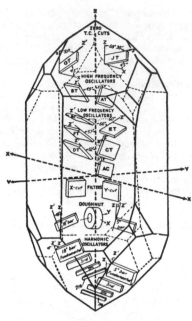

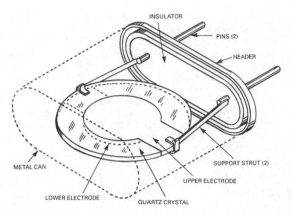

Crystal holder contains a quartz disk cut and polished to oscillate at a specified frequency.

Crystal blank positions for different cuts in natural quartz. The angle of the cut with respect to the natural faces of crystal determines the electrical characteristics of finished crystal blank.

crystal-controlled transmitter A transmitter whose carrier frequency is directly controlled by the electromechanical characteristics of a quartz crystal unit.

crystal-controlled watch An electronic watch whose timing is maintained by a quartz crystal oscillator. It also includes an integrated-circuit frequency divider that can drive a moving hand analog or digital display.

crystal detector A crystal diode or equivalent earlier crystal-catwhisker combination that can rectify a modulated RF signal to obtain the audio or video signal directly. Crystal diodes in microwave receivers combine an incoming RF signal with a local oscillator signal to produce an IF signal.

crystal diode *Semiconductor diode.*

crystal filter A highly selective tuned circuit that has one or more quartz crystals. It is used in IF amplifiers of communication receivers to improve the selectivity.

crystal holder A housing that provides proper support, mechanical protection, and connections for a quartz crystal plate. When the crystal plate is installed, the combination is called a *crystal unit.*

crystal lattice filter A crystal filter that has two matched pairs of series crystals and a higher-frequency matched pair of shunt or lattice crystals. Additional filter sections can be cascaded to improve the passband response.

crystalline A material is single-crystalline when its atoms are arranged in a single continuous array; it is polycrystalline if its atoms are in groups of arrays (crystallites); and it is amorphous if its atoms are not in ordered arrays.

crystalline directions Directions relative to three axes that pass through a common origin that are chosen to describe the symmetry of a crystal.

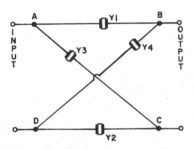

Crystal lattice filter.

crystal microphone A microphone that converts incident sound waves into audio signals by the deformation of a piezoelectric wafer sandwiched between a diaphragm and an electrode. The sound vibrations cause the crystal to generate an output voltage between the diaphragm and electrode connected to the opposing faces of the crystal. The piezoelectric element can be Rochelle salt or ammonium dihydrogen phosphate (ADP). Typically, two crystals are cemented together. Alternatively, barium titanate ceramic can be the transducing element. It has characteristics that are similar to those of the crystal microphone. Both are also called *piezoelectric microphones.* See also *piezoelectricity.*

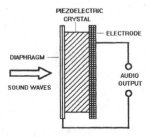

Crystal microphone output voltage is proportional to incident sound pressure that stresses the piezoelectric crystal.

crystal mixer   A mixer that depends on the nonlinear characteristic of a crystal diode to mix two frequencies. In radar receivers it converts the received radar signal to a lower IF value by mixing it with a local oscillator signal.

crystal orientation   The angle, with respect to crystal faces, at which a silicon or other semiconductor crystal is sliced.

crystal oscillator   An oscillator whose AC output frequency is determined by the mechanical properties of a piezoelectric crystal.

crystal oven   A temperature-controlled oven for a crystal to stabilize its temperature and minimize frequency drift.

crystal pickup   A phonograph pickup whose needle movements in the record groove cause deformation of a piezoelectric crystal, thereby generating an audio output voltage between opposite faces of the crystal. The piezoelectric material is generally Rochelle salt or barium titanate. It is also called a crystal cartridge or piezoelectric pickup.

crystal plate   A precisely cut slab of quartz crystal that has been lapped to final dimensions, etched to improve stability and efficiency, and coated with metal on its major surfaces for connecting purposes. Also called quartz plate.

crystal pulling   A method for growing crystal boules in a protective environment. A rod or crystal seed puller with a crystal seed on its end is immersed in a molten solution of the crystal, or *melt,* which is kept in a liquid state by induction heating in a crucible. As the seed puller is withdrawn, the *boule* is formed. See also *Czochralski method.*

crystal rectifier   *Semiconductor diode.*

crystal set   A radio receiver that has a crystal detector stage for demodulation of the received signals, but no amplifier stages.

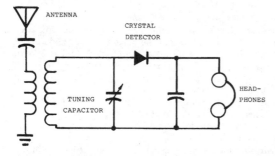

Crystal set needs no power supply to receive strong radio signals if it has a long outdoor antenna.

crystal shutter   A mechanical waveguide or coaxial-cable shorting switch that, when closed, prevents undesired RF energy from reaching and damaging a crystal detector.

crystal slab   A relatively thick piece of crystal from which crystal blanks are cut.

crystal spectrometer   *Bragg spectrometer.*

crystal-stabilized transmitter   A transmitter with automatic frequency control that includes a crystal oscillator which provides the reference frequency.

crystal system   One of the seven main categories of crystals based on the symmetry of their external form or internal structure. The systems are cubic, tetragonal, hexagonal, trigonal, orthorhombic, monoclinic, and triclinic.

crystal unit   *Crystal holder.*

crystal video receiver   A broad-tuning radar or other microwave receiver that consists only of a crystal detector and a video or audio amplifier.

CS   Abbreviation for *chip select.*

C-scale sound level in decibels [dBC]   The sound level in decibels as read when a standard sound-level meter is switched to weighting scale C, which weights frequencies between 70 and 4000 Hz uniformly and discriminates slightly against frequencies above and below this range.

C scan [Carrier System for Controlled Approach of Naval aircraft]   An all-weather instrument landing system on aircraft carriers. Two shipborne transmitters produce a cross-shaped beam whose intersection corresponds to the glide path. Receivers and instrumentation in the aircraft actuate a cockpit display that shows the pilot his exact location with respect to the glide path.

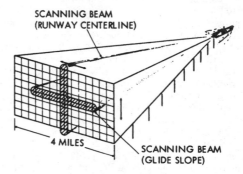

C-scan beams transmitted by aircraft carrier intersect at glide path.

CSMA/CD   Abbreviation for *carrier-sense multiple-access with collision detection.* A local area network (LAN) protocol that permits each station with equal access to the network to "listen" (carrier sense) to determine whether the network is clear for transmission. If two or more stations try to transmit at the same time, each will detect a "collision" and abort its transmission, sending again after waiting a random length of time.

C scope   A radarscope that produces a C display.

CSPDN   Abbreviation for *circuit-switched public data network.*

CSTN   Abbreviation for *color supertwisted nematic.*

CT   Symbol for *center tap.* Used on circuit diagrams.

CT-cut crystal   A quartz crystal cut at an orientation so that its resonant frequency is below 500 kHz.

CTD   Abbreviation for *current-transfer device.*

CTFM sonar   Abbreviation for *continuous-transmission frequency-modulated sonar.*

CTFT   Abbreviation for *color thin-film transistor.*

CTIA   Abbreviation for *Cellular Telecommunication Industry Association.*

CTL   Abbreviation for *complementary-transistor logic.*

CTS   Abbreviation for *clear to send.*

CU   Abbreviation for *crosstalk unit.*

cubical antenna   A full-wavelength antenna array whose elements form the 12 edges of a cube. It is also called a *cubical quad antenna.*

cubical quad antenna   *Cubical antenna.*

cubicle   An enclosure for high-voltage equipment.

cubic semiconductor   A semiconductor device with a crystalline structure whose three axes (X, Y, and Z) are perpendicular to each other, providing equal spacings for all atoms along these directions.

cue circuit   A one-way communication circuit that conveys program control information.

CUJT   Abbreviation for *complementary unijunction transistor*.

Cuk converter   A variation of the buck-boost switching power supply converter that produces low output ripple, usually used where input/output isolation is not required.

cumulative ionization   *Avalanche effect*.

cup core   A core that encloses a coil to provide magnetic shielding. It usually has a powdered iron center post through the coil.

Curie point   The temperature above which a ferromagnetic material becomes substantially nonmagnetic.

current [I or i]   The flow of electrons or holes measured in amperes (A) or in fractions of an ampere [milliamperes (mA), microamperes (μA), nanoamperes (nA), picoamperes (pA)]. Current can be induced by the application of an electric field through a conductor or by changing the electric field across a capacitor (displacement current).

current amplification   The ratio of output signal current to input signal current for an electron tube, transistor, or magnetic amplifier, the multiplier section of a multiplier phototube, or any other amplifying device. It is expressed in decibels by multiplying the common logarithm of the ratio by 20.

current amplifier   An amplifier capable of delivering considerably more signal current than is fed in.

current antinode   A point at which current is a maximum along a transmission line, antenna, or other circuit element that has standing waves. Also called current loop.

current attenuation   The ratio of input signal circuit for a transducer to the current in a specified load impedance connected to the transducer, expressed in decibels.

current-balance relay   A relay that operates when the magnitudes of two current inputs reach a predetermined ratio.

current calibrator   A current source that provides adjustable and accurately known alternating and/or direct currents for calibrating ammeters and other current-measuring instruments.

current-carrying capacity   The maximum current that can be continuously carried without causing permanent deterioration of electrical or mechanical properties of a device or conductor.

current compensation   A means of compensating for stray shunt conductance across the terminals of a constant-current power supply.

current-controlled switch   A semiconductor device whose controlling bias sets the resistance at either a very high or very low value, corresponding to the off and on conditions of a switch.

current cutoff   A negative-resistance circuit in regulated power supplies that reduces load current automatically as load resistance is reduced to minimize overload damage and protect sensitive loads.

current density   The current per unit cross-sectional area of a conductor.

current-differencing amplifier   An alternative to the operational amplifier for many functions except DC voltage amplification. The amplifying section consists of four transistors, three of which operate as common emitters. The amplifier does not need a negative power supply but does require external circuitry. It is also called a *Norton amplifier*.

current drain   The current taken from a voltage source by a load. It is also called *drain*.

current feed   Feed to a point where current is a maximum, as at the center of a half-wave antenna.

current feedback   Feedback introduced in series with the input circuit of an amplifier.

current gain   The ratio of output current to input current under specified conditions for a transistor or other amplifying device.

current generator   A two-terminal circuit element whose terminal current is independent of the voltage between its terminals.

current hogging   The condition whereby one of several parallel components or circuits takes more than its designed share of available current often causing malfunction or damage.

current-leak detector   A safety device that indicates current leakage through insulation and other undesired paths in electronic and electric equipment.

current limiter   A device that restricts the flow of current to a certain amount, regardless of applied voltage.

current limiting   A regulated power-supply-circuit feature that primarily protects against overloads and short circuits rather than supplying constant-current.

current-limiting resistor   A resistor inserted in an electric circuit to limit the flow of current to some predetermined value. It protects tubes and other components during warmup.

current loop   *Current antinode*.

current mirror   A circuit formed by connecting two or more transistors so that the current in one node is duplicated in the other. If the sources of two MOSFETs are tied together, and their gates are connected by one of the drains, the current at the first drain would be duplicated at the second drain.

current-mode control   A control method for switching converters in which a dual-loop control circuit adjusts pulse-width-modulation (PWM) in response to a measured output current.

current-mode logic [CML]   Logic in which unsaturated transistors operate from a constant-current source that is switched at very high speed from one transistor to another.

current node   A point of zero current along a transmission line, antenna, or other circuit element that has standing waves.

current regulator   A device that maintains the output current of a voltage source at a predetermined, essentially constant, value despite changes in load impedance.

current relay   A relay that operates at a specified current value rather than at a specified voltage value.

current saturation   *Anode saturation*.

current-sensing resistor   A resistor placed in series with a load to develop a voltage proportional to load current, as required for regulating load current.

current-sinking logic   Logic in which the output current of a transistor flows back to the preceding stage through any inputs that are low. When all inputs are high, the transistor is saturated so its output is low. It is then capable of sinking the input currents of several additional gates.

current-sourcing logic   Logic in which current flows from the output of the driving gate to the inputs of the driven gate when the output of the driving gate is high. Resistor-transistor logic is an example.

current transformer   An instrument transformer whose primary winding is connected in series with a circuit that carries the current to be measured or controlled. The current is measured across the secondary winding.

cursor   1. In radar technology, a bright radial line generated electronically within the PPI display or a transparent disk overlay with a scribed radial line from the center of the PPI display to its circumference. Either type of cursor can be rotated in either direction to align it with an object or target mark so its bearing can be read accurately from bearing graduations around the display. The disk of the mechanical version can be clear or amber, and it is mounted so that it can be turned by peripheral gearing. 2. In computer technology, it is a small marker generated within the CRT or LCD display to indicate the location where text is to be entered on the monitor display. It is typically a short vertical line or box, black on a white background or a light color against a dark background. It can be positioned where needed either by keyboard keys or the *mouse*.

cursor target bearing   Target bearing as measured by a cursor on a PPI radar display.

curtain array   An antenna array that consists of vertical wire elements stretched between two suspension cables. It can be backed by a second curtain that acts as reflector. The active elements are usually half-wave dipoles.

curtain rhombic antenna   A multiple-wire rhombic antenna that has a constant input impedance over a wide frequency range. Two or more conductors join at the feed and terminating ends but are spaced apart vertically from 1 to 5 ft (30 to 150 cm) at the side poles.

custom integrated circuit   Any analog or digital integrated circuit that is designed and fabricated for a particular function or application, not general-purpose or standard. They can be *full-custom* or *semicustom*. A full-custom IC requires a full set of exclusive masks, but a semicustom device might require only two or three custom masks. See *applications-specific IC.*

cut   1. A section of a crystal that has two parallel major surfaces. Cuts are specified by their orientation with respect to the axes of the natural crystal, such as X cut, Y cut, BT cut, and AT cut. 2. An order to stop an action, turn off a television camera, or disconnect all microphones in a radio studio.

Cutler feed   A resonant cavity that transfers RF energy from the end of a waveguide to the reflector of a radar spinner assembly.

cutoff   The minimum value of bias voltage for a given combination of supply voltages that just stops output current in an electron tube, transistor, or other active device.

cutoff attenuator   An adjustable-length waveguide that varies the attenuation of signals passing through the waveguide.

cutoff bias   The DC bias voltage that must be applied to the grid of an electron tube to stop the flow of anode current or base of a bipolar transistor to stop the flow of collector current.

cutoff field   *Critical field.*

cutoff frequency   The limiting frequency beyond which the attenuation, gain, efficiency, or other performance characteristic of a device begins changing so rapidly that the output can no longer be considered useful. Also called critical frequency.

cutoff wavelength   1. The ratio of the velocity of electromagnetic waves in free space to the cutoff frequency in a uniconductor waveguide. 2. The wavelength that corresponds to the cutoff frequency.

cutout   An electric device that is operated manually or automatically to interrupt current flow, such as a circuit breaker, fuse, or switch.

CVD   Abbreviation for *chemical-vapor deposition.*

CVR   Abbreviation for *crystal video receiver.*

CVSD modulation   Abbreviation for *continuously variable step delta modulation,* a scheme for digital voice modulation.

CW   Abbreviation for *continuous wave.*

CW laser   Abbreviation for *continuous-wave laser.*

cybernetics   A comparative study of the methods of automatic control and communication that are common to people and machines.

cycle   1. One complete sequence of values of an alternating quantity, including a rise to a maximum in one direction, a return to zero, a rise to a maximum in the opposite direction, and a return to zero. The number of cycles occurring in 1 s is called the frequency. 2. A set of operations that is repeated as a unit. 3. To run a machine through an operating cycle.

cycle-matching loran   *Low-frequency loran.*

cycle per minute   [cpm]   A unit of frequency of action.

cycle per second   [cps]   Former unit of frequency, now called hertz and abbreviated Hz.

cycle timer   A timer that opens or closes circuits according to a predetermined time schedule.

cyclically magnetized   Under the influence of a magnetizing force that varies between two specific limits long enough so the magnetic induction has the same value for corresponding points in successive cycles.

cyclic binary code   Any binary code that changes by only 1 bit when going from one number to the number immediately following. The gray code is one example.

cyclic function   A function that assumes a given sequence of values repetitively at an arbitrarily varied rate.

cyclic memory   *Circulating memory.*

cyclic shift   A computer shift in which the digits dropped off at one end of a word are returned at the other end of the word.

cycling   A periodic change of the controlled variable from one value to another in an automatic control system. Also called oscillation.

cycling redundancy check   [CRC]   An error-detection scheme in which the block check character is the remainder after dividing all the serialized bits in a transmission block by a predetermined binary number.

cyclotron   An accelerator whose charged particles are successively accelerated by a constant-frequency alternating

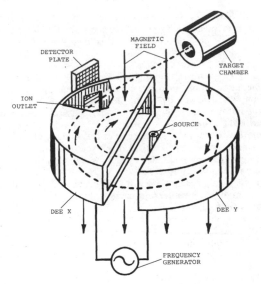

Cyclotron. Positive ions emitted at the source are accelerated between the D-shaped electrodes and bent into a spiral path by the axial magnetic field.

electric field that is synchronized with movement of the particles on spiral paths in a constant magnetic field normal to their path.

cyclotron frequency The frequency at which an electron beam will rotate about an axis when moving at right angles to a uniform magnetic field. This frequency depends upon only the strength of the magnetic field. When rotating in this manner, the electron beam traces the surface of a cone in space.

cyclotron-frequency magnetron A magnetron whose frequency of operation depends on synchronism between the AC electric field and the electrons oscillating in a direction parallel to this field. An example is a split-anode magnetron that has a resonator between the anodes.

cyclotron radiation The electromagnetic radiation emitted by charged particles as they orbit in a magnetic field. At low velocities the radiation is concentrated in a single spectral line, at the cyclotron frequency. At higher velocities the spectral line is spread into a band of frequencies, including harmonics of the cyclotron frequency.

cyclotron resonance A method of accelerating electrons, as for electric propulsion of spacecraft, based on the physical principles of a cyclotron. The electrons of a plasma are trapped by crossed electric and magnetic fields. As the electrons spiral around the magnetic lines of force, they absorb energy continuously from the electric field until they collide with other particles and scatter at high speeds. The only

direction of escape from the trap is through the exhaust nozzle, where they can provide a propulsion thrust.

cyclotron wave A wave associated with the electron beam of a traveling-wave tube.

cylinder A reference to the three-dimensional arrangement of data storage *tracks* formed by a stack of memory disks in a computer hard-disk drive. See also *sector*.

cylindrical antenna An antenna whose hollow cylinders serve as radiating elements.

cylindrical array An electronic scanning antenna that can consist of several hundred columns of vertical dipoles mounted in cylindrical radomes arranged in a circle. It is a beacon interrogator for air traffic control.

cylindrical magnetron An early form of the magnetron-crossed-field electron tube without multiple internal cavities. See *magnetron* and *crossed-field electron tube*.

cylindrical reflector A reflector that is a portion of a cylinder. This cylinder is usually parabolic.

cylindrical wave A wave whose equiphase surfaces form a family of coaxial cylinders.

Czochralski process A process for growing single-crystal boules or ingots of semiconductor material such as silicon,

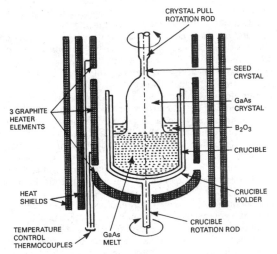

Czochralski process crystal growth equipment has a pressurized crucible, three-heat zones, and a crystal puller.

gallium arsenide, or gallium phosphide. An axial rod with a seed crystal on its end is first immersed in a crucible filled with similar molten crystal. Then the rod is slowly pulled up from the melt while being rotated. The crystal grows out from the seed to form a generally cylindrical boule. The crystal in the crucible is kept in a liquid state by radio-frequency induction.

D

d   Abbreviation for *deci-*.

D   Symbol for *drain*.

D/A   Abbreviation for *digital-to-analog*.

DAC   Abbreviation for *digital-to-analog converter*.

D/A converter   Abbreviation for *digital-to-analog converter*.

daisy-wheel printer   A serial printer in which the printing element is a plastic hub that has a large number of flexible radial spokes, each spoke having one or more different raised printing characters. The wheel is rotated as it is moved horizontally step by step under computer control, and it stops when a desired character is in a desired print position so a hammer can drive that character against an inked ribbon, just as in a typewriter.

damped oscillation   An oscillation whose amplitude decreases with time.

damped wave   A wave in which the amplitudes of successive cycles progressively diminish at the source.

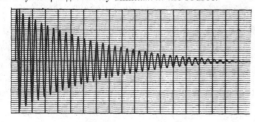

Damped wave.

damper   A diode in the horizontal deflection circuit of a television receiver that makes the sawtooth deflection current decrease smoothly to zero instead of oscillating at zero. The diode conducts each time the polarity is reversed by a current swing below zero.

damping   1. Any action or influence that extracts energy from a vibrating system in order to suppress the vibration or oscillation.   2. Reducing or eliminating reverberation in a room by placing sound-absorbing materials on the walls and ceiling. Also called soundproofing.

damping factor   The ratio of the amplitude of any one of a series of damped oscillations to that of the following one.

D-AMPS [digital AMPS]   A second-generation North American digital cellular system based on a *time-division multiaccess* [TDMA] scheme.

daraf [farad spelled backward]   The unit of elastance, which is the reciprocal of capacitance in farads.

dark current   The current flowing through a photoelectric device in the absence of irradiation.

dark discharge   An invisible electric discharge in a gas.

dark resistance   The resistance of a selenium cell or other photoelectric device in total darkness.

dark space   A region in a glow discharge that produces little or no light.

Darlington pair   A pair of bipolar junction transistors in which the emitter of the first transistor is connected to the base of the second transistor. This configuration provides far higher current gain than a single transistor through direct coupling. The pair can be made on a single die and packaged in a three-terminal transistor case. The pairs are often used in linear ICs, such as operational amplifiers, and in power amplifier output stages. Its most common appli-

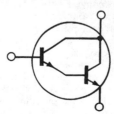

Darlington pair is a dual transistor emitter-follower configuration.

cation is that of an *emitter follower*. The output is taken across a resistor from the emitter of the second transistor to ground. The input resistance at the base of the first transistor is raised to a higher value than that of a single transistor emitter-follower circuit.

DARPA Abbreviation for *Defense Advanced Research Projects Agency;* now called *ARPA* (Advanced Research Projects Agency).

d'Arsonval movement See *permanent magnet, moving-coil meter movement.*

DAS Abbreviation for *data-acquisition system.*

DASD Abbreviation for *direct-access storage device.*

dashpot A device with a piston that moves in a gas- or liquid-filled chamber to absorb energy and delay the opening of electrical contacts. In *magnetic circuit breakers,* it delays the opening of its contacts long enough to allow the passage of harmless transients. See *magnetic circuit breaker time delay.*

data A general term for the words, numbers, letters, and symbols that serve as input for communications and computer processing. It is commonly treated as a collective noun, for use with a singular verb.

data acquisition The process of acquiring data that tracks changes in physical variables such as temperature, pressure, flow rate, and hydrogen ion concentration in factories, process plants, or laboratories. These variables are measured by *thermocouples, strain gauges, flowmeters,* and *pH meters.* This data can be displayed at the measurement site in analog form on meter or chart recorders, or it can be converted to a digital format by an *analog-to-digital converter* [ADC] for local display. The digitized data can then be transmitted over wires with no loss in accuracy to a remote process control room for display, further processing, printout, or storage in computer memory.

data bank One or more collections of data, generally stored on magnetic tapes or disks in a form suitable for computer processing. It is also called a *data base.*

data base *Data bank.*

data communications equipment [DCE] Devices or circuits that provide the functions required to establish, maintain, and terminate a data transmission line, such as those of a modem.

data compression A technique for reducing the number of bits in the original data without information loss for more efficient and faster transmission. The receiver expands the received data bits into the original bit sequence.

data concentrator A device that takes data from several different teleprinter or other slow-speed lines and feeds them to a single higher-speed line. A microprocessor can perform this function.

data converter A converter that changes data from one form to another, as from laser marks on a CD-ROM to magnetic domains on a disk drive.

data domain The aspect of digital systems that is characterized by data flows, data formats, equipment architecture, and state-space concepts.

data element A basic unit of information, such as age, sex, payroll number, geographic location, date, or time.

data encryption The transformation of computer data into a form that is unreadable by nonauthorized recipients. The protection system is based on use of a unique

encryption key assigned to each customer. Data is encrypted with its key at the point of transmission and decrypted at the receiving point.

data-encryption unit [DEU] A computer peripheral device that encrypts blocks of text words with a specified key to produce cipher words. It is reversible for decrypting.

data-flow diagram A graphic representation of a communication system in which data transmitters, data receiver, data storage, and the processes performed on data are shown as nodes, and the logical flow of data is shown as links between nodes.

data format The organization of digital information in a computer storage medium. Typically hierarchical, the format includes specifications for the low-level data recording and high-level file system organization.

data-handling capacity The maximum number of units of information that can be transmitted, received, processed, stored, or otherwise handled by a specific piece of equipment.

data link A wire, radio, or other data-transmission channel used for connecting data-processing equipment to an input terminal, output or display device, or other remotely located data-processing equipment.

data logging The recording and printing of the output values of the sensors and transducers monitoring processes such as the fractional distillation of petroleum or the generation of power in a power station in real time as they occur. Data can be logged in analog form on mechanical printers, but in the latest systems the data is converted from an analog to a digital format for transmission to a remote computer for further processing and computer printout.

data path The path followed through a circuit by a signal being processed. This path can be traced through the circuit stage by stage for fault isolation.

data processing Changing the form, meaning, appearance, location, or other characteristics of data. Processing includes data handling and data reduction.

data processor A circuit or system that processes data by such steps as entering it into the computer, formatting it, modifying or enhancing it, performing programmed calculations, and storing it for further reference. Data processing can be performed by *computers, personal communicators,* and *facsimile machines.*

data set ready [DSR] An RS-232C modem interface control signal that indicates when a terminal is ready for signal transmission.

data signaling rate The data-handling capacity of a set of parallel transmission channels, expressed in bits per second.

data source A device capable of originating signals for a data-transmission system.

data terminal equipment [DTE] A circuit, such as a terminal, that acts as a data source, a data sink, or both.

data terminal ready [DTR] An RS-232C modem interface control signal that indicates to the modem that the data terminals are ready for transmission.

data tone multiple frequency [DTMF] The audio signaling frequency on pushbutton Touch Tone telephones.

data transcription The conversion of data from one recorded form to another, as from magnetic tape to magnetic disks.

data-transmission system A system that transmits data from one instrument to another.

data under voice [DUV] An AT&T digital data service that allows digital signals to travel on the lower portion of the frequency spectrum of existing microwave radio systems. Digital channels initially available handled speeds of 2.4, 4.8, 9.6, and 56 kb/s.

datum line A reference line from which calculations or measurements are made.

dawn chorus *Chorus.*

dB Abbreviation for *decibel.*

dBa Abbreviation for *decibels adjusted.*

dBC Abbreviation for *C-scale sound level in decibels.*

dBf Abbreviation for *decibels above 1 femtowatt.*

dBi A decibel unit for measuring antenna gain. A unit of measurement of the gain relative to an isotropic antenna, or one that radiates equal power in all directions.

dBk Abbreviation for *decibels above 1 kilowatt.*

dBm Abbreviation for *decibels above 1 mW,* a unit for specifying input signal power—1.0 mW across 600 Ω, or 0.775 V RMS (root mean square). Any other voltage level can be converted to dBm by

$$dBm = 20 \times \log \frac{V\ RMS}{0.775}$$

$$= 20 \times \log (V\ RMS) = 2.22$$

dBm0 Signal power measured at a point in a standard test tone level at the same point: dBm0 = dBm − dBr, where dBr is the relative transmission level, or level relative to the point in the system defined as the zero transmission-level point.

dB meter *Decibel meter.*

dBm0p Relative power expressed in dBmp. See *dBm0* and *dBmp.*

dBmp Abbreviation for *decibels above 1 mW psophometrically weighted;* a dBm measurement made with a *psophometric* weighting filter.

dBp Abbreviation for *decibels above 1 picowatt.*

dBrn Abbreviation for *decibels above reference noise,* a relative signal level expressed in decibels above reference noise, where reference noise is 1 pW. Hence, 0 dBrn = 1 pW = −90 dBm.

dBrnC Abbreviation for *decibels above reference noise C,* a dBrn measurement made with a C-message weighting filter. These units are most commonly used in the United States, where psophometric weighting is rarely used.

dBrnC0 Abbreviation for *decibels above reference noise C0,* a noise measurement measured in dBrnC0 referenced to zero transmission level.

DBS Abbreviation for *direct broadcast by satellite.*

dBV Abbreviation for *decibels above 1 volt.*

dBW Abbreviation for *decibels above 1 watt.*

dBx Abbreviation for *decibels above reference coupling.*

DC Abbreviation for *direct current.*

DC amplifier An amplifier capable of amplifying DC voltages and slowly varying voltages. The basic types of DC amplifiers are the direct-coupled amplifier, the carrier amplifier, and combinations of direct-coupled and carrier amplifiers. The most common type of carrier amplifier is the chopper amplifier; when the chopper amplifier is used with a direct-coupled amplifier, the combination is called a chopper-stabilized amplifier.

DCC Abbreviation for *digital color code.*

DC component The average value of a signal. In television it represents the average luminance of the picture being transmitted. In radar it is the level from which the transmitted and received pulses rise.

DC coupling The signal coupling that transmits the zero-frequency term of the Fourier series, which represents the input signal.

DCD Abbreviation for *data-carrier detect.*

DC/DC An abbreviation used to indicate that a converter will change one DC voltage value to a different (higher or lower) DC voltage value.

DC/DC converter An electronic circuit that accepts a direct-current input at one voltage level and converts it to direct-current output at a higher or lower voltage. This is typically accomplished by "chopping" the input DC to convert it to a coarse alternating current, then amplifying and rectifying the AC.

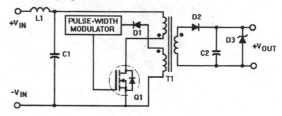

DC/DC converter based on a MOSFET can produce 5, 12, and 15 V DC with an input of 18 to 25 V DC.

DCE Abbreviation for *data communications equipment.*

DC generator A rotating electric machine that converts mechanical power into DC power.

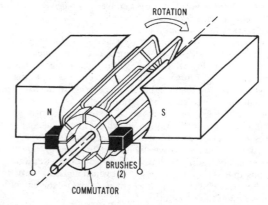

DC generator. When the shaft is rotated by an external source, direct current appears across the brushes.

DC inserter A television transmitter stage that adds to the video signal a DC component known as the pedestal level.

DC magnetic biasing Magnetic biasing by means of direct current in magnetic recording.

DC motor A motor that operates from a DC voltage source and converts electric energy to mechanical energy.

DC offset A direct-current level that can be added to the input signal of an amplifier or other circuit.

DC parametrics The operating characteristics of an integrated circuit or discrete device that can be measured with the device in a static condition. See *parametric tests.*

DC permanent-magnet motor An electric motor suited as an actuator in *closed-loop* control systems because it can be controlled precisely with DC control signals. The speed-torque curve is linear.

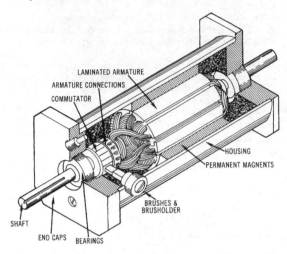

DC permanent-magnet motor is suitable for closed-loop machine control systems.

DC power supply A power supply that provides one or more DC output voltages, such as a DC generator, linear or switching power supply, converter, or dynamotor.

DCR Abbreviation for *direct-conversion receiver.*

DC reinsertion *Clamping.*

DC restoration *Clamping.*

DC restorer A clamping circuit used in television receivers to restore the DC component of the video signal after AC amplification. The resulting DC voltage serves as the bias voltage for the grid of the picture tube, to make average reproduced brightness correspond to the average brightness of the scene being transmitted.

DC self-synchronous system A system for transmitting angular position or motion in which an arrangement of resistors serves as a transmitter that furnishes a receiver with two or more voltages which are functions of transmitter shaft position. The receiver has two or more stationary coils that set up a magnetic field which causes a rotor to take an angular position corresponding to the angular position of the transmitter shaft.

DCTL Abbreviation for *direct-coupled transistor logic.*

DC voltage *Direct voltage.*

DC working volts [DCWV] Obsolete term for *volts DC.*

DCWV Abbreviation for *DC working volts.* Use of VDC is preferred.

DDA Abbreviation for *digital differential analyzer.*

DDC Abbreviation for *direct digital control.*

DDCMP Abbreviation for *digital data communications message protocol.*

DDD Abbreviation for *direct distance dialing.*

DDP Abbreviation for *distributed data processing.*

DDS 1. Abbreviation for *digital data service.* 2. Abbreviation for *direct digital synthesizer.*

deaccentuator A circuit used in an FM receiver to offset the preemphasis of higher audio frequencies introduced at the FM transmitter.

dead band The range of values over which a measured variable can change without affecting the output of a magnetic amplifier or automatic control system.

dead spot 1. A geographic location in which signals from a radio, television, or radar transmitter are received poorly or not at all. 2. A portion of the tuning range of a receiver in which stations are heard poorly or not at all because of improper design of tuning circuits.

dead time 1. The time interval, after a response to one signal or event, during which a system is unable to respond to another. For a radiation counter it is the interval after the start of a count during which the counter is insensitive to further ionizing events. For a transponder it is the interval after the start of a pulse during which no new pulse can be received or produced. Also called insensitive time. 2. The time interval between a change in the input signal to a process-control system and the response to the signal.

dead-time correction A correction applied to an observed counting rate to allow for the probability of the occurrence of events within the dead time. Also called coincidence correction.

dead zone *Dead band.*

deathnium center An imperfection in the arrangement of atoms in a semiconductor crystal. It facilitates the generation and recombination of electron-hole pairs.

de Broglie wavelength The wavelength ascribed by wave or quantum mechanics to a particle having a given momentum. This wavelength is equal to Planck's constant divided by the momentum of the particle.

debugging 1. Removing mistakes from a computer program. 2. The detection and removal of secretly installed listening devices, popularly known as bugs. The basic counterespionage instrument is a sensitive receiver that can be tuned to and homed on a concealed miniature radio transmitter.

debunching A tendency for electrons in a beam to spread out both longitudinally and transversely due to mutual repulsion. It decreases efficiency in velocity-modulation tubes.

Debye effect Selective absorption of electromagnetic waves by a dielectric, due to molecular dipoles.

Debye length The distance at which a given negative particle in a plasma is shielded by surrounding positive particles.

Debye-Sears effect The generation of acoustic waves, consisting of alternate regions of compression and refraction one half-wavelength apart, by a piezoelectric crystal vibrating in a longitudinal mode in a liquid.

decade 1. A group of 10. 2. The interval between any two quantities having the ratio of 10 to 1.

decade box An assembly of precision resistors, coils, or capacitors whose individual values vary in submultiples and multiples of 10. Each section contains 10 equal-value components connected in series. The total value of each section is 10 times that of the preceding section. By appropriately setting the 10-position selector switch for each

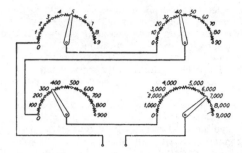

Decade box with precision resistors.

section, the decade box can be set to any desired value within its range.

decade counter   *Decade scaler.*

decade scaler   A scaler that produces 1 output pulse for every 10 input pulses. Also called counter decade, decade counter, and scale-of-ten circuit.

decametric wave   A radio wave between the wavelength limits of 10 and 100 m, corresponding to the high-frequency (HF) range of 3 to 30 MHz.

decay   1. Gradual reduction in the magnitude of a quantity, as of current, magnetic flux, a stored charge, or phosphorescence.   2. *Radioactive decay.*

decay characteristic   *Persistence characteristic.*

Decca   A proprietary *hyperbolic navigation system* operating in the same 70- to 130-kHz frequency band as Loran C, but as a continuous wave rather than pulsed system. A typical chain consists of one master and three secondary stations about 80 mi (130 km) apart.

decelerating electrode   An electrode whose potential provides an electric field to decrease the velocity of the beam electrons in an electron-beam tube.

deceleration time   The time between completion of reading or writing of a magnetic-tape record and actual stopping of the tape.

deci- [d]   A prefix representing 10^{-1}, or one-tenth.

decibel [dB]   A power or voltage measurement unit, referred to another power or voltage. It is computed as

$$dB = 10 \log_{10} P_2/P_1$$

$$dB = 20 \log_{10} V_2/V_1$$

decibel meter   An instrument that directly measures the power level of a signal in decibels above or below an arbitrary reference level. Also called dB meter.

decibels above 1 femtowatt [dBf]   A power level equal to 10 times the common logarithm of the ratio of the given power in watts to 1 fW (10^{-15} W).

decibels above 1 kilowatt [dBk]   A power level equal to 10 times the common logarithm of the ratio of a given power in watts to 1 kW.

decibels above 1 milliwatt [dBm]   A power level equal to 10 times the common logarithm of the ratio of a given power in watts to 0.001 W. A negative value, such as −2.7 dBm, means decibels below 1 mW.

decibels above 1 milliwatt psophometrically weighted [dBmp]   A unit that specifies telephone-channel noise lev-

els measured with the standard international psophometer weighting curve. It more nearly gives the true interfering effect of line noise than does a measuring set having flat frequency response. The power of each interfering tone is compared with the power of an 800-Hz tone creating the same interference during listening tests.

decibels above 1 picowatt [dBp]   A power level equal to 10 times the common logarithm of the ratio of a given power in watts to 1 pW, or 10^{-12} W.

decibels above 1 volt [dBV]   A voltage level equal to 20 times the common logarithm of the ratio of a given voltage in volts to 1 V.

decibels above 1 watt [dBW]   A power level equal to 10 times the common logarithm of the ratio of a given power in watts to 1 W.

decibels above reference coupling [dBx]   A measure of the coupling between two circuits, expressed in relation to a reference value of coupling that gives a specified reading on a specified noise-measuring set when a test tone of 90 dBa is impressed on one circuit.

decibels above reference noise [dBrn]   A unit that shows the relationship between the interfering effect of a noise frequency, or band of noise frequencies, and a fixed amount of noise power commonly called reference noise. A 1-kHz tone that has a power level of −90 dBm was originally selected as the reference noise power. This reference level was later changed to −85 dBm, and the new unit called decibels adjusted, abbreviated dBa.

decibels adjusted [dBa]   A unit that shows the relationship between the interfering effect of a noise frequency, or band of noise frequencies, and a reference noise power level of −85 dBm. This unit replaces dBrn, which was based on a reference noise level of −90 dBm. The new unit gives less weight to low tones and thus more nearly matches the effect of sound on people. Also called adjusted decibel.

decimal-binary conversion   Conversion of a number written in the decimal scale of 10 to the same number written in the binary scale of 2.

decimal-binary switch   A switch that connects a single input lead to appropriate combinations of four output leads (representing 1, 2, 4, and 8) for each of the decimal-numbered settings of its control knob. Thus, for position 7, output leads 1, 2, and 4 would be connected to the input.

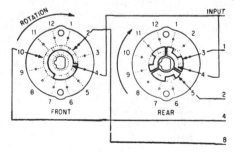

Decimal-binary switch.

decimal code   A code in which each allowable position has 1 of 10 possible states. The conventional decimal number system is a decimal code.

decimal-coded digit   One of 10 arbitrarily selected patterns of 1s and 0s used to represent the decimal digits.

decimal digit   One of the 10 digits, 0, 1, 2, 3, 4, 5, 6, 7, 8, and 9, used in the scale of 10. Two of these digits, 0 and 1, also serve as binary digits in the scale of 2.

decimal notation   A system of notation that uses the scale of 10.

decimal number system   A system of positional notation in which the successive digits relate to successive powers of the base 10.

decimal point   The point that marks the place between integral and fractional powers of 10 in a decimal number.

decimal-to-binary conversion   The mathematical process of converting a number written in the scale of 10 into the same number written in the scale of 2.

decimetric wave   A radio wave between the wavelength limits of 10 and 100 cm, corresponding to the ultrahigh-frequency (UHF) range of 300 to 3000 MHz.

decimillimetric wave   A radio wave between the wavelength limits of 0.1 and 1 mm, corresponding to the frequency range of 300 to 3000 GHz.

decineper   One-tenth of a neper.

decision element   A circuit that performs a logic operation such as AND, OR, NOT, or EXCEPT on one or more binary digits of input information representing "yes" or "no," and expresses the result in its output.

decision table   A table of all possibilities to be considered in a computer problem, along with the action to be taken for each.

deck   *Tape deck*

deck switch   *Gang switch.*

declination   The angle between the horizontal component of the earth's magnetic field and true north.

declinometer   An instrument for measuring the exact direction of a magnetic field to determine magnetic declination.

decode   1. The complement of encode.   2. The generation of an analog sample from the digital character signal that represents the sample.

decoder   1. A matrix network whose combination of inputs produces a single output for converting digital information to analog form.   2. A circuit that responds to a particular coded signal while rejecting others.   3. A device that decodes.   4. A device that unscrambles matrix-encoded signals for quadraphonic sound systems.   5. *Matrix.*   6. *Code converter.*

decommutation   The process of recovering a signal from the composite signal previously created by a commutation process.

decommutator   The section of a telemetering system that extracts analog data from a time-serial train of samples representing a multiplicity of data sources transmitted over a single RF link.

decoupling network   Any combination of resistors, coils, and capacitors placed in power-supply leads or other leads that are common to two or more circuits, to prevent unwanted interstage coupling.

decoy transponder   A transponder that returns a strong signal when triggered directly by a radar pulse. When used for electronic countermeasures, the transponders produce large and misleading target signals on enemy radar screens.

decrement   1. The quantity by which a variable is decreased.   2. *Damping factor.*

decremeter   An instrument for measuring the logarithmic decrement or damping of a wave train.

decrypt   To convert a cryptogram or series of electronic pulses into plain text by electronic means.

decryption   The conversion of an encrypted message to its original form by cryptogear after transmission.

DECT   Abbreviation for *digital-enhanced cordless telephone.*

dedicated   Reserved for a specific use or application, such as a dedicated line or dedicated microprocessor.

dedicated line   A permanently wired line used between two points exclusively for one type of service, such as for data communication or for a radio studio-transmitter link.

dee   A hollow accelerating electrode in a cyclotron, shaped like the letter D.

dee line   A structural member that supports the dee of a cyclotron and acts with the dee to form the resonant circuit.

deemphasis   A process for reducing the relative strength of higher audio frequencies before reproduction, to complement and thereby offset the preemphasis that was previously introduced to help these components override noise or reduce distortion. Used chiefly in frequency- and phase-modulated receivers. It is also called *postemphasis* and *postequalization.*

deemphasis network   An RC filter inserted into a system that restores preemphasized signals to their original form.

deenergize   To disconnect from the source of power.

deep ultraviolet   A region of the ultraviolet spectrum centering on 257 nm suitable for high-definition photolithography for the fabrication of integrated circuits.

deerhorn antenna   A dipole antenna whose ends are swept back to reduce wind resistance when mounted on an airplane.

defect   A chemical or structural irregularity that degrades the crystal structure of the semiconductor material or of the deposited materials that reside on its surface. Defects can be active mobile impurities that impact the electrical device characteristics over time, or inactive particulates that interfere with photolithographic patterning. The most common defects in semiconductor processing are those originating from humans (skin flakes, skin oil, cosmetics, and nasal droplets).

definition   1. The fidelity with which a television or facsimile receiver forms an image.   2. The extent to which the fine-line details of a printed circuit correspond to the master drawing.   3. The fidelity with which details are reproduced in an image.

deflection   The displacement of an electron beam from its straight-line path by an electrostatic or electromagnetic field.

deflection coil   One of the coils in a deflection yoke.

deflection defocusing   Defocusing that becomes greater as deflection is increased in a cathode-ray tube because the beam hits the screen at an increasingly greater slant, and its spot becomes increasingly more elliptical as it approaches the edges of the screen.

deflection electrode   An electrode whose potential provides an electric field that deflects an electron beam.

deflection factor   The reciprocal of the deflection sensitivity in a cathode-ray tube. Deflection factor is expressed in amperes per inch for electromagnetic deflection and volts per inch for electrostatic deflection.

deflection plane A plane perpendicular to the cathode-ray-tube axis containing the deflection center.

deflection sensitivity The displacement of the electron beam at the target or screen of a cathode-ray tube per unit of change in the deflection field. Usually expressed in inches per volt applied between deflection electrodes or inches per ampere in a deflection coil. Deflection sensitivity is the reciprocal of deflection factor.

deflection voltage The voltage applied between a pair of deflection electrodes to produce an electric field.

deflection yoke An assembly of one or more electromagnets that is placed around the neck of an electron-beam tube to produce a magnetic field for deflection of one or more electron beams. Also called yoke.

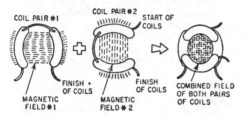

Deflection yoke for a television picture tube contains four separate coils.

deflection-yoke pullback 1. In a color picture tube, the distance between the maximum possible forward position of the yoke and the position of the yoke that gives optimum color purity. 2. In a monochrome picture tube, the maximum distance the yoke can be moved back along the tube axis without producing neck shadow.

defocus To make a beam of X-rays, electrons, light, or other radiation deviate from an accurate focus at the intended viewing or working surface.

degassing The process of driving out and exhausting the gases occluded in the internal parts of an electron tube, generally by heating during evacuation.

degauss To remove, erase, or clear information from a magnetic tape or disk.

degaussing 1. *Demagnetizing*. 2. A method of neutralizing the magnetic field of a ship by placing a cable around its hull and sending a direct current through it with the value needed to neutralize the magnetic effect of the hull. The current adjustment is made at a degaussing station equipped with underwater equipment that indicates when the resultant magnetic field has been sufficiently weakened so it will not actuate a magnetic mine.

degaussing coil A plastic-encased coil, about 12 in (30 cm) in diameter, that can be plugged into a 120-VAC wall outlet and moved slowly toward and away from a color television picture tube to demagnetize adjacent parts.

degaussing control A control that automatically varies the current in degaussing coils as a ship changes heading or rolls and pitches.

degeneration *Negative feedback*.

deglitcher A nonlinear filter or other special circuit that limits the duration of switching transients in digital converters.

degradation *Moderation*.

°C Abbreviation for degree *Celsius*.

°F Abbreviation for degree *Fahrenheit*.

°R Abbreviation for degree *Rankine*.

deionization 1. The recombination of ions in a glow or arc discharge to form neutral atoms and molecules. 2. The neutralization of static electrical charges on nonconductive surfaces by blowing ionized air from an ionizer over the surfaces. See *electrostatic discharge* [ESD] *protection*.

deionization potential The potential at which ionization of the gas in a gas-filled tube ceases and conduction stops.

deionization time The time required for a gas tube to regain its preconduction characteristics after interruption of anode current, so the grid regains control. Also called recontrol time.

delay The amount of time by which an event is retarded.

delay circuit A circuit in which the output signal is delayed by a specified time interval with respect to the input signal.

delay coincidence circuit A coincidence circuit that is actuated by two pulses, one of which is delayed by a specified time interval with respect to the other.

delay distortion Phase distortion in which the rate of change of phase shift with frequency of a circuit or system is not constant over the frequency range required for transmission. Also called envelope delay distortion. It occurs on communication lines because of differences in signal propagation speeds at different frequencies, measured in microseconds of delay relative to the delay at 1700 Hz. It can seriously impair data transmission, but has little effect on voice transmission.

delayed automatic gain control An automatic-gain-control system that does not operate until the signal exceeds a predetermined magnitude. Weaker signals thus receive maximum amplification. Also called biased automatic gain control, delayed automatic volume control, and quiet automatic volume control.

delayed automatic volume control *Delayed automatic gain control*.

delayed sweep A technique for operating an oscilloscope that adds a precise amount of time between the trigger point and the beginning of the oscilloscope's horizontal sweep to permit making special measurements, such as risetime. It also permits the sweep to be triggered anywhere along the *X*-axis baseline.

delay equalizer A corrective network that makes the phase delay or envelope delay of a circuit or system substantially constant over a desired frequency range.

delay line A device that produces a time delay of a signal. A transmission line using either lumped or distributed constants gives a delay determined by the electrical length of the line. In one form, the inductance is increased by winding a helix on a flexible powdered iron core to serve as the inner conductor. High capacitance is achieved by using only a few layers of thin paper to separate the inner conductor from the stranded outer conductor. An ultrasonic delay line gives a delay determined by the length of the path taken by acoustic waves through the medium. Also called artificial delay line.

delay-line cable A special cable in the monochrome channel of a color television receiver to provide a time delay just long enough to make the monochrome and chrominance signals arrive together at the cathode-ray tube.

delay-line memory   *Delay-line storage.*

delay-line storage   A computer storage or memory device consisting of a delay line and means for regenerating and reinserting information into the delay line. Also called delay-line memory.

delay multivibrator   A monostable multivibrator that generates an output pulse a predetermined time after it is triggered by an input pulse.

Dellinger effect   A form of shortwave radio fadeout believed to be caused by rapid shifting of ionosphere layers during solar eruptions.

Delta   A U.S. ballistic missile designed to launch nuclear warheads that has been converted into a launch vehicle for low earth-orbiting (LEO) and geostationary satellites.

delta connection   A combination of three components connected in series to form a triangle like the Greek letter delta. Also called mesh connection.

delta function   A distribution profile large at points close to an origin and zero for all significant distances from the origin.

delta-matched antenna   A single-wire antenna, usually one half-wavelength long, to which the leads of an open-wire transmission line are connected in the shape of a Y. The flared parts of the Y match the transmission line to the antenna. Because the top of the Y is not cut, the matching section has the triangular shape of the Greek letter delta. Also called Y antenna.

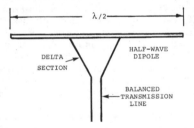

Delta-matched antenna feed method for a half-dipole antenna.

delta-matching transformer   The Y-shaped matching section of a delta-matched antenna.

delta modulation   A pulse-modulation technique in which a continuous analog signal is converted into a serial bit stream which corresponds to changes in analog input levels. It is usually used by devices that are continuously variable-slope delta (CVSD)-modulated.

delta network   A set of three branches connected in series to form a mesh.

delta noise   The noise signal voltage induced in the winding of a ferrite-core storage by partially selected cores.

delta pulse-code modulation   A modulation system that converts audio signals into corresponding trains of digital pulses to give greater freedom from interference during transmission over wire or radio channels.

DELTIC   [DElay-Line-TIme-Compression]   A method of sampling incoming radar, sonar, seismic, speech, or other waveforms along with reference signals, compressing the samples in time, and comparing them by autocorrelation. The time compression greatly reduces the complexity and size of the equipment needed for signal analysis.

demagnetization   Removal of residual magnetism.

demagnetization curve   That part of a magnetic material's hysteresis loop which shows the peak value of residual induction and the way that magnetization is reduced to zero when a demagnetizing force is applied.

demagnetizer   Apparatus or equipment for removing undesired magnetism from an object or material. Magnetic tape and computer disks can be erased with a demagnetizer. The secondary coil of an AC transformer will demagnetize small metal objects.

demagnetizing   Removing magnetism from a ferrous material. Also called degaussing.

demagnetizing force   A magnetizing force applied in the direction that reduces the residual induction in a magnetized object.

demand factor   The ratio of the maximum downward-driven machine power drawn by an electronic system to the total connected load of the system. The power drawn is either the instantaneous value or the value measured over a specified period of time.

demodulation   The process of converting a modulated RF carrier signal to a form that can be heard or displayed. When the carrier is unmodulated, the process is called detection.

demodulator   1. A circuit that removes the modulation signal from a modulated carrier signal. In a receiver the carrier is an RF signal.   2. The functional section of a modem that converts received analog data line signals to digital form for processing by a computer. It is the inverse of a modulator.

demultiplexer   A circuit that separates two or more signals previously combined by a compatible multiplexer and transmitted over a single channel.

demultiplexing circuit   A circuit that separates the signals which have been combined for transmission by multiplex.

dendritic web   A silicon or other crystal grown as a thin, narrow strip several yards long.

dendritic-web process   A process for forming crystalline material into a ribbon as it is drawn vertically from a molten bath to produce photovoltaic material.

densimeter   *Density indicator.*

densitometer   1. An instrument that measures the optical density of a material.   2. An instrument that measures the amount of darkening of film badges to determine the radiation dosage received by the wearer.

density   1. Weight per unit volume.   2. A measure of the light-transmitting properties of an area. It is expressed as the common logarithm of the ratio of incident light to transmitted light.   3. Amount per unit cross-sectional area, as for current, magnetic flux, or electrons in a beam.

density indicator   An instrument that measures the density of a liquid or solid material. One version measures the absorption of gamma rays by the material because this absorption is proportional to density. Also called densimeter.

density modulation   Modulation of an electron beam by making the density of the electrons in the beam vary with time.

density packing   The number of units of useful information that can be stored within a given linear dimension on a single track of a magnetic tape or disk by a single head.

dependent node   A node having one or more incoming branches.

depletion Reduction of the charge-carrier density in a semiconductor below the normal value for a given temperature and doping level.

depletion layer An electric double layer formed at the surface of contact between a metal and a semiconductor having different work functions. Electrons diffuse from the substance having the lower work function toward the other substance, leaving equivalent positive charges at the layer in the first substance. This action occurs because the mobile charge-carrier density is insufficient to neutralize the fixed charge density of donors and acceptors. Also called space-charge layer. Formerly called barrier layer and blocking layer.

depletion-layer capacitance The capacitance of the imaginary capacitor formed by the charges of a depletion layer. This capacitance is a function of reverse voltage.

depletion-mode transistor A "normally on" junction field-effect transistor (JFET), metal-oxide FET (MOS-FET), or metal-semiconductor FET (MESFET) that conducts when its gate and source are at the same potential and its gate bias is zero volts. N-channel FETs require positive bias to increase zero-bias drain current, but they require negative bias to pinch them off. The bias is complementary in P-channel FETs. See also *enhancement-mode transistor*.

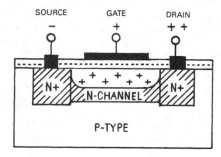

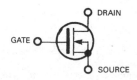

Depletion-mode, N-channel *normally-on* MOSFET conducts whenever the drain is positive with respect to the source. It is turned off with negative gate bias. Its schematic symbol is shown below.

depletion (transition) region The region at a junction between N- and P-type semiconductor materials where the difference potential energies between the two materials creates an energy barrier. This barrier results in an electric field that depletes the semiconductor of free (mobile) charge carriers. The width of the region decreases with external *forward bias* and it increases with external *reverse bias*.

depolarization Prevention of polarization in an electric cell or battery.

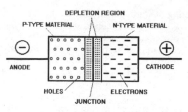

Depletion region on both sides of the PN junction decreases with external forward bias and increases with reverse bias (shown).

deposited-carbon resistor A resistor that has a thin film of carbon deposited on a supporting mandrel.

deposited oxide An oxide layer deposited on a surface by a method such as evaporation or sputtering that does not require a chemical reaction with the substrate.

deposition The process of depositing materials on a substrate. It typically refers to the deposition of thin conducting and insulating films that form MOSFET gates, capacitors, thin-film resistors, and the interconnect traces for an integrated circuit. See *metallization*.

depth finder *depth sounder*.

depth of heating The depth below the surface of a material in which effective dielectric heating can be confined when the applicator electrodes are applied adjacent to one surface only.

depth of modulation The ratio of the difference in field strength of the two lobes of a directional antenna system to the field strength of the greater at a given point in space. Used to determine direction in a radio guidance system.

depth sounder A marine electronic instrument for determining the depth of water beneath a ship by measuring the round-trip time for sound energy from a transducer mounted underwater on the ship's hull to travel to the sea floor and be reflected back to the transducer. Most depth sounders transmit ultrasonic pulses at about 200 kHz. Sound travels about 4800 ft/s (0.0002083 s/ft) in sea water, so a pulse round trip to a depth of 120 ft takes about 1/20 s. It is a dedicated form of sound *sonar equipment*. Its principal components are a display/control head, which includes the display and transmitter/receiver circuit, and the *depth transducer*, a *piezoelectric device*. The trans-

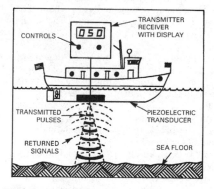

Depth sounder projects ultrasonic pulses down to the sea floor, measures their round-trip transit time, and converts that time to a depth reading in feet or fathoms.

ducer, acting as an antenna, directs the ultrasonic energy in a narrow beam and receives weak echo signals for processing to provide a digital readout or a depth display. The narrow beam makes it necessary for the host ship to remain within 10° of vertical to obtain accurate readings. The output pulse triggers counting circuitry in a digital display and the return signal stops the count, which is translated by the circuitry into a depth reading. The latest systems have power-conserving liquid-crystal displays. It is sometimes spelled *depthsounder,* and it is also called a *depth finder* and a *fathometer.* See also *fishfinder.*

depth transducer An ultrasonic transducer that acts as a speaker to transmit bursts of sound energy and as a microphone to listen for the returning echoes. In *sonar* systems, the energy can be directed downward into the water through an angle from horizontal to vertical, while the beams of other systems are swept through angular sectors. Specialized submarine sonar directs the sound beam upward to determine depth below an ice pack. *Depth finders* direct sound energy vertically downward, but some *fishfinders* can direct the beam through a vertical angle to find and follow schools of fish. The transducers in commercial depth finders and fish finders are made from *piezoelectric* ceramics that oscillate at either 50 or 200 kHz. Water absorbs high frequencies more than low frequencies, so a 50-kHz transducer is more effective in water deeper than 400 ft, but 200 kHz provides better resolution.

derate To reduce the rating of a device to improve reliability or to permit operation at high ambient temperatures.

derivative action A system correction control response whose speed depends on the speed at which the system error increases. It is also called *rate action.*

derived sound system A four-channel sound system that is artificially synthesized from conventional two-channel stereo sound by an adapter, to provide feeds to four loudspeakers for approximating quadraphonic sound.

derived measurements Measurements made directly by formulas and calculation. Examples are alternating current measurements made after the direct measurement of peak-to-peak amplitude. These include *maximum value, average value,* and *root mean square* [RMS] *voltage.*

DESC Abbreviation for *Defense Electronic Supply Center.*

desensitization An automatic-gain-control effect that occurs when a receiver is tuned to one channel and there is a strong signal on a nearby channel. The strength of the desired signal appears to be decreased by the presence of the nearby signal.

despun antenna A satellite antenna that is rotated at a rate equal to and opposite from the rate at which the satellite is spinning for stabilization, so the directional antenna can be pointed continuously at the earth.

destructive breakdown 1. The *punchthrough* of the dielectric of a capacitor caused by exposure to overvoltage. 2. The breakdown of the insulating layer between the gate and channel of a field-effect transistor or CMOS logic gate because of electrostatic discharge (ESD) or the application of excessive voltage.

destructive readout [DRO] The loss of data that can occur in certain semiconductor memory devices when they are being read. Additional circuitry might be required to restore any data lost, if it were needed again.

detection The process of converting an RF carrier signal to a form that can be heard or displayed. If the carrier is unmodulated, the detected result is a DC voltage that will act on a simple diode wavemeter. If the carrier is modulated, the modulation signal is obtained as output, and the process is more often called demodulation.

detectivity A figure of merit representing gain over noise for low-level solid-state radiation detectors like photodiodes.

detector The stage in a receiver where demodulation takes place. In a superheterodyne receiver it is called the second detector, although it is actually a demodulator. The so-called first detector in a superheterodyne is a frequency converter that changes the incoming carrier frequency to the intermediate frequency of the receiver.

detent A mechanism on a multiposition control to hold it firmly in each position. One common type consists of a spring-loaded ball that falls into equally spaced indentations on a plate as the control shaft is rotated.

detune To change the inductance or capacitance of a tuned circuit so its resonant frequency is different from the incoming signal frequency.

detuning stub A quarter-wave stub that matches a coaxial line to a sleeve-stub antenna. The stub detunes the outside of the coaxial feed line while tuning the antenna itself.

DEU Abbreviation for *data-encryption unit.*

deuterated potassium dihydrogen phosphate [DKDP] A ferroelectric electrooptical crystal.

Deutsche Industrie Normenausschus [DIN] A German institute that sets industrial standards. In the audio field, plugs and sockets having DIN geometry are used throughout the world.

deviation 1. The difference between the actual value of a controlled variable and the desired value corresponding to the set point. 2. *Frequency deviation.*

deviation distortion Distortion in an FM receiver caused by inadequate bandwidth, inadequate amplitude-modulation rejection, or inadequate discriminator linearity.

deviation ratio The ratio of the maximum possible frequency deviation to the maximum audio modulating frequency in an FM system.

deviation sensitivity 1. The rate of change of course indication with respect to the change of displacement from the course line in a navigation indicator. 2. The lowest frequency deviation that produces a specified output power in an FM receiver.

device A term for any electronic, electrical, or mechanical component or product whose intended function cannot be performed if it is further divided. Examples of electronic devices include the resistor, capacitor, transformer, relay, switch, diode, transistor, integrated circuit, connector, and LED display.

device under test [DUT] The device (or circuit) that is plugged into or otherwise connected to a special circuit designed for checking its performance characteristics.

Dewar flask A double-walled glass container or bottle with an evacuated space between its walls to minimize the transfer of heat. The inner walls are usually silvered. It is widely used as a container for liquefied cryogenic gases such as helium and nitrogen. The thermos bottle is a miniature Dewar flask.

DF Abbreviation for *direction finder.*

DFG Abbreviation for *diode function generator.*

D flip-flop A delay flip-flop in which the input data is delayed by one clock pulse period. The output is therefore a function of the input that appeared one pulse earlier.

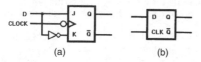

(a)   (b)

D (delay) flip-flop (*a*) formed with a J-K flip-flop and a NOT gate and (*b*) schematic symbol.

DFSK Abbreviation for *double frequency-shift keying.*

DFT Abbreviation for *discrete Fourier transform.*

DGPS Abbreviation for *Differential Global Positioning System.*

DI Abbreviation for *dielectric isolation.*

diac [DIode AC switch] A bidirectional diode that has a symmetrical switching mode. It is triggered to its on state when its breakover voltage is exceeded in either direction by an applied voltage or a trigger spike. Applications include triggering of triacs. Also called trigger diode.

diagram A schematic or other line drawing that explains the operation of a circuit or piece of equipment.

dial 1. A separate scale or other device for indicating the value to which a control is set. 2. A telephone calling device that generates the pulses or tones required for establishing a desired connection.

dial-back security A security feature of some computer modems that lets them hang up and return a call to a prearranged number to avoid unwanted interception of messages.

dial telephone system A telephone system in which connections between customers are established automatically by electronic and mechanical switching systems controlled by rotary pulse-generating dials or pushbutton equivalents.

dial tone A tone in a dial telephone system to indicate that the equipment is ready for the dialing of a number.

dial-up A temporary connection to a host computer, typically over public telephone lines with a *modem.*

diallyl phthalate A high-strength plastic resin that exhibits high compressive strength, resistance to corrosion and chemicals, low water absorption, and dimensional stability. It is used to make IC sockets and terminal blocks and strips.

diamagnetic material A material that has a magnetic permeability less than 1, such as bismuth and antimony. Diamagnetic materials are repelled by a magnet and therefore tend to position themselves at right angles to magnetic lines of force.

diameter equalization Increasing the high-frequency response in proportion to decreasing diameter of a disk recording.

diamond antenna *Rhombic antenna.*

diamond circuit A gate circuit that provides isolation between input and output terminals when in its off state, by operating transistors in their cutoff region. In the on state, the output voltage follows the input voltage as required for gating both analog and digital signals, while the transistors provide current gain to supply output current on demand.

diamond stylus A stylus with a carefully ground diamond as its point.

diaphragm 1. A thin flexible sheet that can be moved by sound waves, as in a microphone, or can produce sound waves when moved, as in a loudspeaker. 2. An adjustable opening used in television cameras to reduce the effective area of a lens to increase the depth of focus. 3. *Iris.*

diathermy Therapeutic use of RF energy to produce heat within some part of the body.

diathermy interference Television interference caused by diathermy equipment. It produces a herringbone pattern in a dark horizontal band across the picture.

dibit A pair of binary digits represented by a single modulation condition in a data-transmission system. Thus, in four-phase modulation a phase shift of 225° is 00, 315° is 01, 45° is 11, and 135° is 10.

dichotomizing search A computer search in which an ordered set of items is divided into two parts, one of which is rejected. The process is repeated on the accepted part until the items with the desired property are found.

dichroic The transmission of different colors by an object or material depending on its inherent properties. The term has been extended to include different frequencies as well as colors. See *dichroic antenna* and *dichroic mirror.*

dichroic antenna An antenna that transmits signals on one output frequency and receives signals on a different input frequency.

dichroic mirror A glass surface coated with a special metal film that reflects certain colors of light while allowing other colors to pass through.

dicing The process of sawing apart or scribing and breaking apart completely processed semiconductor wafers into individual diodes, transistors, or integrated circuit dies or chips, usually for packaging as complete devices.

Dicke radiometer A radiometer-type receiver that detects weak signals in noise by modulating or switching the incoming signal before it is processed by conventional receiver circuits. After amplification and detection, the modulation frequency is recovered. The product of this signal and a reference waveform from the input modulator is smoothed in a low-pass filter, and the filter output drives a display device that indicates the presence of a signal.

die (plural: **dice**) In electronics, the smallest active subdivision of a wafer on which multiples of a device such as a diode, transistor, or integrated circuit are fabricated before being cut or sawed apart (diced) to form an independent unit, typically rectangular in shape. A die is synonymous with a chip, although the term chip generally refers to an integrated-circuit die cut apart from the fabricated wafer.

die bonding The mounting of a semiconductor device (diode, transistor, thyristor, or integrated circuit) die or chip on a leadframe or ceramic substrate by brazing, cementing, or other means to provide mechanical contact along with a thermal path and sometimes to make an electrical contact with its package. Many different bonding methods have been developed to attach individual dies to package substrates: metal alloys (eutectic die attaches, solders, silver-filled glasses), and organic adhesives (such as polyimides,

dielectric

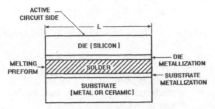

Die bonding process of a silicon die to a substrate, shown schematically.

silicones, and epoxies). The most common technique for bonding a silicon chip or die to a ceramic package or metal can is with a gold-silicon (Au-Si) bond at about 425°C. The silicon diffuses into the gold until the Au-Si eutectic composition is reached. Bonding with soft solder (95% lead, 5% tin) is a lower-temperature (about 300°C) alternative.

dielectric Another term for an insulator or material that has poor electric conductivity. Dielectrics such as air, mica, paper, plastic film, or ceramic separate the metal foil or metallized plates in a capacitor to isolate them electrically and store electric energy. In semiconductor fabrication, dielectric materials such as silicon dioxide provide electrical isolation between adjacent active devices, and metallized interconnects as well as between the gate and channel of MOSFETs.

dielectric absorption The persistence of electric polarization in certain dielectrics after removal of the electric field. The effect can last for several years in certain mixtures of wax that are allowed to harden in a strong electric field. Electrets are based on this effect.

dielectric constant The property of a material that determines how much electrostatic energy can be stored per unit volume when unit voltage is applied. In effect, it is the ratio of the capacitance of a capacitor filled with a given dielectric to that of the same capacitor having only a vacuum as dielectric. Also called permittivity.

dielectric current The current flowing at any instant through a surface of a dielectric that is located in a changing electric field.

dielectric diode A capacitor in which the negative electrode can emit electrons into the normally insulating region between the plates. The charge stored on the capacitor is thus continuously in transit between the electrodes, to give current flow in one direction. Cadmium sulfide crystals can serve as the dielectric.

dielectric dispersion The phenomenon in which the dielectric constant of an insulating material varies with frequency.

dielectric dissipation factor The cotangent of the dielectric phase angle of a dielectric material.

dielectric fatigue The decrease in the ability of some dielectrics to withstand breakdown following long exposure to voltage.

dielectric gas A gas that has a high dielectric constant, such as sulfur hexafluoride. It acts as an insulator in laser cavities and high-powered microwave equipment.

dielectric heating The heating of a nominally insulating material by placing it in a high-frequency electric field. The heat results from internal losses during the rapid reversal of polarization of molecules in the dielectric material.

dielectric hysteresis A lagging of the electric field in a dielectric with respect to the alternating voltage applied to the dielectric. This effect causes a dielectric hysteresis loss comparable to that produced by magnetic hysteresis in a ferrous material.

dielectric insulator layer A layer of nonconducting material laid down between layers of metallization to prevent electrical short-circuiting between them.

dielectric isolation [DI] The electrical isolation of active components in an integrated circuit from each other by an insulator (dielectric material). The active devices are formed in "tubs" of silicon dioxide (glass) dielectric material rather than by reverse-biased PN junctions.

dielectric lens A lens made of dielectric material so that it refracts radio waves in the same manner that an optical lens refracts light waves. Microwave antennas can be equipped with these lenses.

dielectric-lens antenna An aperture antenna whose beam width is determined by the dimensions of a dielectric lens through which the beam passes.

dielectric loss The electric energy that is converted into heat in a dielectric subjected to a varying electric field.

dielectric matching plate A dielectric plate used as an impedance-matching transformer in a waveguide.

dielectric phase angle The angular difference in phase between the sinusoidal alternating voltage applied to a dielectric and the resulting alternating current.

dielectric power factor The cosine of the dielectric phase angle.

dielectric-rod antenna A surface-wave antenna whose end-fire radiation pattern is produced by the propagation of a wave from a cavity down the length of a tapered dielectric rod. If the rod diameter is less than a half wavelength, the wave continues beyond the end of the rod into free space. If the rod is made of polystyrene, it is called a *polyrod;* if it is made of a *ferrite* material, it is called a *ferrod.* The higher the dielectric constant, the thinner the rod can be.

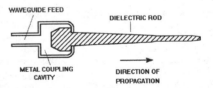

Dielectric rod antenna made from a tapered dielectric rod with a diameter less than a half wavelength propagates the wave in the direction of the rod

dielectric strength The maximum potential gradient a material can withstand without rupture. Usually specified in volts per millimeter of thickness. Also called electric strength.

dielectric test A destructive test on a device, circuit, or product in which a voltage higher than the rated value is applied for a specified time. The objective is to determine the failure voltage of insulating materials to verify their selection for the manufacture of components and products and to determine their *margin of safety.*

dielectric waveguide A waveguide consisting of a dielectric material surrounded by air. Electromagnetic waves travel through the solid dielectric in much the same way they travel through a hollow waveguide.

difference amplifier   *Differential amplifier.*

difference detector   A detector circuit in which the output is a function of the difference between the amplitudes of the two input waveforms.

difference of potential   The voltage between two points.

differential   The difference between levels for turn-on and turn-off operation in a control system.

differential amplifier   An amplifier whose output is proportional to the difference between the voltages applied to its two inputs. Some operational amplifiers can operate in a differential mode. Also called difference amplifier.

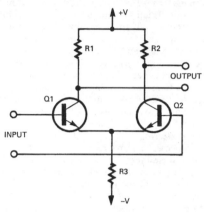

Differential amplifier is a direct-coupled AC and DC signal amplifier that discriminates against unwanted noise voltages.

differential analyzer   An analog computer designed for integrating and solving differential equations.

differential capacitance   The derivative with respect to voltage of a charge characteristic, such as an alternating charge characteristic or a mean charge characteristic, at a given point on the characteristic.

differential capacitor   A two-section variable capacitor having one rotor and two stators so arranged that as capacitance is reduced in one section, it is increased in the other.

differential comparator   A comparator having at least two high-gain differential-amplifier stages, followed by level-shifting and buffering stages, as required for converting a differential input to single-ended output for digital logic applications.

differential delay   The difference between the maximum and minimum frequency delays occurring across a band.

differential discriminator   A discriminator that passes only pulses whose amplitudes are between two predetermined values, neither of which is zero.

differential frequency circuit   A circuit that provides a continuous output frequency equal to the absolute difference between two continuous input frequencies.

differential frequency meter   A circuit that converts the absolute frequency difference between two input signals to a linearly proportional DC output voltage that can drive a meter, recorder, oscilloscope, or other device.

differential gain   The amount that unity is exceeded by the ratio of the output amplitudes of a small high-fre-quency sinewave signal at two stated levels of a low-frequency signal on which it is superimposed in a video transmission system. Differential gain is expressed in percent by multiplying the difference by 100 and expressed in decibels by multiplying the common logarithm of the ratio itself by 20.

differential gain control   *Sensitivity-time control.*

Differential Global Positioning System [DGPS]   A system that compensates for inherent inaccuracies of a position determined by receiving uncorrected GPS-C (degraded commercial) signals. A ground receiving station of known location determines a correction signal based on the error between its known position and its position given by the GPS-C receiver. That correction signal is then transmitted to specialized receivers on ships (or other mobile platforms) within a 500-mi radius so that they can correct their positions as determined by their GPS-C receivers. DGPS permits a ship with a GPS receiver to determine its true position to within 3 m. See also *Wide-Area Augmentation System.*

differential input   An input circuit that rejects voltages that are the same at both input terminals and amplifies the voltage difference between the two input terminals. The circuit can either be balanced or floating and can also be guarded.

differential-input impedance   The impedance between the inverting and noninverting input terminals of a differential amplifier, consisting primarily of resistance and capacitance.

differential keying   Chirp-free break-in keying of a continuous-wave transmitter, achieved by making the oscillator turn on fast before the keyed amplifier stage can pass any signal and turn off fast after the keyed amplifier stage has cut off.

differential linearity   The degree of variation in size of adjacent steps in the output of a digital converter over its operating range.

differentially coherent phase-shift keying [DPSK]   Phase-shift keying in which a particular signal phase can be decoded only by comparison with the phase of the preceding bit.

differential microphone   *Double-button carbon microphone.*

differential-mode electromagnetic interference   Interference that causes the potential of one of two signal leads to change with respect to the other lead.

differential-mode signal   A signal applied between the two ungrounded terminals of a balanced three-terminal system.

differential modulation   Modulation in which the choice of the significant condition for any signal element is dependent on the choice for the previous signal element.

differential null detector   A null indicator whose differential transformer delivers an output voltage that is proportional to the vector difference between two input voltages. One input is a 1-kHz source voltage, and the other is the unknown signal, applied through an amplifier and phase shifter that can be adjusted so the difference between the two voltages is zero.

differential operational amplifier   A dual-input amplifier that can perform mathematical calculations on the difference voltage between the two input signals as organized for the amplifier with external active and passive components.

differential pair The circuit formed by connecting the sources or emitters of two transistors together at a current source. The difference in voltage between the two gates steers the current between the two drains.

differential permeability The slope of the magnetization curve for a magnetic material.

differential phase The difference in output phase of a small high-frequency sinewave signal at two stated levels of a low-frequency signal on which it is superimposed in a video transmission system.

differential phase-shift keying [DPSK] A modulation technique for data transmission in which the frequency remains constant but phase changes will occur from 90°, 180°, and 290° to define the digital information.

differential pressure pickup An instrument that measures the difference in pressure between two pressure sources and translates this difference into a change in inductance, resistance, voltage, or some other electrical quality.

differential probe An oscilloscope accessory that permits the user to make measurements from two points in a circuit without reference to ground. It allows an oscilloscope to be safely grounded without isolation from ground by transformers or optoisolators. It includes a differential amplifier and typically has selectable attenuation ratios. It can make accurate measurements of small signal differences despite high common-mode voltages.

differential pulse-height discriminator *Pulse-height selector.*

differential relay A two-winding relay that operates when the difference between the currents in the two windings reaches a predetermined value.

differential synchro 1. *Synchro differential receiver.* 2. *Synchro differential transmitter.*

differential transducer A transducer that simultaneously senses two separate variables and provides an output proportional to the difference between them.

differential transformer A transformer that joins two or more sources of signals to a common transmission line.

differential transformer transducer *Linear variable differential transformer* [LVDT].

differential voltmeter A voltmeter that measures only the difference between a known and an unknown voltage.

differential winding A winding whose magnetic field opposes that of a nearby winding.

differentiating circuit An analog circuit whose output is proportional to the derivative of its input. In its simplest form, it is an input voltage source in series with a capacitor, a resistor, and a switch. When the switch is closed, current flows, and the differentiated signal is measured as the voltage across the resistor. It is also called a *differentiator.* See also *integrating circuit.*

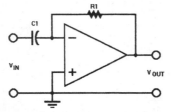

Differentiating circuit is an operational amplifier with a capacitor at the inverting terminal and a resistor in the feedback loop.

differentiating network *Differentiating circuit.*

differentiator 1. A mechanism or electronic circuit whose output is proportional to the derivative of an input. 2. *Differentiating circuit.*

diffracted wave A wave whose front has been changed in direction by an obstacle or other nonhomogeneity in a medium, other than by reflection or refraction.

diffraction The scattering of a beam of electromagnetic radiation (such as light or radio energy) or particles (such as electrons) as it passes through an object; this is caused by constructive interference from differential parts of the object.

diffraction instrument *Diffractometer.*

diffraction pattern The pattern produced on film exposed in an X-ray diffraction camera, consisting of portions of circles having various spacings depending on the material being examined.

diffraction propagation Propagation of electromagnetic waves around objects, or over the horizon, by diffraction. The action occurs because every point in a wavefront generates a spherical front that falls off in intensity away from the forward direction. A continuous series of such actions carries radiation around objects, but with rapidly decreasing intensity.

diffraction scattering Elastic scattering that occurs when inelastic processes remove particles from a beam.

diffraction velocimeter *Laser velocimeter.*

diffused junction A semiconductor junction that has been formed by the diffusion of an impurity within a semiconductor crystal.

diffused-junction rectifier A semiconductor diode in which the PN junction is produced by diffusion.

diffused-junction transistor A transistor in which the emitter and collector electrodes have been formed by diffusion of an impurity metal into the semiconductor wafer without heating.

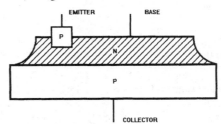

Diffused-junction transistor section view showing the N-type mesa base.

diffused-mesa transistor A diffused-junction transistor in which an N-type impurity is diffused into one side of a P-type wafer. A second PN junction, required for the emitter, is produced by alloying or diffusing a P-type impurity into the newly formed N-type surface. After contacts have been applied, undesired diffused areas are etched away to create a flat-topped peak called a mesa.

diffused resistor An integrated-circuit resistor produced by a diffusion process in a semiconductor substrate.

diffuse reflection Reflection of light, sound, or radio waves from a surface in all directions according to the cosine law.

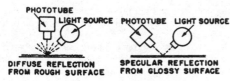

Diffuse-reflection measurement, as compared to direct or specular reflection.

diffuse sound Sound that disperses with uniform energy density in all parts of a room or space.

diffuse transmission Transmission in which all the emergent radiation is observed.

diffuse-transmission density The value of the photographic transmission density obtained when light flux impinges normally on the sample and all the transmitted flux is collected and measured.

diffusion 1. A dispersion of particles, usually in the form of a gas through a unit volume—chamber, room, building, or region. 2. A high-temperature process in which donors or acceptors are distributed in the crystalline lattice structure of semiconductor materials to change its electrical characteristics. The process takes place in a diffusion furnace, usually at temperatures between 850 and 1150°C. The diffusion in acceptors and donors in one semiconductor die will form a PN junction.

diffusion bonding Bonding in which the molecules of one metal diffuse into the crystalline lattice structure of another metal to form a solid solution that is equivalent to an electrical connection.

diffusion capacitance The rate of change of stored minority-carrier charge with the voltage across a semiconductor junction.

diffusion coefficient The constant of proportionality in Fick's law.

diffusion constant The diffusion current density in a homogeneous semiconductor divided by the charge-carrier concentration gradient. The resulting constant is also equal to the product of the drift mobility and the average thermal energy per unit charge of carriers.

diffusion length The average distance that minority carriers diffuse between generation and recombination in a homogeneous semiconductor.

diffusion process A method of producing a junction by diffusing an impurity metal into a semiconductor at a high temperature.

diffusion pump A vacuum pump that creates a stream of rapidly moving gas molecules by heating a fluid in a chamber. The gas molecules purge the chamber of remaining oxygen, nitrogen, and other gas molecules not removed by the mechanical pump. It is normally turned on only after a coarse vacuum has been pumped in the chamber by a mechanical pump to about several hundred millitorr. The mechanical pump is kept in operation to act as a backup for the diffusion pump.

digit A character that stands for zero or for a positive integer smaller than the base of an ordinary number system.

digital Represented by discrete digits, each distinct from the next. A method of representing and manipulating information by switching current on or off.

digital adder An adder that accepts two numbers in digital form and gives their sum in the same digital form.

digital advanced television An alternate term for *high-definition television* (HDTV).

digital attenuator An absorptive microwave attenuator that uses a single current-controlled diode attenuator in combination with a digital driver which provides the correct level of control current for giving the discrete attenuation level corresponding to the binary control code. Each attenuation level is determined by a unique logic gate and resistor.

digital audio tape [DAT] A proprietary system for recording music and voice on tape cassettes in a digital rather than conventional analog format for playback. DAT is said to have performance advantages over both compact disks (CD) and analog tape cassettes. A DAT player can both record and playback digital recordings made on compatible systems and can play back analog tape cassettes but not record on them.

digital circuit A circuit that operates like a switch in that it has only two states: on and off.

digital clock A clock that has a large direct-reading digital display of time in the form 12:15 or 12:15:55 and sometimes a day-month-year display. A digital clock may also provide time-code signals in a variety of forms, such as series or parallel binary-coded decimal, level shift, or modulated carrier.

digital communication A communication system in which data is transmitted in the form of trains of pulse signals, and digital information is transmitted directly from computers, radar, tape readers, teleprinters, and telemetering equipment. Delta pulse-code modulation can convert audio signals to digital signals. The system can also include automatic switching circuits that choose the fastest path from origin to destination.

digital comparator A comparator that accepts two different values of input digital data and determines which is larger, or compares one digital input value against preset upper and lower digital limits and provides pass/fail information.

digital compression A technique for converting analog television signals into digital signals and the subsequent compression of that signal to about one-tenth of its original length.

digital computer A software-programmed computer that processes information in a digital format. Today, it typically refers to a computer whose architecture is based on the *Von Neumann machine*, with a single central processor (CPU) and both volatile semiconductor memory (DRAM) and nonvolatile magnetic-disk memory (floppy disk and/or hard-disk drive). It accepts data in digital format from a keyboard, disk drive, or modem, and solves problems by repeatedly performing addition, subtraction, multiplication, and division at high speed. Its principal display for desktop computers is a cathode-ray tube monitor that can display processed data, calculation results, electronic mail, operator help text, menus, and other text and graphic presentations in color. However, active-matrix liquid crystal (AMLCD) displays are installed on portable, battery-powered or notebook-style personal computers. The principal classes of computers today are *mainframes, workstations, network servers,* and *microcomputers,* more

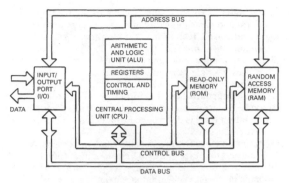

Computer based on Von Neumann's design has a central processing unit (CPU), random-access and read-only memory (ROM and RAM), and input/output (I/O) functions.

generally called desktop *personal computers,* which include *notebook computers.*

digital computer system   The components of the basic computer system today are the computer in a separate case, cathode ray-tube monitor, keyboard, printer, and mouse or other input aid. A computer printer, typically ink-jet or laser today, prints the hard copy. Circuitry in the separate case contains the microprocessor, peripheral input/output (I/O) devices, semiconductor memory on circuit boards, and a power supply. Both volatile (DRAM) and nonvolatile (ROM) memory are included. The case typically contains a *floppy-disk drive* and a *hard-disk drive* (Winchester-style) disk drive. Many systems now include a CD-ROM drive, internal modem, and facsimile circuit boards. Dual speakers might be included for reproducing CD-ROM audio recordings or sound and music from the World Wide Web.

digital converter   A converter that changes one form of digital input data to its equivalent in some other code.

digital data   Information transmitted and received in a coded binary format, such as ASCII.

digital data service [DDS]   An AT&T communication system developed specifically for digital data, using existing local digital lines combined with data-under-voice microwave transmission facilities.

digital differential analyzer [DDA]   A differential analyzer that will synthesize a waveform digitally and perform integration.

digital echo modulation   A modem transmitter design technique in which the line signal is synthesized by generating signal elements in a time sequence. The signal elements can be stored in a digital memory as pulse-code modulation or delta-modulation samples.

digital frequency meter   A frequency meter in which the exact value of the frequency being measured is indicated on a digital display, after the number of cycles in the input signal is counted for a fixed period. The time interval is determined by a master oscillator and frequency dividers.

digital gaussmeter   A gaussmeter that measures magnetic flux density and indicates its value in gauss on a direct-reading digital display.

digital integrated circuit   An integrated circuit that is used primarily for pulse processing, as contrasted to a linear integrated circuit which provides linear amplification of signals.

digitally programmed amplifier   An amplifier whose gain can be controlled by digital signals from a computer or other source.

digitally programmed power supply   A power supply in which the value of the output voltage or current can be controlled by digital signals from a computer or other source.

digital meter   A meter that provides a direct-reading digital display, eliminating the need for reading the value represented by the position of a moving pointer on a dial. The meter includes circuits for sampling a measured analog quantity, converting the instantaneous value to digital form, and presenting it as a continuously updated display that can use light-emitting diodes, liquid crystals, cold-cathode indicators, or other display devices. A three-digit display provides readings up to 999, with or without a floating decimal. A 3½-digit display has a fourth digit position, commonly limited to 1 for reading up to 1999, but some displays give limits of 2999 or 3999. Polarity is sometimes also indicated.

digital micromirror device [DMD]   An active semiconductor device consisting of a matrix of about half a million movable mirrors, each measuring 4×4 μm and mounted independently at two corners above a memory IC chip with microscopic torsion hinges. The computer-generated signals switch the memory cells on the device, causing selected mirrors to tilt toward the television projection screen (digital 1) or away from it (digital 0) to form the moving colored image at any instant of time. When red, blue, or green light is focused on the DMD, each mirror forms a *pixel,* whose color is that of the incident light at that moment. The micromirrors are *micromachined* on a substrate measuring approximately 2.5 cm square. It is intended to replace the cathode ray tube in projection TV systems. Image brightness is related to the frequency of pixel selection during each frame. The device is also known as a *digital micromirror display device.*

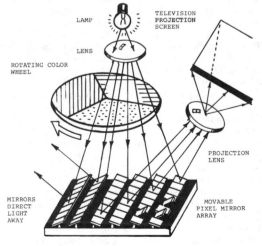

Digital micromirror device (DMD) contains nearly half a million micromirrors that can form moving TV images for projection on a large screen.

digital microwave radio Transmission of voice and data signals in digital form on microwave links, as in the 2-GHz common-carrier bands. Pulse-code modulation is used.

digital modulation A method of placing digital traffic on a microwave system without use of modems, by transmitting the information in the form of discrete phase or frequency states determined by the digital signal.

digital multimeter [DMM] A test instrument capable of measuring at least the five basic electrical variables (AC and DC voltage, AC and DC current, and resistance in ohms) and displaying the results on a digital display, typically liquid crystal (LCD). Some models can also measure and display capacitance values and frequency, test diodes, and perform audible continuity checks. Analog values are converted to digital values for display. Some can generate and transmit the measurement values in ASCII code. Measurement functions in handheld, battery-powered DMMs are switched with either a rotary switch or keypad. The principal classes of DMMs today are AC-line-powered benchtop and battery-powered handheld models.

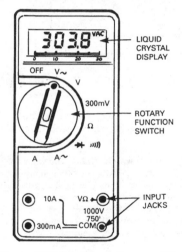

Digital multimeter (DMM): basic models measure AC and DC voltage, AC and DC current, and resistance; many also test diodes and perform audible continuity checks.

digital multiplier A multiplier that accepts two numbers in digital form and gives their product in the same digital form, usually by making repeated additions. The multiplying process is simpler if the numbers are in binary form wherein digits are represented by a 0 or 1.

digital output An output signal consisting of a sequence of discrete quantities coded in an appropriate manner for driving a printer or digital display.

digital panel meter [DPM] An electronic instrument containing analog-to-digital conversion circuitry and a digital display in a case designed for mounting in a panel cutout. Battery or line-powered, it can be modified to display values of a wide range of physical variables such as voltage, current, power, rpm, temperature, or flow rate. The displays are typically liquid crystal (LCD) or light-emitting diode (LED). Some versions also contain circuit boards for the processing of low-level signals from thermocou-

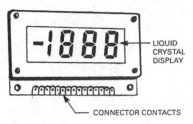

Digital panel meter (DPM): a basic model contains an analog-to-digital converter and a digital display; some also provide digital signal output.

ples, flowmeters, RTDs, and other sensors, as well as provisions for generating ASCII conversions of those measurements for transmission to a central control computer.

digital phase shifter A phase shifter whose control pulse provides a predetermined amount of signal-phase shift, with the polarity of the pulse determining the direction of the shift.

digital plotter A recorder that produces permanent hard copy in the form of a graph from digital input data.

digital printer See *dot-matrix printer, daisy-wheel printer, ink-jet printer* and *laser printer*.

digital readout *Digital display.*

digital recording Magnetic recording in which the information is first coded in a digital form, generally with a binary code that uses two discrete values of residual flux. In nonreturn-to-zero recording, the two values correspond to saturation of the tape in opposite directions. In return-to-zero recording, the tape either is saturated in one direction or is in a neutral or biased state.

digital remote control A battery-powered, handheld, infrared transmitter that controls consumer entertainment products such as television receivers, video cassette recorders, and stereo systems. A keypad contains the function keys that perform channel and volume selection and turn the equipment on and off. The entertainment product being remotely controlled receives coded infrared *commands* from the remote control and internal circuitry carries them out. *Universal digital remote controls* can control many different brands and types of consumer entertainment products.

digital satellite system [DSS] An all-digital direct satellite broadcast system that transmits high-quality television signals to privately owned receivers each equipped with an 18-in (45.72-cm) diameter paraboloidal antenna. The scrambled signals can be decoded by the subscribers' receivers. Up to 175 channels are available with this system. The antenna must be mounted outdoors or in a protective RF-transparent shelter and pointed at the satellite.

digital selective calling [DSC] A nautical version of electronic messaging or Call Waiting. A DSC transmitter sends a brief digitally encoded emergency message automatically over VHF channel 70 to alert a DSC-equipped ship that it has sent a message intended for it.

digital scrambler A scrambler used in suppressed-carrier digital modulation systems to modify the bit pattern in a nearly random manner to minimize interference errors. An inverse descrambling operation is required at the receiving end of the system.

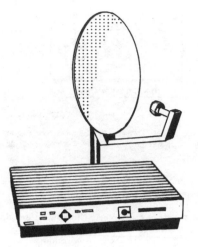

Digital Satellite System (DSS) receiver provides up to 175 channels for home reception. It has an 18-in diameter antenna reflector.

digital service unit [DSU]   The interface between a user's data terminal and a digital data service.

digital set-top converter   A consumer electronics appliance that can be placed on a television receiver to perform typical cable conversion functions, as well as decompress and decode digital signals. Some versions are expected to include keypads that will permit interaction between the consumer and the service provider, such as making requests for pay programs, responding to questionnaires, or participating in live cable programs in progress.

digital signal processing [DSP]   The processing of signals in digital-format technology, typically after conversion from an analog format. DSP techniques offer greater stability and accuracy than analog processing. It is used to analyze, enhance, filter, modulate, or otherwise manipulate signals that originate in analog form: images, sounds, and radar pulses. DSP offers better recovery for weak or fading signals and faster and more efficient processing in systems such as sonar, radar, and telecommunications.

digital simultaneous voice and data [DSVD]   A proposed standard that will allow voice and data communications to be transmitted over the same wire or cable simultaneously. It is intended for use in teleconferencing to allow two people to work on the same document from two different locations while they discuss it as if they were talking over a duplex telephone line.

digital speech communication   The transmission of voice messages in binary or other digital form by wire lines or radio.

digital synchronometer   A time comparator that provides a direct-reading digital display of time with high precision by making accurate comparisons between its own digital clock and high-accuracy time transmissions from WWV or a loran-C station.

digital telemetering   The telemetering of data in digital form over wire or radio links.

digital telephone   A telephone terminal that digitizes a voice signal for transmission and decodes a received digital signal back to a voice signal. It will usually multiplex 64

kilobits per second [kbps (kb/s)] and separate data inputs at multiples of 8 kbps.

digital television converter   A converter used to convert television programs from one system to another, such as for converting 525-line 60-field U.S. broadcasts to 625-line 50-field European PAL or SECAM standards. The video signal is digitized before conversion.

digital thermometer   A thermometer in which the measured value of temperature is shown on a direct-reading digital display.

digital-to-analog [D/A]   Pertaining to systems having analog input and digital output.

digital-to-analog converter [D/A converter or DAC]   A circuit that converts a digital signal input to an analog signal output. It can be a discrete-component module, hybrid circuit, or monolithic integrated circuit.

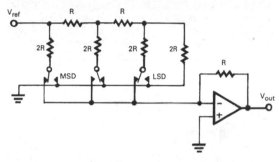

Digital-to-analog converter (DAC) converts input binary-coded signals to an equivalant analog signal. The R-2R (DAC) is illustrated.

digital-to-synchro converter [D/S converter or DSC]   A converter that changes BCD or other digital input data to a three-wire synchro output signal representing corresponding angular data.

digital transducer   A transducer that measures physical quantities and transmits the information as coded digital signals rather than as continuously varying currents or voltages.

digital video disk [DVD]-ROM   An optical storage disk with a 4.7- to 8.5-Gbyte capacity intended as a replacement for the compact-disk (CD) ROM.

digital video disk [DVD]-ROM drive   A drive for the DVD optical storage disk intended as a replacement for the compact-disk (CD) ROM. It is expected to be compatible with both CD-ROM and audio and video CDs. Stand-alone and one-time recordable versions are planned.

digital voltmeter [DVM]   A voltmeter in which the measured value of voltage is shown on a direct-reading digital display, usually with polarity sign as well as decimal point.

digital volt-ohm-milliammeter [DVOM]   A volt-ohm-milliammeter in which a digital display serves in place of an indicating meter.

digital watch   A watch that displays time as segmented digits rather than with moving hands and has crystal-controlled oscillator timing circuitry rather than spring-wound or self-winding mechanical clockwork. It contains analog-to-digital conversion and display-driver circuitry and a display, typically liquid-crystal (LCD), all powered by a miniature button power cell. By contrast, an *electronic* *watch* also has crystal-controlled electronic time-keeping circuitry, but it has a moving-hand time display and a rotating mechanical date display.

digitize   To convert an analog measurement of a quantity into a numerical value.

dilution   Reducing the intensity of a color by adding white.

dimmer   An electric or electronic control for varying the intensity of a lamp or other light source.

DIN   Abbreviation for *Deutsche Industrie Normenausschus.*

DIN jack   A class of five-conductor female connectors for computers, tape recorders, and audio components based on the European DIN standards.

DIN plug   A class of five-conductor male connectors for computers, tape recorders, and audio components based on European DIN standards.

diode   1. A two-terminal semiconductor (rectifying) device that exhibits a nonlinear current-voltage characteristic. It allows current to flow in one direction (*forward bias*), but blocks it in the opposite direction (*reverse bias*). Its terminals are the *anode* and *cathode.* There are two kinds of semiconductor diode: the PN junction, which has an electrical barrier at the interface between the N and P types, and a Schottky diode, whose barrier is formed between metal and semiconductor regions. See *Gunn diode, light-emitting diode* [LED], *semiconductor diode, semiconductor laser,* and *rectifier.* 2. A two-electrode electron tube, now obsolete.

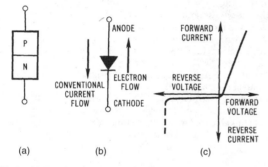

Diode: (*a*) functional diagram, (*b*) schematic showing current and electron flow, and (*c*) characteristic curve.

diode characteristic   A plot of the response curve of a diode when subjected to both forward and reverse voltages within its specified limits..

diode-connected transistor   A bipolar junction transistor whose base and collector are short-circuited together to form a diode. The diode function can also be obtained by short-circuiting transistors in an integrated circuit.

diode demodulator   A demodulator consisting of one or more diodes that provide a rectified output whose average value is proportional to the original modulation. It is also called a *diode detector.*

diode detector   *Diode demodulator.*

diode gate   An AND gate that uses diodes as switching elements.

diode laser   *Semiconductor laser.*

diode limiter   A peak-limiting circuit employing a diode that becomes conductive when signal peaks exceed a predetermined value.

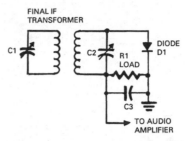

Diode demodulator or diode detector circuit rectifies a radio-frequency signal.

diode logic   Logic circuitry with a diode in each input lead configured so that the bias voltage on each diode can be varied to turn the diodes on or off.

diode matrix   A matrix of diodes used between a set of input connections and a set of output connections. Usually used for code conversion.

diode mixer   A mixer that uses a crystal or semiconductor diode. It is generally small enough to fit directly into an RF transmission line.

diode modulator   A modulator that uses one or more diodes to combine a modulating signal with a carrier signal. Used chiefly for low-level signaling because of its inherently poor efficiency.

diode ring modulator   A matched set of four diodes connected cathode-to-anode as a ring for use as a modulator or demodulator.

diode-transistor logic   [DTL]   Logic in which each input diode of a gate circuit performs an AND or an OR function to control the base current of a bipolar transistor that provides power gain for driving additional gates.

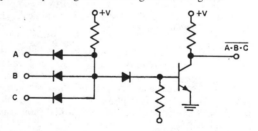

Diode-transistor logic as used for a basic NAND gate.

DIP   Abbreviation for *dual in-line package.*

diplexer   A coupling system that allows two different transmitters to operate simultaneously or separately from the same antenna.

diplex radio transmission   The simultaneous transmission of two signals on a common carrier wave.

diplex reception   Simultaneous reception of two signals that have some feature in common, such as a single receiving antenna or a single carrier frequency.

dip meter   An absorption wavemeter that contains solid-state bipolar or field-effect transistor circuitry.

dipole   *Dipole antenna.*

dipole antenna   An antenna approximately one half-wavelength long, split at its electrical center for connection to a transmission line. The impedance of the antenna is

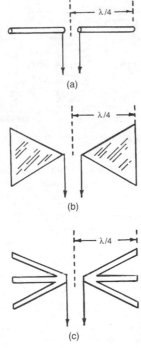

(a)

(b)

(c)

Dipole antennas are broadband radiators: examples are (a) conventional, (b) triangular, and (c) fan.

about 72 Ω. The radiation pattern is a maximum at right angles to the axis of the antenna. Also called a dipole antenna, doublet, doublet antenna, and half-wave dipole.

dipole moment A term that specifies mathematically the field caused by a given distribution of electric or magnetic charges.

dipping sonar An airborne (typically helicopter-transported) sonar system whose transducer is lowered by winch to various levels below the ocean surface for listening to and determining the coordinates of a submerged submarine. It is generally deployed where there is a high probability of the presence of an unidentified submarine, as determined by sound echoes from a cluster of *sonobuoys*. Its transducer is located in a streamlined torpedo-shaped case with stabilizing fins so that it tracks smoothly along the desired underwater course dragged by the helicopter's suspension/umbilical cable.

dip soldering A manufacturing process in which pretinned and fluxed metallic surfaces on components are soldered to printed circuit boards or other insulating substrates by bringing both component and circuit board into contact with molten solder. The component can be leaded or have tabs or exposed metal surfaces such as on leadless surface-mount components.

dipulse Pulse transmission in which the presence of one cycle of a sinewave tone represents a binary 1 and the absence of one cycle represents binary 0.

direct conversion receiver A module for a cable or a satellite *digital* (TV) *set-top box* [DSTB] that allows the user to connect to an RF source and receive the digital output.

direct-coupled amplifier [always spell out, to avoid confusion with DC amplifier] A DC amplifier in which a resistor or a direct connection provides the coupling between stages so the DC component of a signal is preserved. The frequency response starts at zero frequency (DC) and extends to some specified upper limit.

direct-coupled transistor logic [DCTL] Logic in which transistors in gate, flip-flop, and inverter circuits are coupled together directly, without resistors or other coupling components.

direct coupling The coupling of two circuits by means of a nonfrequency-sensitive device such as a wire, resistor, or battery so both direct and alternating current can flow through the coupling path. In a coaxial cavity, direct coupling is achieved by connecting directly to the center conductor in the cavity.

direct current [DC] An electric current that flows in one direction. For direct-current terms, see DC entries.

direct digital control [DDC] Control of a process or machine by digital computer. It is also called *direct numerical control*.

direct distance dialing [DDD] A telephone exchange service that allows a telephone user to dial subscribers outside the local area.

directed branch A branch having an assigned direction.

direct inductive coupling Coupling by inductance that is common to the two circuits. One circuit can be connected directly to a tap on a coil in the other circuit.

directional antenna An antenna that radiates and/or receives most transmitted energy in one specific direction. Consequently, it must be aimed at the intended transmitter/receiver for the most efficient transmission and/or reception.

directional beam Electromagnetic or acoustic energy concentrated in a given direction.

directional characteristic The variation in the behavior of a transducer or other device with respect to direction.

directional coupler A waveguide device that extracts a fixed small fraction of the energy flowing in one direction

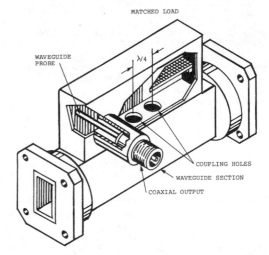

Directional coupler extracts a small measure of the energy flowing in one direction to measure power.

in the waveguide to determine the output power of a microwave system. That energy is used to actuate a *wattmeter*. By the proper use of one or more directional couplers, reflected signal power can be prevented from affecting the accuracy of power measurements. It is typically a short section of a waveguide coupled to the main waveguide by means of two small holes or *apertures* whose centers are a quarter wavelength apart. The short section contains a matched load in one end and a coaxial transition in the other end. The degree of coupling between the main waveguide and the short section is determined by the aperture diameters. See also *bidirectional coupler*.

directional hydrophone   A hydrophone whose response varies significantly with the direction of sound incidence.

directional microphone   A microphone whose response varies significantly with the direction of sound incidence.

directional phase shifter   A passive phase shifter in which the phase change for transmission in one direction differs from that for transmission in the opposite direction.

directional response pattern   *Directivity pattern.*

direction finder [DF]   *Radio direction finder.*

direction-finder deviation   The difference between the observed radio bearing obtained with a direction finder and the true bearing of the transmitter.

direction-finding station   A radio station that has equipment for determining the direction of arrival of radio waves. Two or more such stations working together can determine the location of the transmitter by triangulation.

direction of polarization   The direction of electric lines of force or the electric vector in a polarized wave.

direction of propagation   The direction of time-average energy flow at any point in a homogeneous isotropic medium. For a linearly polarized wave it is the direction of the electric vector. In a uniform waveguide the direction of propagation is often taken along the axis.

directive gain   An antenna rating equal to 4π times the ratio of the radiation intensity in a given direction to the total power radiated by the antenna.

directivity   1. The value of the directive gain of an antenna in the direction of its maximum value. The higher the directivity value, the narrower the beam in which the radiated energy is concentrated. 2. The ratio of the power measured at the forward-wave sampling terminals of a directional coupler, with only a forward wave present in the transmission line, to the power measured at the same terminals when the direction of the forward wave in the line is reversed. The ratio is usually expressed in decibels. For a perfect coupler it would be infinitely high.

directivity factor   1. The ratio of radiated sound intensity at a remote point on the principal axis of a loudspeaker or other transducer to the average intensity of the sound transmitted through a sphere passing through the remote point and concentric with the transducer. The frequency must be stated. 2. The ratio of the square of the voltage produced by sound waves arriving parallel to the principal axis of a microphone or other receiving transducer to the mean square of the voltage that would be produced if sound waves having the same frequency and mean-square pressure were arriving simultaneously from all directions with random phase. The frequency must be stated. Directivity factor in acoustics is equivalent to directivity as applied to antennas.

directivity index   The directivity factor expressed in decibels. It is 10 times the logarithm to the base 10 of the directivity factor. Also called directional gain.

directivity pattern   A graphical or other description of the response of a transducer used for sound emission or reception. It is presented as a function of the direction of the transmitted or incident sound waves in a specified plane and at a specified frequency. Also called beam pattern and directional response pattern.

directivity signal   A spurious signal present in the output of a coupler because the directivity of the coupler is not infinite.

directly ionizing particles   Charged particles having sufficient kinetic energy to produce ionization by collision. Examples include electrons, protons, and alpha particles.

direct memory access [DMA]   A computer feature, set up by the central processing unit, that provides for direct data transfer from a peripheral device to the computer memory or to magnetic disk or tape drives.

direct numerical control [DNC]   *Direct digital control.*

director   A parasitic element placed a fraction of a wavelength ahead of a dipole receiving antenna to increase the gain of the array in the direction of the major lobe. It is usually a rod slightly shorter than the receiving dipole, with no connection to the lead-in.

direct-view storage tube   A cathode-ray tube with a storage grid whose secondary electron emission provides a bright display for long, controllable periods. The tube includes one or more writing guns, a flooding gun, and circuitry for selectable erasing. Its applications have largely been preempted by digital oscilloscopes with semiconductor memory devices that can store images for longer periods more efficiently.

direct voltage   *DC voltage.*

disc   Alternate spelling for *disk*.

discharge   1. The passage of electricity through a gas, usually accompanied by a glow, arc, spark, or corona.   2. To remove a charge from a battery, capacitor, or other electric energy storage device.   3. The conversion of chemical energy to electric energy in a battery.

discone antenna   A biconical antenna with one of its cones spread out to 180° to form a disk. The center conductor of the coaxial line terminates at the center of the disk, and the cable shield terminates at the vertex of the cone. Both the input impedance and radiation pattern remain essentially

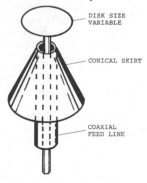

Discone antenna is a variation of the conical antenna, except that the flat disk replaces the upper cone.

constant over a wide frequency range. The disk is normally parallel to the earth, giving an omnidirectional radiation pattern in a horizontal plane.

discrete-component microcircuit   A microcircuit consisting chiefly of separate active and passive components that were manufactured before installation on the microcircuit board or substrate.

discrete device   A separate electronic part, device, or component having only one function per package. Examples include both chip and packaged resistors, capacitors, inductors, diodes, and transistors as well as specialized devices including LEDs, SCRs, phototocells, and pressure transducers.

discrete transistor   A single transistor encapsulated in a suitable package, with pins for external connections. Two transistors are sometimes formed on the same wafer to give thermal matching. A Darlington pair transistor is considered to be discrete.

discrete variable   A quantity that can assume any one of a number of individually distinct or separate values.

discrete word intelligibility   The percent intelligibility obtained when the speech units under consideration are words, usually presented so as to minimize the contextual relation between them.

discrimination   1. The degree of rejection of unwanted signals in a receiver or other equipment containing tuned circuits.   2. *Conditional jump.*

discriminator   A circuit in which magnitude and polarity of the output voltage depend on how an input signal differs from a standard or another signal. Thus a frequency discriminator converts frequency deviations from a carrier frequency into corresponding amplitude variations. A pulse-height discriminator delivers an output voltage only for pulses that exceed a predetermined height. A phase discriminator converts phase variations into corresponding amplitude variations.

discriminator transformer   A transformer designed for use in a stage where FM signals are converted directly to AF signals or in a stage where frequency changes are converted to corresponding voltage changes.

dish   A concave reflector that has a surface which is parabolic or part of a sphere, used in a microwave antenna. The periphery is usually circular.

disk   In electronics, this term generally refers to a memory disk. These include the flexible plastic oxide-coated disks, now packaged in hard plastic cases, called *floppy disks* or *diskettes* and stacked hard (aluminum) disks or platters with metallized magnetic coatings in Winchester-style hard drives. Other media disks are the compact disk (CD) and the compact-disk read-only memory (CD-ROM). Data is recorded on these disks as a series of pits formed by a laser beam. An alternative spelling for the term is *disc,* sometimes used to distinguish *magnetic memory disks* from the laser-generated CD and CD-ROM *discs.*

disk drive   A secondary or mass memory for computers that stores and retrieves digital data on one or more surfaces of disk-shaped magnetic media. The disks rotate under electromagnetic heads. A disk drive is capable of taking digital data from memory and writing it on disks, or reading it from disks and restoring it to memory. The read-write head on a disk drive alters the magnetic domains of the surface media to form small regions that represent ones or zeros. The principal types are *hard-disk drive* and *flexible-* or *floppy-disk drive.*

diskette   *Floppy disk.*

disk file   *Disk storage.*

disk operating system   [DOS]   A software program that manages the activities of a computer system, including its related peripheral equipment. It sets priorities, determines the locations of data in semiconductor and disk memory, and lists all of the files and their status. If this program is not installed in the computer's hard-disk drive at the factory it must be loaded from disks by the owner as the initial preparation for operation. The DOS must be compatible with the computer's microprocessor and with all applications software.

disk-seal tube   An electron tube with disk-shaped electrodes arranged in closely spaced parallel layers, to give low interelectrode capacitance along with high-power output up to 2.5 GHz. The edges of the disk electrodes are fused into and project through the glass or ceramic envelope, to serve as external contacts. Also called lighthouse tube.

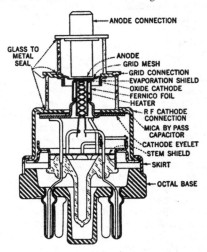

Disk-seal tube construction.

disk storage   The storage of data in digital format on a magnetic disk that can be written to, written over, or erased. It is a peripheral or embedded resource in a digital computer. The randomly accessed disk is rotated by a motor under electromagnetic read/write heads. Examples are the *hard-disk drive* (Winchester-style) and the *floppy or diskette drive.* The CD-ROM, a read-only memory disk, is not considered to be disk storage.

disk-type motor   A permanent-magnet DC motor with a lightweight glass-fiber epoxy (GFE) rotor designed for high-speed, fast response in servosystems where a more conventional rotor would have excessive inertia. The rotor is made by pressing stamped copper coil windings into the disk rotor, which is mounted at right angles to the drive shaft. The magnetic field is produced by opposing sets of discrete cylindrical magnets. The motor case has a flat "pancake" form factor, making it suitable for applications where space is restricted, as in industrial robots. It is also called a *pancake motor* and a *printed-circuit motor,* reflecting an earlier rotor manufacturing process.

dislocation An imperfection in the geometric arrangement of atoms in a crystal.

disperse A data-processing operation in which grouped input items are distributed among a larger number of groups in the output.

dispersion 1. The process of separating radiation into components having different frequencies, energies, velocities, or other characteristics. A prism or diffraction grating disperses white light into its component colors. A magnetic field disperses or sorts electrons according to their velocities. 2. Scattering of microwave radiation by an obstruction. 3. A distribution of finely divided particles in a medium.

dispersive line A delay line that delays each frequency a different length of time. Both quartz and aluminum delay lines can be designed to have this property.

dispersive medium A medium in which the phase velocity of an electromagnetic wave is a function of frequency. A plasma is a dispersive medium, but free space is not because waves of all frequencies travel in space with the velocity of light.

displacement current A hypothetical current assumed to exist in the presence of time-varying electric fields. It was postulated by Maxwell to explain the transfer of current through space between capacitor plates.

displacement gyroscope A gyroscope that senses, measures, and transmits angular displacement data.

displacement law *Radioactive displacement law.*

displacement transducer A transducer that converts a linear or angular movement into an electric signal. Methods of accomplishing this conversion include the use of variable-inductance, variable-resistance, variable-capacitance, and electron-tube or transistorized circuits.

display A visual presentation of output information, as on the screen of a cathode-ray tube or in readable characters of a digital display.

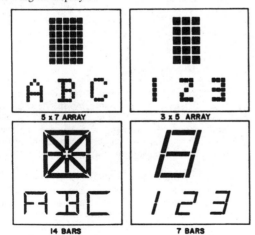

Display styles obtained with various configurations of light-emitting diodes or other light sources.

display loss *Visibility factor.*

display storage tube A storage tube that accepts information introduced as an electric signal to be read at a later time as a visible output.

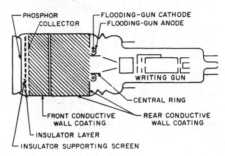

Display storage tube having writing gun for creating characters of display and larger coaxial flooding gun for spraying electrons in broad beam to intensify pattern created by writing gun.

disruptive discharge A sudden and large increase in current through an insulating medium due to complete failure of the medium under electrostatic stress.

dissector tube *Image dissector.*

dissipation An undesired loss of energy, generally by conversion into heat. Thus, the collector dissipation rating in watts for a transistor is the maximum amount of energy that can be lost as heat at the collector electrode without damage to the transistor.

dissipation factor The reciprocal of Q.

dissipation line A length of stainless steel or Nichrome wire used as a noninductive terminating impedance for a rhombic transmitting antenna when several kilowatts of power must be dissipated.

dissociative recombination The capture of an electron by a positive molecular ion. The electron combines with the ion and dissociates it into two neutral atoms.

dissolve The merging of two television camera signals in such a way that as one scene disappears, another slowly appears.

dissonance An unpleasant combination of harmonics heard when certain musical tones are played simultaneously.

distance measuring equipment [DME] A two-way aircraft ranging system that includes an airborne interrogator and a ground-based transponder. The airborne interrogator transmits 3.5 µm, 1 kW pulses at the rate of 30/s on one of 126 channels which are 1 MHz apart in the 1025- to 1150-MHz band. The transponder replies with similar pulses on another channel 63 MHz above or below the interrogating channel. The signal received by the aircraft is compared with its transmitted signal, their time difference is derived, and distance is computed and displayed. There are about 2000 ground stations worldwide. See also *Tacan.*

distance resolution The ability of a radar system to distinguish between two separate targets (aircraft or ships) that are close together. A system with poor distance resolution will combine both targets as a single return on a *plan-position indicator* [PPI] screen. A system with good distance resolution will show separate returns for each target at long range. It is usually high in radars with narrow

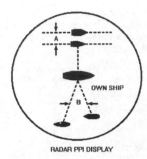

RADAR PPI DISPLAY

Distance resolution (A) is illustrated by the separation of two ships, and bearing resolution (B) by the separation of two ships, for radar on own ship.

beamwidths and high operating frequencies. However, it can be degraded by the presence of rain or fog.

distortion 1. Any undesired change in the waveform of a signal. 2. Any undesired deviation of an image from proportionality with the original scene.

distortion analyzer An analyzer that measures total harmonic distortion of an audio signal by removing the fundamental frequency with a narrow-band rejection filter and measuring the amplitude of the remaining components for comparison with that of the fundamental frequency.

distortion meter An instrument that measures distortion of a sinusoidal signal by removing the fundamental-frequency component and measuring any harmonics that remain.

distress frequency A frequency assigned to distress calls, generally by international agreement. For ships at sea and aircraft over the sea, it is 500 kHz.

distress signal The international signal used when a ship, aircraft, or other vehicle is threatened by grave and imminent danger and requests immediate assistance. In radiotelegraphy, it consists of three dots, three dashes, and three dots (the international code for SOS) transmitted as a single signal in which the dashes are emphasized. In voice radio, it is the word "mayday."

distributed capacitance The capacitance that exists between adjacent turns in a coil or between adjacent conductors in a cable. Also called self-capacitance.

distributed constant A circuit parameter that exists along the entire length of a transmission line. For a transverse electromagnetic wave on a two-conductor transmission line, the distributed constants are series resistance, series inductance, shunt conductance, and shunt capacitance per unit length of line.

distribution coefficients The tristimulus values of monochromatic radiations having equal power.

distribution control *Linearity control.*

disturbance 1. An undesired interference or noise signal affecting radio, television, or facsimile reception. 2. An undesired command signal in a control system.

dither A force having a controlled amplitude and frequency, applied continuously to a device driven by a servomotor so the device is constantly in small-amplitude motion and cannot stick at its null position. Used in some

recorders to make the pen ready to move instantly. Also called buzz.

dither injector *Antistiction oscillator.*

dither-tuned magnetron A magnetron that has a motor-driven tuning plunger which provides rapid repetitive tuning over a narrow band as required for reception of frequency-agile radar signals that have a different frequency for each pulse.

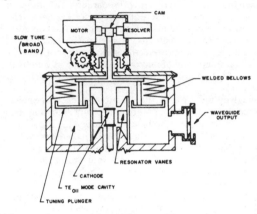

Dither-tuned magnetron in which motor-driven cam moves tuning plunger up and down rapidly in output cavity.

divalent silver-oxide cell A silver-oxide primary cell in which highly active divalent silver oxide replaces the monovalent type as a depolarizing cathode, to increase power output.

divergence The spreading of a cathode-ray stream due to repulsion of like charges (electrons).

divergence loss The portion of the transmission loss in an acoustic system caused by the divergence or spreading of sound rays.

diversity radar A radar that uses two or more transmitters and receivers, each pair operating at a slightly different frequency but sharing a common antenna and video display, to obtain greater effective range and reduce susceptibility to jamming.

diversity receiver A radio receiver designed for space or frequency diversity reception.

diversity reception Radio reception in which the effects of fading are minimized by combining two or more sources of signal energy carrying the same modulation. Space diversity takes advantage of the fact that fading does not occur simultaneously for antennas spaced several wavelengths apart. Frequency diversity takes advantage of the fact that signals differing slightly in frequency do not fade simultaneously.

divided-carrier modulation Modulation in which the carrier is divided into two components that are 90° out of phase, each modulated by a different signal. When these components are added, the frequency is unchanged, but the resulting signal varies in amplitude and phase in accordance with the modulating signals.

divider A circuit or device capable of dividing one quantity or a variable by another or by a fixed number such as 2 or 10.

dividing network   *Crossover network.*

DKDP   Abbreviation for *deuterated potassium dihydrogen phosphate.*

D layer   The lowest layer of ionized air above the earth, occurring in the D region only in the daytime hemisphere. It reflects frequencies below about 50 kHz and partially absorbs higher-frequency waves.

DLC   Abbreviation for *data-link control.*

DLM   Abbreviation for *double-level metal.*

DMA   Abbreviation for *direct memory access.*

DMD   Abbreviation for *digital micromirror display.*

DME   Abbreviation for *distance-measuring equipment.*

DMM   Abbreviation for *digital multimeter.*

DMOS   Abbreviation for *double-diffused metal-oxide semiconductor.*

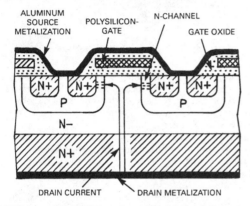

DMOS power MOSFET is made with multiple parallel gate and source cells so that it can handle high currents.

DMOS power MOSFET   A power MOSFET made with the vertical DMOS process. It has multiple silicon gate cells that are formed by an integration process to distribute current and dissipate heat. Cell density can exceed a half million per square inch. It has a common parallel-connected source and a common drain on the under surface of the die.

DNC   Abbreviation for *direct numerical control.*

DOC   Abbreviation for *Department of Communications,* the communications regulatory agency of Canada.

docking   The process of bringing two spacecraft together in space.

document   The source of data for input into a computer.

documentation   The flowcharts, written procedures, and operator instructions produced with a computer program, for guidance in making later modifications and for running the program on a computer.

DOD   Abbreviation for *direct outward dialing.*

Doherty amplifier   A linear RF power amplifier that is divided into two sections whose inputs and outputs are connected by quarter-wave networks. Operating parameters are so adjusted that for all values of input signal voltage up to one-half maximum amplitude, section No. 1 delivers all the power to the load. Above this level, section No. 2 comes into operation. At maximum signal input, both sections are operating at peak efficiency, and each

section is delivering half the total output power to the load.

Dolby system   A noise-reduction system developed by Dr. Ray M. Dolby to reduce hiss and other high-frequency noise originating in magnetic tape. During quiet passages the level of tape noise is comparable to that of the music, whereas during loud passages the noise is masked by the music. The Dolby system provides a predetermined amount of extra amplification for low levels of the higher audio frequencies during recording, with corresponding attenuation during playback to restore the music to its correct level while reducing tape noise. In the A-type professional Dolby version, the audio spectrum is divided into four bands, with the gain in each band automatically varied with signal level. The simplified B-type version developed for home tape recorders and cassette decks uses only one frequency range, extending upward from about 600 Hz.

dolorimeter   An instrument for measuring pain, based on the dol as the unit of pain intensity.

domain   *Magnetic domain.*

dome   An enclosure for a sonar transducer, projector, or hydrophone and associated equipment. It is designed to have minimum effect on sound waves traveling under water.

dominant mode   *Fundamental mode.*

dominant wave   The electromagnetic wave that has the lowest cutoff frequency in a given uniconductor waveguide. It is the only wave that will carry energy when the excitation frequency is between the lowest cutoff frequency and the next higher cutoff frequency.

dominant wavelength   The single wavelength of light that when combined in suitable proportions with a reference standard light matches the color of a given sample.

donor   An impurity from column V of the periodic table which adds a mobile electron to the conduction of silicon, making it more N-type. Commonly used donors are arsenic and phosphorous. See *acceptor.*

donor impurity   *Donor.*

donor level   An intermediate energy level close to the conduction band in the energy diagram of an extrinsic semiconductor.

dopant   An impurity element added to a semiconductor material under precisely controlled conditions to create PN junctions required for transistors and semiconductor diodes. Also called doping agent.

doped junction   A junction produced by adding an impurity to the melt during growing of a semiconductor crystal.

doped-junction transistor   *Grown-junction transistor.*

doping   The process of adding a chemical impurity (dopant) to the crystal structure of a semiconductor when it is in the molten state to modify its electrical properties. For example, adding boron to silicon makes it more P-type. Concentrations can range from a few parts per billion (for resistive semiconductor regions) to a fraction of a percent (for highly conductive regions).

doping agent   *Dopant.*

doping compensation   The addition of donor impurities to a P-type semiconductor or of acceptor impurities to an N-type semiconductor.

Doppler broadening   Frequency spreading that occurs in single-frequency radiation when the radiating atoms, molecules, or nuclei do not all have the same velocity. Each radiating particle can then give rise to a different Doppler shift.

Doppler current meter   An ultrasonic instrument that measures the speed of water. An ultrasonic beam is projected into the water and its volume reverberation signal picked up. The difference between transmitted and received frequencies, as produced by the Doppler effect, is then proportional to water speed.

Doppler effect   The change in the observed frequency of a wave due to relative motion of source and observer. When the distance between source and observer is decreasing, the observed frequency is higher than the source frequency. When the distance is increasing, the observed frequency is lower. The effect occurs for sound waves as well as radio waves.

Doppler frequency   *Doppler shift.*

Doppler radar   A radar based on Doppler shift of an echo due to relative motion of target and radar. This shift in frequency permits differentiation between fixed and moving targets. The velocity of a moving target can be determined with high accuracy by measuring the frequency shift. A Doppler radar can be either continuous wave or pulsed.

Doppler radar altimeter   A radar altimeter based on the Doppler effect.

Doppler radar guidance   Missile guidance based on a Doppler navigation system built into the missile to determine surface velocity and drift angle.

Doppler radar-inertial navigation   An airborne navigation system in which the ground velocity signals developed by a Doppler radar navigation system are used to null accumulative errors of the inertial navigation system.

Doppler radar navigation   An airborne navigation system based on the Doppler effect to determine drift and ground speed. Four beams of pulsed microwave energy are beamed toward the earth along the corners of an imaginary pyramid whose peak is at the aircraft. Echoes from the front-pointing beams undergo upward Doppler shift, and echoes from the rearward beams undergo downward Doppler shift. Similarly, drift causes Doppler shift of echoes from beams on one side with respect to beams on the other side of the aircraft. Comparison of the Doppler shifts in a computer provides complete information for navigation even in zero-visibility weather, at all altitudes, without reference to ground stations.

Doppler shift   The amount of the change in the observed frequency of a wave due to Doppler effect, expressed in hertz. Also called Doppler frequency.

Doppler VOR   A VHF radio range that is compatible with conventional VOR but is much less critical in site requirements. The variable signal that produces azimuth information is developed by feeding an RF signal sequentially to the elements of a ring-shaped antenna array.

doroid   A coil that resembles a half toroid and a removable core segment to simplify the winding process.

DOS   Abbreviation for *disk operating system.*

dosage   *Dose.*

dose   The amount of ionizing radiation delivered to a specified volume, measured in rads, reps, or rems. For X-rays or gamma rays having quantum energies up to 3 MeV, the roentgen unit may be used. Also called absorbed dose and dosage.

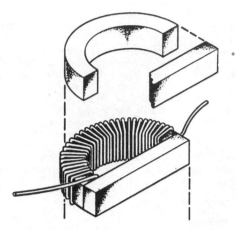

Doroid coil, showing the method of assembling the core.

dose equivalent   The product of absorbed dose in rads and a number of modifying factors due to nonuniform distribution of internally deposited isotopes in radiobiology. The unit of dose equivalent is the rem.

dosemeter   *Dosimeter.*

dose rate   The rate at which ionizing radiation is absorbed by a material. It can be measured as an ionizing dose rate in rads material per second.

dose-rate meter   An instrument that measures radiation dose rate.

dosimeter   An instrument that measures the total dose of nuclear radiation received in a given period. Also called dosemeter.

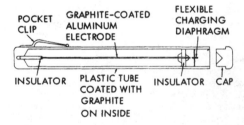

Dosimeter. A central wire and graphite-coated plastic case together form capacitor that is initially charged by DC voltage. Ionizing radiation dissipates the charge. It can measure remaining charge to determine the wearer's dosage.

dosimetry   The measurement of radiation doses.

dot cycle   A mark pulse followed by an equal-length space pulse element in teletypewriter systems.

dot generator   A signal generator that produces a dot pattern on the screen of a three-gun color television picture tube, for use in convergence adjustments. When convergence is out of adjustment, the dots occur in groups of three, one for each of the receiver primary colors. When convergence is correct, the three dots of each group converge to form a single white dot.

dot matrix   A method of generating characters with a matrix of dots.

Dot-matrix characters and numerals produced with 5 × 7 format.

dot-matrix printer A printer that has a printhead with a vertical row of five or seven stiff wires which can be pushed against an inked ribbon under the control of individual solenoids, to make dots on paper. As the row of wires is moved horizontally, the wires are appropriately energized so the resulting 5 × 7 or 7 × 9 dot-matrix pattern forms the desired ASCII characters. In a thermal-matrix printer the character-forming dots are produced by heat rather than impact. It is also called a *matrix printer*.

dot-sequential color television A color television system in which the red, blue, and green primary-color dots are formed in rapid succession along each scanning line.

double-amplitude-modulation multiplier A multiplier in which one variable is amplitude-modulated by a carrier, and the modulated signal is again amplitude-modulated by the other variable. The resulting double-modulated signal is applied to a balanced demodulator to obtain the product of the two variables.

double-balanced mixer A balanced mixer that uses accurately matched components, balanced transformers, and the bridge or ring configuration of diodes to minimize conversion loss and reduce oscillator radiation from the antenna in critical applications using superheterodyne receivers.

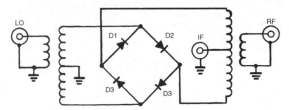

Double-balanced mixer minimizes conversion loss and reduces oscillator radiation from the antenna.

double-break switch A switch that opens a conductor at two different points.

double bridge *Kelvin bridge.*

double-button carbon microphone A carbon microphone with two carbon-filled buttonlike containers, one on each side of the diaphragm, to give twice the resistance change obtainable with a single button. Also called differential microphone.

double-channel duplex A method that provides for simultaneous communication between two stations through the use of two RF channels, one in each direction.

double-channel simplex A method that provides for nonsimultaneous communication between two stations through the use of two RF channels, one in each direction.

double-conversion receiver *Double superheterodyne.*

double-diffused metal-oxide semiconductor [DMOS] A metal-oxide semiconductor manufacturing process in-

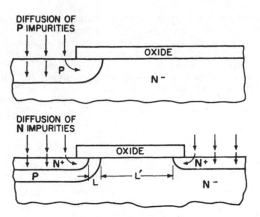

Double-diffused metal-oxide semiconductor process.

volving two-stage diffusion of impurities through a single mask opening. Either depletion or enhancement modes can be produced. Short diffusion times keep channel lengths short, as required for production of diodes and transistors for high-speed logic and microwave applications.

double-diffused transistor A transistor in which two PN junctions are formed in the semiconductor wafer by gaseous diffusion of both P- and N-type impurities. An intrinsic region can also be formed.

double-diode *Duodiode.*

double-layer supertwisted-nematic LCD A variation on twisted-nematic LCDs that aligns the liquid crystal molecules so that they twist from 90° to as much as 270°. The greater twist requires a more nonlinear voltage, allowing more lines to be displayed.

double-level metal [DLM] An integrated-circuit metal interconnect process that employs two vertical levels of metal, separated by a dielectric or insulating layer. See *single-level metal.*

double limiter *Cascade limiter.*

double local oscillator A circuit that generates two RF signals which are accurately spaced several hundred hertz apart and mixes these signals to give an accurate audio frequency for use as a reference.

double-moding Undesirable shifting of a magnetron from one frequency to another at irregular intervals.

double modulation A method of modulation in which a subcarrier is first modulated with the desired intelligence, and the modulated subcarrier is then used to modulate a second carrier having a higher frequency.

double-pole double-throw [DPDT] A six-terminal switch or relay contact arrangement that simultaneously connects one pair of terminals to either of two other pairs of terminals.

double-pole-piece magnetic head A magnetic head with two pole pieces, opposite in polarity and mounted on opposite sides of the magnetic recording medium. Either or both can have an energizing winding.

double-pole single-throw [DPST] A four-terminal switch or relay contact arrangement that simultaneously opens or closes two separate circuits or both sides of the same circuit.

double-precision number *Double-length number.*

double-pulse recording Magnetic recording in which each stored bit consists of two regions magnetized in opposite polarities, with unmagnetized regions on both sides. The order of the polarity determines whether the bit is a 1 or a 0.

doubler *Frequency doubler.*

double-sideband reduced-carrier transmission [DSBRC transmission] Double-sideband transmission in which the useless RF power in the carrier is reduced and the power in the intelligence-carrying sidebands correspondingly increased.

Double-sideband reduced-carrier transmission spectrum, in which carrier power is reduced and power in sidebands increased without exceeding FCC limits for average power output.

double-sideband transmission The transmission of a modulated carrier wave accompanied by both of the sidebands resulting from modulation. The upper sideband corresponds to the sum of the carrier and modulation frequencies, whereas the lower sideband corresponds to the difference between the carrier and modulation frequencies.

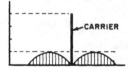

Double-sideband transmission in which carrier is 100% amplitude-modulated, giving upper and lower sidebands that carry useful intelligence.

double-stub tuner A tuner consisting of two stubs, usually ⅜ wavelength apart, connected in parallel with a transmission line. Used for impedance-matching.

doublet *Dipole.*

doublet antenna *Dipole.*

double-throw switch A switch having two operating positions, each providing different circuit connections.

double-track tape recording A magnetic recording in which two adjacent tracks are placed on the tape either for stereophonic reproduction or for doubling playing time by recording the second monophonic track in the opposite direction.

doublet trigger A trigger signal consisting of two pulses spaced a predetermined amount for coding purposes.

double-tuned amplifier An amplifier whose stages are tuned to two different resonant frequencies to obtain wider bandwidth than is possible with single-frequency tuning.

double-tuned detector An FM discriminator whose limiter output transformer has two secondaries, one tuned above the resting frequency and the other tuned an equal amount below. Without modulation, both diodes conduct equally at the resting frequency, and the audio output is zero. Signal frequency deviation makes one diode conduct more than the other, giving audio output.

double-vee antenna A variation of the open dipole consisting of a pair of rods, each bent in the shape of a V and

mounted horizontally. The transmission line is fed at the apex. It is suitable for receiving the FM broadcast band.

double zepp antenna A straight, full-wavelength active element with maximum radiation at right angles to the antenna. See also *zepp antenna.*

downconverter A converter that changes an incoming modulated or unmodulated carrier frequency to a lower frequency which is within the tuning range of a receiver or radio test set.

down counter A pulse counter that starts at its maximum count and decreases one count at a time to zero. Thus, a modulus 16 binary down counter decreases in 16 steps from 1111 to 0000.

down-lead *Lead-in.*

downlink The radio or optical transmission path downward from a communication satellite to the earth or an aircraft, or from an aircraft to the earth. The upward path is the uplink.

download The process of copying a file of data, graphics, or software from one computer to another for reading or printout. The source can be the Internet, a mainframe or server computer in a network, or between any two compatible personal computers.

downrange Any area along the flight course of a missile. Downrange tracking stations report on missile flight behavior and receive telemetered data from the missile.

downward modulation Modulation in which the instantaneous amplitude of the modulated wave is never greater than the amplitude of the unmodulated carrier.

Dow oscillator *Electron-coupled oscillator.*

DPDT Abbreviation for *double-pole double-throw.*

DPM Abbreviation for *digital panel meter.*

DPSK Abbreviation for *differential phase-shift keying.*

DPST Abbreviation for *double-pole single-throw.*

DQPSK Abbreviation for *differential-quadrature phase-shift keying.*

drain One of the three regions that form a field-effect transistor. Majority carriers that originate at the source and traverse the channel are collected at the drain to complete the current path. The flow between the source and the drain is controlled by the voltage bias applied to the gate. The drain is analogous to the collector of a bipolar junction transistor and the anode of an electron tube. See also *channel, field-effect transistor* [FET], *gate,* and *source.*

drain wire A bare metallic conductor in contact with foil shielding of a signal cable to provide a low-resistance path to ground.

DRAM Abbreviation for *dynamic random-access memory.*

D region The lowest region of the *ionosphere,* about 25 to 56 mi (40 to 90 km) above the earth. It causes most of the attenuation of radio waves in the region of 1 to 100 MHz, so it contributes very little to shortwave radio propagation. It absorbs energy from radio waves below about 10 MHz during daylight hours. It is below the *E region.*

dress The careful arrangement of connecting wires in a circuit to prevent undesirable coupling and feedback.

drift 1. A slow change in some characteristic of a device such as frequency, balance current, direction (as in a gyro), or desired course of travel. Temperature variations are a common cause of frequency drift and unbalance in circuits. 2. The movement of current carriers in a semiconductor under the influence of an applied voltage. 3. Gradually developing changes in the offset voltage and both offset currents of an operational amplifier.

drift mobility The average drift velocity of carriers per unit electric field in a homogeneous semiconductor. Also called mobility.

drift space A space in an electron tube that is substantially free of externally applied alternating fields where repositioning of electrons takes place. In a klystron the velocity-modulated electrons form bunches in this space.

drift velocity The average velocity of an electron that is moving under the influence of an electric field. Drift velocity corresponds to the net current in an electron tube or a semiconductor device.

drip loop A downward loop formed in an antenna lead-in or other cable just before it enters a building. Water drips off at the bottom of the loop.

drive *Excitation.*

drive control *Horizontal drive control.*

driven array An antenna array consisting of many of driven elements, usually half-wave dipoles, fed in phase or out of phase from a common source. Examples include broadside, collinear, and end-fire arrays.

driven element An antenna element that is directly connected to the transmission line.

drive pulse A pulsed magnetomotive force applied to a magnetic cell from one or more sources.

driver 1. The amplifier stage preceding the output stage in a receiver or transmitter. It is also called a driver stage. 2. The part of a loudspeaker that converts electric energy into acoustic energy. 3. An amplifier for powering high-current loads. 4. A transistor whose output can drive a relay, solenoid, motor, or other high-current device. 5. *Bus drivers* that rapidly charge and discharge capacitance.

driver stage *Driver.*

driving-point admittance The complex ratio of alternating current to applied alternating voltage for an electron tube, network, or other transducer.

driving-point function A response function for which the variables are measured at the same port.

driving-point impedance The complex ratio of applied alternating voltage to the resulting alternating current in an electron tube, network, or other transducer.

driving signal A signal that times horizontal or vertical scanning at a television transmitter. Driving signals are usually provided by a central sync generator at the transmitter.

DRO Abbreviation for *destructive readout.*

drop-in An unwanted character, digit, or bit formed accidentally on a magnetic recording surface.

dropout A reduction in output signal level during reproduction of recorded data, sufficient to cause a processing error.

dropout count The number of dropouts detected in a given length of magnetic tape.

dropout current The maximum current at which a relay or other magnetically operated device will release to its deenergized position.

dropout error The loss of a recorded bit or any other error occurring in recorded magnetic tape because of foreign particles on or in the magnetic coating or defects in the backing.

dropout voltage The maximum voltage at which a relay or other magnetically operated device will release to its deenergized position.

dropping resistor A resistor used in series with a load to decrease the voltage applied to the load.

dry circuit A relay circuit that has open-circuit voltages which are very low and closed-circuit currents which are extremely small, so there is no arcing to abrade the contacts. As a result, an insulating film can develop that prevents closing of the circuit when the contacts are brought together mechanically by the relay.

dry contact A contact that does not break or make current.

dry reed relay A reed relay with a glass-encapsulated reed contact capsule that does not contain mercury. See *reed relay* and *mercury-wetted reed switch.*

dry reed switch A reed switch whose glass-encapsulated reed contact capsule does not contain mercury. See *reed switch* and *mercury-wetted reed switch.*

D/S Abbreviation for *digital-to-synchro.*

DSBRC transmission Abbreviation for *double-sideband reduced-carrier transmission.*

DSC Abbreviations for *Digital Selective Calling* and *digital-to-synchro converter.*

D/S converter Abbreviation for *digital-to-synchro converter.*

DSP Abbreviation for *digital signal processing.*

DSR Abbreviation for *data set ready.*

DSS Abbreviation for *digital satellite system.*

DSTB Abbreviation for *digital set-top box.*

DSTN Abbreviation for *dual-scan super twisted nematic display.*

DSU Abbreviation for *digital service unit.*

DSVD Abbreviation for *digital simultaneous voice and data.*

DT-cut crystal A quartz crystal cut to give a resonant frequency below about 500 kHz.

DTE Abbreviation for *data-terminal equipment.*

D3-D3 channel bank A specific generation of AT&T channel PCM terminals that multiplex 24 voice channels into a 1.544-MHz digital bit stream. The specification associated with D3 channel banks is the basis for all PCM device specifications.

DTL Abbreviation for *diode-transistor logic.*

DTL/TTL An abbreviation indicating that a circuit will operate with inputs and/or outputs that are either diode-transistor or transistor-transistor logic.

DTMF Abbreviation for *dual-tone multifrequency.*

DTR Abbreviation for *data terminal ready.*

DTX Abbreviation for *discontinuous transmission.*

dual-channel amplifier An AF amplifier with two separate amplifiers for the two channels of a stereophonic sound system, usually operating from a common power supply mounted on the same chassis.

dual-diversity receiver A diversity radio receiver whose two antennas feed separate RF systems with mixing occurring after the converter. An automatic selection system can connect the output to that channel which is stronger at each instant.

dual in-line package [DIP] A flat rectangular, leaded package for circuit board mounting of integrated circuits, relays, resistor networks, and other miniature components. DIPs for ICs can have up to 60 leads or pins that are in parallel rows at right angles to the body. A plastic DIP is made by molding epoxy resin on the stamped leadframe that contains the device or chip. Ceramic DIPs are made by brazing metal pins or leads to the sides of ceramic rectangular substrates and bonding on a separate ceramic cover. See *CERDIP*.

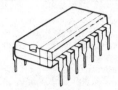

Dual in-line package for integrated circuit.

dual laser A gas laser that has Brewster windows and concave mirrors at opposite ends, the mirrors having different reflectivities to produce two different visible or infrared wavelengths from a helium-neon laser beam.

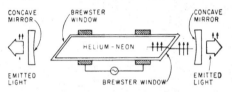

Dual laser containing helium-neon gas.

dual-mode transducer A transducer inserted into a circular waveguide to select between two types of polarization.

dual modulation The process of modulating a common carrier wave or subcarrier with two different types of modulation, each conveying separate information.

dual-scan supertwisted nematic display [DSTN] A *liquid-crystal display* [LCD] for notebook and subnotebook computers that is a compromise between the *active matrix LCD* [AMLCD] *thin-film technology* [TFT]-type display, offering the brightest and fastest display for portable computers, and the less-expensive, lower-power *supertwisted nematic* [STN] display.

dual-tone multifrequency [DTMF] The *tone dialing* system based on producing two nonharmonic related frequencies simultaneously to identify the number dialed. Eight frequencies have been assigned to the four rows and four columns of a typical keypad.

dual-trace oscilloscope An oscilloscope with two separate vertical input circuits that permit two waveforms to be observed simultaneously on the same CRT or computer monitor display.

dub To transfer recorded material from one recording to another, with or without the addition of new sounds, background music, or sound effects.

duct 1. An atmospheric condition that makes possible abnormally long-range radio and radar signal propagation in the troposphere. Temperature inversions cause abnormal changes in dielectric constant that make radio waves refract up and down between the two air layers forming the duct or between one air boundary and the ground. Also called tropospheric duct. 2. An enclosed runway for cables.

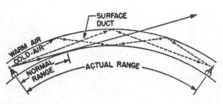

Duct of cold air under warm air makes radio waves follow the curvature of earth to the ship at right.

dumbbell slot A dumbbell-shaped hole in a wall or diaphragm of a waveguide, designed to serve as a slot radiator.

dumb terminal An asynchronous ASCII *video data terminal* [VDT] that is dependent on a communications network computer to perform data protocol. It is called "dumb" because it does not include a microprocessor or *central processing unit* [CPU] for processing data. It is used as a terminal in travel agencies, financial institutions, and airport and train ticket counters.

dummy An artificial address, instruction, or other unit of information inserted into a digital computer solely to fulfill prescribed conditions (such as word length or block length) without affecting operations.

dummy antenna A device that has the impedance characteristic and power-handling capability of an antenna but does not radiate or receive radio waves. It is used for testing transmitters. It is also called an *artificial antenna*.

dummy load A dissipative device used at the end of a transmission line or waveguide to convert transmitted energy into heat, so essentially no energy is radiated outward or reflected back to its source.

dump 1. To withdraw all power from a computer accidentally or intentionally. 2. In digital computer programming, to transfer all or part of the contents of one section of computer memory into another section.

dump check A computer check that usually consists of adding all the digits during dumping and verifying the sum when retransferring.

duodecimal number system A number system based on the equivalent of the decimal number 12.

duplex *Duplex operation.*

duplex channel A communication channel providing simultaneous transmission in both directions.

duplexer A switching component for radar that permits alternate use of the same antenna for both transmitting and receiving. It contains the TR switch that blocks out the receiver when the transmitter is operating. Other forms of duplexers serve in two-way radio communication with a single antenna at lower frequencies.

duplexing *Duplex operation.*

duplex operation The operation of associated transmitting and receiving apparatus concurrently, as in ordinary telephones, without manual switching between talking and listening periods. A separate frequency band is required for each direction of transmission. Also called duplex and duplexing.

duplication check A computer check that requires the results of two independent performances (either concurrently on duplicate equipment or at a later time on the same equipment) in the same operation to be identical.

DUT Abbreviation for *device under test.*

duty cycle The ratio of the pulse width to the period (on time to total time) expressed as a *percentage.* It is the same as *duty factor,* except that term is expressed as a *decimal.*

duty factor The ratio of the pulse width to the period (on time to total time) expressed as a *decimal.* It is the same as *duty cycle,* except that term is expressed as a *percentage.*

DUV Abbreviations for *deep ultraviolet* and *data under voice.*

DVD Abbreviation for *digital video disk.*

DVM Abbreviation for *digital voltmeter.*

DVOM Abbreviation for *digital volt-ohm-milliammeter.*

dwell A controlled time interval or delay during which a specified action occurs, such as the closing of contacts or the maximum lift position of a cam.

DX Abbreviation for distance reception, related to the reception of, or communication with, distant radio stations.

dyadic operation Simultaneous operation on two operands in a computer.

dye laser A liquid laser in which the lasing material is an active organic fluorescent material dissolved in a solvent. One organic dye used for this purpose is anthracene. The dye can be excited by another laser, a flashlamp, or some other pulsed light source, to give either pulsed or continuous tunable output in the visible spectrum. Tuning is achieved by such means as rotating a diffraction grating or varying the optical path length of the lasing cell. The normal wavelength range is 0.3 to 1.2 μm.

dyn Abbreviation for *dyne.*

dynamic accuracy The degree of conformance to the true value when relevant variables are changing with time.

dynamic convergence The process whereby the locus of the point of convergence of electron beams in a color television or other multibeam cathode-ray tube is made to fall on a specified surface during scanning. Without dynamic convergence that varies with beam angle, the locus would be a spherical surface at a constant radius from the center of deflection of the beam.

dynamic focusing The process of varying the focusing electrode voltage for a color picture tube automatically so the electron-beam spots remain in focus as they sweep over the flat surface of the screen. Without dynamic focusing, part of the image would be out of focus at all times.

dynamic headphone A headphone in which a flexible miniature cone or diaphragm is attached to a voice coil positioned in the magnetic field of a permanent magnet. The operation is the same as for a dynamic loudspeaker.

dynamic loudspeaker A loudspeaker whose moving diaphragm is attached to a current-carrying voice coil that interacts with a constant magnetic field to give the in-and-out motion required for the production of sound waves. These waves correspond to the audio-frequency current flowing through the voice coil. In a permanent-magnet loudspeaker the constant magnetic field is produced by a permanent magnet, and in an excited-field loudspeaker it is produced by a field coil. Also called dynamic speaker and moving-coil loudspeaker.

dynamic microphone A moving-conductor microphone whose flexible diaphragm is attached to a coil positioned in the fixed magnetic field of a permanent magnet. When sound waves move the diaphragm back and forth, the attached voice coil cuts magnetic lines of force. The desired AF voltage is thus induced in the coil. It is also called a *moving-coil microphone.*

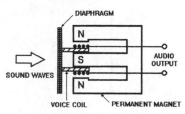

Dynamic microphone voice coil is moved in the permanent magnet field by the incident sound wave to induce a voltage from the coils.

dynamic noise suppressor An audio frequency filter circuit that automatically adjusts its bandpass limits according to signal level, generally by means of reactance tubes. At low signal levels, when noise becomes more noticeable, the circuit reduces the low-frequency response and sometimes it also reduces the high-frequency response.

dynamic pickup A pickup in which the electric output is due to motion of a coil or conductor in a constant magnetic field. In a moving-coil phono pickup, the coil is moved by the needle that follows the grooves of a record. Also called dynamic reproducer and moving-coil pickup.

dynamic random-access memory [DRAM] A volatile semiconductor read-write memory that requires periodic refreshing to preserve the charges on its capacitive memory cells that retain data. DRAMs are the primary memories of most computers, and have the lowest cost per bit of any semiconductor memory. DRAMS containing a million bits (megabits) of data are now being produced.

dynamic range The ratio of the maximum specified signal level capability of a system or component to its noise level. It is usually expressed in decibels.

dynamic response The response of a switching device or circuit to a rapidly changing waveform such as a step function. Dynamic response is usually measured in microseconds. See *response time.*

dynamic sensitivity The alternating component of phototube anode current divided by the alternating component of incident radiant flux.

dynamic speaker *Dynamic loudspeaker.*

dynamometer An electric generator or motor that measures the torque of a rotating shaft.

dyne [dyn] The CGS unit of force. One dyne is the force that gives an acceleration of 1 centimeter per second per second to a mass of 1 gram.

dynode An electrode whose primary function is secondary emission of electrons. Multiplier phototubes and some types of television camera tubes have dynodes.

E

e 1. Symbol for the base of the system of natural or napierian logarithms, with the approximate value of 2.71828. 2. Symbol for the instantaneous value of an alternating voltage.

E 1. Symbol for *emitter* on transistor circuit diagrams. 2. Abbreviation for *exa-*.

E 1. Symbol for *electric field strength.* 2. Symbol for *voltage.*

EAE Abbreviation for *extended arithmetic element.*

Early effect The reduction in the effective base width of a bipolar transistor when the width of the collector-base PN junction is increased by increasing the collector-base voltage. The effect was discovered by J. M. Early.

EAROM Abbreviation for *electrically alterable read-only memory.*

earphone A small, lightweight electroacoustic transducer that fits inside the ear, to function the same as a headphone or telephone receiver. It is used chiefly with hearing aids.

earphone coupler A shaped cavity for testing earphones. A microphone is mounted inside to measure pressures developed in the cavity.

earth British term for *ground.*

earth current *Telluric current.*

earthed British term for *grounded.*

earthshine Faint illumination of the dark side of the crescent moon, caused by reflection of sunlight from the earth.

earth-stabilized vehicle A space vehicle that is stabilized so one axis points toward the center of the earth.

earth-surface potential [ESP] A voltage in the earth's surface caused by fluctuations of the earth's magnetic field.

EBCDIC Abbreviation for [Extended Binary-Coded Decimal Interchange Code]

E-beam Abbreviation for *electron beam.*

E-beam generator A generator of a fine electron beam that can draw integrated-circuit elements directly on a photoresist-coated semiconductor wafer or expose the wafer through a mask in IC fabrication. See *electron-beam lithography.*

EBICON [Electron Bombardment Induced CONductivity] A television camera tube that differs from orthicon and vidicon tubes chiefly in the construction of its target.

EBS Abbreviation for *electron-bombarded semiconductor.*

ebullator A cooling system whose a fluorocarbon cooling agent maintains a constant temperature through bubbling, despite severe changes in ambient external temperature. It encloses a klystron oscillator when constant frequency is required.

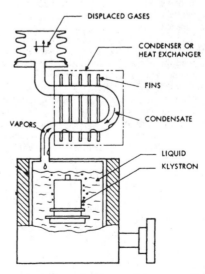

Ebullator used as enclosure for klystron of airport surveillance radar.

140

Eccles-Jordan circuit *Flip-flop circuit* or *bistable multivibrator.*

ECCM Abbreviation for 1. *electronic counter countermeasures* and 2. *electromagnetic counter countermeasures.*

ECDIS Abbreviation for *electronic chart display information system.*

E cell A timing device that converts the current-time integral of an electrical function into an equivalent mass integral (or the converse operation) up to a maximum of several thousand microampere-hours. It consists essentially of a gold central electrode surrounded by an electrolyte, with the silver container serving as the second electrode. It works like a small integrating coulometer in which metal is transferred through the electrolyte. Timing periods range is from seconds to months. Also called electrolytic cell.

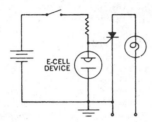

E-cell circuit.

ECG 1. Abbreviation for *electrocardiogram.* 2. Abbreviation for *electrocardiograph.*

echo 1. A wave that has been reflected from impedance mismatches returned with sufficient delay and magnitude to be perceived in some manner as a wave distinct from that directly transmitted, causing a time delay that interferes with speaking and listening. 2. The signal reflected by a radar target, or the trace produced by this signal on the screen of the cathode-ray tube in a radar receiver. It is also called radar echo and return. 3. A sound wave reflected from a hard surface. 4. *Ghost signal.*

echo area *Radar cross section.*

echo box A calibrated high-Q resonant cavity that stores part of the transmitted radar pulse power and gradually feeds this energy into the receiving system after completion of the pulse transmission. It provides an artificial target signal for test and tuning purposes. It is also called a phantom target.

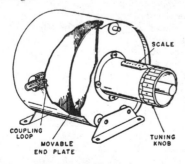

Echo box for coaxial cable feed from 10-cm radar.

echo chamber A reverberant room or enclosure used in a studio to add echo effects to sounds for radio or television programs.

echo depth sounder *Depth sounder.*

echogram The recording or display obtained with ultrasonic pulses and echo-ranging techniques, as in medical applications, ocean-bottom profile measurements, and geological exploration.

echoing area *Backscattering coefficient.*

echo matching Rotating a radar antenna or antenna array to a position at which the two echoes corresponding to the two directions of an echo-splitting radar are equal.

echoplex A method of checking for data-transmission errors by arranging for the computer to transmit its incoming data back to the sending terminal for printout and comparison with the originally transmitted data. It is generally used only at startup or for periodic checks.

echo suppressor 1. A circuit that desensitizes electronic navigation equipment for a fixed period after the reception of one pulse, for the purpose of rejecting delayed pulses arriving from indirect reflection paths. 2. A circuit or device used on a transmission line to prevent a reflected current from returning to the sending end of the line, typically a voice-operated gate that allows simplex communication.

ECL Abbreviation for *emitter-coupled logic.*

ECM Abbreviation for 1. *electronic countermeasures* and 2. *electromagnetic countermeasures.*

ECO Abbreviation for *electron-coupled oscillator.*

E core A transformer core made from E-shaped laminations used in conjunction with I-shaped laminations.

ECS Abbreviation for *electronic chart system.*

EDC Abbreviation for *error detection and correction.*

eddy current A circulating current induced in a conducting material by a varying magnetic field. These currents are undesirable in most instances because they represent loss of energy and cause heat. Laminations are used for the iron cores of transformers, filter chokes, and AC relays to shorten the paths for eddy currents and thus keep eddy-current losses at a minimum. At radio frequencies eddy-current paths must be broken up still more with powdered-iron cores.

eddy-current heating *Induction heating.*

eddy-current loss Energy loss caused by undesired eddy currents circulating in a magnetic core.

EDGE An air-to-ground bomb precisely guided by signals from GPS satellites that have been corrected by a ground station to permit the bomb to hit a target at a slant range of up to 20 mi (32 km) from the target. It permits the aircraft carrying the bomb to remain far enough from the target to avoid possible damage from the detonation as well as to minimize the threat from the enemy's defensive weapons.

edgeboard connector *Card-edge connector.*

edge effect An outward-curving distortion of lines of force near the edges of two parallel metal plates that form a capacitor. A correction must be made for this effect when computing capacitance from the geometry of a structure, or a special guard ring must eliminate the effect.

edgewise bend A bend in a rectangular waveguide made so that the longitudinal axis remains in a plane parallel to the wide side of the waveguide.

Edison effect The emission of electrons from hot bodies. The rate of emission increases rapidly with temperature. Discovered by Edison in 1883, when a current flow was obtained between the filament of an incandescent lamp and an auxiliary electrode inside the lamp. It is also called the Richardson effect.

Edison storage battery An alkaline storage battery that produces an open-circuit voltage of 1.2 V per cell. The active material on the negative plates is an iron alloy, whereas that on the positive plates is nickel oxide. It is also called nickel-iron battery.

EDIF Abbreviation for *electronic design interchange format*.

EDO Abbreviation for 1. *enhanced data output* and 2. *extended data output*.

EDP Abbreviation for *electronic data processing*.

EEPROM or E²PROM Abbreviation for *electrically erasable programmable read-only memory*.

effective acoustic center The point on or near a loudspeaker or other acoustic generator from which spherically divergent sound waves appear to diverge.

effective antenna length The electrical length of an antenna, as distinguished from its physical length.

effective area A directional antenna rating, equal to the square of the wavelength multiplied by the power gain (or directive gain) of an antenna in a specified direction, and the result divided by 4π.

effective current The value of alternating current that will give the same heating effect as the corresponding value of direct current. The effective value is 0.707 times the peak value in the case of sine-wave alternating currents.

effective echoing area of target The area of a hypothetical perfect radar target, perpendicular to the incident beam, that would produce at the receiver a signal equal to that produced by the actual target. For an average aircraft it is generally from 1 to 10 m².

effective height The height of the center of radiation of a transmitting antenna above the effective ground level.

effective isotropically radiated power [EIRP] The product of the net radiated RF power of a transmitter and the gain of the antenna system in one direction relative to an isotropic (omnidirectional) radiator.

effective particle velocity The root-mean-square value of the instantaneous particle velocities at a point. It is also called root-mean-square particle velocity.

effective perceived noise decibel [epndB] A subjective noise unit that includes the effects of tone and duration. It is a Federal Aviation Administration noise-certification requirement for aircraft.

effective radiated power [ERP] The product of antenna input power and antenna power gain, expressed in kilowatts.

effective radius of earth A radius value substituted for the geometric radius to correct for atmospheric refraction when the index of refraction in the atmosphere changes linearly with height. Under conditions of standard refraction the effective radius of the earth is 8500 km, or four-thirds the geometric radius.

effective resistance *High-frequency resistance.*

effective sound pressure The root-mean-square value of the instantaneous sound pressures at a point during a complete cycle, expressed in dynes per square centimeter. It is also called pressure, root-mean-square sound pressure, or sound pressure.

effective thermal resistance The effective temperature rise per unit power dissipation of a designated junction of a semiconductor device, above the temperature of a stated external reference point, under conditions of thermal equilibrium.

effective value *Root-mean-square.*

effective voltage The value of DC voltage that will heat a resistive component to the same temperature as the AC voltage that is being measured.

effective wavelength The wavelength of a monochromatic X-ray that undergoes the same percentage attenuation in a specified filter as the heterogeneous X-ray beam under consideration.

efficiency 1. The ratio of useful output of a device to total input, generally expressed as a percentage. It is normally measured at full rated output power with typical input conditions. 2. The probability that a count will be produced in a counter tube by a specified particle or quantum incident.

EFL 1. Abbreviation for *emitter-follower logic*. 2. Abbreviation for *emitter-function logic*.

EFT Abbreviation for *electronic funds transfer*.

EGA Abbreviation for *enhanced graphics adapter*.

EHF Abbreviation for *extremely high frequency*.

E-H T junction A waveguide junction composed of a combination of E-plane and H-plane T junctions having a common point of intersection with the main waveguide.

E-H tuner An E-H T junction used for impedance transformation, having two arms terminated in adjustable plungers.

EHV Abbreviation for *extra-high voltage*.

EIA Abbreviation for *Electronic Industries Association*.

EIA color code One of the systems of color markings developed by the Electronic Industries Association for specifying electrical values and terminal connections of resistors, capacitors, and other components. It was formerly called RETMA color code and RMA color code.

eight-level code A teletypewriter code that uses eight impulses, in addition to the start and stop impulses, to define a character.

Einstein-de Haas effect The rotation induced in a freely suspended ferromagnetic object when magnetization of the object is reversed.

EIRP Abbreviation for *effective isotropic radiated power*.

EISA Abbreviation for *Extended Industry Standard Architecture*.

EKG 1. Abbreviation for *electrocardiogram*. 2. Abbreviation for *electrocardiograph*.

elastance The reciprocal of capacitance, measured in darafs.

elastic wave *Acoustic wave.*

elastomer A rubberlike material that returns to its original shape and dimensions after it is stretched or deformed.

elastooptical effect The change that is produced in the index of refraction of a material by an internal strain resulting from either stationary or traveling elastic waves.

E layer A layer of ionized air occurring at various heights in the E region of the ionosphere, capable of bending radio waves back to earth. Average height is about 100 km. It is also called the *Kennelly-Heaviside* layer. See also *ionosphere*.

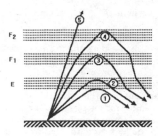

E layer reflects signals along paths 1 and 2. F layers reflect signals along paths 3 and 4. Signals following path 5 pass through all layers into outer space if they are directed at an angle higher than the critical angle.

elbow *Corner.*

electret A permanently polarized piece of dielectric material produced by heating the material and placing it in a strong electric field during cooling. Some barium titanate ceramics, carnauba wax, and mixtures of certain other organic waxes can be polarized in this way. The electric field of an electret corresponds to the magnetic field of a permanent magnet.

electret microphone A capacitor microphone whose diaphragm is a charged dielectric foil electret of polyester or other plastic having a thin layer of gold on its upper surface. When sound waves move the diaphragm, the permanently stored static charges in the electret produce a correspondingly varying AF voltage between the output terminals.

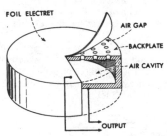

Electret-microphone construction.

electric Containing, producing, arising from, or actuated by electricity. Examples are electric energy, electric lamp, and electric motor. It can be used interchangeably with electrical.

electrical Related to or associated with electricity, but not containing it or having its properties or characteristics. Examples are electrical engineer, electrical handbook, and electrical rating. It can be used interchangeably with electric.

electrical angle The angle that specifies a particular instant in an AC cycle. Usually expressed in degrees. One cycle is equal to 360°; hence a quarter-cycle is 90°. The phase difference between two alternating quantities is expressed as an electrical angle.

electrical axis The *x* axis in a quartz crystal. There are three in a crystal, each parallel to one pair of opposite sides of the hexagon. All pass through and are perpendicular to the optical or *z* axis.

electrical boresight The tracking axis of a radar antenna or highly directional radio antenna, corresponding to the null of a conical-scanning antenna or the maximum of a directional antenna.

electrical center The point that divides a component into two equal electrical values.

electrical degree A unit equal to $\frac{1}{360}$ cycle of an alternating quantity.

electrical distance The distance between two points, expressed in terms of the duration of travel of an electromagnetic wave in free space between the two points. A convenient unit is the light-microsecond, which is approximately 983 ft or 300 m.

electrical interference An electrical or electromagnetic disturbance that interrupts normal electronic equipment functions caused by operating electrical apparatus not intended to radiate electromagnetic energy.

electrical length The length of a conductor expressed in wavelengths, radians, or degrees. Distance in wavelengths is multiplied by 2π (6.2832) to give radians or by 360 to give degrees.

electrical rules check Computer software that determines whether circuit connections shown on a schematic are logical.

electrically alterable read-only memory [EAROM] A read-only memory that can be reprogrammed electrically in the field many times, after the entire memory is erased by applying an appropriate electric field. Term now considered obsolete.

electrically erasable programmable read-only memory [EEPROM or E^2 PROM] A semiconductor, nonvolatile, programmable ROM that can be erased electrically without removing it from its host circuit. It can be selectively rewritten electrically. See also *erasable-programmable ROM* and *flash memory.*

electrically tuned oscillator An oscillator whose frequency is determined by the value of a voltage, current, or power. Electric tuning includes electronic tuning, electrically activated thermal tuning, electromechanical tuning, and tuning methods in which the properties of the medium in a resonant cavity are changed by external electric means. An example is the tuning of a ferrite-filled cavity by changing an external magnetic field.

electrically variable coil An iron-core coil whose inductance can be varied over a wide range by changing a small DC control current.

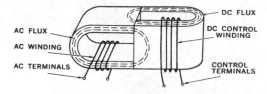

Electrically variable coil having two C cores rotated at 90° to each other, with DC control winding on one half and AC winding on other half.

electrical zero A standard reference position from which rotor angles are measured in synchros and other rotating devices.

electric arc   A discharge of electricity through a gas, normally characterized by a voltage drop approximately equal to the ionization potential of the gas. It is also called an *arc*.

electric cell   *Cell.*

electric charge   *Charge.*

electric circuit   A path or group of interconnected paths capable of carrying electric currents.

electric conduction   The conduction of electricity by means of electrons, ionized atoms, ionized molecules, or semiconductor holes.

electric contact   A physical contact that permits current flow between conducting parts.

electric control   The control of a machine or device by switches, relays, or rheostats, as contrasted to electronic control by transistors or electron tubes.

electric controller   A device that governs in some predetermined manner the electric power delivered to apparatus.

electric coupling   A rotating machine whose torque is transmitted or controlled by electric or magnetic means.

electric dipole   A pair of equal and opposite charges an infinitesimal distance apart.

electric-discharge lamp   A lamp whose light is produced by current flow through a gas or vapor in a sealed glass enclosure. Examples of these lamps include argon glow, mercury-vapor, neon glow, and sodium-vapor.

electric-discharge machining [EDM]   A metal-cutting process in which high-frequency discharges from a negatively charged metal tool remove metal from the work piece by electroerosion. There is no electrolyte, but the work is submerged in oil to flush away eroded particles and to delay each spark until peak energy is built up.

electric displacement   *Electric flux density.*

electric displacement density   *Electric flux density.*

electric doublet   *Dipole.*

electric dynamometer   An electric generator or motor equipped with a display for indicating torque.

electric energy   The integral with respect to time of the instantaneous power input or power output of a circuit or device. The basic unit is the watthour.

electric eye   Slang term for *photocell.*

electric field   1. The region around an electrically charged body in which other charged bodies are acted on by an attracting or repelling force.   2. The electric component of the electromagnetic field associated with radio waves and electrons in motion.

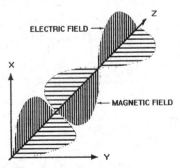

Electric field and magnetic field of an electromagnetic wave propagating in free space.

electric field strength [E]   The magnitude of the electric field vector.

electric field vector   The force on a stationary positive charge per unit charge at a point in an electric field. Usually measured in volts per meter. Also called electric vector.

electric flux   *Electric line of force.*

electric flux density   A vector whose magnitude is equal to the charge per unit area that would appear on one face of a thin metal plate which is placed at a point in an electric field and oriented for maximum charge. The vector is perpendicular to the plate and directed from the negative to the positive face. It is also called electric displacement and electric displacement density.

electric image   An array of electric charges, either stationary or moving, in which the density of charge is proportional at each point to the light values at corresponding points in an optical image to be reproduced.

electricity   A fundamental quantity in nature, consisting of electrons and protons at rest or in motion. Electricity at rest has an electric field that possesses potential energy and can exert force. Electricity in motion (an electric current) has both electric and magnetic fields that possess potential energy and can exert force.

electric lamp   A lamp in which light is produced by electricity. Examples include incandescent, arc, glow, mercury-vapor, and fluorescent lamps.

electric line of force   An imaginary line, each segment of which represents the direction of the electric field at that point. Also called electric flux.

electric motor   *Motor.*

electric noise   Unwanted electric energy in a receiver or transmission system, other than crosstalk. Sources include electric appliances, electric motors, engine ignition, and power lines.

electric precipitation   *Electrostatic precipitation.*

electric propulsion   Propulsion of spacecraft and other vehicles by electrothermal, electrostatic, or plasma techniques, as contrasted to chemical propulsion, which involves direct use of fuel.

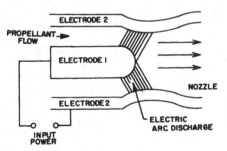

Electric propulsion with an arc jet between two electrodes to heat and accelerate gaseous propellant.

electric reset relay   A relay that remains in the on condition after actuation until reset by applying an independent electric input.

electric shield   A housing, usually aluminum or copper, placed around a circuit to prevent interaction with other circuits by providing a low resistance and a reflecting path to ground for high-frequency radiation.

electric shock *Shock.*

electric strength *Dielectric strength.*

electric vector *Electric field vector.*

electroacoustic The operation of a product or device based on both electricity and acoustics, such as a loudspeaker or microphone.

electroacoustic effect *Acoustoelectric effect.*

electroacoustic transducer A transducer that receives waves from an electric system and delivers waves to an acoustic system, or vice versa.

electrocardiogram [ECG or EKG] A record made by an electrocardiograph. It is also called a cardiogram.

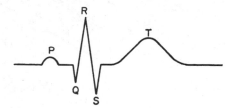

Electrocardiogram showing a typical tracing with electrodes on the surface of body. The initial P wave, produced by electric activation of atrium, is about 90 ms in duration. The peaked QRS complex due to activation of the ventricles lasts about 80 ms. The final slowly varying T wave is related to electric recovery of the ventricles.

electrocardiograph [ECG or EKG] An instrument for recording the waveforms of voltages developed in the chest and lower parts of the human body in synchronism with the action of the heart.

electrochemical machining A metal-cutting process that is the reverse of electroplating. A low DC voltage is applied between the workpiece and a tool having the shape of the desired cut, and an electrolyte is pumped at high pressure through the gap between workpiece and tool. Electrochemical action in the gap erodes metal from the workpiece.

electrode 1. A conducting element that performs one or more of the functions of emitting, collecting, or controlling the movements of electrons or ions in an electron tube, or the movements of electrons or holes in a semiconductor device. 2. A terminal or surface at which electricity passes from one material or medium to another, as at the electrodes of a battery, electrolytic capacitor, or weider. 3. One of the terminals used in dielectric heating or diathermy for applying the high-frequency electric field to the material being heated.

electrodeless discharge A luminous discharge produced by a high-frequency electric field in a gas-filled glass tube with no internal electrodes.

electrodeposition The process of depositing a substance on an electrode by electroplating or electroforming. It is also called electrolytic deposition.

electrode potential 1. The instantaneous voltage of an electrode with respect to the cathode of an electron tube. 2. The voltage existing between an electrode and the solution or electrolyte in which it is immersed.

electrode radiator A metal structure, often with a large area, that is an external extension of an electrode of an electron tube to facilitate the dissipation of heat. See *heatsink*.

electrodermal reaction The change in electric resistance of the skin during emotional stress. This is one of the functional variables measured by polygraphs.

electrodesiccation The destruction of tissue by electric sparks generated at the tip of a small movable electrode.

electrode voltage The voltage between an electrode and the cathode or a specified point on a filamentary cathode of an electron tube.

electrodiagnosis Diagnosis of disease by studying electric activity of parts of the body and responses to stimulation of electrically excitable tissues.

electrodynamic instrument An instrument that depends for its operation on the reaction between the current in one or more movable coils and the current in one or more fixed coils.

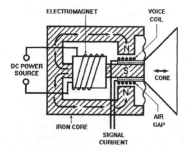

Electrodynamic speaker operation depends on the interaction between the electromagnet and the voice coil.

electrodynamic loudspeaker *Excited-field loudspeaker.*

electrodynamometer A low-frequency, transmission-type power meter that is connected between a source and a load. The current-sensing element is a fixed or stationary coil, and the voltage-sensing element is a movable or potential coil which rotates in the magnetic field of the stationary coil. The torque on the movable coil is opposed by a spring and is proportional to the average value of the instantaneous product of the currents in the two coils. The load current passes through the current-coil terminals, and the voltage terminals are connected across the load. The current through the movable coil is proportional to the load voltage, so its angular rotation is a function of the instantaneous voltage-current product. The pointer then indicates the average load power.

electroflor A material that changes color when electrically activated but does not radiate light.

electrofluiddynamic converter A converter that transforms the dynamic energy of a gaseous fluid into electric energy by passing through an electrostatic field a gas which contains electrically charged particles. The action is similar to that of a Van de Graaff generator, but higher power density can be obtained.

electroforming 1. The electrodeposition of metal on a conducting mold in sufficient thickness to make a desired metal object, such as a complex waveguide structure. The mold is often of graphite-coated wax so it can be removed by melting. 2. Production of a PN junction in a point-contact diode or transmitter by passing a large current pulse through the semiconductor material.

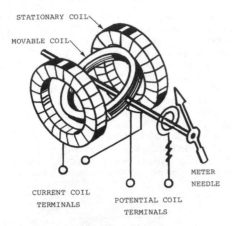

Electrodynamometer power meter is an electromechanical transmission-type meter.

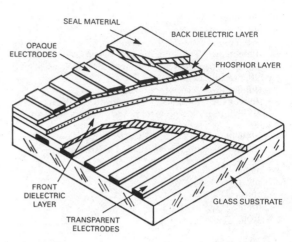

Electroluminescent display is a thin-film panel with a manganese-doped, zinc-sulfide phosphor luminescent layer.

electrogasdynamics Conversion of the kinetic energy of a moving gas to electricity, for such applications as high-voltage electric power generation, air-pollution control, and paint spraying.

electrography A form of electrostatography in which electrostatic images are formed on an insulating medium without the aid of electromagnetic radiation. It includes xeroprinting, where the charged image is permanent, as required for repetitive printing, and electrographic recording, in which the charged image is formed by electric means.

electrojets Powerful natural currents of electricity that flow continuously around the earth's upper atmosphere and participate in the production of auroral light that is visible in the darkened sky of the polar regions.

electroless deposition Chemical deposition of a metal on a material, without electrolytic or electroplating action.

electroluminescence The emission of light caused by stimulating a polycrystalline phosphor with an electric field, a technology for displays and indicators.

electroluminescent display A digital display whose phosphor layer is sandwiched between transparent conductive segments that form characters by electroluminescence when energized by either AC or DC voltages. The phosphor most commonly used is zinc sulfide doped with manganese; this emits yellow light, but colors from red to green can be obtained by filtering.

electroluminescent display screen An electroluminescent surface with an extra semiconductor control layer that permits the storing of images for controllable periods of at least 1 hour. The desired image is projected onto the screen with ultraviolet light, and is completely erased with infrared radiation.

electroluminescent panel A light source consisting of a suitable phosphor placed between sheet-metal electrodes (one of which is essentially transparent) separated by only a few thousandths of an inch (0.001 cm), with an AC voltage applied between the electrodes.

electrolysis The production of chemical changes by passing current from an electrode to an electrolyte, or vice versa, as in electroplating, electroforming, or electropolishing.

electrolyte A liquid, paste, or other conducting medium in which the flow of electric current takes place by migration of ions.

electrolytic capacitor A capacitor whose dielectric is an oxide of aluminum or tantalum. The oxide is formed by electrochemical reaction with those metals. Three styles are wet foil, wet slug, and dry or solid slug. Aluminum capacitors are wet-foil style, but tantalum capacitors can be made in all three styles. Both aluminum and tantalum units can be *polarized* or *nonpolarized*. As a class they exhibit high capacitance per unit volume (high volumetric efficiency). An aluminum capacitor can store five or more times as much charge as an equivalent-size film capacitor; a tantalum capacitor can store three times as much charge as an equivalent aluminum capacitor. See also *aluminum electrolytic capacitor* and *tantalum capacitor*.

electrolytic cell 1. A cell consisting of electrodes separated by an electrolyte. It can be used to store electric energy for use on demand, as in a storage cell; to generate electric energy, as in a dry cell; or to produce a desired electrochemical reaction when electric energy is applied. 2. *E cell.*

electrolytic deposition *Electrodeposition.*

electrolytic iron A pure iron that has excellent magnetic properties, produced by an electrolytic process.

electrolytic recording Electrochemical recording in which the chemical change is made possible by the presence of an electrolyte.

electrolytic switch A switch with two electrodes projecting into a chamber containing a precisely measured quantity of a conductive electrolyte, leaving an air bubble of predetermined width. When the switch is tilted from true horizontal, the bubble shifts position and changes the amount of electrolyte in contact with the electrodes, thereby changing the amount of current passed by the switch. It can be a leveling switch in gyro systems.

electrolytic tough-pitch copper [ETPC] Copper that has a minimum conductivity of 99.9%. It is the material of choice for making power-tube anodes.

electromagnet A magnet consisting of a coil wound around a soft iron or steel core. The core is strongly magnetized when current flows through the coil and almost completely demagnetized when the current is interrupted. It attracts a movable external iron object such as the armature of a relay. In a solenoid, the iron core itself is movable.

electromagnetic A reference to the combined electric and magnetic fields associated with radiation or movements of electrons or other charged particles through conductors or space.

electromagnetic circuit breaker *Magnetic circuit breaker.*

electromagnetic compatibility [EMC] The ability of a device or circuit to function correctly in its intended electromagnetic environment without transmitting unwanted signals to adjacent equipment or receiving unwanted interference from nearby sources.

electromagnetic constant The speed of propagation of electromagnetic waves in a vacuum. The latest measurements, based on microwave techniques give a value of 299,793 km/s.

electromagnetic counter-countermeasures [ECCM] *Electronic counter-countermeasures.*

electromagnetic countermeasures [ECM] *Electronic countermeasures.*

electromagnetic coupling Coupling that exists between circuits or conductors when they are mutually affected by the same electromagnetic field.

electromagnetic deflection Deflection of an electron stream by means of a magnetic field. In a television picture tube, the magnetic fields for horizontal and vertical deflection of the electron beam are produced by sending sawtooth currents through coils in a deflection yoke that goes around the neck of the picture tube.

electromagnetic delay line A delay line consisting simply of a transmission line carrying pulse trains. The delay time generally available is not sufficient for storing a large number of pulses within a reasonable line length.

electromagnetic disturbance An electromagnetic phenomenon, usually impulsive, that is superimposed on a desired signal. The disturbance can be random or periodic.

electromagnetic energy Energy associated with radio waves, heat waves, light waves, X-rays, and other types of electromagnetic radiation.

electromagnetic environment [EME] The electromagnetic fields existing in a given area.

electromagnetic field The field associated with electromagnetic radiation, consisting of a moving electric field and moving magnetic field acting at right angles to each other and at right angles to their direction of motion.

electromagnetic flowmeter A flowmeter that offers no obstruction to liquid flow. Two coils produce an electromagnetic field in the conductive moving fluid. The current induced in the liquid, detected by two electrodes, is directly proportional to the rate of flow.

electromagnetic focusing Focusing the electron beam in a television picture tube by means of a magnetic field parallel to the beam, produced by sending an adjustable value of direct current through a focusing coil mounted on the neck of the tube.

electromagnetic forming *Magnetic forming.*

electromagnetic horn A horn-shaped antenna structure used to provide highly directional radiation characteristics. Signal power is fed to the horn by a waveguide or an exciting dipole or loop at the input end of the horn.

electromagnetic induction The production of a voltage in a coil by a change in the number of magnetic lines of force passing through the coil.

electromagnetic interference [EMI] An electromagnetic disturbance caused by such radiating and transmitting sources as electrostatic discharge (ESD), lightning, radar, radio and TV signals, motors with brushes, and power lines. It can induce unwanted voltages in electronic circuits, damage components, and cause malfunction. Shields, filters, and transient suppressors protect electronics from EMI.

electromagnetic lens An electron lens whose electron beams are focused by an electromagnetic field.

electromagnetic loudspeaker *Magnetic-armature loudspeaker.*

electromagnetic mirror A surface or region capable of reflecting radio waves, such as an ionized layer in the upper atmosphere.

electromagnetic mixing Mixing of molten alloys by exposing the melt to a strong magnetic field while passing direct current between electrodes at opposite ends of the crucible. Stirring action results from interaction of the magnetic field of the current-carrying molten alloy with the external transverse magnetic field.

electromagnetic noise An electromagnetic disturbance that is not sinusoidal.

electromagnetic plane wave A transverse electric wave, transverse electromagnetic wave, or transverse magnetic wave.

electromagnetic prospecting Prospecting for ore bodies by measuring electromagnetic waves.

electromagnetic pulse [EMP] The pulse of electromagnetic radiation generated by a large thermonuclear explosion. Hardening of underground missile sites and control centers includes shielding to prevent the pulses from interfering with communication and electronic equipment.

electromagnetic pump A pump that moves a conductive liquid through a pipe by sending a large current transversely through the liquid. This current reacts with a magnetic field that is at right angles to the pipe and current flow, to move the current-carrying liquid conductor just as a solid conductor is moved in an electric motor.

electromagnetic radiation Radiation associated with a periodically varying electric and magnetic field that is traveling at the speed of light, including radio waves, light waves, X-rays, and gamma radiation.

electromagnetic relay *Electromechanical relay.*

electromagnetic shield A metal screen or enclosure placed around circuits to reduce the effects of both electric and magnetic fields. Electromagnetic fields are caused by motors, generators, relays, or devices whose operation depends on alternating fields. Shielding is achieved by a reflection or absorption of fields. Reflection occurs at the surface, and it is not usually affected by shield thickness. Absorption, however, occurs within the shield and is highly dependent on thickness.

electromagnetic spectrum The total range of wavelengths or frequencies of electromagnetic radiation, extending from the longest radio waves to the shortest known cosmic rays. It is also called a spectrum.

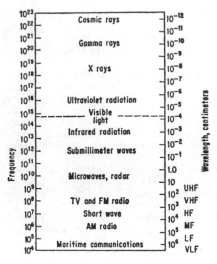

Electromagnetic spectrum.

electromagnetic susceptibility [EMS] The tolerance of circuits and components to all sources of interfering electromagnetic energy.

electromagnetic transducer A transducer whose operation depends on the EMF generated in a conductor or coil when it is moving across perpendicular lines of magnetic flux. An example is the linear velocity transducer.

electromagnetic unit [EMU] A CGS unit based on the assignment of unity to the strength of each of two like magnetic poles that repel each other with a force of 1 dyn at a distance of 1 cm in a vacuum.

electromagnetic vulnerability [EMV] The operation of equipment or systems that can be compromised by the electromagnetic environment.

electromagnetic wave A wave of electromagnetic radiation, characterized by variations of electric and magnetic fields.

electromagnetism Magnetism produced by an electric current rather than by a permanent magnet.

electromechanical transducer A transducer for receiving waves from an electric system and delivering waves to a mechanical system, or vice versa.

electrometer An instrument for measuring voltage without drawing appreciable current. Early electrometers were based on the electrostatic force exerted between bodies that are charged with the voltage to be measured, such as suspended parallel strips of gold leaf. Modern solid-state electrometers can have an input resistance above 10^{14} ohm.

electrometer tube A high-vacuum electron tube that has a high input impedance (low control-electrode conductance) to facilitate measurement of extremely small direct currents or voltages.

electromigration The motion of ions in a metal conductor within an integrated circuit, typically in aluminum surfaces in response to high current passage. It causes voids in the conductor which can grow until current flow is blocked. Its destructive effects are aggravated at high temperature and high current flow, but these effects can be minimized by limiting current densities and alloying the aluminum with copper or titanium.

electromotive force [EMF] The force that tends to produce an electric current in a circuit. It is usually called *voltage.*

electromotive series A listing of metal elements in the order of their electromotive force or the relative ease with which they lose electrons. In effect, it is a table of comparative activities of metals. Each metal in the list of common metals is more negative than the one that precedes it and more positive than the one that follows it. The series is, in part, as follows: sodium, magnesium, beryllium, aluminum, manganese, zinc, chromium, iron, cadmium, cobalt, nickel, tin, lead, antimony, copper, silver, and mercury. It is also called the *list of standard oxidation potentials.*

electromyogram [EMG] The record produced by an electromyograph.

electromyograph [EMG] An instrument for measuring and recording voltages generated by muscles in the body.

electron An elementary atomic particle that carries the smallest negative electric charge (1.6×10^{-19} C). It has $\frac{1}{1837}$ of the mass of the hydrogen atom), is highly mobile, and orbits the nucleus of an atom.

electron beam A stream of electrons that can "write" on phosphor surfaces such as a CRT screen, expose photoresist-coated semiconductor wafers by direct writing or exposure through a mask, or magnify objects by passing through magnetic "lenses." It can also be a cutting tool. See *electron-beam lithography, electron-beam microscope,* and *electron-beam machining.*

electron-beam drilling The drilling of tiny holes in a ferrite, semiconductor, or other material with a sharply focused electron beam to melt and evaporate or sublimate the material in a vacuum.

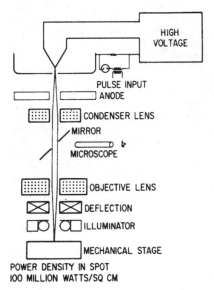

Electron-beam drilling of ferrite wafers on a mechanical stage platform in a vacuum enclosure.

electron-beam lithography   Lithography in which a resist-coated semiconductor wafer is placed in the vacuum chamber of a scanning-beam electron microscope and exposed to an electron beam under digital computer control. It can produce fine lines in very small areas, as required for the production of integrated circuits. After exposure, the wafer is removed from the vacuum chamber for conventional development and other production processes.

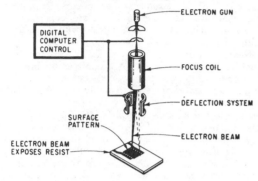

Electron-beam lithography setup.

electron-beam machining   A machining process that takes place in a vacuum. Heat is produced by a focused and controlled electron beam at a sufficiently high temperature to volatilize and thereby remove metal in a desired manner. Drilling and cutting are examples of specific applications.

electron-beam magnetometer   A magnetometer that depends for its operation on the change in intensity or direction of an electron beam which passes through the magnetic field to be measured.

electron-beam tube   An electron tube whose performance depends on the formation and control of one or more electron beams.

electron coupler   A microwave amplifier tube whose electron bunching is produced by an electron beam projected parallel to a magnetic field while it is subjected to a transverse electric field produced by a signal generator. The electron beam traces the surface of a cone as it rotates about the central axis at the signal-generator frequency. A second coupler, known as the output coupler, is required to extract power from the spiraling motions of the electrons and deliver this power to an output load. It is also called a Cuccia coupler.

electron coupling   A method of coupling two circuits inside an electron tube, used principally with multigrid tubes. The electron stream passing between electrodes in one circuit transfers energy to electrodes in the other circuit.

electron device   A device in which conduction is principally by electrons moving through a vacuum, gas, or semiconductor, as in a crystal diode, electron tube, transistor, or selenium rectifier.

electronegative   The property of a negative electric polarity.

electron emission   The liberation of electrons from an electrode into the surrounding space, usually under the influence of heat, light, or a high voltage. It is also called emission.

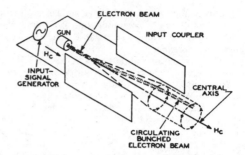

Electron coupler at the input of electron-beam parametric amplifier.

electron flow   A current produced by the movement of free electrons toward a positive terminal. The direction of electron flow is opposite to that of current.

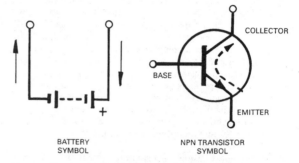

Electron flow in a battery and in an NPN transistor is indicated by arrow.

electron fluence   *Fluence.*

electron gun   An assembly of a cathode, grid, anodes, and other electrodes that can produce an electron beam and then control, focus, and deflect it so that it will converge at a specified location. It is a subassembly in cathode-ray tubes, television camera tubes, and many different kinds of microwave power tubes.

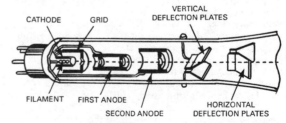

Electron gun for an electrostatically focused cathode-ray tube widely used in oscilloscopes.

electronic   An adjective that refers to a device, component, product, circuit, or system whose function depends on included electronic circuitry.

electronic air cleaner   *Precipitator.*

electronically agile radar   A radar system that has the capability for making rapid changes in its frequency, power output, modulation technique, and beam direction. This is accomplished by computer control. Ground-based, shipboard, or airborne agile radar is likely to have a *phased-array antenna*.

electronically programmable logic device [ELPD] A field-programmable logic device with transistor rather than fuse links for controlling the signal paths in its fixed array of logic blocks and connections. To reprogram an ELPD, the chip is exposed to ultraviolet (UV) light that erases its EPROM section and opens connections made during initial programming. It is then reprogrammed by loading a new program into the EPROM that stores the signal-path data in the transistor cells that are its memory.

electronically tuned oscillator An oscillator whose operating frequency can be changed by changing an electrode voltage or current.

electronic altimeter *Radio altimeter.*

electronic calculator A calculator with an integrated circuit processor that performs calculations and shows results on a digital display. Most basic models provide all four arithmetic operations (+, −, ×, ÷), usually with a floating decimal. More complex models include exponential and trigonometric functions, sometimes with a large choice of programs that can be inserted by changing a prerecorded magnetic program card. Larger desk models, operating from AC power, can provide printed records along with the digital display. Eight-digit displays predominate in pocket-sized battery-operated models. The displays usually include either seven-segment light-emitting diodes or liquid crystals.

electronic chart display information system [ECDIS] A shipboard electronic chart display and information system that integrates data from GPS satellites and the ship's speed log, gyrocompass, and radar, using electronic charts supplied by a national hydrographic office.

electronic chart system [ECS] The data for a navigational chart that has been digitized and stored in a computer for display. The positions displayed by these systems are usually based on measurements made from *GPS satellite* transmissions.

electronic circuit A circuit containing one or more diodes, transistors, integrated circuits, or other devices that provide some function.

electronic clock A clock that determines time from crystal-controlled electronic circuitry rather than spring-wound clockwork or a geared electric motor. It can have hour, minute, and second hands; a digital display in the form of mechanically rotated or positioned numerals; or a digital electronic display such as segmented liquid-crystal (LCD) or light-emitting diode (LED) modules. See also *digital clock.*

electronic commutator An electron-tube or transistor circuit that switches one circuit connection rapidly and in succession to many other circuits, without the wear and noise of mechanical switches. An example is the radial-beam tube, in which a rotating magnetic field causes an electron beam to sweep over one anode after another and produce the desired switching action.

electronic composition Typesetting in which characters are generated by electron or laser beams at speeds above about 6000 words per minute.

electronic control The control of a product, process, or system by electronic circuitry. Examples include automotive engine controls, heating, ventilation and air-conditioning [HVAC] controls, and security system controls. Any product or system controlled by a digital computer is under electronic control.

electronic controller An electronic circuit product or system for controlling a product, appliance, instrument, or system. Examples include programmable controllers, microcontrollers, board-level computers, personal computers, and digital panel meters (DPMs) with set-point circuitry for process control.

electronic counter A circuit or device that contains electronic circuitry which counts. A bistable multivibrator or flip-flop is the simplest form of electronic counter. These circuits can be interconnected to provide additional counting stages.

electronic counter-countermeasures [ECCM] A military offensive or defensive response to enemy efforts to jam radar operation and radio communications with broadband RF generators and various modulation techniques. This response can include measures to overpower the countermeasures, the changing of modulation methods, retuning the transmitters, or efforts to destroy the jamming transmitters and their antennas by weapons. These include missiles that can home on RF radiation or "smart bombs" that can be directed to their targets by lasers. See also *electronic countermeasures* [ECM]. It is also called *electromagnetic counter-countermeasures.*

electronic countermeasures [ECM] Jamming activities that render an enemy's search and fire-control radar ineffective and/or disable radio communications with high-powered, broadband noise. Traveling wave tube amplifiers can generate noise over wide bands of the radar frequency range. In addition, radio and radar transmitters can be attacked by missiles that home on antennas and destroy them. Emissions from enemy radars can be intercepted, delayed, amplified and retransmitted to give enemy operators false target range and bearing information. Passive ECM includes the dispersal of *chaff* to confuse radar missile-guidance systems and the firing of flares to mislead infrared-seeking guided missiles. It is also called *electromagnetic countermeasures.* See also *chaff* and *electronic counter-countermeasures* [ECCM].

electronic data processing [EDP] Automatic data processing with a computer or other electronic equipment.

electronic design interchange format [EDIF] A standardized exchange language for design information.

electronic engineering Engineering that deals with practical applications of electronics.

electronic flash tube *Flash tube.*

electronic fuel injection The forced injection of fuel under pressure into an automobile or truck engine, under electronic control.

electronic funds transfer [EFT] The transfer of funds from a bank account to another account by computer.

electronic game A self-contained version of a video game, with its own microprocessor-controlled screen or other type of display. Miniature, pocket-sized versions are battery-powered and can display animated figures and symbols for game playing on a liquid-crystal panel.

electronic heating Heating by means of RF current produced by a transistorized oscillator or an equivalent RF power source. The two kinds of electronic heating are induction heating for metals and dielectric heating for nonmetals. It is also called high-frequency heating and RF heating.

electronic ignition An ignition system in which transistors or other electronic components replace distributor points to control the firing of spark plugs.

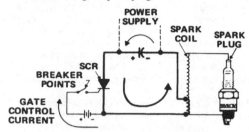

Electronic ignition from a single silicon controlled rectifier (SCR) triggered by breaker points of distributor.

Electronic Industries Association [EIA] A trade association made up chiefly of electronic component and equipment manufacturers. Its functions include standardization of sizes, specifications, and terminology for electronic products. Known as Radio Manufacturers Association (RMA) 1924–1950, Radio-Television Manufacturers Association (RTMA) 1950–1953, and Radio-Electronics-Television Manufacturers Association (RETMA) 1953–1957.

electronic intelligence [ELINT] A worldwide U.S. Air Force network that has fixed stations, specially equipped aircraft, and reconnaissance satellites to monitor and record enemy electromagnetic emissions. These signals are processed to give the nature and deployment of enemy warning and missile guidance radars, fire control, and countermeasures systems.

electronic jamming *Jamming.*

electronic keying Keying accomplished solely by electronic means.

electronic lock A lock that has a magnetically coded key about the size of a credit card. In one version, developed for hotels and motels, the lock code can be changed electronically from a central console as soon as a guest checks out, with simultaneous preparation of new coded keys.

electronic megaphone A megaphone consisting of a microphone, audio amplifier, and horn loudspeaker built as a single unit.

electronic micrometer An electronic instrument for measuring and indicating small linear distances in air or across nonmetallic materials.

electronic motor control The control of electric motors by electronic circuitry. AC motors can be controlled with *triacs* or back-to-back *silicon controlled rectifiers* [SCRs], and DC motors can be controlled with silicon controlled rectifiers. *DC servomotors* for machine tools and robots can be controlled with bipolar or MOSFET power transistors. *Stepping motors* can be stepped with the output of integrated circuit switches. Brushless DC motors can be controlled with *Hall-effect sensors* in their windings and transistorized switching circuitry.

electronic multimeter *Digital multimeter* (DMM).

electronic music Music consisting of tones originating in electronic sound and noise generators used alone or in conjunction with electroacoustic shaping circuits and sound-recording equipment. The resulting sounds might or might not resemble those of conventional musical instruments.

electronic musical instrument A musical instrument in which an audio signal is produced by a pickup or audio oscillator and amplified electronically to feed a loudspeaker, as in an electric guitar, electronic carillon, electronic organ, or electronic piano.

electronic navigation Navigation by means of electronic circuits, including radio, radar, loran, and GPS.

Electronic Numerical Integrator [ENIAC] An early digital computer that was completed in 1946 at the University of Pennsylvania in Philadelphia to compute ballistic trajectories for artillery shells. It contained 17,468 vacuum tubes and could add 5000 numbers in 1 s.

electronic organ A musical instrument that uses electronic circuits to produce music similar to that of a pipe organ.

electronic photometer *Photoelectric photometer.*

electronic piano A piano without a sounding board in which vibrations of each string affect the capacitance of a capacitor microphone and produce audio-frequency signals that are amplified and reproduced by a loudspeaker.

electronic potentiometer A potentiometer circuit that is continuously balanced by an electronic servosystem.

electronic profilometer An electronic instrument for measuring surface roughness. The stylus of a pickup is moved over the surface being examined, and the resulting varying voltage is amplified, rectified, and measured with a meter calibrated to read directly in microinches of deviation from smoothness.

electronic relay An electronic circuit that provides the function of a relay but has no moving parts. A solid-state relay is an example.

electronics 1. A division of electrical engineering that concentrates on circuits and devices which operate at frequencies higher than 50/60 Hz to amplify, oscillate, detect, rectify, and process signals, as distinguished from the generation, transmission, and consumption of electric power. It includes the research, development, design, manufacture, testing, and service of electronic components, circuits, products, and systems. 2. An adjective that modifies studies, documentation, activities, and objects related to the field of electronics.

electronic scanning The steering of streams of electrons or beams of visible light, RF energy, or infrared energy under the control of electronic circuits. Examples include raster scanning in TV transmission tubes and TV receiver CRTs, pattern writing in oscilloscope CRTs, beam steering (both transmitting and receiving) in phased-array antennas, and the direction of laser beams with mirrors and lenses driven by electronic circuits.

electronic security Methods of preventing intruders from obtaining useful information by intercepting radar and navigation signals.

electronic trapping The accumulation of electrons in imperfections in the silicon dioxide of erasable, programmable memories that permits negative charges to build up and delay the erasure of that memory device.

electronics engineer   An engineer whose training includes a degree in electronic engineering from an accredited college or university, a degree in electrical engineering with a major in electronics, or comparable knowledge and experience as required for working with electronic circuits and devices.

electronics industry   The industrial organizations engaged in the design, development, manufacture, and substantial assembly of electronic equipment, systems, assemblies, and components.

electronic speedometer   A speedometer in which a transducer sends speed and distance pulses over wires to the speed and mileage indicators, eliminating the need for a mechanical link involving a flexible shaft.

electronics serviceman   A serviceman who is qualified to repair and maintain electronic equipment.

electronics technician   A technician with both theoretical and practical training in electronics technology who is qualified to work under the direction of an electronics engineer or independently in assembling, testing, and repairing electronic equipment, in factories, laboratories, and private business.

electronic switch   An electronic circuit that functions as a high-speed switch.

electronic thermometer   A thermometer with a sensor, usually a thermistor, that is placed on or near the object being measured. An oral version gives a reading in less than 20 s, as compared to 3 min or more with conventional glass-mercury thermometers.

electronic tube   *Electron tube.*

electronic tuning   Tuning of a transmitter, receiver, or other tuned equipment by changing a control voltage rather than by adjusting or switching components manually.

electronic tuning range   The frequency range of continuous tuning, between two operating points of specified minimum power output, for an electronically tuned oscillator.

electronic tuning sensitivity   The rate of change in oscillator frequency with changes in electrode voltage or current for an electronically tuned oscillator.

electronic viewfinder   A television camera viewfinder that has a small cathode-ray picture tube to show the image being televised.

electronic voltmeter   A voltmeter that uses the rectifying and amplifying properties of electron devices and their associated circuits to obtain desired characteristics, such as high input impedance, wide frequency range, and peak indications. It is called a vacuum-tube voltmeter when its electron devices are vacuum tubes.

electronic warfare [EW]   The use of electronic devices and circuits in active and passive warfare for surveillance, guidance, communications, navigation, fire control, and weapon activation. Radar, sonar, satellites, and radio monitoring provide surveillance. Guidance for missiles and other weapons can be performed by RF and lasers. Communications can be achieved with radios, telephones, fiber optics, and ultrasonics. Navigation can be aided by GPS satellites, inertial guidance, and night-vision scopes. Artillery and missiles can be directed by radar, lasers, night-vision scopes, and forward observers with field telephones. Warheads can be detonated by radar and radio fuses, signals transmitted over wires, and magnetic fields.

electronic watch   A watch based on a quartz crystal or a tuning fork with battery-powered electronic circuits to provide greater accuracy than is possible with conventional spring-type mechanical movements.

electronic wattmeter   A wattmeter that uses two matched electronic voltmeters to give a reading proportional to the product of two voltages, on a scale calibrated to read power values directly. One voltage is that appearing across the load, and the other is obtained across a resistor in series with the line.

electron image   1. An image formed in a stream of electrons. The electron density in a cross section of the stream is at each point proportional to the brightness of the corresponding point in an optical image.   2. A pattern of electric charges on an insulating plate, with the magnitude of the charge at each point being proportional to the brightness of the corresponding point in an optical image.

electron injector   The electron gun that injects a beam of electrons into the vacuum chamber of a mass spectrometer, betatron, or other large electron accelerator.

electron lens   An arrangement of electrodes, with or without magnetic focusing coils that controls the size of a beam of electrons in an electron tube.

electron microscope   A microscope that employs a beam of electrons focused by electromagnetic-coil "lenses" in a vacuum chamber to magnify specimens. The electron beam is formed by a heated filament in an electron gun and is shaped by the coils to magnify the image in a way that is analogous to that of an optical microscope. There are two general types of electron microscope: *transmission* and *scanning*. In the *transmission electron microscope* [TEM] the electron beam passes through a thin specimen to form an image on a fluorescent screen. In the *scanning-electron microscope* [SEM] the electron beam "illuminates" the specimen during scanning to form 3-D pictures of it on a television monitor. See also *scanning electron microscope* [SEM], *scanning tunneling microscope* [STM] and *transmission electron microscope* [TEM].

electron mirror   An electrode or other element that produces total reflection of an electron beam.

electron multiplier   An electron-tube structure that employs secondary electron emission from solid reflecting electrodes (dynodes) to produce current amplification. The electron beam containing the desired signal current is reflected from each dynode surface in turn. At each reflection, an impinging electron releases two or more secondary electrons, so the beam builds up in strength. A typical arrangement of nine dynodes can give an amplification of several million. It is also called a multiplier or a secondary-electron multiplier.

electron-multiplier phototube   *Multiplier phototube.*

electron-multiplier tube   An electron tube that includes an electron multiplier.

electron optics   That branch of electronics concerned with the control of electron beams in a vacuum having electron lenses and electric or magnetic fields, or both.

electron-positron pair   The electron and positron simultaneously created by the process of pair production.

electron radius   The classical value of 2.81777×10^{-13} cm for the radius of an electron. It is obtained by equating the rest-mass energy of the electron to its electrostatic self-energy.

electron rest mass   A physical constant equal to 9.1091×10^{-28} g.

electron shell The arrangement of electrons in a given orbital outside the nucleus of an atom. All electrons in a shell have the same energy level.

electron spin The rotation of an electron about its own axis, contributing to the total angular momentum of the electron.

electron spin resonance The interaction of electric and magnetic fields with the spin of an electron about its own axis.

electron spin resonance spectrometer A spectrometer based on electron paramagnetic resonance.

electron synchrotron A synchrotron designed to accelerate electrons. The electron beam is allowed to strike an internal target, producing high-energy gamma rays that are used outside the machine.

electron trajectory The path of one electron in an electron tube.

electron tube An electron device in which electricity is conducted by electrons moving through a vacuum or gaseous medium within an hermetically sealed envelope. A tube can perform rectification, amplification, modulation, demodulation, oscillation, limiting, and a variety of other functions. Examples include *cathode-ray tubes, receiving tubes,* and *crossed-field tubes.*

electron-tube amplifier An amplifier that has vacuum receiving and/or power amplifying components. See *triode* and *pentode.*

electron-tube coupler A coupler specifically designed to be inserted between an electron tube and an input or output device, as between a magnetron and a transmission line.

electronvolt [eV] A unit of energy equal to the energy acquired by an electron when it passes through a potential difference of 1 V in a vacuum. One electronvolt is equal to 1.60219×10^{-12} erg.

electron-wave tube An electron tube that has mutually interacting streams of electrons with different velocities that cause a signal modulation to change progressively along the length of the electron streams.

electrooptical effect The effect wherein certain transparent dielectrics become doubly refracting when placed in an electric field.

electrooptical material A material that is capable of transforming electrical information into optical information or performing some optical function in response to an electric signal. One example is lead lanthanum zirconate titanate, a transparent ferroelectric ceramic whose optical properties can be changed by an electric field. In lasers, such materials can be used for beam deflection, beam modulation, and *Q* switching.

electrooptical modulator An optical modulator that includes a Kerr cell, an electrooptical crystal, or other signal-controlled electrooptical device to modulate the amplitude, phase, frequency, or direction of a beam of light. With a laser beam, modulating frequencies well into the gigahertz range are possible.

electrooptical shutter A shutter that modulates a beam of light with a Kerr cell.

electrooptics That branch of optics concerned with the effect of an electric field on light rays passing through an electrooptical material. One application is the Kerr cell, in which a signal voltage is applied to the liquid cell to modulate a light beam directed through the cell. The term elec-

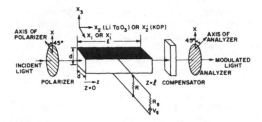

Electrooptical modulator in which the intensity of a light beam is ideally proportional to DC voltage applied to the opposite faces of a ferroelectric or piezoelectric crystal.

trooptics is used interchangeably with the broader term optoelectronics. For consistency and simplicity in this dictionary, most terms relating to combinations of electricity and electronics with optics will be found near the entries for optoelectronics and photoelectric.

electrophoresis The movement of charged particles suspended in a fluid medium, under the influence of an electric field. It is also called cataphoresis.

electrophoretic display A liquid crystal display that has had light-absorbing dye added to the liquid to improve both color and luminance contrast. Individual electrically charged dye particles move when an electric field is applied. If white dye particles are suspended in a black fluid between transparent electrodes, a DC voltage makes the particles deposit on one electrode, and the display appears white when viewed through that electrode. When the polarity of the voltage is reversed, the particles move to the other electrode, and the display appears dark.

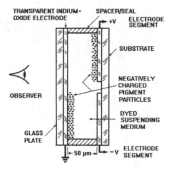

Electrophoretic display depends on the attraction of negatively charged pigment particles.

electrophorus A device that produces electric charges by induction. It consists of a metal plate and a disk of resinous insulating material. In operation, the insulating disk is negatively charged by rubbing it with fur. The metal plate, held by an insulating handle, is placed on the disk so it is charged by induction (bottom surface positive and top surface negative). The top surface is touched with a finger to remove the negative charge. When lifted off, the plate then has a strong positive overall charge.

electrophotography Original name for xerography, as invented by Chester F. Carlson in 1937. Electrophotography now includes both xerography and xeroradiography.

electroplating The electrodeposition of an adherent metal coating on a conductive object for protection, decoration, or other purposes. The object to be plated is placed in an electrolyte and connected to one terminal of a DC voltage source. The metal to be deposited is similarly immersed and connected to the other terminal. Ions of the metal provide transfer of metal as they make up the current flow between the electrodes.

electropolishing The process of producing a smooth, lustrous surface on a metal by making it the anode in an electrolytic solution and preferentially dissolving the minute protuberances.

electropositive The property of a positive electric polarity.

electrorefining The process of dissolving a metal from an impure anode by means of electrodeposition and redepositing it in a purer state on a cathode.

electroresistive effect The change in the resistivity of certain materials with changes in applied voltage. The varistor is based on this effect.

electroscope An instrument for detecting an electric charge by means of the mechanical forces exerted between electrically charged bodies. In one form, two narrow strips of gold leaf suspended in a glass jar spread apart when charged. The angle between the strips is then proportional to the charge.

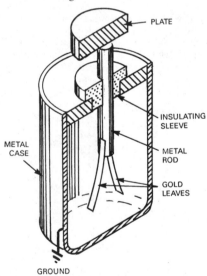

Electroscope demonstrates that an electrostatic charge causes the gold leaves to be repelled in proportion to charge, but they collapse when the electroscope is discharged.

electrosensitive paper A conductive paper that darkens when electric current is sent through it.

electrosensitive recording Recording in which the image is produced by passing electric current through the record sheet.

electrostatic A reference to electricity at rest, such as an electric charge on an object.

electrostatic accelerator *Electrostatic generator.*

electrostatic air cleaner *Precipitator.*

electrostatic cathode-ray tube A cathode-ray tube (CRT) whose electron beam is deflected by electrostatic plates. It is a common component in conventional oscilloscopes.

electrostatic copier A copying machine in which a photosensitive material is electrically charged in the pattern of the original being copied, and the latent image is developed by applying a finely powdered carbon toner that has been oppositely charged. Examples include Xerox and Electrofax copying processes.

electrostatic deflection The deflection of an electron beam by means of an electrostatic field produced by electrodes on opposite sides of the beam. The electron beam is attracted to a positive electrode and repelled by a negative electrode.

electrostatic detector An instrument for determining the presence or absence, polarity, and relative magnitude of electrostatic charges.

electrostatic discharge [ESD] The movement of static electricity from a nonconductive surface that can damage or destroy semiconductors and other circuit components. Static electricity can build on paper, plastic, or other nonconductors and be discharged by human skin (finger) contact. It can also be generated by scuffing shoes on a carpet or by brushing a nonconductor. MOSFETs and CMOS logic ICs are especially vulnerable because it causes internal local heating that melts or fractures the dielectric silicon oxide that insulates gates from other internal structures.

electrostatic discharge [ESD] **protection** Methods for protecting devices and circuits against ESD by the following. 1. Making surfaces on packages and containers for transporting vulnerable devices conductive to prevent or dissipate static buildup. 2. Grounding conductive work surfaces. 3. Requiring that handlers wear grounded, conductive wrist straps and conductive outer garments. 4. Maintaining at least 50% relative humidity and active air ionization in the work area.

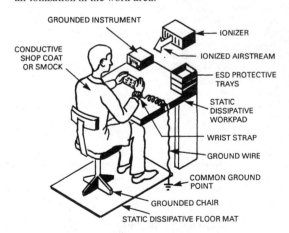

Electrostatic discharge (ESD) control: Protective steps at the workstation ensure that ESD does not damage or destroy ESD-sensitive components and circuits.

electrostatic discharge [ESD] **sensitivity** The vulnerability of a circuit or device to damage or destruction by ESD.

electrostatic discharge [ESD] **simulator** Equipment for simulating the discharge of static electricity from the human body or naturally occurring discharge.

electrostatic field The region around an electrically charged object that will induce an electrical charge on a second object, causing it to experience a force; the voltage gradient between two points at different potentials.

electrostatic fieldmeter An instrument with a noncontact probe or sensor for measuring the electrostatic field produced by a charged body. It provides a measurement of electrostatic field strength or electrostatic voltage at a calibrated distance from a charged body.

electrostatic focusing A method of focusing an electron beam by the action of an electric field, as in the electron gun of a cathode-ray tube.

electrostatic generator A high-voltage generator whose electric charges are generated by friction or induction, then transferred mechanically to an insulated electrode to build up a voltage that can reach 9 MV. Examples include the Van de Graaff generator and the Wimshurst machine.

electrostatic gyroscope A gyroscope with a small beryllium ball that is electrostatically suspended within an array of six electrodes in a vacuum inside a ceramic envelope. External coils provide a rotating magnetic field for spinning the ball at about 30,000 rpm, after which it can take up to 500 days to coast to a stop. Two photoelectric detectors sense the positions of marks on the ball and thereby obtain signals for determining and controlling the spin axis of the ball.

electrostatic headphone A headphone whose moving diaphragm is an extremely thin membrane having a conductive coating. The diaphragm is positioned between two acoustically transparent metal plates to which it is alternately attracted and repelled as the audio signal varies the high DC voltage between the plates.

electrostatic induction The process of charging an object electrically by bringing it near another charged object.

electrostatic instrument A meter that depends for its operation on the forces of attraction and repulsion between electrically charged bodies.

electrostatic lens A lens consisting of coaxial metal cylinders and pierced diaphragms operated at potentials that produce electrostatic focusing of an electron beam directed along the axis of the lens.

electrostatic loudspeaker A loudspeaker in which the mechanical forces are produced by the action of electrostatic fields. In one type, the fields are produced between a thin metal diaphragm and a rigid metal plate. Also called capacitor loudspeaker.

electrostatic machine *Electrostatic generator.*

electrostatic memory *Electrostatic storage.*

electrostatic microphone *Capacitor microphone.*

electrostatic painting A painting process that depends on the particle-attracting property of electrostatic charges. A DC voltage of about 100 kV is applied to a grid of wires through which the paint is sprayed, to charge each particle. The metal objects to be sprayed are connected to the opposite terminal of the high-voltage circuit so that they attract the particles of paint and thereby minimize waste of paint.

electrostatic photomultiplier A photomultiplier whose electrostatic fields cause the electron stream to be reflected off each dynode in turn.

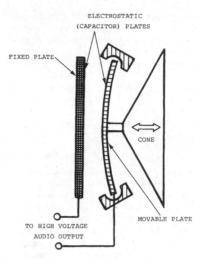

Electrostatic loudspeaker relies on the force of a high-voltage electrostatic field between plates.

electrostatic precipitation The removal of dust, smoke, or other finely divided particles from air by charging the particles with an electric field so that they are attracted to oppositely charged collector electrodes.

electrostatic precipitator *Precipitator.*

electrostatic printer A line printer with high-intensity lamps that project images of characters onto a sensitized drum to form electrostatic patterns that attract ink powder. The images are then transferred to paper and fused just as in electrostatic copiers. More complex versions have character-forming capabilities, as in laser-beam printers, cathode-ray versions like the Videograph, and other charge-forming computer-output printers.

electrostatic propulsion Propulsion of spacecraft or other vehicles by employing electric fields to accelerate charged particles in a desired direction.

electrostatic recording Recording with a signal-controlled electrostatic field.

electrostatic relay A relay whose actuator element consists of nonconducting media separating two or more conductors that change their relative positions because of the mutual attraction or repulsion of electric charges applied to the conductors.

electrostatics The science that deals with electricity at rest such as with charged objects and constant-intensity electric fields.

electrostatic scanning Scanning that involves electrostatic deflection of an electron beam.

electrostatic separator A separator for a finely pulverized mixture that falls through a powerful electric field between two electrodes. Materials with different specific inductive capacitances are deflected by varying amounts and fall into different sorting chutes.

electrostatic shield A grounded metal screen, sheet, or enclosure placed around a device or between two devices to prevent electric fields from acting through the shield. It can prevent interaction between the electric fields of adjacent parts on a chassis.

electrostatic storage   A memory that stores information in the form of the presence or absence of electrostatic charges at specific locations, such as on the screen of a special cathode-ray tube known as a storage tube or cells of dynamic random-access memories.

electrostatic storage tube   *Storage tube.*

electrostatic tweeter   A tweeter loudspeaker in which a flat metal diaphragm is driven directly by a varying high voltage applied between the diaphragm and a fixed metal electrode.

electrostatic unit [ESU]   A CGS unit based on a unit of charge that exerts a force of 1 dyn on another unit charge at a distance of 1 cm in a vacuum.

electrostatic voltmeter   A voltmeter that measures the voltage applied between fixed and movable metal vanes. The resulting electrostatic force deflects the movable vane against the tension of a spring. An attached pointer moving over a scale indicates the voltage of the circuit. It can measure high values of DC voltage.

electrostatography   A generic term covering all processes involving the forming and use of electrostatic charged patterns for recording and reproducing images. The most important application today is in xerography for office copying machines.

electrostriction   The change in dimensions that occurs in some dielectric materials when they are placed in an electric field. The change is independent of the polarity of the electric field. The reverse effect does not take place, whereas in piezoelectricity the effect is reversible. It is also called the electrostrictive effect.

electrostriction transducer   A transducer that depends on production of an elastic strain in certain symmetric crystals when an electric field is applied.

electrostrictive effect   *Electrostriction.*

electrostrictive relay   A relay with an electrostrictive dielectric as the actuator.

electrothermal instrument   An instrument that depends for its operation on the heating effect of a current.

element   1. A distinct functioning part of an electron tube, semiconductor device, antenna array, or other device, contributing directly to its performance.   2. One of the 112 known substances that cannot be divided into simpler substances by chemical means. Scientists believe the total could eventually reach 114. The most recently discovered elements are generally very short-lived nuclear fission products.   3. *Component.*

elemental area   *Picture element.*

elementary charge   The unit charge of electricity corresponding to the charge on a single electron, and equal to about 4.80298×10^{-10} electrostatic unit.

elementary particle   One of more than 60 particles from which all matter is made. The list now includes the photon, two types of neutrino, the electron, the mu meson, two pi mesons, two K particles, the proton, the neutron, the lambda particle, three sigma particles, two xi particles, each with its antiparticle and the boson. It is also called a *fundamental particle.*

elevated duct   A tropospheric radio duct that has both its upper and lower boundaries above the ground.

elevation   1. *Altitude.*   2. *Elevation angle.*

elevation angle   The angle that a radio, radar, or other beam of radiation makes with respect to the horizontal. Also called elevation.

elevation deviation indicator   An indicator that presents visually the relationship between a target and the elevation angle of a radar or radio beam.

elevation indicator   A component that presents visually the angle between a fixed reference point and a target in the same vertical plane.

ELF   Abbreviation for *extremely low frequency.*

ELF radio transmitter   A U.S. Navy communications system that broadcasts *extremely-low-frequency* [ELF] signals simultaneously from two antenna systems buried in the ground in two locations. One antenna array in Wisconsin consists of two perpendicular cables, each 14 mi long. A second one in Michigan consists of three cables: two are 14 mi long, and one is 28 mi long.

ELINT   Abbreviation for *electronic intelligence.*

ellipsometer   An instrument for measuring thickness and other parameters in extremely thin insulating films used in semiconductor manufacturing. It is based on a polarimetric technique in which the change in polarization states caused by reflection of light from a film-covered surface is measured at a fixed angle of incidence. This change in polarization is a function of thickness.

elliptically polarized wave   An electromagnetic wave whose direction of propagation describes an ellipse at certain frequencies.

elliptical polarization   Polarization in which the magnitude of the vector representing the wave varies as the radius of an ellipse while the vector rotates about a point.

elliptical waveguide   A flexible waveguide with an elliptical cross section that can replace a rectangular waveguide.

elliptic-function filter   A multistage LC microwave filter with extremely sharp selectivity, low insertion loss, and low delay distortion. It is installed in IF amplifiers of satellite communication receivers where cascaded bridged-T networks give equalization for a bandwidth of 5 MHz at a 70-MHz IF value.

elongation   The extension of the envelope of a signal caused by delayed arrival of multipath components.

ELPD   Abbreviation for *electronically programmable logic device.*

E-mail   An abbreviation for *electronic mail,* computer-to-computer exchange of typed messages, a service available through commercial providers that offer access to the *Internet.*

embedded computer   Computer circuitry on a single circuit board, located within equipment or an instrument, that performs computation and data processing to support the host equipment applications, which could be measurement, testing, diagnosis, or control.

embedding   The process of molding an insulating plastic around a component or assembly to form a solid block having only the leads or terminals exposed. See also *encapsulating.*

EMC   Abbreviation for *electromagnetic compatibility.*

EME   Abbreviation for *electromagnetic environment.*

emergency alert system   The designated successor to the *emergency broadcast system,* an agreement by U.S. commercial broadcasters to transmit emergency public service information in the event of war or natural disaster. It requires the installation in both portable and line-powered receivers of dedicated integrated circuits capable of responding to a coded signal by turning on power to the receiver automatically, so that the emergency warnings, messages, or instructions can be heard.

emergency broadcast system   A system of broadcast stations and interconnecting facilities authorized by the Federal Communications Commission to operate in a controlled manner during a war, threat of war, state of public peril or disaster, or other national emergency.

emergency position indicating radio beacon [EPIRB]   A small, battery-powered maritime transmitter that broadcasts an emergency signal to alert rescue services. It is primarily intended for operation beyond the range of shore-based VHF transceivers and has a function that is limited to a Mayday emergency. It can be triggered only when a vessel is on fire, in danger of sinking, or has a person on board with a life-threatening injury or disease. Conventional EPIRBs transmit on two recognized VHF emergency frequencies: 121.5 MHz (civilian) and 243.0 MHz (military). EPIRBs usually have water-activated batteries with long shelf-lives. The latest 406-MHz EPIRBs transmit distress signals to orbiting satellites which relay the requests to the National Oceanic and Atmospheric Administration (NOAA). The U.S. Coast Guard is then alerted by NOAA on 406 and 121.5 MHz. Category I EPIRBs are designed and mounted so that they deploy automatically, float free of their shipboard mounting, and activate. Category II EPIRBs must be manually deployed to be activated. See also *COSPAS/SARSAT*.

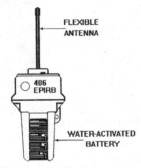

Emergency position-indicating rescue beacon (EPIRB) transmits signals at 406 MHz from a ship in distress to orbiting satellites, which relay the information to shore-based search and rescue stations.

emergency radio channel   Any radio frequency reserved for emergency use, particularly for distress signals. The emergency frequencies for standard channels are 500 kHz and 8.364, 121.5, and 243 MHz.

EMF (or **emf**)   Abbreviation for *electromotive force.*

EMG   1. Abbreviation for *electromyogram.*   2. Abbreviation for *electromyograph.*

EMI   Abbreviation for *electromagnetic interference.*

emission   1. Any radiation of energy by means of electromagnetic waves, as from a radio transmitter.   2. *Electron emission.*

emission bandwidth   The band of frequencies comprising 99% of the total radiated power, extended to include any other discrete frequency at which the power is more than 0.25% of the total radiated power.

emission line   A characteristic frequency of electromagnetic radiation emitted by atoms, molecules, or ions under certain conditions, as when electrons move from one energy level to another.

emission spectrometer   A spectrometer that measures percent concentrations of preselected elements in samples of metals and other materials. When the sample is vaporized by an electric spark or arc, the characteristic wavelengths of light emitted by each element are measured with a diffraction grating and an array of photodetectors.

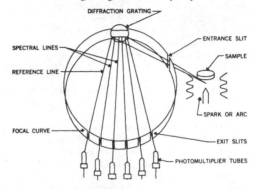

Emission spectrometer.

emission spectrum   The spectrum produced by radiation from any emitting source, such as the spectrum of radiation from an invisible infrared source.

emissivity   The ratio of the radiation emitted by a surface to the radiation emitted by a perfect black-body radiator at the same temperature.

emitron   A television camera tube similar to an iconoscope, made in Great Britain.

emittance   The power radiated per unit area of a radiating surface.

emitter [E]   One of the three doped regions of a bipolar junction transistor. With the emitter-to-base PN junction forward-biased, the emitter injects minority carriers (electrons or holes) into the base region, where they either recombine or diffuse into the collector. The flow of minority carriers from the emitter to the collector under the control of the base-emitter PN junction amplifies signals. See also *base, bipolar transistor,* and *collector.*

emitter bias   A bias voltage applied to the emitter electrode of a transistor.

emitter-coupled logic [ECL]   The fastest bipolar transistor-transistor logic (TTL) family, its transistors are biased into the active region for faster switching and higher power handling. It is widely used as very-high-speed logic in computers. There are several ECL logic families.

emitter follower   *Common collector amplifier.*

emitter-follower logic [EFL]   A form of logic, used in large-scale integration, in which NPN and PNP transistors are combined to give high-performance structures that are similar to complementary-transistor logic but use minimum silicon area.

emitter-function logic [EFL]   A form of large-scale integration that combines the performance advantages of emitter-coupled logic with the compactness of multi-emitter structures. The simplification of gate design reduces propagation delay, power dissipation, and the number of logic levels required.

emitter junction   A transistor junction normally biased in the low-resistance direction to inject minority carriers into a base.

emitter stabilization   The use of a zener diode or similar device to maintain essentially constant emitter voltage in a transistor circuit despite normal variations in base bias.

$E_{m,n}$ mode   British term for *$TM_{m,n}$ mode.*

$E_{m,n}$ wave   British term for *$TM_{m,n}$ wave.*

EMP   Abbreviation for *electromagnetic pulse.*

emphasis   *Preemphasis.*

emphasizer   *Preemphasis network.*

empirical   Based on actual measurement, observation, or experience, rather than on theory.

empty band   Possible energy levels in an atom that do not correspond to the energy of any electron in the given substance in the given state.

EMS   Abbreviation for *electromagnetic susceptibility.*

EMU   Abbreviation for *electromagnetic unit.*

emulation   The modification of an instrument or system by the addition of hardware and/or software so that it performs in the same way as another product. Examples include computer printers with programmable circuitry that permits them to respond to the same input/output signal formats as a printer made by another manufacturer. This term should be distinguished from *simulation.*

emulator   A product or system that includes hardware and/or software that permits it to perform the same way as another product.

EMV   Abbreviation for *electromagnetic vulnerability.*

enable   To activate a circuit or gate by applying a signal or pulse of appropriate form or by removing a suppression signal.

enabling gate   A circuit that initiates the start and determines the length of a generated pulse.

enabling pulse   A pulse that prepares a circuit for some subsequent action.

enameled wire   Wire coated with an insulating layer of baked enamel.

encapsulation   The process of applying a protective coating to a component or assembly by dipping it in a thick insulating plastic fluid or other insulating material that later hardens. Examples of encapsulation materials include RTV silicone, epoxy, and silicon oxide (glass).

encipher   To convert plain text into unintelligible form by means of a cipher system. Its synonyms are encode and encrypt.

enciphered facsimile communication   Communication in which security is accomplished by mixing key pulses, produced by a key generator, with the output of the facsimile converter. Plain text is recovered by subtracting the identical key at the receiving terminal.

enclosure   A housing for loudspeakers or other electronic equipment.

encode   1. To express given information by means of a code.   2. To prepare a routine in machine language for a specific computer.   3. *Encipher.*

encoder   1. An electromechanical component that converts shaft rotation into output pulses that can be counted

to determine shaft revolutions or shaft-angle. The principal types are optical shaft-angle encoders that include light emitters, photodetectors, and a shaft-mounted disk with alternating transparent and opaque patterns. When rotating, the disk "chops" the light beams to produce output codes. The two basic types of optical shaft-angle encoder are the absolute and incremental. They can serve as feedback sensors in closed-loop control systems.   2. A circuit that performs repeated sampling, compression, and analog-to-digital conversion to convert an analog signal to a serial stream of pulse-code modulated (PCM) samples representing the analog signal. See also *absolute optical shaft-angle encoder* and *incremental optical shaft-angle encoder.*

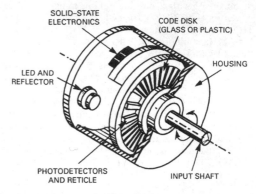

SOLID–STATE ELECTRONICS
CODE DISK (GLASS OR PLASTIC)
HOUSING
LED AND REFLECTOR
PHOTODETECTORS AND RETICLE
INPUT SHAFT

Encoder contains a rotating code disk whose opaque areas intermittently break the light path between a photoemitter and a photosensor, generating pulses that can be converted to a count of shaft revolutions per minute.

encoder disk   See *absolute optical shaft-angle encoder disk* and *incremental optical shaft-angle encoder disk.*

encrypt   *Encipher.*

encryption   The process of scrambling and descrambling video, audio, or data signals with specialized encoding circuits to prevent unauthorized reception. Decoding is done by reversing the encoding process. An example is the encoding of a "play for pay" television program at the source by the cable or satellite service provider that can be decoded by the consumer with the provider's *set-top* box.

encryption key   A string of numbers that must be entered into the encryption algorithm before it can convert the data. Similarly, a *decryption key* converts data back to its original form.

end-around carry   A computer carry signal that is sent directly to the least significant digit place when generated in the most significant digit place.

end effect   The effect of capacitance at the ends of an antenna. It requires that the actual length of a half-wave antenna be about 5% less than a half-wavelength.

end effector   A mechanical tool at the end of a robot arm for performing functions. An electromechanical "hand" can grasp, a suction cup can pick up flat plates, an electromechanical screwdriver can drive screws, a paint sprayer sprays paint, and a welding torch can weld.

end-fed antenna   An antenna that receives its signal at one end. It is also called a voltage-fed antenna because either end of the antenna is a point of maximum voltage. An example is the voltage-fed *zepp antenna*.

end-fed vertical antenna   *Series-fed vertical antenna*.

end-fire array   A linear array whose direction of maximum radiation is along the axis of the array. It can be either unidirectional or bidirectional. The elements of the array are parallel and in the same plane, as in a fishbone antenna.

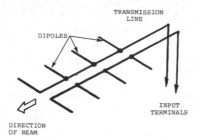

End-fire array radiates the maximum field intensity from the end rather than broadside.

end-of-tape marker   A metal tab, light-reflecting strip, transparent section, or special code placed at the end of the permissible recording area on magnetic tape.

endothermic   A process in which heat is absorbed.

end shield   A shield placed at each end of the cathode of a magnetron to prevent electrons from bombarding the end seals.

energize   To apply rated voltage.

energy   Ability to do work. Energy can be transferred from one form to another, but it cannot be created or destroyed. Electrical energy is measured in watt-seconds, joules, or fractions of joules.

energy band   The sets of discrete but closely adjacent energy levels, equal in number to the number of atoms, that arise from each of the quantum states of the atoms of a substance when the atoms condense to a solid state from a nondegenerate gaseous condition. Also called Bloch band. For a semiconductor, the highest energy level is the conduction band containing only the excess electrons resulting from crystal impurities. The next highest level is the valence band, usually completely filled with electrons. Between these bands is the forbidden band, which is wider for an insulating material than for a semiconductor. It does not exist in a conducting material.

energy beam   An intense beam of light, electrons, or other nuclear particles, that can cut, drill, form, weld, or otherwise process metals, ceramics, and other materials.

energy conservation law   Energy can be neither created nor destroyed, only changed from one form to another.

energy dependence   The characteristic response of a radiation detector to a given range of radiation energies or wavelengths, as compared to the response of a standard open-air chamber.

energy diagram   *Energy-level diagram*.

energy gap   The energy range between the bottom of the conduction band and the top of the valence band in a semiconductor.

energy level   A constant-energy state for particles in an atom. Only a limited number of electrons can exist at each energy level. An electron radiates energy when moving to a lower energy level, and absorbs energy when moving to a higher energy level.

energy-level diagram   A diagram in which the energy levels of the particles of a quantized system are indicated by distances of horizontal lines from a zero energy level. It is also called an energy diagram.

energy product curve   A curve obtained by plotting the product of the values of magnetic induction B and magnetic field strength H for each point on the demagnetization curve of a permanent-magnet material.

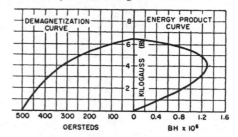

Energy product curve at right is drawn by obtaining values of B and H from the demagnetization curve at left, multiplying them for each point, then plotting the BH products against B.

engineer   A person who has had the training and experience required to perform professional duties in a branch of engineering. An engineer generally designs, constructs, operates, or supervises, whereas a scientist seeks to uncover new knowledge, new principles, or new materials.

engineering   A profession in which a knowledge of the mathematical and physical sciences, gained by study, experience, and practice, is applied with judgment to the utilization of the materials and forces of nature.

engineering workstation   See *workstation*.

enhanced-carrier demodulation   An amplitude demodulation system in which a synchronized local carrier of proper phase is fed into the demodulator to reduce demodulation distortion.

enhanced graphics adapter [EGA]   A graphics adapter that supplies a TTL level RGB signal for a personal computer color monitor. The signal allows up to 64 colors to be displayed.

enhanced small device interface [ESDI]   A standard interface for hard-disk drives that assures the compatibility of drives from different manufacturers.

enhanced video connection [EVC] **standard**   A connection standard promoted by the Video Electronics Standards Association (VESA).

enhancement-depletion circuit   An integrated circuit that contains both enhancement-mode and depletion-mode FETs.

enhancement-mode FET   A reference to a mode of a field-effect transistor (MOSFET or MESFET) in which a specific bias is necessary to cause source-to-drain current flow because the device is fully *depleted*. An enhancement-mode FET is called "normally off." An N-channel FET requires a positive bias, and a P-channel

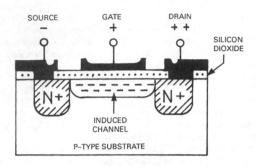

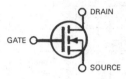

Enhancement mode "normally off" N-channel MOSFET transistor requires a positive gate bias to cause source-to-drain current flow.

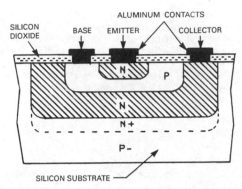

Epitaxial-diffused junction transistor in an integrated circuit has its emitter, base, and collector contacts on the upper surface of the silicon chip.

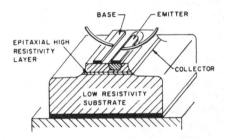

Epitaxial diffused-mesa transistor construction.

FET requires a negative bias to conduct. See also *channel, CMOS, depletion-mode FET, MOSFET,* and *MESFET.*

ENIAC Abbreviation for *Electronic Numerical Integrator.*

entropy The transmission efficiency of an information-handling system, expressed as the logarithm of the number of possible equivalent messages that can be sent by selection from a particular set of symbols.

envelope 1. The glass or metal housing of an electron tube, also called a bulb. 2. A curve drawn to pass through the peaks of a graph showing the waveform of a modulated RF carrier signal.

envelope delay distortion *Delay distortion.*

ephemeris A table that gives the calculated positions of a satellite or celestial body at regular intervals of time.

ephemeris time The fundamental standard of time, based on the orbital motion of the earth about the sun. The ephemeris second is equal 1/31556925.9747 of the tropical year defined by the mean motion of the sun in longitude at noon on January 1, 1900.

EPI Abbreviation for *electronic position indicator.*

EPIRB Abbreviation for *emergency position indicating radio beacon.*

epitaxial diffused-junction transistor A junction transistor produced by growing a thin high-purity layer of semiconductor material on a heavily doped region of the same type.

epitaxial diffused-mesa transistor A diffused-mesa transistor with a thin high-resistivity epitaxial layer that is deposited on the substrate to serve as the collector.

epitaxial junction A doped epitaxial layer of one conductivity type, grown on an epitaxial substrate having the opposite conductivity to produce a PN junction.

epitaxial layer A semiconductor layer with the same crystalline orientation as the substrate on which it is grown. An example of the process is the condensing silicon atoms on a silicon substrate at 1200°C; at this temperature, the

atoms are mobile and able to take up the orientation of the substrate lattice.

epitaxy The controlled growth of a single-crystal layer, called an *epilayer* on a single-crystal (crystalline) substrate. In *homoepitaxy* (silicon layers on a silicon substrate) the epilayer duplicates the substrate's crystallographic structure (orientation of the substrate's lattice). In *heteroepitaxy* (silicon-on-sapphire), the deposited epilayer has a crystalline structure which differs from that of its substrate. It permits the formation of a crystal layer on the substrate that has fewer defects and/or different properties than the host substrate.

epndB Abbreviation for *effective perceived noise decibel.*

epoxy A plastic resin that exhibits high-strength adhesion, low shrinkage, and excellent insulating properties that is used as an adhesive and for encapsulating components and circuits. It can be used to bond the layers of glass-fiber cloth to make glass-fiber epoxy (GFE) circuit boards, and it is the most widely used plastic resin for encapsulating commercial-grade discrete semiconductor devices and integrated circuits, as well as such components as resistors, capacitors, inductors, and relays. See *epoxy glass.*

epoxy glass A high-strength plastic suitable for high-temperature applications. A mixture of epoxy and glass fibers, it offers a low coefficient of thermal expansion and low water absorption. It is the base material of military and industrial-grade conventional and surface-mount circuit boards and cards and laminate that is a suitable substrate for microwave stripline circuitry, plug-in memory modules, and radio-frequency components. It will withstand operating temperatures of 400 to 500°F. It is also called *glass-fiber epoxy* [GFE].

EPPI   Abbreviation for *expanded plan-position indicator.*

EPR   Abbreviation for *electron paramagnetic resonance.*

EPROM   Abbreviation for *erasable programmable read-only memory.*

EPU   Abbreviation for *extended processing unit.*

equal-energy source   A light source for which the time rate of emission of energy per unit of wavelength is constant throughout the visible spectrum.

equal-energy white   The light produced by a source that radiates equal energy at all visible wavelengths.

equalization   The effect of all frequency-discriminative techniques employed in transmitting, recording, amplifying, or other signal processing to obtain a desired overall frequency response. Also called frequency-response equalization.

equalization curve   A curve showing the frequency response needed in a high-fidelity sound-reproducing system to compensate for preemphasis introduced at the broadcasting or recording studio.

equalizer   An electrical network whose phase delay or gain varies with frequency to compensate for an undesired amplitude or phase characteristic in a frequency-dependent transmission line.

equalizing pulse   One of the pulses occurring just before and just after the vertical synchronizing pulses in a television signal and serving to minimize the effect of line-frequency pulses on interlace. The equalizing pulses occur at twice the line frequency and make each vertical deflection start at the correct instant for proper interlace.

equally tempered scale   A series of notes selected by dividing the octave into 12 equal intervals.

equatorial mounting   A mounting for a radio-"telescope" antenna that provides for continuous motion about an axis parallel to the earth's axis of rotation. This allows the antenna to be directed constantly at any point in the sky with a constant-speed motor that compensates for the earth's rotation.

Equatorial orbit of a satellite around earth: the satellite remains over equatorial regions despite the rotation of earth.

equatorial orbit   An orbit in the plane of the earth's equator.

equiangular spiral antenna   A frequency-independent broadband antenna, cut from sheet metal, that radiates a very broad circularly polarized beam on both sides of its surface. This bidirectional radiation pattern is its chief limitation.

Equiangular spiral antenna, cut from sheet metal.

equilibrium   A body that is in a state of rest (static). The simplest condition occurs when a body is acted on by only two forces, which are equal and opposite.

equilibrium orbit   *Stable orbit.*

equiphase surface   Any wave surface over which the field vectors at the same instant are in phase or 180° out-of-phase.

equiphase zone   The region in space within which a difference in phase of two radio signals is indistinguishable.

equipment   One or more assemblies capable of performing a complete function.

equipotential   The property of having the same potential at all points.

equipotential cathode   *Indirectly heated cathode.*

equipotential line   An imaginary line in space or in a medium that possesses the same potential at all points.

equipotential surface   An imaginary surface in space, or in a medium, on which all points have the same potential.

equisignal localizer   An aircraft guidance localizer in which the localizer on-course line is centered in a zone of equal amplitude of two transmitted signals. Deviations from this zone are detectable as unbalance in the levels of the two signals. It is also called an equisignal radio-range beacon or a tone localizer.

equisignal radio-range beacon   *Equisignal localizer.*

equisignal sector   *Equisignal zone.*

equisignal zone   The region in space within which the difference in amplitude of two radio signals (usually emitted by a single station) is indistinguishable.

equivalent binary digits   The number of binary digits that is equivalent to a given number of decimal digits or other characters. When a decimal number is converted into a binary number, the number of binary digits needed is in general about 3.3 times the number of decimal digits. In coded decimal notation, the number of binary digits needed is ordinarily 4 times the number of decimal digits.

equivalent circuit   A circuit that is electrically equivalent to a more complex circuit or device.

equivalent loudness level   *Loudness level.*

equivalent noise temperature   The absolute temperature at which a perfect resistor would generate the same noise as an actual resistor having the same resistance value.

equivalent roentgen   *Roentgen equivalent physical.*

equivalent series inductance [ESL]   The effective inductance in series with an "ideal" capacitor caused by parasitic effects in circuits being rapidly switched. See *parasitic effects.*

equivalent series resistance [ESR] The effective resistance in series with an "ideal" capacitor caused by internal leads or the positions of internal elements.

equivalent sound absorption The area of perfectly absorbing surface that will absorb sound energy at the same rate as the given surface or object under the same conditions. The acoustical unit of equivalent absorption is the sabin.

equivalent stopping power 1. *Relative stopping power.* 2. *Stopping equivalent.*

erasable programmable read-only memory [EPROM] A nonvolatile semiconductor ROM memory that can be erased by exposure to ultraviolet (UV) light and rewritten electrically. The chip is covered with a quartz window permitting it to be exposed to UV light for erasure. The window is recovered with opaque material before the EPROM is reprogrammed. See also *EEPROM* and *OTO EPROM.*

erasable storage A storage whose data can be altered at any time.

erase 1. To remove recorded material from magnetic tape by passing the tape through a strong, constant magnetic field (DC erase) or through a high-frequency alternating magnetic field (AC erase). 2. To change all the binary digits in a digital computer memory to binary zeros. 3. To eliminate previously stored information in a charge-storage tube by charging or discharging all storage elements.

erase oscillator The oscillator in a magnetic recorder that provides the high-frequency signal needed to erase a recording on magnetic tape. The bias oscillator usually serves also as the erase oscillator.

erasing head A magnetic head that obliterates material previously recorded on magnetic tape.

ERC Abbreviation for *electrical rules check.*

E region The region of the ionosphere extending from about 60 to 90 mi (95 to 160 km) above the earth, between the D and F regions.

erg The absolute CGS unit of energy and work. It is the work done when a force of 1 dyn is applied through a distance of 1 cm. One foot-pound is equal to 13.56×10^6 ergs.

ergometer An instrument for measuring the amount of work performed by specific muscles in the body under controlled conditions.

ergonomics The science and technology related to the design and manufacture of tools, controls, equipment, and furniture best suited for long-term human use in homes, offices, factories, ships, aircraft, and vehicles. The term means "measurement of work." It considers all aspects of the environment such as lighting, temperature, humidity, and ambient noise as well as human physiological limitations and fatigue. Its objectives are to maximize human work performance safely. A well-designed workstation will include provisions for long-term comfort, protection against excessive noise or other potential dangers, and provisions for variations in operator size and weight. Tools and controls are designed to fit the hands and/or feet comfortably and will be made of appropriate materials in sizes, shapes, and colors that promote safety and efficient use. It is also called *human factors engineering.*

Erlang A unit of communication traffic density, equal to the average number of simultaneous calls originated during a specific hourly period.

ERP Abbreviation for *effective radiated power.*

error 1. The difference between the true value and a calculated or observed value. 2. *Malfunction.*

error amplifier An operational or differential amplifier that controls feedback in a closed-loop circuit or system. It produces an error voltage when the sensed output differs from the reference voltage. In switching power supplies, it adjusts the pulse-width modulator to correct the sensed output voltage. See *reference amplifier.*

error code A specific character entered into a disk, tape, or memory to indicate that a conscious error was made in the associated block of data. Machines reading the error code can be programmed to throw out the entire block automatically.

error-correcting code A code in which each data signal conforms to rules so that errors in received signals can be automatically detected and some or all of the errors corrected automatically.

error-detecting code *Self-checking code.*

error detection and correction [EDC] A process for storing digital data that has first been given a digital tag derived from that data. When the data is retrieved, the tag is used to check the data for correction.

error rate The ratio of the number of erroneous characters to the total number of characters received in a data-transmission system.

error-rate damping A type of damping in which servo control is accomplished by two voltages: one proportional to the error, and the other proportional to the rate at which the error changes. It is also called a *proportional plus derivative control.*

error signal A voltage or current proportional to the difference between the desired and the actual response in a *closed-loop system.*

error voltage A voltage, usually obtained from a tachometer whose output is proportional to the difference between the angular positions of the input and output shafts of a servosystem. This voltage acts on the system to produce a motion that tends to reduce the error in position. It is also called an *error signal.*

ERTS Abbreviation for *earth resources technology satellite.*

Esaki diode *Tunnel diode.*

escape velocity The minimum velocity that will enable an object to escape from the surface of a planet or other body without further propulsion. The escape velocity for the earth is 25,020 mi/h (40,266 km/h); for the moon, 5364 mi/h (8633 km/h).

ESCA Abbreviation for *electron spectroscopy for chemical analysis.*

E scope A radarscope that produces an E display.

escutcheon An ornamental plate around a dial, window, control knob, or other panel-mounted part.

ESD Abbreviation for *electrostatic discharge.*

ESDI Abbreviation for *enhanced small-device interface.*

ESP Abbreviation for *earth-surface potential.*

ESU Abbreviation for *electrostatic unit.*

etalon An optical device consisting of two spaced, parallel, and partially silvered glass plates. When placed in the cavity of a laser, the etalon is tilted at an angle that pro-

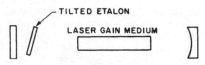

Etalon in single-frequency laser.

vides enough reflection to prevent oscillation at all unwanted axial modes of the long cavity.

etch 1. The removal of copper lamination bonded to a circuit board substrate by acid etch with a subtractive mask and etch process that forms the conductors, lands, and pads on a printed circuit board. 2. The removal of semiconductor material, oxide, or thin metal films by acid or plasma (ion) bombardment such as nitride and oxide etch.

etchant An acid that removes unwanted materials such as metal or oxides from a substrate.

etched printed circuit A printed circuit formed by chemical etching or chemical and electrolytic removal of unwanted portions of a layer of conductive, typically copper foil material, bonded to an insulating base.

etching The act of removing unwanted portions of a metal or semiconductor by chemical or electrolytic action.

ETDMA Abbreviation for *enhanced time-division multi-access*.

Ethernet A local area network (LAN) software standard that determines system operation. It is organized with software, circuit cards, and adapters in each computer or terminal in the network.

ETPC Abbreviation for *electrolytic tough pitch copper*.

ETSI Abbreviation for *European Telecommunications Standards Institute*.

Ettingshausen effect When a metal strip is placed with its plane perpendicular to a magnetic field and an electric current is sent longitudinally through the strip, corresponding points on opposite edges of the strip will have different temperatures.

ETV Abbreviation for *educational television*.

E unit A method of designating radar signal-to-noise ratio. E-1 is a 1 to 1 ratio, barely perceptible; E-2 is 2 to 1, weak; E-3 is 4 to 1, good; E-4 is 8 to 1, strong; E-5 is 16 to 1, very strong or saturating.

EUT Abbreviation for *equipment under test*.

eutectic The liquid alloy composition with the lowest freezing point. For lead-tin solder, the eutectic is 63% tin and 37% solder, giving a freezing point of 361°F.

eV Abbreviation for *electronvolt*.

evacuate To remove gases and vapors from an enclosure. Also called exhaust.

evacuation The removal of gases and vapors from the envelope of an electron tube during manufacture.

evanescent mode The mode of oscillation in which the amplitude diminishes along a waveguide without change of phase.

evaporant The material that is to be deposited as a thin film by evaporation.

evaporated dielectric film A film of dielectric material deposited by vacuum evaporation for passivation of components with active and passive films. It is also used as a capacitor dielectric in thin-film capacitors.

evaporation The process of vaporizing a material by heat in a vacuum to produce a thin film.

evaporative-cooled *Vaporization-cooled*.

EVC Abbreviation for *enhanced video connection standard*.

E vector A vector that represents the electric field of an electromagnetic wave. In free space it is perpendicular to the H vector and the direction of propagation.

even harmonic A harmonic that is an even multiple of the fundamental frequency.

even parity check A method of detecting when bits are dropped by adding 1 bit whenever a character is represented by an odd number of bit patterns. All characters are then represented by an even number of bits, and failure to do so is thus a parity error.

EVR Abbreviation for *electronic video recording*.

EW Abbreviation for *electronic warfare*.

E wave British term for *transverse magnetic wave* (TM wave).

exa- [abbreviated E] A prefix representing 10^{18}.

EXCEPT A logic operator that has the property that if P and Q are two statements, then the statement P EXCEPT Q is true only when P alone is true. It is false for the other three combinations (P false Q false, P false Q true, and P true Q true). The EXCEPT operator is equivalent to AND NOT.

EXCEPT gate A gate that produces an output pulse only for a pulse on one or more input lines and the absence of a pulse on one or more other lines.

excess electron An electron in excess of the number needed to complete the bond structure in a semiconductor, generally resulting from donor impurities.

excess-three code A number code in which the decimal digit n is represented by the 4-bit binary equivalent of $n + 3$, so decimal digits 0 through 9 become 0011, 0100, 0101, 0110, 0111, 1000, 1001, 1010, 1011, and 1100, respectively.

excitation 1. The signal voltage that is applied to the control electrode of an electron tube. 2. Application of signal power to a transmitting antenna. 3. The transfer of a nuclear system from its ground state to an excited state by adding energy. 4. The application of voltage to field coils to produce a magnetic field, as required for the operation of an excited-field loudspeaker or a generator.

excitation winding The magnetic amplifier winding that applies a unidirectional magnetomotive force to the core.

excited-field loudspeaker *Electrodynamic loudspeaker*.

exciter 1. The part of a directional transmitting antenna system that is directly connected to the transmitter. 2. A crystal oscillator or self-excited oscillator that generates the carrier frequency of a transmitter. 3. A small auxiliary generator that provides field current for an AC generator. 4. A loop or probe that extends into a resonant cavity or waveguide. 5. *Exciter lamp.*

exciting current *Magnetizing current.*

exciton The combination of an electron and a hole in a semiconductor that is in an excited state. In attracting the electron, the hole acts as a positive charge.

exclusion principle *Pauli exclusion principle.*

EXCLUSIVE-OR gate A logic gate whose output is 1 if any one but not more than one of its inputs is 1. The output is 0 if more than one input is 1 or all inputs are 0. Also called XOR gate.

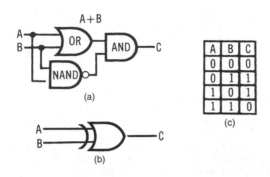

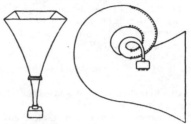

A	B	C
0	0	0
0	1	1
1	0	1
1	1	0

(c)

XOR

EXCLUSIVE OR gate: (*a*) logic symbol, (*b*) schematic symbol, and (*c*) truth table.

execute   To perform the operations in the stored program of a computer.

executive routine   A digital computer routine designed to process and control other routines.

exhaust   *Evacuate.*

exhaust tube   A glass or metal tube through which air from the envelope of an electron tube is evacuated.

exosphere   A region of the atmosphere above the thermosphere, beginning at roughly 600 km, where the density is so low that an upward-traveling molecule makes no further collisions until it falls back to the base of the exosphere. At roughly 10,000 km the exosphere merges into the interplanetary medium. The exosphere overlaps the magnetosphere.

exothermic   A condition of radiating heat or other energy. All radioactive processes are exothermic. It is also called *exoergic.*

expanded sweep   A cathode-ray sweep in which the movement of the electron beam across the screen is speeded up during a selected portion of the sweep time.

expander   The part of a compandor that is used at the receiving end of a circuit to return the compressed signal to its original form. It attenuates weak signals and amplifies strong signals.

expert systems   Computer programs written on a basis of a database gathered from the practical experience of experts in specific fields or disciplines. Examples include medical diagnostic programs, weather prediction (hurricane and tornado) programs, and earthquake or volcano eruption programs. All are based on personal experience, empirical data, readings from test instruments, history, and other factors.

exploring coil   A small coil used to measure a magnetic field or to detect changes produced in a magnetic field by a hidden object. The coil is connected to an indicating instrument either directly or through an amplifier. It is also called a magnetic test coil or a search coil.

exponential absorption   The removal of particles or photons from a beam at an exponential rate as the beam passes through matter.

exponential amplifier   An amplifier capable of supplying an output signal proportional to the exponential of the input signal.

exponential antenna   A television receiving antenna that has a series of active elements mounted parallel to each other, with element lengths adjusted so their ends form two natural logarithmic curves. The antenna provides high gain over the VHF and UHF television bands.

exponential decay   Decay of radiation, charge, signal strength, or some other quantity at an exponential rate.

exponential horn   A horn whose cross-sectional area increases exponentially with axial distance.

Exponential horns, straight and curled.

exponential transmission line   A two-conductor transmission line whose characteristic impedance varies exponentially with electrical length along the line.

exposure dose   A measure of X-ray or gamma radiation at a point, based on its ability to produce ionization. The unit of exposure dose is the roentgen.

exposure-dose rate   The exposure dose per unit time. The unit of exposure-dose rate is the roentgen per unit time.

exposure meter   An instrument that measures the intensity of light reflected from an object for the purpose of determining proper camera exposure. Modern exposure meters consist of a photovoltaic cell connected to an indicating meter.

extended binary-coded decimal interchange code [EBCDIC]   A computer code in which 8 binary bits represent a single character, giving a possible maximum of 256 characters. It is the BCD code extended to 8 binary bits. EBCDIC-encoded messages are returned to the sending station for verification of data integrity.

extended data output [EDO]   A reference to a *dynamic random access memory* [DRAM] format that offers faster speed, (approximately 60 ns) than the 70 to 80 ns of conventional DRAMs.

Extended Industry Standard Architecture [EISA]   An extension for Intel 80386 microprocessors (and later processors) of the ISA bus for PC-compatible computers that provides compatibility with earlier ISA systems. See *Industry Standard Architecture.*

extragalactic radio source   A discrete source of radio signals outside known galaxies or quasars. More than 10,000 such sources have been detected with radio telescopes, at frequencies ranging from 10 MHz to 100 GHz.

extra-high voltage [EHV]   A voltage above 345 kV for electric power transmission.

extraordinary component   The component of light that is plane-polarized and passes through a Nicol prism. The ordinary component is totally reflected by the prism.

extraterrestrial noise   Cosmic, solar, radio, and other electromagnetic noise from sources not related to the earth.

extremely high frequency [EHF]   A Federal Communications Commission designation for a frequency in the band from 30 to 300 GHz, corresponding to a millimetric wave between 1 and 10 mm.

extremely low frequency [ELF]   A frequency below 300 Hz in the radio spectrum, corresponding to a wavelength above 1000 km.

extrinsic properties   The properties of a semiconductor as modified by impurities or imperfections within the crystal.

extrinsic semiconductor   A semiconductor whose electrical properties are dependent on impurities added to the semiconductor crystal, in contrast to an intrinsic semiconductor whose properties are characteristic of an ideal pure crystal.

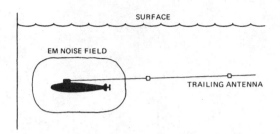

Extremely low frequency signals, in range of 30 to 100 Hz, reach a submerged nuclear submarine trailing long antenna.

F

f Abbreviation for *femto-*.

f Symbol for *frequency*.

F 1. Symbol for *fuse*. 2. Abbreviation for *Fahrenheit*. 3. Abbreviation for *farad*.

°F Abbreviation for degree Fahrenheit.

F– *A–*.

F+ *A+*.

fA Abbreviation for *femtoampere*.

FAA Abbreviation for *Federal Aviation Administration*.

FACCH Abbreviation for *fast associated control channel*.

fab A slang (colloquial) term for the first part or "front end" of semiconductor manufacturing. See *front-end fabrication*.

face bond A bond that is made directly between the bonding pad of a face-down semiconductor chip and a mating contact on a mounting substrate.

faceplate The transparent or semitransparent glass front of a cathode-ray tube, through which the image is viewed or projected. The inner surface of the face is coated with fluorescent chemicals that emit light when hit by an electron beam. Also called face.

facsimile [fax] A system for transmitting and receiving text and/or graphics over the public telephone system. An electric signal representing the image of the *subject copy* is generated by optical scanning in a *scanner* and then is encoded before being transmitted by modem over the telephone lines to a compatible fax receiver or *recorder,* typically a machine capable of both transmitting and receiving. The speed of fax modems is typically 14.4 kb/s. The receiving fax machine decodes the signal and converts it into a format for operating a printer to print out a *facsimile copy* based on the received signal.

facsimile copy The printout of a fax machine on thermal or plain paper. See also *subject copy*.

facsimile machine A modern personal or business machine that combines the dual functions of fax scanning and transmission and receiving and printout in a single integrated unit. In a common scanning method, the *subject copy* is inserted in the machine so it passes around a drum rotating at constant angular velocity. An optical carriage with a photocell is mounted on a traverse drive belt that advances a distance equal to the height of a specified rectilinear area or horizontal strip across the width of the subject copy. The photocell generates a stream of pixels encountered in a helical-scan track. The output of the photocell or cells is converted by a *modem* for transmission over the public dial-up telephone line to a compatible fax machine that contains a mode for converting the data back to a signal that can be printed out on plain or thermal paper. Fax machines are typically assigned an exclusive telephone number so they can operate 24 hours a day without interfering with voice telephone service. Some machines can also serve as *scanners* or document digitizers for converting text and graphics to a digital format for processing or storage by computer, and some can also serve as office copying machines.

facsimile signal The picture signal produced by scanning the subject copy in a facsimile transmitter.

facsimile transmission The transmission of digital data produced by scanning fixed graphic material, including photographs, for reproduction in record form.

fade To change signal strength gradually.

fadeout A gradual and temporary loss of a received radio or television signal caused by magnetic storms, atmospheric disturbances, or other conditions along the transmission path. A blackout is a fadeout that can last several hours or more at a particular frequency.

fader A multiple-unit volume control for the gradual changeover from one microphone, audio channel, or television camera to another. In each changeover, the level is held essentially constant because one section of the fader increases signal level in its channel while the other fader section reduces signal level a corresponding amount.

fading   Variations in the field strength of a radio signal, usually gradual, that are caused by changes in the transmission medium.

fading margin   An attenuation allowance made in radio system planning so that anticipated fading will still keep the signal above a specified minimum signal-to-noise ratio.

Fahnestock clip   A spring-type terminal for making a temporary electrical connection.

Fahrenheit [F]   A temperature scale in which the freezing point of water is 32° and the boiling point of water is 212° at normal atmospheric pressure. Absolute zero is –459.6°F. To change degrees Fahrenheit to degrees Celsius, subtract 32 and then multiply the result by ⅝. Fahrenheit and Celsius scales agree at one point: –40°F = –40°C. Multiply degrees Celsius by ⅝ and add 32 to get degrees Fahrenheit.

fail-safe control   A control so designed that control-circuit failure cannot cause a dangerous condition under any circumstances.

fairlead   A plastic, wood, or metal tube with a funnel-shaped end through which a trailing antenna is reeled on an aircraft.

fall time   The time for the voltage at the trailing edge of a pulse to fall from 90 to 10% of the original negative or positive amplitude. It is the opposite of *rise time.*

false echo   A misleading extra echo indication on the plan-position indicator of a ship's radar set. On rivers, shore structures can cause false echoes. A target can also be hit by a side lobe of the beam, and the echo reflected back into the main beam path by a mast, stack, or tower.

fan beam   1. A radio beam having an elliptically shaped cross section in which the ratio of the major to minor axes usually exceeds 3 to 1. The beam is broad in the vertical plane and narrow in the horizontal plane.   2. A radar beam with a fan shape.

fan dipole   See *dipole antenna.*

fan-in   The number of inputs that can be connected to a digital logic gate.

fan-in circuit   A circuit having many inputs feeding to a common point.

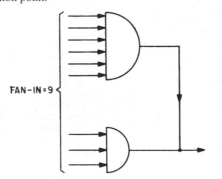

Fan-in circuit for logic gates.

fan marker   *Fan marker beacon.*

fan marker beacon   A VHF radio facility with a vertically directed fan beam intersecting an airway to provide a fix. When used near an airport as part of an instrument land-ing system, it is called a boundary, middle, or outer marker beacon.

fanning beam   A radio beam that is swept back and forth repeatedly over a limited arc.

fan-out   The number of parallel compatible logic outputs that can be driven from an integrated circuit without exceeding the device specifications.

fan-out circuit   A circuit with a single output point that feeds many branches.

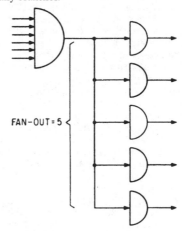

Fan-out circuit with a fan-in circuit for logic gates.

farad [F]   The SI unit of capacitance. A capacitor has a capacitance of 1 F when a voltage change of 1 V/s across it produces a current of 1 A. The farad is too large a unit for practical work, so two smaller units are generally used. The microfarad is equal to one-millionth of a farad, and the picofarad is equal to one-millionth of a microfarad.

faraday   A unit of quantity of electricity equal to 96,500 C.

Faraday cage   See *Faraday shield.*

Faraday dark space   The relatively nonluminous region that separates the negative glow from the positive column in a cold-cathode glow-discharge tube.

Faraday effect   When a plane-polarized beam of light passes through certain transparent substances in a direction parallel to the lines of a strong magnetic field, the plane of polarization is rotated a certain amount. The same effect governs the action of a ferrite rotator in a waveguide.

Faraday rotation   The rotation of a signal's polarization caused by its passage through the ionized E and F layers of the upper atmosphere.

Faraday rotation isolator   An isolator for circular wave-guides that passes waves traveling in one direction but not

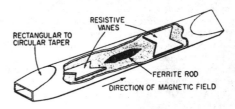

Faraday rotation isolator.

in the other direction. A ferrite rod is positioned between resistive vanes to rotate the plane of polarization of the waves 45°. With appropriate rectangular-to-circular transitions, the isolator will work with rectangular waveguides.

Faraday rotator phase shifter A latching, reciprocal phase shifter that includes a Faraday rotator section and nonreciprocal quarter-wave plates at each end. It includes a square ferrite-filled waveguide. An axial coil is wound around the waveguide, and its magnetic field is completed outside the waveguide wall with ferrimagnetic yokes. Quarter-wave plates at each end convert energy in the rectangular enclosure either to right- or left-hand circularly polarized energy, depending on the direction of propagation. The energy phase is changed by a variable axial magnetic field supplied by the coil around the waveguide. See *phase shifter*.

Faraday shield 1. An electrostatic shield placed between the primary and secondary windings of a transformer to reduce coupling capacitance and common-mode noise. 2. An electrostatic shield made of open metal-wire mesh. 3. A plastic bag with both outer surfaces metallized to protect delicate, ESD-sensitive components from damage or destruction from ESD. The shield provides electrostatic shielding while passing electromagnetic waves. It need not be grounded. See *electrostatic shield*.

Faraday's law The voltage induced in a circuit is proportional to the rate at which the magnetic flux linkages of the circuit are changing. Also called law of electromagnetic induction.

faradic current An intermittent and nonsymmetrical alternating current similar to that obtained from the secondary winding of an induction coil.

far-end crosstalk Crosstalk that travels along the disturbed circuit in the same direction as desired signals in that circuit. When it occurs at carrier telephone repeater stations, the output signals of one repeater go out also over the output line for the other repeater.

far field The radiation field in the Fraunhofer region surrounding a transmitting antenna.

far region *Fraunhofer region.*

far zone *Fraunhofer region.*

fast Fourier transform [FFT] A Fourier transform method for calculating the frequency spectrum, in both magnitude and angle, for any function of time by means of special operating programs that speed machine computation of complex Fourier series.

fathometer *Depth sounder.*

fault 1. A defect such as an open circuit, short circuit, or inadvertent ground in a component, circuit, or transmission line. 2. A defect in a semiconductor device that can cause the device to fail during normal operation. It is usually the result of a processing defect.

fault mode current The current in a circuit under fault conditions, such as during arcbacks and load short-circuits.

fault simulation This term usually means computer simulation of processing defects in integrated-circuit fabrication. The circuit description is modified (faulted) to correspond to a processing defect, and the simulation is rerun to determine whether the test program would find this defect.

FAX (or **fax**) Abbreviation for *facsimile*.

fc Abbreviation for *footcandle*.

FC-AL Abbreviation for *fiber-channel arbitrated loop*.

FCC Abbreviation for *Federal Communications Commission*.

FDCC Abbreviation for *forward control channel*.

FDDI Abbreviation for *fiber distributed data interface*.

FDM Abbreviation for *frequency-division multiplex*.

FDMA Abbreviation for *frequency-division multiple-access*.

FDX Abbreviation for *full-duplex transmission*.

FEC Abbreviation for *forward error correction*.

FED Abbreviation for *field-emitter display*.

Federal Aviation Administration [FAA] An agency created by Congress in 1958, with full authority over both military and civilian airspace requirements.

Federal Communications Commission [FCC] A U.S. government agency responsible for the following: 1. The policy governing allocation of radiated emissions (frequencies). 2. Establishing owner eligibility for commercial radio and TV stations and their licensing. 3. The qualification and licensing of amateur and professional radio and radar operators and electronic technicians.

feed 1. To supply a signal to the input of a circuit, transmission line, or antenna. 2. That part of a radar antenna connected to or mounted on the end of the transmission line which radiates RF energy to the reflector or receives energy therefrom.

feedback The return of a portion of the output of a circuit or device to its input. With positive feedback, the signal fed back is in phase with the input and increases amplification but can cause oscillation. With negative feedback, the signal fed back is 180° out of phase with the input and decreases amplification but stabilizes circuit performance and tends to minimize noise and distortion.

feedback amplifier An amplifier with a passive network that returns a portion of the output signal to its input in a way that changes the performance characteristics of the amplifier.

feedback controller A circuit or element in a *closed-loop system* that controls the motor or other actuator in a closed loop system to correct an *error signal*.

feedback control loop A control loop circuit that senses any deviation from the desired output level and changes the drive waveforms of the power switch to compensate for those changes.

feedback control system *Closed loop control system.*

feedback oscillator An oscillating circuit with an amplifier whose output is fed back in phase with the input. Oscillation is maintained at a frequency determined by the values of the reactive components in the amplifier and the feedback circuits.

feedback path The transmission path from the loop output signal to the loop feedback signal in a feedback control loop.

feedback regulator A feedback control system that tries to maintain a prescribed relationship between certain system signals and other predetermined quantities. Some system signals in a regulator are adjustable reference signals. Under certain methods of operation, a feedback regulator can also be a servomechanism.

feedback resistor The resistor in parallel with an operational amplifier to couple output to input and act as a voltage control.

feedback signal *Primary feedback.*

feedback winding   A winding that makes feedback connections in a magnetic amplifier.

feeder   A transmission line between a transmitter and an antenna.

feed forward   A method for improving line regulation by directly sensing the input voltage of a circuit.

feed-forward control   Process control in which changes are detected at the process input, and an anticipating correction signal is applied before process output is affected.

feed-forward network   An external distortion-reducing network for use with a class C amplifier in communication-satellite service. Signal samples are taken before and after the distorting amplifier to obtain an error signal. This error signal is processed and combined beyond the amplifier output, to provide an undistorted signal by cancellation of the common error terms.

feed line   A wire or cable for connecting an antenna to the transmitter and receiver. See *transmission line*.

feedthrough   A conductor that connects patterns on opposite sides of a printed-circuit board. It is also called a *via* connection.

feedthrough capacitor   A feedthrough insulator that provides a desired value of capacitance between the feedthrough conductor and the metal chassis or panel through which the conductor is passing. It serves as a bypass in UHF circuits.

feedthrough insulator   *Feedthrough terminal.*

feedthrough terminal   An insulating sleeve with a coaxial conductor, made to mount in a hole formed in a metal panel, case, or firewall to pass electrical power or signals through a partition. It is also called a *feedthrough insulator*.

feet per second [ft/s]   A unit for specifying the speed of sound through a medium. In air at standard sea-level conditions, the speed of sound is approximately 1080 ft/s (330 m/s), and 4800 ft/s (1460 m/s) in water.

female connector   A connector that has one or more contacts set into recessed openings. Jacks, sockets, and wall outlets are examples of female connectors.

femto- [f]   A prefix representing 10^{-13}, which is 0.000 000 000 000 001, or one-thousandth of a millionth of a millionth.

femtoampere [fA]   A unit of current equal to 10^{-13} A.

femtosecond [fs]   A unit of time equal to 10^{-13} second.

femtovolt [fV]   A unit of voltage equal to 10^{-13} V.

femtowatt [fW]   A unit of power equal to 10^{-13} W.

FEP   An abbreviation for *fluorinated ethylene propylene*.

Fermat's principle   An electromagnetic wave will take the path of least travel time when propagating between two points.

fernico   An iron-nickel cobalt alloy whose coefficient of expansion is similar to that of glass, making it suitable for making glass-to-metal seals for power tubes and other vacuum chambers. See also *Kovar*.

ferric   Containing a trivalent compound of iron.

ferric oxide [Fe_2O_3]   A magnetic iron oxide (red) used as a coating on magnetic recording tapes.

ferrimagnetic limiter   A power limiter that replaces TR tubes in microwave systems. It uses ferrimagnetic material that exhibits nonlinear properties, such as ferrite or garnet.

ferrimagnetic material   A ferrite material that exhibits ferrimagnetism, such as yttrium-iron garnet polycrystalline materials.

ferrimagnetism   A type of magnetism in which the magnetic moments of neighboring ions tend to align antiparallel to each other. The moments are of different magnitudes, however, so there can be a large resultant magnetization. It is observed in the ferrites and similar compounds.

ferristor   A miniature two-winding saturable reactor that operates at a high carrier frequency and can be connected as a coincidence gate, current discriminator, free-running multivibrator, oscillator, or ring counter.

ferrite   A powdered, compressed, and sintered magnetic material that has high resistivity, consisting chiefly of ferric oxide combined with one or more other metals. The high resistance makes eddy current losses extremely low at high frequencies. Examples of ferrite compositions include nickel ferrite, nickel-cobalt ferrite, manganese-magnesium ferrite, yttrium-iron garnet, and single-crystal yttrium-iron garnet. It is also called ferrospinel.

ferrite bead   A bead made of ferrite powder, placed on a connecting wire in a high-frequency circuit to introduce inductance for suppressing parasitic oscillations.

ferrite circulator   A combination of two dual-mode transducers and a 45° ferrite rotator that control and switch microwave energy in a rectangular waveguide.

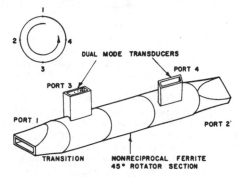

Ferrite circulator with four rectangular ports.

ferrite-core memory   A magnetic memory consisting of read-in and read-out wires threaded through a matrix of tiny toroidal cores molded from a square-loop ferrite. Some cores of this type are only 0.018 in (0.5 mm) in diameter. It is also called a core memory and a core plane. It is now an obsolete technique.

ferrite isolator   An isolator that passes energy in one direction through a waveguide while making possible the absorption of energy from the opposite direction. A ferrite rod, centered on the axis of a short length of circular waveguide, is located between rectangular-waveguide end sections displaced 45° with respect to each other. A signal in the desired direction is rotated 45° by the ferrite rod to make it correct for the output waveguide. A backward signal is rotated 45° in the wrong direction, and its energy is absorbed by resistance cards.

ferrite limiter   A passive low-power microwave limiter that has an insertion loss of less than 1 dB when it is operating in its linear range, with minimum phase distortion. The input signal is coupled to a single-crystal sample of

ferrite phase-differential circulator

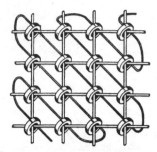

Ferrite-core memory construction. Technology is now obsolete.

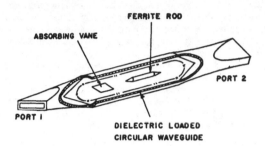

Ferrite isolator.

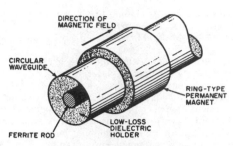

Ferrite rotator in circular waveguide.

either yttrium-iron garnet or lithium ferrite, which is biased to resonance by a magnetic field. They are used to protect sensitive receivers from burnout and blocking by a strong interfering signal.

ferrite phase-differential circulator A combination microwave duplexer and load isolator that functions as a switching device between a high-power radar magnetron, a radar receiver, and a radar antenna. It consists of a T junction, a ferrite section, and a short-slot hybrid coupler with a dual adapter that connects a matched load and the antenna to the hybrid. The ferrite section consists of a double waveguide unit, with a wall between the two waveguide lines. Ferrite slabs are placed in the waveguide lines to produce the desired differential phase shift.

ferrite phase shifter A microwave radar component that alters the frequency phase of the signal passing through it. A matrix of phase shifters permits a phased-array radar beam to be electronically scanned. The most common are ferrite phase shifters enclosed in a waveguide. Alternatives are microstrip versions that include ferrimagnetic materials (ferrites and garnets or ceramic materials with magnetic properties. A phase shifter can be either *reciprocal* or *nonreciprocal*. The toroidal phase shifter is the most common nonreciprocal type; Reggia-Spenser and Faraday rotator phase shifters are the most popular reciprocal units.

ferrite-rod antenna An antenna that consists of a coil wound on a rod of ferrite for radio receivers. The coil generally serves as the tuning inductance for the first stage of the receiver. It is also called a ferrod or loopstick antenna.

ferrite rotator A gyrator consisting of a ferrite cylinder surrounded by a ring-type permanent magnet, inserted in

a waveguide to rotate the plane of polarization of the electromagnetic wave passing through the waveguide. The dimensions of the ferrite cylinder and the magnetization of the ferrite determine the amount of rotation.

ferrite switch A ferrite device that blocks the flow of energy through a waveguide by rotating the electric field vector 90°. The switch is energized by sending direct current through its magnetizing coil. The rotated energy is then reflected from a reactive mismatch or absorbed in a resistive card.

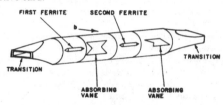

Ferrite switch.

ferrite-tuned oscillator An oscillator with the resonant characteristic of a ferrite-loaded cavity that is changed by varying the ambient magnetic field, to permit electronic tuning. In one example, the tuning range is 500 to 1300 MHz.

ferroacoustic storage A delay-line storage medium consisting of a thin tube of magnetostrictive material, a central conductor passing through the tube, and an ultrasonic

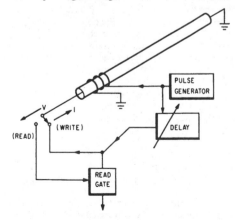

Ferroacoustic-storage principle. At top is thin tube of magnetostrictive material with a central conductor wire.

driving transducer at one end of the tube. To write, an ultrasonic pulse is first sent down the line from the transducer. After a delay corresponding to the time required for the ultrasonic pulse to reach the desired storage point on the line, a short current pulse is applied to the central conductor. This alters the magnetic state of the line at that point. For readout, an ultrasonic pulse is again sent down the line; after a delay corresponding to the location of the desired stored bit, the gate is opened momentarily to provide access to the voltage pulse produced if a bit is stored at that point when the ultrasonic pulse passes it.

ferrod   *Ferrite-rod antenna.*

ferrodynamic instrument   An electrodynamic instrument whose forces are materially augmented by the presence of ferromagnetic material.

ferroelectric converter   A converter that transforms thermal energy into electric energy by using the change in the dielectric constant of a ferroelectric material when heated beyond its Curie temperature. A large capacitor with a ferroelectric dielectric, such as barium strontium titanate, initially at its Curie temperature, is charged by a battery through a diode, then heated beyond the Curie temperature. The dielectric constant then drops, and capacitance drops correspondingly, so that capacitor voltage goes up. After this voltage is discharged through a load, the capacitor is cooled and the cycle is repeated. Solar radiation on spacecraft could provide the required heat.

ferroelectric material   A nonlinear dielectric material whose electric dipoles line up spontaneously by mutual interaction, just as magnetic dipoles line up in a magnetic material. Examples of ferroelectric materials include barium titanate, potassium dihydrogen phosphate, and Rochelle salt. They are used in ceramic capacitors, acoustic transducers, and dielectric amplifiers.

ferroelectric shutter   A shutter consisting of a slab of ferroelectric crystal located between polarizers whose planes are at right angles. The shutter opens to pass light when activated by a pulse of up to 100 V.

ferrofluid   A colloidal suspension of ultramicroscopic magnetic particles in a liquid that acts as a lubricant, damping agent, or heat-transfer medium in a moving object. Permanent magnets or electromagnetics hold the ferrofluid in position even when it is under pressure as in rotary seals. It transfers heat from high-frequency speakers (tweeters) and some medium-frequency, hi-fi speakers.

ferromagnetic amplifier   A parametric amplifier based on the nonlinear behavior of ferromagnetic resonance at high RF power levels. In one version, microwave pumping power is supplied to a garnet or other ferromagnetic crys-

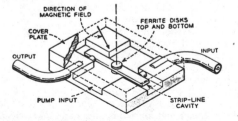

Ferromagnetic amplifier for 4.5-GHz input and output signals, with a 9-GHz pump input.

tal mounted in a cavity containing a stripline. A permanent magnet provides sufficient field strength to produce gyromagnetic resonance in the garnet at the pumping frequency. The input signal is applied to the crystal through the stripline, and the amplified output signal is extracted from the other end of the stripline.

ferromagnetic material   A magnetic material that has a permeability considerably greater than the permeability of a vacuum and which varies with the magnetizing force. The various forms of iron, steel, cobalt, nickel, and their alloys are examples.

ferromagnetic resonance   A resonant condition occurring when the apparent permeability of a magnetic material at microwave frequencies reaches a sharp maximum. This resonance occurs in the presence of a steady transverse magnetic field when the microwave frequency equals the precession frequency of the electron orbits in the atoms of the magnetic material. The resonance frequency depends on the strength of the transverse field.

ferromagnetics   The technology of the storage of information and control of pulse sequences by means of the magnetic polarization properties of materials.

ferrometer   An instrument that can make permeability and hysteresis tests of iron and steel.

ferroresonant circuit   A resonant circuit with a saturable reactor that provides nonlinear characteristics. Tuning is accomplished by varying circuit voltage or current.

ferroresonant transformer   A transformer designed to behave as a tuned circuit by resonating at a specific frequency. The output of the transformer is nearly impervious to variations in input voltage, but if operated at other than the ferroresonant frequency, it has high losses.

ferrospinel   *Ferrite.*

ferrous   Containing a divalent compound of iron.

FES   Abbreviation for *field-emission spectroscopy.*

FET   Abbreviation for *field-effect transistor.*

FET resistor   A field-effect transistor whose gate is tied to the drain; the resultant structure serves as a resistance load for another transistor.

FFT   Abbreviation for *fast Fourier transform.*

FHP motor   Abbreviation for *fractional-horsepower motor.*

fiber bundle   A flexible bundle of glass or other transparent fibers, parallel to each other, that will transmit an image or signals from one end of the bundle to the other.

fiber-channel arbitrated loop [FC-AL]   A serial interface for computer networks that provides wider bandwidth than the parallel *Small Computer Systems Interface* [SCSI].

fiber distributed data interface [FDDI]   A network that transmits data over optical fiber cable in *non-return-to-zero, invert-on-ones* (NRZI) format at data rates of 100 megabits per second.

Fiberglas   Trademark of Owens-Corning Fiberglas Corp. for its glass fiber materials.

fiberoptic cable   Optical fiber or fibers with a protective sheath or jacket suitable for transmitting visible and ultraviolet light for *fiberoptic communication.*

fiberoptic communication   Communication carried on by the transmission of coded or modulated optical signals, generally in the infrared region. Examples include television transmission and computer data transmission, as in a local area network. They can be short-, medium-, or long-haul systems.

fiberoptics The technique for transmitting light through long, thin, flexible fibers of glass, plastic, or other transparent materials. Bundles of parallel fibers can transmit complete images.

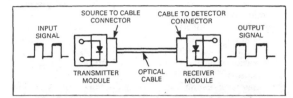

Fiberoptics: electrical signal is converted to a modulated light beam for transmission through an optical fiber. The received light beam is then demodulated electronically to obtain an electrical output signal.

fiberscope An arrangement of parallel glass fibers with an objective lens on one end and an eyepiece at the other end. The assembly can be bent as required to view objects that are inaccessible for direct viewing.

fiber waveguide *Optical waveguide.*

fidelity The degree to which a system accurately reproduces at its output the essential characteristics of the signal impressed on its input.

field 1. One of the equal parts into which a frame is divided in interlaced scanning for television. A field includes one complete scanning operation from top to bottom of the picture and back again. In the present U.S. television broadcasting system there are two fields per frame, with each field taking ¹⁄₆₀ s and including 262.5 lines. 2. A region containing electric or magnetic lines of force, or both. 3. The area covered by a lens. 4. A set of characters treated as a unit for computer processing, such as a name or amount field. 5. An operating location for equipment.

field coil A coil that produces a constant-strength magnetic field in an electric motor, generator, loudspeaker, or other electromagnetic device.

field effect The change produced by an electric field on the equilibrium balance of free electrons and holes in a semiconductor material.

field-effect device A semiconductor device whose properties are determined largely by the effect of an electric field on a region within the semiconductor.

field-effect phototransistor A field-effect transistor that responds to modulated light as the input signal. The action is similar to that of a gate signal voltage.

field-effect transistor [FET] A unipolar, voltage-controlled transistor whose operation requires the movement of only one kind of charge carrier: electrons or holes. It has three electrodes: *source, drain,* and *gate.* FET current moves in a conductive *channel* between the source and the drain under the gate region; N-doped FETs conduct with positive gate bias, and P-doped FETs conduct with negative gate bias. Its drain characteristic is similar to those of a pentode plate. The two kinds are junction FET (JFET) and metal-oxide FET (MOSFET). See *CMOS, HCMOS, JFET, MESFET,* and *MOSFET.*

field-effect varistor A passive two-terminal nonlinear semiconductor device that maintains constant current over a wide voltage range.

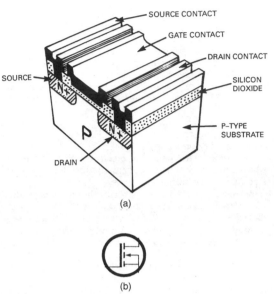

Field-effect transistor: (*a*) an N-channel, enhancement-mode MOSFET is one example. (*b*) Its schematic symbol.

field emission The liberation of electrons from an unheated solid or liquid by a strong electric field at the surface.

field-emitter cathode An unheated metal or semiconductor cathode that emits electrons when exposed to a strong electronic field.

field-emitter display [FED] A flat-panel display for computer monitors and instrument displays that depends on the formation of hundreds of electron beams to replace the single beam for pixel activation. It is intended as a *cathode-ray tube* [CRT] replacement.

field-enhanced photoelectric emission The increased photoelectric emission resulting from the action of a strong electric field on the emitter.

field-enhanced secondary emission The increased secondary emission resulting from the action of a strong electric field on the emitter.

field-free emission current The electron current emitted by a cathode when the electric field at the surface of the cathode is zero.

field frequency The number of fields transmitted per second in television. In the United States, it is 60 fields per second. The field frequency is equal to the frame frequency multiplied by the number of fields that make up one frame. Also called field repetition rate.

field intensity *Field strength.*

field ion emission microscope A high-magnification microscope with an intense electric field that is applied to a sharp metal point to make movements of atoms visible on the point.

field-ion microscope A specialized microscope for the study of atomic structure in the nanometer size range. It is basically a vacuum tube containing a refrigerant in an inner chamber and ionized helium in an outer chamber. When high voltage is impressed on a metal specimen etched to an extremely fine point, ionized helium atoms

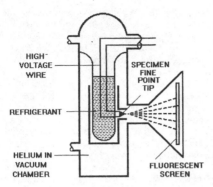

Field ion microscope, a vacuum tube without lenses, is used to study the structure of metals. Ionized helium from the specimen tip hits the fluorescent screen to reveal the atomic structure of the specimen.

projected from the point strike an internal fluorescent screen to form lighted patterns of the atomic structure of the specimen magnified more than a million times. It contains no optical or electronic lens.

field-neutralizing coil   A coil that is placed around the faceplate of a color television picture tube. Direct current is sent through this coil to produce a constant magnetic field that offsets the effect of the earth's magnetic field on the electron beams.

field-neutralizing magnet   A permanent magnet mounted near the edge of the faceplate of a color picture tube to serve the same function as a field-neutralizing coil. Also called rim magnet.

field of force   A region in space in which force is exerted on electric charges by other stationary or moving charges.

field pattern   *Radiation pattern.*

field period   The time required to transmit one television field, equal to ⅟₆₀ s in the United States.

field pole   A structure of magnetic material for mounting a field coil of a loudspeaker, motor, generator, or other electromagnetic device.

field-programmable gate array [FPGA]   A gate array based on static RAM (SRAM) transistors in place of fuses to control the signal paths in an array of gates. Each time the system is powered, the program is downloaded into the FPGA to determine which transistors are on or off. Some have *antifuse* links that melt together, forming permanent interconnections when programming pulses are applied.

field-programmable logic array [FPLA]   A programmed logic array with internal connections of the logic gates that can be programmed once in the field by passing high current through fusible links, with avalanche-induced migration to short base-emitter junctions at desired interconnections. Also called programmable logic array.

field quantum   The fundamental field particle that is the result of quantizing a field.

field repetition rate   *Field frequency.*

field-sequential color television   A color television system whose individual red, blue, and green primary colors are associated with successive fields.

field strength   The strength of an electric, magnetic, or electromagnetic field at a point. For electromagnetic radiation, it is generally expressed in volts, millivolts, or microvolts per meter of effective antenna height. It is also called field intensity.

field-strength meter   A calibrated radio receiver that can be used to measure the field strength of radiated electromagnetic energy from a radio transmitter.

FIFO   Abbreviation for *first-in first-out.*

figure-eight radiation pattern   A radiation pattern that has equal broad lobes 180° apart, resembling the numeral 8.

figure of merit   A performance rating that governs the choice of a device for a particular application. Thus the figure of merit of a magnetic amplifier is the ratio of usable power gain to the control time constant.

filament   1. A coil of tungsten wire that acts as the heating/lighting element in an incandescent lamp. It reaches white heat when a current is passed through it.   2. The coil of tungsten or other high-resistance wire that acts as a heating coil for an adjacent cathode in an electron tube when current is passed through it. The heat drives electrons from the coated or filled cathode to form an electron beam. Early thermionic tubes had their filaments coated with electron-rich materials so that they could function both as heater and cathode.

filament current   The current supplied to the filament of an electron tube for heating purposes.

filament emission   Liberation of electrons from a heated filament wire in an electron tube.

filament resistance   The resistance in ohms of the filament of an electron tube. For metal filaments, the resistance increases with temperature, so the hot resistance is many times the cold resistance.

filament transformer   A small transformer originally made to supply filament or heater current for one or more electron tubes.

filament-type cathode   *Filament.*

filament voltage   The voltage applied to the terminals of the filament in an electron tube.

filament winding   The secondary winding of a power transformer that furnishes AC heater or filament voltage for one or more electron tubes.

file   A collection of related records on magnetic tape or disks for computer memory.

file server   A computer, typically a mainframe or minicomputer, with a large internal memory capable of storing programs, data bases, or other data that can be shared by less-powerful computers in a network.

file transfer protocol [FTP]   A set of rules for transferring data files from one computer to another.

filled band   An energy band in which each energy level is occupied by an electron.

film capacitor   *Plastic-film capacitor.*

film resistor   A fixed resistor whose resistance element is a thin layer of conductive material on an insulated form. The film can be carbon, metal (tin oxide or tantalum nitride) or cermet.

filter   A selective device that transmits a desired range of matter or energy while substantially attenuating all other ranges. Thus an electric filter is a network that transmits alternating currents of desired frequencies while substantially attenuating all other frequencies. An acoustic filter transmits only desired sound frequencies. An optical filter

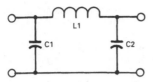

Filter: a basic resistance-capacitance-inductive (RCL) π (pi) input filter can "smooth" ripples in raw DC.

transmits desired wavelength ranges in the visible, ultraviolet, and infrared spectrums.

filter capacitor A capacitor for a power-supply filter system that provides a low-reactance path for alternating currents and acts to suppress ripple currents without affecting direct currents. Electrolytic capacitors are generally used for this purpose.

filter choke An iron-core coil in a power-supply filter system that will pass direct current while offering high impedance to pulsating or alternating currents.

filter response curves Plots of signal amplitude versus frequency for filters. The shapes of the curves indicate filter characteristics, e.g., low-pass, high-pass, bandpass, and band-reject.

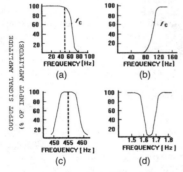

Filter response curves for four basic filters: (*a*) low-pass, (*b*) high-pass, (*c*) bandpass, and (*d*) band-reject.

final amplifier The transmitter stage that feeds the antenna.

finder An optical or electronic device that shows the field of action covered by a television camera.

fine chrominance primary The chrominance primary that is associated with the greater transmission bandwidth in the two-primary U.S. system of color television. The fine chrominance primary is the I signal, and has frequency components up to 1.5 MHz. The coarse chrominance primary is the Q signal, and has a bandwidth of only 0.5 MHz.

finger A software tool that can determine if other users are logged onto the *Internet*. It also serves as an E-mail address directory.

finished crystal blank The finished crystal product after the completion of all processes, including application of electrodes to the faces of the crystal.

finite Having fixed and definite limits of magnitudes.

finite clipping Clipping in which the threshold level is large but below the peak input signal amplitude.

fire-control radar A specialized radar system for tracking identifiable targets in the air, on land, or on water to deter-

mine their range and bearing to permit precise aiming of artillery, rapid-fire defensive guns, and missiles, as well as for setting the fuses on shells or missiles. This class of radar obtains its contacts in a handoff from longer range *search radars,* generally after the targets have been identified as aggressive or threats.

firing 1. The transition from the unsaturated to the saturated state of a saturable reactor. 2. The gas ionization that initiates current flow in a gas discharge tube. 3. Excitation of a magnetron or TR tube by a pulse. 4. Exposure of a material to high temperature, to oxidize and vaporize organic binders as required for achieving desired properties. A thick-film production process.

firmware A term that applies to a read-only memory device that permanently stores computer code, which can include data conversion tables, applications programs, subroutines, and high-level language programs. The storage device can be a stand-alone PROM, EPROM, or EEPROM, or it can be integrated into a *microcontroller.* See also *software.*

first breakdown The normal avalanche or Zener breakdown of a semiconductor device, as distinct from second breakdown.

first detector *Mixer.*

first Fresnel zone The Fresnel zone that is centered on the line-of-sight path between two microwave antennas and bounded by all paths whose lengths are one half-wavelength longer than the direct path.

first harmonic *Fundamental frequency.*

first-in first-out [FIFO] A method of establishing the order in which data are taken from a computer memory or products are taken from inventory. In a FIFO memory, input data propagates automatically toward the output terminals. When data is removed from the output, other data in the memory moves down the line automatically to fill the vacant locations.

first-order servo A servo that has zero static error but a finite steady following error for a velocity input.

first quantum number *Main quantum number.*

fir-tree antenna A vertical array of horizontal dipoles fed by transposed two-wire line and backed by a reflector array.

fishfinder A specialized form of electronic depth sounder for finding schools of fish. They have either liquid-crystal displays (LCDs) or cathode-ray tubes (CRTs) that present a graphic profile of the ocean floor under the ship, showing fish as well as underwater vegetation and structures. (This contrasts with the single digital readout found on most depth sounders.) Every fishfinder has a display/control head and a transducer that must be matched in frequency to work. The display/control head has three parts: display, transmitter/receiver, and signal-processing software. Other circuits can add temperature, speed, and position information. Some models provide a three-dimensional image of the ocean floor by splitting the transducer beam into sections. Others have steerable scanning transducers that allow the user to locate fish more precisely.

fishpole antenna *Whip antenna.*

fix In navigation, the determination of a position of known latitude and longitude. It is typically marked on an aeronautical or nautical chart for further reference in plotting an ongoing course. It can be plotted from visual observa-

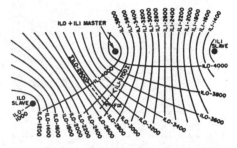

Fix obtained from one master and two slave loran stations. Pair 1L0 gave a time difference of 2900 μs and pair 1L1 gave a time difference of 3700 μs, so the location is at intersection of loran lines 1L0-2900 and 1L1-3700.

tion of known landmarks by triangulation, celestial navigation techniques, loran, or GPS NAVSTAR satellites.

fixed attenuator An *attenuator* for microwave systems consisting of a waveguide section with a resistive card, tapered at both ends, fixed to the center of the broad inside wall in the middle of the waveguide section. The contours of the card are trimmed to obtain the desired attenuation value. It is also called a *fixed resistive-card* attenuator. See also *flap attenuator* and *rotary-vane attenuator*.

fixed bias A constant value of bias voltage, independent of signal strength. For an electron tube or transistors it can be provided by a battery or other DC voltage source. It can also be obtained from a voltage divider connected across the voltage source.

fixed capacitor A capacitor that has a definite capacitance value which cannot be adjusted.

fixed contact A relatively immovable contact that is engaged and disengaged by a moving contact to make and break a circuit, as in a switch or relay.

fixed-frequency transmitter A transmitter designed for operation on a single carrier frequency.

fixed-point arithmetic 1. A method of calculation in which the computer does not consider the location of the decimal or radix point because it is given a fixed position. 2. A type of arithmetic in which the operands and results of all arithmetic operations must be properly scaled so as to have a magnitude between certain fixed values.

fixed-point calculation A computer calculation in which a fixed location of the decimal point or binary point in each number is used or assumed.

fixed-point system A point system of positional notation in which the location of the point is assumed to remain fixed with respect to one end of the numerical expressions.

fixed resistor A resistor that has a fixed resistance value.

fixed station A permanent earth station that is a gateway or base station in satellite communications.

fixed word length The length of a computer machine word that always contains the same number of characters or digits.

fixture A portable or fixed tool that properly aligns and locates parts temporarily for further work such as soldering, welding, or riveting to assure that the assemblies are uniform and meet drawings and specifications. For example, a stainless steel plate with cutouts to accommodate

components in their intended positions can be placed over a hybrid circuit substrate for insertion of the components prior to resoldering to assure their correct positioning. The fixture remains in place until the fixture and substrate have passed through a furnace to bond the components. This term is generally distinguished from a *jig,* which is a portable tool for placement over a part to guide hole drilling.

fL Abbreviation for *footlambert.*

flap attenuator An *attenuator* for microwave systems consisting of a waveguide section with a tapered resistive card that can be moved in and out of a slot cut down the center of the broad wall of the waveguide section. The hinge arrangement allows the card penetration, and hence the

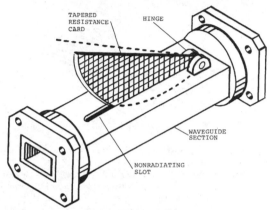

Flap attenuator has a hinged and tapered resistive card that can be moved in or out of the waveguide slot to control attenuation.

attenuation, to be varied from 0 to some maximum value, typically 30 dB. Because the longitudinal slot is centered on the broad wall, no microwave *energy* is radiated. It is also called a *variable resistive-card attenuator.* See also *fixed attenuator* and *rotary vane attenuator.*

flashback voltage The peak inverse voltage at which ionization occurs in a gas tube.

flash magnetization Magnetization of a ferromagnetic object by a current impulse of such short duration that magnetization does not penetrate beyond a shallow surface layer of the material. It is sometimes used in electromagnetic crack detectors.

flash memory An erasable, reprogrammable semiconductor read-only memory (ROM) that permits a data block or groups blocks to be erased and reprogrammed without erasing the entire memory. A variation on the EEPROM electrically erasable memory, some devices can store 1 megabit. It can upgrade system software, act as a portable main computer memory, and replace disk drives and field-programmable memory. See also *EEPROM* and *EPROM.*

flashover An electric discharge around or over the surface of an insulator.

flash tube A gas discharge tube in a photoflash unit to produce high-intensity, short-duration flashes of light for photography. Also called electronic flash tube and photoflash tube.

flat cable A cable made of round or rectangular, parallel copper wires arranged in a plane and laminated or molded into a ribbon of flexible insulating plastic.

flat-conductor cable A cable made of wide, flat conductors arranged side by side in a plane and protected by ribbons of insulating plastic.

flatpack A square or rectangular electronic component package for circuit board mounting of semiconductor devices, resistor networks, or resistor-capacitor networks. Hermetically sealed flatpacks typically have leads projecting from the two opposing sides of its case, in the same plane as the case. Plastic flatpacks usually have leads projecting from all four sides of the case in the same plane as the case (quad flatpack).

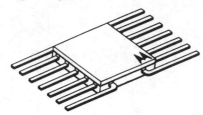

Flatpack with 14 leads.

flat response Uniform amplification or reproduction of a specified band of frequencies.

flat television display A television picture device whose thickness dimension is small with respect to its length and height, typically less than 2 in or 50 mm. These displays, made with active-matrix liquid-crystal (AMLC) technology have replaced cathode-ray tubes in battery-powered portable television sets. Neon gas-discharge flat displays have also been developed as alternatives to the CRT.

flat-top antenna An antenna that has two or more lengths of wire parallel to each other, in a plane parallel to the ground, each fed at or near its midpoint.

flat-top response *Bandpass response.*

flattopping A term for the distorted audio signal produced by a single-sideband transmitter if its microphone is set at too high an output level. The peaks of the voice waveform are cut off by the transmitter because of over-modulation. See also *clipping.*

F layer One of the layers of ionized air that occur at various heights in the F region of the ionosphere, capable of reflecting radio waves back to earth at frequencies up to about 50 MHz. In the daytime hemisphere there are normally two F layers, called the F1 and F2 layers.

F1 layer The lower of the two ionized layers normally existing in the F region in the daytime hemisphere. It is usually somewhere between 90 and 150 mi (145 and 240 km) above the earth, and it chiefly affects frequencies from about 1.5 to about 25 MHz. See also ionosphere.

F2 layer The single ionized layer normally existing in the F region in the nighttime hemisphere. In the daytime hemisphere it is the higher of the two F layers. See also ionosphere.

Fleming's rule 1. *Left-hand rule.* 2. *Right-hand rule.*

Fletcher-Munson curves Equal-loudness curves for human response to pure tones, plotted against frequency. They show the average sound intensity needed to produce a

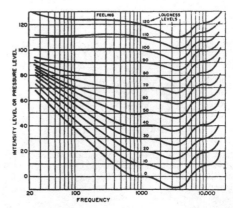

Fletcher-Munson curves of equal-loudness contours for 13 different loudness levels in phons. Frequency scale is in hertz, and intensity level is in decibels.

given loudness sensation throughout the audio-frequency range.

flexible (or **flex**) **circuit** A form of printed circuit board that is fabricated on a flexible rather than rigid substrate material capable of being bent 180° with a 1-in radius without sustaining damage.

flexible coupling 1. A coupling designed to allow a limited angular movement between the axes of two waveguides. 2. A coupling that connects two shafts end to end and permits rotation despite minor shaft misalignment.

flexible waveguide A waveguide that can be bent or twisted without appreciably changing its electrical properties.

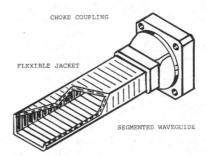

Flexible waveguide simplifies the interconnection between rigidly mounted waveguide sections with low losses.

flicker A visual sensation produced by periodic fluctuations in light at rates ranging from a few cycles per second to a few tens of cycles per second. In NTSC television, interlaced scanning eliminates the flicker that might otherwise be noticed at 30 frames per second.

flicker effect Random variations in the output current of an electron tube that has an oxide-coated cathode, caused by random changes in cathode emission.

flicker noise Electric noise produced by the flicker effect in an electron-tube circuit. It is low-frequency noise that exceeds shot noise. The noise also occurs in semiconductors as a result of the trapping of charges at a semiconductor surface, but it is insignificant at most semiconductor operating frequencies.

flip chip A semiconductor die with all terminations on one side in the form of solder pads or bump contacts. After the surface of the chip has been passivated or otherwise treated, it is flipped over for attaching to a matching substrate on which interconnecting thin films and possibly also thin-film components have previously been deposited. All connections are then made simultaneously by applying heat or a combination of ultrasonic energy and pressure.

flip-flop circuit A two-stage multivibrator circuit that has two stable states. In one state, the first stage is conducting and the second is cut off. In the other state, the second stage is conducting and the first stage is cut off. A trigger signal changes the circuit from one state to the other, and the next trigger signal changes it back to the first state. For counting and scaling purposes, a flip-flop can deliver one output pulse for each two input pulses. It is also known as a bistable multivibrator, Eccles-Jordan circuit, or trigger circuit. See also *D flip-flop, J-K flip-flop, R-S flip-flop* and *T flip-flop.*

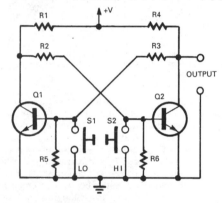

Flip-flop circuit is a bistable multivibrator; also known as an Eccles-Jordan, or binary circuit. It can function as a simple digital memory.

FLIR An abbreviation for *forward-looking infrared unit.*

floating The condition wherein a device or circuit is not grounded and not tied to an established voltage supply.

floating address *Symbolic address.*

floating-average-position action Floating action in which there is a predetermined relation between deviation of the controlled variable and the rate of change of the time-average position of a final control element that is moved periodically from one of two fixed positions to the other.

floating battery A storage battery connected permanently in parallel with another power source. The battery normally handles only small charging or discharging currents, but it can take over the entire load upon failure of the main supply.

floating-carrier modulation *Controlled-carrier modulation.*

floating charge Application of a constant voltage to a storage battery, sufficient to maintain an approximately constant state of charge while the battery is idle or on light duty.

floating grid An electron-tube grid that is not connected to a circuit. The grid assumes a negative potential with respect to the cathode, due to electrons hitting the grid wires, and the tube is then sensitive to external effects

such as movement of a hand near the envelope. Also called free grid.

floating input An amplifier or other circuit in which no input terminal is connected to circuit ground.

floating junction A transistor junction through which the average current is zero.

floating output A circuit output that is not grounded or referenced to another output.

floating point Pertaining to a number system in which the location of the point does not remain fixed with respect to one end of the numerals.

floating-point arithmetic A method of calculation that automatically accounts for the location of the decimal or radix point.

floating-point calculation A computer calculation in which provisions are made for varying the location of the decimal point (if base 10) or binary point (if base 2).

floating-point routine A computer routine that permits floating-point operation for a specific problem.

floating-point system A point system of positional notation in which the position of the point is regularly recalculated and can be moved. A floating-point system usually locates the point by expressing a power of the base, and involves the use of two sets of digits. For floating decimal notation the base is 10, so 6,200,000 would be 6.2, 6. For floating binary notation the base is 2, so 88 would be 11, 3.

float switch A switch actuated by a float at the surface of a liquid.

flood To direct a large-area flow of electrons toward a storage assembly in a charge storage tube.

floodlighting Covering a wide area with radar waves.

floppy disk (or **diskette**) A magnetic memory medium for personal computers made as a double-sided flexible Mylar disk that has been coated with a ferrite magnetic compound. Each side is organized as concentric circles, called *tracks,* and each track is divided into *sectors.* There are 512 bytes of information in each sector of the standard 3.5-in diameter double-density disk. It is contained within a rigid square plastic protective jacket with a spring-loaded shutter. One read/write head of the *floppy-disk drive* contacts each side of the disk to write and read data when the shutter is opened. The 3.5-in disk has a formatted capacity of 1.44 Mbytes. The earlier 5.25-in floppy disk is now obsolete.

floppy-disk drive An electromechanical computer peripheral component for reading and writing data to a *floppy*

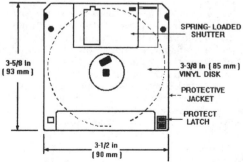

Floppy disk is a magnetically coated vinyl disk within a plastic protective jacket. The reverse side of a 3½ in disk with 1.44 Mb capacity is shown.

disk. It accepts the jacketed floppy disk and its drive motor spins the disk within its jacket when it is reading or writing data to the disk. Its read/write heads move in and out to contact both sides of the disk.

flow   The movement of electric charges, gases, liquids, or other materials or quantities.

flowchart   A graphical representation of a program or routine for a digital computer. Also called control diagram.

flow control   The procedure for regulating the flow of data between two devices to prevent the loss of data once a device's buffer has reached its capacity.

flowmeter   An instrument that measures and indicates the rate of flow of a liquid or gas.

flow soldering   *Wave soldering.*

flow transmitter   A device that measures the flow of liquids in pipe lines and converts the results into proportional electric signals that can be transmitted to distant receivers or controllers.

fluctuation noise   *Random noise.*

fluorescence   Emission of light or other electromagnetic radiation by a material exposed to another type of radiation or a beam of particles, with the luminescence ceasing within about 10^{-8} s after irradiation is stopped. Certain minerals fluoresce in characteristic colors during exposure to ultraviolet radiation. A cathode-ray screen fluoresces when hit by the electron beam in the tube. Other materials give off characteristic X-rays when irradiated by higher-frequency X-rays.

fluorescent lamp   A lamp whose illumination depends on the ionization of mercury vapor. Electrons emitted by the electrode at one end of a sealed glass tube collide with the electrons in the outer rings of mercury atoms to produce ultraviolet light, which irradiates the phosphor coating on the inner walls of the tube to produce visible light. It is also called a *fluorescent tube.*

fluorescent screen   A sheet of material coated with a fluorescent substance that emits visible light when irradiated with ionizing radiation such as X-rays or electron beams. In a cathode-ray tube the fluorescent screen is a coating on the inside surface of the tube face.

fluorescent tube   *Fluorescent lamp.*

fluorinated ethylene propylene   [FEP]   A thermoplastic, melt-extrudable form of ethylene polypropylene that exhibits outstanding electrical properties and resistance to chemicals and heat. It is also known by the DuPont trade name of *Teflon.*

fluorine   A yellowish gas element with a pungent odor that is denser than air and extremely poisonous, with an atomic number of 9 and an atomic weight of 18.99. With hydrogen it forms hydrofluoric acid, an important etchant of glass and of silicon oxides in integrated circuit manufacturing.

fluorocarbon resin   General term for a family of plastics that has excellent electrical insulating qualities and relatively high service temperatures. Examples include polychlorotrifluoroethylene resin, marketed as Kel-F, and polytetrafluoroethylene resin, marketed as Teflon and Fluon.

fluorod   A rod made from silver-activated phosphate glass for solid-state dosimeters. Under irradiation, the rod absorbs ultraviolet light and emits orange fluorescent light. Measurement of the intensity of the emitted light with a photomultiplier gives a measure of the absorbed dose of radiation.

fluorography   Photography of an image produced on a fluorescent screen.

fluorometer   An instrument that measures the intensity of X-rays and other radiation by measuring the intensity of the fluorescence produced.

fluorometry   Measurement of the intensity and color of fluorescent radiation.

fluoroscope   An instrument that includes an *X-ray tube* and a *fluorescent screen* for the observation of the contents of opaque containers placed between the tube and the screen. It can be used by physicians for rapid observation of the internal organs and bones of trauma patients in emergency rooms when urgency does not permit conventional X-ray photography. It is also included in the monitors that screen packages and luggage for contraband and weapons at airport security stations. The X-ray radiation forms a visible image on the screen.

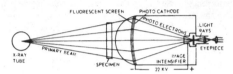

Fluoroscope with an image intensifier for the inspection of welds, honeycomb assemblies, thick plastic or rubber products, and explosive charges.

fluoroscopy   The use of a fluoroscope for X-ray examination.

flush antenna   An aircraft antenna that has no projections beyond the surface of the aircraft. Examples include ditch antennas and slot antennas.

flutter   1. Distortion that occurs in sound reproduction as a result of undesired speed variations during the recording, duplicating, or reproducing process. The variations in speed and hence pitch occur at a much higher rate than for wow. 2. A fast-changing variation in received signal strength. It can be caused by antenna movements in a high wind or by interaction with a signal on another frequency.

flutter echo   A radar echo that consists of a rapid succession of reflected pulses resulting from a single transmitted pulse.

flux   1. The electric or magnetic lines of force in a region. 2. The rate of flow of particles or photons through a unit area. 3. The product of the number of particles per unit volume and their average velocity. 4. A chemical that removes oxide films from the surfaces of metals in preparation for soldering, brazing, or welding. Rosin is widely used as a flux for soldering electronic circuits components to circuit boards.

flux gate   A detector that produces an electric signal whose magnitude and phase are proportional to the magnitude and direction of the external magnetic field acting along its axis. A flux gate consists of three magnetic cores that have appropriate excitation windings and load windings to give perfect balance in the absence of external magnetic fields.

flux-gate compass   An electronic compass that measures the earth's magnetic field with coils of thin wire rather than a rotating compass card. Any deviation from a correct reading can be corrected and the signals can be averaged electronically to provide steady and accurate heading information for ships or aircraft.

flux-gate magnetometer A magnetometer that determines the strength of an external magnetic field by measuring the degree of saturation of its core.

flux guide A shaped piece of magnetic material that guides electromagnetic flux in desired paths in induction heating. The guide can either direct flux to preferred locations or prevent the flux from spreading beyond definite regions.

flux leakage Magnetic flux that does not pass through an air gap or other part of a magnetic circuit where it is required.

flux linkage The product of the number of turns in a coil and the number of magnetic lines of force passing through the turns.

fluxmeter An instrument for measuring magnetic flux. It is usually calibrated to read either in maxwells or webers.

flyback *Retrace.*

flyback converter A *buck-boost* switching power supply with a single switching transistor that eliminates the output inductor. Energy is stored in the transformer primary during the first half of the switching period (when the transistor is conducting), but during the second half (*flyback* period) when the transistor is off, the energy is transferred to the transformer secondary and load. It produces a negative output from a positive input.

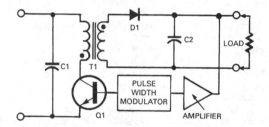

Flyback converter is a single-transistor switching power supply circuit.

flyback power supply A high-voltage power supply that produces the DC voltage of about 10 to 25 kV required for the second anode of a cathode-ray tube in a television receiver or oscilloscope. The sudden reversal of horizontal deflection-coil current in the horizontal output transformer during each flyback induces a voltage pulse that is increased to the required higher value by autotransformer action, then rectified and filtered. Also called kickback power supply.

flyback transformer *Horizontal output transformer.*

fly-by-wire system An aircraft flight-control system that uses electric wiring instead of mechanical or hydraulic linkages to control the actuators for the ailerons, flaps, and other control surfaces of an aircraft. Wiring makes multiple redundancy feasible, for greater combat survivability. Control signals are provided by a computer whose inputs are flight sensors, preflight programs, and manual actions by the crew. It is also used in manned and unmanned spacecraft.

flywheel effect The ability of a resonant circuit to maintain oscillation at an essentially constant frequency when fed with short pulses of energy at constant frequency and phase.

FM 1. Abbreviation for *frequency-modulated.* 2. Abbreviation for *frequency modulation.*

FM/AM Amplitude modulation of a carrier by subcarriers that are frequency-modulated by information.

FM/AM multiplier A multiplier whose frequency deviation from the central frequency of a carrier is proportional to one variable, and its amplitude is proportional to the other variable. The frequency-amplitude-modulated carrier is then consecutively demodulated for FM and for AM. The final output is proportional to the product of the two variables.

FM cyclotron *Synchrocyclotron.*

FM/FM Frequency modulation of a carrier by subcarriers that are frequency-modulated by the information to be transmitted.

FMLB Abbreviation for *first make/last break,* a reference to the sequence in which connector contacts are made and broken.

FOCC Abbreviation for *forward control channel.*

FM pickup *Variable-capacitance pickup.*

FM/PM Phase modulation of a carrier by subcarriers that are frequency-modulated by information.

FM/PM telemetering A telemetering system that uses several frequency-modulated subcarriers to phase-modulate the main RF carrier.

f number A lens rating obtained by dividing the focal length of the lens by the effective maximum diameter of the lens. The lower the f number, the shorter the exposure required, or the lower the illumination needed for satisfactory results with a television or still picture camera. An f number of 3.5, for example, is usually expressed as f/3.5.

foamed plastic A plastic resin that has been expanded into a multicellular structure with low density and relatively high strength.

focal length The distance between the optical center of a lens and the television camera screen or photographic camera film when the camera is focused on a distant object.

focal spot The small area on the target of an X-ray tube that gives off X-rays when hit by the electron stream.

focus 1. The point of convergence for rays of light or electrons of a beam that converges to form a minimum-diameter spot. 2. To move a lens or adjust a voltage or current to obtain a focus.

focus control A control that adjusts spot size at the screen of a cathode-ray tube to give the sharpest possible image. It can vary the current through a focusing coil or change the position of a permanent magnet.

focus-defocus mode A mode of storage of binary digits in which the writing beam of a cathode-ray storage tube is initially focused. For one type of binary digit it remains focused, and for the other type it is suddenly defocused to a small concentric circular area, in the time interval before the beam is cut off and moved to the next position.

focusing 1. The process of controlling convergence or divergence of the electron paths within one or more beams to obtain a desired image or current density distribution in the beam. 2. The process of moving an optical lens toward or away from a screen or film to obtain the sharpest possible image of a desired object.

focusing anode An anode in a cathode-ray tube that changes the size of the electron beam at the screen. Varying the voltage on this anode alters the paths of electrons in the beam and changes the position at which they cross or focus.

focusing coil A coil that produces a magnetic field parallel to an electron beam, for the purpose of focusing the beam. The coil is usually mounted on the neck of a cathode-ray picture tube, and it carries a direct current whose value can be adjusted by a focus control rheostat.

focusing electrode An electrode to which a potential is applied to control the cross-sectional area of the electron beam in a cathode-ray tube.

focusing grid A focusing electrode.

focusing magnet A permanent magnet that produces a magnetic field for focusing an electron beam.

foil In electronics 1. The copper foil bonded to the surface of a glass-fiber epoxy (GFE), paper phenolic, or other circuit board material. The foil is etched or mechanically milled by a subtractive process to form the circuit board conductive traces and pads. 2. The pure etched aluminum foil that forms the plates in wet aluminum electrolytic capacitors. 3. The tantalum foil in wet tantalum capacitors.

foil pattern The pattern (either negative or positive) of the traces and pads to be etched or milled on a circuit board. These patterns can made by photographing scale drawings or they can be computer-generated with appropriate software.

foldback current limiting A method for protecting power supplies against overload by reducing the output current as the load approaches the short-circuit condition so that internal power dissipation is minimized under short-circuit conditions.

folded-dipole antenna A pair of half-wave dipoles in shunt with their ends connected to increase the center impedance to 300 Ω so it will match a 300-Ω two-wire transmission line without a *balun*. It has greater bandwidth than a single half-wave open dipole, which has an impedance of 70 Ω. It is suitable for television and FM reception.

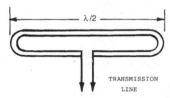

Folded-dipole antenna has a higher impedance than an open dipole, making it easier to match to a two-wire line.

folded horn An acoustic horn whose the path from throat to mouth is folded or curled to give the longest possible path in a given volume.

foot [ft] A unit of length, equal to 0.3048 m. The SI unit of length, the meter, is preferred for scientific measurements.

footcandle [fc] A former unit of illumination replaced by the lumen per square foot. The SI unit of illumination, the lux, is preferred.

footlambert [fL] A unit of luminance. The SI unit of luminance, the candela per square meter, is preferred.

footprint A slang or colloquial term referring to the form factor and area occupied by a component on a circuit board or substrate.

forbidden band An energy band that excludes electrons in a given material.

force-balance transducer A transducer with a sensing member whose output is amplified and fed back to a device that returns the sensing member to its rest position.

The feedback signal then also serves as the output of the transducer.

forced oscillation The oscillation of some physical quantity of a system when external periodic forces determine the period of the oscillation. Also called forced vibration.

forced vibration *Forced oscillation.*

forcing The application of control impulses that are larger than warranted by the error in a system, to achieve a greater rate of correction.

foreground processing The processing of high-priority programs by a computer, often by interrupting background processing of lower-priority programs.

fork oscillator An oscillator that has a tuning fork as the frequency-determining element.

formal logic A discipline that investigates propositions by methods which abstract the contents of the propositions and deal only with their logic forms.

format The predetermined arrangement of information on a form, page, file, or message.

form factor 1. The ratio of the effective value of an alternating quantity to the average value during a half-cycle. The form factor is about 1.11 for a pure sine wave and equal to the ratio of the readings obtained for a given AC quantity on root-mean-square and rectifier-type meters. 2. A factor that takes into account the shape of a coil when computing its inductance. Also called shape factor.

forming The application of voltage to an electrolytic capacitor to form an oxide dielectric that produces a desired permanent change in electrical characteristics as a part of the manufacturing process.

form-wound coil A coil that is formed or bent into an irregular shape, as in a CRT deflection yoke.

FORTRAN [FORmula TRANslation] One of several specific procedure-oriented programming languages that can be used on a computer to translate into machine language a program whose steps are written in relatively simple language.

forty-five rpm record A 7-in-diameter (17-cm-diameter) disk recorded and reproduced at a speed of 45 rpm, with a center hole 1.5 in (3.8 cm) in diameter and grooves designed for a stylus having a point radius of 1 mil (0.0254 mm). It is now an obsolete recording technology.

forward bias A voltage applied across a rectifying diode PN junction with a polarity that provides a low resistance conducting path. By contrast, reverse bias causes the PN junction to block normal current.

forward coupler A directional coupler that samples incident power.

forward control channel [FDCC] A control channel used from a land station to a mobile station in a *cellular mobile telephone system.*

forward converter Another term for a *buck-derived* switching power supply similar to the *flyback* circuit with a single switching transistor. Unlike in the flyback converter, energy is transferred to the transformer secondary while the transistor switch is conducting, and is stored in an output inductor.

forward current The current that flows through a rectifying junction in the conducting direction.

forward direction The direction of least resistance to current flow through a rectifier or semiconductor diode.

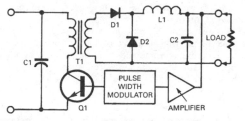

Forward converter is a single-transistor switching power supply circuit.

forward error correction [FEC]   One of two AMTOR communication modes in which each character is sent twice. The receiving station checks for errors in mark/space ratio. If an error is detected, a space is printed to show that an incorrect character was received. It is also called *mode B*.

forward-looking infrared unit [FLIR]   A U.S. military active infrared detection system that includes a laser transmitter and receiver which permit the system to determine the range and bearing of a target as well as detect, recognize, identify, and determine its bearing by sensing its radiation. FLIR systems include discrete detector arrays to scan object space and generate video signals. The signal processor amplifies the electrical signal and sends coded target information to the display. The IR detector must be cooled cryogenically to a specified operating temperature, such as 30 K for a mercury-doped germanium (Ge : Hg) detector for the highest performance. See also *passive infrared* system.

forward path   The transmission path from the loop actuating signal to the loop output signal in a feedback control loop.

forward recovery time   The time required for the forward current of a semiconductor diode to reach a specified value after a forward bias is instantaneously applied.

forward-scatter propagation   *Scatter propagation.*

forward voice channel [FVC]   A voice channel used from a *land station* to a *mobile station* in a *cellular mobile telephone system.*

forward wave   A wave whose group velocity is in the same direction as the electron-stream motion in a traveling-wave tube.

forward-wave crossed-field amplifier [FWA]   A forward-wave crossed-field amplifier, also called the M-type forward-wave amplifier. See *crossed-field electron tube.*

Foster-Seeley discriminator   *Phase-shift discriminator.*

Foucault current   A current induced in the interior of conductors by variations of magnetic flux.

foundry   A factory that manufactures semiconductor devices from a polished blank wafer of silicon or gallium arsenide semiconductor through all masking and processing steps including wafer testing. It is also called *wafer fab.*

foundry service   A business set up to produce custom large-scale integrated circuit wafers from the customers' designs, but where fabrication is performed according to the vendor's design rules and standards.

four-channel sound system   *Quadraphonic sound system.*

four-course radio range   A radio range that beams on-course signals in four different directions for aircraft guidance. The Adcock radio range is an example.

four-horn feed   A cluster of four rectangular horn antennas that form the radiating and receiving elements of a parabolic or lens-type radar antenna, to define the four quadrants of coverage.

Fourier analysis   The process of determining the amplitude, frequency, phase, coherence functions, correlation functions, power spectra, transfer functions, and other functions of each sinusoidal component in a given waveform.

Fourier analyzer   A digital spectrum analyzer that provides pushbutton or other switch selection of averaging, coherence function, correlation, power spectrum, and other mathematical operations involved in calculating Fourier transforms of time-varying signal voltages for such applications as identification of underwater sounds, vibration analysis, oil prospecting, and brain-wave analysis.

Fourier series   A mathematical expression that permits any periodic function to be represented as a combination of sine and cosine terms which are integral multiples of a fundamental frequency.

Fourier transform   A mathematical expression relating the energy in a transient to that in a continuous-energy spectrum of adjacent frequency components.

four-layer device   A PNPN semiconductor device that has four layers of alternating P- and N-type material to give three PN junctions. A silicon controlled rectifier (SCR) is an example.

four-layer diode   A semiconductor diode that has three junctions. Terminal connections are made to the two outer layers which form the junctions. A Shockley diode is an example.

four-layer transistor   A junction transistor that has four conductivity regions but only three terminals. A thyristor is an example.

four-level laser   A laser whose lowest level for a laser transition is an excited state rather than the ground level. Less energy is ordinarily required to obtain the necessary population inversion in a four-level laser because the terminal level can be initially almost empty.

four-of-eight code   An 8-bit code with 4 bits that are always 1 and 4 bits that are always 0, to give high immunity to errors. It is in Touch-Tone telephones.

4PDT   Abbreviation for *four-pole double-throw.*

four-phase modulation   Modulation that encodes data on a carrier frequency as a succession of phase shifts which will be 45, 135, 225, or 315°. Each phase shift contains 2 bits of information called dibits, as follows: 225° represents 00, 315° is 01, 45° is 11, and 135° is 10. It is in some modems for data transmission over wire lines.

four-pole double-throw [4PDT]   A 12-terminal switch or relay contact arrangement that simultaneously connects two pairs of terminals to either of two other pairs of terminals. It can switch sets of stereo loudspeakers.

four-pole single-throw [4PST]   An eight-terminal switch or relay contact arrangement that simultaneously opens or closes two separate pairs of circuits.

4PST   Abbreviation for *four-pole single-throw.*

four-quadrant multiplier   An analog multiplier integrated circuit that can multiply, divide, square, and extract the square root of analog inputs that are either negative or positive. The inputs represent arithmetical values in the four quadrants defined by the Cartesian coordinates. There are also one- and two-quadrant multipliers. See *analog multiplier.*

four-tape sort   A merge-sort that sorts input data on two tapes into complete sequences alternately on two output tapes. The output tapes are used for input for the next run. The process is then repeated until the data is in one sequence on one output tape.

four-track tape   A magnetic tape that records two tracks for each direction of travel, to double the amount of stereo music that can be recorded on a given length of ¼-in (0.635-cm) tape.

four-wire circuit   The part of a telephone system that operates with two pairs of wires—one pair for the transmit path (generally from the microphone) and the other pair for the receive path (generally from the receiver).

four-wire repeater   A repeater that provides amplification in opposite directions on two transmission paths.

Fowler-Nordheim tunneling   A quantum-mechanical process in which electrons tunnel through a thin dielectric from (or to) a floating gate to (or from) a conducting channel. It is the erase mechanism in flash memories and the program and erase mechanism in EEPROMs.

fox message   A standard message for testing teletypewriter circuits and machines that includes all the alphamerics: THE QUICK BROWN FOX JUMPED OVER A LAZY DOG'S BACK 1234567890.

FPGA   Abbreviation for *field-programmable gate array.*

FPLA   Abbreviation for *field-programmable logic array.*

FPS   Abbreviation for *front panel setting.*

FQFP   A reference to a type of integrated circuit package.

fractional-horsepower motor [FHP motor]   Any motor built into a frame smaller than that for a motor with open construction and a continuous rating of 1 hp at 1800 rpm.

frame   1. One complete coverage of a television picture. In the NTSC system a frame contains 525 horizontal scanning lines, repeated at the rate of 30 frames per second. Each frame is scanned in two interlaced fields, each covering 262.5 lines.   2. A single complete picture on motion-picture film. For 35-mm film the standard rate of projection is 24 frames per second. This means that a special projector is required to convert this to 30 frames per second for U.S. television.   3. A rectangular area representing the size of copy handled by a facsimile system. The width of a facsimile frame is the available line width, and the length is determined by the service requirements.

frame frequency   The number of times per second that the frame is completely scanned in television.

frame of reference   A set of lines or surfaces that are references for coordinates defining a moving or stationary point.

frame period   A time interval equal to the reciprocal of the frame frequency. In the NTSC system the frame period is 1⁄30 s.

framer   A device for adjusting facsimile equipment so that the start and end of a recorded line are the same as on the corresponding line of the subject copy.

framing   1. Adjusting a television picture to a desired position on the screen of the picture tube.   2. Adjusting a facsimile picture to a desired position in the direction of line progression. Also called phasing.

framing control   1. A control that adjusts the centering, width, or height of the image on a television receiver screen.   2. A control that shifts a received facsimile picture horizontally.

Franklin antenna   A broadside linear antenna consisting of a collinear dipole array of half-wavelength dipoles whose feed line is connected between the ends of one pair of dipoles. Quarter-wave stubs are connected between the ends of the additional adjacent pair of dipoles.

Franklin antenna is a six-element collinear dipole array.

Fraunhofer diffraction   Diffraction that occurs when radiation passes the edges of one or more apertures.

Fraunhofer lines   Dark lines in the spectrum of sunlight, as obtained with a spectroscope. They are due to absorption of certain wavelengths by gases and vapors in the solar atmosphere.

Fraunhofer region   The region where energy flow from an antenna acts as if it were coming from a point source located in the vicinity of the antenna. It is beyond the Fresnel region, and begins at a distance equal to about twice the square of antenna length divided by the wavelength. It is also called the far region or the far zone.

free-air ionization chamber   An air-filled ionization chamber with a sharply defined beam of radiation that passes between the electrodes without striking them or other internal parts of the equipment. The observed ionization current is then caused entirely by ions and electrons resulting from the action of radiation on the air. It is used for X-ray dosimetry. It is also called an open-air ionization chamber.

free electron   An electron that is not constrained to remain in a particular atom. It is therefore able to move freely in matter or a vacuum when acted on by external electric or magnetic fields.

free-electron laser [FEL]   A source of coherent radiation that operates from the gigahertz band to the soft X-ray bands with high efficiency, high power, and tunable frequency. Its electrons travel near the speed of light. The higher the electron velocity, the shorter the output radiation wavelength. An external spatially periodic pump field is imposed on the electron beam causing the relativistic electrons to oscillate transversely and thus emit radiation. FEL output is coherent because the electrons are longitudinally bunched while oscillating transversely in the pump field.

free field   A field whose boundary effects are negligible over the region of interest. An object in a free sound field will have the same disturbing effect as a boundary unless the acoustic impedance of the object matches the acoustic impedance of the medium.

free-field emission   The electron emission that occurs when the electric field at the surface of an emitter is zero.

free-field room   *Anechoic room.*

free gyro   A gyroscope mounted in two or more gimbal rings so that it is free to maintain a fixed orientation in space, with no external means for changing its normal precession.

free-running Operating without external synchronizing pulses, as in a free-running multivibrator.

free-running frequency The frequency of operation of a normally synchronized oscillator that operates in the absence of a synchronizing signal.

free-running multivibrator *A stable multivibrator.*

free-space antenna pattern An antenna radiation pattern obtained when the antenna is mounted so that it is not affected by nearby buildings, trees, hills, or earth.

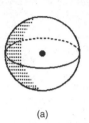 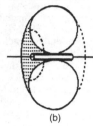

(a)　　　　　(b)

Free-space antenna patterns for (*a*) an isotropic antenna and (*b*) a dipole antenna.

free-space field intensity The field intensity that would exist at a point in the absence of waves reflected from the earth or other reflecting objects.

free-space loss The theoretical radiation loss, depending only on frequency and distance, that would occur if all variable factors were disregarded when transmitting energy between two antennas.

free-space radiation pattern The radiation pattern that an antenna would have in free space.

free vibration *Free oscillation.*

free wave *Free progressive wave.*

F region The region of the ionosphere above about 90 mi (145 km). It includes the F1 and F2 layers in the daytime hemisphere, and a single F layer at night, all capable of reflecting radio waves back to earth at frequencies up to about 50 MHz.

frequency [f] The number of complete cycles per unit of time for a periodic quantity such as alternating current, sound waves, or radio waves. A frequency of 1 cycle/s is 1 hertz (1 Hz). Other frequency units are kilohertz (kHz), megahertz (MHz), gigahertz (GHz), and terahertz (THz). Frequency in megahertz is equal to 300 divided by wavelength in meters.

frequency agility The ability to shift the frequency of a radar transmitter rapidly and continually to avoid jamming by the enemy, to reduce mutual interference with friendly sources, to enhance echoes from targets, or to provide the required patterns of electronic countermeasures or electronic counter-countermeasures radiation.

frequency allocation Assignment of available frequencies in the radio spectrum to specific stations for specific purposes, to maximize the use of frequencies with minimum interference between stations. Allocations in the United States are made by the Federal Communications Commission.

frequency band A continuous range of frequencies extending between two limiting frequencies. A band can include many channels; for example, the broadcast band extends from 535 to 1605 kHz. The frequency and wavelength limits for all bands in the radio spectrum are given in the entry for *band*.

frequency bridge A bridge in which the balance varies with frequency in a known manner, such as occurs in the Wien bridge. It is used to measure frequency.

frequency calibrator An instrument that generates a highly accurate signal at one or more fixed frequencies, for use in calibrating other frequency sources.

frequency changer *Frequency converter.*

frequency characteristic *Amplitude-frequency response.*

frequency compensation *Compensation.*

frequency converter A circuit, device, or machine that changes an alternating current from one frequency to another, with or without a change in voltage or number of phases. In a superheterodyne receiver, the oscillator and mixer-first detector stages together serve as the frequency converter. Also called frequency changer.

frequency counter An instrument for measuring frequencies. A crystal oscillator and digital counting circuit can measure both frequencies from low to RF and period. Some frequency counters can also measure risetime and pulse width.

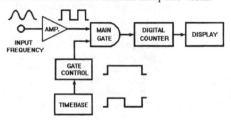

Frequency counter block diagram. The number of cycles in the signal being measured is counted for the time determined by the time base.

frequency cutoff The frequency at which the current gain of a transistor drops 3 dB below the low-frequency gain value.

frequency departure *Frequency drift.*

frequency deviation 1. The peak difference between the instantaneous frequency of a frequency-modulated wave and the carrier frequency. Also called deviation. 2. *Frequency drift.*

frequency-deviation meter An instrument that indicates the number of cycles a broadcast transmitter has drifted from its assigned carrier frequency.

frequency discriminator A discriminator circuit that delivers an output voltage which is proportional to the deviations of a signal from a predetermined frequency value. Used in frequency-modulated receivers and automatic frequency-control circuits. Examples include the double-tuned detector, locked-in oscillator, and ratio detector.

frequency distortion Nonlinear distortion in which the relative magnitudes of the different frequency components of a wave are changed during transmission or amplification. Frequency distortion occurs when an audio amplifier cannot amplify equally well all the frequencies present in the input signal. It is also called amplitude distortion, amplitude-frequency distortion, or waveform-amplitude distortion.

frequency diversity Diversity reception for carrier frequencies separated 500 Hz or more and having the same modulation. It is based on the knowledge that fading does not occur simultaneously on different frequencies. The receiver minimizes the effects of fading by selecting at each instant the frequency that has the higher signal strength.

frequency-diversity radar A radar that transmits on two or more frequencies simultaneously, with means for combining the resulting echoes or selecting the strongest of the echoes from the target at each instant.

frequency-division multiaccess [FDMA] A satellite multiplex communication system that uses an analog method of modulating telephone signals. The satellite repeaters can accept several signals at different frequencies within the repeater bandwidth.

frequency-division multiplex [FDM] A multiplex system for transmitting two or more signals over a common path by switching to a different frequency band for each signal.

frequency-domain reflectometer A tuned reflectometer for measuring reflection coefficients and impedance of waveguides over a wide frequency range, by sweeping a band of frequencies and analyzing the reflected returns.

frequency doubler An amplifier stage whose resonant anode circuit is tuned to the second harmonic of the input frequency. The output frequency is then twice the input frequency. Also called doubler.

frequency-doubling transponder A transponder that doubles the frequency of the signal before retransmission.

frequency drift A gradual change in the frequency of an oscillator or transmitter due to temperature or other changes in the circuit components that determine frequency. Also called frequency departure and frequency deviation.

frequency-exchange signaling Signaling in which the change from one signaling condition to another is accompanied by decay in amplitude of one or more frequencies and buildup in amplitude of one or more other frequencies.

frequency frogging Interchanging of frequency allocations for carrier channels to prevent singing, reduce crosstalk, and reduce the need for equalization. Modulators in each repeater translate a low-frequency group to a high-frequency group, and vice versa.

frequency-hour One frequency is transmitted for 1 hour as in international shortwave broadcasting.

frequency hysteresis Failure of an oscillator to change frequency smoothly when continuously tuned because of a long transmission line between oscillator and load.

frequency-independent antenna An antenna whose operation is essentially independent of frequency for all frequencies above a lower limit.

frequency interlace Interlace of interfering signal frequencies with the spectrum of harmonics of scanning frequencies in television, to minimize the effect of interfering signals by altering the appearance of their pattern on successive scans.

frequency keying Keying in which the carrier frequency is shifted alternately between two predetermined values.

frequency lock A method of recovering, in a single-sideband suppressed-carrier receiver of a power-line carrier communication system, the exact modulating frequency that is applied to a single-sideband transmitter.

frequency meter An instrument for measuring the frequency of an alternating current.

frequency-modulated jamming Jamming in which a constant-amplitude RF signal is varied above and below a center frequency to produce a signal covering a wide band of frequencies.

frequency-modulated laser A helium-neon or other laser whose ultrasonic modulation cell impresses a frequency-modulated video signal on the output beam of the laser.

frequency-modulated radar A continuous-wave radar whose carrier frequency is alternately increased and decreased at a predetermined rate. The frequency of the beat between the returning echo and the wave transmitted at the instant of echo arrival is proportional to range.

frequency-modulated radio altimeter An aircraft altimeter that gives accurate absolute altitude indications from a few feet above the surface up to a limit of about 5000 ft (1.5 km). The frequency of the downward-radiated radio wave is varied back and forth continuously at some cyclic rate. Some of the energy from the transmitter is also fed to the input of the receiver for mixing with the energy reflected from the surface. The beat between the direct and reflected waves, equal to the change in transmitter frequency during the time of travel of the wave to the surface and back, is a direct indication of altitude. Also called low-altitude radio altimeter and terrain-clearance indicator.

frequency-modulated scanning sonar A scanning sonar whose transmitter frequency is electronically decreased at a constant rate for several seconds, then suddenly increased to its original value, and the process repeated. The tone of the echo is then related to target position. The chief drawback of this sonar is Doppler error when there is relative motion of transmitter and target.

frequency modulation [FM] Modulation in which the instantaneous frequency of the modulated wave differs from the carrier frequency by an amount proportional to the instantaneous value of the modulating wave. The amplitude of the modulated wave is constant. A frequency-modulation broadcast system is practically immune to atmospheric and human interference.

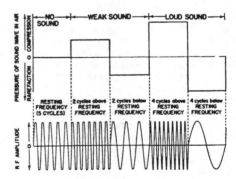

Frequency modulation diagram showing how the carrier or resting frequency varies with the strength of audio modulation.

frequency-modulation broadcast band The band of frequencies extending from 88 to 108 MHz, used for frequency-modulation radio broadcasting in the United States.

frequency-modulation broadcast channel A frequency band 200 kHz wide in the frequency-modulation broadcast band, designated by its center frequency. Assigned center frequencies begin at 88.1 MHz and continue in 100 successive steps of 0.2 MHz up to 107.9 MHz in the United States.

frequency-modulation receiver A radio receiver that receives frequency-modulated waves and delivers corresponding sound waves.

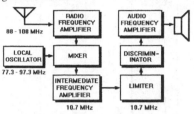

Frequency modulation receiver block diagram is typical for reception in the 88- to 100-MHz FM band.

frequency-modulation recording Data recording that converts the analog signal into frequency deviations above and below a center frequency.

frequency-modulation transmitter A radio transmitter that transmits a frequency-modulated wave.

frequency-modulation tuner A tuner containing an RF amplifier, converter, IF amplifier, and demodulator for frequency-modulated signals, that feeds a low-level AF signal to a separate AF amplifier and loudspeaker.

frequency modulator A circuit or device that produces frequency modulation.

frequency monitor An instrument that indicates the amount of deviation of the carrier frequency of a transmitter from its assigned value.

frequency multiplier A harmonic conversion transducer whose output frequency signal is an exact integral multiple of the input frequency. It is also called multiplier.

frequency-offset transponder A transponder that changes the signal frequency by a fixed amount before retransmission.

frequency overlap The portion of a 6-MHz television channel that is common to both the monochrome and chrominance signals in a color television system. Frequency overlap is a form of bandsharing.

frequency pulling A change in the frequency of an oscillator due to a change in load impedance.

frequency pulsing Oscillator operation at two different frequencies alternately. The buildup of oscillation at one frequency creates conditions favorable for oscillation at the second frequency, and vice versa.

frequency pushing A change in the operating frequency of an oscillator, resulting from changes in bias voltage within the specified bias voltage limits for the circuit.

frequency range The useful range of frequencies over which a transmission system can be operated when combined with different circuits under a variety of operating conditions. In contrast, bandwidth is a measure of useful frequency range with fixed circuits and fixed operating conditions.

frequency-regulated power supply An AC power supply whose dominant parameters are the range and regulation of the output frequency. The output voltage is typically regulated. A power oscillator operating from a DC power supply usually generates the output frequency.

frequency regulator A circuit that maintains the frequency of an AC generator at a predetermined value.

frequency relay A relay that operates only at a predetermined frequency, such as a resonant-reed relay.

frequency response *Amplitude-frequency response.*

frequency-response equalization *Equalization.*

frequency run A series of tests made to determine the amplitude-frequency response characteristic of a transmission line, circuit, or device.

frequency-scan antenna A radar antenna whose scanning is produced in one dimension by frequency variation, as in a phased-array antenna.

frequency scanning The scanning of an oscillator output frequency back and forth over a desired frequency band.

frequency-selective ringing The use of two or more different frequencies of ringing current in a telephone system, in combination with ringers that are mechanically or electrically tuned to operate for only one of the ringing frequencies.

frequency-selective surface An array of passive resonant elements that acts as a reflector over its resonant frequency band but passes RF energy essentially without attenuation at other frequencies. It provides dual-frequency capability for microwave tracking and radar systems.

frequency separation The difference between a repeater input frequency (received) and the output frequency (transmission). Repeaters on the 10-m amateur band normally are separated in frequency by 100 kHz. See also *offset.*

frequency-separation multiplier A multiplier that has each of its variables split into low-frequency and high-frequency parts. These parts are multiplied separately and the results are added to give the required product. This system makes possible high accuracy and broad bandwidth.

frequency separator The circuit that separates the horizontal and vertical synchronizing pulses in a monochrome or color television receiver.

frequency shift A change in the frequency of a radio transmitter or oscillator.

frequency-shift converter A circuit that converts a received frequency-shift signal to an amplitude-modulated or a DC signal.

frequency-shift keying [FSK] A form of frequency modulation in which the modulating wave shifts the output frequency between predetermined values corresponding to the frequencies of correlated sources. When frequency-shift keying is used for code transmission, operation of the keyer shifts the carrier frequency back and forth between two distinct frequencies to designate mark and space. When frequency-shift keying is used for facsimile transmission, one carrier frequency represents picture black, and another, generally 800 Hz away, represents picture white.

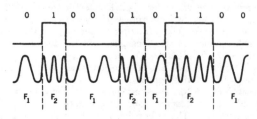

Frequency-shift keying.

frequency-slope modulation   Modulation that sweeps the carrier signal periodically over the entire width of the band, much as in chirp radar. Modulation of the carrier with a voice or other communication signal changes the bandwidth of the system without affecting the uniform distribution of energy over the band. The desired information can be recovered from any part of the system bandwidth; this permits filtering out parts of the band that have interference without losing desired information.

frequency splitting   The undesirable rapid shifting of the output frequency of a magnetron, caused by alternation between two modes of operation.

frequency stability   The ability of an oscillator to maintain a desired frequency. It is expressed as percent deviation from the assigned frequency value.

frequency stabilization   The process of controlling the center frequency of an oscillator so it does not differ from that of a reference source by more than a prescribed amount.

frequency standard   A stable oscillator, usually controlled by a crystal, tuning fork, or atomic clock, primarily for frequency calibration.

frequency swing   1. The instantaneous departure of the frequency of an emitted wave from the center frequency during frequency modulation.   2. The difference between the maximum and minimum design values of the instantaneous frequency in a frequency-modulation system.

frequency synthesizer   A circuit that provides a choice of many different frequencies by setting 10-position knobs or equivalent rows of 10 pushbuttons to the exact frequency value desired. The instrument can cover part or all of the entire radio spectrum from DC up into the gigahertz range and provide a choice of waveforms that usually includes pulse, sine, square, and triangle. The output frequencies are derived from one or more precision quartz crystal oscillators. One is usually 1 or 5 MHz, acting with harmonic generators, dividers, multipliers, and mixers to provide direct or indirect frequency synthesis.

frequency tolerance   The extent to which the carrier frequency of a transmitter may be permitted to depart from the assigned frequency.

frequency-to-voltage converter   [F/V converter]   A converter that provides an analog output voltage which is proportional to the frequency or repetition rate of the input signal derived from a flowmeter, tachometer, or other AC generating device. In a process control system, this analog output voltage can be fed back to regulate the process. F/V converters are also used for stabilization and linearization of voltage-controlled oscillators.

frequency tripler   An amplifier or other device that delivers output voltage at a frequency equal to three times the input frequency.

frequency-type telemeter   A telemeter that uses the frequency of a signal as the translating means.

Fresnel diffraction   Diffraction of microwaves that permits reception behind obstacles with no focusing performed by the obstacles. The Fresnel diffraction region of an antenna is the near-field region where the radiation pattern varies with distance.

Fresnel lens   A thin lens constructed with stepped setbacks so it has the optical properties of a much thicker lens.

Fresnel region   The region between the near field of an antenna and the Fraunhofer region. The boundary between

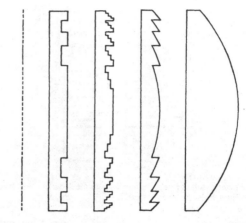

Fresnel-lens configurations. Left to right: simple Fresnel-zone plate, phase-reversing half-period zone plate, quarter-period zone plate, optical Fresnel lens, and equivalent conventional optical lens.

the two is generally considered to be at a radius equal to twice the square of antenna length divided by wavelength.

Fresnel zone   One of the conical zones that exist between microwave transmitting and receiving antennas due to cancellation of some part of the wavefront by other parts that travel different distances. The boundary of the first Fresnel zone includes all paths a half wavelength longer than the line-of-sight path. The outer boundaries of the second, third, and fourth Fresnel zones are formed by paths 1, 1½, and 2 wavelengths longer than the direct path.

Fresnel-zone plate   A thin plastic disk that has alternate radiopaque and radiolucent rings spaced so that millimeter-wave incident radiation is brought to a focus at a point beyond the zone plate, just as an optical lens brings light waves to a focus. The action differs from that of a Fresnel lens in that diffraction at the angular apertures and subsequent interference of the diffracted radiation give the focusing action.

Fresnel-zone plate: black rings are radiopaque grooves in a plastic sheet.

Fresnel-zone reflector A stepped reflector that reflects incident electromagnetic radiation to a point focus. The distance between steps, in the direction of the incident radiation, is approximately half a wavelength at the receiving frequency.

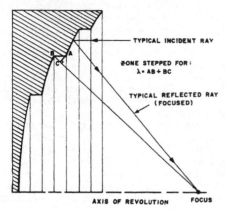

Fresnel-zone reflector.

friction bonding Soldering of a semiconductor chip to a substrate by vibrating the chip back and forth under pressure to create friction that breaks up oxide layers and helps alloy the mating terminals.

fringe area An area just beyond the limits of the reliable service area of a television transmitter in which signals are weak and erratic. High-gain directional receiving antennas and high-sensitivity receivers are generally required for satisfactory fringe-area reception.

fringe effect The extension of the electrostatic field of an air capacitor outside the space between its plates.

fringe howl A howl or squeal heard when some circuit in a radio receiver is on the verge of oscillation.

frit Powdered glass with a chemical binder that is applied selectively to semiconductor devices and packages and fired to form hermetic glass seals.

front contact The stationary contact of the normally open contacts on a relay.

front end The tuner of a television receiver, containing one or more RF amplifier stages, the local oscillator, and the mixer, along with all channel-tuning circuits.

front-end fabrication The fabrication of semiconductor devices to the level of completed and tested wafers. The term is synonymous with wafer fab.

front loading Placing material in front of a loudspeaker to change the acoustic impedance and thereby alter the radiation pattern. One method of achieving front loading is by placing an inverted exponential horn in front, to reduce the size of the opening through which acoustic energy emerges, and thereby improve the low-frequency radiation pattern.

front porch The portion of a composite picture signal that lies between the leading edge of the horizontal blanking pulse and the leading edge of the corresponding sync pulse. The duration of the front porch is 1.27 μs in the standard U.S. television signal.

fs Abbreviation for *femtosecond*.

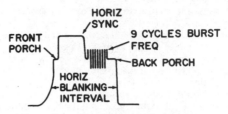

Front porch of a horizontal sync pulse for color telecast, and back porch with 3.58-MHz color-burst frequency.

FS Abbreviation for *full scale*.

FSF Abbreviation for *frequency scaling factor*.

F scope A radarscope that produces an F display.

FSK Abbreviation for *frequency-shift keying*.

FSR Abbreviation for *full-scale range*.

FSS Abbreviation for *fixed satellite service*.

ft Abbreviation for *foot*.

FTP Abbreviation for *file transfer protocol*.

ft/s Abbreviation for *foot per second*.

Fuchs antenna A long-wire, voltage-fed radiator, an even number of wavelengths. One end is connected directly to the transmitter or tuning unit without a transmission line. It is suitable for transmitting or receiving at high frequencies or VHF.

fuel cell A system that converts chemical energy directly into electricity when it is supplied with hydrogen-rich fuel such as hydrazine, kerosene, or hydrogen gas and oxygen. The cell consists of a closed chamber containing an electrolyte, usually phosphoric acid, sandwiched between two electrodes, the anode and the cathode. Both are porous plates containing a small amount of a catalyst, such as platinum. The hydrogen-rich fuel is fed into the cell from an external tank and injected into the anode, and oxygen (air) is injected into the cathode. Within the cell, hydrogen molecules separate into two positive hydrogen ions and two electrons. The electrons flow in the external circuit, providing energy for the load. The hydrogen ions are drawn toward the cathode and combine with oxygen and electrons at the cathode to form water, which is then drained off. Fuel cells can be stacked and connected in series to provide increased power. Although their efficiency is low, they have been practical in spacecraft, and they are under consideration as stand-by power sources for commercial power plants.

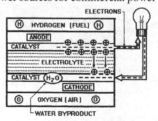

Fuel cell converts chemical energy directly into electricity when fed with reactants—hydrogen rich fuel and oxygen. It does not require recharging.

full adder *Adder*.

full-bridge converter A circuit that operates as a forward converter. It includes a bridge circuit consisting of four switching transistors to drive the transformer primary.

full break-in [QSK]   A method of amateur radio operation that permits operators to hear signals between code characters. This allows another amateur radio operator to break into the communication without waiting for the transmitting station to finish.

full-duplex operation   A mode of radio or wire communication permitting the simultaneous transmission of information between two locations in both directions simultaneously. It is also called *full duplex transmission.*

full-duplex transmission [FDX]   *Full-duplex operation.*

full load   The greatest load that a circuit or equipment is designed to carry under specified conditions. Any additional load is an *overload.*

full scale [FS]   The maximum reading on a particular scale for a specific range of a measuring instrument.

full-wave bridge   A bridge arrangement of four diode or tube rectifiers that provides full-wave rectification of the secondary voltage of the power transformer. A center tap is not needed on the secondary winding of the transformer. The output voltage contains both halves of each input cycle. See also *bridge rectifier.*

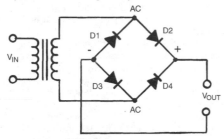

Full-wave bridge provides low-ripple DC from an AC input.

full-wave control   Phase control that acts on both halves of each AC cycle for varying load power over the full range from zero to the full-wave maximum value. The output voltage is usually applied in its AC form, but a rectifying stage can be added to provide a pulsating DC output voltage. The control element can be a silicon controlled rectifier or other solid-state power control device.

full-wave rectification   Rectification in which output current flows in the same direction during both half-cycles of the alternating input voltage.

full-wave rectifier   A double-element rectifier that provides full-wave rectification. One element functions during positive half-cycles and the other during negative half-cycles.

full-wave vibrator   A vibrator with an armature that moves back and forth between two fixed contacts, to change the direction of direct-current flow through a transformer at regular intervals and thereby permit voltage step-up by the transformer. It has largely been replaced by electronic DC-to-DC converters based on multivibrators.

fully active homing   Homing in which the missile generates its own radar signals and carries a computer that provides guidance signals for lock-on to give a collision course.

fully ionized plasma   Plasma in which all the neutral particles have lost at least one electron.

function   A quantity whose value depends upon the value of one or more other quantities.

functional language   A programming language based on expressions to be evaluated.

function digit   A computer instruction-word digit that determines the arithmetic or logic operation to be performed.

function generator   1. An analog computer circuit that indicates the value of a given function as the independent variable is increased.   2. A signal generator that delivers a choice of a number of different waveforms, with provisions for varying the frequency over a wide range.

function key   A special key on a keyboard to control a mechanical function, initiate a specific computer operation, or transmit a signal that would otherwise require multiple key strokes.

function multiplier   An analog computer circuit that accepts the changing values of two functions and produces the changing value of their product as the independent variable is changed.

function switch   A network that has many inputs and outputs connected so that input signals expressed in one code will produce output signals which are a function of the input information but in a different code.

fundamental   *Fundamental frequency.*

fundamental component   *Fundamental frequency.*

fundamental field particle   The field quantum that is the result of quantizing a field.

fundamental frequency   1. The lowest frequency component of a complex vibration, sound, or electric signal. It is the basis for harmonic analysis of a wave. The fundamental frequency is the reciprocal of the period of a wave. Also called first harmonic, fundamental, and fundamental component. The frequency that is twice the fundamental frequency is called the second harmonic.   2. The first order or lowest frequency of an intended mode of vibration for a quartz plate or other vibrating object. Also called fundamental mode.

fundamental-frequency magnetic modulator   A magnetic modulator whose output is at the fundamental frequency of the supply.

fundamental mode   1. The waveguide mode that has the lowest critical frequency. Also called dominant mode and principal mode.   2. *Fundamental frequency.*

fundamental particle   *Elementary particle.*

fundamental tone   The component tone of lowest pitch in a complex tone.

fundamental unit   An arbitrarily defined unit that is the basis of a system of units. All other units in the system are derived from a set of fundamental units. The dimensions of any physical quantity may be expressed as combinations of fundamental units.

fundamental wavelength   The wavelength corresponding to the fundamental frequency.

fungiproofing   Application of a protective chemical coating that inhibits growth of fungi on electronic equipment in humid tropical regions.

fuse   1. A thermal device that protects electrical and/or electronic circuits when it is placed in series between the load and the power line by opening the circuit in the presence of a predetermined overcurrent. It contains a metal wire or foil strip that burns out or *blows* when the current exceeds the rated value of the fuse. It must be

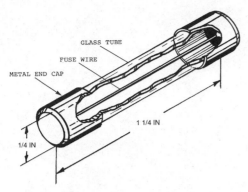

GLASS TUBE
FUSE WIRE
METAL END CAP
1 1/4 IN
1/4 IN

Fuse. The standard replaceable fuse can be rated as fast-acting or delayed response.

removed and replaced to restore the protected circuit's function. For electronics applications, most fuses have tubular glass bodies, metal end caps, and a coaxial thermal element. Standard American fuses are 1¼ in long by ¼ in diameter. Standard voltage ratings are 125- and 250-V AC and standard current ratings are ⅕ to 15 A. European standard fuses are 0.8 in long (20 mm) by 0.2 in (5 mm) diameter. Response times are measurable in ms. Common classes are *fast-acting* and *slow-blow* or *time-delay*. Some are made as leaded devices to be soldered on a circuit board; they must be unsoldered for replacement. *Thermal protector fuses* protect heat-producing appliances by opening in response to excessive heat rather than overcurrent.   2. A device or circuit for detonating bombs, missiles, projectiles, or explosive charges. It is also spelled *fuze.*

fuse alarm   A lamp that indicates the presence and location of a blown fuse.

fuse holder   A mechanical device that provides the electrical connections for a standard fuse to the protected circuit and clamps it in position to maintain secure contacts while also permitting easy fuse removal for the replacement of a blown fuse. A common example is a plastic cylinder threaded at one end for mounting in a hole through a panel of the host equipment so it extends behind the panel. A removable spring-loaded cap clamps the fuse in position but is easily removed for fuse replacement. Terminals permit wires to be soldered to the protected circuit. When convenient accessibility to the fuse is not required, two spring clips can be mounted on the circuit board to clamp the fuse end caps to complete the circuit. A *fuse block* is an assembly of two spring clips on a small insulating plastic block.

fused junction   *Alloy junction.*

fuse wire   Wire made from an alloy that melts at a relatively low temperature and overheats to this temperature when carrying a particular value of overload current.

fusible resistor   A resistor that protects a circuit against overload. Its resistance limits current flow to protect against surges when power is first applied to a circuit. It opens the circuit when current drain exceeds its design limits.

fuzzy logic   A logic based on conditional rather than absolute true or false statements. The results obtained with fuzzy reasoning are less definite than those obtained with strict classical logic, but they apply to a larger universe of cases.

f V   Abbreviation for *femtovolt.*

F/V   Abbreviation for frequency-to-voltage, usually used in connection with converters.

FVC   Abbreviation for *forward voice channel.*

F/V converter   Abbreviation for *frequency-to-voltage converter.*

F/V-V/F   An abbreviation that indicates a converter will convert in both directions, from frequency-to-voltage and from voltage-to-frequency.

f W   Abbreviation for *femtowatt.*

G

g   Abbreviation for *gram*.

G   1. Symbol for *conductance*. 2. Symbol for *gate*. 3. Abbreviation for *gauss*. 4. Symbol for *generator*. 5. Abbreviation for *giga-*. 6. Abbreviation for *gravitational force*. 7. Symbol for *grid*.

G$_m$ Symbol for *transconductance*.

GaAlAs   Symbol for *gallium aluminum arsenide*.

GaAs   Symbol for *gallium arsenide*.

GaAsFET   Symbol for *gallium arsenide field-effect transistor*.

GaAsP   Symbol for *gallium arsenide phosphide*.

gadolinium [Gd]   A rare-earth metallic element. It has an atomic number of 64.

gage   A measuring device or instrument. It is also spelled *gauge*.

gain   1. The increase or decrease of signal amplitude after amplification.   2. An increase or decrease in the effective power radiated by an antenna in a specific direction. It is also an increase or decrease in received signal strength from a certain direction. It is typically expressed in decibels; an increase is a positive number, and a decrease is a negative number.

GaInAs   Symbol for *gallium indium arsenide*.

gain-bandwidth product   A figure of merit for amplifiers, based on their gain and the bandwidth, as measured under specified conditions.

gain control   A device for adjusting the gain of a circuit, system, or component.

gain margin   The amount of increase in gain that would cause oscillation in a feedback control system.

GaInP   Symbol for *gallium indium phosphide*.

gain-time control   *Sensitivity-time control*.

gain tracking error   The variation of gain from a constant level determined at 0 dBm input level when measuring the dependence of gain on signal level by comparing the output signal to the input signal over a range of input signals.

galactic noise   Radio-frequency noise that originates in space from all celestial bodies except the sun.

galena   A crystalline form of lead sulfide (PbS) that was the detector in crystal detectors receivers.

Galileo spacecraft   A 2½-ton unmanned spacecraft launched by NASA in October 1989 for the determination of the atmosphere of Jupiter and its moons. It contained a 746 lb probe capsule that detached from Galileo in July 1995 and plunged into Jupiter's atmosphere in December. Galileo, which has traveled at about 107,000 mph, is expected to complete its mission of orbiting Jupiter by 1998.

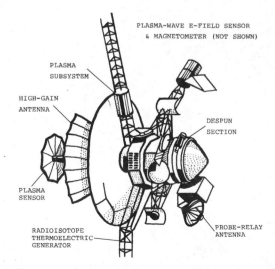

PLASMA-WAVE E-FIELD SENSOR
& MAGNETOMETER (NOT SHOWN)

PLASMA SUBSYSTEM

HIGH-GAIN ANTENNA

DESPUN SECTION

PLASMA SENSOR

RADIOISOTOPE THERMOELECTRIC GENERATOR

PROBE-RELAY ANTENNA

Galileo spacecraft was designed to orbit and explore the planet Jupiter.

gallium [Ga]   A metallic element with an atomic number of 31 and an atomic weight of 69.72. It is an *acceptor* impurity in germanium and silicon semiconductor devices.

gallium aluminum arsenide [GaAlAs]   A semiconductor material used to make light-emitting diodes that produce a red light. See also *aluminum gallium arsenide.*

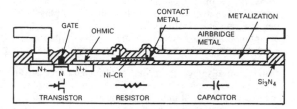

Gallium arsenide integrated-circuit section view shows how active and passive components are formed on a GaAs substrate.

gallium arsenide [GaAs]   A binary compound semiconductor material formed from Group III material gallium and Group V compound arsenic. It has higher electron mobility than silicon, so higher-speed logic ICs can be made from it. Electrons travel twice as fast in GaAs than in silicon. It is suitable for making discrete optoelectronic devices [light-emitting diodes (LEDs) and laser diodes], microwave transistors (MESFETs), and integrated circuits (MMICs and optoelectronic ICs).

gallium arsenide FET   A high-frequency, voltage-controlled current amplifier similar to a silicon MOS-FET. See also *metal semiconductor field-effect transistor* [MESFET].

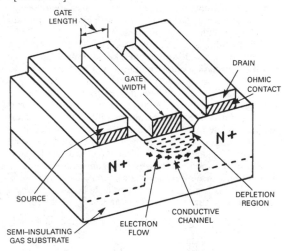

Gallium arsenide FET transistor, called a MESFET, is similar in construction to a silicon MOSFET.

gallium-arsenide laser   A laser that emits light at right angles to a junction region in gallium arsenide, at a wavelength of 9000 A. It can be modulated directly at microwave frequencies.

gallium arsenide phosphide [GaAsP]   A trinary compound semiconductor material used to make light-emitting diodes.

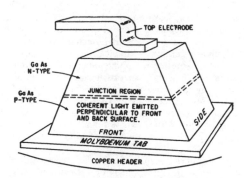

Gallium-arsenide laser mounted above a parabolic reflector. The greater transparency of N-type material, facing outward, increases the laser's light output.

The color of the visible light produced is red or amber, depending on the proportions of arsenide and phosphide. The wavelength range is 650 to 670 nm.

gallium indium arsenide [GaInAs]   A trinary compound semiconductor material used as a cathode surface in high-sensitivity multiplier phototubes for detecting very low light levels, as in astronomy and medicine.

gallium indium phosphide [GaInP]   A semiconductor material used to make light-emitting diodes that produce a yellow light.

gallium nitride [GaN]   A binary compound semiconductor material used to make light-emitting diodes that emit blue light in the 450 to 500 nm wavelength range.

gallium phosphide [GaP]   A binary compound semiconductor material used to make light-emitting diodes to give either green or red light in the wavelength range of 690 to 790 nm.

galvanic   The property of electricity flowing as a result of chemical action.

galvanic cell   An electrolytic cell that is capable of producing electric energy by electrochemical action.

galvanometer   An instrument for indicating or measuring a small electric current. The measurement is made from mechanical motion derived from electromagnetic or electrodynamic forces produced by the current.

game theory   The mathematical process of selecting an optimum strategy for competing with an opponent's strategy.

gamma   1. A numerical indication of the degree of contrast in a television or photographic image. It is equal to the slope of the straight-line portion of the *H*-and-*D* curve for the emulsion or screen.   2. A unit of magnetic field strength, equal to 10 microoersteds or 0.00001 oersted.

gamma-absorption gage   *Gamma gage.*

gamma emitter   An atom whose radioactive decay process involves the emission of gamma rays.

gamma ferric oxide   The ferromagnetic form of ferric oxide, used as a coating on magnetic tape.

gamma radiation   Radiation of gamma rays.

gamma radiography   Radiography with gamma rays.

gamma rate   The rate of gamma-ray exposure that can be tolerated by a device without failure. An indication of radiation hardness. Gamma rate is the ionizing dose rate measured in rads per second (rads/s).

gamma rays A quantum of electromagnetic radiation emitted by a nucleus as a result of a quantum transition between two energy levels of a nucleus. They are high-energy photons above 50 kiloelectronvolts (keV), emitted as short-wavelength electromagnetic radiation from most nuclei undergoing spontaneous nuclear disintegration or fission. They are more penetrating than alpha and beta particles and are not affected by magnetic fields. They have no definite range, but can damage or destroy semiconductor devices. They can be found in outer space, near nuclear reactors, and as by-products of a nuclear weapon explosion.

gamma-ray source A quantity of radioactive material that emits gamma radiation and is in a form convenient for radiology.

gamma-ray spectrometer An instrument that measures the energy distribution of gamma rays.

gamma-ray thickness gage A thickness gage that depends on a radioactive source and a radiation detector to measure the amount of radiation transmitted by a moving sheet of material. Absorption (and hence transmission) of radiation by the material is proportional to its thickness or weight.

GaN Abbreviation for *gallium nitride*.

ganged capacitor *Ganged tuning capacitor*.

ganged switch Two or more sets of switch contacts mounted on a common shaft that can be switched by a single control knob. See *rotary switch*.

ganged tuning Simultaneous tuning of two or more circuits with a single control knob.

ganged tuning capacitor A radio receiver tuning capacitor with two or more sections. Alternate plates are mounted on a common shaft so that they can be interleaved simultaneously by twisting a single knob to change their effective capacitance value, which changes the tuning frequency.

ganged volume control A combination of two or more volume controls, one for each channel of a stereophonic sound system, mounted on a common shaft to permit changing simultaneously the volume of all loudspeakers without changing their balance with respect to each other.

ganging A mechanical means of operating two or more controls with one control knob.

GaP Abbreviation for *gallium phosphide*.

gap length The physical distance between adjacent surfaces of the poles of a longitudinal magnetic recording head.

gap loading The electronic gap admittance that results from the movement of electrons in the gap of a microwave tube. The three types are: multipactor gap loading, primary transit-angle gap loading, and secondary-electron gap loading.

gas, 2-D electron A gaseous concentration of electrons formed by electrostatic forces in pure gallium arsenide (GaAs) at the junction with pure aluminum gallium arsenide (AlGaAs) in a heterostructure FET.

gas capacitor A capacitor that consists of two or more electrodes separated by a gas, other than air, which serves as a dielectric.

gas detector A gas-sensing semiconductor or other device whose resistance decreases in the presence of a deoxidizing gas like carbon monoxide, hydrogen, methane, propane, and volatile oils.

gas diode A tube that has a heated cathode and an anode in an envelope which contains a small amount of an inert gas or vapor. When the anode is made sufficiently positive, the electrons flowing to it collide with gas atoms and ionize them. As a result, anode current is much greater than that for a comparable vacuum diode.

gas discharge Conduction of electricity in a gas due to the movement of ions produced by collisions between electrons and gas molecules.

gas-discharge display An alphanumeric display that depends on the ionization of neon gas in a sealed envelope containing opposing electrodes in the form of diodes. The electrodes, in the form of transparent conductive dots or segments, can provide an alphanumeric readout. A dot matrix can display all of the ASCII characters; seven segments can display all numbers and some letters; 14 segments can display all numbers and letters. About 160 V is required to ionize the gas. Traces of radioactive materials are added to enhance ionization at cold ambient temperatures. See also *gas-plasma display*.

gas-discharge lamp *Discharge lamp*.

gas doping The introduction of impurity atoms into a semiconductor material by epitaxial growth, with streams of gas that are mixed before being fed into the reactor vessel.

gasdynamic laser A gas laser that converts thermal energy directly into coherent radiation at an efficiency high enough to offer promise of wireless power transmission. The gas is usually a mixture of nitrogen and carbon dioxide with a catalyst of helium or water vapor, heated to about 2000 K by an arc jet, shock tube, or other means. The wavelength is the standard CO_2 laser transition of 10.6 μm in the infrared spectrum and can be pulsed or continuous wave.

gas etching The removal of material from a semiconductor circuit by reaction with a gas that forms a volatile compound. The etching of silicon slices with hydrogen chloride at 1000°C or higher is an example.

gas-filled cable A coaxial or other cable that contains gas under pressure that insulates and keeps out moisture.

gas-filled radiation-counter tube A gas tube used to detect radiation by means of gas ionization.

gas-filled rectifier *Cold-cathode rectifier*.

gas laser A laser whose active medium is a discharge in a gas contained in a glass or quartz tube with a Brewster-angle window at each end. The gas can be excited by a high-frequency oscillator or direct-current flow between electrodes inside the tube. The function of the discharge is to pump the medium, to obtain a population inversion. Operation can be pulsed or continuous. Basic types are: 1. atomic (neural) gas lasers, such as helium-neon lasers; 2. ion lasers, such as metal vapor lasers; 3. molecular

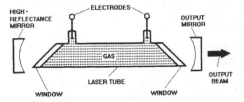

Gas laser: A glass tube sealed at both ends by windows at Brewster's angle with an electrode at each end. Two concave mirrors (one partially transmitting) form the optical cavity. An electric discharge between the electrodes excites the gas to start laser action.

gas lasers, such as carbon dioxide lasers; 4. chemical lasers that have population inversion for lasing produced by a chemical reaction between gases.

gas lens A lens that guides laser light with the variations in the refractive indexes of the contained gases. Made in the shape of a long pipe, it can confine a laser beam to the axis of the tube. The lens is suitable for long-range communications. The refractive index in the tube can be varied by confining the beam in a helical heating element aligned along the axis of the tube to concentrate the gas in that region. Also, different gases can be injected so that they flow in opposite directions in the pipe.

gas-plasma display A flat-panel display for television and computer monitors based on matrixes of diode gas-plasma cells. A display with 786 by 576 pixels (for a total of 442,368) accommodates both the NTSC and PAL analog TV broadcast standards. It is also called a *plasma display*.

gas scattering The scattering of electrons or other particles in a beam by residual gas in the vacuum system.

gassy tube An electron tube that has been incompletely or incorrectly evacuated so that residual gas remains in the envelope. This could be caused by the outgassing of inadequately prepared internal components such as grids or plates, a leak in the envelope, or a faulty *pinchoff*. None of these conditions can be corrected by *gettering*. A gassy tube, even if it is functional at the time the condition is discovered, will eventually fail. Internal gas can be determined by holding an operating Tesla coil against the glass envelope and looking for an internal glow. It is also called a *soft tube*.

gas tube An electron tube in which the contained gas or vapor performs the primary role in the operation of the tube.

gas X-ray tube An X-ray tube with a cold cathode that emits electrons by positive-ion bombardment when the applied cathode-anode voltage is sufficiently high.

gate 1. A digital logic circuit with one or more inputs and an output organized so that an output occurs when specific conditions are met at the inputs. The simplest is a NOT gate whose output is always opposite to its input. An AND gate, with two or more inputs, delivers an output only when all of its inputs are one or zero simultaneously. An OR gate provides an output when its inputs are either one or zero. 2. The control electrode of a FET, MESFET, or HEMT. A voltage applied to a gate regulates the conducting properties of the *channel* region beneath the gate. A MESFET (metal semiconductor FET) gate is in close contact with the semiconductor. A MOSFET (metal-oxide semiconductor FET) gate is separated from the semiconductor by a thin oxide layer, typically 100 to 1000 Å thick. 3. A combination of transistors that form logic functions such as NOT, NAND, or NOR. See also *channel, drain, FET, MESFET, MOSFET,* and *source*.

gate array A semicustom integrated circuit premanufactured as matrix of nondedicated gates. It is completed to custom specifications with one or more layers of metal that uniquely define the gate interconnections for each application. All gates, drains, and sources are accessible, and all metal mask levels except the final ones are fixed and predefined.

gate circuit A circuit that admits and amplifies or passes a signal only when a gating pulse is present.

gate-controlled rectifier A three-terminal semiconductor device that controls the unidirectional current flow

between the rectifier terminals by a signal applied to a third terminal called the gate. The silicon controlled rectifier is an example.

gate-controlled switch [GCS] A semiconductor device that can be switched from its nonconducting or off state to its conducting or on state by applying a negative pulse to its gate terminal and turned off at any time by applying reverse drive to the gate. (Once a silicon controlled rectifier is turned on, it can be turned off only by interrupting its cathode-anode current.)

gate current The alternating or pulsating direct current that flows through the gate winding of a magnetic amplifier.

gated-beam tube A pentode electron tube that has special electrodes which form a sheet-shaped beam of electrons. This beam can be deflected away from the anode by a relatively small voltage applied to a control electrode, thus giving extremely sharp cutoff of anode current. It is an obsolete device.

gated buffer A low-impedance inverting driver circuit that can function as a line driver in multivibrators or for pulse differentiation.

gated sweep A radar sweep whose duration and starting time are controlled to exclude undesired echoes from the screen.

gate equivalent circuit A unit of measure for specifying relative complexity of digital circuits, equal to the number of individual logic gates that would have to be interconnected to perform the same function as the digital circuit under evaluation.

gate generator A circuit that generates gate pulses. In one form it consists of a multivibrator that has one stable and one unstable position.

gate length The distance between the source and the drain of a MOS transistor measured on the photomask plate. It is also called the *patterned* or *drawn* length. When determined from the actual transistor characteristics, it is called *effective gate length*.

gate pulse A pulse that triggers a gate circuit so it will pass a signal.

gate turnoff [GTO] A type of silicon controlled rectifier that can be turned on by a pulse of gate current and turned off by applying a pulsed negative bias between gate and cathode terminals. Used for power-switching applications at power-line and higher frequencies.

gate-turnoff switch An all-diffused three-junction semiconductor switching device that can be turned on or off from its gate input terminal.

gate voltage The voltage across the terminals of the gate winding in a magnetic amplifier.

gating The process of selecting those portions of a wave that exist during one or more selected time intervals or that have magnitudes between selected limits. It is achieved by applying a pulsed voltage to a normally cut-off electron tube, transistor, or magnetic amplifier to make the device conductive (open the gate) for the duration of the pulse. The opposite action (the device is cut off for the duration of the pulse) blacks out a television cathode-ray tube during retraces and is called *blanking*.

gauss [G; plural is gauss] The CGS electromagnetic unit of magnetic induction B. The tesla, an SI unit, is preferred.

Gaussian distribution A distribution of random variables comparable to that found in nature, characterized by a symmetrical and continuous distribution decreasing gradually to zero on either side of the most probable value. Also called normal distribution.

Gaussian noise Noise that has a frequency distribution which follows the Gaussian curve.

Gaussian noise generator A signal generator that produces a random noise signal whose frequency components have a Gaussian distribution centered on a predetermined frequency value.

Gaussian well A potential well whose value varies according to a Gaussian distribution.

gaussmeter A magnetometer with a scale graduated in gauss or kilogauss. One version measures the magnetic field strength between the pole faces of magnets. For magnetrons, gaussmeter readings range from 1.2 to 9.6 kG.

Gb 1. Abbreviation for *gigabit*. 2. Abbreviation for *gilbert*.

GBIC Abbreviation for *gigabit interface converter*.

GCA Abbreviation for *ground-controlled approach*.

GCI Abbreviation for *ground-controlled interception*.

GCS Abbreviation for *gate-controlled switch*.

Geiger counter A radiation counter based on a Geiger counter tube in appropriate circuits for detecting and counting ionizing particles such as cosmic-ray particles. Each particle ionizes the gas in the tube so that the total ionization per event is independent of the energy of the ionizing particle. Ionization results in electron flow from the tube wall to the center electrode, producing one output voltage pulse for each particle. It is also called a *Geiger-Mueller counter* and a *Geiger counter tube*.

Geiger counter tube A radiation-counter tube operated in the Geiger region. It typically consists of a gas-filled cylindrical metal chamber containing a fine-wire anode at its axis. It is also called a Geiger counter, Geiger-Mueller counter tube, or Geiger-Mueller tube.

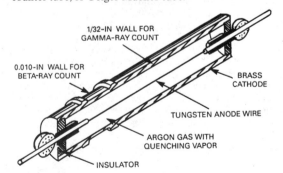

1/32-IN WALL FOR GAMMA-RAY COUNT

0.010-IN WALL FOR BETA-RAY COUNT

BRASS CATHODE

TUNGSTEN ANODE WIRE

ARGON GAS WITH QUENCHING VAPOR

INSULATOR

Geiger counter tube detects the presence of beta particles and gamma rays that ionize the argon gas, shorting the axial anode wire to the cathode, producing a count.

Geiger-Mueller counter [G-M counter] *Geiger counter.*
Geiger-Mueller counter tube *Geiger counter tube.*
Geiger-Mueller region *Geiger region.*
Geiger-Mueller threshold *Geiger threshold.*
Geiger-Mueller tube *Geiger counter tube.*
Geiger-Nuttall law The logarithm of the decay constant of an alpha emitter is linearly related to the logarithm of the range of the alpha particles emitted by it.

Geiger region The range of applied voltage in which the charge collected per isolated count is independent of the charge liberated by the initial ionizing event of a radiation-counter tube. It is also called a *Geiger-Mueller region*.

Geiger threshold The lowest applied voltage at which the charge collected per isolated tube count in a radiation-counter tube is substantially independent of the nature of the initial ionizing event. It is also called a *Geiger-Mueller threshold*.

Geissler tube An experimental two-electrode discharge tube that shows the luminous effects of electric discharges through various gases at low pressures. The electrodes are at opposite ends of the tube as in modern neon signs.

gelled cell A lead-acid cell containing a nonspillable gelled electrolyte for portable use. It can be recharged many times, like an ordinary storage battery.

generalized sort A sort program that will accept the introduction of parameters at run time and does not generate a program.

general-purpose interface bus [GPIB] An interface for the computer control of scientific, medical, and process-control equipment or test instruments. It enables measurements to be processed by the computer, and instructs the controlled equipment how to correct spurious deviations from expected results.

generation rate The time rate of creation of electron-hole pairs in a semiconductor.

geodimeter An optoelectronic distance-measuring instrument. The time required for a modulated light beam to travel from the master unit to a distant mirror and return is measured and converted into distance. In one version, a Kerr cell provides 10-MHz modulation of the light beam.

geographic information system [GIS] A system that handles a computer-based mapping and charting system whose complexity can vary with the user's requirements.

geomagnetic electrokinetograph An instrument that can be suspended from the side of a ship to measure the direction and speed of ocean currents while the ship is under way. Electrodes suspended in the conductive sea water are connected to a potentiometer recorder to measure the voltage induced in the moving conductive sea water by the magnetic field of the earth. Two runs at right angles to each other are generally made to determine the direction and rate of drift caused by local ocean currents.

geomagnetism The magnetic phenomena associated with the earth and its atmosphere.

geomagnetically induced current [GIC] Electric current flowing in metal objects, such as pipelines or power transmission lines that has been induced by *geomagnetism*.

geometric distortion An aberration that causes a reproduced picture to be geometrically dissimilar to the perspective plane projection of the original scene.

geometric disturbance A dramatic change in the earth's magnetic field that occurs over a short time interval.

geometric factor The ratio of the change in a navigation coordinate to the change in distance, taken in the direction of maximum navigation-coordinate change.

geometric horizon The locus of points on the earth where straight lines from a point of reference in space are tangent to the earth's surface.

geometric inertial navigation Inertial navigation whose computations are based on the outputs of accelerometers that are maintained in alignment with predetermined geographic directions, such as local vertical and geographic north.

geometric mean The square root of the product of two quantities.

geometry factor The average solid angle at a radiation source that is subtended by the aperture or sensitive volume of a radiation detector, divided by the complete solid angle. It also denotes counting yield or counting efficiency.

geophone A transducer for seismology that responds to motion of the ground at a location on or below the surface of the earth.

geophysical prospecting Prospecting that involves the use of acoustic, electric, and electronic equipment for measuring the physical properties of the earth as influenced by underground mineral and oil deposits.

geophysics A branch of science that deals with factors affecting the structure of the earth.

georef [GEOgraphic REFerence] Abbreviation for *world geographic reference system.*

georef grid [GEOgraphic REFerence grid] The grid system used on U.S. Air Force aeronautical charts for identifying the location of any point or area in the world. The world chart is divided into 24 parallel north-south strips 15° wide lettered from A through Z (I and O omitted), beginning at the 180th meridian, and into 12 parallel east-west strips 15° wide lettered from A through M (I omitted), beginning at the South Pole. Each quadrangle is subdivided into 15 lettered units eastward and 15 lettered units northward. These are lettered from A through Q (I and O omitted). Each 1° quadrangle is subdivided into 60 numbered minute units. Minute units may be subdivided further into decimal parts.

geostationary orbit A circular satellite orbit that is 35,900 km (22,300 mi) above the equator. The satellite revolves around the earth in exactly the same time it takes the earth to rotate on its axis; thus the satellite appears stationary over one location near the equator. The satellite moves from west to east, and makes one rotation every 24 h in synchronism with the earth's rotation. It is also known as *geosynchronous orbit.*

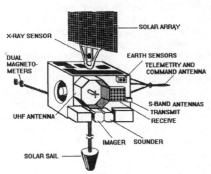

Geostationary Orbiting Earth Satellite (GOES-8) monitors the weather over the Atlantic Ocean and eastern United States. The sail stabilizes it against constant solar radiation.

Geostationary Orbiting Earth Satellite [GOES] A high-altitude *geostationary* orbiting earth satellite that pro-

vides information about the paths of major storms. It has an on-board camera capable of continuously observing and monitoring the 48 contiguous U.S. states every 15 min. It weighs 4640 lb (2100 kg) and has a large cubical casing 7 ft (2.1 m) on a side with a single-wing, dual-panel solar array that extends 20 ft (6 m). A 10-ft (3-m) cone-shaped solar sail at the end of a 58-ft (17.2-m) boom stabilizes the satellite by balancing the pressure exerted by constant solar radiation. A sounder monitors the infrared radiation wavelengths that reach the satellite from earth's atmosphere. This information is used to form heat and humidity profiles at different altitudes. These profiles can indicate the presence and temperature of water vapor, carbon dioxide, and other gases. Wind speed and direction can also be determined from this data. GOES 8, one of a series of five of these satellites, was launched in 1994.

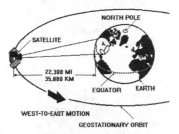

Geostationary satellite orbit is synchronized with the earth's rotation so it remains nearly stationary above a location near the equator.

geostationary satellite A satellite in geostationary orbit. See *geostationary orbit.*

geosynchronous orbit An orbit with the same period as that of the earth revolving about its axis. Thus the satellite rotational speed and the earth's speed of rotation are in synchronism.

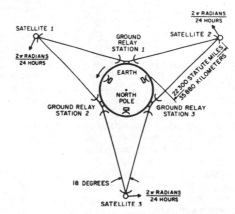

Geosynchronous orbit of communication satellites.

geosynchronous satellite *Geostationary satellite.*

germanium [Ge] A brittle, grayish white metallic element with an atomic number of 32 and an atomic weight of 72.95. It is in the Group 1V A of the periodic table, with silicon and carbon. It has important semiconductor properties and was widely used in early diodes and transistors, but it has been largely replaced by silicon for most commercial semiconductor devices.

germanium diode   A semiconductor diode that uses a germanium crystal pellet as its rectifying element.

germanium transistor   A transistor made from the semiconductor material germanium.

getter   A special metal strip mounted on a "U"-shaped frame that is welded inside the vacuum-tube envelope of a tube during manufacture and vaporized after the tube has been evacuated. When the vaporized metal condenses, it absorbs residual gases. Metals used as getters include barium, calcium, magnesium, potassium, sodium, and strontium. Some getters leave a silvery film on the inside of the glass envelope.

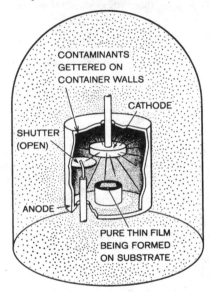

Getter sputtering equipment.

gettering   Removal of residual gas from an electron tube, after evacuation, by evaporation of a getter. An ionizing electron discharge can be passed through the gas during the gettering process, to accelerate absorption of the residual gas by the getter.

getter ion pump   An ion pump whose ionized gas molecules are attracted or propelled to a getter containing material that collects and holds the molecules, thereby improving the vacuum.

getter sputtering   The deposition of high-purity thin films at ordinary vacuum levels with a getter to remove contaminants remaining in the vacuum. Ionized argon can sputter metal from the cathode when the high DC voltage is applied, and the resulting metal atoms carry contaminants from the vacuum to the walls of the container. A shutter is then swung aside so pure metal can travel down from the cathode to the substrate and form the desired film.

GeV   Abbreviation for *gigaelectronvolt.*

GFE   Abbreviation for *glass-fiber epoxy.*

ghost   *Ghost image.*

ghost image   1. An undesired duplicate image at the right of the desired image on a television receiver. It is caused by multipath effect: a reflected signal traveling over a longer path arrives slightly later than the desired signal. Hills and large buildings on either side of the direct path between the transmitter and receiver are the most common causes of ghost images. 2. An undesired or unknown echo indication on the screen of a radar indicator.

ghost pulse   *Ghost signal.*

ghost signal   1. The reflection-path signal that produces a ghost image on a television receiver. Also called echo. 2. Any signal appearing on a loran display at other than the basic repetition rate being observed. Also called ghost pulse.

GHz   Abbreviation for *gigahertz.*

giant air shower   *Extensive shower.*

GIC   Abbreviation for *geomagnetically induced current.*

GIF   Abbreviation for *Graphics Interchange Format.*

giga- [G]   A prefix representing 10^9, which is 1,000,000,000, or a billion. It was formerly called a kilomega-.

gigabit [Gb]   One thousand megabits, or 10^9 bits.

gigabit 1394   An IEEE 1394 serial bus that operates at gigabit data rates.

gigabit ATM   An *asynchronous transfer mode* LAN that operates at gigabit data rates.

Gigabit Ethernet   An *Ethernet* LAN that operates at gigabit data rates.

gigabit interface converter [GBIC]   A 20-position single-cable connector.

Gigabit Token Ring   A *Token Ring* LAN that operates at gigabit data rates.

gigacycle   One thousand megacycles per second (10^9 or 1,000,000,000 cycles per second). It was formerly called a kilomegacycle. It is now called gigahertz.

gigaelectronvolt [GeV]   A unit of energy equal to 10^9 eV. It was formerly called a billion electronvolt (BeV) in the United States and France.

gigahertz [GHz]   One thousand megahertz, or 10^9 Hz. It was formerly called a kilomegacycle.

gigawatt [GW]   One thousand megawatts, or 10^9 W.

gigohm [GΩ]   One thousand megohms, or 10^9 Ω.

gilbert [Gb]   The CGS unit of magnetomotive force. The magnetomotive force in gilberts is equal to the line integral of the magnetic field strength in oersteds around the magnetic circuit. One gilbert is equivalent to 0.7956 ampere-turn. One gilbert per centimeter is equal to 1 oersted. The SI unit, the ampere-turn, is preferred.

gimbal   A mounting that has two mutually perpendicular and intersecting axes of rotation. A body mounted on gimbals is free to incline in any direction.

gimbal lock   Alignment and locking of gimbals that are normally at right angles to each other in a two-axis gyroscope. This catastrophic malfunction occurs when the precession angle reaches 90°, usually as a result of excessive angular motion of the aircraft or missile in which the gyroscope is mounted.

GIS   Abbreviation for *Geographic Information System.*

glass-ambient seal   *Glassivated hermetic seal.*

glass-bonded mica   An insulating material made by compressing a mixture of powdered glass and powdered natural or synthetic mica at high temperature.

glass capacitor   A capacitor with a glass dielectric and protective glass housing that are fused to give monolithic construction with low losses and high stability.

glass fiber   A glass thread less than 0.001 in thick that in woven form is an acoustic, electric, or thermal insulating material and is a reinforcing material in laminated plastics. Also used as a light and light signal conductor in fiber optics.

glass-fiber epoxy   See *epoxy glass.*

glassivated hermetic seal   A hermetic seal used for encapsulating semiconductor and resistor chips. It is formed by pyrolytic deposition of glass followed by densifying to provide an encapsulating layer that will withstand extremely hostile ambient conditions. It is also called a *glass-ambient seal, glassivation,* and *glass-passivated seal.*

glassivation   *Glassivated hermetic seal.*

glass laser   A solid-state laser whose glass serves as the host for laser ions of such materials as erbium, holmium, neodymium, and ytterbium. Neodymium glass lasers and neodymium-ytterbium glass lasers operate at 1.06 μm and neodymium-ytterbium-erbium glass lasers operate at 1.54 μm. It is also called an *amorphous laser.*

glass-passivated seal   *Glassivated hermetic seal.*

glass-to-metal seal   An airtight seal between glass and metal parts of an electron tube made by fusing together glass and a special metal alloy such as kovar that have nearly the same temperature coefficients of expansion.

G line   A single dielectric-coated round wire for transmitting microwave energy by means of surface waves.

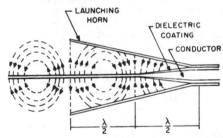

G line, showing horn feed from coaxial cable at left.

glint   A pulse-to-pulse variation in the amplitude and apparent origin of a reflected radar signal, due to reflection of the radar beam from different surfaces of a rapidly moving target. Also called glitter.

glitch   1. An undesired transient voltage spike occurring on a signal being processed. In a digital-to-analog converter, a glitch can occur at a major carry, such as when switching from 0111111111 to 1000000000 because there is an interim condition in which all bits are 0.   2. A minor technical problem arising in electronic equipment.

glitter   *Glint.*

Global Navigation Orbiting Satellite System [GLONASS]   A worldwide satellite radio navigation system being developed in Russia. It includes 24 satellites in 25,500-km orbits and is similar to the U.S. GPS.

global positioning system [GPS]   An accurate real-time navigation system developed for U.S. military ships and aircraft but available for use by civilian ships and aircraft. A specialized receiver is required to obtain the signals

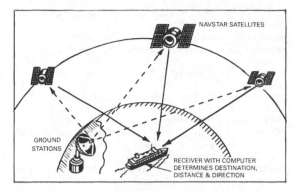

Global positioning system (GPS) is a navigation aid for ships and aircraft. The receiver measures the positions of nearby orbiting Navstar satellites and calculates an accurate latitude and longitude "fix."

from the NAVSTAR satellites. It has also been adapted as an aid in terrain surveys and for tracking police and emergency vehicles as well as bus, truck, and taxi fleets. The system is scheduled for completion in 1993 with 24 satellites. See *NAVSTAR satellite.*

Global System for Mobile [GSM] **communications**   A second-generation cellular mobile telephone scheme that is a combination of frequency-division and time-division multiple access (FDMA and TDMA). Widely used in Europe, it is compatible with the *Integrated Service Digital Network* [ISDN].

global variable   A variable in a computer program that can be shared by any object or subroutine within the program.

GLONASS   Abbreviation for *Global Navigation Orbiting Satellite System.*

glove box   A sealed box with gloves attached to and passing through openings in the box. It can be used for handling certain radioactive materials safely. It is also used in delicate production processes such as the assembly and encapsulation of semiconductor devices and integrated circuits. The interior of the box can be maintained above atmospheric pressure, as in a clean room, to prevent contamination by foreign particles in the air of the room.

glow discharge   A discharge of electricity through a gas in an electron tube, characterized by a cathode glow and a voltage drop in the vicinity of the cathode that is much higher than the ionization voltage of the gas.

glow-discharge cold-cathode tube   *Glow-discharge tube.*

glow-discharge rectifier   A glow-discharge tube used as a rectifier.

glow-discharge tube   A gas tube that depends for its operation on the properties of a glow discharge. Also called glow-discharge cold-cathode tube and glow tube.

glow lamp   A two-electrode electron tube whose light is produced by a negative glow close to the negative electrode when voltage is applied between the two electrodes. The envelope contains a small quantity of an inert gas such as neon or argon. With an AC voltage, both electrodes appear to glow because each is negative half the time. Neon glow lamps, which have an orange-red glow, are the most common examples. Less-used argon glow lamps have a blue-violet glow. Wattages of glow lamps range from 0.04 to 3 W.

glow switch An electron tube that contains contacts which are operated thermally by a glow discharge. It is a starter in some fluorescent lamp circuits. The heat of the glow discharge causes a bimetallic strip to close the contacts.

glow tube *Glow-discharge tube.*

glow voltage The voltage at which a glow discharge begins in a gas tube.

G-M counter Abbreviation for *Geiger-Mueller counter.*

GMSK Abbreviation for *gaussian minimum-shift keying*

GMT Abbreviation for *Greenwich mean time.*

gnd Abbreviation for *ground.*

gnomonic projection An aeronautical chart made by projecting the surface of a sphere onto a plane touching the sphere at one point. The projection is made by radials from the center of the sphere. It is used in determining great-circle courses for aircraft and for tracking aircraft electronically.

GOES Abbreviation for *Geostationary Orbiting Earth Satellite.*

GΩ Abbreviation for *gigohm.*

Golay cell A radiometer that measures the increase in pressure in a gas chamber as temperature rises when radiation is absorbed.

gold [Au] A precious-metal element with an atomic number of 79 and an atomic weight of 196.97. It is electroplated on electronic components that must withstand severe corrosive conditions, such as those that exist in the tropics.

gold doping A technique for controlling the lifetime of minority carriers in a diffused-mesa transistor and transistor-transistor (TTL) logic. Gold is diffused into the base and collector regions to reduce storage time in transistor circuits.

gold-leaf electroscope An electroscope whose two suspended strips are gold foil or leaf.

goniometer An instrument for measuring angles, calculating and solving mathematical problems or electrical functions, and establishing directional phase differences between two transmitted or received signals. In one form it has two fixed windings mounted 90° to each other, along with a rotatable third winding. In another form it can measure the angles between the reflecting surfaces of a crystal or prism. An X-ray version measures the angular positions of the axes of a quartz crystal.

go/no-go test A test that is based on the measurement of one or more parameters but can have only one of two possible results, to pass or reject the device under test.

Gopher A menu-based service to assist in finding information on the *Internet* that is presented in the form of tables of contents.

GPI Abbreviation for *ground-position indicator.*

GPIB Abbreviation for *general-purpose interface bus.*

GPS Abbreviation for *global positioning system.*

graded filter A power-supply filter whose connections to the output stage of a receiver or amplifier are made at or near the filter input to obtain maximum DC voltage. Ripple is less important at this stage because it has low gain, and there are no subsequent stages that might accentuate the ripple.

graded-index fiber A *multimode optical fiber* made with many layers of different refractive indexes with the highest in the center. It has an upper limit of about 500 MHz. See also *step-index fiber.*

graded insulation A method for insulating high-voltage components such as pulse transformers. The insulation to ground is reduced more or less uniformly from the high-potential end of the winding to the ground or low-potential end.

graded junction *Rate-grown junction.*

graded-junction transistor *Rate-grown transistor.*

graded seal A glass-to-metal seal in an electron tube consisting of successive bands of glass with differing temperature coefficients starting with the bulb or envelope. The glasses are melted together in a progression of temperature coefficients so that the glass which will actually contact the metal in the seal will have a temperature coefficient closely matching that of the metal.

grain A small particle of metallic silver remaining in a photographic emulsion after developing and fixing. These grains together form the dark areas of a photographic image.

graininess Visible coarseness in a photographic image under specified conditions, due to silver grains.

gram [g] One-thousandth of a kilogram.

gramophone British term for *phonograph.*

gram-rad A unit of integral absorbed dose of radiation, equal to 100 ergs/g.

graph A line drawing that shows the relation between two variable quantities.

graphic equalizer A circuit in a high-fidelity audio system that permits the frequency response of the system to be adjusted manually in selected bands of the audio spectrum. A typical circuit includes five linear potentiometers, one for each band (0 to 100 Hz, 100 to 400 Hz, 400 Hz to 1 kHz, 1 to 4 kHz, and 4 to 10 kHz) arranged horizontally so that their wipers move vertically.

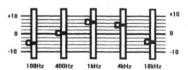

Graphic equalizer permits the adjustment of the high and low tones in an audio system by frequency band.

graphic terminal A cathode-ray-tube or other kind of computer terminal that is capable of producing some form of line drawing based on data being processed by or stored in a computer.

graphical language A programming language that expresses programs in a graphical form resembling flowcharts.

graphical user interface [GUI] The capability for human intervention in the formation of graphics on a computer display with a combination of window displays, menus, icons, and a mouse or trackball.

Graphics Interchange Format [GIF] A method for converting graphic images into computer files. The image is differentiated into pixels, and each pixel is described by a string of numbers. A compressed data format minimizes the computer data required to represent an image. Mathematical code replaces repetitive data to save computer memory space and reduce the time required to transmit the file over a computer network. It was developed by Compuserve Corp.

graphite An amorphous form of *carbon* that can serve as a monolithic resistor in some applications. It is also used to make DC motor brushes. See *carbon*.

graticule A network of horizontal and vertical lines mounted over the face of an oscilloscope screen that permits the values of variables on each axis to be estimated once values have been assigned to the graduations. It typically has an aspect ratio of 10 graduations wide and 8 graduations high with a centerline in both dimensions.

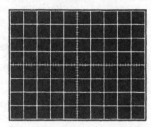

Graticule on the face of an oscilloscope CRT helps in the estimation of the values of variables on the vertical axis.

grating 1. An arrangement of fine parallel wires in waveguides that pass only a certain type of wave. 2. An arrangement of crossed metal ribs or wires that acts as a reflector for a microwave antenna and offers minimum wind resistance. 3. *Diffraction grating.*

FRAME OF GRATING
INSULATING CROSS-BARS SUPPORTING GRATING WIRES
GRATING WIRES

Gratings for circular waveguides. A grating of radial wires blocks all transverse electric waves, and the other two gratings block certain kinds of transverse magnetic waves.

grating converter A wave converter that consists of a double grating positioned just ahead of a coaxial sheet grating in a circular waveguide. One grating conforms to the pattern of the arriving wave and the other conforms to the pattern of the converted wave.

grating plate A transparent plate with scribed rulings that produces an optical diffraction pattern.

grating reflector An openwork metal structure that provides a good reflecting surface for microwave antennas.

gravitational force [G] The gravitational pull of the earth, or the comparable force required to accelerate or decelerate a freely movable body at the rate of approximately 32.16 feet per second per second (9.8 meters per second per second).

gravitational wave A hypothetical wave that travels at the speed of light and causes intense fluctuation in electron density in the F region of the earth's atmosphere. This wave propagates the effect of gravitational attraction.

graybody A body whose spectral emissivity remains constant through the spectrum and is less than that of a blackbody radiator at the same temperature.

Gray code A modified binary code in which sequential numbers are represented by expressions that differ only in one bit, to minimize errors as follows:

Decimal	Binary	Gray
0	000	000
1	001	001
2	010	011
3	011	010
4	100	110
5	101	111
6	110	101
7	111	100

Gray-code disk for converting shaft angle to cyclic or inverted Gray code.

gray filter *Neutral-density filter.*

gray line A band around the earth that separates daylight from darkness, indicating a transition between day and night.

gray scale A series of achromatic tones with varying proportions of white and black, to give a full range of grays between white and black. A gray scale is usually divided into 10 steps.

great circle path Either one of two paths between two points on the surface of the earth. One of the great circle paths is the shortest distance between two points.

green gain control A variable resistor used in the matrix of a three-gun color television receiver to adjust the intensity of the green primary signal.

green gun The electron gun whose beam strikes phosphor dots emitting the green primary color in a three-gun color television picture tube.

green laser A gas laser that uses mercury and argon to generate a green line at 5225 Å, corresponding to the wavelength that is most readily transmitted through sea water. Antisubmarine warfare is a possible application.

green restorer The DC restorer for the green channel of a three-gun color television picture tube circuit.

Green's function A kernel in which the integral operator is the inverse of a differential operator.

green video voltage The signal voltage output from the green section of a color television camera, or the signal voltage between the receiver matrix and the green-gun grid of a three-gun color television picture tube.

Greenwich mean time [GMT]   The former name for universal time (UT).

Grenz ray   Long wavelength X-rays (1 to 10 Å) produced by special low-voltage X-ray tubes operating from 5- to 15-kV power supplies. These weak X-rays are suitable for skin therapy and radiography of biological specimens.

Grenz tube   A low-voltage X-ray tube that has a special glass window capable of transmitting X-ray wavelengths ranging from 1 to 10 Å. These rays are blocked by ordinary glass.

grid   1. An electrode located between the cathode and anode of an electron tube with one or more openings through which electrons or ions can pass under certain conditions. A grid controls the flow of electrons from cathode to anode.   2. A network of equally spaced lines forming squares, for determining permissible locations of holes on a printed-circuit board or a chassis.   3. *Potter-Bucky grid.*

grid bearing   A bearing that has a reference line of grid north.

grid control   The control of anode current of an electron tube by varying the voltage of the control grid with respect to the cathode.

grid current   Electron flow to a positive grid in an electron tube.

gridded tube   A high-power, high-frequency, grid-controlled vacuum tube that provides wideband linear amplification.

grid detection   Detection in the grid circuit of a vacuum tube, as in a grid-leak detector.

grid dissipation   The power lost as heat at the grid of an electron tube.

grid-drive characteristic   A relation between electric or light output of an electron tube and the control-electrode voltage, as measured from cutoff.

grid driving power   The average product of the instantaneous value of the grid current and the alternating component of the grid voltage of an electron tube over a complete cycle.

grid emission   Electron or ion emission from a grid of an electron tube.

grid-glow tube   A glow-discharge tube with one or more control electrodes that initiate but do not limit the anode current except under certain operating conditions.

grid limiting   Limiting action achieved by placing a high-value resistor in series with the grid of a vacuum tube. The voltage drop across this resistor increases with input signal strength, giving a varying negative grid bias that serves to level input signals which are above a certain value.

grid modulation   Modulation produced by feeding the modulating signal to the control-grid circuit of any electron tube in which the carrier is present.

ground [gnd]   1. A conducting path, intentional or accidental, between an electric circuit or equipment and the earth, or some conducting body serving in place of the earth. Also called earth (British).   2. The lowest energy state of a nucleus, atom, or molecule. All other states are excited.

ground absorption   Energy loss caused by dissipation of radio waves in the ground during transmission.

ground clamp   A clamp that connects a grounding conductor to a grounded object.

ground clutter   Clutter on a ground or airborne radar caused by the reflection of signals from the ground or objects on the ground. It is also called ground return.

ground control   Control of an aircraft or missile in flight by a person on the ground.

ground-controlled approach [GCA]   An airport ground radar system that provides information to aircraft making approaches for landings. It consists of an airport surveillance radar for guiding the aircraft to the start of the final approach path and a precision approach radar for showing the exact position of the aircraft on its final approach path.

ground-controlled interception [GCI]   A radar system that permits a controller at a ground or ship radar to direct an aircraft by radio to make an interception of another aircraft.

ground distance   The mean sea-level great-circle component of distance from one point to another.

grounded   Connected to earth or to some conducting body that serves in place of the earth. Also called earthed (British).

grounded-base amplifier   See *common-base amplifier.*

grounded-cathode amplifier   See *common-cathode amplifier.*

grounded-collector amplifier   See *common-collector amplifier.*

grounded-drain amplifier   See *common-drain amplifier.*

grounded-emitter amplifier   See *common-emitter amplifier.*

grounded-gate amplifier   See *common-gate amplifier.*

grounded-grid amplifier   See *common-grid amplifier.*

grounded-source amplifier   See *common-source amplifier.*

ground environment   1. The entire complement of equipment installed on the ground to make up a communication or electronic system, facility, or station.   2. The environment that surrounds and affects a system or equipment operating on the ground.

ground-equalizer coil   A coil with relatively low inductance that is placed in one or more of the circuits connected to the grounding points of an antenna to distribute the current to the various points in a desired manner.

ground fault   Accidental grounding of a conductor.

ground-fault interrupter [GFI]   A fast-acting circuit breaker that also senses very small ground-fault currents such as might flow through the body of a person standing on damp ground while touching a hot AC line wire. The interrupter limits the time the current can flow through the fault by tripping the circuit breaker in as little as 0.025 s; this limits the total energy flow through the human body to a safe value. A typical trip current setting for homes is 5 mA.

ground-fault interrupter IC   A monolithic integrated circuit version of a *ground fault interrupter* circuit. It contains Zener diodes and an operational amplifier and can detect when a short-circuit or fault closes a magnetic path between external coils. The AC coupling through the external coils triggers oscillations which opens a relay when their voltage exceeds a threshold value, which, in turn, opens a power-line relay.

grounding outlet   An outlet that has, in addition to the current-carrying contacts, one grounded contact that can be used for grounding portable appliances and equipment.

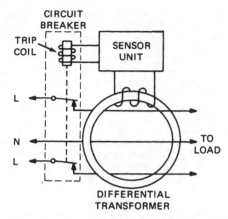

Ground-fault interrupter with a differential transformer. Currents in line wires L are normally equal and opposite, so sensor output is zero. When a fault path occurs from one line to a grounded neutral N, sensor reacts to the ground-fault current and trips a circuit breaker.

grounding plate An electrically grounded metal plate on which a person stands to discharge static electricity picked up by his body, or a similar plate buried in the ground to act as a ground rod.

ground loop A condition when two or more system components share a common electrical ground line and unwanted (spurious) voltages are unintentionally induced. A circuit that has more than one ground point connected to earth ground, with the points differing enough in ground potential produces a circulating current in the ground system. These undesirable circulating currents can be avoided by connecting the DC distribution system to earth ground with only one wire.

ground noise The residual system noise in the absence of the signal in recording and reproducing.

ground plane A grounding plate, above-ground counterpoise, or arrangement of buried radial wires required with a ground-mounted antenna that depends on the earth as the return path for radiated RF energy.

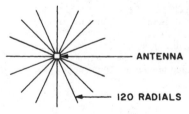

Ground plane consisting of buried radial wires at least one-quarter wavelength long at the lowest frequency.

ground-plane antenna An omnidirectional quarter-wave vertical radiator mounted above a *ground plane*. The vertical radiator is a monopole that is one-quarter wavelength of the operating frequency. It has several radials slightly longer than one-quarter wavelength, extending horizontally. It can be fed by an unbalanced coaxial cable with the energized conductor connected to the bottom of the vertical radiator and the ground shield of the cable connected to the horizontal radials. Each of the horizontal radials is joined at the end nearest the vertical radiator. They might droop because they are only supported at their ends. The antenna can produce vertically polarized waves at frequencies above 7 MHz and as high as 300 MHz.

ground-position indicator [GPI] An air-position indicator that makes allowances for drift, to give the actual position with respect to a fixed ground point.

ground potential Zero voltage with respect to the ground or to a chassis that acts as a ground connection.

ground-reflected wave A radio wave that is reflected from the ground somewhere along the transmission path.

ground return 1. An echo received from the ground by an airborne radar set. 2. The use of the earth or a chassis as the return path for a transmission line. 3. The use of a chassis as a return path for a circuit. 4. *Ground clutter.*

ground rod A copper or bronze rod that is driven into the earth to provide a good ground connection.

groundscatter propagation Multihop ionospheric radio propagation along other than the great-circle path between transmitting and receiving stations. Radiation from the transmitter is first reflected back to earth from the ionosphere, then scattered in many directions from the earth's surface. Also called non-great-circle propagation.

ground speed The speed of a vehicle along its track. For an aircraft, it is the speed relative to the earth's surface.

ground-speed recorder A recorder that makes a permanent record of the horizontal speed of a moving object with respect to the earth.

ground state The lowest energy state of a nucleus, atom, or molecule. Also called normal state.

ground vector A vector that represents the track and ground speed of an aircraft.

ground wave 1. A radio wave that is propagated over the earth and is ordinarily affected by the presence of the ground and the troposphere. The ground wave includes all components of a radio wave over the earth, except ionospheric and tropospheric waves. The ground wave is refracted because of variations in the dielectric constant of the troposphere, including the condition known as a surface duct. Also called surface wave. 2. One of the waves formed in the ground by an explosion. It can be a longitudinal wave (compression), transverse wave (shear), or surface wave (similar to water ripples), induced by direct ground shock of a ground or subsurface burst or by blast transmitted through the air.

ground wire A conductor that connects electric equipment to a ground rod or other grounded object.

grouped-frequency operation The use of different frequency bands for channels in opposite directions in a two-wire carrier system.

grouping 1. Periodic error in the spacing of recorded lines in a facsimile system. 2. Nonuniform spacing between the grooves of a disk recording.

group velocity The velocity of propagation of the envelope of a plane wave occupying a frequency band that has an approximately constant envelope delay from phase velocity in a medium whose phase velocity varies with frequency.

growing The growth of semiconductor crystals by slow crystallization from a melt. See also Czochralski process.

grown-diffused transistor A junction transistor whose final junctions are formed by diffusion of impurities near a grown junction.

grown junction A junction produced by changing the types and amounts of donor and acceptor impurities that are added during the growth of a semiconductor crystal from a melt.

growth curve A curve showing how a quantity increases with time.

GSI Abbreviation for *grand-scale integration.*

GSM Abbreviation for *Global System for Mobile communications.*

GTO Abbreviation for *gate turnoff.*

guard band Rings of doped semiconductor material that surround transistors in integrated circuits to isolate them from adjacent transistors to prevent mutual interference; N-type material surrounds P-type material, and P-type material surrounds N-type material, forming an effective back-biased PN junction.

guarded input An amplifier or other circuit whose ungrounded input terminal is electrically shielded and isolated to minimize pickup of interference.

guard ring A ring-shaped auxiliary electrode in an electron tube or other device to modify the electric field or reduce insulator leakage. In a counter tube or ionization chamber a guard ring can also define the sensitive volume.

GUI Abbreviation for *graphical user interface.*

Gudden-Pohl effect The momentary illumination produced when an electric field is applied to a phosphor previously excited by ultraviolet radiation.

guidance beam The RF or infrared beam that is aimed directly at a guided missile for transmitting control instructions to the missile. In contrast, the tracking beam in antimissile warfare is aimed directly at the target at all times, rather than at the missile sent up to intercept the target. See also *beam rider.*

guidance system Any system that controls the path of a missile from launch to target. Examples are homing guidance and beam-rider guidance.

guide *Waveguide.*

guided Controlled or controllable as to direction by preset mechanisms, radio commands, or built-in self-reacting circuits in vehicles.

guided missile An unmanned missile that is guided entirely or substantially all the way to its target by internal or external means.

guided wave A wave whose energy is concentrated near a boundary or between substantially parallel boundaries separating materials of different properties and whose direction of propagation is effectively parallel to these boundaries. Waveguides transmit guided waves.

guide wavelength The wavelength in a waveguide.

Guillemin line A network or artificial transmission line used in high-level pulse modulation to generate a nearly square pulse, with steep rise and fall. In radar sets it controls pulse width.

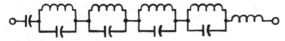

Guillemin line in the modulator for a high-power, pulse-modulated radar produces nearly rectangular pulses.

gullwing The bent form of the input-output leads on a surface-mounted semiconductor package that permits the package to expand or contract after the package is soldered to a circuit board.

gun 1. *Electron gun.* 2. *Soldering gun.*

gun killer An adapter that can be inserted between the socket and base of a color picture tube, to permit turning off independently any one or all of the three guns (red, green, and blue) in the tube during repair or adjustment procedures.

Gunn amplifier A microwave amplifier based on a Gunn diode that functions as a negative-resistance amplifier when placed across the terminals of a microwave source. The reflected power is then greater than the incident RF power. A nonreciprocal device, such as a ferrite circulator, separates the incident wave from the amplified reflected wave.

Gunn diode A two-terminal semiconductor device that exploits the Gunn effect to produce microwave oscillation or amplify an applied microwave signal. The frequency of oscillation depends on domain transit time and can exceed 50 GHz. Its operation is in the transit-time mode. A Gunn diode is a transferred-electron diode.

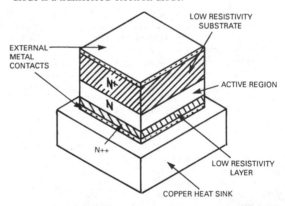

Gunn diode self-oscillates when mounted in an appropriate tunable cavity to produce a microwave-frequency output.

Gunn effect An effect discovered by J. B. Gunn in 1963; microwave oscillation occurs in a small block of N-type gallium arsenide when a constant DC voltage above a critical value is applied to contacts on opposite faces. Generated frequencies range from 500 MHz to well over 50 GHz, depending on the dimensions of the block and other factors.

Gunn oscillator An oscillator that includes a Gunn diode to generate a frequency that can exceed 50 GHz. It can have mechanical, varactor, or YIG tuning to meet the requirements of applications such as local oscillators, low-power transmitters, and microwave laboratory equipment.

guy wire A wire that holds a pole or tower in an upright position.

GW Abbreviation for *gigawatt.*

gyrator A device that causes a reversal of signal polarity for one direction of propagation but not for the other. It is a linear, passive, two-port electric circuit element that is effectively a half wavelength longer for one transmission direction than the other. There are microwave and optical gyrators.

gyrator filter   A highly selective active filter that includes a gyrator which is terminated in a capacitor so that it has an inductive input impedance. The resulting synthetic inductor can be tuned with another capacitor. Gyrators for filters can be constructed in monolithic integrated-circuit form.

gyro   *Gyroscope.*

gyrocompass   A compass with a gyroscope to provide a reference direction.

gyro flux-gate compass   A compass with a flux gate, horizontally stabilized by a gyroscope, that senses the horizontal component of the earth's magnetic field. Because it is fixed with respect to the aircraft, the compass reacts to each change in heading with a change in current. This current is amplified and used to actuate the dial of a master indicator. Also called flux-gate compass.

gyrofrequency   The natural frequency of rotation of a charged particle under the influence of a constant magnetic field, such as the magnetic field of the earth.

gyro horizon   *Artificial horizon.*

gyroklystron   A *gyrotron* amplifier having two or more resonant cavities. See also *klystron* and *gyrotron.*

gyromagnetic   A reference to the magnetic properties of rotating electric charges, such as spinning electrons moving within atoms.

gyromagnetic coupler   A coupler with a single-crystal YIG resonator that provides coupling at the required low signal levels between two crossed stripline resonant circuits. It is used in signal limiters and electronically tunable filters.

gyromagnetic effect   The rotation induced in a body by a change in its magnetization, or the magnetization resulting from a rotation.

gyromagnetic ratio   The ratio of the magnetic moment to the angular momentum for a charged particle moving in a closed orbit.

gyromonotron   A gyrotron oscillator having a single resonant cavity. See also *gyrotron* and *M-type crossed-field tube.*

gyroscope   A wheel mounted and driven at high speed so that its spinning axis is free to rotate about either of two

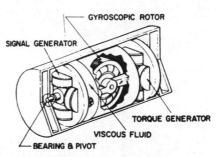

Gyroscope for airplanes, missiles, and spacecraft. A mechanical rotor driven at high speed provides the positional reference.

other axes perpendicular to itself and to each other. It maintains a stable equilibrium as required for autopilots. Also called gyro.

gyroscopic horizon   A gyroscopic instrument that indicates the lateral and longitudinal attitude of an aircraft by simulating the natural horizon.

gyro sight   A gun sight equipped with a gyroscope.

gyro stabilizer   A stabilizer with a gyroscope that compensates for the roll and pitch of a ship.

gyrotron   A cyclotron-resonance maser whose structure includes a cathode, a collector, and a circular waveguide of a gradually varying diameter. Electrons emitted by the cathode are accelerated by an electric field and guided by a static magnetic field. This nonuniform induction field causes the electrons to spiral and bunch before giving up their energy to the microwave field within the circular waveguide in a cyclotron maser interaction. In continuous-wave operation, gyrotron oscillators (*gyromonotrons*) have attained 212 kW at 28 GHz and pulsed power of 1 MW at 100 GHz. See also *gyroklystron* and *gyro-TWT.*

gyro-TWT   A gyrotron traveling-wave tube. See *gyrotron* and *traveling-wave tube.*

G – Y signal   The green-minus-luminance color-difference signal used in color television. It is combined with the luminance signal in a receiver to give the green color-primary signal.

H

h Abbreviation for *hour*.

h Symbol for *Planck's constant*.

H 1. Symbol for *heater*. 2. Abbreviation for *henry*.

H Symbol for *magnetic field strength*.

H² Symbol for *deuterium*.

H³ Symbol for *tritium*.

HAAT Abbreviation for antenna *height above the average terrain*.

halation An area of glow surrounding a bright spot on a fluorescent screen, caused by phosphor scattering or multiple reflections at front and back surfaces of the glass faceplate.

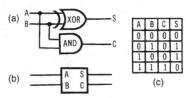

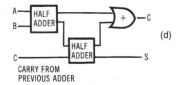

Half-adder: (*a*) logic symbol, (*b*) schematic symbol, (*c*) truth table, and (*d*) two half-adders forming a full adder.

half-adder logic A binary circuit that has two input and two output channels for binary signals. Two half-adders can be combined to give one adder circuit.

half-bridge A bridge circuit that has power sources in each of the two bridge arms which replace the single power source of a conventional bridge.

half-bridge converter A switching power supply topology that operates as a forward converter but has a bridge circuit consisting of two switching transistors to drive the transformer primary.

half-cycle The time interval corresponding to half a cycle or 180° at the operating frequency of a circuit or device.

half-cycle transmission A data-transmission and control system that uses synchronized sources of 60-Hz power at the transmitting and receiving ends. Either of two receiver relays can be actuated by choosing the appropriate half-cycle polarity of the 60-Hz transmitter power supply.

half-duplex A wired or wireless transmission system that permits communication in only one direction at a time. Examples are citizens band (CB) radio and a mobile marine transceiver with "push-to-talk" transmitter switches.

half-life The average time required for half the atoms of a sample of a radioactive substance to lose their radioactivity by decaying into stable atoms.

half-power frequency One of the two values of frequency, on the sides of an amplifier response curve, at which the voltage is 70.7% of a midband or other reference value.

half-power width An angular rating for antenna beam width. In a plane containing the direction of the maximum of the radiation lobe, it is the full angle between the two directions in that plane in which the radiation intensity is half the maximum value of the lobe.

half-rhombic antenna A long-wire antenna array that has sides which are several wavelengths long at the lowest frequency of operation. The radiating sides form half of a rhombic antenna, with the lines connected to opposite ends. When terminated in a resistor, the radiation pattern is unidirectional; when unterminated, it is bidirectional.

half-shift register   A logic circuit that consists of a gated input storage element, with or without an inverter.

half-step   *Semitone.*

halftone characteristic   A relation between the density of the recorded copy and the density of the subject copy in a facsimile system.

half-value layer   *Half-value thickness.*

half-value period   *Half-life.*

half-wave   1. In describing antennas and transmission lines, an adjective meaning *half-wavelength,* e.g., having an electrical length equal to half of the operating wavelength.   2. In describing rectifiers, power supplies, and test instruments, an adjective meaning *half-cycle,* a reference to half of one cycle at the operating frequency of the circuit or device.

half-wave antenna   An antenna whose electrical length is typically half of the operating wavelength being transmitted or received. It is also called a *half-wavelength antenna.*

half-wave bridge   A rectifier circuit consisting of a transformer with a single diode in the secondary circuit that provides positive or negative DC pulses at a rate determined by the input frequency. Ripple can be reduced with the addition of a filter capacitor across the load.

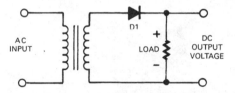

Half-wave bridge is a simple, single-diode AC rectifier circuit that produces a pulsed-DC output.

half-wave control   Phase control that acts on only half of each AC cycle, for varying load power from zero to half the full-wave maximum value or from half-power to full-power. The output is a pulsating DC voltage. The control element can be a silicon controlled rectifier or other solid-state power control device or a thyratron or other controllable gas tube.

half-wave dipole   *Dipole.*

half-wavelength   The distance corresponding to an electrical length of half a wavelength at the operating frequency of a transmission line, antenna element, or other device.

half-wavelength dipole antenna   *Dipole antenna.*

half-wave line   A transmission line whose electrical length is half the wavelength of the signal being transmitted.

half-wave rectification   Rectification in which current flows only during alternate half-cycles.

half-wave rectifier   A rectifier that provides half-wave rectification.

Hall coefficient   The constant of proportionality in the relation of the transverse electric field (Hall field) to the product of current density and magnetic flux density in an electric conductor. The sign of the majority carrier can be inferred from the sign of the Hall coefficient.

Hall constant   The constant of proportionality in the equation for a current-carrying conductor in a magnetic field.

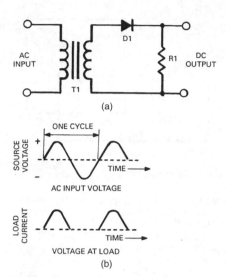

(a)

ONE CYCLE

AC INPUT VOLTAGE

VOLTAGE AT LOAD

(b)

Half-wave rectifier circuit (*a*) and its input and output waveforms (*b*).

The constant is equal to the transverse electric field (Hall field) divided by the product of the current density and the magnetic field strength. The sign of the majority carrier can be inferred from the sign of the Hall constant.

Hall effect   A galvanomagnetic effect in which a transverse DC voltage is developed across a current-carrying conductor inserted in a magnetic field so that its direction is perpendicular to the direction of current flow in the conductor. It can occur in metals, crystals, and semiconductors. The magnitude of the voltage provides a direct estimate of the concentration of charge carriers in the conductor. It is also called the *Hall-effect voltage.*

Hall-effect gaussmeter   A gaussmeter with a *Hall-effect transducer* [HET] on the end of a pair of flexible wires that can be inserted between the poles of a magnet to measure its field strength. A battery or other DC source provides current through the transducer, and a voltmeter in the instrument's case is calibrated to give a voltage that is proportional to magnetic field strength.

Hall-effect gyrator   A gyrator in which the Hall effect reverses the polarity of a signal when input and output are interchanged.

Hall-effect switch   A magnetically actuated momentary keyswitch that contains a *Hall-effect transducer* [HET], transistor amplifier with a trigger circuit integrated on a silicon chip, and a small permanent magnet. When the IC is powered by a DC current and the key is depressed, the IC is moved into the permanent magnetic field against spring pressure, and a Hall-effect voltage is produced. The keyswitch returns to its off position when the key is released.

Hall-effect transducer   [HET]   A transducer, typically made from a chip of doped semiconductor, that will produce a Hall-effect voltage across its opposing faces when it is placed in a magnetic field and a current is passed through it perpendicular to that magnetic field. The Hall voltage developed across the faces of the chip is propor-

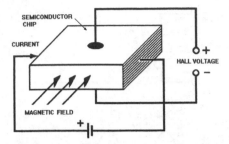

Hall-effect transducer (HET) develops a voltage between the top and bottom surfaces of a silicon chip in a transverse magnetic field.

tional to the magnetic field strength. It is also called a *Hall generator* and a *Hall-effect sensor*.

Hall-effect voltage *Hall effect.*

Hall generator *Hall-effect transducer* [HET].

Hall mobility The product of conductivity and the Hall constant for a conductor or semiconductor. It is a measure of the mobility of the electrons or holes in a semiconductor.

Hall voltage The no-load voltage developed across a semiconductor or metal conductor due to the Hall effect when a specified value of control current flows in the presence of a specified magnetic field.

ham Slang for *radio amateur.*

Hamming code An error-detecting and correcting code used in data transmission. When used with a 7-bit block, there would be information bits at positions 3, 5, 6, and 7 and parity check bits at 1, 2, and 4. This code detects all single and double errors in the block.

handheld personal computer [HPC] A class of small battery-powered personal computers with limited performance capabilities, but they typically include such mobile communications functions as fax and E-mail. Data in memory can be transferred to a large compatible PC for storage or further processing. HPCs include *liquid-crystal displays* and small complete *keyboards.*

handoff The transfer of a cellular telephone call between cells when transmission quality drops below a specified connection threshold.

handset The component of a telephone that includes the microphone at one end with the earphone at the other end, spaced to span the distance between the mouth and the ear.

handshaking The process by which predetermined characters are exchanged by receiving and transmitting communications equipment to establish synchronization. A similar process is used in the transfer of data in computer systems.

hard disk A rigid magnetic memory disk for the storage of data in a computer hard-disk drive. The aluminum disk, coated on both sides with a metallic magnetic material, can store data. A *read/write head* can read and write data from the disk when it is rotating at high speed. The most common version is the *Winchester-drive* disk that cannot be removed from the drive for data exchange or storage. Four or more disks can be stacked on a common spindle, each with its own read/write head to provide exceptionally high data-storage density. More than 2 Gbyte can be stored on a 5¼ in drive. It is also called a *platter.* See also *hard-disk drive* and *Winchester drive.*

hard-disk drive An electromechanical, digital data memory drive for computers that reads and writes data on one or more *hard disks* or *platters* on a common spindle which rotates at high speed. The randomly accessed data is written and read by one *read/write head* for each hard-disk surface; the heads "fly" at microinch heights over the coated disks to read and write data without making contact with the disk surface. Most disk drives are based on the *Winchester drive* design. The data tracks stacked on top of each other form *cylinders.* Some personal computer hard drives now have capacities of more than 2 Gbyte. Multiple drives can be networked to provide very high digital mass storage capacity suitable for *mainframe* and *server* computers. See also *cylinder, hard disk, read-write inductive head, magnetic read/write head,* and *Winchester drive.*

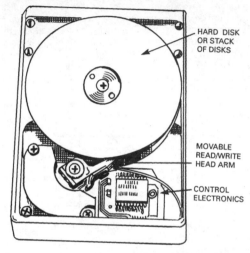

Disk drive: a compact form of the Winchester hard-disk drive, it can have up to six disks or platters. The read/write head "flies" above the disk surface.

hard-drawn copper wire Copper wire that has been drawn to size through several dies without being annealed, to increase its hardness and tensile strength.

hardened circuit A circuit made with components whose tolerance to radiation from a nuclear explosion has been increased by various radiation-hardening procedures.

hard-limited repeater A repeater in communication satellites that clips the composite input signal at a very low level, then amplifies, filters frequency-translates, and retransmits the resulting signal. The repeater output power level remains constant regardless of the input power level, whereas in a linear repeater it varies with input power.

hard limiter A limiter that exhibits negligible variations in output once the range of limiting action is reached.

hardness 1. The penetrating ability of X-rays. The shorter the wavelength, the harder the rays and the greater their penetrating ability. 2. The degree of evacuation in an X-ray or other vacuum tube. The harder the tube, the better its vacuum.

hard-wired circuit A circuit that cannot be altered without unsoldering and resoldering or unwrapping and rewrapping wired connections. See also *wire wrapping.*

hard X-ray   An X-ray that has high penetrating power.

Harm   Acronym for *homing-antiradiation missile*. An antiradiation air-to-ground missile that homes on radiation emitted by an enemy radar to destroy or disable it.

harmonic   A sinusoidal component of a periodic wave, with a frequency that is an integral multiple of the fundamental frequency. The frequency of the second harmonic is twice that of the fundamental frequency or first harmonic. Also called harmonic component and harmonic frequency.

harmonic analysis   1. Any method for identifying and evaluating the harmonics that make up a complex waveform of voltage, current, or some other varying quantity. 2. The expression of a given function as a series of sine and cosine terms that are approximately equal to the given function, such as a Fourier series.

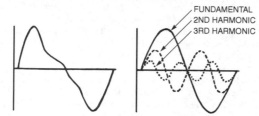

Harmonic analysis of a complex wave showing the fundamental and harmonic components.

harmonic analyzer   An analyzer that measures the strength of each frequency component in a complex wave. Also called harmonic wave analyzer.

harmonic antenna   An antenna whose electrical length is an integral multiple of a half-wavelength at the operating frequency of the transmitter or receiver.

harmonic attenuation   Attenuation of an undesired harmonic component in the output of a transmitter, as with the use of a pi network whose shunt reactances are tuned to have zero impedance at the harmonic frequency to be suppressed.

harmonic component   *Harmonic.*

harmonic content   The components that remain after the fundamental frequency has been removed from a complex wave.

harmonic conversion transducer   A conversion transducer whose output signal frequency is a multiple or submultiple of the input frequency, as in frequency dividers and frequency multipliers.

harmonic distortion   Nonlinear distortion in which undesired harmonics of a sinusoidal input signal are generated because of circuit nonlinearity.

harmonic filter   A filter that is tuned to suppress an undesired harmonic in a circuit.

harmonic frequency   *Harmonic.*

harmonic generator   A generator operated so that it generates strong harmonics along with the fundamental frequency.

harmonic interference   Interference caused by the presence of harmonics in the output of a radio station.

harmonic-mode crystal unit   *Overtone crystal unit.*

harmonic wave analyzer   *Harmonic analyzer.*

harness   An assembly of insulated wires of various lengths, bent to a pattern and bound together before installation in equipment with cord or plastic cable ties.

harp antenna   A variation of the *ground-plane antenna* that consists of a number of vertical radiators which are connected at their bottom ends. The longest radiator is one-quarter wavelength long. Its range is determined by the longest and shortest vertical radiators, and its pattern is omnidirectional.

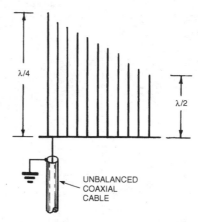

Harp antenna has an array of vertical radiators connected at their bottom ends.

hartley   A unit of information content, equal to the designation of 1 of 10 possible and equally likely values or states of anything used to store or convey information. One hartley is equal to $\log_2 10$ bits or 3.323 bits.

Hartley oscillator   An inductive-capacitive (LC) oscillator whose frequency-determining resonant tank contains a tapped coil. The collector and base are at opposite ends of the tuned circuit to provide the necessary 180° phase shift, and feedback occurs through mutual inductance between the two parts of the coil. Developed as a vacuum-tube circuit, it will function with bipolar junction or field-effect transistor amplifiers. It can generate sine-wave frequencies up to UHF.

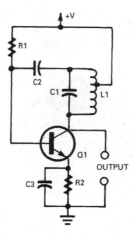

Hartley oscillator has a tapped coil and a tuning capacitor that couples the collector and base circuits.

Hartley's law The total number of bits of information that can be transmitted over a channel in a given time is proportional to the product of channel bandwidth and transmission time.

Hay bridge A four-arm bridge, similar to the Maxwell-Wien bridge, for measuring inductances with large values of Q. It can also be used for determining the incremental inductance of iron-cored reactors. Its balance is frequency-dependent.

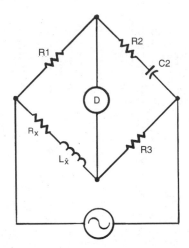

Hay bridge can measure inductances with high Q values.

Haystack antenna A large and powerful radar-type antenna erected by the U.S. Air Force in Massachusetts for microwave research, including satellite relay communication, long-range tracking of spacecraft, and radio astronomy. It has two reflectors arranged according to the Cassegrainian principle of optical telescopes, housed in a metal-frame radome about 50 m in diameter. Movement is controlled by a computer for exact tracking of orbiting spacecraft or radio stars.

HBT Abbreviation for *heterostructure bipolar transistor*.

HCMOS Abbreviation for *high-speed silicon gate CMOS*.

H display A radar *B display* modified to include an indication of the angle of elevation. The target appears as two closely spaced bright spots that merge to form a short line. The slope of the line is proportional to the sine of the target's angle of elevation.

HDLC Abbreviation for *high-level data-link control*.

head 1. The photoelectric unit that converts the sound track on motion-picture film into corresponding audio signals in a motion-picture projector. 2. *Cutter.* 3. *Magnetic head.*

head amplifier 1. An amplifier that is mounted close to the head which serves as its signal source, for amplifying the weak signal before it is fed through a cable to the main amplifier. 2. British term for the video amplifier that is mounted close to the pickup tube in a television camera.

header 1. A glass or metal mounting plate or base with hermetically sealed feed-through pins that support the functional elements of an electron tube (anode, cathode, grids, or a complete electron gun assembly). After the internal elements are welded to the pins projecting into the header, the lip of the header is mated with a glass or metal bulb to form the complete tube. Sealing is done in a glass lathe or metal-envelope sealing machine. The header includes a coaxial glass or metal exhaust stem for evacuating the tube on a vacuum pump. 2. An array of pins molded in a plastic block for use as a connector or interface between a cable and a circuit board.

heading The horizontal direction of a vehicle, expressed as an angle between a reference line and the line extending in the direction the vehicle is headed, usually measured clockwise from the reference line. Heading is the instantaneous actual direction in which an aircraft, ship, or other vehicle is pointed, whereas course is the intended direction of travel. Also called relative heading.

heading marker A line of light produced on a radar plan-position indicator (PPI) display at the instant when the rotating radar antenna is at the position corresponding to the ship's heading, to indicate whether objects ahead of the ship are located to the port or starboard.

headlight antenna A radar antenna small enough to be housed within the thickness of an airplane wing in the manner of an automobile headlight, yet it is capable of producing a beam that can be directed much like a searchlight.

headphone An electroacoustic transducer that is held against the ears by a clamp passing over the head, for private listening to the audio output of a communication, radio, or television receiver or other source of AF signals. It is also called a headset.

headset *Headphone.*

head-up display [HUD] A display, principally for combat aircraft, that projects illuminated navigational or weapons-aiming information onto the windscreen so that the pilot need not look away from a forward view. The information can be presented in a plan position or perspective format complete with moving icons for ease of interpretation.

hearing aid A miniature, portable sound amplifier for persons with impaired hearing, consisting of a microphone, audio amplifier, earphone, and power cell.

hearing loss The ratio of the threshold of audibility of a particular human ear to the normal threshold, expressed in decibels.

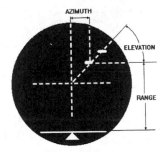

H display presents bearing, range, and altitude. The target appears as a double dot with range on the vertical axis and bearing on the horizontal axis.

heart pacer   *Pacemaker.*

heat   *Thermal radiation.*

heater   An electric heating element for supplying heat to an indirectly heated cathode in an electron tube.

heater current   The current flowing through a heater for an indirectly heated cathode in an electron tube.

heater-type cathode   *Indirectly heated cathode.*

heater voltage   The voltage applied to the terminals of the heater in an electron tube.

heat loss   Power loss that is due to the conversion of electric energy into heat.

heat pipe   A sealed metal tube that has an inner lining of wicklike capillary material and a small amount of fluid in a partial vacuum. When one end of the pipe becomes hot, the fluid boils, and the molecules of steam move along the pipe at high speed. They condense at the other end of the pipe, giving up latent heat, and the capillary lining draws the liquid back to the hot end for a repetition of the heat-transferring cycle. It is also constructed in flattened or plate-shaped form.

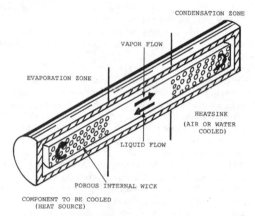

CONDENSATION ZONE

VAPOR FLOW

EVAPORATION ZONE

HEATSINK
(AIR OR WATER
COOLED)

LIQUID FLOW

POROUS INTERNAL WICK

COMPONENT TO BE COOLED
(HEAT SOURCE)

Heat pipe with an internal porous wick conducts heat by capillary action in a closed-evaporation condensation cycle.

heat run   A series of temperature measurements made on an electric or electronic device during operating tests under various conditions.

heat-seal   To bond or weld a flexible plastic film or sheet of material to itself or another material by heat alone. Dielectric heating is frequently used for this purpose.

heat shield   A layer of material that gives protection from heat. It is positioned on the front of a reentry capsule.

heat-shrinkable plastic   A plastic material that can be shrink-fitted over terminals and other objects of varying size and shape primarily for insulating. Commonly used in tubing form, but also available as tape, sheets, and caps. After it is placed on the object to be covered, the material is heated with a hot-air blower to a critical temperature of about 100°C, to make it shrink up to about 50% and conform to the shape of the object when it cools.

heat shunt   A heatsink placed in contact with the lead of a delicate component to prevent overheating during soldering.

heatsink   Metal heat radiators designed to remove heat from components, particularly power transistors, by thermal conduction, convection, or radiation. Heat is conducted away from the internal junction of a semiconductor device to a surface where it can be dissipated. It can be a finned aluminum extrusion, finned copper, or aluminum stampings located under the device case, or aluminum extrusions with radiating fins that clip on the device case.

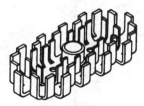

Heatsink for a power transistor is made of aluminum.

heat transfer   The transfer of heat from one location to another by conduction through fluids or solid materials, by convection involving actual movement of a heated fluid or gas, or by radiation of electromagnetic heat waves.

hecto-   A prefix representing 10^2, or hundreds.

hectometric wave   A radio wave between the wavelength limits of 100 and 1000 m, corresponding to the medium-frequency (MF) range of 300 to 3000 kHz.

height   The vertical distance to a target from a horizontal plane passing through the location of the height computer.

height control   The television receiver control that adjusts picture height.

height effect   *Antenna effect.*

height finder   A radar set that measures and determines the height of an airborne object.

height marker   A form of calibration marker included in some radar displays.

height-range indicator   A radarscope that gives both height and range of a target.

Heisenberg uncertainty principle   The simultaneous precise determination of velocity (or any related property) of a particle and its position is impossible. The smaller the particle, the greater the degree of uncertainty.

Heising modulation   *Constant-current modulation.*

helical antenna   A helical radiator connected at right angles to a metal or metal-mesh reflecting disk but insulated from it. It produces a circularly polarized wave that can rotate either clockwise or counterclockwise. The disk diameter is sized to be equal to the lowest operating wavelength, and the helix is formed to a radius about 15% of the center-frequency wavelength with turn spacing about a quarter-wavelength long. It is also called a *helix antenna.*

helical scanning   1. A method of facsimile scanning in which a single-turn helix rotates against a stationary bar to give horizontal movement of an elemental area.   2. A method of radar scanning in which the antenna beam rotates continuously about the vertical axis while the ele-

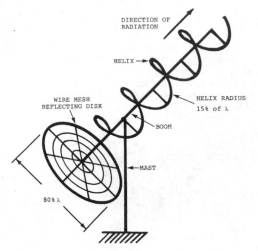

Helical antenna for satellite communication radiates waves in the axial mode.

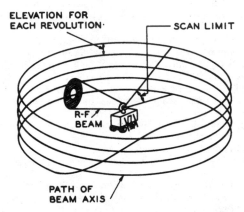

Helical scanning.

vation angle changes slowly from horizontal to vertical. Thus a point on the radar beam describes a distorted helix.

helicon wave   An energy-carrying transverse electromagnetic wave that is supported in solid material by free carriers in the presence of a magnetic field.

Helitow   A version of the wire-guided TOW missile modified for firing from helicopters. The system includes controls, displays, digital guidance, a height precision rate-compensated sight, and a SPRITE-detector-based forward-looking infrared detector (FLIR) with a dedicated missile tracking video channel.

helitron oscillator   An electrostatically focused low-noise backward-wave oscillator. The microwave output signal frequency can rapidly be swept over a wide range by varying the voltage applied between the cathode and the associated RF circuit.

helium [He]   A light, inert gas element with an atomic number of 2 and an atomic weight of 4.00. In electronics it is used to test for leaks in vacuum envelopes and as a pro-

tective gas for heliarc welding because it does not support combustion.

helium-cadmium laser   A metal-vapor ion laser whose cadmium vapor, produced by heat or other means, migrates through a high-voltage glow discharge in helium, generating a continuous laser beam at wavelengths in the ultraviolet and blue parts of the spectrum (about 0.3 to 0.5 μm).

helium cryostat   A cryostat whose liquefied helium-3 isotope is combined with adiabatic demagnetization to produce a temperature of 0.3 K.

helium-neon laser [HeNe laser]   An atomic gas laser based on a combination of helium and neon gases. One application is reading bar code labels in point-of-sale terminals on checkout counters. Its operation is normally at three distinct wavelengths: 0.6328, 1.15, and 3.39 μm.

helium spectrometer   A small-mass spectrometer that detects the presence of helium in a vacuum system. For leak detection, a jet of helium is applied to suspected leaks in the outer surface of the system, and the output indicator of the spectrometer is watched to determine the exact point at which helium enters.

helix   A spread-out single-layer coil of wire, either wound around a supporting cylinder or made of stiff enough wire to be self-supporting.

helix antenna   *Helical antenna.*

helix recorder   A recorder that uses helical scanning.

helix waveguide   A waveguide that consists of closely wound turns of insulated copper wire covered with a lossy jacket.

Helmholtz resonator   An acoustic enclosure that has a small opening dimensioned so that the enclosure resonates at a single frequency determined by the geometry of the resonator.

HELWEPS   A U.S. Navy shipboard high-energy laser weapon based on chemical laser technology. It consists of a megawatt laser, a 5-ft-diameter reflecting mirror, and additional low-power guidance lasers. Its applications range from defensive countermeasures to offensive amphibious support. It has an effective range of 10 km.

HEMT   Abbreviation for *high-electron mobility transistor.*

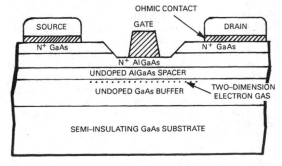

HEMT is a high-frequency, gallium arsenide MESFET made from layers of gallium arsenide and aluminum gallium arsenide.

HEM wave   Abbreviation for *hybrid electromagnetic wave.*

HeNe laser   Abbreviation for *helium-neon laser.*

henry [H; plural: henrys]   The SI unit of inductance or mutual inductance. The inductance of a circuit is 1 H when a current change of 1 A/s induces 1 V.

heptode A seven-electrode electron tube that contains an anode, a cathode, a control electrode, and four additional electrodes which are ordinarily grids.

hermaphroditic connector A connector made with both mating parts exactly alike at their mating surfaces. A bifurcated connector is an example.

hermetic Permanently sealed by fusion, soldering, or other means, to prevent the passage of air, moisture vapor, and all other gases.

hermetically sealed crystal unit A crystal unit sealed in its glass or metal holder, usually by soldering, for protection against all external conditions except vibration and temperature.

hermetically sealed relay A relay that is permanently sealed in its metal, glass, or ceramic housing by fusion or soldering.

hermetic seal A seal that prevents passage of air, water vapor, and all other gases.

herringbone pattern An interference pattern sometimes seen on television receiver screens, consisting of a horizontal band of closely spaced V- or S-shaped lines.

hertz [Hz] The SI unit of frequency, equal to 1 cycle per second.

Hertz antenna An ungrounded half-wave antenna.

Hertzian wave *Radio wave.*

Hertz vector A vector that specifies the electromagnetic field of a radio wave.

HET Abbreviation for *Hall-effect transducer.*

heterodyne To mix two AC signals of different frequencies in a nonlinear device to produce two new frequencies, corresponding respectively to the sum of and the difference between the two original frequencies. This action is the basis of all superheterodyne receivers.

heterodyne conversion transducer *Converter.*

heterodyne detector A detector whose modulated carrier frequency is combined with the signal of a local oscillator that has a slightly different frequency, to provide an audio-frequency beat signal which can be heard with a loudspeaker or headphones. It is used chiefly for code reception.

heterodyne frequency Either of the two new frequencies resulting from heterodyne action. One is the sum of the two input frequencies; the other is the difference between the input frequencies.

heterodyne frequency meter A frequency meter with a known adjustable frequency is heterodyned with an unknown frequency until a zero beat is obtained. Alternatively, the unknown frequency can be heterodyned with a fixed known frequency to produce a lower-frequency signal, usually in the audio-frequency range, whose value is measured by other means. Also called heterodyne wavemeter.

heterodyne harmonic analyzer A harmonic analyzer that produces a complex input voltage which is mixed with the output of a variable-frequency oscillator, and the magnitude of the sum or difference frequency for each input harmonic is measured with a meter.

heterodyne interference *Heterodyne whistle.*

heterodyne oscillator A separate variable-frequency oscillator that produces the second frequency required in a heterodyne detector for code reception.

heterodyne reception Radio reception whose incoming RF signal is combined with a locally generated RF signal of different frequency, followed by detection. The resulting beat frequency can be audible or at a higher intermediate frequency, as in a superheterodyne receiver. Also called beat reception.

heterodyne repeater A radio repeater whose received radio signals are converted to an intermediate frequency, amplified, and reconverted to a new frequency band for transmission over the next repeater section.

heterodyne wavemeter *Heterodyne frequency meter.*

heterodyne whistle A steady high-pitched audio tone heard in an amplitude-modulation radio receiver occurring when two signals that differ slightly in carrier frequency enter the receiver and heterodyne to produce an audio beat. It is also called *heterodyne interference.*

heterojunction An abrupt transition or junction between two different semiconductor materials or two parts of the same semiconductor material that are doped to be P- and N-type materials. Examples include conventional silicon diodes and rectifiers, or gallium arsenide grown on a substrate of silicon. See *PN junction.*

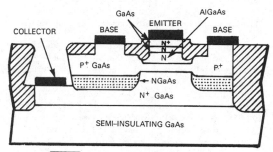

Heterojunction bipolar transistor (HBT) is a high-frequency gallium arsenide transistor.

heterojunction bipolar transistor [HBT or HJBT] A bipolar transistor made from layers of gallium arsenide (GaAs) and aluminum gallium arsenide (AlGaAs).

heterostructure A semiconductor structure containing at least two adjacent layers of different chemical composition, but of similar crystalline structure.

heuristic Related to trial-and-error exploratory methods of problem solving in which solutions are discovered by evaluation of the progress made toward the final result.

heuristic routine A computer routine for attacking a problem by a trial-and-error approach, frequently involving the act of learning.

hexadecimal notation A notation in the scale of 16, based on decimal digits 0 to 9 and six more digits that can be represented by A, B, C, D, E, and F.

HF Abbreviation for *high frequency.*

HGC Abbreviation for the *Hercules graphic card,* a circuit board for personal computer color monitors.

HgTe Symbol for *mercuric telluride.*

hi-fi *High fidelity.*

high band The television band extending from 174 to 216 MHz, which includes channels 7 to 13.

high boost *High-frequency compensation.*

Decimal	Binary	Octal	Hexa-decimal
0	0	0	0
1	1	1	1
2	10	2	2
3	11	3	3
4	100	4	4
5	101	5	5
6	110	6	6
7	111	7	7
8	1000	10	8
9	1001	11	9
10	1010	12	A
11	1011	13	B
12	1100	14	C
13	1101	15	D
14	1110	16	E
5	1111	17	F

Hexadecimal notation with decimal and binary equivalents.

high definition   The television equivalent of high fidelity, in which the reproduced image contains such a large number of accurately reproduced elements that picture details approximate those of the original scene.

high-definition television [HDTV]   Technology to achieve an approximate 2:1 improvement over existing NSTC, SECAM, or other television formats in both horizontal and vertical resolution, improved and more natural color rendition, a wider CRT aspect ratio of at least 5:3, and stereophonic digital audio. See also *NTSC* and *SECAM*.

high-electron mobility transistor [HEMT]   A heterostructure junction FET that is a variation of the MESFET. The term was originated by Fujitsu.

high fidelity   Fidelity of audio reproduction of such high quality that listeners hear almost exactly what they would have heard if they had been present at the original performance. Also called hi-fi.

high-fidelity receiver   A radio receiver that reproduces audio frequencies with high fidelity, to duplicate faithfully the original sound picked up by the microphone.

high frequency [HF]   A Federal Communications Commission designation for a frequency in the band from 3 to 30 MHz, corresponding to a decametric wave between 10 and 100 m.

high-frequency (HF) **band**   A band of frequencies extending from 0.003 to 0.030 GHz in accordance with IEEE Standard 521-1976.

high-frequency bias   A sinusoidal signal that is mixed with a data signal being recorded on magnetic tape, to increase the linearity and dynamic range of the recorded signal. The bias frequency is usually three to four times the highest information frequency to be recorded.

high-frequency compensation   Increasing the amplification at high frequencies with respect to that at low and middle frequencies in a given band, such as in a video band or an audio band. Also called high boost.

high-frequency trimmer   A trimmer capacitor that controls the calibration of a tuning circuit at the high-frequency end of a tuning range in a superheterodyne receiver.

high level   The more positive of the two logic levels or states in a binary digital logic system. When both high and low levels are negative, the opposite connotation can be applied. The larger negative voltage (the less positive voltage) is assumed to be the high level or high state.

high-level data-link control [HLDLC]   A CCITT standard data communication line protocol. See *CCITT*.

high-level detector   A power detector or linear detector with a voltage-current characteristic that is essentially a straight line or two intersecting straight lines extending up to high input voltage levels.

high-level language   An applications-oriented computer programming language, to be distinguished from a machine-oriented programming language. Examples include Basic, C, C++, Cobol, Java, and Pascal.

high-level modulation   Modulation produced in the anode circuit of the last stage of a transmitter, where the power level approximates that at the output of the system.

highlight   A bright area in a television image.

high-mu tube   A vacuum tube that has a high amplification factor.

high-order digit   A digit that occupies the most significant position in a positional notation system. In conventional notation it is the digit at the left.

high-pass filter   A filter that transmits all frequencies above a given cutoff frequency and substantially attenuates all others. See *filter characteristics*.

high-performance parallel interface [HIPPI]   An interface standard for computer networks. See also *FC-AL* and *ISDN*.

high-potting   Testing a circuit or product with a high voltage (potential), generally as a production operation.

high Q   A characteristic of a component that has a high ratio of reactance to effective resistance so that its Q factor is high.

high-recombination-rate contact   A semiconductor-to-semiconductor or metal-to-semiconductor contact where thermal-equilibrium carrier densities are maintained substantially independent of current density.

high-speed printer   *Line printer.*

high-speed regulator   A power-supply regulator without an output capacitor to permit rapid programming of output voltage changes and/or quick response as a current source.

high tension   1. High voltage measured in thousands of volts.   2. British term for *anode voltage,* of about several hundred volts.

high-threshold logic [HTL]   Logic characterized by higher supply voltages (about 15 V), higher noise immunity, and higher threshold voltages than other logic families. It was obtained by adding a Zener-diode voltage drop to the normal diode voltage drop of diode-transistor logic circuits. An obsolete technology.

HIPPI   Abbreviation for *high-performance parallel interface.*

hiss   Random noise in the audio-frequency range, similar to prolonged sibilant sounds.

histogram   A graphic presentation in the form of vertical bars of varying height in which the height of the bar represents the quantity in a specified category. For example, bars could represent the dollar value of products produced by an electronics firm in 1 year or earnings for that firm for a period such as 10 years.

hit   An exact match between two items of data.

H$_{m,n}$ mode   British term for *TE$_{m,n}$ mode*.

hog-trough linear array   A sectoral antenna horn fed by a horizontal array of dipoles about 26 ft (8 m) long used as an IFF interrogator in air-traffic control, [ATC]. It is usually mounted on top of an air-traffic control radar antenna. Its beam is 2.3° wide in azimuth and 42° wide in elevation. This array is used in conjunction with a stationary omnidirectional antenna in an ATC system.

H$_{m,n}$ wave   British term for *TE$_{m,n}$ wave*.

H network   An attenuation network composed of five branches. Two are connected in series between an input terminal and an output terminal. Two are connected in series between another input terminal and output terminal. The fifth is connected from the junction point of the first two branches to the junction point of the second two branches. Also called H pad.

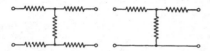

H network has five impedance branches; T network at right has only three branches.

hold   1. To maintain storage elements at equilibrium voltages in a charge-storage tube by electron bombardment. 2. To retain information in a computer memory for further use after it has been first used.   3. A condition in which an integrator, sample-and-hold amplifier, or other charge-storing circuit maintains constant output after the input signal has been removed.   4. A designed stop in an operation or test, such as a delay in the countdown for launching a missile or in completing a telephone call.

hold control   A manual control that changes the frequency of the horizontal or vertical sweep oscillator in a television receiver so that the frequency more nearly corresponds to that of the incoming synchronizing pulses. The two controls are called the horizontal hold control and the vertical hold control. Also called speed control.

holding current   The minimum current required to maintain a switching device in a closed or conducting state after it is energized or triggered.

hold time   The time when pressure is applied to a welded joint after welding current ceases to flow.

hold-up time   The time a power supply output voltage remains within specifications following the loss of input power. In general, off-line *switching power supplies* offer longer hold-up times. Its value depends on the energy stored in a high-voltage capacitor. The output load can be decreased or the line voltage increased to lengthen hold-up time.

hole   A mobile electron vacancy in a semiconductor that acts like a positive electron charge (1.6×10^{-19} C) with a positive mass. Unoccupied locations among the electrons that are bound in their orbits. In the presence of an electric field, holes move in a direction opposite that of electrons, thus producing current. Holes are induced in a semiconductor by adding small quantities of an acceptor dopant to the host crystal.

hole conduction   Conduction occurring in a semiconductor when electrons move into holes under the influence of an applied voltage to create new holes. The apparent movement of those holes is toward the more negative terminal and is hence equivalent to a flow of positive charges in that direction.

hole-electron pair   A positive (hole) paired with a negative (electron) charge carrier.

hole injection   The production of holes in an N-type semiconductor when voltage is applied to a sharp metal point in contact with the surface of the material.

hole mobility   The ability of a hole to travel readily through a semiconductor.

hologram   The special photographic plate used in holography. When this negative is developed and illuminated from behind by a coherent gas-laser beam, it produces a three-dimensional image in space.

holographic bar-code reader   A variation on the earlier *universal product code* or UPC bar-code reader for identification of products in retail stores. Its holographic scanner permits it to read the UPC on the bottom and sides of any coded item passed at most angles over the scanner window. The system, located below the checkout counter, includes a low-power laser and a motor-driven rotating holographic scanning disk with different pie-shaped holographic lenses. Each facet contains a different hologram with a unique combination of focal length, skew angle, and elevation to deflect the laser beam in different paths. On at least one of these sweeps the laser beam will reach and be reflected back from the UPC bars on the package. A detector converts the optical signals to electrical pulses. A computer then compares these signals against a database of product names, sizes, and prices stored in memory. A receipt is printed out while updating the store's inventory. See also *Universal Product Code* [UPC] and *bar-code reader*.

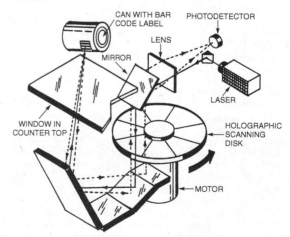

Holographic bar-code reader can read the UPC code on the bottom or side of any item passed over the scanner window.

holographic memory   A memory whose information is stored in the form of holographic images on thermoplastic or other recording films.

holography   Three-dimensional photography based on the use of laser light with photographic plates. The resulting

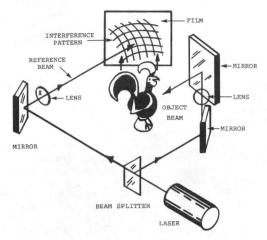

Holography. Laser light beam is split so that one part—the object beam—is directed by mirrors onto the object, and the second part—the reference beam—is directed by mirror and lens to the photographic film. Reflected light from the object beam converges with the reference beam to form an interference pattern that is captured by the film that stores the hologram.

images can be viewed in three dimensions without special glasses. The coherent light output of the laser beam is channeled to illuminate both the subject and the photographic plate during the production of the hologram. The result is an interference pattern on the photographic plate that bears no resemblance to the original until developed and illuminated from behind by a similar coherent laser beam.

holomicrography   The use of holography to produce three-dimensional images with various types of microscopes.

home   1. To fly toward a radiation-emitting source guided by the radiated waves.   2. To travel to a target with the guidance of heat radiation, laser beams, radar echoes, radio waves, sound waves, or other phenomena originating in or reflected from the target.   3. The normal or starting position for a stepping relay.

home mobile station   A *mobile station* (cellular telephone) that is subscribed in its home *cellular mobile telephone system.*

homing   1. The process of approaching a desired point by maintaining constant some indicated navigation parameter other than altitude.   2. The use of radiation from a target to establish a collision course in missile guidance. The three types of homing are active, semiactive, and passive.   3. Flying toward a radio or radar transmitter by receiving the transmitted radiation for navigation guidance.   4. Returning to the starting position, as in a stepping relay or tuning motor.

homing antenna   A directional antenna array used by an aircraft or missile to fly directly to a target that is emitting or reflecting radio or radar waves.

homing beacon   A radio beacon, either airborne or on the ground, that is a target for an aircraft if it is equipped with a radio compass or homing receiver. It is also called a *radio homing beacon.*

homing device   1. A transmitter, receiver, or adapter for homing aircraft or by aircraft for homing purposes.   2. A

circuit in a guided missile or aircraft that homes it on a target.   3. A control device that automatically starts in the correct direction of motion or rotation to achieve a desired change, as in a remote-control tuning motor for a television receiver. A nonhoming device might go first to the end of its travel in the wrong direction.

homing guidance   A guidance system for a missile that directs itself to a target with self-contained circuitry that reacts to a particular characteristic emission of the target. Those emissions include heat, light, sound, a reflected radar echo or other electromagnetic radiation. Homing guidance can be active, semiactive, or passive.

homogeneous radiation   Radiation that has an extremely narrow band of frequencies or a beam of monoenergetic particles of a single type, so that all components of the radiation are alike.

homojunction   A PN junction that is formed in a single type of semiconductor material.

homopolar machine   *Homopolar generator.*

hood   1. An opaque shield placed above or around the screen of a cathode-ray tube or other display device to eliminate extraneous light.   2. A protective covering, usually ventilated to carry away dust, fumes, and gases, to provide a safe working position for handling dangerous chemicals or radioactive materials.

hookswitch   A switch that connects the telephone circuit to the subscriber loop. The name is derived from older telephones where the switch was activated by lifting the receiver off and onto the hook on the side of the phone.

hookup   An arrangement of circuits and apparatus for a particular purpose.

hookup wire   Tinned and insulated solid or stranded soft-drawn copper wire for making low-power circuit connections. The size is No. 12 to 32 AWG (American wire gage).

hop   A single reflection of a radio wave from the ionosphere back to the earth in traveling from one point to another. The term is modified by adjectives such as single-hop, double-hop, and multihop. The number of hops is called the order of reflection.

horizon   The apparent junction of earth and sky as seen from a transmitting antenna site. The horizon bounds that part of the earth's surface which is reached by the direct wave of a radio station.

horizon sensor   A passive infrared device that detects the thermal discontinuity between the earth and space, to establish a stable vertical reference for control of the attitude or orientation of a missile or satellite in space. Thermistors serve as the infrared detectors.

horizontal amplifier   An amplifier for signals that deflect the electron beam horizontally in a cathode-ray tube.

horizontal blanking   Blanking of a television picture tube during the horizontal retrace.

horizontal blanking pulse   The rectangular pulse that forms the pedestal of the composite television signal between active horizontal lines. This pulse causes the beam current of the picture tube to be cut off during retrace. It is also called line-frequency blanking pulse.

horizontal centering control   The centering control provided in a television receiver or cathode-ray oscilloscope to shift the position of the entire image horizontally in either direction on the screen.

horizontal convergence control The control that adjusts the amplitude of the horizontal dynamic convergence voltage in a color television receiver.

horizontal definition *Horizontal resolution.*

horizontal deflection electrodes The pair of electrodes that moves the electron beam horizontally from side to side on the fluorescent screen of a cathode-ray tube by electrostatic deflection.

horizontal deflection oscillator The oscillator that produces, under control of the horizontal synchronizing signals, the sawtooth voltage waveform that is amplified to feed the horizontal deflection coils on the picture tube of a television receiver. It is also called horizontal oscillator.

horizontal drive control The control in a television receiver, usually at the rear of the set, that adjusts the output of the horizontal oscillator. Also called drive control.

horizontal flyback Flyback of the electron beam of a television picture tube that returns from the end of one scanning line to the beginning of the next line. It is also called horizontal retrace and line flyback.

horizontal frequency *Line frequency.*

horizontal hold control The hold control that changes the free-running period of the horizontal deflection oscillator in a television receiver so that the picture remains steady in the horizontal direction.

horizontal linearity control A linearity control that permits narrowing or expanding the width of the left-hand half of a television receiver image to give linearity in the horizontal direction so circular objects appear as true circles. It is usually mounted at the rear of the receiver.

horizontal line frequency *Line frequency.*

horizontally polarized wave A linearly polarized wave whose electric field vector is horizontal.

horizontal oscillator *Horizontal deflection oscillator.*

horizontal output stage The television receiver stage that feeds the horizontal deflection coils of the picture tube through the horizontal output transformer. It can also include a part of the second-anode power supply for the picture tube.

horizontal output transformer A transformer in a television receiver that provides the horizontal deflection voltage, the high voltage for the second-anode power supply of the picture tube, and the filament voltage for the high-voltage rectifier. It is also called a flyback transformer and horizontal sweep transformer.

horizontal polarization Transmission of radio waves so that the electric lines of force are horizontal and the magnetic lines of force are vertical. With this polarization, transmitting and receiving dipole antennas are placed in a horizontal plane. The NTSC television system favors horizontal polarization, whereas the British television system favors vertical polarization.

horizontal resolution The number of individual picture elements or dots that can be distinguished in a horizontal scanning line of a television or facsimile image. They are determined by observing the wedge of fine vertical lines on a test pattern. Also called horizontal definition.

horizontal retrace *Horizontal flyback.*

horizontal scanning Rotation of a radar antenna in bearing entirely around the horizon or in a sector of the horizontal plane.

horizontal sweep The sweep of the electron beam from left to right across the screen of a cathode-ray tube.

horizontal sweep transformer *Horizontal output transformer.*

horizontal synchronizing pulse The rectangular pulse transmitted at the end of each line in a television system to keep the receiver in line-by-line synchronism with the transmitter. It is also called a line synchronizing pulse.

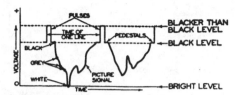

Horizontal synchronizing pulse occurs at the end of each line of picture signal.

horizon tracker A circuit for establishing a vertical reference in a navigation system by precisely tracking the visible horizon.

horn A flared pipe with a loudspeaker driver that improves the radiation of sound and achieves desired directional characteristics. The cross-sectional area increases progressively from the throat to the mouth. It is also called an acoustic horn.   2. An electromechanical or air-actuated signaling device.

horn antenna A microwave antenna made by flaring out the end of a circular or rectangular waveguide into a horn-shape for radiating radio waves directly into space. In a rectangular horn antenna, either one or both transverse dimensions increase linearly from the small end or throat to the mouth.

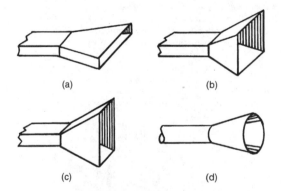

Horn antennas or radiators for waveguides: (*a*) H-plane, (*b*) pyramidal, (*c*) E-plane, and (*d*) conical.

horn feed A horn antenna that feeds a parabolic reflector in a radar antenna system.

horn loudspeaker A loudspeaker whose radiating element is coupled to the air or another medium with a horn.

horn mouth The end of a horn that has the larger cross-sectional area.

horn throat The end of a horn that has the smaller cross-sectional area.

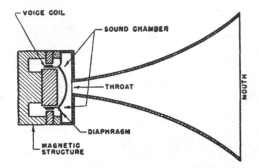

Horn loudspeaker.

horsepower [hp]   A unit of power, equal to 746 W. Use of the SI unit of power, the watt, is preferred.

horseshoe magnet   A permanent magnet or electromagnet whose core is horseshoe-shaped or has parallel sides like a U, to bring the two poles near each other.

host system   1. The system in which a device or circuit is embedded.   2. A computer system that contains emulation circuitry or an accessory card to imitate another system.

hot   1. Highly radioactive.   2. *Energized.*

hot-air soldering   Soldering with a narrow blast of air whose temperature is closely controlled at the value required for soldering individual joints on printed-circuit boards.

hot carrier   A carrier that can be either an electron or a hole which has relatively high energy with respect to the carriers normally found in majority-carrier devices such as thin-film transistors. Hot carriers are injected either by emission over a potential barrier that exists at a metal-semiconductor junction or by tunneling through an extremely thin insulating layer.

hot-carrier diode   *Schottky barrier diode.*

hot-cathode vacuum gage   A vacuum gage with a heated filament (cathode), grid helix, and anode. Electrons are accelerated from the cathode to the grid. Those not collected by the grid move toward the anode where they collide with gas molecules, creating positive ions. Ions attracted to the negative anode cause ion current, which is proportional to pressure (vacuum). See also *cold-cathode vacuum gage* and *thermocouple vacuum gage.*

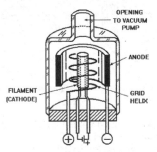

Hot-cathode vacuum gage can measure pressure (vacuum) of 10^{-8} to 10^{-3} torr.

hot electron   An electron in excess of the thermal-equilibrium number. For metals it has an energy greater than the Fermi level. For semiconductors, the energy must be a definite amount above that of the edge of the conduction band. A hot electron (or hot hole) can be generated by photoexcitation, quantum-mechanical tunneling, minority-carrier injection across a forward-biased PN junction, high-field acceleration in nonmetallic materials, and Schottky emission over a forward-biased metal-semiconductor junction.

hot-electron injection   In semiconductor electrically erasable memories, the injection of electrons into the memory cell's floating gate by a vertical electric field. The electrons' excess energy is acquired from a high-source-to-drain channel electric field.

hot-wire ammeter   An ammeter for measuring alternating or direct current by sending it through a fine wire. The resulting expansion or sag of the wire due to heat deflects the meter pointer. It is also called a *thermal ammeter.*

hot-wire instrument   An instrument that depends for its operation on the expansion by heat of a current-carrying wire.

hot-wire microphone   A velocity microphone that depends for its operation on the change in resistance of a hot wire as the wire is cooled or heated by varying particle velocities in a sound wave.

hour [h]   A unit of time, equal to 3600 s. Use of the SI unit of time, the second, is preferred.

housekeeping   The series of computer operations required prior to a processing run, such as the setting up of constants, variables, limits, storage locations, and any other preliminary steps that do not contribute directly to the solution of a problem.

howl   An undesirable prolonged sound produced by a radio receiver or AF amplifier system, caused by electric or acoustic feedback.

howling   *Acoustic feedback.*

hp   Abbreviation for *horsepower.*

H pad   *H network.*

h parameter   One of a set of four transistor equivalent-circuit parameters that conveniently specify transistor performance for small voltages and currents in a particular circuit. Also called hybrid parameter.

HPC   Abbreviation for *handheld personal computer.*

HSSDC   Abbreviation for *high-speed serial data connection.*

H system   A radar navigation system that has two ground radar beacons in conjunction with airborne equipment which gives the direction and distance to each beacon.

HTL   Abbreviation for *high-threshold logic.*

hue   The name of a color, such as red, yellow, green, blue, or purple, corresponding to the dominant wavelength. White, black, and gray are not considered hues.

hue control   A control that varies the phase of the chrominance signals with respect to that of the burst signal in a color television receiver, to change the hues in the image. Also called phase control.

hum   1. An electric disturbance occurring at the power-supply frequency or its harmonics, 60 or 120 Hz in the United States and 50 or 100 Hz in other countries. An example is ripple.   2. A sound produced by an iron core of a transformer because of loose laminations or magnetostrictive effects. The frequency of the sound is twice the power-line frequency.

human engineering *Human factors engineering.* See *ergonomics.*

human factors engineering *Human engineering.* See *ergonomics.*

hum bar A dark horizontal band extending across a television picture, caused by excessive hum in the video signal applied to the input of the picture tube.

hum-bucking coil A coil wound on the field coil of an excited-field loudspeaker and connected in series opposition with the voice coil so that hum voltage induced in the voice coil is canceled by that induced in the hum-bucking coil.

humidity detector A detector that opens or closes a switch when the amount of moisture in the atmosphere reaches a preset value. The sensing element might be moisture-absorbing paper, stretched human hairs, stretched nylon fibers, or any other material with a characteristic that changes with humidity in a known and repeatable manner.

hum modulation Modulation of an RF signal or detected AF signal by hum. This hum is heard in a radio receiver only when a station is tuned in.

hum slug A copper ring placed around the core of an excited-field loudspeaker adjacent to the voice coil to serve as a single shorted turn for hum currents induced by the field coil.

hunting An undesirable oscillation of an automatic control system. The controlled variable swings on both sides of the desired value.

Hurter-and-Driffield curve *H-and-D curve.*

Huygens principle Every point on an advancing wavefront acts as a source that sends out new waves. The combined effect is propagation of the wave as a whole.

HVDC Abbreviation for *high-voltage direct current.*

H vector A vector that represents the magnetic field of an electromagnetic wave. In free space it is perpendicular to the E vector and the direction of propagation.

H wave British term for *transverse electric wave* (TE wave).

hybrid 1. A product with two or more different characteristics or types of structure. 2. *Hybrid junction.*

hybrid circuit See *hybrid integrated circuit.*

hybrid coil *Hybrid transformer.*

hybrid computer A computer that includes both a programmable digital computer and an analog computer, used primarily in testing and simulation, such as wind-tunnel testing of aircraft models and water-tank testing of ship models.

hybrid electromagnetic wave [HEM wave] An electromagnetic wave that has components of both the electric and magnetic field vectors in the direction of propagation.

hybrid integrated circuit 1. A microcircuit made by bonding discrete active components such as diodes, transistors, integrated circuits, and passive components to a ceramic substrate that has metallized conductors and pads. Resistors can be formed on the substrate by masking,

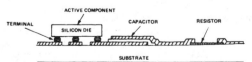

Hybrid integrated circuit constructed on insulating substrate.

metal deposition, and firing; capacitors can be formed by successive metal and dielectric deposition steps and firing. Inductors for microwave hybrids can also be formed by metal deposition and firing. Active and passive components are typically in chip form for surface mounting. Hybrids can be digital or analog (RF and microwave). See *monolithic integrated circuit.*

hybrid junction A transformer, resistor, or waveguide circuit or device that has four pairs of terminals so arranged that a signal entering at one terminal pair will divide and emerge from the two adjacent terminal pairs but will be unable to reach the opposite terminal pair. Also called bridge hybrid and hybrid.

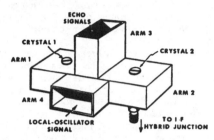

Hybrid junction for rectangular waveguides, serving as a balanced mixer for radar receiver.

hybrid microcircuit A circuit in which one type of microcircuit is combined with some other type of microcircuit or with discrete components.

hybrid parameter *H parameter.*

hybrid ring A doughnut-shaped waveguide that acts as a hybrid junction for four waveguide sections. Coaxial lines can substitute for waveguides under certain conditions.

hybrid set Two or more transformers interconnected to form a hybrid junction. It is also called a *transformer hybrid.*

hybrid T A microwave hybrid junction composed of an E-H T junction with internal matching elements. It is reflectionless for a wave propagating into the junction from any arm when the other three arms are match-terminated.

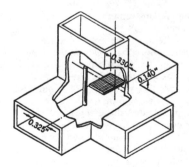

Hybrid T for waveguide, showing the use of a post and iris inside for matching.

hydrogen [H] A gaseous element that has the simplest known atom, consisting of only one proton and one electron. It has an atomic number of 1 and an atomic weight of

1.01. Its three isotopes are ordinary or light hydrogen, deuterium, and tritium.

hydrogen laser   A molecular gas laser whose hydrogen generates coherent wavelengths near 0.6 μm in the vacuum ultraviolet region. Peak output power above 100 kW has been obtained.

hydrogen line   Monochromatic 1.42-GHz radiation from atoms of hydrogen. It is the only known single-frequency natural radiation in the RF portion of the electromagnetic spectrum. This line originates from cool hydrogen in the galaxy and allows radio astronomers to determine directly the velocity properties of any observed cloud of cool hydrogen by means of the Doppler effect.

hydrogen maser   A gas maser whose hydrogen gas is the basis for providing an output signal that has a high degree of stability and spectral purity. The resonator frequency is 1.420 405 751 786 4 GHz.

hydrogen-oxygen fuel cell   A fuel cell that has hydrogen fuel fed into the anode side of the cell under pressure and forced into the pores of the electrode where it reacts with the catalytic agent and the electrolyte to release electrons to the electrode. These electrons flow through an external circuit to perform useful work, and then return to the cathode side of the cell. The by-product of this reaction is water.

hydrogen thyratron   A hydrogen-filled rectifier tube that provides high peak currents at high anode voltages for magnetron pulse circuits. Hydrogen as a replacement for mercury improves the thyratron's thermal stability.

hydrometer   A direct-reading instrument that has a graduated float whose position in a liquid is determined by the density or specific gravity of the liquid. It can measure the state of charge of storage batteries.

hydronic radiation   A form of underwater electromagnetic radiation for communication under water. Voice communication has been achieved between scuba divers 250 m apart, and signals have been transmitted 50 km.

hydrophone   An electroacoustic transducer that responds to waterborne sound waves and delivers essentially equivalent electric waves. It can detect the approach of submarines and other ships.

hydrophotometer   A photometer that measures light transmission through a fixed distance in sea water.

hydrostatic pressure   *Static pressure.*

hygrometer   An instrument that measures the humidity of the atmosphere.

hygroscopic   Tending to absorb moisture.

hyperacoustic zone   The region in the upper atmosphere, between 60 and 100 mi (100 and 160 km) above the earth, where the distance between the rarefied air molecules roughly equals the wavelength of sound. Sound is transmitted with less volume than at lower levels. Sound waves cannot be propagated above this zone.

hyperbolic horn   A horn whose equivalent cross-sectional radius increases according to a hyperbolic law.

hyperbolic navigation system   Any system of radio navigation in which the navigating aircraft receives synchronized signals from at least three known points, as in loran. The time difference between signals received from any two stations determines a line of position in the form of a hyperbola that has the two transmitters at its foci. Crossing this line of position with another line of position, obtained with at least one other transmitting station, establishes a fix.

hyperfrequency wave   Unofficial designation for microwaves that have wavelengths in the range from 1 cm to 1 m.

hypersonic   A velocity greater than five times the speed of sound in air (greater than Mach 5).

hypersonics   The branch of acoustics that deals with acoustic waves and vibrations at frequencies above 500 MHz. Ultrasonics covers lower frequencies down to about 20 kHz.

hypertext   Key words that appear in text on computer screens that, when highlighted, will bring up additional information on the subject in the form of files, such as those found on the *Internet.*

hysteresigraph   An instrument that automatically measures and draws the hysteresis curve for a specimen of magnetic material.

hysteresis   1. The inability of a magnetic material to retrace the response curve during demagnetization that it traced during magnetization.   2. In displays, the tendency for a segment to remain either *on* or *off* when switched, a problem often solved by maintaining a sustaining bias on the segments to keep all lighted pixels glowing without lighting any that should be off.   3. An effect in oscillators in which a given set of operating conditions can result in multiple values of output power and/or frequency.

hysteresis curve   A curve that shows the steady-state relation between the magnetic induction in a material and the steady-state alternating magnetic intensity that produces it.

hysteresis distortion   Distortion that occurs in circuits which contain magnetic components, due to nonlinearity caused by hysteresis.

hysteresis error   The maximum separation due to hysteresis between upscale-going and downscale-going indications of a measured variable.

hysteresis heater   An induction heater with a ferrous charge or charge container that is heated principally by hysteresis losses caused by varying magnetic flux in the magnetic material. In normal induction heating, the heat is due to eddy-current losses.

hysteresis loop   A curve that shows, for each value of magnetizing force, two values of the magnetic flux density in a cyclically magnetized material: one when the magnetizing force is increasing, the other when it is decreasing.

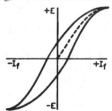

Hysteresis loop.

hysteresis loss   The power loss in an iron-core transformer or other AC device because of magnetic hysteresis.

hysteresis motor   A small synchronous motor for light constant-speed duty, as for tape drives. It starts because of the hysteresis losses induced in its hardened steel secondary member by the revolving field of the primary.

Hz   Abbreviation for *hertz.* It was formerly called *cycle per second.*

I [from intensity]   Symbol for *current*.

IAGC   Abbreviation for *instantaneous automatic gain control*.

IAT   Abbreviation for *international atomic time*.

IC   Abbreviation for *integrated circuit*.

ICAO   Abbreviation for *International Civil Aviation Organization*.

ICBM   Abbreviation for *intercontinental ballistic missile*.

I channel   The 1.5-MHz-wide channel in the NTSC color television system is for transmitting cyan-orange color information. The signals in this channel are known as I signals. See also Q channel.

ICI   Abbreviation for *International Commission on Illumination*.

ICN   Abbreviation for *idle-channel noise*.

icon   In computer technology, graphical symbols, typically line drawings or silhouettes displayed in boxes near the edges of the computer screen that represent, by analogy, a function performed by resident software. Examples include a wastebasket to represent a text or graphic deletion function, a filing cabinet to represent the file function, and a life ring to represent the user-help function.

iconoscope   A television camera tube with a beam of high-velocity electrons that scans a photoemissive mosaic capable of storing an electric charge pattern corresponding to an optical image focused on the mosaic. The mosaic consists of globules of light-sensitive material on a mica sheet that has a conducting film on its back surface. A small value of capacitance exists between each globule and the metal film. Each globule emits electrons in proportion to the light incident on it, producing charges on the capacitances between globules. The electron beam discharges each capacitance in turn during mosaic scanning. The resulting variations in the current taken from the metal film form the desired output signal. This is obsolete technology.

ICW   Abbreviation for *interrupted continuous wave*.

IDE   Abbreviation for *integrated device electronics*.

I demodulator   The demodulator whose chrominance signal and the color-burst oscillator signal are combined to recover the I signal in a color television receiver.

identification   The process of determining the identity of a specific displayed radar target or determining which of several returns represents a specific target in radar.

identification, friend or foe [IFF]   A beacon system that identifies a friendly aircraft when it shows up on radar. A coded interrogation signal triggers the beacon carried by friendly aircraft, initiating the automatic transmission of a properly coded identification signal. The related air traffic radar beacon system is now used worldwide to track aircraft and determine their altitudes for air traffic control purposes. Interrogation occurs at 1.03 GHz and all replies are at 1.09 GHz.

IDP   Abbreviation for *integrated data processing*.

IDTV   Abbreviation for *improved-definition TV*.

IEC   Abbreviation for *International Electrotechnical Commission*.

IEEE   Abbreviation for *Institute of Electrical and Electronics Engineers*.

IEEE 1394 Serial Bus   A serial interface for computer networks.

IF   Abbreviation for *intermediate frequency*.

IF amplifier   The section of a superheterodyne receiver that amplifies signals after they have been converted to the fixed IF value by the frequency converter. It is located between the frequency converter and second detector.

IFF   Abbreviation for *identification, friend or foe*.

IF harmonic interference   Interference due to acceptance of harmonics of an IF signal by RF circuits in a superheterodyne receiver.

IFR   Abbreviation for *instrument flight rules*.

IFR conditions   Weather conditions below the minimum specified for visual flight rules where instrument flight rules apply.

IF rejection   The ability of an FM radio receiver to reject signals of government and commercial stations operating at or near the IF value of the receiver.

IF response ratio   The ratio of the field strength at a specified frequency in the IF band to the field strength at the desired frequency. Each field is applied in turn, under specified conditions, to produce equal outputs.

IF signal   A modulated or continuous-wave signal whose frequency is the IF value of a superheterodyne receiver. It is produced by frequency conversion before demodulation. This value is usually 455 kHz for a broadcast-band radio receiver, 10.7 MHz for an FM radio receiver, approximately 45 MHz for a television receiver picture channel, and 4.5 MHz for a television receiver sound channel.

IF stage   One of the stages in the IF amplifier of a superheterodyne receiver.

IF transformer   The transformer at the input and/or output of an IF amplifier stage in a superheterodyne receiver that provides coupling and selectivity.

IGBT   Abbreviation for *insulated-gate bipolar transistor.*

IGFET   Abbreviation for *insulated-gate field-effect transistor,* now known as *metal-oxide semiconductor field-effect transistor* [MOSFET].

ignition interference   Interference caused by the spark discharges in automotive or other ignition systems.

ignore   A computer instruction that indicates that no action should be taken by a computer.

IHF   Abbreviation for *Institute of High Fidelity.*

IHF power   *Music power.*

IHF standard   One of the standards adopted by the Institute of High Fidelity for specifying performance of audio amplifiers, FM tuners, and other high-fidelity equipment.

I²L   Abbreviation for *integrated injection logic.*

ILF   Abbreviation for *infralow frequency.*

illegal character   A character or combination of bits that is not accepted as a valid representation by a computer or a specific routine. Illegal characters are commonly detected and used as an indication of a machine malfunction.

illuminant C   The reference white of color television. It closely matches average daylight.

illumination   1. The geometric distribution of power that reaches various parts of a dish reflector in an antenna system.   2. The power distribution to elements of an antenna array.   3. The density of the luminous flux falling on a surface. It is equal to the flux divided by the area of the surface when the latter is uniformly illuminated. The SI unit is the lux. It is also called *luminous flux density.*

illumination control   A photoelectric control that turns on room lights when outdoor illumination decreases below a predetermined level.

illumination meter   An instrument for measuring the illumination on a surface directly in lux (lumen per square meter). Also called illuminometer.

illumination sensitivity   The signal output current divided by the incident illumination on a camera tube or phototube.

illuminometer   *Illumination meter.*

ILS   Abbreviation for *instrument landing system.*

IM   Abbreviation for *intermodulation.*

image   1. An optical counterpart of an object, as a real image or virtual image.   2. A fictitious electrical counterpart of an object, as an electric image or image antenna.   3. The scene reproduced by a television or facsimile receiver.

image admittance   The reciprocal of image impedance.

image antenna   A fictitious electrical counterpart of an actual antenna, acting mathematically as if it existed in the ground directly under the real antenna and served as the direct source of the wave that is reflected from the ground by the actual antenna.

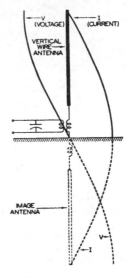

Image antenna for simple Marconi vertical-wire antenna, with voltage and current distribution curves.

image converter   A converter that uses optical fibers to change the form of an image, for more convenient recording and display or the coding of secure messages.

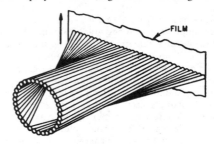

Image converter for converting a circular input pattern to a line pattern for recording on upward-moving film by optical fibers.

image-converter tube   An electron tube that converts an image obtained in one region of the electromagnetic spectrum to an image in another region. For example, an image captured in ultraviolet (UV) or near infrared (IR) can be converted to a visible-light image. Its construction is similar to that of an *image intensifier tube.* It is a component in active and passive infrared imaging systems. See *active night-vision system* and *FLIR.*

image distortion Failure of the reproduced image in a television receiver to appear the same as that scanned by the television camera.

image enhancement A method of improving color television pictures by comparing each video line, element by element, with the preceding and following lines. Any differences between vertically aligned elements are added to the middle-line element in the proper phase to enhance picture outlines and contrast.

image frequency An undesired carrier frequency that differs from the frequency to which a superheterodyne receiver is tuned by twice the intermediate frequency. If the oscillator frequency is higher than that of the desired incoming carrier, the image frequency will be lower by a value equal to the oscillator frequency. If the receiver has poor selectivity, a strong image-frequency signal can get through its tuning circuits to beat with the local oscillator and produce the correct intermediate frequency.

image-frequency rejection ratio The ratio of the response of a superheterodyne receiver at the desired frequency to the response at the image frequency.

image iconoscope A camera tube that projects an optical image on a semitransparent photocathode. The resulting electron image emitted from the other side of the photocathode is focused on a separate storage target. The target is scanned on the same side by a high-velocity electron beam, neutralizing the elemental charges in sequence to produce the camera output signal at the target. Its sensitivity is much higher than that of an iconoscope.

image impedance One of the impedances that, when connected to the input and output of a transducer, will make the impedances in both directions equal at the input terminals and at the output terminals. The load impedance and the equivalent internal impedance of the transducer are then images of each other. This condition gives maximum power transfer. When the two image impedances are equal, their value is the same as the characteristic impedance.

image-intensifier tube A vacuum tube that admits low-level light and intensifies it by electronic multiplication

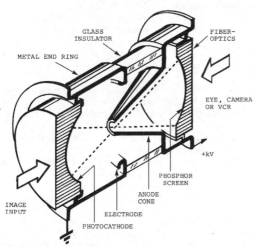

Image-intensifier tube has fiberoptic faceplates on its input and output ends and an electron lens that inverts the image.

and scanning to produce an image of the scene that can be viewed directly in real time at the eyepiece. It gains sufficient light from natural background night lighting and does not need supplementary illumination from infrared projectors or lasers, and is therefore passive. Faint light collected from the object being observed is collected by a glass lens and then is passed through a lens formed from bundles of optical fibers to a gallium arsenide (GaAs) photocathode surface. This construction extends its operation into the near-infrared band and cascades the emission of electrons which are accelerated by an electric field. It is an integral part of military night-vision monoculars and binoculars. The most modern versions require only two, rather than three, amplification stages. See *night-vision goggles* and *night-vision scope*.

image interference Interference that occurs in a superheterodyne receiver when a station broadcasting on the image frequency is received along with the desired station. Image interference can occur when the receiver circuits do not have sufficient selectivity to reject the image-frequency signal. For a standard radio broadcast receiver, the image frequency would be 910 kHz higher than that of the desired station.

image-interference ratio A superheterodyne receiver rating that indicates the effectiveness of the preselector in rejecting signals at the image frequency.

image isocon A camera tube similar to an image orthicon but responsive to much lower light levels, including near darkness.

image orthicon A camera tube that produces an electron image with a photoemitting surface. The image is focused on one side of a separate storage target that is scanned on its opposite side by an electron beam, usually consisting of low-velocity electrons. It has such high sensitivity that images can be picked up even in semidarkness. Both color and monochrome versions are widely used in television broadcasting studios.

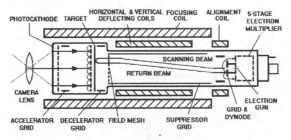

Image-orthicon camera tube with a focusing and deflection system.

image processing The use of computers and mathematical algorithms to analyze, enhance, and interpret digitized television images.

image rejection Suppression of signals at the image frequency in a superheterodyne receiver.

image response A form of interference of received signals produced when the mixer of a superheterodyne receiver responds to a signal equal to the local oscillator (LO) frequency minus the intermediate frequency (IF) when the desired signal frequency is equal to the LO plus the IF. It also occurs when the mixer responds to a signal frequency equal to the LO plus the IF when the desired signal is the LO minus the IF.

image-storage array   A solid-state panel or silicon chip with image-sensing elements that can be a metal-oxide semiconductor or a charge-coupled light-sensitive device which can be manufactured in a high-density configuration.

image-storage tube   A storage tube that has information introduced to it by means of radiation, usually light. It is read later as a visible output.

image tube   A general category of vacuum tubes that includes *image converter tubes* and *image intensifier tubes.*

IMF   Abbreviation for *instantaneous frequency measurement.*

immittance [IMpedance and adMITTANCE]   A term that denotes both impedance and admittance, as commonly applied to transmission lines, networks, and certain kinds of measuring instruments.

impact ionization   Ionization produced by the impact of a high-energy charge carrier on an atom of semiconductor material. The effect is an increase in the number of charge carriers.

impact-noise analyzer   An analyzer that is paired with a sound-level meter to evaluate the characteristics of impact-type sounds and electric noise impulses that cannot be measured accurately with a noise meter alone. Applications include assessment of possible damage to hearing by punch presses, pile drivers, and pneumatic drills.

impact printer   A line printer that has one or more character fonts, a ribbon or other inking device, a paper transport, and some means of impacting desired characters or character elements on the paper. Impact printers include chain and drum printers that print on-the-fly on either the front or back of the paper. Some printers use multicopy forms that have encapsulated inks on the paper itself. Operating speeds can be up to 2000 lines per minute. Dot-matrix printers can also be impact printers, with characters formed by impact of stiff wires selected by solenoid action from a 5×7 or 7×9 matrix.

IMPATT amplifier   A diode amplifier based on an IMPATT diode. Its operating frequency range is from about 5 to 100 GHz, primarily in the C and X bands, with power output up to about 20 W continuous wave or 100 W pulsed.

IMPATT diode [IMPact Avalanche Transit-Time diode]   A solid-state microwave diode that has a negative-resistance characteristic produced by a combination of impact avalanche breakdown and charge-carrier transit-time effects in a thin chip which is usually gallium arsenide or silicon. When suitably mounted in a tuned cavity or waveguide, the diode can either be an oscillator or an amplifier operating in the gigahertz range.

impedance [Z]   The total opposition offered by a component or circuit to the flow of an alternating or varying current. Impedance Z is expressed in ohms, and is a combination of resistance R and reactance X, computed as the square root of the sum of the squares of R and X. Impedance is also computed as $Z = E/I$, where E is the applied AC voltage and I is the resulting alternating current flow in the circuit. In computations, impedance is handled as a complex ratio of voltage to current.

impedance bridge   A bridge that can measure or compare impedances that might contain capacitance, inductance, and resistance.

impedance characteristic   A graph that plots the impedance of a circuit against frequency.

impedance coupling   Coupling of two signal circuits with an impedance, usually a choke. It is used in audio amplifiers when high gain and limited bandpass are required.

impedance feedback   Use of a passive impedance network to provide feedback from the output terminal of an operational amplifier to the input summing junction.

impedance match   The condition in which the impedance of a connected load is equal to the internal impedance of the source or the surge impedance of a transmission line. This gives maximum transfer of energy from source to load, minimum reflection, and minimum distortion.

impedance-matching network   A network of two or more resistors, coils, and/or capacitors to couple two circuits so that the impedance of each circuit will be equal to the impedance into which it looks.

impedance-matching transformer   A transformer that obtains an impedance match between a given signal source and a load that has a different impedance than the source.

impedance triangle   A diagram that consists of a right-angle triangle, whose sides are proportional to the resistance and reactance, respectively, of an AC circuit. The hypotenuse represents the impedance of the circuit. The cosine of the angle between the resistance and the impedance lines is equal to the power factor of the circuit.

impedor   A synonym for impedance to describe a circuit element that has impedance.

imperfection   Any deviation in structure from that of an ideal crystal.

implanted atom   An atom introduced into semiconductor material by ion implantation.

implanted device   1. A heart pacemaker or other medical electronic device that is surgically placed in the body.   2. A resistor or other device that is fabricated within a silicon or other semiconducting substrate by ion implantation.

implode   To burst inward.

implosion   The inward collapse of an evacuated container, such as the glass envelope of a cathode-ray tube.

impregnated coil   A coil in that has the spaces between the turns of insulated wire filled with an insulating varnish or plastic material.

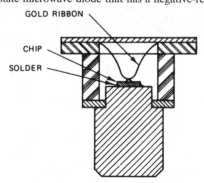

GOLD RIBBON

CHIP

SOLDER

IMPATT diode chip in a hermetically sealed package. The copper stud on the bottom serves as a terminal and a heatsink. The other terminal is at the top.

improvement threshold The upper limit value of carrier-to-noise ratio. Below that limit the signal-to-noise ratio decreases more rapidly than the carrier-to-noise ratio.

impulse 1. A pulse so short that it can be regarded mathematically as having infinitesimally short duration. 2. A pulse, generally with a waveform that rises rapidly to a sharply peaked maximum and falls rapidly to zero.

impulse bonding A variation of stitch bonding in which connections are made directly by impulse welding of the interconnecting wire to terminal pads formed on active or passive areas of the substrate or directly over plated-through holes.

impulse excitation *Pulse excitation.*

impulse generator *Surge generator.*

impulse noise Noise characterized by transient short-duration disturbances distributed essentially uniformly over the useful passband of a transmission system.

impulse relay A relay that stores the energy of a short pulse, to operate the relay after the pulse ends.

impulse solenoid A solenoid that operates on pulse power, at speeds up to several hundred strokes per second. Applications include tape drives and punches, high-speed sorting gates, and shutter drives.

impulse test *Pulse test.*

impurity A material such as boron, phosphorus, or arsenic added in small quantities to a crystal to produce an excess of electrons (donor impurity) or holes (acceptor impurity). It is also called a dopant.

impurity level An energy level caused by the presence of impurity atoms.

impurity profile A plot of the variations in concentration of a dopant in a semiconductor wafer such as one made of silicon or gallium arsenide.

impurity scattering Scattering of electrons by impurity atoms in the crystal.

impurity semiconductor A semiconductor whose properties are caused by impurity levels produced by foreign atoms.

in Abbreviation for *inch.*

InAs Symbol for *indium arsenide.*

in-band signaling The transmission of signaling tones within the channel normally reserved for voice transmission.

incandescence Emission of visible radiation by a heated object, such as a lamp filament heated by electric current.

incandescent lamp An electric lamp whose light is produced by sending electric current through a filament of resistance material to heat it to incandescence.

inch [in] A unit of length, equal to 2.54 cm. The use of the SI unit of length, the centimeter, is preferred.

inch per second [abbreviated in/s] A magnetic tape speed rating. Commonly used speed values include 1⅞, 3¾, 7½, 15, and 30 in/s (4.76, 9.5, 19, 38, and 76 cm/s).

inching *jogging.*

incidence angle The angle between an approaching beam of radiation and the perpendicular (normal) to the surface that is in the path of the beam.

incidental amplitude modulation Amplitude modulation that results unintentionally from the process of frequency modulation and/or phase modulation.

incidental frequency modulation Frequency modulation that results unintentionally from the process of amplitude modulation.

incidental phase modulation Phase modulation that results unintentionally from the process of amplitude modulation.

incidental radiation device A device that radiates radio-frequency energy during normal operation, although not intentionally designed to generate such energy.

incident light The direct light that falls on a surface.

incident wave 1. A wave that impinges on a discontinuity or on a medium that has different propagation characteristics. 2. A current or voltage wave that is traveling through a transmission line in the direction from source to load.

inclination The angle that a line, surface, vector, or aircraft makes with the horizontal.

inclined synchronous orbit A nonequatorial, hence non-stationary, synchronous and circular orbit.

inclinometer 1. An instrument that measures the direction of the earth's magnetic force with relation to the plane of the horizon. 2. An instrument that measures the attitude of an aircraft with respect to the horizontal.

incoherent scattering Scattering of particles or photons, in which the scattering elements act independently of one another, so that there are no definite phase relationships between the different parts of the scattered beam.

incoherent waves Waves that have no fixed phase relationship.

increductor A variable inductance that has a saturable core, used in some high-frequency circuits.

increment A small change in the value of a variable.

incremental frequency shift A method of superimposing incremental intelligence on another intelligence by shifting the center frequency of an oscillator a predetermined amount.

incremental permeability The ratio of a small cyclic change in magnetic induction to the corresponding cyclic change in magnetizing force when the average magnetic induction is greater than zero.

incremental printer A printer that prints sequentially, character by character, on each line. An example is a computer-controlled ink-jet printer.

incremental sensitivity The smallest change in a quantity being measured that can be detected by a particular instrument.

incremental shaft-angle optical encoder An electromechanical encoder with an internal light-emitting diode (LED), photodetectors, and incremental shaft-mounted code disk. When the code disk rotates, the light beam is "chopped" and the pattern is converted by photodetectors into two or three square-wave outputs with pulses equal to the number of lines on the disk. Shaft speed and position can be determined with respect to a previous reference angle by counting pulses. See also *absolute shaft-angle optical encoder* and *incremental shaft-angle encoder disk.*

incremental shaft-angle optical encoder disk A glass or plastic disk with a pattern of equally spaced radial opaque and clear segments that is mounted on the rotating shaft of an incremental shaft-angle optical encoder. See also *incremental shaft-angle optical encoder* and *absolute shaft-angle optical encoder disk.*

incremental tuner A television tuner with its antenna, RF amplifier, and RF oscillator tuning coils continuous or in

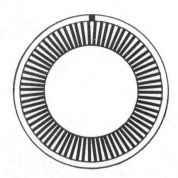

Incremental shaft-angle optical encoder disk.

small sections connected in series. Rotary switches make connections to the required portions of the total inductance necessary for a given channel, or short-circuit all of an inductance except that required for a given channel.

independent-sideband modulation   Modulation in which the upper and lower sidebands carry entirely different information signals. The carrier can be either transmitted or suppressed.

independent variable   The independent quantity or condition that, through the action of the control system's automatic controller, directs the change in the controlled variable according to a predetermined relationship.

indexing   The process of establishing memory addresses in a computer by adding the value in an address field of an instruction to a value stored in a specified index register.

index of refraction   The ratio of the velocity of a wave in a vacuum to that in a specified medium.

index register   A computer register whose contents automatically modify addresses incorporated in instructions just prior to their execution. The original instruction remains intact and unmodified in the memory.

indicator   A cathode-ray tube or other device that presents information transmitted or relayed from some other source, as from a radar receiver.

indirect address   An address in a computer instruction that indicates a location where the address of the referenced operand is to be found.

indirectly controlled variable   The quantity or condition that is controlled by its relation to the controlled variable and is not directly measured for control in a feedback control system.

indirectly heated cathode   A cathode whose heat is supplied by an independent heater element in a thermionic tube. As a result, this cathode has the same potential on its entire surface. By contrast, the potential along a directly heated filament varies from one end to the other. Also called an equipotential cathode, heater-type cathode, and unipotential cathode.

indirectly ionizing particles   Uncharged particles that can directly liberate ionizing particles or initiate a nuclear transformation. Examples include neutrons and photons.

indirect scanning   Scanning in which a narrow beam of light is moved across the area being televised. It is used in flying-spot scanning of films where the light transmitted by each illuminated elemental area is picked up in turn by one or more photosensors.

indium [In]   A metallic element with an atomic number of 49 and an atomic weight of 114.82. It is an important dopant for semiconductors and is in the same group of the periodic table as boron, aluminum, and gallium. See *indium antimonide.*

indium antimonide [InSb]   A binary alloy semiconductor formed from indium and antimony that offers very high electron mobility. It is used to make infrared detectors, infrared filters, magnetoresistors, and Hall-effect devices. Tunnel diodes, transistors and laser diodes have also been made with InSb.

indium antimonide detector   A photovoltaic infrared detector that operates at liquid nitrogen temperatures to give high sensitivity at a wavelength of about 5 μm in the infrared region. It is used in missile guidance systems, target-recognition systems, and for monitoring laser radiation.

indium arsenide [InAs]   An intermetallic compound that has semiconductor properties. It is used in Hall-effect sensors.

indium phosphide [InP]   An intermetallic compound that has semiconductor properties.

induced charge   An electrostatic charge produced on an object by an electric field.

induced current   A current produced in a conductor by a time-varying electromagnetic field, as in induction heating.

induced electron emission   Electron emission produced by impinging X-rays on a sample of a material being analyzed. The energy range of the resulting emitted electrons depends on the chemical composition. It is used in spectrometers.

induced voltage   A voltage produced in a circuit by a change in the number of magnetic lines of force passing through a coil in the circuit.

inductance [L]   The property of a circuit or circuit element that opposes a change in current flow. Inductance causes current changes to lag behind voltage changes. Inductance is measured in henrys, millihenrys, and microhenrys.

inductance bridge   An instrument, similar to a Wheatstone bridge, that can measure an unknown inductance by comparing it with a known inductance.

inductance-capacitance [LC]   A circuit containing both inductance and capacitance, as provided by coils and capacitors.

inductance standard   *Standard inductor.*

inductance switch   A cryogenic two-level variable inductor that can be switched from one level to the other by a control current.

induction   The process of producing a voltage, electrostatic charge, or magnetic field in an object by interaction with lines of force.

induction brazing   An electric brazing process that produces heat by induced current flowing through the resistance of the joint being brazed.

induction coil   A device for changing direct current into high-voltage alternating current. Its primary coil contains relatively few turns of heavy wire, and its secondary coil, wound over the primary, contains many turns of fine wire. Interruption of the direct current in the primary by a vibrating-contact arrangement induces a high voltage in the secondary.

induction compass   A compass whose indications depend on the current generated in a coil revolving in the magnetic field of the earth.

induction-conduction heater A heating device whose electric current is conducted through a charge but restricted by induction to a preferred path.

induction coupling A coupling whose torque is transmitted by the interaction of the magnetic field produced by magnetic poles on one rotating member and induced currents in the other rotating member.

induction field 1. The portion of the electromagnetic field of a transmitting antenna that acts as if it were permanently associated with the antenna. The radiation field leaves the transmitting antenna and travels through space as radio waves. 2. The electromagnetic field of a coil carrying alternating current, responsible for the voltage induced by that coil in itself or in a nearby coil.

induction furnace A furnace whose electric energy is transformed into heat by electromagnetic induction.

induction hardening A process of hardening a ferrous alloy by induction heating.

induction heater A generator and associated equipment for induction heating. At low frequencies, down to power-line frequencies, rotating equipment generates the power. At higher frequencies, more suitable for surface heating, electronic generators generate the power.

induction heating Heating of a conducting material by placing the material in a varying electromagnetic field. The resulting eddy currents induced in the material produce heat, just as current flowing through a resistor produces heat. The frequencies used range from 60 to over 500,000 Hz, depending on the size and shape of the object to be heated. Also called eddy-current heating.

induction motor An AC motor with a primary winding on one member (usually the stator) connected to the power source, and a polyphase secondary winding or a squirrel-cage secondary winding on the other member (usually the rotor) that carries induced current.

induction potentiometer A resolver-type synchro that delivers a polarized voltage whose magnitude is directly proportional to angular displacement from a reference position and whose phase indicates the direction of shaft rotation from the reference position. The linear range can be as great as 85°.

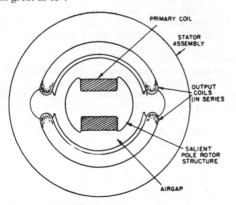

Induction potentiometer.

inductive 1. Related to inductance. 2. A reference to the inducing of a voltage through mutual inductance.

3. Related to the inducing of an electric charge by electrostatic induction.

inductive circuit A circuit that contains a higher value of inductive reactance than capacitive reactance.

inductive coupling Coupling of two circuits by means of the mutual inductance provided by a transformer. Coupling by the self-inductance common to two circuits is called direct inductive coupling. It is also called a transformer coupling.

inductive diaphragm A resonant window in a waveguide that provides the equivalent of inductive reactance at the frequency being transmitted.

inductive feedback Feedback of energy from the output circuit of an amplifier to its input through an inductance or by inductive coupling.

inductive kick The voltage induced in an iron-core coil when coil current is suddenly interrupted. The induced voltage can be many times greater than the applied voltage.

inductive load A load that is predominantly inductive, so the alternating load current lags behind the alternating voltage of the load. It is also called lagging load.

inductive post A metal post or screw extending across a waveguide parallel to the E field, to add inductive susceptance in parallel with the waveguide for tuning or matching purposes.

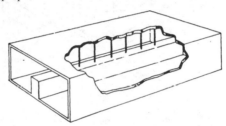

Inductive posts in a filter for a ridge waveguide.

inductive reactance $[X_L]$ Reactance due to the inductance of a coil or circuit. Inductive reactance is measured in ohms and is equal to $6.28fL$, where f is the frequency in hertz and L is the inductance in henrys.

inductive tuning Tuning that is performed by a variable inductance.

inductive window A conducting diaphragm extending into a waveguide from one or both sidewalls of the waveguide that has the effect of an inductive susceptance in parallel with the waveguide.

inductometer A calibrated variable inductance, generally used with an inductance bridge.

inductor *Coil.*

inductor generator An AC generator whose fixed windings and flux linkages are varied by the rotation of a toothed ferromagnetic rotor. It can generate high power at frequencies up to several thousand hertz for induction heating.

Industry Standard Architecture [ISA] The name given to the bus architecture developed by IBM for the PC AT version of its personal computer. See also *Extended Industry Standard Architecture.*

inertance The acoustical equivalent of inductance.

inert gas A chemically inert gas such as argon, helium, krypton, neon, and xenon.

inertial-celestial guidance An inertial guidance system with the capability for checking its position automatically from time to time by means of celestial guidance.

inertial-gravitational guidance system A guidance system that is independent of all outside information except gravity.

inertial guidance A self-contained electronic guidance system that automatically follows a given course toward a ground target whose precise geographical position has been set into the computer of the system along with the starting point or a check-point position. It responds to inertial effects resulting from each change of course or speed and makes appropriate corrections under control of its built-in computer. It is independent of interference and is not subject to jamming. It is used in guided missiles and submarines.

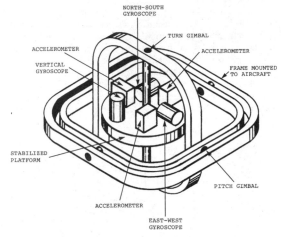

Inertial guidance is based on gyroscopes and accelerometers on a stable platform free to move about all three axes. Changes in direction and velocity are computed to determine the distance and direction of travel.

inertial navigation Inertial guidance for navigation in aircraft, ships, spacecraft, and submarines. It provides dead-reckoning navigation based on automatic calculation of direction, distance, and speed as derived from gyroscopic and acceleration measurements.

inertial space A frame of reference designed with respect to fixed stars.

inertia switch A switch that is actuated by an abrupt change in the velocity of the host on which it is mounted.

infinite attenuation Attenuation so great that a voltage applied to the input terminals of the filter produces no output voltage. The term specifies a frequency at which infinite attenuation would be produced by a filter if its coils and capacitors had zero loss.

infinite baffle A loudspeaker baffle that prevents acoustic energy from traveling from the front of the loudspeaker diaphragm to the back. It is usually a large sealed enclosure with the loudspeaker mounted in a hole cut into one side.

infinite clipping Clipping characterized by a very small threshold level, resulting in an output waveform that is essentially rectangular.

infinite-impedance detector A detector with an input circuit that has infinite impedance.

infinite line A hypothetical transmission line whose characteristics correspond to those of an ordinary line that is infinitely long.

infinite resolution The ability to provide a stepless, continuous change in value or change in output over the entire range of a device. An example is the change in resistance of a potentiometer whose contact arm moves over a gapless plastic-coated metal resistance element.

infinity [∞] 1. An indefinitely large number or amount. 2. Any number larger than the maximum number that a computer can store in any register. When such a number is calculated, the computer usually stops and signals an alarm indicating an overflow.

inflection point A point at which a curve takes a definite change in direction.

information processing The processing of data that represents information, such as index entries.

information retrieval Recovery of specific information from stored data or from a collection of documents, generally with the aid of a computer.

information science The science that deals with the properties and control of information, including its storage, retrieval, and dissemination.

information superhighway A metaphor for a two-way digital data network whose carrying capacity is virtually limitless. All forms of digital information—text, voice, audio, video, computer code, photographs, graphics, and motion pictures—can be transmitted over telephone wires, coaxial cable, and fiberoptic cable. The *Internet* is considered to be an information superhighway.

information system Any means for communicating knowledge from one person to another, such as by simple verbal communication, and completely computerized methods of storing, searching, and retrieval of information. The conventional printed book is the best known, most convenient, and most effective of all information systems.

information theory The mathematical theory concerned with information rate, channel width, distortion, noise, and other factors affecting the transmission of information.

infradyne receiver A superheterodyne receiver in which the intermediate frequency is higher than the signal frequency, to obtain high selectivity.

infralow frequency [ILF] A frequency in the band from 300 to 3000 Hz in the radio spectrum, corresponding to a wavelength between 100 and 1000 km.

infrared [IR] *Infrared radiation.*

infrared absorption spectrum A spectrum produced by molecular absorption of infrared radiation.

infrared beacon A source of infrared radiation that can establish a geographic reference point. A bearing can be determined from it.

infrared camera A camera that uses high-resolution scanning techniques in combination with infrared detectors to obtain thermographs showing images created by the infrared energy which is naturally radiated by all objects at all temperatures above absolute zero.

infrared communication set The components required to operate a two-way electronic system based on infrared radiation as the media to carry intelligence.

infrared detector A device that determines the presence and/or bearing of an object by measuring infrared radiation from its surfaces. Photon-type infrared detectors have high sensitivity and fast time constants, but are sensitive only to a narrow range of wavelengths in the near-infrared region and require cryogenic operating temperatures. Thermal-type infrared detectors absorb infrared energy, causing a temperature rise that changes an electrical property such as resistance.

infrared-emitting diode [IRED] An infrared-emitting diode that has maximum emission in the near-infrared region, typically at 0.9 μm for PN gallium arsenide. It is used in industrial controls, light modulators, logic circuits, optical switching, position encoders, and tape readers.

infrared guidance system A missile guidance system that includes an infrared detector with amplifiers and control units to detect and home on a heat-emitting enemy target.

infrared homing Homing that permits a target to be tracked by the infrared radiation that it emits.

infrared image converter An electron tube that converts an invisible infrared-illuminated scene into a visible image on a fluorescent screen. An infrared lens focuses the desired scene on a photocathode at the input end of the tube. The resulting stream of electrons emitted from the back of the photocathode is proportional to the illumination at each point. It passes through an electron lens that focuses the stream on a fluorescent screen at the other end of the tube to give a visible image.

infrared imagery Imagery produced electronically by sensing infrared electromagnetic radiations emitted or reflected from a target surface.

infrared optics Lenses, prisms, and other optical elements suitable for use with infrared radiation.

infrared polarizer A polarizer that consists of a thin film of pyrolytic graphite. It will provide up to 98% polarization of incident light at 4 μm and beyond in the infrared region.

infrared radiation Electromagnetic radiation in the infrared spectrum, ranging in wavelength from about 0.75 to 1000 μm.

infrared receiver A device that intercepts and/or demodulates infrared radiations which carry intelligence.

infrared remote control *Digital remote control.*

infrared scanner A scanner based on mechanical or electrical techniques that provides line-by-line scanning of the field of view by one or more infrared detectors. In one arrangement, a motor-driven scanning mirror is simultaneously tilted by a motor-driven cam to give both horizontal and vertical scanning.

infrared spectrum Emitted energy at wavelengths between 0.75 and 1000 μm are defined as *infrared radiation.* (Wavelengths between 100 and 1000 μm are millimeter waves). Infrared is electromagnetic radiation generated by vibration and rotation of the atoms and molecules within any material at temperatures above absolute zero (0 K or –273°C). If the radiating source is hotter than about 1000 K, the emitted energy can be seen by the human eye as red light at about 0.75 μm. The spectrum can be subdivided into four divisions: near infrared (NIR, 0.7 to 3 μm); middle infrared (MIR, 3 to 6 μm), far infrared (FIR, 6 to 15 μm); and extreme infrared (XIR, 15 to 1000 μm). The first three divisions include spectral intervals called *atmospheric infrared windows* in which the earth's atmosphere is relatively transparent. The 8- to 12-μm band is best for thermal imaging at ranges beyond 900 m, but

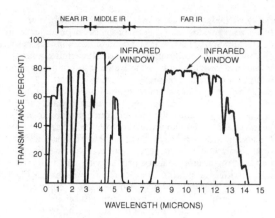

Infrared spectrum extends between visible light (left) and the microwave region (right), starting at about 100 μm. Two infrared "windows" in the atmosphere are shown.

the shorter wavelengths of 3 to 5 μm are suitable for short-range thermal imaging.

infrared transmitter A transmitter that emits energy in the infrared spectrum. It may be modulated with intelligence signals.

infrared-transparent material An optical material that transmits infrared radiation. Examples include sodium chloride (0.25 to 16 μm), cesium iodide (1 to 50 μm), and high-density polyethylene (16 to 300 μm).

infrasonic frequency A frequency below the audible range.

infrasonic voltmeter A voltmeter that measures voltages accurately at infrasonic frequencies, generally for both sine and square waves. It can also measure higher frequencies: one commercial model has 3% accuracy at 0.05 to 30,000 Hz.

inherent regulation A form of system regulation that restores it to equilibrium after an upset without the intercession of a compensating control element.

inhibit To prevent an action.

inhibiting input A gate input that blocks an output which might otherwise occur in a computer.

inhibition gate A gate circuit placed in parallel with the circuit being controlled, for use as a switch.

inhibit pulse A drive pulse that tends to prevent flux reversal of a magnetic cell by certain specified drive pulses.

in-house Relating to an operation produced or carried on within a plant or organization, rather than done elsewhere under contract.

initialize To set various counters, switches, and addresses in a computer to zero, or to other starting values, at the beginning of a routine or at prescribed points in a routine.

initial permeability The normal permeability that exists when both the magnetizing force and the magnetic induction approach zero.

injection 1. A process that makes the minority carrier density in a semiconductor region rise above the equilibrium value. It can be produced by a forward bias on the rectifying barrier or by irradiation with light or other penetrating radiation. 2. Placing a spacecraft in orbit.

injection laser *Laser diode.*

injection laser diode *Laser diode.*

ink-jet cartridge   A container in an *ink-jet printer* that is filled with ink for spraying as droplets on paper to form characters and graphics. It can contain black ink, primary-colored ink, or combinations of these. It includes a micro-miniature nozzle that is activated by computer and computer or printer memory.

ink-jet printer   A nonimpact printer for personal computers that prints by forming and projecting droplets of ink electrostatically onto plain paper from a nozzle on an *ink-jet cartridge* under computer control. The characters are formed in a dot-matrix format as the cartridge, mounted on a traverse drive belt, is scanned across the width of the paper. Primary colors can be printed with primary-colored ink cartridges, and more complex colors and a nearly black hue can be obtained by overprinting various color combinations. Some models can print with 720 by 360 dots per inch (dpi) resolution, and some can print up to 10 pages per minute (ppm) in black and up to 2 ppm in color. Some ink-jet printers contain microprocessors and memory to store images before printing them. There are versions for both desktop and notebook computers. It is also called a *bubble-jet printer.*

in-line electron guns   An assembly of three electron guns in the horizontal plane that produces red, green, and blue in color TV or computer CRTs. Their electron beams strike vertical phosphor stripes through a slotted mask. Sony Trinitron CRTs contain these guns.

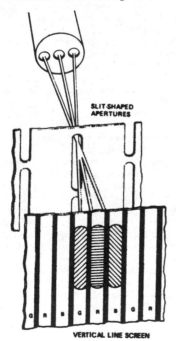

In-line electron guns.

in-line heads   Two magnetic-tape heads mounted so that their gaps are in exact vertical alignment. This is now the standard arrangement for stereophonic tape players. They are also called stacked heads.

INMARSAT   Abbreviation for *International Maritime Satellite Organization.*

inorganic liquid laser   A liquid laser that includes an inorganic liquid such as neodymium-selenium oxychloride or neodymium-doped phosphorus chloride as the active material. These liquids are highly toxic and corrosive, but do not require refrigeration or pulsed operation. Also called neodymium liquid laser.

InP   Abbreviation for *indium phosphide.*

in phase   Characteristic waveforms that are of the same frequency and pass through corresponding values at the same instant.

in-phase rejection   *Common-mode rejection.*

in-phase signal   *Common-mode signal.*

in-port   The entrance for a network.

input   1. The power or signal fed into an electronic device, or the terminals to which the power or signal is applied.   2. The data that is to be processed by a computer.   3. To feed signals or data into a computer or other piece of equipment.

input block   A section of internal memory in a computer that is generally reserved for the receiving and processing of input information.

input capacitance   The short-circuit transfer capacitance that exists between the input terminal and all other terminals of an electron tube (except the output terminal) connected together. This quantity is equal to the sum of the interelectrode capacitances between the input electrode and all other electrodes except the output electrode.

input equipment   The equipment that feeds data into a computer.

input gap   An interaction gap that initiates a variation in an electron stream. In a velocity-modulated tube this gap is in the buncher resonator.

input impedance   The impedance that exists between the input terminals of an amplifier or transmission line when the source is disconnected.

input level   The ratio in decibels of audio input signal power to a reference power level of 1 mW when the signal is working into a given impedance. It is expressed as dBm.

input/output [I/O]   A reference to the connectors or procedures for transmitting information into and out of a computer.

input/output limited   A reference to a computer whose speed of data processing is limited by the speed of the input and/or output equipment.

input/output pads   Areas of conductive metallization on the periphery of an integrated circuit where wires are bonded for the input of signals and power, and the output of signals.

input register   A register that accepts input information for a computer at one speed and supplies the information to the central processing unit at another speed, usually much greater.

input resonator   *Buncher resonator.*

input transient   A spike or step change in the input line to a circuit, typically a voltage.

input transformer   A transformer that provides a correct impedance match between a signal source and the input of a circuit or device.

inrush current   The maximum instantaneous input current drawn by a circuit when it is first turned on.

INS   Abbreviation for *ion neutralization spectroscopy*.

in/s   Abbreviation for *inch per second*.

InSb   Symbol for *indium antimonide*.

insensitive time   *Dead time*.

insert earphone   A small earphone that fits partially inside the ear.

insertion gain   The ratio of the power delivered to a part of the system following an inserted amplifier to the power delivered to that same part before insertion of the amplifier. It is expressed in decibels.

insertion head   A mechanism that inserts a leaded component in a printed-wiring board holes. The mechanism can also include automatic tools for cutting, forming, and clinching the leads of each component.

insertion loss   The loss in load power due to the insertion of a component or device at some point in a transmission system. It is expressed as the ratio in decibels of the power received at the load before insertion of the apparatus to the power received at the load after insertion.

insertion voltage gain   The complex ratio of the alternating component of voltage across the output terminals of a system when an amplifier is inserted between the source and output to the output voltage when the source is connected directly to the output termination.

inside spider   A flexible device placed inside a voice coil to center it accurately with respect to the pole pieces of a dynamic loudspeaker.

instantaneous automatic gain control [IAGC]   The portion of a radar system that automatically adjusts the gain of an amplifier for each pulse, to obtain a substantially constant output-pulse peak amplitude with different input-pulse peak amplitudes. The adjustment is fast enough to act during the time a pulse is passing through the amplifier.

instantaneous companding   Companding in which the effective gain variations are made in response to instantaneous values of the signal wave.

instantaneous particle velocity   The total particle velocity at a point minus the steady velocity at that point.

instantaneous peak power   An audio amplifier power rating that has little practical value because it gives a highly inflated picture of amplifier capability.

instantaneous power output   The rate at which energy is delivered to a load at a particular instant.

instantaneous recording   A recording intended for direct reproduction without further processing.

instantaneous sampling   The process of obtaining a sequence of instantaneous values of a wave.

instantaneous sound pressure   The total instantaneous pressure at a point minus the static pressure that exists at that point when no sound waves are present. Its commonly used unit is the microbar.

instantaneous speech power   The rate of sound energy radiation from a speech source at any given instant.

instantaneous value   The value of a sinusoidal or otherwise varying quantity at a particular instant.

instantaneous volume velocity   The total instantaneous volume velocity at a point minus the static volume velocity at that point.

instant-on switch   A switch that applies a reduced filament voltage to all tubes in a television receiver continuously, so the picture appears almost instantaneously after the set is turned on. The switch inserts a voltage-reducing choke in series with the primary of the power transformer and opens the high-voltage secondary winding to reduce filament voltage to about half the normal value.

Institute of Electrical and Electronics Engineers [IEEE]   A nonprofit professional organization of engineers and scientists, formed by combining the American Institute of Electrical Engineers and the Institute of Radio Engineers.

Institute of High Fidelity [IHF]   An organization established to develop performance specifications for audio amplifiers, FM tuners, and other high-fidelity equipment.

instruction code   An artificial language for describing or expressing the instructions that can be carried out by a digital computer. Each instruction word usually contains a part that specifies the operation to be performed and one or more addresses that identify a particular location in memory or serve some other purpose.

instruction counter   A counter that indicates the location of the next computer instruction to be interpreted.

instruction register   The register that temporarily stores the instruction currently running in a computer.

instruction time   The time required to carry out an instruction that has a specified number of addresses in a particular computer. A typical instruction time is 30,000 instructions per second.

instrument   1. A device for measuring and sometimes recording and controlling the value of a quantity under observation. The term "instrument" is usually applied to combinations of a meter with associated solid-state or tube circuits.   2. *Meter*.

instrument approach   An approach to a landing that is made by navigation instruments and radio guidance, with visual reference to the landing area only after the aircraft breaks through the overcast.

instrument approach system   An aircraft navigation system that furnishes guidance in the vertical and horizontal planes to aircraft during descent from an initial-approach altitude to a point near the landing area. Completion of a landing requires guidance to touchdown by visual or other means.

instrumentation amplifier   An amplifier that accepts a voltage signal as an input and produces a linearly scaled version of this signal at the output. It is a closed-loop fixed-gain amplifier, usually differential, and has high input impedance, low drift, and high common-mode rejection over a wide range of frequencies. Widely used for amplifying millivolt-level signals from strain-gage bridges, thermocouples, and other types of transducers.

instrument flight rules [IFR]   Regulations governing flying when weather conditions are below the minimum for visual flight rules.

instrument landing   An aircraft landing made with the aid of an instrument landing system.

instrument landing approach   An aircraft landing approach without visual reference to the ground. It is made with the aid of aircraft instruments, ground-based electronic devices, or communication systems.

instrument landing system [ILS]   A radio system that provides an aircraft with the directional, longitudinal, and vertical guidance electronics necessary for landing. It usually employs UHF ground transmitters and fixed directional antennas to define a beam that laterally localizes the runway extension and defines a slope plane at some angle between 2 and 5° leading to the optimum point of touch-down on the runway.

instrument shunt　A resistor connected in parallel with an ammeter to extend its current range.

instrument takeoff　A takeoff that depends on aircraft instruments or other aids, without visual reference to the ground.

instrument transformer　A transformer that transfers primary current, voltage, or phase values to the secondary circuit with sufficient accuracy to permit connecting an instrument to the secondary rather than the primary. Used so only low currents or low voltages are brought to the instrument. With a current transformer, the primary winding is inserted into the circuit carrying the current to be measured or controlled. With a voltage transformer, the primary winding is connected across the circuit whose voltage is to be measured or controlled.

insulated　Separated from other conducting surfaces by a nonconducting material.

insulated-gate bipolar transistor [IGBT]　A four-layer discrete power semiconductor device that combines the characteristics of a power MOSFET and a thyristor. A PNP transistor drives an N-channel MOSFET in a pseudo-Darlington pair. An internal JFET conducts most of the voltage, permitting the internal MOSFET to have a lower voltage rating and a lower drain-to-source "on" resistance [R/DS (on)] than similarly rated MOSFETs. This permits it to be used in circuits rated above 300 V. It has a cross section similar to a power MOSFET except for its P+ substrate, which causes the IGBT to function in a way that is more similar to a bipolar transistor than to a power MOSFET. It is also called a COMFET, GEMFET, and IGT. In a switching power supply, it can convert DC from a battery to AC to drive the motor of an electric vehicle.

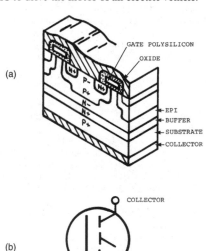

Insulated-gate bipolar transistor (*a*) is similar to a power MOSFET, but it is a minority-carrier device. (*b*) Its schematic symbol.

insulated-gate field-effect transistor　See *metal-oxide semiconductor field-effect transistor.*

insulation　A material that has high electric resistance and is therefore suitable for separating adjacent conductors in an electric circuit or preventing possible future contact between conductors.

insulation resistance　The electric resistance between two conductors separated by an insulating material.

insulator　1. A material that exhibits poor conduction of electricity and heat. Insulators separate and isolate conductors to prevent electrical short circuits and protect personnel against shock or burns from powered circuits. 2. A substance whose normal energy band is full and separated from the first excitation band by a forbidden band that can be penetrated only by an electron with an energy of several electronvolts. Examples of insulators include silicon dioxide (glass), silicon nitride, rubber, plastics, ceramics, and wood.

integer　A whole number.

integral action　A control action whose rate of change of the correcting force is proportional to the deviation.

integrated circuit [IC]　A monolithic semiconductor device that contains many active components (diodes and transistors) and passive components (resistors, capacitors, and inductors) which function as a complete circuit.

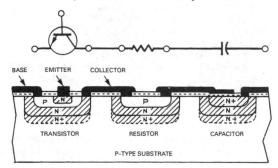

Integrated circuit: a section of a bipolar IC showing the formation of an NPN transistor, P-type resistor, and MOS-type capacitor on the same substrate.

integrated-circuit capacitor　A capacitor that can be produced in a silicon substrate by conventional semiconductor production processes. A junction capacitor exploits the capacitance of a reverse-biased PN junction, which can be formed at the same time as the emitter or collector junctions of transistors. For a metal-oxide semiconductor capacitor, an N+ region is diffused into the silicon to form the bottom electrode on which a controlled thickness of silicon oxide is formed to serve as dielectric. Maximum capacitance values are limited to a few hundred picofarads.

integrated-circuit package [IC]　A monolithic semiconductor device that contains many active components (diodes and transistors) and passive components (resistors, capacitors, and inductors) which function as a complete circuit.

integrated-circuit resistor　A resistor that can be made in or on an integrated-circuit substrate as part of the manufacturing process. Film resistors are deposited on the surface of the substrate. Semiconductor resistors, which exploit the bulk resistivity of the semiconductor, can be bulk, diffused, or ion-implanted.

integrated component A component incorporated into or deposited on the substrate of an integrated circuit.

integrated device electronics [IDE] A standard interface for computer hard-disk drives that improves the interface transfer rate by making the controller part of the hard disk, thus eliminating it as a separate component.

integrated injection logic [I²L] A digital logic in which minority carriers (holes) in the N layer are generated by a forward-biased PN junction called the injector. I²L transistors are inverted from the conventional bipolar logic transistor, and the logic is organized for multiple outputs from a single input rather than the more conventional multiple inputs, but only one output.

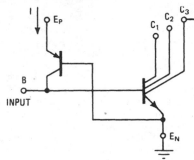

Integrated injection logic for a complete gate fits into the area of one ordinary transistor on an integrated-circuit substrate.

integrated optical circuit [IOC] An integrated circuit capable of operating at optical wavelengths, that can include such functions as lasers, modulators, photodetectors, switches, and optical-fiber conductors.

integrated optical device A device that functions at optical wavelengths and can be constructed on a semiconducting substrate such as that used for integrated circuits.

integrated optics Optics involving the use of components that can be constructed simultaneously in a single substrate for operation as a system at optical wavelengths. Examples include optical waveguide components, surface lasers, and optical communication systems.

Integrated Services Digital Network [ISDN] A communications network that can carry digitized voice and data multiplexed into the public dial-up telephone network. It can communicate at rates up to 128 kb/s, but it requires special telephone lines, communications hardware, and software. The user must pay a provider a fee for this service.

integrating accelerometer A transducer that measures velocity and/or distance by the time integration of measured acceleration. When installed in a missile, it can be preset to cut off fuel flow when the required speed is reached.

integrating circuit An analog circuit that integrates its input signal. In its simplest form it consists of a resistor, capacitor, and switch in series with a voltage source. When the switch is closed, current flows in the circuit, and the integrated signal is measured as an output voltage across the capacitor. It is also called an *integrator*. See also *differentiating circuit*.

integrating filter A filter that builds up a cumulative charge on an output capacitor when subjected to successive pulses of applied voltage.

integrating gyroscope A gyroscope that senses the rate of angular displacement and measures and transmits the time integral of this rate.

integrating ionization chamber An ionization chamber whose collected charge is stored in a capacitor for subsequent measurement.

integrating meter An instrument that totalizes electric energy or some other quantity consumed over a period of time.

integrating timer A timer that totalizes a number of small time intervals.

integrator A circuit or device whose output is the integral of its input with respect to time.

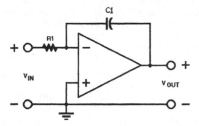

Integrator is an operational amplifier with a capacitor in its feedback loop.

intelligence Data, information, or messages that are to be transmitted or stored.

intelligence signal Any signal that conveys information, such as code, facsimile diagrams and photographs, music, television scenes, and spoken words.

intelligent cell 1. In cellular telephony, circuitry that monitors calls, determines where the *cellular telephone* is located, and finds a way to deliver service to that mobile unit. 2. A cell that coexists comfortably and indestructibly with the interference in the cell. It increases capacity and improves the performance of voice and data transmissions. It is also called the *power-delivery intelligent cell* and the *processing-gain intelligent cell*.

intelligent terminal A computer terminal that has a microcomputer or microprocessor which provides text editing or other forms of processing for keyboarded information before the information is fed over wire lines to a large central computer. It is also called a smart terminal.

intelligibility *Articulation.*

INTELSAT Abbreviation for *International Telephone Satellite Organization.*

intensifier electrode An electrode that increases the velocity of electrons in a beam near the end of their trajectory after deflection of the beam.

intensifier ring A metallic ring-shaped coating on the inside of the glass envelope of a cathode-ray tube near the fluorescent screen. When a high positive voltage is applied to this ring, it increases the velocity of the electrons in the beam and thereby increases the intensity of the picture on the screen.

intensify To increase the brilliance of an image on the screen of a cathode-ray tube.

intensifying screen A thin screen coated with a substance that fluoresces readily under the influence of X-rays. It is placed next to the emulsion of X-ray film to increase the effect of X-rays on the film. In industrial radiography a sheet of thin lead is used for this purpose. The secondary electrons and X-rays emitted by the lead produce the intensifying action.

intensitometer   An instrument for determining relative X-ray intensities during radiography, to control exposure time.

intensity   The strength or amount of a quantity, as of current, magnetization, radiation, or radioactivity. The symbol *I* for current is derived from this word.

intensity control   *Brightness control.*

intensity level   An acoustics term that specifies the relation of one sound intensity to another. The intensity level is expressed in decibels and is equal to 10 times the common logarithm of the ratio of the intensities.

intensity modulation   The modulation of electron-beam intensity in a cathode-ray tube in accordance with the magnitude of the received signal. The luminance of the trace on the screen then varies with signal strength. It is also called *Z*-axis modulation.

interaction-circuit phase velocity   The phase velocity of a wave traveling through the interaction gap of a traveling-wave tube in the absence of electron flow.

interaction crosstalk   Crosstalk that results from mutual coupling between two paths by means of a third path.

interaction gap   An interaction space between electrodes in a microwave tube.

interaction space   A region of an electron tube where electrons interact with an alternating electromagnetic field.

interactive processing   Computer processing that permits the user to modify the operation appropriately while observing results at critical steps.

interactive terminal   A computer terminal designed for two-way communication between the operator and a computer. A video display terminal is one example.

interactive video   A video system, typically computer-based, that permits the user to communicate with the video system to select material to be viewed and the method of viewing.

intercarrier beat   An interference pattern that appears on television pictures when the 4.5-MHz beat frequency of an intercarrier sound system gets through the video amplifier to the video input circuit of the picture tube.

intercarrier noise suppressor   *Noise suppressor.*

intercarrier sound system   A television receiver arrangement in which the television picture carrier and the associated sound carrier are amplified together by the video IF amplifier and passed through the second detector. It gives the conventional video signal plus a frequency-modulated sound signal whose center frequency is the 4.5-MHz difference between the two carrier frequencies. The new 4.5-MHz sound signal is then separated from the video signal for further amplification before going to the frequency-modulation detector stage.

intercept   1. To meet or interrupt the course of a moving vessel, aircraft, or missile.   2. To tap or tune to a telephone or radio message not intended for the listener.

interchannel interference   Interference produced in a common channel by signals from one or more other channels.

intercom   *Intercommunication system.*

intercommunication system   An audio-frequency amplifier system that provides two-way voice communication between two or more locations which are usually in the same structure. Each station contains a dynamic loudspeaker that also serves as a microphone. The amplifier

can be at a central station, or each station can have its own amplifier. Connections between stations can be made by wires or carrier signals traveling over electric wiring in the building. It is also widely used on ships and large aircraft. Also called an intercom.

interconnection diagram   A diagram that shows the external connections required between two or more devices, circuits, or other equipment.

intercontinental ballistic missile   [ICBM]   A missile that flies a ballistic trajectory after guided powered flight, usually over ranges in excess of 4000 mi (6400 km).

intercontinental missile   A missile designed for travel from one continent to another, such as an intercontinental ballistic missile.

interdigital structure   A pair of comb-shaped electrodes with multiple digits or "fingers" that are interleaved to form a serpentine path. In a *traveling-wave tube* [TWT], it functions as a delay line to bring the traveling RF wave into contact with the electron beam many times during its transit from the cathode to the catcher electrode. Energy is transferred from the electron beam to the RF wave at these *interaction nodes*. These structures can also be deposited as thin metallic layers on piezoelectric substrates to function as transducers on *surface acoustic wave* [SAW] devices. When chemically milled as stand-alone structures in microminiature actuators, they can transfer energy to resonant structures. This structure is also the form of power transistor gate and drain structures to increase the length of those interactive surfaces. It is also called an *interdigitated structure.*

interdigital transducer   Two interleaved comb-shaped metallic patterns applied to a piezoelectric substrate such as quartz or lithium niobate for converting microwave voltages to surface acoustic waves or vice versa.

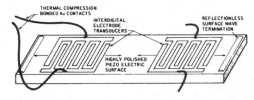

Interdigital transducers of a delay line for frequencies from 10 MHz to 2 GHz in a surface-acoustic-wave (SAW) device.

interdigitated structure   *Interdigital structure.*

interface   A shared boundary or junction. 1. The connector-to-connector junction between a computer and an external power or signal source. 2. The junction in a semiconductor device between semiconductor materials doped with different impurities, such as P-type silicon and N-type silicon, in a *PN diode* or *bipolar-junction transistor* [BJT].

interface card   A circuit board that converts a computer input/output bus into some standard I/O configuration such as 8- or 16-bit parallel, BCD, RS-232C, or IEEE-488.

interface connection   *Feedthrough.*

interference   1. Any undesired energy that tends to interfere with the reception of desired signals. Manmade interference is generated by improperly operating electric devices, with the resulting interference signals either radiated through space as electromagnetic waves or traveling over power lines. Radiated interference can also be caused by atmospheric phenomena such as lightning. Radio transmitters themselves can interfere with each other in certain locations.   2. The systematic alternate reinforcement and attenuation of two or more coherent waves when they are superimposed.

interference eliminator   *Interference filter.*

interference filter   1. A filter that attenuates manmade interference signals entering a receiver through its power line. It is also called an interference eliminator.   2. A filter that attenuates unwanted carrier-frequency signals in the tuned circuits of a receiver.

interference generator   A generator designed to produce RF signals that are amplitude-modulated or frequency-modulated by random frequencies which have erratic amplitudes, to simulate atmospheric static.

interference guard band   One of the two bands of frequencies that border the authorized communication band and frequency tolerance of a station, provided to minimize interference between stations on adjacent channels.

interference pattern   A plot showing the interaction between incident electromagnetic energy and a nontransmissive obstacle. Examples include the patterns formed when light passes through a grating or the redistribution of RF energy when the transmitter is close to the earth's surface.

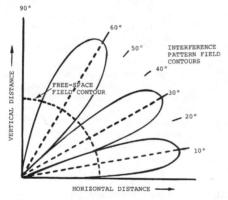

Interference pattern shows the contours or lobes of constant field strength of a transmitter.

interference threshold   The minimum signal-to-noise ratio required for essentially error-free message transmission and reception.

interferometer   An instrument that measures very small distances by splitting a light beam into two parts directed over separate paths and then reuniting the beams. The difference in the path lengths is displayed as a light interference pattern. One application is measuring very thin films.

interlace   *Interlaced scanning.*

interlaced scanning   A scanning process in which the distance from center to center of successively scanned lines is two or more times the nominal line width, so that adjacent lines belong to different fields. In NSTC television, double interlace is used, the 262.5 alternate lines are scanned in one field and the remaining 262.5 lines in the next field. It minimizes flicker of the picture. It is also called *interlace, interlacing,* and *line interlace.*

interlaced stacked rhombic array   A combination of rhombic antennas for radio communication over long distances at high and very high frequencies.

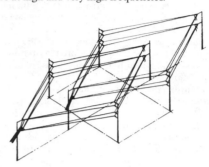

Interlaced stacked rhombic array.

interlacing   *Interlaced scanning.*

interleave   To alternate between parts of two or more sequences of data or operations, such as running two jobs simultaneously on a computer by switching control back and forth between two programs.

interlock circuit   A circuit in which one action cannot occur until one or more other actions have first taken place. The interlocking action is generally obtained with relays.

interlock relay   A relay composed of two or more coils, each with its own armature and associated contacts, so arranged that movement of one armature or the energizing of its coil is dependent on the position of the other armature.

interlock switch   A switch designed for mounting on a door, drawer, or cover in such a way that it opens automatically when the door or other part is opened.

intermediate frequency [IF]   The frequency produced by mixing the received signal with that of the local oscillator in a superheterodyne receiver.

intermediate-range ballistic missile [IRBM]   A ballistic missile that has a range between about 1700 and 3500 mi (2700 and 5600 km).

intermediate state   A state of partial superconductivity that occurs when a magnetic field of appropriate strength is applied to a superconducting material below its critical temperature (the temperature below which it would be completely superconducting if no magnetic field were present).

intermediate subcarrier   A carrier that can be modulated by one or more subcarriers and used to modulate another carrier.

intermetallic compound   A semiconductor that consists only of metallic atoms held together by metallic bonds, to give a basic crystal structure consisting of two different metallic elements. An intermetallic compound is semiconducting when the two metals together contribute just enough electrons to fill the valence band, as in bismuth telluride, gallium arsenide, gallium phosphide, indium antimonide, indium arsenide, indium phosphide, and mercuric telluride.

intermittent defect   A defect that is not continuously present.

intermittent-duty rating An output rating based on the operation of a device for specified intervals of time rather than continuous duty.

intermittent-duty relay A relay that must be deenergized at intervals to avoid overheating of its coil.

intermittent reception A radio receiver fault that occurs when the receiver operates normally for a time and then becomes defective for a time. The process repeats itself at regular or irregular intervals.

intermittent service area The geographic area surrounding the primary service area of a broadcast station. The ground wave is received but it is subject to some interference and fading.

intermodulation [IM] The modulation of the components of a complex wave by each other, producing new waves whose frequencies are equal to the sums and differences of integral multiples of the component frequencies of the original complex wave.

intermodulation distortion Nonlinear distortion characterized by the appearance of output frequencies equal to the sums and differences of integral multiples of the input frequency components. Harmonic components also present in the output are usually not included as part of the intermodulation distortion.

intermodulation interference Interference that occurs when the signals from two undesired stations differ by exactly the IF value of a superheterodyne receiver and both signals are able to pass through the preselector because of poor selectivity. The undesired signals combine in the mixer stage to give an undesired IF signal that interacts with the desired IF signal.

internal resistance The resistance of a voltage source, acting in series with the source.

international atomic time [IAT] Time based on atomic clocks operating in conformity with the definition of the second as the SI unit of time.

International Civil Aviation Organization [ICAO] A United Nations agency that establishes worldwide signal standards for navigation aids used by civil aviation.

International Commission on Illumination [ICI] British version of Commission Internationale de l'Eclairage.

International Electrotechnical Commission [IEC] An organization based in Geneva, Switzerland, that sets standards for electronic products and components which are adopted by the safety standards agencies of many countries. IEC does not perform tests or have enforcement power.

International Maritime Satellite Organization [INMARSAT] An international organization that provides a global fixed and mobile satellite communications network.

international Morse code The code universally used for radiotelegraphy.

International Radio Consultative Committee [CCIR] An international organization concerned with the establishment of radio and television broadcasting standards throughout the world.

international radio silence A 3-min period of radio silence on the international distress frequency of 500 kHz only, commencing 15 and 45 min after each hour. During that time radio stations can listen on that frequency for distress signals from ships and aircraft.

international standard atmosphere A standardized atmosphere, adopted internationally, for comparing the performance of aircraft and missiles. It is a pressure of 1013.2 mbar at a mean sea-level temperature of 15°C, and a lapse rate of 6.5°C/km of altitude up to 11 km, above which the temperature is assumed constant at −56.5°C.

International System of Units [SI, for Système International d'Unités] A system of units, adopted internationally in 1960, that has the ampere, candela, kelvin, kilogram, meter, mole, and second as the seven base units from which other units are derived.

International Telecommunications Union [ITU] An international civil organization established to provide standardized communication procedures, including frequency allocations and radio regulations, on a worldwide basis.

International Telegraph and Telephone Consultative Committee [CCITT] A United Nations advisory committee established to recommend international standards for data communication.

International Telephone Satellite Organization [INTELSAT] An international organization that provides a global fixed and mobile satellite communications network, primarily for point-to-point communication.

international unit A unit accepted internationally between 1908 and 1950 as a legal standard.

Internet A worldwide collection of interconnected computer networks now accessible to private citizens with the requisite computer and software support.

Internet protocol [IP] Rules governing how packets of information are transmitted over the *Internet* and interpreted as a message.

interphase transformer An autotransformer or set of mutually coupled reactors specified when two or more high-power rectifiers are operated in parallel and have out-of-phase ripple voltages.

interpreter A computer executive routine that translates and executes a program in high-level language into machine language or code, one line at a time. See also *assembler* and *compiler*.

interrecord gap An unrecorded space left between records on magnetic tape to permit tape stop-start operations without reading or recording errors.

interrogator A radio, radar, or sonar transmitter that elicits a reply from a *transponder*. For example, a coded radio signal transmitted to commercial aircraft in flight will elicit a coded reply from an on-board transponder that identifies the aircraft on an air-traffic control (ATC) plan-position indicator (PPI) radarscope for air-traffic control purposes. See also *identification, friend or foe* [IFF].

interrogator-responsor A transmitter and receiver combined, for sending out pulses to interrogate a radar beacon and for receiving and displaying the resulting replies.

interrupt To disrupt temporarily the normal operation of a routine, as by a special signal from the computer.

interrupter An electric, electronic, or mechanical device that periodically interrupts the flow of a direct current to produce pulses.

interrupting capacity The highest current that a device can interrupt at its rated voltage.

interstage Between stages.

interstage transformer A transformer that provides coupling between two stages.

interstation noise suppressor *Noise suppressor.*

interval The spacing in pitch or frequency between two sounds. The frequency interval is the ratio of the frequencies or the logarithm of this ratio.

intervalometer An electric timing device for measuring the interval between events, the interval between an event and a reference time, or the predetermined intervals between a series of actions, such as the firing of rockets.

interval timer An instrument for signaling the expiration of a predetermined time. It can also actuate a switch at the end of the time interval. It is also called a timer.

intrinsic-barrier diode A PIN diode that has a thin region of intrinsic material separating the P- and the N-type regions.

intrinsic carrier density The equilibrium density of holes and free electrons in an intrinsic semiconductor material.

intrinsic characteristic A characteristic of a material itself, independent of impurities.

intrinsic coercive force The magnetizing force required to reduce to zero the intrinsic induction of a magnetic material that is in a symmetrically and cyclically magnetized condition.

intrinsic coercivity The maximum value of the intrinsic coercive force, corresponding to the saturation flux density for the material.

intrinsic condition Conduction that results from the movement of only those holes and free electrons which are present in the parent semiconductor material (not produced by impurity elements).

intrinsic flux The product of the intrinsic flux density and the cross-sectional area in a uniformly magnetized sample of magnetic material.

intrinsic flux density *Intrinsic induction.*

intrinsic hysteresis loop A curve that shows the relation between intrinsic flux density and magnetizing field strength when the magnetizing field is cycled between equal negative and positive values. Hysteresis is indicated by the lack of coincidence between the ascending and descending branches of the loop.

intrinsic induction The additional magnetic induction that exists in a given magnetic medium, above that which would exist at the same location for the same magnetizing force if the medium were a vacuum. It is also called *intrinsic flux density.*

intrinsic-junction transistor A four-layer transistor that has an I-type semiconductor layer between the base and collector layers, as in PNIP, NPIN, PNIN, and NPIP transistors.

intrinsic layer A layer of semiconductor material whose properties are essentially those of the pure undoped material.

intrinsic mobility The mobility of the electrons in an intrinsic semiconductor.

intrinsic noise Noise caused by a device or transmission path, independent of modulation.

intrinsic permeability The ratio of intrinsic induction to the corresponding magnetizing force.

intrinsic property A property of a semiconductor that is characteristic of the ideal crystal.

intrinsic region A semiconductor region whose current flow consists of approximately equal numbers of electrons and positive holes.

intrinsic semiconductor A semiconductor whose electrical properties are essentially characteristic of the pure ideal crystal. Current flow consists of both electrons and positive

holes in approximately equal numbers; hence the material is neither N- nor P-type. Also called I-type semiconductor.

intrinsic temperature range The temperature range in which the charge-carrier concentration of a semiconductor is substantially the same as that of an ideal crystal.

intrusion alarm A photoelectric, capacitance-controlled, electric, acoustic, or other system for setting off an alarm that announces the presence of an intruder at the boundaries of a protected area or inside that area.

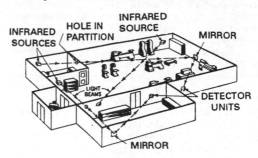

Intrusion alarm system for small factory consisting of three photoelectric systems.

Invar Trademark for an alloy of nickel and iron that contains about 36% nickel. It remains essentially constant in length over a wide range of temperature. It is used for tuning forks and microwave cavities.

inverse conical-scan jamming An electronic countermeasure technique that provides azimuth deception for an enemy tracking radar by sensing its scan-rotation frequency and retransmitting the incident signal with inverse gain modulation of the pulse amplitude.

inverse current The current that results from an inverse voltage in a contact rectifier or semiconductor device.

inverse direction The direction of greater resistance in a rectifier, going from the positive electrode to the negative electrode. It is the opposite of the conducting direction.

inverse electrode current The current that flows through an electrode of an electron tube in the direction opposite to that for which the tube is designed. Thus, for inverse anode current the electrons would flow from the anode to the cathode.

inverse feedback *Negative feedback.*

inverse-feedback filter A tuned-filter circuit connected to the output of a high-selectivity amplifier that has negative feedback. The filter is adjusted so the negative-feedback output is zero at the desired resonant frequency, but increases rapidly to reduce the amplification as the signal frequency departs from this value.

inverse function The function that would be obtained if the dependent and independent variables of a given function were interchanged.

inverse limiter A limiter whose output is constant for instantaneous input values within a specified range. Above and below that range it is linear or responds to some other prescribed function of the input. It is used to remove the low-level portions of signals from an output wave, such as the annoying effects of crosstalk.

inverse-parallel connection A connection of two rectifying elements so that the cathode of the first is connected to the anode of the second, and the anode of the first is connected to the cathode of the second.

inverse peak voltage 1. The peak value of the voltage that exists across a rectifier tube or X-ray tube during the half-cycle in which current does not flow. 2. The maximum instantaneous voltage value that a rectifier tube or X-ray tube can withstand in the inverse direction (with anode negative) without breaking down and becoming conductive.

inverse photoelectric effect The transformation of the kinetic energy of a moving electron into radiant energy at impact, as in the production of X-rays.

inverse piezoelectric effect The contraction or expansion of a piezoelectric crystal under the influence of an electric field, as in crystal headphones.

inverse-square law When electromagnetic, thermal, or nuclear radiation from a point source is emitted uniformly in all directions, the amount received per unit area at any given distance from the source, assuming no absorption, is inversely proportional to the square of that distance.

inverse voltage An effective value of voltage that exists across a rectifier tube or semiconductor rectifier during the half-cycle in which the anode is negative and current does not normally flow.

inversion 1. The process of scrambling speech for secrecy by beating the voice signal with a fixed higher audio frequency and using only the difference frequencies. The original low audio frequencies then become high audio frequencies, and vice versa. 2. A shallow layer of air whose temperature increases with altitude (temperature normally decreases with altitude). The resulting rapid change in air density causes radio waves to bend.

inversion layer A surface layer of doped semiconductor material that has changed to the opposite conductivity from that of adjacent regions. It results from surface ions, surface passivation materials, and induced electric fields.

invert To change to an opposite state.

inverted-cone antenna An antenna that consists of wires which form a cone whose apex is at ground level. The upper ends of the wires are supported by crosswires attached to a circular arrangement of poles. It is a broadband, omnidirectional, vertically polarized radiator, much like a discone antenna.

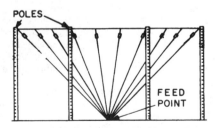

Inverted-cone antenna for a high-frequency communication system.

inverted file A data file whose normal sequence has been reversed.

inverted-L antenna An antenna that consists of a long horizontal wire with the vertical lead-in wire connected to one end. It is also called an *L antenna.*

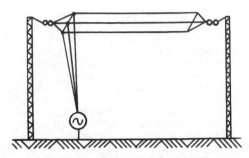

Inverted L antenna for low frequencies requires two masts to support the insulated top section, but does not require base or guy insulators.

inverted magnetron A *magnetron* whose internal geometry is an inversion of conventional magnetron coaxial cathode geometry. Its cathode is outside its anode. See also *crossed-field electron tube.*

inverted speech *Scrambled speech.*

inverter 1. A circuit or device for converting a low DC voltage to a much higher AC voltage with an oscillator or chopper. It is followed by a step-up transformer. When a rectifier is added to give a DC output voltage, the combination becomes a converter. 2. A circuit that takes in a positive-going pulse and delivers a negative-going pulse, or vice versa. 3. *Inverting amplifier.*

inverter transformer A transformer (usually with four windings) combined with power transistors to convert direct current at low voltage into alternating current at higher voltages.

inverting amplifier An operational amplifier whose inverting or negative input is held near ground by negative feedback regardless of the magnitudes of the two input voltages. The output voltage change is inverted with respect to the input voltage change. It is also called an inverter.

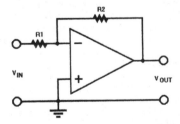

Inverting amplifier showing the feedback loop around the operational amplifier.

inverting function A logic circuit that inverts the input signal so the output is out of phase with the input. The inverting action is indicated by a circle on the output terminal of the amplifier symbol.

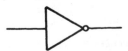

Inverting function symbol.

inverting terminal   The negative input terminal of an operational amplifier. A positive-going voltage at the inverting terminal gives a negative-going output voltage.

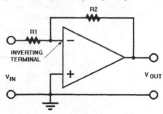

Inverting terminal shown on the schematic symbol for an operational amplifier.

I/O   Abbreviation for *input/output*.

IOC   Abbreviation for *integrated optical circuit*.

ion   A charged atom or group of atoms. A negative ion has gained one or more extra electrons. By contrast, a positive ion has lost one or more electrons.

ion backscattering   Large-angle elastic scattering of monoenergetic ions in a beam directed at a metallized film on silicon or some other thin multilayer system. Ions are detected by a silicon surface-barrier detector that produces a pulse proportional to the energy of the backscattered ion. It is used in nondestructive determination of the depth distribution of atoms after metallization.

ion beam   A beam of ions drawn from a single source by a high voltage in a vacuum.

ion-beam scanning   The process of analyzing the mass spectrum of an ion beam in a mass spectrometer either by changing the electric or magnetic fields of the mass spectrometer or by moving a probe.

ion burn   Deactivation and discoloration of a small area of phosphor at the center of the screen of a magnetically deflected cathode-ray tube, due to bombardment by heavy negative ions. An ion trap eliminates the effect by preventing the ions from leaving the electron gun.

ion chamber   *Ionization chamber.*

ion charging   Dynamic decay caused by ions striking the storage surface of a charge-storage tube.

ion gun   *Ion source.*

ionic focusing   *Gas focusing.*

ionic-heated cathode   A hot cathode that is heated primarily by ionic bombardment of the emitting surface.

ion implantation   The controlled doping of a semiconductor wafer with energetic ion beams of boron, phosphorous, or arsenic in regions defined by oxide masking. Ions penetrate the surface and slow to rest within the wafer. Resistors, diodes, MOSFETs, and GaAs ICs are ion-implantated. A more precise method than *diffusion*, it is especially effective for shallow (<1 μm) distributions. Implantation is done at room temperature, and any resulting crystal lattice damage can usually be corrected by annealing at about 700°C.

ion-implanted resistor   An integrated-circuit resistor produced on a semiconductor surface by ion implantation of impurities.

ionization   A process by which a neutral atom or molecule loses or gains electrons, thereby acquiring a net charge and becoming an ion. Ionization can be produced by collisions of particles, by radiation, and by other means.

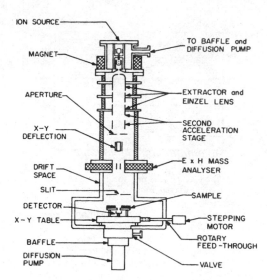

Ion-implantation equipment in a vacuum chamber.

ionization by collision   Ionization produced by collisions of high-velocity electrons or ions with neutral atoms or molecules.

ionization current   *Gas current.*

ionization gage   *Vacuum gage.*

ionization path   The trail of ion pairs produced by an ionizing particle in its passage through matter. It is also called an ionization track.

ionization potential   The amount of energy, expressed in electronvolts, required to remove an electron from a neutral atom.

ionization spectrometer   *Bragg spectrometer.*

ionization time   The time interval between the initiation of conditions for conduction in a gas tube and the establishment of conduction at some stated value of tube voltage drop.

ionization track   *Ionization path.*

ionization voltage   The energy per unit charge, usually expressed in volts, required to remove an electron from a particular kind of atom to an infinite distance. It is also called electron binding energy.

ionized layer   One of the atmospheric layers, such as the E layer or F layer, that reflect radio waves back to earth under certain conditions. See also *ionosphere*.

ionizing energy   The average energy lost by an ionizing particle when producing an ion pair in a gas. For air the ionizing energy is about 32 eV.

ionizing event   An event that produces one or more ions.

ionizing particle   A particle that produces ion pairs directly when it passes through a substance. The kinetic energy of the particle must be considerably greater than the ionizing energy of the medium.

ionizing radiation   Any radiation that is capable of producing ions when it passes through a gaseous or solid material.

ionosphere   A region in the earth's outer atmosphere where ions and electrons are present in quantities sufficient to affect the propagation of radio waves. It begins

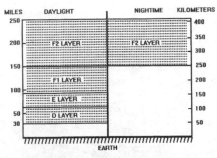

Ionosphere. The height of the principal layers in miles and kilometers. The D, E, and F1 layers dissipate after sunset.

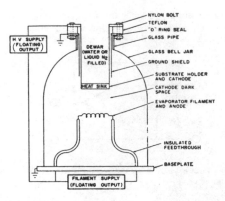

Ion plating. Most of the substrate can be cooled externally as shown, to eliminate heat that might change the properties of certain substrates, such as piezoelectric substrates for transducers.

about 30 mi (50 km) above the earth and extends above 250 mi (400 km), with the height depending on the season of the year and time of day. The chief regions of the ionosphere and their approximate heights are D region: 30 to 60 mi (50 to 100 km), E region: 60 to 90 mi (100 to 150 km), F region: 90 to 250 mi (150 to 400 km).

ionospheric error The total systematic and random error that results from the reception of a navigation signal after ionospheric reflections. It can be due to variations in transmission paths, nonuniform height of the ionosphere, or nonuniform propagation within the ionosphere.

ionospheric scatter A form of scatter propagation in which radio waves are scattered by the lower E layer of the ionosphere to permit communication over distances of from 600 to 1400 mi (1000 to 2250 km) when using the frequency range of about 25 to 100 MHz.

ionospheric storm A turbulence in the F region of the ionosphere, usually due to a sudden burst of radiation from the sun. It is accompanied by a decrease in the density of ionization and an increase in the virtual height of the region. The higher frequencies in the band from 3 to 30 MHz are most affected by the resulting radio blackouts.

ionospheric wave *Sky wave.*

ion pair A positive ion and an equal-charge negative ion, usually an electron, that are produced by radiation on a neutral atom or molecule.

ion plating A method of depositing a thin metallic film on a ceramic or other substrate by first using ion bombardment in a vacuum to clean the surface of the substrate. The polarity is then reversed, so the substrate becomes the cathode in a high-voltage system in which the desired film material is deposited by evaporation.

ion pump A vacuum pump whose residual gas molecules are first ionized and then attracted or propelled by electric charges into an auxiliary pump or an ion trap.

ion repeller An electrode that produces a potential barrier against ions in a charge-storage tube.

ion sheath A film of positive ions that forms on or near an electrode surface in a gas tube and limits the control action.

ion source A device that produces, focuses, accelerates, and emits gas ions as a narrow beam. It is also called an ion gun.

ion spot 1. A dark spot formed near the center of the screen of a cathode-ray tube due to ion burn. 2. A spurious signal that results from bombardment of the target or photocathode of a camera tube or image tube by ions.

ion trap An arrangement that prevents ions in the electron beam of a cathode-ray tube from bombarding the screen and producing an ion spot. Usually a part or all of the electron gun is tilted, and an external permanent magnet bends the electron beam so it will pass through the tiny output aperture of the electron gun. The ions are heavier and hence less affected by the magnetic field. They are trapped harmlessly inside the gun.

ion-trap magnet One or more small permanent magnets with pole pieces placed around the neck of a television picture tube to provide a magnetic field for ion-trap action in the electron gun. It is also called a beam bender.

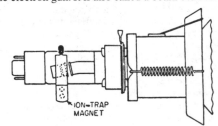

Ion-trap magnet on the neck of a picture tube.

ion yield The number of ion pairs produced per incident particle or quantum.

IP Abbreviation for *Internet protocol.*

IPO Abbreviation for *input-process-output.*

I-phase carrier A carrier phase separated by 57° from the color subcarrier in a color television receiver.

IR Abbreviation for *infrared.*

iraser [InfraRed mASER] *Infrared maser.*

IRBM Abbreviation for *intermediate-range ballistic missile.*

irdome [InfraRed DOME] A dome that protects an infrared detector and its optical elements. It is made from quartz, silicon, germanium, sapphire, calcium aluminate, or other material that offers high transparency to infrared energy.

IR drop The voltage drop produced across a resistance R by the flow of current I through the resistance. It is also called a resistance drop.

iridium 1. An element with an atomic number of 77. 2. A telecommunications system proposed by Motorola Corp. that would launch a fleet of 66 satellites into low polar orbits (LEO) and be linked in a network to earth stations. It would provide telephone voice, digital data, fax (facsimile), and paging by relaying messages from a transmitter anywhere on earth to the nearest satellite for relay to other satellites until they are downloaded to the earth station closest to the message destination. Each three-sided satellite would transmit in the 1.610- to 1.6266-GHz band. Eleven satellites would be arranged in each of six orbital planes passing through the poles. Originally intended to be a constellation of 77 satellites, the project was named after iridium, an element with an atomic number of 77. It would bypass long-distance telephone companies' services.

IRIG Abbreviation for *Inter-Range Instrumentation Group.*

iris An impedance-matching "window" in a waveguide made from metal plates joined at their edges, perpendicular to the guide walls. An opening parallel to the narrow walls is inductive; one parallel to the wide walls is capacitive. The magnitude of reactance depends on the opening width. It is also called a *matching diaphragm.*

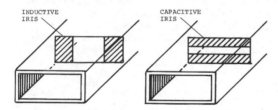

Iris examples in rectangular waveguides: inductive and capacitive.

iron [Fe] A metallic element with an atomic number of 26 and an atomic weight of 55.85.

iron-cobalt alloy An iron alloy that contains up to 65% cobalt and has high flux densities at low magnetizing currents. Its chief drawbacks are high cost, relatively low permeability, and high hysteresis loss.

iron-core choke *Iron-core coil.*

iron-core coil A coil whose solid or laminated iron or other magnetic material forms part or all of the magnetic circuit linking its winding. It is also called an iron-core choke.

iron-core transformer A transformer whose laminations of iron or other magnetic material make up part or all of the path for magnetic lines of force that link the transformer windings.

iron loss *Core loss.*

iron-nickel alloy An iron alloy that contains 20 to 80% nickel. It has high permeability and low hysteresis losses at low flux densities, and it is more readily rolled into thin laminations than silicon steels.

iron-vane instrument An alternating current measuring instrument that has an iron-vane movable element.

irradiance *Radiant flux density.*

irradiation The exposure of a material, object, or patient to X-rays, gamma rays, ultraviolet rays, or other ionizing radiation.

ISA Abbreviation for *industry standard architecture.*

ISAM Abbreviation for *indexed sequential-access method.*

ISDN Abbreviation for *integrated services digital network.*

I signal The in-phase component of the chrominance signal in color television. It has a bandwidth of 0 to 1.5 MHz. It consists of $+0.74(R - Y)$ and $-0.27(B - Y)$, where Y is the luminance signal, R is the red camera signal, and B is the blue camera signal.

island effect The restriction of emission from the cathode of an electron tube to certain small areas of the cathode when the grid voltage is lower than a certain value.

isobar A line that connects points with the same value of a quantity, such as a barometric pressure line on a meteorological chart.

isochronous A characteristic of a fixed frequency or periodicity.

isocon image tube An improved version of the *image orthicon* tube in which the return electrons are isolated from the scattered electrons to improve the *signal-to-noise* [SN] *ratio.* The reflected electrons are isolated, and only the scattered electrons are received by the electron multiplier. The scattered part of the return beam is proportional to the charges on the target. Several electrons are scattered for each electron that impacts the target to neutralize the positive charges. The higher the secondary emission, the greater the scatter gain.

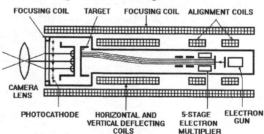

Isocon image tube has a separation section that isolates the scattered and reflected return beam components.

isoelectronic A reference to atoms that have the same number of electrons outside the nucleus of the atom.

isoelectronic trap A bound state induced in a semiconductor by adding an impurity atom. The trap can bind either an electron or a hole. In some light-emitting diodes, the impurity atom introduced by doping determines the color of the light produced.

isolation amplifier A unity-gain amplifier that provides total isolation between input and output signal channels. It is used in industrial applications where millivolt signals must be transmitted in the presence of dangerously high voltages. It is also used with medical electronic equipment to improve patient safety by interrupting ground loops and leakage current paths.

isolation diffusion   Diffusion that produces the back-to-back junctions required for isolating active devices from each other in an integrated circuit.

isolation network   A network inserted into a circuit or transmission line to prevent interaction.

isolation transformer   A transformer inserted into a circuit to separate one section of the circuit from undesired influences of other sections. It is usually made with a 1 to 1 ratio of primary turns to secondary turns to eliminate a direct connection without changing voltages.

isolator   A passive attenuator whose loss in one direction is much greater than that in the opposite direction. A ferrite isolator for waveguides is an example.

isomagnetic line   A line passing through points that have equal magnetic force but not necessarily the same deviation from vertical.

isophotometer   A direct-recording photometer that automatically scans and measures optical density of all points in a film transparency or plate and plots the measured density values in a quantitative two-dimensional isodensity tracing of the scanned areas.

isopotential path   A line passing through points that have the same potential or field strength.

isosynchronous   A serial communications technique that is a hybrid of *synchronous* and *asynchronous* serial protocols. It retains the clock interconnect of the synchronous protocol, but it does not generate synch characters. A start bit is generated as in the synchronous protocol.

isothermal   A term meaning without temperature change.

isotropic antenna   A theoretical antenna capable of radiating signal energy in all directions with a spherical radiation pattern from a point source. It is used as a standard. It is also called a *unipole*.

isotropic dielectric   A dielectric whose dielectric constant is independent of the direction of the applied electric field.

isotropic medium   A medium whose properties are the same in all directions.

isotropic radiator   A radiator that sends out energy equally in all directions.

I²L   Abbreviation for *integrated injection logic*.

I²R loss   *Copper loss.*

iterative array   An array of a large number of interconnected identical processing modules. When combined with appropriate driver and control circuits in a computer, they permit a large number of simultaneous parallel operations. The technique permits the use of slower, less expensive circuits for individual operations while still providing the required fast total execution time for a problem.

iterative filter   A four-terminal filter that provides iterative impedance.

iterative impedance   The impedance that, when connected to one pair of terminals of a four-terminal transducer, will cause the same impedance to appear between the other two terminals. The iterative impedance of a uniform transmission line is the same as the characteristic impedance. When a four-terminal transducer is symmetrical, the iterative impedances for the two pairs of terminals are equal and the same as the image impedances and the characteristic impedance.

iterative process   A mathematical process for calculating a desired result by a repeating cycle of operations that yields a result closer and closer to the desired result.

ITU   Abbreviation for *International Telecommunications Union.*

ITU-TS   An abbreviation for *International Telecommunications Union, Telecommunications Standardization Sector,* an advisory committee established by the United Nations to recommend worldwide standards for data transmission. It was formerly known as *CCITT.*

I-type semiconductor   *Intrinsic semiconductor.*

J

j   A complex operator that is mathematically equivalent to the square root of −1.

J   Abbreviation for *joule*.

jack   A female connecting device into which a plug can be inserted to make circuit connections. The jack can also have contacts that open or close to perform switching functions when the plug is inserted or removed.

jack box   A box that mounts over and protects one or more jacks and sometimes one or more switches.

jacket   A plastic, rubber, or other covering over the insulation, core, or sheath of a cable for mechanical protection.

jammer   A transmitter that jams radio or radar transmissions.

jamming   Radiation or reradiation of electromagnetic waves so that they impair the usefulness of a specific segment of the radio spectrum being used by the enemy for communication or radar. It is also called electronic jamming.

J antenna   A vertical, end-fed antenna whose half-wavelength radiator with a parallel quarter-wavelength section is connected to it by a common-base conductor. It provides a horizontal omnidirectional radiation pattern and does not require a ground connection. It is useful in VHF and UHF frequencies above 7 MHz.

Janus antenna array   An array that provides both forward and backward beams, for airborne Doppler navigation systems. It was named after the Roman god who guarded an entrance and had a face on each side of his head.

Janus technique   A technique for generating a Doppler signal for an airborne navigation radar that is used to measure basic ground speed and drift angle. Microwave energy in the X band is radiated forward and backward from the aircraft toward the earth. The backscattered energy from both beams is detected, and the echo frequency of the aft beam is subtracted from that of the fore beam. The resulting low audio frequency is then converted to distance units to measure ground speed.

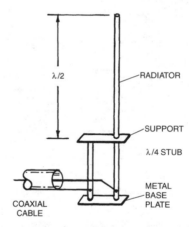

J antenna is a half-wave antenna with a parallel quarter-wave stub connected at the bottom.

J carrier system   A carrier system that provides 12 telephone channels in a bandwidth of approximately 140 kHz.

JCL   Abbreviation for *job control language*.

J display   A radar display whose time base is a circle so that the target echo appears as an outward radial deflection from the time base. It offers better range accuracy than an *A display*.

JEDEC   Abbreviation for *Joint Electronic Devices Engineering Council*.

JEIDA   Abbreviation for the *Japan Electronic Industry Development Association*.

JFET   Abbreviation for *junction field-effect transistor*.

JI   Abbreviation for *junction isolation*.

JIT   Abbreviation for *just in time*.

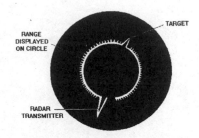

J display shows the target as a radial pip appearing on the range circle to provide increased range accuracy, but no bearing information is given.

jitter   A form of analog signal distortion caused by rapid, spurious variations in the signal from a reference timing position. In the transmission of data, it can cause errors, particularly at high speeds. Jitter can be in amplitude, time, frequency, or phase.

J-K flip-flop   A flip-flop that changes the state of its output if both the J and K inputs are 1 (high) when a clock pulse arrives. If only the J input is 1, a clock pulse drives the output to 1. If only the K input is 1, a clock pulse drives the output to 0. The choice of the letters JK for this flip-flop was purely arbitrary.

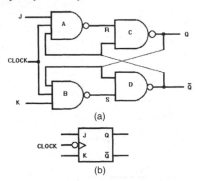

J-K flip-flop (*a*) formed with an R-S flip-flop and (*b*) its schematic symbol.

Johnson noise   *Thermal noise.*

Joint Electronic Devices Engineering Council [JEDEC] A U.S. group established to solve industrywide electronic problems.

Joint Photographic Experts Group Standard [JPEG]   A standard adopted by the International Standards Organization (ISO) for compressing still television images with little or no data loss. It can be used for moving images only if it treats each frame as a separate entity, repeating each detail in each frame.

Joint Tactical Information Distribution System [JTIDS] A decentralized, spread-spectrum data communications and navigation system developed for the U.S. military services. It uses wide bandwidth *phase coding, frequency hopping, and time-division multiple access*. It operates in the 960- to 215-MHz band to permit aircraft and ships to determine their positions accurately, and it includes an independent time synchronization system.

Josephson effect   The passage of paired electrons through a weak connection (Josephson junction) between superconductors, as in the tunnel passage of paired electrons through a thin dielectric layer separating two superconductors.

Josephson junction   A superfast switch suitable as a digital logic or memory element when immersed in liquefied cryogenic gas. Two lead, niobium, or other superconducting-alloy electrodes are separated by a very thin insulating layer. Current flowing in a control conductor above another insulating layer creates a magnetic field that modulates electron-tunneling current through the thin insulation between the two electrodes. The control current allows very fast switching between electron pairs and single-electron tunneling. Its speed is the result of direct current flow without a voltage drop across the junction. It is also called a *tunneling cryotron*.

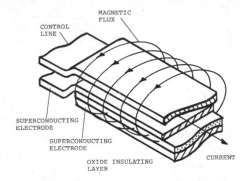

Josephson junction can switch current about four times faster than a transistor.

joule [J]   The unit of energy or work in the International System of Units (SI). It is the work done when a force of 1 newton moves the point of application 1 m in the direction of the force (1 watt = 1 joule per second or 1 J/s).

Joule effect   1. The heating effect produced by the flow of current through a resistance.   2. *Magnetostriction.*

Joule's law   The rate at which heat is produced in a circuit with a constant resistance. It is proportional to the square of the current.

joystick   A two-axis displacement control operated by a lever or ball, for *XY* positioning of a device or an electron beam. It is now a popular accessory for electronic games based on microprocessors.

JPEG   Abbreviation for *Joint Photographic Experts Group Standard.*

JTIDS   Abbreviation for *Joint Tactical Information Distribution System.*

jump   A digital-computer programming instruction that conditionally or unconditionally specifies the location of the next instruction and directs the computer to that instruction. It can alter the normal sequence of the computer.

jumper   A short length of conductor that makes a connection between two points or terminals in a circuit or provides a path around a break in a circuit.

jump table   A software table that gives the location in computer memory of functions that are frequently needed by the program being run.

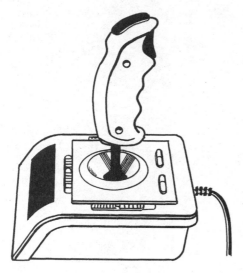

Joystick provides control of images on a computer screen.

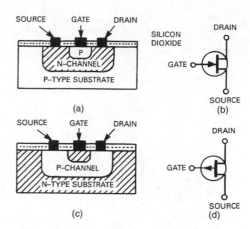

Junction field-effect transistor (JFET): N-channel version (*a*) and its symbol (*b*), and P-channel version (*c*) and its symbol (*d*).

junction 1. A region of transition between two different semiconducting regions in a semiconductor device, such as a PN junction, or between a metal and a semiconductor. The four types of junctions are alloy, diffused, electrochemical, and grown. 2. A fitting that joins a branch waveguide at an angle to a main waveguide, as in a T junction. It is also called a waveguide junction.

junction box An enclosure for connecting wires or cables to form joints. It provides mechanical protection for the joints.

junction capacitor An integrated-circuit capacitor that makes use of the capacitance of a reverse-biased PN junction. It can be formed at the same time as the emitter or collector junctions of transistors. Maximum capacitance values are limited to a few hundred picofarads.

junction diode *PN-junction diode.*

junction field-effect transistor [JFET] A unipolar, three-layer transistor whose operation depends on the movement of only one carrier—either electrons or holes. A "normally on" device, current flows in the *channel* whenever the *drain* voltage of an N-channel JFET is positive with respect to the *source* or that of a P-channel JFET is negative with respect to the source. Current in a JFET channel is controlled by the *gate* voltage. It must be negative with respect to the source in an N-channel JFET or positive in a P-channel JFET to "pinch off" current flow.

Gate voltage is applied to the channel across a PN junction, in contrast to its application across an insulator in a MOSFET. See *field-effect transistor* and *MOSFET.*

junction isolation A fabrication technique that electrically isolates active elements in an integrated circuit from each other by PN junctions. Typically N-type epitaxial layers are grown on P-type substrates, and P-type isolation wells are diffused around each area that is to be electrically isolated from the other circuitry. See *dielectric isolation.*

junction point *Branch point.*

junction transistor A bipolar transistor whose central base region is between the emitter and collector regions and is separated from them by PN junctions. Major categories of junction transistors include grown-junction, alloy-junction, diffusion (such as mesa and planar), and epitaxial.

justify 1. To adjust the printing positions of characters on a page so full lines have the desired length and both left and right margins are flush. 2. To shift the contents of a data storage register so either the most significant or least significant digit is at some specified position in the register.

just-in-time [JIT] A manufacturing term for the practice of requiring vendors to deliver products (components) to the factory exactly when specified prior to their use in production to permit inventory to be kept at minimum levels.

just scale A musical scale formed by taking three consecutive triads, each having the ratio 4 to 5 to 6 or 10 to 12 to 13, with the highest note of one triad serving as the lowest note of the next.

K

k   Abbreviation for *kilo-*.

K   1. Symbol for *cathode*.   2. Abbreviation for *kelvin* (plural kelvins); here K is used without the degree sign and is separated by a space from the temperature value, as in 273.16 K.   3. Unofficial abbreviation for *kilohm* (1000 Ω); here the K follows the value, with no space between, as in 10K for 10,000 Ω.   4. Symbol for *relay*.

kA   Abbreviation for *kiloampere*.

karaoke   A multimedia entertainment and educational system invented in Japan for nightclub entertainment. The basic system consists of a microphone and audio system, source of background music [typically compact disks (CDs) or CD-ROMs], and closed-circuit television. Patrons can sing or perform to recorded music, and the composite performance can be recorded on video tape. It is also used for educational and training purposes.

Karnaugh map   A truth table that has been rearranged to show a geometrical pattern of functional relationships for gating configurations. With this map, essential gating requirements can be recognized in their simplest form.

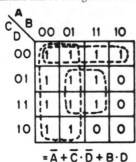

$$= \overline{A} + \overline{C} \cdot \overline{D} + B \cdot D$$

Karnaugh map, with dotted lines enclosing 1s that represent gating requirements of expression below map.

kb   Abbreviation for *kilobit*.

K band   A band of frequencies extending from 18.0 to 27.0 GHz, corresponding to wavelengths of 1.7 and 1.1 cm, respectively, in accordance with IEEE Standard 521-1976. This band corresponds to parts of both the J band (10.0 to 20.0 GHz) and K band (20.0 to 40.0 GHz) of the U.S. Military Joint Chiefs of Staff (JCS) triservice frequency designations (1970).

K_a band   A band of frequencies extending from 27.0 to 40.0 GHz, corresponding to wavelengths of 1.1 and 0.75 cm, respectively, in accordance with IEEE Standard 521-1976. This band corresponds to part of the K band (20.0 to 40.0 GHz) of the U.S. Military Joint Chiefs of Staff (JCS) triservice frequency designations (1970).

K_u band   A band of frequencies extending from 12.0 to 18.0 GHz, corresponding to wavelengths of 2.5 and 1.67 cm, respectively, in accordance with IEEE Standard 521-1976. This band corresponds to part of the J band (10.0 to 20.0 GHz) of the U.S. Military Joint Chiefs of Staff (JCS) triservice frequency designations (1970).

kBd   Abbreviation for *kilobaud*.

kb/s   Abbreviation for *kilobit per second*.

kbyte   Abbreviation for *kilobyte*.

kbyte/s   Abbreviation for *kilobyte per second*.

kc   Abbreviation for an obsolete term *kilocycle*, now called kilohertz and abbreviated kHz.

K carrier system   A carrier system that provides 12 telephone channels in a bandwidth of approximately 60 kHz.

KCL   Abbreviation for *Kirchoff's current law*.

KDP crystal   Abbreviation for *potassium dihydrogen phosphate crystal*.

keep-alive circuit   A circuit used in TR, anti-TR, or other gas-discharge tubes to produce residual ionization and reduce the initiation time of the main discharge.

keeper   A bar of iron or steel placed across the poles of a permanent magnet to complete the magnetic circuit when the magnet is not in use. It avoids the self-demagnetizing effect of leakage lines. It is also called a *magnet keeper*.

kelvin [K, always written without the degree symbol °] The unit of temperature in the International System of Units (SI), equal to the fraction ½₂₇₃.₁₆ of the thermodynamic temperature of the triple point of water, formerly called degree Kelvin. The kelvin temperature scale is based on Celsius degrees, but with the entire scale shifted so 0 K is at absolute zero. In this scale, water freezes at 273.16 K and boils at 373.16 K. Add 273.16 to any Celsius (formerly centigrade) value to get the corresponding value in kelvins.

Kelvin balance An ammeter that measures the current sent through two coils in series, one fixed just below the other, which is attached to one arm of a balance. The resulting force between the coils is then balanced against the force of gravity acting on a known weight at the other end of the balance arm.

Kelvin bridge A seven-arm bridge that can compare the four-terminal resistances of two four-terminal resistors or networks. It is also called a double bridge and a Thomson bridge.

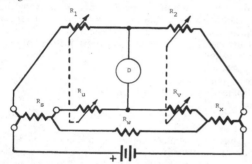

Kelvin bridge has four terminals that can measure resistance as low as 100 μΩ with accuracy better than 1%. Two pairs of resistors are usually ganged.

Kennelly-Heaviside layer *E. layer.*

Kerr cell A cell that contains a pair of electrodes in a dielectric liquid such as nitrobenzene. The dielectric becomes doubly refracting when under electric stress. If crossed Nicol prisms or Polaroid filters are placed before and after the Kerr cell, no light passes through the combination when no voltage is applied to the cell. When a signal voltage is applied to the cell, the plane-polarized light that enters the cell becomes elliptically polarized to an extent dependent on the voltage, and a proportional amount of light passes through the second prism. It can modulate a beam of light or serve as a high-speed camera shutter.

Kerr magneto-optical effect Rotation of the plane of polarization of plane-polarized light when reflected from the polished pole surface of a strong magnet. The angle of rotation is proportional to the magnetizing force. It is also called the magneto-optical effect.

keV Abbreviation for *kiloelectronvolt.*

key 1. A hand-operated switch for transmitting code signals. Also called signaling key. 2. A special lever-type switch for opening or closing a circuit only as long as the handle is depressed. It is also called a switching key. 3. A projection that slides in a mating slot to achieve correct alignment of two parts being put together.

keyboard A matrix of keyswitches that permits an operator to enter information manually into the central processing unit of a computer.

keyed automatic gain control Automatic gain control in which an AGC transistor in a television receiver is biased to cutoff and is unblocked only when the peaks of positive horizontal sync pulses act on its base. This technique prevents the AGC voltage from being affected by noise pulses that occur between sync pulses.

keyed rainbow generator A rainbow generator that has facilities for generating 3.85-MHz colorburst pulses, for making crossover adjustments and for general color television receiver troubleshooting.

keying The forming of signals, such as for telegraph transmission, by modulating a DC or other carrier between discrete values of some characteristic.

keying frequency The maximum number of times per second that a black line signal occurs when scanning the subject copy in facsimile.

keystone distortion Camera-tube distortion evident because the length of a horizontal scan line is linearly related to its vertical displacement. It occurs when the electron beam in the camera tube scans the image plate at an acute angle. A system that has keystone distortion distorts a rectangular pattern into a trapezoidal pattern. The distortion is normally corrected by special transmitter circuits.

keyswitch A switch that is operated by depressing a key on the keyboard of a data-entry terminal. The switch can have mechanical contacts, sealed magnetic reeds, a saturable core, a change in capacitance, a Hall-effect transducer, or some other type of transducer that generates the output for encoding.

key system A miniature PABX that accepts 4 to 10 lines and can direct them to as many as 30 telephone sets.

keyword-in-context index [KWIC index] A computer-generated listing of titles of documents, produced on a printer, with the keywords lined up vertically in a fixed position within the title and arranged in alphabetic order.

kg Abbreviation for *kilogram.*

kG Abbreviation for *kilogauss.*

kHz Abbreviation for *kilohertz.*

kickback The voltage developed across an inductance when current flow is cut off and the magnetic field collapses.

kickback power supply *Flyback power supply.*

kickpipe A metal or plastic pipe used to protect cables from mechanical injury when brought through a floor or deck.

Kikuchi lines A pattern that consists of pairs of white and dark parallel lines, obtained when an electron beam is scattered (diffracted) by a crystalline solid. The pattern gives information on the structure of the crystal.

kilo- [k] A prefix representing 10^3, or 1000.

kiloampere [kA] One thousand amperes.

kilobaud [kBd] One thousand bauds.

kilobit [kb] One thousand bits. In digital memories it is 1024 bits.

kilobit per second [kb/s] One thousand bits per second, used in specifying the modulation rate of a digital transmission system.

kilobyte [kbyte] One thousand bytes. In digital memories it is 8192 bits.

kilobyte per second [kbyte/s] One thousand bytes per second, a rating used in specifying the speed of a digital transmission system.

kilocurie [kCi] One thousand curies.

kilocycle [abbreviated kc] An obsolete term for one thousand cycles per second. It is now called *kilohertz* and abbreviated kHz.

kiloelectronvolt [keV] One thousand electronvolts. It is the energy acquired by an electron that has been accelerated through a voltage difference of 1000 volts.

kilogauss [kG; plural **kilogauss**] One thousand gauss.

kilogram [kg] The unit of mass in the International System of Units (SI).

kilohertz [kHz] One thousand hertz. It was formerly called a *kilocycle*.

kilohm [kΩ: often abbreviated K on diagrams] One thousand ohms. Thus 15K on a diagram means 15,000 Ω or ohms.

kilolumen [klm] One thousand lumens.

kilometer [km] One thousand meters, or 3280 ft.

kilometric wave A radio wave between the wavelength limits of 1 and 10 km, corresponding to the low-frequency (LF) range of 30 to 300 kHz.

kiloton One thousand tons.

kilovolt [kV] One thousand volts.

kilovoltage A voltage of the order of thousands of volts, such as the voltage applied to an X-ray tube.

kilovoltampere [kVA] One thousand voltamperes.

kilovoltmeter A voltmeter whose scale is calibrated to indicate voltage in kilovolts.

kilovolt peak [kV P] The maximum value in kilovolts of the positive peak of an applied voltage waveform.

kilovolt peak-to-peak [kV P-P] The voltage in kilovolts as measured between the maximum positive and negative peaks of the applied voltage waveform.

kilowatt [kW] One thousand watts.

kilowatthour [kWh] One thousand watthours.

kinetic energy Energy associated with motion.

Kirchoff's current law [KCL] A law commonly used in the analysis and solution of networks which states that at each instant of time, the algebraic sum of the instantaneous values of all currents flowing toward a point is equal to the simultaneous values of all the currents flowing away from the point. See also *Kirchoff's voltage law.*

Kirchoff's voltage law [KVL] A law commonly used in the analysis and solution of networks which states that at each instant of time, the algebraic sum of voltage rise is equal to the algebraic sum of voltage drops, both being taken in the same direction around the closed loop. See also *Kirchoff's current law.*

K line One of the characteristic lines in the X-ray spectrum of an atom. It is produced by excitation of the electrons of the K shell.

klm Abbreviation for *kilolumen.*

klystron An electron tube that periodically bunches electrons with electric fields. The resulting velocity-modulated electron beam is fed into a cavity resonator to sustain oscillations within the cavity at a desired microwave frequency. It can function as an oscillator or amplifier in UHF applications such as microwave relay and radar transmitters and receivers.

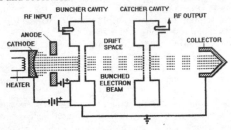

Klystron two-cavity amplifier provides high microwave power output.

klystron frequency multiplier A two-cavity klystron whose output cavity is tuned to a multiple of the fundamental frequency.

klystron repeater A microwave repeater that consists of a klystron inserted directly into a waveguide. Incoming waves velocity-modulate the electron stream emitted by the cathode of the tube. A second cavity converts the velocity-modulated beam back into waves, but with greatly increased amplitude, and feeds them into the output waveguide.

km Abbreviation for *kilometer.*

kmc Abbreviation for the obsolete term *kilomegacycles* (1000 megacycles). It has been replaced by the term *gigahertz* [GHz].

knee The curve that joins two relatively straight portions of a characteristic curve.

knockout A partially punched-out disk about ¾ in (2 cm) in diameter that is located on the sides or bottom of electrical equipment cabinets or boxes. It can be readily removed with a hammer and screwdriver blow for the passage of a power cable and the attachment of a clamp at a location permitting the shortest cable run.

knot [kt] A unit of speed equal to 1 nautical mile per hour or 1.15 mi/h (1.85 km/h).

knowledge base In device and circuit manufacture, the accumulated knowledge of device or circuit design based on the experience of experts that is stored in computer memory as a reference for decision making in automated fabrication processes.

kΩ Abbreviation for *kilohm,* equal to 1000 Ω or ohms.

Kooman antenna An array of half-wavelength open dipoles stacked vertically a half wavelength apart that operate in phase. It has an insulated wire mesh or horizontal metal bar reflector behind and parallel with the array.

Kovar Trademark for an iron-nickel-cobalt alloy with a temperature coefficient similar to glass that is used in making metal-to-glass seals.

krypton [Kr] An inert gaseous element with an atomic number of 36 and an atomic weight of 83.80.

krypton discharge tube A cold-cathode discharge tube that contains krypton gas, used primarily as a high-voltage switching device. It has a short delay time and a very fast rise time.

K shell   The innermost layer of electrons surrounding the atomic nucleus; it has electrons characterized by the principal quantum number 1.

kt   Abbreviation for *knot*.

kV   Abbreviation for *kilovolt*.

kVA   Abbreviation for *kilovoltampere*.

KVL   Abbreviation for *Kirchoff's voltage law*.

kV P   Abbreviation for *kilovolt peak*.

kV P-P   Abbreviation for *kilovolt peak-to-peak*.

kW   Abbreviation for *kilowatt*.

kWh   Abbreviation for *kilowatthour*.

KWIC index   Abbreviation for *keyword-in-context index*.

L

L 1. Abbreviation for *lambert.* 2. Symbol for *coil.* 3. Abbreviation for *inductance.*

L Mathematical symbol for *inductance.*

laboratory power supply A self-contained power supply that converts an AC line voltage to a regulated DC voltage which can be varied over a specified range. Designed primarily for use with equipment being developed or tested in a laboratory, it typically includes meters for reading current and voltage and control knobs.

labyrinth A loudspeaker enclosure that has air chambers at the rear which absorb rearward-radiated acoustic energy, to minimize acoustic standing waves.

LADAR [LAser Detection and Ranging] *Laser radar.*

ladder adder A digital-to-analog converter that has a resistive ladder network and high-speed low-resistance transistor switches to produce an analog output which is proportional to the value of a binary or binary-coded-decimal digital input.

ladder attenuator An attenuator with a series of symmetrical sections whose impedance remains essentially constant in both directions as the amount of attenuation is varied.

ladder network A network composed of a sequence of H, L, T, or pi networks connected in tandem. Some versions include piezoelectric ceramic elements as impedances for ladder filters. It is used in narrow-bandpass filters and analog-to-digital converters.

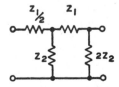

Ladder network, showing ratios of impedance values.

lag 1. The difference in time between two events or values considered together. It is expressed in degrees when comparing alternating quantities; thus the current through a perfect coil lags the applied voltage by 90°. 2. A persistence of the electric-charge image in a camera tube for a few frames after a scene change. 3. The time delay between an initiating action and the desired effect.

lagging current An alternating current that reaches its maximum value of 90° behind the voltage which produces it. A lagging current flows only in a circuit that is predominantly inductive.

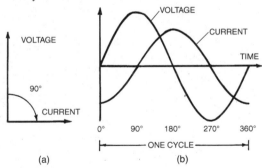

Lag of current behind voltage is 90° in a pure inductive circuit: (*a*) vector diagram and (*b*) waveforms.

lagging load *Inductive load.*

lambda [λ] A Greek letter that designates wavelength in meters.

lambert [L] A CGS unit of luminance equal to $1/\pi$ candela/cm^2. It corresponds to the uniform luminance of a perfectly diffusing surface that is emitting or reflecting light at the rate of 1 lumen/cm^2. The SI unit of luminance, the candela per square meter, is preferred.

Lamb wave An electromagnetic wave propagated over the surface of a solid whose thickness is comparable to the wavelength of the wave.

laminate A product made by bonding together two or more layers of materials.

laminated core An iron or special steel core for a coil, transformer, armature, or other electromagnetic device. It is made of laminations stamped from sheet iron or steel. The laminations are typically insulated from each other by the application of varnish. Laminated construction minimizes the effect of eddy currents.

laminated plastic A plastic material that is made by applying heat and pressure to sheets of filler materials which have been impregnated with a thermosetting resin.

lamination One of the thin punchings of iron or steel that forms a laminated core for a magnetic circuit.

lamp bank A number of incandescent lamps connected together in parallel or series as a resistance load for full-load tests of electric equipment.

lamp cord Two twisted or parallel insulated wires, typically No. 18 or No. 20 AWG, used chiefly for connecting electric equipment to wall outlets.

LAN Abbreviation for *local area network.*

landing beam A radio beam, highly directional in both elevation and azimuth, that slants upward from the landing surface of an airport. It is produced by a landing beacon and serves as the glide path in an instrument landing system for aircraft.

landline The national and international terrestrial telephone network that also includes microwave, satellite, and undersea cable links.

land mobile service Radio service between base and mobile stations or between mobile stations operating on land.

land return *Ground clutter.*

land station A station other than a *mobile station* in a *cellular mobile telephone system* that links to the *mobile station.*

Langmuir dark space A nonluminous region that surrounds a negatively charged probe inserted into the positive column of a glow discharge tube.

Langmuir probe A small metallic conductor inserted within a plasma to sample the plasma current.

L antenna *Inverted-L antenna.*

lanthanide One of the rare-earth elements with atomic numbers 57 to 71, inclusive. All have chemical properties similar to those of lanthanum.

lanthanum [La] A rare-earth element with an atomic number of 57 and an atomic weight of 138.91.

lanthanum-doped lead zirconate-lead titanate *Lead lanthanum zirconate titanate.*

lapel microphone A small microphone that can be attached to a lapel or pocket on the clothing of the speaker to permit free movement while speaking.

Laplace's law The strength of the magnetic field at a given point, due to an element of a current-carrying conductor, is directly proportional to the strength of the current and the projected length of the element and inversely proportional to the square of the distance of the element from the point in question.

Laplace transform A special case of a Fourier transform.

large-scale integration [LSI] Integrated circuits containing between 100 and 5000 gate equivalents, or 1000 to 16,000 bits of memory.

Larmor frequency The angular frequency of precession of a charged particle rotating in a magnetic field. The frequency value is proportional to the strength of the magnetic field and the gyro-magnetic ratio.

Larmor orbit The circular motion of a charged particle in a uniform magnetic field. Although the motion of the particle is unimpeded along the magnetic field, motion perpendicular to the field is always accompanied by a force perpendicular to the direction of motion and the field; the path is therefore helical.

laryngaphone A microphone that is clamped against the throat of a speaker, to pick up voice vibrations directly without responding to background noise.

LASA [Large Aperture Seismic Array] An array of 525 seismometers that covers an area 200 km in diameter in eastern Montana; each instrument is buried in a 60-m-deep hole to minimize effects of local noise. The array can be steered electronically to pick up earth disturbances in an area of interest, such as underground nuclear explosions.

LASCR Abbreviation for *light-activated silicon controlled rectifier.*

LASCS Abbreviation for *light-activated silicon controlled switch.*

lase To generate coherent electromagnetic waves, as in a laser.

laser [Light Amplification by Stimulated Emission of Radiation] An active electron device that converts input power into a very narrow, intense beam of coherent visible or infrared light. The input power excites the atoms of an optical resonator to a higher energy level, and the resonator forces the excited atoms to radiate in phase. The four basic types of laser are gas, liquid, semiconductor, and solid. Major applications include cutting, drilling, heating, welding, and other machining operations; distance measurement, alignment, and other surveying applications; missile guidance, ranging and tracking of moving targets, and other military applications; recording and holography; and communication.

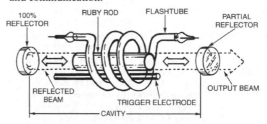

Laser. A ruby laser is shown, with a simplified cavity diagram.

laser-acoustic delay A delay that involves interaction of a laser beam and a transparent acoustic delay line to provide a variable time delay for RF and microwave signals.

laser altimeter A continuous-wave gas laser whose output light beam is modulated by up to three radio frequencies before it is directed downward from an aircraft to scan the earth. The laser light reflected from the terrain is picked up by a telescope mirror system that is coaxial with the transmitting optics, sensed by a photomultiplier, and phase-compared with the transmitted signal to obtain round-trip propagation time and altitude.

laser diode   A semiconductor diode that produces a coherent infrared beam at one frequency. It is typically made from gallium arsenide (GaAs) or GaAs doped with various other materials such as indium and aluminum. These diodes are made with structures that concentrate their emissions along a narrow path. See also *semiconductor laser.*

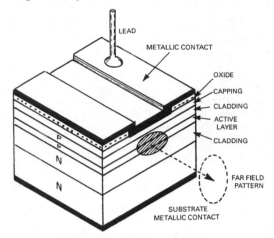

Laser diode, a solid-state laser intended for continuous-wave operation in fiber-optic transmission systems.

laser drill   A drill that concentrates light from a ruby laser to generate intense heat for drilling holes as small as 0.0001 in (2.5 μm) in diameter in gem stones, tungsten carbide, and other hard materials. Hole size is easily changed.

laser flash tube   A high-power air- or water-cooled xenon flash tube that produces high-intensity flashes for laser pumping applications.

laser fusion   The use of an intense beam of laser light to heat a small pellet of deuterium and tritium to a temperature of about 100,000,000°C, as required for initiating a fusion reaction. One major problem has been the development of a laser with sufficiently high power in the visible light spectrum to achieve the required energy transfer within the pellet at the critical instant when it is hit by the beam.

laser guidance   Guidance that continuously illuminates a target with a laser beam from an aircraft or other location so missiles, bombs, or projectiles equipped with suitable seeker heads can home in on the laser energy reflected by the target. Laser and television-guided bombs have been used successfully on North Vietnam and Iraq targets. They are called "smart bombs."

laser-guided bomb   A bomb that carries a guidance receiver for homing on a target which is illuminated by a laser beam from the bomb-carrying aircraft or a spotter aircraft.

laser gyro   A gyro with two laser beams that travel in opposite directions over a ring-shaped path formed by three or more mirrors. When the position of the ring is changed in inertial space, the clockwise and counterclockwise paths have different lengths and produce a frequency difference proportional to the rate at which the gyro position changes. Rotation is thus measured without the need for a spinning mass. It is also called a ring laser.

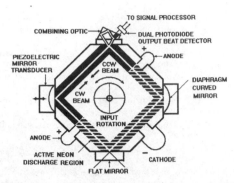

Laser gyro. The area between the cathode and the anodes is the active region where a voltage discharge provides the excited neon atoms for lasing. The two opposing laser beams combine to form interference fringes.

laser interferometer   An interferometer that has a laser as its light source for measuring displacements with a linear resolution of the order of 0.000001 in (25 nm) and angular resolution of about 0.1 arc-second, at ranges up to about 5 m. It can also control step-and-repeat cameras for fabricating and positioning photomasks for the production of integrated circuits.

laser intrusion-detector   A photoelectric intrusion-detector with a laser light source that produces an extremely narrow and essentially invisible beam around the perimeter of the area being guarded. The narrow and intense beam of a laser can be projected long distances, as required for protecting entire airports or large military areas.

laser jamming   An electronic countermeasure. A continuous-wave laser that directs jamming energy back to a hostile laser receiver to prevent it from interfering with laser rangefinders, radars, and tracking equipment during a military operation.

laser memory   A computer memory with a controlled laser beam that acts on individual and extremely small areas of photosensitive or other surfaces for storage and subsequent readout of digital data or other types of information.

laser printer   A nonimpact desktop printer for personal computers that prints text and graphics by an electrostatic process similar to xerography. A low-power laser diode receives digital signals serially that represent the scanned elements of the page (characters or image) from the host computer or internal printer memory. The laser converts the signals to light pulses which are directed at a rotating scanning prism mirror. The light is then focused through a cylindrical lens onto the charged surface of a rotating photoconductor drum. Where light strikes the drum, charge is removed, creating a latent image on the drum's surface. A dry powder called *toner* is attracted only to the charged areas. Transfer and detach-charge coronas apply an electrostatic charge to plain paper driven by rollers across the surface of the drum. After the toner pattern is transferred to the paper, heated rollers fuse it to the paper. Any remaining toner is removed from the drum before it is recharged to begin the cycle again. Some printers can print up to 12 pages per minute (ppm) with a resolution of 600 by 600 dots per inch (dpi). Others can combine the functions of printing with facsimile/modem, copying, and scanning.

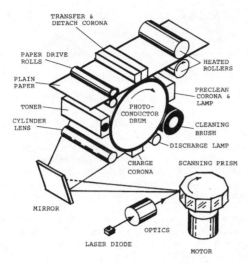

Laser printer. Toner is attracted electrostatically to the charged areas, and heated rollers fuse the toner to plain paper to print text and graphics.

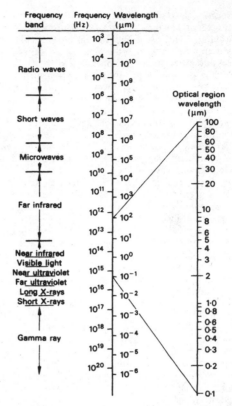

Laser spectrum covers entire optical region from wavelengths of about 0.1 μm in far ultraviolet to about 100 μm in far infrared.

laser radar [LADAR]   A military night-vision and ranging system based on the principles of radar that uses the measured time of flight of pulsed laser energy to produce an image. The returned light energy beam is scanned and detected with components that are similar to those used in imaging infrared sensors.

laser radiation detector   A photodetector that responds primarily to the coherent visible, infrared, or ultraviolet light of a laser beam. One application is warning tank and other military vehicle operators that they are being illuminated by a laser-target designator and are therefore vulnerable to laser-guided weapons.

laser rangefinder   A simplified version of laser radar, used primarily by artillery forward observers, battalion observation posts, tanks, helicopters, and other military vehicles for accurately measuring target distance. A digital counter determines elapsed time between a transmitted laser pulse and the return pulse reflected off the target. Multiplying one-way travel time by the velocity of light gives the range.

laser scriber   A laser-cutting setup replacement for a diamond scriber for dicing thin slabs of silicon, gallium arsenide, and other semiconductor materials used in the production of semiconductor diodes, transistors, and integrated circuits. The laser beam vaporizes grooves in the upper active face of the wafer, after which a breaking operation separates the chips along the grooves. It can also scribe sapphire and ceramic substrates.

laser seismometer   A laser interferometer system that detects seismic strains in the earth by measuring changes in distance between two granite piers located at opposite ends of an evacuated pipe. A helium-neon or other laser beam makes a round trip through the pipe. Movements as small as 80 nm (one-eighth the wavelength of the 632.8-nm helium-neon laser radiation) can be detected.

laser spectrum   The spectrum that includes all optical wavelengths, ranging from infrared through visible light to ultraviolet, in which coherent radiation can be produced by various types of lasers.

laser spooler   A circuit, generally on a circuit board, that permits a laser printer to be shared by two or more computers.

laser threshold   The minimum pumping energy required to initiate lasing action in a laser.

laser transit   A transit with a laser mounted over the sighting telescope to project a clearly visible narrow beam onto a small target at the survey site. It can provide precise positioning of dredges for underwater excavations and alignment of tooling jigs for wing panels of large aircraft.

laser-triggered switch   A high-voltage, high-power switch that consists of a spark gap which is triggered into conduction by a laser beam. It can switch voltages ranging from a few kilovolts to well over 3 MV in electric power systems.

laser velocimeter   A velocity-measuring instrument with a continuous-wave laser that sends a beam of coherent light at an object moving at right angles to the beam. The needlelike diffraction lobes reflected by the moving object sweep past the optical grating in the receiver, thereby generating in a photomultiplier a series of impulses from which velocity can be determined and read out. In another version, the Doppler shift of scattered light from a moving target is measured by detecting with a photodetector the beat signal between scattered and unscattered light. The beat frequency is proportional to target velocity. It is also called a diffraction velocimeter and an optical diffraction velocimeter.

laser welding   The use of a laser beam to spot weld the thin foils and fine wires used in the manufacture of integrated circuits and other microelectronic devices.

lasing   The process of generating radiation anywhere in the laser spectrum at a frequency that is characteristic of the laser material by pumping or exciting electrons into higher energy states.

latching circulator   A switchable circulator that requires no holding current to maintain a desired microwave path after switching action is accomplished by an electromagnet or permanent magnet acting on a ferrite core. It is used in striplines and waveguides.

latching phase shifter   A phase shifter that requires no holding current to maintain a desired phase shift after the device is switched magnetically to the new value. It is used in striplines and waveguides.

latching reed relay   A reed-type relay whose holding coil keeps the relay contacts closed until they are released either by a reverse-polarity pulse in that coil or by a pulse applied to a separate release coil.

latching relay   A relay with contacts that lock in the energized or deenergized positions or both until reset manually or electrically. The latching action is produced by a mechanical latch. In contrast, in a lock-up relay the lock-up is accomplished magnetically or electrically.

latch-in relay   A relay that maintains its contacts in the last position assumed, even without coil energization.

latchup   An undesirable condition in which either a PNPN or an NPNP thyristor turns to an ON state, thereby bypassing or short-circuiting the device which cannot be turned off by the gate.

latency   The time a digital computer takes to deliver information from its memory. It can be the time spent waiting for the desired location on a magnetic disk to appear under a reading head. In a serial storage system, latency is the access time minus the word time.

latent image   A stored image, as in the form of charges on a mosaic of small capacitances.

lateral transistor   A transistor whose emitter- and collector-base junctions are formed in separate areas, with the current flowing between the junctions in a plane parallel to the surface.

lattice   1. A pattern of identifiable intersecting lines of position laid down in fixed positions with respect to the transmitters that establish the pattern for a navigation system. 2. An orderly arrangement of atoms in a crystalline material.

lattice constant   The distance over which a crystal lattice repeats itself. In Group III–V compound semiconductors, the lattice constant is twice the distance between atoms in the crystal or the square root of 8 times the distance between the atoms in the crystal: 0.59 nm in indium phosphide and 0.57 nm in gallium arsenide (GaAs).

lattice imperfection   A deviation from a perfect homogeneous lattice in a crystal.

lattice network   A network composed of four branches connected in series to form a mesh. Two nonadjacent junction points serve as input terminals, and the remaining two junction points serve as output terminals.

lattice scattering   Scattering of electrons by collisions with vibrating atoms in a crystal lattice, reducing the mobility of

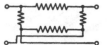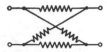

Lattice network as drawn in two different ways.

charge carriers in the crystal and thereby affecting its conductivity.

lattice-wound coil   *Honeycomb coil.*

LAU   Abbreviation for *lobe access unit.*

launcher   Any device that holds, supports, and sometimes directs a missile or spacecraft during launching.

launching   The process of transferring energy from a coaxial cable or transmission line to a waveguide.

launching guidance   Navigation control of a missile during launching.

launch window   An interval of time during which a spacecraft must be launched to reach a desired objective.

lava   A natural fired stone that consists chiefly of magnesium silicate, used for insulators.

law of electric charges   Like charges repel; unlike charges attract.

law of electromagnetic induction   *Faraday's law.*

law of electromagnetic systems   An electromagnetic system tends to change its configuration so that the flux of magnetic induction will be a maximum.

law of electrostatic attraction   *Coulomb's law*

law of induced current   *Lenz's law.*

law of magnetism   Like poles repel; unlike poles attract.

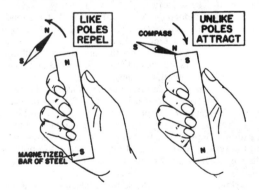

Law of magnetism.

laws of electric networks   *Kirchhoff's laws.*

lay   1. The manner in which the turns of a coil are positioned with respect to each other. In a perfect lay, the wires are staggered with respect to the wires of the previous layer so the coil occupies minimum space. 2. The axial length of a turn of a helix in a cable that has twisted strands.

layer winding   A coil-winding method that lays out adjacent turns evenly side by side along the length of the coil form. Any number of additional layers can be wound over the first, usually with sheets of insulating material between the layers.

layout   A diagram that indicates the positions of various parts on a chassis or panel.

lazy H antenna   An antenna array that has two or more dipoles stacked one above the other to obtain greater directivity.

L band   A band of frequencies extending from 1.0 to 2.0 GHz, corresponding to wavelengths of 30 and 15 cm, respectively, in accordance with IEEE Standard 521-1976. It was widely referred to as the *twenty-centimeter band* because 20 cm is its approximate midband wavelength. It corresponds with the D band (1.0 to 2.0 GHz) of the U.S. Military Joint Chiefs of Staff (JCS) triservice frequency designations (1970).

LBEN   Abbreviation for *low-byte enable.*

lb/in²   Abbreviation for *pounds per square inch.*

LC   Abbreviation for *inductance-capacitance.*

L-carrier system   A telephone carrier system that occupies a frequency band which extends from 68 kHz to over 8 MHz. The system is used with coaxial cable systems and microwave and tropospheric scatter radio systems.

L cathode   A dispenser cathode that has a porous tungsten body covered with a layer of barium and oxygen atoms. Evaporated electron-emitting barium is replaced continuously from a compound located in a chamber behind the tungsten.

LCC   Abbreviation for *leadless ceramic chip carrier.*

LCD   Abbreviation for *liquid crystal display.*

LC product   The product of the inductance value L in henrys and the capacitance value C in farads.

LCR   Abbreviation for *level crossing rate.*

L/C ratio   The ratio of inductance L in henrys to capacitance C in farads for a resonant circuit.

LDCC   Abbreviation for *leaded ceramic chip carrier.*

LDE   Abbreviation for *long-delayed echo.*

LDMOS   An abbreviation for *laterally diffused metal-oxide silicon* technology.

LDR   Abbreviation for *light-dependent resistor.*

lead [pronounced led; Pb]   A soft gray metallic element used as a shielding material in nuclear work and alloyed with tin in solder. It has an atomic number of 82.

lead [pronounced leed]   1. The angle by which one alternating quantity leads another in time, expressed in degrees or radians. The current through a perfect capacitor leads the applied voltage by 90°.   2. The distance between a moving target and the point at which a gun or missile is aimed.   3. A wire that connects together two points in a circuit.

lead-acid cell   The cell in a secondary battery whose electrodes are grids of lead that contain lead oxides which change in composition during charging and discharging. The electrolyte is dilute sulfuric acid. Leakproof versions have a lead grid rolled like a bandage, sealed into a container with a carefully measured amount of acid.

leader   An unused or blank length of tape at the beginning of a reel, used primarily for threading purposes.

leadframe   A stamped or etched metal frame of Kovar or other metal that includes a surface for mounting a die or chip and the input/output leads. After the chip or die is bonded to the leadframe and wire bonds are made, the central part of the leadframe is molded in epoxy to form a SIP, DIP, or SOT package. The leads are then bent downward. See *dual-in-line package.*

lead-in   A single wire that connects a single-terminal outside antenna to a receiver or transmitter. Dipoles and other two-terminal antennas use transmission lines for this purpose. It is also called a *down-lead.*

leading current   An alternating current that reaches its maximum value of 90° ahead of the voltage which produces it. A leading current flows in any circuit that is predominantly capacitive.

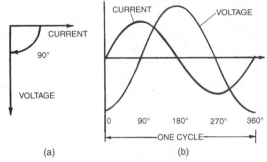

Lead of current ahead of voltage is 90° in a pure capacitive circuit: (*a*) vector diagram and (*b*) waveforms.

leading-edge pulse time   The time for the instantaneous amplitude of a pulse to reach a stated fraction of the peak pulse amplitude.

leading ghost   A ghost displaced to the left of the image on a television receiver screen.

leading load   A load that is predominantly capacitive, so its current leads the alternating voltage applied to the load.

lead-in insulator   A tubular ceramic insulator with an axial hole for the passage of a lead-in wire from an antenna to the transmitter. The insulator is press-fit in a hole formed in a transmitter case, cabinet, or bulkhead.

lead lanthanum zirconate titanate [PLZT]   A ferroelectric, ceramic, electrooptical material whose optical properties can be changed by an electric field or by being placed in tension or compression. It is used in a variety of optoelectronic memory and display devices and is also called lanthanum-doped lead zirconate-lead titanate.

leadless ceramic chip carrier [LCC]   A flat ceramic package for surface mounting of very-large-scale integrated (VLSI) circuits on circuit boards without holes. Its typical number of input/output pins is 20 to 68.

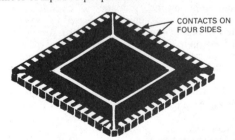

Leadless ceramic chip carrier is a surface-mount package for multipin integrated circuits.

lead-on-chip [LOC] **package**   A molded-epoxy DIP package for very large scale ICs that have their bonding pads arranged down the center of the chip.

lead sulfide [PbS]   The mineral galena that served as a crystal detector in the early days of radio and is now used as an infrared detector.

lead-sulfide cell   A cell that detects infrared radiation. Either its generated voltage or change of resistance can be used as a measure of the intensity of the radiation.

lead telluride [PbTe]   A compound of lead for the manufacture of infrared detectors.

lead zirconate titanate [PZT]   A ferroelectric, ceramic, electrooptical material that has lower optical transparency than PLZT but its other properties are similar.

leakage   Undesired and gradual escape or entry of a quantity, such as escape of electromagnetic radiation through joints in shielding, flow of electricity over or through an insulating material, and flow of magnetic lines of force beyond the region in which useful work is performed.

leakage current   1. The unwanted current flowing from input to output or input to the case of an isolated circuit. 2. The unwanted flow of current through or over the surface of an insulating material or insulator. 3. The unwanted current that occurs in a reverse-biased semiconductor because of structural defects. 4. The current flow within a capacitor because of partial breakdown of the dielectric.

leakage flux   Magnetic lines of force that go beyond their intended path and do not serve their intended purpose.

leakage inductance   Self-inductance caused by leakage flux in a transformer.

leakage power   The RF power transmitted through an ionized TR or pre-TR tube.

leakage radiation   Radiation from an object other than the intended radiating system. A common example is electromagnetic radiation that escapes through joints or defects in shielding.

leakage reactance   Inductive reactance caused by leakage flux that links only the primary winding of a transformer.

leakage resistance   The resistance of the path over which leakage current flows. It is normally a high value.

leak detector   An instrument that finds small holes or cracks in the walls of a vessel. The helium spectrometer is an example.

leaky-wave antenna   A wideband microwave antenna that radiates a narrow beam whose direction varies with frequency. It is fundamentally a perforated waveguide, thin enough to permit flush mounting for aircraft and missile radar applications. Scanning is achieved by sweeping the frequency between two limits at the desired scanning rate.

leaky waveguide   A waveguide that has a narrow longitudinal slot through which energy leaks out continuously.

leased line   A telephone line reserved for the exclusive use of leasing customers. It is also called a *private line*.

least significant bit [LSB]   The bit that carries the lowest value or weight in binary notation for a numeral. For example, when 13 is represented by binary 1101, the 1 at the right is the least significant bit.

least significant character   The character in the rightmost position in a number or word.

Lecher line   *Lecher wires.*

Lecher-line oscillator   A Hartley oscillator that has Lecher wires as a tank circuit.

Lecher wires   Two parallel wires that are several wavelengths long and a small fraction of a wavelength apart that measure

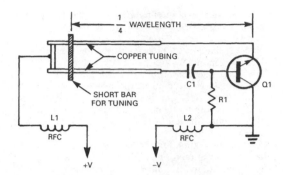

Lecher-line oscillator circuit.

the wavelength of a microwave source which is connected to one end of the wires. A sliding shorting bar is moved along the wires, the positions of standing-wave nodes are noted, and the distance between nodes is then used to determine wavelength or frequency. It is also called a *Lecher line*.

Leclanche cell   A primary dry cell that has a cylindrical zinc container as its negative electrode and manganese dioxide as its positive element, with an electrical connection through a carbon electrode. It has a nominal voltage of 1.5 V, and an energy density of 85 Wh/kg or 165 Wh/L. It is also called a *carbon-zinc cell*.

LED   Abbreviation for *light-emitting diode.*

Leduc effect   If a magnetic field is applied at right angles to the direction of a temperature gradient in a conductor, a temperature difference is produced at right angles both to the direction of the temperature gradient and the direction of the magnetic field.

LEED   Abbreviation for *low-energy electron diffraction.*

left-hand polarized wave   An elliptically polarized transverse electromagnetic wave in which the rotation of the electric field vector is counterclockwise for an observer looking in the direction of propagation. It is also called a counterclockwise polarized wave.

left-hand rule   1. For a current-carrying wire: If the fingers of the left hand are placed around the wire so that the thumb points in the direction of electron flow, the fingers will be pointing in the direction of the magnetic field produced by the wire. For conventional current flow (the opposite of electron flow), the right hand is used.   2. For a movable current-carrying wire or an electron beam in a magnetic field: If the thumb, first, and second fingers of the

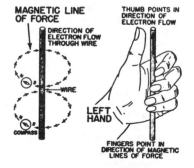

Left-hand rule for electron flow.

left hand are extended at right angles to one another, with the first finger representing the direction of magnetic lines of force and the second finger representing the direction of electron flow, the thumb will be pointing in the direction of motion of the wire or beam. It is also called Fleming's rule.

left-hand taper A taper that has a greater resistance in the counterclockwise half of the operating range of a rheostat or potentiometer (looking from the shaft end) than in the clockwise half.

left-justify 1. To adjust the printing positions of characters on a page so the left margin is lined up. 2. To shift the contents of a register so the left or most significant digit is at some specified position.

left signal The output of a microphone placed to pick up the intensity, time, and location of sounds originating predominantly to the listener's left of the center of the performing area when making a stereo recording or broadcast.

left stereo channel The left signal as electrically reproduced in stereo FM broadcasts or stereo records.

length 1. The number of subunits of data, usually digits or characters, that can be simultaneously stored or processed by a given computer device. 2. The time that data is delayed during transmission, usually expressed in microseconds. Thus the length of a delay line is rated in microseconds.

lens 1. A dielectric or metallic structure that is transparent to radio waves and can bend those waves to produce a desired radiation pattern. 2. One or more disks of precisely contoured glass or other transparent material that can focus light rays and form images by refraction. 3. An arrangement of electrodes that produces an electric field which focuses electrons into a beam. 4. An arrangement of coils or permanent magnets that produces a magnetic field which focuses electrons. 5. A structure that concentrates sound waves by refraction.

lens antenna A microwave antenna that has a dielectric lens placed in front of the dipole or horn radiator to concentrate the radiated energy into a narrow beam. It can also focus received energy on the receiving dipole or horn.

Lenz's law The current induced in a circuit due to its motion in a magnetic field or due to a change in its magnetic flux in a direction to exert a mechanical force opposing the motion or to oppose the change in flux. It is also called the law of induced current.

LEO Abbreviation for *low earth orbit,* the low-altitude orbit of a satellite orbiting the earth. See also *medium earth orbit* [MEO].

letter An alphabet character that represents sounds of a spoken language.

level 1. The difference between a quantity and an arbitrarily specified reference quantity, usually expressed as the logarithm of the ratio of the quantities. In audio and communications a common reference level for power is 1 mW, and power levels are expressed in dBm (decibels above 1 mW). Thus 0.01 mW is −20 dBm, 1 mW is 0 dBm, and 1 W is 30 dBm. 2. A specified position on an amplitude scale applied to a signal waveform, such as reference white level and reference black level in a standard television signal. 3. A charge value that can be stored in a given storage element of a charge-storage tube and distinguished in the output from other charge values. 4. Volume of sound. 5. A single bank of contacts, as on a stepping relay.

level above threshold The pressure level of a sound in decibels above its threshold of audibility for the individual observer. It is also called the sensation level.

level converter An amplifier that converts nonstandard positive or negative logic input voltages to standard DTL or other logic levels.

level indicator 1. An instrument that indicates liquid level. 2. An indicator that shows the audio voltage level at which a recording is being made. It can be a volume-unit meter, neon lamp, or cathode-ray tuning indicator.

level shifting Changing the logic level at the interface between two different semiconductor logic systems.

lever switch A switch that has a lever-shaped operating handle.

Leyden jar An early form of capacitor that consists simply of metal foil applied to the inner and outer surfaces of a glass jar. The terminals are a central brass knob and the outer foil.

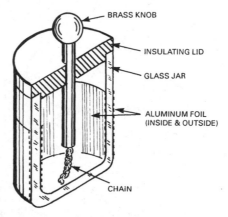

BRASS KNOB

INSULATING LID

GLASS JAR

ALUMINUM FOIL (INSIDE & OUTSIDE)

CHAIN

Leyden jar is an early form of capacitor that demonstrated the storage and discharge of electrical charges.

LF Abbreviation for *low frequency.*

LFOV Abbreviation for *limited field-of-view radar.*

LGA Abbreviation for *land grid array,* a connector socket configuration.

library A collection of commonly used gate or logic-level "building blocks" including NOR and NAND gates and flip-flops, grouped for easy access and manipulation on computer graphics workstation displays.

LIDAR [Laser Infrared raDAR] A meteorological instrument with a ruby laser that generates intense infrared pulses in beam widths as small as 30 seconds of arc, for measuring atmospheric conditions. Reflections and scattering effects of clouds, smog layers, and some atmospheric discontinuities are measured by radar techniques. LIDAR can also track weather balloons, smoke puffs, and rocket trails.

lie detector *Polygraph.*

life test A test of reliability in which a device, circuit, or system is operated under accelerated conditions over a specified period of time to simulate its life expectancy.

lifetime *Mean life.*

LIF Abbreviation for *low insertion force.*

LIFO Abbreviation for *last-in first-out.*

light Electromagnetic radiation that has wavelengths capable of causing the sensation of vision. It ranges from 4000 Å or 0.4 μm (far violet) to 7700 Å or 0.77 μm (far red). The velocity of light is the same as that of radio waves: 186,282.3960 mi/s (299,792.4562 km/s). Invisible infrared radiation produced by lasers is sometimes also called light.

light-activated silicon controlled rectifier [LASCR] A silicon-controlled rectifier that has a glass window to admit incident light that replaces, or adds to the action of an electric gate current in providing switching action. It is also called a photo-SCR and a photothyristor.

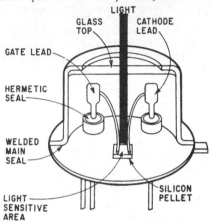

Light-activated silicon controlled rectifier.

light-activated silicon controlled switch [LASCS] A semiconductor device that has four layers of silicon alternately doped with acceptor and donor impurities, as in a light-activated silicon controlled rectifier. All four of the P and N layers are made accessible by terminals. When a light beam hits the active light-sensitive surface, the photons generate electron-hole pairs that turn the device turn on. Removal of light does not reverse the response. The switch can be turned off only by removing or reversing its positive bias.

light amplifier An amplifier whose input and output signals consist of light.

light carrier injection A method for introducing the carrier in a facsimile system by periodic variation of the scanner light beam. Its average amplitude is varied by the density changes of the subject copy. It is also called light modulation.

light chopper A rotating fan or other mechanical device that interrupts a light beam aimed at a photosensor to permit AC amplification of the photosensor output and to make its output independent of strong, steady, ambient illumination.

light-controlled oscillator An oscillator whose output frequency varies with incident light.

light current An electrical current that flows in semiconductor devices such as diodes and transistors because of incident light on the device's PN junction or junctions.

light-dependent resistor [LDR] *Photoconductive cell.*

light detector *Photodetector.*

light-emitting diode [LED] A semiconductor PN junction diode that, when forward-biased, emits light at a wave-

length that is a function of its material and dopants. The crystal structure of gallium arsenide (GaAs) permits non-visible infrared (IR) emission. Phosphorous added to GaAs to form gallium arsenide phosphide (GaAsP) shifts the emission toward the visible red region. Gallium phosphide (GaP) emits in the red, yellow, and green regions. Aluminum gallium arsenide (AlGaAs) on GaAsp produces brighter (high-intensity) red and yellow light emission. GaP on GaP emits high-intensity green.

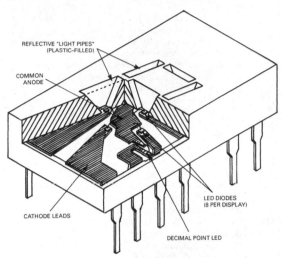

Light-emitting diode (LED) display: seven illuminated segments and a decimal point are formed by plastic-filled light pipes.

light-emitting diode (LED) display A display consisting of an array of light-emitting diode (LED) dies mounted on a rigid insulating substrate or metal leadframe with external connections for individually illuminating each LED. A typical display has a molded, opaque plastic, dual-in-line (DIP) package with clear plastic "light pipe" slots or holes over each LED arranged to form illuminated bar or dot patterns. It can form illuminated numbers, letters or other characters when the appropriate dies are energized. It can be multiplexed to minimize external circuitry. It might contain decoding and driving circuitry. See also *decoder/driver, light-emitting diode LED,* and *multiplex.*

light flux *Luminous flux.*

light-gating cathode-ray tube A cathode-ray tube whose electron beam varies the transmission or reflection properties of a screen that is positioned in the beam of an external light source. It can project a large display on a movie-type screen. An example is the dark-trace tube.

light gun A light pen mounted in a gun-type housing.

lighthouse tube *Disk-seal tube.*

light-microsecond The distance a light wave travels in free space in 0.000001 second. It is used as a unit of electrical distance, equal to approximately 983 ft (300 m).

light modulation *Light carrier injection.*

light modulator The combination of a source of light, an appropriate optical system, and a means for varying the resulting light beam to produce an optical sound track on motion-picture film.

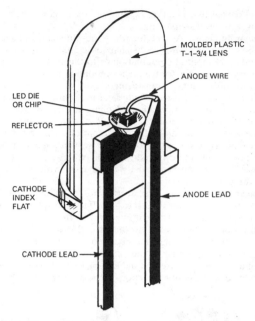

Labels on figure:
MOLDED PLASTIC T–1–3/4 LENS
ANODE WIRE
LED DIE OR CHIP
REFLECTOR
CATHODE INDEX FLAT
ANODE LEAD
CATHODE LEAD

Light-emitting diode (LED) lamp: reflector and plastic lens of radial-leaded LED focus light into a conical beam.

light-negative   The property of negative photoconductivity, hence decreasing in conductivity (increasing in resistance) when exposed to light. Selenium can exhibit this property.

lightning   An electric discharge between clouds or a cloud and the earth.

lightning arrester   A device that provides a discharge path to ground for lightning which strikes an antenna or transmission line. It generally contains a spark gap that has high resistance at normal circuit voltages. This gap breaks down to become a low resistance when acted on by the high-voltage surge of lightning. Also called arrester.

lightning generator   A high-voltage power supply that generates surge voltages which resemble lightning, for testing insulators and other high-voltage components.

lightning rod   A grounded metal rod that projects above the highest point on a structure; it has a sharp point where a high charge density builds up to facilitate leakage of charged particles to ground, minimizing the destructive lightning discharges during a storm.

light-operated switch   A switch operated by a beam or pulse of light, such as a light-activated silicon controlled rectifier.

light pen   A tiny photocell or photomultiplier, mounted with or without a fiber or plastic light pipe in a pen-shaped housing. It is held against a cathode-ray screen to detect the instant at which the electron beam goes through a particular location during its scanning sweep. It can make measurements electronically from the screen or change the nature of the display.

light pipe   A flexible or rigid transparent glass or plastic rod or a bundle of transparent glass or plastic optical fibers that will permit light to pass from one end to the other with minimum loss.

light-positive   Exhibiting positive photoconductivity, hence increasing in conductivity (decreasing in resistance) when exposed to light. Selenium ordinarily has this property.

light ray   A beam of light that has a small cross section.

light relay   *Photoelectric relay.*

light-sensitive cell   *Photodetector.*

light-sensitive detector   *Photodetector.*

light-sensitive resistor   *Photoconductive cell.*

light year   The distance traveled by light through space in 1 year, equal to approximately 5.88×10^{12} mi (9.46×10^{12} km).

limen   *Threshold.*

limit bridge   A form of Wheatstone bridge used for rapid routine electrical tests of manufactured products. No attempt is made to balance the bridge for each test; instead, all products that produce deflections within limits corresponding to permissible tolerance are passed.

limited field-of-view radar   [LFOV]   *Limited-scan array radar.*

limited-scan array radar   A radar whose antenna beam scanning is limited to about ± 10 degrees for such applications as controlling the approach of aircraft from the ground and locating the trajectories of enemy artillery or mortar shells. Its multifunction array antenna is simpler than that of a wide-angle, beam-scanning radar because it does not require as many phase shifters. It is also known as a *limited-field-of-view* [LFOV] *radar.* See *multifunction radar* and *phase shifter.*

limited stability   *Conditional stability.*

limiter   A circuit that limits the amplitude of its output signal to some predetermined threshold level. It can act on positive or negative swings or on both. It is located after the IF amplifier in some FM receivers to remove all amplitude variations from the frequency-modulated signal. For infinite limiting, the threshold level is very small, and the output has an essentially rectangular waveform. It is also called an amplitude limiter, automatic peak limiter, clipper, peak clipper, and peak limiter.

limiting   A desired or undesired amplitude-limiting action performed on a signal by a limiter. It is also called *clipping.*

limiting velocity   The maximum attainable velocity of an electron traveling through a solid material determined by the composition of that material.

limit switch   A switch that cuts off power automatically at or near the limit of travel of a moving object which is electrically controlled.

linac [LINear ACcelerator]   *Linear accelerator.*

LiNbO$_3$ Symbol for *lithium niobate.*

Lindemann glass   A lithium borate-beryllium oxide glass that contains no element higher in atomic number than oxygen. It is used as window material for low-voltage X-ray tubes because it will pass X-rays of extremely long wavelength, such as grenz rays.

line   1. A transmission or power line.   2. A production line for mass-production assembly of electronic equipment.   3. The path covered by the electron beam of a television picture tube in one sweep from left to right across the screen.   4. One horizontal scanning element in a facsimile system.   5. *Trace.*

line amplifier   An audio amplifier that feeds a program to a transmission line at a specified signal level. It is also called a program amplifier.

linear 1. A relationship in which one function is directly proportional to another function providing a straight sloped line when plotted. 2. A circuit whose output varies in direct proportion to its input as in a *linear device*. See *nonlinear*.

linear absorption coefficient The fractional decrease in intensity of a beam of photons or particles per unit distance traversed.

linear accelerator An accelerator that has ring-shaped electrodes arranged in a straight line. When the electrode potentials are properly varied in amplitude at an ultrahigh frequency, particles passing through the electrodes receive successive increments of energy and are accelerated along an essentially linear path. It is also called a linac and a linear electron accelerator.

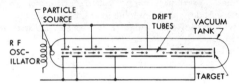

Linear accelerator diagram showing its operating principle.

linear actuator An actuator that converts electric energy into linear mechanical motion.

linear amplifier An amplifier whose changes in output current are directly proportional to changes in applied input voltage.

linear array *Collinear array.*

linear backward-wave oscillator A backward-wave oscillator that generates the required magnetic field by current flow through an electrode called a sole, located adjacent to the cathode.

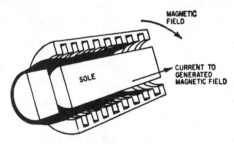

Linear backward-wave oscillator, showing a current-carrying sole at the right of the cathode.

linear-beam tube [O-type] A microwave electron tube that operates with the DC magnetic field parallel to the DC electric field. The DC magnetic field only focuses the electron beam. Tubes of this type include the *two-cavity* and *reflex klystron,* the helix *traveling-wave tube* [TWT], the *coupled cavity TWT,* the *forward-wave amplifier* [FWA], and the *backward-wave amplifier* and *oscillator.*

linear control A rheostat or potentiometer that has uniform distribution of resistance along the entire length of its resistance element.

linear device An amplifying analog device with both linear input and output relationship versus a nonlinear digital device that is either completely on or completely off over its input signal range.

linear detector A detector whose output signal voltage is directly proportional to the changes in input carrier amplitude for amplitude modulation or to the changes in input carrier frequency for frequency modulation.

linear differential transformer An electromechanical transducer that converts physical motion into an output voltage whose amplitude and phase are proportional to position. In one version, a movable iron core is positioned between two windings. Displacement of the core from its null position causes the voltage in one winding to increase, while simultaneously reducing the voltage in the other winding. The difference between the two voltages varies with linear position.

linear electric motor An electric motor that has, in effect, been split and unrolled into two flat sheets; the motion between rotor and stator is linear rather than rotary. Either the stator or the rotor can be extended to form a track along which the other member moves much like a train, provided the current in the energizing coils is switched on and off in a sequence that induces currents able to produce a force acting in one direction on the moving member.

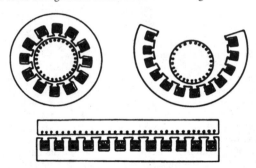

Linear electric motor is, in effect, an ordinary induction motor that has been cut and unrolled.

linear electron accelerator *Linear accelerator.*

linear feedback control system A feedback control system with linear relationships between the pertinent measures of the system signals.

linear integrated circuit An integrated circuit whose output is an amplified linear version of its input or whose output is a predetermined variation of its input. Examples include the operational amplifier, comparator, voltage reference, and analog multiplier. It is contrasted with a digital integrated circuit that is used primarily for processing binary data. See *analog integrated circuit.*

linearity 1. The condition in which the change in the value of one quantity is directly proportional to the change in the value of another quantity. 2. Uniformity of distribution of the lines and elements of an image on a television picture tube so that straight lines in a scene are straight in the image.

linearity control A television receiver control that varies the amount of correction applied to the sawtooth scanning wave to provide the desired linear scanning of lines; lines appear straight, and round objects appear as true circles. Separate linearity controls, known as the horizontal linearity and the vertical linearity controls, are usually provided for the horizontal and vertical sweep oscillators. It is also called distribution control.

linearly polarized wave A transverse electromagnetic wave whose electric field vector is always along a fixed line.

linear modulation Modulation in which the amplitude of the modulation envelope (or the deviation from the resting frequency) is directly proportional to the amplitude of the intelligence signal at all modulation frequencies.

linear power amplifier A power amplifier whose signal output voltage is directly proportional to the signal input voltage.

linear predictive code [LPC] **modulation** A digital voice modulation scheme.

linear pulse amplifier A pulse amplifier in which the peak amplitude of the output pulses is directly proportional to the peak amplitude of the corresponding input pulses.

linear rectifier A rectifier whose output current or voltage contains a wave which has a form identical to that of the envelope of an impressed signal wave.

linear-regulated power supply An analog power supply whose *regulator* circuit provides a linear response to load changes, keeping the power-supply output voltage constant.

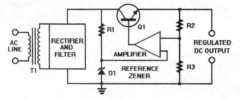

Linear-regulated power supply circuit provides a constant voltage output.

linear scan A radar scan whose beam is oscillated back and forth over a fixed angle in a given plane, as in sector scanning.

linear stopping power The energy loss per unit distance when a charged particle passes through a medium.

linear sweep A cathode-ray sweep in which the beam moves at constant velocity from one side of the screen to the other, then suddenly snaps back to the starting side.

linear taper A potentiometer taper that gives the same change in resistance with rotation over its entire range.

linear time base A time base that moves the electron beam of a cathode-ray tube at a constant speed along its horizontal time scale.

linear transducer A transducer whose output is directly proportional to its input.

linear variable-differential transformer [LVDT] A transformer whose diaphragm or other transducer sensing element moves an armature linearly inside the coils of a differential transformer to change the output voltage by changing the inductances of the coils in equal but opposite amounts. Its applications include measurement of acceleration, force, pressure.

line balance An electrical condition in which the conductors of a transmission line have the same electrical characteristics with respect to each other, other conductors, and ground.

line-balance converter *Balun.*

line conditioning The addition of compensating reactances to a data-transmission line to reduce amplitude and phase delays over certain bands of frequencies.

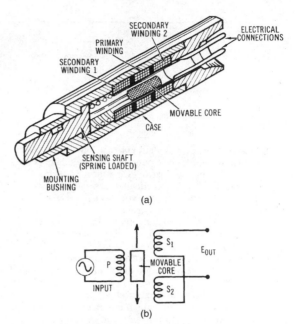

(a)

(b)

Linear variable differential transformer (LVDT): (*a*) section view and (*b*) schematic diagram.

line cord A two-wire cord that terminates in a two-prong plug at one end and is connected permanently to a radio receiver or other appliance at the other end. It makes connections to a source of power and is also called a power cord. The wire alone is commonly called a lamp cord. A third wire and prong are included in some line cords to make a safety connection to ground.

line-cord resistor An insulated wirewound resistor and two regular wires incorporated into a line cord.

line coupling The coupling capacitors, line-tuning circuits, and lead-in circuits that together provide a connection between power or telephone lines and the transmitter-receiver assembly of a carrier communication system.

line diffuser An oscillator within a television monitor or receiver that produces small vertical oscillations of the spot on the screen to make the line structure of the image less noticeable at short viewing distances.

line driver An integrated circuit that acts as the interface between logic circuits and a two-wire transmission line.

line drop The voltage drop that exists between two points on a power line or transmission line, due to the impedance of the line.

line equalizer An equalizer that contains inductance and/or capacitance; it is inserted in a transmission line to modify the frequency response of the line.

line feed A signal that causes a line printer to feed paper up to the next printing line.

line filter 1. A filter inserted between a power line and a receiver, transmitter, or other unit of electric equipment to prevent passage of noise signals through the power line in either direction. It is also called a power-line filter. 2. A filter inserted in a transmission or high-voltage power line for carrier communication purposes.

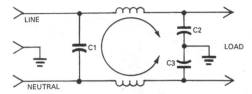

Line filter for use between equipment and power line to trap RF interference.

line finder 1. A device that automatically advances the platen of a line printer or typewriter a given number of lines to the position of the next field in a printed form. 2. A switching device that automatically seeks and locates an idle telephone or telegraph circuit to the destination being called.

line flyback   *Horizontal flyback.*

line-focus tube   An X-ray tube whose focal spot is roughly a line. This gives an essentially square beam of X-rays at one angle of reflection from the target.

line frequency   The number of times per second that the scanning spot sweeps across the screen in a horizontal direction in a television system. In the NTSC system, with 525 lines and 30 complete pictures per second, it is 15,750 sweeps per second, including the horizontal scans made during the vertical return intervals. It is also called horizontal frequency and horizontal line frequency.

line-frequency blanking pulse   *Horizontal blanking pulse.*

line hydrophone   A directional hydrophone that consists of one straight-line element, an array of suitably phased elements mounted in line, or the acoustic equivalent of such an array.

line impedance   The impedance, measured across the terminals of a transmission line.

line interlace   *Interlaced scanning.*

line-length (loop length) compensation   Changes in the transmit and receive path gains within a telephone to compensate for different signal levels at the ends of different line lengths. A short line close to the central office will attenuate signals less, reducing the need for gain. Loop current is generally an indication of line length in compensation circuits.

line microphone   A highly directional microphone that consists of a single straight-line element or an array of small parallel tubes of different lengths, with one end of each abutting a microphone element to give highly directional characteristics. It is also called a machine-gun microphone.

line noise   Noise originating in a transmission line, from such causes as poor joints and inductive interference from power lines.

line of force   An imaginary line in an electric or magnetic field, each segment of which represents the direction of the field at that point.

line of position [LOP]   The intersection of two surfaces of position, to establish a position or fix in navigation.

line of propagation   The path taken by a radio wave through space.

line of sight   The straight, unobstructed path between two points.

line-of-sight distance   The distance from a transmitter to the horizon, normally representing the range limit of a radio or radar station. It is usually under 200 mi (320 km). Under certain conditions, atmospheric refraction can extend the range.

line-of-sight path   The direct, essentially straight, path taken by a radio wave from a transmitting antenna to a receiving antenna.

line-of-sight stabilization   The stabilization of a radar antenna mounted on a ship or aircraft to compensate for roll and pitch by changing the elevation angle of the antenna. For horizontal scanning, the beam would be aimed at the horizon at all times.

line pad   A pad inserted between a program amplifier and a transmission line, to isolate the amplifier from impedance variations of the line.

line printer   An impact or nonimpact printer that requires a full line of data and a print command signal from a computer before an entire line can be printed. The characters are not necessarily printed simultaneously. Speeds can be up to 2000 lines per minute for impact printers and more than 10,000 lines per minute for nonimpact printers. It is also called a high-speed printer.

line regulation   The maximum change in the output voltage or current of a regulated power supply for a specified change in AC line voltage, such as from 105 to 125 V. Line regulation can be expressed as a percentage of the output or the absolute value of the change.

line-sequential color television   A color television system that generates an entire line in one color, with the colors changing from line to line in a red, blue, and green sequence.

line spectrum   The spectrum of electromagnetic radiation emitted spontaneously from bound electrons as they jump from high to low energy levels in an atom. Each different jump in energy level has its own frequency, and these frequencies make up the line spectrum of the atom involved.

lines per minute [LPM]   A rating for the operating speed of a line printer.

line-stabilized oscillator   An oscillator with a section of high-Q transmission line that functions as a frequency-controlling element.

line stretcher   A section of waveguide or rigid coaxial line whose physical length is varied to change its electrical length. A telescoping mechanism commonly achieves this purpose.

line synchronizing pulse   *Horizontal synchronizing pulse.*

line transformer   A transformer inserted in a system for such purposes as isolation, impedance matching, or additional circuit derivation.

line trap   A filter that consists of a series inductance shunted by a tuning capacitor. It is inserted in series with the power or telephone line of a carrier-current system to minimize the effects of variations in line attenuation and to reduce carrier energy loss.

line triggering   Triggering of an oscilloscope or other device by pulses derived from the power-line frequency.

line turnaround   The reversal of transmission direction from a sending unit to a receiving unit or vice versa when using a *half-duplex circuit.*

line-type modulator   The most common form of modulator for pulsing a magnetron transmitter. It typically

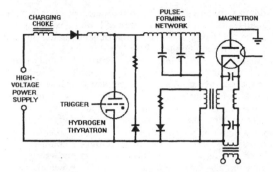

Line-type modulator pulses the cathode of the magnetron in the radar transmitter.

includes a high-voltage supply. The pulse-forming network (PFN) produces a rectangular pulse at the cathode of the magnetron cathode, with the voltage and current sufficient to cause magnetron oscillation. The PFN is charged between pulses, and a trigger fires the thyratron, shorting the input to the PFN, causing a pulse to appear at the transformer. See also *active-switch modulator* and *magnetic modulator*.

line voltage   The voltage provided by a power line at the point of use. In the United States it is usually between 110 and 125 V at outlets in homes, with 120 V as an average.

line-voltage regulator   A regulator that counteracts variations in power-line voltage to provide an essentially constant voltage for the connected load.

link   1. A radio transmitter-receiver system that connects two locations.   2. A flat strip that acts as a removable connector between two terminal screws.

link coupling   Coupling that consists of two coils connected together by a short length of transmission line, with each coil inductively coupled to the coil of a separate tuned circuit.

link neutralization   Neutralization by link coupling between the output and input tuned circuits.

lin-log amplifier   An amplifier whose automatic-gain-control circuit operates linearly for low-amplitude input signals, but logarithmically for high-amplitude signals.

lin-log receiver   A radar receiver that has a linear amplitude response for small-amplitude signals and a logarithmic response for large-amplitude signals.

lip microphone   A contact microphone located against the upper lip of a speaker. An acoustic balancing arrangement cancels sounds originating at a distance. It is useful where noise level is extremely high, as in military tanks. Sound waves from the person's lips enter through only one aperture and act on the microphone. Sound waves from a distance enter through both apertures and act on opposite sides of the diaphragm simultaneously so their effects cancel.

lip-sync   Synchronization of sound and picture so the facial movements of speech coincide with the sounds.

liquid cooling   The use of circulating liquid to cool (remove heat) from electronic components.

liquid crystal   An organic compound that has a liquid phase and a molecular structure similar to that of a solid crystal. The liquid is normally transparent, but it becomes translucent (almost opaque) in localized areas in which the alignment of the molecules is disturbed by applying an electric field with shaped electrodes. Liquid crystals have three phases: nematic, smectic, and cholesteric; the nematic phase, in which the elongated molecules are lined up in one direction but are not in layers, is the most common in liquid-crystal displays.

liquid crystal display [LCD]   A digital display that consists essentially of two sheets of glass separated by a thin layer of sealed-in liquid crystal material. The outer surface of each glass sheet has a transparent conductive coating such as tin oxide or indium oxide, with the viewing-side coating etched into character-forming segments that have leads extending to the edges of the display. The liquid is normally transparent. A voltage applied between front and back electrodes disrupts the orderly arrangement of the molecules, darkening the liquid enough to form visible characters although no light is generated. Power drain is negligible, making liquid crystal displays ideal for digital watches. For other applications, edge or back lighting can improve brightness and permit viewing in darkness.

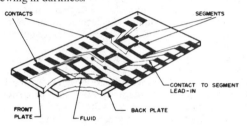

Liquid crystal display. Conductive segments are transparent with liquid darkening behind each energized electrode to form dark visible characters on a bright background. The liquid layer is generally less than 0.025 mm thick. Contrast can be improved by making the back plate reflective.

liquid laser   A laser whose active lasing material is a liquid. The normal wavelength range is 0.3 to 1.2 μm. Examples include chelate, dye, and inorganic liquid lasers.

liquid-level gage   A gage that measures the level of liquid in a tank by sensing elements inside or outside the tank.

liquid-metal mirror   A parabolic mirror suitable for an optical telescope or as a laser light collector in astrophysical experiments that is formed by spinning liquid mercury in a shallow circular pan with a paraboloidal cross section. A parabolic surface is formed when the pan is spun by a drive motor. The curvature can be altered to focus the antenna precisely by adjusting motor speed. Because mercury is a liquid, the mirror can only point straight up. Practical versions are supported by air bearings.

liquid-phase epitaxy [LPE]   A method for growing a semiconductor film on a semiconductor wafer by passing it through the surface of a molten element or compound. The melt temperature is controlled to permit the crystalline growth of the film on the substrate.

Lissajous figure   The pattern that appears on an oscilloscope screen when sine waves are applied simultaneously to both horizontal and vertical deflection plates.

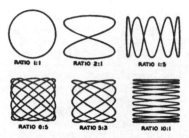

Lissajous figures for sinewaves with various frequency ratios.

list processing language A programming language that is widely used in artificial intelligence research to manipulate categories or lists of items.

LiTaO₃ Symbol for *lithium tantalate.*

lithium [Li] An alkali metal element that has characteristics similar to those of sodium. It can be used in the cathodes of phototubes because it gives a high response at the extreme violet end of the light spectrum. It has an atomic number of 3 and an atomic weight of 6.939.

lithium niobate [LiNbO₃] A ferroelectric piezoelectric crystal that is transparent in the visible and infrared range from 0.38 to 5 μm. Applications include storage of holograms and its use in wideband crystal filters for communication equipment.

lithium power cell A power cell based on lithium electrochemistry that offers higher watt-hours per unit volume and higher voltage than other primary systems—as much as three times the energy output and twice the voltage output of zinc alkaline manganese cells. The cells are classified by their electrolytes: organic liquids, inorganic liquids, or solids. The most popular lithium chemistries are lithium dioxide (Li-SO₂), thionyl chloride (Li-SOCl₂), solid-state iodine, (Li-I₂), manganese dioxide (Li-MnO₂), carbon monofluoride (Li-CF$_x$), and copper oxide (Li-CuO). For example, Li-SOCl₂ offers a nominal voltage of 3.6 V with 11.0 Wh/in³; Li-SO₂ offers a nominal 3.0 V with 7.4 Wh/in³.

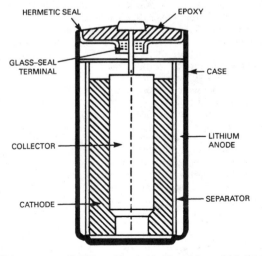

Lithium power cell: lithium thionyl-chloride cell has a high 3.6-V output and long life.

lithium tantalate [LiTaO₃] A piezoelectric crystal grown at high temperature from a melt of lithium oxide and tantalum pentoxide. It is used for wideband crystal filters in communication equipment.

lithography The transfer of a pattern or image from one medium to another, as from a mask to a sensitized (photoresist-coated) semiconductor wafer. If visible light causes the transfer, it is *photolithography. Microlithography* is the process where dimensions are measured in micrometers. See *aligner, electron-beam lithography, mask, stepper,* and *X-ray lithography.*

litz wire Wire that consists of a number of separately insulated strands woven together so each strand successively takes up all possible positions in the cross section of the entire conductor. It reduces skin effect and RF resistance and can be used for winding coils.

live 1. Broadcast directly at the time of production, instead of from recorded or filmed program material. 2. *Energized.*

live chassis A radio, television, or other chassis that has a direct chassis connection to one side of the AC line. For safety, a live chassis must be completely enclosed by an insulating cabinet.

live end The end of a radio studio that gives almost complete reflection of sound waves.

live room A room that has minimum sound-absorbing material.

lm Abbreviation for *lumen.*

LM Abbreviation for *line monitor.*

LMDS Abbreviation for *local multipoint distribution.*

lm/ft² Abbreviation for *lumen per square foot.*

lm/m² Abbreviation for *lumen per square meter.*

lm·s Abbreviation for *lumen second.*

lm/W Abbreviation for *lumen per watt.*

ln Abbreviation for *natural logarithm.*

L network A network composed of two branches in series, with the free ends connected to one pair of terminals. The junction point and one free end are connected to another pair of terminals.

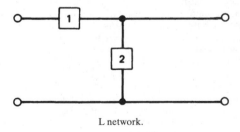

L network.

LO Abbreviation for *local oscillator.*

load 1. The device that receives the useful signal output of an amplifier, oscillator, or other signal source. 2. A device that consumes electric power. 3. The amount of electric power that is drawn from a power line, generator, or other power source. 4. The material to be heated by an induction heater or dielectric heater. 5. To enter data into the storage or working registers of a computer. 6. To place a reel, disk, cartridge, or some other type of recording media into a machine that extracts the stored data or the audio or video content.

load cell A piezoelectric crystal for measuring pressure. Pressure is applied to the piezoelectric crystal, and the resulting voltage across the crystal is measured. It is also used for measuring tension and other forces.

load circuit The complete circuit required to transfer power from a source to a load, such as the coupling network, leads, and load material connected to the output terminals of an induction heater.

loaded antenna An antenna that has extra inductance in series to increase its electrical length.

loaded impedance The impedance at the input of a transducer when the output is connected to its normal load.

loaded line A line that contains loading coils.

loaded Q The Q of a circuit or a device under working conditions.

load factor The ratio of average electric load to peak load, usually calculated over a 1-h period.

load impedance The complex impedance presented to a transducer by its load.

load impedance diagram A diagram that shows how the performance of an oscillator is affected by variations in load impedance.

loading The addition of inductance to a transmission line to improve its transmission characteristics throughout a given frequency band.

loading coil 1. An iron-core coil connected into a telephone line or cable at regular intervals to lessen the effect of line capacitance and reduce distortion. It is also called a Pupin coil and a telephone loading coil. 2. A coil inserted in series with a radio antenna to increase its electrical length and thereby lower its resonant frequency.

loading disk A circular metal piece mounted at the top of a vertical antenna to increase its natural wavelength.

loading error The error introduced when more than negligible current is drawn from the output of a device. In potentiometers the loading error varies with the position of the wiper and the current drawn.

load line A straight line drawn across a series of tube or transistor characteristic curves to show how output signal current will change with input signal voltage for a specified load resistance.

load matching Matching the load circuit impedance to that of the source to give maximum transfer of energy, as desired in induction and dielectric heating.

load-matching network A network for load matching in induction and dielectric heating.

load-matching switch A switch in a load-matching network that compensates for a sudden change in load characteristics, such as that which occurs when passing through the Curie point of the load material in induction and dielectric heating.

load regulation The maximum change in the output voltage or current of a regulated power supply for a specified change in load conditions. Load regulation can be expressed as a percentage of the output or as the absolute value of the change.

load sharing The operation of two computers in duplex so they can share the load of the system during peak hours. At other times, one computer handles the entire load, and the other one serves as backup in case of trouble.

load transfer switch A switch that connects a generator or power source optionally to either of two load circuits.

lobe One of the three-dimensional portions of the radiation pattern of a directional antenna. The direction of maximum radiation coincides with the axis of the major lobe. All other lobes in the pattern are called minor lobes.

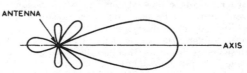

Lobe configuration for one plane of a directional antenna.

lobe access unit [LAU] In a token-ring *local area network,* a device that permits the connection of additional workstations to a single *lobe* in the network.

lobe switch A switch for systematically shifting the radiation pattern of an antenna.

lobe switching A method for determining the exact direction to a target by periodically shifting the beam of a radar antenna slightly to the left and to the right of the dead-ahead position by electronic or mechanical means. While comparing received signal strengths, the entire antenna is turned until equal signals are received from both lobes. The antenna is then accurately aimed, without an impracticably narrow beam width involved. It is used in radio direction finding.

LOC Abbreviation for *lead-on-chip package.*

local area network [LAN] A data communications network capable of transmitting at moderate to high data rates (100 kb/s to 50 Mb/s) within a limited geographic region that can be as small as a single building or as large as a campus-style cluster of buildings. It uses its own switching equipment and does not depend on common carrier circuits, but it can have *gateways* or *bridges* to other public or private networks.

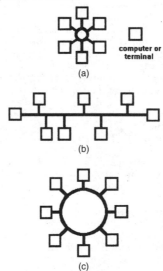

Local area network (LAN) topology: (*a*) star, (*b*) bus, and (*c*) ring.

local channel A standard broadcast channel that permits several stations to operate with power that does not exceed 250 W.

localizer A radio facility that provides signals for the lateral guidance of aircraft with respect to a runway centerline.

localizer on-course line A line in a vertical plane that passes through a localizer, on either side of which the received indications have opposite sense.

localizer sector The sector included between two radial equisignal localizer lines that have the same specified difference in depth of modulation.

local multipoint distribution [LMDS] A short-range television transmission system similar to multichannel multipoint distribution (MMDS), but operating in a different frequency band and able to carry up to 49 channels. Its parabolic receiving antenna is 6 in (150 mm) in diameter, and it can be mounted on a window sill. The signal is less subject to interference than MMDS because it is frequency-modulated. See also *multichannel multipoint distribution* [MMDS] and *satellite master antenna television receiving* [SMATV].

local oscillator [LO] The oscillator in a superheterodyne receiver; its output is mixed with the incoming modulated RF carrier signal in the mixer to give the lower frequency needed to produce the IF signal.

location A place that can be uniquely specified for the storage of data in a computer system.

locked-in line A telephone line that remains established after the caller has hung up. Automatic circuits for this purpose can be installed at police and fire stations for tracing anonymous callers of bomb threats, false fire alarms, and other nuisance calls, as well as legitimate calls for help.

locked oscillator A sinewave oscillator whose frequency can be locked by an external signal to the control frequency divided by an integer. It can be a frequency divider.

locked-oscillator detector A type of discriminator that does not react to amplitude modulation so it requires no limiter preceding it. The circuit has three tank circuits; each circuit is tuned to the signal resting frequency and located so that average current changes only with signal frequency. It acts as an FM detector.

locked-oscillator quadrature-grid FM detector An FM detector that functions as a directly driven quadrature-grid detector for strong signals and as a locked-oscillator detector for relatively weak signals. Some television receivers include it.

locked-rotor current The current drawn by a stalled electric motor.

lock-in The synchronizing of two oscillators by coupling them together or by applying sync pulses so their frequencies are equal or the ratio of their frequencies is an integral number.

lock-in amplifier An amplifier that uses some form of automatic synchronization with an external reference signal to detect and measure very weak electromagnetic radiation at radio or optical wavelengths in the presence of very high noise levels. For optoelectronic applications the low-level light signal can be chopped for drift-free AC amplification, with the reference signal generated in synchronism with the chopping wheel. Other versions for radio frequencies use phase-sensitive detectors in combination with a signal chopper and an external reference frequency source. Locking on the very weak repetitive signal rejects noise and other interfering frequencies.

lock-in range The frequency range over which an oscillator can be synchronized by a synchronization signal.

lock-on The instant at which a radar begins to track its target automatically.

lock-up relay A relay that locks in its energized position either by permanent magnetic biasing (which can be released only by applying a reverse magnetic pulse) or by auxiliary contacts (which keep its coil energized until the circuit is interrupted). By contrast, in a latching relay the lock-up is accomplished mechanically.

lodar A direction finder that determines the direction of arrival of loran signals, free of night effect, by observing the separately distinguishable ground and sky-wave loran signals on a cathode-ray oscilloscope and positioning a loop antenna to obtain a null indication of the component selected to be most suitable. It is also called *lorad*.

LOFAR A submarine detection system based on autocorrelation techniques for long-range analysis of patterned sound picked up at the low-frequency end of the sound spectrum by underwater hydrophones of the Caesar submarine detection system.

Loftin-White circuit A direct-coupled amplifier circuit.

logamp Abbreviation for *logarithmic amplifier*.

logarithm [log] The power to which a number, called the base, must be raised to equal the original number. The common system of logarithms has 10 as a base: here, the logarithm of 1000 to the base 10 is 3 because 10^3 is 1000. Another commonly used base is 2.71828, designated by e and known as the hyperbolic, Napierian, or natural logarithm; here the notation $\log_e N$ is abbreviated as $\ln N$.

logarithmic amplifier [logamp] An amplifier whose output signal is a logarithmic function of the input signal.

logarithmic decrement The natural logarithm of the ratio of the amplitude of one oscillation to that of the next which has the same polarity, when no external forces are applied to maintain the oscillation.

logarithmic diode A diode that has an accurate semilogarithmic relationship between current and voltage over wide forward dynamic ranges. It is used in circuit applications such as dividing, multiplying, logarithmic conversions, and signal compression.

logarithmic multiplier A multiplier in which each variable is applied to a logarithmic function generator. The outputs are added together and applied to an exponential function generator, to obtain an output proportional to the product of two inputs.

logarithmic scale A scale whose graduations are spaced logarithmically rather than linearly.

Logarithmic scale gives a constant percentage accuracy of readings for all pointer positions.

logger A recorder that automatically scans measured quantities at specified times and prints or logs their values on a chart.

logic The translation of formal logic into functional electronic circuits that can be represented by a system of symbols (e.g., AND, OR, and NOT). The functions are performed by switching circuits or *gates* with only two states—ON or OFF or OPEN or CLOSED—making it possible to use binary numbers in solving problems. Each logic symbol can be represented as a *truth table* that gives the output for all possible input conditions. Gate circuits are the basis for digital computers.

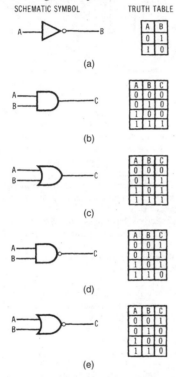

Logic gates. The basic five are shown schematically, with their truth tables: (*a*) NOT, (*b*) AND, (*c*) OR, (*d*) NAND, and (*e*) NOR.

logic analysis Determination of the sequence of logic steps required during a computer run to produce the desired output files from the input files.

logic analyzer An analyzer that locates trouble in digital systems by displaying digital signal levels simultaneously for a number of locations at a predetermined instant of time, which is usually determined by counting clock ticks. The logic analyzer can generate a trigger that stops data collection at the desired instant and displays the pulse signals on a cathode-ray tube screen for analysis of circuit or program errors, isolation of glitch sources, detection of illegal states, mapping of data flow, and diagnosis of other computer problems.

logic comparison The operation of comparing two items in a computer and producing a 1 output if they are equal or alike and a 0 output if they are not alike.

logic design The design of the complete system of logic elements required for a specific application in a digital computer or other digital system.

logic diagram A diagram that represents the logic elements of a computer and their interconnections without necessarily showing construction or engineering details.

logic element The smallest building block that can be represented by an operator in an appropriate system of symbolic logic for a computer data-processing system. Typical logic elements are the AND gate and the flip-flop.

logic function A means of expressing a definite state or condition in magnetic amplifier, relay, and computer circuits. Examples include: (a) the AND function, where an output is produced only when the correct number of input signals is present and combined; (b) the OR function, where an output is obtained when any one of a number of input signals is applied; (c) the NOT function, where the output obtained is the inverse of the input signal.

logic level One of the two voltages whose values have been arbitrarily chosen to represent the binary numbers 1 and 0 in a particular digital system. The magnitude and polarity of the voltage levels must be specified for a particular application to avoid confusion because the level used for 1 can be either higher or lower than that for 0. Terms such as positive logic, negative logic, normal logic, and inverse logic are ambiguous because they can be construed differently for NPN than for PNP devices.

logic operation A nonarithmetic operation in a computer, such as comparing, selecting, making references, matching, sorting, and merging, where yes/no decisions are involved.

logic sum A computer addition in which the result is 1 when either one or both input variables is a 1, and the result is 0 when the input variables are both 0.

logic swing The voltage difference between the logic levels used for 1 and 0. The magnitude of the swing is chosen arbitrarily for a particular system and is usually well under 10 V.

logic switch A diode matrix or other switching arrangement that is capable of directing an input signal to one of several outputs.

logic symbol A graphic symbol that represents the means for performing some specified simple computer operation, such as NOT, AND, OR, NAND, and NOR.

log-periodic antenna A broadband antenna whose electrical lengths and element spacings are chosen so the bidirectional radiation pattern, impedance, and other properties of the antenna are repeated at a number of other frequencies that are equally spaced when plotted on a logarithmic scale.

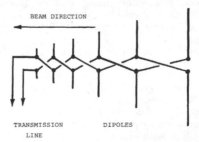

Log-periodic dipole array antenna propagates radio energy in the backfire direction, as shown.

log-periodic dipole array A broadband antenna array whose dipole lengths and spacings increase with distance from a source, with the transmission lines being transposed between adjacent dipole elements. The radiation pattern is unidirectional in the backfire direction, toward the source.

log-periodic folded-dipole array A unidirectional broadband antenna whose elements are arranged as in a log-periodic dipole array, but with all the folded dipoles connected in series with the transmission line rather than in shunt. A phasing strip in each folded dipole is adjusted experimentally to produce a good backfire beam.

Log-periodic folded-dipole array.

log-periodic folded-monopole array A unidirectional broadband array that is essentially half of a log-periodic folded-dipole array, fed against a ground plane.

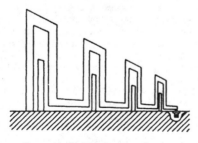

Log-periodic folded-monopole array.

log-periodic folded-slot array A unidirectional broadband antenna that consists of a single metal sheet from which slots are cut in the pattern of a log-periodic folded-dipole array.

loktal base A receiving tube base that has a grooved center post which locks firmly in a corresponding eight-pin loktal socket. The tube pins are sealed directly into the glass envelope. It is also called a loctal base.

loktal tube An electron tube that has a loktal base.

lone electron An electron that is alone on an energy level.

long-delayed echo [LDE] An echo of a shortwave radio signal that is heard with a delay ranging from 2 to 30 s, for which no explanation can be found. Since radio waves travel

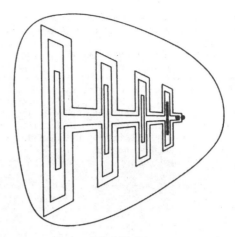

Log-periodic folded-slot array.

at the speed of light, about 186,000 mi/s or 300,000 km/s, one complete passage around the earth takes only about 0.14 s, and even a trip to the moon and back takes less than 2 s.

longitudinal current A current that flows in the same direction in both wires of a pair and uses the earth or other conductors for a return path.

longitudinal magnetization Magnetization of a magnetic recording medium in a direction essentially parallel to the line of travel.

longitudinal parity Parity associated with bits recorded on one track in a data block, to indicate whether the number of recorded bits in the block is even or odd.

longitudinal wave A wave whose direction of displacement at each point of the medium is perpendicular to the wavefront.

long-line effect An effect that occurs when an oscillator is coupled to a transmission line with a bad mismatch. Two or more frequencies might be equally suitable for oscillation, and the oscillator jumps from one of these frequencies to another as its load changes.

long-path communication Amateur radio communication that results from pointing beam antennas in the directions indicated by the longer great-circle path between stations.

long-persistence screen A fluorescent screen that contains phosphorescent compounds which increase the decay time, so a pattern can be seen for several seconds after it is produced by the electron beam.

long-play record [LP record] A 10- or 12-in (25.4- or 30.5-cm) phonograph record that plays at a speed of 33⅓ rpm, and has closely spaced grooves to give playing times up to 30 min for one 12-in side. It is played with a 1-mil stylus whose point radius is 0.001 in (0.0254 mm).

long wave An electromagnetic wave that has a wavelength longer than the longest broadcast band wavelength of about 545 m, corresponding to frequencies below about 550 kHz.

long-wire antenna An antenna whose length is several times greater than its operating wavelength, to give a directional radiation pattern.

loop 1. A curved conductor that connects the ends of a coaxial line or other transmission line and projects into a resonant cavity for coupling purposes. 2. A closed curve on a graph, such as a hysteresis loop. 3. A closed path or circuit over which a signal can circulate, as in a feedback control system. 4. A block of computer code carried out by a computer repetitively until criteria specified in the program are met. 5. The loop formed by the two subscriber wires (tip and ring) connected to the telephone at one end, and the central office (or PBX) at the other end. It is a floating system, not referred to ground, or AC power.

loop actuating signal   The signal derived by mixing the loop input signal with the loop feedback signal of a control system.

loop antenna   An antenna that consists of one or more complete turns of a conductor, usually tuned to resonance by a variable capacitor connected to the terminals of the loop. The radiation pattern is bidirectional, with maximum radiation or pickup in the plane of the loop and minimum radiation at right angles to the loop. It is also called a loop.

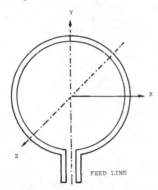

Loop antenna that is small with respect to wavelength will have the same current and phase in all parts of the loop.

loopback   A type of diagnostic test in which the transmitted signal is returned to the sending device after passing through all or part of a communications link or network.

loop control   *Photoelectric loop control.*

loop coupling   The use of a partial or complete loop of wire to couple with the magnetic field of a tuned circuit, tuned cavity, or other tuned device, to feed in or extract RF power.

loop current   The DC current flowing in the subscriber loop, typically provided by the central office or PBX, and typically from 20 to 120 mA.

loop difference signal   A type of loop actuating signal that is produced at a summing point of a feedback control loop when a particular loop input signal is applied to that summing point.

loop feedback signal   The signal derived as a function of the loop output signal and fed back to the mixing point for control purposes.

loop gain   The product of the gain values acting on a signal passing around a closed-loop path. In a feedback control system it is the forward gain multiplied by the feedback

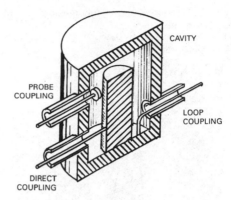

Loop coupling for microwave signals in a coaxial cavity compared with probe and direct coupling methods.

network gain. In a repeater, carrier terminal, or complete closed system the loop gain is the maximum gain that can be used without causing oscillation or singing, and the loop gain here can therefore be less than the product of the gain values.

loopstick antenna   *Ferrite-rod antenna.*

loop test   A telephone or telegraph line test that is made by connecting a faulty line to good lines, to form a loop in which measurements can be made to determine the position of the fault.

LOP   Abbreviation for *line of position.*

loran [LOng-RAnge Navigation]   A long-distance radio navigation system for aircraft and ships; it transmits synchronized pulses simultaneously from widely spaced transmitting stations. Hyperbolic lines of position are determined by measuring the difference in the time of arrival of these pulses. The intersection of two of these lines of position, obtained from either three or four stations, gives a position fix. Standard loran operates on frequencies between 1.8 and 2.0 MHz. The two original versions of loran, both now obsolete, were known as loran A and loran B.

loran C   An extremely accurate, long-range system of navigation, similar to loran B, that gives accuracy comparable to that obtained by celestial observations with a sextant. It operates in the 90- to 110-kHz frequency band. The receiver measures pulse spacing in loran fashion to obtain a rough indication of a position, and it measures precisely the relative phase of the RF carriers in the master and slave pulse envelopes. All older loran stations have been converted to loran C.

loran chain   A chain of four or more loran stations; it forms three or more pairs of stations for loran navigation.

loran chart   A marine or aeronautical navigation chart that shows loran lines of position superimposed on topographic detail.

loran D   A tactical loran system that uses the coordinate converter of low-frequency loran C. It can operate in conjunction with inertial systems on aircraft, independently of ground facilities and without radiating RF energy which could reveal the aircraft's location. The inertial system errors that build up progressively with time can be periodically corrected by loran station transmissions at randomly selected brief intervals.

loran fix A fix obtained by determining the intersection of two loran lines of position.

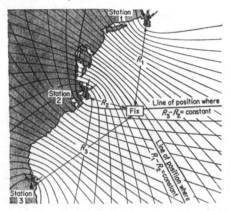

Loran fix on loran chart for the East Coast of the United States.

loran indicator An indicator that displays the pulse signals from two loran ground stations simultaneously and shows the time difference between reception of the two signals.

loran line A line of position on a loran chart. Each line is the locus of points whose distances from two fixed stations differ by a constant amount.

loran set A receiving set or indicator that displays the pulses from loran transmitting stations.

loran station A transmitting station in a loran system.

loran triplet A combination of three loran stations, with one of the stations forming a pair with each of the other stations.

Lorentz force A force that acts on a charged particle in motion in an electric and magnetic field.

Lorentz force equation The equation that relates the force on a charged particle to its motion in an electric and magnetic field.

Lorenz instrument landing system A continuous-wave instrument landing system used in continental Europe and Great Britain; it consists of a runway localizing beacon and two radio marker beacons.

loss 1. Power that is dissipated in a device or system without doing useful work. 2. *Transmission loss.*

Lossev effect Radiation that results from recombination of charge carriers injected in a PN or PIN junction which is biased in the forward direction.

loss factor The power factor of a material multiplied by its dielectric constant. The loss factor varies with frequency and determines the amount of heat generated in a material.

loss modulation *Absorption modulation.*

loss tangent A measure of the amount of power lost as heat when a dielectric or semiconductor material is subjected to a high-frequency electric or electromagnetic field.

lossy The property of certain dielectric materials to cause losses in a circuit. They can be introduced intentionally.

lossy line A transmission line that has intentionally high attenuation per unit length. High loss can be achieved with Nichrome wire as the center conductor.

loudness The intensity characteristic of an auditory sensation. The loudness of sound can be described on a scale extending from soft to loud.

loudness contour A curve that shows the related values of sound pressure level and frequency required to produce a given loudness sensation for a typical listener.

loudness control A combination volume and tone control that boosts bass frequencies when the control is set for low volume. It compensates automatically for the reduced response of the ear to low frequencies at low volume levels. Some loudness controls also provide the same automatic compensation for treble frequencies. It is also called a compensated volume control.

loudness level The sound pressure level, in decibels relative to 0.0002 μbar of a pure 1-kHz tone, that is judged by listeners to be equivalent in loudness to the sound under consideration. It is also called equivalent loudness level. One decibel of change under these conditions is called a phon. See also *Fletcher-Munson curves.*

loudspeaker An electroacoustic transducer that converts audio-frequency electric power into acoustic power and radiates the acoustic power effectively for some distance in air. It is also called a speaker.

loudspeaker dividing network *Crossover network.*

loudspeaker impedance The impedance rating of the voice coil of a loudspeaker. It corresponds to the impedance of the amplifier output terminals to which the loudspeaker should be connected to obtain its rated performance. Common impedance values are 4, 8, and 16 Ω.

loudspeaker system A combination of one or more loudspeakers with associated baffles, horns, and dividing networks, arranged to work together as a coupling between the driving electric circuit and the air, or between the circuit and another acoustic medium.

louver 1. An arrangement of concentric or parallel slats or equivalent grille members that conceals and protects a loudspeaker while allowing sound waves to pass. It can be molded integrally with a plastic radio cabinet. 2. An arrangement of fixed or adjustable slotlike openings provided in a cabinet for ventilation.

low band The band that includes television channels 2 to 6, extending from 54 to 88 MHz.

low-energy electron diffraction instrument An instrument that uses slow electron diffraction in an ultrahigh vacuum to examine the surface structure of solids. The resulting diffraction patterns are visible on a fluorescent screen.

lower sideband The sideband that contains all frequencies below the carrier-frequency value which are produced by an amplitude-modulation process.

lowest useful high frequency The lowest high frequency that is effective at a specified time for ionospheric propagation of radio waves between two specified points. The frequency value is determined by such factors as absorption, transmitter power, antenna gain, receiver characteristics, type of service, and noise conditions.

low frequency [LF] A Federal Communications Commission designation for a frequency in the band from 30 to 300 kHz, corresponding to a kilometric wave between 1 and 10 km.

low-frequency compensation Compensation that extends the frequency range of a broadband amplifier to lower frequencies.

low-frequency induction heater An induction heater whose current flow at the commercial power-line frequency is induced in the charge to be heated.

low-frequency padder A trimmer capacitor connected in series with the oscillator tuning coil of a superheterodyne receiver. It is adjusted during alignment to calibrate the circuit at the low-frequency end of the tuning range.

low level The less positive of the two logic levels or states in a digital logic system. When both high and low levels are negative, the opposite connotation can be used: the smaller negative voltage (the more positive voltage) is assumed to be the low level or low state.

low-level language A programming language written in the machine code of the host computer.

low-level modulation Modulation produced at a point in a system where the power level is low compared with the power level at the output of the system.

low-level radio-frequency signal A radio-frequency signal that has insufficient power to fire a TR, ATR, or pre-TR tube.

low-loss insulator An insulator that has negligible loss at high radio frequencies.

low-loss line A transmission line that has low power dissipation per unit length.

low-noise amplifier An amplifier whose background noise is very weak or inaudible in the absence of desired signals.

low-order digit A digit that occupies the least significant position in a positional notation system. In conventional notation it is the digit at the right.

low-order position The rightmost position in a number or word.

low-pass filter A filter that transmits alternating currents below a given cutoff frequency and substantially attenuates all other currents.

low tension British term for *low voltage,* as generally applied to filament and heater voltages of tubes.

low vacuum A low-quality vacuum in which so much gas or vapor is still present that ionization can occur in an electron tube.

low-velocity scanning The scanning of a target with electrons whose velocity is less than the minimum velocity needed to give a secondary-emission ratio of unity. It is used in the emitron, image orthicon, and vidicon camera tubes.

lox An abbreviation for liquid oxygen, a cryogenic fuel that is liquid below −183°C.

L pad A volume control that has essentially the same impedance at all settings. It consists mainly of an L network that has both elements adjusted simultaneously.

LPC Abbreviation for *linear predictive code.*

LPM Abbreviation for *lines per minute.*

LP record Abbreviation for *long-play record.*

LRC Abbreviation for *longitudinal redundancy check.*

LSA diode [Limited Space-charge Accumulation diode] A microwave diode whose space charge is developed in the semiconductor by the applied electric field. It is dissipated during each cycle before it builds up appreciably, thereby limiting transit time and increasing the maximum frequency of oscillation.

LSA mode One of the three operating modes of a transferred-electron diode, in which the entire structure becomes a negative resistance during a portion of each operating cycle. The frequency of oscillation is determined by the

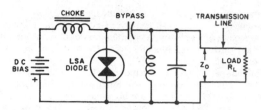

LSA diode in an oscillator circuit.

surrounding circuit and is independent of the transit time of the charge carriers. The other two modes are the quenched-domain and transit-time modes.

LSB Abbreviation for *least significant bit.*

L scope A radarscope that produces an L display.

LSI Abbreviation for *large-scale integration.*

lubber line A line placed on a compass or cathode-ray tube display parallel to the longitudinal axis of a ship or aircraft, as a reference for determining the heading.

Lucite A Du Pont trademark for its transparent acrylic resins.

lug A stamped metal strip terminal to which wires can be soldered. It is also called a soldering lug.

lumen [lm] The SI unit of luminous flux equal to the flux on a unit surface when all points are at unit distance from a uniform point source of 1 candela.

lumen per square foot [lm/ft²] A unit of illumination and luminous exitance. The SI unit of luminous exitance, the lumen per square meter, is preferred.

lumen per square meter [lm/m²] The SI unit of luminous exitance.

lumen per watt [lm/W] The SI unit of luminous efficacy.

lumen second [lm·s] The SI unit of quantity of light, equal to the quantity of light delivered in 1 s by a flux of 1 lm.

lumerg [LUMen-ERG] A unit of luminous flux, equal to 1 erg of radiant energy emitted by a source that has a luminous efficacy of 1 lm/W.

luminance The luminous intensity of any surface in a given direction per unit of projected area of the surface, as viewed from that direction. The SI unit of luminance is the candela per square meter. It was formerly called *brightness.*

luminance carrier *Picture carrier.*

luminance channel A path intended primarily for the luminance signal in a color television system.

luminance flicker Flicker that results only from the fluctuation of luminance.

luminance primary One of the three transmission primaries whose amount determines the luminance of a color in a color television system.

luminance signal The color television signal that has exclusive control of the luminance of the picture. It is made up of 0.30 red, 0.59 green, and 0.11 blue and is capable of producing a complete monochrome picture. It is also called the *Y signal.*

luminescence Emission of light by a material at lower than incandescent temperatures, as a result of chemical or electrical action, exposure to certain types of radiation, or other nonthermal processes.

luminescence threshold The lowest frequency of radiation that is capable of exciting a luminescent material. It is also called the *threshold of luminescence.*

luminescent   Capable of exhibiting luminescence.

luminescent screen   The screen in a cathode-ray tube; the screen becomes luminous when bombarded by an electron beam, and it maintains its luminosity for a time measurable in seconds.

luminophor   A luminescent material that converts part of the absorbed primary energy into emitted luminescent radiation.

luminosity   The ratio of luminous flux to the corresponding radiant flux at a particular wavelength. It is expressed in lumens per watt.

luminosity coefficients   The constant multipliers for the respective tristimulus values of any color, such that the sum of the three products is the luminance of the color.

luminous efficacy   The total luminous flux divided by the total radiant flux. The SI unit of luminous efficacy is the lumen per watt. It was formerly called *luminous efficiency*.

luminous efficiency   *Luminous efficacy*.

luminous emittance   *Luminous exitance*.

luminous exitance   The average luminous flux that leaves a surface per unit area, for which the SI unit is the lumen per square meter. It is also called *luminous emittance*.

luminous flux   The total visible energy produced by a source per unit time. It corresponds to the time rate of flow of light and is usually measured in lumens. It is also called *light flux*.

luminous flux density   The luminous flux per unit area of a surface. When it is referring to the luminous flux falling on a surface, it is called *illumination*. When it is referring to the luminous flux leaving a surface, it is called *luminous exitance*.

luminous intensity   The luminous flux emitted by a source in an infinitesimal solid angle, divided by the solid angle.

luminous sensitivity   The output current of a phototube or camera tube divided by the incident luminous flux.

lumped constant   A single constant that is electrically equivalent to the total of that kind of distributed constant existing in a coil or circuit.

lumped impedance   An impedance concentrated in a single component rather than distributed throughout the length of a transmission line.

Luneberg lens   An artificial lens that focuses radiated electromagnetic energy at ultrahigh frequencies, to increase the gain of an antenna. The lens has a spherically symmetrical but nonuniform distribution of dielectric constant and index of refraction. It is designed to bend divergent rays from the feed so they are parallel as they emerge.

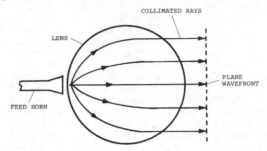

Luneberg lens is a sphere whose index of refraction is variable. It will collimate the rays from a RF feed source placed anywhere around its circumference.

lux [x]   The SI unit of illumination, equal to 1 lm/m^2.

Luxemburg effect   Cross-modulation between two radio signals during their passage through the ionosphere, due to the nonlinearity of the propagation characteristics of free charges in space. Because of this effect, the programs of a powerful station can be heard when a receiver is tuned to a weaker station on a different frequency.

LVDT   Abbreviation for *linear variable-differential transformer*.

Lw   Symbol for *lawrencium*.

lx   Abbreviation for *lux*.

Lyman alpha radiation   Radiation at a wavelength of 1215 Å, associated with one of the spectral lines of hydrogen in the Lyman series.

Lyman series   A group of spectral lines in the ultraviolet spectrum of hydrogen with the wavelength range of 912 to 1215 Å.

M

m 1. Abbreviation for *meter* (metric unit of length). 2. Abbreviation for *milli-*.

M 1. Abbreviation for *mega-*. 2. Unofficial abbreviation for *megohm* (1,000,000 Ω); here the M follows the value, with no space between, as in 5M for 5,000,000 Ω.

M Symbol for *mutual inductance.*

MA Abbreviation for 1. *megampere.* 2. *memory address.*

MAC Abbreviation for *media access controller.*

machine check An automatic check in a computer, or a programmed check of machine functions.

machine code *Machine language.*

machine error A deviation from correctness in computer-processed data, caused by equipment failure.

machine instruction An instruction written in a *machine language* that a compiler can recognize and execute without translation.

machine language 1. Information in a form that can be handled by a computer, as on floppy or hard disk. 2. Characters or instructions expressed in a form that a computer can process without conversion, translation, or programmed interpretation. In machine language, the value of each bit in each instruction in the program must be written out in binary, octal, or hexadecimal digital form. It is also called *machine code.*

machine learning The ability of a machine to improve its performance, based on past performance.

machine translation *Mechanical translation.*

machine word The standard number of characters that a computer regularly handles in each transfer. For example, a machine can regularly handle numbers or instructions in units of 36 binary digits.

Mach number [pronounced mock] The ratio of flight speed to the speed of sound in the medium in which the object moves. At sea level, Mach 1 is approximately 741 mi/h (1192 km/h) at 32°F (0°C) in dry air, but at 30,000 ft (9.14 km) altitude it is about 675 mi/h (1086 km/h).

macro *Macroinstruction.*

macrocell A group of logic gates on a semicustom gate array integrated circuit organized to perform higher-level functions than simple gates (e.g., flip-flop, half adder). See *gate array.*

macroinstruction A computer instruction written in a simple higher-level language or even as a single symbol that is equivalent to one or more ordinary machine instructions. A macroinstruction is expanded into machine instructions by the assembler software of a computer, making it unnecessary for programmers to write out frequently occurring instruction sequences. It is also called a *macro.*

macrometeorite Any meteoroid particle larger than a pea.

macroprogramming The process of writing a computer program in terms of macroinstructions.

macroscopic Large enough to be read by the unaided eye.

macroscopic cross section The cross section per unit volume.

macrosonics The technology of sound at signal amplitudes so large that linear approximations are not valid, as in the use of ultrasonics for cleaning or drilling.

MACSAT Abbreviation for *multiple access communications satellite.*

MAD Abbreviation for *magnetic anomaly detector.*

madar An abbreviation for *malfunction, analysis, detection, and recording,* a self-diagnostic system to detect malfunctions in avionics equipment.

magamp Abbreviation for *magnetic amplifier.*

Magellan A NASA spacecraft, designed to probe the surface of the planet Venus with radar. It was launched in 1986. It has since completed its radar probe and was then assigned to orbiting the planet to gather information for mapping the Venusian gravity fields, whose strength varies from point to point as do those of earth.

maglev Abbreviation for a *magnetically levitated* transportation system, an experimental concept for ground transportation. It includes the suspension, guidance, and propulsion of a mass transport vehicle by magnetic forces rather than driven wheels. There would be no physical contact between the maglev car and its guideway, making it possible for it to travel faster than a vehicle with a wheeled suspension. Roadbed maintenance would be virtually eliminated. One concept is a superconducting maglev system and the other is an electromagnetic system. See *superconductor*.

magnet An object that produces a magnetic field outside itself. It has the property of attracting other magnetic objects such as iron, and attracting or repelling other magnets. A permanent magnet produces a permanent magnetic field. By contrast an electromagnet possesses magnetic properties only when current is flowing through its windings.

magnet charger A charger that restores or establishes the field strength of a permanent magnet by applying a strong magnetic field produced by a surge of direct current through a large electromagnet.

magnetic airborne detector *Magnetic anomaly detector.*

magnetically operated solid-state switch A combination of a Hall generator, trigger circuit, and an amplifier in a single integrated circuit. It can be opened or closed at up to 10,000 operations per second by moving a magnet or otherwise changing a magnetic field. There are no contacts to wear out so there is no contact bounce.

magnetic amplifier [magamp] An amplifier based on the nonlinear properties of saturable reactors, alone or in combination with other circuit elements, to provide amplification. The DC or AC input signal is applied to a control winding to change the degree of saturation of the core, thereby producing a larger change in alternating current in the output winding.

magnetic anisotropy The dependence of the magnetic properties of some materials on direction.

magnetic anomaly A variation in the expected and normal pattern of a magnetic field, as in the magnetic field of the earth. It can be caused by any large ferrous mass, such as a submerged submarine.

magnetic anomaly detector [MAD] An airborne magnetometer that detects submerged submarines. A direct current passes through a coil wound on a high-permeability core and balances out the effect of the earth's magnetic field on the core. An alternating current saturates the core an equal amount on both positive and negative swings. The magnetic field of a submarine causes unequal swings, thereby producing an output signal. In Navy Orion P3 antisubmarine patrol planes the magnetic sensor is located in a stinger-shaped housing that projects from the tail of the plane.

magnetic-armature loudspeaker A loudspeaker whose diaphragm is driven by a ferromagnetic armature that is alternately attracted and repelled by interaction between the field of a permanent magnet and the field produced in the armature by a coil that carries audiofrequency currents. It is also called an electromagnetic loudspeaker, a magnetic loudspeaker, and a moving-armature loudspeaker.

magnetic bearing 1. The angle in the horizontal plane between the direction of magnetic north and an aircraft or vessel's course. It is measured clockwise from mag-

netic north. 2. A bearing that supports a shaft in free space by magnetic fields which are generally produced by a combination of electromagnets and permanent magnets.

magnetic bias A steady magnetic field applied to the magnetic circuit of a relay or other magnetic device.

magnetic biasing The biasing of a magnetic recording medium during recording by superposing an additional magnetic field on that of the signal being recorded. It provides a linear relationship between the amplitude of the signal and the remanent flux density in the recording medium.

magnetic biasing coil A winding on a saturable reactor that establishes a basic magnetization of the core in both polarity and magnitude.

magnetic blowout A permanent magnet or electromagnet that produces a magnetic field which lengthens the arc between opening contacts of a switch or circuit breaker, thereby helping to extinguish the arc.

magnetic bottle A magnetic field that confines a stream of plasma to minimum volume to produce the pinch effect in an electron tube and in controlled thermonuclear fusion experiments.

magnetic brake A friction brake controlled by electromagnetic means.

magnetic bubble A cylindrical stable (nonvolatile) region of magnetization produced in a thin-film magnetic material by an external magnetic field. The direction of magnetization is perpendicular to the plane of the material. It is the basis for magnetic bubble memory.

magnetic-bubble memory *Bubble memory.*

magnetic card A card that has a magnetic stripe on which data can be stored by selective magnetization, much as on computer magnetic tape. Applications include credit cards for on- or off-line point-of-sale terminals, cards for obtaining cash outside banking hours at self-service automatic teller machines (ATM) and monthly commuter tickets in which each use changes the recorded information appropriately to represent the number of rides remaining.

magnetic cartridge *Variable-reluctance pickup.*

magnetic cell One unit of a magnetic memory capable of storing 1 bit of information as a 0 or a 1 state.

magnetic character A character printed with magnetic ink, as on bank checks, for reading by both machines and humans.

magnetic-character reader A character reader that reads special type fonts printed in magnetic ink, such as those used on bank checks. It feeds the character data directly to a computer for processing.

magnetic circuit A complete closed path for magnetic lines of force; it has a reluctance that limits the amount of magnetic flux which can be sent through the circuit by the magnetomotive force.

magnetic circuit breaker A *circuit breaker* whose contacts open automatically in response to an increase in magnetic attraction from a *solenoid* caused by *overload* current in the protected circuit. It is also called an *electromagnetic circuit breaker*. See also *thermal circuit breaker*.

magnetic circuit breaker time delay A *dashpot* mechanism in a *magnetic circuit breaker* that prevents an undesirable response to low-level current transients due to a brief *overload* in the protected circuit. The action of the dashpot delays contact closure.

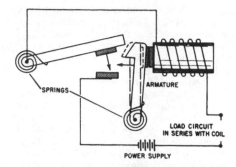

Magnetic circuit breaker whose armature normally locks contacts in a closed position. Overload current through solenoid attracts the armature, and the spring at left opens contacts to break the circuit.

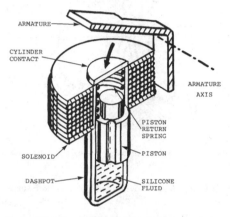

Magnetic circuit breaker time delay. The piston in the dashpot fluid delays the breaker response to allow low-level nuisance transients to pass.

magnetic clutch A clutch in which motion is transmitted from one rotating shaft to another by the attraction between magnetized poles.

magnetic contactor A contactor actuated by electromagnetic means.

magnetic core A ferrous metal core in a coil or transformer that provides a better path than air for magnetic flux, thereby increasing the inductance of the coil and increasing the coupling between the windings of a transformer.

magnetic current sheath The sheath that develops at the surface of a plasma immersed in a magnetic field. The space charge can be either positive or negative. A net current flows along the sheath surface, perpendicular to the magnetic field.

magnetic damping Damping of a mechanical motion by the reaction between a magnetic field and the current generated by the motion of a coil through the magnetic field.

magnetic declination The angle between true north (geographical) and magnetic north (the direction of the compass needle). The angle is different for different locations and it varies from year to year.

magnetic deflection Deflection of an electron beam by a magnetic field, as in a television picture tube.

magnetic delay line A delay line that stores data in a computer; it consists essentially of a metallic medium along which the velocity of propagation of magnetic energy is small compared to the speed of light. Storage is accomplished by recirculation of wave patterns that contain information, usually in binary form.

magnetic dip The angle that the magnetic field of the earth makes with the horizontal at a particular location. It is also called magnetic inclination.

magnetic dipole An elementary dipole associated with nuclear particles. It consists of two equal magnetic poles that have opposite polarity, are closely spaced together, and are so small that directive properties are independent of size and shape. It is also called a magnetic doublet.

magnetic discriminator A magnetic amplifier with a transformer and other components that can sense the polarity and magnitude of coded pulses. It produces output control voltages that could make rudder corrections during the flight of a guided missile.

magnetic disk A disk that has a magnetizable coating on which binary data or other kinds of information can be recorded for storage and readout. It is widely used as read/write memory in computers.

magnetic domain A movable magnetized area in a thin-film magnetic material. Its presence or absence can designate a 1 or 0. Either state is stable (nonvolatile), so is not affected by power shutdown in a computer. The direction of magnetization in a domain is parallel to the plane of the material and is about 100 times as great as that in a magnetic bubble. It is also called a domain.

magnetic-domain memory *Domain-tip memory.*

magnetic doublet *Magnetic dipole.*

magnetic electron multiplier An electron multiplier in which the paths of the emitted secondary electrons are controlled by an applied magnetic field.

magnetic field Any space or region in which a magnetic force is exerted on moving electric charges. The magnetic field can be produced by a current-carrying coil or conductor, by a permanent magnet, or by the earth itself.

magnetic field strength [H] The magnitude of the magnetic field vector. It is expressed in amperes per meter in SI units. Formerly expressed in oersteds in CGS units, it is also called a magnetizing force.

magnetic film memory A magnetic memory that consists essentially of a magnetic film and associated read-in and read-out writing, producible with packing densities of several thousand elements per square inch by vacuum deposition, electroplating, chemical etching, and/or other integrated-circuit production techniques. Rotational switching of the magnetic field of an anisotropic magnetic material such as permalloy can be used, instead of changing the magnetic field strength of an element.

magnetic flaw detector A flaw detector in which a ferrous object is magnetized with an electromagnet or permanent magnet and sprayed with an ink that contains fine iron particles. Surface or near-surface flaws then appear as black lines. They are easier to see if the surface being examined is first painted white.

magnetic flowmeter A flowmeter that depends on the presence of magnetic constituents in a liquid or slurry. It must be initially calibrated for the amount of magnetic material present per unit volume.

magnetic fluid A suspension of iron particles or colloidal ferrite particles in a carrier fluid. The colloidal suspension can be controlled in position, location, shape, specific gravity, surface, trajectory, or velocity by a magnetic field. By contrast, in magnetic clutch fluids the iron particles chain together and solidify when a magnetic field is applied.

magnetic fluid clutch A friction clutch that is engaged by magnetizing a liquid suspension of powdered iron located between pole-pieces mounted on the input and output shafts.

magnetic flux The magnetic lines of force produced by a magnet.

magnetic flux density [B] The number of magnetic lines of force per unit area at right angles to the line. Its SI unit is the tesla. The gauss was formerly used as the CGS unit of magnetic flux density. It is also called magnetic induction.

magnetic focusing Focusing an electron beam through the action of a magnetic field.

magnetic forming The forming of metal into desired shapes with strong magnetic fields that push the metal against a forming die. The magnetic field is produced by charging a large capacitor bank, then dumping the stored energy into an induction coil in less than a millionth of a second. It is also called electromagnetic forming.

magnetic gap A nonmagnetic section in a magnetic circuit, such as an air gap.

magnetic hardness comparator An instrument that compares the hardness of steel parts. A sample part that has the desired hardness is placed in one coil, and the parts to be tested are inserted one by one into a similar coil. If the two coils then have the same magnetic properties as displayed on a cathode-ray oscilloscope screen, the parts in the two coils have the same hardness.

magnetic head The electromagnet for reading, recording, or erasing signals on a magnetic disk or tape.

magnetic heading A heading in which the direction of the reference line is magnetic north.

magnetic hysteresis Internal friction that occurs between the molecules of a magnetic material when subjected to a varying magnetic field. It results in heat loss and makes the magnetic induction dependent on the previous state of magnetization of the material.

magnetic inclination *Magnetic dip.*

magnetic induction 1. The process of generating or inducing currents or voltages in conductors with a magnetic field. In general, the magnetic field must be changing, or there must be relative motion between the conductor and the field. 2. *Magnetic flux density.*

magnetic ink Ink that contains magnetic particles to permit printed characters to be read by either a magnetic character reader or a human.

magnetic-ink character recognition [MICR] A check-processing system. After conventional printing of blank checks, bank and customer identifying numerals and special characters are imprinted in magnetic ink to specifications established by the American Banking Association. After the check is cashed, up to ten numerals and four special characters are added to provide the amount of the check and any special control data required for automatic processing by computer.

magnetic leakage The expansion of magnetic flux outside the path along which it can do useful work.

magnetic lens A lens that has an arrangement of electromagnets or permanent magnets to produce magnetic fields which focus a beam of charged particles.

magnetic levitation The use of magnetic forces for stable suspension of a ground vehicle above or below a suitable guideway. One concept is ferromagnetic attraction; AC or DC magnets on the vehicle ride below a ferromagnetic rail to provide attraction forces for suspension from the rail. Another concept is superconductive induction; suspension is provided by a repulsive-force interaction between superconducting magnet coils of the vehicle and currents induced in the guideway conductors. Levitated vehicles are generally propelled with linear electric motors or some other form of propulsion for high-speed ground transportation.

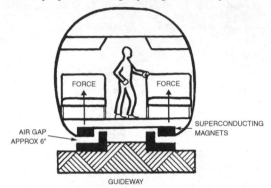

Magnetic levitation vehicle (maglev) is suspended above the guideway by the repulsive force of the superconducting magnets.

magnetic line of force An imaginary line, each segment of which represents the direction of the magnetic flux at that point.

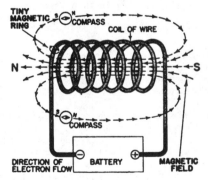

Magnetic lines of force around current-carrying coil make up the magnetic field of coil.

magnetic loudspeaker *Magnetic-armature loudspeaker.*

magnetic material A material that has a permeability considerably greater than that of air or a vacuum. Ferromagnetic materials are strongly magnetic; paramagnetic materials are feebly magnetic.

magnetic memory *Magnetic storage.*

magnetic meridian A horizontal line oriented, at each point on the surface of the earth, in the direction of the horizontal component of the magnetic field of the earth at that point.

magnetic microphone   *Variable-reluctance microphone.*

magnetic mine   An underwater mine that is detonated when the hull of a passing vessel changes the magnetic field at the mine.

magnetic modulator   A cathode pulse modulator for radar magnetrons that is based on the saturation characteristics of inductors. It does not require a thyratron tube or switching device. The inductors transfer their energy resonantly through parallel capacitors in a pi network to the magnetron cathode. See also *line-type modulator* and *active-switch modulator.*

magnetic moment   The moment of a magnetic dipole. A magnetic moment is associated with the intrinsic spin of a particle and the orbital motion of a particle in a system.

magnetic north   The direction indicated by the north-seeking element of a magnetic compass when influenced only by the earth's magnetic field. Because magnetic meridians often follow zigzag lines, the compass needle at any given place does not necessarily point to the magnetic pole.

magnetic pickup   *Variable-reluctance pickup.*

magnetic pole   1. One of the two poles of a magnet near which the magnetic intensity is greatest. These poles are known as the north and south poles.   2. Either of two locations on the surface of the earth toward which a compass needle points. The north magnetic pole is near the geographic North Pole and attracts the south pole of a compass needle.

magnetic reaction analyzer   An analyzer with a Hall detector that measures magnetic field intensities directly with high resolution, making it possible to locate tiny defects during magnetic inspection of welds and other materials.

magnetic reading head   A magnetic head that transforms magnetic variations in magnetic tape or disk into corresponding voltage or current variations.

magnetic recorder   *Magnetic tape recorder.*

magnetic recording   Recording by a signal-controlled magnetic field.

magnetic recording head   A magnetic head that transforms electric variations into magnetic variations for storage on magnetic media.

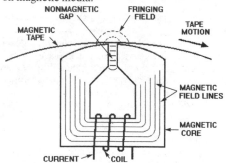

Magnetic recording head for a tape recorder has a nonmagnetic gap.

magnetic recording medium   A magnetizable material used in a magnetic recorder to retain the magnetic varia-

tions imparted during the recording process. It can be a tape, card, or disk.

magnetic reed relay   *Reed relay.*

magnetic reed switch   *Reed switch.*

magnetic reproducing head   A magnetic head that converts magnetic variations on magnetic media into electric variations.

magnetic-resonance imaging [MRI]   A medical diagnostic imaging technique that detects the variation in density and relaxation time of hydrogen nuclei throughout a patient's body when it is subjected to a strong magnetic field. The patient is placed in a tunnel surrounded by electromagnets. The two-dimensional displays on a monitor of sections taken at right angles to the patient (slices) are constructed by computer data processing based on reception of signals from the resonant hydrogen atoms.

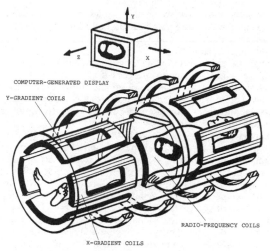

Magnetic-resonance imaging (MRI). Signals are given off by atoms that resonate after they have been subjected to intense magnetic fields and RF pulses. A computer forms a single video image of a "splice" of the subject from the data collected.

magnetic-resonance spectrum   A spectrum produced by varying the RF electromagnetic field that is superimposed on a steady or slowly varying magnetic field about which the atoms of a material precess, to make molecules change their magnetic quantum numbers as they absorb or emit quanta of radio waves. It is the basis for magnetic resonance imaging (MRI) used in medical diagnosis.

magnetic rigidity   A measure of the momentum of a particle, equal to the product of the magnetic intensity perpendicular to the path of the particle and the resultant radius of curvature of the path of the particle.

magnetics   The branch of science that deals with magnetic phenomena.

magnetic saturation   The maximum possible magnetization of a magnetic substance.

magnetic separator   An apparatus for separating powdered magnetic ores from nonmagnetic ores. An electromagnet deflects magnetic materials from the path taken by nonmagnetic materials.

magnetic shield   An enclosure made from high-permeability magnetic material that protects instruments and electronic assemblies from the effects of stray magnetic fields.

magnetic shunt   A piece of magnetic material that diverts an adjustable amount of magnetic flux around an air gap, usually for calibration purposes.

magnetic sound track   A magnetic stripe on motion-picture film that forms a magnetic track for recording the sound accompaniment of the film.

magnetic spectrograph   A spectrograph based on the action of a constant magnetic field on the paths of electrons or other charged particles. It separates particles that have different velocities.

magnetic storage   Storage that represents information by varying degrees of magnetization of a magnetic material. The material can be magnetic disks, plates, or tapes. It is also called magnetic memory.

magnetic storm   A storm that causes rapid and erratic changes in the strength of the magnetic fields of the earth, affecting both radio and wire communications. It is caused by sun-spot activity.

magnetic-stripe credit card   A credit card that has one or more magnetic stripes which contain the data required for establishing credit at an on- or off-line point-of-sale terminal. Added features include secret personal identification codes and encryption of on-line encoded data to minimize fraud and abuse of cards.

magnetic tape   A plastic tape that is coated or impregnated with magnetizable iron oxide particles.

magnetic-tape player   A machine capable of playing back magnetic tapes but not recording. It is a function included in some AM/FM battery-powered portable audio systems, and in *tape decks* in home and automotive stereo systems. See also *magnetic tape recorder.*

magnetic-tape reader   A machine that plays back coded information from a magnetic tape, usually containing digital data for use in computer operation or for the operation of numerically controlled machines. It is to be distinguished from a *magnetic-tape player* which plays back voice and music in the audio-frequency band.

magnetic-tape recorder   A machine capable of recording and playing back audio-frequency signals on a *magnetic tape.* In recording, it converts the signal to magnetic variations in the medium and in playing back it converts those magnetic variations back into audio-frequency sig-

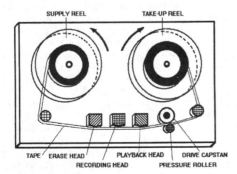

Magnetic-tape recorder organization. Tape moves from left to right.

nals. Typical units include an amplifier and a loudspeaker. Portable units are battery or AC-to-DC-transformer powered. A complete system is included in *telephone-answering machines.* See also *magnetic-tape player.*

magnetic-tape unit   A computer peripheral unit that consists of a tape transport, reading and recording heads, and associated electric and electronic equipment.

magnetic test coil   *Exploring coil.*

magnetic unit   A unit used in measuring magnetic quantities. The SI magnetic units are the ampere-turn, ampere per meter, tesla, and weber.

magnetic-vane meter   An AC meter with a metal-vane moving element pivoted inside a coil.

magnetic variometer   An instrument that measures differences in a magnetic field with respect to space or time, as contrasted to a magnetometer, which measures the absolute value of the intensity and/or direction of a magnetic field.

magnetic wave   A magnetostatic wave that does not depend on the motion of magnetic fields. A surface magnetic wave is an example.

magnetic-wave device   A device that depends on magnetoelastic or magnetostatic wave propagation through or on the surface of a magnetic or dielectric material. Applications are comparable to those for other types of microwave delay lines.

magnetism   A property possessed by iron, steel, and certain other magnetic materials. These materials can produce or conduct magnetic lines of force capable of interacting with electric fields or other magnetic fields.

magnetization   1. The degree to which a particular object is magnetized.   2. The process of magnetizing a magnetic material.

magnetization curve   *B-H curve.*

magnetizing current   The current that flows through the primary winding of a power transformer when no loads are connected to the secondary winding. This current establishes the magnetic field in the core and furnishes energy for the no-load power losses in the core. It is also called exciting current.

magnetizing force   *Magnetic field strength.*

magnet keeper   *Keeper.*

magnetoelastic energy   The energy associated with the change in dimensions of a ferromagnetic material during magnetization.

magnetoelectric   The generation of voltages by magnetic techniques, as in an ordinary generator.

magnetoelectric effect   One of the effects observed when a material carrying an electric current is placed in a transverse magnetic field. These effects include the Hall effect, Ettingshausen effect, Nernst effect, and magnetoresistance.

magnetoelectric generator   *Magneto.*

magnetoelectric transducer   A transducer that measures the voltage generated by the movement of a conductor in a magnetic field.

magnetograph   An instrument that analyzes light spectra to measure the intensity of a magnetic field in hot gases, such as those of the sun or other stars.

magnetohydrodynamics [MHD]   The study of the effects of magnetic fields on superheated ionized gases and conducting fluids. It is the subject of research on controlled fusion for nuclear power plants. It has been proposed as a method for generating electric power at higher efficiencies than steam generators.

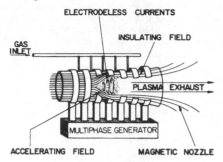

ELECTRODELESS CURRENTS

INSULATING FIELD

GAS INLET

PLASMA EXHAUST

MULTIPHASE GENERATOR

ACCELERATING FIELD   MAGNETIC NOZZLE

Magnetohydrodynamic system.

magnetoionic wave component   Either of the two elliptically polarized wave components into which a linearly polarized wave incident on the ionosphere is separated because of the earth's magnetic field.

magnetometer   An instrument that measures the magnitude and direction of a magnetic field, such as the earth's magnetic field.

magnetomotive force   The force that produces a magnetic field. It is the total magnetizing force acting around a complete closed magnetic circuit, and it corresponds to voltage (electromotive force) in an electric circuit. If due to current in a coil, magnetomotive force is proportional to ampere-turns. The CGS unit of magnetomotive force is the gilbert, equal to about 0.8 ampere-turn. The corresponding SI unit of magnetomotive force, the ampere-turn, is now preferred.

magneton   *Bohr magneton.*

magnetooptical effect   *Kerr magnetooptical effect.*

magnetooptical laser   A laser whose continuous magnetic field contributes to the generation of coherent radiation.

magnetooptical material   A material whose optical properties can be changed by applying a magnetic field: for example, the Faraday effect can be applied in certain transparent materials to deflect a laser beam.

magnetooptical modulator   An arrangement for modulating a beam of light by passing it through a single crystal of yttrium-iron garnet. It provides intensity modulation by causing a magnetic field to produce optical rotation.

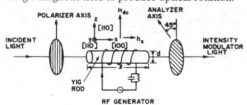

POLARIZER AXIS   ANALYZER AXIS

INCIDENT LIGHT   INTENSITY MODULATOR LIGHT

YIG ROD

RF GENERATOR

Magnetooptical modulator in which the modulation of a light beam is proportional to the magnetization in the direction of light propagation through a YIG crystal.

magnetooptics   The branch of physics concerned with the effect of magnetic fields on light.

magnetopause   The boundary between the magnetosphere of the earth or another planet and the interplanetary plasma or solar wind of the solar atmosphere.

magnetophotophoresis   The movement of dust and other particles under the combined influences of a magnetic field and radiant energy such as light.

magnetoplasmadynamics   The generation of electricity by shooting a beam of ionized gas through a magnetic field, to give the same effect as moving copper bars near a magnet.

magnetoresistance   The change in resistance associated with a change in magnetization in some materials.

magnetoresistance multiplier   An analog multiplier in which thin-film magnetoresistors are mounted in air gaps in the magnetic core of the multiplier. An input signal applied to a coil wound on the core of the multiplier produces push-pull resistance swings in the two magnetoresistors, thereby unbalancing the bridge in which they are connected. It is in control systems and analog computers for measuring power and providing such functions as division, multiplication, squaring, and square-rooting.

magnetoresistive head   A disk-drive read head whose operation depends on the *magnetoresistive effect.* Certain metals, when exposed to a magnetic field, change their resistance to the flow of current. The read element of a magnetoresistive head exploits this property for reading high-density magnetic disks. It has made possible the storage of 1.3 billion bits/in^2. See also *inductive thin-film recording heads.*

magnetoresistor   A resistor whose resistance value changes with the strength of the applied magnetic field. They are generally used in pairs in a bridge circuit, with a permanent magnet providing a magnetic bias. Current flow through series-opposing coils wound on the magnet poles reduces the flux at one resistor while increasing it at the other, to unbalance the bridge and give an output current.

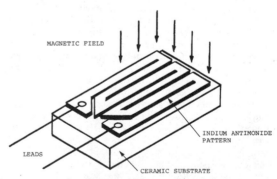

MAGNETIC FIELD

INDIUM ANTIMONIDE PATTERN

LEADS

CERAMIC SUBSTRATE

Magnetoresistor is made as an etched indium antimonide serpentine pattern on a ceramic substrate.

magnetosphere   A region that extends to heights of several earth radii above the earth. It is caused by the earth's magnetic field. It includes the Van Allen radiation belt and consists of trapped particles, chiefly electrons and protons, that spiral about the magnetic lines from pole to pole.

magnetostatic   Magnetic properties that do not depend upon the motion of magnetic fields.

magnetostriction   The change in the dimensions of a ferromagnetic object when the object is placed in a magnetic field. It is also called Joule effect.

magnetostriction hydrophone A magnetostriction microphone that responds to waterborne sound waves.

magnetostriction loudspeaker A loudspeaker whose mechanical displacement is derived from the deformation of a material that has magnetostrictive properties.

magnetostriction microphone A microphone that depends for its operation on the generation of an electromotive force by the deformation of a material that has magnetostrictive properties. Ultrasonic and underwater sound generation are its chief applications.

magnetostriction oscillator An oscillator in which the anode circuit is inductively coupled to the grid circuit through a magnetostrictive element. Its frequency of oscillation is determined by the magnetomechanical characteristics of the coupling element.

magnetostriction transducer A sonar transducer that changes an alternating current to sound energy at the same frequency and forms the sound energy into a beam. A large number of nickel or nickel alloy tubes and coils are connected in a series-parallel arrangement. One end of each tube is attached to a diaphragm that is in contact with the sea water. The transducer also acts as a microphone for returning echoes.

magnetostrictive The property of a material to change dimensions when placed in a magnetic field.

magnetostrictive delay line A delay line made of nickel or other magnetostrictive material. The amount of delay is determined by a shock wave traveling through the length of the line at the speed of sound.

magnetostrictive relay A relay that functions as a result of dimensional changes occurring in a magnetic material subjected to a magnetic field.

magnetostrictive resonator A ferromagnetic rod that can be made to vibrate at one or more definite resonant frequencies by applying an alternating magnetic field.

magnetostrictor A device for converting electric oscillations to mechanical oscillations by employing the property of magnetostriction. The device consists of a bar of magnetic material, anchored at one point and surrounded by a coil that carries the oscillating current. For maximum energy, the system must be driven at or near its natural frequency. It has ultrasonic applications.

magnetothermoelectric single crystal A crystal basis for a solid-state refrigerator. When the crystal carries an electric current in the presence of an intensely strong transverse magnetic field of over 100 kG, it is capable of cooling to temperatures of 100°C below room temperature.

magnetotransistor A lateral bipolar transistor with two collectors that is designed as a magnetic-field sensor. Two types: vertical magnetotransistors (VMTs) depend on vertically flowing carriers for their magnetic operation; lateral magnetotransistors (LMTs) depend on the lateral flow of carriers for magnetic response. When the magnetic field is applied to both types of device, the Lorentz force acts on flowing carriers, causing deflection.

magnetron A microwave-frequency power oscillator consisting of a multicavity cylindrical anode and a coaxial cathode within a vacuum tube. A permanent magnetic field is completed through the tube parallel to the axial cathode from external paired or single C-shaped magnets. When powered, a DC electric field forms between the anode and the cathode, at right angles to the magnetic field. The result-

ing crossed field between the cathode and the anode causes electrons emitted from the cathode to move in curved trajectories. Oscillations occur in the anode's resonant cavities which form a slow-wave structure when the DC voltage is adjusted, so that the average rotational velocity of the electrons corresponds to the phase velocity of the field in the slow-wave structure. It is also called a *cylindrical magnetron*, a *traveling-wave magnetron*, and a *conventional magnetron*. See also *crossed-field electron tubes.*

magnetron amplifier A magnetron that functions as an amplifier.

magnetron oscillator *magnetron.*

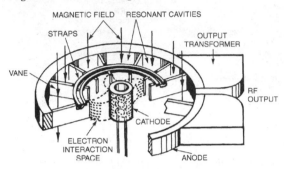

Magnetron oscillation occurs when the cathode DC voltage causes the electrons in the axial magnetic field to rotate at a velocity that corresponds to the phase velocity of the field in the anode cavity structure.

magnetron strapping *strapping.*

magnet steel Steel that has high retentivity and usually contains some combination of tungsten, cobalt, chromium, and manganese with steel. It is permanent magnet material.

magnet wire Copper wire with a varnish insulation in any of the sizes commonly used for winding the coils of transformers, relays, and other electromagnetic devices.

mAh Abbreviation for *milliampere-hour.*

main A line that brings power from a generator, converter, or service-cutoff switch to the main distribution center for power lines inside a building. It is the British term for *power line.*

mainframe A term that now designates a large computer system compared with a workstation, personal computer, or minicomputer. It is capable of performing massive data-processing tasks such as telephone switching or bank transactions.

main gap The conduction path between a principal cathode and a principal anode of a glow-discharge tube.

main lobe The radiated field from an antenna that contains the maximum radiation in one specific direction. It is also called the *major lobe.*

main quantum number A positive integer that specifies the size of an electron orbit.

mains British term for *power line.*

main memory The general-purpose memory of a computer; it provides fast access to stored instructions and data being processed. It typically consists of dynamic random-access memories (DRAM).

maintenance 1. Preventive or corrective measures to keep equipment in satisfactory operating condition.

2. Operating and check runs to keep computer programs and files up to date and free of errors.

main terminal In a bidirectional thyristor, the terminal marked 1. The other terminal, through which the principal current flows, is marked 2.

major apex face One of the three large sloping faces extending to the apex or pointed end of a natural quartz crystal. The other three smaller sloping faces are the minor apex faces.

major cycle The time interval between successive appearances of a given memory position in a serial-access computer memory.

majority carrier The type of carrier that constitutes more than half the total number of carriers in a semiconductor device. Majority carriers can be either holes or electrons. The other type of carrier is then known as the minority carrier. In N-type material, electrons are the majority carriers. In P-type material, holes are the majority carriers.

majority emitter An electrode from which a flow of majority carriers enters the interelectrode region of a transistor.

majority logic Logic in which the output depends on the state of the majority of inputs.

major lobe The radiation lobe that contains the direction of maximum radiation.

make-before-break contacts Double-throw contacts arranged so that the moving contact establishes a new circuit before interrupting the old one.

make contact A normally open stationary contact on a relay. Its circuit is closed when the relay is energized.

management information system [MIS] A computer-based information system that receives data on materials cost, overhead, operating expense, sales, and profit and loss, and processes that information at regular intervals. It provides summary reports for management guidance and decision making on demand. The reports can be visual and graphic displays on a monitor or they can be printouts.

Manchester code A computer code in which a binary 1 is represented by a transition from the 1 to the 0 level in the middle of a bit, and a binary 0 is a transition from the 0 to the 1 level in the middle of a bit. In this code, one bit time is the longest interval permitted without a transition. It is also called *phase coding*.

Manchester code for binary data. The space between each pair of dashed vertical lines represents one bit time.

manganese [Mn] A metallic element with an atomic number of 25 and an atomic weight of 54.94.

manganin An alloy that contains 84% copper, 12% manganese, and 4% nickel for making precision wirewound resistors because of its low temperature coefficient of resistance.

Manhattan project The project by the War Department lasting from August 1942 to August 1946 that developed the atomic bomb.

manipulated variable The quantity or condition that is varied by the controller, to change the value of the controlled variable in a feedback control system.

manipulator A mechanism, typically consisting of a series of segments, jointed or sliding relative to one another, for grasping and moving objects. It usually has at least three degrees of freedom. The joints can be powered by electric, hydraulic, or pneumatic motors. It can be remotely controlled in real time as in a *telecheric* or by programmed computer as in a *robot*. It is also called an *arm*.

manmade interference Electromagnetic interference that results from normal or abnormal operation of electric or electronic equipment, such as harmonic or spurious signals from transmitters. Noise produced by sparking is generally called static rather than interference.

manmade static High-frequency noise signals created by sparking in an electric circuit. When picked up by radio receivers, manmade static causes buzzing and crashing sounds.

map-matching guidance Guidance of an aircraft or missile by a radar map previously recorded by a reconnaissance flight over the terrain of the route compared with radar echoes received during the guided flight.

Marconi antenna An antenna that is connected to ground at one end through the receiver or transmitter input coil and suitable tuning reactances.

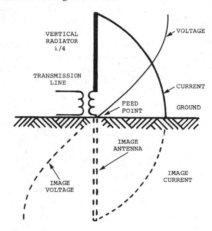

Marconi antenna is a vertical radiator about a quarter-wavelength long that is grounded at one end. The current and voltage distribution are shown.

marker generator 1. An RF generator that injects one or more frequency-identifying pips on the pattern produced by a sweep generator on a cathode-ray oscilloscope screen. It is used for adjusting response curves of tuned circuits, as when aligning FM and television receivers. 2. An RF generator that generates pulses which have precise amplitude, shape, duration, and recurrence characteristics, as required for producing reference indexes on a radarscope for such quantities as target range, bearing and elevation.

mark-space ratio The ratio of the duration of a single pulse to the interval between two successive recurrent pulses.

MARS [Military Affiliate Radio System] A worldwide network of radio stations operated by amateurs both on and off military installations, sponsored jointly by the U.S. Air Force, Army, and Navy to provide an auxiliary and emergency alternate communication system.

Mars Global Surveyor A spacecraft that is to go into orbit around Mars and use a battery of six instruments to scan the planet's surface for a full Martian year (about two earth years), seeking evidence of past or present water and life forms.

Mars Pathfinder Lander A spacecraft that will fly directly to Mars and land on its surface to transmit images of the Martian terrain with a color TV camera. It will monitor the Martian weather and deploy a small roving unit that will explore the terrain and sample rocks and soil.

maser [microwave amplification by stimulated emission] A microwave amplifier that amplifies by stimulating atoms or molecules to an unstable higher energy level. A microwave input signal interacts with the atoms or molecules to stimulate the emission of excess energy at the same frequency and phase as the stimulating wave, thus providing coherent amplification at a wavelength determined by the dimensions of the cavity or resonant structure. The application of external energy required for amplification or oscillation is called *pumping*. The radiated energy greatly exceeds the energy level of the pumping signal. There are three types: *gas, solid state,* and *traveling wave.* Ammonia atoms or molecules are the parametric material in beam-type gas maser oscillators. Solid-state masers depend on the electrons of parametric atoms or molecules. There are two- and three-level solid-state masers. Masers can function as low-noise preamplifiers for very weak signals in radio astronomy or long-distance radar, and as time and frequency standards. The stimulated emission principles of the maser apply in the visible light and infrared regions, where the equivalent device is the *laser.*

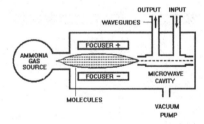

Maser that uses ammonia as its parametric material.

mask 1. A transparent (glass or quartz) plate covered with opaque patterns that define the size and shape of integrated-circuit elements at one level of fabrication. Examples include wells, emitters, gates, drains, and channels. Selected areas of a *photoresist*-coated wafer are exposed to ultraviolet light to define the areas to be etched. Opaque mask areas can be formed from deposited chrome, iron oxide, or silicon. As many as 20 successively registered masks might be used to make an IC. 2. A transparent (glass or plastic) plate with opaque emulsion forming a stencil for defining conductors, pads, ground planes, and contacts in printed-circuit board manufacture. Light passing through the mask exposes the pattern onto a photoresist-coated circuit board. Exposed photoresist is chemically removed, and the pattern that remains defines the copper-clad areas to be removed by acid etch. See *lithographic.*

masking 1. The amount the threshold of audibility of a sound is raised by the presence of another sound. The unit customarily used is the decibel. It is also called audio masking and aural masking. 2. A programmed procedure for eliminating radar coverage in areas where such transmissions can be useful to the enemy for navigation purposes, by weakening the beam in those directions, or by using additional transmitters on the same frequency at suitable sites to interfere with homing. 3. Applying a covering or coating on a semiconductor surface to provide a masked area for selective deposition or etching. 4. The use of tones, noise, music, or other sounds to hide or mask a clear signal for secrecy purposes. The masking signal must be available at the receiving terminal for subtraction, leaving the desired signal.

mask programming A factory method for custom finishing semiconductor read-only memories (ROM), by applying one or more metallized film layers by mask-and-etch processing with a custom mask. By contrast, *field-programmable* ROMs such as a PROM, EPROM, EEPROM, and flash are field-programmed with special equipment that organizes the memory cells by selective voltage application.

mask-programmed read-only memory [ROM] A read-only memory in which a mask produces the metallized interconnection pattern corresponding to the desired permanently stored program or data.

mass The quantity of matter in a body. Mass is a measure of inertia and determines resistance to acceleration independently of gravitational force, whereas weight is the force exerted by a body under the influence of gravity at a particular location.

mass conversion factor *Atomic mass conversion factor.*

mass-energy equation The equation developed by Albert Einstein for interconversion of mass and energy, written as $E = mc^2$, where E is the energy in ergs, m is mass in grams, and c is the velocity of light in centimeters per second.

mass memory A large-capacity magnetic memory disk or memory system consisting of multiple parallel Winchester-style hard disks for mainframe computers or supercomputers.

mass migration Metallic erosion that occurs when a high current density is passed through a conductor that has a small cross-sectional area, such as the conductors in integrated circuits.

mass spectrograph A mass spectrometer that provides a permanent record of the mass spectrum lines of a material on a photographic plate.

mass spectrometer A spectrometer that analyzes a substance in terms of the ratios of mass to charge of its components. A gas or a compound in the vapor state is bombarded by electrons, and the resulting ions are accelerated and separated according to their mass-to-charge ratios, in the most common type, combined electric and magnetic fields deflect the ions of the substance and focus each type in turn on an output electrode for detection and measurement. In another type, sorting of ions is based on the time of flight of the ions through a drift tube during acceleration by electric fields.

mass storage *Mass memory.*

mast A vertical metal pole that acts as an antenna or antenna support.

master 1. The negative metal counterpart of a disk recording, produced by electroforming as one step in the production of phonograph records. 2. *Master station.*

master-antenna television [MATV]   An antenna system that consists of an antenna array which is capable of receiving available broadcast signals and amplifying them as required for distribution over coaxial cables to a number of individual television receivers that are normally within a single home, apartment, hotel, or other building.

master brightness control   A variable resistor that adjusts simultaneously the grid bias on all guns of a three-gun color picture tube.

master clock   The electronic or electric source of standard timing signals, often called clock pulses, required for sequencing the operation of a computer.

master control   The control console that contains the main program controls for a radio or television transmitter or network.

master gain control   1. A variable resistor or potentiometer in a stereo amplifier that controls the gain of both audio channels simultaneously.   2. A control in a radio, television, or recording studio that changes the overall audio output level without affecting the mixer controls which determine the balance of the microphones and other sound sources. It can fade out or fade in the sound volume.

master oscillator   An oscillator that establishes the carrier frequency of the output of an amplifier or transmitter.

master oscillator-power amplifier [MOPA]   An oscillator state followed by an RF power amplifier stage that serves also as a buffer.

master routine   *Routine.*

master-slave flip-flop   A combination of two flip-flops, one of which (the master) receives its information on the leading edge of a clock pulse. The other (the slave or output flip-flop) receives its information on the trailing edge of the pulse. It prevents false triggering when two or more gate inputs are applied almost simultaneously.

master station   The reference station of a synchronized group of radio stations to which the emissions of other stations of the group are referred. In a loran system it is the A or master station.

master synchronization pulse   A pulse, distinguished from other telemetering pulses by amplitude and/or duration, that indicates the end of a sequence of pulses.

match   A data-processing operation similar to a merge, except that instead of producing a sequence of items made up from the input sequences, the sequences are matched against each other on the basis of some key.

matched diodes   Two diodes that have exactly the same outline dimensions and electrical characteristics. One can have forward polarity and the other reverse polarity, or both can have the same polarity.

matched load   A load with an impedance value that results in maximum absorption of energy from the signal source.

matched power gain   The power gain obtained when the impedance of a load is matched to the effective output impedance of the amplifier to which it is connected.

matched termination   A termination that produces no reflected wave at any transverse section of a waveguide or other transmission line.

matched transmission line   A transmission line that has a matched termination.

matched waveguide   A waveguide that has a matched termination.

matching   Connecting two circuits or parts together so that their impedances are equal or are equalized by a coupling device, to give maximum transfer of energy.

matching device   A device that matches unequal impedances, such as the output transformer of a radio receiver.

matching diaphragm   A diaphragm that consists of a slit in a thin sheet of metal, placed transversely across a waveguide for matching purposes. The orientation of the slit with respect to the long dimension of the waveguide determines whether the diaphragm acts as a capacitive or inductive reactance.

matching stub   A short length of two-wire transmission line connected at the antenna or receiver end of a regular transmission line to add inductive or capacitive reactance for matching purposes. The short-circuiting conductor can be a slider that can be moved along the stub.

matching transformer   A transformer inserted between unequal impedances for matching purposes, to give maximum transfer of energy.

match-terminated   Terminated in a load equal to the characteristic impedance of the transmission line.

math coprocessor   A circuit that processes floating-point arithmetic. Advanced microprocessors incorporate this function on the chip.

matrix   1. A computer logic network that consists of a rectangular array of intersections of input/output leads, with diodes, fuses, antifuses, or other circuit elements connected at some of these intersections. The network usually functions as an encoder or decoder.   2. The section of a color television transmitter that transforms the red, green, and blue camera signals into color-difference signals and combines them with the chrominance subcarrier. It is also called a color coder, color encoder, or encoder.   3. The section of a color television receiver that transforms the color-difference signals into the red, green, and blue signals needed to drive the color picture tube. It is also called a color decoder or decoder.   4. A set of mathematical elements arranged in rows and columns for solving certain types of problems.

matrixing   The process of performing a code conversion with a matrix, as in converting color television signal components from one form to another.

matrix printer   *Dot-matrix printer.*

matrix sound system   A quadraphonic sound system whose four input channels are combined into two channels by a coding process for recording or for stereo FM broadcasting. They are decoded back into four channels for playback of recordings or for quadraphonic stereo reception. The matrix system can be referred to as a 4-2-4 system, in contrast to discrete quadraphonic sound, which is a 4-4-4 system because four channels are used throughout. Matrix examples include the SQ system developed by CBS and the QS system developed by Sansui.

MATV   Abbreviation for *master-antenna television.*

MAU   Abbreviation for *multistation access unit.*

maximum average forward current   The highest average forward current that can flow through a diode for a given junction temperature.

maximum output   The greatest average output power delivered to the rated load of a receiver or amplifier, regardless of distortion.

maximum sound pressure   The maximum absolute value of the instantaneous sound pressure at a point during any given cycle. It is expressed in microbars.

maximum undistorted output   The greatest average output power into the rated load of an amplifier whose distortion does not exceed a specified limit when the input is sinusoidal. It is also called maximum useful output.

maximum usable frequency [MUF]   The upper limit of the frequencies that can be transmitted at a specified time for point-to-point radio transmission that depends on propagation by reflection from the ionized layers of the ionosphere. Higher frequencies can be transmitted only by sporadic and scattered reflections.

maximum useful output   *Maximum undistorted output.*

maxwell [Mx]   The CGS unit of magnetic flux. It is equal to 1 G/cm^2, or to one magnetic line of force. The SI unit, the weber, is now preferred.

Maxwell-Boltzmann law   A law that gives the distribution of velocities among the molecules of a perfect steady-state gas.

Maxwell-Boltzmann statistics   Statistics that represent the distribution of particles among the various possible energy levels at such high temperatures that a large number of energy levels are excited.

Maxwellian distribution   The velocity distribution of the molecules of a gas in thermal equilibrium.

Maxwell inductance bridge   A four-arm AC bridge that compares inductances. Bridge balance is independent of frequency.

Maxwell mutual-inductance bridge   An AC bridge that measures mutual inductance in terms of self-inductance. Bridge balance is independent of frequency.

Maxwell triangle   The equilateral-triangle form of a chromaticity diagram in which the primary colors are represented at the corners of the triangle.

Maxwell-Wien bridge   A four-arm AC bridge widely used for accurate inductance measurements. It depends on a capacitance standard rather than a resistor. Bridge balance is independent of frequency.

mayday [French m'aider]   The international radiotelephone distress signal for ships, aircraft, and spacecraft.

Mb   Abbreviation for *megabit.*

mbar   Abbreviation for *millibar.*

Mbar   Abbreviation for *megabar.*

MBE   Abbreviation for *molecular-beam epitaxy.*

MBGA   Abbreviation for *micro ball-grid array.*

Mb/s   Abbreviation for *megabit per second.*

Mbyte   Abbreviation for *megabyte* (one million bytes).

Mbyte/s   Abbreviation for *megabyte per second.*

Mc   Abbreviation for the obsolete term *megacycle.* It has been replaced by megahertz, abbreviated as MHz.

MCC   Abbreviation for *Microelectronics and Computer Technology Corporation.*

mcd   Abbreviation for *milicandela.*

MCGA   Abbreviation for *multicolor graphics array.*

MCIF   Abbreviation for *Miniature Card Implementer's Forum,* a standards organization for memory card manufacturers.

MCM   Abbreviation for *multichip module.*

MCT   Abbreviation for *MOS-controlled thyristor.*

MCU   Abbreviation for *microcontroller* or *microcomputer.*

MCW   Abbreviation for *modulated continuous wave.*

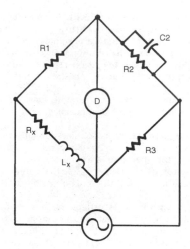

Maxwell-Wien bridge measures inductance accurately with a capacitance standard.

MDAC   Abbreviation for *multiplying digital-to-analog converter.*

m-derived filter   A filter whose design is based on a constant *m* which is derived from the constant K. See *constant-k filter.*

m-derived section   A T or pi network section designed so that when two or more sections are joined in a filter unit, their impedances are matched at all frequencies, although the sections can have different resonant frequencies.

mean carrier frequency   The average carrier frequency of a transmitter, corresponding to the resting frequency in a frequency-modulation system.

mean free path   1. The average distance that a particle travels between successive collisions.   2. The average distance that sound waves travel between successive reflections in an enclosure.

mean free time   The average time between successive collisions of a particle.

mean lethal dose   *Median lethal dose.*

mean life   The average time that an atom or other system exists in a particular form. For a radionuclide, the mean life is the reciprocal of the disintegration constant. Mean life is 1.443 times the radioactive half-life. In a semiconductor, mean life is the time required for injected excess carriers to recombine with others of the opposite sign. It is also called average life or lifetime.

mean pulse time   The arithmetic mean of the leading-edge pulse time and the trailing-edge pulse time.

mean range   The range that is exceeded by half the particles under consideration.

mean time between failures [MTBF]   A measure of quality expressed in hours of life, stating the relative reliability of a component, circuit, or system. MTBF data is either based on data derived from actual operating conditions or calculated from data in military standard handbook MIL-HDBK-217E.

measurand   In instrumentation, the variable that is being measured. Examples include temperature, pressure, flow rate, frequency, voltage, current, and power.

measurement The determination of the magnitude, amount, or other parameter of a characteristic or quantity.

mechanical axis One of the Y axes in a quartz crystal. There are three, each perpendicular to one pair of opposite sides of the hexagon.

mechanical filter A filter that consists of shaped metal rods which act as coupled mechanical resonators when used with piezoelectric or magnetostrictive input and output transducers. It is used in IF amplifiers of highly selective receivers.

mechanical joint A joint that is made by clamping one conductor to another conductor or to a terminal mechanically, without soldering them.

mechanical modulator A device that varies a carrier wave by moving some part of a circuit element. Examples include motor-driven capacitor plates and motor-driven choppers.

mechanical scanning A scanning method that employs a beam of light controlled by a rotating scanning disk, rotating mirror, or other mechanical device to break up a scene or image into a rapid succession of narrow lines, as required for conversion into electric pulses.

mechatronics A method for designing subsystems of electromechanical products by the simultaneous simulation of electronic, mechanical, and hydraulic elements to ensure optimum system performance. Computer-simulated components are modeled as graphical representations for the creation of schematics which become simulator input. The schematics are used by the simulator to determine which models are needed and how they are to be interconnected.

median The value at the halfway point in a series, wherein there is the same number of values above as below the median.

medical electronics A branch of electronics that develops electronic instruments and equipment for such medical applications as diagnosis, therapy, research, anesthesia control, cardiac control, and surgery.

medium frequency [MF] A Federal Communications Commission designation for a frequency in the band from 300 to 3000 kHz, corresponding to a hectometric wave between 100 and 1000 m.

medium-scale integration [MSI] A term generally applied to integrated circuits with 100 to 1000 transistors or 10 to 100 equivalent gates. MSU also applies to memory devices with less than 1024 bits of memory. See *LSI, SSI,* and *VLSI.*

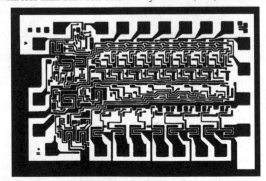

Medium-scale integration: The metallization mask has connections for about 70 equivalent gates of a scaler.

mega- [M] A prefix representing 10^6, or 1,000,000.

megabar [Mbar] An absolute unit of pressure equal to 1,000,000 bars. One megabar is almost exactly equal to normal atmospheric pressure.

megabit [Mb or Mbit] Approximately one million bits, or 1,048,576 bits.

megabit per second [Mb/s] One million bits per second for specifying the modulation rate of a digital transmission system.

megabyte [MB or Mbyte] 1024 kbytes, or 1,048,576 bytes or 8,388,608 bits.

megabyte per second [Mbyte/s] One million bytes per second, for specifying the speed of a digital transmission system.

megacurie [MCi] One million curies.

megacycle [Mc] One million cycles per second. It is now called megahertz and abbreviated MHz.

megaelectronvolt [MeV] One million electronvolts.

megagauss [MG] One million gauss.

megagauss physics The production, measurement, and application of megagauss fields, as produced by discharge of capacitor banks or explosive flux-compression techniques.

megahertz [MHz] One million hertz. It was formerly called a *megacycle.*

megampere [MA] One million amperes.

megarad A dose of radiation equal to 10^6 rads. It is sometimes stated as megarad (Si) or megarad (SiO), indicating the equivalent material absorbing the radiations. See *rad.*

megavolt [MV] One million volts.

megavoltampere [MVA] One million voltamperes.

megawatt [MW] One million watts.

megawatthour [MWh] One million watthours.

Megger A trademark of James G. Biddle Co. for its high-range ohmmeter, which has a hand-driven DC generator as its voltage source. It can measure insulation resistance values in megohms.

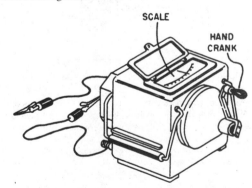

Megger. An instrument for measuring high resistance values.

megohm [MΩ] One million ohms.

megohm-farad A rating for insulation resistance of energy-storage capacitors.

megohmmeter An ohmmeter that measures resistance in megohms, gigohms, and teraohms. High test voltages are required, sometimes more than 1 kV for the highest ranges.

Meissner effect When a superconductor is cooled below the temperature required for superconductivity, the material appears to become perfectly diamagnetic. The induced magnetization opposes the applied magnetic field so effectively that there is no magnetic field in the material.

mel   A unit of pitch. By definition, a simple 1-kHz tone 40 dB above a listener's threshold produces a pitch of 1000 mels. The pitch of any sound that is judged by the listener to be *n* times that of the 1-mel tone is *n* mels.

memory   A device that stores information in electrical, magnetic, or optical form. In computers, memories accept and hold binary numbers only. Among the many forms of memory are charge-coupled devices (CCD), magnetic bubble memories (MBM), random-access memories (RAM), read-only memories (ROM), floppy disks, and Winchester hard disks.

memory capacity   *Storage capacity.*

memory card   A solid-state memory device packaged as a plug-in card module. There are three kinds: flash, dynamic random-access [DRAM], and static random access [SRAM]. Both DRAM and SRAM cards require a power source. (It can be a battery for SRAM cards.) Flash cards do not require power. The modules are typically packaged in PCMCIA standard Type I, II, and III cases. The cards are components in digital cameras, personal digital assistants [PDAs], pagers, and cellular telephones. See also *memory module.*

memory cell   A single storage element of a memory, together with associated circuits for storing and reading out 1 bit of information.

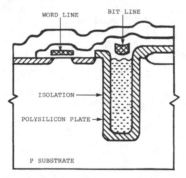

Memory cell structure based on a 0.7-μm CMOS process.

memory integrated circuit   A memory that consists of cell matrixes that include address selection and amplification circuits. It stores information as binary numbers. See *volatile memory, nonvolatile memory, field-programmable memory, erasable memory, random-access memory,* and *read-only memory.*

memory module   An assembly of four to six semiconductor integrated circuit memory devices mounted on a circuit board for insertion onto an existing computer circuit board to supplement the computer's memory. It is typically organized as a plug-in, single-in-line (SIP) component.

MEMS   Abbreviation for *microelectromechanical systems.*

menu   In computer technology, command options available to the computer user in the form of either a row of icons or a written list.

MEO   Abbreviation for *medium earth orbit,* the medium altitude orbit of a satellite orbiting the earth. See also *low earth orbit* [LEO].

mercuric telluride   [HgTe]   An intermetallic compound that has characteristics similar to those of indium antimonide.

mercury   [Hg]   A silvery white liquid metal that becomes a solid at −40°C. It is used in mercury switches and electron tubes because the vapor of mercury ionizes readily and conducts electricity. The green line of mercury 198 very closely approaches pure monochromatic light. Its atomic number is 80.

mercury arc   An electric discharge through ionized mercury vapor; it gives off a brilliant bluish green light that contains strong ultraviolet radiation.

mercury battery   A battery consisting of mercury cells.

mercury cell   A *primary* low-voltage power cell, typically made as a *button cell* for powering watches, hearing aids, and other low-power portable circuits. It has a nominal voltage of 1.3 V and energy density of 100 Wh/Kg or 470 Wh/L. It is being replaced by other button cells such as silver-oxide and zinc/air because mercury is toxic. It is also called the *zinc-mercuric-oxide cell.*

Mercury cell with flat pellet structure. Other versions use cylindrical structures.

mercury delay line   An acoustic delay line with mercury as the medium for sound transmission.

mercury relay   A relay containing mercury that when moved by a magnetic plunger connects the relay contacts together.

mercury switch   A switch that is closed by tilting the switch body so that a large globule of mercury moves across the contacts to bridge them.

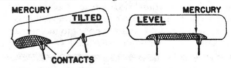

Mercury-switch construction and operation.

mercury thermostatic switch   A thermostatic switch in which heat causes mercury to expand and complete a circuit between contacts that project into the mercury column.

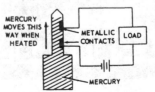

Mercury thermostatic switch.

mercury-vapor lamp A lamp whose light is produced by an electric arc between two electrodes in an ionized mercury-vapor atmosphere. It gives off a blue-green light that is rich in ultraviolet radiation.

mercury-vapor rectifier *Mercury-arc rectifier.*

mercury-vapor tube A gas tube whose active gas is mercury vapor.

mercury-wetted reed relay A reed relay in which the contacts are covered with a mercury film, just as in a mercury-wetted reed switch.

mercury-wetted reed switch A reed switch capsule that contains a pool of mercury at one end and is normally operated vertically. The mercury keeps the contacts on the reeds covered with a mercury film by capillary action. Each operation of the switch renews this mercury film contact, thereby increasing the operating life of the switch many times. In another version, mercury-wetted contacts are obtained without a mercury pool, permitting mounting in any position.

merge To collate two or more similarly ordered sets of values into one set that has the same ordered sequence.

mesa The raised area that remains when semiconductor material is etched away for access to regions beneath the surface.

mesa diffusion A method of growing PN junctions by creating a single base region over the entire surface of a semiconductor slice, then etching away valleys between emitters, to leave islands or mesas of processed material for use as transistor elements.

mesa transistor A transistor whose germanium or silicon wafer is etched down in stages so the base and emitter regions appear as physical plateaus above the collector region.

MESFET Abbreviation for *metal semiconductor field-effect transistor.*

mesosphere A region of decreasing temperature immediately above the stratosphere, extending roughly from an altitude of 60 to 90 km.

message An oral or written communication made in any plain or secret language or code; it has a definite beginning or end. In a computer, it can be a group of words transported or processed as a unit in binary or other coded form.

metal-backed screen *Aluminized screen.*

metal-ceramic *Cermet.*

metal detector An electronic circuit that detects concealed metal objects, such as guns, knives, or buried pipe lines, generally by radiating a high-frequency electromagnetic field and detecting the change produced in that field by the ferrous or nonferrous metal object being sought. It is also called an *electronic locator,* a metal locator, or a radio metal locator.

metal-film resistor A resistor whose film of a metal, metal oxide, or alloy is deposited onto an insulating substrate to form an integrated circuit or discrete resistor.

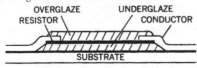

Metal-film resistor on insulating substrate.

metal-glaze resistor A thick-film resistor made by mixing fine metal particles and powdered glass and placing it on a ceramic substrate by dipping, brushing, or spraying, before firing it to produce a glaze. The resistance value can be adjusted or increased by grinding a spiral groove. The metal particles are generally equal parts of palladium and silver. The cermet resistor is an example.

metal-insulator semiconductor [MIS] A semiconductor construction in which an insulating layer, generally a fraction of a micrometer thick, is deposited on the semiconducting substrate before the pattern of metal contacts is applied. It can produce a field-effect region at the surface of the semiconductor material. Applications include capacitors, diodes, field-effect transistors, luminescent diodes, microstrip devices, varactors, and other semiconductor devices.

metallic antenna lens A lens that consists of contoured parallel metal surfaces placed in front of an antenna to focus the beam. The vanes change the phase velocity in proportion to the distance traveled between metal surfaces at each part of the lens.

metallic insulator A shorted quarter-wave section of a transmission line that acts as an extremely high impedance at a frequency corresponding to its quarter-wavelength. It can act as a mechanical support.

metallization 1. The process of depositing a thin film of conductive metal onto an appropriately masked semiconductor substrate to obtain the desired pattern for cell or gate interconnections. Metal layers are typically 1 to 2 μm thick. 2. A thin film of metal sputtered on the masked surface of a ceramic substrate of a resistor network, resistance-capacitance networks, or hybrid circuit to form conductive traces in a set pattern to connect elements of deposited resistors or capacitors to external pins.

metallized resistor A resistor made by depositing a thin film of high-resistance metal on a glass surface, ceramic rod, or tube.

metal locator *Metal detector.*

metal master *Original master.*

metal negative *Original master.*

metal-nitride-oxide semiconductor [MNOS] A semiconductor structure that has a double insulating layer (instead of the usual silicon dioxide gate insulator found in metal-oxide semiconductor structures). Typically, a layer of silicon dioxide (SiO_2) is nearest the silicon substrate, with a layer of silicon nitride (Si_3N_4) over it. The ability of the double insulating layer to store charges makes it useful in memory transistor arrays, capacitors, and other semiconductor devices.

metal-organic vapor-phase epitaxy [MOVPE] The formation of an epitaxial layer by chemical reaction between gaseous compounds of the layer constituents. For indium and gallium deposition the reacting gases are alkyls, and in phosphorous and arsenic they are hydrides.

metal-oxide resistor A metal-film resistor whose metal oxide such as tin oxide is deposited as a film on an insulating substrate.

metal-oxide semiconductor [MOS] A metal-insulator-semiconductor structure in an insulating layer of an oxide of the substrate material. For a silicon substrate, the insulating layer is silicon dioxide (SiO_2). Field-effect transistors, capacitors, resistors, and other semiconductor devices are made with this structure. MOS processes include CMOS, DMOS, NMOS, PMOS.

metal-oxide semiconductor capacitor [MOS capacitor] An integrated-circuit capacitor with its N⁺ region diffused into the silicon substrate to form the bottom electrode. A controlled thickness of silicon oxide is formed on this, to serve as the dielectric. The top electrode is a layer of metal that is deposited at the same time as the interconnections for the integrated circuit. Maximum capacitance values are limited to a few hundred picofarads.

metal-oxide semiconductor field-effect transistor [MOS-FET] A voltage-driven field-effect transistor with an oxide-insulated metal or polycrystalline gate. A unipolar transistor, its operation depends on the movement of one majority carrier (electrons in N-type, holes in P-type), not both as does the bipolar junction transistor. The MOSFET controlling gate voltage is applied to the channel region across the insulating layer, rather than across a PN junction as in the JFET. The gate is insulated with silicon oxide (SiO₂) or silicon nitride (SiN). A MOSFET can be P- or N-channel, and it can operate in either the depletion or enhancement mode. MOSFETs are included in CMOS and BiCMOS digital logic. A MOSFET offers low power drain because of insulation between its source and drain, simpler processing than for a JFET, economical use of silicon, and easier interconnection on a chip. It is also called an insulated-gate field-effect transistor (IGFET). See *field-effect transistor* [FET], *junction field-effect transistor* [JFET], and *metal semiconductor field-effect transistor* [MESFET].

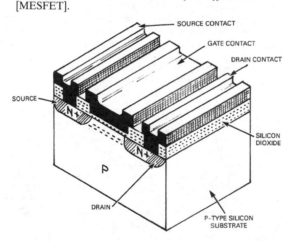

Metal-oxide semiconductor field-effect transistor (MOSFET): N-channel enhancement-mode MOSFET.

metal-oxide semiconductor resistor [MOS resistor] A metal-oxide semiconductor field-effect transistor that replaces a resistor in an integrated circuit.

metal-oxide semiconductor transistor [MOS transistor or MOST] *Metal-oxide semiconductor field-effect transistor.*

metal-oxide silicon device A diode, capacitor, or other semiconductor device with a metallic oxide such as silicon dioxide that serves as an insulating layer.

metal-oxide varistor [MOV] A variable resistor, *varistor,* for protecting AC-powered circuitry from transient spikes or overvoltage. Molded from zinc-oxide and furnace fired to form a monolithic block. MOVs have

symmetrical voltage-current curves that suppress bidirectional transients by a clipping. Equivalent to two back-to-back Zener diodes, a MOV can absorb up to 10 kJ. It can be packaged as a radial-leaded disk or block with terminal posts. See *transient voltage suppressor* (TVS) and *surgector.*

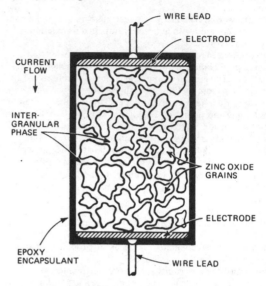

Metal-oxide varistor construction.

metal semiconductor field-effect transistor [MESFET] A field-effect transistor with its metal gate formed directly on its semiconductor channel, which is effectively a Schottky diode. It differs from a MOSFET whose gate is separated from its channel by an insulating oxide layer. It is made from group III–V materials, such as gallium arsenide (GaAs) or indium phosphide (InP).

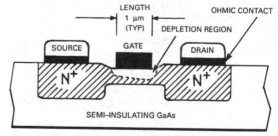

Metal semiconductor field-effect (MESFET) transistor made from gallium arsenide has a structure similar to a silicon MOSFET.

metal-tank mercury-arc rectifier A mercury-arc rectifier whose anodes and the mercury cathode are enclosed in a metal container or chamber.

metal transistor can A standardized metal can for *hermetic* sealing of semiconductor devices and miniature relays.

meteoric scatter A form of scatter propagation in which meteor trails scatter radio waves back to earth. Two radio

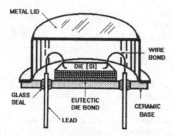

Metal transistor can standardized for the packaging of military and some industrial semiconductor devices.

links, working in opposite directions, are used. Any message transmitted on one link is sent back on the other, so the sender can verify satisfactory reception. The sender transmits only the first character of the message until it is received and returned as a result of scatter by a meteor. The desired message, previously recorded on magnetic tape, is then transmitted at high speed as a burst that can last from a fraction of a minute to several minutes, depending on the size and course of the meteor.

meteorograph An instrument that measures and records meteorological data such as air pressure, temperature, and humidity. When carried aloft, it is also called an aerograph. When used with a radio transmitter, it is called a radiosonde.

meter 1. A device that measures the value of a quantity under observation. The term meter is usually applied to an indicating instrument alone, such as a voltmeter or an ohmmeter. 2. [m] The basic unit of length in the metric system, equal to 39.37 in or 3.28 ft. A meter is equal to 100 cm or 1000 mm. A meter was defined in 1960 as 1,650,763.73 wavelengths of orange-red line of krypton 86.

meter-kilogram-second-ampere unit [MKSA unit] A practical absolute electrical unit based on the meter, kilogram, second, and ampere as fundamental units.

meter-kilogram-second unit [MKS unit] An absolute unit based on the meter, kilogram, and second as fundamental units.

meter-type relay A relay with a meter movement that has a contact-bearing pointer which moves toward or away from a fixed contact mounted on the meter scale. It is also called a contact-making meter or an instrument-type relay.

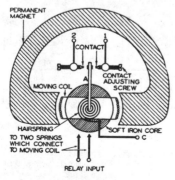

Meter-type relay.

metric sabin A unit of sound absorption for a surface, equivalent to 1 m^2 of perfectly absorbing surface.

metric wave A radio wave between the wavelength limits of 1 and 10 m, corresponding to the very high-frequency (VHF) range of 30 to 300 MHz.

metrology The science of measurement.

MeV Abbreviation for *megalectronvolt.*

MEW Abbreviation for *microwave early warning.*

MF Abbreviation for a *medium frequency* and *multiple feedback.*

MFC Abbreviation for *multifrequency code.*

MF-TDMA Abbreviation for *multiple-feedback, time-division multiaccess.*

mG Abbreviation for *milligauss.*

MG Abbreviation for *megagauss.*

MGD Abbreviation for *MOS gate driver.*

mH Abbreviation for *millihenry.*

MHD Abbreviation for *magnetohydrodynamics.*

mho The former unit of conductance and admittance, now replaced by the siemens as the SI unit of conductance. Both are the reciprocal of the ohm. A resistance or impedance of 1 Ω is equal to a conductance or admittance of 1 siemens.

MHz Abbreviation for *megahertz.*

mi Abbreviation for *mile.*

MIC Abbreviation for *microwave integrated circuit.*

mica A transparent mineral that splits readily into thin sheets which have excellent insulating and heat-resisting qualities. It serves as the dielectric in mica capacitors and as electrode spacers in electron tubes.

mica capacitor A capacitor whose dielectric is mica.

Michelson interferometer An interferometer that uses a half-reflecting dielectric sheet, a fixed sheet-metal reflector, and a movable sheet-metal reflector to give accurate frequency measurements at millimeter wavelengths. When the two reflectors are the same distances from the 45° half-reflecting sheet the path lengths are identical, and the two

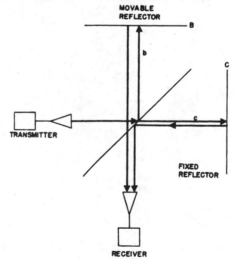

Michelson interferometer with a 45° half-reflecting dielectric sheet at its center and two metal-sheet reflectors for frequency measurements at 35 GHz. The open arrows are antenna symbols.

waves arrive at the receiver in phase to give a power maximum. As the movable reflector is moved toward or away from the half-reflecting sheet, alternate maxima and minima of detected power are observed at the receiver.

MICR Abbreviation for *magnetic-ink character recognition*.

micro *Microcomputer*.

micro- 1. [μ] A prefix that represents 10^{-6} or one-millionth. 2. A prefix that indicates smallness, as in microwave. 3. A prefix that indicates extreme sensitivity, as in microradiometer.

microactuator Any of a class of microminiature actuators fabricated by *micromachining* techniques. These include *micromotors, microvalves* and *microresonators*.

microammeter An ammeter whose scale is calibrated to indicate current values in microamperes.

microampere [μA] One-millionth of an ampere.

micro ball-grid array [MBGA] A miniature *ball grid array* [BGA] package for surface mounting ICs.

microbar [μbar] One-millionth of a bar. The microbar was the unit of pressure formerly used in acoustics. The *newton per square meter* is now the SI unit of pressure.

microcell The geographical area, generally circular, in which a single cellular telephone transmitter/receiver station is effective.

microchannel plate A mosaic of several million fine glass tubes with metallized ends that multiplies electrons in *image-converter* and *image-intensifier tubes* when a high voltage is applied across the ends. Electron gains in excess of 100,000 are achieved. The plates are included in *night-vision binoculars, night-vision scopes,* and *oscilloscope* CRTs where they increase beam current delivered to the phosphor. It is also called a *microchannel plate array* [MCP].

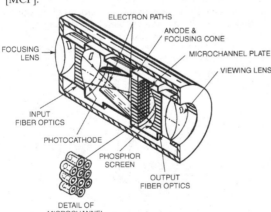

Microchannel plate is a mosaic of microminiature glass tubes for multiplying electrons in an image-intensifier tube.

microcircuit *Integrated circuit* or *hybrid circuit*.

microcode The lowest level of instructions that directly control the interaction of a microprocessor's computing elements or machine instructions built permanently into controller circuitry.

microcomputer 1. A computer with a central processing unit (CPU) that is a *microprocessor*. Examples include the personal computer (PC) and engineering workstation.

2. The term might also refer to a microcontroller. See *microcontroller* (MCU).

microcontroller [MCU] A monolithic integrated circuit with a complete central processing unit (CPU) and enough semiconductor memory (RAM, ROM, EPROM, or EEPROM) and input-output (I/O) capability to be considered as equivalent to a "computer on a chip." Because of its limited memory, a MCU is usually used in control rather than data-processing applications.

microdensitometer A high-sensitivity densitometer used in spectroscopy to detect spectrum lines too faint on a negative to be seen by the human eye.

microelectromechanical systems [MEMS] A class of microminiature machines, actuators, and sensors fabricated by micromachining, primarily from silicon. Practical products made as MEMs now include accelerometers, fuel injectors, ink-jet nozzles, and pressure sensors. See *micromachining*.

microelectronics The technology of constructing and using electronic circuits and devices in extremely small packages through the use of integrated-circuit or other special manufacturing techniques. It is also called microminiaturization.

microfarad [μF] One-millionth of a farad.

microhenry [μH; plural microhenrys] One-millionth of a henry.

microhertz [μHz] One-millionth of a hertz.

microhm [μΩ] One-millionth of an ohm.

microlithography The process of writing thin-line patterns on the surface of a semiconductor wafer to form an integrated circuit. A precise form of lithography, it can define micrometer-wide features on a photoresist-coated substrate. The photoresist is exposed to ultraviolet light, electron beams, or X-rays through a *mask*. See *lithography* and *mask*.

microlock A phase-locking loop system for transmitting and receiving information by radio with reduced bandwidth. It is employed in tracking satellites and telemetering data to ground stations at line-of-sight distances as great as 3000 mi (4800 km).

microlock network A network of radio stations that uses microlock equipment to track missiles and satellites.

microlux [μlx] One-millionth of a lux.

micromachining The fabrication of microminiature sensors and actuators with micrometer-scale dimension integrated circuit fabrication methods. These include photolithographic masking, deposition, and chemical etching. Among the devices developed as R&D experiments

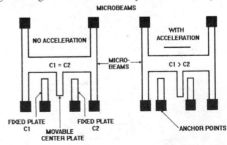

Micromachined capacitive accelerometer has two fixed outer plates and a movable center plate. When accelerated, the center plate is deflected, changing the capacitance of the accelerometer in proportion to the acceleration.

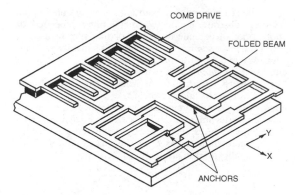

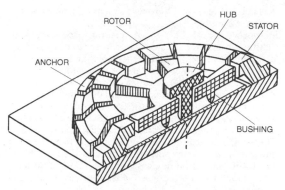

Micromachined linear resonator consists of a pair of folded beams that are vibrated in the *X* direction by an electrostatically driven comb structure. Lateral or *Y*-direction motion is restrained by the geometry of the folded beams.

Micromotor made by micromachining techniques is driven by electrostatic rather than electromagnetic forces.

are electric motors and even a microminiature vehicle. Practical products now in production include accelerometers, pressure sensors, fuel injectors, and ink-jet nozzles. Most are made from silicon or aluminum deposited on silicon, but diamonds have been used as a raw material. The processes are classified as either *bulk* or *surface*. See also *bulk micromachining* and *surface micromachining*.

micrometeorite   A fine dust particle, composed mainly of iron and silicates as in larger meteorites, distributed throughout the solar system. It is generally too small to do structural damage to spacecraft, but it has an erosive effect on exposed surfaces and is a potential hazard to unprotected astronauts.

micrometer [μm]   One-millionth of a meter, a unit now used for specifying wavelengths of light. Visible light is in the range from about 0.4 μm for purple to 0.75 μm for red. The angstrom was also used for specifying wavelengths of light: 1 Å is equal to 10^{-4} μm, so 1 μm is equal to 10,000 Å. It is also called a micron.

micromho [μmho]   One-millionth of a mho.

micromicro- [μμ]   A prefix that represents one-millionth of a millionth, or 10^{-12}. It is now called *pico-*.

microminiaturization   The process of converting discrete or hybrid circuits to integrated circuits.

micromotor   A subminiature motor made by *micromachining* processes that is driven electrostatically rather than electromagnetically. Experimental versions have diameters of 0.1 to 0.2 mm and are about 4 to 6 μm high. It is expected to find applications in intrusive medical instruments.

micron [μ]   One-millionth of a meter. This term for specifying wavelengths of light and thin-layer thicknesses was abrogated by international agreement in 1967, and *micrometer* is now preferred.

microoptics   The technology for constructing and applying optical devices manufactured in extremely small packages by using integrated-circuit or other special manufacturing techniques.

microphone   An electroacoustic transducer that responds to sound waves and delivers essentially equivalent electric waves. It is also called a mike (slang).

microphone boom   An overhead extension arm that supports a microphone within range of the sound to be picked up but outside the range of a television camera.

microphone button   A button-shaped telescoping container that is filled with carbon particles and serves as the resistance element of a carbon microphone.

microphone cable   A special shielded cable for connecting a microphone to an audio amplifier.

microphone mixer   A mixer that feeds two or more microphones into the input of an AF amplifier. Separate controls permit adjusting the output level of each microphone.

microphone preamplifier   An AF amplifier that amplifies the output of a microphone before the signal is sent through a transmission line to the main AF amplifier. The preamplifier can be built into the microphone housing or stand.

microphone shield   A protective foamed plastic covering that protects a microphone diaphragm from condensed moisture originating from the operator's breath when in use, or protects it from rain, fog, or sleet.

microphone stand   A stand that supports a microphone in a desired position above the floor or on a table.

microphone transformer   An iron-core transformer that couples certain types of microphones to a microphone preamplifier, to a transmission line, or to the main AF amplifier.

microphonic   Vulnerable to vibration that produces microphonics. A tube in a radio receiver is microphonic if a pinging sound is heard from the loudspeaker when the side of the tube is tapped with a finger.

microphonics   Noise caused by mechanical vibration of the elements of an electron tube, component, or system. The vibration causes modulation of the signal currents flowing through or controlled by the vibrating device. It can be heard as noise in an AF system and seen as an undesirable interference pattern in facsimile and television images.

microphotometer   A photometer that provides highly accurate illumination measurements. In one form, the changes in illumination are picked up by a phototube or photosensor and converted into current variations that are amplified.

micropower circuit   A circuit whose individual component power consumption is down to microwatts or less, reducing heat generation so greatly that high packaging densities become feasible. Microwatt circuits and nanowatt circuits are examples.

microprocessor [MPU] 1. A central processing unit (CPU) fabricated on a single large integrated circuit chip, containing the basic arithmetic, logic, and control elements of a computer that are required for processing data. 2. An IC that accepts coded instructions, executes the instructions received, and delivers signals that describe its internal status. The instructions can be entered or stored internally. The MPU must be supplemented by other peripheral chips such as input/output (I/O) logic devices and semiconductor memory to meet the criteria of a computer. MPUs are classified as 4-bit, 8-bit, 16-bit, and 32-bit devices with 16-bit and 32-bit units now made by CMOS processing. See *microcomputer* and *microcontroller.*

micropump A microminiature pump fabricated by micromachining techniques, one of a class of *microactuators.* The diaphragms can be moved by embedded piezoelectric films, electrostatic forces, or by thermal expansion. It is expected to find applications in biomedical research.

microrad One-millionth of a rad.

microradiography The radiography of small objects that have details too fine to be seen by the unaided eye, with optical enlargement of the resulting negative.

microradiometer A radiometer that measures weak radiant power, in which a thermopile is supported on and connected directly to the moving coil of a galvanometer.

microreflectometer A reflectometer that measures the reflectance of very small areas of an image.

microrem [μrem] One-millionth of a rem.

microroentgen [μR] One-millionth of a roentgen.

microsecond [μs] One-millionth of a second.

microseismograph A microseismometer that has recording facilities.

microsensor A microminiature sensor fabricated by micromachining techniques, one of a class of *microactuators.* One example consists of a pair of folded beams that are set in vibrational motion by an electrostatically driven comb structure. It is expected to find applications in biomedical research.

microsiemens [μS] One-millionth of a siemens.

microstrip line A controlled-impedance microwave frequency transmission line made as conductive metal traces or conductors on one side of a dielectric substrate with a conductive ground plane bonded to the other side. It can

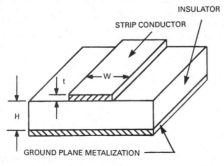

Microstrip transmission line for microwave frequencies has an insulating substrate between its strip conductor and conductive ground plane.

be made as foil bonded to both sides of a glass-fiber epoxy (GFE) substrate, conductive metal films deposited on both sides of ceramic dielectric, or thin film formed on both sides of a semiconductor chip. It is a microwave analogy for a two-wire transmission line.

microswitch *Snap-action switch.*

Micro Switch Trademark of Micro Switch Division of Honeywell for its line of switches.

microvalve A microminiature valve fabricated by micromachining techniques, one of a class of *microactuators.* The diaphragms can be moved by embedded piezoelectric films, electrostatic forces, or by thermal expansion. It has found applications in automotive electronics.

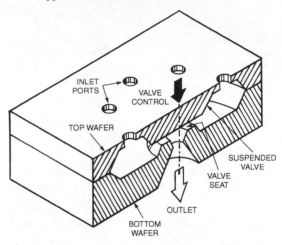

Microvalve made by micromachining techniques has a diaphragm that moves perpendicular to its base substrate by the action of an embedded piezoelectric film, electrostatic forces, or thermal expansion.

microvolt [μV] One-millionth of a volt.

microvoltmeter A voltmeter whose scale is calibrated to indicate voltage values in microvolts.

microvolts per meter A measure of the intensity of the signal produced by a radio transmitter at a given point. It is equal to the signal strength in microvolts at the receiving antenna divided by the effective height of the antenna in meters. Stronger signals are expressed in millivolts per meter.

microwatt [μW] One-millionth of a watt.

microwatt circuit A micropower circuit whose individual components consume only a few microwatts. They are typically made with planar diffusion technology and deposited thin-film technology on the same silicon wafer.

microwave A reference to *microwave frequencies* or *microwave spectrum.*

microwave acoustics The branch of acoustics that deals with acoustic waves traveling through or on the surface of a solid material at microwave frequencies. It is also called *microsonics.*

microwave altimeter An altimeter that operates in the microwave spectrum. A 13.9-GHz version in the Skylab satellite vehicle detected variations of up to 110 km in the surface of the ocean, corresponding to gravity-altering changes in ocean-floor topology.

microwave antenna A combination of an open-end waveguide and a parabolic reflector or horn, to receive and transmit microwave signal beams at microwave repeater stations.

microwave beacon system A system that includes an *interrogator* and an automatic *transponder* that will respond to a microwave interrogation signal and transmit a coded reply that will indicate the position of the transponder on a radar plan-position indicator (PPI) display. This system is used for air-traffic control of commercial aircraft and is useful in search and rescue efforts for aircraft that have been forced down under emergency conditions in remote locations. See also *identification, friend or foe* [IFF] and *interrogator*.

microwave circuit A circuit that operates at microwave frequencies.

microwave circulator *Circulator.*

microwave counter A counter that measures frequency in the microwave spectrum by counting cycles for a precise interval of time and converting the count to the corresponding frequency value in gigahertz on a digital display.

microwave delay line A small solid-state passive device that introduces controllable time delay into a microwave signal line.

microwave detector A device that reacts to the presence of microwave radiation by changing its resistance or some other electrical parameter. One basic type is the crystal detector. Another type is the bolometer, which depends on the temperature-sensitive resistance characteristic of a barretter or thermistor.

microwave diathermy Diathermy operated at frequencies in the microwave spectrum, generally at 2.45 GHz, to heat deep tissues for medical purposes.

microwave diode A diode intended primarily for operation at microwave frequencies. Examples include BARITT, Gunn, IMPATT, LSA, TRAPATT, and varactor diodes. All these microwave diodes convert DC energy directly to microwave energy.

microwave filter A filter that consists of resonant cavity sections or other elements built into a microwave transmission line to pass desired frequencies while rejecting or absorbing other frequencies.

microwave frequencies *Microwave spectrum.*

microwave gyrator *Gyrator.*

microwave heating The heating of food with electromagnetic energy in or just below the microwave spectrum for cooking, dehydration, sterilization, thawing, and other purposes. The operating frequencies most often used are 915 MHz and 2.45 GHz, in the bands assigned for industrial, scientific, and medical purposes.

microwave integrated circuit [MMIC] An analog integrated circuit that contains components capable of amplifying signals at frequencies of 1 GHz or higher. Some silicon MMICs contain bipolar transistors, but gallium arsenide (GaAs) or GaAs on a silicon substrate MMIC, typically based on MESFETs, are capable of operating above 4 GHz. Analog MMICs conserve space and power in airborne phased-array radars.

microwave interferometer An instrument that measures the free-electron density of an ionized gas by measuring transmission of extremely high-frequency radio waves through the gas.

microwave landing system [MLS] A proposed microwave frequency version of the *instrument landing system* [ILS] that will take advantage of greater antenna directivity and a wider available frequency spectrum.

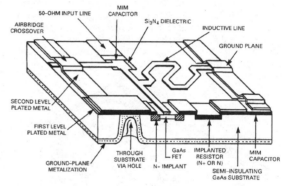

Microwave monolithic integrated circuit (MMIC) has active and passive components formed on a common substrate.

microwave monolithic IC [MMIC] An integrated circuit that operates in the microwave frequency region containing diodes, microwave transistors (MESFETs), resistors, capacitors, and transmission lines on a single chip, typically

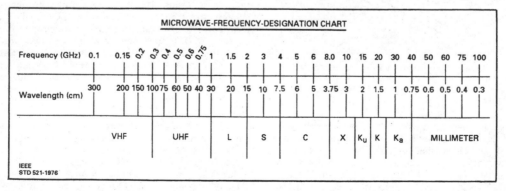

Microwave frequency designation chart.

gallium arsenide (GaAs). MMICs are replacements for microwave hybrid circuits and function as amplifiers, attenuators, or switches.

microwave multiplier phototube A multiplier phototube that can demodulate light beams which are modulated at microwave frequencies. Transmission secondary-emission dynodes amplify the photoelectrons emitted by the photocathode. The resulting bunched photoelectrons inside the tube are focused by an axial magnetic field created by a helix. This electron beam excites an RF wave on the helix that is coupled through the glass envelope of the tube to another short length of helix connected to the output coaxial connector.

microwave oven An appliance that heats food rapidly with microwave energy. It is suitable for both thawing frozen food and cooking most homogenous foods. A *magnetron* tube oscillating at 2.45 GHz radiates microwave energy that is directed through a waveguide into the oven cavity, where the food is cooked within seconds or minutes. The distribution of RF energy within the oven cavity is improved with a rotating fanlike microwave reflector called a *mode stirrer*. Typical power ratings of small ovens are 400 to 500 W while large ovens have ratings of 700 to 800 W. Materials with high moisture content, such as food, will absorb microwave energy. The food molecules align themselves with the microwave energy, and the rapidly changing RF polarity creates friction, causing heating.

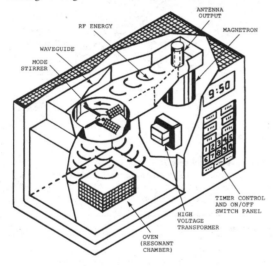

Microwave oven magnetron generates RF power at 2.45 GHz, and its output probe transmits that energy through a waveguide to the resonant oven chamber. The energy is distributed evenly by a rotating mode stirrer.

microwave radiometer A radiometer in which a microwave receiver detects microwave thermal radiation and similar weak wideband signals that resemble noise and are obscured by receiver noise. Most designs are based on the original Dicke radiometer, generally with a latching ferrite circulator to switch back and forth between an internal reference source and the microwave power source being measured.

microwave radiometry Radiometry that depends primarily on measurement of microwave wavelengths which emanate from beneath a surface. The measurements are thus independent of surface irregularities or variations in surface temperatures. One application involves determining the orientation of the line that joins a space vehicle with the center of mass of the earth, the moon, or the sun.

microwave refractometer An instrument that measures the refractivity of the atmosphere, which is proportional to the dielectric constant of the atmosphere. The instrument includes two precision microwave transmission cavities, one hermetically sealed to serve as a reference. Travel time of a microwave signal through each cavity is measured by associated electronic circuits and converted into the refractivity value of the air being sampled.

microwave repeater A radio repeater that forms highly directional radio beams at microwave frequencies to link towers spaced up to 50 mi (80 km) apart. Each tower has a receiver and transmitter for each direction in which signals are picked up, amplified, and passed on.

microwaves *Microwave spectrum.*

microwave spectroscopy Spectroscopy that concentrates on determining of the selective absorption of microwaves at various frequencies by solid or gaseous materials. Its applications include the study of atomic, crystalline, and molecular structures.

microwave spectrum A spectrum of wavelengths between the shortwave region and the far-infrared region, commonly considered to include wavelengths from about 30 to 0.3 cm (1 to 100 GHz). It is also called *microwave frequency.*

microwave system A radio system based on microwave propagation.

microwave transistor A transistor that operates at microwave frequencies. Both bipolar and field-effect transistors can be made to operate efficiently in this part of the frequency spectrum. Microwave transistors are planar NPN silicon or gallium arsenide FETs, with dimensions in the micrometer range.

microwave tube An electron tube that operates at wavelengths in the range of about 30 to 0.3 cm.

MIDI Abbreviation for *Musical Instrument Digital Interface.*

migration 1. The uncontrolled movement of certain metals, particularly silver, from one location to another, usually with associated undesirable effects such as oxidation or corrosion. It can cause serious problems on printed-circuit boards that have gold-plated silver contacts; here the silver migrates through the protective gold plating. 2. The movement of charges through a semiconductor material by diffusion or drift of charge carriers or ionized atoms. 3. The movement of crystal defects through a semiconductor crystal under the influence of high temperature, strain, or a continuously applied electric field.

mike Slang for *microphone.*

mil 1. A unit of angular measurement. A true mil is the angle determined by an arc whose length is one-thousandth of the radius. For practical purposes, the mil is considered to be $\frac{1}{6400}$ (instead of $\frac{1}{6283}$) of 360°. 2. One-thousandth of an inch, 0.00254 cm, 25.4 µm, or 25.4×10^4 Å.

MIL Abbreviation for Military Specification, which replaces JAN specifications.

mile [mi]   A unit of length, equal to 5280 ft or 1.609344 km.

Miller code   A code used internally in some computers, in which a binary 1 is represented by a transition in the middle of a bit (either up or down), and a binary 0 is represented by no transition following a binary 1. A transition between bits represents successive 0s. In this code, the longest period possible without a transition is two bit times.

Miller code for binary data. The space between each pair of dashed vertical lines represents one bit time.

Miller effect   The increase in the effective grid-cathode capacitance of a vacuum tube due to the charge induced electrostatically on the grid by the anode through the grid-anode capacitance.

Miller integrator   A resistor-capacitor charging network that has a high-gain amplifier which parallels the capacitor. Used to produce a linear time-base voltage.

milli- [m]   A prefix that represents 10^{-3}, or one-thousandth.

milliammeter   An ammeter whose scale is calibrated to indicate current values in milliamperes.

milliampere [mA]   One-thousandth of an ampere.

milliampere-hour [mAh]   One-thousandth of an ampere-hour.

millibar [mbar]   One-thousandth of a bar.

millibarn [mb]   One-thousandth of a barn.

millicandela [mcd]   One-thousandth of a candela.

milligauss [mG; plural: **milligauss**]   One-thousandth of a gauss.

milligaussmeter   A gaussmeter that measures magnetic field strength in milligauss.

millihenry [mH; plural millihenrys]   One-thousandth of a henry.

millilumen [mlm]   One-thousandth of a lumen.

millimeter [mm]   One-thousandth of a meter.

millimeter band   A band of frequencies extending from 40.0 GHz (0.75 cm wavelength) to 300 GHz (1 mm wavelength) in accordance with IEEE Standard 521-1976. It was so named because wavelengths are more conveniently measured in millimeters (mm) than centimeters (cm). It corresponds with both the L band (40.0 to 60.0 GHz) and M band (60.0 to 100.0 GHz) in the U.S. Military Joint Chiefs of Staff (JCS) triservice frequency designations (1970).

millimeter of mercury   A unit of pressure for vacuum equipment, approximately equal to 1/760 atmosphere, or 1 torr.

millimeter-wave amplifier   An amplifier that amplifies millimeter waves.

millimeter-wave tube   A tube capable of operating as an oscillator or amplifier at wavelengths of a few millimeters.

millimetric wave   A radio wave between the wavelength limits of 1 and 10 mm, corresponding to the extremely high-frequency (EHF) range of 30 to 300 GHz.

milliohm [mΩ]   One-thousandth of an ohm.

milliohmmeter   An ohmmeter whose scale is calibrated to indicate resistance values in milliohms.

millirad [mrd]   One-thousandth of a rad.

milliradian [mrad]   One-thousandth of a radian, equal to 0.0572957795°.

millirem [mrem]   One-thousandth of a rem.

milliroentgen [mR]   One-thousandth of a roentgen.

millisecond [ms]   One-thousandth of a second.

millisiemens [mS]   One-thousandth of a siemens.

millisone   A unit of loudness equal to 0.001 sone.

millitorr   One-thousandth of a torr.

millivolt [mV]   One-thousandth of a volt.

millivoltage   A voltage whose magnitude is most conveniently expressed in millivolts.

millivoltmeter   A voltmeter whose scale is calibrated to indicate voltage values in millivolts.

millivolts per meter   A rating often used for signal intensities greater than 1000 μV/m.

milliwatt [mW]   One-thousandth of a watt.

Millman's theorem   Any number of constant-current sources that are directly connected in parallel can be converted into a single current source whose total generator current is the algebraic sum of the individual source currents, and whose total internal resistance is the result of combining the individual source resistances in parallel.

Milstar   Abbreviation for military strategic and tactical relay, a proposed network to replace FLTSATCOM and AFSATCOM with virtually unjammable survivable communications satellites. Its primary mode will be EHF, but its UHF capability will allow it to use existing air- and ground-based terminals.

MIM capacitor   A capacitor formed by first depositing a metal layer on a substrate, and then adding an insulating layer and finally a second metal layer. MIM capacitors are formed on some ICs during the processing of other active and passive components or on hybrid circuit substrates.

MIMIC   Abbreviation for *microwave/millimeter wave monolithic integrated circuit* program. A program initiated by the U.S. Department of Defense to encourage the economic manufacture of gallium arsenide (GaAs) microwave integrated circuits for defense applications; the GaAs microwave industry's equivalent of the VHSIC program. See *VHSIC.*

min   Abbreviation for *minute.*

Miniature Card Implementer's Forum [MCIF]   An association that promotes the MCIA memory card as a standard.

miniaturization   Reduction in the size and weight of a system, package, or component with the use of small parts arranged for maximum utilization of space.

miniaturize   To redesign a component or piece of electronic equipment to fit in less space.

minicomputer   A general-purpose stored-program digital computer, capable of performing essentially the same arithmetic, logic, and input/output operations as a large computer. Smaller in size, it has a smaller primary memory, shorter word lengths (usually a 32-bit word size as compared to up to 64 bits per word in large computers), and lower price. A minicomputer is generally operated by a single operator or control system at a time, with no time-sharing.

minimum-access programming   The programming of a digital computer so that minimum waiting time is required to obtain information out of the memory.

minimum resolvable temperature [MRT]   The minimum temperature difference necessary to resolve objects seen in thermal-imaging systems.

minor apex face One of the three smaller sloping faces near but not touching the apex of a natural quartz crystal. The three larger sloping faces are the major apex faces.

minor cycle The time required for the transmission or transfer of one machine word, including the space between words, in a digital computer that uses serial transmission.

minor face One of the three longer sides of a natural hexagonal quartz crystal.

minority carrier The type of carrier that constitutes less than half of the total number of carriers in a semiconductor device. The minority carriers are holes in an N-type semiconductor and electrons in a P-type semiconductor device.

minority-carrier diffusion length The average distance a minority carrier, usually a hole, travels in a solid before combining with a majority carrier, usually an electron.

minority emitter An electrode from which a flow of minority carriers enters the interelectrode region of a transistor.

minor lobe Any lobe except the major lobe of a radiation pattern. It is also called a *secondary lobe* or *side lobe*.

minus zone The bit positions in a computer code that represent the algebraic minus sign.

minute [min] A unit of time, equal to 60 s.

Minuteman II A U.S. Air Force ballistic missile that was designed to launch nuclear warheads but has been converted to launch low earth-orbiting [LEO] satellites.

MIPS Acronym for *million instructions per second,* a measure of performance for computers.

mirror reflection *Direct reflection.*

mirror-reflection echo A radar echo that undergoes multiple reflections, as by reflection from the side of an aircraft carrier or other large flat surface, before being reflected from a nearby target.

MIS 1. Abbreviation for *management information system.* 2. Abbreviation for *metal-insulator semiconductor.*

mismatch The condition in which the impedance of a source does not match or equal the impedance of the connected load or transmission line.

mismatch factor *Reflection coefficient.*

missile An object that is dropped, propelled, or otherwise projected through air or water toward a target. Although a missile is normally intended to destroy or damage a target, it can also serve for photoreconnaissance, detection and meteorographic measurements. When the trajectory of a missile can be changed after launching, by remote control or by an internal control system, it is known as a guided missile.

missile acquisition The process of identifying a missile on a search radar screen and tracking it to determine its range, bearing, and elevation. This information is needed to aim and fire defensive weapons. The acquisition of a target is also called *lock-on.*

missile decoy A vehicle that simulates a missile in flight, attracting enemy radar and drawing fire, to increase the odds for penetration by manned bombers or other weapon systems.

missile plume The region of electromagnetic and other disturbances that follow a missile during reentry and make the missile more readily detectable.

missile range A marked-off course or area for test missile flight under observation.

missing-pulse detector A detector circuit that detects the absence of a pulse by comparing the energy of incoming pulses with that of preceding presumably normal pulses or by comparing the energy content of each pulse with a preset DC reference voltage.

mission The assigned task or objective of a person or group, or of a military, naval, or aerospace operation.

MITI Abbreviation for *Ministry of International Trade and Industry,* a Japanese government agency.

mixed highs The high-frequency signal components that are intended to be reproduced achromatically (without color) in a color television picture.

mixer 1. A device that has two or more inputs, usually adjustable, and a common output. It combines separate audio or video signals linearly in desired proportions, to produce an output signal. 2. The stage in a superheterodyne receiver where the incoming modulated RF signal is combined with the signal of a local RF oscillator to produce a modulated IF signal. Crystal diodes are widely used as mixers in radar and other microwave equipment. They are also called the first detector and the mixer-first detector. The mixer and oscillator together form the converter. 3. A nonlinear device in which two light beams are combined to form new beams that have frequencies equal to the sum or the difference of the input wavelengths. The difference frequency between two light beams can be so low that it falls in the spectrum of radio waves rather than that of light waves.

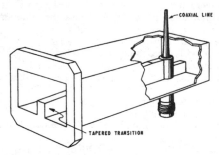

Mixer: the modulated microwave carrier signal in the ridge waveguide is mixed with the oscillator signal in the crystal diode inside the coaxial line to provide an IF output.

mixer-first detector *Mixer.*

mixer transistor A transistor that performs only the frequency-conversion function of a converter in a superheterodyne receiver. It is supplied with voltage or power by a separate local oscillator.

mixing Combining two or more signals, such as the outputs of several microphones.

mixing amplifier An amplifier that has inputs for the application of two or more different signals and a common output which delivers a composite signal.

MKSA unit Abbreviation for *meter-kilogram-second-ampere unit.*

MKS unit Abbreviation for *meter-kilogram-second unit.*

mlm Abbreviation for *millilumen.*

MLS Abbreviation for *microwave landing system.*

mm Abbreviation for *millimeter.*

MMDS Abbreviation for *multichannel multipoint distribution.*

MMIC Abbreviation for *microwave integrated circuit.*

MNOS Abbreviation for *metal-nitride-oxide semiconductor.*

MNP International standards for data compression, such as MNP5.

mobile cellular telephone *Cellular mobile telephone.*

mobile phone A telephone that is permanently installed in a vehicle and depends on the host vehicle's battery for its power. It is also called a *car phone.*

mobile station A radio station intended for operation while its host vehicle, ship, or aircraft is in motion or during halts at unspecified points.

mobile telephone service A service that provides radiotelephone communication from a mobile vehicle to a regular telephone or to another similarly equipped vehicle.

mobile telephone switching office [MTSO] The station that switches signals from a *cellular mobile telephone* to the *landline* or to another cellular mobile telephone. It operates on a trunk-to-trunk basis, and its interconnection scheme is similar to a *private-branch exchange* [PBX].

mobile transmitter A radio transmitter designed for installation in a vessel, vehicle, or aircraft, and normally operated while in motion.

mobile unit A truck or other vehicle equipped with television studio equipment for television pickups at remote locations. Picture and sound signals are usually sent back to the main transmitter by a microwave transmitter and antenna on the truck.

mobility The velocity of a charged particle in response to an applied electric field. Units are $cm^2/V \cdot s$.

mode 1. A state of a vibrating system that corresponds to a particular field pattern and one of the possible resonant frequencies of the system. The three common modes of vibration in a quartz plate are the extensional, flexural, and shear modes. 2. A form of propagation of guided waves that is characterized by a particular field pattern in a plane transverse to the direction of propagation. The field pattern is independent of position along the axis of the waveguide. For uniconductor waveguides the field pattern of a particular mode of propagation is also independent of frequency. The TE mode is the transverse electric mode, and the TM mode is the transverse magnetic mode. Also called transmission mode. 3. One of several alternative methods of operating a system or device.

mode changer *Mode transducer.*

mode filter A selective device that passes energy along a waveguide in one or more modes of propagation and substantially reduces energy carried by other modes.

mode jump A sudden and irregular change in the oscillation frequency and power output of a magnetron, due to a change in the mode of operation from one pulse to the next.

mode locking Locking of the internal cavity modes of a pulsed laser in proper phase and amplitude to divide each output pulse into a train of extremely sharp and equally spaced pulses.

modem [MOdulator-DEModulator] A combination modulator and demodulator located at each end of a telephone line to convert binary digital information to audio tone signals suitable for transmission over the line, and vice versa. With a two-wire line, transmission can be in only one direction at a time (simplex or half-duplex). With a four-wire line, simultaneous transmission is possible in both directions (full duplex), with only one modem at each end. Facsimile modems operate at rates up to 14.4 Kb/s and data modems operate at rates up to 33.6 Kb/s.

mode purity 1. The ratio of power present in the forward-traveling wave of a desired mode to the total power present in the forward-traveling waves of all modes. 2. The extent to which an ATR tube in its mount is free from undesirable mode conversion.

mode separation The frequency difference between resonator modes of oscillation in a microwave oscillator.

mode shift A change in the mode of magnetron operation during the interval of a pulse.

mode skip Failure of a magnetron to fire on successive pulses.

MODFET Abbreviation for *modulation-doped field-effect transistor.*

modifier A quantity that can alter the address of an operand in a computer, such as the cycle index.

modify 1. To alter the address of the operand in a computer instruction. 2. To alter a computer subroutine according to a defined parameter.

modular construction A device, circuit, or system that has been packaged in a case having standardized outline dimensions so that it is directly interchangeable with other products in the same style package. Input, output, and power connections or pins have identical functions to permit easy removal and replacement in the event of their failure or internal circuit design change. Examples include solid-state relays, hybrid circuits, plug-in computer memory cards, and power supplies. However, the characteristics and ratings of products sharing the same style package need not be identical.

modulate To vary the amplitude, frequency, or phase of a wave or the velocity of the electrons in an electron beam in some characteristic manner.

modulated amplifier The amplifier stage in a transmitter where the modulating signal is introduced to modulate the carrier.

modulated carrier An RF carrier whose amplitude or frequency has been varied in accordance with the intelligence to be conveyed.

modulated continuous wave [MCW] A form of emission in which the carrier is modulated by a constant AF tone. In telegraphic service, the carrier is keyed.

modulated light Light that has been made to vary in intensity in accordance with variations in an audio, facsimile, or code signal.

modulated oscillator An oscillator whose input signal varies the output frequency.

modulated stage The RF stage to which the modulator is coupled and in which the carrier wave is modulated by an audio, video, code, or other intelligence signal.

modulated wave A carrier wave whose amplitude, frequency, or phase varies in accordance with the value of the intelligence signal being transmitted.

modulating-anode klystron A klystron that has an electrode between the cathode and its drift-tube section which can turn the electron beam on and off for pulse generation.

modulating signal A signal that causes a variation of some characteristic of a carrier.

modulation The process for varying some characteristic of one wave in accordance with another wave. In radio broadcasting some stations use amplitude modulation; others use frequency modulation. In television, the picture portion of the program is amplitude modulated, and the sound portion is frequency modulated. Other types of modulation include phase, pulse-amplitude, pulse-code, pulse-duration, pulse-frequency, pulse-position, and pulse-time modulation.

modulation capability The maximum percentage modulation that is possible without objectionable distortion.

modulation-demodulation amplifier A unidirectional amplifier with an amplitude modulator that has conversion gain followed by a demodulator. It is applied in wideband microwave systems.

modulation depth The difference in brightness between black and white in a CRT display.

modulation-doped field-effect transistor [MODFET] A transistor whose structure is similar to that of a MOSFET. MODFET is the term used by the University of Illinois, Cornell University, and Honeywell Inc. It has the same meaning as HEMT, TEGFET, or SDHT.

modulation envelope A curve drawn through the peaks of a graph that shows the waveform of a modulated signal. The modulation envelope represents the waveform of the intelligence carried by the signal.

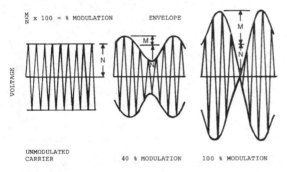

Modulation envelope and modulation percentage.

modulation factor The ratio of the peak variation in the modulation of a transmitter to the maximum variation for which the transmitter was designed. It can be expressed as a percentage. In amplitude modulation, the modulation factor is the ratio of half the difference between the maximum and minimum amplitudes of an amplitude-modulated wave to the average amplitude. For frequency modulation, it is the ratio of the actual frequency swing to the frequency swing required for 100% modulation, expressed in percentage.

modulation frequency The rate of variation of a frequency-modulated signal about the carrier frequency.

modulation index The ratio of the frequency deviation to the frequency of the modulating wave in a frequency-modulation system when a sinusoidal modulating wave is applied.

modulation meter An instrument that measures the modulation factor of a modulated wave train at a transmitter. The readings are usually expressed as a percent.

modulation noise Noise that is caused by the modulating signal, making the noise level a function of the strength of the signal.

modulation percentage The percentage value obtained by multiplying the modulation factor by 100.

modulation transfer function [MTF] The curve that expresses the luminance contrast between black and white lines on a graphic-alphanumeric display screen as the number of lines increases.

modulator A transmitter circuit or device that varies the amplitude, frequency, phase, or other characteristic of a carrier signal in accordance with the waveform of a modulating signal which contains useful information. The carrier can also be a direct current, pulse train, light beam, laser beam, or other transmission medium.

module A standardized, uniform, interchangeable package for electronics that simplifies maintenance, replacement, and storage. A module can be a closed case or a uniformly sized circuit board. Examples include low-voltage power supplies and ATR boxes for aircraft avionics.

module generation The automatic construction of major functions of an integrated circuit by specifying parameters controlling the structure and/or performance of that function. Examples are RAM, ROM, PLA, datapath, state machine, registers, multiplexers, and standard logic families.

modulo *n* check A computer check system in which each number to be operated on is associated with a check number. This check number is equal to the remainder obtained when the number operated on is divided by *n*. Thus, in a modulo 4 check, the remainder serving as the check number will be 0, 1, 2, or 3.

mΩ Abbreviation for *milliohm*.

MΩ Abbreviation for *megohm*.

moire A spurious pattern in a reproduced television picture, resulting from interference beats between two sets of periodic structures in the image. Moires can be produced, for example, by interference between patterns in the original subject and that of the target grid in an image orthicon, or between subject patterns and the line and phosphor-dot patterns of a color picture tube.

mol Abbreviation for *mole*.

MOL Abbreviation for *manned orbiting laboratory*.

molded component An electronic component that has been encased in a molded plastic insulating material such as epoxy to keep out dust and moisture.

mole [mol] The SI unit for amount of substance. One mole is that amount of a substance that contains as many elementary entities as there are atoms in 0.012 kg of carbon 12. When the mole is used, the type of particle (atom, molecule, ion, electron, and so forth) or group of particles must be specified.

molecular-beam epitaxy [MBE] A technique for forming or growing thin films on substrates in ultra-high-vacuum chambers by directing one or more beams of the requisite elements or compounds (atoms or molecules) at the surface of the target crystalline substrate. This film growth technique can produce high-quality crystalline semiconductors, metals, and insulation layers with tight dimensional control. It makes possible some electrical and optical properties not obtainable in natural crystals.

molecular gas laser A gas laser that can operate continuously at output powers in the kilowatt range, with a coherent output at desired frequencies in the entire gas laser spectrum from 0.15 to 773.5 µm. Laser action results from molecular transitions. Pulsed operation is also feasible.

Pumping can be chemical, electric, or optical. The carbon dioxide (CO_2) laser is the most common and most important. Molecular hydrogen lasers operate in the vacuum ultraviolet region, near 0.16 µm.

molecular microwave amplifier A solid-state amplifier whose operation is based on the interaction between uncharged molecular matter and the microwave field. Examples include the maser and the parametric amplifier.

molecular pump A vacuum pump that exhausts the molecules of the gas and carries them away by the friction between the molecules and a rapidly revolving disk or drum.

molecule The smallest particle into which an element or compound can be divided and still retain the chemical properties of the element or compound in mass.

mole fraction The number of atoms of a certain isotope of an element, expressed as a fraction of the total number of atoms of that element, that are present in an isotopic mixture.

molybdenum [Mo] A metallic element, sometimes used for electrodes of electron tubes. Its atomic number is 42.

molybdenum permalloy A high-permeability alloy that consists chiefly of nickel, molybdenum, and iron. A typical formulation has 4% molybdenum, 79% nickel, and 17% iron.

MOMBE Abbreviation for *metal-organic molecular-beam epitaxy.*

momentary-contact switch A switch that returns from the operated condition to its normal circuit condition when the actuating force is removed.

monatomic layer A coating that consists of a single layer of atoms.

monaural *Monophonic.*

monitor 1. An instrument that measures continuously or at intervals a condition which must be kept within prescribed limits, such as the image picked up by a television camera, the sound picked up by a microphone at a radio or television studio, a variable quantity in an automatic process control system, the transmissions in a communication channel or band, or the position of an aircraft in flight. 2. A person who watches a monitor.

monitor head An additional playback head on some tape recorders, to permit playing back the recorded sounds off the tape while the recording is being made.

monitoring Using a monitor.

monitoring amplifier A power amplifier used primarily for evaluation and supervision of a program.

monitoring antenna An antenna that picks up the RF output of a transmitter at the transmitter site for overall monitoring purposes.

mono *Monostable multivibrator.*

monochromatic The property of having only one color, corresponding to a negligibly small region of the spectrum.

monochromatic radiation Electromagnetic radiation that has a single wavelength, or photons that all have the same energy. Although no radiation is strictly monochromatic, it can have an extremely narrow band of wavelengths, as does sodium light and the spectrum of mercury 198.

monochromatic sensitivity The response of a device to light of a given color.

monochromator An instrument that isolates a narrow portion of the spectrum for analysis, transmission, or other purposes, as in a spectrometer.

monochrome The property of having only one chromaticity. This is achromatic in black and white television, involving only shades of gray between black and white.

monochrome bandwidth The video bandwidth of the monochrome channel or the monochrome signal in color television.

monochrome channel Any path intended to carry the monochrome signal in a color television system. This path can also carry other signals, such as the chrominance signal.

monochrome display adapter [MDA] A circuit board for personal computer color monitors that produces a TTL monochrome video signal for text-only display. It allows only three degrees of intensity: off, on, and intensified. Characters are produced on a 7×9 dot matrix within a 9×14 cell.

monochrome signal 1. A signal wave that controls luminance values in monochrome television. 2. The portion of a signal wave that has major control of the luminance values in a color television system, regardless of whether the picture is displayed in color or in monochrome.

monochrome television Television that reproduces a picture in monochrome: only shades of gray between black and white. It is also called black and white television.

monochrome transmission Transmission of a signal wave that controls the luminance values in a television picture but not the chromaticity values. The result is a monochrome picture.

monolithic Constructed from a single crystal or other single piece of material.

monolithic ceramic capacitor *Ceramic monolithic multilayer capacitor* (MLC).

monolithic circuit An alternate term for an *integrated circuit* (IC). A circuit fabricated on a single chip of semiconductor material.

monolithic crystal filter A filter whose single quartz plate provides the functions of multiple resonators.

monolithic integrated circuit *Integrated circuit.*

monolithic multilayer capacitor [MLC] A ceramic capacitor with multiple metallized dielectric layers having a volume of only about 1 mm^3 but a large equivalent surface area. It is formed by depositing metal films on paper-thin "green" ceramic strips, and then stacking, compressing, and firing the strips to form the monolithic chip. MLCs can have different dielectric properties and a range of capacitive values. Metallized end surfaces permit the MLC to be soldered to surface-mount circuit boards. Its dimensions have been standardized internationally as number codes related to the outside dimensions in millinches. Examples include 1210 to 6560.

monophonic A reference to sound that is transmitted, recorded, or heard over a single path. Binaural and stereophonic pertain to sounds arriving over two paths, to give the effect of auditory perspective. It is also known as monaural sound.

monophonic recorded tape Magnetic tape that has a single recording track, for use in a monophonic sound system.

monophonic recorder A magnetic-tape recorder that has a single-channel audio system which consists of a microphone, amplifier, and one recording head. It produces a single recorded track on magnetic tape.

monophonic sound system   A sound-reproducing system in which one or more microphones feed a single channel that terminates in one loudspeaker system.

monopole antenna   An antenna mounted on an imaging ground plane, to produce a radiation pattern approximating that of a dipole.

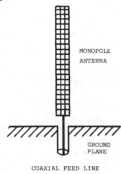

MONOPOLE
ANTENNA

GROUND
PLANE

COAXIAL FEED LINE

Monopole antenna is half of a dipole operated in conjunction with its image in a conducting ground plane perpendicular to it.

monopulse radar   Radar that obtains directional information with high precision with a receiving antenna system that has two or more partially overlapping lobes in the radiation patterns. Sum and difference channels in the receiver compare the amplitudes of the antenna outputs (sum-and-difference monopulse radar) or compare the phases of the antenna outputs (phase-sensing monopulse radar).

monoscope   A signal-generating electron-beam tube that produces a picture signal by scanning an electrode which has a predetermined pattern of secondary-emission response over its surface. This fixed image is printed on the electrode during manufacture of the tube, to give a useful test pattern for testing and adjusting television equipment.

monostable   Having the property of only one stable state.

monostable circuit   A circuit that has only one stable condition, to which it returns in a predetermined time interval after being triggered. A Schmitt-trigger circuit is an example.

monostable multivibrator   A multivibrator with one stable state and one unstable state. A trigger signal is required to drive the unit into the unstable state, where it remains for a predetermined time before returning to the stable state. It is also called a mono or one-shot multivibrator.

monostable timer   A monostable multivibrator that has controls for adjusting the width or time duration of the output pulse.

monostatic radar   A conventional radar whose transmitter and receiver are at the same location and share the same antenna. By contrast, in bistatic radar the receiver and transmitter are some distance apart and use different antennas.

monostatic reflector   A reflector that reflects energy only along the line of the incident ray. A corner reflector is an example.

monotonicity   In an analog-to-digital converter, the condition in which there is an increasing output for every increasing value of input voltage over the full operating range. The derivative of the output with respect to the input is therefore always positive.

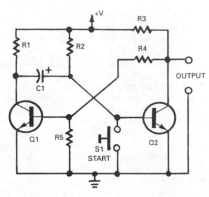

Monostable multivibrator or one-shot stays in a stable state until a trigger pulse at its input terminal initiates a change.

moon   A natural celestial body that orbits as a satellite about the earth. Its mean diameter is 2160 mi (3476 km), its mean distance from the earth is 238,857 mi (384,403 km), and its surface gravity is about one-sixth of that on the earth.

Moore code   An equal-length message-transmitting code that has 70 possible characters with four marking intervals and four spacing intervals. If a received character has other than eight intervals, an error signal is printed, or repetition is automatically requested. A seven-unit version that has 35 characters uses three marking and four spacing intervals.

MOPA   Abbreviation for *master oscillator-power amplifier.*

Morse code   A message-transmitting code that consists of dot and dash signals. The international Morse code is universally used for radiotelegraphy. The American Morse code applies only for wire telegraphy.

A ·—	T —
B —···	U ··—
C —·—·	V ···—
D —··	W ·——
E ·	X —··—
F ··—·	Y —·——
G ——·	Z ——··
H ····	**PUNCTUATION**
I ··	PERIOD ·—·—·—
J ·———	INTERROGATION ··——··
K —·—	BREAK —···—
L ·—··	END OF MESSAGE ·—·—·
M ——	END OF TRANSMISSION ···—·—
N —·	**NUMBERS**
O ———	1 ·———— 6 —····
P ·——·	2 ··——— 7 ——···
Q ——·—	3 ···—— 8 ———··
R ·—·	4 ····— 9 ————·
S ···	5 ····· 0 —————

Morse code (international) for use in radio communications.

Morse-code generator   An automatic code generator that is operated by a typewriter keyboard. Depression of a key causes a corresponding international Morse-code character to be generated, followed by a letter space.

MOS   Abbreviation for *metal-oxide semiconductor.*

mosaic A light-sensitive surface used in television camera tubes; it consists of a thin mica sheet coated on one side with a large number of tiny photosensitive silver-cesium globules, insulated from each other. The picture is optically focused on the mosaic, and the resulting charges on the globules are scanned by the electron beam of the camera tube.

MOS capacitor Abbreviation for *metal-oxide semiconductor capacitor.*

MOS-controlled thyristor [MCT] A power semiconductor device that combines a MOS transistor as the gate and a thyristor as the power source. This composite device has the lowest forward voltage drop of any voltage-controlled power source, including power MOSFETs and IGBTs.

MOS EAROM An electrically alterable read-only memory that made with metal-oxide semiconductor technology.

MOSFET Abbreviation for *metal-oxide semiconductor field-effect transistor.*

MOS gate driver [MGD] An integrated circuit for driving high-voltage MOSFETs or IGBTs with ratings to 600 V in such applications as power supplies, motor speed controls, UPS (uninterrupted power-supply) systems, and amplifiers. It can also drive dual-transistor forward converters.

MOS LSI Large-scale integration made with metal-oxide semiconductor technology.

MOS PROM A programmable read-only memory made with metal-oxide semiconductor technology.

MOS RAM A random-access memory made with metal-oxide semiconductor technology.

MOS resistor Abbreviation for *metal-oxide semiconductor resistor.*

MOS ROM A read-only memory made with metal-oxide semiconductor technology.

MOS transistor Abbreviation for *metal-oxide semiconductor transistor.*

most significant bit [MSB] The bit that carries the highest value or weight in binary notation for a numeral.

motherboard A printed circuit board assembly for interconnecting arrays of plug-in electronic "daughter" boards. See also *backplane.*

mother crystal A raw piezoelectric crystal as found in nature or grown artificially.

Motion Picture Experts Group standard [MPEG] A television standard for picture compression that uses interframe coding and accepts some loss of detail within a frame to obtain somewhat higher transmission speed than the JPEG standard.

motion-picture pickup The use of a television camera to pick up scenes directly from motion-picture film.

motor A machine that converts electric energy into mechanical energy by utilizing forces exerted by magnetic fields produced by current flow through conductors.

motorboating Oscillation in a system or component, usually indicated by a succession of pulses occurring at a very low audio-frequency rate. It is generally caused by excessive positive feedback, such as through a common power supply. When a loudspeaker is connected to the system, as in a radio, the pulses produce a put-put sound like that of a motorboat.

motor control *Electronic motor control.*

motor effect The repulsion force exerted between adjacent conductors carrying currents in opposite directions.

motor-generator set A motor and generator that are coupled mechanically for changing one power-source voltage to other desired voltages or frequencies.

mount 1. A shock or vibration isolator, usually consisting of one elastic member and one or more relatively inelastic members that supports the equipment to be isolated. 2. The flange or other means for connecting a switching tube, or tube and cavity to a waveguide.

mountain effect The effect of rough terrain on radio-wave propagation, causing reflections that produce errors in radio direction-finder indications.

mouse A computer peripheral whose motion on a horizontal plane causes the cursor on the computer monitor's screen to move accordingly.

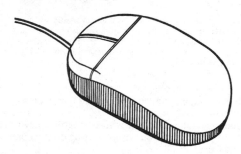

Mouse is a popular input device for a personal computer.

MOV Abbreviation for *metal-oxide varistor.*

movable contact The relay contact that is mechanically displaced to engage or disengage one or more stationary contacts.

moving-armature loudspeaker *Magnetic-armature loudspeaker.*

moving-coil galvanometer A galvanometer that measures current sent through a coil suspended or pivoted in a fixed magnetic field.

moving-coil hydrophone A moving-coil microphone that responds to waterborne sound waves.

moving-coil loudspeaker *Dynamic loudspeaker.*

moving-coil meter A meter whose pivoted coil is the moving element.

moving-coil microphone *Dynamic microphone.*

moving-coil pickup *Dynamic pickup.*

moving-conductor hydrophone A moving-conductor microphone that responds to waterborne sound waves.

moving-conductor loudspeaker A loudspeaker whose mechanical forces result from reactions between a steady magnetic field and the magnetic field produced by current flow through a moving conductor.

moving-conductor microphone A microphone whose electric output results from motion of a conductor or coil in a magnetic field. Examples include dynamic or velocity microphones.

moving-iron instrument An instrument that depends on current in one or more fixed coils acting on one or more pieces of soft iron, at least one of which is movable.

moving load A section of waveguide with a plunger that controls the position of a sliding, tapered, low-reflection load, to adjust the phase of residual reflection from the load.

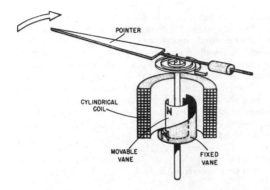

Moving-iron instrument whose fixed and rotatable soft iron vanes are similarly magnetized by a coil that carries current being measured. Like poles repel, causing the meter pointer to move against the force of the torsion spring. The meter will measure either alternating or direct current.

moving-magnet instrument   An instrument that depends on the action of a movable permanent magnet in aligning itself in the resultant field produced either by a fixed permanent magnet and an adjacent coil or coils carrying current or by two or more current-carrying coils whose axes are displaced by a fixed angle.

moving-magnet magnetometer   A magnetometer that depends for its operation on the torques acting on a system of one or more permanent magnets which can turn in the field to be measured.

moving-target indicator [MTI]   A radarscope that limits the display of radar information to moving targets. Echos caused by reflections from stationary objects are canceled by a memory circuit.

MOVPE   Abbreviation for *metal-organic vapor-phase epitaxy.*

MPC-2   An industry standard for the compatibility of multimedia computer software that permits the integration of a CD-ROM drive, sound card, speakers, and microphone.

MPEG   Abbreviation for *Motion Picture Experts Group.*

MPEG-2   An international standard for digital information exchange between personal computers and many new consumer entertainment services. See *MPEG.*

MPU   Abbreviation for *microprocessing unit.*

MQFF   Abbreviation for *metric pitch quad flat package.*

MQR   Abbreviation for *multiquantum-well structure.*

mR   Abbreviation for *milliroentgen.*

MR   Abbreviation for *memory read.*

mrad   Abbreviation for *milliradian.*

mrd   Abbreviation for *millirad.*

mrem   Abbreviation for *millirem.*

MRT   Abbreviation for *minimum resolvable temperature.*

ms   Abbreviation for *millisecond.*

mS   Abbreviation for *millisiemens.*

MS   Designation for Military Standard.

MSB   Abbreviation for *most significant bit.*

M scope   A radarscope that produces an M display.

M shell   The third layer of electrons about the nucleus of an atom; the layer has electrons characterized by the principal quantum number 3.

MSI   Abbreviation for *medium-scale integration.*

MSMT package [micro SMT]   A concept for the minimal packaging of semiconductor devices for surface-mount applications while they are still in wafer form. The packages are only slightly larger than the bare dies or chips, permitting very high density population of circuit boards. They are smaller than either the DIP and SMT cases for comparable products.

MSPS   Abbreviation for *million samples per second,* a measurement of time used predominately in referencing data acquisition and digital signal processing (DSP).

MSS   Abbreviation for *mobile satellite service.*

M-Star   A system of 72 interactive satellites intended to speed up communications through private corporate networks. The satellites will circle the earth in low-altitude orbits. It is sponsored by Motorola Inc. See also *Iridium.*

MSW   Abbreviation for *magnetostatic wave.*

MTBF   Abbreviation for *mean time between failures.*

MTI   Abbreviation for *moving-target indicator.*

MTL   Abbreviation for *merged-transistor logic.*

MTSO   Abbreviation for *mobile telephone switching office.*

MTTR   Abbreviation for *mean time to repair.*

M-type backward-wave oscillator   See *carcinotron* and *crossed-field tubes.*

M-type microwave electron tubes   See *crossed-field tubes.*

mu or μ law   A companding-encoding law commonly used in the United States.

μ [Greek letter mu]   1. Symbol for *amplification factor.* 2. Abbreviation for *micro-.*

μA   Abbreviation for *microampere.*

μbar   Abbreviation for *microbar.*

μC   Abbreviation for *microcomputer.*

μCi   Abbreviation for *microcurie.*

Mueller bridge   A bridge that measures with high precision the resistance values of three- and four-terminal resistors, particularly resistance thermometers.

μF   Abbreviation for *microfarad.*

MUF   Abbreviation for *maximum usable frequency.*

μ-factor   The ratio of the magnitude of an infinitesimal change in the voltage at the jth electrode of an n-terminal electron tube to the magnitude of an infinitesimal change in the voltage at the lth electrode, with the current to the mth electrode remaining unchanged.

μH   Abbreviation for *microhenry.*

μHz   Abbreviation for *microhertz.*

multianode tube   An electron tube that has two or more main anodes and a single cathode. This term is used chiefly for pool-cathode tubes.

multiaperture device   A magnetic core that has two or more openings through which windings are threaded. In one arrangement, used as a variable-gain adaptive component, flux can be switched around the small apertures without disturbing the flux around the main aperture.

multiband antenna   An antenna that can satisfactorily receive on more than one frequency band.

multicellular horn   1. A cluster of horn antennas that have mouths which lie in a common surface. The horns are fed from openings spaced one wavelength apart in one face of a common waveguide, to provide a desired directional radiation pattern for the radiated energy.   2. A combination of individual horn loudspeakers that have individual driver units or are joined in groups to a com-

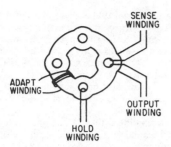

Multiaperture device.

mon driver unit. Subdivision of a large horn into smaller horns makes it easier to control the pressure and phase of the acoustic waves across the mouths of the horns.

multichannel analyzer   *Pulse-height analyzer.*

multichannel multipoint distribution [MMDS]   A short-range, amplitude-modulated microwave-frequency signal distribution system that broadcasts from a transmitting tower capable of carrying up to 33 channels to rooftop grating-style antennas. The receiving antennas must be in the direct line of sight of the transmitter. See also *local multipoint distribution* [LMDS] and *satellite master-antenna television* [SMATV].

multichip module   1. Multiple integrated circuits on a small printed-circuit board. The combination of integrated circuits and substrate is then molded in plastic. Some products are available in standard plastic quad flatpack body sizes with leads for direct mounting on a printed-circuit board.   2. A small printed-circuit board with one or more DIP-packaged integrated circuits mounted on it, made as a daughter board to plug into a mother board to add extra circuitry, particularly memory. See also *single-in-line-memory module.*

multicolor graphics array   A display device that provides a multicolor graphic display such as a *liquid-crystal display* [LCD], *plasma panel,* or *cathode-ray tube* [CRT].

multicoupler   A device that connects several receivers to one antenna and properly matches the impedances of the receivers to the antenna.

multicurrent-range diode   A diode that has a nearly constant coefficient over a wide range of operating terms. A silicon planar epitaxial diode is an example.

multidrop line   A single communications circuit that interconnects many stations, each of which contains terminal devices.

multielectrode tube   An electron tube that contains more than three electrodes associated with a single electron stream.

multifrequency transmitter   A radio transmitter capable of operating on two or more selectable frequencies, one at a time, using preset adjustments.

multifunction array radar   A radar system that combines both search and track functions. It typically includes a phased-array antenna with high-power RF phase shifters under the control of digital signal processors. Agile antenna beam scanning permits the dual functions if a compromise frequency between the optimum for searching and the optimum for tracking is determined. See *search radar, tracking radar,* and *Patriot.*

multigrid tube   An electron tube that has more than one grid electrode.

multigun tube   A cathode-ray tube that has more than one electron gun.

multihole coupler   A directional coupler that has two rows of graduated-diameter holes are spaced a quarter-wavelength apart on the broad waveguide faces of the coupler. Power flowing in one direction through the primary waveguide couples through the holes and excites waves in only one direction in the secondary waveguide. Because of the spacing between the holes, waves traveling in the reverse direction are out of phase and cancel.

multihop transmission   Radio-frequency transmission in which radio waves are reflected and refracted between the earth and ionosphere several times in their path of travel to a receiver far beyond the direct transmission range. It is also called multiple-hop transmission.

multijunction cell   A photovoltaic cell in which two or more different materials are positioned over each other to form one semiconducting unit.

multilayer ceramic capacitor [MLC]   *Ceramic monolithic multilayer capacitor.*

multilayer dielectric reflector   A sequence of thin layers of transparent material of controlled thicknesses and indexes of refraction. High reflectivity is established at certain wavelengths by constructive interference of the Fresnel reflections from the dielectric interfaces. These reflectors establish the optical cavity of a laser.

multilayer printed-circuit board   A circuit board manufactured by stacking and properly registering two or more partially cured circuit boards with copper traces and bonding them together to form a monolithic unit by heat and pressure. Lands are vertically aligned to permit drilling and through-hole plating processes. It permits high component density in a small volume, particularly in military and aerospace circuitry. It is also called a multilayer printed-wiring board.

multimedia   A reference to the many forms of media that can be processed for display, listening, and, sometimes, interaction on personal computers. These include video, 2-D and 3-D graphics, audio, fax/modem, telephone, and videophone.

multimeter   1. A meter that has a number of measuring functions, usually with several scale ranges for each function.   2. *Volt-ohm-milliammeter.* See *digital multimeter.*

multimode   The ability of a fiberoptic cable, transmission line, or waveguide to propagate signals in two or more *modes.*

multimode graded fiber   Optical fiber that has a graded core which will support more than one optical transmission mode.

multimode optical cable   Optical cable capable of transmitting light signals in more than one optical transmission mode.

multimode optical fiber   An optical fiber that can support the propagation of multiple transmission modes.

multimode radar   Radar that has two or more functions such as simultaneous ground mapping; terrain-following maneuvers during air strikes; and identification, tracking, and rangefinding of fixed or moving targets.

multimode step-index fiber   Optical fiber that has a stepped core which will support more than one optical transmission mode.

multimode waveguide A waveguide that propagates more than one mode at its operating frequency.

multipath cancellation Effectively complete cancellation of radio signals because of the relative amplitude and phase differences of components arriving over different paths.

multipath reception Reception in which the transmitter signals arrive at a receiving antenna over two or more paths, one direct and the others reflected from buildings or other obstacles. One result is ghosts in television pictures.

multipath transmission The propagation phenomenon that results in signals reaching a radio receiving antenna by two or more paths, causing distortion in radio and ghost images in television. At least one of the paths involves reflection from some object. It is also called multipath.

multiple *Parallel.*

multiple-access communications satellite [MACSAT] An experimental class of communications satellites placed in low-earth orbit (LEO).

multiple-address code A computer instruction code that specifies more than one address or storage location. The instruction can give the locations of the operands, the destination of the result, and the location of the next instruction.

multiple-beam klystron A klystron that has a number of electron guns, each of which produces its own beam which interacts with its own sequence of RF gaps. All beams are ultimately dissipated in a common collector. Each beam section thus acts as a conventional klystron, with the beams contributing to the power output in a rigidly phase-locked manner that cannot be achieved readily with separate klystrons.

multiple-break contacts Contacts so arranged that a circuit is interrupted in two or more places when the contacts open.

multiple-carrier time-division multiaccess [MC-TDMA] A communications system sharing multiple low-burst-rate TDMA carriers. It trades off on the peak power disadvantage of single-carrier TDMA and the traffic management inflexibility of FDMA.

multiple circuit A circuit in which two or more identical or different circuits are connected in parallel to a common signal or power source.

multiple-contact switch *Selector switch.*

multiple course One of a family of lines of position defined by a navigation system, any one of which can be selected as a course line.

multiple decay *Branching.*

multiple disintegration *Branching.*

multiple electron-beam lithography An advanced form of electron-beam lithography in which a single electron beam is split up into hundreds of beams so that the system can simultaneously etch the circuitry for all of the integrated circuits on a wafer. The single beam is split when it reaches a metal screen with many holes, each acting as a separate lens to direct an electron stream to the silicon wafer below it. See *electron-beam lithography.*

multiple-hop transmission *Multihop transmission.*

multiple modulation Modulation in which the modulated wave from one process becomes the modulating wave for the next. Thus, in an amplitude-modulated pulse-position-modulation system, one or more signals position-modulate their respective pulse subcarriers, which are spaced in time and are used to amplitude-modulate a carrier.

multiple plug A device that is inserted into a single receptacle or wall outlet to act as more than one receptacle.

multiple scattering Scattering in which the final displacement is the vector sum of many small displacements. Multiple scattering is greater than single or plural scattering.

multiple sound track A group of sound tracks printed adjacently on a common base, independent in character but having a common time relationship, such as for stereophonic sound recording.

multiple-speed floating action Floating action in which a final control element is moved at two or more speeds, each corresponding to a definite range of values of deviation.

multiple-track range A range system that has two closely spaced synchronized pulse stations. The indicator in the aircraft has several predetermined time difference settings by which a number of approximately parallel tracks can be flown.

multiple-trip echo An echo returned from a target so distant that the time required for the radar pulse to go out to the target and return to the receiver is longer than the interval between two successive pulses. These echoes show up at false ranges, indentifiable because they change when the pulse repetition rate is changed. It is called a *second-trip echo* when the echo arrives between the second and third pulses.

multiple tube counts Spurious counts induced by previous tube counts in a radiation-counter tube.

multiple-tuned antenna A low-frequency antenna that has a horizontal section connected to a multiplicity of tuned vertical sections.

multiple-unit tube An electron tube that contains within one envelope two or more groups of electrodes associated with independent electron streams, such as a duodiode, duotriode, diode-pentode, duodiode-triode, duodiode-pentode, and triode-pentode.

multiplex To transmit two or more signals simultaneously on a single wire, bus, or channel. 1. In telecommunications, to switch signals in a timed sequence so that many

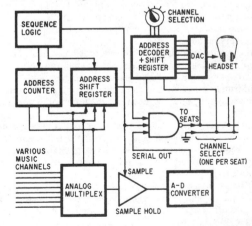

Multiplex for distributing eight different music channels to each seat of a commercial aircraft over a single pair of wires. An A/D converter sends out serial words corresponding to samples of a channel, in sequence. A 3-bit address code is added to 8 bits of analog information. At each seat an address decoder is linked to a channel selection switch with a D/A converter operating only for digital words corresponding to selected channel.

different receptors share a single wire or bus to conserve wiring. 2. In LED and LCD displays, to time-share a common bus to permit the operation of multiple display segments or dots with a minimum of wiring at a speed high enough that the display does not appear to flicker.

multiplex channel A carrier channel that provides two or more services simultaneously in the same or opposite directions.

multiplexer [MUX] A circuit in a signal transmission system that divides signal channel into two or more subchannels either by splitting the frequency band into narrower bands (*frequency-division multiplexing* [FDM]) or by allotting a common channel to several different transmitting devices, one at a time (*time-division multiplexing* [TDM]).

multiplexing A process of transmitting more than one signal over a single wire or cable. There are two types: parallel and serial processing. Parallel processing frequency-shares the bandwidth of a channel, with each multiple input sharing part of the available bandwidth. Serial processing time-shares multiple signals. In serial processing, signal speed is so fast that it is possible to multiplex four different coded signals through a single decoder-driver and have them appear on four different displays without a flicker.

multiplex radio transmission The simultaneous transmission of two or more signals by radio, using a common carrier wave.

multiplex transmission *Multiplex.*

multiplication An increase in current flow through a semiconductor due to increased carrier activity.

multiplication point A mixing point whose output is obtained by multiplication of its inputs in a feedback control system.

multiplier 1. An analog circuit whose output is the arithmetic product of two input signals. Some integrated-circuit multipliers can perform multiplication in all four quadrants, as well as perform division, squaring, and the determination of square roots. Some digital multipliers are also important in digital signal processing (DSP). There are analog and digital multipliers. 2. A resistor in series with a meter to permit voltage readings in a specified range. An array of multipliers with different values will permit the measurement of a wide range of voltages. See also *electron multiplier* and *frequency multiplier.*

multiplier accumulator Circuitry dedicated to multiplying numbers and adding. The very fast units are typically able to multiply two numbers and add the product of previously accumulated results in a single clock period.

multiplier phototube *Photomultiplier.*

multiplier resistor *Multiplier.*

multiplying factor The number by which the reading of a given meter must be multiplied to obtain the true value.

multipoint line A single communications line or circuit interconnecting several stations that typically requires a polling mechanism to address each connected terminal.

multipolar The property of having more than one pair of magnetic poles.

multipole moment A measure of the charge, current, and magnet distributions of a system of particles.

multiport network A network that has more than one port.

multiposition action Automatic control action in which a final control element is moved to one of three or more predetermined positions, each corresponding to a definite value or range of values of the controlled variable.

multiprocessing The simultaneous or interleaved execution of two or more programs by a computer.

multiprogramming Programming that allows two or more arithmetic or logic operations to be performed by a computer either simultaneously or on a time-sharing basis.

multiquantum-well structure [MQW] A device made as a stack of quantum-well layers, each of which quantum-mechanically traps electrons or holes. These multilayered structures are typically made of very thin (<20 μm) alternating layers of gallium arsenide (GaAs) and aluminum gallium arsenide (AlGaAs). MQW structures are formed in solid-state lasers and heterojunction FETs similar to HEMTs, SDHTs, TEGFETs, and MODFETs.

multisegment magnetron A magnetron with an anode divided into more than two segments, usually by slots parallel to its axis.

multistation access unit [MAU] In token-ring *local area networks* [LANs], it is a wiring concentrator with up to eight stations.

multithreaded processor A computer processor that is divided into a large number of "virtual" processors which can simultaneously process tasks by making new requests for data while waiting for earlier ones to be satisfied. After requesting data to complete a task, it carries out partial tasks while waiting for all the data to become available. It puts aside the first task and then proceeds to the next, sending out a second request for data. The process continues until a data request is filled. Then the relevant task is resumed and the information is used.

multitrace oscilloscope An oscilloscope that has a single-beam cathode-ray tube that is time-shared by two or more input signals.

multitrack recording system A recording system that provides two or more recording paths on a medium, carrying related or unrelated recordings that have a common time relationship.

multiturn potentiometer A precision wire-wound potentiometer whose resistance element is formed into a helix, generally with from 2 to 10 turns. The actuating shaft must be rotated a corresponding number of turns to cover the full resistance range. The wiper arm slides along the shaft as its contact moves along the helix.

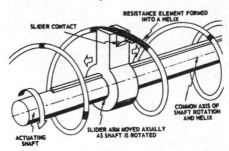

Multiturn potentiometer.

multivibrator A relaxation oscillator circuit that has two transistors or other active elements coupled so that one is cut off when the other is conducting. An *astable multivibra-*

tor is free-running. The frequency of spontaneous transition between these two states is determined by the time constants of the coupling elements and sometimes by an external voltage. A *monostable multivibrator* needs a trigger signal to drive the circuit into its unstable state, and the circuit constants determine the time for returning to the stable state. The third basic type of multivibrator is the *flip-flop* or *bistable* multivibrator. An external trigger signal is required for each transition. It is also called an Eccles-Jordan circuit.

μlx Abbreviation for *microlux*.

μm Abbreviation for *micrometer*.

μmho Abbreviation for *micromho*.

Munsell chroma *Chroma*.

Munsell value The dimension, in the Munsell system of object-color specification, that indicates the apparent luminous transmittance or reflectance of the object on a scale which has approximately equal perceptual steps under the usual conditions of observation.

μΩ Abbreviation for *microhm*.

μP Abbreviation for *microprocessor*.

μR Abbreviation for *microroentgen*.

μrd Abbreviation for *microrad*.

μrem Abbreviation for *microrem*.

Murray loop test A method of localizing a fault in a cable by replacing two arms of a Wheatstone bridge with a loop formed by the cable under test and a good cable connected to the far end of the defective cable.

μs Abbreviation for *microsecond*.

μS Abbreviation for *microsiemens*.

MUSA [multiple-unit steerable antenna] An electrically steerable receiving antenna whose directional pattern can be rotated by varying the phases of the contributions of the individual units. These units are usually stationary rhombic antennas.

musical echo A flutter echo that is periodic and has a flutter whose frequency is in the audio range.

Musical Instrument Digital Interface [MIDI] The creation of sound by digitally programming a music synthesizer integrated circuit that creates an analog signal which is amplified and played through a speaker. An enhancement is *wavetable MIDI* sound that improves the realism of MIDI sound.

music power An output capability rating for high-fidelity audio amplifiers; it takes into account such factors as transient response, power supply regulation, and the RMS power capability. The music power rating can be as much as 20% higher than the RMS power rating of an amplifier. It is also called IHF power because the measurement procedure has been standardized by the Institute of High Fidelity (IHF).

muting Silencing, or reducing in volume.

mutual conductance *Transconductance*.

mutual-conductance meter *Transconductance meter*.

mutual inductance A measure of the amount of inductive coupling that exists between two coils. It is related to the flux linkages produced in one coil by current in the other coil. It is measured in henrys, millihenrys, and microhenrys, the same as for inductance.

mutual interference Interference that affects two or more pieces of equipment because of interactions such as those caused by harmonics and other spurious emissions of transmitters and spurious responses and images of receivers.

MUX Abbreviation for *multiplexer*.

μV Abbreviation for *microvolt*.

μW Abbreviation for *microwatt*.

mV Abbreviation for *millivolt*.

MV Abbreviation for *megavolt*.

MVA Abbreviation for *megavoltampere*.

MVBR Abbreviation for *multivibrator*.

mW Abbreviation for *milliwatt*.

MW Abbreviation for *megawatt*.

MWh Abbreviation for *megawatthour*.

MWR Abbreviation for *memory write*.

Mx Abbreviation for *maxwell*.

Mylar A DuPont tradename for *polyester* or *polyethylene terephthalate plastic* resin. It is widely used as a dielectric film in capacitors and as a base for magnetic tape.

Mylar capacitor A film-type *polyester* capacitor.

myriametric wave A radio wave between the wavelength limits of 10 and 100 km, corresponding to the very low-frequency (VLF) range of 3 to 30 kHz. (The FCC designated all frequencies below 30 kHz as VLF.)

N

n   Abbreviation for *nano-*.

N   1. Abbreviation for *negative*.   2. Abbreviation for *newton*.   3. Symbol for *nitrogen*.

N   1. Symbol for *neutron number*.   2. Symbol for number of turns.

nA   Abbreviation for *nanoampere*.

NA   Abbreviation for *numerical aperture*.

NAB   Abbreviation for National Association of Broadcasters.

NAB curve   The standard playback equalization curve adopted by the National Association of Broadcasters for disk recordings.

NAND   A logic operator which has the property that if $P, Q, R, \ldots$ are statements, then the NAND of $P, Q, R, \ldots$ is true if and only if at least one statement is false; the NAND of $P, Q, R, \ldots$ is false if and only if all statements are true.

NAND gate   An AND gate followed by an inverter, to give an output logic state of 0 when all input signals are 1.

nano-   [n]   A prefix that represents 10^{-9}, which is 0.000000001 or one-thousandth of a millionth.

nanoampere   [nA]   One-thousandth of a microampere, or 10^{-9} A.

nanocircuit   A flip-flop or other logic circuit that operates at speeds above 20 MHz, in the nanosecond range.

nanofarad   [nF]   One-thousandth of a microfarad, or 10^{-9} F. A unit not widely used in the United States.

nanohenry   [nH]   One-thousandth of a microhenry, or 10^{-9} H.

nanometer   [nm]   One-thousandth of a micrometer, or 10^{-9} m.

nanosecond   [ns]   One-thousandth of a microsecond, or 10^{-9} s. Light travels approximately 1 ft (0.3 m) in 1 ns.

nanovolt   [nV]   One-thousandth of a microvolt, or 10^{-9} V.

nanovoltmeter   A voltmeter that is sufficiently sensitive to measure thousandths of a microvolt.

nanowatt   [nW]   One-thousandth of a microwatt, or 10^{-9} W.

narrow-band axis   The direction of the phasor that represents the coarse chrominance primary in color television; it has a bandwidth extending from 0 to 0.5 MHz.

***N*-ary code**   A code that uses N distinct types of elements. As an example, with pulse-code modulation there would be N distinct pulse height levels.

NASA   Abbreviation for *National Aeronautics and Space Administration*.

National Aeronautics and Space Administration   [NASA] A civilian agency established in 1958 to control aeronautical and space research and exploration activities sponsored by the United States, except those associated with military and defense systems.

National Electrical Code   A set of regulations that governs construction and installation of electric wiring and apparatus in the United States, established by the American National Board of Fire Underwriters for safety purposes.

National Electrical Manufacturers Association   [NEMA] An organization of manufacturers of electric products.

National Institute of Standards and Technology   [NIST]   A U.S. government agency with the responsibility for maintaining time, weight, length, and other basic measurement standards for the United States. It also performs basic research to support U.S. commercial and industrial manufacturing and services. It was formerly called the U.S. Bureau of Standards.

National Television System Committee   [NTSC]   A committee organized in 1940 by representatives of U.S. companies and organizations interested in television. It formulated television standards for black and white television in 1940–1941 and for color television in 1950–1953 that were approved by the Federal Communications Commission. Countries using NTSC standards include the United States, Canada, Japan, and Mexico.

NATO identification system [NIS]   The International version of the military *identification friend or foe* [IFF] system.

natural frequency   1. A frequency at which a body or system will oscillate freely.   2. The lowest resonant frequency of an antenna, circuit, or component.

natural logarithm [ln]   A logarithm that has a base 2.71828.

natural period   The period of a free oscillation of a body or system.

natural radioactivity   Radioactivity exhibited by naturally occurring radionuclides.

natural wavelength   The wavelength that corresponds to the natural frequency of an antenna or circuit.

NAU   Abbreviation for *network addressable unit.*

nautical mile   A measure of distance equal to 1 minute of arc on the earth's surface. The United States has adopted the International Nautical Mile, equal to 1852 m or 6076.11549 ft, which is approximately 1.15 mi.

Nautilus   A long-range, high-power, laser-based defense system against short-range rockets, developed by the U.S. Army. The laser beam will destroy the rocket by melting away its metal skin.

navigation aid   A device or system that provides a navigator with some or all of such navigation data as present position, heading, speed, location of fixed objects and other craft, right-left steering directions or automatic steering control, and altitude.

navigation beacon   A light, radio beacon, or radar beacon that provides navigation aid to aircraft and ships.

navigation instrument   *Navigation aid.*

navigation satellite   An earth-orbiting satellite from which radio measurements can be made under all weather conditions, to give the position of a ship or aircraft. See *NAVSTAR, Global Positioning System.*

navigation way point   A point on the surface of the ocean that can be established in terms of latitude and longitude which lies along the intended course of a ship where course changes occur or where the course is checked. Similarly, a point along the flight path of an aircraft (that can also have the third dimension of altitude) where the flight path is changed or the progress of the flight plan is checked.

NAVSTAR [NAVigation System using Time And Ranging]   A global system of up to 24 navigation satellites developed to provide instantaneous and highly accurate worldwide three-dimensional location by air, sea, and land vehicles equipped with suitable receivers.

NAVSTAR GPS   Abbreviation for *NAVSTAR Global Positioning System.*

NAVSTAR satellite [NAVigation System using Time and Ranging]   One of the 24 continuously orbiting satellites in the global positioning system (GPS). See *Global Positioning System.*

NC   1. Symbol for *no connection.*   2. Symbol for *normally closed,* used with reference to relay and switch contacts.   3. Abbreviation for *numerical control.*

N-channel   A conduction channel formed by electrons in an N-type semiconductor, as in an N-type field-effect transistor.

N-channel MOS [NMOS]   A metal-oxide semiconductor process for the fabrication of field-effect transistors in which electrons are the dominant charge carriers in the semiconductor channel, making the channel N-type. NMOS transistors run at least twice as fast as PMOS (P-channel MOS) devices because the mobility of electrons is higher than that of holes. In P-channel MOS, holes are the dominant charge carriers in the channel. See also *P-channel MOS.*

NDRO   Abbreviation for *nondestructive readout.*

Nd:YAG laser   Abbreviation for *neodymium-doped yttrium-aluminum garnet laser.*

NEAR   Abbreviation for *Near Earth Asteroid Rendezvous Spacecraft.*

Near Earth Asteroid Rendezvous Spacecraft [NEAR]   A space probe launched in 1996 for a three-year mission to orbit the asteroid Eros. It contains five scientific instruments. The probe will measure the mass, density, and composition of Eros.

near-end crosstalk   Interference that can occur at carrier telephone repeater stations when output signals of one repeater leak into the same end of the other repeater.

near field   1. The acoustic radiation field that is close to the loudspeaker or other acoustic source.   2. The electromagnetic field that exists in the near region, within a distance of one wavelength from a transmitting antenna.

near infrared   That portion of the infrared spectrum that contains the shortest wavelengths, adjacent to the 0.7-μm red end of the visible spectrum.

near region   The region immediately surrounding a transmitting antenna, extending out to a distance of one wavelength, in which the strength of the induction field varies inversely with the square of the distance. Also called near zone.

near zone   *Near region.*

neck   The small tubular part of the envelope of a cathode-ray tube, extending from the funnel to the base and housing the electron gun.

NEDT   Abbreviation for *noise-equivalent differential temperature.*

needle   *Stylus.*

needle drag   *Stylus drag.*

needle force   *Stylus force.*

needle scratch   *Surface noise.*

negative   1. A terminal or electrode that has more electrons than normal. Electrons flow out of the negative terminal of a voltage source.   2. A designation used to describe an opposite character to positive, as in negative feedback, negative image, negative resistance, and negative transmission.

negative bias   A base or gate bias voltage that makes the control base or gate of a transistor negative with respect to its emitter or source.

negative charge   An electric charge in which the object in question has more electrons than protons.

negative conductance   A property of certain semiconductors in which the local current density decreases whenever the local electric field exceeds a threshold level. The effect occurs in silicon transit-time devices, gallium arsenide Gunn devices, and in other semiconductor crystals that have a limited-space-charge mode of oscillation.

negative electricity   A negative charge, such as that produced in a resin object by rubbing with wool.

negative electron An electron, as distinguished from a positive electron or positron.

negative electron affinity A characteristic of some semiconductor surfaces that permits their use as photoemitters, secondary emitters, and cold-cathode emitters when they are mounted in a vacuum. With this characteristic, an electron that has energy above a minimum level encounters no work function barrier at the surface of the semiconductor and can escape into the vacuum much like other emitter materials.

negative feedback Feedback in which a portion of the output of a circuit, device, or machine is fed back 180° out of phase with the input signal. Negative feedback decreases amplification, to stabilize the amplification with respect to time or frequency, and to reduce distortion and noise. It is also called degeneration, inverse feedback, or stabilized feedback.

negative glow The luminous glow in a glow-discharge cold-cathode tube; it occurs between the cathode dark space and the Faraday dark space.

negative image An image in which dark areas are bright and bright areas are dark. It is also called a reversed image.

negative impedance An impedance that exhibits a voltage drop across itself when the current through it increases.

negative input–positive output [NIPO] A logic circuit that accepts a negative-going input pulse and delivers a positive-going output pulse.

negative ion An atom that has more electrons than normal and therefore has a negative charge.

negative logic Digital logic in which the more negative logic level represents 0. This logic level term is not recommended for use because it can be construed differently for NPN devices than for PNP devices. To avoid confusion, the magnitude and polarity of the voltage levels for 1 and 0 in a specific logic circuit should always be specified.

negative modulation 1. Modulation in which an increase in brightness corresponds to a decrease in amplitude-modulated transmitter power. U.S. television transmitters and some facsimile systems employ it. 2. Modulation in which an increase in brightness corresponds to a decrease in the frequency of a frequency-modulated facsimile transmitter.

negative picture phase The video signal phase in which the signal voltage swings in a negative direction for an increase in brilliance.

negative plate The internal plate structure that is connected to the negative terminal of a storage battery. Electrons flow from the negative terminal through the external load circuit to the positive terminal.

negative proton *Antiproton.*

negative reactance A negative inductance characterized by the response that its reactance increases with frequency, or a negative capacitance whose reactance decreases as frequency increases. It is therefore necessary to measure it at more than one frequency to identify a negative reactance.

negative resistance A resistance whose value increases with increasing current, causing the voltage drop across it to decrease. This is a characteristic of some semiconductor devices and circuits.

negative-resistance oscillator An oscillator whose parallel-tuned resonant circuit is connected to a two-terminal negative-resistance device.

negative-resistance region An operating region in which the current decreases when the applied voltage is increased.

negative temperature coefficient [NTC] The characteristic of a material that causes its resistance to decrease with temperature. Carbon has a negative coefficient of resistance, but all metals and most metal alloys have *positive temperature coefficients of resistance* [PTC].

negative terminal The terminal of a battery or other voltage source that has more electrons than normal. Electrons flow from the negative terminal through the external circuit to the positive terminal.

negative thermion *Thermoelectron.*

negative-transconductance oscillator An oscillator whose output is coupled back to the input without phase shift. The phase condition for oscillation is satisfied by the negative transconductance of the device.

negative transmission *Negative modulation.*

negative-true logic A logic system in which the voltage representing a logical 1 has a lower or more negative value than the one representing a logical 0. Most parallel I/O (input/output) buses use negative-true logic because of the common organization of logic circuits.

NEMA Abbreviation for *National Electrical Manufacturers Association.*

nematic liquid crystal A liquid-crystal material whose elongated molecules are parallel to each other but are not in layers. See also *supertwisted nematic liquid crystal* and *twisted nematic liquid crystal.*

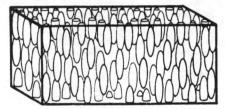

Nematic liquid crystals showing arrangement of elongated molecules.

neodymium [Nd] A rare-earth element with an atomic number of 60.

neodymium amplifier A light amplifier whose amplification is provided by the action of pulsed flash tubes on neodymium-doped glass rods.

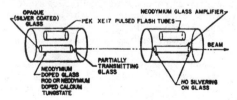

Neodymium amplifier in straight-through operation.

neodymium-doped yttrium-aluminum garnet laser [Nd: YAG laser] A crystalline solid laser whose YAG crystal is doped with neodymium. It is also called a YAG laser.

neodymium glass laser An amorphous solid laser whose glass is doped with neodymium. Its characteristics are

neodymium liquid laser

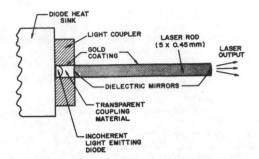

Neodymium-doped yttrium-aluminum garnet laser end-pumped by a single light-emitting diode.

comparable to those of a pulsed ruby laser, but its wavelength of radiation is outside the visible range.

neodymium liquid laser   *Inorganic liquid laser.*

neon [Ne]   An inert gas with an atomic number of 10 and an atomic weight of 20.18. It is used in lamps, tubes, and alphanumeric display panels that are electrically diodes. When a voltage of about 160 V is placed across the electrodes, the gas is ionized, and it emits its characteristic red-orange glow.

neon glow lamp   A glow lamp that contains neon gas. It produces a characteristic red-orange glow. The watt ratings of standard sizes range from 0.04 to 3 W. It is also called a *neon lamp*.

Neon glow lamp with screw base (left) and bayonet base (right).

neon indicator   A neon glow lamp used as a visual indicator of voltage.

neon lamp   *Neon glow lamp.*

neon oscillator   An oscillator circuit that consists of a neon glow lamp, a capacitor, and sometimes also a resistor.

Neoprene   A trademark of Du Pont for polychloroprene.

NEP   Abbreviation for *noise equivalent power.*

neper [Np]   A unit that expresses the ratio of two voltages, two currents, or two power values as a logarithm. The number of nepers is the natural logarithm of the square root of the ratio of the two values being compared. The neper has a base of 2.71828, whereas the decibel has the common-logarithm base of 10. One neper is equal to 8.686 dB.

nephelometer   An instrument that measures the size and number of dust particles suspended in a medium, based on directing a beam of light through the particles to a photomultiplier.

Nernst bridge   A four-arm bridge that contains capacitors instead of resistors, for measuring capacitance values at high frequencies.

Nernst effect   If heat flows through a strip of metal whose surface is perpendicular to a magnetic field, a voltage is developed between opposite edges of the strip.

Nernst-Ettinghausen effect   A thermomagnetic effect that occurs in certain pure crystals. A temperature difference is

produced in a direction perpendicular to a longitudinal electric current and an applied magnetic field.

net   1. Many communication stations equipped to communicate with each other, often on a definite time schedule and in a definite sequence.   2. Many computers able to interact with each other because of a common address and common communication links. An example is the *Internet.*

netlist   A description of electronic schematic design symbols and their interconnection points that is stored in computers or workstations for logical connection in computer-aided circuit design. Netlist transfer is the most common way of moving designs from one design system or tool to another.

network   1. Connected components such as resistors, diodes, capacitors, and even transistors that perform certain functions such as filtering or biasing. An active network also contains a source of energy, whereas a passive energy does not.   2. In telecommunications, a group of devices such as radio transceivers that are organized to form a large system.   3. A group of computers or terminals coupled together to form a system such as a local area network (LAN) or wide area network (WAN).   4. A group of radio and television stations that are interconnected by cable or microwave link so that they can broadcast the same program simultaneously.

network analyzer   1. A computer in which networks are set up to simulate power-line systems or other physical systems to obtain solutions to various problems before actual construction.   2. An RF analyzer that makes accurate measurements of magnitude, phase, and group delay of either of two signals with respect to a reference signal. It can also measure impedance and admittance over a wide frequency range.

network bridge   1. A circuit located within a local area network (LAN), but not on a customer's premises, that bridges together the voice signals from several locations in an audio teleconference.   2. In a local area network (LAN), a transition between networks with different protocols.

network constant   One of the values of resistance, inductance, mutual inductance, or capacitance that make up a network.

neural network   An electronic circuit that simulates or mimics the functions of a network of living neurons and axons such as those found in the brain. It contains adjustable elements with connections that fan out to an interconnected network of similar elements. Programming the adjustable elements can "teach" the network to perform specific functions.

neuroelectricity   A current or voltage generated in the nervous system.

neuroengineering   A field of research concerned with the building of *artificial neural networks* [ANNs] for the performance of specified activities. The objectives of this engineering specialty includes the design of electronic and photonic circuitry based on models of neural networks and the writing of applications software.

neuron   A living nerve cell.

neutral   The property of having the same number of electrons as protons; in a normal condition there is no electric charge.

neutral-density filter   An optical filter that reduces the intensity of light without appreciably changing its color. It is also called a *gray filter.*

neutral ground   A ground connection made to the neutral conductor or neutral point of a power line, transformer, motor, generator, or other device connected to the line. The connection can be made directly or through a grounding device that has resistance and/or impedance.

neutralize   To stop regeneration in an amplifier stage by neutralization.

neutralizing voltage   The AC voltage that is fed from the collector circuit to the base circuit of an amplifier, or the inverse. It is 180° out of phase with and equal in amplitude to the AC voltage that is transferred between these circuits over an undesired path, usually through the base-to-collector transistor capacitance.

neutral zone   *Dead band.*

neutron   An elementary nuclear particle that has a zero charge and a mass number of 1, making its mass approximately the same as that of a proton. Ionization is produced by the products of neutron collisions.

neutron fluence   A measure of radiation hardness in integrated circuits.

newton [N]   The SI unit of force, equal to the force that produces an acceleration of 1 m/s^2 to a mass of 1 kg. One newton is equal to 10^5 dyn.

newton per square meter [N/m^2]   The SI unit of pressure or stress. A pressure of 1 N/m^2 is equal to 1 pascal.

nF   Abbreviation for *nanofarad.*

NF   Abbreviation for *noise figure.*

nH   Abbreviation for *nanohenry.*

nibble   A term for a half of a byte (four bits).

nicad battery   Abbreviation for *nickel-cadmium battery.*

NiCd battery   Abbreviation for *nickel-cadmium battery.*

Nichrome   A trademark of Driver-Harris Co. for an alloy of nickel, chromium, and sometimes iron. The alloy has high electric resistance and the ability to withstand high temperatures for long periods of time. It is used to make wirewound resistors and electric heating elements.

nickel [Ni]   A hard, silver-white, malleable metal element with an atomic number of 28 and an atomic weight of 58.71. In electronics it is plated on other metals because of its resistance to oxidation, and is alloyed with other metals to form materials with special thermal characteristics. It is also used in the manufacture of secondary, rechargeable cells and batteries (nickel-cadmium (Ni-Cd) and nickel-metal hydroxide (Ni-MH). It has *magnetostrictive* properties. Arrays of nickel tubes can form a magnetostrictive transducer capable of emitting powerful ultrasonic signals for *sonar* when energized, and it can convert sound echoes back to electrical signals.

nickel-cadmium battery [NiCd battery or nicad battery]   A sealed secondary battery that has a nickel anode, a cadmium cathode, and an alkaline electrolyte. It is widely used to power cordless appliances, transceivers, cellular telephones and computers.. Without recharging, it can serve as a primary battery. It has nominal voltage of 1.25 V per cell.

nickel-cadmium cell [NiCd cell]   A secondary cell with a nominal 1.25 V per cell.

nickel-metal hydroxide [Ni-MH] **battery**   A rechargeable battery technology with higher cell capacity than nickel cadmium (Ni-Cd) technology. Nominal cell voltage is 1.2 V. Nickel is the positive electrode and a hydrogen-absorbing alloy is the negative electrode.

Nicol prism   A prism made by cementing together two pieces of transparent crystalline Iceland spar with Canada balsam. It produces plane-polarized light from ordinary unpolarized light by eliminating the ordinary component of the original light by total reflection at the cementing layer. Only the extraordinary component passes.

night effect   A polarization error that occurs in radio direction finders at night.

night-vision binoculars   A binocular night-vision aid consisting of two parallel *image intensifier tubes,* each equipped with an objective lens and an ocular lens for focusing and enlarging the image. A high-voltage power supply is required to operate the tubes. The binoculars are useful in the low natural night light provided by sky glow, stars, moon, or cloud reflection. Available light can be magnified as much as 30,000 times. Helmet-mounted military versions permit safer flying of aircraft or driving of vehicles at night. The helmet mounted versions are called *night-vision goggles* and are designated as ANVIS for *airborne night-vision imaging system.* See also *image-intensifier tube* and *night-vision scope.*

night-vision goggles   *Night-vision binoculars.*

night-vision scope   A monocular *night-vision telescope* consisting of a *light-intensifier tube* equipped with objective and ocular lenses for focusing and enlarging the image. A high-voltage power supply is required to operate the instrument. Typically, a battery-powered multiplier circuit is located within the scope housing. It functions at night in low natural light—sky glow, moonlight, and cloud reflection. Available light can be magnified up to 30,000 times. Military versions can be mounted on sniper rifles and machine guns. It is also called a *nightscope, night-vision telescope, sniperscope, snooperscope,* and *starlight scope.* See also *image intensifier tube* and *night-vision binoculars.*

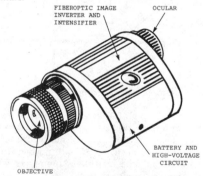

Night-vision scope amplifies natural light from the night sky 30 to 50,000 times to form an image of an object or scene in darkness.

NI junction   A junction between N-type material and intrinsic material in a semiconductor.

niobium [Nb]   A metallic element with an atomic number of 41 and an atomic weight of 92.9.

Ni-MH   Abbreviation for *nickel-metal hydride.*

NIPO   Abbreviation for *negative input-positive output.*

NIS   Abbreviation for *NATO identification system.*

NIST   Abbreviation for *National Institute of Standards and Technology.* Formerly the National Bureau of Standards.

NIST-7 The seventh generation of atomic clock built by NIST. It is 10 times more accurate than its predecessor NBS-6. It uses cesium atoms that move in a tube in the clock in a predictable motion at a specific rate by which time can be measured. The standard for one second is the amount of time taken for cesium atoms to vibrate 9,192,631,770 times. It depends on laser beams rather than magnetic fields to control the stream of electrons.

nit [nt] A name given to the SI unit of luminance, the candela per square meter.

nitrogen [N] A colorless, odorless, gas element with an atomic number of 7 and an atomic weight of 14.01. It forms about 78% of the earth's atmosphere. Nitrogen is used in electronics as a protective gas in welding and as a fill for coaxial cables and waveguides to keep out moisture because it is not a very active element and does not support combustion.

nm Abbreviation for *nanometer*.

N/m² Abbreviation for *newton per square meter*.

NMOS Abbreviation for *N-channel MOS*.

NMOS RAM A random-access memory that is made with N-channel metal-oxide semiconductor technology.

NMR Abbreviation for *nuclear magnetic resonance*.

NMRR Abbreviation for *normal-mode rejection ratio*.

NO Symbol for *normally open*.

noble metal A metal, such as gold, silver, or platinum, that has high resistance to corrosion and oxidation. It is used to make thin-film circuits, metal-film resistors, and other metal-film devices.

nodal diagram A diagram that shows the order and mode of waves propagated in a waveguide.

node 1. A point on a transmission line at which either a current or voltage standing wave has an essentially zero amplitude. 2. An interconnection or junction point in a network where one or more terminals are located.

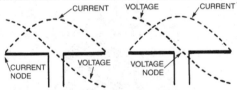

Nodes of current and voltage for a half-wave dipole.

noise Unwanted high-frequency, low-level impulses caused by natural phenomena such as lightning or man-made sources such as the arcing of motor brushes, relay contact openings, transformers, and unshielded radio transmitters. It can cause errors in communications systems and computer readouts.

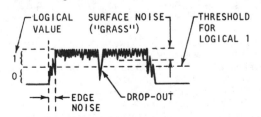

Noise as it can occur in a connection with logic levels 1 and 0. Edge noise and dropouts can cause errors.

noise analyzer An instrument that determines the amplitudes of the frequency components in a noise signal.

noise behind the signal *Modulation noise.*

noise-canceling microphone *Close-talking microphone.*

noise cancellation technology A technique for canceling the effects of sound emitted from industrial, commercial, and consumer appliances and machines, automobiles, aircraft, and other sources by generating antinoise signals. Computer-based circuitry analyzes the composition of noise and generates sounds that are 180° out of phase from the original sounds.

noise-current generator A current generator whose output is a random function of time.

noise diode A diode designed for operation at saturation in a noise-generating circuit.

noise dosimeter An instrument that determines if the total daily exposure of a person to audible noise is within government-established limits. One version has a detachable microphone that can be positioned anywhere on the body. It might have a digital display in addition to a sealed-in meter that reads the maximum measured level.

noise equivalent differential temperature [NEDT] The temperature difference required at the input of the detector to produce a peak signal-to-noise (RMS) ratio of unity at the detector preamplifier output.

noise equivalent power [NEP] The power that would be present at a given point in a circuit due to noise alone if the useful signal were removed without changing operating conditions.

noise factor The ratio of the total noise power per unit bandwidth at the output of a system to the portion of the noise power that is due to the input termination at the standard noise temperature of 290 K.

noise figure [NF] The noise factor expressed in decibels. Noise figure is equal to $10 \log_{10}$ (noise factor).

noise filter A filter that is inserted in an AC power line to block noise interference which would otherwise travel through the power line in either direction and affect the operation of receivers.

noise generator An instrument that generates one or more types of noise signals, covering a specified portion of the frequency spectrum. The noise generated could include pink or white noise and various industry-standardized noise signals. The noise can be essentially random, such as occurs in nature, or it can be pseudorandom, in which noise patterns of known content and duration occur at a fixed repetition rate.

noise immunity The weakest signal that can be accepted by a circuit without getting lost in internally generated circuit noise.

noise jamming A brute-force electronic countermeasure technique that produces clutter over the entire cathode-ray tube display of an enemy tracking radar. The disclosure to the enemy of jamming action or the ability of some missiles to home on the noise jammer are two drawbacks to this kind of jamming.

noise level The level of electric or acoustic noise at a particular location. It is expressed in decibels with respect to a specified reference level. The value of the noise is integrated over a specified frequency range.

noise limiter A limiter circuit that cuts off all noise peaks that are stronger than the highest peak in the desired signal being received, thereby reducing the effects of atmospheric or man-made interference. It is also called an automatic noise limiter.

noise-measuring set   *Circuit-noise meter.*

noise pollution   The condition in which the loudness and other characteristics of a noise together make that noise undesirable in the human environment. Factors affecting the degree of noise pollution include the responses of different people to various sounds, exposure duration, time of day, and the amount of pain or physical damage to human ears caused by the noise.

noise power ratio   The ratio in decibels of the noise level in a telephone measuring channel that has its baseband fully noise-loaded to the level in that channel when all the baseband except the measuring channel is noise-loaded.

noise quieting   The ability of a receiver to reduce noise background in the presence of a desired signal. It is expressed in decibels.

noise reduction   A process whereby the average transmission of the sound track of a sound motion-picture print, averaged across the track, is decreased for signals of low level and increased for signals of high level. Because background noise introduced by the sound track is less at low transmission, this process reduces film noise during soft passages.

noise source   A device that generates a random noise signal for test purposes. Common noise sources are a temperature-limited diode operated at cathode saturation, an electron multiplier, a crystal diode with positive bias, a nonoscillating reflex klystron, and a nonoscillating magnetron.

noise suppressor   A circuit that blocks the AF amplifier of a radio receiver automatically when no carrier is being received, to eliminate background noise. It is also called an intercarrier noise suppressor, interstation noise suppressor, and squelch circuit.

noise temperature   The temperature at which the thermal noise power of a passive system per unit bandwidth is equal to the noise at the actual terminals. The standard reference temperature for noise measurements is 290 K.

noise-voltage generator   A voltage generator whose output is a random function of time.

no-load loss   The power loss of a device that is operated at rated voltage and frequency but is not supplying power to a load.

nominal band   A frequency band equal in width to that between zero frequency and the maximum modulating frequency in facsimile.

nominal frequency   The specified frequency value of a crystal unit, as distinguished from the actual frequency measured during operation.

nominal line width   The average separation between centers of adjacent scanning or recording lines.

nomograph   A chart that has three or more scales on which equations can be solved graphically by placing a straightedge on two known values and reading the answer where the straightedge crosses the scale for the unknown value.

nonbridging   Switching action in which the movable contact leaves one fixed contact before touching the next.

noncoherent MTI   A moving-target indicator system that uses ground clutter in place of a coherent reference oscil-

lator as the reference signal. A moving target can be detected only when ground clutter exists at the same range and bearing as the target.

noncoherent radiation   Radiation that has are no definite phase relations between different points in a cross section of the beam.

nonconductor   An insulating material.

noncontacting piston   *Choke piston.*

noncontacting plunger   *Choke piston.*

noncorrosive flux   Flux that is free from acid and other substances that might cause corrosion when used in soldering.

nondegenerate gas   A gas formed by a system of particles whose concentration is sufficiently weak so the Maxwell-Boltzmann law applies. Examples are molecules or atoms of a body in the gaseous state, electrons emitted by a hot cathode, electrons and ions in a cloud or a plasma, electrons supplied to a conduction band by donor levels in an N-type semiconductor, and holes resulting from the passage of electrons from the normal band to an impurity band of acceptor levels in a P-type semiconductor.

nondestructive breakdown   The breakdown of the barrier between the gate and channel of a field-effect transistor without causing failure of the device. In a junction field-effect transistor, avalanche breakdown occurs at the PN junction.

nondestructive readout [NDRO]   A computer memory in which the storage media that gives the readout of the desired data is self-reversible, eliminating the need for special circuits to restore the stored data.

nondestructive testing   Testing by techniques that do not damage or destroy the items under test.

nondirectional   *Omnidirectional.*

nondirectional antenna   *Omnidirectional antenna.*

nondirectional beacon   An omnidirectional beacon or other radio transmitter facility designed for use with an airborne direction finder to provide a line of position.

nondirectional microphone   *Omnidirectional microphone.*

nonerasable storage   A computer storage medium that cannot be erased and reused, such as factory-programmed read-only memory (ROM) or field-programmable memory (PROM).

nonferrous   A material that does not contain iron.

non-great-circle propagation   *Groundscatter propagation.*

nonimpact printer   A line printer whose characters are produced electrically, electronically, or optically rather than mechanically. Only a single copy can be produced. Examples include electrostatic, ink-jet, laser-beam, and thermal-matrix printers. Operating speeds are much higher than for impact printers and can be well over 10,000 lines per minute for laser printers.

noninduced current   A current that flows through a winding of a transformer not due to a voltage induced in the transformer. The filament current in a bifilar winding is an example.

noninduced voltage   An alternating or direct voltage that is applied uniformly to an entire winding so that no appreciable potential difference is induced along the winding. The filament voltage supplied to a load through a bifilar winding is an example.

noninductive   The property of having negligible or zero inductance.

noninductive capacitor A capacitor constructed so that it has practically no inductance. Foil layers are staggered during winding so an entire layer of foil projects at either end for contact-making purposes. All currents then flow laterally rather than spirally around the capacitor.

noninductive circuit A circuit that has practically no inductance.

noninductive load A load that has practically no inductance. It can consist entirely of resistance or may be capacitive.

noninductive winding A winding constructed so that the magnetic field of one turn or section cancels the field of the next adjacent turn or section. The winding is hairpin or bifilar.

noninverting amplifier An operational amplifier whose input signal is applied to the ungrounded positive input terminal to give a gain greater than unity and cause the output voltage to change in phase with the input voltage.

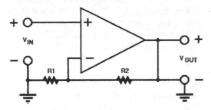

Noninverting amplifier accepts the signal on a noninverting terminal, and output and input voltages are in phase.

nonlinear Not directly proportional.

nonlinear amplifier An amplifier whose output is not directly related to its input.

nonlinear capacitor A capacitor that has a nonlinear charge characteristic or a reversible capacitance that varies with bias voltage.

nonlinear circuit A circuit whose characteristics cannot be specified by linear differential equations in which time is the independent variable.

nonlinear detection Detection based on the nonlinear device characteristic, such as occurs in square-law detection.

nonlinear distortion Distortion of the output of a system or component because it does not have a linear relation to the input. Amplitude, harmonic, and intermodulation distortions are examples of nonlinear distortion.

nonlinear element An element in which an increase in applied voltage does not produce a proportional increase in current.

nonlinear feedback control system A feedback control system in which the relationships between the input and output signals are not linear.

nonlinearity The deviation of any functional relationship from direct proportionality.

nonlinear material A material that exhibits physical changes that are not proportional to the intensity of the energy or force to which it is subjected. For example, a piezoelectric material that does not provide an electrical output that is proportional to the pressure across its opposing faces.

nonlinear network A network not specifiable by linear differential equations in which time is the independent variable.

nonlinear taper Nonuniform distribution of resistance throughout the element of a potentiometer or rheostat.

nonmagnetic Not magnetizable, and hence not affected by magnetic fields. Examples are air, glass, paper, and wood. All of them have a magnetic permeability of 1, the same as a vacuum.

nonmagnetic steel A steel alloy that contains about 12% manganese and sometimes a small quantity of nickel. It is practically nonmagnetic at ordinary temperatures. Some stainless-steel alloys are nonmagnetic.

nonmicrophonic *Antimicrophonic.*

nonpolarized electrolytic capacitor An electrolytic capacitor in which the dielectric film is formed adjacent to both metal electrodes, to give the same construction in both directions of current flow. It can be formed by placing two polarized capacitors back-to-back.

nonpolarized relay *Neutral relay.*

nonquantized system A system of particles whose energies are assumed to be capable of varying in a continuous manner. The number of microscopic states of the system, defined by the positions and velocities of the particles at a given instant, is then infinite.

nonradiative transition A change from one energy level to another in an atom without absorption or emission of radiation. The necessary energy can be supplied or carried away by vibrations in a solid material or by motions of atoms or electrons in a plasma.

nonresonant line A transmission line on which there are no standing waves.

non-return-to-zero code [NRZ code] A communications code in which a binary 1 is represented by one bit time at the 1 level and a binary 0 is represented by one bit time at the 0 level. This permits storing about twice as much data as can be stored with the return-to-zero code.

Non-return-to-zero code for binary data. The space between each pair of dashed vertical lines represents one bit time.

non-return-to-zero-inverted code [NRZI code] A communications code in which the modulation switches from one state to the opposite state to represent a binary 0 and remains unchanged to represent a binary 1.

nonsinusoidal wave A wave whose form differs from that of a sine wave and therefore contains harmonics.

nonspecular reflection Reflection from a rough surface, producing diffraction and scattering of waves.

nonstorage camera tube A television camera tube whose picture signal is at each instant proportional to the intensity of the illumination on the corresponding elemental area of the scene at that instant.

nonstorage display The display of nonstored information in a storage tube without appreciably affecting the stored information.

nonsynchronous Not related in phase, frequency, or speed to other quantities in a device or circuit. It is also called *asynchronous.*

nonsynchronous transmission　A data-transmission process that does not control the unit intervals within a block or a group of data signals with a clock.

nonsynchronous vibrator　A vibrator that interrupts a DC circuit at a frequency unrelated to other circuit constants, and does not rectify the resulting stepped-up alternating voltage.

nonthermal radiation　Electromagnetic radiation emitted by accelerated charged particles that are not in thermal equilibrium. Aurora light and fluorescent lights are examples.

nonvolatile memory　A memory device that retains its data when power is removed. Examples of nonvolatile semiconductor memories include the ROM (read-only memory), PROM (programmable ROM), EPROM (erasable PROM), EEPROM (electrically erasable PROM), and the flash memory. Ferrite-core memories and magnetic-bubble memories (MBM) are nonsemiconductor nonvolatile memories.

nonrecurring engineering [NRE] **charge**　A one-time charge for design engineering, tooling, and setup in the production of custom-made products. For semiconductor devices this includes photomask development.

NOR　A logic operator which has the property that if *P, Q, R,* . . . are statements, then the NOR of *P, Q, R,* . . . is true if and only if all statements are false, false if and only if at least one statement is true.

NORAD　The joint North American Air Defense Command that coordinates operations of various services in defense of continental United States and regions of the North American continent.

NOR device　A logic device whose output is 1 only when all its control signals are 0.

NOR gate　A gate whose output represents 1 only when all the input signals represent 0. The output is thus inverted from that of an OR gate.

normal　1. The perpendicular to a line or surface at the point of contact.　2. The expected or regular value of a quantity.

normal distribution　*Gaussian distribution.*

normal induction　The limiting induction, either positive or negative, in a magnetic material that is under the influence of a magnetizing force varying between two specific limits.

normalize　To multiply all quantities by a constant so they fall within the operating ranges of a computer or the scale ranges of a graph.

normalized admittance　The reciprocal of the normalized impedance.

normalized impedance　An impedance divided by the characteristic impedance of a transmission line or waveguide.

normal logic　A logic level term whose use is not recommended because it can be construed differently for NPN devices than for PNP devices. To avoid confusion, the magnitude and polarity of the voltage levels for 1 and 0 in a specific logic circuit should always be specified.

normally closed [NC]　A term applied to relay or switch contacts that are connected to complete a circuit when the relay or switch is not energized.

normally open [NO]　A term applied to relay or switch contacts that are connected to break a circuit when the relay or switch is not energized.

normal mode　A characteristic distribution of vibration amplitudes among the parts of a system; each part is vibrating freely at the same natural frequency and phase.

normal-mode rejection ratio [NMRR]　The ability of an amplifier to reject spurious signals at the power-line frequency or at harmonics of the line frequency.

normal permeability　The ratio of the normal induction B of a magnetic material to the corresponding magnetic intensity H.

normal position　The deenergized position of the contacts of a relay.

normal state　*Ground state.*

normal threshold of audibility　The minimum sound pressure level at the entrance to the ear, at a specified frequency, that produces an auditory sensation in a large percentage of normal persons in the age group from 18 to 30 years.

north magnetic pole　The magnetic pole located approximately at 71°N latitude and 96°W longitude, about 1140 nautical miles south of the North Pole.

north pole　The pole of a magnet whose magnetic lines of force are considered to be leaving the magnet.

North Pole　A geographical point on the earth that is one end of the axis about which the earth revolves. It is almost directly beneath the star Polaris.

Norton amplifier　*Current-differencing amplifier.*

Norton's theorem　An electrical engineering theorem which states that any linear two-terminal AC network will behave in a manner equivalent to an AC-current source in parallel with a single impedance. In the DC case, the Norton source becomes a constant DC current, and the impedance becomes a resistance. See also *Thevenin's theorem.*

NOT　A logic operator exhibiting the property that if *P* is a statement, then the NOT of *P* is true if *P* is false, false if *P* is true.

notation　A method of representing numbers in which a number is expressed as a sum of coefficients multiplied by successive powers of a chosen base. Examples of notation systems and their bases are binary—2; ternary—3; quaternary—4; quinary—5; decimal—10; duodecimal—12; hexadecimal—16; duotricenary—32; and biquinary—2, 5.

notch antenna　A microwave antenna whose radiation pattern is determined by the size and shape of a notch or slot in a radiating surface.

notch filter　A band-rejection filter that produces a sharp notch in its frequency response curve of a system. In television transmitters it provides attenuation at the low-frequency end of the channel to prevent possible interference with the sound carrier of the next lower channel.

notching relay　A relay whose operation depends on successive input pulses.

note　A conventional sign that indicates the pitch, duration, or both of a tone sensation. It is also the sensation itself or the vibration causing the sensation. The word serves when no distinction is desired between the symbol, the sensation, and the physical stimulus.

notebook computer　A small, battery-powered, portable, personal computer packaged in a hinged case with a keyboard located in its lower half and a flat-panel display located inside its cover. The display is typically a monochrome or color liquid-crystal flat-panel with diagonal dimensions up to about 10.2 in (26 cm). Some contain an internal hard-disk drive and a floppy-disk drive or CD-

Notebook computer is a battery-powered personal computer with many of the features of a desktop model. An LCD display replaces the CRT monitor.

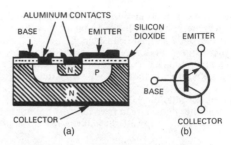

NPN transistor (*a*) has a P-type silicon base and an N-type silicon emitter and collector; (*b*) its schematic symbol.

ROM drive, and at least 8 Mbytes of random-access memory (RAM). Some models have self-contained battery chargers. Its weight is typically 4 to 8 lb (2 to 4 kg). It is also called a *laptop computer*. See also *subnotebook computer*.

notice to mariners [NTM]   A notice sent to mariners by mail or radio that specifies changes to be made in printed charts or computer databases to update charts to reflect changes in the positions of buoys, changes in water depth, the presence of new maritime hazards, or the establishment of restricted areas.

NOT gate   A gate whose output is 1 only when its single input is 0 and vice versa. It is also called an *inverter*.

NOT logic   A logic circuit in which a 0 input will produce a 1 output. Conversely, if the input is 1, there will be a 0 output.

noval base   An electron-tube base that has positions for nine pins which extend directly through the glass envelope. The spacing between pins 1 and 9 is greater than the spacings between the other pins, for orienting purposes.

Np   Abbreviation for *neper*.

NPIN transistor   An intrinsic-junction transistor whose intrinsic region is sandwiched between the P-type base and N-type collector layers.

NPIP transistor   An intrinsic-junction transistor whose intrinsic region is between two P regions.

N⁺-type semiconductor   An N-type semiconductor whose the excess conduction electron concentration is very large.

NPN   Negative-positive-negative, in referring to a semiconductor in which a layer with P-type conductivity is located between two layers that have N-type conductivity.

NPN transistor   A two-junction bipolar transistor formed with an N-type silicon collector and emitter and a P-type base. To conduct, its emitter must be negative and the collector must be positive with respect to the base. See *bipolar transistor* and *PNP transistor*.

NPO material   A temperature-compensating dielectric material that has an ultrastable temperature coefficient as applied to ceramic capacitors. The term is derived from negative-positive-zero.

N quadrant   One of the two quadrants in which the N signal of an A-N radio range is heard.

N region   The region in a semiconductor where conduction electron density exceeds hole density.

NRZ code   Abbreviation for *nonreturn-to-zero code*.

NRZI code   Abbreviation for *nonreturn-to-zero-inverted code*.

ns   Abbreviation for *nanosecond*.

N signal   A dash-dot signal heard in either a bisignal zone or an N quadrant of a radio range.

NSS   Abbreviation for *network and switching subsystem*.

nt   Abbreviation for *nit*.

NTC   Abbreviation for *negative temperature coefficient*.

NTM   Abbreviation for *notice to mariners*.

NTSC   Abbreviation for *National Television System Committee*.

N-type conductivity   The conductivity associated with conduction electrons in a semiconductor.

N-type negative resistance   A current-stable negative resistance in which a given current value can correspond to only one possible voltage value. Examples include four-layer diodes, silicon unijunction transistors, and gas discharge tubes.

N-type semiconductor   A semiconductor whose density of holes in the valence band is exceeded by the density of electrons in the conduction band. N-type material is created by adding donor impurities (doping)—materials such as arsenic or phosphorous—to the crystal structure of silicon.

nuclear   A reference to the nucleus of an atom.

nuclear fusion   A thermonuclear reaction in which the nuclei of an element of low atomic weight unite under extremely high temperature and pressure to form a nucleus of a heavier atom. The associated loss in mass is released as energy. This reaction takes place in the hydrogen bomb, where hydrogen nuclei combine to form helium nuclei.

nuclear magnetic resonance [NMR]   Resonance encountered in energy transfers between an RF magnetic field and a nucleus placed in a constant magnetic field that is sufficiently strong to decouple the nucleus from its orbital electrons. The amount of energy absorbed by the atoms at resonance is a clue to identification of the atoms involved.

nuclear magnetic resonance spectrometer   A spectrometer in which nuclear magnetic resonance is used for the analysis of protons and nuclei and for the study of changes in chemical and physical quantities over wide frequency ranges.

nuclear magnetometer *Nuclear resonance magnetometer.*

nucleus [plural: nuclei] The central part of an atom; it possesses a positive charge and contains nearly all the mass of the atom. The nucleus consists of protons and neutrons, together known as nucleons, except for the hydrogen nucleus, which consists only of one proton. It is also called the *atomic nucleus.*

nuclide A species of atom characterized by the number of protons and neutrons and energy content in the nucleus, or alternatively by the atomic and mass numbers and atomic mass. To be regarded as a distinct nuclide, the atom must be capable of existing for a measurable lifetime, generally greater than 10^{-10} s. Also called nuclear species.

null A position of minimum or zero indication or strength, such as that between adjacent lobes of an antenna radiation pattern.

null-balance system A measuring system whose input quantity is measured by producing a null with a corresponding calibrated balancing voltage, current, or other parameter.

null detector *Null indicator.*

null indicator A galvanometer or other device that indicates when voltage or current is zero. It can determine when a bridge circuit is in balance. It is also called a null detector.

null method A method of measurement in which the measuring circuit is balanced to bring the pointer of the indicating instrument to zero, as in a Wheatstone bridge. The settings of the balancing controls are then read.

null modem A device that connects two data terminal devices directly by emulating physically the connections of a data-communications device.

number A designation that represents a quantity, a position in a sequence, a direction, or some other magnitude. It can be one or more digits, a word, a sequence of pulses, or any equivalent designation. As an example, fourteen, 14, XIV, and 1110 all represent the same number. It is also called a *numeral.*

numeral 1. A conventional symbol that represents a number, such as the Arabic digits 0 to 9. 2. *Number.*

numeric 1. A number or numeral. 2. *Numerical.*

numerical aperture [NA] A number that expresses the light-gathering power of an optical fiber. It is equal to the sine of the acceptance angle and the half angle of the *acceptance cone.* It is a function of the *indexes of refraction* of the core and cladding. Typical values range from 0.20 to 0.27.

numerical control [NC] A control system for machine tools and industrial processes in which numerical values corresponding to the desired positions of tools or controls are recorded on punched or magnetic tapes in such a way that they can provide automatic control of the operation. Computers or programmable controllers maximize the operating efficiency of numerical control.

numerical display A display that gives the exact numerical value of a quantity in readable form.

nutating feed An oscillating antenna feed for producing an oscillating deflection of a tracking radar beam; the plane of polarization remains fixed.

nutation A periodic variation in the inclination of the axis of a spinning gyroscope to the vertical, between certain limiting angles.

nutation field The time-variant three-dimensional field pattern of a directional or beam-producing antenna that has a nutating feed.

nutator A mechanical or electric device that moves a radar beam in a circular, conical, spiral, or other manner periodically to obtain greater air surveillance than could be obtained with a stationary beam.

nV Abbreviation for *nanovolt.*

nW Abbreviation for *nanowatt.*

nibble A slang term used to represent some fraction of a byte. When a byte is 8 bits, a nibble is half a byte or 4 bits. A nibble is generally assumed to be 4 bits, so there are 4 nibbles in a 16-bit byte. This term is discouraged unless its meaning and that of byte are precisely specified.

nylon An easily machinable and moldable plastic resin that offers high tensile strength, high impact resistance, and high resistance to corrosion. It also forms low-friction surfaces. It is used to make insulators and the insulating parts of electronic components.

Nyquist criterion A parameter in servomechanism theory that corresponds to the open-loop harmonic response function.

Nyquist diagram A diagram for determining the stability of a control system. It is a closed polar plot of the loop transfer function of the system.

O

object code The machine language instructions actually executed by a computer system. It is produced by a compiler from an applications program written in a higher-level language. It is the form to which all programs are reduced before they are run on a processor.

objective lens The first lens through which rays pass in an optical or electronic lens system.

object-oriented programming In computer science, a program writing technique in which data representing objects or elements of the system and code defining the legal operations on the object data are packaged together. This programming permits concepts used in the solution of a problem or the performance of a function to be transferred to another function.

oblique-incidence transmission The transmission of a radio wave at a slant with respect to the ionosphere and back to earth, as in long-distance radio communication.

obstacle gain The increase in signal strength obtained over a long radio-communication path when a mountain obstacle or range of hills is located about halfway between the transmitting and receiving antennas. This obstacle gain offsets some of the path losses normally expected.

occluded gas Gas absorbed in a material, as in the electrodes, supports, leads, and insulation of a vacuum tube.

occultation The interference in the reception of a signal from a spacecraft to earth as a result of blockage by planet, its rings, or atmosphere.

OCR 1. Abbreviation for *optical character recognition.* 2. Abbreviation for *optical character reader.*

octal A base-8 number representation system based on the numbers 0 through 7. It is used to create machine code programs and is useful in visualizing bit patterns.

octal base An electron-tube base that has a central aligning key and positions for eight equally spaced pins. Pins not needed for a particular tube are omitted without the positions of the remaining pins being changed.

octal notation A notation based on the scale of eight. Here the number 235 means 5 times 1, plus 3 times 8, plus 2 times 8 squared, which is equal to 157 in decimal notation.

octal socket A socket for an octal tube.

octal tube An electron tube with an octal base.

octantal error An error that occurs in a measured bearing when spaced antennas in an electronic navigation system are used. The error varies sinusoidally through 360°, with four positive and four negative maximums.

octave The interval between any two frequencies that has a ratio of 2 to 1. Thus, going one octave higher means doubling the frequency. Going one octave lower means changing to one-half the original frequency.

octave-band noise analyzer An analyzer with filters that permit measurements of sound levels in different one-octave bands centered on ANSI preferred frequencies. For evaluating speech interference levels, the most important octave bands are 0.6 to 1.2, 1.2 to 2.4, and 2.4 to 4.8 kHz.

octave-band oscillator An oscillator that can be tuned over a frequency range of 2 to 1 so that its highest frequency is twice its lowest frequency.

octave-band pressure level The band pressure level for a frequency band that corresponds to a specified octave of sound. The location of an octave-band pressure level on a frequency scale is usually specified as the geometric mean of the upper and lower frequencies of the octave.

octave filter A bandpass filter whose upper cutoff frequency is twice its lower cutoff frequency.

octet An 8-bit byte.

odd-even check A forbidden-combination check in which an extra digit is carried along with each word to indicate whether the total number of ls in the word is odd or even.

odd-even nuclei Nuclei that have an odd number of protons and an even number of neutrons.

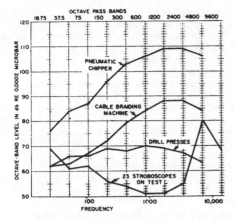

Octave-band noise analyzer measurements of pressure levels encountered in industry at frequencies up to 10,000 Hz. To prevent damage to hearing, level should not exceed 85 dB for daily exposure over a period of years in any band from 300 to 4800 Hz.

odd harmonic   A harmonic that is an odd multiple of the fundamental frequency.

odd-odd nuclei   Nuclei that have an odd number of protons and an odd number of neutrons.

Oe   Abbreviation for *oersted.*

OEM   Abbreviation for *original equipment manufacturer.*

oersted [Oe]   The unit of magnetic field strength *H* in the CGS system until 1930 when it was replaced by the gauss. Both terms have now been replaced by ampere per meter as the SI unit.

off   A term used (in connection with "on") to designate the inoperative state of a device or one of two possible conditions in a circuit.

off-center dipole   A rotating dipole mounted in a parabolic reflector at an angle to the axis of rotation, to give conical scanning.

off-center PPI display   A plan-position-indicator display in which the zero position of the time base is not at the center of the display. The off-center arrangement permits enlargement of the display for a selected portion of the radar service area.

off-hook   The condition that occurs when a telephone handset is removed from its cradle, thus connecting it to the phone system and permitting loop current to flow. The central office detects the DC current as an indication that the phone is not available for ringing.

off-line   A reference to peripheral equipment or devices not in direct communication with the central processing unit of a computer.

off-line power supply   A power supply, either linear or switching, that operates directly from the utility AC power line. The input voltage is rectified and filtered. It can be a stand-alone power supply, or it can be built as an integral part of its host product such as a radio or television receiver.

off-line storage   A storage device or system not under the control of the central processing unit.

off-scale   Beyond the normal indicating range, as when excessive current drives the pointer of a meter swing beyond the right-hand limit of the printed scale.

offset   1. In a process control system, the steady-state difference between the desired control point and that actually obtained. Offset is an inherent characteristic of positioning controller action.   2. In a linear amplifier, the change in input voltage that is required to produce zero output voltage in the absence of an input signal.   3. In a digital circuit, the DC voltage on which a signal is impressed.   4. In a DC amplifier, the small and relatively constant temperature-dependent voltage that exists between the two input terminals.   5. The amount of unbalance between the two halves of a symmetrical circuit.

offset current   The direct current that appears as an error at either input terminal of a DC amplifier when the input current source is disconnected.

offset voltage   The DC voltage that remains between the input terminals of a DC amplifier when the output voltage is zero.

off-the-shelf   Available for immediate shipment.

OFHC   Abbreviation for *oxygen-free high-conductivity copper.*

O guide   A surface-wave transmission line that consists of a hollow cylindrical structure made of a thin dielectric sheet.

ohm [Ω]   The SI unit of resistance and impedance. The electric resistance between two points of a conductor when a constant potential difference of 1 V, applied to these points, produces in the conductor a current of 1 A, the conductor not being the seat of any electromotive force.

ohm-centimeter   A unit of resistivity. The resistivity of a sample in ohm-centimeters is equal to its resistance R in ohms multiplied by its cross-sectional area A in square centimeters, and the result is divided by the sample length L in centimeters (resistivity = RA/L).

ohmic contact   A purely resistive contact between two surfaces or materials.

ohmic heating   1. Heating produced by sending current through a resistance.   2. Heating of plasma with a pulsed voltage to accelerate plasma electrons and thereby cause more heat-producing collisions with plasma ions.

ohmic value   The resistance in ohms.

ohmmeter   An instrument that measures electric resistance. Its scale can be graduated in ohms or megohms.

ohm-per-square   The resistance of any square area, measured between parallel sides, of thin films of resistive materials.

ohm-per-volt [Ω/V]   A sensitivity rating for measuring instruments obtained by dividing the resistance of the instrument in ohms at a particular range by the full-scale voltage value at that range. The higher the ohm-per-volt rating, the more sensitive the meter, and the less current it will draw from a circuit during a measurement.

Ohm's law   The current I in a circuit is directly proportional to the total voltage E in the circuit and inversely proportional to the total resistance R of the circuit. The law is expressed in three forms: $E = IR, I = E/R, R = E/I.$

oil diffusion pump   A diffusion pump that is similar to a mercury-vapor vacuum pump but oil replaces mercury vapor.

olfactronics   The science that deals with the detection and identification of odors. Highly sensitive electronic circuits are generally required for positive identification of predetermined odors, such as those given off by drugs, gunpowder, and other explosives.

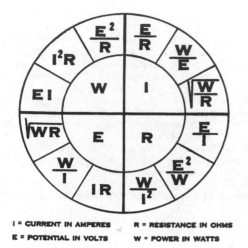

I = CURRENT IN AMPERES   R = RESISTANCE IN OHMS
E = POTENTIAL IN VOLTS   W = POWER IN WATTS

Ohm's law and related expressions for power.

omega A Greek letter; in its capital form Ω it represents the word ohms; in its lowercase form ω as a letter symbol for a value equal to 6.28 times frequency ($\omega = 2\pi f$).

Omega A hyperbolic phase-comparison navigation system suitable for aircraft use. It has eight stations located around the world and operates in the 10- to 4-kHz (VLF) band, using a fixed transmission pattern. It gives position accuracy within 2.3 to 4.6 mi (3.7 to 7.4 km).

Ω [Greek letter omega] Symbol for *ohm*.

Ω/V Symbol for *ohm per volt*.

omni *VHF omnirange.*

omnibearing line One of an infinite number of straight lines radiating from the geographical location of a VHF omnirange.

omnidirectional Radiating or receiving equally well in all directions. It is also called nondirectional.

omnidirectional antenna An antenna that has an essentially circular radiation pattern in azimuth and a directional pattern in elevation. It is also called a nondirectional antenna.

omnidirectional beacon A beacon that radiates radio signals equally well in all directions.

omnidirectional hydrophone A hydrophone that responds equally well in all directions to waterborne sound waves.

omnidirectional microphone A microphone whose response is essentially independent of the direction of sound incidence. It is also called a nondirectional microphone or an astatic microphone.

omnidirectional range A radio facility that provides bearing information to or from its location at all bearings within its service area, as directional guidance for pilots. It is also called an omnirange.

omnidistance The distance between a vehicle and an omnibearing-distance facility.

omnirange *Omnidirectional range.*

on A term used (in connection with "off") to designate the operating state of a device or one of two possible conditions in a circuit.

on-course curvature The rate of change of the indicated navigation course with respect to distance along the course line or path.

on-course signal A signal indicating that the aircraft receiving it is on course, following a guiding radio beam.

one-many function switch A function switch in which only one input is excited at a time, and each input produces a combination of outputs.

one's complement A number modified so that addition of the modifying number to its original value, plus 1, will equal an even power of 2. A one's complement is obtained mathematically by subtracting the original value from a string of 1s, and electronically by inverting the states of all bits in the number. Thus, the one's complement of 010101 (representing 21) is 101010 (representing 42).

one's complement arithmetic A binary arithmetic system in which negative numbers are created by inverting individual bits in the binary representation of the positive number.

one-shot multivibrator *Monostable multivibrator.*

one state A state of a magnetic cell in which the magnetic flux through a specified cross-sectional area has a positive value, when determined from an arbitrarily specified direction for positive flux. The opposite state in which the magnetic flux has a negative value, when similarly determined, is called a zero state.

one-time-only memory [OTO] An ultraviolet-erasable programmable read-only memory (EPROM) in a plastic, opaque or "windowless" dual-in-line package (DIP) that does not permit it to be reprogrammed. It is used where the product or system program is fixed and not subject to change, and it costs less than the reprogrammable part. See *erasable programmable read-only memory.*

one-to-partial-select ratio The ratio of a 1 output to a partial-select output from a magnetic cell.

O network A network composed of four impedance branches connected in series to form a closed circuit. Two adjacent points serve as input terminals, and the remaining two junction points serve as output terminals.

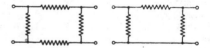

O network has four impedance branches; pi network at right has only three branches.

one-way repeater A repeater that amplifies signals traveling in one direction over wire lines or through space as radio waves.

on-hook The condition that occurs when a telephone handset is positioned in its cradle, opening the DC path and disconnecting it from the phone system so that no loop current flows. The central office considers an on-hook phone to be available for ringing.

on-line A reference to peripheral equipment under direct control of the central processing unit of a computer, permitting input data to be processed as it is received, and output data to be transmitted on a real-time basis.

on-line data reduction The processing of information as rapidly as it is received by the computing system.

on/off control A simple control system in which the device being controlled is either full on or full off, with no intermediate operating positions.

on/off keying Keying in which the modulated wave is transmitted to form the mark signal and suppressed to form the space signal.

on/off switch A switch that turns a receiver or other equipment on or off. It is combined with a volume control in radio and television receivers. It is also called power switch.

on period That part of an operating cycle in which an electron tube, transistor, or other active device is conducting.

on the air Transmitting a radio signal.

on the beam Following a radio beam.

opacimeter *Turbidimeter.*

opacity The ability of a substance to block the transmission of radiant energy. Opacity is the reciprocal of transmission.

op-amp Abbreviation for *operational amplifier.*

opaque A composition that prevents the passage of electromagnetic waves or atomic particles.

opaque plasma A plasma through which an electromagnetic wave cannot propagate and is, therefore, either absorbed or reflected. In general, a plasma is opaque for frequencies below the plasma frequency.

opcom Abbreviation for *optical communication.*

OPDAR Abbreviation for *optical radar.*

open A break in a path for electric current.

open-air ionization chamber *Free-air ionization chamber.*

open-architecture A system whose characteristics comply with industry standards, permitting compatibility and interconnection with other systems that also comply with those standards. An example is personal computers that have the same operating system such as MS-DOS or Unix.

open circuit An electric circuit that has been broken so there is no complete path for current flow.

open-circuit jack A jack that has no circuit-shorting contacts. The circuit can be closed only by the connections made to the plug that is inserted into the jack.

open-circuit parameter One of a set of four transistor equivalent-circuit parameters that specify transistor performance when the input and output currents are chosen as the independent variables. When these parameters are being measured, either the input or output circuit must be open for alternating current.

open-circuit voltage The voltage at the terminals of a source when no appreciable current is flowing.

open collector An output structure in certain bipolar logic families. The output is characterized by an active transistor pulldown for taking the output to a low-level voltage, and no pullup device. Resistive pullups are generally added to provide the high-level output voltage. Open-collector devices are useful when several devices are bused together on one I/O bus such as IEEE-488.

open-ended An object or circuit capable of being extended or expanded.

open loop A signal path that has no feedback.

open-loop gain The gain of an amplifier as measured without feedback.

open-loop control system A control system that has no means for comparing the output with the input for control purposes.

open subroutine A computer subroutine that is inserted directly into the linear operational sequence, in contrast to a closed subroutine, which must be entered by a jump. An

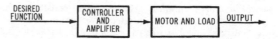

Open-loop control system has no feedback loop and depends on the accuracy of the input pulses.

open subroutine must be recopied at each point that it is needed in a routine.

open-wire feeder *Open-wire transmission line.*

open-wire transmission line A transmission line that consists of two spaced parallel wires supported by insulators at the proper distance to give a desired value of surge impedance. This line acts as a pure resistance when properly terminated. It is also called open-wire feeder.

operand A quantity that is to be operated on by a computer.

operate time The total elapsed time from the instant a relay coil is energized until the contacts have opened or firmly closed.

operating angle The electrical angle (portion of the base voltage cycle) during which collector current flows in a transistor amplifier. Operating angles for three types of amplifiers are class A—360°, class B—180° to 360°, class C—less than 180°.

operating frequency The frequency at which a device or circuit operates.

operating point A point on the family of characteristic curves for an electron tube, transistor, or other active device that corresponds to the average electrode voltages or currents in the absence of a signal.

operating power The power that is actually supplied to the antenna of a transmitter.

operating system [OS] An integrated collection of supervisory programs and subroutines that controls the execution of computer programs and performs special system functions such as formatting floppy and hard disks.

operation The action specified by a single computer instruction.

operational Immediately usable.

operational amplifier [op amp] A voltage feedback, high-gain amplifier whose gain and response characteristics are determined by external components such as resistors, capacitors, or diodes. Now widely available in integrated-circuit form, op amps are the basic circuits in many different monolithic and hybrid linear circuits. Op amps are the front ends of many different sensors.

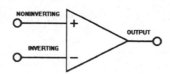

Operational amplifier schematic symbol.

operational power supply A regulated power supply whose control amplifier design has been optimized for signal-processing applications. It will generally have one or more operational amplifiers and a power amplifier in addition to conventional power-supply circuits.

operational programming The process of controlling the output voltage of an operational power supply with voltage, current, or other signals that are operated on by the power supply in a predetermined fashion.

operational trigger A circuit that provides trigger action in response to very small changes in input by combining some of the features of a Schmitt trigger with those of an operational amplifier, which has positive feedback.

operation code 1. The list of operation parts occurring in an instruction code, together with the names of the corresponding operations. 2. *Operation part.*

operation number A number that indicates the position of an operation or its equivalent subroutine in the sequence which forms a problem routine for a computer.

operation part The part of a computer instruction that usually specifies the kind of operation to be performed but not the location of the operands. Also called operation code.

operations research [OR] The use of analytic methods, adapted from mathematics, to provide criteria for decisions concerning the actions of people, commercial or military organizations, and machines.

operation time The time after simultaneous application of all electrode voltages for a current to reach a stated fraction of its final value.

operator 1. A person whose duties include operation, adjustment, and maintenance of a piece of equipment, such as a computer or transmitter. 2. A symbol that represents a mathematical operation to be performed.

opposition The condition in which the phase difference between two periodic quantities that have the same frequency is 180°, corresponding to one half-cycle.

optic 1. A reference to the eye. 2. A reference to the lenses, prisms, and mirrors of a camera, microscope, or other conventional optical instrument.

optical A reference to visible or near-visible light. The extreme limits of the optical spectrum are about 1000 Å (0.1 μm or 3×10^{15} Hz) in the far ultraviolet and 300,000 Å (30 μm or 10^{13} Hz) in the far infrared.

optical amplifier An optoelectronic amplifier whose electric input signal is converted to light, amplified as light, then converted back to an electric signal for the output.

optical axis 1. The straight line that passes through the centers of curvature of the surfaces of a lens. Light rays passing along this direction are neither refracted nor reflected. 2. In a quartz crystal, the Z axis is the optical axis that runs lengthwise through the mother crystal from apex to apex.

optical cable One or more *optical fibers* in a protective sheath or jacket suitable for the transmission of light signals. It is also called a *fiber-optic cable.*

optical character reader [OCR] A character reader that scans printed or handwritten characters with some form of photoelectric system and compares combinations of light and dark areas with character data stored in a computer or other recognition unit to find a match. The corresponding machine-readable code for the identified character is then fed into a computer or other data-processing machine. It is also called a *photoelectric character reader.*

optical character recognition [OCR] Automatic recognition of printed characters by optical methods that are gen-

ABCDEFGH IJKLMNOP

Optical character recognition font developed by American National Standards Institute.

erally performed with a scanning light beam and a light-sensitive device.

optical communication [op-com] Communication over short distances with beams of visible, infrared, or ultraviolet radiation, or over much longer distances with laser beams.

optical computer A computer based on various combinations of holography, lasers, and mass-storage memories for such applications as ultrahigh-speed signal processing, image deblurring, and character recognition. Specific applications include fingerprint identification and the sharpening of images obtained with coherent side-looking synthetic-aperture radar. A typical system consists of an optical input device, image scanner, image analyzer, bandwidth compressor, and digitizer feeding a digital computer. It has applications in machine processing, enhancement, and analysis of aerial, radiographic, and other images.

optical computing A form of data processing that is based largely on the concepts of Fourier optics and holography. It has applications in processing images obtained from acoustic, electronic, microwave, optical, or other sources. Applications also include analysis and enhancement of aerial and radiographic images.

optical contact bond A bond that depends on the property of two highly polished plane metal surfaces to adhere with great force when pressed together. The bond is used for the cold welding of parts that must transfer acoustic waves at microwave frequencies. The bond must be a small fraction of the wavelength of the energy being transferred.

optical coupler *Optoisolator.*

optical coupling Coupling between two circuits by a light beam or light pipe terminated with transducers at opposite ends, to isolate the circuits electrically.

optical density *Photographic transmission density.*

optical diffraction velocimeter *Laser velocimeter.*

optical encoder An encoder that converts positional information into corresponding digital data by interrupting light beams directed on photoelectric devices.

optical fiber A transparent fiber made of glass or plastic capable of transmitting modulated analog or digital light signals in the visible and infrared regions in a fiber-optic system. Glass fiber is preferred for transmission distances greater than 1 km because of lower losses. A single fiber in a protective sheath can be an optical cable. See also *single-mode fiber* and *multimode fiber.*

optical-fiber cable *Optical waveguide.*

optical filter A filter that consists of a pane of glass or other selectively transparent material which transmits only certain wavelength ranges in the visible, ultraviolet, and infrared spectrums.

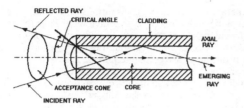

Optical fiber diagram illustrating the key fiberoptic terminology.

optical heterodyning The process of combining an incoming laser signal beam with a laser local oscillator beam by focusing both beams on the surface of a radiation detector. A current at the difference frequency of the two laser beams will be generated by the nonlinear process of photodetection for conventional amplification.

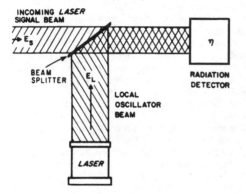

Optical heterodyning.

optical isolator *Optoisolator.*

optically coupled isolator *Optoisolator.*

optical maser *Laser.*

optical memory A computer memory based on optical techniques that generally depend on an addressable laser beam, a storage medium which responds to the beam for writing and perhaps for erasing, and a detector which reacts to the altered character of the medium when the beam reads out stored data.

optical modulator A device for modulating or chopping a beam of radiant energy. Tuning fork, motor-driven, or vibrator devices will work at frequencies up to several thousand hertz, and electrooptical devices will work for frequencies into the gigahertz range.

optical pumping The process of altering the number of atoms or atomic systems in a set of energy levels by absorption of incident light, thereby raising the energy levels. In a semiconductor laser, the semiconductor material is irradiated with photons that have sufficient energy to produce the electron-hole pairs required for population inversion and lasing action.

optical reader A computer data-entry machine that converts printed characters, bar or line codes, and pencil-shaded areas into a computer-input code format. When scanning occurs, the optical image is converted into an electric signal that is unique for each character, and this signal in turn is recognized and converted into the corre-

sponding computer code. Examples include bar code, mark-sense, and optical character readers.

optical relay An optoisolator whose output device is a light-sensitive switch that provides the same on and off operations as the contacts of a relay. It has an infrared emitting LED (IRED) as its input device and a photosensitive diode, transistor, SCR, IC or other device at its output.

optical resonance Luminescence caused by the exciting and emitted radiation that have essentially the same frequencies.

optical scanner A scanner that moves a beam of light across printed or written characters, symbols, or bar codes, measures the reflected light photoelectrically, and converts the resulting signals into corresponding digital representations.

optical shaft-angle encoder See *absolute shaft-angle encoder* and *incremental shaft-angle encoder.*

optical sound recorder *Photographic sound recorder.*

optical sound reproducer *Photographic sound reproducer.*

optical sound track A sound track that consists of a variable-width or variable-density photographic recording of sound signals, placed at one side of the frames of a motion-picture film.

optical star tracker A star tracker whose optoelectronic equipment locks onto the light of a single star.

optical storage Storage of large amounts of data in permanent form on photographic film or its equivalent, for nondestructive readout by laser light source and photodetector. It can store fixed data in such applications as airborne navigation computers and machine-tool or process-control computers. Optical storage is generally in the form of a rotating disk optical detector with reading heads for each recorded circle of data on the disk or with heads that are moved to the desired track for readout.

optical twinning A defect that occurs in natural quartz crystals. Those small regions of unusable material are discarded when a piezoelectric crystal is cut.

optical waveguide A waveguide whose light-transmitting material such as a glass or plastic fiber transmits information from point to point at wavelengths in the ultraviolet, visible-light, or infrared regions of the spectrum. It is also called a fiber waveguide or an optical-fiber cable.

optical window The spectral region between 3000 and 20,000 Å (0.3 and 2 μm in wavelength). Visible and near-visible radiation will pass through the atmosphere of the earth in this region.

optics The branch of science concerned with the phenomena of light and vision. It is related to the electromagnetic spectrum between microwaves and X-rays. This range includes ultraviolet, visible, and infrared radiation.

optimum bunching The bunching condition that produces maximum power at the desired frequency in an output gap of a klystron microwave tube.

optimum coupling *Critical coupling.*

optimum load The load impedance value for maximum transfer of power from source to load.

optimum programming Computer programming based on instructions and data that are stored to minimize access time.

optimum working frequency The most effective frequency at a specified time for ionospheric propagation of radio waves between two specified points.

optoacoustic modulator A modulator whose acoustic wave interacts with a coherent light beam in a medium such as a single crystal of lead molybdate. It can deflect and modulate a laser beam.

optocoupler A solid-state device that provides high electrical isolation by converting the input signal to light emission and reconverting it to an electrical signal. It consists of a photoemitter (such as a light- or infrared-emitting diode—LED or IRED) and a photodetector (such as a phototransistor). Optocouplers prevent the transmission of unwanted noise and provide coupling between systems operating at different voltage levels. It is also called an optoisolator or photocoupler.

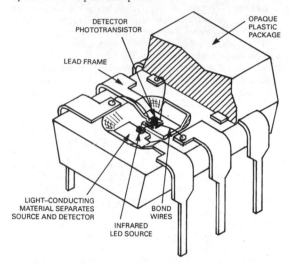

Optocoupler or optoisolator provides an isolated optical link between two different circuits with an infrared LED transmitter and phototransistor receiver.

optoelectronic amplifier An amplifier whose input and output signals and the method of amplification can be either electronic or optical.

optoelectronic device Devices that are responsive to or emit or modulate light waves. Today most are solid state. Examples include light-emitting diodes (LEDs), photodiodes, phototransistors, laser diodes, and optocouplers.

optoelectronic isolator *Optoisolator.*

optoelectronics The branch of electronics that is concerned with solid-state and other electronic devices for

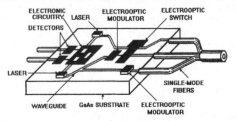

Optoelectronic integrated circuit contains several different optoelectronic functions on a single substrate.

generating, modulating, transmitting, and sensing electromagnetic radiation in the ultraviolet, visible light, and infrared bands of the spectrum. Examples of optoelectronic devices include cathode-ray tubes, electroluminescent displays, holographic equipment, image converters, lasers, light-emitting diodes, light modulators, liquid crystal displays, optical memories, optical waveguides, optoisolators, photodetectors, solar cells, television camera tubes, and electrooptical devices such as Kerr cells. Although sometimes used interchangeably with electrooptics, optoelectronics is a much broader term.

optoisolator See *optocoupler.*

optophone A photoelectric device that converts ordinary printed characters into characteristic sounds which can be recognized by a blind person.

OR 1. A logic operator whose output is a true logic 1 if at least one of the inputs is a true logic 1. 2. Abbreviation for *operations research.*

orbit 1. The path described by a body in its revolution about another body, as by a manmade satellite revolving about the earth. 2. To revolve about another body.

orbital 1. A reference to electrons outside the nucleus of an atom. 2. An energy state or wave function from which the probability of finding an electron at a particular point can be calculated.

orbital electron An electron that is moving in an orbit around the nucleus of an atom.

orbital-electron capture Electron capture in which the electron usually comes from an orbit of the atom or molecule that contains the transforming nucleus.

orbital quantum number A number equal to the angular momentum of an electron in its orbital motion around a nucleus. This number can have all whole values from zero to $n-1$, where n designates the main quantum number.

orbital velocity The velocity of a body in orbit. An earth satellite achieves a stationary orbit at an orbital velocity of about 1.9 mi/s (3 km/s), corresponding to an altitude of 22,300 mi (35,700 km).

orbiting astronomical observatory [OAO] A satellite launched in orbit around the earth primarily for astronomical measurements. One version carries instrumentation to obtain ultraviolet absorption data of intergalactic gas clouds for mapping the sky in the ultraviolet region.

orbiting geophysical observatory An instrumentation satellite that collects and transmits to earth data on cosmic rays, magnetic fields in space, interplanetary dust particles, and other geophysical data.

orbiting solar observatory An astronomical satellite launched in a solar orbit for telemetering to earth continuous measurements of solar phenomena.

order 1. The highest power of the variable in the denominator of a filter's transfer action. It is equal to the number of poles, and in a *switched-capacitor filter,* to the number of op amps (operational amplifiers); it determines the steepness of the rolloff. 2. The number of vibrations or half-period variations of a field along diameters of a circular waveguide or along the wider transverse axis of a rectangular waveguide. 3. A sequence of items or events.

ordinary-wave component One of the two components into which a radio wave entering the ionosphere is divided under the influence of the earth's magnetic field. The ordinary-wave component has characteristics similar to those expected in the absence of a magnetic field. It is also called an *O-wave component*. The other component is the extraordinary-wave component.

ordinate The value that specifies distance in a vertical direction on a graph.

organic semiconductor An organic material that has unusually high conductivity. It is often enhanced by the presence of certain gases, and has other properties commonly associated with semiconductors. An example is anthracene.

OR gate A multiple-input gate circuit whose output is energized when any one or more of the inputs is in a prescribed state.

orientation 1. The relationship between the length, width, and thickness directions of a quartz plate and the rectangular axes of the mother crystal. 2. The physical positioning of a directional antenna or other device that has directional characteristics.

origin The intersection of the reference axes on a graph.

original equipment manufacturer [OEM] A manufacturer who produces equipment in its final form: the manufacturer generally assembles components and subassemblies obtained from other sources.

OR logic A logic circuit organized so that any input will produce an output.

orthicon *Image orthicon.*

orthoferrite A magnetic material that contains a combination of ferrites and rare-earth elements. It forms magnetic domains in bubble memories.

orthogonal mode In a magnetic device, a mode of operation characterized by two mutually perpendicular directions of excitation.

OS Abbreviation for *operating system.*

oscillation 1. A periodic change in a variable, as in the amplitude of an alternating current or the swing of a pendulum. 2. *Cycling.* 3. *Vibration.*

oscillator 1. A circuit that generates alternating current at a frequency determined by the values of its components. The oscillations are produced by positive feedback with an electron tube, transistor, magnetic amplifier, or other amplifying device. Oscillations are maintained by drawing power from a battery or other source of power. For stable frequency characteristics, the circuit can include a crystal, tuning fork, or other essentially unvarying source of vibrations. 2. The stage of a superheterodyne receiver that generates an RF signal of the correct frequency to mix with the incoming signal and produce the IF value of the receiver. 3. The stage of a transmitter that generates the carrier frequency of the station or some fraction of the carrier frequency.

oscillator coil The RF transformer in the oscillator circuit of a superheterodyne receiver or in other oscillator circuits that provides the feedback required for oscillation.

oscillator harmonic interference Interference that occurs in a superheterodyne receiver when an undesired carrier signal beats with a harmonic of the local oscillator to produce the correct IF value.

oscillator-mixer-first detector *Converter.*

oscillator padder An adjustable capacitor connected in series with the oscillator tank circuit of a superheterodyne receiver that permits adjusting the tracking between the

oscillator and preselector at the low-frequency end of the tuning dial.

oscillator radiation The field strength produced at a distance by the local oscillator of a television or radio receiver.

oscillatory circuit A circuit in which oscillations can be generated or sustained.

oscillatory surge A surge that includes both positive and negative polarity values. A unidirectional surge is a pulse.

oscillograph A recorder that produces a permanent record of the instantaneous values of one or more varying electrical quantities. In a cathode-ray oscillograph the record is produced, usually on photographic film, by the electron beam of a cathode-ray tube. In other types of oscillographs, the trace is made by a light beam, pen, heated stylus, or other means of writing directly on plain or treated chart paper.

oscilloscope *Cathode-ray oscilloscope.*

oscilloscope tube A cathode-ray tube that produces a visible pattern which is a graphic representation of electric signals.

OSI Abbreviation for *open system interconnection.*

osmium [Os] A hard metallic element of the platinum group with an atomic number of 76. When alloyed with iridium, it is used to make styli for disk recorders and phonographs.

OSS Abbreviation for *operation subsystem.*

OTO EPROM Abbreviation for *one-time-only erasable read-only memory.*

O-type microwave electron tubes See *linear-beam tubes.*

outage A failure in an electric power system. Because an outage lasting only a fraction of a second can cause errors in data processing, many computers are equipped with uninterruptible power supplies as regulators for line-voltage transients and as backups for complete power-line failures.

outdiffusion Diffusion of impurity atoms out of a semiconductor surface by applying heat.

outdoor antenna A receiving antenna erected outside a building, usually in an elevated location.

outer marker A marker approximately 5 mi (8 km) from the approach end of the runway in an instrument landing system for aircraft. It provides a fix along the localizer course line.

outer-shell electron *Conduction electron.*

outer space The space beyond the atmosphere of the earth.

outgassing The baking out of an electron tube during evacuation to remove residual gases occluded in the tube elements.

outlet A power-line termination from which electric power can be obtained by inserting the plug of a line cord. It is also called a convenience receptacle or receptacle.

out-of-phase The property of waveforms that are of the same frequency but do not have corresponding values at the same instant.

output 1. Output terminals of a device. 2. The result of signal or data processing seen at the output terminals of the circuit or computer. 3. The useful energy from a battery or power supply.

output block A portion of the internal memory of a computer that is reserved for receiving, processing, and transmitting data to be transferred out.

output capacitance   The short-circuit transfer capacitance between the output terminal and all other terminals of an electron tube or transistor, except the input terminal, connected together.

output impedance   The impedance presented by a source to a load. For maximum power output, the output impedance should match the load impedance.

output indicator   A meter or other device that is connected to a radio receiver to indicate variations in output signal strength for alignment and other purposes. It need not indicate the exact value of the output.

output-limited   Restricted by the need to await completion of a computer output operation, as in process control or data processing.

output meter   An AC voltmeter connected to the output of a receiver or amplifier to measure output signal strength in volume units or decibels.

output power   The power delivered by a system or component to its load.

output register   The computer register or memory that holds processed data until it can be fed to an output device or line.

output resonator   The resonant cavity that is excited by density modulation of the electron beam in a klystron and delivers useful energy to an external circuit. It is also called the catcher or catcher resonator.

output stage   The final stage in electronic equipment. In a radio receiver, it feeds the loudspeaker directly or through an output transformer. In an AF amplifier, it feeds one or more loudspeakers, the recording head of a sound recorder, a transmission line, or any other load. In a transmitter it feeds the transmitting antenna.

output winding   A winding other than the feedback winding of a saturable reactor through which power is delivered to the load.

overall electric efficiency   The ratio of the power absorbed by the load to the total power drawn from the supply lines in induction and dielectric heating, stated as a percentage.

overall sound-pressure level   The sound-pressure level measured over the entire frequency range of interest, such as from 20 to 20,000 Hz.

overbunching   The bunching condition produced by continuation of the bunching process beyond optimum bunching in a velocity-modulation tube such as a klystron.

overcoupling   The condition in which two resonant circuits are tuned to the same frequency but coupled so closely that two response peaks are obtained. It can obtain broadband response with substantially uniform impedance.

overcurrent protection   Protection against abnormally high currents in a circuit, component, power supply, or load. It is generally achieved by limiting the duration and magnitude of the abnormally high current.

overdamping   Damping greater than that required for critical damping.

overdriven amplifier   An amplifier whose input signal waveform is intentionally distorted by driving the base past cutoff or into collector-current saturation.

overflow   1. The condition that arises when the result of an arithmetic operation exceeds the capacity of the number representation in a digital computer.   2. The carry digit arising from this condition.

overload   A load greater than that for which a device is designed. It can cause overheating of the power-handling components and distortion in signal circuits.

overload capacity   The current, voltage, or power level beyond which permanent damage will occur to the device or circuit.

overload level   The level at which operation of a system ceases to be satisfactory because of signal distortion, overheating, or other effects.

overload protection   Protection against excessive current by a device that automatically interrupts current flow when an overload occurs. Examples are fuses and circuit breakers.

overload relay   A relay that operates when current flow in a circuit exceeds the normal value for that circuit, to provide overload protection.

overmodulation   Amplitude modulation greater than 100%, causing distortion because the carrier voltage is reduced to zero during those parts of each cycle.

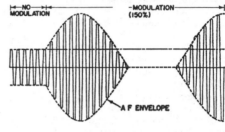

Overmodulation.

over-radiation alarm   A radiation detector that trips an alarm when a predetermined level of radioactivity is reached.

override   To cancel the influence of an automatic control by means of a manual control.

overshoot   A transient change in output voltage that exceeds specified accuracy limits. It usually occurs when power is turned on or off, or with the introduction of a step change in output load or input line.

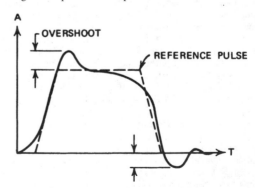

Overshoot occurring with square-wave input.

over-the-horizon communication   *Scatter propagation.*

over-the-horizon radar   A long-range radar whose transmitted and reflected beams are bounced off the ionosphere layers to achieve ranges far beyond the line of sight. It can detect and identify launches of ballistic missiles. The missiles change the reflective characteristics of the ionosphere and as a result, alter the transmitted radar signals. Thus the returned signals have different characteristics.

overthrow distortion   Distortion that occurs when the maximum amplitude of a signal wavefront exceeds the steady-state amplitude of the signal wave. It is also called overshoot distortion.

overtone   A component of a complex tone that has a pitch higher than that of the fundamental pitch. The term overtone is synonymous with harmonic. The nth harmonic is called the $(n-1)$st overtone.

overtone crystal unit   A crystal unit in which the quartz plate operates at a higher order than the fundamental. It is also called the harmonic-mode crystal unit.

overtravel   Movement beyond the operating position in a limit switch.

overvoltage   A voltage higher than the normal or predetermined limiting value.

overvoltage crowbar   A circuit that monitors the output of a power supply and prevents the output voltage from exceeding a preset voltage under any failure condition. It places a low resistance (crowbar) across the output terminals when an overvoltage occurs. A silicon controlled rectifier can be used as the crowbar.

overvoltage protection [OVP]   Circuit protection against excessive voltage provided by one or more overvoltage protective devices built into the circuit such as a metaloxide varistor (MOV), transient voltage suppressing diode (TVS), or gas-discharge tube, or combinations of these. The protective device or devices dissipate or shunt electrical impulse energy at a voltage low enough to ensure the survival of the circuit.

overvoltage protective device   A device designed to dissipate or shunt electrical impulse energy that exceeds a preset voltage level. Examples include the Zener diode, MOV, and transient voltage suppression (TVS) diode.

overvoltage relay   A relay that operates when the voltage applied to its coil reaches a predetermined value.

overvoltage test   A test made at a voltage well above rated operating voltage.

overwrite   To place new data in a memory location so that any data previously stored in that location is destroyed. This is a property found in computer disks, tape cassettes, and read/write (random-access memories) SRAM and DRAM.

OVP   Abbreviation for *overvoltage protection.*

O-wave component   *Ordinary-wave component.*

Owen bridge   A four-arm AC bridge that measures self-inductance in terms of capacitance and resistance. Bridge balance is independent of frequency.

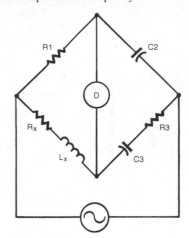

Owen bridge can measure a wide range of inductance values in terms of resistance and capacitance.

oxide-coated cathode   A cathode that has been coated with oxides of alkaline-earth metals to improve electron emission at moderate temperatures.

oxide isolation   Isolation of the elements of an integrated circuit by forming a layer of silicon oxide around each element.

oxide passivation   Passivation of a semiconductor surface by producing a layer of an insulating oxide on the surface.

oxygen [O]   A colorless, odorless, tasteless, gas element with an atomic number of 8 and an atomic weight of 15.99. It is the most abundant element on earth and the second most abundant free gas in the earth's atmosphere (next to nitrogen), forming about 21% of its volume. It is a very active element and it will combine with nearly all other elements. It is essential to all life processes and for combustion. Oxides of aluminum, tantalum, and silicon are important materials in electronics.

oxygen-free high-conductivity copper [OFHC]   Pure copper that has 100% conductivity, used for the construction of high-power electron tubes because it does not release appreciable gas when hot.

ozone-producing radiation   Ultraviolet radiation shorter in wavelength than about 220 nm, at which oxygen is decomposed to produce ozone.

P

p Abbreviation for *pico-*.

P 1. Symbol for *primary winding,* used on circuit diagrams to identify the primary winding of a transformer. 2. Abbreviation for *peak.* 3. Abbreviation for *positive.*

P 1. Symbol for *permeance.* 2. Symbol for *power.*

pA Abbreviation for *picoampere.*

Pa Abbreviation for *pascal.*

PABX Abbreviation for *private automatic branch exchange.*

pacemaker A pulsed, battery-operated oscillator implanted in the body to deliver electric impulses to the muscles of the lower heart, either at a fixed rate or in response to a sensor that detects when the patient's pulse rate slows or ceases. Several different power sources are used. The most common and least costly is a mercury-zinc, lithium, or other primary cell that requires replacement surgically under local anesthetic at intervals of 2 to 6 years. It is also called a cardiac pacemaker and a heart pacer.

pack To combine several different brief fields of information into one machine word for a digital computer.

packaged magnetron An integral structure consisting of a magnetron, its permanent magnetic circuit, and its output matching device.

packaging The process of enclosing a device or circuit in a protective package. It could be a dual-in-line (DIP) package, single-in-line package (SIP), can, module, or case.

packaging density The number of components per unit volume in a system or subsystem.

packed data Information that has been compressed to make optimal use of memory. Four BCD (binary-coded decimal) digits can be packed in a 16-bit memory location.

packet A group of bits transmitted as a block on a *packet-switched network.*

packet protocols In digital telecommunications, methods for sending digital data from one location to another. A packet, as defined by the international X.25 standard, includes bits that (1) define the start and end of a packet, (2) are positioned in front of the data to identify the sender and receiver and supply other control information, (3) make up the data itself, and (4) check to enable the receiver to determine whether the bits have been transmitted correctly. In the X.25 method, data can be transmitted in both directions at the same time. In addition to the data, a packet contains acknowledgments of successful (or faulty) receipt of packets from the other end.

packet-switched radio network A radio network in which a computer divides large data files into short segments or *packets* for transmission. The packets are exchanged by sending and receiving stations, and they are reassembled before reaching their final destination, which could be a computer or facsimile (fax) machine. A channel is occupied only during the transmission of the packet.

packet transmission The transmission of standardized packets of data over transmission lines in a fraction of a second by high-speed switching computers that have the message packets stored in fast-access memory. The system provides automatic checking for transmission errors.

packing density The number of units of digital information per unit length or unit area of a recording or storage medium for a digital computer, such as the number of binary digits of polarized spots per inch of magnetic tape length.

pad 1. An arrangement of fixed resistors that reduces the strength of an RF or AF signal a desired fixed amount without introducing appreciable distortion. It is also called a fixed attenuator. The corresponding adjustable arrangement is called an attenuator. 2. *Terminal area.*

padder A trimmer capacitor inserted in series with the oscillator tuning circuit of a superheterodyne receiver to control calibration at the low-frequency end of a tuning range.

page The contents of one section of a computer memory, generally a subdivision of a program that is moved as a block into the main computer memory.

page A Portion of a program or data to be moved in and out of the main memory as needed.

pager A pocket-sized radio receiver that receives signals indicating that a specified telephone number has been dialed. It emits an audible and/or visual or tactile signal to page the person assigned to that number. Some versions also display the calling telephone number and brief alphanumeric messages on a liquid-crystal display (LCD). It is also called a *beeper* (slang). See also *paging system*.

Pager informs a user that a personal telephone call has been received. Some models include a display for phone numbers and brief messages, in addition to giving an audible alarm.

page printer A printer for a computer that composes and organizes an entire page of text. It provides a printout in final page format, including allowance of space for illustrations.

paging system A radio common-carrier service that has one or more central stations which transmit beeps plus signal display on a selective calling basis to small battery-operated pagers carried by each person to be paged. In a beep-plus phone display system, the person hears both the alerting beep and the message.

pair The two associated conductors that form part of a communication channel.

pairing A faulty interlace in a television picture in which the lines of one field do not fall exactly between those of the preceding field. When this defect is serious, the lines of alternate fields tend to pair up and fall on one another, cutting the vertical resolution in half.

PAL Abbreviation for *phase-alternation line*.

palladium [Pd] A metallic element with an atomic number of 46 and an atomic weight of 106.4 It is used for plating ferrous metals.

Palmer scan A combination of a circular or raster-type radar antenna scan with a conical scan. The beam swings around the horizon concurrently with the conical scan.

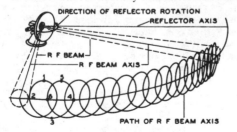

Palmer scan.

palmtop computer A portable personal microcomputer, generally weighing less than 1 lb (0.5 kg), for taking notes, listing appointments, and transmitting facsimiles. Most versions have miniature keyboards. See also *personal communications device* [PCD] and *electronic organizer.*

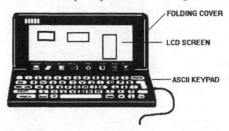

Palmtop or subnotebook computer performs such computer functions as word processing, spreadsheet preparation, scheduling, and the retrieval of personal data.

PAM Abbreviation for *pulse-amplitude modulation*.

PAM/FM Pulse-amplitude modulation of subcarriers, used to frequency-modulate a carrier.

pan To tilt or otherwise move a television camera vertically and horizontally to keep it trained on a moving object or secure a panoramic effect.

pancake coil A coil whose turns are arranged in a flat spiral.

pancake motor A DC servomotor that has a large diameter and thin cross section. It can provide controlled torque for small angular movements, such as for driving scanning antennas or robots. It is also called a *torquer*.

panel meter A meter designed for mounting on a panel, usually in a round or rectangular hole.

panning Moving a television camera across a field of view.

panoramic adapter An adapter used with a search receiver to provide a visual presentation, on an oscilloscope screen, of a band of frequencies extending above and below the center frequency to which the search receiver is tuned.

panoramic display A display that simultaneously shows the relative amplitudes of all signals received at different frequencies.

panoramic indicator An indicator connected to a radio receiver that shows signals received over a band of frequencies centered about the specific frequency to which the receiver is tuned.

panoramic receiver A radio receiver that permits continuous observation, on a cathode-ray-tube screen, of the presence and relative strength of all signals in a wide frequency band through which the receiver is periodically tuned.

paper capacitor A fixed capacitor that consists of two strips of metal foil separated by oiled or waxed paper or other insulating material and rolled together in compact tubular form. The foil strips are staggered so one projects from each end of the roll, and the connecting wires are attached to the projecting foil strips. It is not normally used in electronic circuits.

PAR Abbreviation for *precision approach radar*.

parabolic antenna An antenna that includes a *parabolic reflector* to concentrate the transmitted or received RF energy into a parallel beam. It is a common configuration for terrestrial microwave transmission and receiving antennas as well as satellite-transmission receiving antennas. Some parabolic antennas are sections of complete parabolas that form the transmitted beam into horizontal or vertical fan shapes. It is also called a *dish antenna*.

parabolic microphone   A microphone at the focal point of a parabolic sound reflector to give improved sensitivity and directivity.

parabolic reflector   A reflector whose inner surface is shaped by rotating a parabola about its axis. When a microwave transmitting dipole, horn, or other antenna is placed at its focal point, the reflector concentrates the radiation into a parallel beam. For reception, incoming radiation is reflected to the receiving antenna at the focal point. The reflector can be made from wire screen or sheet metal.

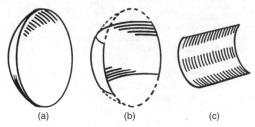

Parabolic reflector forms include: (a) full paraboloid, (b) cut paraboloid, and (c) parabolic cylinder.

parallax   The apparent displacement of the position of an object caused by a shift in the point of observation. Thus, the pointer of a meter will appear to be at different positions on the scale, depending on the angle from which the meter is read.

parallel circuit   A circuit in which the same voltage is applied to all components and the current divides among the components according to their resistances or impedances.

parallel cut   A Y cut in a quartz crystal.

parallel feed   *Shunt feed.*

parallel input/output [I/O]   The fastest, simplest method for interconnecting to data-processing devices with a minimum of circuitry. Data is transferred in a bit-parallel format, with the width of the interconnect bus generally equal to the computer memory width, in bits. Although 16-bit buses are common, 32-bit buses will simultaneously handle the transmission of four characters.

parallel interface [Centronics type]   The protocol for 36-pin, byte-serial interface connectors now widely accepted as the standard in computer-to-printer communications.

parallelism   1. In computing, the ability to perform a large number of operations simultaneously but independently, such as switching all the logic gates in an entire array. 2. In *neural networks,* the ability to affect all the neurons simultaneously, but dependently, because the network architecture causes feedback between elements.

parallel memory   Computer memory in which all bits, characters, or words are essentially equally available.

parallel operation   The flow of information through a part or all of a computer over two or more lines or channels simultaneously.

parallel-plate waveguide   A waveguide that consists of two metal strips whose width is large compared to the spacing between them. For the dominant transverse electromagnetic wave, this waveguide has an infinite cutoff wavelength, and guide wavelength is equal to free-space wavelength.

parallel processing   Processing of two or more programs simultaneously in a computer that has more than one active processor.

parallel-resistance equivalence   Resistor clusters configured in shunt or parallel have equivalent resistance values that are smaller than the lowest resistor value in the cluster. If two resistors of equal value are in parallel, the equivalent value is the product of the two resistor values divided by their sum. Thus, two 10-Ω resistors in parallel have the equivalent value of 5 Ω.

parallel resonance   Resonance in a parallel resonant circuit. The inductive and capacitive reactances are equal at the frequency of the applied voltage. The impedance of the parallel resonant circuit is then a maximum, so maximum signal voltage is developed across it. It is also called antiresonance.

parallel resonant circuit   A resonant circuit in which the capacitor and coil are in parallel with the applied AC voltage. It is also called antiresonant circuit.

parallel-T network   *Twin-T network.*

parallel transfer   Computer data transfer in which the characters of an element of information are transferred simultaneously over a set of parallel paths.

parallel transmission   A transmission mode in which bits are sent simultaneously over separate, typically unidirectional, transmission lines.

parallel-wire line   A transmission line that consists of two parallel wires.

parallel-wire resonator   A resonator that consists of a length of parallel-wire transmission line short-circuited at one end.

paramagnetic   Having a magnetic permeability greater than that of a vacuum and essentially independent of the magnetizing force. In ferromagnetic materials, the permeability varies with magnetizing force.

paramagnetic amplifier   *Maser.*

paramagnetic resonance   Resonance observable in a paramagnetic material as a peak in the energy absorption spectrum at a frequency related to the strength of the applied magnetic field and the gyromagnetic ratio. The phenomenon is useful in studying the energy states of nuclei, atoms, molecules, and crystal lattices.

paramagnetism   Magnetism that involves a permeability only slightly greater than unity.

parameter   A quantity to which arbitrary values can be assigned, such as the value of a transistor or tube characteristic, or the value of a circuit component. A parameter is usually not changed during a given set of conditions.

parameterized cell   Integrated circuit building blocks such as adders, barrel shifters, or multiplexers, that can be slightly modified to meet particular requirements for speed, power consumption, or bit length.

parametric amplifier [paramp]   A microwave amplifier that includes an electron tube or solid-state device whose reactance can be varied periodically by an AC voltage at a pumping frequency.

parametric device   A device whose operation depends essentially on the variation of some parameter with time. As an example, the parameter can be a reactance that is varied by an AC control voltage.

parametric frequency converter   A frequency converter that depends on the variation of the reactance parame-

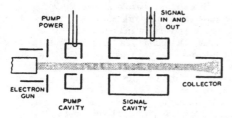

Parametric amplifier based on an electron-tube.

ter of an energy-storage element to obtain frequency conversion.

parametric modulator A modulator based on the variation in the reactance parameter of an energy-storage element to produce modulation.

parametric test A test that measures dc conditions of an integrated circuit, such as maximum current, leakage, and output drive.

paramp Abbreviation for *parametric amplifier.*

parasitic Undesired capacitance and inductance that occurs between the elements in active devices such as transistors or passive devices such as resistors or capacitors that limit frequency response. Parasitics in semiconductor devices are minimized by zone-isolation wafer-fabrication techniques.

parasitic element An antenna element that serves as part of a directional antenna array but has no direct connection to the receiver or transmitter. A parasitic element reflects or reradiates the energy that reaches it in a phase relationship that gives the desired radiation pattern. It is also called a passive element.

parasitic oscillation An undesired self-sustaining oscillation or a self-generated transient impulse in an oscillator or amplifier circuit, generally at a frequency above or below the correct operating frequency.

parasitic suppressor A suppressor, usually in the form of a coil and resistor in parallel, inserted in a circuit to suppress parasitic high-frequency oscillations.

PARD Abbreviation for *periodic and random deviation.*

parity An error-detection method used in input/output (I/O) procedures where noise can cause data errors. Parity is determined by counting the number of ones in the data word. Odd parity sets the parity bit so that the total number of ones sent is odd. Even parity sets the parity bit for an even number.

parity bit A 0 or 1 bit that is added to an existing character to ensure that the total number of 1 bits in the data field is either even or odd for parity check purposes. See *parity check.*

parity check The addition of a noninformation bit to a data block to be transmitted to ensure that the number of 1s is either odd or even, according to a predetermined requirement.

parsec [PARallax-SECond] A unit of distance for interstellar space equal to 3.26 light-years, 206,000 astronomical units, or 19.15×10^{12} mi (30.82×10^{12} km).

part An article that is an element of a subassembly, is not normally useful by itself, and is not amenable to further disassembly for maintenance purposes. The term is used chiefly for structural members in electronic equipment. Examples include circuit boards, knobs, fas-

teners, and cases. Transistors, resistors, capacitors, coils, switches, relays, transformers, and similar items that have distinct electrical characteristics are usually called components.

partial A sound-sensation component that is distinguishable as a simple tone, cannot be further analyzed by the ear, and contributes to the character of the complex sound. The frequency of a partial can be higher or lower than the basic frequency and might or might not be an integral multiple or submultiple of the basic frequency.

partial node The points, lines, or surfaces in a standing-wave system where some characteristic of the wave field has a minimum amplitude that differs from zero.

particle Any very small part of matter, such as an atom, electron, proton, molecule, neutron, alpha particle, or beta particle.

particle accelerator *Accelerator.*

particulate A characteristic of having the form of separate, very small particles.

partition noise Noise that arises in an electron tube when the electron beam is divided between two or more electrodes, as between screen grid and anode in a pentode.

parts per million [PPM] A unit for specifying the precision with which a frequency, voltage, or other parameter is generated, measured, or controlled.

party-line carrier system A single-frequency carrier telephone system whose carrier energy is transmitted directly to all other carrier terminals of the same channel.

pascal [Pa] The SI unit of pressure or stress. A pressure of 1 Pa is equal to 1 N/m^2.

Paschen's law The sparking potential between two parallel-plate electrodes in a gas is proportional to the product of gas pressure and electrode spacing.

pass A complete cycle of reading, processing, and writing in a computer.

passband A frequency band in which the attenuation of a filter is essentially zero.

pass element An active element, such as a transistor or tube, installed in series with the load of a regulated DC power supply and controlled by an amplifier so that it varies its series resistance as required to maintain a constant DC output voltage.

passband filter A filter that allows only the band of frequencies to pass without attenuation. Frequencies above and below the passband are rejected.

passivate See *passivation.*

passivated transistor A transistor that has been protected against premature failure by passivation.

passivation The deposition of a layer of insulating material such as silicon dioxide or silicon nitride over a semiconductor wafer or a region of a passive device to stabilize and protect its surface against moisture, contamination, and mechanical damage. In semiconductors it reduces reverse-current leakage, increases breakdown voltage, and raises the power dissipation rating. It is also a method for encapsulating chip-type, surface-mount resistors.

passive communication satellite A satellite that reflects communication signals between stations, without providing amplification. An example was the Echo satellite.

passive component An electronic component that is not powered and cannot amplify, oscillate, or convert the form of its input energy. Examples include resistors, capacitors, coils, and transformers.

passive corner reflector   A corner reflector that reflects energy from a distant transmitting antenna. It improves the reflection of radar signals from objects that would not otherwise be good radar targets, such as fiberglass boats.

passive detection   The detection of a target or other object by means that do not reveal the position of the detecting instrument.

passive double reflector   A combination of two passive reflectors positioned to bend a microwave beam over the top of a mountain or ridge, generally without appreciably changing the general direction of the beam.

passive electronic countermeasures   Passive techniques for overcoming *electronic jamming,* such as determining the frequency or modulation schemes of the jamming signals and retuning or changing the modulation of friendly transmitters to minimize or eliminate the interference. The ability to perform these evasive adjustments is called *agility.* See also *antijamming* [AJ] and *spread-spectrum modulation.*

passive element   *Parasitic element.*

passive filter   A filter made from passive components with no active elements.

passive guidance   Guidance of a vehicle by preset or inertial devices without reliance on external signals or observations.

passive homing   Homing that depends only on energy emanating naturally from the target, as in the form of infrared radiation, light, sound, electromagnetic radiation, ionization of air, or air pollution by exhaust gases.

passive infrared detection system   An infrared receiving system that does not require infrared illumination. It includes an optical receiver, cryogenically cooled detector array, signal-processing electronics, and a display (typically a cathode-ray tube [CRT]). The system can sense and display a pattern of infrared radiation which has characteristics that permit it to be recognized as an aircraft, armored vehicle, tank, or ship. It can determine bearing, but not range. The target's IR emission depends on its temperature, emissivity, and viewing angle. The optical receiver acts like an antenna in a radar system by collecting the radiation and sending it on to the detector. Some military systems can be mounted in aircraft or on ground vehicles. These systems can also be used to study wildlife or changes in the environment. See also *active infrared detection system* and *FLIR.*

passive infrared tracking   Tracking by an infrared detector that responds to the normal infrared radiation emitted by all objects. No infrared light source is used, and tracking can therefore be done without conveying information to the target being tracked.

passive jamming   The use of passive reflectors to return spurious and confusing signals to enemy radars. An example is chaff.

passive-matrix display   A liquid-crystal video display developed for *notebook computers* that can display text and images in color. It was developed to replace the *cathode-ray tube* monitor in portable, battery-powered computers. Screen size is from about 10 to 12 in measured diagonally. It is a lower-cost alternative to the *active-matrix display,* but its colors are not as bright and contrast is lower. It is also called a *dual-scan display* and a *passive-matrix screen.*

passive-matrix screen   *Passive matrix display.*

passive network   A network that has no source of energy and hence no active gain elements.

passive radar   Detection of an object at a distance by picking up the microwave electromagnetic energy that any object normally radiates when it is above absolute zero in temperature. Passive radar requires an apparent temperature difference between the object and its surroundings. Radio astronomy techniques in receivers distinguish from noise a desired signal whose level can be less that 1 pW.

passive sonar   Sonar restricted to underwater listening equipment, with no transmission of location-revealing acoustic pulses. It can determine the bearing of the objects it detects but not the range. See also *active sonar.*

passive surveillance   In military and law enforcement electronics, surveillance carried out by receive-only techniques to determine location, direction, range, identification, and transmission parameters (frequency, bandwidth, signal modulation) of possible threats or targets. A technique that prevents radar systems from being attacked by antiradiation missiles, and in submarine sonar, a technique to prevent acoustic torpedoes from homing in on the sonar transducer. It is also used to monitor all sources of active radiation or acoustic energy to ensure that they do not reveal their presence to the enemy's surveillance equipment.

passive transducer   A transducer that contains no internal source of power.

paste solder   Finely powdered solder metal combined with a flux.

PA system   Abbreviation for *public-address system.*

patch   1. A temporary connection between jacks or other terminations on a patchboard. 2. A section of coding inserted into a computer routine to correct a mistake or alter the routine.

patchboard   A board or panel that has many jacks at which circuits are terminated. Short cables called patchcords are plugged into the jacks to connect various circuits temporarily as required in broadcast, communication, and computers. Patchboards for analog or hybrid computers are designed for quick removal without disturbing the patches made on the board, to permit plugging in a patchboard already set up for the next problem.

patchcord   A cord terminated with plugs at each end, to connect two jacks on a patchboard.

patent   A document that confers on inventors for a term of years the exclusive right to make, use, and sell their inventions in practical form.

path   In printed-circuit technology, a conductive strip connecting pads on the surface of one- or two-sided boards or inside a multilayer board, formed by print and etch subtractive methods on copper-foil laminated surfaces, or by plating with additive methods.

path control system   A *closed-loop position control* system that provides for guiding the load or *end effector* to a desired position along a specified arc, trajectory, or path, usually under computer control, as in a *robot.*

path length   The length of a magnetic flux line in a core.

Patriot   A U.S. Army ground-to-air missile that is 17 ft 5 in long with a diameter of 16 in. With a range in excess of 50 mi, it is launched at 2000 mi/h and propelled by solid rocket fuel. It is radar-guided close to the target. Its own on-board radar locks on that target and explodes its warhead on contact or nearby. A system consists of eight unmanned launchers, an electrical generator truck, and a command station. Launchers carry four Patriots.

pattern recognition   The matching of patterns formed by actual images or objects with those stored electronically in computer memory. Silhouettes or masks of objects or scenes to be recognized are stored in computer memory and compared with actual objects or scenes viewed in real time by a television camera. Simple systems can perform automatic parts sorting as a go/no go operation as objects move by on a conveyor belt beneath a television camera. Any part that does not have the correct silhouette or profile is automatically removed by a rejection mechanism. In more advanced systems, the TV camera scans an actual scene and compares it with significant photographic visual elements stored in computer memory. It has been demonstrated as a navigation aid in slow-flying guided missiles. Lakes, rivers, shorelines, and mountain ranges are high-contrast visual elements.

Pauli-Fermi principle   Each level of a quantized system can include one, two, or no electrons. If there are two electrons, they must have spins in opposite directions.

payload   The total weight of instrumentation and/or passengers carried by a space vehicle for the intended mission of the flight. Fuel, navigation equipment, and control equipment are not considered parts of the payload.

P band   A band of radio frequencies that extended from 225 to 390 MHz in the obsolete U.S. military designations.

PBN   Abbreviation for *pyrolytic boron nitride.*

PbS   Symbol for *lead sulfide.*

PBS   Abbreviation for *Public Broadcasting Service.*

PbTe   Symbol for *lead telluride.*

PBX   Abbreviation for *private branch exchange.*

PC   Abbreviation for *personal computer* (usually an IBM PC or compatible), *programmable controller, printed circuit,* and *program counter.*

PCB   Abbreviation for *printed-circuit board.*

PCD   Abbreviation for *personal communications device.*

P-channel   A conduction channel formed by holes in a P-type semiconductor, as in a PMOS field-effect transistor.

P-channel MOS [PMOS]   A metal-oxide semiconductor manufacturing process in which selective diffusion of a P-type dopant forms closely spaced source and drain regions within a silicon substrate, with the conducting channel consisting of holes. In contrast, the channel consists of electrons in the N-channel MOS process.

PCM   Abbreviation for *pulse-code modulation.*

PCMCIA   Abbreviation for *Personal Computer Memory Card International Association.*

PCMCIA card   A memory or I/O card conforming to the PCMCIA and JEIDA standards for insertion into a microcomputer. The three categories are: Type I (3.33 mm), Type II (5 mm), and Type III (10.5 mm). A slot that will accept Type III cards is compatible with Type I and Type II cards. A PCMCIA slot is designed to accept peripherals conforming to the PCMCIA standards.

PCM/FM   Pulse-code modulation on frequency modulation.

PCN   Abbreviation for *personal communications network.*

PCS   Abbreviation for *personal communications services.*

PDA   Abbreviation for *personal digital assistant.*

PDBM   Abbreviation for *pulse-delay binary modulation.*

PDC   Abbreviation for *personal digital cellular system.*

PDIP   Abbreviation for *plastic dual-in-line package.*

P display   *Plan-position indicator.*

PDM   Abbreviation for *pulse-duration modulation.*

PDM/FM   Pulse-duration modulation on frequency modulation.

PDM/PM   Pulse-duration modulation on phase modulation.

peak [P]   The maximum instantaneous value of a quantity. It is also called a crest.

peak amplitude   The maximum amplitude of an alternating quantity, measured from its zero value.

peak clipper   *Limiter.*

peak detector   A detector whose output voltage approximates the true peak value of an applied signal. The detector tracks the signal in its sample mode and preserves the highest input signal in its hold mode.

peak electrode current   The maximum instantaneous current that flows through an electrode.

peak energy density   The maximum absolute value of the instantaneous energy density in a specified time interval.

peak envelope power [PEP]   The average RF power supplied by a transmitter during one RF cycle at the highest peak-to-peak values of the modulation envelope. It is equal to the input power indicated by an ammeter and voltmeter when the output amplifier is driven by a continuous RF signal that has the peak amplitude which the amplifier can amplify within allowable distortion limits. For single-sideband transmitters, modulation must be present during measurement because the carrier is suppressed.

peaker   A small fixed or adjustable inductance that resonates with stray and distributed capacitances in a broadband amplifier to increase the gain at the higher frequencies.

peak field strength   *Peak magnetizing force.*

peak flux density   The maximum flux density in a magnetic material in a specified cyclically magnetized condition.

peaking   Increasing the response of a circuit at a desired frequency or band of frequencies.

peaking circuit   A circuit that improves the high-frequency response of a broadband amplifier. In shunt peaking, a small coil is placed in series with the collector load. In series peaking, the coil is placed in series with the base of the following stage. It is present in video amplifiers, often with both types of peaking in the same stage. The circuit converts an input signal to a more peaked waveform.

peaking coil   A small coil placed in a circuit to resonate with the distributed capacitance of the circuit at a frequency for which peak response is desired. A video amplifier near the cutoff frequency is an example.

peaking control   An adjustable resistor-capacitor circuit that controls the wave shape of the horizontal oscillator output pulses, as required to give a linear sweep.

peaking transformer   A transformer that has the number of ampere-turns in the primary needed to produce many times the normal flux density values in the core. The flux changes rapidly from one direction of saturation to the other twice per cycle, inducing a highly peaked voltage pulse in a secondary winding.

peak inverse voltage [PIV]   The maximum rated value of an AC voltage acting in the direction opposite to that in which a device is designed to pass current.

peak level The maximum instantaneous level that occurs during a specified time interval, such as the peak sound pressure level in acoustics.

peak limiter *Limiter.*

peak load The maximum instantaneous load or the maximum average load over a designated interval of time.

peak magnetizing force The upper or lower limiting value of magnetizing force associated with a cyclically magnetized condition.

peak particle velocity The maximum absolute value of the instantaneous particle velocity in a specified time interval.

peak power The maximum instantaneous pulsed power of a transmitted radar pulse. The resting time of a radar transmitter is long compared to its operating time, so the average power is low compared to the peak power.

peak power output The output power of a radio transmitter as averaged over one carrier cycle at the maximum amplitude that can occur for any combination of transmitted signals.

peak pulse amplitude The maximum absolute peak value of a pulse, excluding spikes and other unwanted portions.

peak pulse power The power at the maximum of a pulse of power, excluding spikes.

peak response The maximum response of a device or system to an input stimulus.

peaks Momentary high-volume levels during a radio program; they drive the volume indicator at the studio or transmitter upward.

peak signal level The maximum instantaneous signal power or voltage at a specified point in a facsimile system.

peak sound pressure The maximum absolute value of the instantaneous sound pressure in a specified time interval. The SI unit of sound pressure is the newton per square meter.

peak speech power The maximum value of the instantaneous speech power within the time interval considered.

peak-to-peak [P-P] From a positive to a negative peak in an alternating quantity.

peak-to-peak amplitude The sum of the extreme swings of an alternating quantity in positive and negative directions from its zero value. For a sinusoidal waveform, the peak amplitude in either direction is half the peak-to-peak amplitude.

peak-to-peak voltmeter A voltmeter that measures the voltage difference between the positive and negative peaks of a voltage. Two peak-reading voltmeters connected in series opposition can accomplish this purpose.

peak value The maximum instantaneous value of a varying current, voltage, or power during the time interval under consideration. For a sine wave, it is equal to 1.414 times the effective value. It is also called crest value.

peak voltmeter A voltmeter that reads peak values of an alternating voltage.

PECL Abbreviation for *positive emitter-coupled logic.*

pedestal 1. The structure that supports a radar antenna. 2. A flat-topped pulse that elevates the base level for another pulse. 3. *Blanking level.*

pedestal level *Blanking level.*

Peltier effect The emission of heat at one junction of a thermocouple and the absorption of heat at the other junction when electric current is sent through a thermo-

couple. It was discovered by the French physicist, Jean Peltier, in 1834. It is the inverse of the *Seebeck effect.* See also *thermoelectric cooler.*

pencil beam A narrow radar beam that has an essentially circular cross section.

pencil-beam antenna A unidirectional antenna designed so that cross sections of its major lobe are approximately circular.

pencil tube A long, thin disk-seal UHF oscillator or amplifier tube.

penetration depth The nominal depth below the surface of a conductor within which current is concentrated by the skin effect during induction heating. The higher the frequency, the less the penetration.

Pentium An Intel Corporation tradename for a 500 series, 32-bit microprocessor that is a fast *central processing unit* (CPU) for personal computers, servers, and supercomputers. Each device has 5.5 million transistors and it can add 70 million numbers in 1 s. Versions are available with speeds up to 200 MHz.

pentode A five-electrode electron tube that contains an anode, a cathode, a control electrode, and two additional electrodes which are ordinarily grids.

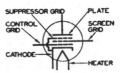

Pentode symbol.

PEP Abbreviation for *peak envelope power.*

perceived noise level The sound-pressure level of a reference sound that is judged as noisy as a given sound. The value is commonly expressed in PNdB.

percent harmonic distortion A measure of the harmonic distortion in a system or component. It is equal to 100 times the ratio of the square root of the sum of the squares of the root-mean-square voltages of each of the individual harmonic frequencies, to the root-mean-square voltage of the fundamental. Current values can replace voltage values.

percent modulation The modulation factor expressed as a percentage.

percent ripple The ratio of the effective value of the ripple voltage to the average value of the total voltage, expressed as a percentage.

perceptron A network of artificial neurons that has pattern-recognizing capabilities.

perfboard A contraction of the term "perforated board," for a blank phenolic laminated-paper circuit board with rows and columns of closely spaced holes for inserting component leads for circuit prototyping. The 0.043-in diameter holes are spaced 0.10 in apart on centers to conform to the 0.10-in standard center-to-center pin spacing of DIP packages, circuit board relays, coils, transformers, and other components. Some also contain solder-plated ground or power buses.

perfect dielectric A dielectric in which all the energy required to establish an electric field in the dielectric is returned to the electric system when the field is removed. A vacuum is the only known perfect dielectric.

perigee   The point on an elliptical orbit at which a satellite is farthest from the earth.

period   The time required for one complete cycle of a regularly repeated series of events.

periodic   The property of having a repetition rate.

periodic and random deviation [PARD]   The noise and ripple voltage that is superimposed on the output of the power supply. It is typically specified at full load and expressed as peak-to-peak or RMS volts over a given bandwidth.

periodic damping   Damping in which a pointer or other moving object oscillates about a new position before coming to rest.

periodic duty   Intermittent duty in which the load conditions are regularly recurrent.

periodic electromagnetic wave   A wave whose electric field vector is repeated in detail at a fixed point after the lapse of a time known as the period.

periodic law   Certain properties of the elements are periodic functions of their atomic numbers. When the elements are arranged in the order of their atomic numbers, these properties recur in regular cycles.

periodic line   A transmission line that has successive identical sections, with nonuniform electrical properties within each section. An example is a loaded line that has uniformly spaced loading coils.

periodic quantity   An oscillating quantity whose values recur at equal increments of time, space, or some other independent variable.

periodic rating   A rating that defines the load which can be carried for specified alternate periods of load and rest.

periodic table   A table in which the elements are arranged according to the periodic law, so elements with similar characteristics are logically grouped together.

periodic wave   A wave that repeats itself at regular intervals, such as a sine wave.

periodic waveguide   A waveguide that has discontinuities at spaced intervals.

peripheral   1. In computer systems, any functional unit not packaged in the computer enclosure that is part of the system. Examples include monitor, printer, keyboard, mouse, and sometimes the modem.   2. Integrated-circuit devices that support a microprocessor to give it a complete set of functions. Examples include input/output (I/O) devices, additional memory devices, and, in some systems, a coprocessor.

permalloy   A magnetic alloy that has high permeability; it usually contains iron, nickel, and small quantities of other metals. It can shield components, tubes, and equipment from stray magnetic fields.

permanent echo   A signal reflected from an object that is fixed with respect to a radar site.

permanent magnet [PM]   A piece of hardened steel or other magnetic material that has been strongly magnetized and retains its magnetism indefinitely.

permanent-magnet centering   Centering of the image on the screen of a television picture tube by magnetic fields produced by permanent magnets mounted around the neck of the tube.

permanent-magnet dynamic loudspeaker   *Permanent-magnet loudspeaker.*

permanent-magnet erasing head   An erasing head that uses the fields of one or more permanent magnets for erasing magnetic tape.

permanent-magnet focusing   Focusing of the electron beam in a television picture tube by the magnetic field produced by one or more permanent magnets mounted around the neck of the tube.

permanent-magnet loudspeaker [PM loudspeaker]   A moving-conductor loudspeaker in which the steady magnetic field is produced by a permanent magnet. It is also called a permanent-magnet dynamic loudspeaker.

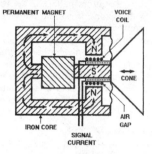

Permanent-magnet loudspeaker operation depends on the interaction between the magnet and the voice coil.

permanent-magnet, moving-coil meter   A DC meter movement that consists of a small coil of wire supported on jeweled bearings between the poles of a permanent magnet. Spiral springs serve as connections to the coil and keep the coil and its attached pointer at the zero position on the meter scale. When the direct current to be measured is sent through the coil, its magnetic field interacts with that of the permanent magnet to produce rotation of the coil. It was formerly known as a d'Arsonval movement.

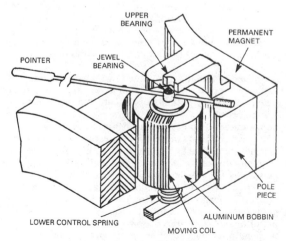

Permanent-magnet moving-coil (D'Arsonval) meter movement can be organized as as an analog voltage, current or power meter.

permanent-magnet moving-iron instrument A meter that depends for its operation on a movable iron vane which aligns itself in the resultant field of a permanent magnet and an adjacent current-carrying coil.

permanent-magnet second-harmonic self-synchronous system A remote indicating arrangement that consists of a transmitter unit and one or more receiver units. All units have permanent-magnet rotors and toroidal stators that use saturable ferromagnetic cores and are excited with alternating current from a common external source. The coils are tapped at three or more equally spaced intervals, and the corresponding taps are connected together to transmit voltages that consist principally of the second harmonic of the excitation voltage. The rotors of the receiver units will assume the same angular position as that of the transmitter rotor.

permanent-magnet stepper motor A stepper motor with a rotor that includes a strong permanent magnet. Each stator coil is energized independently in sequence. The rotor aligns itself with the stator coil that is energized, thus rotating 90° steps if there are four stator coils. For smaller steps, more stator coils or speed-reducing gears are used.

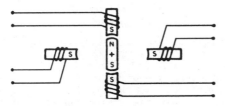

Permanent-magnet stepper motor. Energizing stators in a clockwise sequence steps a permanent-magnet rotor 90° clockwise.

permanent-split capacitor motor [PSC motor] A capacitor motor with a starting capacitor and an auxiliary winding that remains in the circuit for both starting and running. It is also called a capacitor start-run motor.

permeability A measure of the ability of any given material to act as a path for magnetic lines of force, compared with air. The permeability of air is assumed as 1.

permeability tuner A television or radio tuner with a tuning dial that moves the powdered iron cores of coils in the tuning circuits.

permeability tuning The tuning of a resonant circuit by moving a ferrite core in or out of a coil, thereby changing the effective permeability of the core and the inductance of the circuit.

permeameter An instrument that measures the magnetic flux or flux density produced in a test specimen of ferromagnetic material by a given magnetic intensity, to permit computation of the magnetic permeability of the material.

permeance [P] A characteristic of a portion of a magnetic circuit, equal to magnetic flux divided by magnetomotive force. Permeance is the reciprocal of reluctance.

permittance Obsolete term for *capacitance.*

permittivity *Dielectric constant.*

perpendicular magnetization In magnetic recording, magnetization of the recording medium in a direction perpendicular to the line of travel and parallel to the smallest cross-sectional dimension of the medium.

Pershing An Army surface-to-surface guided missile powered by a solid propellant. It is capable of carrying a nuclear warhead for ranges up to 400 nautical miles (740 km).

persistence A measure of the length of time that the screen of a cathode-ray tube remains luminescent after excitation is removed. Long-persistence screens are included in PPI radar displays. Medium-persistence screens are widely used in television receivers. Short-persistence screens are used in cathode-ray oscilloscopes and some types of radar displays. The last number in the type designation of a cathode-ray tube indicates its persistence, on a scale ranging from 1 for short persistence to 7 for long persistence.

persistence characteristic The relation between luminance and time after excitation of a luminescent screen. It is also called the decay characteristic.

persistence of vision The ability of the eye to retain the impression of an image for a short time after the image has disappeared. This characteristic enables the eye to fill in the dark intervals between successive images in movies and television and give the illusion of motion.

persistent current A magnetically induced current that flows undiminished in a superconducting material or circuit. This current, in turn, produces a persistent magnetic field.

persistent-image device An optoelectronic amplifier capable of retaining an image for a definite length of time.

personal communications device [PCD] A palm-sized, limited-capability, battery-powered personal computer intended primarily for taking notes, paging, and facsimile transmission.

personal communications services [PCS] A short-range personal digital cellular telephone communications service intended primarily for urban areas. Antennas with 9-in diameters are mounted on the third or fourth floors of downtown buildings. A base station in or on the building routes the call, which is handed off from one base station to the next by controllers as the person with the telephone moves. The signal travels to a public switching telephone network that sends the call to the intended receiver.

personal computer [PC] A computer based on a microprocessor central processing unit (CPU) intended for personal use in the home or office. A minimum PC includes a monitor (black-and-white or color), and it can have one or two floppy-disk drives and one hard drive. Up to 32 megabytes of memory is included in standard models. Options include a printer, a data modem, facsimile (fax) modem, and a CD ROM drive. Typically a desk-

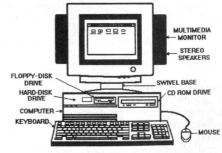

Personal computer can include a CD-ROM drive, stereo speakers, fax, and modem, in addition to hard- and floppy-disk drives.

top unit, the term also applies to smaller and lighter portable *laptop* and *notebook* models.

personal digital assistant [PDA] Any of a number of small, battery-powered, limited-duty computers for taking notes, scheduling work, and, in some models, communication. They typically have small but complete standard keyboards and monocolor liquid-crystal displays (LCDs). See *personal communications device* [PCD], *subnotebook computer* and *palmtop computer.*

Personal Digital Cellular [PDC] **system** The standard digital cellular system in Japan, based on a time-division multiaccess (TDMA) scheme.

PERT [Program Evaluation and Review Technique] A management control tool for defining, integrating, and inter-relating what must be done to accomplish a desired objective on time. A computer can compare current progress against planned objectives and give management the information needed for corrections, planning, and decision making.

perturbation A change in a known system.

perturbation theory The study of the effect of small changes on the behavior of a system.

PESIS Abbreviation for *photoelectron spectroscopy of inner-shell electrons.*

PESOS Abbreviation for *photoelectron spectroscopy of outer-shell electrons.*

PET An abbreviation for *positron-emission tomography.*

peta- [P] A prefix that represents 10^{15}.

PET scanner *Positron-emission tomography scanner.*

pF Abbreviation for *picofarad.*

PF Abbreviation for *power factor.*

PFM Abbreviation for *pulse-frequency modulation.*

PGA Abbreviation for *pin-grid array.*

PGBM Abbreviation for *pulse-gated binary modulation.*

ph Abbreviation for *phot.*

phantom circuit A communication circuit derived from two other communication circuits or from one other circuit and ground, with no additional wire lines.

phantom target *Echo box.*

phase [φ] The position of a point on the waveform of an alternating or other periodic quantity with respect to the start of the cycle. It is expressed in degrees, with 360° representing one complete cycle.

phase-alteration line [PAL] A 625-line, 50-field color television system originally developed in Germany. The hue and saturation information are carried by quadrature modulation, but with one of the two modulations switched 180° from line to line at the transmitter. A delay line in the receiver restores the correct phase of the two modulations by delaying one modulation for the duration of the line. Other countries now using PAL include Austria, Belgium, Brazil, Denmark, Finland, the Netherlands, Sweden, Switzerland, and the United Kingdom (Great Britain). It is also called the *phase-alternation line system.*

phase-amplitude modulation multiplier A multiplier whose carrier phase is made proportional to one variable, and its amplitude is made proportional to the other variable. This signal is applied to a detector whose averaged output is proportional to the product of the two variables. The detector can be a balanced demodulator or a synchronous detector.

phase-angle meter *Phasemeter.*

phase-angle voltmeter A voltmeter that provides both a direct reading of the phase angle and the magnitude of an AC voltage.

phase comparator A comparator that accepts two RF input signals of the same frequency and provides two video outputs which are proportional, respectively, to the sine and cosine of the phase difference between the two inputs.

phase-comparison tracking system A tracking system that provides target trajectory information by continuous-wave phase-comparison techniques.

phase conjugacy A fundamental requirement for retrodirectivity in antenna array. Each element in the array must have an outgoing wave that is delayed exactly as much as the incoming wave was advanced.

phase constant A rating for a line or medium through which a plane wave of a given frequency is being transmitted. It is the imaginary part of the propagation constant, and it is the space rate of decrease of phase of a field component (or of the voltage or current) in the direction of propagation, in radians per unit length. The real part of the propagation constant is the attenuation constant.

phase control 1. A control that changes the phase angle at which the AC line voltage fires a thyratron, ignitron, silicon controlled rectifier, or other control device. It is also called phase-shift control. 2. *Hue control.*

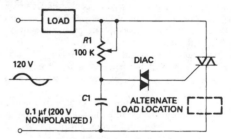

Phase-control circuit with a diac-triggered triac.

phase converter A converter that changes the number of phases in an AC power source without changing the frequency.

phased-array antenna An antenna array whose radiation pattern is determined by the phase relationships of the signals that excite the radiating elements. With adjustable phase shifters operating under computer control, the

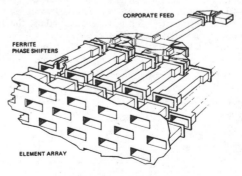

Phased-array antenna.

beam can be scanned in bearing or elevation without mechanical movements.

phased-array radar   *Array radar.*

phase delay   The very short time delay that occurs when a single-frequency wave is transferred from one point to another in a system.

phase detector   A circuit that provides a DC output voltage which is related to the phase difference between an oscillator signal and a reference signal. It controls the oscillator to keep it in synchronism with the reference signal. Color television receivers maintain the subcarrier oscillator in synchronization with the colorburst reference signal with a phase detector. It is also called a phase discriminator or a phase-to-voltage converter.

phase deviation   The peak difference between the instantaneous angle of a modulated wave and the angle of the sine-wave carrier in phase modulation.

phase difference   The time in electrical degrees by which one wave leads or lags another.

phase discriminator   *Phase detector.*

phase distortion   *Phase-frequency distortion.*

phase equalizer   A network that compensates for phase-frequency distortion within a specified frequency band.

phase focusing   An automatic action that helps to keep the electrons of a multicavity magnetron in phase with the rotating field. Lagging electrons receive energy from the radial component of the gap field to reduce the phase lag, and leading electrons give up energy to the gap field to reduce the phase lead.

phase-frequency distortion   A distortion that occurs because phase shift is not proportional to frequency over the frequency range required for transmission. It is also called phase distortion.

phase generator   An instrument that accepts single-phase input signals over a given frequency range, or generates its own signal. It provides continuous shifting of the phase of this signal with one or more calibrated dials.

phase inverter   A circuit or device that changes the phase of a signal by 180°, as required for feeding a push-pull amplifier stage without using a coupling transformer, or for changing the polarity of a pulse.

phase jitter   Abrupt spurious variations in analog signals on the telephone line generally caused by power and communications apparatus on that line that shifts the signal phase back and forth.

phase lock   The technique for making the phase of an oscillator signal follow exactly the phase of a reference signal, by comparing the phases between the two signals and using the resultant difference signal to adjust the frequency of the reference oscillator.

phase-locked loop   [PLL]   A circuit that consists essentially of a phase detector that compares the frequency of a

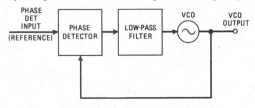

Phase-locked loop block diagram.

voltage-controlled oscillator with that of an incoming carrier signal or reference-frequency generator. The output of the phase detector, after passing through a loop filter, is fed back to the voltage-controlled oscillator to keep it exactly in phase with the incoming or reference frequency. Color television, telemetry, and many other receivers include a PLL.

phasemeter   An instrument that measures the difference in phase between two alternating quantities which have the same frequency. It is also called a phase-angle meter.

phase-modulated transmitter   A transmitter that transmits a phase-modulated wave.

phase modulation   [PM or φM]   Angle modulation in which the phase (expressed as an angle in degrees) of a carrier varies with the amplitude of the modulating signal wave.

phase modulator   A modulator that provides phase modulation of a carrier signal.

phase multiplier   A circuit that multiplies the frequency of signals for phase comparison so that phase differences can be measured to higher resolution.

phaseout   The discontinuation of a major project or operation according to a gradual schedule.

phase-plane analysis   Nonlinear analysis that provides a dynamic graphic time portrait of a simple circuit which has variable inductance and/or capacitance. It is applied in parametric-amplifier circuits.

phase-propagation ratio   The propagation ratio divided by its magnitude.

phase quadrature   *Quadrature.*

phaser   1. A microwave ferrite phase shifter based on a longitudinal magnetic field along one or more rods of ferrite in a waveguide.   2. A device for adjusting facsimile equipment so the recorded elemental area bears the same relation to the record sheet as the corresponding transmitted elemental area bears to the subject copy in the direction of the scanning line.

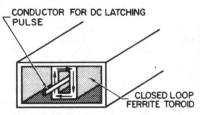

Phaser with a rectangular ferrite toroid in a waveguide.

phase resolution   The minimum phase change that can be distinguished by a given system.

phase reversal   A change of 180° or one half-cycle in phase.

phase-reversal keying   [PRK]   Keying by ±90° phase deviation of the carrier.

phase-reversal modulation   A form of pulse modulation in which reversal of signal phase distinguishes between the two binary states in data transmission.

phase-reversal switch   A switch in a stereophonic sound system that reverses the connections to one loudspeaker so its acoustic output is reversed 180° in phase.

phase-sensing monopulse radar   A monopulse radar whose receiving antenna has two or more apertures separated by several wavelengths, each with its own feed. The apertures give identical radiation patterns. Phase comparison of the arriving signals gives the desired directional information with high precision.

phase-sensitive amplifier   A servoamplifier whose output signal polarity or phase is dependent on the phase relationship between an input and a reference voltage.

phase-sequence relay   A relay that responds to the order in which the voltages or currents in a polyphase system reach maximum positive values.

phase-shaped antenna   *Shaped-beam antenna.*

phase shift   1. A change in the phase relationship between two alternating quantities.   2. The phase relationship between a scattered wave and the incident wave associated with a particle or photon that undergoes scattering.

phase-shift bridge   A mutual-inductance bridge that measures the ratio of two voltages in both magnitude and phase.

phase-shift circuit   A network that provides a voltage component which is shifted in phase with respect to a reference voltage.

phase-shift control   *Phase control.*

phase-shift discriminator   A discriminator that includes two similarly connected diodes and requires a limiter in its input to remove amplitude variations from the frequency- or phase-modulated input signal. The diodes are fed by a transformer that is tuned to the center frequency. When the frequency of the input signal swings away from this center frequency, one diode receives a stronger signal than the other. The net output of the diodes is then proportional to the frequency displacement. It is also called a Foster-Seeley discriminator.

phase shifter   A component in high-frequency, *phased-array* radar systems that causes a phase shift in transmitted frequency because of the active element within the shifter. The simultaneous phase change in a matrix of phase shifters will change the direction of the radar beam electronically. The selection of a suitable shifter for specific antenna system depends on its operating frequency and the RF power to be transmitted. At frequencies above S-band, *ferrite waveguide phase shifters* exhibit lower insertion loss than semiconductor diode units. But *diode phase shifters* predominate in systems that operate at lower microwave frequencies where RF powers are low and there are system size and weight constraints. Phase shifting permits the transmitting and receiving beams to be scanned.

phase-shift keying [PSK]   A modulation system for data transmission in some modems. In its simplest form, the binary modulating signal produces the 0 and 180° phases of the carrier for representing either mark and space or binary 1 and 0. For higher data rates over wire lines, four or eight different phase shifts can occur. The system requires an accurate and stable reference phase at the receiver to distinguish between the various phases employed. With dif-

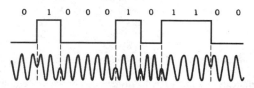

Phase-shift keying.

ferentially coherent phase-shift keying, this reference requirement is eliminated by encoding data in terms of phase changes and detecting these by comparison with the phase of the preceding bit.

phase-shift microphone   A microphone with phase-shift networks that produces directional properties.

phase-shift omnidirectional radio range   An omnidirectional radio range that indicates the azimuthal position of an aircraft by two carrier waves. One is continuously changed in phase. The two waves are in phase only along a reference line that is usually north.

phase-shift oscillator   An oscillator with a network that provides a phase shift of 180° per stage connected between the output and input of an amplifying device.

phase simulator   A precision test instrument that generates reference and data signals at the same frequency but precisely separated in phase.

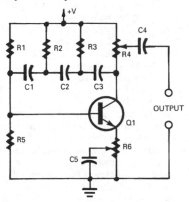

Phase-shift oscillator with an RC network that provides the required 180° phase shift for a transistor amplifier.

phase-splitter   A circuit that takes a single input signal voltage and produces two output signal voltages 180° apart in phase.

phase-to-voltage converter   *Phase detector.*

phase velocity   The velocity of travel of a point that has a certain phase in an electromagnetic wave in the direction of propagation. In a waveguide the phase velocity can be greater than the wave velocity.

phasing   *Framing.*

phasing capacitor   A capacitor for a crystal filter circuit that neutralizes part the capacitance of the crystal holder.

phasing link   A delay line that connects together the bays of a stacked antenna so the signals from all bays are in phase at the transmission line.

phasor   A quantity expressed in complex form, with or without time variation. A phasor can represent a vector, but a vector does not involve a complex plane and hence is not a phasor.

PHEMT   Abbreviation for *pseudomorphic high-electron-mobility transistor.*

phenol formaldehyde   A wear-resistant, rigid plastic that can be molded to precise dimensions. It has excellent electrical insulation properties and retains its properties to ambient temperatures of 250°C. It is used to make knobs, handles, and other electronic hardware. It is also called *Bakelite.*

phenolic A thermosetting plastic material available in many combinations of phenol and formaldehyde, often with added fillers. It provides a broad range of physical, electrical, chemical, and molding properties.

φ [Greek letter phi] Symbol for *phase*.

phenolic laminate A rigid circuit-board material made by laminating paper with phenolic. Available in different thicknesses, the material is immune to common solvents, light in weight, and easily machined. It is widely used to make circuit boards for consumer entertainment products and appliances. See also *perfboard*.

φM Symbol for *phase modulation*.

pH indicator An instrument that measures and indicates the hydrogen ion concentration of a solution on a scale of pH values from 0 to 14. The number 7 indicates a neutral solution, lower numbers indicate acidity, and higher numbers indicate alkalinity.

Phoenix An air-to-air solid-propellant guided missile powered by a solid propellant that has both radar and infrared acquisition, a speed of about Mach 5, and a range of about 400 nautical miles (740 km).

phon The unit of loudness level of a sound. It is numerically equal to the sound pressure level, in decibels relative to 0.0002 μbar, of a pure 1-kHz tone that is judged by listeners to be equivalent in loudness to the sound under consideration.

phone 1. *Headphone.* 2. *Telephone.*

phone jack A female-type audio connector originally developed for telephone switchboards but adapted for consumer entertainment audio systems. The most common jacks are for mating with ¼-in, ⅛-in (*mini*), and ³⁄₃₂-in (*submini*) diameter phone plugs. It can have contacts for two-conductor (*monaural*) and three-conductor (*stereo*) plugs. These jacks are made for panel or chassis mounting and in-line wire or cable connection. See *phone plug.*

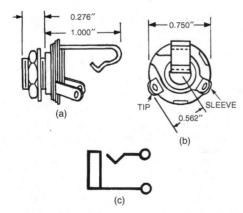

Phone jack for two conductors (monaural applications): (*a*) side view, (*b*) end view, and (*c*) schematic symbol.

phone patch A device that connects an amateur or citizens band transceiver temporarily to a telephone system.

phone plug A male-type audio connector originally developed for telephone switchboards but adapted for consumer entertainment audio systems. The most common are for mating with ¼-in, ⅛-in (*mini*), and ³⁄₃₂-in (*sub-*

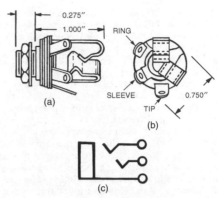

Phone jack for three conductors (stereo applications): (*a*) side view, (*b*) end view, and (*c*) schematic symbol.

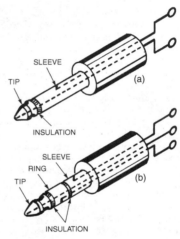

Phone plugs with ¼-in diameter shanks: (*a*) monaural applications, and (*b*) stereo applications.

mini) inside-diameter phone jacks. It can have internal conductors for two-conductor (*monaural*) and three-conductor (*stereo*) plugs. See *phone jack.*

phonetic alphabet A list of standard words to help in the positive identification of letters in a voice message transmitted by radio or telephone.

A	Alfa	J	Juliet	S	Sierra
B	Bravo	K	Kilo	T	Tango
C	Charlie	L	Lima	U	Uniform
D	Delta	M	Mike	V	Victor
E	Echo	N	November	W	Whiskey
F	Foxtrot	O	Oscar	X	X-ray
G	Golf	P	Papa	Y	Yankee
H	Hotel	Q	Quebec	Z	Zulu
I	India	R	Romeo		

phonocardiogram A graphic recording of the sounds of the heart.

phonocardiograph An instrument that provides a graphic record of heart murmurs and other sounds.

phono cartridge *Phonograph pickup.*

phonograph An instrument that converts the sound groove variations of a phonograph record into sound waves. In an electric phonograph, the needle movements in the record grooves are converted into audio-frequency currents and amplified sufficiently for reproduction by a loudspeaker. In a mechanical phonograph, the needle actuates a sound-producing diaphragm directly. The turntable on which the record is placed may be driven by an electric or a spring motor. It is also called a gramophone (British).

phonograph record A shellac-composition or vinyl plastic disk, usually 7, 10, or 12 in (17.8, 25.4, or 30.5 cm) in diameter, on which sounds have been recorded as modulations in grooves. Common speeds used are 16⅔, 33⅓, 45, and 78 rpm. Also called record.

phono jack A female-type audio connector, also called an *RCA jack*. It was developed for phonograph players but adapted for consumer entertainment audio systems. A low-cost product, it is made in many styles for panel or chassis mounting, and in-line wire or cable connection. See also *phono plug*.

phonon A unit of thermal energy in a crystal lattice, equal in value to the product of Planck's constant and the thermal vibration frequency.

phono pickup *Phonograph pickup*.

phono plug A male-type audio connector, also called an *RCA plug*. It was developed for phonograph players but adapted for consumer entertainment audio systems. A low-cost product, it is made in many styles. See also *phono plug*.

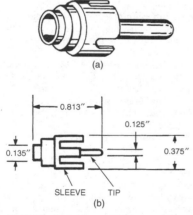

(a)

(b)

Phono plug (or RCA plug): (*a*) general view and (*b*) side view, with dimensions.

phosphor Any material with phosphorescent, fluorescent, or luminescent properties.

phosphor bronze An alloy of copper, tin, and phosphorus, for making contact springs in switches and relays.

phosphor-dot faceplate The glass faceplate on which the trios of color phosphor dots are applied in a shadow-mask three-gun color television picture tube.

phosphorescence A form of luminescence in which the emission of light continues more than 10^{-8} s after excitation by radiation that has a shorter wavelength, such as by electrons, ultraviolet light, or X-rays. When emission of light occurs only during excitation, the result is fluorescence. It is also called afterglow.

phosphorogen A substance that promotes phosphorescence in another substance, as manganese does in zinc sulfide.

phosphorus [P] A chemical element with an atomic number of 15 and an atomic weight of 30.97. It is commonly used as a *donor* doping element for silicon to form an N-type semiconductor.

phot [ph] The CGS unit of illumination, equal to 1 lm/cm². The SI unit (now preferred) is the lux, equal to 1 lm/m².

photocathode A photosensitive surface that emits electrons when exposed to light or other suitable radiation. Phototubes, television camera tubes, and other light-sensitive devices include photocathodes.

photocell A solid-state photosensitive electron device whose current-voltage characteristic is a function of incident radiation. Examples include photoconductive cells, phototransistors, and photovoltaic cells. It is also called an electric eye (slang) and a photoelectric cell. A phototube is not a photocell.

photochemical activity Chemical changes caused by radiant energy, such as light.

photochromic compound A chemical compound that changes in color when exposed to visible or near-visible radiant energy. The effect is reversible. It produces very high-density microimages.

photochromic glass A glass that darkens when exposed to light, but regains its original transparency a few minutes after light is removed. The rate of clearing increases with temperature.

photoconduction A process by which the conductance of a material is changed by incident electromagnetic radiation, such as visible, infrared, or ultraviolet light. An increase in radiation intensity increases the conductance (decreases the resistance).

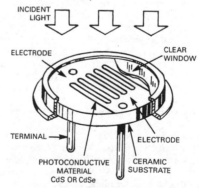

Photoconductive cell: resistance changes when it is exposed to light. Resistive elements are cadmium sulfide or cadmium selenide.

photoconductive cell A photocell whose resistance varies with the illumination on the cell. The selenium cell is an example. When made from a semiconductor material such as cadmium sulfide or cadmium selenide, it gives a good response to infrared radiation. Cooling with liquid air or gas improves the infrared response of many photoconductive detectors. It is also called a light-dependent resistor, a light-sensitive resistor, or a photoresistor.

photoconductivity Conductivity that varies with illumination.

photoconductor A semiconductor whose conductivity varies with illumination.

photocoupler *Optoisolator* or *optocoupler*.

photocurrent An electric current that varies with illumination.

photodarlington A Darlington amplifier whose input transistor that is a phototransistor. The action is essentially the same as if a photodiode were illuminating a standard Darlington amplifier in the same package. It is included in some optoisolators.

photodetector A detector that responds to radiant energy. Examples include photoconductive cells, photodiodes, photoresistors, photoswitches, phototransistors, phototubes, and photovoltaic cells. It is also called a *light detector, light-sensitive cell, light-sensitive detector,* or *photosensor*.

photodiode A semiconductor diode in which the reverse current varies with illumination. Examples include the alloy-junction and the grown-junction photocells.

photodiode parametric amplifier A photodetector arrangement for modulated laser beams. A diode serves simultaneously as a photodiode and a varactor diode, with parametric amplification raising the apparent equivalent resistance of the photodetector. The pumping frequency of the parametric amplifier portion is equal to the sum of the resonant frequencies of the two tank circuits.

photodissociation The removal of one or more atoms from a molecule by the absorption of a quantum of electromagnetic or photon energy.

photoelastic effect The change in the optical properties of a dielectric when subjected to mechanical stress.

photoelectric A property related to the electrical effects of light, such as the emission of electrons, generation of a voltage, or a change in resistance when exposed to light.

photoelectric absorption The absorption of photons in the photoelectric effect.

photoelectric autocollimator An instrument that automatically produces precise electric error signals which are functions of the magnitude and direction of an angular displacement. Accuracy is achieved with a photodetector at the location of the human eye in the basic optical instrument.

photoelectric cathode A cathode that functions primarily by photoelectric emission.

photoelectric cell *Photocell.*

photoelectric character reader *Optical character reader.*

photoelectric color comparator *Color comparator.*

photoelectric colorimeter A colorimeter with a phototube or photocell, a set of color filters, an amplifier, and an indicating meter for quantitative determination of color. It can determine the constituents of a liquid in which the color varies with the constituents in a known manner.

photoelectric color register control A photoelectric control system that functions as a longitudinal position regulator for a moving material or web. It maintains a preset relationship between repetitive register marks when printing successive colors.

photoelectric constant A quantity that, when multiplied by the frequency of the radiation which is causing emission of electrons, gives the voltage absorbed by the escaping photoelectron. The constant is equal to h/e,

where h is Planck's constant and e is the electronic charge.

photoelectric control Control of a circuit or unit of equipment in response to a change in incident light.

photoelectric counter A photoelectrically actuated device that records the number of times a given light path is intercepted by an object.

photoelectric current A current of electrons emitted from the cathode of a phototube under the influence of light.

photoelectric cutoff register controller A photoelectric control system that functions as a longitudinal position regulator to maintain the position of the point of cutoff with respect to a repetitive pattern on a moving material.

photoelectric densitometer An electronic instrument that measures the density or opacity of a film or other material. A beam of light is directed through the material, and the amount of light transmitted is measured with a photocell and meter.

photoelectric directional counter A photoelectrically actuated device that records the number of times a given light path is intercepted by an object moving in a given direction.

photoelectric effect The emission of electrons from a body because of visible, infrared, or ultraviolet radiant energy. The energy of a photon is absorbed for each electron emitted.

photoelectric emission The emission of electrons by certain materials upon exposure to radiation in and near the visible region of the spectrum.

photoelectric flame-failure detector A photoelectric control that cuts off fuel flow when the fuel-consuming flame is extinguished.

photoelectric intrusion-detector A burglar-alarm system that is activated by the interruption of a light beam by an intruder who reduces the illumination on a photosensor and closes an alarm circuit.

photoelectric lighting controller A photoelectric relay actuated by a change in illumination to control the illumination in a given area or at a given point.

photoelectric liquid-level indicator A level indicator in which rising liquid interrupts the light beam of a photoelectric control system.

photoelectric material A material that will emit electrons when exposed to radiant energy in a vacuum. Examples are barium, cesium, lithium, potassium, rubidium, sodium, and strontium.

photoelectric opacimeter *Photoelectric turbidimeter.*

photoelectric photometer A photometer with a photocell, phototransistor, or phototube that measures the intensity of light. It is also called an electronic photometer.

photoelectric pickup A pickup that converts a change in light to an electric signal.

photoelectric pinhole detector A photoelectric control system that detects minute holes in an opaque material.

photoelectric pulse generator A pulse generator with apertures on a shaft-mounted disk or drum that alternately transmits and interrupts light falling on a photocell. The output pulse frequency derived from the photocell is directly proportional to shaft speed.

photoelectric register control A system for the control of an industrial process in which a change in the amount of light reflected from a moving surface triggers a response, such as process shutdown. It contains a uniform light source, photosensors, an amplifier, and one or more relays. Special register marks on a moving web of paper, fabric, or metal can signal the system that the material supply is running out and shut down the process until it is resupplied.

photoelectric relay   A relay combined with a photocell and amplifier arranged so changes in incident light on the phototube make the relay contacts open or close. It is also called a *light relay*.

photoelectric sensitivity   The ratio of photoelectric emission current to incident radiant energy.

photoelectric timer   A timer that is actuated by incident light for the control of some activity, such as turning a light on or off. The control circuit that includes a photoelectric device and amplifier could, for example, be activated by sunrise or sunset.

photoelectric transducer   A transducer that converts changes in light energy to changes in electric energy.

photoelectric tristimulus colorimeter   A colorimeter that uses three or more combinations of light sources, filters, and phototubes to measure colors with high accuracy.

photoelectric tube   *Phototube.*

photoelectric turbidimeter   A photoelectric instrument that determines the turbidity of relatively clear solutions. It is also called a *photoelectric opacimeter*.

photoelectric work function   The energy required to transfer electrons from a given metal to a vacuum or other adjacent medium during photoelectric emission. It is expressed in electronvolts.

photoelectric yield   *Photoelectric sensitivity.*

photoelectromagnetic effect   When light falls on a flat surface of an intermetallic semiconductor located in a magnetic field that is parallel to the surface, excess hole-electron pairs are created. Those carriers diffuse in the direction of the light but are deflected by the magnetic field to give a current flow through the semiconductor that is at right angles to both the light rays and the magnetic field.

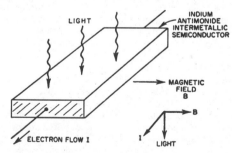

Photoelectromagnetic effect.

photoelectron   An electron emitted by the photoelectric effect.

photoemissive   The property of materials to emit electrons upon exposure to radiation in and near the visible region of the spectrum.

photoemitter   A material that emits electrons when it is illuminated sufficiently. With a photocathode, the emitted electrons are collected in a vacuum. With an internal photoemitter, such as a photoconductive or photovoltaic device, the electrons are detected by measuring the change

they produce in the resistance or some other property of the material. In a negative-electron-affinity photoemitter, high efficiency of light detection is obtained because the electrons are generally deep within the material and are diffused to the surface.

photoemitter cathode   An unheated cathode that emits electrons when it is exposed to light. Electron emission can occur at the surface that is exposed to incident light, as in conventional tubes. It can also be from the opposite side, as in a transmission photocathode.

photoflash tube   *Flash tube.*

photoflash unit   A portable electronic light source for photography. It consists of a capacitor-discharge power source, a flash tube, a battery for charging the capacitor, and sometimes a high-voltage pulse generator to trigger the flash.

photogenerator   A semiconductor-junction device capable of generating or emitting light when pulsed.

photogoniometer   A goniometer based on a phototube or photocell as a sensing device for studying X-ray spectra and X-ray diffraction effects in crystals.

photographic transmission density   The common logarithm of opacity. A film that transmits 100% of the light has a density of 0, and a film transmitting 10% has a density of 1. It is also called optical density.

photoionization   The removal of one or more electrons from an atom or molecule by absorption of a photon of visible or ultraviolet light. It is also called atomic photoelectric effect.

photo-island grid   The photosensitive surface of an image dissector tube for television cameras.

photoisolator   *Optoisolator.*

photolithography   Lithographic techniques that depend on ultraviolet (UV) light to expose photoresist-sensitized semiconductor materials through a mask. See *lithography, electron-beam lithography,* and *X-ray lithography.*

photoluminescence   Luminescence stimulated by visible, infrared, or ultraviolet radiation.

photomagnetic effect   The direct effect of light on the magnetic susceptibility of certain substances.

photomagnetoelectric effect   The generation of a voltage when a semiconductor material is positioned in a magnetic field and one face is illuminated.

photomask   A film or glass negative that has many high-resolution images for the production of semiconductor devices and integrated circuits.

photometer   An instrument that measures the intensity of a light source or the amount of illumination on a surface.

photometry   Measurement of luminous flux and related quantities, such as illuminance, luminance, luminosity, and luminous intensity.

photomixer   A phototransistor that detects optical beats in the superheterodyne receiver of an optical communication system based on coherent laser sources.

photomultiplier counter   A scintillation counter that has a built-in multiplier phototube.

photomultiplier tube   A glass or metal electron tube with a photocathode and discrete or continuous dynodes in front of an anode. Electrons from the photocathode are reflected sequentially from each dynode, where secondary emission adds electrons to the stream at each reflection. The typical photocathode diameter is ¼ to 2 in. It usually has from 5 to 16 dynodes. Typical current amplification is 1

× 10^6 DC; power requirements are from 1 to 3 kV. It is also called a *multiplier phototube* and an *electron-multiplier phototube.*

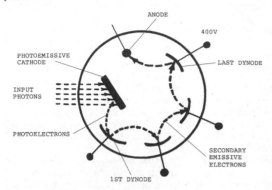

ANODE

400V

PHOTOEMISSIVE CATHODE

LAST DYNODE

INPUT PHOTONS

PHOTOELECTRONS

SECONDARY EMISSIVE ELECTRONS

1ST DYNODE

Photomultiplier tube provides high gain from the additional secondary electrons emitted at each reflection from a dynode.

photon A quantum of electromagnetic radiation, equal to Planck's constant multiplied by the frequency in hertz. Electromagnetic radiation can be considered as photons of light, X-rays, gamma rays, or radio waves.

photon coupling Coupling of two circuits by photons passing through a light pipe.

photonegative The property of negative photoconductivity, hence decreasing in conductivity (increasing in resistance) under the action of light. Selenium can exhibit this property.

photon emission spectrum The relative numbers of optical photons emitted by a scintillator material per unit wavelength as a function of wavelength. The emission spectrum can also be given in alternative units such as wave number, photon energy, or frequency.

photonephelometer An instrument that measures the clarity of a liquid.

photon flux The total amount of luminous flux arriving at the photocathode of a multiplier phototube per unit time, expressed in photons per second.

photopositive The property of positive photoconductivity, hence increasing in conductivity (decreasing in resistance) under the action of light. Selenium has this property.

photoresist A light-sensitive organic compound whose composition is altered by exposure to ultraviolet (UV) light. It is applied to a surface of a semiconductor wafer (typically as a liquid by spinning) to achieve a coating of even thickness. It is then cured before being exposed through a mask for selective material removal. Photoresists can be negative or positive. The molecular structure of *positive photoresist* breaks down on exposure to light, permitting removal of the exposed areas with solvents. By contrast, in *negative photoresist,* the molecular structure of the exposed region is made impervious to solvents which remove areas *not* exposed to UV. After developing, washing, and drying, photoresist defines a pattern that resists subsequent chemical etching to permit the exposed surfaces of the wafer to be etched by chemical action.

photoresistor *Photoconductive cell.*

photo-SCR *Light-activated silicon controlled rectifier.*

photosensitive *Light-sensitive.*

photosensitive recording Recording by the exposure of a photosensitive surface to a signal-controlled light beam or spot.

photosensor *Photodetector.*

photosphere The apparent surface of the sun or of a star from which light is seen to radiate.

phototelegraphy *Facsimile.*

photothermoelectric effect The generation of the voltage in a heat-transmitting semiconductor when it is exposed to light.

photothyristor *Light-activated silicon controlled rectifier.*

phototransistor A junction transistor that might have only collector and emitter leads or also a base lead, with the base exposed to light through a tiny lens in the housing. Collector current increases with light intensity, as a result of amplification of base current by the transistor structure.

phototube An electron tube whose output signal is related to the total radiation that is producing photoelectric emission from the photocathode. The photocathode surface can be chosen for maximum response to a particular part of the visible, infrared, or ultraviolet spectrum. A phototube is not a photoelectric cell. It is also called an electric eye (slang) and a photoelectric tube.

photovoltaic cell *Solar cell.*

photovoltaic effect The generation of a voltage in a *photovoltaic* or *solar cell,* typically a form of silicon *PN diode,* when light or other electromagnetic radiation is absorbed near the *PN junction.* Because the effect produces a current, no battery is required as in *photoconduction.*

phthalocyanine Q switching Laser *Q* switching in which a solution of metal-organic compounds known as phthalocyanines is placed in a cell between an uncoated ruby laser crystal and a high-reflectivity mirror. When the incident ruby light reaches a certain level, the solution suddenly becomes almost perfectly transparent to this light, permitting the release of all the energy stored in the ruby as a giant pulse. The solution then returns to its absorbing state, in readiness for formation of another pulse.

physical optics The branch of optics that considers light as a form of wave motion in which the energy of the light is propagated by wavefronts rather than by rays.

pi The Greek letter π, which designates the ratio of the circumference of a circle to its diameter. A complete circle contains 2π radians. For most purposes sufficient accuracy is obtained with 3.14159265359, 3.14159, or simply 3.1416.

pickoff A device that converts mechanical motion into a proportional electric signal.

pickup 1. A device that converts a sound, scene, measurable quantity, or other form of intelligence into corresponding electric signals, as in a microphone or television camera. A pickup is a transducer only when energy conversion is also involved, as in a microphone. In a telemetering system the end instrument is a pickup. 2. The minimum current, voltage, power, or other value at which a relay will complete its intended function. 3. Interference from a nearby circuit or system. A potentiometer in an automatic pilot that detects the motion of the airplane around the gyro and initiates corrective adjustments.

pickup arm   A pivoted arm that holds a phonograph pickup cartridge. It is also called a tone arm.

pickup cartridge   A cartridge that contains the electromechanical translating elements and the reproducing stylus of a phonograph pickup.

pickup current   The current value at which a magnetically operated device starts to operate. It is also called pull-in current.

pickup spectral characteristic   The spectral response of a pickup that converts radiation to electric signals, as measured at the output terminals of the pickup tube.

pickup tube   *Camera tube.*

pickup value   The minimum voltage, current, or power at which the contacts of a previously deenergized relay will always assume their energized position. It is also called pull-in value.

pickup voltage   The voltage value at which a magnetically operated device starts to operate.

pico- [p; pronounced pie-ko]   A prefix that represents 10^{-12}, which is 0.000000000001, or one-millionth of a millionth. It was formerly called micromicro-.

picoammeter   An ammeter whose scale is calibrated to indicate current values in picoamperes.

picoampere [pA]   One-millionth of a microampere.

picofarad [pF]   One-millionth of a microfarad. It is also called a "puff."

picosecond [ps]   One-millionth of a microsecond.

picowatt [pW]   One-millionth of a microwatt, or 10^{-12} W.

pictorial wiring diagram   A wiring diagram that contains actual sketches of circuit components and shows clearly all connections between the parts.

picture brightness   The brightness of the highlights of a television picture, usually expressed in candelas per square meter.

picture carrier   A carrier frequency located 1.25 MHz above the lower frequency limit of a standard NTSC television signal. In color television, this carrier transmits luminance information; the chrominance subcarrier, which is 3.579545 MHz higher, transmits the color information. The sound carrier, 5.75 MHz above the picture carrier, transmits the sound information for both black and white and color television. It is also called the luminance carrier.

picture element   The smallest subdivision of a television or facsimile image. In a color television receiver it is one color phosphor dot. In a black and white television receiver it is a square segment of a scanning line whose dimension is equal to the nominal line width.

picture frequency   1. A frequency that results from scanning copy in a facsimile system.   2. *Frame frequency.*

picture inversion   Reversal of black and white shades in the recorded copy in a facsimile system.

picture line-amplifier output   The junction between the television studio facility and the line feeding a relay transmitter, a visual transmitter, or a network.

picture line standard   The number of horizontal lines in a complete television image. The NTSC standard is 525 lines.

picture monitor   A cathode-ray tube and associated circuits, arranged to view a television picture or its signal characteristics at station facilities.

picture signal   The signal that results from the scanning process in television or facsimile.

picture-signal amplitude   The difference between the white peak and the blanking level of a television signal.

picture-signal polarity   The polarity of the signal voltage that represents a dark area of a scene with respect to the signal voltage representing a light area. It is expressed as black negative or black positive.

picture size   The useful viewing area on the screen of a television receiver, in square inches.

picture synchronizing pulse   *Vertical synchronizing pulse.*

picture transmission   The transmission, over wires or by radio, of a picture that has a gradation of shade values.

picture transmitter   *Visual transmitter.*

picture tube   *Television picture tube.*

Pierce oscillator   An oscillator that has a piezoelectric crystal connected between the collector and base of a transistor. It is basically a Colpitts oscillator, with voltage division provided by the base-to-emitter and collector-to-emitter capacitances of the circuit.

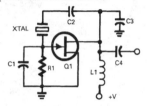

Pierce oscillator is a crystal-controlled oscillator similar to the Colpitts oscillator.

pie winding   A coil winding that is divided into sections called pies to reduce the distributed capacitance of the coil.

Pie winding with three pie sections.

piezodielectric   The property of a material to change its dielectric constant when a mechanical force is applied.

piezoelectric   The property of a material to generate a voltage when mechanical force is applied, or to produce a mechanical force when a voltage is applied, as in a piezoelectric crystal.

piezoelectric axis   One of the axes in a crystal in which either tension or compression will generate a voltage.

piezoelectric ceramic   A ceramic material that has piezoelectric properties similar to those of some natural crystals.

piezoelectric crystal   A crystal that has piezoelectric properties. Crystal loudspeakers and crystal microphones are made from this material.

piezoelectric effect   The generation of a voltage between opposite faces of a piezoelectric crystal as a result of strain due to pressure or twisting, and the reverse effect in which application of a voltage to opposite faces causes deformation to occur at the frequency of the applied voltage. The deformation can produce ultrasonic waves that are useful in ultrasonic cleaning and depth finders.

piezoelectric gage   A pressure-measuring gage that depends on a piezoelectric material to develop a voltage when sub-

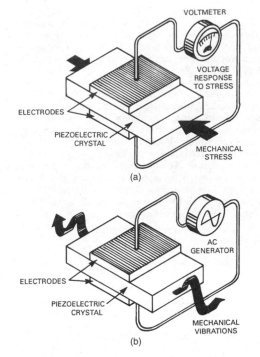

(a)

(b)

Piezoelectric effect: voltage is produced across the electrodes when the crystal is stressed (*a*), and high-frequency oscillations are produced when alternating current is applied across the electrodes (*b*).

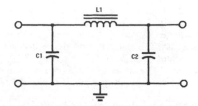

Pi filter reduces AC ripple in DC power supplies.

pillbox antenna   A microwave antenna that consists of a cylindrical parabolic reflector enclosed by two plates perpendicular to the cylinder. The plates are spaced to permit the propagation of only one mode in the desired direction of polarization. It is fed on the focal line.

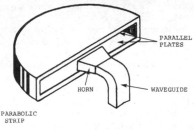

Pillbox antenna distributes the radiation along a long narrow opening, making it a line source of radiation.

jected to pressure. It can measure the pressures that result from underwater explosions.

piezoelectricity   Electric energy resulting from the piezoelectric effect.

piezoelectric loudspeaker   *Crystal loudspeaker.*

piezoelectric microphone   *Crystal microphone.*

piezoelectric pickup   *Crystal pickup.*

piezoelectric transducer   A transducer whose output voltage is produced by deformation of a crystal or ceramic material that has piezoelectric properties.

piezoelectric vibrator   An element cut from piezoelectric material, usually in the form of a plate, bar, or ring, with electrodes attached to the element to excite one of its resonant frequencies.

piezoid   *Finished crystal blank.*

piezooptical effect   The change produced in the index of refraction of a light-transmitting material by externally applied stress.

pi filter   A filter that has a series element and two parallel elements connected in the shape of the Greek letter pi.

pigment   A solid that will naturally reflect some photons of light while absorbing photons of other wavelengths, without producing appreciable luminescence.

pigtail   A short, flexible wire, usually stranded or braided, connected between a stationary terminal and a terminal that has a limited range of motion, as in relay armatures.

pigtail splice   A splice made by twisting together the bared ends of parallel conductors.

pilot   A single frequency, sent over a transmission system to measure or control its characteristics.

pilot channel   A narrow channel over which a single frequency is transmitted to operate an alarm or automatic control.

pilot circuit   The portion of a control circuit that carries the controlling signal from the master switch to the controller.

pilot lamp   A small lamp that indicates a circuit is energized. It is also called a pilot light. When used to illuminates a dial, it is called a dial lamp.

pilot light   *Pilot lamp.*

pilot regulator   A regulator that maintains a constant level at the receiving end of a carrier-derived circuit despite variations in the attenuation of the transmission line. The regulator usually monitors the resistance of a pilot wire that is exposed to essentially the same temperatures as the transmission circuit being regulated.

pilot spark   A weak spark that ionizes the air in preparation for a greater spark discharge.

pilot subcarrier   A subcarrier that serves as a control signal for the reception of stereo FM broadcasts.

pi mode   The mode of operation of a magnetron in which the phases of the fields of successive anode openings facing the interaction space differ by π radians.

pin   A metal lead or terminal that projects from an active or passive component package. It plugs into a hole in a circuit board or a socket for the transmission of input/output (I/O) signals or to receive power. Pins are typically spaced (pitched) 0.10 in-on-centers in DIP packages and 0.05 in-on-centers in SOIC packages.

PIN [Positive-Intrinsic-Negative]   A semiconductor structure that has a high-resistance intrinsic region between low-resistance P- and N-type regions. Microwave diodes, photodiodes, switching diodes, and voltage-dependent variable resistors are made with this structure.

PIN diode A junction diode whose heavily doped P and N regions are separated by a relatively thick layer of high resistivity intrinsic (I) layer. It can switch microwave transmission lines and function as a microwave limiter, replacing a TR tube in systems where peak powers are less than 100 kW. It can also act as a variable microwave attenuator and as an electronically controlled, rapid-acting phase shifter for microwave phased-array systems.

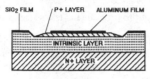

PIN diode has many applications at microwave frequencies. The intrinsic (I) layer is high-resistivity P-type silicon.

pinch A pressed glass stem that supports the internal leads of electron tubes.

pinch effect 1. Constriction of ionized gas to a narrow thread in the center of a straight or doughnut-shaped electron tube through which a heavy current is passed. It is also called rheostriction. 2. Constriction, and sometimes momentary rupture, of molten metal through which a heavy current is flowing.

pinchoff The equivalent of collector cutoff in a field-effect transistor.

pinchoff voltage (V_p or $-V_p$) A gate-to-source bias voltage (V_{GS}) applied to reduce drain-to-source current to zero in a field-effect transistor (FET). In a *depletion-mode* N-channel FET the pinchoff voltage is negative, but for a depletion-mode P-channel FET the pinchoff voltage is positive. See *enhancement mode* and *depletion mode*.

pincushion corrector A television receiver circuit that compensates for pincushion distortion. The horizontal pincushion corrector uses the vertical sawtooth voltage to vary the load on the horizontal sweep system at the vertical rate and thereby straighten the sides of the picture. The vertical corrector circuit uses a parabolic voltage at the horizontal sweep rate to make the picture straight at the top and bottom of the screen.

pincushion distortion Distortion in which all four sides of a received television picture are concave (curve inward).

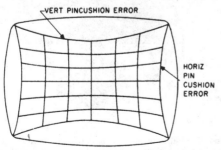

Pincushion distortion.

pin-grid array [PGA] A package for integrated circuits with more than 68 input/output pins. Its array of pins is on the bottom of the package. The pins are inserted in matching holes on a printed circuit board or socket. There can be as many as 600 input/output pins on a PGA, but the typical number is 68 to 144.

pine-tree array An array of horizontal dipole antennas arranged to form a radiating curtain. It is backed up with reflectors.

pi network A network that has three impedance branches connected in series with each other to form a closed circuit. The three junction points form an input terminal, an output terminal, and a common input and output terminal.

ping A sonic or ultrasonic pulse sent out by an active echo-ranging sonar.

ping analyzer An analyzer that stores a selected reflected sonar return or ping, analyzes the waveform, and displays the results on a storage-type cathode-ray tube to aid in the identification of the underwater target.

pinhole detector A photoelectric device that detects extremely small holes and other defects in moving sheets of material.

pin jack A small jack that mates with a plug whose thickness is comparable to that of an ordinary pin.

pink noise Noise whose intensity is inversely proportional to frequency over a specified range, to give constant energy per octave.

Pioneer 6 A 140-lb spacecraft launched on a journey through the solar system on December 16, 1965. It was still transmitting a useful signal on December 16, 1995, making it the oldest operating satellite. It had circled the sun more than 35 times, and had traveled more than 18 billion miles.

pip The target echo indication on a radarscope. It can be a bright spot of light, as on a PPI display, or a sharply peaked pulse, as on an A display.

piped program A radio or television program sent over commercial transmission lines.

pipelined operations A mode of operation for displays when a stream of data such as pixels passes through a set of processing circuits.

pipelining The overlapping of multistep computer processing instructions. For example, the first step of a two-step instruction is processed simultaneously with the second step of the previous instruction.

pi point The frequency at which the insertion phase shift of a device is an integral multiple of π radians or 180°.

Pirani gage A vacuum gage based on the principle that the temperature and resistance of a heated filament vary with the pressure of the surrounding gas. If the amount of gas to conduct away the heat from the filament is insufficient, the filament becomes hotter and its resistance increases.

piston A sliding metal structure used in waveguides and cavities installed for tuning or for reflecting most of the incident energy. It is also called a plunger and a waveguide plunger.

piston action Movement of the entire diaphragm of a loudspeaker as a unit when driven at low audio frequencies.

piston attenuator A microwave attenuator inserted in a waveguide to introduce an amount of attenuation that can be varied by moving an output coupling device along its longitudinal axis.

pistonphone   A small chamber equipped with a reciprocating piston that has a measurable displacement. It establishes a known sound pressure in the chamber for testing microphones.

pitch   1. The attribute of auditory sensation that depends primarily on the frequency of the sound stimulus, but also on the sound pressure and the waveform of the stimulus. Pitch determines the position of sound on a musical scale whose standard pitch is 440 Hz for the tone A. With this standard, middle C is 261.6 Hz.   2. The center-to-center spacing between pads, rows of bumps, pins, posts, and leads on an integrated circuit or circuit board.   3. Rotation of an aircraft, missile, or ship around its traverse axis.

PIV   Abbreviation of *peak inverse voltage*.

pixel   Acronym for *picture element*. The smallest resolvable and addressable segment (in terms of X and Y coordinates) on a video display screen. It can be the pattern made by the red, green, and blue (RGB) electron guns on a color television or computer screen, or the smallest electrode on a liquid-crystal display panel.

PLA   Abbreviation for *programmable logic array*.

place   A position that corresponds to a given power of the base in positional notation. A digit located in any particular place is a coefficient of a corresponding power of the base. Places are usually numbered from right to left. Zero is the place at the right if there is no decimal, binary, or other point, or the column immediately to the left of the point if there is one.

planar-array antenna   An array antenna whose radiating element centers are all in the same plane.

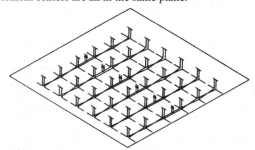

Planar-array antenna.

planar diode   A diode that has planar electrodes in parallel planes.

planar epitaxial passivated diode   A diode that has an oxide-passivated planar structure built into a high-resistivity epitaxial layer grown on a low-resistivity silicon substrate. This construction gives high conductance, fast recovery time, low leakage, and low capacitance.

planar junction transistor   A junction transistor similar to a diffused-junction transistor. Localized penetration of the impurity is achieved by coating parts of the wafer surface with an oxide such as silicon dioxide. This process is known as surface passivation.

planar process   A silicon transistor manufacturing process in which a fractional-micron-thick oxide layer is grown on a silicon substrate. A series of etching and diffusion steps is

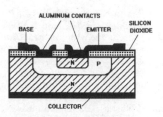

Planar junction NPN transistor showing the positions of the base, emitter, and collector and their external contacts formed in this process.

then carried out to produce the transistor inside the silicon substrate.

Planck's constant [h]   A universal constant equal to 6.626 $\times 10^{-27}$ erg · s. It is the proportionality factor that, when multiplied by the frequency of a photon, gives the energy of the photon.

Planck's law   The fundamental law of quantum theory. It states that energy transfers associated with radiation are proportional to the frequency of the radiation. The proportionality factor is Planck's constant.

plane-earth factor   The ratio of the electric field strength that would result from propagation over an imperfectly conducting plane earth to that which would result from propagation over a perfectly conducting plane.

plane of polarization   The plane that contains the electric field vector and the direction of propagation of a plane-polarized wave. In a horizontally polarized wave this plane is horizontal.

plane-polarized light   Light whose electric vectors of all components of the radiation are in the same fixed plane.

plane-polarized wave   An electromagnetic wave whose electric field vector at all times lies in a fixed plane that contains the direction of propagation through a homogeneous isotropic medium.

plane wave   A wave whose equiphase surfaces form a family of parallel planes.

plan-position indicator [PPI]   A polar-coordinate radar display on which targets appear as bright spots at the same relative locations as they would on a circular map of the area being scanned. The radar antenna is represented as a point at the center of the map. The sweep line is radial from the center of the screen, and it rotates synchronously with the

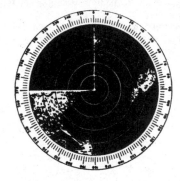

Plan-position-indicator presentation obtained from a search radar showing the location of the aircraft. The radar antenna is pointing at 270°.

radar antenna. The radial distance on the screen from the center to the spot representing the target is an indication of range. The angle measured clockwise from true north (usually at the top of the screen) is an indication of bearing.

plasma Any mixture of particles that contains approximately equal numbers of positive and negative particles along with neutral particles, so the mixture is electrically neutral. The term is generally applied to a gas that is sufficiently ionized to be conductive and affected by magnetic fields. A true plasma, which is completely ionized and has no neutral particles, is produced by temperatures above 20,000 K. In solids, plasma can exist either as electrons and positively charged donors or as holes and negatively charged acceptors.

plasma confinement The use of a magnetic field to confine a gas plasma in the central region of an evacuated chamber long enough to develop the high temperature needed for a controlled-fusion reaction.

plasma display A display panel typically formed as transparent metallized electrodes deposited on opposing faces of parallel sheets of glass that have been sealed at the edges to form a thinly spaced cavity. A gas such as neon is entrapped in the cavity before it is hermetically sealed. The application of voltage across pairs of opposing electrodes causes the gas to ionize in the region and create illuminated spots or segments. The electrodes, in effect, form a matrix of gas diodes that, when activated by voltage, can form alphanumeric characters—dot matrix or segmented. When neon is ionized, it produces a characteristic red-orange color. The displays can be multiplexed under computer control to conserve electrical connections. A plasma display differs from a vacuum-fluorescent display which is electrically a triode and yields characteristic blue-green characters.

plasma rocket engine An electric propulsion system for spacecraft, based on cyclotron resonance in a plasma that is trapped by crossed electric and magnetic fields. It is also called a coaxial plasma accelerator or an electromagnetic rocket engine.

plasma sheath An envelope of ionized gas that surrounds a spacecraft or other body moving through an atmosphere at hypersonic velocities. The plasma sheath affects transmission, reception, and diffraction of radio waves.

plastic One of a large and varied group of organic materials that can be formed into shape by flow at some stage, usually by applying heat alone or with pressure. Plastics are classified as thermoplastic or thermosetting.

plastic-film capacitor A capacitor made with alternate layers of metal foil (usually aluminum) and thin films of plastic dielectric rolled into compact tubular form, just as for a paper capacitor. The foil strips are staggered to project alternately from the row ends, and leads are soldered to the foil. Plastic films in common use include Mylar (a polyester), polycarbonate, polypropylene, polystyrene, and polysulfone. All but polystyrene can be metallized to eliminate the foil strips. It is also called a film capacitor.

plastic leaded-chip carrier [PLCC] A leaded quad integrated-circuit package intended as a replacement for the plastic dual-in-line package (DIP) in conventional circuit board assembly.

plastic quad leaded flatpack [PQFP] A type of square plastic integrated-circuit package that has input/output leads projecting from all four sides.

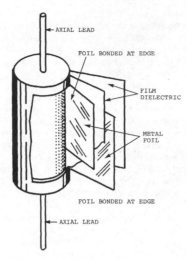

Plastic-film capacitor dielectric is plastic film that is either metallized or sandwiched between metal foil.

plate 1. One of the conducting surfaces in a capacitor. 2. One of the electrodes in a storage battery. 3. [P] *Anode.*

plate circuit *Anode circuit.*

plate current *Anode current.*

plated crystal A crystal unit whose electrodes are metal films deposited directly on the quartz surfaces.

plate detection *Anode detection.*

plated-through hole An interface connection formed by electrodeposition of copper on the walls of a hole in a printed circuit board.

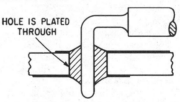

Plated-through hole in a printed circuit board.

platform In the computer industry, a term meaning a complete integrated computer system, such as a desktop or notebook personal computer.

platform stabilization Radar antenna stabilization in which a platform is pivoted so it can be driven to a horizontal position regardless of the motion of the vehicle. Gyroscope signals actuate drive motors that maintain this horizontal position as the vehicle pitches and rolls.

platinum [Pt] A heavy, almost white metal that resists the action of practically all acids, is capable of withstanding high temperatures, and is little affected by sparks. It has an atomic number of 78.

platter Slang term for *hard-memory disk.*

playback Reproduction of a recording.

playback head A head that converts a changing magnetic field on magnetic tape into corresponding electric signals. It is also called a reproduce head.

playback loss *Translation loss.*

PLCC Abbreviation for *plastic leaded-chip carrier.*

PLD Abbreviation for *programmable logic device.*

Plexiglas Trademark of Rohm & Haas Co. for its clear acrylic molding powders and clear acrylic sheets.

PLL Abbreviation for *phase-locked loop.*

PL/M A high-level programming language developed primarily for microcomputers and microprocessors.

PLMN Abbreviation for *public land-mobile network.*

POF Abbreviation for *plastic optical fiber.*

plotter An instrument with an automatically controlled pen or pencil that traces the curves of one or more variables as functions of one or more other variables.

plug The male half of a connector that is normally movable and generally attached to a cable or removable subassembly. A plug is inserted into a jack, outlet, receptacle, or socket.

plugging The braking of an electric motor by reversing its connections, so it tends to turn in the opposite direction. The circuit is opened automatically when the motor stops, so the motor does not actually reverse.

plug-in unit A component or subassembly that has plug-in terminals so all connections can be made simultaneously by pushing the unit into a suitable socket.

Plumbicon Trademark of N. V. Philips for its line of television camera tubes.

plumbing A slang term for the pipelike waveguide circuit elements in microwave radio and radar equipment.

plunger *Piston.*

PLZT Abbreviation for *lead lanthanum zirconate titanate.*

PM 1. Abbreviation for *permanent magnet.* 2. Abbreviation for *phase modulation.*

PM loudspeaker Abbreviation for *permanent-magnet loudspeaker.*

PMOS Abbreviation for *P-channel MOS.*

PMOS RAM A random-access memory made with P-channel metal-oxide semiconductor technology.

PN Positive-negative, a reference to an interface between P- and N-type conductivities in a semiconductor.

PN boundary A surface in a PN junction at which the donor and acceptor concentrations are equal.

PNdB A rating for perceived noise level in decibels. The maximum allowable noise made by an aircraft taking off is specified this way; a typical limit is 112 PNdB.

PN diode A diode, typically made of silicon or gallium arsenide, that has a single PN junction. It is also called a *PN-junction diode.*

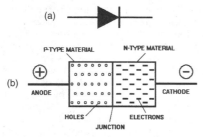

PN diode: (*a*) schematic symbol and (*b*) functional diagram, showing a forward-bias condition.

PNIN transistor An intrinsic-junction transistor that has its intrinsic region between N regions.

PNIP transistor An intrinsic-junction transistor that has its intrinsic region sandwiched between the P-type base and the N-type collector.

PN junction The basic structure formed by the intimate contact of P-type and N-type semiconductors. A PN junction will conduct electric current with one polarity of applied voltage (forward bias), but will not conduct with the opposite polarity (reverse bias). Diodes and rectifiers are PN junctions, and there is a PN junction in all bipolar junction transistors.

PNP Positive-negative-positive, used in referring to a semiconductor in which a layer with N-type conductivity is located between two layers that have P-type conductivity.

PNPN device A device that consists of four alternate layers of P- and N-type semiconductor material. The device always blocks in one direction only when the positive end is positively biased. PNPN devices include four-layer diodes, gate-turnoff silicon controlled rectifiers, and silicon controlled switches. Its function depends on how many of the layers are brought out to terminals.

PNP transistor A semiconductor bipolar junction transistor with a P-type collector and emitter and an N-type base. Current amplification in the device is caused by injection of holes from the emitter into the base and their subsequent collection in the collector. See *NPN transistor.*

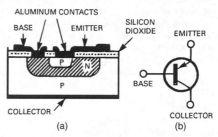

PNP bipolar transistor has an N-type silicon base (*a*) and P-type silicon emitter and collector (*b*).

Pockel's effect The change produced in the refractive properties of certain transparent piezoelectric crystals when an electric field is applied. One application is in the readout of stored electrostatic patterns.

point The character or space that separates the integral and fractional parts of a numerical expression in positional notation, such as the binary point in binary notation and the decimal point in decimal notation. It is also called a base point or radix point.

point contact A pressure contact between a semiconductor die and a metallic point.

point-contact diode A semiconductor diode that depends on a point contact to provide rectifying action. It can function as a detector in RF and microwave circuits. An example is a Schottky diode.

point-contact rectifier A diode whose stray capacitance is minimized with a point contact as a junction, so the flow of current is essentially radial away from the junction. It has high-frequency applications.

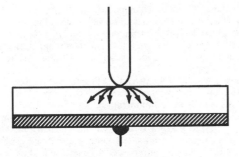

Point-contact rectifier.

point-contact transistor A transistor that has a base electrode and two or more point contacts located near each other on the surface of an N-type semiconductor. Pressure of the points creates a small volume of P-type material under each point to produce the necessary junctions for a PNP transistor. This is an obsolete technology.

pointer The needle-shaped stamping that moves over the scale of a moving-coil meter.

point-of-sale terminal [POS] **terminal** A replacement for the cash register in retail business that combines the cash register function with an on-line data terminal able to perform electronic accounting and carry out credit-card transactions. Data from the terminal is sent to a central computer to keep inventory continuously updated.

point source A radiation source whose dimensions are small compared with the distance from which the radiation is used.

point-to-point communication Radio communication between two fixed stations.

point-to-point wiring A method for assembling circuits on perforated circuit board for constructing prototype circuits. Typically, insulated hookup wires are soldered to exposed pins on active and passive components to complete the circuit. It is a lower-cost but more labor-intensive alternative to construction with a printed-circuit board (PCB).

polar A reference to, measured from, or having a pole, such as the poles of the earth or of a magnet.

polar coordinates A system of coordinates in which the location of a point is specified by its distance from a fixed point and the angle that the line from this fixed point to the given point makes with a fixed reference line.

polar diagram A diagram based on polar coordinates to show the magnitude of a quantity in some or all directions from a point. Examples include directivity patterns and radiation patterns.

polar grid *Grid north.*

polarimeter An instrument that measures the state of polarization of polarized light.

Polaris A Navy surface-to-surface intermediate-range ballistic missile designed to be launched from submarines and surface ships for accurate bombardment of small target areas with conventional or nuclear warheads at ranges up to 2500 nautical miles (4600 km). It uses inertial guidance, with true north, the ship's position, the ship's speed,

and the position of the target being fed into a computer in the missile at the instant of firing.

polarity The characteristic that an object exhibits with opposite properties within itself, such as opposite charges (positive and negative in a battery) or opposing magnetic poles (north and south in a magnet).

polarity of picture signal The polarity of the black portion of a picture signal with respect to the white portion. In a black-negative picture, the potential corresponding to the black areas of the picture is negative with respect to the potential corresponding to the white areas of the picture. In a black-positive picture the potential corresponding to the black areas of the picture is positive.

polarity-reversing switch A switch that interchanges the connections to a device.

polarization 1. The direction of the electric field as radiated from a transmitting antenna. Horizontal polarization is standard for television in the United States, and vertical polarization is standard in Great Britain. 2. Vibration of the vectors of a beam of light or other electromagnetic radiation in a particular direction or manner, as in plane-polarized light. 3. A chemical change occurring in dry cells during use, increasing the internal resistance of the cell and shortening its useful life. 4. A displacement of bound charges in a dielectric when placed in an electric field.

polarization beam splitter A prism that separates light beams passing through it according to polarization.

polarization diversity A communication technique that doubles the number of available channels by using two signals per channel. Each signal has its own distinctive polarization. One signal can have vertical polarization and the other horizontal polarization, or one can be circularly polarized in a clockwise direction, while the other is counterclockwise.

polarization-diversity reception Diversity reception that involves the use of a horizontal dipole and a vertical dipole at the same location. The individual receiver outputs are combined, as in space-diversity reception. The arrangement counteracts the changes in polarization of a received radio wave during fading.

polarization error An error in a radio direction-finder indication due to changes in the polarization of the received wave as atmospheric conditions change. The error is generally greatest at night.

polarization modulation A form of modulation used in optical communication.

polarization receiving factor The ratio of the power received by an antenna from a given plane wave of arbitrary polarization to the power received by the same antenna from a plane wave of the same power density and direction of propagation. It is determined when the state of polarization has been adjusted for maximum received power.

polarized electrolytic capacitor A standard electrolytic capacitor whose dielectric film is formed adjacent to only one metal electrode. Its impedance to the flow of current is then greater in one direction than in the other.

polarized light Light that vibrates in only one plane.

polarized meter A meter that has a zero-center scale, with the direction of deflection of the pointer depending on the polarity of the voltage or the direction of the current being measured.

polarized plug   A plug that can be inserted into its receptacle only when in a predetermined position.

polarized receptacle   A receptacle for use with a polarized plug, to ensure that the grounded side of an AC line or the positive side of a DC line is always connected to the same terminal on a piece of equipment.

polarized relay   A relay in which the direction of movement of the armature depends on the direction of the energizing current in the relay coil.

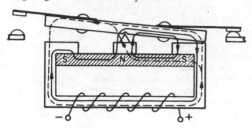

Polarized relay: When a coil is momentarily energized with reverse polarity at a sufficiently high level, the armature moves to the other position and is held there by a permanent magnet.

polarizer   A Nicol prism or other device for polarizing light.

polarizing current   The direct current that is sent through the coil of an iron-core component to establish a reference value of magnetic flux.

polar modulation   Amplitude modulation in which the positive excursions of the carrier are modulated by one signal and the negative excursions by another signal.

Polaroid   Trademark of Polaroid Corp. for a plastic sheet material that produces plane-polarized light.

polar orbit   A satellite orbit running north and south so the satellite vehicle orbits over both the North and South Poles. United States polar-orbit satellites are launched from Vandenberg Air Force Base, California. A satellite in a polar orbit sees a different view on each orbit because the earth is rotating beneath it.

Polar orbit of earth satellite. Earth is rotating continuously about its polar axis, so each orbit traverses a different part of earth.

pole   1. A region in a magnet that has polarity: north pole or south pole.   2. An output terminal on a switch; a double-pole switch that has two output terminals.   3. The roots of the polynomial expression in the denominator of a filter's transfer function; the greater the number of poles, the more complex the circuit and the steeper its rolloff.

pole face   The end of a magnetic core that faces the air gap in which the magnetic field performs useful work.

pole piece   A piece of magnetic material that forms one end of an electromagnet or permanent magnet, shaped to control the distribution of the magnetic flux in the adjacent air gap.

police radio   Two-way radio communication equipment installed in police cars for car-to-headquarters and sometimes car-to-car communications.

Polish notation   A notation system for digital-computer logic, developed by J. Lukasiewicz, in which arithmetic and logic expressions are written without parentheses. In Polish notation, each operator precedes its operands; thus, $(a + b)c$ could be written as $+abc$. It is also called *prefix notation*.

polling   A technique for communicating with and identifying a device on a computer input/output bus that requires service. The processor can periodically interrogate each peripheral device to determine the device's status.

polybutadiene   An improved form of synthetic rubber, reduced from butadiene.

polycarbonate   A thermosetting plastic resin that is virtually unbreakable and weather resistant. It can be molded to make instrument cases and ground to make optical lenses. Light in weight and easily machined, it can be formed by compression and cemented with solvents. As a film it is used as a capacitor dielectric and as the base material for compact disks (CDs). It can also serve as a glass substitute for making protective windows for instrument displays. It is also known by the General Electric tradename of *Lexan*.

polychloroprene   A rubberlike compound used for jacketing on wire and cable that is subject to rough usage, moisture, oil, greases, solvents, and chemicals. It is called neoprene in the wire and cable industry.

polychlorotrifluoroethylene resin   A fluorocarbon resin that has high dielectric strength, widely used for electrical insulation.

polycrystalline cell   A photovoltaic cell made of crystalline semiconductor compounds (two or more different atoms).

polydirectional microphone   A microphone that has provisions for changing its directional characteristics.

polyester   A plastic resin with high dielectric and mechanical strength and resistance to chemicals and moisture. It retains these properties over the temperature range of −7 to 105°C. It is a widely used dielectric in film capacitors. It can be strengthened by adding glass fibers for manufacturing insulators and terminal blocks. It is also known by the DuPont tradename *Mylar* and as *polyethylene terepthalate*.

polyester film   A plastic film made from a polyester, such as Mylar. It is used as a backing for magnetic tape to obtain high strength and resistance to humidity change, and used as a dielectric film in metallized and foil capacitors.

polyethylene   A transparent, water-resistant plastic resin. Tough and flexible, it has excellent insulating properties at ultra-high frequencies. It is used as an insulating material in coaxial cables. In film form, it is used as a capacitor dielectric and as the base material for transparent envelopes and parts-storage bags.

polyethylene terepthalate   *Polyester.*

polygon-type delay line   A delay line made from strain-free fused silica shaped as a polygon, with the surfaces designed to reflect ultrasonic beams many times.

polygraph An instrument that indicates or records one or more functional variables of a body that can change when a person undergoes the emotional stress associated with a lie. Those variables include blood pressure, heart rate, and skin resistance. It is also called lie detector.

polygraphy The science of using a lie detector.

polyimide A tough high-temperature plastic used in film and sheet form for multilayer printed-circuit boards, hybrid-circuit substrates, film capacitor dielectrics, and as a base for flexible circuits and tape cable.

polymethyl methacrylate substrate A support layer of plastic material used to make optical video disks.

polyolefin A plastic insulation for wires in either irradiated or nonirradiated form. It offers performance comparable to that of Teflon in extreme environmental conditions encountered at high altitudes in missiles and spacecraft.

polyphase A reference to two or more phases of an AC power line.

polyphenylene sulfide [PPS] A high-strength, chemical-resistant plastic resin for molding insulating connector bodies, headers and other electronic components. It is typically strengthened by mixing in glass fibers.

polyplexer A radar unit that combines the functions of duplexing and lobe switching.

polypropylene A thermoplastic resin that offers high tensile and dielectric strength, low water absorption, and excellent chemical resistance. In film form it is used as a film-capacitor dielectric. It is made by the polymerization of propylene.

polyrod antenna A microwave antenna that consists of a parallel arrangement of rods made from polystyrene or other good dielectric material. When excited at one end, as from the end of a waveguide, the rods will radiate from their other ends.

polystyrene A thermoplastic resin with excellent electrical properties. It is easily machined, highly resistant to chemicals, and dimensionally stable. It is used to make microwave waveguides and dielectric rod antennas. In film form, it is used as a film-capacitor dielectric.

polystyrene capacitor A capacitor with polystyrene film as a dielectric between rolled strips of metal foil.

polysulfone A tough, rigid, high-strength thermoplastic resin that retains its characteristics over temperatures from 150 to 300°F. It is used to make insulating parts of electrical and electronic components.

polytetrafluoroethylene resin A fluorocarbon resin that has high dielectric strength and a slippery feel, widely used for electrical insulation. The standard designation of the Society of the Plastics Industry is TFE-fluorocarbon resin. Its trademark names include Teflon (DuPont) and Fluon (Imperial Chemical Industries, England).

polyurethane A tough, durable, elastomeric plastic resin that withstands abrasion and ultraviolet radiation. It is resistant to common solvents, and it retains its properties over temperatures from −80 to 250°F. It has replaced natural rubber as a jacket and insulation in many electronics products.

polyvinyl chloride [PVC] A corrosion-resistant plastic resin that is easy to mold, light in weight, and has excellent dielectric properties. It is impact resistant, has low water absorption, and is easily machined, bonded, and heat sealed. It is used as wire and cable insulation and jacketing.

popcorn noise Noise produced by erratic jumps of bias current between two levels at random intervals in operational amplifiers and other semiconductor devices.

population inversion The condition in which a higher energy state in an atomic system is more heavily populated with electrons than a lower energy state of the same system. This condition must be established to create lasing action in a laser.

porcelain A fired ceramic material used to make insulators and capacitors.

porcelain capacitor A fixed capacitor whose dielectric is a high grade of porcelain molecularly fused to alternate layers of fine silver electrodes to form a monolithic unit that requires no case or hermetic seal. It has been essentially replaced by ceramic multilayer capacitors.

port 1. An opening in a waveguide component, through which energy can be fed or withdrawn, or measurements made. 2. An opening in a base-reflex enclosure for a loudspeaker, designed and positioned to improve bass response. 3. An entrance or exit for a network or computer.

positional crosstalk The variation in the path followed by any one electron beam as the result of a change impressed on any other beam in a multibeam cathode-ray tube.

positional notation A system of notation in which the significance of each digit of a number depends on its position in the sequence. Successive digits are interpreted as coefficients of successive powers of an integer called the base.

position control system A *closed-loop control system* that controls the movement of an object or load from its initial position to a specified end position. It typically has two feedback loops: a velocity loop containing a tachometer as the sensor, and a *position loop* with either a resolver or an encoder as the sensor. See also *velocity control system*.

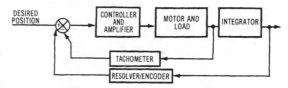

Position control system block diagram shows that it has two feedback loops.

position fix The intersection of two plotted bearing lines on a map, chart or display.

positioning action Automatic control action in which there is a predetermined relation between the value of a controlled variable and the position of a final control element.

position sensor A device that measures a position and converts that measurement into a form convenient for transmission. It is also called a position transducer.

positive bias A bias on the base or gate of a transistor that is positive with respect to the emitter or source.

positive charge The charge that exists in a body which has fewer electrons than normal.

positive column The luminous glow, often striated, that occurs between the Faraday dark space and the anode in a glow-discharge tube. The positive column is a plasma and has equal numbers of electrons and positive ions.

positive electricity   The positive charge that is produced in a glass object by rubbing it with silk.

positive electrode   The electrode that serves as the anode in a primary cell when the cell is discharging. It is connected to the positive terminal of the cell. Electrons flow through the external circuit to the positive electrode.

positive electron   *Positron.*

positive emitter-coupled logic   Standard ECL logic circuits that run from a positive power supply. See *emitter-coupled logic.*

positive feedback   Feedback in which a portion of the output of a circuit or device is fed back in phase with the input, to increase the total amplification. Excessive positive feedback causes instability and distortion. When positive feedback is sufficiently high, oscillation occurs. In a sound system, excessive transfer of acoustic energy back from the loudspeaker to the microphone causes howling.

positive ghost   A ghost image that has the same tonal variations as the primary television image.

positive-going   Increasing in a positive direction.

positive image   A picture as normally seen on a television picture tube; it has the same rendition of light and shade as in the original scene being televised.

positive ion   An atom with fewer electrons than normal and therefore has a positive charge.

positive ion emission   The thermionic emission of positive ions from a cathode in an electron tube due to the loss of cathode material or impurities in the cathode.

positive logic   Digital logic whose more positive logic level represents 1. The use of this logic level term is not recommended because it can be construed differently for NPN devices than for PNP devices. To avoid confusion, the magnitude and polarity of the voltage levels for 1 and 0 in a specific logic circuit always should be specified.

positive magnetostriction   Magnetostriction in which the application of a magnetic field causes the expansion of a material.

positive modulation   Modulation in which an increase in brightness of the image being televised causes an increase in the transmitted power of an amplitude-modulated (AM) transmitter or an increase in the frequency of a frequency-modulated (FM) transmitter. It is also called *positive transmission.*

positive ray   A stream of positively charged atoms or molecules produced by a suitable combination of ionizing agents, accelerating fields, and limiting apertures.

positive temperature coefficient [PTC]   The characteristic of materials that causes its resistance to increase with temperature. All metals and most metal alloys have positive temperature coefficients of resistance, but carbon has a *negative temperature coefficient of resistance* [NTC].

positive terminal   The terminal of a battery or other voltage source toward which electrons flow through the external circuit.

positive transmission   *Positive modulation.*

positive zero   The zero value reached by counting down from a positive number in the binary system.

positron [POSItive elecTRON]   A nuclear particle that has the mass of an electron and a positive charge which is exactly equal in magnitude to the negative charge of an electron. Positrons are formed in the beta decay of many radionuclides. It is also called an antielectron and a positive electron.

positron-emission tomography [PET]   A medical imaging technique based on the detection of trace amounts of radioactive isotopes in solution that have been injected into the patient's body. The solution emits *positrons* that collide with electrons and annihilate each other, releasing two *gamma rays.* The rays move in opposite directions to leave the body and strike the radiation detectors within the *PET scanner.* The gamma-ray detectors emit flashes of energy (scintillation) that are converted into electrical signals. A computer records the location of each energy flash and plots the source of radiation within the patient's body. That data is translated into a color-coded image showing isotope concentration proportional to activity level for diagnostic purposes. See also *single-photon-emission computed tomography* [SPECT].

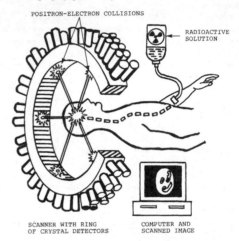

POSITRON-ELECTRON COLLISIONS

RADIOACTIVE SOLUTION

SCANNER WITH RING OF CRYSTAL DETECTORS

COMPUTER AND SCANNED IMAGE

Positron-emission tomography (PET): gamma rays emitted from a radioactive solution injected into the patient are detected by sensors positioned around his body. A video image of the patient's internal organs is computer generated from the scanned detector signals.

positron-emission tomography [PET] **scanner**   Medical imaging equipment consisting of an array of crystal radiation detectors mounted in a ring that surrounds a patient's body to detect radiation from injected isotopes for conversion into an image of the patient's inner tissues. Its output is sent to a computer that constructs an image and displays it on a monitor. It is also called a *PET scanner.*

post   *Waveguide post.*

postacceleration   Acceleration of beam electrons after deflection in an electron-beam tube.

postdeflection focus   The focusing of a beam of electrons after deflection, as in the chromatron.

postemphasis   *Deemphasis.*

postequalization   *Deemphasis.*

POS terminal   Abbreviation for *point-of-sale terminal.*

postforming   The forming, bending, or shaping of cured laminated sheets by applying heat rapidly and pressing the sheet over a mold.

postprocessing Steps taken to determine an answer or solution after the completion of data processing.

postregulation A power supply output circuit with a linear regulator that improves line-load regulation and reduces ripple and noise. In *pulse-width-modulated switching supplies,* postregulation can add expense and degrade efficiency.

pot Slang term for *potentiometer.*

potassium [K] An alkali metal that has photosensitive characteristics, used to make the cathodes of phototubes when maximum response is desired to blue light. It has a moderate thermal neutron-absorption cross section. It has an atomic number of 19 and an atomic weight of 39.10.

potassium dihydrogen phosphate crystal [KDP crystal] A piezoelectric crystal. It is used to make sonar transducers.

pot core A ferrite magnetic core that has the shape of a pot, with a magnetic post in the center and a magnetic plate as a cover. The coils for a choke or transformer are wound on the center post.

potential The degree of electrification as referred to some standard, such as the earth. The potential at a point is the amount of work required to bring a unit quantity of electricity from infinity to that point. The words potential and voltage are synonymous.

potential barrier A region within a semiconductor device where the voltage can be high enough to oppose moving electric charges and might even cause them to reverse their direction. This phenomenon occurs in PN-junction diodes and bipolar transistors.

potential difference The voltage that exists between two points, more often called voltage.

potential energy Energy due to the position of a particle or piece of matter with respect to other particles.

potential gradient The difference in the values of the voltage per unit length along a conductor or through a dielectric.

potential transformer *Voltage transformer.*

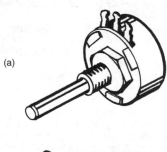

(a)

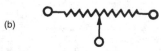

(b)

Potentiometer. (*a*) This volume control or control potentiometer is one of many different kinds of variable resistor; (*b*) its schematic symbol.

potentiometer 1. A variable resistor that has a continuously adjustable sliding contact which is generally mounted on a rotating shaft, and functions as a voltage divider. It is also called a pot (slang). 2. An instrument for measuring a voltage by balancing it against a known voltage.

potentiometric controller A controller that operates on the null balance principle. An error signal is produced by balancing the sensor signal against a set-point voltage in the input circuit. The error signal is amplified to keep the load at a desired temperature or other parameter.

potting A process for protecting a component or assembly by mounting the component or assembly in a can and pouring in an insulating compound such as room-temperature vulcanizing RTV silicone rubber.

pounds per square inch [lb/in^2 or psi] A unit of pressure that specifies air, hydraulic, and steam pressure and force.

powdered iron core *Ferrite core.*

powder metallurgy The production of magnetic cores, permanent magnets, and other molded metal objects by compressing finely powdered metals in molding dies, then sintering them to a temperature below the fusion point of the metal.

power 1. The time rate of doing *work* or the speed in which work is done. 2. [P] The rate at which electric energy is fed to or taken from a device, measured in watts. Power is a definite quantity, whereas level is a relative quantity. 3. The result obtained when a number is multiplied by itself a particular number of times. Thus 125 is the third power of 5.

power amplification *Power gain.*

power amplifier An AF or RF amplifier that delivers maximum output power to a loudspeaker or other load, rather than maximum voltage gain, for a given percent distortion.

power-amplifier stage 1. An AF amplifier stage that is capable of handling considerable AF power without distortion. 2. An RF amplifier stage that serves primarily to increase the power of the carrier signal in a transmitter.

power attenuation *Power loss.*

power bandwidth The frequency range for which half the rated power of an audio amplifier is available at rated distortion. This specification is a measure of the power available at the critical high and low frequencies.

power bipolar transistor A bipolar junction transistor capable of handling currents of 1 A or more without damage or destruction. Some can handle current as high as 100 A, and others can handle voltage as high as 1200 V. It finds applications in switching regulated power supplies and motor controls.

power converter A converter that changes DC power to AC power.

power cord *Line cord.*

power density The amount of power per unit area in a radiated microwave or other electromagnetic field, usually expressed in watts per square centimeter. A power density of 10 mW/cm^2 is generally regarded as the maximum safe dosage for the constant exposure of radar operating and maintenance personnel.

power detector A detector circuit that will handle strong input signals without objectionable distortion.

power divider A device that produces a desired distribution of power at a branch point in a waveguide system.

power dump The removal of all power from a computer, accidentally or intentionally.

power factor [PF] The ratio of active power to apparent power. As a percentage rating, it is equal to the resistance of a part or circuit divided by the impedance at the operating frequency, with the result multiplied by 100. A pure resistor has a power factor of 100%. A pure capacitor has a power factor of 0% leading, and a pure coil has a power factor of 0% lagging. Power factor is equal to the cosine of the phase angle between the current and voltage when both are sinusoidal.

power-factor correction Addition of capacitors to an inductive circuit to increase the power factor. The capacitors offset part or all of the inductive reactance, making the total circuit current more nearly in phase with the applied voltage.

power-factor meter A direct-reading instrument that measures power factor.

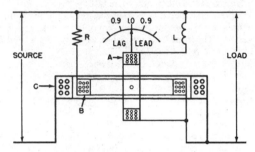

Power-factor meter with a crossed-coil construction.

power frequency The frequency at which electric power is generated and distributed. Most power companies in the continental United States are interconnected and operated at exactly the same frequency, which is usually maintained between 59.98 and 60.02 Hz. When the time indicated by electric clocks in this power grid differs more than 3 s from the standard time signals of WWV, steps are taken to bring back the entire power grid to correct time.

power gain 1. The ratio of the power delivered by a transducer to the power absorbed by the input circuit of the transducer. The power gain in decibels is 10 times the logarithm of the ratio of the power values. It is also called power amplification. 2. An antenna rating equal to 12.56 times the ratio of the radiation intensity in a given direction to the total power delivered to the antenna.

power integrated circuit A class of integrated circuits constructed specifically to handle high voltages and currents because of special active-element insulation built into the device. Some combine analog small-signal logic with a power MOSFET switch on the same chip.

power level The amount of power being transmitted past any point in an electric system. When expressed in decibels, it is equal to 10 times the logarithm to the base 10 of the ratio of the given power to a reference power value. It is also expressed in volume units.

power-level indicator An AC voltmeter calibrated to read AF power levels directly in decibels or volume units.

power line Two or more wires conducting electric power from one location to another.

power-line carrier A carrier frequency, generally below 600 kHz, that transmits control signals or information over power lines.

power-line filter *Line filter.*

power-line interference Interference caused by radiation from high-voltage power lines. The interference is generally noticeable only when the receiving antenna is within a few hundred feet of the line.

power loss The ratio of the power absorbed by the input circuit of a transducer to the power delivered to a specified load. It is expressed in decibels and is also called power attenuation.

power MOSFET A MOS field-effect transistor capable of handling more than 1 A without damage or destruction. Some MOSFETs can handle currents as high as 100 A, and others can handle voltage as high as 1200 V. It finds applications in high-frequency switching-regulated power supplies and motor controls.

power output The AC power in watts delivered by an amplifier to a load.

power output tube *Power tube.*

power pack A power supply unit that converts the available power line or battery voltage to the voltage values required by a unit of electronic equipment.

power rating The power available at the output terminals of a component or unit of equipment that is operated according to the manufacturer's specifications.

power relay 1. A relay that functions at a predetermined value of input power. 2. The final relay in a sequence of relays controlling a load or a magnetic contactor.

power semiconductor A semiconductor device that is capable of dissipating power in excess of 1 W in normal operation. It can handle currents of thousands of amperes or voltages up into thousands of volts, at frequencies up to 10 kHz.

power spectrum level The power level for the acoustic power in a band 1 Hz wide, centered at a specified frequency.

power supply A power line, generator, battery, power pack, or other source of power for electronic equipment.

power switch *on/off switch.*

power transformer An iron-core transformer that has a primary winding which is connected to an AC power line and one or more secondary windings which provide different alternating voltage values.

power transistor A MOSFET or bipolar junction transistor capable of handling currents of 1 A or more without self-damage or self-destruction.

power tube An electron tube capable of handling more current and power than an ordinary voltage-amplifier tube. It is used in the last stage of an AF amplifier or in high-power stages of an RF amplifier. It is also called a power-amplifier tube or a power output tube.

power winding A saturable reactor winding that receives power from a local source.

Poynting's vector A vector that represents the direction and amount of energy flow at a point in a wave at a given instant of time.

P-P Abbreviation for *peak-to-peak.*

PPBM Abbreviation for *pulse-polarization binary modulation.*

PPI Abbreviation for *plan-position indicator.*

PPI repeater A radar display that duplicates at a remote location the plan-position-indicator display at the main radar console.

P+-type semiconductor   A P-type semiconductor that has an excess mobile hole concentration.

PPM   1. Abbreviation for *parts per million.*   2. Abbreviation for *pulse-position modulation.*

PPPI   Abbreviation for *precision plan-position indicator.*

PPS   Abbreviation for *polyphenylene sulfide.*

pps   Abbreviation for *pulses per second.*

PQFP   Abbreviation for *plastic quad-leaded flatpack.*

practical system   A system of electrical units in which the units are convenient multiples or submultiples of CGS units. Practical units are the ampere, coulomb, farad, henry, joule, ohm, volt, watt, and watthour.

preamble   The portion of a commercial radiotelegraph message that is sent first: it contains the message number, office of origin, date, and other numerical data not part of the original message.

preamp   Abbreviation for *preamplifier.*

preamplifier [preamp]   An amplifier that is connected to a low-level signal source to provide gain and impedance matching so the signal can be further processed without appreciable degradation of the signal-to-noise ratio. A preamplifier can also perform equalizing and mixing.

precession   A change in the orientation of the axis of a gyroscope or other rotating body.

precipitation attenuation   Attenuation of radio waves by passage through regions of precipitation in the atmosphere.

precipitation clutter   Clutter caused by rain or other precipitation within the range of a radar.

precipitation static   Static interference due to the discharge of large charges built up on an aircraft or other vehicle by rain, sleet, snow, or electrically charged clouds.

precipitator   An electronic apparatus for removing smoke, dust, oil mist, or other small particles from air. A high direct voltage, of the order of 10 kV, is obtained from a high-voltage rectifier and applied to a fine wire mesh through which the air is drawn by a fan. Particles in the air are charged by this screen and are then drawn through a system of parallel charged plates that attract the particles and remove them from the air. It is also called an air cleaner, an electronic air cleaner, an electrostatic air cleaner, or an electrostatic precipitator.

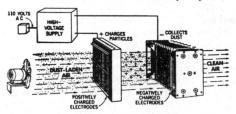

Precipitator operating principles.

precision   The quality of being exactly defined or stated. A six-place table has greater precision than a four-place table. The accuracy of either table would be reduced by errors in compilation or printing, however.

precision guided munitions   Aerial bombs that are guided to their targets by add-on, remotely controlled guidance packages that can steer the bomb during its free-fall descent to the target. The guidance package is controlled by an infrared seeker that searches for and tracks a laser beam. This method minimizes the possibility of jamming, more likely with RF or microwave links. The attacking aircraft is equipped with a laser that can illuminate the target precisely. However, that beam can be attenuated by rain or fog, causing the bomb to go out of control. Bombs with these add-on guidance packages are called *smart bombs.*

preconduction current   The low value of anode current flowing in a thyratron or other grid-controlled gas tube prior to the start of conduction.

predetection   Before detection. In predetection recording, the modulated carrier is recorded as received or as taken from an IF stage of the receiver.

predistortion   *Preemphasis.*

preemphasis   The first part of a process for increasing the strength of some frequency components with respect to others, to help these components override noise or reduce distortion. It emphasizes the higher audio frequencies in frequency- and phase-modulated transmitters and in sound recording systems. It is also called accentuation, emphasis, predistortion, and preequalization. The original relations are restored by the complementary process of deemphasis before reproduction of the sounds.

preemphasis network   An *RC* filter inserted in a system to emphasize one range of frequencies with respect to another. It is also called an emphasizer.

preemption   The act of interrupting a lower-priority transmission or process to accommodate a higher-priority user.

preequalization   *Preemphasis.*

preferential recombination   Recombination that takes place immediately after an ion pair is formed.

preferred numbers   A series of numbers adopted by EIA and the U.S. Department of Defense for use as nominal values of resistors and capacitors, to reduce the number of different sizes that must be kept in stock for replacements.

preferred transistor type   A transistor type recommended to designers of electronic equipment for general use, to minimize the number of transistors required for stock supply.

prefix   A combining form used with a unit of measure to indicate a larger or smaller quantity as a power of 10.

prefix notation   *Polish notation.*

preform   Material that has been formed into a convenient shape and size for further processing, such as a solder ring.

P region   The region in a semiconductor where hole density exceeds conduction electron density.

preoscillation current   *Starting current.*

prerecord   To record program material before it is required for broadcasting, transmission, or other use.

prerecorded tape   *Recorded tape.*

preregulator   A separate regulator circuit that increases the power-handling capability of a regulated power supply by minimizing power dissipation in the series regulator elements. Silicon controlled rectifiers or triacs have replaced the thyratrons in older tube circuits.

prescaler   A scaler that extends the upper frequency limit of a counter by dividing the input frequency by a precise amount, generally 10 or 100.

preselector   A tuned RF amplifier stage installed ahead of the frequency converter in a superheterodyne receiver to increase the selectivity and sensitivity of the receiver.

presence The impression, as created by a recording or radio receiver, that the original music or voice source is in the room.

preset To establish an initial value or condition, generally by setting one or more controls in advance.

preset guidance Missile guidance in which the path of a missile is determined by controls that are set before launching.

preset parameter A parameter whose value is not changed during the running of a computer subroutine.

press 1. To mold a phonograph record from a stamper. 2. *Pinch*.

pressed cathode A dispenser cathode made by compacting and heating a mixture of tungsten-molybdenum alloy and barium-calcium aluminate, in a ratio such that evaporated electron-emitting surface barium is continuously replenished from within the cathode body.

pressure The *force* per unit area or force divided by the area over which it is distributed. It is typically expressed in pounds per square inch, or *psi*.

pressure altimeter An altimeter that measures and indicates altitude by differences in atmospheric pressure.

pressure hydrophone A pressure microphone that responds to waterborne sound waves.

pressure microphone A microphone whose output varies with the instantaneous pressure produced by a sound wave acting on a diaphragm. Examples are capacitor, carbon, crystal, and dynamic microphones.

pressure pickup A device that converts changes in the pressure of a gas or liquid into corresponding changes in some more readily measurable quantity such as inductance or resistance.

pressure spectrum level The effective sound pressure level for the sound energy contained within a band 1 Hz wide, centered at a specified frequency.

pressure switch A switch that is actuated by a change in pressure of a gas or liquid.

pressure transducer A transducer that includes a potentiometer, strain gage, variable-reluctance device, crystal, or other electric device for converting pressure to a proportional electric output.

pressurization Use of an inert gas or dry air, at a pressure several pounds above atmospheric pressure, inside the waveguide system of a radar or in a sealed coaxial line. Pressurization prevents corrosion by keeping out moisture and minimizes high-voltage breakdown at high altitudes.

prestore To store a quantity in an available computer location before it is required in a routine.

pretersonics The branch of electronics concerned with acoustic waves at microwave frequencies (above 500 MHz), traveling on or in piezoelectric or other solid substrates. It is also called acoustoelectronics.

pretravel The distance or angle through which the actuator of a switch moves from the free to the operating position.

pre-TR tube [pre-Transmit-Receive tube] A gas-filled RF switching tube that protects the TR tube from excessively high power and protects the radar receiver from frequencies other than the fundamental.

preventive maintenance A procedure of inspecting, testing, and reconditioning a system at regular intervals, according to specific instructions intended to prevent failures in service or retard deterioration.

PRF Abbreviation for *pulse repetition frequency*.

primaries The colors of constant chromaticity and variable amount that, when mixed in proper proportions, produce or specify other colors.

primary *Primary winding*.

primary battery A battery that consists of one or more primary cells.

primary carrier flow The current flow that is responsible for the major properties of a semiconductor device. It is also called primary flow.

primary cell A cell that delivers electric current as a result of an electrochemical reaction which is not efficiently reversible.

primary color A color that cannot be matched by any combination of other primary colors. In color television, the three primary colors emitted by the phosphors in the color picture tube are red, green, and blue.

primary-color unit The area within a color cell in a color picture tube that is occupied by one primary color.

primary current The current flowing through the primary winding of a transformer.

primary dark space A narrow nonluminous region that appears between the cathode and the cathode glow of some gas-discharge tubes.

primary detector *Sensor*.

primary electron An electron emitted directly by a material rather than as a result of a collision.

primary element *Sensor*.

primary emission Electron emission directly caused by the temperature of a surface, irradiation of a surface, or the application of an electric field to a surface.

primary feedback Feedback that is obtained from the controlled variable and compared with the reference input to obtain the actuating signal for a feedback control system. It is also called a feedback signal.

primary flow *Primary carrier flow*.

primary frequency standard The national standard of frequency as maintained by the National Institute of Standards and Technology, Washington, D.C. The operating frequency of a radio station is determined by comparison with multiples of this standard frequency as broadcast by station WWV.

primary fuel cell A fuel cell in which the fuel and oxidant are continuously consumed.

primary ionization 1. The ionization produced by primary particles in a collision, as contrasted to the total ionization, which includes the secondary ionization produced by delta rays. 2. The total ionization produced in a counter tube by incident radiation without gas amplification.

primary ionizing event *Initial ionizing event*.

primary ion pair An ion pair produced directly by the causative primary particle or photon. An ion cluster is a group of ion pairs produced at or near the site of a primary ionizing event; it includes the primary ion pair and any secondary ion pairs formed.

primary radiation Radiation that arrives directly from its source without interaction with matter.

primary radiator The portion of an antenna system from which energy leaves the transmission system. The distribution of the energy can be subsequently modified by other parts of the antenna system.

primary service area   The geographical area in which the ground wave of a broadcast station is not subject to objectionable interference or fading.

primary skip zone   The area around a radio transmitter that is beyond the ground-wave range but not far enough out for good skip-distance reception. Radio reception is not reliable in the primary skip zone.

primary standard   A unit directly defined and established by some authority, against which all secondary standards are calibrated.

primary transit-angle gap loading   The electronic gap admittance that results from the traversal of the gap by an initially unmodulated electron stream.

primary voltage   The voltage applied to the terminals of the primary winding of a transformer.

primary winding   The transformer winding that receives signal energy or AC power from a source. It is also called the primary.

prime contractor   A contractor who has a direct contract for an entire project. A prime contractor can in turn assign parts of the work to subcontractors.

priming speed   The rate of priming successive storage elements in a charge-storage tube.

principal axis   A reference direction for angular coordinates, used in describing the directional characteristics of a transducer. It is usually an axis of structural symmetry or the direction of maximum response.

principal E plane   A plane that contains the direction of maximum radiation and the electric vector.

principal H plane   A plane that contains the direction of maximum radiation and the magnetic vector. The electric vector is everywhere perpendicular to the H plane.

printed-circuit motor   A DC permanent-magnet electric motor with a lightweight, low-inertia armature made from copper-foil laminated GFE. It has conductors that are electrically equivalent to copper wire windings etched on a round, flat circuit board by the subtractive process. The case of the motor has a flat disk form, so it is also called a *pancake motor*. Originally made for rapid-response servosystems such as those in machine tools, sewing machines, and robots, the design has since been superseded by motors with armatures made by a punched copper-laminate process. See *disk-type DC motor*.

printed-wiring board   A conductive pattern formed on one or both sides of an insulating base by etching, plating, or stamping. It is also called a printed-circuit board.

printed-wiring connector   A connector that makes interconnections between printed-wiring terminations and conventional external wiring.

printout   The printed output of a computer as produced by a printer.

print-through   Transfer of signals from one recorded layer of magnetic tape to the next on a reel.

priority interrupt   An interrupt structure in computers organized so that devices with higher priority can interrupt the servicing of devices with lower priority.

private automatic branch exchange [PABX]   A customer-owned, switchable private branch exchange that provides internal and external station-to-station dialing. All telephone connections are made by remotely controlled switches.

private branch exchange [PBX]   A class of telephone service that typically provides the same service as PABX—a private telephone switch located on the customer's premises that gives the customer access to a public telecommunications exchange.

private line   A telephone line that does not go through the telephone service provider's central office and is reserved for exclusive use by a single customer. See also *leased line*.

PRK   Abbreviation for *phase-reversal keying*.

probability multiplier   *Coincidence multiplier.*

probability of collision   The probability that an electron will collide with an atom or molecule when moving through a distance of 1 cm.

probability of ionization   The ratio of the number of collisions followed by ionization to the total number of collisions in a gas during a specified time.

probability theory   A measure of the likelihood of occurrence of a chance event. The theory predicts behavior of a group rather than single items.

probable error   The amount of error that is most likely to occur during a measurement. Half the results will have a greater error than this value, and the other half will have less error.

probe   1. A metal rod that projects into but is insulated from a waveguide or resonant cavity. It provides coupling to an external circuit for injection or extraction of energy. When a probe is movable in a slot, it can measure the standing-wave ratio. It is also called a *waveguide probe*. 2. A large test prod that has interconnected components or circuits built into its handle.   3. An unmanned, instrumented vehicle sent into the upper atmosphere or space to gather environmental information.

probe microphone   A small microphone that measures the sound pressure at a point without significantly altering the sound field in the neighborhood of the point.

process   To assemble, compile, generate, interpret, compute, and otherwise act on information in a computer.

process control   Automatic control of a complex industrial process.

process control monitor [PCM]   A pattern of active and passive devices formed on a semiconductor wafer during processing that can be used to test the quality of the other devices on the substrate. A monitor can include field-effect transistors, diodes, load resistors, or circuits such as ring oscillators. Data taken from the PCM provides the critical link between the device design and the modeling and fabrication process. This data is instrumental in grading the quality of finished wafers. There can be 40, 50, or more PCM patterns on each wafer. DC parametric testing is done on the FETs, Schottky diodes, ohmic contacts, implanted layers, and isolation; AC testing is done on ring oscillators, typically formed as interconnected NOR gates.

processing gain   The ratio of the bandwidth of the spread-spectrum signal to the bandwidth of the unspread signal.

processor   *Microprocessor.*

product demodulator   A demodulator whose output is the product of an amplitude-modulated carrier input voltage and a local oscillator signal voltage at the carrier frequency. With proper filtering, the output is proportional to the original modulation.

production control   The procedure for planning, routing, scheduling, dispatching, and expediting the flow of materials, parts, subassemblies, and assemblies within a plant, from the raw state to the finished product, in an orderly and efficient manner.

production model   A model in its final mechanical and electrical form, made by the production tools, jigs, fixtures, and methods to be used in producing subsequent units.

product modulator   A modulator whose output is proportional to the product of the carrier and modulating-signal voltages. The carrier is then normally suppressed.

profile chart   A vertical cross-sectional drawing of a microwave path between two stations, indicating terrain, obstructions, and antenna height requirements.

program   1. A precise sequence of coded instructions used by a digital computer in solving a problem. A complete program includes plans for transcription of data and effective use of results.   2. A sequence of audio signals alone, or audio and video signals, transmitted for entertainment or information.

program circuit   A telephone circuit that has been equalized to handle a wider range of frequencies than is required for ordinary speech signals, for use in transmitting musical programs to the stations of a radio network.

program control   A controller whose desired value is automatically changed periodically in accordance with a predetermined program.

program counter   A register that provides the address of the next instruction to be fetched from memory in a computer. The register is normally incremented one step automatically after each instruction fetched.

program failure alarm   A signal-actuated electronic relay circuit that gives a visual and aural alarm when the program fails on the line being monitored. A time delay is provided to prevent the relay from giving a false alarm during intentional short periods of silence.

program generator   A program that permits a computer to write other programs automatically.

program level   The level of the program signal in an audio system, expressed in volume units.

programmable calculator   An electronic calculator that includes provision for changing its internal program, usually by inserting a new magnetic card on which the desired calculating program has been stored. Some calculators permit new programs to be keyed in manually.

programmable controller   [PC]   An industrial controller based on digital logic that can be programmed in the field to perform tasks previously performed by banks of electromechanical relays. It is assigned to timing and sequencing functions and can operate simple robots, chemical processes, and assembly-line functions.

programmable counter   A counter that divides an input frequency by a number which can be programmed into decades of synchronous down counters. These decades, with additional decoding and control logic, give the equivalent of a divide-by-*N* counter system.

programmable decade resistor   A decade box designed so the value of its resistance can be remotely controlled by programming logic as required for the control of the load, time constant, gain, and other parameters of circuits in automatic test equipment and automatic controls.

programmable-gain amplifier   An amplifier that can be operated under direct control of a digital computer or controlled by autoranging techniques, for signal-scaling under control of a digital input.

programmable logic array   [PLA]   A premanufactured array of gates whose transistors can be organized to perform any combinational-logic function. PLAs are programmed by computers.

programmable logic device   A logic device that forms only a part of a complete computing system. Examples include content-addressable memories, erasable programmable read-only memories, field-programmable logic arrays, microprocessors, programmable logic arrays, random-access memories, random logic, and read-only memories.

programmable power supply   A power supply whose output voltage can be changed by digital control signals. It can include provisions for manual selection of a choice of preset output voltages.

programmable read-only memory   [PROM]   A semiconductor read-only memory that can be written to or programmed only once, typically in the field with special equipment that burns out fusible links in a network of logic devices to set a specific memory location to a desired logic level, thus establishing the program. PROMs can be made with bipolar or CMOS technology.

programmable unijunction transistor   [PUT]   A PNPN device (thyristor) with an anode gate. For a constant gate voltage, the device remains off or nonconducting until the anode voltage exceeds the gate voltage by a predetermined amount, at which point it turns on. The action is similar to that of a silicon controlled switch.

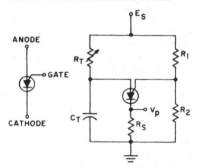

Programmable unijunction transistor symbol and an example of its use in a relaxation oscillator.

programmed check   1. A computer check in which a sample problem with a known answer, selected to have programming similar to that of the next problem to be run, is put through the computer.   2. A series of self-checking tests inserted in the computer program for a problem.

programmed electron-beam welding   Electron-beam welding controlled by switching the beam in sequence from connection to connection on an integrated circuit, in a vacuum, by digital control data stored in a computer. Similar programming can be used for removing unwanted portions of a thin film.

programmer   1. A person who prepares sequences of instructions for a computer without necessarily converting them into the detailed codes.   2. A device that controls the motion of a missile in accordance with a predetermined plan.

programming   Preparing a sequence of operating instructions for a computer.

programming language The language used by a programmer to write a program for a computer.

program parameter A parameter that can have different values during the course of a given program in a digital computer.

program register The register in the control unit of a digital computer that stores the current instruction of the program and controls the operation of the computer during the execution of that instruction.

program-sensitive error A computer error that occurs only when a unique combination of program steps is executed.

progressive scanning In a cathode-ray tube display, the rectilinear process in which the distance from center to center of successively scanned lines is equal to the nominal line width (one field makes up the frame, and all lines are scanned in sequence). It is also called sequential scanning.

progressive wave A wave that is propagated freely in a medium.

progressive-wave antenna *Traveling-wave antenna.*

projection cathode-ray tube A television cathode-ray tube that produces an intensely bright but relatively small image which can be projected onto a large viewing screen by an optical system.

projection optics A system of mirrors and lenses that projects the image onto a screen in projection television. The Schmidt system is an example. It is also called *reflective optics.*

projection television receiver A television receiver equipped with a system of lenses and mirrors that project on a large-size screen an intensely bright image formed on the face of a projection cathode-ray tube.

projector 1. A horn that projects sound essentially in one direction from a loudspeaker. 2. A machine that projects film images onto a screen. 3. *Underwater sound projector.*

projector efficiency *Transmitting efficiency.*

PROM Abbreviation for *programmable read-only memory.*

propagation The travel of electromagnetic waves or sound waves through a medium, or the travel of a sudden electric disturbance along a transmission line. It is also called wave propagation.

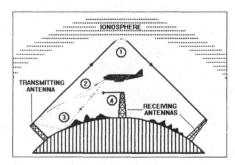

Propagation. Radio waves can travel by four different modes: (1) sky wave, (2) direct wave, (3) reflected wave, and (4) surface wave.

propagation anomaly A change in propagation characteristics due to a discontinuity in the medium of propagation.

propagation constant A rating for a line or medium through which a plane wave of a given frequency is being transmitted. It is a complex quantity; the real part is the attenuation constant in nepers per unit length, and the imaginary part is the phase constant in radians per unit length.

propagation delay The time required for a signal to pass through a complete operating circuit. It is generally measured in nanoseconds.

propagation factor *Propagation ratio.*

propagation loss The attenuation of signals passing between two points on a transmission path.

propagation ratio The ratio, for a wave propagating from one point to another, of the complex electric field strength at the second point to that at the first point. It is also called propagation factor. The field strength is a vector with a magnitude less than 1, and it is the attenuation ratio.

propagation time delay The time required for a wave to travel between two points on a transmission path.

propagation velocity The velocity of propagation of radio waves is equal to the velocity of light which is 2.998×10^8 m/s or 299.8 m/μs. This corresponds to 983.6 ft/μs, 0.1863 mi/μs, or 0.1618 nautical mi/μs.

proportional band The range of values of the controlled variable that will cause a controller to operate over its full range.

proportional control Control in which the amount of corrective action is proportional to the amount of error.

proportional counter A radiation counter that consists of a proportional counter tube and its associated circuits.

proportional counter tube A radiation-counter tube operated at voltages high enough to produce ionization by collision and adjusted so the total ionization per count is proportional to the ionization produced by the initial ionizing event.

proportional ionization chamber An ionization chamber in which the initial ionization current is amplified by electron multiplication in a region of high electric field strength, as it is in a proportional counter. It measures ionization currents or charges over a period of time, rather than counting.

proportional plus derivative control *Error-rate damping.*

proportional-position action Control action in which there is a continuous linear relation between the value of the controlled variable and the position of a final control element.

proportional region The range of applied voltage for a radiation-counter tube in which the charge collected per isolated count is proportional to the charge liberated by the initial ionizing event.

proportional response *Rate control.*

protective resistance A resistor placed in series with a gas tube or other device to limit current flow to a safe value.

protector tube A glow-discharge cold-cathode tube that becomes conductive at a predetermined voltage, to protect a circuit against overvoltage. An example is the surge-voltage protector (SVP) tube.

protocol A set of conventions for the transfer of information between computer devices. The simplest protocols define only the hardware configuration, while more complex protocols define timing, data formats, error detection and correction techniques, and software structures.

protocol converter A circuit that changes one computer or communications protocol into another.

proton An elementary particle that has a positive charge equal in magnitude to the negative charge of the electron. The atomic number of an element indicates the number of protons in the nucleus of each atom of that element. The rest mass of a proton is 1.67×10^{-24} g, or 1836.13 times that of an electron.

proton microscope   A microscope that is similar to the electron microscope but uses protons instead of electrons as the charged particles.

proton scattering microscope   A microscope that produces protons in a cold-cathode discharge which are accelerated and focused on a crystal in a vacuum chamber. Protons reflected from the crystal strike a fluorescent screen to give a visual and photographable display that is related to the structure of the target crystal. It is used for the analysis and orientation of crystals.

prototype   A model suitable for use in the complete evaluation of the form, design, and performance of a product.

proximity detector   A sensing device that produces an electric signal when approached by an object or when approaching an object.

proximity effect   The redistribution of current in a conductor due to the presence of another current-carrying conductor.

proximity fuze   A fuze that detonates a warhead when the target is within some specified region near the fuze. A microwave source can be the activating element.

proximity switch   A switch that is operated when the actuating device is moved near it, without physical contact. An example is a reed switch that operates when a permanent magnet is brought near it.

proximity warning indicator   An airborne anticollision system that gives the host aircraft a warning of another aircraft approaching on a possible collision course.

PRR   Abbreviation for *pulse repetition rate.*

ps   Abbreviation for *picosecond.*

PSC motor   Abbreviation for *permanent-split capacitor motor.*

pseudocode   An arbitrary code, independent of the hardware of a computer, that must be translated into computer code before it can direct the computer.

pseudomorphic high-electron mobility transistor-[PHEMT]   An RF gallium-arsenide (GaAs) power transistor made with a special epitaxial layer grown on the GaAs to optimize it for lower voltage and higher efficiency for cellular telephones and RF modems.

pseudorandom sequence   A sequence of codes or numbers that appears to be random and satisfactorily meets the randomness requirements of a specific application. The sequence might be of finite length and eventually repeat itself.

psi   Abbreviation for *pounds per square inch.*

PSK   Abbreviation for *phase-shift keying.*

psophometer   An instrument that measures noise in electric circuits. When connected across a 600-ohm resistance in the circuit under test, the instrument gives a reading that by definition is equal to half the psophometric electromotive force actually existing in the circuit.

psophometric electromotive force   The true noise voltage that exists in a circuit.

psophometric voltage   The noise voltage as actually measured in a circuit under specified conditions.

psophometric weighting   A frequency weighting similar to the C-message weighting that is used as the standard for the European telephone system testing.

PSPDN   Abbreviation for *packet-switched public data network.*

PSRR   Abbreviation for *power-supply rejection ratio.*

psychogalvanometer   An instrument that tests physiological reaction by determining how skin resistance changes when a voltage is applied to electrodes in contact with the skin.

psychosomatograph   An instrument that records muscular action currents or physical movements during tests of mental-physical coordination.

PTC   Abbreviation for *positive temperature coefficient.*

PTM   Abbreviation for *pulse-time modulation.*

P-type conductivity   The conductivity associated with holes in a semiconductor, which are equivalent to positive charges.

P-type semiconductor   A form of semiconductor material whose electron density in the conduction band is exceeded by the density of holes in the valence band. P-type material is formed by the addition of *acceptor* impurities. An example is doping silicon with boron. See also *acceptors, dopants, donors,* and *impurities.*

public-address amplifier   An AF amplifier that provides sufficient output power to loudspeakers for adequate sound coverage at public gatherings.

public-address system [PA system]   A complete system for amplifying sounds and providing adequate volume for large public gatherings.

Public Broadcasting Service [PBS]   The national network of public television stations.

public television   Broadcasting by noncommercial television stations, supported mainly by viewers, foundation grants, and government funds. Commercials are not allowed, but large corporations can be credited for underwriting production costs of special programs.

puff   British abbreviation (slang) for *picofarad.*

pull-in current   *Pickup current.*

pulling   An effect that forces the frequency of an oscillator to change from a desired value. It is caused by undesired coupling to another frequency source or the influence of changes in the oscillator load impedance.

pulling figure   The total frequency change of an oscillator when the phase angle of the reflection coefficient of the load impedance varies through 360°, the absolute value of this reflection coefficient being constant at 0.20.

pull-in value   *Pickup value.*

pulsar signal   A pulsed radio signal that comes from essentially a point source in interstellar space.

pulsatance   Angular velocity in radians, equal to 2π times frequency in hertz.

pulsating current   A direct current that increases and decreases in magnitude.

pulsating DC   The output from a rectifier before it is filtered. The polarity of a pulsating DC source does not change with time but its amplitude does.

pulsation welding   Resistance welding in which the current is applied in timed pulses rather than continuously. It improves the transfer of surface heat to the water-cooled welding electrodes and thereby increases electrode life.

pulse   A momentary, sharp change in a current, voltage, or other quantity that is normally constant. A pulse is characterized by a rise and a decay and has a finite duration.

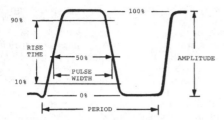

Pulse parameters are illustrated in this diagram.

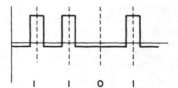

Pulse code representing a binary number.

pulse amplifier An amplifier designed specifically to amplify electric pulses without appreciably changing their waveforms.

pulse amplitude The peak, average, effective, instantaneous, or other magnitude of a pulse, usually with respect to the normally constant value. The exact meaning should be specified when giving a numerical value.

pulse-amplitude modulation [PAM] The process of amplitude modulating a pulse carrier waveform.

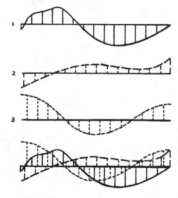

Pulse-amplitude modulation (PAM). Three voice signals (1, 2, and 3) are sampled at different times so they can be transmitted in the correct time sequence.

pulse analyzer An instrument that measures pulse widths and repetition rates and displays the waveform of a pulse on a cathode-ray tube screen.

pulse-averaging discriminator A telemetering subcarrier discriminator that uses resistive and capacitive tuning components to give an output which is proportional to average pulse width.

pulse bandwidth The bandwidth limits of the amplitude of a pulse. Outside that limit it is below a prescribed fraction of the peak amplitude.

pulse carrier A pulse train functioning as a carrier.

pulse code A code that consists of various combinations of pulses, such as the Morse code, Baudot code, and the binary code of computers.

pulse code modulation [PCM] A modulation process in which the signal is sampled periodically by the process of *pulse-amplitude modulation* [PAM]. This is performed by an electronic circuit acting like a stepping switch. Then each PAM sample is quantized or coded. The *coder* measures the height or amplitude of each sample and converts

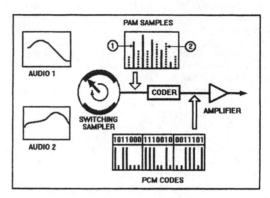

Pulse-code modulation (PCM). Signals are sampled sequentially and the samples are binary-coded for transmission.

it into a binary number, such as 1011000 or 1110010. Thus, the transmitted pulses (ones) all have the same value. At the receiving end, the binary code is converted back into PAM samples.

pulse coder *Coder.*

pulse-compression radar A radar system whose transmitted signal is linearly frequency-modulated or otherwise spread out in time to reduce the peak power that must be handled by the transmitter. Signal amplitude is kept constant. The receiver includes a linear filter to compress the signal and thereby reconstitute a short pulse for the radar display.

pulse counter A device that indicates or records the total number of pulses received during a time interval.

pulse dialer Telephone system apparatus that generates pulse trains corresponding to digits or characters for impulse or loop-disconnect dialing.

pulsed altimeter A radar altimeter that emits pulses of RF energy.

pulsed Doppler radar A method for obtaining range information from a continuous wave (CW) Doppler radar. The received pulses are small segments of the CW returns. A fixed target produces uniform pulses, but pulses from a moving target vary in amplitude periodically. This is caused by phase coherence. Each time a fixed target echo returns, it is mixed with a voltage that has undergone the same difference in phase since the instant of its transmission.

pulse decay time The interval of time required for the trailing edge of a pulse to decay from 90 to 10% of the peak pulse amplitude.

pulse decoder A decoder that extracts useful information from a pulse-coded signal.

pulse-delay binary modulation [PDBM] A form of pulse modulation in which a delayed pulse represents a binary 1 and an undelayed pulse represents a binary 0. It is used in optical data communication.

pulse-delay network A network that consists of two or more components such as resistors, coils, and capacitors that delay the passage of a pulse.

pulse demoder A circuit that responds only to pulse signals which have a specified spacing between pulses. It is also called a constant-delay discriminator.

pulse discriminator A discriminator circuit that responds only to a pulse which has a particular duration or amplitude.

pulsed laser A laser that produces a pulse of coherent light at fixed time intervals, as required for ranging and tracking applications. It can also permit higher output power than can be obtained with continuous laser operation.

pulsed maser *Two-level maser.*

pulsed oscillator An oscillator that generates a carrier-frequency pulse or a train of carrier-frequency pulses, as the result of self-generated or externally applied pulses.

pulse dot soldering iron A soldering iron that provides heat to the tip for a precisely controlled time interval, as required for making a good soldered joint without overheating adjacent parts.

pulse droop A distortion of an otherwise essentially flat-topped rectangular pulse, characterized by a decline of the pulse top.

pulsed ruby laser A laser that has a ruby as its active material. The extremely high pumping power required is obtained by discharging a bank of capacitors through a special high-intensity flash tube, giving a coherent beam with a duration of about 0.5 ms.

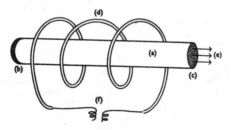

Pulsed ruby laser showing ruby rod (*a*), heavily silvered flat end (*b*), lightly silvered flat output end (*c*), exciting flash tube (*d*), coherent output beam (*e*), and leads to capacitor bank (*f*).

pulse duration The time interval between the first and last instants at which the instantaneous amplitude reaches a stated fraction of the peak pulse amplitude. It is also called pulse length or pulse width.

pulse-duration coder *Coder.*

pulse-duration discriminator A circuit in which the sense and magnitude of the output are a function of the deviation of the pulse duration from a reference.

pulse-duration error An error caused by pulse duration, which makes certain targets appear longer or thicker than they actually are in the direction of the radar beam.

pulse-duration modulation [PDM] A form of pulse-time modulation in which the duration of each pulse is varied. It is also called pulse-width modulation.

pulse duty factor The ratio of average pulse duration to average pulse spacing. This is equivalent to the product of average pulse duration and pulse repetition rate.

pulse equalizer A circuit that produces output pulses of uniform size and shape, in response to input pulses which can vary in size and shape.

pulse excitation A method of producing oscillator current in which the duration of the impressed voltage in the circuit is relatively short compared with the duration of the current produced. It is also called impulse excitation.

pulse-forming line A continuous line or ladder network that has parameters which give a specified shape to the modulator pulse in a radar modulator.

pulse-forming network A network that shapes the leading and/or trailing edge of a pulse.

pulse-frequency modulation [PFM] A form of pulse-time modulation in which the pulse repetition rate is the characteristic varied. A more precise term for pulse-frequency modulation would be pulse repetition-rate modulation.

pulse-frequency spectrum *Pulse spectrum.*

pulse-gated binary modulation [PGBM] A form of pulse modulation in which a binary 1 is represented by a pulse and binary 0 by the absence of a pulse. It has applications in optical data communication.

pulse generator A generator that produces repetitive pulses or single signal-initiated pulses.

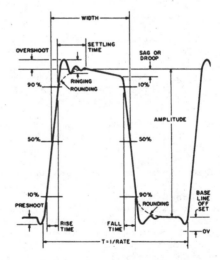

Pulse-generator output specifications.

pulse-height analyzer An instrument capable of indicating the number of occurrences of pulses falling within each of one or more specified amplitude ranges.

pulse-height discriminator A circuit that produces a specified output pulse when and only when it receives an input pulse whose amplitude exceeds an assigned value. It is also called an amplitude discriminator.

pulse-height selector A circuit that produces a specified output pulse only when it receives an input pulse whose amplitude lies between two assigned values. It is also called an amplitude selector or a differential pulse-height discriminator.

pulse interleaving A process of combining pulses from two or more sources in time-division multiplex for transmission over a common path.

pulse interrogation The triggering of a transponder by a pulse or pulse mode.

pulse interval *Pulse spacing.*

pulse jitter A relatively small variation of the pulse spacing in a pulse train. The jitter can be random or systematic, depending on its origin. It is generally not coherent with any pulse modulation imposed.

pulse length *Pulse duration.*

pulse mode A finite sequence of pulses in a prearranged pattern for selecting and isolating a communication channel.

pulse-mode multiplex A process or device for selecting channels with pulse modes. This process permits two or more channels to use the same carrier frequency.

pulse moder A device for producing a pulse mode.

pulse-modulated radar Radar that transmits in a series of discrete pulses.

pulse-modulated waves Recurrent wave trains in which the duration of the trains is short compared with the interval between them.

pulse modulation Modulation in which the amplitude or time of some characteristic of a pulse carrier is varied by the modulating wave. Examples include pulse-amplitude and pulse-time modulations.

pulse modulator A device that applies pulses to the circuit element where modulation occurs.

pulse multiplex *pulse-mode multiplex.*

pulse navigation system A navigation system that depends on the time required for a pulse of RF energy to travel a given distance. Examples include loran and radar.

pulse-numbers modulation Modulation in which the pulse density per unit time of a pulse carrier is varied in accordance with a modulating wave by making systematic omissions without changing the phase or amplitude of the transmitted pulses. As an example, omission of every other pulse could correspond to zero modulation; reinserting some or all pulses then corresponds to positive modulation, and omission of more than every other pulse corresponds to negative modulation.

pulse packet The volume of space occupied by a single radar pulse. It is defined by the angular width of the beam, the duration of the pulse, and the distance of the pulse from the antenna.

pulse-phase modulation *Pulse-position modulation.*

pulse-polarization binary modulation [PPBM] A form of pulse modulation in which a right circularly polarized pulse represents a binary 1 and an oppositely polarized pulse represents a binary 0. It has applications in optical data communication.

pulse-position modulation [PPM] A form of pulse-time modulation in which the position in time of a pulse is varied. It is also called pulse-phase modulation.

pulser A generator that produces high-voltage, short-duration pulses, as required by a pulsed microwave oscillator or a radar transmitter. In a vacuum-tube pulser, the pulse is produced by discharging a capacitor through the load. In a line-type pulser, an unterminated transmission line is charged through a high impedance and discharged through the load.

pulse radar Radar that transmits high-power pulses which are spaced far apart in comparison with the duration of each pulse. The receiver is active for the reception of echoes in the interval following each pulse.

pulse-rate telemetering Telemetering in which the number of pulses per unit time is proportional to the magnitude of the measured quantity.

pulse recurrence interval The time, usually expressed in microseconds, between pulses in radar and loran.

pulse recurrence rate *Pulse repetition rate.*

pulse regeneration The process of restoring pulses to their original relative timings, forms, and magnitudes.

pulse regenerator A device or circuit capable of restoring or regenerating a pulse train to its original characteristics.

pulse repeater A device that receives signal pulses from one circuit and transmits corresponding pulses into another circuit. It can also change the frequency and waveform of the pulses and perform other functions.

pulse-repetition frequency [PRF] *Pulse-repetition rate.*

pulse-repetition rate [PRR] The number of times per second that a pulse is transmitted. In radar the pulse-repetition rate is usually between 400 and 3000 pulses per second. It is also called the pulse-recurrence rate or pulse-repetition frequency.

pulse reply The transmission of a pulse or pulse mode by a transponder as the result of an interrogation.

pulse resolution The minimum time separation between input pulses that will permit a circuit or component to respond properly.

pulse rise time The interval of time required for the leading edge of a pulse to rise from 10 to 90% of the peak pulse amplitude.

pulse scaler A scaler that produces an output signal when a prescribed number of input pulses has been received.

pulse selector A circuit or device for selecting the proper pulse from a sequence of telemetering pulses.

pulse separation The time interval between the trailing edge of one pulse and the leading edge of the succeeding pulse.

pulse shaper A transducer that changes one or more characteristics of a pulse, such as a pulse regenerator or pulse stretcher.

pulse shaping Intentionally changing the shape of a pulse.

pulse spacing The interval between the corresponding pulse times of two consecutive pulses. It is also called the pulse interval.

pulse-spacing analyzer An analyzer that converts the spacing between two adjacent pulses to a proportional voltage. Applications include measuring the periodicity of neural activities.

pulse-spacing modulation A form of pulse-time modulation in which the pulse spacing is varied. It is also called pulse-interval modulation.

pulse spectrum The frequency distribution of the sinusoidal components of a pulse in relative amplitude and in relative phase. It is also called the pulse-frequency spectrum.

pulses per second [pps]   The number of pulses per second in a varying DC quantity. The unit for an AC quantity is hertz.

pulse spike   An unwanted pulse of relatively short duration superimposed on a main pulse.

pulse spike amplitude   The peak pulse amplitude of a pulse spike.

pulse stretcher   A pulse shaper that produces an output pulse whose duration is greater than that of the input pulse and whose amplitude is proportional to the peak amplitude of the input pulse.

pulse synthesizer   A circuit that supplies pulses which are missing from a sequence due to interference or other causes.

pulse test   An insulation test in which the applied voltage is a pulse that has a specified wave shape. It is also called an impulse test.

pulse tilt   A distortion in an otherwise essentially flat-topped rectangular pulse, characterized by either a decline or a rise of the pulse top.

pulse-time modulation [PTM]   A form of modulation in which the time of occurrence of some characteristic of a pulse carrier is varied from the unmodulated value. Examples include pulse-duration, pulse-frequency, pulse-position, and pulse-spacing modulation.

pulse train   A sequence of pulses that have similar characteristics.

pulse-train spectrum   The frequency distribution of the sinusoidal components of a pulse train, in amplitude and in phase angle.

pulse transformer   A transformer capable of operating over a wide range of frequencies. It can transfer nonsinusoidal pulses without materially changing their waveforms.

pulse transmitter   A pulse-modulated transmitter whose peak power-output capabilities are usually large with respect to the average power-output rating.

pulse-type altimeter   *Radar altimeter.*

pulse-type scanning sonar   A scanning sonar that transmits a pulse of sound power in all directions simultaneously. The volume of surrounding water is then scanned rapidly for viewing of all echoes on a PPI screen.

pulse-type telemeter   A telemeter that employs characteristics of intermittent electric signals other than their frequency as the means of conveying information.

pulse valley   The portion of a pulse between two specified maxima.

pulse width   *Pulse duration.*

pulse-width modulation [PWM]   1. A form of analog control in which the duration of the conduction time of the output transistor or transistors in a switching-regulated power supply is varied by modulating the bias on the gate or base of the transistors in response to changes in the load. This keeps the output voltage of the power supply constant over varying operating conditions.   2. In radar, the modulation of the transmitter by rapidly turning it on and off to allow time for the return of the echo signal.

pulsing circuit   A circuit that provides abrupt changes of voltage or current in some characteristic pattern.

pumping   A method of exciting electrons to a higher energy state in a laser. Pumping can be done optically with a light source, electrically with a discharge in a gas or with current in a semiconductor, or chemically by a reaction between two gases. Pumping produces the electron population inversion that is required to initiate and sustain laser action. In a maser, pumping is achieved with a microwave signal that differs in frequency from the output signal.

pumping band   A group of energy states to which ions in the ground state are initially excited when pumping radiation is applied to a laser medium. The pumping band is generally higher in energy than the levels that are to be inverted.

pumping frequency   The pumping frequency provided in a laser, maser, quadrupole amplifier, or other amplifier that requires high-frequency excitation.

pumping radiation   Light applied to the sides or end of a laser crystal to excite the ions to the pumping band.

pumping voltage   A reverse voltage delivered to the varactor of a parametric amplifier at a multiple of the RF signal frequency.

punch-through   A destructive discharge through a dielectric layer of a semiconductor device, typically gate to channel in a MOSFET transistor or CMOS logic circuit caused by electrostatic discharge (ESD) or a transient overvoltage. See *electrostatic discharge* [ESD].

puncture   *Breakdown.*

puncture voltage   The voltage at which a test specimen is electrically punctured.

pure tone   *Simple tone.*

purity   1. The degree to which a primary color is pure and not mixed with the other two primary colors used in color television.   2. A ratio of distances on the CIE chromaticity diagram that compares a sample color with a reference standard light.

purity coil   A coil mounted on the neck of a color picture tube, to produce the magnetic field needed for adjusting color purity. The direct current through the coil is adjusted to a value that makes the magnetic field orient the three individual electron beams so each strikes only its assigned color of phosphor dots.

purity control   A potentiometer or rheostat that adjusts the direct current through the purity coil.

purple boundary   A straight line drawn between the ends of the spectrum locus on the CIE chromaticity diagram.

purple plague   Failure of a gold-wire bond on aluminum because of the formation of porous intermetallic compounds that are mechanically weak and electrically nonconductive. It is most likely to occur at temperatures above 200°C. The compound has a purple color.

pushbutton control   Control of machines, missiles, and other complex equipment entirely by relays and automatic electric or electronic control circuits, once a human operator has operated a pushbutton switch.

pushbutton dialing   The dialing of a desired telephone number by pushing buttons in a particular sequence to initiate automatic switching in telephone exchanges, as with a TouchTone telephone.

pushbutton switch   A master switch that is operated by finger pressure on the end of an operating button.

pushbutton tuner   A device that automatically tunes a radio receiver or other piece of equipment to a desired frequency when the button assigned to that frequency is pressed.

pushing   A change in the resonant frequency of a circuit due to changes in the applied voltages.

pushing figure   The amount of change in oscillator frequency produced by a specified change in oscillator electrode current, excluding thermal effects.

push-pull amplifier A balanced amplifier that includes two similar transistors working in phase opposition.

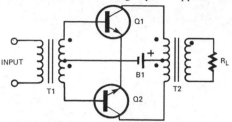

Push-pull amplifier example shown is transformer coupled, class B.

push-pull converter A power supply topology typically configured as a forward converter but with two switching transistors and a center-tapped transformer. The switching of the transistors on and off alternately by pulse-width modulation is very efficient.

push-pull currents *Balanced currents.*

push-pull microphone A microphone that includes two identical microphone elements which are actuated 180° out of phase by a sound wave, as in a double-button carbon microphone.

push-pull oscillator A balanced oscillator that has two different transistors in phase opposition.

push-pull-parallel amplifier A push-pull amplifier that has two or more tubes or transistors in parallel in each half of the circuit, to obtain higher output power.

push-pull transformer An AF transformer that has a center-tapped winding, for use in a push-pull amplifier.

push-pull voltages *Balanced voltages.*

push-push amplifier An amplifier that employs two similar transistors with bases or gates connected in phase opposition and collectors or drain connected in parallel to a common load. It can function as a frequency multiplier to emphasize even-order harmonics in a high-fidelity sound system.

push-push currents Currents flowing in the two conductors of a balanced line that, at every point along the line, are equal in magnitude and in the same direction.

push-push voltages Voltages (relative to ground) on the two conductors of a balanced line that, at every point along the line, are equal in magnitude and have the same polarity.

push-to-talk switch A switch mounted directly on a microphone or handset to provide a convenient means for switching two-way radiotelephone equipment, intercom equipment, or electronic dictating equipment to the talk position. The switch must be released for listening.

pushup list A computer list whose next item to be retrieved is the oldest item placed in the list. It corresponds to first-in, first-out.

PUT Abbreviation for *programmable unijunction transistor.*

PVC Abbreviation for *polyvinyl chloride.*

pW Abbreviation for *picowatt.*

PWM Abbreviation for *pulse-width modulation.*

pylon antenna A vertical antenna constructed of one or more sheet metal cylinders, each cylinder with a lengthwise slot.

pyramidal horn antenna An aperture antenna that has an outward-flaring rectangular horn. Its dimensions determine the shape of the radiation pattern.

pyranometer An instrument that measures the intensity of the radiation received from any part of the sky.

pyrheliometer An instrument that measures the total intensity of solar radiation.

pyroconductivity Electric conductivity that develops in solids only at high temperature. The solids are essentially nonconductive at atmospheric temperatures.

pyroelectric effect The development of charges in certain crystals when they are unequally heated or cooled.

pyrolytic boron nitride [PBN] A high-temperature material suitable for making furnace crucibles.

pyromagnetic A reference to the interaction of heat and magnetism.

pyrometer An instrument that measures temperatures by electric means, especially temperatures beyond the range of mercury thermometers. Examples include the radiation, resistance, and thermoelectric pyrometers.

Pythagorean scale A musical scale arranged so that frequency intervals are represented by the ratios of integral powers of the numbers 2 and 3.

PZT Abbreviation for *lead zirconate titanate.*

Q

Q A figure of merit for an energy-storing device, tuned circuit, or resonant system. It is equal to reactance divided by resistance. The Q of a capacitor, coil, circuit, or system thus determines the rate of decay of stored energy; the higher the Q, the longer it takes for the energy to be released. It is also called Q factor and quality factor.

QAM Abbreviation for *quadrature amplitude modulation.*

Q antenna A dipole that is matched to its transmission line by stub-matching.

Q channel The 0.5-MHz-wide band in the NTSC color television system for transmitting green-magenta color information. The I channel is 1.3 MHz wide, and the luminance channel is 3.579545 MHz wide.

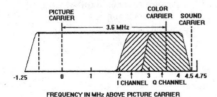

Q-channel bandwidth of 0.5 MHz is centered on the color carrier, and the I-channel bandwidth is 1.3 MHz in the NTSC system.

Q demodulator The demodulator that combines the chrominance signal and 90° phase-shifted signal of the color-burst oscillator to recover the Q signal in a color television receiver.

Q (quality) factor The ratio of the center frequency to the bandwidth. A measure of the steepness of a filter's amplitude response.

QFP Abbreviation for *quad flatpack.*

Q meter An instrument that measures the Q of a circuit or circuit element by determining the ratio of reactance to resistance. It is also called a *quality-factor meter.*

Q multiplier A filter that gives a sharp response peak or a deep rejection notch at a particular frequency, equivalent to boosting the Q of a tuned circuit at that frequency.

QPSK Abbreviation for *quaternary phase-shift keying.*

QRP An international radio Q signal that means "reduce your power."

Q signal 1. The quadrature component of the chrominance signal in color television; it has a bandwidth of 0.5 MHz. It consists of +0.48(R − Y) and +0.41(B − Y), where Y is the luminance signal, R is the red camera signal, and B is the blue camera signal. 2. A three-letter abbreviation starting with Q, used in the International List of Abbreviations for radiotelegraphy to represent complete sentences.

QSK Letters that stand for *full break in* in amateur radio operation.

QSL card A card sent by one radio amateur to another to verify radio communication with each other.

QSOP Abbreviation for *quad small-outline package.*

Q spoiling A method of preventing laser action by quenching while a large population excess is being pumped up, so a more powerful pulse of light is obtained when the laser is triggered by Q switching.

Q switching A technique for keeping the Q of the cavity of a laser at a low value while an ion population inversion is being built up, then suddenly switching the Q to a high value just before instability occurs. The technique gives a very high rate of stimulated emission. Switching action can be achieved with Kerr cells, rotating reflecting prisms, or thin foils of gold inserted between the laser crystal and a high-reflectivity end plate. Phthalocyanine molecules in solution have served as repeatable Q switching elements for ruby lasers.

quad An adjective meaning four. As applied in electronics, it means four separate functions in a single package. An example is four independent, electrically isolated, discrete devices or integrated circuits on the same substrate. Examples include diodes, resistors, operational amplifiers, and NOR gates.

quad flatpack [QFP] An epoxy-molded flat package for large-scale ICs with leads projecting from all four sides; hence, the term "quad" for quadrant. A typical QPF will have 64 leads with 16 projecting from each side. There is also a *thin quad flatpack* [TQFP].

quadrant 1. A 90° sector of a circle. 2. The fourth part of something, as one of the quadrants in a four-course radio range.

quadrant electrometer An electrometer for measuring voltages and charges by means of electrostatic forces between a suspended metal plate and a surrounding metal cylinder that is divided into four insulated parts connected oppositely in pairs. The voltage to be measured is applied between the two pairs of quadrants.

quadraphonic sound system A four-channel sound system that has speakers which are normally placed in the four corners of a room. The front pair provides stereophonic sound; the rear pair provides the sound that would normally be heard after reflection from the walls of the auditorium or other large room in which the recording was made. Listeners within the area of the four speakers then enjoy reproduction approximating that of a seat in the original concert hall. In a discrete quadraphonic system the four channels are kept acoustically separate throughout the recording and playback processes. In the QS or SQ matrix system, the four microphone channels are converted into two channels by a coding process before recording or broadcasting. They are then decoded back into four channels for playback. In a derived quadraphonic system, the four channels are synthesized from a conventional two-channel stereo source by an adapter. It is also called a four-channel sound system.

quadrature Separated in phase by 90° or one quarter-cycle. Also called phase quadrature.

quadrature amplifier An amplifier that shifts the phase of a signal 90°. In a color television receiver it amplifies the 3.58-MHz chrominance subcarrier and shifts its phase 90° for the Q demodulator.

quadrature amplitude modulation [QAM] Quadrature modulation in which some form of amplitude modulation is applied for both digital inputs.

quadrature component The reactive component of a current or voltage due to inductive or capacitive reactance in a circuit.

quadrature filter A filter that eliminates the quadrature components of signals in systems where information is contained in the quadrature modulation components of the carrier.

quadrature modulation Modulation in which two carrier components, differing in phase by 90°, are each modulated by a different signal.

quadrature-phase subcarrier signal The portion of the chrominance signal that leads or lags the in-phase portion by 90°.

quadricorrelator A circuit that is added to the automatic-phase control loop in a color television receiver to obtain improved performance under severe interference conditions.

quadruple diversity The simultaneous combining of four received signals with space, frequency, or other diversity reception techniques.

quadrupole amplifier A low-noise parametric amplifier that consists of an electron-beam tube whose quadrupole fields act on the fast cyclotron wave of the electron beam to produce a high amplification at frequencies in the range of 400 to 800 MHz. The cyclotron frequency is approximately equal to the frequency of the signal to be amplified, and the pumping frequency is about twice this value.

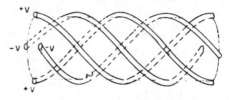

Quadrupole amplifiers have a four-wire twisted quadrupole structure that applies transverse DC electric fields to an electron beam.

quadrupole moment A term for specifying mathematically the field caused by a given distribution of electric or magnetic charges.

quality factor Q.

quality-factor meter Q *meter*.

quantity A positive or negative real number for numerical data.

quantity of electricity 1. The amount of electric charge stored in a capacitor, measured in coulombs or similar units. 2. The amount of current flowing through a circuit in a given time, measured in coulombs. One coulomb is one ampere flowing for one second.

quantity of radiation The total radiated energy passing through a unit area per unit of time. Expressed in ergs per square centimeter or wattseconds per square centimeter.

quantization Division of the range of values of a wave into a finite number of smaller subranges. Each subrange is represented by an assigned or quantized value.

quantization distortion Inherent distortion introduced in the process of quantization. It is also called quantization noise.

quantization level One of the subrange values obtained by quantization.

quantization noise *Quantization distortion.*

quantize To restrict a variable to a discrete number of possible values. The age of a person is usually quantized as a whole number of years.

quantized pulse modulation Pulse modulation that involves quantization, such as pulse-numbers and pulse-code modulation.

quantized system A system of particles whose energies can have only discrete values.

quantizer A device that measures the magnitude of a time-varying quantity in multiples of some fixed unit or quantum, at a specified instant or specified repetition rate. It delivers a proportional response which is usually in pulse code or digital form. The amplitude of the response signal is at each instant proportional to the number of quanta measured. The action of a quantizer is essentially the same as that of an analog-to-digital converter.

quantizing The process of representing any value between certain limits by the nearest of a limited number of values selected to cover the range. It is used in pulse-code modulation.

quantizing encoder An encoder that converts voltages to digital form. The voltage corresponding to the contents of a register is obtained from the output of a digital-voltage decoder. The error between this voltage and the input voltage is quantized to the nearest power of 2, and subtracted from the contents of the register. Each time this process is repeated, the error is diminished.

quantizing noise Noise that is correlated with a signal generally associated with the quantizing error introduced by analog-to-digital and digital-to-analog conversions in digital transmission systems.

quantum [plural **quanta**] The smallest quantity of energy that can be associated with a given phenomenon. The quantum of electromagnetic radiation is the photon.

quantum efficiency The average number of electrons photoelectrically emitted from a photocathode per incident photon of a given wavelength in a phototube.

quantum electronics The branch of electronics that studies the various energy states of matter, such as the motions within atoms or groups of atoms and various phenomena in crystals. Examples of practical applications include the atomic hydrogen maser and the cesium atomic-beam resonator.

quantum mechanics The study of atomic structure and related phenomena in terms of quantities that can actually be measured.

quantum number A number assigned to one of the various values of a quantized quantity in its discrete range. As an example, the principal quantum number of an electron determines its energy level with respect to the minimum-energy level or ground state that has a quantum number of 1.

quantum state One of the states in which an atom can exist permanently or momentarily.

quantum theory A theory that atoms or molecules emit or absorb energy by a process which takes place in a series of steps, each step being the emission or absorption of an amount of energy called the quantum. For light or other radiation, the quantum is the photon. The energy of the photon is equal to the frequency of the radiation in hertz multiplied by Planck's constant, which is 6.626×10^{-27} erg s.

quantum voltage The voltage through which an electron must be accelerated to acquire the energy corresponding to a particular quantum.

quantum well A semiconductor device structure formed from alternating thin layers of Group III–V materials with differing bandgap properties fabricated to obtain physical properties not found in bulk quantities of either constituent material. An example is a quantum well formed from gallium arsenide (GaAs) and aluminum gallium arsenide (AlGaAs). GaAs, with its smaller bandgap, is sandwiched between layers of AlGaAs with a larger bandgap. This heterostructure quantum-mechanically confines the electrons to the smaller-bandgap material. Multiple quantum-well structures can also be formed.

quantum yield The number of photon-induced reactions of a specified type per photon absorbed.

quarter-phase *Two-phase.*

quarter-square multiplier *Four-quadrant multiplier.*

quarter-wave The property of having an electrical length of one quarter-wavelength.

quarter-wave antenna An antenna whose electrical length is equal to one quarter-wavelength of the signal to be transmitted or received.

quarter-wave attenuator An arrangement of two wire gratings spaced an odd number of quarter-wavelengths apart in a waveguide that attenuates waves traveling through in one direction. The wave reflected from the first grating is canceled by that reflected from the second. Thus, all energy reaching the attenuator is either transmitted through the gratings or absorbed by them, with no resultant reflection.

quarter-wavelength The distance that corresponds to an electrical length of a quarter of a wavelength at the operating frequency of a transmission line or antenna element.

quarter-wave line *Quarter-wave stub.*

quarter-wave plate A plate of mica or other doubly refracting crystal material thick enough to introduce a phase difference of one quarter-cycle between the ordinary and extraordinary components of light passing through.

quarter-wave stub A section of transmission line that is one quarter-wavelength long at the fundamental frequency being transmitted. When shorted at the far end, it has a high impedance at the fundamental frequency and at all odd harmonics. It has a low impedance for all even harmonics. It is also called a quarter-wave line and a quarter-wave transmission line.

quarter-wave termination A nonreflecting waveguide termination that consists of an energy-absorbing wire grating or semiconducting film stretched across the waveguide one quarter-wavelength from a metal-plate termination. The wave reflected by the grating is canceled by the wave reflected from the plate.

quarter-wave transformer A section of transmission line (approximately one quarter-wavelength long) that can match a transmission line to an antenna or load.

quarter-wave transmission line *Quarter-wave stub.*

quarter-wavelength plate A transparent plate that changes the phase of a light beam passing through it by a quarter of a wavelength.

quarter-wavelength vertical antenna *Ground-plane antenna.*

quartz A natural or artificially grown piezoelectric crystal composed of silicon dioxide. Thin slabs or plates are carefully cut from it and ground to serve as a crystal plate for controlling the frequency of an oscillator.

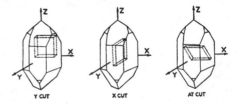

Quartz in natural form, as cut for three types of crystal plates.

quartz delay line An acoustic delay line with quartz as the medium of sound transmission.

quartz-fiber electroscope An electroscope whose gold-plated quartz fiber serves the same function as the gold leaf of a conventional electroscope.

quartz lamp A mercury-vapor lamp that has a transparent envelope made from quartz instead of glass. Quartz resists heat, permitting higher currents, and it passes ultraviolet rays that are absorbed by ordinary glass.

quartz plate *Crystal plate.*

quartz pressure gage A pressure gage that includes a highly stable quartz crystal resonator whose frequency changes directly with applied pressure.

quartz thermometer A thermometer whose response is based on the sensitivity of the resonant frequency of a quartz crystal to changes in temperature.

quartz watch An electronic watch that keeps time under the control of a quartz crystal operating in a battery-powered oscillator circuit. The time can be indicated conventionally by hands or by a digital display, typically liquid crystal today.

quasar A quasi-stellar source of radio signals near the limit of the observable universe that is receding farther into space at about half the speed of light and radiating energy beyond anything explainable by nuclear or other known phenomena.

quasi-bistable circuit An astable circuit that is triggered at a rate which is high compared to its own natural frequency.

quasi-conductor A conductor that has a Q much less than unity.

quasi-dielectric A dielectric that has a Q greater than unity.

quasi-ferroelectric ceramic A ceramic whose behavior differs from that of true ferroelectric and dielectric materials. Examples include some compositions of lead lanthanum zirconate titanate [PLZT].

quasi-linear feedback control system A feedback control system with a relationship between the input and output signals that is substantially linear despite the existence of nonlinear elements within the system.

quasi-monostable circuit A monostable circuit that is triggered at a rate which is high compared to its own natural frequency.

quasi-optical A reference to properties similar to light waves, such as the line-of-sight range limitation.

quasi-passive satellite A passive satellite that has special reflector-type antennas which provide signal enhancement by concentrating the received signal and reflecting it back to the receiving station in a narrow beam.

quasi-resonant power supply A resonant power supply whose power switch controls the repetition of the resonant cycles, making energy transfer discontinuous.

quasi-single-sideband The transmission of parts of both sidebands to simulate single-sideband transmission.

quaternary compound semiconductor A semiconductor made from a compound of four elements such as indium, gallium, arsenic, and phosphorous to form indium gallium arsenide phosphide (InGaAsP). Quaternary compounds are grown epitaxially on top of binary compound substrates.

quaternary phase-shift keying [QPSK] Modulation of a microwave carrier with two parallel streams of non-return-to-zero data in such a way that the data is transmitted as 90° phase shifts of the carrier. This gives twice the message channel capacity of binary phase-shift keying in the same bandwidth.

quaternary signaling Signaling in which information is represented by presence and absence or plus and minus variations of four discrete levels of one parameter of the signaling medium.

quenched-domain Gunn diode A Gunn diode whose frequency of oscillation is increased with a circuit that quenches the domain before it reaches the anode. This starts a new operating cycle earlier than for a free-running Gunn diode.

quenched-domain mode One of the three operating modes of a transferred-electron diode in which the formation and extinction of space-charge domains are controlled by the surrounding circuit. The other two modes are the LSA mode and the transit-time mode.

quenched spark gap A spark gap that has provisions for rapid deionization. One form consists of many small gaps between electrodes that have relatively large mass and are good radiators of heat. The electrodes cool the gaps rapidly and thereby stop conduction.

quenched spark-gap converter A spark-gap generator that uses the oscillatory discharge of a capacitor through a coil and a quenched spark gap as a source of RF power. The spark gap usually consists of closely spaced gaps operating in series to give good quenching action.

quenching 1. The process of terminating a discharge in a gas-filled radiation-counter tube by inhibiting reignition. 2. Cooling suddenly, as in heat-treating metals.

quenching circuit A circuit that diminishes, suppresses, or reverses the voltage applied to a counter tube to inhibit multiple discharges from an ionizing event.

quenching frequency The frequency at which the oscillations in a superregenerative receiver are suppressed or quenched.

queue In computer operation, a list of processes to be executed in sequential order, information blocks to be processed in sequential order, or a mixture of the two.

quiescent Without an input signal.

quiescent-carrier modulation A system of modulation in which the carrier is suppressed during intervals when no modulation is applied.

quiescent-carrier telephony A radiotelephone system in which the carrier is suppressed when there are no voice signals.

quiescent current The electrode current that corresponds to the electrode bias voltage.

quiescent push-pull amplifier A push-pull amplifier in which the control grids are biased so negatively that little anode current flows when there is no signal. There is no noise when tuning between stations, but tone quality is poor for weak signals.

quiescent value The voltage or current value for an electron-tube electrode when no signals are present.

quiet automatic volume control *Delayed automatic gain control.*

quieting sensitivity The least signal input for which the output signal-to-noise ratio does not exceed a specified limit in FM receivers.

quiet tuning A tuning circuit that silences the output of a radio receiver until it is accurately tuned to an incoming carrier wave.

quinary code A code based on five possible combinations for representing digits. An example is biquinary notation.

Q value *Disintegration energy.*

R

R 1. Abbreviation for *resistor.* 2. Abbreviation for *Rankine.* 3. Abbreviation for *roentgen.* 4. Abbreviation for *resistance.*

R Mathematical symbol for *resistance.*

°**R** Abbreviation for degree *Rankine.*

*R*ₒₙ A symbol for *on resistance,* the outward resistance of a power switching device when it is forward-biased to the fully "on" or conducting state.

rabbit ears A V-shaped indoor dipole television antenna whose arms are adjustable in angle and usually in length. The antenna can be attached to the television set or mounted on a base for use on top of the set.

race A transient condition in which two or more memory elements are changing state simultaneously in an asynchronous computer circuit.

race condition 1. A situation that occurs when inputs to a semiconductor device gate traverse parallel but different circuit paths. Differing path delays can result in unpredictable signal arrival times at a gate and uncertain transition time for the output of the gate. 2. An ambiguous condition that occurs in control counters when one flip-flop changes to its next state before a second one has had sufficient time to latch.

RACES Abbreviation for *Radio Amateur Civil Emergency Service.*

raceway A channel that holds and protects wires, cables, or busbars.

rack 1. A standardized steel cabinet that holds 19-in (48.26-cm) panels of various heights; mounted on the panels are radio receivers, amplifiers, and units of electronic test equipment. Mounting holes for the panels, usually drilled and threaded for 10-32 machine screws, are spaced apart in multiples of ½ and ⅝ in (1.27 and 1.5875 cm) to accommodate notch spacings of correspondingly standardized panels. Originally designed to hold relay panels in telephone exchanges. 2. A straight bar with gear teeth that engages with a drive gear or pinion for producing straight-line motion.

rad 1. An acronym for *radiation-absorbed dose.* 2. Abbreviation for *radian.*

radar [RAdio Detecting And Ranging] A system based on transmitted and reflected RF energy for detecting and locating objects, measuring distance or altitude, navigating, homing, and bombing. In detecting and ranging, the time interval between transmission of the energy and reception of the reflected energy establishes the range of an object in the transmitted beam's path. In primary radar, the return signal is produced by reflection of the transmitted energy from the target. In secondary radar, the transmitted energy triggers a responder beacon that sends an entirely new signal back to the radar set.

radar aid *Radar navigation aid.*

radar altimeter A radio altimeter for aircraft to give accurate absolute altitude indications at altitudes above the common 5000-ft (1,524-m) limit of FM radar altimeters. Pulse-type radar equipment sends a pulse straight down and measures its total time of travel to the surface and back to the aircraft. Its lowest useful altitude is about 250 ft (76.2 m). It is also called a *pulsed altimeter.*

radar altitude An absolute altitude as determined by a radar altimeter, also called radio altitude.

radar astronomy The study of astronomical bodies and the earth's atmosphere by radar techniques, including tracking of meteors and the reflection of radar pulses from the moon and the planets.

radar band See *microwave frequencies.*

radar beacon A radar receiver-transmitter that transmits a strong coded radar signal whenever its radar receiver is triggered by an interrogating radar on an aircraft or ship. The coded beacon reply can be used by navigators to determine their own positions in terms of bearing and range from the beacon.

radar beam   The movable beam of RF energy produced by a radar transmitting antenna. Its shape is commonly defined as the loci of all points at which the power has decreased to one-half that at the center of the beam.

radar boresight target   A target located by a survey at a known bearing, elevation, and distance from a radar. It is used for collimation and orientation of the radar antenna system.

radar calibration   The process of determining the extent and accuracy of the radar coverage of a given aircraft-warning or tactical air control radar installation. Calibration includes computing the theoretical coverage, making flights to check coverage, and preparing calibration charts, diagrams, and overlays.

radar cell   The volume in space that extends outward for one radar pulse length from the transmitter, with a cross section corresponding to the angular width of the radar beam as defined by points of half-power intensity.

radar chart   A special map used in radar navigation. For radar-equipped ships, radar charts emphasize the coastline, hills, large buildings, and other objects that give prominent radar echoes to ships well off shore. For air navigation, radar charts similarly show outlines of cities, rivers, lakes, bridges, railroads, and other objects that appear on airborne radar screens.

radar clutter   *Clutter.*

radar command guidance   A missile guidance system in which radar equipment at the launching site determines the positions of both target and missile continuously, computes the missile course corrections required, and transmits these by radio to the missile as commands.

radar contact   Recognition and identification of an echo on a radar screen. An aircraft is said to be in radar contact when its radar echo can be seen and identified on a PPI display.

radar control   Control of an aircraft, guided missile, or gun battery by radar.

radar control area   The area or airspace in which radar control of aircraft or guided missiles is exercised.

radar cross section   The area intercepting the power which, if scattered equally in all directions, would produce a received radar echo equal to that actually obtained from the target. It is also called *echo area.*

radar display   The pattern that represents the output data of a radar system, generally produced on the screen of a cathode-ray tube or a liquid-crystal digitized display.

radar drift   The drift of an aircraft as determined by a timed series of bearings taken on a fixed radar target.

radar echo   *Echo.*

radar equation   An equation that relates the transmitted and received powers and antenna gains of a primary radar system to the echo area and distance of the radar target.

radar fence   A network of radar warning stations maintained as a barrier against surprise attack.

radar fire control   Weapon fire control by radar.

radar fix   A position determined by radar.

radar handoff   Transferring radar identification and control over an aircraft from one controller to another without interrupting surveillance. This procedure is used in commercial aircraft flight control.

radar homing   1. Homing on the source of a radar beam. 2. Homing in which a missile-borne radar locks onto a target and guides the missile to that target.

radar horizon   The lowest elevation angle at which a radar can operate effectively at a particular location, taking into account the terrain in the vicinity and the curvature of the earth.

radar illumination   Illumination of a target by a radar external to a missile, to produce echo signals suitable for homing use by receiving equipment in the missile.

radar indicator   A cathode-ray tube or digitized liquid-crystal display and associated equipment that provides a visual indication of the echo signals picked up by a radar set.

radarman   See *radar operator.*

radar mile   The time required for a radar pulse to travel to a target 1 statute mile (1.609 km) away and return, equal to 10.75 μs.

radar modulator   A modulator that varies the amplitude, phase, frequency, pulse repetition rate, and/or pulse duration of a signal generated by a radar transmitter to which it is directly connected.

radar nautical mile   The time interval of approximately 12.367 μs that is required for the RF energy of a radar pulse to travel 1 nautical mile (1.852 km) and return.

radar navigation   Navigation by radar equipment and radar navigation aids.

radar navigation aids   Modifications to conventional navigational aids to enhance their effectiveness for radar navigation at night or in the fog. An example is the addition of *corner reflectors* to the superstructures of navigational buoys and towers to enhance their ability to reflect radar beams.

radar navigator   A person trained in radar navigation.

radar net   A network of radar stations set up to detect aircraft entering a defined airspace. It is also called a radar screen.

radar operator   1. A person who operates a radar set. 2. An aircraft crew member who operates radar equipment, such as a radar navigator or radar bombardier.

radar performance figure   The ratio of the pulse power of a radar transmitter to the power of the minimum signal detectable by the receiver.

radar picket   A ship or aircraft equipped with early-warning radar and operating at a distance from the area being protected, to extend the range of radar detection.

radar plot   A ship's course plotted directly on a radar plan-position indicator [PPI], rather than by visual sightings and plotting in pencil directly on a navigational chart. When done with reference to a navigational chart and depth sounder, it permits the mariner to pilot a ship safely through waters that are deep enough to avoid running aground while also avoiding collisions with other ships or obstructions. The course can be plotted directly on the face of the PPI with an erasable china-marking pencil.

radar prediction   A graphic representation of what can be expected to show on a radar screen when an actual radar scan is made. These can be made with scaled topographic models scanned with ultrasonic waves or they can be simulated with computer graphics.

radar probing   The process of obtaining data on distant objects in space by radar signals. Echoes have been received from signals sent to the planet Venus, 28,000,000 mi (45,000,000 km) away.

radar range   The maximum distance at which a radar set is ordinarily effective in detecting objects. It is usually assumed to be the distance at which a radar set can detect a specified object at least 50% of the time. The range in free space varies directly with receiver power sensitivity and target echo area, but it varies as the square of antenna gain and as the fourth power of transmitted power.

radar range marker A mark or line that is scribed or electronically formed on the face of a PPI display that indicates the range to the object detected.

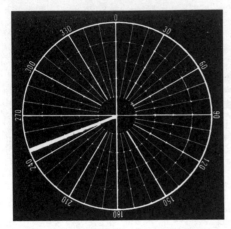

Radar range marker generates circles electronically on a plan-position-indicator display, and bearing lines are scribed on a transparent plastic graticule mounted over the face of the radarscope.

radar receiver A high-sensitivity radio receiver that amplifies and demodulates radar echo signals and feeds them to a radar display.

radar reconnaissance The use of radar to obtain information on location and strength of enemy forces and/or obtain terrain information.

radar reflector A reflector that reflects or deflects radar waves. See also *corner reflector*.

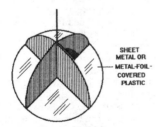

SHEET METAL OR METAL-FOIL-COVERED PLASTIC

Radar reflector for boats with fiberglass hulls will enhance their radar returns on larger nearby radar-equipped ships.

radar repeater A cathode-ray tube monitor that reproduces the visible intelligence of a radar display at a remote position. When used with a selector switch, the visible intelligence of any one of several radar systems can be reproduced.

radar resolution The resolution of a radar system.

radar scan The circular, spiral, rectangular, or other motion of a radar beam as it searches for a target.

radarscope A term that originally referred to the cathode-ray tube (CRT) display of radar information but now also refers to such forms of radar display as digitized, raster-scanned presentations available on liquid-crystal displays for smallcraft navigation.

radar system A complete assembly of radar equipment for detecting and ranging, consisting essentially of a transmitter, antenna, receiver, and indicator.

radar shadow A region shielded from radar illumination by an intervening reflecting or absorbing medium such as a hill or mountain.

radar signal analysis Analysis of radar reflections from an aircraft, ballistic missile, or orbiting-satellite to obtain as much information as possible about their size, shape, and purpose.

radar signal simulator Equipment whose electric output can be applied directly to the indicator of a radar set to produce artificial radar echo indications.

radar silence A period of time during which radar transmission is stopped, generally for security reasons.

radarsonde 1. An electronic system for automatically measuring and transmitting high-altitude meteorological data from a balloon, kite, or rocket by pulse-modulated radio waves when triggered by a radar signal. 2. A system in which radar techniques are used to determine the range, elevation, and bearing of a radar target carried aloft by a radiosonde.

radar station The ground, air, or sea location of a radar system that transmits or receives signals.

radar surveillance The use of one or more radar sets to locate distant targets and provide approach information. Surveillance can include facilities for identifying a target as friend or foe and means for relaying data to appropriate information and/or control centers.

radar surveying Surveying in which airborne radar measures accurately the distance between two ground radio beacons positioned along a baseline. This eliminates the need for measuring distance along the baseline in inaccessible or extremely rough terrain.

radar target An object being tracked or watched on a radar display or radarscope.

radar telescope A large radar antenna and associated equipment used for radar astronomy.

radar theodolite A theodolite based on radar to obtain azimuth, elevation, and slant range to a reflecting target, for surveying or other purposes.

radar track command guidance Command guidance that uses two radars external to a missile: one tracks the target and one tracks the missile. The radar receiver outputs are fed to a computer, and the output of the computer is in turn fed into a data transmitter that transmits flight information to the missile for correcting its flight path to the target.

radar tracking Tracking a moving object by radar.

radar trainer A trainer for teaching radar techniques and operations by simulating various radar target displays.

radar transmitter The transmitter section of a radar system.

radar video data processor A radar system that provides uncluttered displays for air traffic control by converting the input signals into digital data that can be synchronized with a system range clock. Associated circuits permit collecting and displaying weak target signals surrounded by noise.

radar wave A transmitted or reflected radio wave originating from a radar transmitter.

rad-hard Abbreviation for *radiation-hardened*.

radiac [RAdioactivity Detection, Identification, And Computation] 1. Detection, identification, and measurement of the intensity of nuclear radiation in an area. 2. *Radiac set.*

radiac set A complete radioactivity detecting, identifying, and measuring system. It is also called a radiac, a radiac instrument, and a radiacmeter.

radial One of a number of radial lines of position defined by an azimuthal radio navigation facility, and identified in terms of the bearing (usually magnetic) of all points on that line from the facility.

radial field A field of force directed toward or away from a point in space.

radial-leaded component An electronic component whose twin leads project in parallel from the body of the components, permitting direct insertion into holes in a conventional circuit board. Examples include light-emitting diodes, some aluminum electrolytic capacitors, and dipped tantalum capacitors. See also *axial-leaded component.*

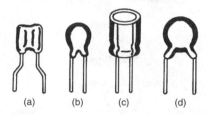

Radial-leaded component examples are (*a*) monolithic ceramic capacitor, (*b*) dipped tantalum capacitor, (*c*) aluminum electrolytic capacitor, and (*d*) ceramic disk capacitor.

radian [rad] The angle that intercepts an arc whose length is equal to its radius. A complete circle contains 2π rad. One radian is 57.29579°, and 1° is 0.01745 rad.

radiance The radiant flux per unit solid angle, per unit of projected area of the source. The usual unit is the watt per steradian per square meter. This is the radiant analog of luminance.

radianlength The distance between points in a sinusoidal wave that differ in phase by an angle of 1 radian. One radianlength is equal to the wavelength divided by 2π.

radian per second A unit of angular velocity.

radiansphere The boundary between the near and far fields of a small antenna. It is a spherical surface whose radius is equal to the wavelength divided by 6.28.

radiant Emitted or transmitted along radii, as from a point source.

radiant energy Energy transmitted in the form of electromagnetic radiation, such as radio waves, heat waves, and light waves.

radiant flux The time rate of flow of radiant energy.

radiant flux density The amount of radiant power per unit area that flows across or onto a surface. It is also called irradiance.

radiant gain The ratio of emitted radiant flux to incident radiant flux at specified ports in an optoelectronic device.

radiant intensity The energy emitted per unit time, per unit solid angle about the direction considered. It is expressed in watts per steradian.

radiant sensitivity The signal output current of a camera tube or phototube divided by the incident radiant flux at a given wavelength.

radiate To send out energy, such as electromagnetic waves, into space.

radiated interference Electromagnetic interference caused by radiated noise and other undesired signals from power lines or energized electric equipment.

radiated noise Electromagnetic energy that produces undesired noise in receiving equipment.

radiated power The total power emitted by a transmitting antenna.

radiating circuit A circuit capable of sending electromagnetic waves into space, such as the antenna circuit of a radio transmitter.

radiating element A basic subdivision of an antenna that in itself is capable of radiating or receiving RF energy.

radiating guide A waveguide that radiates energy into free space through slots, gaps, or horns.

radiation The emission of energy in any part of the electromagnetic spectrum from X-rays to radio waves. However, the term is usually understood by most persons to mean *nuclear radiation:* alpha, beta, and gamma rays.

radiation-absorbed dose [rad] A unit that specifies the amount of energy transferred to a material by ionizing radiation. One rad is the radiation dose that deposits an energy density of 100 ergs per gram in the absorbing material. The material must be specified because the energy differs with each material; for example, 1 rad of silicon = 100 ergs/g of silicon. Gallium arsenide devices typically have total dose hardness from 10^7 to 10^8 rads and dose rate hardness of 10^8 to 10^{11} rads/s.

radiation absorber An insulating material in sheet form; it has a conductive backing and is used as dielectric and reflecting elements for absorbing unwanted RF energy.

radiation belt *Van Allen belt.*

radiation burn A burn caused by overexposure to radiant energy.

radiation counter An instrument that detects or measures nuclear radiation by counting the resultant ionizing events. Examples include Geiger and scintillation counters.

radiation-counter tube *Counter tube.*

radiation damage Permanent damage to a semiconductor device can be caused by radiation of neutral particles and charged particles. This includes displacement damage, charge-transfer damage, and mechanical damage. Neutron and gamma-ray damage, for example, decreases transconductance, drain saturation current, and pinchoff voltage in MESFETs; neutron damage decreases power output and small-signal gain in MESFET amplifiers.

radiation danger zone A zone within which the maximum permissible constant dose rate is exceeded.

radiation detector A device for converting radiant energy to a form more suitable for observation.

radiation dose The total amount of ionizing radiation absorbed by material or tissues. It is commonly expressed in rads.

radiation efficiency The ratio of the power radiated to the total power supplied to an antenna at a given frequency.

radiation excitation *Radiation ionization.*

radiation field The electromagnetic field that breaks away from a transmitting antenna and radiates outward into space as electromagnetic waves. The other type of electromagnetic field associated with an energized antenna is the induction field.

radiation-hardened circuit An integrated circuit with modified circuit elements that have been made with isolation techniques to prevent those elements from being shorted out when exposed to heavy radiation. Conventional circuits will short out as a result of that exposure because the radiation generates electrical current inside the semiconductor material. Environments exposed to differing very high radiation levels are classified as (1) high total dose, (2) high dose rate, and (3) SEU (single-event upset). Circuits requiring in excess of 1000 krads Si are considered *strategic rad-hard,* while those requiring up to 50 krads Si are considered *tactical rad-hard.*

radiation hardening Improving the ability of a device, circuit, or equipment to withstand nuclear or other radiation. The techniques apply chiefly to dielectric and semiconductor materials.

radiation hardness The ability of a device or circuit to withstand the effects of radiation from space, nuclear reactors, nuclear weapons, and even alpha particles emitted from packaging ceramics. Neutrons and gamma-ray radiation are considered to be the most damaging. Both have high penetration and no charge polarity. Radiation hardness is measured in *dose rate, fluence,* and *total dose.*

radiation hazard A health hazard that arises from exposure to ionizing radiation.

radiation hazards meter An instrument that detects and measures electromagnetic radiation at power density levels which are potentially hazardous to human life processes. They are levels produced by high-energy radio sources in transmitters, industrial processing equipment, and microwave ovens.

radiation intensity The power radiated from an antenna per unit solid angle in a given direction.

radiation ionization Ionization of the atoms or molecules of a gas or vapor by electromagnetic radiation. It is also called *radiation excitation.*

radiation length The mean path length required to reduce the energy of relativistic charged particles by the factor $1/e$ or 0.368 as they pass through matter. The radiation length for relativistic electrons in air is 0.5 cm in lead.

radiation lobe The portion of a radiation pattern that is bounded by one or two cones of nulls.

radiation loss The portion of the transmission loss that is due to radiation of RF power from a transmission system.

radiation monitor A radiation detector that measures continuously the level of ionizing radiation and sometimes actuates an alarm when a preset danger level is exceeded.

radiation pattern A graphical representation of the radiation of an antenna as a function of direction. It is also called field pattern.

radiation potential The voltage corresponding to the energy in electronvolts required to excite an atom or molecule and cause emission of one of its characteristic radiation frequencies.

radiation pressure The extremely small pressure exerted on a surface by electromagnetic radiation, or the larger pressure exerted on a surface or interface by a sound wave.

radiation pyrometer A pyrometer that focuses the radiant power from the object or source to be measured on a thermocouple, thermopile, bolometer, or other suitable detector that provides electric output for an indicating instrument. It is also called a radiation thermometer.

radiation resistance The total radiated power of an antenna divided by the square of the effective antenna current measured at the point where power is supplied to the antenna.

radiation shield A shield or wall of lead or other material that effectively absorbs nuclear radiation.

radiation warning symbol A standard symbol used on posters displayed in locations where radiation hazards exist. The symbol consists of a magenta trefoil printed on a yellow background.

Radiation warning symbol.

radiation window A window that is transparent to alpha, beta, gamma, and/or X-rays. It protects the item that it covers from foreign matter.

radiator 1. The part of an antenna or transmission line that radiates electromagnetic waves either directly into space or against a reflector for focusing or directing. 2. A body that emits radiant energy.

radio 1. The transmission of signals through space by electromagnetic waves. The term is also applied to the transmission of audio and code signals, although television and radar also depend on electromagnetic waves. 2. *Radio receiver.*

radio- 1. A prefix that denotes radioactivity or a relationship to it, as in radiocarbon. 2. A prefix that denotes the use of radiant energy, particularly radio waves.

radioactive The property of exhibiting radioactivity. The source can be natural or manmade.

radioactive decay The spontaneous transformation of a nuclide into one or more different nuclides. The process involves (*a*) the emission from the nucleus of alpha particles, electrons, positrons, and gamma rays, (*b*) the nuclear capture or ejection of orbital electrons, or (*c*) fission. The rate of radioactive decay is expressed in terms of the half-life of the nuclide.

radioactive element An element that disintegrates spontaneously, giving off various rays and particles. Examples include promethium, radium, thorium, and uranium.

radioactive emanation A radioactive gas given off by certain radioactive elements. Thus, radium gives off radon, thorium gives off thoron, and actinium gives off actinon.

radioactive half-life The time required for a particular radioisotope to decrease to half its initial value.

radioactive isotope *Radioisotope.*

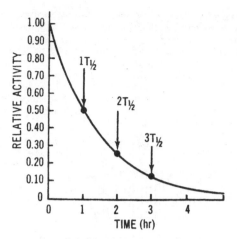

Radioactive-decay curve with its exponential form. The rate of disintegration is proportional to the number of atoms present at a given time. The curve shown is for radioactive material that decays to half its original value (one half-life) in 1 h, and to half of the new value (two half-lives) in the next hour.

radioactive material A material that has one or more constituents that exhibit significant radioactivity.

radioactive nuclide *Radionuclide.*

radioactive source Any quantity of radioactive material intended for use as a source of ionizing radiation.

radioactive standard A sample of radioactive material, usually with a long half-life, in which the number and type of radioactive atoms at a definite reference time is known. It is used for calibrating radiation-measuring equipment.

radioactivity Spontaneous nuclear disintegration, a property possessed by elements such as radium, uranium, thorium, and their products. Alpha or beta particles and sometimes gamma rays are emitted by disintegration of the nuclei of atoms.

radio aid *Radio navigation aid.*

radio altimeter An absolute altimeter that depends on the reflection of radio waves from the earth for the determination of altitude, as in an FM radar altimeter and a pulsed radar altimeter.

Radio Amateur Civil Emergency Service [RACES] An amateur radio service that provides communications during civil emergencies brought on by earthquake, fire, flood, hurricane, or tornado.

radio astronomy The study of radio waves emitted by astronomical bodies.

radio beacon A nondirectional radio transmitting station in a fixed geographic location that emits a characteristic signal from which bearing information can be obtained by a radio direction finder on a ship or aircraft. Some operate continuously. Others transmit only in response to an interrogation signal and can also provide range information.

radio beam A concentrated stream of RF energy as in radio ranges and microwave radio relays. A radar beam is a radio beam transmitted for a specific application.

radio bearing A bearing taken with respect to a radio transmitter, obtained with a radio direction finder.

radiobiology That branch of biology which deals with the effects of radiation on living tissue.

radio blackout *Radio fadeout.*

radio broadcast A program broadcast from a radio transmitter for general reception.

radio broadcasting Radio transmission intended for general reception.

radio broadcast station A station that transmits radio programs in the broadcast band, intended to be received by the general public.

radio capacity In cellular telephony, a measure of traffic capacity. It is defined as $m = M/K$, the number of channels per sector where M is the total number of frequency channels, K is the cell reuse factor, and S is the number of sectors.

radio channel A band of frequencies of a width sufficient to permit its use for radio communication. The width of a channel depends on the type of transmission and the tolerance for the frequency of emission.

radio circuit 1. An arrangement of components and connecting wires for radio purposes. 2. *Radio communication circuit.*

radio clock A clock that keeps precise time and date because its movement is controlled by radio signals transmitted from the U.S. Department of Commerce atomic clock in Boulder, Colorado. A built-in microcontroller processes the time signals and adjusts the clock's hands. Some are able to change from standard to daylight savings time automatically. These clocks are commercial products for sale to the general public.

radio command A command issued by radio to initiate or terminate some activity on a remote platform or vehicle. Examples include commands given to a missile for course correction or commands to a spacecraft to fire its rockets for a course change or turn on instruments to begin monitoring phenomena. However, household garage-door openers issue coded radio commands.

radio common carrier The official designation for such miscellaneous radio services as paging and portable telephones.

radio communication Communication by radio waves, such as by radio facsimile, radiotelegraph, radiotelephone, and radioteletypewriter.

radio control The control of stationary or moving objects by signals transmitted through space by radio.

radio deception The use of radio to deceive the enemy, as by sending false dispatches or using enemy call signs.

radio detection The detection of the presence of an object by radiolocation without precise determination of its position.

radio determination The determination of position, or the obtaining of information relating to position, by the propagation properties of radio waves.

radio direction finder [RDF] A radio aid to navigation that uses a rotatable loop or other highly directional antenna arrangement to determine the direction of arrival of a radio signal. Examples include aural-null and automatic direction finders. Also called direction finder.

radio distress signal The letters SOS transmitted without letter spaces in Morse code ($\cdots - - - \cdots$) or the spoken word "mayday," transmitted on one of the international distress frequencies.

radioelement An element tagged with one or more radioisotopes.

radio engineering   The field of engineering that deals with the generation, transmission, and reception of radio waves and with the design, manufacture, and testing of associated equipment.

radio facsimile   Facsimile communication by radio.

radio fadeout   A sudden and abnormal increase in ionization in the lower layers of the ionosphere, causing increased absorption of radio waves passing through these regions. Signals at receivers then fade out or disappear. The fadeout occurs suddenly and may last up to 1 h. Frequencies from about 3 to 10 MHz are most affected, but only where part or all of the signal path is in daylight. Transmission on frequencies below about 100 kHz is usually simultaneously improved. Also called blackout and radio blackout.

radio field strength   The effective value of the electric or magnetic field strength at a point due to the passage of radio waves of a specified frequency. It is expressed as the electric field intensity in microvolts or millivolts per meter.

radio field-to-noise ratio   The ratio of radio field strength to that of noise at a given location.

radio fix   1. Determination of the position of the source of radio signals by obtaining cross bearings on the transmitter with two or more radio direction finders in different locations, then computing the position by triangulation. 2. Determination of the position of a vessel or aircraft equipped with direction-finding equipment by obtaining radio bearings on two or more transmitting stations of known location and computing the position by triangulation.   3. Determination of position of an aircraft in flight by identification of a radio beacon or by locating the intersection of two radio beams.

radio frequency [RF]   A frequency at which coherent electromagnetic radiation of energy is useful for communication. Radio frequencies are designated as follows: very low frequency, below 30 kHz; low frequency, 30–300 kHz; medium frequency, 300–3000 kHz; high frequency, 3–30 MHz; very high frequency, 30–300 MHz; ultrahigh frequency, 300–3000 MHz; superhigh frequency, 3–30 GHz; extremely high frequency, 30–300 GHz. (For entries starting with radio-frequency, see RF entries.)

radio-frequency interference [RFI]   Interference with radio reception that occurs at or near the frequency of a desired signal. It can be caused by natural sources, such as lightning, or manmade sources, such as arc welders, diathermy machines, DC motor commutation, or the opening of relay contacts. It is also known as *RF interference*.

radio galaxy   A galaxy, consisting of billions of stars, that emits radio signals of varying strength from essentially its entire volume in the sky, thousands of light-years away.

radiogoniometer   A goniometer used as part of a radio direction finder. In the Bellini-Tosi system, two loop antennas positioned at right angles to each other are connected to two field coils in the radiogoniometer. Bearings are obtained by a rotatable search coil that is inductively coupled to the field coils.

radiogoniometry   The science of determining the direction of arrival of radio waves.

radiogram   A message transmitted by radio.

radio guard   A military ship, aircraft, or radio station designated to listen for and record radio transmissions and handle message traffic on one or more designated frequencies.

radio guidance system   A guidance system that uses radio signals to guide a flight-borne missile or other vehicle from a ground station.

radio-guided bomb   An aerial bomb guided by radio control from outside the missile.

radio homing beacon   *Homing beacon.*

radio horizon   The locus of points at which direct rays from a transmitter become tangential to the surface of the earth. The distance to the radio horizon is affected by atmospheric refraction.

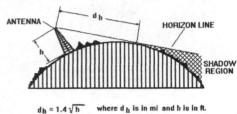

$$d_h = 1.4 \sqrt{h} \quad \text{where } d_h \text{ is in mi and } h \text{ is in ft.}$$

Radio horizon. "Line-of-sight" radio signal propagation from an antenna of height h is limited to the horizon distance d_h, assuming that the receiver is close to the earth's surface.

radio interference   Interference with reception of a desired radio signal by an undesired radio signal or by a radio disturbance.

radio interferometer   An interferometer that operates at radio frequencies, used for radio astronomy and in satellite tracking.

radio landing beam   A radio beam for vertical guidance of aircraft during descent to a landing surface.

radio line of position   A line of position obtained with a radio direction finder.

radio link   A radio system that provides a communication or control channel between two specific points.

radiolocation   The determination of direction, position, or motion of an object by exploiting the known properties of radio waves.

radio log   A log of radio messages sent and received, together with other pertinent information, maintained by radio operators.

radiological   A reference to nuclear radiation, radioactivity, and atomic weapons.

radiological defense   Defense against the effects of radioactivity from atomic weapons, including detection and measurement of radioactivity, protection of persons from radioactivity, and decontamination of areas and equipment.

radiological dose   The total amount of ionizing radiation absorbed by an individual exposed to any radiating source.

radioluminescence   Luminescence produced by radiant energy, as by X-rays, radioactive emissions, alpha particles, or electrons.

radio marker beacon   *Marker.*

radio metal locator   *Metal detector.*

radiometallography   Examination of the crystalline structure and other characteristics of metals and alloys with X-ray equipment.

radiometeorograph   *Radiosonde.*

radiometeorology   The branch of meteorology that covers the propagation of radio energy through the atmosphere and the use of radio and radar equipment in meteorology.

radiometer An instrument that measures radiant energy. Examples include the bolometer, microradiometer, microwave radiometer, and thermopile.

radiometry Measurement of quantities associated with radiant energy.

radio multiplexing 1. Dividing a radio channel into a number of voice or code channels through frequency division or time division. 2. Connecting two or more transmitters or receivers to the same antenna through appropriate coupling networks.

radio navigation Navigation by radio signals, with such equipment as radio direction finders, radio ranges, radio beacons, loran, and GPS.

radio navigation aid A navigation aid that uses radio signals, as contrasted to a radar navigation aid.

radio navigation guidance The guidance or control of a guided missile along a course established by external radio transmitters.

radio net A net of radio stations established for communication purposes.

radio noise Noise that occurs in the radio spectrum.

radionuclide A substance that exhibits radioactivity. It is also called a radioactive nuclide.

radio operator A person who operates radio transmitting and receiving equipment.

radiopaque Not appreciably penetrable by X-rays or other forms of radiation.

radiopaque obstacle An obstacle that creates a communication blackout between a spacecraft and the earth.

radiophare *Radio beacon.*

radiophone *Radiotelephone.*

radiophoto 1. A photograph transmitted by radio to a facsimile receiver. 2. *Facsimile.*

radiophotoluminescence Luminescence exhibited by minerals such as fluorite and kunzite as a result of irradiation with beta and gamma rays followed by exposure to light.

radio propagation prediction A prediction of the quality or nature of radio propagation as influenced by such factors as sunspots and seasonal changes; the prediction is published periodically by the National Institute of Standards and Technology (NIST).

radio range A radio transmitting facility that radiates signals from which aircraft can determine their bearings with respect to the transmitting site. The A-N radio range provides four courses. It is also called a radio-range beacon and range.

radio-range beacon *Radio range.*

radio-range leg One of the courses or beams in a four-course radio range.

radio-range monitor An instrument that automatically monitors the signal from a radio-range beacon. It gives a warning to attendants when the transmitter deviates a specified amount from its correct bearings and transmits a distinctive warning to approaching planes when trouble exists at the beacon.

radio receiver A receiver that converts radio waves into intelligible sounds or other perceptible signals. It is also called a radio, radio set, or receiver.

radio reception Reception of messages, programs, or other intelligence by radio.

radio relay system *Radio repeater.*

radio repeater A repeater that acts as an intermediate station in transmitting radio communication signals or radio programs from one fixed station to another. It extends the reliable range of the originating station. A microwave repeater is an example. It is also called a *radio relay system* or a *relay system.*

radio scanner *Scanning radio.*

radio scattering *Scattering.*

radio signal A signal transmitted by radio.

radio silence A period during which transmissions by a radio station are stopped. It can permit reception of signals from other stations or permit reception of weak distress signals.

radio sky The sky as it would appear if human eyes were sensitive to radio waves instead of to light.

radiosonde [pronounced radio sond] A meteorograph combined with a radio transmitter. When carried aloft by a balloon, it transmits radio signals that can be recorded at a ground station and interpreted in terms of the pressure, temperature, and humidity at regular intervals during the ascent. It is also called a radiometeorograph. The equipment descends by parachute when the balloon bursts.

radio source A region in the sky from which radio waves are received by radio-astronomy equipment.

radio spectrum The entire range of frequencies in which useful radio waves can be produced. It extends from the audio range to about 300 GHz. The radio spectrum is divided into eight bands (see band or radio frequency). It is also called the RF spectrum.

radio star A discrete radio source in the sky. It does not usually correspond to a known optical star and should therefore be called a radio source. The sun is the only star definitely known to emit radio waves.

radio station A station equipped to engage in radio communication or radio broadcasting.

radio-station interference Interference caused by radio stations other than that from which reception is desired.

radio sun The sun as defined by its electromagnetic radiation in the radio portion of the spectrum.

radio system A complete radio equipment installation that provides multichannel communication between two points.

radiotelegraph A reference to telegraphy over radio channels.

radiotelegraph transmitter A radio transmitter that is capable of handling code signals (type A1 and B emissions).

radiotelegraphy Telegraphy transmitted by radio waves rather than wire. The international Morse code is generally used in radiotelegraphy.

radiotelephone 1. A reference to telephony over radio channels. 2. A radio transmitter and radio receiver that together permit two-way telephone communication by radio. It is also called a radiophone.

radiotelephone distress call The word "mayday," corresponding to the French pronunciation of m'aider, spoken under the same conditions that the signal SOS would be transmitted in code by radiotelegraphy.

radiotelephone transmitter A radio transmitter capable of handling AF modulation, such as voice and music.

radiotelephony Two-way voice communication (telephony) based on radio without connecting wires between stations.

radio telescope A system consisting of an antenna for collecting celestial radio signals and a receiver for detecting and recording them. The antenna is analogous to the objective lens or mirror of an optical telescope, while the

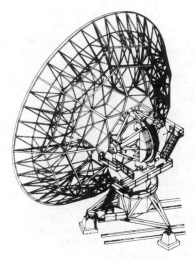

Radio telescope antenna with an 82 ft (25 m) diameter.

receiver/recorder is analogous to the eye-brain combination, photographic plate, or video recorder. The appearance of the sky at radio wavelengths differs from its optical appearance because the sun is less significant, while the Milky Way radiates strong signals. The rest of the sky is dotted with radio sources almost entirely unrelated to objects visible to the unaided eye. The radio window extends from a few millimeters to tens of meters in wavelength, permitting many kinds of antennas to be useful.

radioteletype [RTTY] A narrow-band, direct-printing form of radio telegraphy intended for automatic copying with teleprinters connected to radio receivers and transmitters. The radio signals are sent from one teleprinter to another.

radioteletypewriter A teletypewriter and the associated equipment for operation over a radio channel rather than wires.

radiothermoluminescence Luminescence exhibited by certain vitreous and crystalline substances as a result of irradiation with beta and gamma rays followed by heating.

radiotracer *Radioactive tracer.*

radio transmission The transmission of signals through space at radio frequencies by radiated electromagnetic waves.

radio transmitter A transmitter that produces RF power for transmission through space in the form of radio waves.

radio tube *Electron tube.*

radio wave An electromagnetic wave produced by reversal of current in a conductor at a frequency in the range from about 10 kHz to 3000 GHz. Radio waves travel through space at approximately the speed of light (299,792.458 km/s or about 186,000 mi/s).

radio wave propagation The transfer of energy through space by electromagnetic radiation at radio frequencies.

radio window A band of frequencies extending from about 6 MHz to 30 GHz that permits radiation from the outer universe to enter and travel through the atmosphere of the earth.

radium [Ra] A highly radioactive metallic element that gives off alpha, beta, and gamma rays, with an atomic number of 88.

radix *Base.*

radix notation A positional notation in which the successive digits are interpreted as coefficients of successive integral powers of a number called the radix or base. The represented number is equal to the sum of this power series. Thus, 5762 is the sum of the power series $5 \times 10^3 + 7 \times 10^2 + 6 \times 10^1 + 2 \times 10^0$, where 10 is the base.

radix point *Point.*

RADNOS [transposition of NO RADio plus S] A radio fadeout encountered chiefly in arctic regions, considered to be caused by solar explosions, sunspots, or the aurora borealis.

radome [RAdar DOME] A protective housing for a radar antenna, made from dielectric material (usually plastic laminate) that is transparent to RF energy.

radon [Rn] A heavy gaseous radioactive element with an atomic number of 86 and atomic weight of 222. It is a daughter of radium in the uranium radioactive series.

rad per unit time A unit of absorbed dose rate.

railing The jamming of radar transmissions by transmitters at a pulse rate of 50 to 150 kHz. It causes images resembling fence railings to appear on radar screens.

railroad radio service A radio communication service related to the operation and maintenance of a railroad common carrier.

rail voltage British term for *supply voltage.*

rain attenuation Attenuation of radio waves when passing through moisture-bearing cloud formations or areas where rain is falling. The attenuation increases with the density of the moisture in the transmission path.

RAM Abbreviation for *random-access memory.*

Raman scattering Scattering of light by the molecules of transparent gases, liquids, and solids. It results from a change in the frequency in the incident radiation because of interaction of this radiation with the molecules. It is used in studying molecular structure of materials.

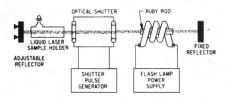

Raman scattering occurs in a sample liquid with laser action when powerful pulses of light are beamed through a liquid by a ruby laser.

Raman spectrometer A spectrometer in which a continuous-wave laser serves as the excitation source. A helium-neon laser operating at 0.6328 μm is commonly used. It is also called a laser Raman spectrometer.

RAMARK [RAdar MARKer] A fixed radar beacon that emits radar waves continuously to provide a bearing indication to radar-equipped ships and aircraft.

ramp generator A circuit that generates a sweep voltage which increases linearly in value during one cycle of sweep, then returns to zero suddenly to start the next cycle.

Ramsauer effect The low attenuation of slow-moving electrons by inert gases.

Rand Corporation [Research ANd Development] A nongovernmental nonprofit organization engaged in research for the welfare and national security of the United States.

random-access discrete-address system A radio-communication service in which a large group of users share a broad band of channels simultaneously. Voice modulation is converted into digital form, and the resulting pulses are transmitted in sequence, each at a different carrier frequency and a different assigned instant in time. Each receiver in a given service has an assigned combination of channel frequencies and time slots.

random-access memory [RAM] A memory that is organized so that any address location can be written in or read in any sequence. Digital bits can be stored temporarily and can be rapidly changed as desired. Random access means that there is ready access to any storage location in the memory. A more precise term would be read/write memory or RWM. Most semiconductor memory devices are randomly accessible as opposed to serially accessible as in a magnetic or video tape or charge-coupled device (CCD). In normal use, the term implies a dynamic random-access memory (DRAM), a volatile semiconductor memory that must be regularly refreshed to maintain all data that is stored in its "cells." The DRAM is the most widely used read/write memory in computers. See *dynamic random-access memory* [DRAM] and *static random-access memory* [SRAM].

random error An error that can be predicted only on a statistical basis.

random logic Logic based on combinations of simple gates that provide functions which can be programmed either during manufacture or in the field.

random noise Noise characterized by a large number of overlapping transient disturbances occurring at random, such as thermal or shot noise. It can be produced intentionally for test purposes or for jamming enemy transmissions.

random-noise testing Testing with a complex wave whose randomly varying frequencies and amplitudes are applied to a mechanical shake table. The signal can be obtained experimentally, as from a missile telemetering record, or can be generated electronically.

random number A number formed by a set of digits in which each successive digit is equally likely to be any of n digits to the base n. Random numbers are obtained entirely by chance.

random variable A discrete or continuous variable that can assume any one of a number of values, each of which has a fixed probability.

random-wire antenna A multiband wire antenna of any convenient length that is connected directly to a transmitter or impedance-matching network without a feed line.

range 1. The distance from a radar set or weapon to a target. 2. The distance capability of an aircraft, missile, gun, radar, or radio transmitter. 3. The difference between the maximums and minimums of a variable quantity. 4. A line defined by two fixed landmarks, for missile, vehicle, and other test purposes. 5. A line of bearing defined by a radio range. 6. *Radio range.*

range-amplitude display A radar display on which a time base provides the range scale and echoes appear as deflections normal to the base. The base is usually a straight line, as in the A display, or a circle, as in the J display.

range circle A radar range marker in the form of a circle.

range gate A gate circuit that selects radar echoes within a small range interval.

range-height indicator display A radar display that presents visually the scalar distance between a reference point and a target, along with the vertical distance between a reference plane and the target.

range mark *Radar range marker.*

range marker generator A signal generator that generates the signal required for the production of radar range markers on a radarscope. Its action is initiated by the sync pulse that starts the time base.

range rate The rate at which the distance from the measuring equipment to the signal source being tracked is changing with respect to time.

range resolution *Distance resolution.*

range selector A control that selects the range scale on a radar indicator.

range target A reflective target at a precisely known range from a radar antenna for radar range system alignment.

ranging Determining distance.

rank The number of independent cut-sets that can be selected in a network. The rank is equal to the number of nodes minus the number of separate parts.

Rankine [R] An absolute temperature scale based on Fahrenheit degrees but with the entire scale shifted so 0°R is at absolute zero. In this scale, water freezes at 459.6°R and boils at 639.6°R. Add 427.6 to a Fahrenheit value to obtain the corresponding Rankine value.

Raphael bridge A type of slide-wire Wheatstone bridge for locating faults in transmission lines.

rapid scanning Scanning with a narrow radar beam at the rate of 10 sweeps per second or more, as required for tracking fast-moving targets.

rapid-start fluorescent lamp A fluorescent lamp with a ballast that has a low-voltage winding which is continuously connected to the cathode heaters of the lamp, to initiate and maintain a gaseous discharge.

rare earth An element that has an atomic number in the range from 57 to 71 inclusive. The rare earths are cerium, dysprosium, erbium, europium, gadolinium, holmium, lanthanum, lutecium, neodymium, praseodymium, promethium, samarium, terbium, thulium, and ytterbium.

rare-earth chelate laser *Chelate laser.*

rare-earth magnet A permanent magnet formed from compounds of cobalt and one or two of the following: cerium, lanthanum, mischmetal, praseodymium, samarium, or yttrium. One example, the samarium-cobalt magnet, has many applications.

raster A predetermined pattern of scanning lines that provides substantially uniform coverage of an area. In television the raster is seen as closely spaced parallel lines, most evident when there is no picture.

raster burn A change in the characteristics of the scanned area on the target of a camera tube, resulting in a spurious signal when a larger or tilted raster is scanned.

raster data Graphic data stored as a series of cells, each having X-Y coordinates and a value. It is a simple but inefficient means for storing geographic information.

rate control Control of the rate of change of the independent variable in an automatic control system. It is also called proportional response or throttling control.

rated accuracy   The advertised accuracy of a manufactured instrument.

rated coil current   The steady-state coil current at which a relay is designed to operate.

rated coil voltage   The coil voltage at which a relay is designed to operate.

rated contact current   The current that contacts are designed to carry for their rated life.

rated output   The output power, voltage, current, or other value at which a machine, device, or circuit is designed to operate under specified normal conditions.

rate-grown junction   A grown junction produced by varying the rate of semiconductor crystal growth periodically and using a melt that contains both N- and P-type impurities. The two types of impurities alternately predominate. It is also called a graded junction.

rate-grown transistor   A junction transistor in which both impurities (such as gallium and antimony) are placed in the melt at the same time and the temperature is suddenly raised and lowered to produce the alternate P- and N-type layers of rate-grown junctions. It is also called a graded-junction transistor.

rate gyroscope   A spinning or vibratory gyroscope that measures the rate of change of direction of a platform.

rate meter   *Counting-rate meter.*

rate-of-climb indicator   An instrument that indicates the rate of climb or descent of an aircraft.

rate of closure   The speed at which two airborne aircraft or other moving objects close the distance between them. With aircraft approaching each other, the rate of closure is the sum of their speeds.

rate of decay   The time rate at which the sound pressure level, velocity level, or sound-energy density level is decreasing at a given point and at a given time. The practical unit is the decibel per second.

rate-of-turn control   A gyroscopic instrument that furnishes a rate-of-turn signal to an automatic pilot system in an aircraft.

rate signal   A signal proportional to the time derivative of a specified variable.

rate transmitter   A transmitter in a missile being launched that, with a ground receiver, will indicate the rate of speed increase.

rating   A designation of an operating limit for a machine, apparatus, or device used under specified conditions.

ratio   The value obtained when one quantity is divided by another of the same kind, to indicate their relative proportions.

ratio arms   Two adjacent arms of a Wheatstone bridge designed so they can be set to provide a variety of indicated resistance ratios.

ratio control   Control in which a predetermined ratio between two physical quantities is maintained.

ratio detector   An FM detector circuit with two diodes that requires no limiter at its input. The audio output is determined by the ratio of two developed IF voltages whose relative amplitudes are a function of frequency.

ratio meter   A meter that measures the quotient of two electrical quantities. The deflection of the meter pointer is proportional to the ratio of the currents flowing through two coils.

rationalized unit   A unit in a system of measurement that is designed to minimize occurrence of the constant 4π in equations.

ratio of transformation   The ratio of the secondary voltage of a transformer to the primary voltage under no-load conditions, or the corresponding ratio of currents in a current transformer.

ratio-type telemeter   A telemeter that uses the phase or magnitude relations of two or more electrical quantities as the translating means.

rat race   *Hybrid ring.*

raw data   Unprocessed data that might or might not be in machine-readable form.

rawin   [RAdar WINd or RAdio WINd; pronounced ray win] 1. Determination of wind direction and velocity by radar or by radio direction-finding in conjunction with a radiosonde, radiosonde balloon, or a balloon carrying a radar reflector. 2. Wind information gathered with radar tracking or radio direction-finding in connection with a specially equipped balloon.

rawinsonde   A radiosonde used in rawin.

ray   *Beam.*

Rayleigh cycle   A cycle of magnetization that does not extend beyond the initial portion of the magnetization curve, between zero and the upward bend. In this region the permeability is low and there is little hysteresis.

Rayleigh disk   An acoustic radiometer that measures particle velocity. A thin disk suspended by its edge from a fine fiber tends to take a position perpendicular to the horizontal component of sound particle velocity.

Rayleigh distribution   A mathematical statement of a natural distribution of random variables.

Rayleigh line   A spectrum line in scattered radiation that has the same frequency as the corresponding incident radiation. It arises from ordinary or Rayleigh scattering.

Rayleigh scattering   Selective scattering of light by very small particles suspended in air, as by dust.

Rayleigh wave   A type of wave that can be propagated near the surface of a solid and is characterized by elliptical motion of particles.

RC   Abbreviation for *resistance-capacitance.*

RC amplifier   *Resistance-coupled amplifier.*

RC circuit   Abbreviation for *resistance-capacitance circuit.*

RC constant   The time constant of a resistance-capacitance circuit, equal in seconds to the resistance value in ohms multiplied by the capacitance value in farads.

RC coupling   *Resistance coupling.*

RC differentiator   Abbreviation for *resistance-capacitance differentiator.*

RC filter   Abbreviation for *resistance-capacitance filter.*

RCG circuit   Abbreviation for *reverberation-controlled gain circuit.*

RCM   Abbreviation for *radar countermeasure.*

RC oscillator   Abbreviation for *resistance-capacitance oscillator.*

RCTL   Abbreviation for *resistor-capacitor-transistor logic.*

$R_{DS(on)}$ A symbol for the static resistance between the drain and source of a forward-biased power MOSFET at a specified drain current and gate voltage.

RDF   Abbreviation for *radio direction finder.*

reactance [X]   The opposition offered to the flow of alternating current by pure inductance or capacitance in a circuit, expressed in ohms. It is the component of impedance that is not due to resistance. Inductive reactance is due to inductance; capacitive reactance is due to capacitance.

reactance control circuit   The color television receiver circuit that converts the DC correction voltage from the phase detector into a capacitive reactance change which maintains the 3.58-MHz oscillator at the correct frequency and phase.

reactance frequency divider   A frequency divider that uses a nonlinear coil or capacitor to generate subharmonics of a sinusoidal source.

reactance frequency multiplier   A frequency multiplier that uses a nonlinear coil or capacitor to generate harmonics of a sinusoidal source.

reactance modulator   A modulator whose reactance can be varied in accordance with the instantaneous amplitude of the modulating voltage. This is a circuit that produces phase or frequency modulation.

reaction   1. An action in which one or more substances are changed into one or more new substances, as in a nuclear reaction.   2. British term for *positive feedback*.

reaction cavity   A cavity that can be mounted on a side or end of a waveguide, for automatic or manual frequency-control applications. Tuning is achieved with a micrometer head that controls the position of the plunger in the cavity.

reaction motor   A synchronous motor whose rotor contains salient poles but has no windings and no permanent magnets.

reactivation   Application of a higher voltage than normal to the thoriated cathode of an electron tube for a few seconds, to bring a fresh layer of thorium atoms to the filament surface and thereby improve electron emission.

reactive   A reference to inductive or capacitive reactance.

reactive attenuator   An attenuator that absorbs very little energy.

reactive factor   The ratio of reactive power to apparent power.

reactive-factor meter   A meter that measures and indicates reactive factor.

reactive load   A load that has inductive or capacitive reactance.

reactive power   The power value obtained by multiplying together the effective value of current in amperes, the effective value of voltage in volts, and the sine of the angular phase difference between current and voltage. It is also called reactive voltamperes and wattless power. The unit of reactive power is the var.

reactive voltampere   *Voltampere reactive.*

reactive voltampere-hour   *Voltampere-hour reactive.*

reactive voltampere meter   *Varmeter.*

reactive voltamperes   *Reactive power.*

reactor   A device that introduces either inductive or capacitive reactance into a circuit, such as a coil or capacitor.

read   1. To acquire information, usually from some form of memory in a computer.   2. To generate an output corresponding to the pattern stored in a charge-storage tube.   3. To understand clearly, as in radio communication.

read in   To sense one form of information and transmit this information to an internal memory of a computer.

reading   1. The indication shown by an instrument.   2. To observe the readings of one or more instruments.

reading rate   The number of characters, words, fields, or blocks that can be sensed by an input reading device per unit of time.

reading speed   The rate of reading successive storage elements in a charge-storage tube.

read number   The number of times that a storage element is read without rewriting in a charge-storage tube.

read-only memory [ROM]   A nonvolatile semiconductor memory in which the binary information located at each address is fixed and cannot be changed. It permanently stores information that is used repeatedly, such as tables of data and characters for electronic displays. The factory-programmed ROM consists of a mosaic of undifferentiated cells that are permanently factory-programmed for dedicated (applications-specific) tasks. See also *programmable ROM* [PROM], *erasable ROM,* [EPROM], *electrically erasable ROM* [EEPROM], and *flash memory.*

readout   The presentation of output information by a cathode-ray tube or liquid-crystal display, or printer.

read pulse   A pulse that causes information to be acquired from a magnetic cell or cells.

read time   The time interval between the instant at which information is called for from computer memory and the instant at which delivery is completed in a computer. It is also called access time.

read/write memory   *Dynamic random-access memory* (DRAM) or *static random-access memory* (SRAM).

ready-to-receive signal   A signal sent back to a facsimile transmitter to indicate that a facsimile receiver is ready to accept the transmission.

real power   The component of apparent power that represents true work. It is expressed in watts and can be calculated as the product of voltage and the in-phase component of alternating current or voltamperes multiplied by the power factor.

real time   The performance of a computation during the time of a related physical process, so the results are available for guiding the physical process. It is typical of industrial control.

real-time clock   A circuit that measures time at a rate consistent with the task being performed.

real-time data   Data presented in usable form at essentially the same time the event occurs. The delay in presenting the data must be small enough to allow a corrective action to be taken if required.

real-time delay   An essentially negligible delay time, generally of the order of a few nanoseconds and controllable. It is used in antenna arrays to obtain a desired radiation pattern by connecting appropriate delay lines between the elements of the array.

real-time operation   Computer data processing that is fast enough to be able to process information about events as they occur, as opposed to batch processing that occurs at a time unrelated to the actual events. The computer gathers data from sensors, performs calculations, and responds fast enough to keep pace with a process. Examples are automotive engine control, antilock brakes, and digitally controlled machine tools and robots.

rear projection   A projection television system in which the picture is projected on a ground-glass screen for viewing from the opposite side of the screen.

rebroadcast   Repetition of a radio or television program at a later time.

recalescent point   The temperature at which there is a sudden liberation of heat as a heated metal is cooled.

receiver The complete equipment required for receiving modulated radio waves and converting them into the original intelligence, such as into sounds or pictures. It also produces useful information in a radar receiver.

receiver bandwidth The frequency range between the half-power points on the frequency-response curves of a receiver.

receiver gating The application of operating voltages to one or more stages of a receiver only during that part of a cycle of operation when reception is desired.

receiver noise figure The ratio of noise in a given receiver to that in a theoretically perfect receiver.

receiver radiation Radiation of interfering electromagnetic fields by the oscillator of a receiver.

receiver synchro *Synchro receiver.*

receiving antenna An antenna that converts electromagnetic waves to modulated RF currents.

receiving set *Radio receiver.*

receiving station A station that receives radio signals or messages.

receptacle *Outlet.*

rechargeable battery *Secondary battery.*

recharger A DC power supply that recharges nickel-cadmium or other rechargeable batteries for calculators, portable appliances, instruments, and other battery-operated devices.

reciprocal counter A counter that measures a time interval between two events and computes the reciprocal of the measured value. The reciprocal of the period of a signal is its frequency, and the reciprocal of the time for an object to travel between two points is its speed.

reciprocal-energy theorem A theorem of Rayleigh: If an electromotive force E_1 in one branch of a circuit produces a current I_2 in any other branch, and if an electromotive force E_2 inserted in this other branch produces a current I_1 in the first branch, then $I_1E_1 = I_2E_2$. This is closely related to the reciprocity theorem.

reciprocal ferrite switch A ferrite switch that can be inserted in a waveguide to switch an input signal to either of two output waveguides. Switching is done by a Faraday rotator when acted on by an external magnetic field.

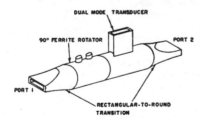

DUAL MODE TRANSDUCER

90° FERRITE ROTATOR   PORT 2

PORT I

RECTANGULAR-TO-ROUND TRANSITION

Reciprocal ferrite switch.

reciprocal impedance Two impedances Z_1 and Z_2 are said to be reciprocal impedances with respect to an impedance Z (invariably a resistance) if they satisfy the equation $Z_1Z_2 = Z^2$.

reciprocity theorem If a voltage located at one point in a network produces a current at any other point in the network, the same voltage acting at the second point will produce the same current at the first point.

reclosing relay A relay that functions to reclose a circuit automatically under certain conditions.

recognition differential The signal strength above noise level that gives a 50% probability of detection of an aural signal.

recombination The process in which free-moving and spatially separated electrons and holes are recaptured by atoms, stopping their movement and current-carrying capability, resulting in zero net charge.

recombination radiation The radiation emitted in semiconductors when electrons in the conduction band recombine with holes in the valence band. If an actual population inversion is achieved between portions of the valence and conduction bands, or between adjacent localized states of acceptors or donors near these respective bands, stimulated emission and laser amplification or oscillation can result.

recombination rate The time rate at which free electrons and holes recombine at the surface or within the volume of a semiconductor.

recombination velocity The normal component of the electron or hole current density on a semiconductor surface divided by the excess electron or hole charge density at the surface.

recommutation Failure of load current to be completely commutated from one ignition to another within the required time, with the result that current is commutated back to the original tube.

reconditioned-carrier receiver A receiver in which the carrier is separated from the sidebands to eliminate amplitude variations and noise, then added at an increased level to one sideband to obtain a relatively undistorted output. It is used with single-sideband transmitters.

reconnaissance satellite An earth satellite that provides strategic information, as by television, photography, or radio link.

reconstituted conductive material Conductive material formed by compressing finely divided particles.

reconstituted mica Mica sheets or shaped objects made by breaking up scrap natural mica, combining it with a binder, and pressing it into forms suitable as an insulating material.

recontrol time *Deionization time.*

record 1. To preserve for later reproduction or reference. 2. A group of related facts or fields of information treated as a unit. 3. Compact disk.

record changer A record player that plays two or more records automatically in succession.

record density *Character density.*

recorded program A radio or television program that depends on compact disks, electric transcriptions, magnetic tapes, or other means of reproduction.

recorded tape 1. A recording that is commercially available on magnetic tape, also called prerecorded tape. 2. Any magnetic tape that has been recorded.

recorder An instrument that makes a permanent record of varying electrical quantities or signals. A common industrial version records one or more quantities as a function of another variable, usually time. Other types include the cathode-ray oscilloscope, facsimile recorder, kinescope recorder, magnetic-tape recorder, and video tape recorder.

record gap A gap that indicates the end of a record on magnetic tape.

record head *Recording head.*

recording 1. Any process for preserving signals, sounds, data, or other information for future reference or reproduction, such as disk recording, facsimile, compact-disk recording, magnetic-tape or wire recording, and photographic recording. 2. The end product of a recording process, such as the recorded audio or video magnetic tape, disk, or record sheet.

recording blank   *Recording disk.*

recording channel   The independent recording tracks on a recording medium.

recording characteristic   A graph that shows the intentional attenuation of bass frequencies and accentuation of treble frequencies involved in making a disk or tape recording.

recording density   The number of bits per unit length in a single linear data recording track.

recording disk   An unrecorded or blank disk for recording sounds by a stylus or a laser.

recording head   A magnetic head used only for recording.

recording lamp   A lamp whose intensity can be varied at an AF rate for exposing variable-density sound tracks on motion-picture film and for exposing paper or film in photographic facsimile recording.

recording level   The amplifier output level required to drive a particular recorder.

recording noise   Noise that is introduced during a recording process.

recording-playback head   A magnetic head for both recording and reproduction.

recording storage tube   A type of cathode-ray tube capable of storing the electric equivalent of an image as an electrostatic charge pattern on a storage surface. There is no visual display, but the stored information can be read out at any desired later time as an electric output signal.

recording stylus   A tool that inscribes the grooves in a mechanical recording medium.

record length   The number of characters required for all the information in a record.

record medium   The physical medium on which a facsimile recorder forms an image of the subject copy.

record player   A motor-driven turntable that with a phonograph pickup obtains AF signals from a phonograph record. These signals must be fed into an AF amplifier for additional amplification before they can be reproduced as sound waves by a loudspeaker. In an electric phonograph the amplifier and loudspeaker are combined with the record player.

recovery package   An instrumentation package carried by a missile or satellite and designed for ejection before impact. The package usually also contains a transmitter that emits signals after ejection, to aid in recovery.

recovery time   1. The time required for the control electrode of a gas tube or surge voltage protector to regain control after anode-current interruption. 2. The time required for a fired TR or pre-TR tube to deionize to such a level that the attenuation of a low-level RF signal transmitted through the tube is decreased to a specified value. 3. The time required for a fired ATR tube to deionize to such a level that the normalized conductance and susceptance of the tube in its mount are within specified ranges. 4. The minimum time from the start of a counted pulse to the instant a succeeding pulse can attain a specific percentage of the maximum value of the counted pulse in a Geiger counter.   5. The interval required, after a sudden decrease in input signal amplitude to a system or compo-

nent, to attain a specified percentage (usually 63%) of the ultimate change in amplification or attenuation due to this decrease.   6. The time required for a radar receiver to recover to half sensitivity after the end of the transmitted pulse, so it can receive a return echo.

rectangular coordinate   A coordinate with respect to one of three mutually perpendicular axes for specifying the location of a point in space.

rectangular horn antenna   A horn antenna that has a rectangular cross section, with one or both transverse dimensions increasing linearly from the small end or throat to the mouth positioned at the end of a rectangular waveguide to radiate radio waves directly into space.

rectangular picture tube   A television picture tube that has an essentially rectangular faceplate and screen. The typical aspect ratio is 3:5.

rectangular wave   A periodic wave that alternately and suddenly changes from one to the other of two fixed values. When the duty cycle is 50% (the two fixed values have equal times), the rectangular wave becomes a square wave.

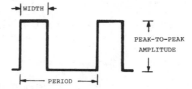

Rectangular wave, unlike a square wave, can have a pulse width that is not equal to half the period.

rectangular waveguide   A waveguide that has a rectangular cross section.

rectification   The process of converting an alternating current (AC) to a unidirectional current (DC).

rectification factor   The change in the average current of an electrode divided by the change in the amplitude of the alternating sinusoidal voltage applied to the same electrode under the condition that the direct voltages of this and other electrodes remain constant.

rectified current   The direct current that results from the process of rectification.

rectifier   A device that converts alternating current into a current with a large unidirectional component. Examples are a gas tube, metallic rectifier, semiconductor diode, and vacuum tube.

rectifier junction   A semiconductor junction that has greater conductivity in one direction than in the other.

rectifier stack   A stacked assembly of semiconductor rectifier disks or wafers.

rectify   To convert an alternating current (AC) into a unidirectional current (DC).

rectilineal compliance   The mechanical compliance that opposes a change in the applied force. An example is the springiness that opposes a force acting on the diaphragm of a loudspeaker or microphone.

rectilinear   Following a straight line.

rectilinear scanning   The process of scanning an area in a predetermined sequence of narrow, straight parallel strips.

recurrence rate   *Repetition rate.*

recycling   Returning to an original condition, as to 0 or 1 in a counting circuit.

Redeye   A man-transportable shoulder-fired guided missile with infrared guidance, for destroying a low-flying enemy aircraft.

red gun   The electron gun whose beam strikes phosphor dots emitting the red primary color in a three-gun color television picture tube.

redistribution   The alteration of charges on an area of a storage surface in a charge-storage tube or television camera tube by secondary electrons from any other area of the surface.

red restorer   The DC restorer for the red channel of a three-gun color television picture tube circuit.

reduced-instruction set computer [RISC]   A computer processor that rapidly executes a small set of basic instructions by leaving out seldom used instructions from the processor. The circuitry saved by not allowing more complex instructions can be assigned as additional registers or to other functions. See *complex-instruction-set computer*.

redundancy   Any deliberate duplication or partial duplication of circuitry or information to sustain circuit or computer operation in the event of a partial equipment failure.

redundancy check   A forbidden-combination check that is based on redundant digits called check digits to detect errors made by a computer.

redundant digit   A digit that is not necessary for an actual computation but reveals a malfunction in a digital computer.

redundant operation   The practice of powering an electronic system with a parallel group of equally rated power supply modules whose total output under full load is greater than that required for full system operation. It increases system reliability by permitting the system to remain operational if one module fails. The other supplies pick up the load until the faulty module can be replaced, usually without system downtime.

red video voltage   The signal voltage output from the red section of a color television camera, or the signal voltage between the matrix and the grid of the red gun in a three-gun color television picture tube.

reed relay   A relay that has contacts mounted on magnetic reeds sealed into a length of small glass tubing. An actuat-

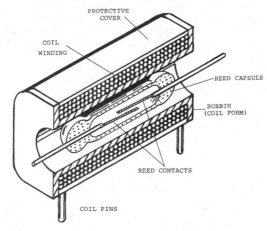

Reed relay contacts are opened or closed by energizing a solenoid coil.

ing coil is wound around the tubing or wound on an auxiliary ferrite-core structure to provide the magnetic field required for relay operation. The contacts can be dry or mercury-wetted. It is also called a magnetic reed relay.

reed switch   A switch that has contacts at the ends of ferromagnetic reeds sealed in a glass tube designed for actuation by an external magnetic field. The contacts can be dry or mercury-wetted. A reed switch is the contact assembly of a reed relay. It is also called a magnetic reed switch.

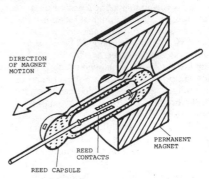

Reed switch contacts are opened or closed by moving the reed capsule into or out of the permanent magnet field.

reentrant   An object with one or more sections directed inward, as in certain types of cavity resonators.

reference black level   The picture signal level in television that corresponds to a specified maximum limit for black peaks.

reference coupling   The coupling between two circuits that gives a reading of 0 dBa (adjusted decibels) on a specified noise-measuring set connected to the disturbed circuit when a test tone of 90 dBa is impressed on the disturbing circuit. It is expressed in decibels above reference coupling, abbreviated dBx.

reference dipole   A straight half-wave dipole tuned and matched for a given frequency. It is a unit of comparison in antenna measurement work.

reference direction   The direction used as a reference for angular measurements.

reference input   An independently established signal used as a standard of comparison in a feedback control system.

reference input element   A feedback control system element that establishes the relationship between the reference input and the command.

reference level   The level designated as a basis of comparison when determining the level of an AF signal in decibels or volume units. A common reference value in voltage, current, and power designations is 6 mW for 0 dB. For sound loudness, the reference level is usually the threshold of hearing. For communication receivers, the common level is 60 μW.

reference line   A line from which angular measurements are made.

reference noise   The power level used as a basis of comparison when designating noise power in dBm. The reference typically is 1 pW (–90 dBm) at 1 kHz.

reference phase The phase of the color-burst signal voltage in a color television receiver, or the phase of the master-oscillator voltage in a color television transmitter.

reference pressure The pressure standard in the sealed chamber of a differential pressure gage. It is usually 1 atmosphere.

reference stimulus A reference quantity that is applied to a telemetering system, such as a reference phase, pressure, or voltage, for calibrating purposes.

reference temperature A standard temperature source used as a reference for calibration of high-temperature measuring devices.

reference time An instant near the beginning of switching that is chosen as a reference for time measurements in a digital computer. It can be the first instant at which the instantaneous value of the drive pulse, the voltage response of the magnetic cell, or the integrated voltage response reaches a specified fraction of its peak pulse amplitude.

reference voltage An AC voltage for comparison purposes, usually to identify an in-phase or out-of-phase condition in an AC circuit.

reference volume The audio volume level that gives a reading of 0 VU on a standard volume indicator. The sensitivity of the volume indicator is adjusted so that reference volume or 0 VU is read when the instrument is connected across a 600-Ω resistance to which is delivered a power of 1 mW at 1 kHz.

reference white 1. The light from a nonselective diffuse reflector that receives the normal illumination of a scene. 2. The standard white color reference for specifying all other colors. The reference white in color television approximates direct sunlight or sky light that has a color temperature of 6500 K. Primary colors are specified in units such that one unit of each primary will combine to produce reference white.

reference white level The picture signal level that corresponds to a specified maximum limit for white peaks.

reflectance *Reflection factor.*

reflected impedance The impedance value that appears to exist across the primary of a transformer when an impedance is connected as a load across the secondary.

reflected resistance The resistance value that appears to exist across the primary of a transformer when a resistive load is across the secondary.

reflected wave A wave reflected from a surface, discontinuity, or junction of two different media, such as the sky wave in radio, the echo wave from a radar target, or the wave that travels back to the source end of a mismatched transmission line.

reflecting curtain A vertical array of half-wave reflecting antennas, generally located one quarter-wavelength behind a radiating curtain of dipoles to form a high-gain antenna.

reflecting target A target that reflects radar waves.

reflection The return or change in direction of light, sound, radar, or radio waves striking a surface or traveling from one medium into another. If the reflecting surface is smooth enough so that each incident ray gives rise to a reflected ray in the same plane, the effect is known as direct, regular, specular, and mirror reflection. If the surface is so rough that reflected rays are distributed in all directions according to the cosine law, the effect is called diffuse reflection.

Reflections of electromagnetic energy can occur at a mismatch in a transmission line, causing standing waves.

reflection coefficient The ratio of some quantity associated with a reflected wave to the corresponding quantity in the incident wave at a given point, for a given frequency and mode of transmission. It is also called a mismatch factor, reflection factor, or transition factor.

reflection Doppler A system that uses Doppler frequency shift to measure position and/or velocity of an object not carrying a transponder.

reflection electron diffraction [RED or RHEED] A process for examining crystal surface structures. Instruments are mounted within the chambers of molecular-beam epitaxy (MBE) systems to make these in-process examinations of epitaxial processes.

reflection error An error in bearing indication due to wave energy that reaches a navigation receiver by undesired reflections.

reflection factor 1. The ratio of the load current that would be delivered by a generator to a particular load without matching to the load current obtained when generator and load impedances are matched. 2. The ratio of the total luminous flux reflected by a given surface to the incident flux, also called reflectance. 3. The ratio of electrons reflected to electrons entering a reflector space, as in a reflex klystron. 4. *Reflection coefficient.*

reflection grating A wire grating placed in a waveguide to reflect one desired wave while allowing one or more other waves to pass freely.

reflection interval The time interval between the transmission of a radar pulse or wave and the reception of the reflected wave at the point of transmission. Multiplying this time interval in microseconds by 492 gives target distance in feet; multiplying the time interval by 0.0931 gives target distance in standard miles. Multiplying microseconds by 0.15 gives target distance in kilometers.

reflection law The angle of incidence is equal to the angle of reflection.

reflection loss 1. The portion of the transition loss that is due to the reflection of power at the discontinuity. 2. The ratio in decibels of the power arriving at a discontinuity to the difference between the incident power and the reflected power.

reflections Radio waves that have been reflected from a building, hill, or other conductive or semiconductive surface during their travel to a television receiving antenna. The resulting longer travel time causes ghost images on the screen.

reflection seismograph A seismograph for prospecting for underground oil deposits. A dynamite detonation or other explosive force near the surface produces sound waves that travel down to the oil strata and are there reflected back to the surface. Measurements of arrival times of the waves at seismographs give data for calculating the depth and extent of the underground oil pool.

reflective optics *Projection optics.*

reflectivity The fraction of incident radiant energy that is reflected from a uniformly irradiated surface.

reflectometer 1. A directional coupler that measures the power flowing in both directions through a waveguide, as a means of determining the reflection coefficient or standing-wave ratio. 2. A photoelectric instrument that measures the optical reflectance of a reflecting surface.

reflector   1. A single rod, system of rods, metal screen, or metal sheet positioned behind an antenna to increase its directivity.   2. A metal sheet or screen that acts as a mirror to change the direction of a microwave radio beam.   3. An electrode in a reflex klystron or other electron tube that reflects a beam of electrons.   4. *Repeller.*

reflector element   A single rod or other parasitic element serving as a reflector in an antenna array.

reflector space   The space in a reflex klystron that follows the buncher space and is terminated by the reflector.

reflector voltage   The voltage between the reflector electrode and the cathode in a reflex klystron.

reflex bunching   The bunching that occurs in an electron stream which has been made to reverse its direction in the drift space.

reflex circuit   A circuit whose signal is amplified twice by the same amplifier tube or tubes, once as an IF signal before detection and once as an AF signal after detection.

reflex klystron   A single-cavity vacuum-tube oscillator that can generate 10 to 500 mW of microwave power in the 1- to 25-GHz range. Electrons from the cathode that have been velocity-modulated by the cavity gap voltage are turned back by the reflector voltage and pass through the cavity gap again in bunches, giving up their energy to the electromagnetic field in the cavity to sustain oscillations. It is used as a low-power microwave source for laboratory measurements and experiments and as a local oscillator in radar systems.

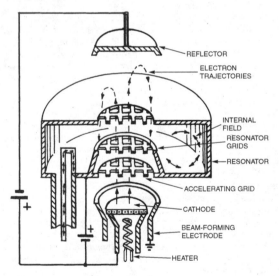

Reflex klystron is a low-power RF oscillator that can produce 10 to 500 mW at frequencies of 1 to 25 GHz.

reflow soldering   The application of heat to parts that have previously been coated with solder, to form a joint by reflowing the solder.

refracted wave   The portion of an incident wave that travels from one medium into a second medium.

refraction   The bending of a heat, light, radar, radio, or sound wave as it passes obliquely from one medium to another in which the velocity of propagation is different.

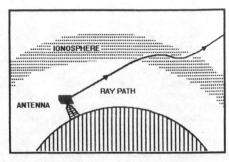

Refraction in transionospheric propagation causes the ray path to bend.

refraction error   A radio bearing error due to bending of one or more wave paths by undesired refraction.

refraction loss   The portion of the transmission loss that is due to refraction resulting from nonuniformity of the medium.

refractive index   The ratio of the phase velocity of a wave in free space to that in a given medium. The refractive index of air at sea level is 1.00029.

refractive modulus   The excess over unity of the modified index of refraction in the troposphere, expressed in millionths and computed as $(n + h/a - 1)10^6$, where n is the index of refraction at a height h above sea level and a is the radius of the earth.

refractivity   The refractive index minus 1.

refractometer   1. An instrument that measures the refractive index of a liquid or solid, usually by measuring the critical angle at which total reflection occurs.   2. An instrument that measures the refractivity of the atmosphere, which is proportional to the dielectric constant of the air being measured. One version has two capacitors, one hermetically sealed to serve as a reference and the other open to the air being sampled. Capacitance varies with dielectric constant, so a comparison of the two capacitances gives the difference in the dielectric constants. This can be readily converted into the refractivity of the air being sampled.

refractory metals   Metals such as tungsten, nickel, molybdenum, tantalum, and titanium that have high melting points and bulk resistivity lower than their thin-film resistivity.

refrangible   Capable of being refracted.

refresh cycle   The frequency of refreshing the capacitor-based memory cells of dynamic RAM (DRAM) memories to preserve their logic state by compensating for continuous charge leakage.

refresh rate   The number of times each second that the information displayed on a nonpermanent display is rewritten. An example is the number of times a cathode-ray tube image must rewritten or reenergized to remain visible and flicker-free.

regenerate   1. To restore pulses to their original shape.   2. To restore stored information to its original form in a storage tube to counteract fading and disturbances.

regeneration   1. Replacement or restoration of charges in a charge-storage tube to overcome decay effects, including loss of charge by reading.   2. *Positive feedback.*

regeneration control   A variable capacitor, variable inductor, potentiometer, or rheostat in a regenerative receiver that controls the amount of feedback and thereby keeps regeneration within useful limits.

regeneration period   The time interval in which the screen of a cathode-ray storage tube is scanned by the beam to regenerate the charge distribution that represents the stored information. It is also called the scan period.

regenerative amplifier   An amplifier that uses positive feedback to give increased gain and selectivity.

regenerative braking   *Dynamic braking.*

regenerative detector   A detector circuit in which RF energy is fed back from the collector circuit to the base circuit to give positive feedback at the carrier frequency, thereby increasing the amplification and sensitivity of the circuit.

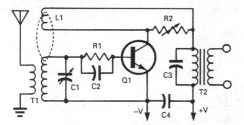

Regenerative-detector circuit.

regenerative receiver   A radio receiver that includes a regenerative detector.

regenerative repeater   A repeater that performs pulse regeneration to restore the original shape of a pulse signal in teletypewriter and other code circuits. Each code element is replaced by a new code element that has specified timing, waveform, and magnitude.

Reggia-Spencer phase shifter   A reciprocal phase shifter that has a bar of ferromagnetic material located axially within a section of waveguide. A solenoid wound around the waveguide produces a longitudinal magnetic field when it is energized by current. The magnetic field causes a variation in the permeability of the material and produces a variation in the RF energy propagation constant, so that phase shift can be controlled by drive current.

regional channel   A standard radio broadcast channel in which several stations can operate with powers not in excess of 5 kW.

register   1. A circuit that holds information in binary format to be processed or transferred. A flip-flop circuit is the simplest form of register as well as the simplest form of memory. Registers are important circuits in computers and are capable of holding one or more words.   2. In the graphic arts and printing trades, the accurate matching or superimposition of two or more images. In printed-circuit processing it is the accurate alignment or positioning of trace or foil patterns on the opposite sides of a two-sided board or on all the layers of a multilayer board. In television receivers it is the accurate superimposition of images from the three color electron guns in a color television tube.

register control   Automatic control of the position of a printed design with respect to reference marks or some other part of the design, as in photoelectric register control.

register file   A small area of computer memory where several data elements, or registers, can be accessed simultaneously, rather than individually.

register length   The number of characters that can be stored by a register in a computer.

register mark   A mark or line printed or otherwise impressed on a web of material for use as a reference to maintain register.

registration   Exact superimposition of all three color images on the screen of a color television receiver, or superimposition of all colors of a design on a printed sheet. It is also called a *register.*

regular reflection   *Direct reflection.*

regulated power supply   A power supply that contains components for maintaining essentially constant output voltage or output current under changing load conditions.

regulating system   *Automatic control system.*

regulation   1. The ability of a power supply to maintain an output voltage within specified limits despite variations in line-power input and output load.   2. The process of holding such quantities as speed, temperature, voltage, or position constant in a system. This is done by methods, typically electronic, that automatically correct errors with feedback to the system. Regulation is based on feedback but control is not.   3. It can be quantized as a percent by determining the change in output voltage that occurs between no load and full load in a transformer, generator, or power supply. That change can be divided by the rated full-load value and multiplying the result by 100 to obtain regulation as a percent.

regulator   A device that maintains a desired quantity at a predetermined value or varies it according to a predetermined plan.

reignition   A process for generating multiple counts within a radiation-counter tube by atoms or molecules excited or ionized in the discharge accompanying a tube count.

reignition voltage   The voltage that is just sufficient to reestablish conduction if applied to a gas tube during the deionization period. The value of this voltage varies inversely with time during the deionization period. It is also called restriking voltage.

Reinartz crystal oscillator   A Reinartz oscillator whose frequency is stabilized by a crystal. The oscillator's resonant frequency is tuned to half the crystal frequency. The resulting regeneration at the crystal frequency improves efficiency and prevents uncontrolled oscillation at other frequencies.

Reinartz oscillator   An inductive-capacitive (LC) oscillator. Regenerative feedback occurs between the collector and emitter of a common-base circuit. The frequency is determined by the LC resonant tank circuit, which is inductively coupled to the collector and emitter coils. It can generate sine-wave frequencies up to UHF. See *Reinartz crystal-controlled oscillator.*

reinserter   *DC restorer.*

reinsertion of carrier   Combining a locally generated carrier signal with an incoming suppressed-carrier signal.

rejection band   The band of frequencies below the cutoff frequency in a uniconductor waveguide.

rejector   *Trap.*

rejector circuit   *Band-elimination filter.*

rel   A unit of reluctance, equal to 1 ampere-turn per magnetic line of force.

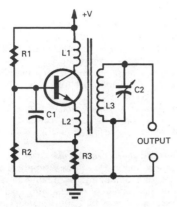

Reinartz oscillator oscillates because of feedback between collector and emitter.

relative address   A designation for the position of a memory location in a computer routine or subroutine. Relative addresses are translated into absolute addresses by adding a reference address, such as that at which the first word of the routine is stored.

relative bearing   A bearing in which the heading of the vehicle serves as the reference line.

relative coding   Computer coding in which all addresses refer to an arbitrarily selected position or in which all addresses are represented symbolically.

relative damping   The ratio of damping torque at a given angular velocity of a moving element in an instrument to the damping torque that would produce critical damping at that same angular velocity.

relative humidity   The ratio of the amount of water vapor present in air to the amount that would saturate it at a given temperature.

relative luminosity   The ratio of the value of the luminosity at a particular wavelength to the value at the wavelength of maximum luminosity.

relative permeability   *Specific permeability.*

relative plateau slope   The percent change in counting rate per unit change of applied voltage near the midpoint of the plateau of a radiation-counter tube.

relative refractive index   The ratio of the refractive indexes of two media.

relative response   The ratio, usually expressed in decibels, of the response under some particular conditions to the response under reference conditions.

relative specific ionization   The specific ionization for a particle of a given medium, relative either to that for the same particle and energy in a standard medium or the same particle and medium at a specified energy.

relative target bearing   The bearing of a radar target expressed relative to the heading of a ship or aircraft.

relative time delay   The difference in time delay encountered by the audio signal and the composite picture signal or between the components of the picture signal traveling over a television relay system.

relative velocity   The time rate of change of a position vector of a point with respect to a reference frame.

relaxation generator   *Relaxation oscillator.*

relaxation inverter   An inverter with a relaxation oscillator circuit that converts DC power to AC power.

relaxation oscillator   An oscillator whose fundamental frequency is determined by the time of charging or discharging a capacitor or coil through a resistor, producing waveforms that may be rectangular or sawtooth. It is also called a *relaxation generator.*

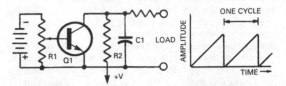

Relaxation oscillator circuit and its sawtooth output waveform.

relaxation time   1. The time constant required for an abrupt change of magnetizing force to make the magnetic induction reach a specified percent of its new value.   2. The travel time of an electron in a metal before it is scattered and loses its momentum.

relay   1. A device that is operated by a variation in the conditions in one electric circuit and makes or breaks one or more connections in the same or another electric circuit. The most common types are electromagnetic, reed, solid-state, and thermal relays.   2. A microwave or other radio system that passes a signal from one radio communication link to another.

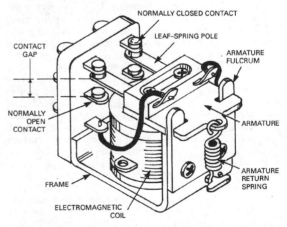

Relay: a two-pole, general-purpose electromechanical relay for switching circuits by remote control.

relay armature   The movable iron part of a relay.

relay broadcast station   A station licensed to transmit, from points where wire facilities are not available, programs for broadcast by one or more broadcast stations.

relay channel   The band of frequencies for transmitting a single television relay signal, including the guard bands.

relay coil   One or more windings on a common form, used with an iron core to form a relay electromagnet.

relay contact   One of the pair of contacts that are closed or opened by the movement of the armature of a relay.

relay chatter   The low-frequency sound produced by a relay when its coil is energized by alternating current or unfiltered rectified current.

relay magnet   The electromagnet that attracts the armature of a relay when energized.

relay rack   *Equipment rack.*

relay receiver   A receiver that accepts a television or microwave relay input signal and delivers a television or microwave relay output signal to the transmitter portion of a repeater station.

relay station   *Repeater station.*

relay system   1. An assembly of relays for switching purposes.   2. *Radio repeater.*

release   A mechanical, electromagnetic, or other arrangement of parts for holding or freeing a device or mechanism as required.

release time   The total elapsed time from the instant that relay coil current starts to drop until the make contacts have opened or the break contacts have closed.

reliability   The probability that a device will perform its purpose adequately for the period of time intended under the operating conditions encountered.

reliability engineering   A field of engineering that deals with the prevention and correction of malfunctions in equipment.

reliability index   A quantitative figure of merit related to the reliability of a piece of equipment, such as the number of failures per 1000 operations or the number of failures in a specified number of operating hours.

reliability test   A test designed specifically to evaluate the level and uniformity of reliability of equipment under various environmental conditions.

reluctance   A measure of the opposition presented to magnetic flux in a magnetic circuit. Reluctance is the reciprocal of permeance, and is therefore equal to magnetomotive force divided by magnetic flux. The unit of reluctance is the rel, equal to 1 ampere-turn per magnetic line of force.

reluctance microphone   *Variable-reluctance microphone.*

reluctance motor   A synchronous motor, similar in construction to an induction motor, in which the member carrying the secondary circuit has salient poles but no DC excitation. It starts as an induction motor but operates normally at synchronous speed.

reluctance pickup   *Variable-reluctance pickup.*

reluctivity   The ratio of the magnetic intensity in a region to the magnetic induction in the same region. Reluctivity is the reciprocal of magnetic permeability.

rem   Abbreviation for *roentgen equivalent man.*

remanence   The magnetic flux density that remains in a magnetic circuit after the removal of an applied magnetomotive force. If the magnetic circuit has an air gap, the remanence will be less than the residual flux density.

remanent charge   The charge that remains in a ferroelectric device when the applied voltage is removed.

remanent induction   The induction that remains in a magnetic material when the magnetomotive force around the magnetic circuit is zero.

Remendur   A high-performance magnetic material that can have a remanence as high as 21.5 kG. It is a malleable, ductile, cobalt-iron-vanadium alloy.

remodulation   Transferring modulation from one carrier to another.

remodulator   A circuit that converts amplitude modulation to audio frequency-shift modulation for transmission of facsimile signals over a voice-frequency radio channel.

remote control   Control of equipment from a distance over wires or by light, radio, sound, ultrasonic waves, or other means.

remote cutoff   The characteristic in which a large negative bias is required for complete cutoff of output current in an electron tube or other amplifying device.

remote indicator   1. An indicator at a distance from the data-gathering sensing element, with data being transmitted to the indicator mechanically, electrically over wires, or by light, radio, or sound waves.   2. *Repeater.*

remote line   A program transmission line between a remote-pickup point and a broadcast studio or transmitter site.

remotely piloted vehicle [RPV]   A robot aircraft, controlled over a two-wave radio link from a ground station or mother aircraft that can be many miles away. Electronic guidance is generally supplemented by remote-control television cameras feeding monitor receivers at the control station.

remote metering   *Telemetering.*

remote pickup   Picking up a radio or television program at a remote location and transmitting it to the studio or transmitter over wire lines or a shortwave or microwave radio link.

remote programming   Control of the output voltage or current of a regulated power supply by a remotely varied resistance or voltage.

remote sensing   1. The acquisition of information by sensors that are not in physical contact with the phenomenon or object under study. Examples are the measurement of radiation, temperature, pressure, and other variables with sensors that might be packaged in compact cases with a battery-powered radio-frequency transmitter. The habits and physiology of animals in the wild can be monitored by remote sensing, as can the activity levels of volcanoes.   2. The use of sense leads to connect the power supply to the load and provide feedback to the voltage regulation circuits of the supply. This circuitry compensates for voltage losses resulting from long cables to a load.

removable media drive   Any of a class memory drives with removable data storage media, intended to back up the *hard drives* of personal computers and perhaps eventually replace the *floppy-disk drive.* Capacities are typically in the range of 100 to 130 Mbytes. See also *zip drive.*

REN   Abbreviation for *ringer-equivalence number.*

rendezvous radar   Radar designed for use in orbital rendezvous and docking in space.

rep   1. Abbreviation for *roentgen equivalent physical.*   2. Abbreviation for *representative.*

repeatability   1. A measure of the variation in the readings of an instrument such as a precision potentiometer when identical tests are made under fixed conditions. Also called reproducibility.   2. The ability of a voltage regulator or voltage reference tube to attain the same voltage drop at a stated time after the beginning of any conducting period.

repeater   An amplifier and associated equipment that processes weak signals and retransmits stronger signals without reshaping their waveforms. It can be a one-way or two-way repeater. Repeaters are used in telephone lines, undersea cable, and fiber-optic cables to overcome the effects of signal attenuation in the transmission media.

There are also radio-frequency repeaters, especially for amplifying weak signals in the microwave region and permitting those signals to be relayed past the horizon for long-distance transmission.

repeater station A station that contains one or more repeaters. It is also called a *relay station*.

repeating timer A timer that continues repeating its operating cycle until excitation is removed.

repeat-point tuning *Double-spot tuning.*

repeller An electrode whose primary function is to reverse the direction of an electron stream in an electron tube. It is also termed a *reflector*.

repertory dialer Electronic equipment that stores a repertory of telephone numbers and dials them automatically on request, as is used in telephone solicitation.

repetition frequency *Repetition rate.*

repetition rate The rate at which recurrent signals are produced or transmitted. It is also called recurrence rate and repetition frequency.

repetitive error The maximum deviation of the controlled variable from the average value upon successive return to specified operating conditions following specified deviation in an automatic control system.

reply An RF signal or combination of signals transmitted by a transponder in response to an interrogation. It is also called response.

reproduce head *Playback head.*

reproducibility *Repeatability.*

reproducing stylus *Stylus.*

reproduction speed The area of copy recorded per unit of time by a facsimile receiver; facsimile feed rate.

repulsion A mechanical force that tends to separate bodies which have electric charges or like magnetic polarity. Adjacent conductors that have currents flowing in opposite directions can repel each other.

repulsion-induction motor A repulsion motor that has a squirrel-cage winding in the rotor in addition to the repulsion-motor winding.

repulsion motor An AC motor that has stator windings connected directly to the source of AC power and rotor windings connected to a commutator. Brushes on the commutator are short-circuited and are positioned to produce the rotating magnetic field required for starting and running. The speed of this type of motor varies considerably as the load is changed.

repulsion-start induction motor An AC motor that starts as a repulsion motor. At a predetermined speed the commutator bars are short-circuited to give the equivalent of a squirrel-cage winding for operation as an induction motor with constant-speed characteristics.

request to send [RTS] An RS-232C control signal between a modem and the user's computer or facsimile (fax) machine that initiates the data transmission sequence on a communication line.

reradiation Undesirable radiation of signals generated locally in a radio receiver, causing interference or revealing the location of the receiver.

rerecording The process of making a recording by reproducing a recorded sound source and recording this reproduction.

rerecording system A system of reproducers mixers, amplifiers, and recorders that combine or modify various sound recordings to provide a final sound record.

rerun point A point in a computer program at which all information is available for rerunning the last-run portion of a problem when an error is detected. Several such points are usually provided for in a program, to eliminate the need for rerunning the entire problem.

rerun routine A routine used after a computer malfunction, a coding error, or an operating mistake to reconstitute a routine from the last previous rerun point.

research Scientific investigation aimed at discovering and applying new facts, techniques, and natural laws.

research and development The conception, design, and first creation of experimental or prototype operational devices.

reserve battery A battery that has long shelf life in an unenergized state. Methods of activating the battery include heat, addition of an electrolyte or water, or use of mechanical shock or other means for breaking a glass ampule or container to allow electrolyte into the battery.

reserve cell A cell in a reserve battery.

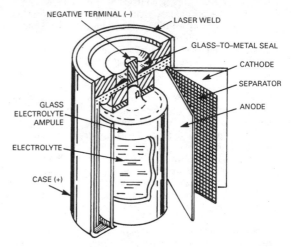

Reserve cell: a lithium-thionyl chloride power cell is activated instantly by breaking a glass electrolyte ampule to activate the cell to full performance after long storage periods.

reset *Clear.*

reset control circuit The magnetic amplifier circuit that resets the flux in the core of the saturable reactor.

reset pulse A pulse that resets an electronic counter to zero or to some predetermined position.

reset rate The number of corrections made per minute by a control system.

reset switch A switch that restores normal operation to a control system after a corrective action.

resettability The ability of the tuning element of an oscillator to retune the oscillator to the same operating frequency for the same set of input conditions.

reshaping circuit A circuit that changes the waveform of a current, such as a limiter or clipper.

residual charge The charge that remains on the plates of a capacitor after an initial discharge.

residual discharge A discharge of the residual charge remaining after the initial discharge of a capacitor.

residual error The sum of random errors and uncorrected systematic errors.

residual field   The magnetic field left in an iron core after excitation has been removed.

residual flux density   The magnetic flux density at which the magnetizing force is zero when the material is in a symmetrically and cyclically magnetized condition. It is also called residual induction, residual magnetic induction, and residual magnetism.

residual gap   The length of the magnetic air gap between the armature and the center of the core face of an energized relay.

residual gas   The small amount of gas that remains in a vacuum tube after the best possible exhaustion by vacuum pumps. Much of this residual gas is removed by the getter.

residual induction   *Residual flux density.*

residual magnetic induction   *Residual flux density.*

residual magnetism   *Residual flux density.*

residual modulation   *Carrier noise level.*

residual resistance   The portion of the electric resistance of a metal that is independent of temperature.

residual resistivity   The constant minimum value of electric resistivity of a metal with decreasing temperature, reached near absolute zero. The residual resistivity is not a characteristic of a particular metal, but varies with impurities in the sample. The purer the specimen, the lower its residual resistivity.

resin   A natural or synthetic organic ingredient in plastics, adhesives, and surface coatings.

resist   See *photoresist.*

resistance   [R]   The opposition that a device or material offers to the flow of direct current, measured in ohms, kilohms, or megohms. In AC circuits, resistance is the real component of impedance.

resistance box   A box that contains a number of precision resistors connected to panel terminals or contacts so that a desired resistance value can be obtained by withdrawing plugs (as in a post office bridge) or by setting multicontact switches. Its usual form is a decade box, in which individual resistance values vary in submultiples and multiples of 10.

resistance braking   *Dynamic braking.*

resistance bridge   *Wheatstone bridge.*

resistance-bridge controller   A controller that operates on the null balance principle. Balancing the bridge produces an error signal that is amplified to keep the load at a desired temperature.

resistance-capacitance   [RC]   Containing both resistance and capacitance, as provided by resistors and capacitors.

resistance-capacitance circuit   [RC circuit]   A circuit that contains resistors and capacitors which determine the time constant. The time constant in seconds is equal to the product of resistance in ohms and capacitance in farads.

resistance-capacitance-coupled amplifier   An amplifier with a capacitor that provides a path for signal currents from one stage to the next, with resistors connected from each side of the capacitor to the power supply or to ground. It can amplify AC signals but cannot handle small changes in direct currents. It is also called an RC amplifier.

resistance-capacitance coupling   Coupling in which capacitors act as the input and output impedances of the circuits being coupled. A coupling capacitor is generally located between the resistors to transfer the signal from one stage to the next. It is also called RC coupling or resistance-capacitance coupling.

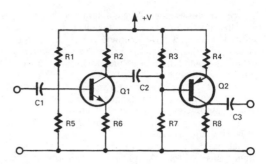

Resistance-capacitance-coupled amplifier that includes both NPN-PNP transistor stages.

resistance-capacitance differentiator   [RC differentiator]   A resistance-capacitance circuit that produces an output voltage whose amplitude is proportional to the rate of change of the input voltage. A square-wave input thus produces sharp output voltage spikes.

resistance-capacitance filter   [RC filter]   A filter that contains only resistance and capacitance elements (no inductance). It reduces ripple in rectifier-type power supplies.

resistance-capacitance oscillator   [RC oscillator]   An oscillator whose frequency is determined by its resistance-capacitance elements.

resistance-coupled amplifier   An amplifier with resistance coupling between stages.

resistance decade   A decade box that contains an assembly of precision resistors whose individual values are related by submultiples or multiples of 10. Each section or decade contains 10 equal-value resistors connected in series, with provisions for making a connection to any junction. The decades are similarly connected in series, with each decade having resistors 10 times larger in value than those in the next lower decade. A resistance decade can thus be set to any desired value within its range, in steps that are multiples of the smallest resistance value.

resistance drop   1. The voltage drop that occurs between two points on a conductor due to the flow of current through the resistance of the conductor. Multiplying the resistance in ohms by the current in amperes gives the voltage drop in volts.   2. *IR drop.*

resistance element   An element of resistive material in the form of a grid, ribbon, or wire, used singly or built into groups to form a resistor for heating purposes, as in an electric soldering iron.

resistance furnace   An electric furnace whose heat is developed by the passage of current through a suitable internal resistance that can be the charge itself, a resistor embedded in the charge, or a resistor surrounding the charge.

resistance hybrid   A hybrid junction that consists entirely of resistors. Operation is essentially independent of frequency up to several hundred megahertz, but attenuation along desired paths is greater than with a hybrid transformer.

resistance-inductance-capacitance circuit   [RLC circuit]   A circuit that contains inductance, capacitance, and resistance.

resistance junction   *Resistance hybrid.*

resistance loss   Power loss due to current flowing through resistance. Its value in watts is equal to the resistance in ohms multiplied by the square of the current in amperes.

resistance magnetometer A magnetometer that depends for its operation on variations in the electric resistance of a material immersed in the magnetic field to be measured.

resistance material A material that has sufficiently high resistance per unit length or volume to permit its use in the manufacture of resistors.

resistance pad A pad that uses only resistances to provide attenuation without altering frequency response.

resistance pyrometer A pyrometer whose heat-sensing element is a length of wire with a resistance that varies greatly with temperature.

resistance standard *Standard resistor.*

resistance-start motor A split-phase motor that has a resistance connected in series with the auxiliary winding. The auxiliary circuit is opened when the motor attains a predetermined speed.

resistance strain gage A strain gage that consists of a strip of material which is cemented to the part under test, and changes its resistance with elongation or compression.

resistance wire Wire made from a metal or alloy that has high resistance per unit length, such as Nichrome. Wire-wound resistors and heating elements are wound from resistance wire.

resistive conductor A conductor used primarily because it has high resistance per unit length.

resistive coupling *Resistance coupling.*

resistive transducer A sensor or transducer whose operation is based on a change in resistance brought about by the measurand. These include the translation of such displacements as fluid velocity, force, load, mechanical strain, pressure, and temperature into electrical outputs.

resistivity The resistance in ohms that a unit volume of a material offers to the flow of current. Resistivity is the reciprocal of the conductivity of a material and is measured in ohm-centimeters. The resistivity of a wire sample in ohm-centimeters is equal to the resistance R in ohms multiplied by the cross-sectional area A in square centimeters and the result divided by the sample length L in centimeters (resistivity = RA/L). Resistivity is also expressed in ohms per circular mil foot, which is the resistance of a sample that is 1 circular mil in cross section and 1 ft long. It is also called specific resistance.

resistor [R] A component made to provide a definite amount of resistance. In circuits it limits current flow or provides a voltage drop. The principal types of discrete resistors are carbon-composition, carbon film, wirewound, metal film, and cermet film. There are also thin-film and thick-film (cermet) resistor chips and resistor networks.

resistor-capacitor-transistor logic [RCTL] Resistor-transistor logic in which capacitors are added to increase switching speed.

resistor color code A method of marking the value in ohms on a resistor by bands of colors as specified in the EIA color code.

resistor core The insulating support on which a resistor element is wound or otherwise placed. It is also called a mandrel.

resistor element The portion of a resistor that provides resistance. It can be pure metal, an alloy, a metallic coating, a carbon-cement mixture, or a plastic that contains finely powdered metal.

resistor network Resistive elements deposited on a ceramic substrate and interconnected by conductive metal

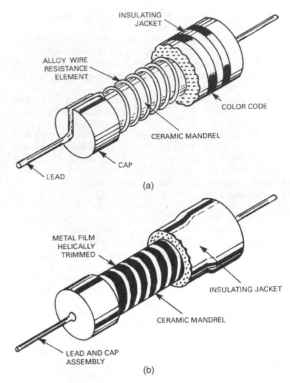

Resistor: axial-leaded wirewound unit (*a*) and axial-leaded metal-film unit (*b*).

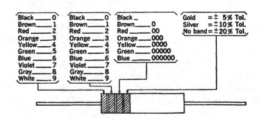

Resistor color code.

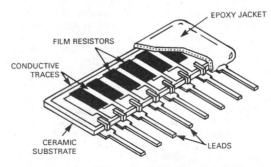

Resistor network has thick- or thin-film resistive elements deposited on a ceramic substrate and connected by conductive traces.

traces. Networks are packaged as multiple resistors in dual-in-line (DIP) and single-in line (SIP) packages.

resistor-transistor logic [RTL] Logic that includes both resistors and transistors, but the transistors only invert the output.

resolution The ability to delineate, detail, or distinguish between nearly equal values of a quantity. In television, it is a measure of sharpness of a television image in both vertical and horizontal axes. TV resolution measurements are usually made with a test pattern that includes test wedges (a collection of black and white lines that converge at the center of the pattern), calibrated in lines per picture height. See *modulation-transfer function*. In radar and sonar, it is the ability to discriminate between two distant targets measured as a function of distance from the antenna or transducer and angular separation of the targets. It is also called *resolving power*.

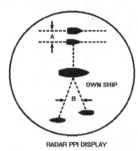

RADAR PPI DISPLAY

Resolution for radar. The minimum separation distance between two targets that permits them to be distinguished in range and the minimum angle between them that permits them to be distinguished in bearing on the radar display.

resolution chart *Test pattern*.

resolution in bearing The angle by which two objects at the same range must be separated in bearing to be distinguishable on a radar display.

resolution in range The distance by which two objects at the same bearing must be separated in range to be distinguishable on a radar display.

resolution sensitivity The minimum change of measured variable that actuates an automatic control system.

resolution time The minimum time interval at which two successive voltage pulses can be registered by a counter.

resolution wedge A group of gradually converging lines on a test pattern for measuring resolution in television.

resolver A synchro whose rotor is mechanically driven to translate rotor angle into electric information corresponding to the sine and cosine of rotor angle. It can interchange rectangular and polar coordinates. It is also called a sine-cosine generator or a synchro resolver.

resolving power 1. The ability of a mass spectrometer to separate adjacent mass spectrum lines. 2. The reciprocal of the beam width of a unidirectional antenna, measured in degrees. 3. The ability of a radar system to form distinguishable images. 4. *Resolution*.

resolving time The minimum time interval between two distinct events that will permit them to be counted or otherwise detected by a particular circuit or device.

resonance [noun; also used as adjective in place of resonant] 1. The condition existing in a circuit when the inductive reactance cancels the capacitive reactance. 2. The condition existing in a body when the frequency of an applied vibration equals the natural frequency of the body. A body vibrates most readily at its resonant frequency. It is also called velocity resonance.

resonance bridge A four-arm AC bridge that measures inductance, capacitance, or frequency. The inductor and the capacitor, which can be either in series or in parallel, are tuned to resonance at the frequency of the source before the bridge is balanced.

resonance characteristic *Resonance curve*.

resonance current step-up The ability of a parallel resonant circuit to circulate a current through its coil and capacitor that is many times greater than the current fed into the circuit.

resonance curve An amplitude-frequency response curve that shows the current or voltage response of a tuned circuit to frequencies at and near the resonant frequency. It is also called resonance characteristic.

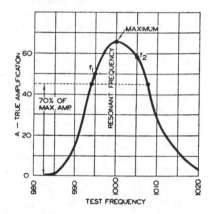

Resonance curve at 1000 kHz for typical RF amplifier.

resonance frequency 1. The frequency at which the inductive reactance of a given resonant circuit is equal to the capacitive reactance, so resonance exists. 2. The frequency at which a quartz crystal, loudspeaker diaphragm, or other object will vibrate readily.

resonance indicator A device that indicates when a circuit is tuned to resonance. It can be a voltmeter, neon lamp, or headphones.

resonance transformer A high-voltage transformer in which the secondary circuit is tuned to the frequency of the power supply.

resonant A reference to resonance. The term resonance is used as an adjective in place of resonant.

resonant cavity *Cavity resonator*.

resonant-cavity maser A maser whose paramagnetic active material is placed in a cavity resonator.

resonant chamber *Cavity resonator*.

resonant-chamber switch A waveguide switch with a tuned cavity in each waveguide branch that functions as a switch contact. Detuning of a cavity blocks the flow of energy in the associated waveguide.

resonant charging choke A choke that resonates with the effective capacitance of a pulse-forming network in a modulator to produce oscillation at the resonant frequency.

resonant circuit A circuit containing inductance and capacitance values that give resonance at an operating frequency. The frequency in hertz at which resonance occurs is $1/2\pi\sqrt{LC}$, where L is in henrys and C is in farads. When the capacitance and inductance are in series, the combination has a lower impedance at resonance than either alone. When in parallel, the combination has a higher impedance than either alone.

resonant diaphragm A diaphragm that has no reactance at a specified frequency.

resonant element *Cavity resonator.*

resonant gap The small internal gap where the electric field of a TR tube is concentrated.

resonant iris A resonant window in a circular waveguide, so called because of its resemblance to an optical iris.

resonant line A transmission line that has values of distributed inductance and distributed capacitance to make the line resonant at the frequency it is handling. Parallel resonance exists when the line is an odd number of quarter-wavelengths long and is short-circuited at the load, and series resonance exists for the same line when it is open at the load end.

resonant-line oscillator An oscillator in which one or more sections of transmission line serve as resonant circuits.

resonant-line tuner A television tuner in which resonant lines are used to tune the antenna, RF amplifier, and RF oscillator circuits. Tuning is achieved by moving shorting contacts that change the electrical lengths of the lines.

resonant power supply A power supply whose energy is stored in an inductor and capacitor and is exchanged at a rate determined by the relative impedances of the components. The inductor current and capacitor voltage are sinusoidal waveforms that shape the input current or voltage before being applied to a power transformer.

resonant resistance The resistance of a resonant circuit or resonant line at the resonance frequency.

resonant tank A series or parallel circuit formed by an inductor and capacitor that achieves resonance when the inductive reactance cancels the capacitive reactance at a specified frequency.

resonant voltage step-up The ability of a coil and a capacitor in a series resonant circuit to deliver a voltage several times greater than the input voltage of the circuit.

resonant window A parallel combination of inductive and capacitive diaphragms in a waveguide structure that provides transmission at the resonant frequency and reflection at other frequencies.

resonant-window switch A waveguide switch that has a resonant window in each waveguide branch serve as switch contacts. Detuning of the window blocks the flow of energy in the associated waveguide.

resonate To reach resonance, as by tuning.

resonating piezoid A piezoid (finished crystal blank) used as a resonator or oscillator rather than as a transducer.

resonator A device that exhibits resonance at a particular frequency, such as an acoustic resonator or cavity resonator.

resonator grid A grid that is attached to a cavity resonator in a velocity-modulated tube to provide coupling between the resonator and the electron beam.

resonator mode The operating mode for which an electron stream introduces a negative conductance into the coupled circuit of an oscillator.

resonator wavemeter Any resonant circuit that determines wavelength, such as a cavity-resonator frequency meter.

responder The transmitter section of a radar beacon.

response 1. A quantitative expression of the output of a device or system as a function of the input. 2. *Amplitude-frequency response.* 3. *Reply.*

response characteristic *Amplitude-frequency response.*

responser *Responsor.*

response time The time required for the output of a control system or element to reach a specified fraction of its new value after the application of a step input or disturbance. It is usually given in seconds. For magnetic amplifiers the response time is often specified in cycles of the power-line frequency. For indicating instruments it is the time for the pointer to come to rest at its new position.

responsor The receiving section of an interrogator-responsor also called a responser.

resting frequency *Carrier frequency.*

restorer *DC restorer.*

restoring spring The spring that moves the armature of a relay away from the magnet core when the relay is de-energized.

restricted radiation device A device whose RF energy is intentionally generated and conducted along wires or radiated, but the total electromagnetic field does not exceed 15 µV/m at a distance in feet equal to 15,700 divided by the frequency in kilohertz (distance in meters equal to 4785 divided by the frequency in kilohertz).

restriking voltage *Reignition voltage.*

resultant A force that combines the effects of two or more forces acting on an object.

retarding field An electric or magnetic field that slows up electrons traveling through an interelectrode space in an electron tube.

retarding-field oscillator An oscillator that employs an electron tube whose electrons oscillate back and forth through a grid that is maintained positive with respect to both the cathode and anode. The frequency depends on the electron transit time and can also be a function of associated circuit parameters. The field in the region of the grid exerts a retarding effect that draws electrons back after they pass through the grid in either direction. It is also called a positive-grid oscillator. The Barkhausen-Kurz and Gill-Morell oscillators are examples.

retarding-field tube An electron tube in a retarding-field oscillator.

retentivity The property of a magnetic material that is measured by the residual flux density corresponding to the saturation induction for the material.

reticle An optical photomask of fine lines that defines individual boundaries of components on integrated circuits, which are stepped and repeated across the wafer for simultaneous multiple IC processing.

retrace The return of the electron beam to its starting point in a cathode-ray tube after a sweep. It is also called flyback.

retrace blanking Blanking a television picture tube during vertical retrace intervals to prevent retrace lines from showing on the screen. Voltage pulses for blanking are derived from a vertical sweep oscillator or vertical deflection circuits, and are applied to the control grid of the picture tube.

retrace interval   The interval of time for the return of the blanked scanning beam of a television picture tube or camera tube to the starting point of a line or field. It is about 7 μs for horizontal retrace and 500 to 750 μs for vertical retrace in NTSC TV broadcasting. It is also called retrace period, retrace time, return interval, return period, or return time.

retrace line   The line traced by the electron beam in a cathode-ray tube in going from the end of one line or field to the start of the next line or field. It is also called *return line*.

retrace period   *Retrace interval.*

retrace time   *Retrace interval.*

retransmission unit   A control unit that is an intermediate station for feeding one radio receiver-transmitter unit automatically to another receiver-transmitter unit for two-way communication.

retroaction   British term for *positive feedback*.

retroreflector   A device that reflects radiation back on a path parallel to the incident rays over a wide range of retroreflector orientation as in laser surveying. One version, the corner reflector, is an efficient radar target.

return   *Echo.*

return-beam mode   A camera-tube operating mode in which the output current is derived from that portion of the scanning beam not accepted by the target.

return-beam vidicon   A vidicon whose electron beam that scans the target is bent back from the target to an electron multiplier and anode which surround the electron gun. Light falling on the target surface changes the resistance of the surface, and thereby modulates the energy in the return beam.

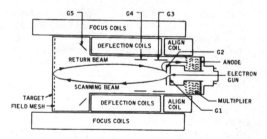

Return-beam vidicon construction.

return interval   *Retrace interval.*

return line   *Retrace line.*

return loss   1. The difference between the power incident upon a discontinuity in a transmission system and the power reflected from the discontinuity.   2. The ratio in decibels of the power incident upon a discontinuity to the power reflected from the discontinuity.

return period   *Retrace interval.*

return time   *Retrace interval.*

return to bias   Magnetization of magnetic tape to saturation in a direction called minus, representing binary 0. Binary 1 signals are recorded by magnetizing the tape in the opposite direction. After each binary 1 pulse the tape returns to the minus (bias) condition. This method requires a clock to read 0s.

return-to-zero code [RZ code]   A communications code in which a binary 0 is represented by one bit time at the 0 level, and a binary 1 is pulsed so that it reaches the 1 level for only half a bit time. The signal therefore returns to 0 (or stays at 0) after each bit. With this code, only half as much data can be stored in a given area or distance as with nonreturn-to-zero code.

return wire   The ground wire, common wire, or negative wire of a DC power circuit.

reverberation   The persistence of sound at a given point after direct reception from the source has stopped. In air it can be caused by repeated reflections from a small number of boundaries or free decay of normal modes of vibration that were excited by the sound source. In water it can be caused by scattering from a large number of inhomogeneities in the medium or reflections from bounding surfaces.

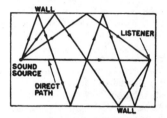

Reverberation paths for sound waves in a room.

reverberation absorption coefficient   The term for the sound absorption coefficient when the distribution of incident sound is completely random.

reverberation chamber   An enclosure that has had all surfaces made as sound-reflective as possible. Acoustic measurements are made in this kind of room. It is also called a reverberation room.

reverberation-controlled gain circuit [RCG circuit]   A circuit for underwater sound equipment that varies the gain of the receiving amplifier in proportion to the strength of undesired reverberations associated with the desired echo.

reverberation reflection coefficient   The term for the sound reflection coefficient when the distribution of incident sound is completely random.

reverberation room   *Reverberation chamber.*

reverberation time   The time in seconds required for the average sound-energy density at a given frequency to reduce to one-millionth of its initial steady-state value after the sound source has been stopped. This corresponds to a decrease of 60 dB.

reverberation-time meter   An instrument that measures the reverberation time of an enclosure.

reverberation transmission coefficient   The term for the sound transmission coefficient when the distribution of incident sound is completely random.

reverberation unit   A circuit that generates reverberation synthetically in a sound system. In one version, audio signal energy is converted into torsional motion of coil springs by a magnetic driver. A magnetic pickup at the other end of the springs converts the motions back into audio signals after a time delay, to give the effect of listening in a reverberant hall.

reverse bias A bias voltage applied to a diode or a semiconductor junction with polarity that permits little or no current to flow. It is the opposite of forward bias.

reverse-blocking thyristor A thyristor that switches only for positive anode-to-cathode voltages. It blocks current flow for negative voltages.

reverse-conducting thyristor A thyristor that switches only for positive anode-to-cathode voltages. It conducts large currents at negative voltages comparable in magnitude to the on-state voltages.

reverse coupler A directional coupler that samples reflected power.

reverse current The small value of direct current that flows when a semiconductor diode is reverse-biased.

reverse-current relay A relay that operates whenever current flows through the relay coil opposite to the normal direction.

reversed image 1. A mirror image in which the right and left sides of the picture are interchanged. 2. *Negative image.*

reverse direction The direction of higher resistance to steady DC flow through a semiconductor junction or device.

reverse emission The flow of electrons in the reverse direction (from anode to cathode) in a vacuum tube during that part of a cycle in which the anode is negative with respect to the cathode. The action is similar to arcback in gas tubes. It is also called back emission.

reverse Polish notation A logic used in some calculators that permits entry of a mathematical problem from left to right exactly as it is normally written.

reverse resistance The resistance of a rectifier diode as measured at a specified value of reverse voltage or reverse current.

reverse voice channel [RVC] A voice channel used from a *mobile station* to a *land station* in a *cellular mobile telephone system.*

reverse voltage A voltage applied to a rectifier diode with the opposite of normal polarity.

reversible counter A counter capable of counting either from left to right or right to left, usually without change of the count state at reversal.

reversible permeability The term for incremental permeability when the change in magnetic induction is exceedingly small.

reversible transducer A transducer whose loss is independent of the direction of transmission.

reversing motor A motor whose direction of rotation can be reversed by changing electric connections or by other means while the motor is running at full speed. The motor will then come to a stop, reverse, and attain full speed in the opposite direction.

reversing switch A switch intended to reverse the connections of one part of a circuit.

rewind 1. The components on a magnetic-tape recorder that return the tape to the supply reel at high speed. 2. To return an audio or video magnetic tape in a cassette to its starting position.

rewrite The process of restoring a memory device to its state prior to reading.

RF (or rf) Abbreviation for *radio frequency.*

RF alternator A rotating alternator that produces high power at frequencies above power-line values, but generally lower than 100 kHz. It has applications in high-frequency heating.

RF amplification Amplification of a radio signal at the same carrier frequency at which it travels through space.

RF amplifier An amplifier that increases the voltage or power of RF signals at the carrier frequency. In a superheterodyne receiver, an RF amplifier can be positioned ahead of the converter.

RFC Symbol for *RF choke.*

RF cable See *coaxial cable.*

RF choke [RFC] An RF coil designed specifically to block the flow of RF current while passing lower frequencies or direct current.

RF coil A coil that has one continuous untapped winding, specifically designed to furnish inductive reactance for tuning purposes in a circuit carrying RF current.

RF converter A power source for producing electric power at a frequency above about 10 kHz.

RF current An alternating current that has a frequency higher than about 10 kHz.

RF energy Alternating-current energy at any frequency in the radio spectrum between about 10 kHz and 300 GHz.

RF generator A generator capable of supplying sufficient RF energy at the required frequency for induction or dielectric heating.

RF harmonic A harmonic of a carrier frequency. The frequency of the second RF harmonic is twice the carrier frequency.

RF head A radar transmitter and part of a radar receiver contained in a package for ready installation and removal in a system.

RFI Abbreviation for *RF interference.*

RF indicator An indicator that shows the presence of RF energy at or near its own resonant frequency. It usually consists of a coil or parallel line connected to an incandescent lamp.

RF interference [RFI] Interference caused by RF sources of energy outside a system—considered to be a factor in electromagnetic interference (EMI).

RF intermodulation distortion Intermodulation distortion that originates in the RF stages of a receiver.

RF oscillator An oscillator that generates alternating current at radio frequencies.

RF pattern A fine herringbone interference pattern that occurs in a television picture because of high-frequency interference.

RF power probe A probe that extracts RF energy from a transmission system.

RF power supply A high-voltage power supply that steps up the output of an RF oscillator with an air-core transformer to the high voltage required for the second anode of a cathode-ray tube. It is then rectified to provide the required high DC voltage.

RF pulse An RF carrier that is amplitude-modulated by a pulse. The amplitude of the modulated carrier is zero before and after the pulse.

RF resistance *High-frequency resistance.*

RF response The response of a receiver to radio frequencies within and outside the channel being received.

RF signal generator A test instrument that generates the radio frequencies required for alignment and servicing of radio, television, VCRs, and other electronic equipment.

RF spectrum *Radio spectrum.*

RF stage A single stage of RF amplification.

RF transformer   A transformer that has a tapped winding or two or more windings which furnish inductive reactance and/or transfer RF energy from one circuit to another by a magnetic field. It can have an air core or some form of ferrite core.

RF transmission line   A transmission line designed primarily to conduct RF energy that consists of two or more conductors supported in a fixed spatial relationship along their own length.

RHEED   Abbreviation for *reflected high-energy electron diffraction.*

rhenium [Re]   A metallic element with an atomic number of 75 and an atomic weight of 186.2.

rheo-   Prefix meaning a flow of current.

rheography   Monitoring of the impedance between two electrodes placed on the skin.

rheostat   A variable resistor whose value can be changed readily with a control knob, to control the current in a circuit. A rheostat has one fixed terminal and one terminal that is connected to the sliding or rolling contact. It is also called a *variable resistor.* A potentiometer is a rheostat that has an additional fixed terminal at the other end of the resistance element.

Rhm   Abbreviation for *roentgen-per-hour-at-one-meter.*

rhodium [Rh]   A metallic element with an atomic number of 45 and an atomic weight of 102.91.

rhombic antenna   An antenna system consisting of four long-wire antennas strung between four masts in a diamond arrangement with the signal feed introduced at one apex. If the apex opposite the feed is open (resonant antenna), it has a bidirectional response between the two apexes. But if the open end is terminated with a matching resistor (nonresonant antenna), it is unidirectional toward the terminated apex. The system has a gain of 20 to 40 times that of a dipole.

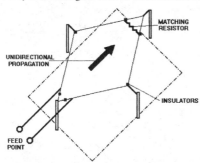

Rhombic antenna is unidirectional if a matching resistor is placed across both antenna ends, but it is bidirectional if the ends are left open.

rhometal   A high-resistivity magnetic alloy that has an initial permeability of 250 to 2000.

rho-theta navigation   *Omnibearing-distance navigation.*

rhumbatron   *Cavity resonator.*

RIAA curve   1. Recording Industry Association of America curve that represents standard recording characteristics for long-play records.   2. The corresponding equalization curve for playback of microgroove records.

ribbon cable   A flat cable made of multiple, round, insulated wires positioned side by side and extruded or bonded together by a process that forms a flexible ribbon.

ribbon microphone   A velocity microphone whose moving element is a thin corrugated metal ribbon mounted between the poles of permanent magnets. The ribbon cuts magnetic lines of force as it is moved back and forth in proportion to the velocity of air particles in a sound wave. As a result, an AF output voltage is induced in the ribbon.

Richardson effect   *Edison effect.*

Richardson equation   An equation that gives the density of thermionic emission at saturation current in terms of the absolute temperature of the filament or cathode of an electron tube.

ride gain   To control the volume range of an AF circuit while watching a volume indicator.

ridge waveguide   A circular or rectangular waveguide that has one or more longitudinal internal ridges which serve primarily to increase transmission bandwidth by lowering the cutoff frequency.

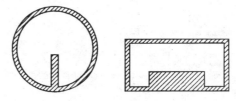

Ridge waveguides.

ridge-waveguide termination   A component that serves as a perfect zero-impedance load when connected to the end of a ridge waveguide, so no signal is reflected back through the waveguide.

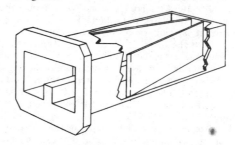

Ridge-waveguide termination.

Rieke diagram   A chart that shows contours of constant power output and constant frequency for a microwave oscillator, drawn on a Smith chart or other polar diagram whose coordinates represent the components of the complex reflection coefficient at the oscillator load.

rig   Slang term for a complete system such as a complete amateur radio station.

Righi-Leduc effect   Development of a difference in temperature between the two edges of a strip of metal in which heat is flowing longitudinally, when the plane of the strip is perpendicular to magnetic lines of force.

right-hand polarized wave   An elliptically polarized transverse electromagnetic wave. The rotation of the electric field vector is clockwise for an observer looking in the

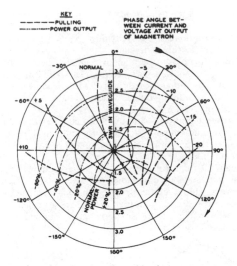

KEY
----- —PULLING
-------—POWER OUTPUT

PHASE ANGLE BET-
WEEN CURRENT AND
VOLTAGE AT OUTPUT
OF MAGNETRON

Rieke diagram showing the effect of load on a magnetron with a normal frequency of 9 GHz. Values on dashed arcs give the amount of frequency-pulling in megahertz. An ideal load is a normal-power curve, passing through SWR = 1 at the center of diagram. The best operating region is for a load phase angle of 0 to −30° because pulling from a normal 9 GHz value is least here.

direction of propagation. It is also called a clockwise polarized wave.

right-hand rule 1. For a current-carrying wire: If the fingers of the right hand are placed around the wire so that the thumb points in the direction of current flow, the fingers will be pointing in the direction of the magnetic field produced by the wire. For electron flow (the opposite of current flow), the left-hand rule applies. 2. For a movable current-carrying wire or an electron beam in a magnetic field: If the thumb, first, and second fingers of the right hand are extended at right angles to one another, with the first finger representing the direction of magnetic lines of force and the second finger representing the direction of current flow, the thumb will be pointing in the direction of motion of the wire or beam. It is also called *Fleming's rule*.

right-hand taper The resistive taper in the resistive element of a potentiometer or rheostat that causes its resistance value to change nonlinearly when the control knob is turned to the right, or clockwise. See also *left-hand taper*.

right-justify 1. To adjust the printing positions of characters on a page so the right margin is lined up. 2. To shift the contents of a register so the right or least significant digit is at some specified position.

right signal The output of a microphone placed to pick up the intensity, time, and location of sounds originating predominantly to the listener's right of the center of the performing area, when making a stereo recording or broadcast.

right stereo channel The right signal as electrically reproduced in reception of stereo FM broadcasts or stereo records.

rim drive A phonograph or sound recorder drive whose rubber-covered drive wheel is in contact with the inside of the rim of the turntable.

ring-around An undesired triggering of a transponder by its own transmitter, or triggering at all bearings so as to give a ring presentation on a radar PPI display.

ring counter A loop of binary scalers or other bistable units so connected that only one scaler is in a specified state at any given time. As input signals are counted, the position of the one specified state moves in an ordered sequence around the loop.

ringer-equivalence number [REN] An indication of the impedance or loading factor of a telephone bell or ringer circuit. A REN of 1.0 equals about 8 kΩ. Telephone operating companies typically permit a maximum of 5.0 REN (1.6 kΩ) on an individual subscriber line. A minimum REN of 0.2 (40 kΩ) is typically required.

ring head A magnetic head whose magnetic material forms an enclosure that has one or more air gaps. The magnetic recording medium bridges one of these gaps, and is in contact with or in close proximity to the pole pieces on one side only.

ringing An oscillatory transient that occurs in the output of a system as a result of a sudden change in input. In television it might produce a series of closely spaced images or a black line immediately to the right of a white object.

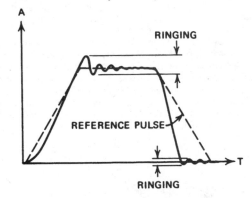

Ringing as it appears on an oscilloscope.

ringing time 1. The time required for the output of an oscillatory circuit to decrease to a predetermined level after its input power is removed. 2. The time between the termination of a transmitted radar pulse and the instant at which the reradiated power from an echo box falls below the minimum required to produce an indication. It is a measure of overall radar performance.

ring laser *Laser gyro.*

ring modulator A modulator that has four diode elements connected in series to form a ring around which current

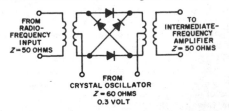

Ring modulator circuit.

flows readily in one direction. Input and output connections are made to the four nodal points of the ring. It can be a balanced modulator, demodulator, or phase detector.

ring oscillator Two or more pairs of transistors operating as push-pull oscillators around a ring, usually with alternate successive pairs of bases and collectors connected to tank circuits. Adjacent transistors around the ring operate in phase opposition.

right-plane circuit A slow-wave structure that consists of circular rings or slotted pipes supported by one or more radial planes. It is a structure in high-power, high-frequency traveling-wave tubes.

ring scaler A scaler whose asymmetrical condition is passed along to the next transistor in line, with the last transistor feeding back to the first to complete the ring.

ring-seal tube An electron tube whose grid and anode are radially symmetrical, with the grid connected to a metal ring sealed into the glass envelope. The construction permits insertion of the tube in a coaxial chamber, with the grid connected to the outer cylinder and the anode to the inner conductor for operation as a grounded-grid amplifier.

ring time The time during which the output of a radar echo box remains above a specified level. The interval starts when a pulse is transmitted, and it is usually considered to end when the energy reradiated by the echo box falls below the minimum needed for an indication on the radar screen.

ring wire One of the two wires connecting the central office to a customer's telephone. It is so named because of the ring portion of the plugs used by operators to make the connection in older equipment. The ring wire is normally negative with respect to the tip wire, the second wire from the central office. The ring wire normally has red insulation, and the tip wire has green insulation. See *tip.*

riometer [Relative Ionospheric Opacity METER] An instrument that measures changes in ionospheric absorption of electromagnetic waves by determining and recording the level of extraterrestrial cosmic radio noise.

ripple 1. The AC component of the output of a DC signal. The term typically refers to the residual 50- to 60-Hz AC component in the output of a DC power supply that arises as a result of incomplete or inadequate filtering. The amount of filtering depends on the ripple frequency and the load resistance. As load resistance declines, more filtering is required. 2. Deviation of the frequency response curve from a generally flat, monotonously rising, or falling trend.

ripple counter A counter that consists of flip-flops in series. When the first flip-flop changes state it affects the second, which in turn affects the third, and so on, until the last in the series is changed.

ripple factor The ratio of the effective value of the AC component of a pulsating DC voltage to the average value.

ripple filter A low-pass filter that reduces ripple while freely passing the direct current obtained from a rectifier or DC generator.

ripple frequency The frequency of the ripple present in the output of a DC source.

ripple voltage The periodic AC component imposed on the output voltage of a power supply.

RISC Abbreviation for *reduced-instruction-set computer.*

rise time The time required for a signal pulse to rise from 10 to 90% of its final steady value. It is a measure of the steepness of the wavefront.

rising-sun magnetron anode A multicavity magnetron anode whose resonators have two different resonant frequencies that are arranged alternately for mode separation. The cavities appear as alternating long and short radial slots around the perimeter of the anode structure, resembling the rays of the sun.

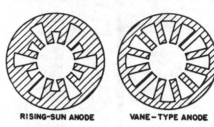

Rising-sun magnetron anode compared with a vane-type magnetron anode.

RJ-11/14 The standard for the common modular telephone jack.

RLC circuit Abbreviation for *resistance-inductance-capacitance circuit.*

R-meter An ionization instrument calibrated to indicate the intensity of gamma rays, X-rays, and other ionizing radiation in roentgens.

RMI Abbreviation for *radio magnetic indicator.*

r/min Abbreviation for *revolutions per minute.*

R – Y signal The red-minus-luminance color-difference signal used in color television. It is combined with the luminance signal in a receiver to give the red color-primary signal.

RMS Abbreviation for *root-mean-square.*

RMS power per channel The stereo power amplifier rating now required in the United States by the Federal Trade Commission. To be meaningful, the load impedance, total harmonic distortion, bandwidth, and other amplifier parameters must also be specified.

roamer A *mobile station* (telephone) that operates in a *cellular mobile telephone system* other than the one from which service is subscribed.

robot A reprogrammable, multifunction manipulator designed to move material, parts, tools, or specialized devices through variable programmed motions for the performance of a variety of tasks. Most true robots are stationary industrial robots in factories that perform assignments from heavy-duty materials handling and spot welding to spray painting. Another class of light-duty robots perform such tasks as picking and placing electronic components on circuit boards and inspecting and testing of finished circuit boards. There are four principal geometric configurations for industrial robots: (1) articulated, revolute, or jointed with low shoulder; (2) articulated revolute, or jointed with high shoulder; (3) polar coordinate or spherical; and (4) Cartesian or rectangular coordinate. The manipulator to which tools are mounted is often referred to as the *arm.* Most industrial robots are driven by DC electric motors, but both hydraulic and pneumatic power is in use. Scientific robots can carry out tests and experiments in distant or inaccessible locations such as deep space, planetary surfaces, or the ocean floor. Some robots

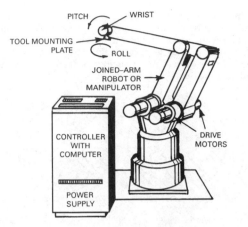

Robot is a programmable mechanical arm or manipulator capable of moving in three dimensions to perform useful work.

have been equipped with wheels or tracks for self-propulsion or have synchronized "legs" for walking. There are also self-propelled service or personal robots that can perform routine delivery or cleaning functions in the office or home. A true robot is distinguished from another class of robotlike machines called telecherics that can perform work in hazardous locations by remote control. They are always under the control of a human operator through signals transmitted over a flexible umbilical cable. See *telecheric*.

Rochelle-salt crystal A crystal of sodium potassium tartrate that has a pronounced piezoelectric effect. It is used in crystal microphones and crystal pickups.

rocking Back-and-forth rotation of the tuning control in a superheterodyne receiver while adjusting the oscillator padder near the low-frequency end of the tuning dial, to obtain more accurate alignment.

roentgen [R] The international unit of exposure dose for X-rays and gamma rays. One roentgen of radiation will ionize dry air sufficiently to produce 1 electrostatic unit of electricity per 1.293 mg of air.

roentgen-equivalent man [rem] A unit of ionizing radiation, equal to the amount that produces the same damage to a person as 1 roentgen of high-voltage X-rays.

roentgen-equivalent physical [rep] A unit of ionizing radiation, equal to the amount that causes absorption of 93 ergs of energy per gram of soft tissue. It is also called equivalent roentgen.

roentgen meter A meter for measuring the cumulative quantity of X-rays or gamma rays, without reference to time.

roentgenography Radiography by X-rays.

roentgen-per-hour-at-one-meter [Rhm] A unit of gamma-ray source strength, corresponding to a dose rate of 1 R/h at a distance of 1 m in air.

roentgen-rate meter An electrically operated instrument that measures radioactivity and is calibrated in roentgens per unit time or any multiple of that unit.

roentgen ray *X-ray*.

roger 1. A code word used in communication, meaning that a message has been received and understood. 2. An expression of agreement.

Roget spiral A helix of wire that contracts in length when a current is sent through, because of the mutual attraction between adjacent turns.

roll Slow upward or downward movement of the entire image on the screen of a television receiver, due to a lack of vertical synchronization.

roll-and-pitch control A control for automatic pilots and remote-attitude indicators that consists of a gyroscope which provides signals for controlling an aircraft about its lateral and longitudinal axes. Some controls also provide a visual presentation of the aircraft's attitude on a panel instrument for the pilot.

rollback *Rerun*.

rolloff Gradually increasing attenuation as frequency is changed in either direction beyond the flat portion of the amplitude-frequency response characteristic of a filter, system, or component.

ROM Abbreviation for *read-only memory*.

roof filter A low-pass filter in carrier telephone systems that limits the frequency response of the equipment to frequencies needed for normal transmission. It blocks unwanted higher frequencies induced in the circuit by external sources. A roof filter improves runaround crosstalk suppression and minimizes high-frequency singing.

room-temperature vulcanizing [RTV] **silicone** A resin for potting or encapsulating electronic circuits and sealing wires and cable against moisture, salt spray, and environmental contaminants that cures at normal room temperature to become an elastic, rubberlike composition.

rooter amplifier A nonlinear amplifier whose negative feedback makes the output voltage vary as the square root or some other root of the input voltage. It is used in television transmitter video amplifiers for gamma correction to compensate for camera-tube characteristics.

root-mean-square [RMS] 1. The square root of the average of the squares of a series of related values. 2. The effective value of an alternating current, corresponding to the DC value that will produce the same heating effect. The RMS value is computed as the square root of the average of the squares of the instantaneous amplitudes for one complete cycle. For a sine wave, the RMS value is 0.707 times the peak value. Unless otherwise specified, alternating quantities are assumed to be RMS values. Another name is effective value.

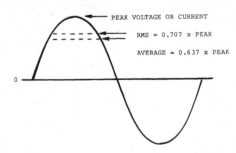

Root-mean-square (RMS) value of a sine wave compared with its peak and average values.

root-mean-square particle velocity *Effective particle velocity.*
root-mean-square sound pressure *Effective sound pressure.*

rosin-core solder Solder made in a tubular or other hollow form with its inner space filled with rosin flux to serve as a noncorrosive flux for soldering joints.

rotary actuator A device that converts electric or pneumatic energy into controlled rotary force. One type consists of an electric motor, gear box, and limit switches.

rotary amplifier *Rotating magnetic amplifier.*

rotary beam antenna A highly directional short-wave antenna system mounted on a mast so that it can be rotated to any desired position, either manually or by an electric motor drive.

rotary converter *Dynamotor.*

rotary coupler *Rotating joint.*

rotary dial A telephone calling device that generates pulses by manual rotation and release of a dial. The number of pulses is determined by how far the dial is rotated before being released.

rotary gap *Rotary spark gap.*

rotary joint *Rotating joint.*

rotary solenoid A solenoid whose armature is rotated when actuated. The rotary stroke is usually converted to linear motion to give a longer stroke than is possible with a conventional plunger-type solenoid. The stroke can range from 25 to 95°.

rotary stepping relay *Stepping relay.*

rotary stepping switch *Stepping relay.*

rotary switch A switch that is operated by rotating its shaft with an attached knob.

rotary-vane attenuator A variable attenuator that consists of three circular waveguide sections—two fixed and one rotatable—each containing a resistive card. Input and output transitions at the ends permit connections to standard waveguide. The attenuation is controlled by rotation of the center section. Minimum attenuation occurs when all three cards lie in the same plane, and maximum attenuation occurs when the card in the rotatable section is at 90° with respect to the other cards. See also *flap attenuator.*

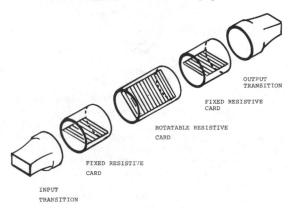

Rotary-vane attenuator. Attenuation is controlled by the rotation of the resistive card in the center section.

rotary variable differential transducer [RVDT] An inductive transducer that operates on the same principle as the linear variable differential transducer (LVDT). It can measure angular displacements and torque.

rotatable loop antenna A loop antenna that can be rotated in azimuth, for use in direction-finding.

rotatable-loop radio compass An automatic direction finder with a loop antenna that is rotated manually to determine the relative bearing between an aircraft or ship and a transmitter.

rotating joint A joint that permits one section of a transmission line or waveguide to rotate continuously with respect to another while passing RF energy. Rotary coupler and rotary joint are other terms for this fitting.

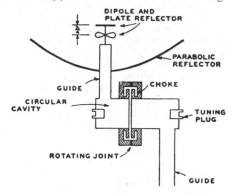

Rotating joint in waveguide feed to radar antenna.

rotating magnetic amplifier A prime-mover-driven DC generator whose power output can be controlled by small field input power. It produces power gain as high as 10,000. It can be a power source for the DC drive motors of large radar antennas as well as for other motor drives in automatic control systems. Examples include the amplidyne and metadyne. It is also called a rotary amplifier or a rotating amplifier.

rotating radio beacon A radio transmitter arranged to radiate a concentrated beam that rotates in a horizontal plane at constant speed and transmits different signals in different directions so ships and aircraft can determine their bearings without directional receiving equipment.

rotating-type scanning sonar Scanning sonar with an electric drive to obtain a rotating receiving-beam pattern. Stationary transducer units arranged in a circle are connected in succession to the receiver by a commutator.

rotation wave *Shear wave.*

rotator A device that rotates the plane of polarization, such as a twist in a rectangular waveguide.

rotoflector An elliptically shaped rotating radar reflector that reflects a vertically directed radar beam at right angles so it radiates horizontally.

rotor The rotating member of a machine or device, such as the rotating armature of a motor or generator, or the rotating plates of a variable capacitor.

rotor plate One of the rotating plates of a variable capacitor, usually directly connected to the metal frame.

rounding error The error that results from round-off of a quantity in a computer or in calculations. It is also called *round-off error.*

round off To change a more precise quantity to a less precise one by dropping certain less significant digits and applying some rule for changing the last significant retained digit.

round-off error *Rounding error.*

round-the-world echo A signal that occurs every ½ s when a radio wave repeatedly encircles the earth at its speed of 186,000 mi/s (300,000 km/s) during unusual backscatter propagation modes.

routine A set of instructions arranged in proper sequence to cause a computer to perform a desired operation, such as the solution of a mathematical problem.

r parameter A transistor parameter relating to resistivity.

rpm Abbreviation for revolutions per minute.

rps Abbreviation for revolutions per second.

RPV Abbreviation for *remotely piloted vehicle*.

r/s SI abbreviation for revolutions per second.

R scope A radarscope that produces an R display.

R-S flip-flop A flip-flop that has two logic inputs, designated R for reset and S for set, with only one of the inputs at high level or 1 at a time. Its operating speed is determined by the charging time of capacitors in the clock steering network connected between the R and S inputs. If the S input is at the 1 level when a clock pulse arrives, the flip-flop and its Q output go to the 1 state. A 1 at the R input gives reset to the 0 or off state. Set-reset flip-flop and S-R flip-flop are other terms for it.

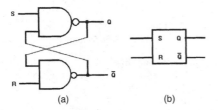

R-S flip-flop (*a*) formed with NAND gates and (*b*) its schematic symbol.

RS-232C interface The EIA standard applicable to the 25-pin interconnection of data terminal and data communications equipment for serial binary data interchange. See also *computer connector*.

RS-423 interface The EIA standard applicable to unbalanced digital interface circuits.

RS-449 The EIA standard that specifies the functional and mechanical characteristics of the general-purpose 37-pin and 9-pin interface between data terminal and data communications equipment for serial binary data interchange. It complies with EIA electrical interface standards RS-422 and RS-423.

RST flip-flop An R-S flip-flop that has an additional trigger input (labeled T) which can be energized to make the flip-flop change state. If the flip-flop is off, a pulse on S or T will turn it on; a pulse on the R input will cause no change. If the flip-flop is on, a pulse on the R or T input will turn it off; a pulse on the S input will cause no change. Also called set-reset-trigger flip-flop and SRT flip-flop.

RTL Abbreviation for *resistor-transistor logic*.

RTS Abbreviation for *request to send*.

RTV Abbreviation for *room-temperature vulcanizing*.

RTTY Abbreviation for *radioteletype*.

rubidium [Rb] A photosensitive metallic element with an atomic number of 37 and an atomic weight of 85.47.

rubidium magnetometer A highly sensitive magnetometer that combines the spin precession principle with optical pumping and monitoring for detecting and recording variations as small as 0.01 gamma (0.1 microoersted) in the total magnetic field intensity of the earth. It can locate and map buried archeological sites and perform airborne or oceanographic magnetic surveys.

rubidium-vapor frequency standard An atomic frequency standard in which the frequency is established by a gas cell that contains rubidium vapor and a neutral buffer gas. It is a secondary frequency standard because the rubidium gas cell is dependent on the gas mixture and pressure; hence it must be calibrated initially.

ruby A single crystal of aluminum oxide with a small fraction of its aluminum atoms replaced by chromium atoms that serve as the source of characteristic red fluorescence under irradiation.

ruby laser A crystalline solid laser in which optical pumping is applied to a rod-shaped ruby crystal with a flash tube to produce an intense and extremely narrow beam of coherent red light.

run One complete performance of a program or routine on a computer.

runaround crosstalk Crosstalk resulting from coupling between the high-level end of one repeater and the low-level end of another repeater, as at a carrier telephone repeater station.

runway localizing beacon A small radio range that provides accurate directional guidance along the runway of an airport and for some distance beyond, for instrument landings.

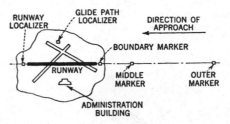

Runway localizing beacon for instrument landings at the opposite end of the runway from the marker beacons.

ruthenium [Ru] A metallic element with an atomic number of 44 and an atomic weight of 101.07.

Rutherford scattering Scattering of moving particles at various angles as a result of interaction with atoms of a solid material.

RVC Abbreviation for *reverse voice channel*.

RVDT Abbreviation for *rotary variable differential transducer*.

RZ code Abbreviation for *return-to-zero code*.

s Abbreviation for *second*.

S 1. Symbol for *secondary winding,* used on circuit diagrams to identify the secondary winding of a transformer. 2. Abbreviation for *siemens*. 3. Symbol for *source* on circuit diagrams that contain field-effect transistors.

sabin A unit of sound absorption for a surface, equivalent to 1 ft^2 of perfectly absorbing surface. It is also called the square-foot unit of absorption. The metric sabin has an area of 1 m^2.

Sabine absorption The sound absorption defined by the equation in which reverberation time in seconds is $0.049V/A$, where V is the volume of the room in cubic feet and A is the total Sabine absorption in sabins.

Sabine coefficient The Sabine absorption of a sound-absorptive surface divided by the area of the surface.

SAC [pronounced as a word] Abbreviation for *Strategic Air Command.*

SACCH Abbreviation for *slow associated control channel.*

Safety of Life at Sea Convention [SOLAS] An international convention, in effect since 1974, specifying that all commercial vessels displacing more than 1100 tons carry charts.

safety factor The amount of load, above the normal operating rating, that a machine can handle without failure.

sag An alternating-current power-line undervoltage condition that lasts more than $\frac{1}{60}$ of a second. A long-term sag is a *brownout.*

SAG Abbreviation for *self-aligned gate technology.*

SAGE [SemiAutomatic Ground Environment] An air defense system whose air surveillance data is processed for transmission to computers at direction centers. Here the data is further processed, evaluated, and analyzed automatically to produce weapon assignment and guidance orders.

Saint Elmo's fire A visible electric discharge sometimes seen on the mast of a ship, on metal towers, and on pro-jecting parts of aircraft, due to concentration of the atmospheric electric field at such projecting parts.

salient pole A structure of magnetic material for mounting a field coil of a generator or motor.

SAM Abbreviation for *surface-to-air missile.*

samarium [Sm] A rare-earth element with an atomic number of 62 and an atomic weight of 150.35.

samarium-cobalt magnet A rare-earth permanent magnet that is more efficient, has lower leakage and greater resistance to demagnetization, and can be magnetized to higher levels than conventional permanent magnets. These characteristics permit reduction of magnet size and weight for a given application. These magnets are in electronic watches, magnetrons, meters, and traveling-wave tubes.

sample-and-hold amplifier A circuit consisting of an operational amplifier and capacitor that temporarily holds the changing voltage values obtained from a sensor, trans-ducer, or other signal source to give an analog-to-digital converter circuit enough time to complete conversion so that it can provide a stable, accurate readout of the changing value on a measurement instrument, such as a *digital panel meter* [DPM].

sampled data Data that is obtained at discrete rather than continuous intervals.

sample size The number of units in a sample.

sampling Selecting a small statistically determined part of the total group under consideration for tests that infer the value of one or several characteristics of the entire group.

sampling action Control action in which the difference between the set point and the value of the controlled variable is measured, and correction is made only at intermittent intervals.

sampling gate A gate circuit that extracts information from the input waveform only when it is activated by a selector pulse.

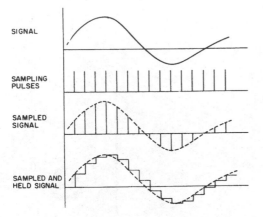

Sample-and-hold process using a train of periodic sampling pulses, as performed by a sample-and-hold amplifier.

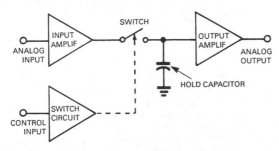

Sample-and-hold amplifier temporarily holds voltage values to permit more accurate, slow-speed sampling rate analog-to-digital conversion.

sampling interval   The time between samples in a sampling or sample-and-hold system.

sampling multiplier   *Averaging multiplier.*

sampling oscilloscope   An instrument whose fast and repetitive signals are slowed down for conventional display on a cathode-ray oscilloscope. An amplitude sample of the signal is selected by a strobe pulse at an instant of time, widened, amplified, and displayed as a bright dot. The next time the signal occurs, the same process is repeated, but the strobe is automatically delayed slightly longer so it produces a sample of an adjacent portion of the signal. This process is repeated many times until a reproduction of the original signal is traced in dots.

sampling plan   A plan that states sample sizes and the criteria for accepting, rejecting, or taking another sample during inspection of a group of items.

sampling rate   The rate with which the amplitude of an analog signal is sampled by a coding circuit. According to the *Nyquist sampling theorem,* if a band-limited signal is sampled at regular intervals and at a rate equal to or greater than the highest frequency of interest, the sample contains all the information of the original signal. The frequency band of interest in telephony ranges from 300 to 3400 Hz, so a sampling rate of 8 kHz is acceptable for transmitting frequencies from 0 to 4000 Hz.

sampling switch   *Commutator switch.*

sampling theorem   Equispaced data that exhibit two or more points per cycle of highest frequency which permits the reconstruction of band-limited functions. It is a process in information theory.

sapphire   A pure variety of gem corundum that occurs in nature and is also produced synthetically. Tips of phonograph needles are made from it because it has a hardness of 9 and takes a fine polish.

sapphire substrate   Synthetic sapphire that is used as a passive insulating base. Silicon can be grown on the sapphire and then etched away selectively to form such solid-state devices as silicon-on-sapphire field-effect transistors and integrated circuits.

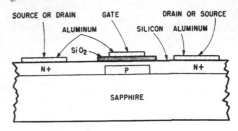

Sapphire substrate in a silicon-on-sapphire field-effect transistor.

SAR   Abbreviation for   1. *synthetic-aperture radar* and 2. *successive-approximation register.*

saser   *Sound laser.*

satellite   An unmanned space vehicle or platform designed to orbit the earth to perform specific functions, such as military surveillance, weather mapping, the transmission of navigational signals, long-term monitoring of environmental phenomena, or relaying telecommunications between widely separated terrestrial transmitters and receivers. Satellites are placed either in high-altitude *geostationary orbit* or *low-altitude orbit* [LAO].

satellite communication   Communication with an active or passive satellite to extend the range of a radio, television, or other transmitter by returning signals to earth from an orbiting satellite. An active satellite with a solar-cell power supply can produce about 500 W of output power.

satellite master antenna television [SMATV]   A local television satellite TV transmission system whose receiving antenna, typically a 6-ft diameter parabolic dish, is located on a building in a city or urban area amid many other buildings, so that it can send satellite programs directly over cables to customers in adjacent buildings. See also *multichannel multipoint distribution* [MMDS] and *local multipoint distribution* [LMDS].

satellite reconnaissance   Strategic reconnaissance obtained from data gathered by a satellite.

saturable-core magnetometer   A magnetometer that depends for its operation on the changes in permeability of a ferromagnetic core as a function of the magnetic field to be measured.

saturable reactor   An iron-core reactor that has an additional control winding which carries direct current. Its value is adjusted to change the degree of saturation of the core, thereby changing the reactance that the AC winding offers to the flow of alternating current. With appropriate external circuits, a saturable reactor can serve as a magnetic amplifier.

saturable-reactor-controlled oscillator An oscillator that has a saturable reactor in its tuning circuit to control its output frequency.

saturable transformer A saturable reactor that has additional windings to provide voltage transformation or isolation from the AC supply.

saturated color A pure color, not contaminated by white.

saturated diode A diode that is passing the maximum possible current. Thus further increases in applied voltage have no effect on current.

saturated logic Logic based on transistors that are allowed to saturate during normal operation. Examples include diode-transistor and transistor-transistor logic.

saturating signal A radar signal whose amplitude is greater than the dynamic range of the receiving system.

saturation 1. The condition in which a further increase in one variable produces no further increase in the resultant effect. 2. The condition occurring when a transistor is driven so hard that it becomes biased in the forward direction (the collector becomes positive with respect to the base, for example, in a PNP-type transistor). In a switching application under saturation conditions, the charge stored in the base region prevents the transistor from turning off quickly. 3. *Anode saturation.* 4. *Color saturation.* 6. *Magnetic saturation.* 6. *Temperature saturation.*

saturation current The maximum possible current that can be obtained as the voltage applied to a device is increased. In a gas tube, it is the current at which the applied voltage is sufficient to attract all ions. In a semiconductor diode, it is that part of the steady-state reverse current that flows as a result of the transport of thermally generated minority carriers across the junction within the regions adjacent to the junction.

saturation curve A curve that shows the way in which a quantity such as current or magnetic flux reaches saturation.

saturation flux density *Saturation induction.*

saturation induction The maximum intrinsic induction possible in a material. It is also called saturation flux density.

saturation magnetostriction The value of magnetostriction that would be reached if the applied magnetizing force were increased indefinitely.

saturation reactance The reactance of the gate winding of a magnetic amplifier during the saturation interval.

saturation region An operating region in which an increase in the actuating component produces no further increase in the output effect.

saturation voltage The minimum voltage needed to produce saturation current.

Saturn A booster rocket for launching large spacecraft such as a space shuttle.

SAW Abbreviation for *surface acoustic wave.*

sawtooth current A current that has a sawtooth waveform.

sawtooth generator A circuit whose output voltage has a sawtooth waveform. It produces sweep voltages for cathode-ray tubes and signal generators.

sawtooth voltage A voltage that has a sawtooth waveform.

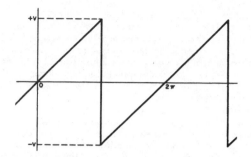

Sawtooth waveform as it would appear on an oscilloscope.

sawtooth waveform A waveform characterized by a slow rise time and a sharp fall, resembling the tooth of a saw.

sb Abbreviation for *stilb.*

S band A band of frequencies extending from 2.0 to 4.0 GHz, corresponding to wavelengths of 15 and 7.5 cm, respectively, in accordance with IEEE Standard 521-1976 frequency designations. It was once referred to as the *ten-centimeter band* because 10 cm is its approximate midband wavelength. This band corresponds to both the E band (2.0 to 3.0 GHz) and the F band (3.0 to 4.0 GHz) of the U.S. Military Joint Chiefs of Staff (JCS) frequency designations (1970).

SBC Abbreviation for *single-board computer.*

SC Abbreviation for *suppressed carrier.*

scalar function A scalar quantity that has a definite value for each value of some other scalar quantity. Thus the resistance of a given conductor is a scalar function of the temperature of the conductor.

scalar quantity A quantity that has only magnitude but not direction, such as resistance, time, or temperature.

scale 1. A series of markings for reading the value of a quantity or setting. 2. A series of musical notes arranged from low to high by a specified scheme of intervals suitable for musical purposes. 3. To change the magnitudes of the units in which a problem is expressed to bring all magnitudes within the capacity of a computer.

scale division That part of a scale between two adjacent scale marks.

scale factor The factor for the multiplication of the reading of an instrument or the solution of a problem to give the true final value when a corresponding scale factor initially brings the magnitude within the range of the instrument or computer.

scale-of-eight circuit A counting circuit that recycles at every eighth pulse.

scale-of-ten circuit *Decade scaler.*

scale-of-two circuit *Binary scaler.*

scaler A circuit that produces an output pulse when a prescribed number of input pulses is received. A single binary scaler stage delivers an output pulse for every 2 input pulses, and a decade scaler delivers an output pulse for every 10 input pulses. It is also called a *counter* or *scaling circuit.*

scaling Counting pulses with a scaler when the pulses occur too fast for direct counting by conventional means.

scaling circuit *Scaler.*

scaling factor The number of input pulses per output pulse of a scaling circuit. It is also called scaling ratio.

scaling ratio   *Scaling factor.*

scan   1. To examine an area or a region in space point by point in an ordered sequence, as when converting a scene or image to an electric signal or when radar is monitoring an airspace for detection, navigation, or traffic control purposes.   2. One complete circular, up-and-down, or left-to-right sweep of the radar, light, or other beam or device in making a scan.

scan axis   The axis that acts as a reference for specifying target displacement in radar or sonar.

scan converter   A cathode-ray tube that is capable of storing radar, television, and data displays for nondestructive readout over prolonged periods of time. Applications include buildup of repetitive signals submerged in noise, conversion of video displays from one scan mode to another, and special overlay and merging effects for television and radar displays.

scanned array   A phased-array radar antenna with phase shifters that operates under computer control to scan the beam in bearing and elevation without mechanical movements.

scanner   1. A radar antenna and reflector assembly that oscillates back and forth about a center position during its search, then moves in any required plane during the tracking of a target.   2. That part of a facsimile transmitter which systematically translates the densities of the elemental areas of the subject copy into corresponding electric signals.   3. A device that automatically samples, measures, or checks a number of quantities or conditions in sequence, as in process control.   4. A specialized radio receiver that scans back and forth over a range of frequencies in quick succession so that the listener can sample the output and switch to a frequency of interest.   5. A personal computer peripheral that includes a scanning head and a circuit board for digitizing text, drawings, and photos so that they can be displayed on the computer monitor and/or printed-out. The scanning head scans shades of gray or color optoelectronically and transmits its output in dots per inch (dpi) to the computer.

scanning   The process of examining an area, a region in space, or a portion of the radio spectrum point by point in an ordered sequence.

scanning beam   A beam of light, an electron beam, or a radar beam that scans.

scanning electron microscope [SEM]   An electron microscope whose operation is based on the ability of electromagnetic condensing lenses to form electrons into a fine beam that can be swept across a specimen, causing it to release a shower of electrons. The electrons spray out in all directions like light rays striking an object, and they are detected by a signal detector circuit and sent to a video monitor. The specimen might have a thin precious-metal film deposited on it to enhance the electron emission process. Specimens have a 3-D appearance, and can be magnified up to 10,000 times. It is used to view integrated circuits, microminiature objects, biological specimens such as bacteria, and viruses, and minute features of plants, insects, and animals. See also *electron microscope, scanning tunneling microscope* [STM], and *transmission electron microscope* [TEM].

scanning hydrographic operational airborne lidar survey [SHOALS]   A system for recording laser depth soundings, intended for use aboard helicopters flying over relatively clear bodies of water.

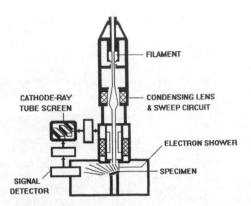

Scanning electron microscope illuminates the specimen with electrons, and the image is converted to a scanned image.

scanning line   The single, continuous narrow strip covered during one scan of a television image or one scan of the subject copy or record sheet in facsimile or digitizing text for computers.

scanning loss   The reduction in sensitivity when scanning across a radar target as compared with the sensitivity obtained when the radar beam is directed constantly at the target. Scanning loss is due to the change in antenna position during the interval in which the signal travels from the antenna to the target and back.

scanning radio   A radio receiver that automatically scans across public service, emergency service, or other radio bands and stops at the first preselected station which is on the air. Some versions use a crystal for each channel of interest; other versions use a frequency synthesizer that permits choosing from thousands of digitally derived frequencies. It is also called a radio scanner.

scanning sonar   Sonar that shows all targets of interest simultaneously, as on a radar PPI display or sector display. The sound pulse can be transmitted in all directions simultaneously and picked up by a rotating receiving transducer. Alternatively, it can be transmitted and received in only one direction at a time by a scanning transducer.

scanning speed   *Spot speed.*

scanning spot   The area that is viewed instantaneously by the pickup system of a facsimile scanner or a television camera.

scanning switch   *Commutator switch.*

scanning transducer   A multielement sonar transducer whose elements are arranged in a circle and electronically switched in sequence to give the equivalent of scanning without mechanical movement.

scanning tunneling microscope [STM]   An electron microscope with a very sharp tip that scans a surface to gather data which is converted by computer software into high-resolution, 3-D images. It can perform atomic-scale manipulation, grow thin crystalline films, and apply strong magnetic fields to low-temperature samples. It is operated in an ultra-high vacuum chamber to eliminate contaminant atoms and is cooled by liquid helium so that very low temperatures slow the movement of atoms, permitting them to be both imaged and manipulated.

scan period *Regeneration period.*

scan rate The angular velocity of a radar beam, generally in the range of 4 to 12 rpm.

SCAP Abbreviation for *silicon capacitive absolute pressure sensor,* a device designed to measure barometric or manifold absolute pressure.

scatterband The total bandwidth occupied by the frequency spread of numerous interrogations that have the same nominal radio frequency in a pulse system.

scattered radiation Radiation that has been changed in direction during its passage through a substance. It can also be increased in wavelength.

scattering 1. The change in direction of a particle or photon because of a collision with another particle or a system. 2. Diffusion of acoustic waves because of nonuniformity of the transmitting medium. 3. Diffusion of electromagnetic waves in a random manner by air masses in the upper atmosphere, permitting long-range reception, as in scatter propagation. It is also called radio scattering.

scattering loss That part of the transmission loss caused by scattering within the medium or roughness of the reflecting surface.

scattering parameter *S parameter.*

scatter propagation Transmission of radio waves far beyond line-of-sight distances with high power and a large transmitting antenna that beams the signal upward into the atmosphere and a similar large receiving antenna that picks up the small portion of the signal which is scattered by the atmosphere. In ionospheric scatter, the phenomenon occurs in the lower E layer of the ionosphere. In tropospheric scatter, the phenomenon is entirely in the earth's lower atmosphere, from ground level to about 30,000 ft (10 km). In meteoric scatter, the trail of a passing meteor scatters the radio waves back to earth. It is also called over-the-horizon communication, or forward-scatter propagation.

SCD Abbreviation for *source control drawing.*

schematic capture The symbols representing functional blocks or individual circuit elements of an electronic schematic are placed on a graphical design workstation and interconnected by graphical representations of wires, cables, and connectors for reproduction, parts listing, and manufacture. It is a form of computer-aided design.

Schering bridge A four-arm AC bridge that measures capacitance and dissipation factor. The bridge balance is independent of frequency.

schlieren photography An optical system designed to photograph changes in gas density due to sound waves, shock waves, or turbulence in wind tunnels. A knife edge at the focal point transmits or cuts off a light beam when the refraction of the intervening gas varies with density.

Schmitt trigger A bistable trigger circuit that converts an AC input signal into a square-wave output signal by switching action. It is triggered at a predetermined point in each positive and negative swing of the input signal.

Schottky barrier An electric potential step at the interface between a metal and a semiconductor in a solid-state device. It usually repels electric charge carriers in the semiconductor from the interface. The gate of a MESFET is actually a Schottky barrier diode.

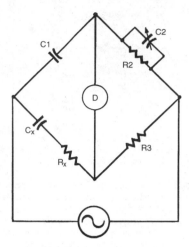

Schering bridge can measure capacitance and dissipation factors.

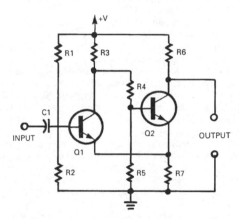

Schmitt trigger circuit accepts sine waves or pulses and produces square waves.

Schottky barrier diode A semiconductor diode formed by contact between a semiconductor layer and a metal contact; it has a nonlinear rectifying characteristic. Hot carriers (electrons for N-type material or holes for P-type material) are emitted from the Schottky barrier of the semiconductor, and move to the metal coating that is the diode base. Majority carriers predominate but there is essentially no injection or storage of minority carriers to limit switching speeds. It is also called a hot-carrier diode or a Schottky diode.

Schottky-clamped transistor A logic circuit that includes a Schottky barrier diode to avoid saturation, and thereby speed up switching action. This construction is used in some transistor-transistor logic (TTL) and integrated injection logic (I^2L).

Schottky diode *Schottky barrier diode.*

Schottky effect An increase in anode current of a thermionic tube beyond that predicted by the Richardson equation. It is due to a lowering of the work function of the cathode when an electric field is produced at the surface of

Schottky emission

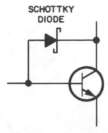

SCHOTTKY
DIODE

Schottky-clamped transistor with a Schottky barrier diode connected between the base and collector.

the cathode by the anode. The same effect removes electrons from the surface of a semiconductor material when an electric field is applied.

Schottky emission Injection of an electron over a semiconductor potential barrier into a region where allowed energy levels are available.

Schottky I²L An integrated injection logic circuit that includes Schottky barrier diodes for clamping purposes.

Schottky noise *Shot noise.*

Schottky power diode A Schottky barrier diode whose heat dissipation is spread over a large area of silicon, but whose diode capacitance is kept low by a small metal-semiconductor contact area.

Schottky TTL gate A transistor-transistor logic gate with Schottky barrier diodes connected across each transistor to reduce the number of charge carriers in the bases when the transistors are on. It then takes less time to deplete the base region of carriers so the transistors can be turned off at high switching speeds.

Schottky TTL memory A random-access memory whose cells include two inverted transistors in a cross-coupled flip-flop circuit, with emitter-base resistances serving as load.

Schroedinger equation A wave equation that states the relation between wave function, particle mass, total energy, potential energy, and Planck's constant.

Schroedinger wave function A wave function that determines the state of a system and satisfies the Schroedinger equation. For a particle or photon, the square of the wave function is proportional to the probability that the particle will be at a particular point at a particular time.

SCI Abbreviation for *scalable coherent interface.*

scintillation 1. A flash of light (optical photons) produced in a phosphor by an ionizing particle or photon. 2. A rapid apparent displacement of a target indication from its mean position on a radar display. One cause is the shifting of the effective reflection point on the target. It is also called target glint or target scintillation. 3. Rapid random fading of microwave radio signals, caused by fluctuations in the refraction of the atmosphere. With optical communication links, the same fluctuations or scintillations can be produced by air turbulence.

scintillation counter A counter whose scintillations are produced in a fluorescent material by an ionizing radiation that is detected and counted by a multiplier phototube and associated circuits. It has wide use in medical research, nuclear research, and prospecting for radioactive ores.

scintillation spectrometer A scintillation counter adapted to the study of energy distributions.

scintillator material A material that emits optical photons in response to ionizing radiation. The five major classes of scintillator materials are (*a*) inorganic crystals such as sodium iodide, thallium single crystals, and zinc sulfide-silver screens; (*b*) organic crystals such as anthracene and transstilbene; (*c*) liquid, plastic, or glass solution scintillators; (*d*) gaseous scintillators.

SCL Abbreviation for *space-charge layer.*

scope 1. *Cathode-ray oscilloscope.* 2. *Radarscope.*

SCPI An abbreviation for *standard commands for programmable instruments.*

SCR Abbreviation for *silicon controlled rectifier.*

SCR bridge An arrangement of four silicon controlled rectifiers to form a *full-wave bridge.*

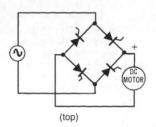

(top)

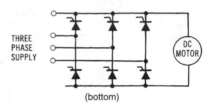

(bottom)

SCR bridge: single-phase (top) and three-phase (bottom).

scrambled speech Speech that has been made unintelligible by inversion, such as for privacy of telephone calls. At the receiving end, the signals are inverted again to restore the original sounds.

scrambler A circuit that divides speech frequencies into several ranges by filters, then inverts and displaces the frequencies in each range so the resulting reproduced sounds are unintelligible. In the simplest system the entire speech frequency range is combined with the output of a fixed-frequency oscillator, and difference frequencies become the inverted signal. The process is reversed at the receiving apparatus to restore intelligible speech. Scramblers that provide secrecy for teletypewriter and code messages generally use cryptology techniques involving pseudorandom key generators, with similar equipment for unscrambling the received messages before or after printout. It is also called a *speech scrambler.*

scratchpad memory A small fast-access memory in a computer for the storage of frequently needed instructions and codes or for holding subtotals until required for final results.

screen 1. The surface on which a television, radar, X-ray, or cathode-ray oscilloscope image is made visible for viewing. It can be a fluorescent screen that has a phosphor layer which converts the energy of an electron beam to visible light, or a translucent or opaque screen on which the optical image is projected. It can also be a liquid-crystal display.   2. *Screen grid.*

screen grid   A grid placed between a control grid and an anode of an electron tube and usually maintained at a fixed positive potential, to reduce the electrostatic influence of the anode in the space between the screen grid and the cathode. It is also called a screen.

screen-grid modulation   Modulation produced by introducing the modulating signal into the screen-grid circuit of a multigrid tube in which the carrier is present.

screen-grid tube   An electron tube that has a screen grid.

scribe and break   The procedure for separating a processed wafer into individual device dies or IC chips. Narrow channels between individual devices are mechanically weakened by scratching them with a diamond tip (scribe), sawing with a diamond blade, or burning with a laser. The wafer is mechanically stressed and broken apart along the channels (called scribe lines), thereby separating the individual chips or dice.

scribing   The process of scratching lines on a semiconductor wafer. Later application of pressure breaks the wafer along these lines to form desired sizes of chips for semiconductor devices and integrated circuits.

scroll bar   In computer science, a graphical construct, consisting of a rectangular box and a position marker, intended for use with a mouse that positions a part of an image, table, or text in a viewing area on the computer monitor.

SCS   Abbreviation for *silicon controlled switch.*

SCSI   Abbreviation for *small computer systems interface.*

SCU   Abbreviation for *subscriber channel unit.*

S-curve   A project management concept for tracking the progress of a specific technology as a function of research and development effort or time. The plot of technological progress with respect to time has been found to form an approximate S shape. When a new technology is introduced progress is slow, but it accelerates as that technology becomes accepted by customers and the industry. However, when the technology matures its progress slows, flattening the curve.

S/D   Abbreviation for synchro-to-digital converter.

SDC   Abbreviation for *synchro-to-digital converter.*

SDCCH   Abbreviation for *stand-alone dedicated control channel.*

SDH   Abbreviation for *synchronous digital hierarchy.*

SDLC   Abbreviation for *synchronous data-link control.*

SDR   Abbreviation for *signal-to-distortion ratio.*

SDRAM   Abbreviation for *synchronous dynamic random-access memory.*

sea clutter   Clutter on a radar display due to reflection of signals from the sea. It is also called sea return and wave clutter.

seal   A joint between two pieces of glass, two pieces of metal, or glass and metal. For electron tubes the joint must be hermetically secure.

sealed crystal unit   A crystal unit whose the quartz plate is sealed in its holder, usually by a gasket under pressure, for protection against humidity or a contaminated atmosphere.

sealed tube   An electron tube that is hermetically sealed.

sealing off   Final closing of the envelope of an electron tube after evacuation.

search   1. To explore a region with radar.   2. To examine a set of items for those which have a desired property.

search coil   *Exploring coil.*

searchlighting   Projecting a radar or sonar beam continuously at a target instead of scanning the area containing the target.

searchlight sonar   A sonar system whose directional transducer concentrates the outgoing pulse of sound energy into a narrow beam and receives the echo reflected from an underwater target. The bearing of a target is determined by aiming the transducer for maximum echo strength.

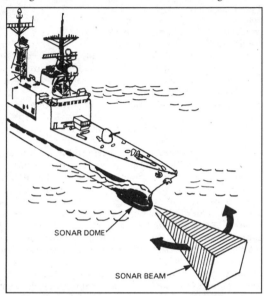

Searchlight sonar sweeps in bearing and depth.

search radar   A radar system that examines a hemispherical volume of space around the antenna with a continuously rotating mechanical antenna or a continuously swept phased-array antenna to locate targets on land, sea, or in the air. These radars typically operate in the S or L frequency bands and hand off information to higher-frequency, higher-definition, shorter-range radars for target tracking or weapon fire control. It is also called a *surveillance radar* when long-range targets are being monitored, as in the civilian air-traffic control system. See also *tracking radar.*

search receiver   A radio receiver that can be tuned over a wide frequency range for detecting and measuring RF signals transmitted by the enemy.

sea return   *Sea clutter.*

Sea Sparrow   U.S. Navy shipboard surface-to-air and surface-to-surface guided missile. Armed with an 85-lb conventional warhead and powered by solid fuel, it is usually guided to its target by reflected radar energy from the target that has been illuminated by shipboard radar. It is smaller than the *Sparrow.* See also *Sparrow.*

SECAM [derived from French for sequential with memory] A 625-line 50-field color television system originally developed in France. In this system, one color signal is transmitted on one line and the other color signal on the next line. Delay and switching circuits in the receiver combine these signals so the three types of color video information can be displayed simultaneously. Other countries now using this system include the Czech Republic, Germany, Lebanon, and Russia.

second [s]   The SI unit of time, based on the time of transition between two specific energy levels in cesium 133. This redefinition of the unit of time was approved at the Twelfth General Conference on Weights and Measures. A second was formerly $\frac{1}{86,400}$ of a mean solar day.   2. A unit of angle, equal to $\frac{1}{3600}°$.

secondary   *Secondary winding.*

secondary battery   *Rechargeable battery.*

secondary cell   *Rechargeable cell.*

secondary electron   1. An electron emitted as a result of bombardment of a material by an incident electron.   2. An electron whose motion is due to a transfer of momentum from primary radiation.

secondary-electron conduction   Movement of charges by free secondary electrons traveling in interparticle spaces of low-density materials under the influence of an externally applied electric field.

secondary-electron-conduction camera tube   A camera tube in which an electron image, generated by a photocathode, is focused on a target that has a backplate and a secondary-electron-conduction layer which provides charge amplification and storage. Image amplification is high enough for dimly illuminated scenes.

secondary-electron multiplier   *Electron multiplier.*

secondary emission   The emission of electrons from a solid or liquid as a result of bombardment by electrons or other charged particles.

secondary-emission ratio   The average number of electrons emitted from a surface per incident primary electron.

secondary grid emission   Electron emission from a grid as a direct result of bombardment of the grid surface by electrons or other charged particles.

secondary lobe   *Minor lobe.*

secondary memory   Storage that is not an integral part of the computer but is directly linked to and controlled by the computer.

secondary parameter   An additional rating or characteristic needed to evaluate the operation of a product beyond its normal limits, such as its temperature coefficient.

secondary service area   The area served by the sky wave of a broadcast station and not subject to objectionable interference. The signal is subject to intermittent variations in intensity.

secondary standard   1. A unit, as of length, capacitance, or weight, used as a standard of comparison in individual countries or localities, but checked against the one primary standard in existence somewhere. 2. A unit defined as a specified multiple or submultiple of a primary standard, such as the centimeter.

secondary surveillance radar [SSR]   The system for air-traffic control that identifies and tracks civil and military aircraft worldwide. It is a component in the FAA's *Air Traffic Control Radar Beacon System* [ATCRBS]. An SSR is typically combined with an air-traffic control radar, with its antenna mounted on top of the ATC radar antenna. The

ground station interrogates the aircraft's transponder with a narrow fan-shaped beam at 1030 MHz, and it receives replies at 1090 MHz for determining the aircraft's range and bearing. The airborne transponder returns a train of pulses that identify the aircraft and report its altitude. It was developed from the military *identification friend or foe* [IFF] system.

secondary voltage   The voltage across the secondary winding of a transformer.

secondary winding   A transformer winding that receives energy by electromagnetic induction from the primary winding. A transformer can have several secondary windings, and they can provide AC voltages that are higher, lower, or the same as that applied to the primary winding. It is also called a secondary.

secondary X-ray   Any X-ray given off by a material when irradiated by X-rays. The frequency of the secondary rays is characteristic of the material.

second breakdown   Destructive breakdown in a bipolar transistor or GTO in which structural imperfections cause localized current concentrations and uncontrollable generation and multiplication of current carriers causing heating. The reaction occurs so suddenly that the thermal time constant of the collector regions is exceeded and the transistor is irreversibly damaged. It is most likely to occur during turnoff of inductive loads. In contrast, primary or avalanche breakdown is the normal sustaining mode of a transistor and does not cause permanent damage.

second-channel attenuation   *Selectance.*

second-channel interference   Interference in which the extraneous power originates from a signal of assigned type in a channel two channels removed from the desired channel.

second detector   The detector that separates the intelligence signal from the IF signal in a superheterodyne receiver.

second-harmonic magnetic modulator   A magnetic modulator whose output frequency is twice the power-supply frequency.

second sound   A type of heat wave that carries energy but moves at very nearly the speed of sound. The heat waves are produced by turning a heater on and off. The phenomenon has been observed in photoconducting crystals of cadmium sulfide.

second-time-around echo   A radar or sonar echo received after an interval exceeding the pulse recurrence interval.

section   Each individual transmission span in a radio relay system. A system has one more section than it has repeaters.

sectionalized vertical antenna   A vertical antenna that is insulated at one or more points along its length. Reactances or driving voltages are applied across the insulated points to modify the radiation pattern in the vertical plane.

sector   In a computer magnetic memory disk, the radial division of concentric *tracks* into sections shaped like pie slices. There are 80 tracks per side and 18 sectors per track on a 3.5-in high-density floppy disk. See also *cylinder.*

sectoral horn   An electromagnetic horn that has two opposite sides parallel while the other two sides diverge.

sector display   A display that shows only a sector of the total service area of a radar system. The sector is usually selectable.

sector scan A radar scan through a limited angle, as distinguished from complete rotation.

secular variation The slow variation in the strength of the magnetic field of the earth, requiring many years for a complete cycle.

secure voice Voice communication that is scrambled or coded.

security classification The classification assigned to defense information or material to denote the degree of danger to the nation that would result from unauthorized disclosure. The usual classifications are confidential, secret, and top secret.

security clearance A clearance that permits a person to have access to classified material or information up to and including a given security classification, provided the person can establish a need-to-know.

Seebeck effect Development of a voltage due to differences in temperature between two junctions of dissimilar metals in the same circuit. It was discovered by J. T. Seebeck, German physicist, in 1821, and is also called the *thermoelectric effect.*

seed A small single crystal of semiconductor material used to start the growth of a large single crystal or boule from which semiconductor wafers are cut.

seeker A missile or other device that finds its target by seeking heat, light, radio waves, sound, or other radiation emitted by the target.

seeker antenna An antenna that is mounted in the forward part of a missile for seeking RF emissions from a target. It is a part of the closed-loop airborne guidance system, which also includes a missile radome, RF receiver, gimbal, and autopilot. It can be mechanically or electronically scanned [phased array].

segment 1. A part of a digital-computer routine short enough to be stored entirely in the internal memory yet containing all the coding necessary to call in and jump automatically to other segments of the routine. 2. An element in a light-emitting diode or liquid-crystal display, as in a seven-segment display.

seismic detector A microphone that detects acoustic waves transmitted through the earth.

seismic surveying A petroleum exploration technique based on variations in the rate of propagation of shock waves in layered subsurface media.

seismograph An instrument that records the time, direction, and intensity of earthquakes or of earth shocks produced by explosions during geophysical prospecting.

seismometer An instrument that measures earth movements or earth shocks.

selectance The reciprocal of the ratio of the sensitivity of a receiver tuned to a specified channel to its sensitivity at another channel separated by a specified number of channels from the one to which the receiver is tuned. It is generally expressed as a voltage or field-strength ratio. It is also called adjacent-channel attenuation or second-channel attenuation.

selection check A check made by a digital computer to verify that the correct register or peripheral is selected for performance of the next instruction.

selective absorption Absorption of radiation as some function of frequency.

selective calling system A radio communication system whose central station transmits a coded call that activates only the receiver to which that code is assigned.

selective diffusion Doping of isolated regions of a semiconductor material to produce individual components in an integrated circuit.

selective epitaxial growth Growth of an epitaxial layer in a semiconductor material after selective masking of the surface with an oxide. Alternatively, the epitaxial layer can be formed on the entire surface and later removed from the regions where it is not required.

selective fading Fading that differs at different frequencies in a frequency band occupied by a modulated wave, causing distortion that varies from instant to instant.

selective jamming Jamming that occurs in only a single radio channel.

selective transmission The transmission of electromagnetic energy at wavelengths other than those reflected or absorbed in a given system.

selectively doped heterostructure transistor [SDHT] A selectively doped heterostructure field-effect transistor. The acronym was originated at AT&T Bell Laboratories.

selectivity The characteristic of a receiver that determines its ability to separate a desired signal frequency from all other signal frequencies.

selectivity control A control that adjusts the selectivity of a radio receiver.

selector An automatic or other device for making connections to any one of a number of circuits, such as a selector relay or selector switch.

selector pulse A pulse that identifies for selection one event in a series of events.

selector relay A relay capable of selecting one circuit automatically from a number of circuits.

selector switch A manually operated multiposition switch. It is also called a *multiple-contact switch.*

selenium [Se] A nonmetallic element that has photosensitive properties. Its resistance varies inversely with illumination. It is also used as a rectifying layer in metallic rectifiers and its atomic number is 34.

selenium rectifier A metallic rectifier that has a thin layer of selenium deposited on one side of an aluminum plate and a conductive metal coating deposited on the selenium. Electrons flow more freely in the direction from the metallic coating to the selenium than in the opposite direction, thus giving rectifying action.

self-absorption Absorption of radiation by the material that emits the radiation, reducing the radiation level against which further shielding must be provided.

self-adapting Capable of changing performance characteristics automatically in response to the environment.

self-aligned gate [SAG] **technology** In the fabrication of gallium arsenide (GaAs) MESFETs, the process of gate metallization on the wafer before the ion implantation is performed, ensuring that the gate is properly aligned over the uniformly doped interaction region.

self-bias Base or gate bias provided automatically by the flow of electrode currents through a resistor in the emitter or base circuit or a bipolar transistor or source or gate circuit of a FET transistor. The resulting voltage drop across the resistor serves as the base or gate bias.

self-cleaning contact   *Wiping contact.*

self-demagnetization   The process by which a magnetized sample of magnetic material tends to demagnetize itself by virtue of the opposing fields created by its own magnetization. Self-demagnetization inhibits the successful recording of short wavelengths or sharp transitions in a recorded signal.

self-discharge   The loss of useful capacity of a cell or battery ion storage because of internal chemical reactions. These could be chemical evaporation or the electrolyte slowly reacting with the anode even when the cell is open circuit.

self-energy   The energy equivalent of the rest mass of a particle. The self-energy of an electron is 511 keV.

self-excited   Operating without an external source of power for excitation.

self-excited oscillator   An oscillator that depends on its own resonant circuits for initiation of oscillation and frequency determination.

self-focused picture tube   A television picture tube that has automatic electrostatic focus incorporated into the design of the electron gun.

self-generating transducer   A transducer that does not require external electric excitation to provide specified output signals.

self-guided   Directed only by built-in self-reacting devices, as in a homing missile.

self-healing capacitor   A capacitor that repairs itself after breakdown caused by excessive voltage. Air capacitors, some wet electrolytic capacitors, and some metallized plastic film capacitors have this characteristic.

self-impedance   The impedance at a pair of terminals of an antenna array or of a network when all other elements or terminal pairs are open-circuited.

self-inductance   Inductance that produces an induced voltage in the same circuit as a result of a change in current flow.

self-induction   The production of a voltage in a circuit by a varying current in that same circuit.

self-instructed carry   A carry in which information goes to succeeding locations automatically as soon as it is generated.

self-locking nut   A nut that has an inherent locking action, so it cannot readily be loosened by vibration. The nut includes a plastic washer.

self-luminous light source   A light source that consists of a radioactive nuclide such as tritium, firmly incorporated in solid and/or inactive materials or sealed in a protective envelope strong enough to prevent leakage of radioactive materials to the atmosphere. The nuclide incorporates or is surrounded by a phosphor that gives off light continuously in the presence of the radioactivity.

self-phased array   *Adaptive array.*

self-pulse modulation   Modulation by an internally generated pulse, as in a blocking oscillator.

self-quenched counter tube   A radiation-counter tube in which reignition of the discharge is inhibited by gas or other internal means.

self-quenched detector   A superregenerative detector that has a time constant of the grid leak and grid capacitor which is sufficiently large to cause intermittent oscillation above audio frequencies. This stops regeneration just before it spills over into a squealing condition.

self-rectifying tube   A hot-cathode X-ray tube that has AC anode voltage, but current flows in only one direction as long as the anode is kept cool.

self-scattering   Scattering of radiation by the material that emits the radiation, increasing the measured activity over that expected for a weightless sample.

self-starting synchronous motor   A synchronous motor provided with the equivalent of a squirrel-cage winding, to permit starting as an induction motor.

self-steering microwave array   An antenna array with electronic circuitry that senses the phase of incoming pilot signals and positions the antenna beam in their direction of arrival. It is used in earth-satellite communication.

self-supporting antenna tower   An antenna tower that requires no guy wires.

self-synchronous device   *Synchro.*

self-wiping contact   *Wiping contact.*

selsyn   [SELf SYNchronous]   *Synchro.*

selsyn generator   *Synchro transmitter.*

selsyn motor   *Synchro receiver.*

selsyn receiver   *Synchro receiver.*

selsyn system   *Synchro system.*

selsyn transmitter   *Synchro transmitter.*

SEM   Abbreviation for *scanning electron microscope.*

Sematech   An acronym for *Semiconductor Manufacturing Technology Research Consortium,* a consortium of American semiconductor manufacturers dedicated to the goal of restoring America's manufacturing leadership in semiconductors. Located in Austin, Texas, half of its annual funding is provided by its member companies and half by the federal government. Research results are transferred to member firms and to the U.S. government for both commercial and military applications.

semiactive homing   Homing of a missile that does not include the transmitter that illuminates the target. The missile contains only the receiver for energy reflected from the target.

semiactive tracking system   A tracking system that tracks a signal source normally aboard the target for other purposes, or a system that uses a ground transmitter to illuminate the target but requires no special equipment on the missile.

semiconductor   A class of materials, such as silicon and gallium arsenide, whose electrical properties lie between those of conductors (e.g., copper and aluminum) and insulators (e.g., glass and rubber). A material that exhibits relatively high resistance in a pure state and much lower resistance when it contains small amounts of certain impurities. It also denotes electronic devices made from semiconductor materials.

semiconductor capacitor   A reverse-biased PN junction device serving as a capacitor.

semiconductor detector   A nuclear-particle detector with a semiconductor sensing element. Examples include the lithium-drifted germanium and the lithium-drifted silicon detectors, both of which will detect gamma rays as well as all other types of particle.

semiconductor diode   A two-electrode semiconductor device that makes use of the rectifying properties of a junction between P- and N-type material in a semiconductor. Examples are a junction diode, or the rectifying properties of a sharp wire point in contact with a semiconductor material, as in a point-contact diode. It is also called a *crystal diode,* a *crystal rectifier,* or a *diode.*

semiconductor diode phase shifter A phase shifter that depends on the junctions of *PIN diodes* as control elements in high-power digital *phase shifters.* The PIN diode can be forward or reverse biased. The intrinsic region in the diode behaves as a lossy dielectric at microwave frequencies. The impedance presented between the network terminals can be varied with external reactive tuning elements. There can be from 10 to 16 PIN diodes per stripline module.

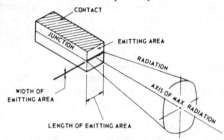

Semiconductor laser based on a PN junction. A pulsed power supply is required to produce high peak currents for injecting electrons across the junction to produce lasing action.

semiconductor laser A laser that produces stimulated emission of coherent light at a PN junction when electrons and holes are driven into the junction by carrier injection, electron-beam excitation, impact ionization, optical excitation, or other means. The most common semiconductor material used to make lasers is gallium arsenide: other materials include cadmium sulfide, lead sulfide, lead selenide, lead telluride, zinc oxide, and zinc sulfide. Output wavelengths are in the range from 0.33 to 31.2 μm. It is also called a diode laser and a laser diode.

semiconductor rectifier *Metallic rectifier.*

semiconductor relay A semiconductor device that provides the equivalent of electromagnetic relay action, such as a silicon controlled switch.

Semiconductor Research Corporation A consortium of more than 60 corporations and U.S. government agencies planning and executing applied research programs at leading U.S. universities to strengthen the competitive ability of the U.S. semiconductor industry.

semiconductor switch A transistor circuit that provides switching action comparable to that of a relay or switch.

semicustom integrated circuit An analog, digital, or mixed-signal integrated circuit that is premanufactured as an array of nondedicated gates and/or transistors that are, in later steps, connected by one or more masking steps to dedicate the device to a specific application. Device completion is done with computer-aided design (CAD) from data stored in computer memory. Two types are the gate array and the standard cell.

semiduplex operation Operation of a communication circuit with one end duplex and the other end simplex. When used in mobile systems, the base station is usually duplex and the mobile stations are simplex.

semiremote control Remote control of a radio transmitter by devices connected to but not an integral part of the transmitter.

semitone The interval between two sounds whose basic frequency ratio is approximately equal to the twelfth root of 2. The interval, in equally tempered semitones, between any two frequencies is 12 times the logarithm to the base 2 (or 39.86 times the logarithm to the base 10) of the frequency ratio. It is also called half-step.

semitransparent photocathode A photocathode in a television camera tube or phototube that emits electrons from one side when the opposite side is exposed to infrared or visible light.

sending Transmitting, as Morse code.

sending-end impedance The input impedance of a transmission line.

sensation level *Level above threshold.*

sense 1. The relation of a change in the indication of a radio navigation facility to the change in the navigation parameter being indicated. 2. To resolve a 180° ambiguity in a reading. 3. To determine the arrangement or position of a device or the value of a quantity. 4. To read punched holes in tape or cards.

sense amplifier An amplifier that detects bipolar differential-input signals from a semiconductor memory and acts as the interface between the memory and logic sections of a computer.

sense antenna An auxiliary antenna attached to a directional receiving antenna to resolve a 180° ambiguity in the directional indication. It is also called a sensing antenna.

sense indicator A flight instrument that determines whether an aircraft is flying toward or away from a VHF omnirange.

sense line An output line that connects a power supply with its load in a *remote-sensing* feedback loop to route the remote voltage (at the load) back to the power supply control feedback loop. See *remote sensing.*

sensing The process of determining the sense of an indication.

sensing antenna *Sense antenna.*

sensing element *Sensor.*

sensitive relay A relay that will operate at small currents, usually below 10 mA.

sensitivity A figure of merit that expresses the ability of a circuit or device to respond to an input quantity. Expressed as divisions per volt or ohms per volt for a measuring instrument, as spot displacement per volt of deflection voltage or ampere of deflection current for a cathode-ray tube, as output current per unit incident radiation density for a camera tube or other photoelectric device, and as microvolts of input signal when specifying minimum signal strength to which a receiver will respond.

sensitivity control A control that adjusts the amplification of RF amplifier stages in a receiver.

sensitivity-time control [STC] An automatic control circuit that changes the gain of a receiver at regular intervals to obtain desired relative output levels from two or more sequential and unequal input signals. In a loran receiver it keeps output signal amplitude essentially constant as the receiver is tuned between input signals of different strength. In a radar receiver it reduces the gain after transmission of a pulse so nearby echo signals do not overload the system, then gradually restores the gain to the maximum value required for more distant targets. It is also called amplitude balance control, anti-clutter gain control, differential gain control, gain-time control, swept gain control, temporal gain control, or time-varied gain control.

sensitometer An instrument that measures the sensitivity of light-sensitive materials.

sensitometry The measurement of the light-response characteristics of photographic film under specified conditions of exposure and development.

sensor A device that senses a change in a physical or chemical quantity and provides an electrical output. Examples include response to a specific physical stimulus such as light, sound, heat, pressure, liquid flow rate, magnetic field, or radio frequency. It might also sense the presence of a chemical vapor. Sensors include the photocell, phototransistor, strain gage, and thermistor. A piezoelectric transducer can also be a sensor.

separately excited Obtaining excitation from a source other than the machine or device itself.

separation The degree, expressed in decibels, of isolation between the left and right stereo channels. A similar rating applies for front and rear quadraphonic channels. The greater the isolation in decibels, the better the separation.

separation circuit A circuit that sorts signals according to amplitude, frequency, or some other characteristic.

separation filter A filter that separates one band of frequencies from another, as in carrier systems.

separation loss The loss in output that occurs when the surface of the coating on magnetic tape fails to make perfect contact with the surfaces of the record or reproduce head.

separator 1. A circuit that separates one type of signal from another by clipping, differentiating, or integrating action. 2. A porous insulating sheet between the plates of a battery.

septate waveguide A waveguide that contains one or more septa placed across the waveguide to control the transmission of microwave power.

septinary number A number whose quantity as represented by each figure is based on a radix of 7.

septum [plural **septa**] A metal plate placed across a waveguide and attached to the walls by conductive joints. The septum usually has one or more windows or irises that give inductive, capacitive, or resistive characteristics.

sequence 1. An arrangement of items or events according to a specified set of rules. 2. To arrange items according to a specified set of rules.

sequence control The automatic control of a series of operations in a predetermined order.

sequencer A device or circuit that determines the order in which a number of actions occur.

sequence relay A relay that opens or closes two or more sets of contacts in a predetermined sequence.

sequence weld timer A timer that controls the sequence and duration of each part of a complete resistance-welding cycle.

sequential color television A color television system in which the primary color components of a picture are transmitted one after the other. The three basic types are the line-, dot-, and field-sequential color television systems. It is also called a sequential system.

sequential control Control of a digital computer so that instructions are fed into the computer in a given sequence during the running of a problem.

sequential interlace Television interlace in which the raster lines of one field fall directly under the corresponding lines of the preceding field.

sequential lobing A radar direction-finding technique in which returned signals are received with two partly overlapping antenna lobes whose output is compared in phase or power to obtain a sequential measure of angular displacement of the echo source from the antenna centerline.

sequential logic Logic whose outputs are dependent on the input states, delays encountered in the logic path, the presence of a discrete timing interval, and the previous state of the logic array. In contrast, combinatorial logic depends only on input states and delays.

sequential scanning *Progressive scanning.*

sequential system *Sequential color television.*

Sergeant A U.S. army surface-to-surface guided missile that is controlled by inertial guidance. Its range is about 75 nautical miles (139 km). It can carry either conventional or nuclear warheads.

serial A reference to to time-sequential transmission of, storage of, or logic operations on the parts of a word in a digital computer, using the same facilities for successive parts.

serial access A memory characteristic wherein all the bits of a byte or word are entered sequentially at a single input or retrieved sequentially from a single output.

serial adder A computer logic circuit that adds two binary numbers in pairs of bits, starting with the least significant bits and handling the carries while progressing step by step to the most significant bits.

serial by bit Digital-computer memory in which the individual bits that make up a computer word appear in time sequence.

serial by character A digital-computer memory in which the characters for coded-decimal or other nonbinary numbers appear in time sequence.

serial by word Digital-computer memory in which the words within a given group appear one after the other in time sequence.

serial digital computer A digital computer that handles the digits serially, although the bits that comprise a digit might be handled either serially or in parallel.

serial input/output [I/O] A type of interconnection in which information is transferred one bit at a time. The most common serial I/O hardware schemes are the RS-232C and current loop.

serial line Internet protocol/point-to-point protocol Two protocols that allow dial-in access to the *Internet* through a serial *modem* link.

serial operation The flow of information through a computer in time sequence, only one digit, word, line, or channel at a time.

serial-parallel A combination of serial and parallel, such as serial by character and parallel by bits comprising the character.

serial printer An electromechanical printer that has type bars, type balls, daisy wheels, or other mechanisms for moving a desired character into printing position and producing an image of that character on paper with a ribbon or other means. Either the paper or the printing device moves step by step to successive positions for printing one character at a time. It is also called a *character printer.*

serial programming Programming in which only one operation is scheduled at one time.

serial storage architecture [SSA] A serial interface for computer networks.

serial transfer Transfer of the characters of an element of information in sequence over a single path in a digital computer.

series 1. An arrangement of circuit components end to end to form a single path for current. 2. The indicated sum of a set of terms in a mathematical expression, as in an alternating series or an arithmetic series.

series circuit A circuit that has all parts connected end to end to provide a single path for current.

series coil The coil that carries the main current in a rotating machine or other device. The shunt coil is connected across the line and usually carries only a small current.

series connection A connection that forms a series circuit.

series element A two-terminal element connected to complete the only path existing between two nodes of a network. Any mesh including one series element must include all the other series elements of the mesh.

series excitation A motor or generator characteristic for obtaining field excitation by allowing the armature current to flow through the field winding.

series-fed vertical antenna A vertical antenna that is insulated from the ground and energized at its base. It is also called an end-fed vertical antenna.

series feed The application of direct voltage to the anode of a tube, the collector of a transistor, or the drain of a FET through the load that is carrying the output signal current. See also *shunt feed*.

series loading Loading in which reactances are inserted in series with the conductors.

series motor A commutator-type motor that has armature and field windings in series. Its characteristics are high starting torque, variation of speed with load, and dangerously high speed on no-load. It is also called a series-wound motor.

series-parallel switch A switch that changes the connections of lamps or other devices from series to parallel, or vice versa.

series peaking The use of a peaking coil and resistor in series as the load for a video amplifier to produce peaking at some desired frequency in the passband. It can compensate for previous loss of gain at the high-frequency end of the passband.

series regulator A transistor in linear power supplies that is connected in series with the load to achieve a constant voltage across the load. Feedback action on the transistor changes its voltage drop, as required, to maintain the constant DC output voltage. It can be part of the stand-alone power supply or part of the host equipment. See *linear regulator*.

series resonance Resonance in a series resonant circuit when the inductive and capacitive reactances are equal at the frequency of the applied voltage. The reactances then cancel each other, reducing the impedance of the circuit to a minimum purely resistive value. Signal current is then a maximum, and the signal voltage developed across either the coil or capacitor can be several times the voltage applied to the combination.

series resonant circuit A resonant circuit that has a capacitor and coil in series with the applied AC voltage.

series T junction A junction in which the impedance of the branch waveguide is predominantly in series with the impedance of the main waveguide.

series-wound motor *Series motor.*

serrated pulse A pulse that has notches or sawtooth indentations in its waveform.

serrated rotor plate *Slotted rotor plate.*

serrated vertical pulse A vertical synchronizing pulse that is broken up by five notches which extend down to the black level of a television signal. It gives six component pulses, each lasting about 0.4 line to keep the horizontal sweep circuits in step during the vertical sync pulse interval.

serrodyne A phase modulator based on the transit-time modulation of a traveling-wave tube or klystron.

server A computer, generally a mainframe or minicomputer/workstation in a network, with large memory capacity assigned to the storage of programs and data not stored in the memories of typical network computers or terminals.

service area The area that is effectively served by a given radio or television transmitter, navigation aid, or other type of transmitter. It is also called coverage.

service band A band of frequencies allocated to a given class of radio service.

service life The length of time that a battery or other active device will provide specified performance under specified conditions of use.

service test A test made under simulated or actual conditions of use to determine the characteristics, capabilities, and limitations of a product.

serving A covering, such as thread or tape, that protects a winding from mechanical damage.

servo *Servomotor.*

servoamplifier An amplifier in a servosystem.

servomechanism A mechanical component in a *servosystem*.

servomotor The electric, hydraulic, or other type of motor that serves as the final control element in a servosystem. It receives power from the amplifier element and drives the load with a linear or rotary motion. It is also called a servo.

servo multiplier An electromechanical multiplier that has one variable which positions one or more ganged potentiometers across which the other variable voltages are applied.

servosystem A *closed-loop control system* for the control of speed, position, or both in machine tools, robots, military fire-control systems, and other electromechanical systems driven by electrical, hydraulic, or pneumatic motors.

sesqui-sideband transmission Transmission of a carrier modulated by one full sideband and half of the other sideband.

set 1. A radio or television receiver. 2. A combination of units, assemblies, and parts connected or otherwise joined together to perform an operational function, such as a radar system. 3. To place a memory device such as a flip-flop in a prescribed state, such as placing it in the 0 state or in the 1 state.

set point The value selected that is to be maintained by an automatic controller.

set-point control   A digital technique for stopping an output device at a series of programmed points called set points, entered manually or automatically under the control of a computer or tape reader.

set pulse   A drive pulse that sets a one-shot multivibrator.

set-reset flip-flop   *RS flip-flop.*

set-reset-trigger flip-flop   *RST flip-flop.*

setscrew   A small headless machine screw with a point at one end and a recessed hexagonal socket or a slot at the other end. It holds knobs or gears on a shaft.

settling time   1. The time elapsed between the application of a perfect step input to an operational amplifier and the time when amplifier output has entered and remained within a specified error band that is usually symmetrical about the final value. Settling time includes the time required for the amplifier to slew from the initial value, recover from the slew-rate-limited overload, and settle within the error band.   2. *Correction time.*

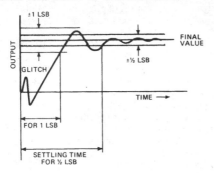

Settling the time for digital-to-analog converter is specified as the time required for the output to settle within a range of ½ or 1 least significant bit of its final value after switching.

set-top box   A circuit that is connected to a television receiver or video cassette recorder (VCR) to decode cable TV programs encoded by the cable TV service provider. It can also compensate for inadequate tuning or extend the tuning range of the TV set. It might also provide for interaction with the program or permit direct selection of "pay for play" cable programming with a keypad.

setup   The ratio between reference black level and reference white level in television, both measured from blanking level. It is expressed as a percentage.

SEU   Abbreviation for *single-event upset.*

sexadecimal notation   *Hexadecimal notation.*

sferics [coined from atmospherics]   *Atmospheric interference.*

sferics set   An electronic system that detects, analyzes, and determines the position of electromagnetic disturbances generated by any atmospheric phenomena.

S/H   Abbreviation for *sample-and-hold.*

shaded-pole motor   A single-phase induction motor that has one or more auxiliary short-circuited windings acting on only a portion of the magnetic circuit. Generally the winding is a closed copper ring imbedded in the face of a pole. The shaded pole provides the required rotating field for starting purposes.

shading   A variation in brightness over the area of a reproduced television picture, caused by spurious signals gener-

ated in a television camera tube during the retrace intervals. Those spurious signals are generally due to redistribution of secondary electrons over the mosaic in a storage-type camera tube, and vary from scene to scene as background illumination changes.

shading coil   *Shading ring.*

shading generator   One of the signal generators in a television transmitter that generates waveforms that are 180° out of phase with the undesired shading signals produced by a television camera. An operator watches the picture on a monitor and adjusts the controls of the shading generators as required to give essentially uniform scene brightness.

shading ring   1. A heavy copper ring placed around the central pole piece of an electrodynamic loudspeaker to serve as a shorted turn that suppresses the hum voltage produced by the field coil.   2. The copper ring in a shaded-pole motor that produces a rotating magnetic field for starting purposes. It is also clamped around part of the core of an AC relay to prevent contact chatter. It is also called a shading coil.

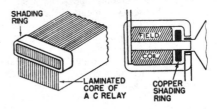

Shading ring on a relay and on an excited-field loudspeaker.

shadow factor   The ratio of the electric field strength resulting from propagation of waves over a sphere to that resulting from propagation over a plane under comparable conditions.

shadow mask   A thin perforated metal mask mounted just behind the phosphor-dot faceplate in a three-gun color picture tube. The holes in the mask are positioned to ensure that each of the three electron beams strikes only its intended color phosphor dot. It is also called an aperture mask.

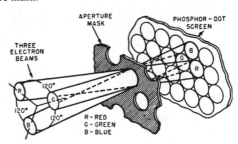

Shadow-mask details showing positions of color phosphor dots with respect to apertures in mask.

shadow-mask color picture tube   A three-gun color picture tube that has a shadow mask.

shadow RAM   A semiconductor random-access memory that copies (shadows) data taken from any memory location so that, if necessary, the data can be reinstated.

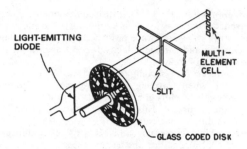

Shaft-position encoder with a photoelectric readout of transparent and opaque segments produced photographically on a glass disk, to give a binary output directly.

shadow region The region in which received field strength is so reduced by some obstruction that effective reception of signals or radar detection of objects is normally improbable.

shaft-position encoder An analog-to-digital converter whose exact angular shaft position is sensed and converted to digital form. In one type, a continuous transducer such as a potentiometer, resolver, or synchro converts the rotary motion first to a proportional electric quantity or time interval. This in turn is converted to digital form by a voltage or time coder. In another type, conversion to digital form is accomplished directly by a disk, mounted on the shaft, that contains various combinations of opaque segments from which a unique digital output is derived for each increment of shaft position. See also *encoder*.

shaker An electromagnetic table capable of imparting known and usually controlled vibratory acceleration to a given object. It is also called a shake table. It tests the ability of products and systems to withstand shock and vibration.

shake table *Shaker.*

Shannon limit The best possible signal-to-noise ratio that can be obtained with the best modulation technique, based on Shannon's theorem relating channel capacity to signal-to-noise ratio.

shape coding Special shapes for control knobs to permit radar operators to set controls and monitor their positions by sense of touch under reduced lighting conditions.

shaped-beam antenna A unidirectional antenna whose major lobe differs significantly from that provided by an aperture which gives uniform phase. It is also called a phase-shaped antenna.

shape factor *Form factor.*

shaping network *Corrective network.*

sharp tuning The property of high selectivity, and therefore the ability to respond only to a desired narrow range of frequencies.

sheath 1. The metal wall of a waveguide. 2. A protective outside jacket on a cable. 3. A space charge formed by ions near an electrode in a gas tube.

sheet grating A grating consisting of thin longitudinal metal sheets that extend inside a waveguide a distance of about one wavelength. It suppresses undesired modes of propagation.

sheet resistivity The resistivity of a sheet of material, measured in ohms per square. The rating depends only on the

Shape coding of control knobs for touch recognition in reduced light.

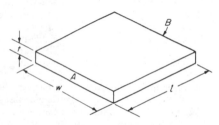

Sheet resistivity is measured between surfaces A and B of square sheet of material.

resistivity and thickness of the material. It is independent of the size of the square.

shelf life The length of time that any active or passive component can remain as effective as a newly produced product and not show signs of deterioration in either performance or appearance. Product life can be shortened by exposure to high or low temperature for prolonged periods (causing freezing or boiling of internal liquids), the presence of airborne contaminants such as sulfur dioxide or salt spray, high humidity, ultraviolet radiation, breakdown of chemical composition, or evaporation of volatile chemical solvents.

shell A group of electrons that form part of the outer structure of an atom and have a common energy level.

SHF Abbreviation for *superhigh frequency*.

shield A metallic housing placed around a circuit or component to suppress the effect of an electric or magnetic field within or beyond definite regions.

shielded-arc welding Arc welding in which the metal electrode is coated with a flux that produces an envelope of protective inert gas.

shielded cable A cable that has a conducting envelope around its insulated conductors. Braided copper wire is most often used for this purpose. See also *shielded wire*.

shielded line *Shielded transmission line.*

shielded pair   A two-wire transmission line surrounded by a metallic sheath.

shielded room   A room that has been made free from electromagnetic interference by the installation of appropriate shielding to the floor, walls, and ceiling and by suppressing interference which might enter over power lines.

shielded transmission line   A transmission line whose elements essentially confine propagated electromagnetic energy to a finite space inside a conducting sheath. It prevents the line from radiating radio waves.

shielded wire   An insulated wire covered with a metal shield, usually made of tinned, braided copper wire. See also *shielded cable*.

shielded X-ray tube   An X-ray tube enclosed in a grounded metal container except for a small window through which X-rays emerge.

shielding   A conductive covering or enclosure that blocks the passage of radio-frequency interference (RFI), electrostatic discharge (ESD), or magnetic fields. RFI and ESD can be blocked by metal screening, punched metal, or sheet metal, and magnetic fields can be blocked by sheets of Mumetal.

shielding metal   A metal that has high magnetic saturation and good attenuation for electromagnetic and electrostatic fields, to protect them from stray effects. Examples include nickel alloys such as Mumetal, nickel-iron alloys, silicon iron or soft cold-rolled steel coated with ferrite powders, and sandwiches of two different shielding metals or alloys.

shift   Displacement of an ordered set of characters one or more places to the left or right in a digital computer. If the characters are the digits of a numerical expression, a shift could be equivalent to multiplying by a power of the base.

shift pulse   A drive pulse that initiates shifting of characters in the register of a digital computer.

shift register   A computer circuit that converts a sequence of input signals into a parallel binary number or vice versa, by moving stored characters to the right or left.

shift-register memory   A memory in which data is entered at one end and must be shifted stage by stage through the entire memory before becoming available again. The data can then be either recirculated or removed.

Shillelagh   An Army gun-launched missile that has a shaped-charge warhead which destroys such hard targets as tanks, bunkers, and buildings. The missile is guided to its target from its airborne or ground assault vehicle by an infrared command link.

shimming   Adjustment of the strength of a magnetic field by thin spacers, shims of soft iron, or compensating coils.

ships' station identification number [SSID]   A ship's identification number for tracking purposes on file with the U.S. Coast Guard and the FCC.

SHOALS   Abbreviation for *scanning hydrographic operational airborne lidar survey*.

shock excitation   Excitation produced by a voltage or current variation of relatively short duration. It initiates oscillation in the resonant circuit of an oscillator.

Shockley diode   A PNPN device that switches rapidly into its conducting state when a critical voltage is reached. Conduction continues until the anode voltage drops below a specified minimum value that is called the turnoff voltage. In its blocking state, the diode impedance is very high.

shock mount   A mount for sensitive equipment to reduce or prevent the transmission of shock motion to the equipment.

shock wave   A sound wave produced by a sudden change in pressure and particle velocity. It can travel faster than the velocity of sound.

shore effect   The bending of radio waves toward the shoreline when traveling over water near a shoreline, caused by the slightly greater velocity of radio waves over water than over land. This effect can cause errors in a radio direction finder's indications.

shore-to-ship communication   Radio communication between a shore station and a ship at sea.

short   *Short-circuit.*

short-baseline system   A trajectory-measuring system that uses a baseline whose length is very small compared with the distance to the object being tracked.

short circuit   A low-resistance connection across a voltage source or between both sides of a circuit or line, usually accidental and usually resulting in excessive current flow that may cause damage. It is also called a *short* (slang).

short-circuit impedance   The impedance of a network when a specified pair or group of its terminals is short-circuited.

short-circuit parameter   One of a set of four transistor equivalent-circuit parameters that specify transistor performance when the input and output voltages are chosen as the independent variables. Two of the four measurements require short-circuiting of the input for alternating current.

short-path communication   Radio communication resulting from pointing a beam antenna in the direction indicated by the shorter of the two great circle paths.

short-range radar   Radar whose maximum line-of-sight range, for a reflecting target that has $1m^2$ of area perpendicular to the beam, is between 5 and 15 mi (8 and 24 km).

short-time rating   A rating that defines the load which a machine, apparatus, or device can carry for a specified short time.

shortwave [SW]   A general term applied to a wavelength shorter than 200 m, corresponding to frequencies higher than the highest broadcast-band frequency.

shortwave antenna   An antenna that receives frequencies above the broadcast band, in the range from about 1.6 to 30 MHz.

shortwave converter   A converter positioned between a receiver and its antenna system to convert incoming high-frequency signals to a lower carrier frequency to which the receiver can be tuned. A converter usually contains a local oscillator and a mixer, as in a superheterodyne receiver.

shortwave receiver   A radio receiver that tunes in stations in the range from about 1.6 to 30 MHz or some portion of that range.

shortwave transmitter   A radio transmitter that radiates shortwaves, generally for communication purposes or for international broadcasting.

shot effect   *Shot noise.*

shot noise   Noise voltage developed in a thermionic tube because of the random variations in the number and velocity of electrons emitted by the heated cathode. The effect causes sputtering or popping sounds in radio receivers and snow effects in television pictures. It is also called Schottky noise or shot effect.

shunt   1. A precision low-value resistor placed across the terminals of an ammeter to increase its range by allowing a known fraction of the circuit current to bypass the meter.   2. A piece of iron that provides a parallel path for magnetic flux around an air gap in a magnetic circuit.   3. To place one part in parallel with another.   4. *Parallel.*

shunted monochrome   A color television technique in which the luminance or monochrome signal is shunted around the chrominance modulator or chrominance demodulator.

shunt-excited   The connection of field windings across the armature terminals, as in a DC generator.

shunt-fed vertical antenna   A vertical antenna connected to ground at the base and energized at a point suitably positioned above the grounding point.

shunt feed   The application of direct operating voltage to the anode of a tube, the collector of a transistor, or drain of a FET through a choke coil that is parallel to and therefore separated from the signal circuit. Thus, only the signal current flows through the load. It is also called *parallel feed.* See also *series feed.*

shunt loading   Loading in which reactances are connected between the conductors of a transmission line.

shunt neutralization   *Inductive neutralization.*

shunt peaking   The use of a peaking coil in a parallel circuit branch that connects the output load of one stage to the input load of the following stage. It compensates for high-frequency loss caused by the distributed capacitances of the two stages.

shunt regulator   A transistor in a linear power supply that acts as a regulator when connected in parallel with the load to permit a constant voltage across the load. It can be part of the stand-alone power supply or part of the host equipment circuit.

shunt T junction   A waveguide junction in which the impedance of the branch guide is predominantly in parallel with the impedance of the main waveguide.

shunt-wound   The connection of an armature and field windings in parallel, as in a DC generator or motor.

shutter   A device that prevents light from reaching the light-sensitive surface of a television camera or other light-sensitive object except during the desired period of exposure.

SI   Abbreviation for *Système International d'Unités,* the French equivalent of *International System of Units.*

SIC   Symbol for *silicon carbide.*

SID   Abbreviation for *system identification.*

sideband   A band of frequencies on each side of the carrier frequency of a modulated radio signal, produced by modulation. The upper sideband contains the frequencies that are the sums of the carrier and modulation frequencies, and the lower sideband contains the difference frequencies.

sideband attenuation   Attenuation in which the transmitted relative amplitude of some sideband component of a modulated signal is smaller than that produced by the modulation process.

sideband interference   *Adjacent-channel interference.*

sideband power   The power contained in the sidebands. A receiver responds to sideband power rather than carrier power when it is receiving a modulated wave.

sideband-reference glide slope   A modification of a null-reference instrument landing system. The upper or sideband antenna is replaced by two antennas that are lower in height and fed out of phase, so a null is produced at the desired glide-slope angle. It reduces unwanted reflections of signals from rough terrain near the approach end of the runway.

sideband splash   *Adjacent-channel interference.*

sideband suppression   Removal of the energy of one sideband from the spectrum of a modulated carrier.

side frequency   One of the frequencies of a sideband.

side lobe   *Minor lobe.*

side-lobe echo   A radar echo due to a side lobe of the radar beam.

side-looking airborne radar   [SLAR]   A high-resolution airborne radar that has antennas aimed to the right and left of the flight path. It can provide high-resolution strip maps with photographlike detail. Another use is mapping unfriendly territory while flying along its perimeter at altitudes up to about 70,000 ft (21 km).

side-looking sonar   Sonar whose beam is wide vertically but narrow in the horizontal plane. It gives a map of acoustic scattering from the ocean bottom on a display or recording paper as the sonar is towed at a desired position above the bottom or mounted on the hull of a moving ship.

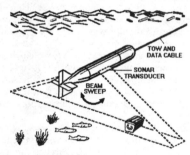

Side-looking sonar scans the sea floor as it is towed by a surface vessel.

sidetone   The sound fed back to the receiver as a result of speaking into a microphone, a consequence of the 2-to-4 wire conversion system. It is important that the speaker be able to speak properly into the handset to avoid this problem.

Sidewinder   An air-to-air guided missile powered by solid fuel and guided to the target by heat-seaking sensors that track an enemy aircraft's afterburners or missile exhaust outlets.

siemens   [S]   The SI unit of conductance, replacing the mho. A resistance or impedance of 1 Ω is equal to a conductance or admittance of 1 S.

SIGINT   Abbreviation for *signal intelligence.*

sign   A symbol (+ or −) used with a numerical quantity to indicate whether the quantity is above or below zero.

signal   Any variation in an electrical current, visible or nonvisible light, audible or ultrasonic energy that conveys information. Signals can be coded in frequency, phase, or amplitude to separate them from unwanted noise.

signal averaging   Analysis of a repetitive voltage waveform, often buried in noise, by slicing it into small segments under control of a sampling circuit and storing the sampled amplitudes. Each subsequent signal slice is added to the store for its point in the cycle, to reinforce the signal of interest at an arithmetic rate while noise and other signals are added at an RMS rate. The result is continuous improvement in signal-to-noise ratio up to the limit of the instrumentation and the stability of the repetitive waveform.

signal comparator   A circuit that correlates information from two or more signals.

signal conditioner   A circuit that shapes or adapts a signal to the requirements of a data-transmission line.

signal contrast   The ratio in decibels between the white and the black signals in facsimile.

signal data converter   A circuit that converts a data-modulated signal from one form to another.

signal distance   The number of digit positions in which the corresponding digits of two binary words of the same length are different.

signal frequency shift   The numerical difference between the frequencies corresponding to white and black signals at any point in a frequency-shift facsimile system.

signal generator   A test instrument that can be set to generate an unmodulated or tone-modulated sinusoidal RF signal voltage or an AF voltage at a known frequency and output level. It is used for aligning or servicing electronic equipment. Signal generators that are capable of producing square and triangular waveforms are generally called function generators.

signaling   1. The act of indicating to the receiving end of a communications circuit that intelligence is to be transmitted.   2. The transmission of control or status information between telephone switching systems that takes the form of dedicated bits or channels of information inserted on trunks with voice data.

signaling channel   A tone channel for signaling purposes.

signaling communication   One-way communication from a base station to a mobile receiver to actuate a signaling device in the mobile unit or to communicate information to the desired mobile unit.

signaling key   *Key.*

signal intelligence [SIGINT]   A combination of communication and electronic intelligence.

signal level   The difference between the level of a signal at a point in a transmission system and the level of an arbitrarily specified reference signal. For audio signals the difference is conveniently expressed as a ratio in decibels since the reference level is then 0 dB.

signal light   A light specifically designed for the transmission of code messages by visible light rays that are interrupted or deflected by electric or mechanical means.

signal output current   The absolute value of the difference between output current and dark current in a camera tube or phototube.

signal plate   The metal plate that backs up the mica sheet which contains the mosaic in one type of cathode-ray television camera tube. The capacitance that exists between this plate and each globule of the mosaic is acted on by the electron beam to produce the television signal.

signal processing   The processing of signals from physical sensors; radio-frequency, X-ray, ultraviolet, sonic, and ultrasonic transmitters; and even natural sources to filter noise,

improve the signal-to-noise ratio, amplify, or otherwise improve signals for reception, retransmission, or conversion to another format. Filters and analog-to-digital, and digital-to-analog converters are examples of signal processors.

signal regeneration   The process of demodulating a received signal to recover its baseband data (thus removing received noise, but creating bit errors) and remodulating the baseband signal into a carrier for retransmission.

signal-separation filter   A bandpass filter that selects the desired subcarrier channel from a composite FM signal.

signal-shaping network   A network inserted in a circuit, usually at the receiving end, to improve the waveform of the code signals transmitted.

signal strength   The strength of the signal produced by a radio transmitter at a particular location, usually expressed as microvolts or millivolts per meter of effective receiving antenna height.

signal-strength meter   A meter that is connected to the automatic volume control circuit of a communication receiver and calibrated in decibels or arbitrary S units to read the strength of a received signal. It is also called an S meter or S-unit meter.

signal-to-distortion ratio [S/D]   The ratio of the input signal to the level of all components that are present when the input signal (usually a 1.020-kHz sinusoid) is eliminated from the output signal, usually by filter.

signal-to-noise [S/N]   A measure of useful signal strength, generally expressed as a ratio.

signal-to-noise ratio [SNR]   The ratio of the amplitude of a desired signal at any point to the amplitude of noise signals at that same point. It is expressed in decibels. The peak value relates to pulse noise, and the RMS value relates to random noise.

signal tracer   A test instrument that facilitates signal-tracing.

signal-tracing   A servicing technique in which the progress of a signal is traced through each stage of a receiver to locate the faulty stage.

signal voltage   The effective (RMS) voltage value of a signal.

signal wave   A wave whose characteristics permit some intelligence, message, or effect to be conveyed. It is also called a *signal.*

signal-wave envelope   The contour of a signal wave that consists of a series of RF cycles.

signal winding   The control winding to which the control signals are applied in a saturable reactor.

signature   The characteristic pattern of radio emission as displayed by detection and classification equipment.

sign digit   A character that designates the algebraic sign of a number in a digital computer. It is usually a single bit (0 or 1) in a preallocated sign position.

significant digit   A digit that appears in the coefficient of a number when the number is written as a coefficient between $1.000\ldots$ and $9.999\ldots$ times a power of 10. Thus 0.009407, which is equal to 9.407×10^{-3}, has four significant digits.

sign position   The position at which the sign of a number is located.

silent zone   *Skip zone.*

silica   *Silicon dioxide.*

silica gel   A chemically inert and highly hygroscopic form of hydrated silica that can absorb water and vapors of solvents in coaxial lines, waveguides, and other nonevacuated enclosures.

silicide A metallically conducting compound formed of a transition metal such as cobalt, chromium, molybdenum, nickel, and silicon.

silicide resistor A thin-film resistor made with a silicide of molybdenum or chromium, deposited by DC sputtering in an integrated circuit when radiation hardness or high resistance values are required. Silicide films in use include $MoSi_2$, $CrSi_2$, and Si-Cr.

silicon [Si] A nonmetallic element that is abundantly available in the form of sand and SiO_2 (glass). It is the 14th element (atomic number 14) in the periodic table, with an atomic weight of 28.09. It has a diamond crystal lattice, a density of 2.328 g/cm₃, and a melting point of 1415°C. It can be processed at moderate temperatures, and its natural oxide (SiO_2), is stable. It can improve the magnetic properties of iron or steel for making magnetic cores.

silicon anodization An anodizing process that isolates the elements of an integrated circuit. It improves performance by lowering the capacitance between elements and increasing the gain and speed of transistors. Applications include I^2L and T^2L circuits.

silicon capacitive absolute pressure sensor [SCAP] A device designed to measure barometric or manifold absolute pressure.

silicon capacitor A capacitor that has a pure silicon crystal slab as the dielectric. A silicon capacitor can have high Q at frequencies up to 5 GHz in low-voltage circuits. When the crystal is grown to have a P zone, a depletion zone, and an N zone, the capacitance varies with the externally applied bias voltage, as in a varactor. The higher the bias voltage of this voltage-controlled capacitor, the lower its capacitance.

silicon carbide [SiC] A semiconductor material used to make light-emitting diodes that produce blue light.

silicon controlled rectifier [SCR] A four-layer, three-terminal (anode, cathode, and gate) PNPN semiconductor device that is normally an open switch in both directions. When a pulse is applied to the gate electrode, anode-to-cathode current is initiated as in a conventional rectifier. Once turned on, it cannot be turned off by removing the gate voltage. But it can be turned off by removing the anode voltage or waiting until the waveform across the device passes the zero level.

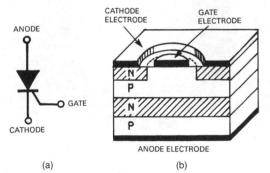

(a)   (b)

Silicon controlled rectifier (SCR) is controlled by its gate (*a*). It has four layers and three electrodes (*b*).

silicon controlled switch [SCS] A four-terminal switching device that has four semiconductor layers, all of which are accessible. It can function as a silicon controlled rectifier, complementary silicon controlled rectifier, gate-turnoff switch, or conventional silicon transistor.

silicon-diffused epitaxial mesa transistor A silicon transistor that has high voltage and power ratings combined with low storage time and low saturation voltage.

silicon dioxide [SiO_2] A crystalline material that has excellent insulating properties. It is an insulating layer in some semiconductor devices. It can be selectively etched and doped to produce a variety of components in integrated circuits. It is also called silica. Quartz is a form of silicon dioxide.

silicone A polymeric organosilicon compound that has excellent insulating, lubricating, water-resisting, and heat-resisting properties. Silicone grease improves heat transfer from a power transistor or other semiconductor device to a heatsink.

silicon epitaxial planar transistor A silicon transistor made by epitaxial processing combined with gold doping by a process that permits making thousands of transistors on one silicon wafer at one time.

silicon gate A gate made from polycrystalline silicon rather than a metal layer. It is used in MOS and CMOS technology.

silicon-gate-controlled AC switch A thyristor that can be gate-triggered from a blocking to a conducting state for either polarity of the applied voltage. It is used for phase-control applications such as lamp dimming and temperature control.

silicon-gate transistor A field-effect transistor whose gate is a film of silicon.

silicon image sensor A solid-state television camera that has a charge-coupled-device as a replacement for a vidicon or other camera tube. The image is focused on an array that can consist of 163,840 picture elements, charging each element in proportion to the light falling on it. The charges are removed from the elements electronically and passed on to the camera output as standard television signals.

silicon imaging device A solid-state industrial television camera that uses charge-coupled-device technology to form individual light-sensitive elements. One version has an array of 512 × 320 elements (a total of 163,840 elements) producing standard 525-line video output within a 3-MHz bandwidth.

silicon nitride [Si_3N_4] A material deposited as a primary passivating layer in some semiconductor devices and integrated circuits to protect the active devices from ionic contamination and mechanical damage.

silicon-on-insulator [SOI] A composite structure consisting of an active layer of silicon deposited on an insulating substrate. Examples of insulators include sapphire (as in silicon-on-sapphire), silicon dioxide, silicon nitride, and even an insulating form of silicon. The ICs formed in the active silicon layer might have additional radiation hardness, speed, and higher operating temperatures.

silicon-on-sapphire [SOS] A form of silicon-on-insulator in which a layer of silicon is epitaxially grown on a synthetic sapphire wafer. Specific regions etched away between individual transistors form CMOS logic. Each MOS transistor is totally isolated from the other transistors. See *silicon-on-insulator.*

silicon rectifier   A semiconductor rectifier whose rectifying action is provided by an alloy junction formed in a high-purity silicon wafer.

silicon resistor   A resistor that uses silicon semiconductor material as a resistance element to obtain a positive temperature coefficient of resistance which does not appreciably change with temperature. It is a temperature-sensing element.

silicon solar cell   A solar cell that consists of P and N silicon layers placed one above the other to form a PN junction where radiant energy is converted into electricity. Theoretical maximum efficiency is 22%, and actual efficiencies better than 11% have been achieved. It is a power source for satellite instrumentation, portable radios, and calculators.

silicon steel   An alloy steel that contains 3 to 5% silicon and has desirable magnetic qualities for iron cores of transformers and other AC devices.

Silsbee effect   The ability of an electric current to destroy superconductivity by the magnetic field that it generates, without raising the cryogenic temperature.

silver [Ag]   A precious-metal element that has better electric conductivity than copper. It is used to make contact points of relays and switches and as a plating on electronic components because it does not readily corrode. It has an atomic number of 47 and an atomic weight of 107.87.

silvered mica capacitor   A mica capacitor whose mica dielectric sheets are coated with a silver film to eliminate the need for interleaving conductive metal foil electrodes between the mica layers.

silver-oxide cell   A primary low-voltage power cell that is typically made as a *button cell* for powering watches, timers, hearing aids, and other low-power portable circuits. It has a nominal voltage of 1.6 V and an energy density of 120 Wh/kg or 500 Wh/L. It is also called a *zinc-silver-oxide cell* and a *silver-zinc cell*.

silver solder   A solder composed of silver, copper, and zinc with a melting point that is lower than silver but higher than lead-tin solder.

silver-zinc cell   *Silver-oxide cell.*

silver-zinc battery   A battery consisting of silver-zinc cells.

SIM   Abbreviation for *subscriber identity module.*

SIMM   Abbreviation for *single-in-line-memory module.*

simple harmonic wave   A wave whose amplitude at any point is a simple harmonic function of time.

simple sound source   A source that radiates sound uniformly in all directions under free-field conditions.

simple tone   1. A sound wave whose instantaneous sound pressure is a simple sinusoidal function of time.   2. A sound sensation characterized by singleness of pitch. It is also called a pure tone.

simplex   *Simplex operation.*

simplex operation   A method of radio operation in which communication between two stations takes place in only one direction at a time. This includes ordinary transmit-receive operation, press-to-talk operation, voice-operated carrier, and other forms of manual or automatic switching from transmit to receive. It is also called simplex.

simulation   The use of a computer and software that has been designed to model a device or system. Programs have been written to model electronic devices such as transistors, digital and analog integrated circuits, and complete electronic circuits. Large dynamic systems such as nuclear power plants, submarines, aircraft, automobiles, and robots have been simulated for engineering analysis, performance evaluation, and operator training.

simulator   1. A computer or other equipment that simulates a desired system or condition and shows the effects of various applied changes, such as in flight simulation.   2. Software that can be used with one computer system to make it execute programs written for another computer system. In contrast, an emulator requires added hardware and/or microprograms along with software to achieve this result.

simulcast   A program broadcast simultaneously by two different types of stations, as by radio and television stations or by FM and AM broadcast stations.

simultaneous color television   A color television system in which the phosphors for the three primary colors are excited at the same time, not one after another. The shadow-mask color picture tube gives a simultaneous display.

simultaneous lobing   A radar direction-finding technique in which the signals received by two partly overlapping antenna lobes are compared in phase or power to obtain a measure of the angular displacement of a target from the antenna's center line.

Si₃N₄   Symbol for *silicon nitride.*

sinad ratio   The ratio in decibels of signal-plus-noise-plus-distortion to noise-plus-distortion at the output of a mobile radio receiver for a modulated-signal input.

SINCGARS   Abbreviation for *single-channel ground and airborne radio system.*

sine-cosine encoder   A shaft-position encoder that has a special type of angle-reading code disk which gives an output that is a binary representation of the sine of the shaft angle.

Sine-cosine encoder disk.

sine-cosine generator   *Resolver.*

sine potentiometer   A potentiometer whose DC output voltage is proportional to the sine of the shaft angle. It can serve as a resolver in computer and radar systems.

sine wave   A wave whose amplitude varies as the sine of a linear function of time.

sine-wave clipper   A clipper circuit that cuts off the top of a sine wave so it resembles a square wave.

sine-wave modulation   A form of modulation in which the envelope of the modulated carrier signal has a sine waveform.

sine-wave response   *Amplitude-frequency response.*

singing   An undesired self-sustained oscillation in a system or component, at a frequency in or above the passband of the system or component. It is generally caused by excessive positive feedback.

singing margin   The difference in level, usually expressed in decibels, between the singing point and the operating gain of a system or component.

singing point   The minimum value of gain of a system or component that will result in singing.

single-address instruction   *One-address instruction.*

single-board computer [SBC]   A circuit board that typically contains a microprocessor, ROM and RAM, serial I/O lines, and parallel I/O ports. It can serve as a main processing unit or as a controller for peripheral and I/O devices. It is often embedded in scientific instruments and automated test equipment, and it is used in prototype system development.

single-button carbon microphone   A carbon microphone that has a carbon-filled buttonlike container on only one side of its flexible diaphragm.

single-channel ground and airborne radio system [SINC-GARS]   A U.S. Army line-of-sight, single-channel, frequency hopping unit operating in the 30- to 88-MHz band with discrete channels for hopping and single-channel functions. It is designed to survive in a nuclear environment and includes electronic countermeasures. The system includes manpack and vehicular transceivers, and has 2320 channels.

single-channel simplex   Simplex operation that provides nonsimultaneous radio communication between stations which use the same frequency channel.

single crystal   A crystal, usually artificially grown, in which all parts have the same crystallographic orientation.

single-domain particle   A ferromagnetic particle so small that it can support only one permanently magnetized region in which the magnetic moments of the atoms are ordered.

single-ended   Unbalanced, as when one side of a transmission line or circuit is grounded.

single-ended amplifier   An amplifier that has only one transistor in each stage so that its operation is asymmetric with respect to ground.

single-ended input   An amplifier or other circuit that has one side of its input grounded.

single-ended push-pull amplifier   An amplifier that has two transmission paths designed to operate in a complementary manner which are connected to provide a single unbalanced output. This circuit provides push-pull operation without the need for a transformer.

single-event upset [SEU]   A reference to the response of an IC to a single radiation event, such as an alpha particle or a cosmic ray, which can cause the temporary failure of an integrated circuit. It generally occurs in space applications and can be caused by such events as solar flares.

single-frequency duplex   Duplex carrier communication that provides communication in opposite directions, but not simultaneously, over a single-frequency carrier chan-

nel. The transfer between transmitting and receiving conditions is automatically controlled by the voices of the communicating parties.

single-frequency simplex   Single-frequency carrier communication that has manual rather than automatic switching to change over from transmission to reception.

single-gun color tube   A color television picture tube that has only one electron gun and one electron beam. The beam is sequentially deflected across phosphors for the three primary colors to form each color picture element, as in the chromatron.

single hop   The range of a radio wave that is radiated at a small angle to the horizontal. It penetrates the ionosphere only a small amount before being reflected back to the surface of the earth. The maximum range that can be spanned by single-hop transmission is about 1500 mi (2400 km) for E-layer transmissions and it is obtained when the radio wave is radiated horizontally.

single-in-line memory module [SIMM]   A circuit board module with four to six dual-in-line-packaged memory devices mounted on a substrate made as a small daughter board. It has edge contacts for insertion into edge connectors mounted on a motherboard. A SIMM can add a large amount of memory to a motherboard while taking up very little space because of vertical mounting.

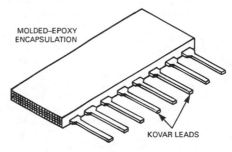

Single in-line package (SIP) resistor network can be mounted vertically on circuit board to save space.

single-in-line package [SIP]   An electronic component package with a thin rectangular case and a row of leads for mounting and connecting projecting from one of its edges. A SIP package saves circuit board space because it is mounted vertically in a row of holes in the circuit board. Resistor networks and memory arrays are SIP-packaged.

single-mode fiber   In fiber-optic communications, an optical fiber with a core diameter so small with respect to the wavelength of light being transmitted that only the lowest transmission mode can be propagated through it.

single-phase   Energized by a single alternating voltage.

single-phase circuit   A circuit energized by a single alternating voltage, applied through two wires.

single-photon emission-computed tomography [SPECT]   A medical imaging technique that provides images of a patient's internal organs and tissue by detecting the presence of trace amounts of commercially available radio isotopes injected in solution into the body. An array of detectors emits electrical signals and a computer records the location of the radio activity and plots the source of radiation, translating that data into an image. Color coding permits easier analysis of the image for diagnostic purposes. It is less versatile but less expensive than *positron-emission tomography*. [PET].

single-polarity pulse *Unidirectional pulse.*

single-pole double-throw [SPDT] A three-terminal switch or relay contact arrangement that connects one terminal to either of two other terminals.

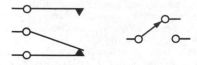

Single-pole double-throw (SPDT) relay and switch contacts.

single-pole single-throw [SPST] A two-terminal switch or relay contact arrangement that opens or closes one circuit.

Single-pole single-throw (SPST) normally open (NO) and normally closed (NC) switch contacts.

single scattering The deflection of a particle from its original path due to one encounter with a single scattering center in the material traversed. Plural scattering involves several successive encounters, and multiple scattering involves many successive encounters.

single-sideband [SSB] A reference to a communication system in which one of the two sidebands for amplitude modulation is suppressed.

single-sideband communication A communication system in which the RF carrier and one sideband are suppressed. Less power is then required at the transmitter for the same effective signal at the receiver, a narrower frequency band can be used, and the signal is less affected by selective fading or manmade interference.

single-sideband converter A converter that is connected to the IF amplifier output of an amplitude-modulation radio receiver to convert the receiver into a single-sideband receiver.

single-sideband filter A bandpass filter that exhibits a slope on one side of the response curve which is greater than that on the other side. It can suppress one sideband and sometimes the carrier frequency.

single-sideband modulation Modulation in which all the components of one sideband are eliminated from an amplitude-modulated wave.

single-sideband receiver A radio receiver that receives single-sideband modulation; it has provisions for restoring the carrier.

single-sideband transmission The transmission of a carrier and substantially only one sideband of modulation frequencies, as in television, where only the upper sideband is transmitted completely for the picture signal. The carrier wave can be either transmitted or suppressed.

single-sideband transmitter A transmitter that transmits one sideband while the other is effectively eliminated.

single-signal receiver A highly selective superheterodyne receiver for code reception; it has a crystal filter in the IF amplifier.

single-stub transformer A shorted section of coaxial line that is connected to a main coaxial line near a discontinuity to provide impedance matching at the discontinuity.

single-stub tuner A section of transmission line terminated by a movable short-circuiting plunger or bar attached to a main transmission line for impedance-matching purposes.

single-sweep oscilloscope An oscilloscope whose sweep must be reset and triggered for each operation to prevent multiple displays of a trace.

single-tone keying Keying in which the modulating wave causes the carrier to be modulated with a single tone for either a mark or a space. The carrier is unmodulated for the other condition.

single track A variable-density or variable-area sound track in which both positive and negative halves of the signal are linearly recorded. It is also called standard track.

single-track recorder A magnetic-tape recorder that records only one track on the tape.

sink 1. A power-consuming device such as the load in a circuit. 2. The region of a Rieke diagram where the rate of change of frequency with respect to phase of the reflection coefficient is maximum for an oscillator. Operation in this region can lead to unsatisfactory performance because of cessation or instability of oscillations.

SINS [Ship's Inertial Navigation System] A navigation system based on a gyroscope-stabilized platform and associated inertial guidance and sonar equipment to provide automatically the ship's position, true north, and true speed of the ship with respect to the ocean bottom. Nuclear-powered submarines are equipped with SINS.

sintering The process of bonding metal or other powders by cold-pressing into the desired shape, then heating it to form a strong cohesive body. Slug-type tantalum capacitors are made by this method.

sinusoidal Varying in proportion to the sine of an angle or time function. Ordinary alternating current is sinusoidal.

sinusoidal electromagnetic wave A wave whose electric field strength is proportional to the sine or cosine of an angle that is a linear function of time or distance.

sinusoidal quantity A quantity that varies according to a sinusoidal function of the independent variable.

SIP Abbreviation for *single in-line package.*

site error An error due to distortion of the radiated field by objects in the vicinity of a radio navigation aid.

skew 1. Deviation of a received facsimile frame from rectangularity due to lack of synchronism between the scanner and the recorder. It is expressed numerically as the tangent of the angle of this deviation. 2. The degree of nonsynchronism of supposedly parallel bits when bit-coded characters are read from magnetic tape.

skiatron *Dark-trace tube.*

skin depth The depth below the surface of a conductor at which the current density has decreased 1 neper below the current density at the surface due to the action of the electromagnetic waves associated with the high-frequency current flowing through the conductor.

skin effect An increase in the effective resistance of a conductor with frequency because of the flow of current through its surface or skin at high frequencies.

skip distance The minimum distance that radio waves can be transmitted between two points on the earth by reflection from the ionosphere, at a specified time and frequency.

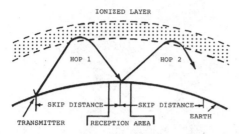

Skip distance is the distance radio waves travel between the transmitter and earth after being reflected from an ionospheric layer.

The skip distance thus includes the maximum ground-wave range and the width of the skip zone. In multihop transmission, the distance of each succeeding hop from earth to ionosphere and back is also the skip distance.

skip zone A ring-shaped area around a radio transmitter, surrounding the ground-wave reception region, within which no radio signals are received. The outer edge of the skip zone is at the minimum distance for reception of sky-wave signals, and the inner edge is at the maximum limit for reception of ground-wave signals. It is also called silent zone and zone of silence.

sky wave A radio wave that travels upward into space and might or might not be returned to earth by reflection from the ionosphere. It is also called an ionospheric wave.

sky-wave correction A correction for sky-wave propagation errors, applied to measured position data. The amount of the correction is based on an assumed position and an assumed ionosphere height.

sky-wave station error The error in station synchronization in sky-wave-synchronized loran that is caused by the effect of ionospheric variations on the time of transmission of the synchronizing signal from one station to the other.

slab A relatively thick crystal cut that is the source of blanks obtained in subsequent transverse cutting.

slant distance The distance between two points not at the same elevation. It is also called slant range.

slant range *Slant distance.*

SLAR Abbreviation for *side-looking airborne radar.*

slave operation Operation of two or more stabilized power supplies or other devices so that coordinated control of the overall supply is achieved by controlling only the master unit.

slave station A navigation station that has some characteristic of its emission controlled by a master station, such as the B station in loran.

SLC Abbreviation for *straight-line capacitance.*

SLDC Abbreviation for *synchronous data link control.*

sleeve 1. The cylindrical contact that is farthest from the tip of a phone plug. 2. Insulating tubing placed over wires or components.

sleeve antenna A single vertical half-wave radiator whose lower half is a metallic sleeve through which the concentric feed line runs. The upper radiating portion, one quarter-wavelength long is connected to the center of the line. It is also called a coaxial antenna.

sleeve stub An antenna that consists of half a sleeve-dipole antenna projecting from a large metal surface.

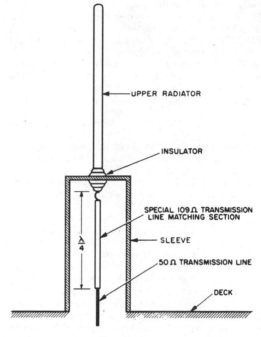

Sleeve antenna.

slewing The movement of a radar antenna or sonar transducer rapidly in a horizontal or vertical direction, or both.

slewing motor A motor that drives a radar antenna or sonar transducer at high speed for slewing to pick up or track a target.

slew rate The maximum rate at which the output voltage of an operational amplifier changes for a square-wave or step-signal input. The rate is specified in volts per microsecond.

SLF Abbreviation for *straight-line frequency.*

SLIC 1. Acronym for *subscriber line interface circuit.* 2. Acronym for *standard (function) linear and (mixed-signal) IC.*

slice 1. In semiconductor technology, to cut a crystal ingot or boule into thin slices or wafers with a saw. After grinding and polishing, the slice becomes a wafer. 2. Another term for *wafer.*

SLICE An acronym for *Simulation Language for Integrated Circuit Emphasis* a BASIC-like language program for the design of the analog sections of mixed-signal (analog and digital) integrated circuits.

slider A sliding movable contact. See also *wiper.*

slide-rule dial A dial with a pointer that moves in a straight line over long straight scales that resemble the scales of a slide rule.

slide switch A switch that is actuated by sliding a button, bar, or knob.

slide-wire bridge A bridge circuit in which the resistance in one or more branches is controlled by the position of a sliding contact on a length of resistance wire stretched along a linear scale.

slide-wire rheostat A rheostat with a sliding contact that rides over a long single-layer coil of resistance wire.

sliding contact   *Wiping contact.*

slip   The difference between synchronous and operating speeds of an induction machine.

SLIP/PPP   Abbreviation for *serial line Internet protocol/point-to-point protocol.*

slip ring   A conducting ring mounted on but insulated from a rotating shaft. It is used with a stationary brush to join fixed and moving parts of a circuit.

slope   1. The projection of a flight path in the vertical plane.   2. The degree of deviation of an essentially straight portion of a characteristic curve from the horizontal or vertical.

slope detector   A discriminator that uses a single tuned circuit and single diode to react to differences in frequency. Its operation is on one of the slopes of the response curve for the tuned circuit. It is seldom used in FM receivers because the linear portion of the curve is too short for large-signal operation.

slope deviation   The difference between the projection in the vertical plane of the actual path of movement of an aircraft and the planned slope for the aircraft. It is expressed in terms of angular or linear measurement.

slope equalizer   An equalizer with an amplifier that makes the attenuation of a section of transmission line constant over the frequency band being transmitted.

slope filter   A filter that has a response which rises or falls with frequency over a given frequency range.

slot antenna   An antenna formed by cutting one or more narrow slots in a large metal surface fed by a coaxial line or waveguide. For unidirectional radiation, the rear of the metal sheet is boxed in, or the slot is energized directly by a waveguide. In another version, diagonal slots are cut into a length of waveguide at precisely spaced intervals.

slot array   An antenna array that consists of many slot antennas, energized separately to give a desired radiation pattern.

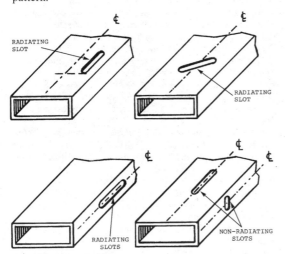

Slot array radiators in a rectangular waveguide, showing those that radiate and those that do not.

slot coupling   The coupling between a coaxial cable and a waveguide by two coincident narrow slots, one in a waveguide wall and the other in the sheath of the coaxial cable.

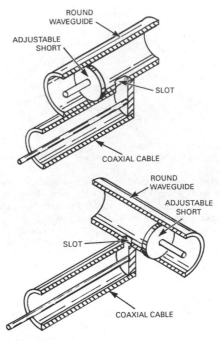

Slot coupling between coaxial cables and circular waveguides.

slot mask   A color picture tube mask that has vertical slots positioned in front of a screen on which color phosphors are arranged in vertical stripes. It is in the Trinitron and other picture tubes that have in-line electron guns. It is also called an aperture grille.

slotted line   *Slotted section.*

slotted rotor plate   A capacitor rotor plate that has radial slots to permit bending different sections of the plate either inward or outward to adjust the total capacitance of a variable capacitor section during alignment. It is also called a *serrated rotor plate* or a *split rotor plate.*

slotted section   A section of waveguide or shielded transmission line with a shield that is slotted to permit the movement of a traveling probe for examination of standing waves. Also called slotted line and slotted waveguide.

slotted waveguide   *Slotted section.*

slow-acting relay   A time-delay relay in which an interval of several seconds can exist between energizing the coil and pulling the armature. The delay can be obtained electrically by placing a solid copper ring on the core of the relay. It is also called a slow-operate relay.

slow-blow fuse   A fuse that can withstand up to 10 times its normal operating current for a brief period, as required for circuits, motors, and inductive loads that draw a very heavy starting current.

slow-operate relay   *Slow-acting relay.*

slow-release relay   A time-delay relay in which there is an appreciable delay between deenergizing of the coil and release of the armature.

slow-scan television   [SSTV]   Television that scans a complete scene in about 8 s. This reduces transmission channel requirements to that for an audio signal (usually less than 5 kHz), permitting the use of ordinary telephone lines for transmission and reception of sequences of still pictures. A high-persistence cathode-ray tube is required to hold the picture for 8 s. The picture can be cut to 120 lines, giving a much coarser image than with standard 525-line NTSC television.

slow wave   A wave that has a phase velocity less than the velocity of light, as in a ridge waveguide.

slow-wave circuit   A microwave circuit whose phase velocity is much slower than the velocity of light.

slug   1. A heavy copper ring placed on the core of a relay to delay operation of the relay.   2. A movable iron core for a coil.   3. A movable piece of metal or dielectric material for tuning in a waveguide or impedance-matching purposes.   4. A sintered, pressed body of a solid tantalum capacitor.

slug tuner   A waveguide tuner that contains one or more longitudinally adjustable pieces of metal or dielectric.

slug tuning   A means for varying the frequency of a resonant circuit by introducing a slug of material into either the electric or magnetic fields or both, as in permeability tuning.

SLW   Abbreviation for *straight-line wavelength.*

SMA   Abbreviation for a small coaxial connector that is effective up to the transmission frequency of 12.4 GHz.

Small Computer Systems Interface [SCSI]   An ANSI specification for small computer interface in Unix workstations and Macintosh computers for connecting computers to peripherals such as disk and tape drives, CD-ROM drives, and scanners. It limits maximum data rates in megabytes per second in asynchronous and synchronous modes. SCSI-1 is specified as an 8-bit data path at a maximum data rate of 2.5 Mbytes/s per second asynchronous mode and 5 Mbytes/s synchronous.

small-outline integrated circuit [SOIC]   A miniature plastic flat integrated-circuit package for surface mounting. Similar in appearance to a dual-in-line package, it has gull-wing leads with typical lead spacing of 0.05 in.

small-outline transistor [SOT]   A plastic package for surface mounting discrete semiconductor devices.

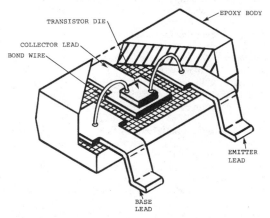

Small-outline transistor (SOT) with its case cut away, showing die and three gull-wing leads for surface mounting.

small-scale integration   An integrated circuit containing less than 100 transistors.

small-signal analysis   Circuit analysis based on such small excursions of current and voltage from their quiescent operating points that linear operation can be assumed.

smart bomb   Slang term for *precision-guided munitions.*

smart terminal   *Intelligent terminal.*

SMATV   Abbreviation for *satellite master-antenna television.*

SMB   Abbreviation for a small coaxial connector that is effective up to the transmission frequency of 10 GHz.

SMD   Abbreviation for *surface-mount device.*

SMDS   Abbreviation for *switched multimegabit data service.*

smear   A television picture defect that causes objects to appear to be extended horizontally beyond their normal boundaries in a blurred or smeared manner. One cause is excessive attenuation of high video frequencies in the television receiver.

smectic material   A liquid crystal material with elongated molecules that are are arranged in layers, and all molecules are parallel.

S meter   *signal-strength meter.*

Smith chart   A special polar diagram that contains constant-resistance circles, constant-reactance circles, circles of constant standing-wave ratio, and radius lines which represent constant line-angle loci. It is useful in solving transmission-line and waveguide problems.

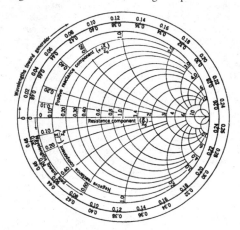

Smith chart.

smoke detector   A photoelectric circuit for actuating an alarm when smoke in an office, home, or other location exceeds a predetermined density.

smooth   To decrease or eliminate rapid fluctuations in measured data.

smoothing choke   An iron-core choke for a power-supply filter circuit that removes ripple.

smoothing factor   A factor that expresses the effectiveness of a filter in smoothing ripple-voltage variations.

SMPS   Abbreviation for *switched-mode power supply.*

SMPTE   Abbreviation for *Society of Motion Picture and Television Engineers.*

SMT   Abbreviation for *surface-mount technology.*

S/N   Abbreviation for *signal-to-noise.*

snap-action switch   A switch that responds to very small movements of its actuating button or lever and changes rapidly and positively from one contact position to the other.

snap-on ammeter An AC ammeter that has a magnetic core in the form of hinged jaws which can be snapped around the current-carrying wire. It is also called a clamp-on ammeter.

snap switch A switch whose contacts are separated or brought together suddenly by the action of a spring placed under tension or compression by the operating knob or lever.

Snell's law When a light wave travels from one medium into another that has a different index of refraction, the product of the sine of the angle of refraction and the refractive index of the refracting medium is equal to the product of the sine of the angle of incidence and the index of refraction of the medium which contains the incident beam.

snooperscope *Night-vision scope.*

snow Small, random white spots produced on a television or radar screen by inherent noise signals that originate in the receiver. It is visible on television screens only when the received signal strength is inadequate, because strong incoming signals will override the noise signals.

SNR Abbreviation for *signal-to-noise ratio.*

snubber A suppression network consisting of a capacitor in series with a resistor in shunt across a thyristor to prevent voltage spikes or stepped voltage from damaging the device when the load is primarily inductive and voltage and current are out of phase. It lowers the circuit's resonant frequency and characteristic impedance.

SOC Abbreviation for *state of charge.*

Society of Motion Picture and Television Engineers [SMPTE] A nonprofit organization established for the advancement of theory and practice related to the production and utilization of motion pictures and television programs.

socket 1. A female connector that accepts a plug to complete an electrical connection. It is usually mounted on the product or component case. It is also known as a *jack.* See also *phone jack* and *phono jack.* 2. *receptacle.* 3. The hollow part of an electric fixture that accepts the light bulb. See also *outlet.*

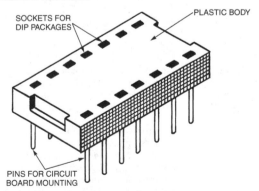

SOCKETS FOR DIP PACKAGES

PLASTIC BODY

PINS FOR CIRCUIT BOARD MOUNTING

Socket for a 14-pin dual-in-line integrated circuit package.

SO DIMM Abbreviation for *small-outline dual-in-line memory module.*

sodium [Na] A metallic alkali element that has a melting point of 97.5°C. In liquid form it can be a coolant for nuclear reactors. It is used on cathodes of phototubes when maximum response is desired at the violet end of the visible spectrum. It has an atomic number of 11 and an atomic weight of 22.99.

sodium-vapor lamp A discharge lamp that contains sodium vapor, used chiefly for highway illumination.

soft magnetic material Magnetic material that is easily demagnetized.

soft start An input circuit that limits the inrush current to a circuit at turn on by presenting a large resistance value at the input. In one technique the resistor is short-circuited by a relay after a preset time interval. In another, the resistor is a thermistor whose resistance value declines with the heating effect of current passage through the thermistor over time, thus placing the full supply voltage across the terminals.

soft superconductor A superconductor whose superconductivity can be destroyed by a weak magnetic field, as low as 25 oersteds. Zirconium and cadmium are examples. A hard superconductor might require over 1000 oersteds, or even as much as 100,000 oersteds for certain alloys.

soft tube 1. An X-ray tube that has a vacuum of about 0.000002 atmosphere. The residual gas is allowed to remain to give less penetrating rays than those of a more completely evacuated tube. 2. *Gassy tube.*

software Programs, routines, languages, and procedures for computer systems that include assemblers, generators, subroutines, compilers, and operating systems. See also *firmware.*

software interrupt The interruption of a user-level program in response to the acknowledgment of a hardware interrupt by the operating system. In high-level language programs, software interrupts can occur only at the end of a program line.

soft X-rays X-rays that have comparatively long wavelengths and poor penetrating power.

SOI Abbreviation for *silicon-on-insulator.*

SOIC Abbreviation for *small outline integrated circuit package.*

SOJ Abbreviation for *small outline "J" leaded package* for semiconductor devices.

solar burst A sudden increase in the RF energy radiated by the sun, generally associated with visible solar flares.

solar cell A silicon photovoltaic cell that converts sunlight directly into electric energy. In satellites it provides power for transmitting equipment. It is also called a *Photovoltaic cell.*

solar-cell array A battery of solar cells mounted so that it can be oriented for exposure to solar radiation. It is a power source for satellites and spacecraft.

solar corpuscle A particle, usually a proton, sprayed out into the solar system by disturbances on the sun. The particles react with the earth's magnetic field to produce ionospheric disturbances.

solar cycle *Sunspot cycle.*

solar flare A violent eruption on the surface of the sun that sends a large cloud of cosmic rays out into space. The magnetic field associated with this cloud is believed to shield the earth from cosmic rays coming from other inter-

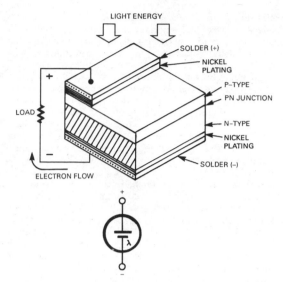

Solar cell generates power when it is exposed to the the sun or light energy source (top); its symbol is shown below.

stellar sources, thus accounting for the sudden decrease in cosmic-ray count on the earth following a solar flare.

solar-flare proton A proton ejected from the sun during solar flares, with energies ranging from below 10 MeV up to several gigaelectron-volts. Cosmic-ray particles from solar flares consist primarily of these protons.

solar flux Radio energy from the sun.

solar flux index A measure of solar activity. The solar flux is a measure of the radio noise at 2800 MHz.

solarimeter An instrument that makes direct readings of the solar radiation intensity from the sun and sky.

solar magnetograph A magnetograph designed primarily to study the powerful magnetic fields of sunspots.

solar pumping The use of sunlight focused directly into a laser rod for pumping to induce lasing action.

solar radio noise Radio noise that originates at the sun and increases greatly in intensity during sunspots and flares. It is generally heard as a hissing noise on shortwave radio receivers.

solar wind A high-speed stream of electrons and hydrogen and helium gas and their nuclei, blowing at speeds up to 700 km/s from the surface of the sun.

SOLAS Abbreviation for the *Safety of Life at Sea Convention.*

solder 1. An alloy that can be melted at a fairly low temperature for joining metals which have much higher melting points. An alloy of lead and tin in approximately equal proportions is the solder most often used for making permanent joints in electronic circuits. 2. To join two metals with solder.

solderability The ability of a metal to be wetted by molten lead-tin solder to form a strong, low-resistance bond with the solder and other prepared metal surfaces.

soldering The process of joining metals by fusion and solidification of an adherent alloy that has a melting point below about 800°F (425°C).

soldering flux A chemical that dissolves oxides from surfaces being soldered. Rosin is widely used for this purpose when soldering electronic circuits because of its noncorrosive qualities.

soldering gun A soldering tool that has as its tip a fast-heating resistance element which serves as a short-circuit across a high-current, low-voltage secondary winding of a step-down transformer built into the unit. A trigger-type switch in a pistol-grip handle permits intermittent use.

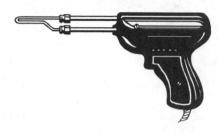

Soldering gun.

soldering iron A tool that applies heat to a joint preparatory to soldering. The source of heat is an internal electric heating element.

soldering paste A soldering flux prepared in the form of a paste.

soldering pencil A small soldering iron, about the size of a pencil, for soldering or unsoldering joints on printed-wiring boards. It has a typical rating of 35 W or less.

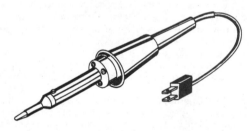

Soldering pencil rated 15 to 40 W is suitable for soldering heat-sensitive miniature components to circuit boards.

solderless connector A connector that clamps wire ends firmly together to provide a good connection without solder. A common form is a cap with tapered internal threads that is twisted over the exposed ends of the wires.

solderless contact *Crimp contact.*

solenoid A coil that surrounds a movable iron core. When the coil is energized by sending alternating or direct current through it, the core is pulled to a central position with respect to the coil. It can convert electric energy into mechanical energy. Other forms of solenoids produce rotary rather than axial movement of the core. The core can be stationary and the coil movable.

solid conductor A conductor that consists of a solid single wire rather than strands.

solid state A reference to the electronic properties of crystalline materials, generally semiconductor—as opposed to vacuum and gas-filled tubes that function by

solid-state amplifier

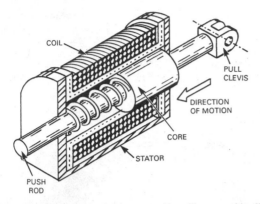

Solenoid with a movable plunger to provide pulling or pushing linear motion.

the flow of electrons through the space, or by flow through ionized gases. Solid state devices interact with light, heat, magnetic fields, mechanical stress, and electric currents.

solid-state amplifier   An amplifier whose active devices are discrete transistors or those constructed as parts of an integrated circuit.

solid-state circuit   A term that today implies a circuit whose active components are semiconductor devices (diodes, transistors, thyristors, and integrated circuits). A circuit that contains no vacuum or gas tubes.

solid-state circuit breaker   A circuit breaker that has a Zener diode, silicon controlled rectifier, or solid-state device connected to sense when load terminal voltage exceeds a safe value. This solid-state device, acting in microseconds, can trip a slower-acting electromechanical circuit breaker that disconnects the load from the power line.

solid-state component   A component whose operation depends on the control of electric or magnetic phenomena in solids, such as a transistor, crystal diode, or integrated circuit.

solid-state DC motor   A DC motor whose brushes have been replaced by two Hall generators located 90° apart on the stator winding. During the rotation of the cylindrical permanent-magnet rotor, the Hall units generate sinusoidal currents that control stator-winding switching by separate solid-state circuits, for precise control of motor speed.

solid-state device   A device other than a conductor whose operation depends on magnetic, electrical, and other properties of solid materials, as opposed to vacuum or gaseous devices.

solid-state dosimeter   A dosimeter that depends on radiophotoluminescence, thermoluminescence, or electric conductivity changes for measurement of radiation dose and dose rate. One example is the fluorod.

solid-state floppy disk card [SSFDC]   A small, flat memory card intended to replace a floppy disk drive. It plugs in a mother board and can easily be removed from personal voice recorders, digital still-picture cameras, and *personal digital assistants* [PDA]. It measures approximately 50 by 40 mm by 0.8 mm thick and includes NAND flash EEPROM. It can be easily interchanged between systems.

solid-state image sensor   *Charge-coupled image sensor.*

solid-state lamp   *Light-emitting diode.*

solid-state laser   A laser made from either a crystalline or amorphous solid material, usually in the form of a rod that is excited by optical pumping. The most common crystalline materials are ruby, neodymium-doped ruby, and neodymium-doped yttrium-aluminum garnet. Solid-state lasers are usually pulsed. Continuous operation is possible, but it is more difficult in glass than in crystals because of the lower thermal conductivity of glass. Solid-state lasers can have either three or four excitation levels. Ruby lasers exhibit three-level operation. Four-level operation is usually obtained with active atoms or ions of an actinide, rare-earth, or transition metal. The normal wavelength range is 0.6 to 3 μm.

solid-state maser   A maser made from a semiconductor material that produces the coherent output beam. Two input waves are required: one wave, called the pumping source, induces upward energy transitions in the active material. The second wave, of lower frequency, causes downward transitions and undergoes amplification as it absorbs photons from the active material. It is typically an amplifier, but it can also serve as an oscillator.

solid-state memory   A computer memory that consists of solid-state memory devices such as DRAM, SRAM, ROM, EPROM, and EEPROM.

solid-state physics   The branch of physics concerned with the structure and properties of solids, including semiconductors.

solid-state relay [SSR]   A relay that has only solid-state components and no moving parts.

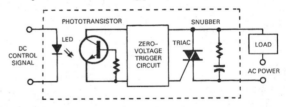

Solid-state relay has an optocoupler input stage, zero-voltage trigger, and a triac as an output power device.

solid-state static alternator   An alternator that converts a square-wave input from a pulse generator into a single- or three-phase AC output voltage. Silicon controlled rectifiers are the current-switching elements. It has no moving parts.

solid-state switch   A switch that has a semiconductor material as the switching element. The switch is triggered by a voltage pulse applied to the anode or gate terminal. Examples include silicon controlled rectifiers, diacs, and triacs.

solid-state television camera   A television camera based on solid-state charge-coupled devices for the image sensor as well as all circuit components.

solid-state thyratron   A power semiconductor device, such as a silicon controlled rectifier, that approximates the extremely fast switching speed and power-handling capability of a gaseous thyratron tube.

solid-state tuner   A tuner that changes the energy absorption of a semiconductor, such as yttrium-iron garnet, by varying the strength of the magnetic field applied to the material, to provide electric tuning.

solid tantalum capacitor An electrolytic capacitor that has an anode made of a porous pellet of tantalum. The dielectric is an extremely thin layer of tantalum pentoxide that is formed by anodization of the exterior and interior surfaces of the pellet. The cathode is a layer of semiconducting manganese dioxide that fills the pores of the anode over the dielectric. The polarity of connections must be observed, just as with other electrolytic capacitors.

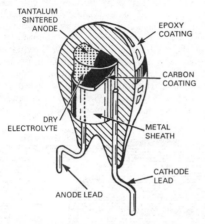

Solid tantalum dipped capacitor with radial leads has a porous tantalum anode slug and dry electrolyte.

Sommerfeld's equation An equation for radio ground-wave propagation that relates field strength at the surface of the earth at any distance from a transmitting antenna to the field strength at unit distance for given ground losses.

sonar [SOund Navigation And Ranging] Equipment for sonic and ultrasonic underwater detection, ranging, sounding, and communication. The most common type is echo-ranging sonar, an active sonar whose sonic or ultrasonic pulse is transmitted, reflected from an object, and received back at the transmitter location. The elapsed time for the pulse gives target range, and the directional characteristics of the transmitting-receiving transducer give target bearing. For underwater communication the pulses can be coded according to international Morse code. Other versions are passive, scanning, and searchlight sonar. The British term is *asdic*. It is also called a *sonar system*.

sonar attack plotter A system that coordinates information from a sonar installation, a ship's gyrocompass, and related devices, and presents graphically the information needed to plan an antisubmarine attack.

sonar beacon An underwater beacon that transmits sonic or ultrasonic signals, to provide bearing information. It can have receiving facilities that permit triggering by an external source.

sonar data computer A computer that calculates two or more factors of sonar data, such as range, bearing, depth, and sound velocity.

sonar dome A streamlined watertight enclosure that provides protection for a sonar transducer, sonar projector, or hydrophone and associated equipment, and offers minimum interference to sound transmission and reception.

sonar modulator A modulator that varies the frequency of a signal generated by a sonar transmitter.

sonar projector An electromechanical device used under water to convert electric energy to sound energy. A crystal, ceramic, or magnetostriction transducer projects the sound energy.

sonar receiver A receiver that intercepts and amplifies the sound signals reflected by an underwater target and displays the accompanying intelligence in useful form. It can also pick up other underwater sounds such as the sounds of whales and porpoises.

sonar receiver-transmitter A single unit of equipment that combines the functions of generating energy for sonic or ultrasonic underwater transmission and receiving the resulting echo signals. It is used primarily for detection and ranging on military ships. Its secondary functions can include underwater communication.

sonar set *Sonar* or *sonar system*.

sonar signal simulator A signal generator that feeds synthetic signals to a sonar receiver or directly to its indicator to give a presentation similar to that which would be observed under actual operating conditions.

sonar train mechanism A mechanism that rotates a sonar projector or transducer. It can also include mechanism for tilting the transducer or projector.

sonar transducer A transducer used under water to convert electric energy to sound energy and sound energy to electric energy.

sonar transducer scanner A circuit or switch that permits sampling the output signals of individual elements or groups of elements in certain types of sonar transducers, to determine the direction of arriving signals.

sonar transmitter A transmitter that generates electric signals of the proper frequency and form for application to a sonar transducer or sonar projector, to produce sound waves of the same frequency in water. The sound waves can be modulated to transmit messages.

sonar window The portion of a sonar dome or sonar transducer that passes sound waves at sonar frequencies with little attenuation, while providing mechanical protection for the transducer.

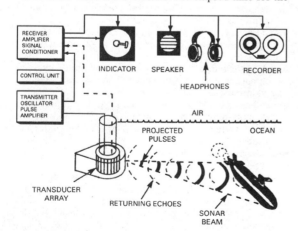

Sonar: an ultrasonic transducer mounted below a ship's waterline detects and then determines the range and bearing of underwater objects such as submarines or mines.

sonde　An instrument that obtains weather data during ascent and descent through the atmosphere, in a form suitable for telemetering to a ground station by radio, as in a radiosonde.

sone　A unit of loudness. By definition, a simple 1-kHz tone that is 40 dB above a listener's threshold of hearing produces a loudness of 1 sone. A sound that is judged by the listener to be n times as loud as that of the 1-sone tone has a loudness of n sones. A millisone is equal to 0.001 sone.

SONET　Acronym for the *Synchronous Optical Network* standard.

sonic　A reference to sound waves and the speed of sound.

sonic barrier　The turbulence encountered by an aircraft as its speed approaches the speed of sound. It is also called a sound barrier or a transonic barrier.

sonic boom　An explosionlike sound heard when a shock wave, generated by an aircraft flying at supersonic speed, reaches the ear.

sonic cleaning　The cleaning of contaminated materials by the action of intense sound from a transducer in the liquid in which the material is immersed. See also *ultrasonic cleaning.*

sonic delay line　*Acoustic delay line.*

sonic drilling　The process of cutting or shaping materials with an abrasive slurry driven by a reciprocating tool attached to an AF electromechanical transducer.

sonic flaw detection　The process of locating imperfections in solid materials by observing internal reflections or a variation in transmission through the materials as a function of sound-path location.

sonic frequency　*Audio frequency.*

sonic mine　*Acoustic mine.*

sonics　The use of sound in any noncommunication process.

sonic speed　*Speed of sound.*

sonic viscometry　Determination of the coefficients of viscosity of liquids or slurries by measurement of the acoustic properties of a transmitted wave or by the reaction of such a medium on a transducer.

sonobuoy　An acoustic receiver and radio transmitter mounted in a buoy that can be dropped from an aircraft by parachute to pick up underwater sounds of a submarine and transmit them to the aircraft. To track a submarine, many buoys are dropped in a pattern that includes the known or suspected location of the submarine, with each buoy transmitting an identifiable signal. An airborne computer then determines the location of the submarine by comparison of the received signals and triangulation of the resulting time-delay data. It is also called a radio sonobuoy.

sonography　The use of sonic or ultrasonic energy for imaging of biological tissues, as in studying the progress of the fetus in pregnancy, determining the size of tumors, and diagnosis of other pathological conditions.

sonoluminescence　Creation of light in liquids by sonically induced cavitation.

SOS　1. The distress signal in radiotelegraphy, consisting of the run-together letters S, O, and S of the international Morse code.　2. Abbreviation for *silicon-on-sapphire.*

SOSUS　Abbreviation for *Sound Surveillance System* or *Underwater Sound System,* a worldwide U.S. Navy system for tracking the ships and submarines of potential enemies. Started in the 1950s, it is now a vast network of underwater microphones that are connected to shore stations by about 30,000 miles of undersea cables.

SOT　Abbreviation for *small-outline transistor.*

sound　1. The sensation of hearing produced when sound waves act on the brain through the auditory organs of the ears. The extreme frequency limits for human hearing are from about 15 to 20,000 Hz, but animals can hear much higher frequencies. It is also called *audio* (slang).　2. A vibration in an elastic medium at any frequency that produces the sensation of hearing. This vibration is propagated by the medium as a sound wave. Frequencies that produce sound are called sound or audio frequencies. Frequencies below the audio range are infrasonic, and frequencies above the audio range are ultrasonic. Sound travels at about 1100 ft/s (335 m/s) in air at sea level and about 4800 ft/s (1463 m/s) in water.

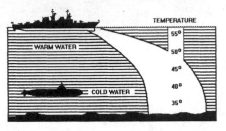

Sound beam from a surface vessel sonar is bent downwards with decreasing water temperature because the speed of sound in water increases with temperature.

sound absorber　A material that absorbs a high percentage of incident sound energy.

sound absorption　The process by which sound energy is diminished in passing through a medium or striking a surface.

sound absorption coefficient　The ratio of sound energy absorbed to that arriving at a surface or medium. It is also called acoustic absorption coefficient or acoustic absorptivity.

sound analyzer　An instrument that measures the levels of the components of a complex sound as a function of frequency.

sound articulation　The percent articulation obtained when speech units are fundamental sounds, usually combined into meaningless syllables.

sound bar　One of the two or more alternate dark and bright horizontal bars that appear in a television picture when AF voltage reaches the video input circuit of the picture tube.

sound barrier　*Sonic barrier.*

sound box　*Acoustic pickup.*

sound carrier　The television carrier that is frequency-modulated by the sound portion of a television program. The unmodulated center frequency of the sound carrier is 4.5 MHz higher than the video carrier frequency for the same channel.

sound channel　1. The series of stages that processes only the sound signal in a television receiver.　2. A layer of sea water that extends from about 700 m down to about 1500 m, in which sound travels at about 1500 m/s, the slowest it can travel in sea water. The density of sea water, which depends upon temperature, pressure, and salinity, is a maximum in the sound channel because salinity and temperature decrease with depth, whereas pressure increases with depth.

sound-effects filter A filter, usually adjustable, that reduces the passband of a system at low and/or high audio frequencies to produce special effects.

sound energy Energy existing in a medium due to sound waves.

sound-energy density Sound energy per unit volume, generally expressed in ergs per cubic centimeter.

sound-energy flux The average rate of flow of sound energy for one period through any specified area, generally expressed in ergs per second.

sound field A region that contains sound waves.

sound film Motion-picture film that has a sound track along one side for reproduction of the sounds which are to accompany the film.

sound-film recorder A device that converts sound signals to a modulated light beam (variable-area or variable-density) to expose film and produce sound images.

sound frequency *Audio frequency.*

sound gate The gate through which film passes in a sound-film projector for conversion of the sound track into AF signals that can be amplified and reproduced.

sound head The section of a sound motion-picture projector that converts the photographic or magnetic sound track to audible sound signals.

sound image The photographic image of a sound, as on a film sound track.

sounding 1. Any penetration of the natural environment, under water or into the atmosphere, for scientific observation and measurement. 2. The determination of the depth of water with lead line or depth finder.

sounding balloon A small free balloon for carrying radiosonde equipment aloft.

sound intensity The sound energy transmitted per unit of time through a unit area, expressed in ergs per second per square centimeter or in watts per square centimeter.

sound laser A device that amplifies sound in much the same way as a laser amplifies light. It is a 1-in long glass rod cooled to near absolute zero to eliminate interference from heat vibrations. Piezoelectric crystals at one end emit sound vibrations when an electric current flows through them. To amplify sound, the atoms in the glass are first pumped with a 2.5-μs sound pulse of selected intensity and frequency. The sound pulse to be amplified is then sent. Phonons from that sound travel through the rod, hitting the excited atoms without being absorbed, thus causing the atoms to release phonons, which multiply and intensify the sound. It is also called a *saser.*

sound level The weighted, sound pressure level at a point in a sound field, determined in the manner specified by the American Standards Association. The meter reading in decibels corresponds to a value of the sound pressure integrated over the audible frequency range with a specified frequency weighting and integration time.

sound-level meter An instrument that measures noise and sound levels in a specified manner. The meter might be calibrated in decibels or volume units. It includes

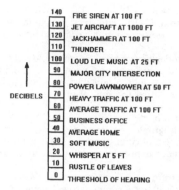

Sound levels are given in approximate decibel values for different noise sources. The threshold of ear pain occurs at about 130 dB.

a microphone, an amplifier, an output meter, and frequency-weighting networks. A volume-unit meter is an example.

sound power The total sound energy radiated by a source per unit time, generally expressed in ergs per second or watts.

sound-powered telephone A telephone operating entirely on current generated by the speaker's voice, with no external power supply. Sound waves cause a diaphragm to move a coil back and forth between the poles of a powerful but small permanent magnet, generating the required AF voltage in the coil.

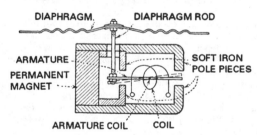

Sound-powered telephone. Diagram shows operating principle.

sound power level A value in decibels equal to 10 times the logarithm to the base 10 of the ratio of radiated sound power to a reference power.

sound pressure *Effective sound pressure.*

sound pressure level [SPL] A value in decibels equal to 20 times the logarithm to the base 10 of the ratio of the pressure of this sound to a reference pressure. Reference pressures in common use are 0.0002 and 1 μbar.

sound probe A probe that explores a sound field without significantly disturbing the field in the region being explored. It can be a small microphone or a small tubular attachment added to a conventional microphone.

sound recorder A recorder that provides a permanent record of sounds, as on optical sound tracks or magnetic tape.

sound recording system A combination of transducing devices and associated equipment suitable for storing sound in a form that can be reproduced reliably.

sound reflection coefficient The ratio of sound energy reflected from a surface to that reaching the surface. It is also called *acoustic reflection coefficient* and *acoustic reflectivity*.

sound-reproducing system A combination of transducing devices and associated equipment for picking up sound at one location and reproducing it at either the same location or some other location, at the same time or at some later time.

sound sensation *Sound.*

sound spectrograph An instrument that records and analyzes the spectral composition of audible sound. Applications include identification of speakers by their voice patterns.

sound spectrum A representation of the amplitudes of the components of a complex sound, arranged as a function of frequency.

sound speed The speed of sound motion-picture film, standardized at 24 frames per second. For 16-mm film this is 36 ft/min (11 m/min), and for 35-mm film it is 90 ft/min (27.4 m/min).

sound stripe A longitudinal stripe of magnetic material placed on some motion-picture films for recording a magnetic sound track.

sound takeoff The point where the sound signal is separated from the video signal in a television receiver for separate IF amplification, demodulation, and AF amplification.

sound track A narrow band, usually along the margin of a sound film, that carries the sound record. It can be a variable-width or variable-density optical track or a magnetic track.

sound transmission coefficient The ratio of transmitted to incident sound energy at an interface in a sound medium. Its value depends on the angle of incidence of the sound. It is also called *acoustic transmission coefficient* or acoustic transmittivity.

sound wave The traveling wave produced in an elastic medium by vibrations in the frequency range of sound. The approximate velocities of sound waves are 1100 ft/s (335 m/s) in air at 0°C, 4800 ft/s (1463 m/s) in water, and 16,400 ft/s (5000 m/s) in steel.

source 1. The power bus that drives a power supply. 2. One of three terminals in a field-effect transistor. Majority carriers (electrons in an N-channel FET or holes in a P-channel FET) originate at the source and flow across the channel to the drain as a result of the electric field applied between the source and the drain. It is analogous to the cathode in a vacuum tube and an emitter in a bipolar junction transistor. See also *channel, drain, FET, MOSFET,* and *gate.*

source-controlled drawing [SCD] A specification for a military semiconductor device that is specific to a program, a vendor, or a customer.

source-follower amplifier *Common-drain amplifier.*

source impedance The impedance presented by a source to a transducer or load circuit.

source language The language in which a problem is programmed for a computer. It must then be translated into an object program in machine language by an *assembler, compiler,* or *interpreter* for use by the computer.

source program A program that is written in source language or code.

source resistance [R_S] A field-effect transistor parameter. It is the change of gate-to-source voltage with respect to a change in drain-to-source current.

south magnetic pole The magnetic pole located approximately at 73°S latitude and 156°E longitude, about 1020 nautical miles (1900 km) north of the South Pole.

south pole The pole of a magnet where magnetic lines of force are assumed to enter.

space 1. The universe, extending from the earth's atmosphere without limit. 2. A nonprinting action in a typewriter or printer.

space attenuation The power loss of a signal traveling in free space, caused by such factors as absorption, reflection, and scattering. It is expressed in decibels.

space charge The net electric charge distributed throughout a volume or space, such as the cloud of electrons in the space near the cathode of a thermionic vacuum tube or phototube.

space-charge-controlled tube A microwave tube that generates RF power by controlling a space-charge-limited current with a grid electrode. A ceramic planar triode tube is an example.

space-charge debunching A process in which the mutual interactions between electrons in a stream spread out the electrons of a bunch.

space-charge density The net electric charge per unit volume.

space-charge effect The repulsion between the electrons emitted by the cathode of a thermionic vacuum tube and the electrons accumulated in the space charge near the cathode, resulting in a reduction in anode current.

space-charge grid A grid, usually positive, that controls the position, area, and magnitude of a potential minimum or of a virtual cathode in a region adjacent to the grid of an electron tube.

space-charge layer *Depletion layer.*

space-charge region A semiconductor region in which the net charge density differs significantly from zero.

space-charge wave An electrostatic wave in a plasma produced by oscillating motions of charges. The waves are longitudinal so that the electric field is in the direction of propagation.

spacecraft A vehicle or platform that travels outside of the earth's atmosphere, either in earth orbit or on a mission into outer space. Examples of spacecraft include communications, surveillance, and weather satellites, space shuttles, and space probes. It can be manned or unmanned.

space current The total current flowing between the cathode and all the other electrodes in a tube, including the anode current and the currents to all other electrodes.

space-diversity reception Radio reception that makes use of two or more antennas located several wavelengths apart, feeding individual receivers whose outputs are combined. The system gives an essentially constant output signal despite fading caused by variable propagation characteristics because fading affects the spaced-out antennas at different instants of time.

space environment The environment encountered by vehicles and living creatures in space, characterized by the absence of air and intense cold and solar radiation.

space factor 1. The ratio of the space occupied by the conductors in a winding to the total cubic content or volume of the winding, or the similar ratio of cross sections. 2. The ratio of the space occupied by iron to the total cubic content of an iron core.

space pattern A geometric pattern on a test chart for measuring geometric distortion in television equipment.

space permeability The factor that expresses the ratio of magnetic induction to magnetizing force in a vacuum. In the CGS electromagnetic system of units, the permeability of a vacuum is arbitrarily taken as unity.

space probe An unmanned instrumented spacecraft that makes measurements of lunar and planetary surfaces and atmospheres, solar winds, and other phenomena in space.

space quadrature A difference of a quarter-wavelength in the position of corresponding points of a wave in space.

space shuttle A reusable earth-orbiting spacecraft, similar in appearance to a fixed-wing airplane. It is launched with rockets and lands as a glider. It contains a cargo bay suitable as a laboratory for space- and electronic-related experiments.

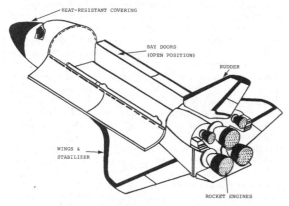

Space shuttle has a cargo bay that can be a laboratory for research to improve the manufacture of electronic components.

Spacetrack A global system of optical, radar, and radiometric sensors for life support connected to a computer facility, used for detecting, tracking, and cataloging all man-made objects that are orbiting the earth.

space wave The component of a ground radio wave that travels more or less directly through space from the transmitting to the receiving antenna. One part of the space wave goes directly from one antenna to the other; another part is reflected off the earth between the antennas.

spade lug An open-ended flat termination for a wire lead, easily slipped under a terminal nut.

S parameter (scattering parameter) One of a set of four parameters for describing two-port networks or devices that do not require the establishment of open-circuit or short-circuit reference planes. The four parameters (input reflection coefficient, output reflection coefficient, forward voltage gain, and reverse voltage gain) are all measured with the device terminated in its characteristic impedance. These dimensionless expressions are obtained from microwave devices or circuits to present a physical interpretation of the network performance. They are related to power gain and mismatch loss. S parameters are obtained by introducing a known signal into the device or circuit.

spark A short-duration electric discharge caused by a sudden breakdown of air or some other dielectric material that separates two terminals. It is accompanied by a momentary flash of light.

spark coil An induction coil that produces spark discharges.

spark gap An arrangement of two electrodes between which a spark might occur. The insulation (usually air) between the electrodes is self-restoring after the passage of the spark. It can be a switching device for protecting equipment against lightning.

spark-gap generator A high-frequency generator with a capacitor that is repeatedly charged to a high voltage and allowed to discharge through a spark gap into an oscillatory circuit. It can generate successive trains of damped high-frequency oscillations.

spark-gap modulation A modulation process that produces one or more pulses of energy by a controlled spark-gap breakdown, for application to the circuit in which modulation takes place.

spark-gap modulator A modulator that uses a controlled spark gap to modulate a carrier.

sparking Intentional or accidental spark discharges, as between the brushes and commutator of a rotating machine, between contacts of a relay or switch, or at any other point at which a current-carrying circuit is broken.

spark spectrum The spectrum produced by a spark discharging through a gas or vapor. With metal electrodes, a spectrum of the metallic vapor is obtained.

spark suppressor A resistor and capacitor in series, connected between a pair of contacts to suppress sparking when the contacts open.

Sparrow An air-to-air or air-to-ground guided missile armed with an 85-lb conventional warhead that is propelled by solid fuel and guided to its target by energy reflected from the enemy target when it is illuminated by the host aircraft's radar. The host aircraft is usually an F-14 or F-15. It is larger than the *Sea Sparrow*. See also *Sea Sparrow*.

spatial filter An optical filter that consists of a very small aperture, such as a pinhole. In laser cavities it can change the beam shape or mode structure, but when outside lasers it can change only beam shape.

spatial frequency The number of black and white line pairs displayed on a screen per degree of visual angle. Spatial frequency is expressed as cycles per degree or as lines per inch for a given viewing distance.

SPC Abbreviation for *statistical process control.*

SPDT Abbreviation for *single-pole double-throw.*

speaker *Loudspeaker.*

SPEC Abbreviation for *Standard Performance Evaluation Corporation,* a nonprofit group sponsored by 24 computer manufacturers.

spec Abbreviation for *specification.*

SPECint92 Abbreviation for a benchmark test to determine the speed of a computer based on simple mathematical operations, generally used in industry and science. See *SPEC.*

SPECT Abbreviation for *single-photon emission computed tomography.*

special-purpose computer A computer that solves a specific type of problem.

specific address *Absolute address.*

specification 1. A technical document defining the characteristics of a specific device, circuit, subassembly, or product to distinguish it from similar products in the same category. 2. A technical document issued by a procurement agency such as the U.S. Department of Defense or a private corporation, listing the criteria that must be used to determine the acceptability of a product or materials in conformity with the description included. For example, military specifications cover resistors, capacitors, integrated circuits, and test procedures.

specific conductivity The conducting characteristic of a material in siemens per cubic centimeter (formerly mhos per cubic centimeter). It is the reciprocal of resistivity.

specific dielectric strength The dielectric strength per millimeter of thickness of an insulating material.

specific electronic charge The ratio of the electronic charge to the rest mass of the electron.

specific emission The rate of emission per unit area.

specific ionization The number of ion pairs formed per unit distance along the track of an ion passing through matter.

specific ionization coefficient The average number of pairs of ions with opposite charges that are produced by electrons which have a specified kinetic energy when traveling a unit distance in a gas at a specified pressure and temperature.

specific permeability The permeability of a substance divided by the permeability of a vacuum. It is also called relative permeability.

specific resistance *Resistivity.*

specific routine A routine expressed in computer coding to solve a particular problem, with addresses of registers and locations specifically stated.

SPECmarks A normalized measure of performance for RISC computers based on a standard set of operations. See *reduced-instruction set computer.*

spectral characteristic The relation between wavelength and some other variable, such as between wavelength and emitted radiant power of a luminescent screen per unit wavelength interval.

spectral color A color that appears in the spectrum of white light. The basic spectral colors are violet, blue, green, yellow, orange, and red.

spectral purity The property of having a single wavelength.

spectral quantum yield The average number of electrons photoelectrically emitted from a photocathode per incident photon of a given wavelength.

spectral radiant intensity The radiant intensity per unit wavelength interval, such as watts per steradian per micrometer.

spectral response *Spectral sensitivity characteristic.*

spectral selectivity The effect of radiation wavelength on the output current of a photoelectric device.

spectral sensitivity characteristic The relation between the radiant sensitivity and the wavelength of the incident radiation of a camera tube or phototube, under specified conditions of irradiation. It is also called spectral response.

spectrograph A spectrometer that provides a permanent record of a spectrum of radiation.

spectrometer An instrument that disperses radiation into its component wavelengths and measures the magnitude of each component.

spectrometry The use of spectrographic techniques for deriving the physical constants of materials.

spectrophotometer An instrument that measures transmission or apparent reflectance of visible light as a function of wavelength, permitting accurate analysis of color or accurate comparison of luminous intensities of two sources at specific wavelengths.

spectrophotometric analysis A method of quantitative analysis based on spectral energy distribution in the absorption spectrum of a substance in solution.

spectroradiometer An instrument that measures the spectral energy distribution of any type of radiation, such as infrared radiation.

spectroscope An instrument that spreads individual wavelengths in radiation to permit observation of the resulting spectrum.

spectroscopy The branch of physical science concerned with the measurement and analysis of visible, infrared, and ultraviolet spectra.

spectrum [plural **spectra**] 1. All the frequencies used for a particular purpose. Thus, the radio spectrum extends from about 10 kHz to about 300 GHz. 2. The result of dispersing an emission (such as light) in accordance with some progressive property, usually its frequency. 3. *Electromagnetic spectrum.*

spectrum analyzer An instrument that measures the amplitudes of the components of a complex waveform throughout the frequency range of the waveform. One version gives the energy at each frequency in the output of a pulsed magnetron. The instruments generally provide a cathode-ray display of amplitude versus frequency.

spectrum line A line recorded by a spectrograph to represent a specific wavelength, atomic mass, or other spectral quantity.

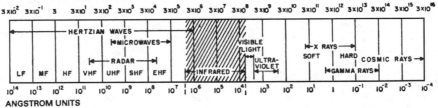

Spectrum of electromagnetic radiation, extending from 0.03 MHz at left to cosmic rays at right. Radio or Hertzian waves overlap longer infrared wavelengths.

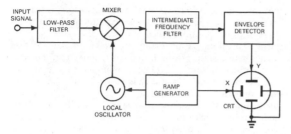

Spectrum analyzer is a test instrument with a narrow-band super-heterodyne receiver that can show voltage versus frequency waveforms on its CRT display.

spectrum locus The locus of points representing the chromaticities of spectrally pure stimuli in a chromaticity diagram.

spectrum signature The spectral characteristics of the transmitter, receiver, and antenna of an electronic system, including emission spectra, antenna patterns, and other characteristics. It is important in studies of electromagnetic interference caused by transmitting and receiving equipment.

specular reflection The visual effect of light from one or more external sources reflected from solid objects. It is direct reflection. Simulation of spectral reflections is a difficult task for a computer because it requires memory-intensive computations.

specular transmission Transmission in which only the emergent radiation parallel to the entrant beam is observed.

specular transmission density The value of the photographic density obtained when the light flux impinges normally on the sample, and only the normal component of the transmitted flux is collected and measured.

speech amplifier An AF amplifier designed specifically for amplification of speech frequencies, as for public-address equipment and radiotelephone systems.

speech clipper A clipper that limits the peaks of speech-frequency signals, as required for increasing the average modulation percentage of a radio-telephone or amateur radio transmitter.

speech compression A method of eliminating certain frequency components of a speech signal to permit its transmission over a narrower frequency band without appreciably affecting intelligibility.

speech digitization The conversion of analog speech waveforms to digital form.

speech frequency *Voice frequency.*

speech inverter *Scrambler.*

speech network A circuit that provides 2-to-4 wire conversion, i.e., connects the microphone and the receiver (or the transmit and receive paths) to the tip and ring phone lines. It also provides sidetone control, and sometimes the DC loop-current interface.

speech power The rate at which sound energy is being radiated by a speech source at a given instant, or the average of the instantaneous values over a given time interval.

speech processing Changing the characteristics of a voice signal to improve communication intelligibility. Methods include audio compression, audio clipping, RF compression, and RF clipping.

speech processor A circuit that increases the average power contained in a speech waveform to improve the readability of a voice signal.

speech recognition system A system that recognizes spoken syllables, words, and phrases. It can be used for the direct entry of voice data into a computer or control system.

speech scrambler *Scrambler.*

speech synthesizer *Voice synthesizer.*

speed 1. A scalar quantity equal to the magnitude of velocity. (Speed does not specify a direction.) 2. The angular velocity of a rotating shaft or device, generally expressed in revolutions per minute. 3. The rate of performance of an act. 4. The aperture of a lens. 5. The exposure time of a shutter. 6. The frequency of a relaxation oscillator.

speed control system A control system that changes the speed of a motor or other drive mechanism, as for a VCR or magnetic-tape recorder. See also *velocity control system.*

speed of light A physical constant equal to 2.997925×10^{10} cm/s. This corresponds to 186,280 statute mi/s, 161,870 nautical mi/s, and 328 yd/μs. All electromagnetic radiation travels at this same speed in free space. The velocity of light is usually considered to be a vector quantity representing the speed of light in a particular direction.

speed of sound The speed at which sound waves travel through a given medium. In air under standard sea-level conditions, sound travels at 1100 ft/s (335 m/s). In water it is about 4800 ft/s (1463 m/s). It is also called sonic speed.

speed regulator A device that maintains the speed of a motor or other device at a predetermined value or varies it in accordance with a predetermined plan.

sphere gap A spark gap between two spherical electrodes with equal diameters.

spherical aberration An image defect caused by the spherical form of an optical or electron lens or mirror, resulting in blurred focus and image distortion.

spherical antenna An antenna that has the shape of a sphere, used chiefly in theoretical studies.

spherical-earth factor The ratio of the electric field strength that would result from propagation over an imperfectly conducting spherical earth to that which would result from propagation over a perfectly conducting plane.

spherical faceplate A television picture tube faceplate that is a portion of a spherical surface.

spherical hyperbola The locus of the points on the surface of a sphere that has a specified constant difference in great-circle distances from two fixed points on the sphere.

spherical wave A wave whose equiphase surfaces form a family of concentric spheres. The direction of travel is always perpendicular to the surfaces of the spheres.

SPI Abbreviation for *serial peripheral interface.*

SPICE Abbreviation for *Simulation Program with Integrated Circuit Emphasis,* a simulation program for modeling transistors electrically. Developed by the University of California at Berkeley, it is available from independent vendors in modified forms for specific applications.

spider A highly flexible perforated or corrugated disk that centers the voice coil of a dynamic loudspeaker with respect to the pole piece without appreciably hindering reciprocating motion of the voice coil and its attached diaphragm.

spike A short-duration transient whose amplitude considerably exceeds the average amplitude of the associated pulse or signal.

spillover The inadvertent transfer of RF energy from the antenna of a continuous wave radar antenna to its receiving antenna. It must be minimized.

spindle A shaft, such as the upward-projecting shaft on a phonograph turntable, used for positioning the record.

spinel A hard crystalline oxide of magnesium and aluminum ($MgAl_2O_4$); it has characteristics which make it suitable as a replacement for sapphire as a substrate for semiconductor material.

spinner A rotating radar antenna and reflector assembly.

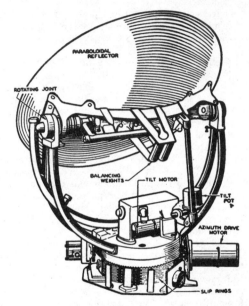

Spinner assembly for a ground radar.

spin-orbit coupling The interaction between intrinsic and orbital angular momentum of a particle.

spin precession magnetometer A magnetometer that applies the proton-free precession principle in quantum electronic instrumentation for measuring and recording geomagnetic phenomena. Examples include the proton magnetometer and the rubidium magnetometer.

spin quantum number A number that gives the angular momentum of the electron considered as a small charged sphere rotating around an axis.

spin stabilization The rotation of a satellite on an axis that is oriented to the earth's axis to give a stabilizing flywheel effect. If solar cells are mounted around the periphery of the satellite, a large number of the cells are exposed to the sun for power generation. Some satellites have mechanisms that spin highly directional antennas at the same speed but opposite in direction from the main bodies of the satellites so that their antennas remain focused on specific receiving station on earth.

spin wave The type of wave that exists within a ferrimagnetic material when all the magnetic moments are precessing uniformly but not in phase.

spiral distortion A distortion in which image rotation varies with distance from the axis of symmetry of the electron optical system of a camera or image tube that uses magnetic focusing. It is also called S distortion.

spiral-four cable *Spiral quad.*

spiral tuner A tuner that has spiral coils and a tuning mechanism which slides a contact along the spiral of each coil. The sliding contacts change the inductance of the coils and thus change the frequency. It is also called a continuous tuner or an inductuner.

SPL Abbreviation for *sound pressure level.*

splash baffle *Arc baffle.*

splatter Distortion due to overmodulation of a transmitter by peak signals of short duration, particularly sounds that contain high-frequency harmonics. It is a form of adjacent-channel interference.

SPLIC Acronym for *special-purpose linear (and mixed-signal) IC.*

splice A joint that connects two lengths of conductor with good mechanical strength and good conductivity.

splicing tape A pressure-sensitive nonmagnetic tape for splicing magnetic tape. It has a hard adhesive that will not ooze and gum up the recording head or cause adjacent layers of tape on the reel to stick together.

split hydrophone A directional hydrophone made so that each transducer or group of transducers produces a separate output voltage.

split integrator A digital-differential-analyzer integrator that can multiply its output or input by a constant.

split-phase current One of two different phases of current obtained from a single-phase AC circuit by reactances.

split-phase motor A single-phase induction motor that has an auxiliary winding connected in parallel with the main winding, but displaced in magnetic position from it. The auxiliary winding produces the required rotating magnetic field for starting. The auxiliary circuit is usually opened when the motor has attained a predetermined speed.

split rotor plate *Slotted rotor plate.*

split-series motor A DC series-connected motor that has one series field winding for each direction of rotation.

split-stator variable capacitor A variable capacitor that has a rotor section which is common to two separate stator sections. It is used in grid and anode tank circuits of transmitters for balancing purposes.

spoiler A rod grating mounted on a parabolic reflector to change the pencil-beam pattern of the reflector to a cosecant-squared pattern. Rotating the reflector and grating 90° with respect to the feed antenna changes from one pattern to the other.

spoking A radar malfunction in which luminous spots continue on the screen for an abnormal length of time and form radial lines, interfering with presentations.

spontaneous emission The spontaneous decay of an excited atom into its ground-state energy level, with the energy carried off by radiation.

sporadic E layer A layer of intense ionization that occurs sporadically within the E layer. It is variable in time of occurrence, height, geographical distribution, penetration frequency, and ionization density. See *E layer.*

sporadic reflection   *Abnormal reflection.*

spot   1. The luminous area produced on the viewing screen of a cathode-ray tube by the electron beam.   2. A commercial announcement of short duration, inserted in programs or broadcast between programs.

spot beam   A radio-frequency beam from a satellite that has a narrow aperture angle to cover a relatively small geographic area.

spot gluing   Applying heat to a glued assembly by dielectric heating to make the glue set in spots that are more or less regularly distributed.

spot size   The cross section of an electron beam at the screen of a cathode-ray tube.

spot welding   1. Resistance welding in which the fusion is limited to a small area.   2. The use of dielectric heating to join together sheets of thermoplastic material at a number of spots.

spread   The range within which the values of a variable quantity occur.

spreader   An insulating crossarm that holds apart the wires of a transmission line or multiple-wire antenna.

spreading resistance   The part of the resistance of a point-contact rectifier that is due to the semiconducting material alone, not including the barrier-layer resistance.

spread-spectrum modulation   A modulation technique that spreads a normally narrow band of transmitted frequencies over a broad band (more than 10 times as wide) with lower energy content to minimize noise and interference. The band is then compressed back to its original narrow frequency band at the receiver. In radar it provides antijamming security protection or low probability of interception.

spring contact   A relay or switch contact mounted on a flat spring, usually of phosphor bronze.

spring-return switch   A switch whose contacts return to their original positions when the operating lever is released. See also *momentary switch.*

sprocket pulse   The pulse generated by the magnetized spot that accompanies every character recorded on magnetic tape. This pulse is used during read operations to regulate the timing of read circuits and provide a count of the number of characters read from tape.

SPST   Abbreviation for *single-pole single-throw.*

spurious count   A count from a radiation counter other than background counts and those directly caused by ionizing radiation.

spurious emission   *Spurious radiation.*

spurious modulation   Undesired modulation that occurs in an oscillator, such as frequency modulation caused by mechanical vibration.

spurious pulse   A pulse in a radiation counter other than one purposely generated or directly caused by ionizing radiation.

spurious radiation   Any emission from a radio transmitter at frequencies outside its frequency band. It is also called spurious emission.

spurious response   Any response of a radio receiver to a frequency which differs from that to which the receiver is tuned.

spurious response ratio   The ratio of the field strength at the frequency that produces a spurious response to the field strength at the desired frequency. Each field is applied in turn to the receiver under specified conditions to produce equal outputs. Image ratio and IF response ratio are special forms of spurious response ratio.

spurious signal   An unwanted signal generated in the equipment itself, such as spurious radiation or undesired shading signals generated in a television camera tube.

spurious tube count   *Spurious count.*

sputtering   A method for depositing a thin layer of metal on a glass, plastic, metal, or other surface in a vacuum. The object to be coated is placed in a large demountable vacuum chamber that has a cathode made of the metal to be sputtered. The chamber is operated under conditions that promote cathode bombardment by positive ions. As a result, extremely small particles of molten metal fall uniformly on the object and produce a thin, conductive metal coating on it. The action occurs to some extent in ordinary electron tubes but is undesirable there because the positive ions knock small particles of the coating off the cathode. It is also called cathode sputtering.

square-foot unit of absorption   *Sabin.*

square-law demodulator   *Square-law detector.*

square-law detection   Detection in which a sinusoidal input gives an output proportional to the square of the input.

square-law detector   A demodulator whose output voltage is proportional to the square of the amplitude-modulated input voltage. It is also called a square-law demodulator.

square-loop ferrite   A ferrite that has an approximately rectangular hysteresis loop.

squareness ratio   The ratio of the flux density at zero magnetizing force to the maximum flux density for a material in a symmetrically cyclically magnetized condition. The ratio alternatively can be based on a magnetizing force halfway between zero and its negative limiting value.

squarer   *Squaring circuit.*

square-rooting circuit   A circuit that produces an output voltage which is proportional to the square root of the input voltage.

square wave   A uniform rectangular waveform whose pulse width is uniformly equal to one-half of the period. It has no even harmonics. It differs from a *rectangular wave,* whose pulse width is not equal to half the period.

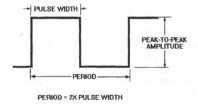

Square wave is a rectangular waveform whose pulse width is half of its period.

square-wave generator   A signal generator that generates a square-wave output voltage.

square-wave modulator   A modulator that delivers a square-wave AF output voltage, generally at a frequency of 1 kHz, for modulating RF signal sources such as klystrons.

square-wave testing The use of a series of related step functions to determine the performance characteristics of a device or system.

squaring circuit 1. A circuit that reshapes a sine or other wave into a square wave. 2. A circuit that contains nonlinear elements and produces an output voltage proportional to the square of the input voltage. It is also called a squarer.

squealing A high-pitched note or squeal produced by a radio receiver that interferes with radio reception. It is caused by interference between stations or by accidental oscillation in the receiver's circuitry.

squelch To quiet a receiver automatically by reducing its gain in response to a specified characteristic of the input, e.g., by reducing gain to suppress background noise when there is no input signal.

squelch circuit *Noise suppressor.*

SQUID [Superconducting QUantum Interference Device] A superconducting ring that couples with one or two junctions. Applications include high-sensitivity magnetometers, near-magnetic-field antennas, and measurement of very small currents or voltages.

squint 1. The angle between the two major lobe axes in a radar lobe-switching antenna. 2. The angular difference between the axis of radar antenna radiation and a selected geometric axis, such as the axis of the reflector. 3. The angle between the full-right and full-left positions of the beam of a conical-scan radar antenna.

squirrel-cage antenna A four-bay stacked array of vertical dipoles mounted on a vertical column. Each bay is balunfed at two points to obtain omni-directional radiation in the horizontal plane.

squirrel-cage induction motor An induction motor that has a secondary circuit consisting of a squirrel-cage winding arranged in slots in the iron core.

squirrel-cage winding A permanently short-circuited winding, usually uninsulated, that has its conductors uniformly distributed around the periphery of the rotor and joined by continuous end rings.

SR Abbreviation for *switch register.*

sr Abbreviation for *steradian.*

SRAM Abbreviation for *static random-access memory.*

SRC Abbreviation for *Semiconductor Research Corporation.*

SR flip-flop *RS flip-flop.*

SRT flip-flop *RST flip-flop.*

SSA Abbreviation for *serial storage architecture.*

SSB Abbreviation for *single-sideband.*

SSB/AM Single-sideband operation with amplitude modulation.

SSFDC Abbreviation for *solid-state floppy-disk card.*

SSI Abbreviation for *small-scale integration.*

SSID Abbreviations for *ships' station identification number.*

SSOP Abbreviation for *shrink small outline package* for semiconductor devices.

SSR 1. Abbreviation for *secondary surveillance radar.* 2. Abbreviation for *solid-state relay.*

SSTV Abbreviation for *slow-scan television.*

stability 1. Freedom from undesired variations. 2. Ability to develop restoring forces that are equal to or greater than the disturbing forces in a control system, so equilibrium is restored. 3. Freedom from undesired oscillation.

stabilization 1. Maintenance of a desired orientation independent of the roll, pitch, and yaw of a ship or aircraft. 2. A treatment of a magnetic material to improve the stability of its magnetic properties.

stabilized feedback *Negative feedback.*

stabilized flight Flight under the control of information obtained from inertia-stabilized references such as gyroscopes.

stabilized platform A platform whose attitude is maintained by two or more gyroscopes and associated servosystems, despite the pitch, roll, and yaw of a vehicle in space, on water, or in water. It supports radar antennas, sonar transducers, and inertial guidance systems.

stable element A navigation instrument or device that maintains a desired orientation independently of the motion of the vehicle.

stack 1. To assign different altitudes by radio to aircraft awaiting their turns to land at an airport. 2. A sonar equipment assembly in the sonar room of a ship.

stacked array Antenna elements that are stacked in an array, one above the other, and connected in phase to increase the gain.

stacked dipoles Two or more dipole antennas arranged above each other on a vertical supporting structure and connected in phase to increase their gain.

stacked heads *In-line heads.*

stage A functional grouping of active and passive components within a larger circuit. The grouping can include integrated circuitry. Examples of stages or functions are the mixer, detector, oscillator, amplifier, and audio output stages of a superheterodyne receiver.

stage-by-stage diagnosis A process of isolating circuit faults by making measurements of voltages at test points or examining the waveforms at those test points in stage-by-stage order, starting from the input end. Power is applied to each stage in succession. This methodical approach will lead to the isolation of the faulty component or components.

staggered circuits Adjacent circuits that are alternately tuned to two different frequencies to obtain broadband response, as in a video IF amplifier.

staggered tuning Alignment of successive tuned circuits to slightly different frequencies, to widen the overall amplitude-frequency response curve.

staggering Adjusting tuned circuits to give staggered tuning.

stagger-tuned amplifier An amplifier that is staggered-tuned to give a wide bandwidth.

staircase generator A signal generator whose output voltage increases in steps. Its output voltage waveform appears on an oscilloscope as a staircase.

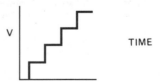

Staircase-generator waveform on an oscilloscope.

STALO [STAble Local Oscillator] A highly stable local RF oscillator for heterodyning signals to produce an intermediate frequency in radar moving-target indicators. Only echoes that have changed slightly in frequency due to reflection from a moving target produce an output signal.

STALO cavity   A cavity resonator that stabilizes the output frequency of a klystron oscillator.

standard   1. A reference that forms a basis for comparison or calibration.   2. A concept that has been established by authority, custom, or agreement, to serve as a model or rule in the measurement of a quantity or the establishment of a practice or procedure.

standard atmosphere   An arbitrary atmosphere for comparing aircraft performance. For official U.S. government use, the standard atmosphere is based on the assumptions that air is a dry perfect gas, the ground temperature is 59°F (15°C), the temperature gradient in the troposphere is 0.003566°F/ft (6.5°C/km), and the temperature in the stratosphere is −67°F (−55°C). The atmospheric pressure at sea level is then 29.92 in Hg (760 mmHg).

standard broadcast band   *Broadcast band.*

standard broadcast channel   The band of frequencies occupied by the carrier and two sidebands of a broadcast signal, with the carrier frequency at the center. Carrier frequencies are spaced 10 kHz apart, starting at 540 kHz and going up to 1,600 kHz.

standard cable   A cable of particular size and construction that serves as a reference for specifying transmission line losses. The standard cable includes a conductor that weighs 20 lb/mi (5.7 kg/km), a loop resistance of 88 Ω/mi (55 Ω/km), a capacitance of 54 nF/mi (34 nF/km), an inductance of 1 mH/mi (0.6 mH/km), and an attenuation constant of 0.103.

standard candle   The unit of candlepower, equal to a specified fraction of the visible light radiated by a group of 45 carbon-filament lamps preserved at the National Institute of Standards and Technology when the lamps are operated at a specified voltage. The standard candle was originally the amount of light radiated by a tallow candle of specified composition and shape.

standard capacitor   A capacitor whose capacitance value is not likely to vary with temperature and is known to a high degree of accuracy. It is also called a *capacitance standard.*

standard cell   1. An integrated circuit produced with computer-aided design. Its graphic elements such as gates and cells or subcircuits (macrocells) are stored in a workstation computer's memory. It makes less efficient use of the semiconductor chip material than hand-crafted, custom-designed ICs. A semicustom or applications-specific integrated circuit (ASIC), unlike a gate array, is not premanufactured, but made in successive steps as are full-custom ICs. It is available as an MSI or LSI. See *semicustom IC.*   2. A primary cell that has a voltage which is accurate and constant enough to be a calibration standard for instruments. The Weston standard cell, for example, has a voltage of 1.018636 VDC at 20°C.

standard deviation   The RMS value of the deviations of a series of like quantities from their mean.

standard-frequency service   A radio communication service for the transmission of standard and specified frequencies of known high accuracy, intended for general reception.

standard-frequency signal   One of the highly accurate signals broadcast by the National Institute of Standards and Technology radio station WWV on 2.5, 5, 10, 15, 20, 25, 30, and 35 MHz at various scheduled times. It is used for testing and calibrating radio equipment throughout the world.

standard IC   An integrated circuit whose design has been widely accepted by customers and is typically manufactured by multiple sources. These devices are generally known by numeric designations without further identification of their prime source (e.g., 709, 7474, 386, 555, 8080).

standard inductor   An inductor (coil) that has a highly stable inductance value with little variation of inductance with current or frequency and a low temperature coefficient. It can have an air core or an iron core. It is a primary standard in laboratories and as a precise working standard for impedance measurements. It is also called an inductance standard.

standard missile   A U.S. Navy two-stage shipboard ship-to-ship or ship-to-air guided missile powered by solid fuel and guided by the launching ship's radar. Its explosive charge is detonated by a radar proximity fuse. It was formerly called the Tartar missile.

standardization   The setting of rules governing the acceptabily of materials or products including such factors as dimensional limits, electrical and mechanical characteristics, style, and format.

standardize   To adjust the exponent and coefficient of a floating-point result in a digital computer so the coefficient lies in the normal range of the computer.

standard microphone   A microphone whose response is accurately known for the condition under which it is to be used.

standard noise temperature   The temperature of 290 K (27°C) for evaluating the noise factor of a signal-transmission system.

standard observer   A hypothetical observer who requires standard amounts of primaries in a color mixture to match every color. The present standard primaries and the standard amounts required to match various wavelengths of the spectrum were established in 1931 by the International Commission on Illumination.

Standard Observer Curve   A curve established by the CIE as the industry standard for relating the total power (radiant flux) emitted from a source to the amount of power to which the eye is sensitive (luminous flux). The curve is on a logarithmic scale, and for reference various wavelengths

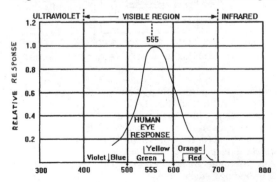

Standard Observer Curve shows the human eye's response to colors in the visible region. Peak eye response occurs at 555 nm in the yellow-green region.

of energy are labeled by color. The eye's response peaks in the yellow-green region. It is also called the *CIE photopic curve.*

standard pitch A musical pitch based on 440 Hz for tone A. With this standard, the frequency of middle C is 261.6 Hz.

standard propagation The propagation of radio waves over a smooth spherical earth of uniform dielectric constant and conductivity, under conditions of standard refraction in the atmosphere.

standard refraction The refraction that would occur in an idealized atmosphere in which the index of refraction decreases uniformly with height at the rate of 39×10^{-6} per kilometer.

standard resistor A resistor that is adjusted with high accuracy to a specified value and is only slightly affected by variations in temperature. It is also called a resistance standard.

standard sea-water conditions Sea water at a static pressure of 1 atmosphere, a temperature of 15°C, and a salinity such that the velocity of propagation is exactly 1500 m/s. Its density is then 1.02338 g/cm³, characteristic acoustic impedance is 153.507 CGS units, and pressure spectrum level of thermal noise is 82.17 dB below 1 µbar. The velocity of sound increases 0.018 m/s per meter of depth.

standards Documents that establish limits on materials and products relating to any or all of the following: dimensional limits and tolerances, weight, composition, electrical and mechanical characteristics, ability to withstand environmental stresses (e.g., temperature excursions, humidity, shock, and vibration), finish, inspection and quality control considerations, packaging, storing, shipping, and labeling.

standard source A light source that consists of a segment of fused thoria immersed in a chamber of molten platinum. When the platinum is at its melting point, the light emitted from the chamber approximates blackbody radiation.

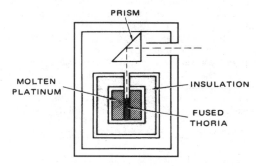

Standard source for radiation energy, as adopted by international agreement.

standard temperature and pressure A temperature of 0°C and a pressure of 760 mmHg.

standard time Mean solar time, based on the transit of the sun over a specified meridian called the time meridian, and adopted for use over an area called a time zone.

standard track *Single track.*

standard tuning frequency The frequency of 440 Hz, corresponding to the note A_4.

standard volume indicator A volume indicator that has the characteristics specified by American National Standards Association.

stand by A request to wait for additional messages to be transmitted a short time later.

standing wave A wave in which the ratio of an instantaneous value at one point to that at any other point is constant with time. A standing wave is produced by two waves with the same frequency traveling in opposite directions, such as a wave and its reflection from a discontinuity.

standing-wave antenna An antenna or antenna system whose current distributions are produced by standing waves of charges on the conductors.

standing-wave detector A detector that can be moved along the length of a transmission line or waveguide to locate the nodes or antinodes of a standing wave.

standing-wave loss factor The ratio of the transmission loss in an unmatched waveguide to that in the same waveguide when matched.

standing-wave meter An indicating instrument for measuring the standing-wave ratio in a transmission line or waveguide. It can include means for finding the locations of nodes and antinodes. The detecting device is generally a bolometer, crystal diode, or thermocouple. It is also called standing-wave-ratio indicator.

standing-wave ratio [SWR] The ratio of the maximum to the minimum amplitudes of corresponding components of a field, voltage, or current along a transmission line or waveguide in the direction of propagation and at a given frequency; or, alternatively, the reciprocal of this ratio.

standing-wave-ratio bridge A bridge that measures the standing-wave ratio in a transmission line, generally to verify the impedance match.

standing-wave-ratio indicator *Standing-wave meter.*

standing-wave system *Stationary-wave system.*

standing-wave voltage ratio [SWVR] The ratio of the maximum to the minimum voltage values along a transmission line.

standoff insulator An insulator that supports a conductor at a safe distance from the surface on which the insulator is mounted.

star network A set of three or more branches with one terminal of each connected at a common node to give the form of a star. It is also called a *star connection.*

starter 1. An auxiliary electrode in a gas tube that initiates conduction. 2. A device used with one type of fluorescent lamp to preheat the cathode and then apply starting voltage to initiate conduction. 3. A controller that starts a motor and brings it up to normal speed.

starting anode An anode that establishes the initial arc in a mercury-arc rectifier.

starting current The value of electron-stream current through an oscillator at which self-sustaining oscillations will start under specified conditions of loading. It is also called *preoscillation current.*

starting rheostat A rheostat that controls the current taken by a motor during starting and acceleration.

starting voltage The minimum voltage that must be applied to a radiation-counter tube to obtain counts with a particular circuit.

star tracker A telescopic instrument, on a missile or other flying platform that locks onto a celestial body and gives guidance to the missile or other platform during flight. A star tracker can be optical or radiometric.

state-variable filter   An active bandpass filter that consists of a summing amplifier followed by two integrators, which are operational amplifiers.

static   1. A hissing, crackling, or other sudden sharp sound that tends to interfere with the reception, comfort level, or enjoyment of the desired signals or sounds. When heard in an ordinary radio receiver, it might be caused by natural electric storms or improperly operating electric apparatus in the vicinity. Crackling sounds heard when listening to long-playing plastic phonograph records are caused by dust particles attracted to the record by surface electric charges built up by friction in rooms with low humidity. Static appears as small white specks or flashes, called snow, on a television picture.   2. Without motion or change.

static characteristic   A relation between a pair of variables such as electrode voltage and electrode current, with all other operating voltages for an integrated circuit, transistor, or other amplifying device maintained constant.

static charge   An electric charge accumulated on an object. See also *electrostatic charge.*

static convergence   Convergence of the three electron beams at an opening in the center of the shadow mask in a color TV picture tube. This is called static convergence because the beams must meet at this point when there are no scanning forces.

static converter   A frequency or voltage converter that includes static switching devices (with no moving parts) such as electron tubes and solid-state devices.

static dump   A dump that is performed at a particular time in a computer run, generally at the end of a run.

static electricity   The transfer of a static charge from one object to another by actual contact or by a spark that bridges an air gap between the objects. See *electrostatic discharge.*

static eliminator   A device that reduces the effect of atmospheric static interference in a radio receiver.

static focus   The focus of the undeflected electron beam in a cathode-ray tube.

static frequency converter   A frequency converter that has no moving parts, such as a solid-state static alternator.

static machine   A machine that generates electric charges, usually by electric induction, to build up high voltages for research purposes. A Van De Graaf generator is an example.

static pressure   The pressure that would exist at a point in a medium with no sound waves present. In acoustics, the commonly used unit is the micro-bar. It is also called hydrostatic pressure.

static read-only memory [SRAM]   A semiconductor volatile memory based on transistor gates that offers high access speeds but is less efficient in its use of silicon than *dynamic RAMs*. Unlike DRAMs, SRAMs do not require refreshing to hold their data, but that data is lost when power is shut off.

static subroutine   A computer subroutine that involves no parameters other than the addresses of the operands.

static switching   Switching of circuits by transistors, silicon controlled rectifiers and other devices that have no moving parts.

static test   A measurement taken under conditions where neither the stimulus nor the environmental conditions fluctuate.

station   1. An assembly line or assembly machine location at which a wiring board or chassis is stopped for insertion of one or more parts.   2. A location where radio, televi-sion, radar, or other electronic equipment is installed.   3. *Broadcast station.*

stationary contact   A contact that is rigidly fastened to the frame of a switch or relay, and does not move during operation.

stationary orbit   A circular, equatorial, and synchronous orbit, in which the satellite appears stationary with respect to any point on the earth's surface because the satellite altitude of 22,300 mi (35,890 km) and orbital velocity of about 1.9 mi/s (3 km/s) keep it in a fixed relation to points on the earth. A stationary orbit must be synchronous, but a synchronous orbit is not necessarily stationary.

stationary satellite   *Geostationary satellite.*

stationary wave   A standing wave in which the amplitudes of the wave components are equal, so the energy flux is zero at all points.

stationary-wave system   An interference pattern charac-terized by stationary nodes and antinodes. It is also called a standing-wave system.

statistical test   A procedure that determines whether observed values or quantities fit a hypothesis well enough for the hypothesis to be accepted.

stator   1. The part of a rotating machine that contains the stationary parts of the magnetic circuit and their associ-ated windings.   2. The stationary set of plates in a vari-able capacitor.

stator plate   One of the fixed plates in a variable capac-itor. Stator plates are insulated from the frame of the capacitor.

statute mile   A unit of distance equal to 5280 ft (1.609344 km).

STC   Abbreviation for *sensitivity-time control.*

steady state   The condition in which circuit values remain essentially constant, after initial transients or fluctuating conditions have disappeared.

steady-state error   The error that remains after transient conditions have disappeared in a control system.

steatite   A dense, nonporous heat-resisting ceramic that consists chiefly of a silicate of magnesium; it has excellent insulating properties, even at high frequencies. It can be molded and fired in various shapes to make sockets, insu-lators, and terminal blocks.

steerable antenna   A directional antenna whose major lobe can be readily shifted in direction.

Stefan-Boltzmann law   The total emitted radiant energy of a blackbody is proportional to the fourth power of its absolute temperature.

Steinmetz formula   An empirical formula for the magnetic hysteresis loss per unit volume of material per magnetiza-tion cycle, specifying that the energy loss in ergs is propor-tional to the 1.6th power of the maximum flux density.

stellar guidance   *Celestial guidance.*

stellar map matching   Guidance in which a map of the stars is matched with the position of the stars as observed through a telescope, to provide guidance for a missile, spacecraft, or other vehicle.

stem   The inward-projecting section of the glass envelope of an electron tube, through which the heavy leads pass that support and make connections to the electrodes.

stenode circuit   An IF amplifier circuit whose crystal filter passes only signals at the exact IF value, giving high selec-tivity.

step 1. One operation in a computer routine. 2. A portion of a step-function waveform, consisting of a single sudden change in amplitude and a period of time at the new amplitude value.

step attenuator An attenuator whose attenuation can be varied in precisely known steps by switches.

step-by-step excitation The successive transitions of an atom to higher levels of excitation.

step-by-step system 1. A control system in which the drive motor moves in discrete steps when the input element is moved continuously. 2. *Strowger system.*

step-down transformer A transformer whose secondary winding AC voltage is lower than its primary winding AC voltage, indicating a voltage step-down.

step function A signal that has zero value before a certain instant of time, and a constant nonzero value immediately after that instant.

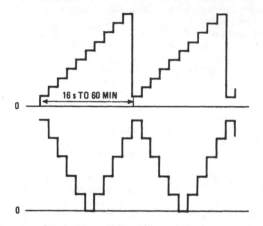

Step functions as viewed on an oscilloscope.

step-function generator A function generator whose output waveform increases and decreases suddenly in steps that might or might not be equal in amplitude.

step-function response The time variation of an output signal when a specified step-function input signal or disturbance is applied.

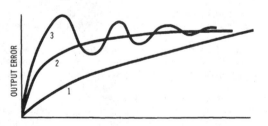

Step-function response in a closed-loop control system is shown graphically as curve 1 (overdamped), curve 2 (critically damped), and curve 3 (underdamped).

step-index fiber A *multimode optical fiber* made of ultrapure quartz clad with a polymer of lower *refractive index* or with a halide-doped, low-refractive-index glass. It has an upper limit of about 30 MHz.

step input The introduction of a *step function* into a *closed-loop system* as a test of its response to an extreme signal. The objective is to determine a step-input command response.

step-input response One of three responses of the load or *end effector* to the introduction of a *step input* into a *closed-loop system* or *servosystem:* overdamped, critically damped, and underdamped. See also *damped wave* and *damping.*

stepper motor A motor that rotates in short and essentially uniform angular movements rather than continuously. Typical steps are 30, 45, and 90°. The angular steps are obtained electromagnetically rather than by ratchet and pawl mechanisms, as in stepping relays. The two basic types are permanent-magnet and variable-reluctance stepper motors.

stepping *Zoning.*

stepping register A register in which an appropriate AC waveform controls the locations of stored data in computers.

step-recovery diode A diode that stores a charge while conducting in the forward direction. When the applied voltage is reversed, the diode conducts for a brief time, up to about 300 ns, until the stored charge is removed, and then abruptly stops conducting. It is in harmonic-generating and pulse-sharpening circuits.

step up To increase the value of electrical quantity.

step-up 1. An increase in the value of an electrical quantity. 2. A reference to an increase in the value of an electrical quantity.

step-up transformer A transformer whose secondary winding AC voltage is higher than its primary winding AC voltage, indicating a voltage step-up.

steradian [sr] The SI unit of solid angle, subtending a spherical surface whose area is equal to the square of the radius. The total solid angle about a point in space is 4π sr.

Sterba-curtain array A stacked array with a curtain reflector, suspended from messenger cables running between two steel towers. The curtain can be parasitic or excited. It is in a high-power transmitter for highly directional long-range communications.

stereo 1. A reference to three-dimensional pickup or reproduction of sound, as achieved with two or more separate audio channels. It is also called stereophonic. 2. *Stereo sound system.*

stereo- A prefix that designates a three-dimensional characteristic.

stereo amplifier An audio-frequency amplifier that has two or more channels, as required for use in a stereo sound system.

stereo broadcasting The broadcast of two sound channels for reproduction by a stereo sound system that has a stereo tuner at its input.

stereo effect The reproduction of sound so that the listener receives the sensation that individual sounds are coming from different locations, as did the original sounds reaching the stereo microphone system.

stereo FM In stereo broadcasting systems in the United States, the main FM carrier is modulated by the sum of the left and right stereo channels, and a subcarrier at 38 kHz is suppressed after being amplitude-modulated by the difference between the left and right channel signals. A 19-

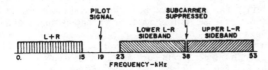

Stereo FM spectrum for one 200-kHz FM channel. The assigned carrier frequency of the station would be at 0 kHz on the frequency scale.

kHz pilot tone is added to indicate at the receivers the presence of a stereo transmission. The additional bandwidth remaining in the channel assigned to the FM station is available for SCA background music service or other services. At the receivers, the 19-kHz tone is doubled and used as a reference for demodulating the subcarrier sidebands. After demodulation, the pairs of modulations are added to get the left channel and subtracted to get the right channel.

stereo microphone system An arrangement of two or more microphones spaced far enough apart to give two different output signals, as required for making a stereo recording or feeding a stereo sound system directly or by radio.

stereophonic *Stereo.*

stereo preamplifier An audio-frequency preamplifier that has two channels, for use in a stereo sound system.

stereo record A single-groove disk record that has V-shaped grooves at 45° to the vertical. Each groove wall has one of the two recorded channels. It is also called a stereo recording.

stereo recorded tape Recorded magnetic tape that has two separate recordings, one for each channel of a stereo sound system. It is also called a stereo recording.

stereo recording 1. *Stereo record.* 2. *Stereo recorded tape.*

stereoscopic A reference to a three-dimensional visual image based on two slightly different images that are integrated by the human brain so they convey depth.

stereoscopic television Television that imparts a three-dimensional appearance to viewed images.

stereo separation The ratio of the electric signal in the right stereo channel to the signal in the left stereo channel when only a right signal is transmitted, or vice versa.

stereo sound system A sound-reproducing system that has a stereo pickup, stereo tape recorder, stereo tuner, or stereo microphone system which feeds two independent audio channels. Each channel terminates in one or more loudspeakers arranged to give listeners the same audio perspective that they would get at the original sound source. It is also called stereo.

stereo subcarrier A subcarrier whose frequency is the second harmonic of the pilot subcarrier frequency for stereo FM broadcasting.

stereo subchannel The band of frequencies from 23 to 53 kHz, containing the stereo FM subcarrier and its associated sidebands.

stereo tape recorder A magnetic-tape recorder that has two stacked playback heads, for the reproduction of stereo recorded tape.

stereo tuner A tuner that has provisions for receiving both channels of a stereo broadcast.

sticking The tendency of a flip-flop or other bi-stable circuit, to stay in one of its two stable states, or switch spontaneously to that state.

stiction [STatic frICTION] Friction that tends to prevent relative motion between two movable parts at their null position. It is often seen in moving-coil meters.

stiffening capacitor A capacitor added to an automobile battery power supply to improve bass response in a car's audio equipment.

stilb [sb] A CGS unit of luminance, equal to 1 cd/cm². The SI unit of luminance, the candela per square meter, is preferred.

stimulated emission A downward transition of electron energy levels under conditions such that a photon of the correct frequency will stimulate another transition and make two photons available. The process repeats to give rapid cumulative buildup of coherent light output, as in lasers. Stimulated emission, population inversion, and light amplification together form the basis for laser operation.

stimulus A signal that affects the controlled variable in a control system.

stitch bonding *Wedge bonding.* See also *wire bonding.*

STM Abbreviations for 1. *scanning tunneling microscope* and 2. *synchronous transfer mode.*

STN Abbreviation for *supertwisted nematic.*

stochastic process A random sampling process in which each and every member of the population has the same opportunity for being selected.

stochastic variable A variable that is dependent on the random variable and is usually measured experimentally.

stoichiometric impurity A crystalline imperfection in a semiconductor.

Stokes' law The wavelength of luminescence excited by radiation is always greater than that of the exciting radiation.

stop The aperture or useful opening of a lens, usually adjustable by a diaphragm.

stop band The band of frequencies that is highly attenuated by a filter or other frequency-sensitive device.

stopping voltage The voltage required to stop an electron emitted by photoelectric or thermionic action.

STP Abbreviation for 1. *shielded twisted pair* (shielded pair of twisted copper wires) and 2. *signaling transfer point.*

storage *Memory.*

storage battery *Battery.*

store To transfer an element of information to memory or storage in a computer for later extraction.

storecasting A Subsidiary Communications Authorization service in which a standard stereo FM broadcast carrier is modulated by an additional signal, usually 60 to 74 kHz above the carrier frequency, for transmission of background music to stores and public buildings.

stored base charge The phenomenon associated with the storage of minority charge carriers in the base region of alloy-junction-type transistors under conditions of saturation.

stored-energy welding Welding with electric energy that is accumulated electrostatically, electromagnetically, or electrochemically at a relatively slow rate and released at the required rate for welding.

stored-program computer A digital computer that can translate or otherwise alter an input program by using internally stored instructions and then executing the rewritten program.

straight dipole   A dipole that consists of a single straight conductor, usually broken at its center for connection to a transmission line.

straight-line capacitance [SLC]   A variable-capacitor characteristic obtained when the rotor plates are shaped so that capacitance varies directly with the angle of rotation.

straight-line frequency [SLF]   A variable-capacitor characteristic obtained when the rotor plates are shaped so the resonant frequency of the tuned circuit containing the capacitor varies directly with the angle of rotation.

straight-line path   The axis of the Fresnel-zone family of paths between two microwave antennas.

straight-line wavelength [SLW]   A variable-capacitor characteristic obtained when the rotor plates are shaped so the wavelength for resonance in the tuned circuit that contains the capacitor varies directly with the angle of rotation.

strain gage   A resistive transducer for measuring mechanical strain. It is essentially a conductor or semiconductor with a small cross-sectional area that is cemented to the surface of the material whose strain is to be measured. The gage elongates or contracts with that surface, and the deformation changes the resistance of the gage. There are five different forms: bare-wire bondable, bondable wire premounted on a paper or plastic carrier base, metal-foil bondable, semiconductor, and deposited-metal (thin-film). The resistance change is converted into a voltage by connecting one or more similar gages as arms of a *strain-gage bridge*.

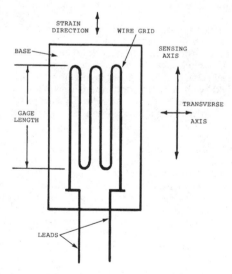

Strain gage. A bondable flat wire grid form of gage is shown.

strain-gage bridge   A Wheatstone bridge that includes one, two, or four similar strain gages in its arms. When the bridge is excited, the resistance change is converted into a voltage for display or recording. The gages of the bridge are cemented to the material to be stressed. In a four-gage

bridge, two gages show increases in resistance and two show decreases, thus giving a larger reading than a single gage.

strain-gage multiplier   A time-varying resistance multiplier that has a strain gage as its time-varying element. One variable controls the strain in the gage, and the other variable controls the current through it. The voltage across the strain gage is proportional to the required product.

strain insulator   An insulator located between sections of a stretched wire or antenna to break up the wire into insulated sections while withstanding the total pull of the wire.

strand   One of the wires or groups of wires in a stranded wire.

stranded conductor   *Stranded wire.*

stranded wire   A conductor composed of a group of wires or a combination of groups of wires, usually twisted together. It is also called *stranded conductor*.

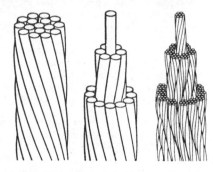

Stranded wire.

strap   *Strapping.*

strapping   Concentric conductive rings in a multicavity magnetron that connect the upper and lower edges of alternate vanes of the resonant cavities that have the same polarity. They suppress undesirable oscillation modes in the magnetron. Typically made from copper, two are brazed to the vanes on each side of the cavities.

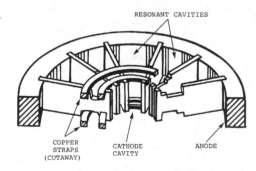

Straps connect alternate anode vanes of a multicavity magnetron together to improve its stability.

stratosphere   A stratum of the earth's atmosphere above the troposphere, extending from about 7 mi (11.25 km) up to about 50 mi (80 km) above the earth. The temperature is essentially constant in the stratosphere.

stray capacitance Undesirable capacitance between circuit wires, wires and the chassis, or components and the chassis of electronic equipment.

stray field Leakage magnetic flux that spreads outward from a coil and does no useful work.

stray radiation Radiation that serves no useful purpose.

streaking A television picture condition indicated by white or black horizontal streaks or smudges that appear to follow images across the screen. The effect is more apparent at the vertical edges of objects where there is an abrupt transition from black to white or white to black. It can be caused by excessive low-frequency response.

streamer An indefinite wavy band that occurs when the gas pressure in a discharge tube is reduced below the value required for a glow discharge through the tube.

striated discharge An electric discharge characterized by alternate light and dark bands in the positive column adjacent to the anode of a glow-discharge tube.

striation technique A technique for making sound waves visible by using their individual abilities to refract light waves.

striking 1. Starting an electric arc by touching the electrodes together momentarily. 2. Electrodeposition of a thin initial film of metal, usually at a high current density.

striking voltage The grid-cathode voltage required to start the flow of anode current in a gas tube.

string A group of data items that are in an ascending or descending sequence according to alphabetic, numerical, or other rules.

string electrometer An electrometer that has a conducting fiber stretched midway between two oppositely charged metal plates. The electrostatic field between the plates displaces the fiber laterally in proportion to the voltage between the plates.

strip To remove insulation from a wire.

strip-chart recorder A recorder with one or more writing pens or other recording devices that trace changes in a measured variable on the surface of a strip chart that is moved at constant speed by a time-clock motor.

stripline A strip transmission line that consists of a flat metal-strip center conductor which is separated from flat metal-strip outer conductors by dielectric strips.

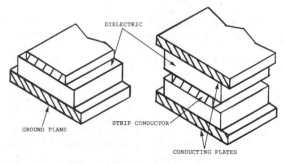

Stripline is effective for short-range microwave transmission when space for conventional coaxial cable or waveguides is restricted.

stripper A hand or motorized tool that removes insulation from wires.

stripping Removal of a metal coating, as when etching away undesired portions of a printed circuit.

strip transmission line A microwave transmission line that consists of a thin, narrow rectangular strip which is separated from a wide ground-plane conductor or mounted between two wide ground-plane conductors. Separation is usually achieved with a low-loss dielectric material, typically alumina ceramic, on which the conductors are formed by printed-circuit techniques.

strobe 1. A pulse that gates the output of a counter, shift register, or other computer circuit, to produce a desired action. 2. A pulse superimposed on a radar image to serve as a marker from which the range of the target can be measured. 3. A line or wedge produced on a radar screen by a jamming signal. 4. *Stroboscope.*

strobe marker A small bright spot or a short gap or other discontinuity produced on the trace of a radar display to indicate the part of the time base that is receiving attention.

stroboscope A controllable intermittent source of intense light, that creates the illusion of slowing down or stopping vibrating or rotating objects. The flashing frequency is adjusted until it corresponds to some multiple of the speed of vibration or rotation of the object under study. It is also called a strobe.

stroboscopic disk A printed disk that has two or more concentric rings, each containing a different number of dark and light segments. When the disk is placed on a phonograph turntable or rotating shaft at a known frequency by a flashing discharge tube, speed can be determined by noting which pattern appears to stand still or rotate slowly.

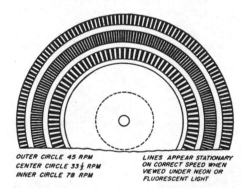

OUTER CIRCLE 45 RPM
CENTER CIRCLE 33⅓ RPM
INNER CIRCLE 78 RPM

LINES APPEAR STATIONARY ON CORRECT SPEED WHEN VIEWED UNDER NEON OR FLUORESCENT LIGHT

Stroboscopic disk for checking speed of phonographs with 120-Hz flashes from glow lamp operating on 60-Hz power.

stroboscopic tachometer A stroboscope that has a scale which reads in flashes per minute or revolutions per minute. The speed of a rotating device is measured by directing the stroboscopic lamp on the device, adjusting the flashing rate until the device appears to be stationary, then reading the speed directly on the scale of the instrument.

stroboscopic tube A cold-cathode gas-filled arc-discharge tube that produces intensely bright flashes of light for use with a stroboscope.

strontium [Sr]   A metallic element that is used in cathodes of phototubes to obtain maximum response to ultraviolet radiation. It has an atomic number of 38.

strontium 90   A radioisotope that has a half-life of about 25 years. It is also called radio strontium.

stub   1. A short section of transmission line, open or shorted at the far end, connected in parallel with a transmission line to match the impedance of the line to that of an antenna or transmitter.   2. A solid projection one quarter-wavelength long that forms an insulating support in a waveguide or cavity.

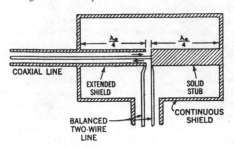

Stub in a cavity to support a conductor during conversion from coaxial line to balanced two-wire line.

stub-matching   The use of a stub to match a transmission line to an antenna or load. Matching depends on the spacing between the two wires of the stub, the position of the shorting bar, and the point at which the transmission line is connected to the stub.

stub-supported line   A transmission line that is supported by short-circuited quarter-wave sections of coaxial line. A stub exactly a quarter-wavelength long acts as an insulator because it has infinite reactance.

stub-tuned filter   A microwave stopband filter that consists of a number of T junctions of different sizes, inserted

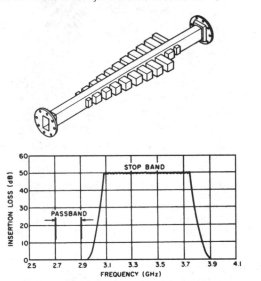

Stub-tuned filter for 2.7–2.9 GHz waveguide, providing 50-dB attenuation between 3.1 and 3.7 GHz.

in a waveguide to produce high attenuation over a band of frequencies. It can suppress undesired frequencies that can be up to the fourth harmonic.

stub tuner   An adjustable shorted stub for adjusting a transmission line for maximum power transfer.

studio   A room in which television or radio programs are produced.

stylus [plural **styli**]   The part of a phonograph pickup that follows the modulations of a record groove and transmits the resulting mechanical motions to the transducer element of the pickup for conversion to corresponding AF signals. It is also called a *needle* and a reproducing stylus.

s-type negative resistance   A voltage-stable negative resistance in which a given current in the operating range can have three different possible values of terminal voltage. Examples include tunnel diodes and the common-emitter input of a point-contact transistor.

styrofoam   A foamed plastic that exhibits low water absorption, is light in weight, and can float. It is also a thermal insulator and can be molded to specific shapes to conform to and support electronic equipment during shipment. It is also available in pellet form for use as a shock-absorbing package fill. However, only conductive styrofoam should be in contact with ESD-sensitive components and circuits.

subassembly   Two or more parts joined together to form a functional entity but which have replaceable components and are subordinate to a higher-level assembly, product, or system. Examples include completed circuit boards, printer heads, speaker cabinets, and tape decks.

subatomic   A reference to particles smaller than atoms, such as electrons, protons, and neutrons.

subaudio   *Infrasonic.*

subcarrier   1. A carrier that is applied as a modulating wave to modulate another carrier.   2. *Chrominance subcarrier.*

subcarrier band   A band associated with a given subcarrier and specified in terms of maximum subcarrier deviation.

subcarrier discriminator   A discriminator that demodulates a telemetering subcarrier frequency.

subcarrier frequencies   *Channels* assigned for specific applications, such as telemetry. There are standardized subcarrier channels for FM/FM telemetry systems. There are 18 IRIG standardized subcarrier bands for telemetry.

subcarrier oscillator   1. The crystal oscillator that operates at the chrominance subcarrier or burst frequency of 3.579545 MHz in a color television receiver. This oscillator, synchronized in frequency and phase with the transmitter master oscillator, furnishes the continuous subcarrier frequency required for demodulators in the receiver.   2. An oscillator for a telemetering system that translates variations in an electrical quantity into variations of a frequency-modulated signal at a subcarrier frequency.

subharmonic   A sinusoidal quantity that has a frequency which is an integral submultiple of the frequency of some other sinusoidal quantity to which it is referred. A third subharmonic would be one-third the fundamental or reference frequency.

subject copy   The text document or graphic that is to be transmitted over the public telephone lines by a *facsimile machine*. See also *facsimile copy*.

subliminal   Below the threshold of conscious responsiveness to a stimulus. Applications include behavior modification that involves audio or video motivational stimuli.

submillimeter band A band of frequencies >300.0 GHz, in accordance with IEEE Standard 521-1976.

submillimeter wavelength A wavelength shorter than 1 mm, corresponding to frequencies above 300 GHz.

subminiaturization Reduction of size and weight of electronic equipment, generally achieved with integrated circuits, printed circuits, modules, and surface-mount components.

subnanosecond Less than 1 ns, or less than one-billionth of a second.

subnanosecond radar A radar that transmits very short pulses with durations of less than 1 ns. It has an effective range resolution less than 3 in (7.5 cm), permitting accurate measurements by radar of ice and snow thickness, surface reflection coefficients, and signal attenuation of sea water.

subnotebook computer A personal computer typically weighing less than 4 lbs (2 kg). It is smaller than a notebook computer, but has a miniature keyboard, small LCD screen, and limited data-processing capability sufficient for note keeping, scheduling, and certain communications functions. It is also called a *personal digital assistant* [PDA].

subrefraction A large-scale positive troposphere refraction effect that determines the trajectories of radio waves and their propagation distance. It can cause diffractive fading due to obstructions, such as mountains, on the path. It is caused by random variations in the refractive index resulting from changes in air pressure, temperature, and water-vapor content. See also *scintillation* and *superrefraction*.

SUBROC [SUBmarine ROCket] A Navy missile that is fired underwater from a conventional torpedo tube but is rocket-powered and most of its trajectory is through the air. The missile reenters the water at supersonic speed and detonates at the preset depth at which an enemy submarine has been located. Its range is about 50 mi (80 km).

subroutine A portion of a routine that causes a computer to carry out a well-defined mathematical or logic operation. At its conclusion, control reverts to the master routine.

subscriber line interface circuit [SLIC] A circuit that performs the 2-to-4 wire conversion, battery feed, line supervision, and common-mode rejection at the central office (CO) or private-branch exchange (PBX) end of the telephone line. Today SLICs are made as integrated circuits.

subset A telephone or other subscriber equipment connected to a communication system.

subsonic Less than the velocity of sound in air, hence less than Mach 1, which is about 738 mi/h (1188 km/h).

substrate 1. In the manufacture of silicon, gallium arsenide or other semiconductors, the wafer on which all active and passive components are formed. 2. In hybrid microcircuitry, a ceramic (e.g., alumina) wafer on which all conductors are formed and, in some hybrids, resistors and capacitors as well. 3. In liquid-crystal displays, glass wafers on which the electrodes are formed. 4. A circuit board on which all components are inserted and/or bonded, typically made of glass-fiber epoxy laminate (GFE).

subsurface wave An electromagnetic wave that has an underwater or underground propagation path. Operating frequencies for communication must generally be below about 35 kHz because of attenuation of higher frequencies by water or earth.

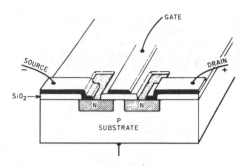

Substrate of field-effect transistor.

subsynchronous Operating at a frequency or speed that is a submultiple of the source frequency.

subsystem An interconnected combination of a set of related components or circuits to form a subdivision of a system. Examples include the computer monitor, printer, keyboard, modem, antenna, power module, and transmit-receive console.

successive approximation register An analog-to-digital conversion method in which the input voltage is compared with the output voltage of a sequentially programmed digital-to-analog (D/A) converter. First, the most significant bit (MSB) of the D/A is turned on and compared with the analog input. If the input is greater than the D/A output, the MSB is left on; otherwise it is turned off. This process is then repeated for all other bits in decreasing order until the least significant bit (LSB) is reached. See also analog-to-digital converter.

Suhl effect An effect that occurs when a strong transverse magnetic field is applied to an N-type semiconducting filament. Holes injected into the filament are deflected to the surface, where they can recombine rapidly with electrons or be withdrawn by a probe. The overall effect is an increase in conductance.

sulfur [S] An element with an atomic number of 16 and an atomic weight of 32.06.

sulfur hexafluoride A dielectric gas that suppresses arcing in high-power radar waveguides.

sulfuric acid A compound of sulfur, hydrogen, and oxygen; it has the chemical formula H_2SO_4. It is the electrolyte in lead-acid batteries.

sum-and-difference monopulse radar A monopulse radar that has a receiving antenna with one aperture and two or more closely spaced feeds. Each feed produces a radiation pattern that is displaced from the antenna boresight axis. Signals arriving off the axis give unequal amplitudes in the two channels, and signals on the axis give equal amplitudes that produce a sharp null in the difference channel and a peak in the sum channel.

sum channel A combination of two stereophonic channels that provide a program which can be recorded or transmitted by a single channel.

summation bridge A bridge that measures such values as temperature, frequency, speed of rotation, time, resistance, and capacitance by adding the original bridge current to the current needed for balance, and presenting the result on an indicator or scale.

summation check A redundancy check that sums groups of digits usually without regard for overflow, and that sum is checked against a previously computed sum to verify the accuracy of a digital computer.

summation network   *Summing network.*

summing amplifier   An amplifier that delivers an output voltage which is proportional to the sum of two or more input voltages or currents.

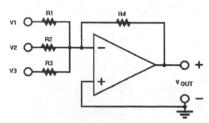

Summing amplifier is an operational amplifier that can provide an output which is the sum of two or more inputs.

summing network   A passive electrical network whose output voltage is proportional to the sum of two or more input voltages. It is also called a summation network.

summing point   A mixing point whose output is obtained by addition, with prescribed signs, of its inputs in a feedback control system. In an operational amplifier circuit, the junction of the input with the feedback network is commonly called the summing point.

S unit   An arbitrary unit of signal strength, that can be used along with decibels on the scale of a signal-strength meter.

S-unit meter   *Signal-strength meter.*

sunlight recorder   A recorder that includes a photosensor, capacitor-charging circuit, and SCR-operated counter to record the integrated value of solar radiation over a period of time.

sunspot   A dark spot on the sun, usually associated with the magnetic storms on the earth that affect radio communication at the lower frequencies.

sunspot cycle   A period of about 11 years in which the number and duration of sunspots and solar flares pass through a cycle of buildup to a maximum value and then drop back to a minimum value. It is also called a solar cycle.

sunspot number   The predicted number of sunspots for a given month.

superconducting   The state of a superconductor when it exhibits superconductivity below a critical temperature.

superconducting generator   A DC generator that in one form consists of a series of flat plates of superconducting lead or niobium arranged in a circle, wired together by superconducting wire, and cooled near absolute zero by an appropriate liquid gas.

superconducting memory   A memory made from multiple cryotrons, thin-film cryotrons, superconducting thin films, or other superconducting memory devices. These operate only under cryogenic conditions and dissipate power only during the read or write operation.

superconducting solenoid   A solenoid that uses superconducting wires to generate strong magnetic fields.

superconducting thin film   A thin film of indium, tin, or other superconducting element, functioning as a cryogenic switching or storage device, as in a thin-film cryotron.

superconducting transition   A transition from a normal state to a superconducting state, occurring at a temperature that depends on the magnetic field as well as on the nature of the material.

superconductive materials   Superconductivity was discovered in mercury and later found in tin and lead. Materials that exhibit the property at low temperature include niobium, titanium, and niobium-tin. Various compounds of barium, bismuth, copper, lanthanum, oxygen, thalium, and yttrium have exhibited the property at higher temperatures. A ceramic compound of thalium barium calcium copper oxide exhibited superconductivity at 125 K.

superconductivity   The property of certain metals, alloys, and ceramics that causes DC electrical resistance to diminish, permitting significant increases in conductivity when cooled below transitional temperatures close to absolute zero or 0 K (−273.16°C or 460°F). The conductor can carry current without loss of energy and generate powerful magnetic fields with coils that do not have magnetic cores. See also *maglev* and *superconductive materials.*

superhet   *Superheterodyne receiver.*

superheterodyne receiver   A receiver that converts all incoming modulated RF carrier signals to a common IF carrier value for additional amplification and selectivity prior to demodulation, with heterodyne action. The output of the IF amplifier is then demodulated in the second detector to give the desired AF signal.

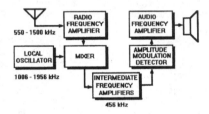

Superheterodyne circuit is the accepted standard for radio reception in the 550- to 1500-kHz broadcast band.

superhigh frequency [SHF]   A Federal Communications Commission designation for a frequency in the band from 3 to 30 GHz, corresponding to a centimetric wave between 1 and 10 cm.

superlattice   A heterostructure composed of thin layers of uniform thickness of different semiconductor materials such as gallium arsenide (GaAs) interspersed with aluminum gallium arsenide (AlGaAs). Each pair of layers has the same thickness.

superposition theorem   The current that flows at a point in a linear network during the simultaneous application of many voltages throughout the network is the sum of the component currents at the point which would be produced by the individual voltages acting separately. Similarly, the voltage between any two points under such conditions is the sum of the voltages that would be produced between these two points by the individual voltages acting separately.

superpower station   A station that broadcasts at extremely high power, generally more than 1 megawatt.

superrefraction   A large-scale negative troposphere refraction effect that governs the trajectories of radio waves and their propagation distance. It can bend the ray paths toward the earth's surface for *trapping* or *ducting*. It is caused by random variations in the refractive index resulting from changes in air pressure, temperature, and water-vapor content. See also *scintillation* and *subrefraction*.

superregeneration Regeneration that breaks up or quenches oscillation at a frequency slightly above the upper audible limit of the human ear by a separate oscillator circuit connected between the base and collector of the amplifier transistor. It prevents the regeneration from exceeding the maximum useful amount.

superregenerative detector A detector that uses superregeneration to obtain extremely high sensitivity with a minimum number of amplifier stages.

superregenerative paramagnetic amplifier A paramagnetic amplifier that gives a much higher gain-bandwidth product than conventional amplifiers, but with a higher noise figure. It can be self-quenched or have separate quenching.

superregenerative receiver A tuned-radio-frequency receiver that has a superregenerative detector.

supersonic Faster than the velocity of sound in air, hence faster than Mach 1, which is about 738 mi/h (1188 km/h). The term applies to airplanes and missiles.

superstructure A semiconductor heterostructure composed of thin layers of different thicknesses to achieve certain desired electrical properties.

supersync signal A combination horizontal and vertical sync signal transmitted at the end of each television scanning line to synchronize the operation of a television receiver with that of the transmitter.

superturnstile antenna A modified turnstile antenna that has wing-shaped dipole elements in pairs mounted at right angles about a common vertical axis. The dipole pairs are fed in quadrature to give substantially omnidirectional radiation over a wide band for FM and television transmitters.

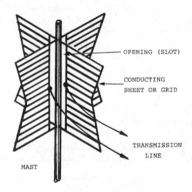

OPENING (SLOT)

CONDUCTING SHEET OR GRID

TRANSMISSION LINE

MAST

Superturnstile antenna is fed in quadrature for TV-signal transmission.

supertwisted nematic [STN] A form of liquid crystal that is used in active-matrix liquid-crystal video displays. See *liquid-crystal display* and *nematic LCDs*.

super VGA A circuit board for a personal computer video monitor that produces higher resolution graphics than the video graphics array (VGA). See also *video graphics array*.

supervisory control system A control system that provides both indication and control for remotely located equipment by electrical means, as by carrier-current channels on power lines.

supply voltage The voltage obtained from a power source for operation of a circuit or device. It is also called rail voltage (British).

suppressed carrier [SC] A carrier that is suppressed at the transmitter. The chrominance subcarrier in a color television transmitter is an example.

suppressed-carrier transmission Transmission in which the carrier component of the modulated wave is suppressed, leaving only the sidebands to be transmitted.

suppressed time delay *Code delay.*

suppressed-zero instrument An indicating or recording instrument whose zero position is below the lower end of the scale markings.

suppression Elimination of any component of an emission, as a particular frequency or group of frequencies in an AF or RF signal.

suppressor A resistor in series with a spark plug or distributor of an gasoline-powered engine to suppress spark interference that might otherwise interfere with radio reception.

suppressor capacitor A capacitor, typically ceramic, that is connected across the sweep circuit primary or secondary winding of a power supply transformer to suppress transient voltage spikes and block them from the power supply.

suppressor grid A grid placed between two positive electrodes in an electron tube primarily to reduce the flow of secondary electrons from one electrode to the other. It is usually placed between the screen grid and the anode.

suppressor-grid modulation Amplitude modulation in which the modulating signal is applied to the suppressor grid of a pentode that is amplifying the carrier signal.

surface acoustic wave [SAW] An acoustic wave that has frequencies up to several gigahertz, traveling on the optically polished surface of a piezoelectric substrate at a velocity which is only about 10^{-5} times that of electromagnetic waves. A surface acoustic wave thus has the slow-travel property of sound while retaining the microwave frequency of its source. It is the basis for delay lines, filters, pulse processors, and other microwave devices and circuits. Piezoelectric substrates in common use include bismuth germanium oxide ($Bi_{12}GeO_{20}$), lithium niobate ($LiNbO_3$), and quartz. It is also called an acoustic surface wave.

surface-acoustic-wave [SAW] **delay line** A delay line whose delay is determined by the distance that a surface acoustic wave travels on a piezoelectric surface.

surface-acoustic-wave filter A filter that consists of a piezoelectric substrate with a polished surface along which surface acoustic waves can propagate. Metallic input and output transducers are deposited at the opposite ends of the active face of the substrate.

surface analyzer An instrument that measures or records irregularities in a surface by moving the stylus of a crystal pickup or similar device over the surface, amplifying the resulting voltage. The output voltage is fed to an indicator or recorder that shows the surface irregularities magnified as much as 50,000 times.

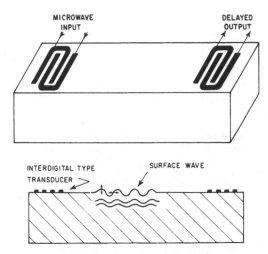

Surface acoustic wave produced by interdigital transducers of a delay line.

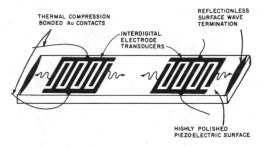

Surface-acoustic-wave delay line.

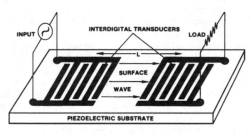

Surface-acoustic-wave filter.

surface barrier   A potential barrier formed at a surface of a semiconductor by the trapping of carriers at the surface. The effective area of the barrier is appreciably larger than for a point-contact transistor.

surface-barrier detector   A semiconductor nuclear-particle detector that has its rectifying junction between an evaporated gold layer and high-resistivity N-type silicon. Its performance is similar to that of a PN junction.

surface-barrier diode   A diode whose thin surface layers, formed either by deposition of metal films or surface diffusion, serve as a rectifying junction.

surface-charge transistor   An integrated-circuit transistor element based on controlling the transfer of stored electric charges along the surface of a semiconductor.

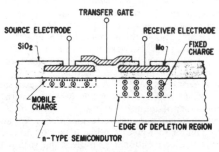

Surface-charge transistor cross section.

surface-contact rectifier   A rectifier whose surface barrier serves as the rectifying contact.

surface density   The quantity per unit area of any substance distributed over a surface.

surface duct   An atmospheric duct whose lower boundary is the surface of the earth.

surface hardening   The hardening of a metallic surface by rapid induction heating and rapid quenching.

surface leakage   The unwanted passage of current over the surface of an insulator.

surface magnetic wave   A magnetostatic wave that can be propagated on the surface of a magnetic material, as on a chip of yttrium-iron garnet (YIG).

surface micromachining   A *micromachining* technique that makes use of deposited or grown layers of aluminum or silicon on a silicon substrate for the fabrication of microminiature mechanical actuators and sensors.

surface-mount device [SMD]   A component in a package with terminals that are stubs or tabs to permit them to be soldered directly to the surface of the circuit board without holes for insertion. Examples include chip capacitors, chip resistors, leadless diodes, trimmers, small-outline packaged transistors (SOT), and integrated circuits (SOIC).

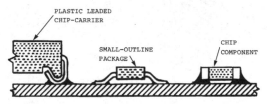

Surface-mount components are leadless so that they can be soldered directly to the surface of the circuit board.

surface-mount technology [SMT]   A manufacturing technology in which leadless components are soldered to lands on the surface of a circuit board without plated-through holes. The term covers all aspects, including the design and manufacture of electronics components, circuit boards without holes, and the tools and equipment needed to pick, place, and temporarily cement the components to the board until all components have been soldered. It also covers cleaning, testing, and quality control.

surface noise   The noise component in the audio output of a phonograph pickup due to irregularities in the contact surface of the groove.

surface-passivated diode   A diode that has been hermetically sealed by a production process which combines surface passivation with glass-to-silicon seals.

surface-passivated transistor   A transistor that has been protected against water, ions, and other environmental conditions by passivation. A protective compound is chemically bonded to the surface of the semiconductor crystal.

surface passivation   A method of coating the surface of a P-type wafer for a diffused-junction transistor with an oxide compound such as silicon oxide, to prevent penetration of the impurity in undesired regions.

surface photoelectric effect   Ejection of an electron from the surface of a solid or liquid by an incident photon whose total energy is absorbed by the material.

surface recombination rate   The time rate for the recombination of free electrons and holes at the surface of a semiconductor.

surface recombination velocity   The velocity with which electrons and holes drift to the surface of a semiconductor and recombine.

surface resistivity   The electric resistance of the surface of an insulator, measured between the opposite sides of a square on the surface. The value in ohms is independent of the size of the square and the thickness of the surface film.

surface-search radar   Shipboard radar that has as its prime function the detection of ships in line-of-sight distance; it gives the range and bearing of each target while maintaining a complete 360° search.

surface-to-air missile [SAM]   A guided missile that is fired at an airborne target from the ground or the deck of a surface ship. Examples include Hawk, SAM-D, Talos, and Standard.

surface-to-surface guided missile   A guided missile that is fired at a surface target from a surface position on land or water. Examples include Pershing, Polaris, and Sergeant.

surface wave   1. A wave that can travel along an interface between two different media without radiation. The interface must be essentially straight in the direction of propagation. The most common interface is that between air and the surface of a metal conductor.   2. *Ground wave.*

surface-wave antenna   An antenna energized so that a surface wave is propagated along its structure. The dielectric-rod antenna is an example.

surface-wave chirp filter   A surface-acoustic-wave filter that has the spacing of its interdigital electrodes graded from one end of the array to the other. It performs pulse compression in chirp radar.

surface waveguide   A waveguide that conducts the electromagnetic field essentially on the outside of the guiding structure.

surface-wave transmission line   A single-conductor transmission line energized so that a surface wave is propagated along the line with satisfactorily low attenuation.

surge   A long-duration overvoltage or overcurrent.

surge admittance   The reciprocal of characteristic impedance.

surge-crest ammeter   A magnetometer with magnetizable links that measures the peak value of transient electric currents.

surge generator   A device that produces high-voltage pulses, usually by charging capacitors in parallel and discharging them in series. It is also called an impulse generator.

surge impedance   *Characteristic impedance.*

surge protector   *Surge suppressor.*

surge suppressor   A device that prevents the maximum voltage or current from exceeding a preset value. Examples include Zener diodes optimized for surge suppression, metal-oxide varistors (MOV), surge voltage protectors (SVP), and thermistors.

surge voltage protector [SVP]   A hermetically sealed gas-discharge tube made with two electrodes properly spaced by insulators and filled with a rare gas. It provides a conductive path for unwanted and excessive transient voltages to prevent damage to components and circuits as well as injury to personnel.

surveillance   Systematic observation of air, surface, or subsurface areas or volumes by visual, electronic, photographic, or other means for intelligence gathering.

susceptance   The imaginary component of admittance.

susceptibility   The ratio of the magnetization of a material to the magnetizing field.

susceptometer   An instrument that measures paramagnetic, diamagnetic, or ferromagnetic susceptibility.

suspension   A fine wire or coil spring that supports the moving element of a meter.

sustained oscillation   Continuous oscillation at a frequency essentially equal to the resonant frequency of the system.

SVP   Abbreviation for *surge voltage protector.*

SW   Abbreviation for *shortwave.*

swamping resistor   A resistor placed in the emitter lead of a transistor circuit to minimize the effects of temperature on the emitter-base junction resistance.

sweep   1. The steady movement of the electron beam across the screen of a cathode-ray tube, producing a steady bright line when no signal is present. The line is straight for a linear sweep and circular for a circular sweep.   2. The steady change in the output frequency of a signal generator from one limit of its range to the other.

sweep amplifier   An amplifier in a television receiver that amplifies the sawtooth output voltage of the sweep oscillator and shape the waveform as required for the deflection circuits.

sweep circuit   The sweep oscillator, sweep amplifier, and other circuitry that produces a deflection voltage or current for a cathode-ray tube (CRT), typically for generating the horizontal timebase or X-axis sweep.

sweep frequency   The rate at which an electron beam is swept back and forth across the screen of a cathode-ray tube.

sweep-frequency reflectometer   A reflectometer that measures standing-wave ratio and insertion loss in decibels over a wide range of frequencies, in either single- or sweep-frequency operation.

sweep generator   A test instrument that generates an RF voltage whose frequency varies back and forth through a given frequency range at a rapid constant rate. It produces an input signal for circuits or devices whose frequency response is to be observed on an oscilloscope. It is also called a sweep oscillator.

sweep oscillator   1. An oscillator that generates a sawtooth voltage which can be amplified to deflect the electron beam of a cathode-ray tube. It is also called a *time-base generator* or a timing-axis oscillator.   2. *Sweep generator.*

sweep voltage   The periodically varying voltage applied to the deflection plates of a cathode-ray tube to give a beam displacement that is a function of time. It is also called time-base voltage.

swept-frequency measurement   The measurement of magnitude and phase parameters of a device, component, or system as a function of frequency.

swept-frequency modulation   *Chirp modulation.*

swept gain control   *Sensitivity-time control.*

swing   The total variation in the frequency or amplitude of a quantity.

swinging   The momentary variation in the frequency of a received radio wave.

swinging choke   An iron-core choke that has a core which can be operated almost at magnetic saturation. The inductance is then a maximum for small currents and swings to a lower value as current increases. It can be the input choke in a power supply filter to provide improved voltage regulation.

switch   A manual or mechanically actuated device for making, breaking, or changing the connections in an electric or electronic circuit.

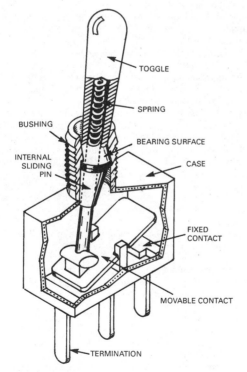

Switch for PC-board mounting has a positive snap action when toggled.

switched-capacitor filter   A filter that includes an operational amplifier and feedback path to obtain selected filtering functions, as opposed to passive inductance-capacitance-resistive filters. It is a form of active filter.

switched multimegabit data service [SMDS]   High-speed packet-switched digital telecommunications system.

Switched Telecommunications Network   A specialized common-carrier communication service owned by telephone companies; it provides both analog and digital information transmission over cable and microwave radio links.

switchhook   *Hookswitch.*

switching diode   A crystal diode that provides essentially the same function as a switch. Below a specified applied voltage it has high resistance corresponding to an open switch, and above that voltage it suddenly changes to the low resistance of a closed switch.

switching frequency   The rate at which the input voltage is switched or "chopped" in a switching or switchmode power supply.

switching losses   Power dissipation when a power switch, typically a power transistor, switches from on to off or off to on.

switching power supply   *Switching-regulated power supply.*

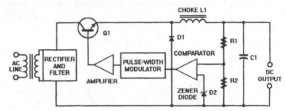

Switching power supply. A simplified schematic for an off-line supply.

switching reactor   A saturable-core reactor that has several input control windings and one or more output windings which essentially duplicate the functions of a relay.

switching-regulated power supply   A power supply containing a *pulse-width modulated* [PWM] *switching regulator* that rectifies AC line voltage and converts it to regulated DC. It can have from one to four power transistors. A sensing circuit detects load changes, and the regulator modulates transistor pulse width to restore the required voltage. High-frequency (20- to 100-kHz) regulators have small, light transformers, so the supply can be lighter and smaller than a comparably rated linear-regulated supply, and its efficiency is higher. It is also called a *switching-power supply* and a *switch-mode power supply.*

switching regulator   A pulse-width-modulator circuit with a closed loop to the load for regulating the output voltage. It is used in switching power supplies and motor controls.

switching time   1. The time interval between the reference time and the last instant at which the instantaneous voltage response of a magnetic cell reaches a stated fraction of its peak value.   2. The time interval between the reference time and the first instant at which the instantaneous integrated voltage response of a magnetic cell reaches a stated fraction of its peak value.

switching transistor   A transistor designed for on/off switching operation.

switching tube A gas tube that switches high-power RF energy in the antenna circuits of radar and other pulsed RF systems. Examples are ATR, pre-TR, and TR tubes.

SWL Abbreviation for *shortwave listener.*

SWR Abbreviation for *standing-wave ratio.*

SWVR Abbreviation for *standing-wave voltage ratio.*

syllabic companding Companding in which the effective gain variations are made at speeds that allow response to the syllables of speech but not to individual cycles of the signal wave.

symbol 1. A design drawn on diagrams to represent a component or identify specific characteristics, quantities, or objects. 2. A letter or abbreviation lettered on diagrams or in equations or text to represent a quantity or unit of measure, to identify an object.

symbolic address A label that identifies a particular word, function, or other information in a computer programming routine, independent of the location of the information within the routine. It is also called a *floating address.*

symbolic coding Coding based on symbols other than actual computer addresses, to make computer programming easier.

symbolic logic Logic in which symbols suitable for calculation express nonnumerical relations for a computer. Boolean algebra is an example.

symbolic programming A method of programming based on arbitrary symbols instead of explicit numerical codes and addresses. The computer then translates the symbols into machine language.

symmetrical avalanche rectifier An avalanche rectifier that can be triggered in either direction. It has a low impedance in the triggered direction.

symmetrical cyclically magnetized condition A condition of a magnetic material when it is cyclically magnetized, and the limits of the applied magnetizing forces are equal and of opposite sign. Thus, the limits of flux density are equal and of opposite sign.

symmetrical relay A relay that has two identical coils which can be interchanged as operate and reset coils.

symmetrical transducer A transducer that permits all possible pairs of specified terminations to be interchanged without affecting transmission because the input and output image impedances are all equal.

symmetrical transistor A junction transistor whose emitter and collector electrodes are identical so that their terminals are interchangeable.

syn (sync) 1. A bit character that synchronizes a time frame in a time-division multiplexer. 2. A synchronous modem sequence that performs bit synchronization or a sequence by a line controller for character synchronization.

sync 1. *Synchronization.* 2. *Synchronize.*

sync compression The reduction in gain applied to the sync signal over any part of its amplitude range with respect to the gain at a specified reference level.

sync generator An electronic generator that supplies synchronizing pulses to television studio and transmitter equipment. It is also called a *sync-signal generator.*

synchro [SYNCHROnous] Any of a number of variations in a small, two-pole *alternator* used in a *synchro system.* It has a *rotor* with a single coil of wire wound on an iron core. The *stator* is a fixed part of the frame. The rotor is energized by alternating current across its two terminals and

the stator has three separate coils, each with a terminal. In a simple synchro system, two synchros are wired together. The unit acting as the transmitter or synchro generator differs slightly from the unit acting as the synchro receiver or synchro motor. It is also called an *autosyn* and a *selsyn.*

synchro angle The angular displacement of a synchro rotor from its electrical zero position.

synchro control transformer A transformer that has its secondary winding on a rotor. When its three input leads are excited by angle-defining voltages, the two output leads deliver an AC voltage that is proportional to the sine of the difference between the electrical input angle and the mechanical rotor angle. The output voltage thus varies sinusoidally with rotor position. It is essentially zero when the mechanical and electrical angles are the same, and it can be used for control purposes.

synchro control transmitter A high-accuracy synchro transmitter that has high-impedance windings.

synchrocyclotron A cyclotron that frequency modulates the radio frequency of its electric field. It is also called an FM cyclotron.

synchro differential receiver A synchro receiver that subtracts one electrical angle from another and delivers the difference as a mechanical angle. One set of three input leads is excited by one set of angle-defining voltages. The other set of three input leads is excited by the other set of angle-defining voltages. The rotor rotates to the difference angle, with a torque proportional to the sine of the difference between the angles. It is also called a differential synchro.

synchro differential transmitter A synchro transmitter that adds a mechanical angle to an electrical angle and delivers the sum as an electrical angle. When its three input leads are excited by the electrical angle-defining voltages, the three output leads deliver voltages that define an angle which is the sum of the electrical input and the mechanical rotor angles. It is also called a differential synchro.

synchro generator *Synchro transmitter.*

synchro motor *Synchro receiver.*

synchronism The condition in which two or more varying quantities have the same speed or reach their peaks at the same instant of time.

synchronization The maintenance of one operation in step with another, as in keeping the electron beam of a television picture tube in step with the electron beam of the television camera tube at the transmitter. It is also called sync.

synchronization error The error caused by imperfect timing of two operations in a navigation system.

synchronization indicator An indicator that presents visually the relationship between two varying quantities or moving objects.

synchronize To produce synchronization. It is also called sync.

synchronized sweep A sweep voltage that is controlled by an AC voltage so that the forward and return traces on a cathode-ray oscilloscope are exactly superimposed and appear as a single trace.

synchronizing The process of maintaining a fixed speed or phase relationship between two varying quantities or moving objects, as between two scanning processes.

synchronizing signal *Sync signal.*

synchronometer An instrument that counts the number of cycles produced by a signal source in a given time interval. It can serve as a master clock when driven by a frequency standard.

synchronous In step or in phase, as applied to two or more circuits, devices, or machines.

synchronous capacitor A synchronous motor that runs without mechanical load and draws a large leading current like a capacitor. It improves the power factor and voltage regulation of an AC power system.

synchronous clock An electric clock driven by a synchronous motor, for operation from an AC power line whose frequency is accurately controlled.

synchronous converter A converter that combines motor and generator windings on one armature and is excited by one magnetic field. It can change AC power to DC power.

synchronous coupling An electric coupling that transmits torque by attraction between magnetic poles on both rotating members.

synchronous data communications A serial input/output (I/O) hardware protocol in which the transmitter and receiver are synchronized to a common clock signal.

synchronous data link control [SLDC] A computer communications protocol.

synchronous demodulator *Synchronous detector.*

synchronous detector A detector that inserts a missing carrier signal in exact synchronism with the original carrier at the transmitter. When the input to the detector consists of two suppressed-carrier signals in phase quadrature, as in the chrominance signal of a color television receiver, the phase of the reinserted carrier can be adjusted to recover either one of the signals. Two synchronous detectors, with carriers differing in phase by 90°, can thus extract the I and Q signals separately from the chrominance signal. It is also called a synchronous demodulator.

synchronous device A device that transfers information at its own rate and not at the convenience of any interconnected device.

Synchronous Digital Hierarchy [SDH] Standards for microwave radio telecommunications based on CCIR, CCIT, and ETSI requirements.

synchronous dynamic RAM [SDRAM] An organization of dynamic random-access memory for personal computers.

synchronous gate A time gate whose output intervals are synchronized with an incoming signal.

synchronous generator *Alternator.*

synchronous inverter *Dynamotor.*

synchronous machine An AC machine whose average speed is proportional to the frequency of the applied or generated voltage.

synchronous modem A modem that uses a derived clocking signal to perform bit synchronization with incoming data.

synchronous motor A synchronous machine that transforms AC electric power into mechanical power with field magnets excited by direct current.

Synchronous Optical Network [SONET] A standard for the transmission of data by means of a fiber-optic network.

synchronous orbit *Geosynchronous orbit.*

synchronous rectifier A rectifier whose contacts are opened and closed at correct instants of time for rectification by a synchronous vibrator or a commutator driven by a synchronous motor.

synchronous satellite *Geostationary satellite.*

synchronous shutdown A characteristic of a power supply that permits the resonance voltage or current cycle to be completed before it shuts down.

synchronous speed A speed value related to the frequency of an AC power line and the number of poles in a synchronous machine. Synchronous speed in revolutions per minute is equal to the frequency in hertz divided by the number of poles, with the result multiplied by 120.

synchronous switch An SCR circuit that controls the operation of ignitrons in such applications as resistance welding.

synchronous transfer An I/O transfer that takes place in a certain amount of time without regard to feedback from the receiving device.

synchronous vibrator An electromagnetic vibrator that simultaneously converts a low direct voltage to a low alternating voltage and rectifies a high alternating voltage obtained from a power transformer to which the low alternating voltage is applied. In power packs, it eliminates the need for a rectifier tube. It is obsolete technology.

synchronous voltage The voltage required to accelerate electrons from rest to a velocity equal to the phase velocity of a wave in the absence of electron flow in a traveling-wave tube.

synchro receiver A *synchro* that provides an angular position related to the applied angle-defining voltages. When two of its input leads are excited by an AC voltage, and the other three input leads are excited by the angle-defining voltages, the rotor rotates to the corresponding angular position. The torque of rotation is proportional to the sine of the difference between the mechanical and electrical angles. It is also called a *receiver synchro, selsyn motor, selsyn receiver,* or *synchro motor.*

synchro resolver *Resolver.*

synchroscope 1. An instrument that indicates whether two periodic quantities are synchronous. The indicator can be a rotating-pointer device, a cathode-ray oscilloscope or liquid-crystal display panel that provides a rotating pattern. The position of the rotating pointer is a measure of the instantaneous phase difference between the quantities. 2. A cathode-ray oscilloscope that shows a short-duration pulse by using a fast sweep which is synchronized with the pulse signal to be observed.

synchro system An electric system for transmitting angular position or motion. The simplest system consists of one *synchro transmitter* or *synchro generator* wired to one *synchro receiver* or *synchro motor.* Any motion of the synchro transmitter shaft changes the stator field in the synchro receiver. The synchro receiver rotor shaft then follows the direction of the changing stator field. More complex systems include *synchro control transformers, synchro differential transmitters,* and *receivers.* It is also called an *autosyn system* and a *selsyn system.*

synchro-to-digital converter [SDC] An electronic circuit that converts synchro or resolver output voltages into parallel binary data that represent angular position.

synchro transmitter A *synchro* that provides voltages related to the angular position of its rotor. When its two input leads are excited by an AC voltage, the magnitudes and polarities of the voltages at the three output leads define the rotor position. It is also called a selsyn genera-

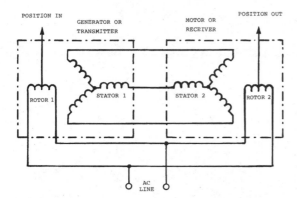

Synchro system. A change in the position of rotor 1 will cause the position of rotor 2 to follow for control purposes or for the transmission of directional information.

tor, selsyn transmitter, synchro generator, transmitter, or transmitter synchro.

synchrotron An accelerator similar to a betatron that has a higher-frequency magnetic field which is applied in synchronism with the orbiting and rapidly accelerating charged particles, to give much higher beam energy than in a betatron. The two types are the electron and proton synchrotrons.

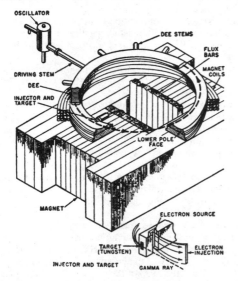

Synchrotron construction, with top half of magnet removed.

synchrotron noise Radio-frequency noise caused by the acceleration of charged particles to high speeds. The particles are guided by a changing magnetic field while they are accelerated many times in a closed path by an RF electric field.

synchrotron radiation Electromagnetic radiation generated by the acceleration of charged relativistic particles, usually electrons, in a magnetic field. It was first encountered in a synchrotron.

synchro zeroing Lining up the zero positions of a synchro system with the zero position of the associated indicator or mechanism being controlled.

sync level The level of the peaks of the synchronizing signal in a television system.

sync limiter A limiter circuit for television that prevents sync pulses from exceeding a predetermined amplitude.

sync pulse One of the pulses that make up a sync signal.

sync separator A circuit that separates synchronizing pulses from the video signal in a television receiver. The signal for the sync separator is usually taken from the collector circuit of the video amplifier.

sync signal A signal transmitted after each line and field to synchronize the scanning process in a television or facsimile receiver with that of the transmitter. The picture, blanking, and sync signals together make up the composite picture signal in a television system. It is also called a synchronizing signal.

sync-signal generator *Sync generator.*

synthetic-aperture radar [SAR] An airborne radar system whose host aircraft moves along a very straight path while it emits microwave pulses continuously at a frequency constant enough to be coherent for a period during which the aircraft travels about 1 km. All echoes returned during this period can then be processed just as if a single antenna as long as the flight path had been used. A long "synthetic" aperture has high resolving power, providing displays with extremely fine detail. It is used in terrain mapping.

synthetic mica A fluor-phlogopite mica made artificially by heating a large batch of raw material in an electric resistance furnace and letting the mica crystallize from the melt during controlled slow cooling.

synthetic quartz A quartz crystal that is grown commercially at high temperature and pressure around a seed of quartz suspended in a solution which contains scraps of natural quartz crystals.

syntony The condition wherein two or more oscillators have exactly the same resonant frequency.

system A combination of several units of equipment integrated to perform a specific function. A personal computer system typically includes the computer, its monitor, a mouse or trackball, a keyboard, and a printer.

systematic errors Errors that have an orderly character and can be corrected by calibration.

system deviation The value of the ultimately controlled variable minus the ideal value in a feedback control system.

Système International d'Unités [SI] French equivalent of *International System of Units.*

system engineering An engineering approach that takes into consideration all the elements related in any way to the equipment under development, including utilization of manpower and the characteristics of each component of the system.

system error The ideal value minus the value of the ultimately controlled variable in a feedback control system.

system noise The output of a system when operating with zero input signal.

systems analysis The analysis of an activity, procedure, method, technique, or business to determine what must be accomplished and how the necessary operations are best accomplished.

T

t Abbreviation for *tonne*.

T 1. Abbreviation for *tera-*. 2. Abbreviation for *tesla*. 3. Symbol for *transformer*. on circuit diagrams. 4. Symbol for *tritium*.

TAB bonding *Tape automated bonding*.

table A collection of data, each item uniquely identified by some label or its relative position in a computer location.

table lookup Obtaining a function value that corresponds to an argument, stated or implied, from a table of function values stored in a computer.

tacan [TACtical Air Navigation] An air navigation system whose single UHF transmitter sends out signals that actuate airborne equipment to provide range and bearing indications with respect to the transmitter location, when interrogated by a transmitter in the aircraft. Each tacan station broadcasts a location-identifying Morse code signal at regular intervals. It is also called tactical air navigation. It is an enhanced military version of *distance-measuring equipment* [DME].

tachometer An instrument that measures angular speed in revolutions per minute. An electric tachometer delivers an output voltage that is proportional to speed.

TACS Abbreviation for *total access communications system*.

tactical air navigation *Tacan*.

tactical missile A guided missile deployed in tactical operations against or in the presence of a hostile force.

tactical radar A radar set deployed in operations against or in the presence of a hostile force.

tactile feel A sudden change in pressure or a click that indicates when a key has been depressed sufficiently on a keyboard as sensed by the fingertips.

tag A label attached to a piece of data in a data-flow computer that specifies where the information is to be used in the program.

tagging *Labeling*.

tail 1. A small pulse that follows the main pulse of a radar transmitter and rises in the same direction. 2. The trailing edge of a pulse.

tailing Excessive prolongation of the decay of a signal.

tail warning radar Radar installed in the tail of an aircraft to warn the pilot that an aircraft is approaching from the rear.

takeoff *Sound takeoff*.

takeup reel The reel that accumulates magnetic tape after the tape is recorded or played by a tape recorder.

talk-back circuit *Interphone*.

talk-down system *Ground-controlled approach*.

talk-listen switch A switch on an intercommunication unit that permits the loudspeaker to function as a microphone when desired.

tangential component A component that acts at right angles to a radius.

tangential wave path The path of propagation of a direct wave that is tangential to the surface of the earth. The path is curved by atmospheric refraction.

tank *Tank circuit*.

tank circuit A circuit capable of storing electric energy over a band of frequencies continuously distributed about the resonant frequency, such as a coil and capacitor in parallel. The selectivity of the circuit is proportional to the Q factor, which is the ratio of the energy stored in the circuit to the energy dissipated. It is also called a *tank*.

tantalum [Ta] A steel-blue, corrosion-resisting metal element with an atomic number of 73 and an atomic weight of 180.95. It has a high melting point and a high specific gravity. The metal is used to make the plates of low-voltage, high-capacitance capacitors. An anodically formed oxide film of the metal has high dielectric strength with a dielectric constant about twice that of aluminium oxide. Tantalum is also used to make trays or "boats" with the high melting point, high strength, and high ductility needed for conveying component parts to be fired or brazed through furnaces.

tantalum capacitor   An electrolytic capacitor whose anode is made from some form of tantalum. Examples include solid tantalum, tantalum-foil electrolytic, and tantalum-slug electrolytic capacitors.

tantalum chip capacitor   A miniaturized solid tantalum capacitor for installation on printed circuit boards and hybrid integrated circuits.

tantalum-foil electrolytic capacitor   An electrolytic capacitor that is made with plain or etched tantalum foil for both electrodes. It contains a weak acid electrolyte.

tantalum-nitride resistor   A thin-film resistor that consists of tantalum nitride deposited on a substrate such as industrial sapphire.

tantalum-oxide capacitor   A capacitor whose dielectric is a film of tantalum oxide which is grown or deposited on substrate material in thick-film or thin-film circuits. It is also called a tantalum capacitor.

tantalum-slug electrolytic capacitor   An electrolytic capacitor made from a sintered slug of tantalum as the anode, in a highly conductive acid electrolyte. Some types can function at operating temperatures as high as 200°C.

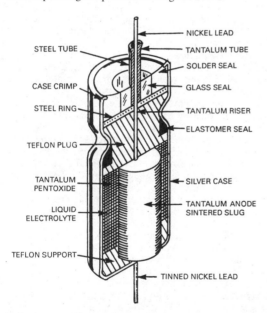

Tantalum-slug, axial-leaded capacitor has a tantalum slug anode and a liquid electrolyte.

T antenna   An antenna that consists of one or more horizontal wires, with a lead-in connection made at the approximate center of each wire.

tap   A connection made at some point other than the ends of a resistor or coil.

tape   1. *Magnetic tape.*   2. *Punched-plastic or paper tape.*

tape-and-ammo-box packaging   A method for taping together components in belts for shipment, storage, and dispensing from rectangular boxes. Components such as capacitors and resistors are uniformly spaced and taped in continuous belts—the axial-leaded components on both sides across their lead ends, and radial-leaded components

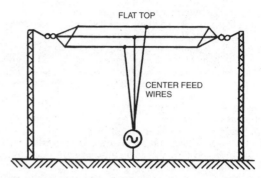

T antenna for low frequencies requires two masts to support the insulated top section, but does not require base or guy insulators.

on one side across their lead ends. The taped belts are then folded to lie flat in a box, much as cartridge belts are placed in ammunition boxes. The components can be cut from the tape by continuous-feed machines prior to automatic insertion. See *also tape-and-reel packaging.*

tape-and-reel packaging   A method for taping together components in belts that are wound on reels for shipment, storage, and dispensing. Components such as capacitors and resistors are uniformly spaced and taped in continuous belts—axial-leaded components on both sides across their lead ends, and radial-leaded components on one side across their lead ends. Small-outline packaged transistors and ICs can be placed in embossed paper carrier strips and covered with tape. The loaded belts are then wound on the reel. They can be cut from the tape by continuous-feed machines prior to automatic insertion. See also *tape-and-ammo-box packaging.*

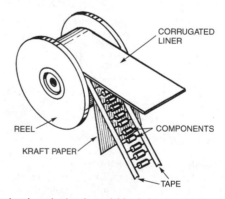

Tape-and-reel packaging for axial-leaded components such as diodes or resistors facilitates their insertion in circuit-board holes by automatic machines.

tape automated bonding [TAB]   An automated method for wiring semiconductor device pads to package leads by means of lead arrays that have been formed by metallizing plastic tape. The leads are aligned with the pads and bonded with heat and pressure.

tape cassette   A package that holds a length of magnetic tape so that the package can be slipped into a tape recorder or video cassette recorder and played without threading the tape. The tape runs back and forth between two reels inside the cassette, called a cartridge.

tape deck A magnetic tape player mounted on a motor board; it includes the tape transport and the bias and erase oscillators but no preamplifier, power amplifier, loudspeaker, or cabinet. It is for stereo and automotive entertainment systems.

tape drive *Tape transport.*

tape eraser *Bulk eraser.*

tape guides Grooved pins of nonmagnetic material mounted at either side of a magnetic recording head assembly to position the magnetic tape on the heads as the tape is being recorded or played.

tape loop A length of magnetic tape that has its ends spliced together to form an endless loop. It is used in message repeater units and some types of tape cartridges to eliminate the need for rewinding the tape.

tape player A machine only for playback of recorded magnetic tapes.

taper The way that resistance is distributed throughout the element of a potentiometer or rheostat. Uniform distribution, with the same resistance per unit length throughout the element is called linear taper. Nonuniform distribution is called nonlinear.

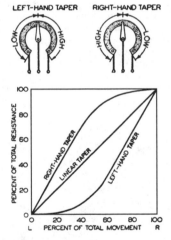

Taper of linear and nonlinear potentiometers.

tape recorder A recorder that records voice, music, and digital data on magnetic tape by selective magnetization of iron oxide particles which form a thin film on the tape. A recorder usually also includes provisions for playing back the recorded material.

tapered transmission line *Tapered waveguide.*

tapered waveguide A waveguide whose physical or electrical characteristic changes continuously with distance along the axis of the waveguide.

tape speed The speed at which magnetic tape moves past the recording head in a tape recorder. Standard speeds are ¹⁵⁄₁₆, 1⅞, 3¾, 7½, 15, and 30 in/s (2.38, 4.76, 9.5, 19, 38, and 76 cm/s). Faster speeds give improved high-frequency response under given conditions.

tape splicer Apparatus that splices magnetic tape automatically or semiautomatically with splicing tape or by fusing the tape with heat.

tape transport The mechanism on the deck of a tape recorder that holds the tape reels, drives the tape past the recording heads, and controls various modes of operation. It is also called a tape drive or tape handler.

tape-wound core A length of ferromagnetic material in tape form, wound so that each turn falls directly over the preceding turn.

tapped control A rheostat or potentiometer that has one or more fixed taps along the resistance element.

target A substance or object exposed to bombardment or irradiation by nuclear particles, electrons, or electromagnetic radiation. In an X-ray tube the target is the anode or anti-cathode, from which X-rays are emitted as a result of electron bombardment. In radar and sonar it is any object capable of reflecting the transmitted beam. In a television camera tube it is the storage surface that is scanned by an electron beam to generate an output signal current which corresponds to the charge-density pattern stored there.

target acquisition The first appearance of a recognizable and useful echo signal from a new target in radar and sonar operation.

target discrimination The ability of a detection or guidance system to distinguish a target from its background or to discriminate between two or more targets that are close together.

target drone A pilotless aircraft controlled by radio from the ground or from a mother ship and used exclusively as a target for antiaircraft weapons.

target fade A momentary reduction in the strength of an echo signal from a radar or sonar target, due to interference or other phenomena. Tracking radar usually includes memory circuits that maintain tracking during this period, to prevent the loss of the target.

target identification Identification of a target to determine whether it is friend or foe.

target noise Statistical variations in a radar echo signal due to the presence on the target of many reflecting elements randomly oriented in space. Target noise can cause scintillation.

target scintillation *Scintillation.*

Tartar An earlier name for a U.S. Navy surface-to-air guided missile intended primarily for defense of destroyers. It is now called the standard missile.

TASI [Time Assignment Speech Interpolation] A method of increasing or even doubling the capacity of a transatlantic telephone-cable circuit by using intervals of silence on each circuit for transmitting information from other channels. The signals for a large group of telephone conversations are split into time segments, and only the segments carrying information are transmitted. They are first labeled so that they can be directed to the correct receiver.

taut-band meter An analog panel meter whose moving element is supported between two extremely fine metal bands under tension. The bands twist when the moving element of the meter rotates under magnetic action and thereby supply return torque while providing electric connections to the moving element.

Tb   Abbreviation for *terabit*.

TC   Abbreviation for *temperature coefficient*.

T carrier   A time-division multiplex system based on digital transmission of pulse-code-modulation-encoded information over cables which contain twisted pairs or over microwave radio links.

TCAS   Abbreviation for *traffic collision and avoidance system*.

T circulator   A circulator that has three identical rectangular waveguides joined asymmetrically to form a T-shaped structure with a ferrite post or wedge at its center. Each port must be separately matched. Power entering any waveguide will emerge from only one adjacent waveguide.

TCP   Abbreviation for *tape carrier package* for semiconductor devices.

TCP/IP   Abbreviation for *transmission control protocol/Internet protocol*.

TC wire bonder   *Thermocompression wire bonder*.

TCXO   Abbreviation for *temperature-compensated crystal oscillator*.

TDD   Abbreviation for *time-division duplexing*.

TDM   Abbreviation for *time-division multiplex*.

TDMA   Abbreviation for *time-division multiaccess*.

TDR   1. Abbreviation for *temperature-dependent resistor*. 2. Abbreviation for *time-domain reflectometer*.

TE   Abbreviation for *transferred electron*.

TEA   Abbreviation for *transferred-electron amplifier*.

TEA laser   Abbreviation for *transversely excited atmospheric-pressure laser*.

tearing   A television picture defect in which groups of horizontal lines are displaced in an irregular manner, caused by inadequate horizontal synchronization.

TED   Abbreviation for *transmission electron-diffraction contrast microscopy*.

Teflon   A DuPont tradename for fluorocarbon resins. See also *tetrafluoroethylene* [TFE *and* FEP].

tele-   Prefix meaning from a distance.

teleammeter   A telemeter that measures and transmits current values to a remote point.

telecast [TELEvision broadCAST]   The transmission of a television program intended for reception by the general public.

telecasting   Broadcasting a television program.

Telecheric   A robotlike machine that can perform tasks such as surveillance, testing, or remote manipulation under the control of a human operator who sends command signals over a flexible cable. Unlike a true robot, it is not under the control of a computer program or other form of program. It can be permanently mounted in one location, or it can be self-propelled with wheels or tracks to move on land or with a propellor to move in water. It can enter locations that are hazardous to humans such as burning buildings, nuclear radiation, toxic chemicals or vapors, or the threat of a bomb explosion. It can also propel itself in the sea where pressures or turbidity make it unsafe for divers. It is a useful bomb-disposal tool.

telecommunication   Any transmission, emission, or reception of signals, writing, images, sounds, or intelligence of any nature by wire, radio, visual, or other electromagnetic systems. The terms telecommunication and communication are used interchangeably.

teleconference   A conference whose the participants are some distance apart but are able to talk to and see each other because of telephone, radio and television links.

telegraph channel   A path suitable for the transmission of telegraph signals between two telegraph stations, either over wires or by radio. It might be one of several channels on a single radio or wire circuit, providing simultaneous transmission in the same frequency range, simultaneous transmission in different frequency ranges, or successive transmission.

telegraph circuit   The complete wire or radio circuit over which signal currents flow between transmitting and receiving apparatus in a telegraph system.

telegraph key   A hand-operated telegraph transmitter, that forms telegraph signals.

telegraph level   The signal power at a specified point in a telegraph circuit when one telegraph channel is in the marking or continuous-tone condition and all other channels are silenced.

telegraph modem   The complete equipment for modulating and demodulating one or more separate telegraph circuits, each circuit containing one or more telegraph channels.

telegraph-modulated wave   A wave obtained by varying the amplitude or frequency of a continuous wave by telegraphic keying.

telegraph repeater   A repeater inserted at intervals in long telegraph lines to amplify weak code signals, with or without reshaping of pulses, and retransmit them automatically over the next section of the line.

telegraph transmitter   A device that controls a source for electric power, to form telegraph signals for radio or wire transmission.

telegraphy   Communication at a distance by code signals that consist of current pulses sent over wires or by radio.

telemeter   *Telemetry*.

telemetry   The methods and equipment required to receive measurement data transmitted over wires or by radio from a remote location. In practice, it is more likely to be data from sensors and transducers on experimental or development aircraft, missiles, or rockets received at a *tracking and control* [TTC] *station*. Typical variables measured are acceleration, velocity, temperature, pressure, flow rate from fuel tanks, stresses or strains, and physiological data from pilots or crew. Typically, a radio carrier signal is multiplexed to transmit data that has been modulated by subcarrier frequencies or channels. The term also applies to the recovery of technical data from a satellite, as distinguished from signals being retransmitted by a satellite functioning as a relay. It is also called *telemetering*.

telemetry antenna   A highly directional antenna, generally mounted on a servo-controlled platform, for tracking aircraft, guided missiles, rockets, or spacecraft from a *telemetry tracking and control* [TTC] *station*.

telemetry channel   A subcarrier frequency assigned to the transmission of one telemetered function in a multiplexed telemetry system.

telemetry system   The equipment and facilities required to carry out *telemetry*. This includes a radio receiver, data processor, and displays at a *telemetry tracking and control* [TTC] *station*. It is also called a *telemetering system*.

telemetry tracking and control [TTC] **station**   A building or shelter containing a *telemetry system*.

telephone

Telemetering antenna is highly directional.

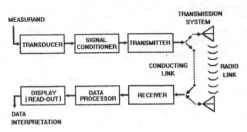

Telemetry system block diagram shows the principal system components.

telephone A unit of telephone communications equipment that is designed for location on a desk or mounting on a wall. It consists of a handset and a transceiver cradle or module. Conventional handsets contain only the microphone and speaker (headphone) in a separate assembly connected by a flexible conductive cord to the transceiver module, which contains the circuitry for converting voice/sound to an electric signal and its inverse. It also contains the necessary switches for entering telephone numbers and the initiation of various telephone services. It also contains the "ringing" circuitry, now an electronic tone generator. In modern units the alphanumeric keypad has replaced the rotary dial. It is also called a *phone* and a *telephone set*. See also *cellular telephone* and *cordless telephone*.

telephone answering machine An electronic circuit that is compatible with the public telephone system and combines both the automatic answering and recording functions. It responds to incoming telephone tones, and triggers a magnetic tape recorder. Some answering machines contain circuitry permitting them to be interrogated from a remote telephone in response to a preset tone code transmitted over the telephone lines, so they will play back recorded telephone calls. The functions of the answering machine and telephone have been combined in some single products.

telephone capacitor A small fixed capacitor connected in parallel with a telephone receiver to bypass higher audio frequencies and thereby reduce noise.

telephone carrier current A carrier current for telephone communication over power lines or to obtain more than one channel on a single pair of wires.

telephone channel A one- or two-way path suitable for the transmission of audio signals between two stations.

telephone circuit The complete circuit over which audio and signaling currents travel in a telephone system between the two telephone subscribers in communication with each other.

telephone current An electric current produced or controlled by the operation of a telephone transmitter.

telephone dial A switch operated by a finger wheel, that makes and breaks a pair of contacts the required number of times for setting up a telephone circuit to the party being called.

telephone exchange A switching center for the interconnection of telephone lines in a given service region.

telephone induction coil A coil in a telephone circuit that matches the impedance of the line to that of a telephone transmitter or receiver.

telephone jack *Phone jack.*

telephone line The conductors extending between telephone subscriber stations and central offices.

telephone loading coil *Loading coil.*

telephone modem A unit of equipment that modulates and demodulates one or more separate telephone circuits, each circuit containing one or more telephone channels. It might include multiplexing and demultiplexing circuits, individual amplifiers, and carrier-frequency sources.

telephone network *Telephone circuit.*

telephone pickup A microphone placed on a telephone set to pick up both voices during a telephone conversation for recording purposes.

telephone plug *Phone plug.*

telephone receiver *Telephone.*

telephone relay An electromechanical relay with multiple contacts, typically 4PDT, that operates with a coil voltage of 5 V with contacts suitable for switching telephone voltages. Today many are miniature flatpack relays intended for printed-circuit board mounting.

telephone repeater A repeater inserted at one or more intermediate points in a long telephone line to amplify telephone signals. It maintains the required current strength in the line.

telephone repeating coil A coil in a telephone circuit for inductively coupling two sections of a line when a direct connection is undesirable.

telephone retardation coil A coil in a telephone circuit that passes direct current while offering appreciable impedance to alternating current.

telephone ring circuit An electronic circuit that responds to incoming calls by emitting a series of tones generated to indicate that the telephone is linked to an incoming call. It has replaced the electromechanical-relay armature that rang a bell mounted on the earliest telephones.

telephone switchboard A switchboard for interconnecting telephone lines and associated circuits.

telephone transmitter *Telephone.*

telephony The transmission of speech and sounds to a distant point for communication purposes.

telephoto lens A lens with a long focal length that will permit a television camera to obtain large images of distant objects.

teleprinter An electric telephone typewriter that can be actuated by signals received over telephone lines and can similarly generate output signals while producing hard copy. It is a data-communication terminal. It might also have a paper-tape punch, paper-tape reader, or magnetic-tape transport.

TELETEX A text communication service between electronic workstations that will gradually replace TELEX when the digital network is introduced.

TELETEXT Broadcast text and graphics for domestic television reception, distinguished from TELETEX.

teletypewriter [TTY] A special electric typewriter that produces coded electric signals which correspond to manually typed characters and automatically types messages when fed with similarly coded signals produced by another machine. The signals can be transmitted directly over telephone wires that connect the machines or used to drive a teletypewriter perforator. This is now obsolete equipment.

teletypewriter exchange service [TWX] A direct-dialing point-to-point service based on teleprinter equipment. The service also permits interfacing with computers. It is also called telex.

televise To pick up a scene with a television camera and convert it into corresponding electric signals for transmission by a television station.

television [TV] A system that converts a succession of visual images into corresponding electric signals and transmits those signals by radio or over cable to distant receivers where the signals can be used to reproduce the original images.

television broadcast band The band extending from 54 to 890 MHz. The 6-MHz channels are assigned to television broadcast stations in the United States. The frequencies are 54 to 72 MHz (channels 2 through 4), 76 to 88 MHz (channels 5 and 6), 174 to 216 MHz (channels 7 through 13), and 470 to 890 MHz (channels 14 through 83).

television camera The pickup unit that converts a scene into corresponding electric signals. Optical lenses focus the scene to be televised on the photosensitive surface of a camera tube. This tube breaks down the visual image into small picture elements and converts the light intensity of each element in turn into a corresponding electric signal.

television camera tube *Camera tube.*

television channel A band of frequencies 6 MHz wide in the television broadcast band, available for assignment to a television broadcast station.

television engineering The field of engineering concerned with the design, manufacture, and testing of equipment required for the transmission and reception of television programs.

television film scanner A motion-picture projector adapted for use with a television camera tube to televise 24 frame per second motion-picture film at the 30 frame per second rate required for television.

television-guided bomb A bomb that carries a small television camera in its nose for guidance. The camera system can be locked on the target before the bomb is dropped, for self-guidance, or the pilot can monitor the camera picture over a microwave relay link and adjust the course of the bomb by remote control after dropping the bomb.

television interference [TVI] Interference produced in television receivers by computers, power tools, amateur radio, and other transmitters.

television picture tube *Picture tube.*

television receiver A receiver that converts incoming television signals into the original scenes along with the associated sounds. It is also called a television set.

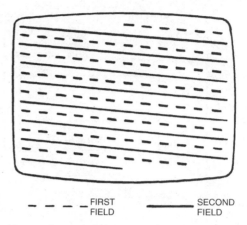

FIRST FIELD — — — —   SECOND FIELD ————

Television receiver scanning sequence for conventional image interlacing.

television reconnaissance Reconnaissance in which television transmits a scene from the reconnoitering vehicle to another location on the surface or in the air.

television recording *Kinescope recording.*

television relay system *Television repeater.*

television repeater A repeater that transmits television signals from point to point by using radio waves in free space as a medium. This transmission is not intended for direct reception by the public. It is also called a television relay system.

television scanning The U.S. NTSC standard requires that 15,750 lines be scanned per second. The vertical scanning rate is 30 ps; there are 525 lines allocated to each *frame* and 262.5 lines to each *field*. Each line is 63.49 µs in duration. The beam for line 1 begins at the top center of the image and proceeds for half a line to the right edge. Retrace to the left occurs, and line 3 follows. Line 5 and successive odd-numbered lines are scanned, for a total of 241.5 lines, ending at the lower right corner. During the vertical retrace interval, 21 lines elapse before the beam is moved to the upper left corner to start scanning line 2. This completes one full field. Then 241.5 successive even-numbered lines are scanned, ending at the middle bottom of the scene. After a 21-line elapse to return the beam to the top center, the second field and a full frame are complete. The sequence then repeats.

television screen The front face of an NTSC television receiver's cathode-ray tube (CRT) has a rectangular form with an aspect ratio of 4 units horizontally and 3 units vertically.

television set *Television receiver.*

television signal The general term for the audio and visual signals that are broadcast together to provide the sounds and pictures of a television program.

television standards for NTSC The NTSC standard for television transmission allows a channel width of 6 MHz. The picture carrier is located 1.25 MHz above the lower boundary of the channel. The aural center frequency is 4.5 MHz above the picture carrier. The transmission is horizontally polarized. The composite picture is amplitude-modulated and the audio signal is frequency-modulated. A total of 525 lines per frame are interlaced two to one, and

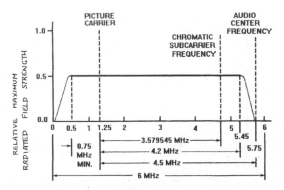

Television transmission standards for the NTSC system showing RF amplitude characteristics.

the scanning sequence is horizontal from left to right and vertical from top to bottom. The horizontal scanning frequency is 15.75 kHz for monochrome or 2/455 times the chrominance subcarrier frequency (about 15.734). The vertical scanning frequency is 60 Hz for monochrome or 2/525 times the horizontal scanning frequency (59.94 Hz) for color.

television transmitter A video and an audio transmitter interconnected together for transmitting a complete television signal.

telex *Teletypewriter exchange service.*

telluric current A natural electric current flowing through the earth. The direction and intensity of this current varies with the earth's magnetic field, auroral and solar activity, and other cosmic phenomena. It is also called earth current.

tellurium [Te] An element with an atomic number of 52 and an atomic weight of 127.60.

telnet A terminal emulation program that allows subscribers to access other computers on the *Internet* interactively.

TEM Abbreviation for *transmission electron microscope.*

TE$_{m,n}$ mode A mode for the propagation of a particular transverse electric wave in a waveguide. It is also called H$_{m,n}$ mode (British).

TE$_{m,n,p}$ mode A mode of wave propagation in a cavity that consists of a hollow metal cylinder closed at its ends. Its transverse field pattern is similar to that of the TE$_{m,n}$ mode in a corresponding cylindrical waveguide. The letter *p* is the number of half-period field variations along the axis. It also applies to closed rectangular cavities.

tempco Abbreviation for *temperature coefficient.*

temperature characteristic The performance of a device as temperature is varied through specified limits.

temperature coefficient [TC or tempco] The amount of change in any performance characteristic of a device as a result of a change in temperature. It is expressed in parts per million (of change) per degree Celsius (ppm/°C).

temperature coefficient of capacitance [TCC] The amount of change that occurs in capacitance of any capacitor or electronic component as a result of a change in temperature, typically measured in parts per million per degree Celsius (ppm/°C).

temperature coefficient of frequency The rate of change of frequency with temperature.

temperature coefficient of linear expansion [TCLE] The amount of change that occurs in any linear dimension of a solid as a result of a change in temperature, typically measured in parts per microinch per degree Celsius.

temperature coefficient of resistance [TCR] The amount of change that occurs in resistivity of any component, usually resistors or variable resistors, as a result of a change in temperature, typically measured in parts per million per degree Celsius (ppm/°C).

temperature-compensated crystal oscillator [TCXO] A crystal oscillator whose temperature-compensation networks have been adjusted to match the crystal characteristics, to give frequency stability over a wide temperature range.

temperature-compensated reference element A special temperature-compensated diode whose stability is comparable to that of standard cells. An example is the passivated alloy silicon diode.

temperature-compensated Zener diode A Zener-diode package that contains one reverse-biased zener PN junction which has a positive temperature coefficient. It is connected in series with one or more forward-biased diodes that have negative temperature coefficients. The result is a Zener voltage that remains essentially constant over a wide temperature range.

temperature-compensating alloy An alloy whose magnetic properties change with temperature. The most common examples are nickel-iron alloys, in which magnetic permeability decreases at a controlled rate with increases in temperature. It is used for shunts in watthour meters, speedometers, tachometers, and voltage regulators, to compensate for temperature changes.

temperature-compensating capacitor A capacitor whose capacitance varies with temperature in a known and predictable manner. In resonant circuits it compensates for changes in the values of other parts with temperature.

temperature-compensating network A network whose components are so chosen that the network characteristics change with temperature in a predetermined manner.

temperature compensation The process of making some characteristic of a circuit or device independent of changes in ambient temperature.

temperature control A control that maintains the temperature of an oven, furnace, or other enclosed space within desired limits.

temperature-controlled crystal unit A crystal unit that contains, in addition to the quartz plate or plates, a heater that maintains the temperature of the quartz plate within specified limits.

temperature correction A correction applied to a measured value to compensate for changes that are due to a temperature that is higher or lower than some standard temperature value.

temperature-dependent resistor [TDR] A nonlinear resistor whose resistance value is some function of the ambient temperature or the temperature rise produced internally by current flow. Examples include barretters, bolometers, and thermistors.

temperature derating Lowering the rating of a device when it is to be used at elevated temperatures.

temperature element The sensing element of a temperature-measuring device or direct-reading temperature indicator.

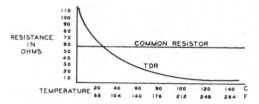

Temperature-dependent negative temperature coefficient resistor characteristic, compared to that for common resistor.

temperature inversion 1. The ocean condition in which surface water is colder than the water below. Because the speed of sound increases with water temperature, a temperature inversion causes a sonar beam to bend in the direction of the colder water. 2. A region in the troposphere at which temperature increases rather than decreases with altitude.

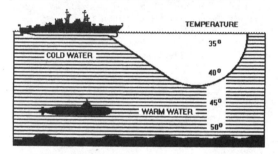

Temperature inversion in the ocean bends a downward-directed sonar beam back toward the surface, because sound travels faster in warm water.

temperature range The total variation in ambient temperature for a given application, expressed in degrees Celsius.

temperature saturation The condition in which the anode current of a thermionic vacuum tube cannot be further increased by increasing the cathode temperature at a given value of anode voltage. The effect is due to the space charge formed near the cathode. It is also called filament saturation or saturation.

TEM wave Abbreviation for *transverse electromagnetic wave*.

tera- [T] A prefix that represents 10^{12}, which is 1,000,000,000,000 or a million million. It is pronounced terra, rhyming with Sarah.

terabit [Tb] One million megabits, equal to 10^{12} bits.

teracycle One million megacycles, or 10^{12} cycles per second. It is now called terahertz and abbreviated THz.

teraelectronvolt [TeV] A unit of energy equal to 10^{12} eV.

terahertz [THz] One million megahertz, or 10^{12} Hz. It was formerly called teracycle.

teraohm [TΩ] One million megaohms, equal to 10^{12} Ω.

teraohmmeter An ohmmeter that has a teraohm range for measuring extremely high insulation resistance values.

terbium [Tb] A rare-earth element with an atomic number of 65.

terminal 1. A screw, soldering lug, or other point to which electric connections can be made. 2. The equipment at the

end of a microwave relay system or other communication channel. 3. One of the electric input or output points of a circuit or component. 4. A device that provides access to a remote computer. See also *video data terminal*.

terminal area The enlarged area of conductor material that surrounds a hole for a lead on a printed circuit. It is also called a land or pad.

terminal board An insulating mounting for terminal connections. It is also called a *terminal strip*.

terminal equipment The equipment at a terminal of a communication channel.

terminal guidance Navigation control of a missile as it approaches its target.

terminal lug A soldering lug attached to a terminal board or at the end of a wire.

terminal pad *Terminal area.*

terminal phase The path of a missile as it approaches its target. For a ballistic missile, the terminal phase is that part of the trajectory between reentry and impact.

terminal strip *Terminal board.*

terminated line A transmission line terminated in a resistance equal to the characteristic impedance of the line, so there are no reflections and no standing waves.

termination The load connected to the output end of a circuit, device, or transmission line.

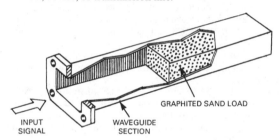

Termination or matched load in a waveguide.

ternary alloy An alloy that contains three metals such as iron, silicon, and aluminum.

ternary compound An alloy that contains three elements. Examples include (1) iron, silicon, and aluminum that offers high resistivity, good magnetic permeability and has magnetic properties approaching those of iron-nickel alloys and (2) aluminum gallium arsenide (AlGaAs), a ternary compound semiconductor grown epitaxially on compound semiconductor substrates.

ternary notation A system of notation with a base of 3 and the characters 0, 1, and 2.

terrain-avoidance radar An airborne radar that provides a display of the terrain ahead of a low-flying airplane, to permit avoidance of obstacles.

terrain echoes *Ground clutter.*

terrain error The navigation error due to distortion of a radiated field by the nonhomogeneous characteristic of the terrain over which the radiation has propagated.

terrain-following radar Airborne radar that provides autopilot control signals as required to maintain a low and constant altitude above the earth. It might also provide a display of the terrain ahead of the plane as required for manual terrain-following by the pilot at night.

tesla [T]   The SI unit of magnetic flux density (magnetic induction). It is equal to 1 Wb/m².

Tesla coil   A high-voltage, high-frequency, air-core transformer that can produce a long spark across a spark gap. Handheld versions are laboratory tools that can determine the presence of air in sealed-off vacuum tubes and evacuated chambers by ionizing the residual gas through the glass envelope. When the spark discharge is held near the envelope, a blue glow can be seen within the evacuated tube or chamber.

test   To check the operation or performance characteristics of a component, equipment, system, or computer program under controlled conditions.

test clip   A spring clip at the end of an insulated wire lead for making temporary connections quickly for test purposes.

test lead   A flexible insulated lead, usually with a test prod at one end, for making tests, connecting instruments to a circuit temporarily, or making other temporary connections.

test pattern   A chart that has various combinations of lines, squares, circles, and graduated shading, transmitted periodically by a television station to check definition, linearity, and contrast for the complete system from camera to receiver. It is also called a resolution chart.

test prod   A metal point attached to an insulating handle and connected to a test lead for conveniently making a temporary connection to a terminal while tests are being made.

test program   *Check routine.*

test record   A magnetic tape record that has recorded frequencies suitable for checking and adjusting audio systems.

test routine   *Check routine.*

test vector   The specification of a set of input signals for the input pins of an integrated circuit and the expected response at the output pins.

tetrad   A group of four pulses that express a digit in the scale of 10 or 16.

tetrafluoroethylene   A plastic resin with unexcelled chemical resistance that is an excellent insulator offering high dielectric strength, very low dissipation factor, very high resistivity, and excellent machinability. It is used as an insulator in coaxial cable and is formed into insulating electrical tape. It is also used to make variable capacitors and electrical standoffs. It is sold under the DuPont tradename of *Teflon.*

tetrode   A four-electrode electron tube that contains an anode, a cathode, a control electrode, and one additional electrode which is ordinarily a grid.

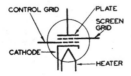

Tetrode symbol. The plate is the anode.

tetrode transistor   A four-electrode transistor.

TeV   Abbreviation for *teraelectronvolt.*

TE wave   Abbreviation for *transverse electric wave.*

TE$_{m,n}$ wave   1. In a circular waveguide, the transverse electric wave for which m is the number of axial planes along which the normal component of the electric vector vanishes, and n is the number of coaxial cylinders (including the boundary of the waveguide) along which the tangential component of the electric vector vanishes. The TE$_{0,1}$ wave is the circular electric wave that has the lowest cutoff frequency, and the TE$_{1,1}$ wave is the dominant wave and has electric lines of force approximately parallel to a diameter of the waveguide.   2. In a rectangular wave-guide, the transverse electric wave for which m is the number of half-period variations of the electric field along the longer transverse dimension, and n is the number of half-period variations of the electric field along the shorter transverse dimension. It is also called the H$_{m,n}$ wave (British), for both circular and rectangular waveguides.

TFE   Abbreviation for *tetrafluoroethylene.*

TFE-fluorocarbon resin   Standard term of Society of the Plastics Industry for polytetrafluoroethylene resin, marketed as Teflon by DuPont and as Fluon by Imperial Chemical Industries in Great Britain.

T flip-flop   A flip-flop circuit that changes state with each application of a trigger or clock pulse to a single input terminal T. It is widely used in counter circuits. It is also called a *toggle.*

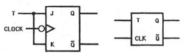

T (toggle) flip-flop (*a*) formed with a J-K flip-flop and (*b*) its schematic symbol.

TFT   Abbreviation for *thin-film transistor* as used in fabricating active-matrix liquid-crystal displays (AMLCD).

thallium [Tl]   An element with an atomic number of 81 and an atomic weight of 204.37. Several thallium isotopes are members of the uranium, actinium, thorium, and neptunium radioactive series.

thallium-activated sodium iodide detector   A gamma-ray detector based on a single crystal of thallium-activated sodium iodide.

thallium oxysulfide   A compound of thallium, oxygen, and sulfur that has photoconductive properties.

thalofide cell   A photoconductive cell whose active light-sensitive material is thallium oxysulfide in a vacuum. It has its maximum response at the red end of the visible spectrum and in the near-infrared region.

THD   Abbreviation for *total harmonic distortion.*

Theater Missile Defense [TMD] **radar**   A U.S. radar system that can be used anywhere in the world for the interception at long range of enemy ballistic missiles.

thermal agitation   Random movements of the free electrons in a conductor; they produce noise signals that might become noticeable when they occur at the input of a high-gain amplifier. It is also called a *thermal effect.*

thermal-agitation voltage   The voltage produced in a circuit by thermal agitation.

thermal ammeter   *Hot-wire ammeter.*

thermal battery   1. A combination of thermal cells.   2. A voltage source that consists of a number of bimetallic junctions connected to produce a voltage when heated by a flame.

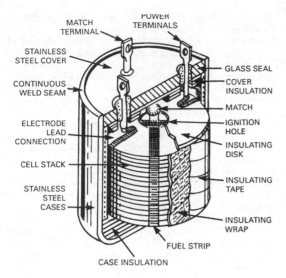

Thermal battery produces full power instantly when internal pyrotechnic materials are ignited electrically to heat the lithium-iron disulfide cells.

thermal cell A reserve cell that is activated by applying heat to melt a solidified electrolyte.

thermal circuit breaker A *circuit breaker* whose contacts open automatically in response to heat caused by *overload* current in the protected circuit. One contact can be a *bimetallic strip*. See also *magnetic circuit breaker*.

thermal compression bonding *Thermocompression bonding.*

thermal conductivity The quantity of heat that passes through a unit volume of a material in unit time when the difference in temperature of the two faces is 1° Celsius.

thermal converter A converter that consists of one or more thermojunctions in thermal contact with an electric heater. The voltage developed at the output terminals by thermoelectric action is then a measure of the input current to its heater. It is also called a thermocouple converter, thermoelectric generator, or a thermoelement.

thermal cutout A heat-sensitive switch that automatically opens the circuit of an electric motor or other electromechanical actuator when the operating temperature exceeds a safe value.

thermal detector A heat detector, such as a bolometer or thermocouple.

thermal drift Drift caused by internal heating of equipment during normal operation or by changes in external ambient temperature.

thermal effect *Thermal agitation.*

thermal flasher A thermoelectric device that opens and closes a circuit automatically at regular intervals. The alternate heating and cooling of a bimetallic strip when it is heated by a resistance element in series with the circuit being controlled causes the action.

thermal imager A laboratory camera or other passive infrared system that gives an infrared image of a scene in which various objects or areas differ in temperature. Temperature difference ranges can be displayed in color.

thermal imaging A passive military night-vision system that depends on the differences in thermal (infrared) emissions of objects being viewed. Unlike *image intensification,* it requires no natural or artificial illumination.

thermal imaging navigation set [TINS] An aircraft system that provides images on a *heads-up* (HUD) display under all-weather, day-night environments. Based on digital signal processing, TINS can display either black-hot or white-hot images. It consists of three units: thermal control unit (TCU), FLIR sensor unit (FSU), and a pod electronics unit (PEU). See *forward-looking infrared* (FLIR) and *thermal imaging.*

thermal inertia The reciprocal of thermal response.

thermal instrument An instrument that depends on the heating effect of an electric current, such as a thermocouple or hot-wire instrument.

thermal ionization The ionization of atoms or molecules by heat, as in a flame.

thermal management The means by which the temperature of a battery system is maintained within a specified range while the battery is charging or discharging.

thermal noise Electric noise produced by thermal agitation of electrons in conductors and semiconductors. This random motion of free electrons increases with temperature. It is also called Johnson noise.

thermal noise generator A generator that depends on the inherent thermal agitation of an electron tube to provide a calibrated noise source.

thermal photograph A photograph made by an image tube or similar device that shows objects on the earth differentiated by their radiations of heat or infrared waves.

thermal photography *Thermography.*

thermal printer A nonimpact printer whose characters are formed by heating selected elements of a 5×7 or 7×9 dot matrix that is in contact with heat-sensitive paper.

thermal protector A temperature-sensing element that is built into a motor or other equipment to interrupt the power when overheating reaches a level that would damage or destroy the machine. The protector can reapply power automatically when the machine has cooled, or it might require manual resetting.

thermal radiation Radiation in the form of heat, emitted by all bodies that are not at absolute zero in temperature. The wavelength range extends from the shortest ultraviolet through visible light to the longest infrared wavelengths. It is also called *heat.*

thermal relay A relay operated by the heat produced by current flow.

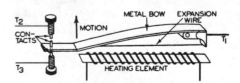

Thermal relay. Basic elements are shown.

thermal runaway A condition that might occur in a power transistor when collector current increases collector junction temperature, reducing collector resistance and allowing a greater current to flow. The increased current increases the heating effect still more. The action might continue until the transistor is destroyed, particularly when the ambient temperature is high.

thermal switch   A temperature-controlled switch.

thermal tuner   A microwave tuner that depends on thermal tuning of a cavity resonator.

thermal tuning   The process of changing the operating frequency of a system with controlled thermal expansion, altering the geometry of the system.

thermal weapon sight [TWS]   A lightweight military thermal imaging device that can be mounted on a weapon or be handheld.

thermel   A thermoelectric device that measures temperature, such as a thermocouple or thermopile.

thermion   A charged particle emitted by a heated body, as by the hot cathode of a thermionic tube.

thermionic   A reference to the emission of electrons as a result of heat.

thermionic cathode   *Hot cathode.*

thermionic converter   A converter that converts heat energy directly into electric energy. In one version, two metal electrodes are separated by a gas at low pressure. When one electrode is heated to about 1100°C, electrons boiled out of it travel through the gas to the other electrode to give an electric current. It is also called a thermionic generator.

thermionic current   A current caused by directed movements of electrons or other thermions, such as the flow of emitted electrons from the cathode to the anode in a thermionic tube.

thermionic detector   A detector that includes a hot-cathode tube.

thermionic diode   A diode electron tube that has a heated cathode.

thermionic emission   The liberation of electrons or ions from a solid or liquid as a result of heat.

thermionic generator   *Thermionic converter.*

thermionic grid emission   The current produced by electrons thermionically emitted from a grid. It is also called primary grid emission.

thermionic tube   *Hot-cathode tube.*

thermionic work function   The energy required to transfer electrons from a given metal to an adjacent medium during thermionic emission, as from a heated filament to a vacuum.

thermistor [THERMal resISTOR]   A bolometer that makes use of the change in resistivity of a semiconductor with a change in temperature. A thermistor has a high negative temperature coefficient of resistance, so its resistance decreases as temperature rises. In critical circuits it compensates for opposite temperature variations in other components. As a

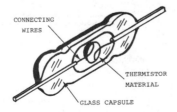

Thermistor bead packaged in a glass capsule.

bolometer it measures temperatures and microwave energy. It can also be a nonlinear circuit element.

thermistor mount   A waveguide mount for the insertion of a thermistor that can measure electromagnetic power.

thermoammeter   An ammeter that is actuated by the voltage generated in a thermocouple through which the current to be measured is sent. It can measure RF currents. It is also called a thermocouple ammeter, a thermocouple instrument, or a thermocouple meter.

thermocline   An interface between warmer and colder water in the ocean, where sound or sonar waves are so sharply bent that submarines can escape detection by hiding under the interface.

thermocompression bonder   *Wire bonder.*

thermocompression wire bonder   *Heat-compression wire bonder.* See also *wire bonder* and *wire bonding.*

thermocouple   A component that consists of two dissimilar conductors welded together at their ends to form a junction. When this junction is heated, the voltage developed across it is proportional to the temperature rise. It measures temperatures, as in a thermoelectric pyrometer, or converts radiant energy into electric energy.

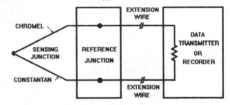

Thermocouple circuit showing extension wires.

thermocouple ammeter   *Thermoammeter.*

thermocouple converter   *Thermal converter.*

thermocouple instrument   *Thermoammeter.*

thermocouple meter   *Thermoammeter.*

thermocouple thermometer   *Thermoelectric thermometer.*

thermocouple vacuum gage   A vacuum gage that depends for its operation on the thermal conduction of the gas present. Pressure is measured as a function of the voltage of a thermocouple whose measuring junction is in thermal contact with a heater that carries a constant current. It has a typical pressure range of 10^{-1} to 10^{-3} torr.

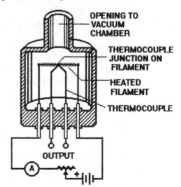

Thermocouple vacuum gage can monitor pressures in the 2 to 10^{-3} torr range.

thermoelectric converter A converter that changes solar or other heat energy to electric energy. It can be a power source on spacecraft.

thermoelectric cooler A module or array for spot-cooling electronic components based on the *Peltier effect*. It is effective for cooling components in high-density circuits. It is typically made from N- and P-doped bismuth telluride semiconductor material. Single-stage modules have couples arranged electrically in series and thermally in parallel. Multistage modules are made by stacking couples thermally in series with decks of electrically insulated and thermally conductive ceramics between them.

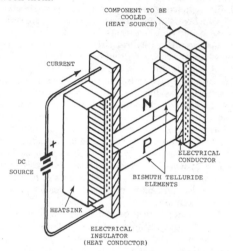

Thermoelectric cooler made from N- and P-doped bismuth-telluride semiconductor couples.

thermoelectric effect *Seebeck effect.*

thermoelectric generator *Thermal converter.*

thermoelectricity Electricity produced by direct action of heat, as by unequal heating of two thermojunctions in the same circuit.

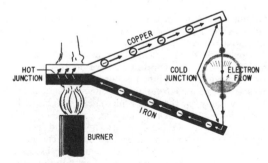

Thermoelectricity principle.

thermoelectric junction *Thermojunction.*

thermoelectric material A material that can convert thermal energy into electric energy or provide refrigeration directly from electric energy. Good thermoelectric materials include lead telluride, germanium telluride, bismuth telluride, and cesium sulfide.

thermoelectric microrefrigerator A refrigeration device based on the Peltier effect for cooling small electronic components such as infrared detectors.

thermoelectric module A device that depends for its operation on the Peltier effect to provide spot cooling of transistors, infrared detectors, and other components when energized by direct current. It can perform precise temperature control of liquids, solids, and gases.

thermoelectric pyrometer A pyrometer whose sensing element is a thermocouple.

thermoelectric series A series of metals arranged in the order of their thermoelectric voltage-generating ratings with respect to some reference metal such as lead.

thermoelectric solar cell A solar cell that first converts the sun's energy into heat with a sheet of metal, and the heat is converted into electricity by a semiconductor material sandwiched between the first metal sheet and a metal collector sheet.

thermoelectric thermometer A thermometer whose thermocouple measuring junction is in thermal contact with the body of the patient. It is also called a thermocouple thermometer.

thermoelectron An electron liberated by heat, as from a heated filament. It is also called a negative thermion.

thermoelement *Thermal converter.*

thermograph 1. A far-infrared image-forming instrument that provides a thermal photograph by scanning a far-infrared image of an object or scene. 2. An instrument that senses, measures, and records the temperature of the atmosphere.

thermography Photography that records radiation in the long-wavelength, far-infrared region, emitted by objects at temperatures ranging from −170 to over 300°F (−112 to 150°C). It is also called *thermal photography*.

thermojunction One of the surfaces of contact between the two conductors of a thermocouple. It is also called a thermoelectric junction.

thermoluminescence Luminescence produced in a material by moderate heat.

thermoluminescent dosimeter A dosimeter based on the principle that when certain irradiated solids are heated, trapped electrons or holes are restored to the ground state, with the resulting emission of light. The amount of this light is measured with a multiplier phototube.

thermomagnetic A reference to the effect of temperature on the magnetic properties of a substance, or to the effect of a magnetic field on the temperature distribution in a conductor.

thermometer An instrument that measures and indicates temperature.

thermomilliammeter A low-range thermoammeter.

thermophone An electroacoustic transducer that produces sound waves which have an accurately known strength by the expansion and contraction of the air adjacent to a conductor whose temperature varies in response to a current input. It is used for calibrating microphones.

thermopile A group of thermocouples that when connected in series gives higher voltage output, or in parallel gives higher current output. It is used for measuring temperature or radiant energy, or converting radiant energy into electric power.

thermoplastic   A plastic that can be softened by heat and rehardened into a solid state by cooling. It can be remelted and remolded many times. Examples are cellulose acetate, cellulose nitrate, methyl methacrylate, polyethylene, polystyrene, and vinyls.

thermoregulator   A high-accuracy or high-sensitivity thermostat. One type consists of a mercury-in-glass thermometer with sealed-in electrodes whose rising and falling column of mercury makes and breaks an electric circuit.

thermorelay   *Thermostat.*

thermosetting   A plastic that solidifies when first heated under pressure and cannot be remelted or remolded without having its original characteristics destroyed. Examples are epoxies, melamines, phenolics, ureas, neoprene, and Teflon TFE.

thermosonic wire bonder   *Ultrasonic wire bonder.* See *wire bonding* and *wire bonder.*

thermosphere   The region of the atmosphere, above the mesosphere where there is strong heating and increasing temperature, resulting from photodissociation and photo-ionization of nitrogen and oxygen atoms. The region extends roughly from 50 to 375 mi (80 to 600 km) altitude.

thermostat   A device that opens or closes a circuit when the temperature deviates from a preset value or range of values. It can actuate the controls of a heating element and produce the required corrective action.

thermostatic switch   A temperature-operated switch that receives its operating energy by thermal conduction or convection from the device being controlled or operated.

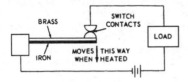

Thermostatic switch with a bimetallic strip (brass and iron).

thermostat materials   Pairs of metals that have widely different coefficients of expansion. When they are joined together, a temperature change makes one material expand more than the other, causing a bend in the shape of the combination that can make or break a circuit as temperature changes. The most effective combinations of materials are nickel and iron and chromium and iron.

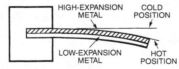

Thermostat materials: differences in coefficients of expansion cause a bimetal strip to bend with increasing heat. The strip can open or close electrical contacts.

Thevenin's theorem   A theorem from electric circuit theory which states that a pair of terminals in a network composed of lumped, linear circuits can be replaced by a voltage source in series with a single impedance. It is also known as the *Helmholtz* or *Helmholtz-Thevenin* theorem. See also *Norton's theorem.*

thick film   A deposition of a film, typically at least one millionth of an inch thick (approximately 10 μm) of conductive or resistive material on an insulating substrate. It is prepared by mixing precious metal powders, powdered glass (frit), and a volatile vehicle to form an ink that is applied to a substrate or mandrel by *silk screening, spraying,* or *wiping.* The substrates or mandrels are furnace-fired to form a hard (often abrasive) film on the substrate. It forms resistive elements on ceramic substrates in chip resistors and resistor networks, dielectric areas for capacitors, and conductive paths on hybrid microcircuits or resistor-capacitor networks. Examples are cermet and proprietary metal-resistive elements.

thick-film capacitor   A capacitor with two overlapping thick-film layers of conducting material that are separated by a deposited dielectric film.

thick-film circuit   A hybrid circuit or resistive-capacitive network on which some or all resistors and capacitors are formed directly on a ceramic substrate as deposited and fired thick conductive or dielectric films. The films are deposited by masking and sputtering, spraying, printing, or combinations of these steps. The active components (diodes, transistors, or integrated circuits) are typically in the form of bare chips or dies that are bonded to the substrate by soldering, brazing, or conductive adhesives. See also *thin-film circuit.*

thick-film resistor   A fixed resistor whose resistance element is a thick film made from particles of noble metals, metal oxides, and glass powders, suspended in organic vehicles or binders.

thin film   A deposition of resistive or conductive material less than one millionth of an inch thick (<10 μm) on an insulating substrate. It is applied on glass, ceramic, or a semiconductor wafer by *sputtering, evaporation,* or *chemical-vapor deposition* [CVD], often through a mask. Examples are nickel-chromium and tin-oxide.

thin-film, active-matrix display [TFAMD] A liquid-crystal display with switching transistors or diodes at each picture element or picture site made as thin films of amorphous or polycrystalline silicon. The active devices set the column bus voltage for pixel storage for the duration of each frame. See also *active-matrix liquid-crystal display* [AMLCD].

thin-film capacitor   A capacitor that is constructed on a substrate by the successive evaporation of thin-film layers of conductive and dielectric films. A thin film of silicon monoxide (SiO) is a typical dielectric.

thin-film circuit   A hybrid microwave or RF microcircuit on which some or all of the resistors, capacitors, and inductors are fabricated from thin conductive and dielectric films, typically less than a few micrometers thick. The films are deposited by masking, sputtering, vacuum evaporation, or combinations of these methods. The active devices such as diodes, transistors or integrated circuits can be bare chips or dies or packaged in surface-mount packages. See also *thick-film circuit.* Some specialized thin-film circuits such as *active-matrix liquid-crystal displays* [AMLCD] have amorphous silicon transistors formed by thin-film deposition.

thin-film cryotron   A circuit intended for operation in a cold environment whose tin or indium thin-film gate element can make the transition from a superconducting to a normal state. A thin-film lead control element, which bridges the gate element but is electrically insulated from it by silicon monoxide insulation, carries the control current. See also *cryotron.*

thin-film ferrite coil An inductor made by depositing a thin flat spiral of gold or other conducting metal on a ferrite substrate. Higher inductance values can be obtained by sandwiching two conductor spirals between three ferrite substrates and connecting the spirals in series.

thin-film inductive read/write head A read/write head for a disk memory drive made by integrated circuit fabrication techniques rather than being wound from wire and assembled from discrete parts. The read element detects changes in the magnetic flux on the surface of the magnetic recording media and converts them into digitally coded signals that are interpreted as data by the drive circuitry.

thin-film magnetoresistor A thin-film resistor whose value can be changed by applying a magnetic field.

thin-film material A material that can be deposited in a desired pattern by a variety of chemical, mechanical, or high-vacuum evaporation techniques. Thin-film resistors are made from gold-platinum or nickel-chromium alloys as well as from tin oxide deposited on a ceramic substrate. Thin-film capacitors are made by depositing alternate layers of a dielectric and a metal; aluminum oxide and silicon oxide are typical dielectrics. Coils are made by depositing pure metal films on ferrite or an alumina substrate.

thin-film microcircuit *Thin-film circuit.*

thin-film resistor A fixed resistor whose resistance element is a metal, alloy, carbon, or other thin-film material.

thin-film technology 1. Thin-film, active-matrix, liquid-crystal display. 2. The application of thin films of conductive materials to substrates. Examples include aluminum sputtered on semiconductor wafers or silver "ink" on ceramic substrates.

thin-film transistor [TFT] In active-matrix liquid-crystal displays (AMLCD), amorphous silicon transistors that control row (address) and column (data) lines on the matrix. The transistors apply a voltage to the LCD material which, along with the display and common electrodes, form capacitors for storing the voltage for the duration of the frame.

thin-window counter tube A counter tube made so that a portion of the envelope has low absorption to permit the entry of short-range radiation.

third harmonic A sinewave component that has three times the fundamental frequency of a complex wave.

Thomson bridge *Kelvin bridge.*

Thomson coefficient The ratio of the voltage existing between two points on a metallic conductor to the difference in temperature of those points.

Thomson cross section *Scattering cross section.*

Thomson effect If a uniform metal conductor is heated at the middle and an external current is passed through it, the heat will be conducted unevenly along the two halves. For example, in copper, cadmium, zinc, and silver, the region where the current is directed from a colder to a hotter part will be cooler than if there were no current, and the region where current is directed from a hotter to a colder part will be warmer. The effect is reversed in such metals as iron and nickel. This effect and the EMF developed are named

for William Thomson, Lord Kelvin. Lead shows no appreciable Thomson effect.

Thomson scattering Scattering of electromagnetic radiation by electrons. The scattering cross section for an electron is 0.657 barn.

Thomson voltage The voltage that exists between two points which are at different temperatures in a conductor.

thoriated-tungsten filament A vacuum tube filament that consists of tungsten mixed with a small quantity of thorium oxide to give improved electron emission.

thorium [Th] A metallic element in the actinide series with an atomic number of 90 and an atomic weight of 232.04. It emits electrons profusely when heated, so it is widely used as a coating for cathodes in electron tubes.

three-gun color picture tube A color television picture tube that has three electron guns which emit electron beams, one for each primary color. Each beam is directed onto phosphor dots that emit only the corresponding primary color. Each gun is controlled by its appropriate primary color signal. The shadow-mask color picture tube is an example.

three-level laser A laser that exhibits three energy levels, one of which is the ground state. Laser action usually occurs between the intermediate and ground states.

three-phase circuit A circuit energized by AC voltages that differ in phase by one-third of a cycle, or 120°.

three-pole switch An arrangement of three single-pole, single-throw switches coupled together to make or break three circuits simultaneously.

three-terminal regulator An integrated-circuit linear regulator packaged in a standard three-terminal transistor package. These devices have a range of current ratings and can be shunt or series. Examples are the industry-standard 7800 series (7805 and 7812).

three-way system A three-unit loudspeaker system that consists of a woofer to handle the lowest frequencies, a midrange unit, and a tweeter for the high frequencies.

threshold 1. The least value of a current, voltage, or other quantity that produces the minimum detectable response. It is also called a limen. 2. The level of pumping at which a laser can go into self-excited oscillation.

threshold current The minimum current value at which a nonself-sustained gas discharge changes to a self-sustained discharge.

threshold effect The inherent suppression of noise in a phase- or frequency-modulated receiver by a carrier whose peak value is only slightly greater than that of the noise.

threshold energy The energy limit, for an incident particle or photon, below which a particular endothermic reaction will not occur or a particular nuclear reaction cannot be observed.

threshold frequency The frequency of incident radiant energy below which there is no photoemissive effect.

threshold gate A computer logic element that has an output of 1 for a minimum sum of input weights, and an output of 0 for a maximum sum of input weights. Threshold gates can be arranged in various combinations to form flip-flops, memory cells, accumulators, and other logic circuits.

thresholding In display technology, the representation of all the gray levels above a certain threshold as 1 (white) and all those below the threshold as 0 (black).

threshold of audibility The minimum effective sound pressure of a specified signal that is capable of evoking an

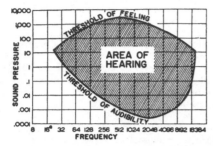

Threshold of audibility, shown as sound pressure in microbars plotted against frequency in hertz. At 1000 Hz, threshold of hearing is 0.0002 μbar, equal to 0 dB when 0.0002 μbar is the reference level.

auditory sensation in a specified fraction of the trials. The threshold can be expressed in decibels relative to 0.0002 μbar or 1 μbar.

threshold of feeling The minimum effective sound pressure of a specified signal that, in a specified fraction of trials, will stimulate the ear until there is the sensation of feeling, discomfort, tickle, or pain. It is expressed in decibels relative to 0.0002 μbar or 1 μbar.

threshold of luminescence *Luminescence threshold.*

threshold sensitivity The smallest amount of a quantity that can be detected by a measuring instrument or automatic control system.

threshold signal The smallest signal that gives a recognizable change in positional information in a navigation system.

threshold value The minimum input that produces a corrective action in an automatic control system.

threshold voltage 1. The lowest gate-to-source voltage of a field-effect transistor (FET) that causes current flow in the source-to-drain channel to start or stop flowing. 2. The lowest voltage at which a specified change in a device or circuit performance occurs.

threshold wavelength The wavelength of the incident radiant energy above which there is no photoemissive effect.

throat The smaller end of a horn or tapered waveguide.

throat microphone A contact microphone that is strapped to the throat of a speaker and reacts to throat vibrations directly rather than to the sound waves they produce.

throttling Control by intermediate steps between full on and full off.

throttling control *Rate control.*

throughput The maximum output of a system, such as lines per minute or pages per minute of a computer line printer.

thulium [Tm] A rare-earth element with an atomic number of 69 and an atomic weight of 168.93.

thumbwheel switch A compact multiposition switch in which finger pressure on the actuating member advances the switch and its indicator to the next contact position.

thump A low-frequency transient disturbance in an audio system.

thyristor A semiconductor switching device whose bistable action depends on PNPN regenerative feedback. A thyristor can be unidirectional or bidirectional, have from two to four terminals, and be triggered from its blocking to its con-

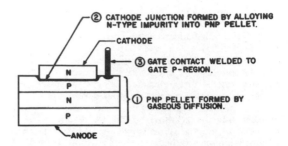

Thyristor PNPN pellet formed by alloy diffusion.

ducting state at a desired point within a single 90° quadrant of the applied AC voltage. The silicon controlled rectifier (SCR) is the most common unidirectional thyristor; others include the gate-turnoff switch (GTO) and light-activated silicon controlled rectifier (LASCR). The triac and silicon bilateral switch are examples of bidirectional thyristors, which can conduct in either direction.

THz Abbreviation for *terahertz.*

TIA Abbreviation for *Telecommunications Industry Association.*

tie-down point One of the frequencies at which a radio receiver is aligned. For the broadcast band, the tie-down points are usually 600 and 1400 kHz.

tie-line A leased communication channel or circuit.

tie point An insulated terminal that permits two or more wires to be connected.

tie wire A wire that connects a number of terminals together.

tight coupling Inductive coupling that links practically all the magnetic flux of one coil to another coil.

tilt The angle that an antenna axis forms with respect to the horizontal.

tilting The forward inclination of the wavefront of radio waves traveling along the ground. The amount depends on the electrical constants of the ground.

tilt stabilization The stabilization of a radar antenna with an additional servomotor that tilts the antenna up or down during scanning, as required to correct for pitch, roll, and yaw of the ship or aircraft.

timbre The attribute of auditory sensation that permits a listener to judge that two sounds similarly presented with the same loudness and pitch are dissimilar. Timbre depends primarily on the spectrum of the stimulus. However, it also depends on the waveform, sound pressure, and frequency location of the spectrum of the stimulus.

time A measure of duration of an event. The fundamental unit of time is the second.

time base The horizontal line formed by the electron beam driven by the sweep-circuit on the screen of a cathode-ray tube.

time-base generator *Sweep oscillator.*

time-base voltage *Sweep voltage.*

time code generator A crystal-controlled pulse generator that produces a train of pulses with various predetermined widths and spacings, from which the time of day and sometimes day of year can be determined. In telemetry and other data-acquisition systems, it provides the precise time of each event.

time constant   The time required for a voltage or current in a circuit to rise to approximately 63% of its steady final value, or to fall to approximately 37% of its initial value. The time constant of a coil that has an inductance L in henrys and resistance R in ohms is L/R. The time constant of a capacitor that has a capacitance C in farads in series with a resistance R in ohms is RC.

time delay   The time required for a signal to travel between two points in a circuit, or for a wave to travel between two points in space.

time-delay circuit   A circuit that delays a signal or action a definite desired period of time.

time-delay fuse   A fuse whose burnout action depends on the time it takes for the overcurrent heat to build up in the fuse and melt the fuse element.

time-delay relay   A relay that provides an appreciable interval of time between the energizing or deenergizing of the coil and the movement of the armature, such as occurs in a slow-acting relay and a slow-release relay.

time discriminator   A circuit in which the sense and magnitude of the output is a function of the time difference between two pulses and their relative time sequence.

time-distribution analyzer   An instrument that indicates the number or rate of occurrence of time intervals falling within one or more specified time-interval ranges. The time interval is delineated by the separation between pulses of a pulse pair.

time-division duplexing [TDD]   A digital transmission method in which two conversations are transmitted over the same carrier frequency by time-sharing the transmissions.

time division multiaccess [TDMA]   A U.S. standard scheme for time sharing digital cellular telephone and satellite communication channels among multiple users assigned to specific transmission *time slots*. Digitized voice and/or data signals are compressed, stored, and processed into the time slots, which are transmitted in *frames*. The transmissions are synchronized to avoid collisions. Receivers extract the user's data bursts from the time slots and demodulate them in milliseconds.

time-division multiplex [TDM]   The transmission of two or more signals over a common path by using a different time interval for each signal.

time-division multiplier   An electronic multiplier whose output is the average of a train of pulses that have their width controlled by one variable and amplitude controlled by the other variable.

time-domain reflectometer [TDR]   An instrument that measures the electrical characteristics of wideband trans-

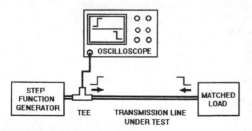

Time-domain reflectometer. It compares reflections from an input voltage step with the original to locate faults in a length of cable or other transmission line.

mission systems, subassemblies, components, and lines by feeding in a voltage step and displaying the superimposed reflected signals on an oscilloscope equipped with a suitable time-base sweep. The display gives the character and location of each pulse-reflecting discontinuity.

time gate   A circuit that gives an output only during chosen time intervals.

time-interval counter   An electronic counter that measures a time interval by counting the number of pulses received from an RF signal generator in that time interval.

time-interval selector   A circuit that produces a specified output pulse when, and only when, the time interval between two pulses lies between specified limits.

time jitter   Variations in the synchronization of the components of a radar system, causing variations in the position of the observed pulse along the time base and reducing the accuracy with which the time of arrival of a pulse can be determined.

time lag   The time between an event aqnd a resultant effect, as between occurrence of a primary ionizing event and its count by a counter.

time-mark generator   A signal generator that produces highly accurate clock pulses which can be superimposed as pips on a cathode-ray screen for timing the events shown on the display.

time modulation   Modulation in which the time of occurrence of a definite part of a waveform is varied in accordance with a modulating signal.

time phase   The property of reaching corresponding peak values at the same instants of time, though not necessarily at the same points in space.

time quadrature   Differing by a time interval corresponding to one-fourth the time of one cycle of the specific frequency.

time response   The time required for the output of a control system to show the effect of the application of a prescribed input signal.

time-shared amplifier   An amplifier that, when combined with a synchronous switch, amplifies signals from different sources, one after another.

time-sharing   The shared use of a circuit or system for two or more purposes during the same overall time interval by allocating small divisions of the total time to each purpose on a fixed schedule, or according to demand.

time signal   An accurate radio signal broadcast at known times each day on several different frequencies by WWV and other stations, for setting clocks.

time-slot assigner circuit [TSAC]   A circuit that determines when a codec will put its 8 bits of data on a PCM bit stream.

time-slot interchange circuit [TSIC]   A device that switches digital highways in PCM-based switching systems; a "digital" crosspoint switch.

time sorter   *Time-distribution analyzer.*

time switch   A clock-controlled switch that opens or closes a circuit at one or more predetermined times.

time-varied gain control   *Sensitivity-time control.*

timing analysis   In integrated-circuit fabrication, the evaluation of integrated-circuit response that accounts for the signal delay times of the circuit elements, but does not consider their logical functionality.

timing-axis oscillator   *Sweep oscillator.*

timing hazard A condition in which spurious data are generated because signals do not travel from one part of the circuit to another fast enough.

timing signal Any signal recorded simultaneously with data on magnetic tape for identifying the exact time of each recorded event.

timing verification In integrated-circuit fabrication, a simulation process in which signals are propagated throughout the entire circuit, and the results are displayed on the workstation screen to verify that the design rules have been met and that the circuit will function at its intended speed.

tin [Sn] A soft, malleable, silver-white metallic element with an atomic number of 50 and an atomic weight of 118.69. It is alloyed with lead to make soft solder for bonding components to circuit boards. The *eutectic* contains 63% tin. Copper wire coated with tin or lead-tin prevents oxidation and makes solder "wetting" easier. Tin oxide is a transparent conductive film that can be deposited on glass mandrels to make film resistors and on glass sheets to form electrodes for liquid-crystal displays (LCDs).

tinned wire Copper wire that has been coated during manufacture with a layer of tin or solder to prevent corrosion and make soldering of connections easier.

tin-oxide resistor A metal-film resistor that consists of tin oxide fused to the surface of a glass or ceramic mandrel. The resistance value is adjusted or increased by cutting or grinding a spiral groove into the film surface.

TINS Abbreviation for *thermal imaging navigation set*.

tinsel cord A flexible cord for headphone and microphone leads. The conductors are strips of thin metal foil or tinsel wound around a strong but flexible central fabric cord.

tip One of the two wires connecting the central office (CO) to a telephone. The name is derived from the tip of the plugs used by operators in older equipment to make the connection. The tip is usually positive with respect to the ring. See *ring*.

tip jack A small single-hole jack for a single-pin contact plug.

Titan An Air Force surface-to-surface intercontinental ballistic missile that has a range of over 6000 mi (9700 km), now employed as a booster for launching spacecraft.

titanium [Ti] A metallic element that has high strength and corrosion resistance with an atomic number of 22.

titration control An electronic control used in chemical processes to regulate acidity or alkalinity.

T junction A waveguide junction with a branch guide that intersects the main guide at right angles.

$TM_{m,n}$ mode A mode that propagates a particular transverse magnetic wave in a waveguide. It is also called the $E_{m,n}$ mode (British).

$TM_{m,n,p}$ mode A mode of wave propagation in a cavity that consists of a hollow metal cylinder closed at its ends. Its transverse field pattern is similar to that of the $TM_{m,n}$ mode in a corresponding cylindrical waveguide for which p is the number of half-period field variations along the axis. It also applies to closed rectangular cavities.

TMN Abbreviation for *telecommunications management network*.

TM wave Abbreviation for *transverse magnetic wave*.

$TM_{m,n}$ wave 1. In a circular waveguide, the transverse magnetic wave for which m is the number of axial planes along which the perpendicular component of the magnetic vector vanishes, and n is the number of coaxial cylinders to which the electric vector is perpendicular. The $TM_{0.1}$ wave is the circular magnetic wave that has the lowest cutoff frequency. 2. In a rectangular waveguide, the transverse magnetic wave for which m is the number of half-period variations of the magnetic field along the longer transverse dimension, and n is the number of half-period variations of magnetic field along the shorter transverse dimension. It is also called an $E_{m,n}$ wave (British) for both circular and rectangular waveguides.

T network A network composed of three branches. One end of each branch is connected to a common junction point, and the three remaining ends are connected to an input terminal, an output terminal, and a common input and output terminal, respectively.

toggle 1. To switch a circuit such as a *flip-flop* to an alternate state. 2. *T flip-flop*.

toggle switch A switch that is operated by the manipulation of a projecting lever which is combined with a spring to provide a snap action for opening or closing a circuit quickly.

token In computer networking, the information package that contains a block of data and a description of its location in the program.

token ring A protocol for a local area network (LAN) that provides for the transfer of messages to specific personal computers in the network.

tolerance A permissible deviation from a specified value, expressed in actual values or more often as a percentage of the nominal value.

toll call A long-distance telephone call for which an individual extra charge is made, based on such factors as distance, length of call, and time of day.

toll line A telephone line or channel that connects different telephone exchanges.

Tomahawk cruise missile A U.S. Navy surface ship- or submarine-launched cruise missile that is 21 ft (6.4 m) long and carries a 1000-lb (450-kg) conventional warhead. Powered by a jet engine, it flies at about 550 mi/h (885 km/h) and has a range of 1150 mi (1850 km). Earlier models had both a long-range terrain-scanning (tercom) guidance system and a near-target video camera guidance system. Advanced models now include a GPS satellite receiver which permits it to hit its target with a stated accuracy ±24 ft (7 m). See *cruise missile*.

TΩ Abbreviation for *teraohm*.

tone 1. A sound wave capable of exciting an auditory sensation that has pitch, or a sound sensation that has pitch. 2. The equality of reproduction of a sound program.

tone arm *Pickup arm*.

tone-burst generator An instrument that produces pulses or bursts of an input frequency provided by an external oscillator. Panel controls are provided for adjusting the number of cycles in the burst and the time interval between bursts.

T1 carrier A pulse-code-modulation (PCM) system operating at 1.544 MHz that can carry 24 individual voice-frequency channels.

tone control A control in an audio amplifier that changes the frequency response to obtain the most pleasing pro-

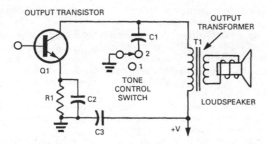

Tone control in a transistorized audio amplifier bypasses higher audio frequencies to ground when switch is in position 2, to emphasize bass notes.

portion of bass to treble. Individual bass and treble controls are included in some amplifiers.

tone generator A signal generator that generates an audio signal suitable for signaling purposes or for testing audio equipment.

tone localizer *Equisignal localizer.*

tone-modulated wave A continuous wave that is modulated by a single audio frequency.

tone ringer The solid-state equivalent of the electromechanical bell that provides the sound when the central office alerts the subscriber that someone is calling. Ringing voltage is typically 80 to 90 V RMS, 20 Hz. See *telephone.*

toner The fine black resinous powder used in copying machines and laser printers to make an electrostatic image readable. The toner is either deposited directly on coated paper or transferred from a charged surface to ordinary paper and then fused to the paper by heating. See *xerography.*

tonne [t] The metric ton, equal to 1000 kg, 2204.623 lb, or 1.1 tons.

TO package A type of package for power transistors, integrated circuits, and other semiconductor devices. The leads are arranged in a circle and project from the base parallel to the axis of the device. The housing is a cylindrical metal can made in standardized sizes. The leads pass out of the housing through glass or other insulating eyelets in its base.

top cap A metal cap positioned at the top of an electron tube and connected to one of the electrodes, usually the control grid. It is an obsolete concept.

top-loaded monopole antenna A vertical antenna that is wider at the top to modify the current distribution and give a more desirable radiation pattern in the vertical

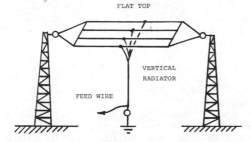

Top-loaded monopole antenna provides a vertical radiation pattern.

plane. A coil can be connected between the enlarged section of the antenna and the remaining structure.

topology An interconnection pattern for electronic components that can be representative of an entire class of circuits such as power supplies or amplifiers.

topside fathometer A fathometer mounted on the deck of a submarine to send sound waves upward. It measures the distance to the surface or to the bottom and top of surface ice.

topside sounder A sounder that sends short radio pulses from a satellite down toward the ionosphere and listens for echoes while sweeping the frequency or changing to different fixed frequencies, to obtain a profile of echo delay versus frequency. This profile is called an ionogram. It can measure ion concentration and electron density in the ionosphere.

toroid A coil or transformer wound on a doughnut-shaped core. The toroidal core gives a maximum magnetic field within itself, with minimum magnetic flux leakage externally.

toroidal phase shifter A nonreciprocal *phase shifter* with a ferromagnetic toroid located within a waveguide section. The toroid is wired to a drive amplifier that can supply either a positive or negative current pulse, which will provide a magnetic field to drive the toroid material into saturation in either a positive or negative state. A complete digital phase shifter contains several lengths of ferrite cores to give differential phase shifts up to 180°. Analog versions must be capable of producing at least 360° of phase shift. See *Reggia-Spencer phase shifter.*

torque A rotational effect on a body that is measured as the product of the *force* present and the radius distance from the axis of rotation to the line of action of the force. The perpendicular distance is referred to as the *lever arm.*

torque-coil magnetometer A magnetometer that depends for its operation on the torque developed by a known current in a coil which can turn in the field to be measured.

torque gradient The amount of torque developed by a synchro per degree of angular difference between transmitter and receiver rotors.

torque motor A motor designed primarily to exert torque while stalled or rotating slowly.

torr The new international standard unit of atmospheric pressure or vacuum. One torr is defined as 1/760 of a standard atmosphere. The torr differs from the earlier millimeter-of-mercury unit by only 1 part in 7 million.

torsion galvanometer A galvanometer that measures the force between the fixed and moving systems by the angle through which the supporting head of the moving system must be rotated to bring the moving system back to its zero position.

torsionmeter An instrument that measures the amount of power transmitted by a rotating shaft. It measures the twisting of the shaft under load or measures the twisting of components mounted on a coupling device inserted between sections of the shaft.

torsion-string galvanometer A sensitive galvanometer whose moving system is suspended by two parallel fibers that tend to twist around each other.

TOS Abbreviation for *tape operating system.*

total dose A measure of radiation hardness of a semiconductor device, measured in *rads.*

total electrode capacitance The capacitance between one electrode and all other electrodes connected together.

total harmonic distortion [THD] The ratio, expressed in percent, of the RMS voltage value for all harmonics present in the output of an audio system to the total RMS voltage at the output, for a pure sinewave input. The lower the percentage figure, the better the audio system.

total ionization 1. The total electric charge on the ions of one sign when the energetic particle that has produced these ions has lost all its kinetic energy. 2. The total number of ion pairs produced by the ionizing particle along its entire path.

touch control A circuit that closes a relay when two metal areas are bridged by a finger or hand.

Touch Tone telephone A telephone that has 12 pushbuttons, each of which produces a distinctive two-frequency musical tone which initiates the proper automatic switching in telephone exchanges, just as the DC pulsing of a rotating dial does.

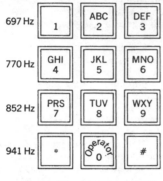

Touch Tone telephone pushbutton arrangement. Each button produces two tone frequencies with values indicated at left and below button pressed.

tourmaline A strongly piezoelectric natural crystal.

tower radiator A tall metal structure that functions as a transmitting antenna.

Townsend avalanche *Avalanche effect.*

Townsend characteristic The current-voltage characteristic curve for a phototube at constant illumination and at voltages below those where glow discharge occurs.

Townsend coefficient The number of ionizing collisions by an electron per centimeter of path length in the direction of the applied electric field in a radiation counter.

Townsend discharge A discharge in a gas at moderate pressure (above about 0.1 mmHg), corresponding to corona. It is free of space charges.

Townsend ionization *Avalanche effect.*

T pad A pad made up of resistors arranged as a T network.

TPI Abbreviation for *tracks per inch.*

TQFP Abbreviation for *thin quad flat package* for integrated circuits.

trace 1. The visible path of a moving spot on the screen of a cathode-ray tube. 2. An extremely small quantity of a substance. 3. An interpretive diagnostic technique that provides an analysis of each executed instruction of a computer and writes it on an output device as each instruction

is executed by the computer. 4. A conductive path on a printed circuit board.

trace concentration A concentration of a substance below the usual limits of chemical detection.

trace interval The time interval for a sweep to trace a desired pattern on the screen of a cathode-ray tube.

tracer 1. A foreign substance, usually radioactive, that is mixed with or attached to a given substance so the distribution or location of the latter can later be determined. 2. A thread of contrasting color woven into the insulation of a wire for identification purposes.

track 1. A path that records one channel of information on a magnetic tape or other magnetic recording medium. The location of the track is determined by the recording equipment rather than by the medium. 2. The horizontal component of the path actually followed by a vehicle, or (marine usage) the intended course. 3. The trace of a moving target on a PPI radar screen or an equivalent plot. 4. To follow the progress of a missile, aircraft, hurricane, or other moving object or action, generally by radar, radio direction finders, infrared, or optical equipment.

trackball A computer control device for moving the cursor on a computer monitor screen. It is basically a ball that is free to rotate in an enclosure surrounded by sensors which translate movements of the ball by the fingertips into cursor motions.

track-command guidance Missile guidance in which the target and missile are tracked by separate radars, and corrective commands are sent to the missile by radio.

track homing The process of following a line of position known to pass through an objective.

tracking 1. The condition in which all tuned circuits in a receiver accurately follow the frequency indicated by the tuning dial over the entire tuning range. 2. A motion given to the major lobe of a radar or radio antenna so that some preassigned moving target in space is always within the major lobe. 3. The following of a groove by a phonograph needle. 4. Maintaining the same ratio of loudness in the two channels of a stereophonic sound system at all settings of the ganged volume control.

tracking beam The beam that is aimed directly at the target at all times in antimissile warfare. Data obtained from this beam is transmitted to the counter-attacking guided missile over what is known as the guidance beam.

tracking element The element in a fire-control system that receives data from the position-finding element and computes the speed and direction of movement of the target and sometimes the rates of change in speed and direction.

tracking filter A bandpass filter whose center frequency follows the average frequency of the input signal.

tracking radar A radar that keeps one or more targets under continuous surveillance so that a more accurate determination can be made of the target's location. Targets of interest detected by a search radar are often "handed off" to and acquired by the tracking radar. Some radars combine both search and track functions by time-sharing the agile beam of a phased-array antenna. See *search radar.*

tracking station A radio, radar, or other station organized to track an object moving through the atmosphere or space.

track in range To adjust the gate of a radar receiver so that it opens at the correct instant to accept the signal from a target that is changing in range.

track made good   The resultant track of an aircraft, represented as a straight line between the departure point and the last point of fix on the surface.

tracks per inch [TPI]   The number of recording tracks produced per inch of radial movement of a magnetic tape recording head on a magnetic disk or other type of recording medium.

traction control   An automotive electronic system which includes sensors that detect when wheel traction is insufficient and the vehicle begins to slip on the road surface. A microcontroller begins to restrict the flow of fuel to the engine, slowing the vehicle, while applying progressive brake pressure to help the driver maintain control.

traffic   The messages transmitted and received over a communication channel.

traffic collision and avoidance system [TCAS]   A system based on *secondary surveillance radar* [SSR]. When a TCAS-equipped aircraft interrogates another nearby TCAS-equipped aircraft, it receives a reply from the other aircraft's SSR *transponder*. The reply permits the interrogating aircraft to determine and display the relative altitude and position of the other aircraft and compute its closing velocity to determine if a collision threat exists. Some versions of TCAS include cockpit displays of complementary escape maneuvers for each converging TCAS-equipped aircraft.

trailer   A bright streak at the right of a dark area or dark line in a television picture, or a dark area or streak at the right of a bright part. It is caused by insufficient gain at low video frequencies.

trailing antenna   An aircraft radio antenna that has one end weighted so it trails free from the aircraft when in flight.

trailing edge   The major portion of the decay of a pulse.

trailing-edge pulse time   The time when the instantaneous amplitude of a pulse last reaches a stated fraction of the peak pulse amplitude.

train   To aim or direct a radar antenna in bearing.

trainer   Equipment used for training operators of radar, sonar, and other electronic equipment by simulating signals received under operating conditions in the field.

trajectory   The path traced through space by a missile or space vehicle.

trajectory-controlled   Guided or directed so its trajectory will follow a predetermined curve, as for a missile.

transaction data   Random and unpredictable new input data for a data-processing system, such as hours worked, quantities shipped, and amounts invoiced.

transadmittance   A specific measure of transfer admittance under a given set of conditions.

transceiver   A radio transmitter and receiver combined in one case with a switch to permit both sending and receiving. Examples include a cellular telephone and mobile radios for the dispatch of emergency vehicles or for marine or aviation communication. It is typically battery-powered and has a short flexible or telescoping antenna.

transconductance $[G_m]$ In transistors, it is the ratio of a change in current with respect to a change in voltage. In a field-effect transistor, it is the ratio of drain-to-source current with respect to a change in gate-to-source voltage. In a vacuum-tube triode, it is the ratio of a change in anode current with respect to a change in control-grid voltage. It can be considered as the amplification factor of the FET divided by drain resistance, or the amplification factor of a triode divided by its anode resistance. It is measured in mhos or siemens and is also called mutual conductance.

transconductance meter   An instrument that indicates the transconductance of a grid-controlled electron tube. It is also called a mutual-conductance meter.

transcribe   1. To record, as to record a radio program by electrical transcriptions or magnetic tape for future rebroadcasting.   2. To copy, with or without translating, from one external computer storage medium to another.

transcriber   The equipment that converts information from one form to another, as for converting computer input data to the medium and language used by the computer.

transducer   A general term for any device or sensor that converts energy from one form to another, as from acoustic energy to electric or mechanical energy. Loudspeakers, microphones, recording heads, and strain gages are examples of transducers.

transducer scanner   A circuit that provides a means of sampling directional signals from individual magnetostriction transducers in a sonar transmitter array. Capacitor plates are arranged radially on a disk rotate with respect to a stationary circular disk that contains matching plates which are connected to the transducer elements, to scan all elements once per revolution of the rotor disk.

transducing piezoid   A piezoid that functions as a transducer.

transfer   To transmit or copy information from one computer to another without changing its form.

transfer check   A check on the accuracy of transfer of a word in a computer, usually made automatically.

transfer function   The mathematical relationship between the output of a control system and its input.

transfer impedance   The ratio of the voltage applied at one pair of terminals of a network to the resultant current at another pair of terminals. All terminals are terminated in a specified manner.

transfer instruction   A computer instruction or signal that specifies the location of the next operation to be performed by the computer.

transfer oscillator   An oscillator that extends the upper frequency limits of an electronic counter. The transfer oscillator mixes the unknown signal with a harmonic of a signal derived internally from a variable-frequency oscillator that is tuned for zero beat. The counter measures the frequency of the variable-frequency oscillator signal, and the counter reading multiplied by the harmonic number gives the unknown frequency.

transferred charge   The net electric charge transferred from one terminal of a capacitor to another via an external circuit.

transferred electron [TE]   A free electron that has been transferred from one minimum to another within a zone. The effective mass of the electron changes to that associated with the new minimum, but the electron does not change its location.

transferred-electron amplifier [TEA]   A diode amplifier, generally based on a transferred-electron diode, made from doped N-type gallium arsenide that provides amplification in the gigahertz range to well over 50 GHz at power outputs typically below 1 W continuous wave. A Gunn amplifier is an example.

transferred-electron device   A semiconductor device, usually a diode, that depends on internal negative resistance caused by transferred electrons in gallium arsenide or indium phosphide at high electric fields. Transit time is minimized, permitting oscillation at frequencies up to several hundred megahertz. Operation can be in the transit-time mode, as in Gunn diodes; in the quenched-domain mode; or in the limited space-charge accumulation (LSA) mode.

transferred-electron diode   A transferred-electron device, generally made from gallium arsenide, that can produce microwave energy directly from a DC input when combined with an appropriate cavity or a microstrip integrated circuit.

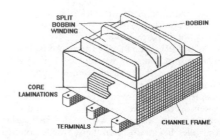

Transformer for circuit board mounting has a low-height profile.

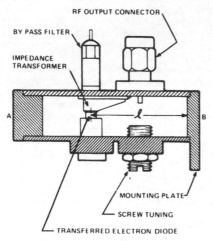

Transferred-electron diode mounted in a tunable waveguide cavity of a microwave oscillator.

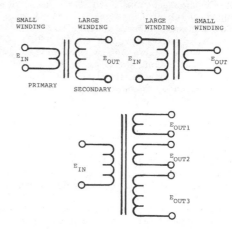

Transformer schematic symbols for (*a*) step-up, (*b*) step-down, and (*c*) multiple winding transformers.

transferred-electron effect   The variation in the effective drift mobility of charge carriers in a semiconductor when significant numbers of electrons are transferred from a low-mobility valley of the conduction band in a zone to a high-mobility valley, or vice versa.

transfer switch   A switch for transferring one or more conductor connections from one circuit to another.

transfer time   The total elapsed time between the breaking of one set of contacts on a relay and the making of another set of contacts, after all contact bounce has ceased.

transformer [T]   A component that consists of two or more coils which are coupled together by magnetic induction. It can transfer electric energy from one or more circuits to one or more other circuits without change in frequency, but usually with changed values of voltage and current.

transformer-coupled amplifier   An audio amplifier whose untuned iron-core transformers provide coupling between stages.

transformer coupling   *Inductive coupling.*

transformer hybrid   *Hybrid set.*

transformer loss   The ratio of the power delivered by an ideal transformer to the power delivered by an actual transformer under specified conditions. It is expressed in decibels.

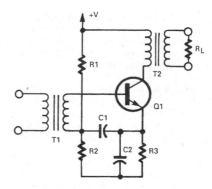

Transformer-coupled transistor amplifier.

transforming section   A length of waveguide or transmission line that has a tapered cross section, used for impedance transformation.

transient   A sudden, very brief spike of high voltage on a power line caused by lightning, electrostatic discharge, or power-line switching. If allowed to enter an electronic circuit it can damage or destroy both active semiconductor and passive components. A fast-response protective device such as a Zener diode or gas-discharge tube in series with the AC line can short-circuit transients before they cause damage. See *transient suppressor.*

transient analyzer An analyzer that generates transients in the form of a succession of equal electric surges of small amplitude and adjustable waveform, applies these transients to a circuit or device under test, and shows the resulting output waveforms on the screen of an oscilloscope.

transient distortion Distortion due to a circuit's inability to amplify transients linearly.

transient motion An oscillatory or other irregular motion that occurs while a quantity is changing to a new steady-state value.

transient oscillation A momentary oscillation that occurs in a circuit during switching.

transient overshoot The maximum value of the overshoot of a quantity as a result of a sudden change in circuit conditions.

transient phenomena Rapidly changing actions that occur in a circuit during the interval between the closing of a switch and the circuit settling to steady-state conditions.

transient recovery time The time required for a circuit output to return to within specified limits following a step change in output load current.

transient response The response of a circuit to a sudden change in an input quantity, such as to a step function.

transient suppressor A device that specifically protects a circuit from destructive voltage surges. Examples include gas-discharge tubes (TVS), metal-oxide varistors (MOV), resistor-capacitor networks, and Zener diodes.

transistance The characteristic that makes possible the control of voltages or currents to accomplish gain or switching action in a circuit. Examples of transistance occur in transistors, diodes, and integrated circuits.

transistor A generic term covering a class of solid-state devices that are capable of amplification and/or switching. The two principal categories are bipolar junction transistor (BJT) and field-effect transistor (FET). BJTs are further classified into NPN and PNP, and FETs are further classified into junction FETs (JFETs) or metal-oxide semiconductor FETs (MOSFETs). All FETs can be further classified as N-channel or P-channel. A transistor can be a discrete device or it can be integrated into an IC. The term also includes the phototransistor and the unijunction transistor. See *bipolar junction transistor, field-effect transistor, phototransistor,* and *unijunction transistor.*

transistor amplifier An amplifier with one or more transistors that provides amplification comparable to that of electron tubes. In a class A transistor amplifier, operation is in the linear region of the collector characteristic. For class B, amplification occurs only during half of each input signal cycle. For class AB, the collector current or voltage is zero for less than half of each input cycle. For class C, collector current or voltage is zero for more than half of each input cycle.

transistor-coupled logic A form of integrated-circuit logic that can be used with common bases and collectors, for multiple-emitter coupling.

transistorized A circuit constructed with transistors in place of electron tubes.

transistorized DC motor 1. A conventional AC motor driven by a transistorized AC/DC converter. 2. A DC motor in which transistors replace the conventional commutator for commutating the current.

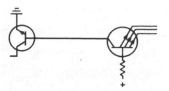
Transistor-coupled logic.

transistor symbol A schematic symbol that represents a transistor in circuit diagrams. The base is represented by a straight line at right angles to its lead. The collector line intersects the base at an angle and has no arrow. The emitter line has an arrow, pointing toward the base for a PNP transistor and pointing away from the base for an NPN transistor.

transistor-transistor logic [TTL] A logic family based on bipolar junction transistors that is characterized by high speed, medium power consumption, and wide application. See also *Schottky logic, low-power Schottky* (LS), and *advanced low-power Schottky* (ALS).

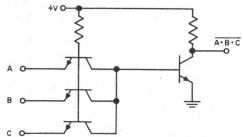
Transistor-transistor logic circuit.

Transit A satellite-based navigation system that includes six satellites. It was developed by the U.S. Navy for ship navigation, but it is available for civilian use. It is scheduled for phase-out by 2000 because of the superiority of the GPS system.

transitional coupling The amount of inductive coupling between two coils that gives the widest passband and flattest response curve without double peaks.

transition coding Color-bar coding of binary data in which the color of the previous bar determines whether the bar being read is a binary 0 or 1. Each consecutive bar has a different color; in one example, a green bar following a white space is a binary 0 bit, and a black to green transition is a binary 1. Transition coding gives higher density of coding and improves accuracy of readout when printing quality is poor.

transition effect A change in the intensity of the secondary radiation associated with a beam of primary radiation as the latter passes from a vacuum into a material medium or from one medium into another.

transition element An element that couples one type of transmission system to another, as for coupling a coaxial cable to a waveguide.

transition factor *Reflection coefficient.*

transition frequency The frequency that corresponds to the intersection of the asymptotes to the constant-

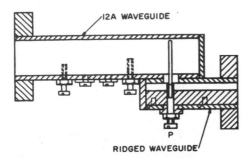

Transition element for coupling a rectangular 12A waveguide to a ridged waveguide is probe P, soldered to center of ridge.

amplitude and constant-velocity parts of the frequency response curve for a recording. This curve is plotted with output voltage ratio in decibels as the ordinate, and the logarithm of the frequency as the abscissa. Below the transition frequency, the level is progressively reduced when making a recording to prevent loud bass notes from distorting the recording. One standard transition frequency value is 500 Hz. It is also called a crossover frequency or a turnover frequency.

transition loss 1. The difference between the power incident upon a transition or discontinuity between two media in a wave propagation system and the power transmitted beyond the discontinuity that would be observed if the medium beyond the discontinuity were match-terminated. 2. The ratio in decibels of the power incident upon a discontinuity to the power transmitted beyond the discontinuity that would be observed if the medium beyond the discontinuity were match-terminated.

transition point A point at which the constants of a circuit change to cause the reflection of a wave being propagated along the circuit.

transition region The region between two homogeneous semiconductors in which the impurity concentration changes.

transit time The time required for an electron or other charge carrier to travel between two electrodes in an electron tube or transistor.

transit-time microwave diode A solid-state microwave diode whose charge-carrier transit time is short enough to permit operation in microwave bands. Bulk diodes (such as Gunn and LSA) and junction diodes (such as BARITT, IMPATT, and TRAPATT) are two major types.

transit-time mode One of the three operating modes of a transferred-electron diode. The space-charge domains are formed at the cathode and travel across the drift region to the anode. The frequency of oscillation is influenced by the dimensions of the drift region. This mode is used in Gunn diodes. The other two modes are the LSA and the quenched-domain modes.

translate To change computer information from one language to another without significantly affecting the meaning.

translation look-aside buffer A buffer used with computer cache memory systems to hold recently completed translations of virtual to physical addresses. See also *physical address* and *virtual address*.

translation loss The amount by which the amplitude of motion of a phonograph stylus differs from the recorded amplitude in a disk record. It is also called playback loss.

translator 1. A computer network or system that has numerous inputs and outputs so connected that when signals representing information expressed in a certain code are applied to the inputs, the output signals will represent the same information in a different code. 2. A combination television receiver and low-power television transmitter that can pick up television signals at one frequency and retransmit them on another frequency to provide reception in areas not served directly by television stations. A translator usually broadcasts on a UHF channel from No. 70 to No. 83.

transliterate To convert the characters of one alphabet to another alphabet.

transmission 1. The process of transferring a signal, message, picture, or other form of intelligence from one location to one or more other locations by fiber-optic cable wire lines, radio, light or infrared beams, or other communication systems. 2. A message, signal, or other form of intelligence that is being transmitted. 3. The ratio of the light flux transmitted by a medium to the light flux incident upon it. Transmission can be either diffuse or specular. It is also called transmittance.

transmission band The frequency range above the cutoff frequency in a waveguide, or the comparable useful frequency range for any other transmission line, circuit, or device.

transmission coefficient The ratio of transmitted to incident energy, or some other quantity at a discontinuity in a transmission medium. For sound waves, it is called the sound transmission coefficient.

transmission control protocol/Internet protocol [TCP/IP] Networking protocol that allows computers to interact with other computers or networks on the *Internet*.

transmission electron microscope [TEM] A microscope in which electron beams replace the visible light of a conventional microscope. Electron beams are formed by electromagnetic coils that emulate the action of glass lenses on light. A filament is the source of a high-voltage electron beam which passes through a series of condensing lenses and then through the specimen. The electrons are first scattered and then focused by a series of magnetic lenses to form an image on a fluorescent screen. The screen can be viewed through an eyepiece. The principle application of the TEM is in medical and genetics applications. See also *electron microscope, scanning tunneling microscope* [STM], *transmission electron microscope* [TEM], and *field-ion microscope*.

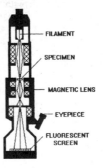

Transmission electron microscope. The electron beam passes through the specimen to form an image that bears little resemblance to the actual object.

transmission grating A diffraction grating produced on a transparent base so radiation is transmitted through the grating instead of being reflected from it.

transmission level The ratio of the signal power at any point in a transmission system to the signal power at some point in the system chosen as a reference point. It is expressed in decibels.

transmission limit A limiting wavelength or frequency above or below which a given type of radiation is not appreciably transmitted by a given medium.

transmission line A waveguide, coaxial cable, fiberoptic cable, twisted pair, or other system of conductors that transfers signal energy efficiently from one location to another.

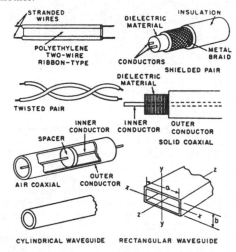

Transmission lines.

transmission-line coupler A coupler that permits the passage of electric energy in both directions between balanced and unbalanced transmission lines.

transmission-line trap An interference trap that can be installed in television receivers to minimize FM and other kinds of interference picked up by the television antenna in the range of 40 to 170 MHz. It consists of a 4⅜-in (11-cm) length of twin-lead-in cable that has a short-circuit at one end and an adjustable ceramic capacitor at the other end, taped to the receiver twin-lead-in cable.

transmission loss 1. The ratio of the power at one point in a transmission system to the power at a point farther along the line, usually expressed in decibels. 2. The actual power that is lost in transmitting a signal from one point to another through a medium or along a line.

transmission measuring set A measuring instrument that consists of a signal source and receiver which have known impedances, to measure the insertion loss or gain of a network or transmission path connected between those impedances.

transmission mode *Mode.*

transmission plane The plane of vibration of polarized light that will pass through a Nicol prism or other polarizer.

transmission primaries The set of three color primaries that correspond to the three independent signals contained in the color television picture signal. The three receiver primaries in the color picture tube form one set. The luminance primary and the two chrominance primaries, known as the Y, I and Q primaries, form another possible set of transmission primaries.

transmission secondary-emission multiplication Electron multiplication in which electrons striking one side of a dynode cause emission of many more electrons from the opposite side of that dynode, with the process building up as the electron stream passes through a series of dynodes.

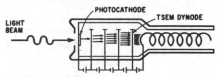

Transmission secondary-emission multiplication in a traveling-wave phototube.

transmission security The aspect of communication security that is concerned with the transmission of messages over wires or by radio.

transmission speed The number of information elements sent per unit time. It can be expressed as bits, characters, word groups, or records per second or per minute.

transmission time The absolute time interval from transmission to reception of a signal.

transmission unit An early signal-level unit now known as the decibel.

transmissivity *Transmittivity.*

transmissometer A photoelectric instrument that measures the visibility of the atmosphere.

transmit To send a message, program, or other information to a person or place by telephone wire, fiber-optic cable, radio, or other means.

transmit negative The transmission of facsimile signals intended for reception as a negative.

transmit positive The transmission of facsimile signals intended for reception as a positive.

transmit-receive [TR] **switch** *TR switch.*

transmit-receive [TR] **tube** *TR tube.*

transmittance *Transmission.*

transmitted-carrier operation Amplitude modulation in which the carrier wave is transmitted.

transmitted wave *Refracted wave.*

transmitter 1. The equipment that generates and amplifies an RF carrier signal, modulating the carrier signal with intelligence, and feeding the modulated carrier to an antenna for radiation into space as electromagnetic waves. 2. In telephony, the microphone that converts sound waves into AF signals. 3. *Synchro transmitter.*

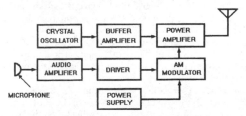

Transmitter block diagram for an AM radio station.

transmitter input polarity The polarity of the part of a television picture signal that represents a dark area of a scene, relative to the polarity of some of the signal which represents a light area.

transmitter synchro *Synchro transmitter.*

transmitting efficiency The ratio of total acoustic power output to electric power input for an electroacoustic transducer.

transmittivity The ratio of the transmitted radiation to radiation arriving perpendicular to the boundary between two media. It is also called *transmissivity*.

transolver A synchro that has a two-phase cylindrical rotor within a three-phase stator to serve as a transmitter or a control transformer with no degradation of accuracy or nulls.

transonic barrier *Sonic barrier.*

transonic speed A speed in the range of about Mach 0.8 to Mach 1.2, corresponding to 600 to 900 mi/h (about 960 to 1450 km/h), at which one or more local points on the body of an aircraft or missile are moving at subsonic speed at the same time that one or more points move at sonic or supersonic speed.

transparent plasma A plasma through which an electromagnetic wave can propagate. In general, a plasma is transparent at frequencies above the plasma frequency.

transpolarizer An electrostatically controlled circuit impedance that can have about 30 discrete and reproducible impedance values. Two capacitors, each having a crystalline ferroelectric dielectric with a nearly rectangular hysteresis loop, are connected in series and act as a single low impedance to an AC sensing signal when both capacitors are polarized in the same direction. Application of 1-μs pulses of appropriate polarity increases the impedance in steps.

transponder A radio frequency or acoustic transceiver that receives a signal (*interrogation*) and automatically transmits a response with a self-generated signal, often encoded and/or at a different frequency from the interrogation signal. See also *beacon* and *IFF.*

transponder dead time The time interval between the start of a pulse and the earliest instant at which a new pulse can be received or produced by a transponder.

transportable phone A mobile telephone that can take its power either from a vehicle's battery or its own battery pack.

transrectification Rectification that occurs in one circuit when an alternating voltage is applied to another circuit.

transversal filter A filter whose frequency transmission properties exhibit a periodic symmetry.

transverse-beam traveling-wave tube A traveling-wave tube in which the direction of motion of the electron beam is transverse to the average direction in which the signal wave moves.

transverse electric wave [TE wave] An electromagnetic wave in which the electric field vector is everywhere perpendicular to the direction of propagation. It is also called an H wave (British).

transverse electromagnetic wave [TEM wave] An electromagnetic wave in which both the electric and magnetic field vectors are everywhere perpendicular to the direction of propagation.

transverse-field traveling-wave tube A traveling-wave tube in which the traveling electric fields that interact with the electrons are essentially perpendicular to the average motion of the electrons.

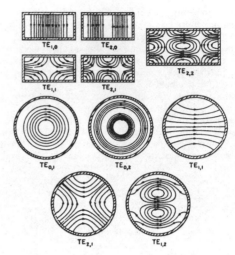

Transverse electric wave modes in rectangular and circular waveguides showing electric fields for each.

transverse-film attenuator An attenuator in the form of a conducting film placed across a waveguide.

transversely excited atmospheric-pressure laser [TEA laser] A gasdynamic laser whose discharge path between electrodes is made extremely short by placing electrodes at opposite sides of the discharge tube. Many pairs of these electrodes are positioned along the length of the tube to provide full excitation. Peak output powers exceeding 1 MW have been obtained.

transverse magnetic wave [TM wave] An electromagnetic wave in which the magnetic field vector is everywhere perpendicular to the direction of propagation. It is also called an *E wave* (British).

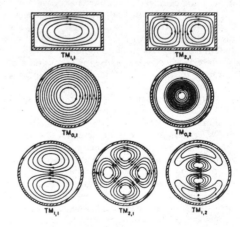

Transverse magnetic wave modes for rectangular and circular waveguides showing magnetic fields for each.

transverse magnetization Magnetization of a magnetic recording medium in a direction perpendicular to the line of travel and parallel to the greatest cross-sectional dimension.

transverse plate   A plate of metal or highly resistant material that closes the end of a waveguide or is an adjustable piston inside the waveguide.

transverse wave   A wave whose direction of displacement at each point of the medium is parallel to the wavefront.

trap   1. A tuned circuit in the RF or IF section of a receiver that rejects undesired frequencies. Traps in television receiver video circuits keep the sound signal out of the picture channel. It is also called a rejector. 2. A semiconductor imperfection that prevents carriers from moving through the material. 3. *Wave trap.*

TRAPATT diode   [TRApped Plasma Avalanche Transit-Time diode]   A solid-state microwave diode whose operating frequency as an oscillator is approximately determined by the thickness of the active layer. It is a transit-time device like the IMPATT diode, but it operates in a different mode; the avalanche zone moves through the drift region, creating a trapped space-charge plasma within the PN junction region.

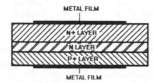

TRAPATT diode is placed in a circuit containing a high-Q resonator and is reverse biased into an avalanche condition.

trapezoidal wave   A square wave that has had one ramp of a sawtooth wave superimposed on it.

trapped flux   Magnetic flux that links with a closed superconducting loop.

trapped mode   A form of propagation in which the energy radiated in the troposphere is almost entirely confined within a duct.

trapped radiation   Radiation from space that has become trapped in the magnetic field of the earth, as in the Van Allen belt.

trapping   A process in which electrons are held at an irregularity in the crystal lattice of a semiconductor until released by thermal agitation.

traveling detector   An RF probe mounted in a slotted-line section of waveguide that, with a detector, can measure standing-wave ratios.

traveling wave   A wave formed by translation of energy along a conductor. The energy is equally divided between current and voltage forms.

traveling-wave accelerator   A plasma engine for space travel in which plasma is accelerated through a tube by a series of coils spaced along the tube and excited by polyphase RF energy.

traveling-wave amplifier   [TWA]   See *traveling-wave amplifier tube.*

traveling-wave amplifier tube   [TWAT]   A traveling-wave tube designed as an amplifier. It can produce pulsed output power in the lower microwave frequencies in excess of 200 W, and it can be designed to operate over the range of 1 to 40 GHz.

traveling-wave antenna   An antenna in which the current distributions are produced by waves of charges propa-gated in only one direction in the conductors. It is also called a progressive-wave antenna.

traveling-wave interaction   The interaction between an electron stream and a slow wave moving through a circuit in approximate synchronism with the velocity of the electrons.

traveling-wave light modulator   A two-conductor transmission line in which some of the dielectric is an electro-optical material capable of modulating a laser beam. The index of refraction of this material, and hence the velocity of light through it, varies with the applied electric field. Suitable dielectric materials include cubic crystals and cuprous chloride and zinc sulfide. The amount of mismatch between the velocity of propagation of light through the crystal and the velocity of the microwave signal being transmitted at a given frequency determines the bandwidth of the modulator.

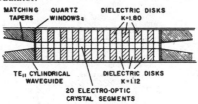

Traveling-wave light modulator with 20 crystal segments to modulate the laser beam.

traveling-wave maser   A ruby maser with a comblike slow-wave structure and some yttrium-iron garnet isolators that gives amplification in the frequency range from about 400 MHz to well over 10 GHz. It operates at the temperature of liquid helium (4.2 K).

traveling-wave parametric amplifier   A parametric amplifier that propagates signal, pump, and difference-frequency waves along a continuous structure which contains nonlinear reactors.

traveling-wave phototube   A traveling-wave tube that has a photocathode and an appropriate window to admit a modulated laser beam. The modulated laser beam causes the emission of a current-modulated photoelectron beam. That beam is accelerated by an electron gun and directed into the helical slow-wave structure of the tube. An alternative arrangement has a transmission-type photocathode and transmission-type dynodes in an electron multiplier that feeds the helical structure. The operation of traveling-wave phototubes is restricted essentially to visible wavelengths of light.

traveling-wave tube   [TWT]   A microwave tube that can function either as an oscillator or as an amplifier. All versions of this tube contain an electron gun, wave-slowing delay line, and target or collector. The traveling radio frequency (RF) wave in the slow wave structure is slowed by a factor of 10 to 20% so that its speed approximately matches that of the electrons in the beam. A high-intensity electron beam formed by the electron gun passes close to the delay line and gives up its energy to it, thereby amplifying the signal or causing oscillation, depending on the tube's structure. The collector dissipates the remaining energy of the beam. The waveguide can be a helix or an interdigitated line. The conventional tube is called a *forward-wave traveling-wave tube.* Its

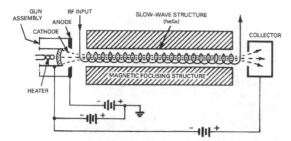

Traveling-wave tube solenoid provides the required longitudinal magnetic field.

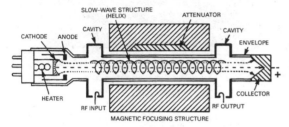

Traveling-wave tube (TWT) amplifier signal is amplified as it travels on a helical delay line by receiving energy from the TWT's electron stream.

dimensions determine its operating frequency range. See also *carcinoton* and *backward-wave oscillator*.

TR box   *TR tube.*

TR cavity   The resonant section of a radar TR tube.

treble   High audio frequencies, such as those reproduced by a tweeter in a sound system.

treble boost   Adjustment of the amplitude-frequency response of a system or component to accentuate the higher audio frequencies.

TRF   Abbreviation for *tuned radio frequency*.

triac [TRIode AC semiconductor switch]   A bidirectional gate-controlled thyristor that provides full-wave control of AC power. With phase control of the gate signal, load current can be varied from about 5 to 95% of full power.

Triac schematic symbol.

triad   A triangular group of three small phosphor dots on the screen of a three-color cathode-ray television tube. Each dot emits one of the three primary colors—red, green, or blue—when scanned by the CRT's electron beam. The set of three phosphor dots makes up a pixel, the smallest picture element on the screen.

triangle generator   A signal generator whose output is a repeating ramp function that has equal positive and nega-

tive rates of change with respect to time. The resulting triangle wave resembles a series of equilateral triangles.

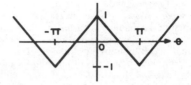

Triangle-generator output waveform as seen on an oscilloscope.

triangulation   The determination of the position of a ship or aircraft by obtaining bearings of the moving object with reference to two fixed radio stations a known distance apart. This gives the values of one side and all angles of a triangle, from which the position can be computed.

tribo-   A prefix meaning related to or resulting from friction.

triboelectricity   Electric charges generated by friction.

triboelectric series   A list of materials that produce an electrostatic charge when rubbed together. They are arranged so that a material has a positive charge when rubbed with a material substance below it in the list and a negative charge when rubbed with a material above it in the list.

triboelectroemanescence   Electron emission that occurs when a metal is fractured or abraded.

triboluminescence   Luminescence produced by friction between two materials.

trichromatic coefficient   *Chromaticity coordinate.*

trickle charge   A continuous charge of a battery at a low rate to maintain the battery in a fully charged condition.

trickle charger   A device that charges a battery at a low rate continuously to keep it fully charged.

triclinic   A reference to a crystal structure that has three unequal axes intersecting at angles, not more than two of which are equal, and not more than one of which is 90°.

tridipole antenna   A horizontally polarized antenna that has three curved dipoles mounted in a horizontal plane to form a circle.

triductor   An arrangement of iron-core transformers and capacitors that triples a power-line frequency. A DC voltage is applied to some of the windings to provide enough premagnetization so the applied AC voltage can saturate the cores and thereby generate the desired third-harmonic output.

trifluorochloroethylene resin   A fluorocarbon base for polychlorotrifluoroethylene resin, marketed as Kel-F.

trigatron   An electronic switch whose conduction is initiated by the breakdown of an auxiliary gap in a gas-filled envelope. The gap between the two main electrodes is normally nonconducting, but it breaks down when a pulse is applied to a trigger electrode.

trigger   1. To initiate a sudden action, as by applying a pulse to a trigger circuit.   2. The pulse that initiates the response of a trigger circuit.   3. *Trigger circuit.*

trigger action   A weak input pulse that initiates main current flow suddenly in a circuit or device.

trigger circuit   1. A circuit or network that exhibits abrupt output changes with a small change in input at a predetermined operating point.   2. A circuit whose action is initiated by an input pulse, as in a radar modulator. 3. *Flip-flop circuit.*

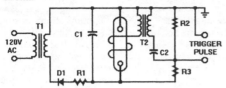

Trigger circuit for an electronic flash lamp suitable for stop-action photography.

trigger diode *Diac.*

triggered blocking oscillator A blocking oscillator that can be reset to its starting condition by a trigger voltage. A parallel-triggered blocking oscillator has less effect on the trigger source than a series-triggered blocking oscillator, but the series-triggered type has less delay.

triggering Initiation of an action in a circuit, which then functions for a predetermined time, as for the duration of one sweep in a cathode-ray tube.

trigger pulse A pulse that starts a cycle of operation.

trigger winding A winding added to a pulse transformer to supply a low-voltage pulse to an external load, usually for synchronizing purposes.

trihedral reflector A corner reflector that has three square or triangular sides which meet at a point. It is an artificial radar reflector and has other applications where signals must be reflected back toward the transmitter over a greater angle than is practical with a plane sheet reflector.

trimmer A variable resistor or variable capacitor. When the term is not modified by the word capacitor, it usually refers to a *trimmer potentiometer.* See also *trimmer potentiometer* and *trimmer capacitor.*

trimmer capacitor A variable or adjustable capacitor most commonly used in RF circuits. The change in capacitance is achieved by mechanically altering the spacing between two metallic plates or by altering the position of a screw with respect to another metallic surface.

trimmer potentiometer A variable, three-terminal resistor for making final adjustments in the resistive values of a circuit for tuning or compensation. A wide range of styles, sizes, and resistance values is available. The most common are single-turn units that can be adjusted with a screwdriver in a slot that turns a wiper over a semicircular resistive element. The resistance values of multiturn trimmers can be adjusted by leadscrew mechanisms that move a wiper along the length of a linear resistive element. It is also called a *set-and-forget variable resistor.*

trimming Fine adjustment of capacitance, inductance, or resistance of a component during manufacture or after installation in a circuit.

Trinitron A color picture tube that has its phosphor screen deposited in narrow vertical stripes instead of dots. The mask in front of the screen is a grille of vertical slots instead of a shadow mask with round holes. A single electron gun emits three beams, one for each primary color, in a horizontal line. The phosphor stripes are very narrow compared to the beams, so each beam spreads across two slots in the mask. The angle at which a beam enters a slot determines which color of phosphor stripe it hits.

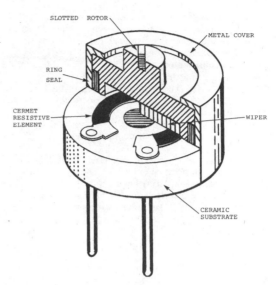

Trimmer potentiometer cutaway, showing the resistive element and rotary wiper.

triode A three-electrode electron tube that contains an anode, a cathode, and a control electrode or grid.

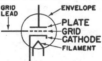

Triode symbol. The plate electrode is usually called the anode.

triode amplifier An amplifier that amplifies with triodes.

triode-hexode converter A triode oscillator and a multigrid mixer in the same tube envelope.

trip action Instability that occurs in a magnetic amplifier because of excessive feedback.

trip coil A coil that opens a circuit breaker or other protective device when coil current exceeds a predetermined value.

triple-conversion receiver A communications receiver that has three different intermediate frequencies, to give higher adjacent-channel selectivity and greater image-frequency suppression.

triple-detection receiver *Double superheterodyne.*

triple-stub transformer A transformer with three stubs that are placed a quarter-wavelength apart on a coaxial line and adjusted in length to compensate for impedance mismatch.

triplet Three radio navigation stations operated as a group for the determination of positions. It is also called a *triad.*

triplexer A dual-duplexer that permits two receivers to function simultaneously and independently in a radar system by disconnecting the receivers during the transmitted pulse.

tripping pulse *Trigger pulse.*

trip value The voltage, current, or power at which a polarized relay will transfer from one contact to another.

tristate logic An output organization in digital logic families that is capable of assuming three output states; two are the normal low-impedance 1 and 0 states, and the third is a high-impedance state that allows many tristate devices to time-share the same wires while allowing only one wire to control the line levels at any given time.

tristimulus colorimeter A colorimeter that measures a color stimulus in terms of tristimulus values.

tristimulus values The amounts of each of the three primary colors that must be combined to establish a match with a given sample color.

tritium [T or H³] The hydrogen isotope that has mass number 3. It is one form of heavy hydrogen; the other is deuterium.

tritium light source A self-luminous lamp filled with radioactive tritium which emits high-energy beta particles during its decay (electrons having a maximum energy of 18.6 keV). These electrons provide excitation for a phosphor coating inside a glass envelope, to give a light source that requires no power supply. Its luminous half-life is about 8 years.

trombone A U-shaped length of waveguide that is adjustable in length.

tropopause The discontinuity that separates the stratosphere from the troposphere.

troposcatter *Tropospheric scatter.*

troposphere The region of the earth's atmosphere that extends from the surface up to about 6 mi (10 km). Temperature generally decreases with altitude, clouds form, and convection in this region.

tropospheric bending Refraction of radio waves by adjacent layers of air masses that have different temperature and humidity characteristics in the troposphere, making possible long-distance transmission of VHF radio waves.

tropospheric duct *Duct.*

tropospheric scatter A form of scatter propagation in which radio waves are scattered by the troposphere. The phenomenon is essentially independent of frequency. It is useful for communication over distances of several hundred kilometers over the entire RF spectrum. It is also called *troposcatter.*

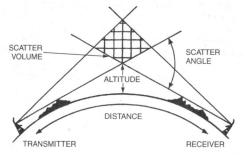

Tropospheric scatter for point-to-point, beyond-the-horizon RF communication at frequencies from 350 MHz to 8 GHz.

tropospheric superrefraction A condition in the troposphere in which radio waves are bent sufficiently to be returned to the earth.

tropospheric wave A radio wave that is propagated by reflection from a region where there is abrupt change in dielectric constant or its gradient in the troposphere.

troubleshooting The practice of locating and repairing faults in equipment after they have occurred.

TR switch [transmit-receive switch] A mechanical switch, electronic device, or circuit that isolates the receiver from the transmitter when a signal is being sent. It generally applies to switches in RF communications transmitters and receivers and low-power radar systems rather than the gas-filled *TR tubes* used in high-power radar systems. Examples of basic switching elements are *ferrite circulators* and *PIN diodes.* PIN switches are easier to apply in coaxial circuitry and at the lower microwave frequencies. Multiple diodes are used when a single diode cannot withstand the required voltage or current.

TR tube [transmit-receive tube] A gas-filled tube that protects the receiver from the transmitter power. It permits low-power RF signals to pass through it with very little attenuation. Higher power causes the gas to ionize and present a short circuit to the RF energy. A typical balanced *duplexer* contains two TR tubes. When the transmitter is on, the TR tubes fire and reflect the RF power to the antenna port of the input hybrid. On reception, signals received by the antenna are passed through the TR tubes and to the receiver port of the output hybrid. It is also called a *TR box* and a *TR switch.*

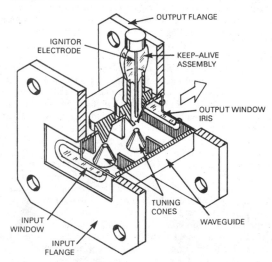

TR tube construction and installation.

true altitude The altitude above mean sea level.

true bearing A bearing given in relation to true geographic north. A magnetic bearing is given in relation to magnetic north, and a relative bearing is given in relation to the lubber line or other axis of a ship or aircraft.

true-bearing rate The rate of change of true bearing.

true-bearing unit A circuit added to a radar set that rotates the PPI display so true north is always at the top of the screen.

true course A course indicated by an angle measured clockwise from true north.

true heading A heading measured with respect to true north.

true north The direction of the North Pole from the observer, or a line showing this direction.

truncate To drop digits at the end of a numerical value. The number 3.14159265 is truncated to five figures in

3.1415, whereas it would be 3.1416 if rounded off to five figures.

truncated paraboloid A radar parabolic reflector that has had part of the top and bottom cut away to broaden the radar beam in the vertical plane.

trunk 1. A path over which information is transferred in a computer. 2. A telephone circuit or channel connecting two central offices or switching entities.

truth table A table that describes a logic function by listing all possible combinations of input values, and indicating for each combination the true output values.

AND			OR			NAND		
B	A	C	B	A	C	B	A	C
0	0	0	0	0	0	0	0	1
0	1	0	0	1	1	0	1	1
1	0	0	1	0	1	1	0	1
1	1	1	1	1	1	1	1	0

Truth tables for three gating functions.

TSAC Abbreviation for *time-slot assigner circuit.*

TSOP Abbreviation for *thin small-outline package* for integrated circuits. *Type I* has leads on the short sides of the rectangular package only, and *Type II* has leads on the long sides of the rectangular package only.

TTC Abbreviation for *telemetry tracking and control station.*

T²L Abbreviation for *transistor-transistor logic.*

TTL Abbreviation for *transistor-transistor logic.*

TTY Abbreviation for *teletypewriter.*

tube coefficients The constants that describe the characteristics of an electron tube, such as amplification factor and transconductance.

tube noise Noise originating in an electron tube, such as that caused by shot noise and thermal agitation.

tube socket A socket that accepts the pins of an electron tube electrically and mechanically.

tube tester A test instrument that measures and indicates the condition of electron tubes for electronic equipment.

tumbling Loss of control in a two-frame free gyroscope, occurring when both frames of reference become coplanar.

tunable-cavity filter A microwave filter that can be tuned by adjusting one or more tuning screws which project into the cavity, or by adjusting the positions of one or more rectangular or circular irises in the cavity or waveguide.

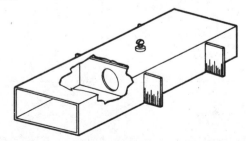

Tunable-cavity filter with two adjustable circular irises and one tuning screw.

tunable echo box An echo box that consists of an adjustable cavity operating in a single mode. It can be calibrated so the setting of the plunger at resonance indicates wavelength.

tunable laser A laser whose output radiation frequency can be tuned over part or all of the ultraviolet, visible, and infrared regions of the spectrum. A flashlamp-excited dye laser with a birefringent tuning element and frequency-doubling quartz crystals is available. Other versions use KDP crystal frequency-doubling stages and mirror-angle tuning. The choice of dye solution affects the tuning range. Applications include research in plasma physics, spectroscopy, chemistry, photoconductivity, biology, and holography.

tunable magnetron A magnetron that can be tuned over a range of frequencies by electronic or mechanical means. Generally the capacitance or inductance of the resonant structure is varied mechanically to achieve tuning.

tune To adjust for resonance at a desired frequency.

tuned amplifier An amplifier whose load is a tuned circuit. Load impedance and amplifier gain then vary with frequency.

tuned antenna An antenna whose inductance and capacitance values provide resonance at the desired operating frequency.

tuned-base oscillator A transistor oscillator whose frequency-determining resonant circuit is located in the base circuit. This is comparable to a tuned-grid electron-tube oscillator.

tuned cavity *Cavity resonator.*

tuned circuit A circuit whose components can be adjusted to make the circuit responsive to a particular frequency in a tuning range. It is also called a tuning circuit.

tuned-circuit oven An electrically heated compartment that accommodates and maintains tuned-circuit elements at an essentially constant temperature to prevent drifting in frequency with changes in temperature.

tuned-collector oscillator A transistor oscillator whose frequency-determining resonant circuit is located in the collector circuit. This is comparable to a tuned-anode electron-tube oscillator.

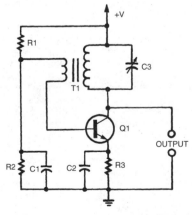

Tuned-collector oscillator oscillates when energy from the tank in the transistor's collector circuit is coupled back to the transistor's base.

tuned dipole A dipole that provides resonance at its operating frequency.

tuned-emitter oscillator A transistor oscillator whose frequency is determined by a tuned resonant circuit in the emitter circuit, coupled to the base to provide the required feedback.

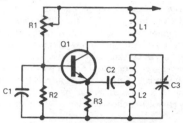

Tuned-emitter oscillator oscillates when energy from an emitter tank circuit is coupled back to the transistor's base.

tuned filter A filter that includes one or more tuned circuits to attenuate or pass signals at the resonant frequency.

tuned radio frequency [TRF] A reference to a receiver in which all RF amplification is carried out at the frequency of the transmitted carrier signal.

tuned radio-frequency amplifier An RF amplifier whose tuned circuits are adjusted to the frequency of the desired transmitted carrier signal.

tuned radio-frequency receiver A radio receiver that consists of numerous amplifier stages which are tuned to resonance at the carrier frequency of the desired signal by a gang capacitor. The amplified signals at the original carrier frequency are fed directly into the detector for demodulation, and the resulting AF signals are amplified by an AF amplifier and reproduced by a loudspeaker. This design has been replaced by superheterodyne receivers that have better selectivity.

tuned radio-frequency stage A stage of amplification that is tunable to the carrier frequency of the signal being received.

tuned transformer A transformer whose associated circuit elements are adjusted to be resonant at the frequency of the alternating current supplied to the primary, thereby causing the secondary voltage to build up to higher values than would otherwise be obtained.

tuner The section of a receiver that contains circuits which can be tuned to accept the carrier frequency of a desired transmitter while rejecting the carrier frequencies of all other stations on the air at that time. A television tuner commonly contains only the RF amplifier, local oscillator and mixer stages, whereas a radio tuner also contains the IF amplifier and second-detector stages.

tungsten [W] A heavy, hard, gray-white metallic chemical element that has an atomic number of 74 and an atomic weight of 183.85. It has a melting point of 3370°C and only fair conductivity. It can be drawn into wire to make filaments for incandescent lamps and wound in coils to make heaters for electron tubes. It is also wound on an insulating core or mandrel to make power resistors, and it is used as an electrode in welding. It has an inverse temperature coefficient of resistance (TCR). It is also called *wolfram*.

tungsten-halogen lamp A tungsten-filament lamp that has halogen gas (related to iodine) added to the normal gas mixture in the envelope to combine with the tungsten evaporated from the filament and prevent blackening. It is used for high-intensity photographic, theater, home and television lighting, and in automobile headlights.

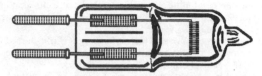

Tungsten-halogen lamp provides higher intensity light than an incandescent lamp and consumes less power.

tuning Adjusting circuits for optimum performance at a desired frequency.

tuning capacitor A variable capacitor used for tuning purposes.

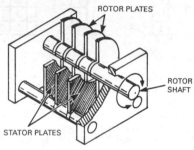

Tuning capacitor with an air dielectric. Rotor plates move within the stator plates to change capacitance value.

tuning circuit *Tuned circuit.*

tuning coil A variable inductance used for tuning purposes.

tuning core A ferrite core that is moved in and out of a coil or transformer to vary the inductance.

tuning fork A U-shaped bar of hard steel, fused quartz, or other elastic material that vibrates at a definite natural frequency when struck or when set in motion by electromagnetic means. It can be a frequency standard.

tuning-fork drive Use of a tuning fork drive for the control of frequency of an oscillator. A high harmonic of the fork frequency is picked up by a coil and amplified to control the frequency of the main oscillator in a transmitter or other equipment.

tuning-fork resonator A tuning fork and associated coils that generate an AC voltage related to the natural vibrating frequency of the fork.

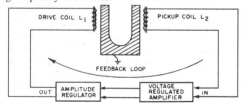

Tuning-fork resonator with a feedback arrangement for continuous oscillation.

tuning indicator A device that indicates when a radio receiver is tuned accurately to a radio station, such as a meter or a cathode-ray tuning indicator. It is connected to a circuit that has a DC voltage which varies with the strength of the incoming carrier signal.

tuning meter   A DC voltmeter or ammeter used as a tuning indicator.

tuning probe   An essentially lossless probe that can be extended through the wall of a waveguide or cavity resonator to project an adjustable distance inside for tuning or impedance matching.

tuning range   The frequency range over which a receiver can be adjusted by a tuning control.

tuning screw   A screw that is inserted into the top or bottom wall of a waveguide and adjusted as to depth of penetration inside for tuning or impedance-matching.

tuning sensitivity   The rate of change of frequency of an oscillator with the position of a mechanical tuner, electric tuning voltage, or other control variable, at a given operating point.

tuning stub   A short length of transmission line, usually shorted at its free end, that is connected to a transmission line for impedance-matching.

tunnel diode   A heavily doped junction diode that has negative resistance in the forward direction over some of its operating range, due to quantum mechanical tunneling. It can be made from a variety of semiconductor materials, including germanium, silicon, gallium arsenide, and indium antimonide, for use as an oscillator or amplifier that can operate well up into microwave frequencies. It is also called an Esaki diode.

tunnel-diode amplifier   A low-noise microwave amplifier that has a tunnel diode as its active element. Applications for the amplifier include radar and telemetry receivers. Its frequency range is from about 200 MHz to well over 10 GHz.

tunnel effect   The piercing of a rectangular potential barrier in a semiconductor by a particle that does not have sufficient energy to go over the barrier. The wave associated with the particle is almost totally reflected on the first slope of the barrier, but a small fraction passes through the barrier.

tunneling   The penetration of a potential barrier in a semiconductor by electrons whose energy is theoretically insufficient to overcome the barrier.

tunneling cryotron   *Josephson junction.*

turbidimeter   An instrument that measures the turbidity of a liquid. A photoelectric turbidimeter measures the amount of light that passes through the liquid. It is also called an *opacimeter*.

turbomolecular pump   A secondary vacuum pump that includes a rapidly rotating turbine for sweeping molecules out of a chamber with the mechanical motion of its vanes against an impeller. It is intended to reach pressures below one millitorr, but it can be used only after rough vacuum pumping by a standard piston pump.

turbulence   The variations in refractive index that accompany microthermal fluctuations in the atmosphere which cause distortion in optical communication systems.

Turing machine   A hypothetical computer that is not limited by its memory capacity.

turn   One complete loop of wire.

turnover cartridge   A phonograph pickup that has two styli and a pivoted mounting which places in playing position the correct stylus for a particular record. There is usually a stylus with 3-mil radius for 78-rpm records and a stylus with 1-mil radius for 45- and 33⅓-rpm records.

turnover frequency   *Transition frequency.*

turns ratio   The ratio of the number of turns in a secondary winding of a transformer to the number of turns in the primary winding.

turnstile antenna   An antenna that consists of one or more layers of crossed horizontal dipoles on a mast, usually energized so the currents in the two dipoles of a pair are equal and in quadrature. It is used with television, FM, and other VHF or UHF transmitters to obtain an essentially omnidirectional horizontally polarized radiation pattern. The superturnstile antenna is a more elaborate version in which the dipole elements are wing-shaped.

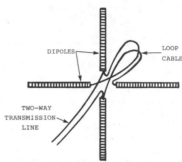

Turnstile antenna is an arrangement of crossed dipoles that radiates circularly polarized waves.

turntable   A motor-driven platter that supports shellac or vinyl recordings while they are being played. It is a consumer electronics product that is a component in stereo systems. Other models were used in studios for making recordings in this obsolete recording process.

turntable rumble   Low-frequency vibration that is mechanically transmitted to a recording or reproducing turntable and superimposed on the reproduction.

turret tuner   A television tuner that has one set of pretuned circuits for each channel, mounted on a drum which is rotated by the channel selector. The rotation of the drum connects each set of tuned circuits in turn to the receiver antenna circuit, RF amplifier, and RF oscillator.

TUV   Abbreviation for *Technische Uberwachungs-Verein Rheinland,* a German organization approved for testing products to the German VDE standards and awarding the *Geprufte Sicherheit* (G. S.) label. Organizations capable of giving TUV approvals are located in the United States.

TV   Abbreviation for *television.*

TVI   Abbreviation for *television interference.*

TWA   Abbreviation for *traveling-wave amplifier.*

TWAT   Abbreviation for *traveling-wave amplifier tube.*

tweaking   Making a small individual adjustment on a component after installing it in a system, such as making a zero correction for an operational amplifier.

tweeter   A loudspeaker that reproduces only the higher audio frequencies, usually those well above 3 kHz. It is usually installed with a crossover network and a woofer.

twin-lead   *Twin-lead cable.*

twin-lead cable   A transmission line that has two parallel conductors separated by insulating material. Line impedance is determined by the diameter and spacing of the conductors and the insulating material and is usually 300 Ω for television receiving antennas. It is also called *balanced transmission line* and *twin-lead*.

twinning   A structural defect that occurs in quartz crystals. The two forms of twinning are electric and optical twinning.

twin-T network   A network that consists of two T networks in parallel, with the capacitor and resistor positions interchanged in one of them. It is also called a parallel-T network.

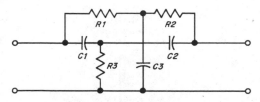

Twin-T network.

twin-triode   Two triode vacuum tubes in a single envelope.

twist   1. The amplitude ratio of a pair of *dual-tone multi-frequency* tones. Because of transmission and equipment variations, a pair of tones that are originally equal in amplitude might arrive with a considerable difference in amplitude. 2. A waveguide section whose cross section is progressively rotated about the longitudinal axis of the waveguide.

twisted pair   A cable composed of two, thin, separately insulated conductors twisted together.

twisted-nematic liquid-crystal display [TNLCD]   A display containing a liquid crystal whose twisted-nematic molecules align on a helical axis in the absence of an electric field, twisting polarized light up to 90°. When the liquid crystal is channeled on its new axis through an exit polarizer, the viewer sees a bright background region. With an electric field across opposing electrodes, the crystals align themselves parallel to the field, and entering polarized light is blocked by the exit polarizer. A dark region appears—typically dots (pixels) or alphanumeric segments. The electrodes can be organized so that the OFF state corresponds to the dark background region and the ON state to bright picture elements. See *liquid-crystal display* [LCD].

twisted-ring counter   A feedback shift register in which the next input digit is the complement of the outgoing digit, giving $2N$ states for N stages. Thus only five stages are needed for a scale-of-10 counter.

two-address programming   Digital computer programming in which each complete instruction includes an operation and specifies the location of two registers, usually one containing an operand and the other containing the result of the operation.

two-dimensional electron gas field-effect transistor [TEGFET]   A field-effect transistor with a two-dimensional electron-gas layer. This acronym was originated by Thomson-CSF.

two-frequency duplex   Carrier communication that provides simultaneous communication in both directions between two stations with different carrier frequencies in the two directions.

two-hole directional coupler   A directional coupler that consists of two parallel coaxial lines in contact, with holes or slots through their contacting walls at two points a quarter-wave apart. These permit the extraction of some of the RF energy traveling in one direction through the main line while rejecting energy traveling in the opposite direction. One end of the secondary line must be terminated in its characteristic impedance.

two-level maser   A solid-state maser that excites all of its molecules simultaneously, permitting only pulsed operation. Each useful interval of oscillation or amplification must be followed by an interval of excitation. A paramagnetic solid material such as cerium ethyl sulfate is used. It is also called a pulsed maser.

two-out-of-five code   An error-detecting code in which each decimal digit from 0 through 9 is represented by a different combination of five binary digits. Two of these are always one kind and three are the other kind, such as two 0s and three 1s.

two-phase   A reference to a phase difference of one quarter-cycle or 90°. It is also called quarter-phase.

two-position action   Automatic control action in which a final control element is moved from one of two fixed positions to the other, with no intermediate positions.

two-receiver radiometer   A radiometer-type receiver that has two independent receiver channels. The IF outputs are multiplied and the product is smoothed in a low-pass filter. A high signal-to-noise ratio is obtained at the output, as required for the detection of extremely weak signals.

two's complement   A binary number formed by interchanging all 1s and 0s of the original binary number and adding 1 to the result. Adding the original number and its two's complement gives a sum that is a power of 2. Thus for 10110 the two's complement is $01001 + 1 = 01010$ and the sum of 10110, 01001, and 1 is $22 + 9 + 1$ or 32 (binary 100000), which is the fifth power of 2.

two-source frequency keying   Keying in which the modulating wave switches the output frequency between predetermined values that correspond to the frequencies of independent sources.

two-tone keying   Keying in which the modulating wave modulates the carrier with one frequency for the marking condition and with a different frequency for the spacing condition.

two-value capacitor motor   A capacitor motor that depends on different values of effective capacitance for starting and running.

two-way repeater   A repeater that amplifies signals coming from either direction.

two-wire circuit   A reference to the two telephone wires connecting the central office to the subscriber's telephone. Commonly referred to as tip and ring, the two wires carry both transmit and receive signals in a different manner.

two-wire repeater   A telephone repeater that provides for transmission in both directions over a two-wire telephone circuit.

TWS   Abbreviation for *thermal weapon sight*.

TWT   Abbreviation for *traveling-wave tube*.

U Abbreviation for *atomic mass unit.*

UAF Abbreviation for *universal active filter.*

UART Abbreviation for *universal asynchronous receiver-transmitter.*

UDLC Abbreviation for *universal data-link control.*

UDLT Abbreviation for *universal digital loop transceiver.*

UHF Abbreviation for *ultrahigh frequency.*

UHV Abbreviation for *ultrahigh voltage.*

UJT Abbreviation for *unijunction transistor.*

UL Abbreviation for Underwriters Laboratories, Inc.

UPS Abbreviation for *ultraviolet photoelectron spectroscopy.*

USART Abbreviation for *Universal synchronous-asynchronous receiver-transmitter.*

ultrafine wire Wire ranging in size from No. 47 through No. 60 American wire gage.

ultrahigh frequency [UHF] **band** A band of frequencies extending from 0.3 to 1.0 GHz (300 to 1000 MHz), corresponding to wavelengths of 100 and 30 cm, respectively, in accordance with IEEE Standard 521-1976. It corresponds to part of the B band (0.25 to 0.5 GHz) and all of the C band (0.5 to 1.0 GHz) in the U.S. Military Joint Chiefs of Staff (JCS) triservice frequency designations (1970).

ultra-high-frequency converter An electronic circuit that converts UHF signals to a lower frequency to permit their reception on a VHF receiver. It converts UHF television signals to VHF signals for reception on VHF television receivers.

ultrahigh voltage [UHV] A voltage above about 1 MV.

ultrashort waves Radio waves shorter than about 10 m in wavelength (about 30 MHz in frequency). Waves shorter than 30 cm (above 1 GHz) are called microwaves.

ultrasonic A reference to signals, equipment, or phenomena involving frequencies just above the range of human hearing, or above about 20 kHz.

ultrasonic bonding The bonding of two identical or dissimilar metals by mechanical pressure combined with a wiping motion produced by ultrasonic vibration. It can attach gold or aluminum wire leads to semiconductor dies or chips. See also *wire bonding* and *wire bonder.*

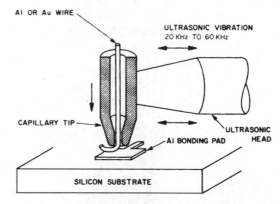

Ultrasonic bonding of a wire lead to an aluminum bonding pad on a silicon semiconductor. The ultrasonic head provides wiping motion that removes oxide films, to give a strong intermolecular bond.

ultrasonic cleaning The cleaning of objects by immersion in a liquid that is acted on by ultrasonic waves.

ultrasonic coagulation The bonding of small particles into large aggregates by the action of ultrasonic waves.

ultrasonic communication Communication through water by keying the sound output of echo-ranging sonar on ships or submarines or with stand-alone ultrasonic projectors.

ultrasonic delay line A delay line that depends on the propagation time of sound through a medium such as fused

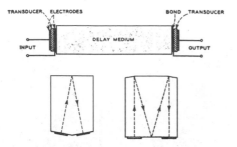

Ultrasonic delay line and methods for placing the transducers to increase the path length and delay time by reflecting the signal beam.

quartz, barium titanate, or mercury to obtain a time delay of a signal.

ultrasonic detector A mechanical, electric, thermal, or optical device that detects and measures ultrasonic waves. It is typically a piezoelectric transducer.

ultrasonic diagnosis The use of ultrasonic echo techniques to obtain a visual image of the interior of the human body for medical diagnostic purposes.

ultrasonic drill A drill that has a magnetostrictive transducer attached to a tapered cone which serves as a velocity transformer. With an appropriate tool at the end of the transformer, holes of many different shapes can be drilled in hard, brittle materials such as tungsten carbide and gems.

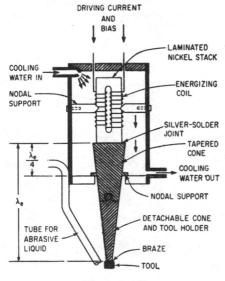

Ultrasonic drill.

ultrasonic equipment Equipment that generates AC energy at frequencies above about 20 kHz to excite or drive an electromechanical transducer for the production and transmission of ultrasonic energy for industrial, scientific, or medical purposes.

ultrasonic flaw detector An ultrasonic generator and detector for determining the distance to a wave-reflecting internal crack or other flaw in a solid object.

ultrasonic frequency A frequency above the audio-frequency range. The term is commonly applied to elastic waves propagated in gases, liquids, or solids.

ultrasonic generator A generator that consists of an oscillator which drives an electroacoustic transducer to produce acoustic waves above about 20 kHz.

ultrasonic holography The use of an interference pattern between two ultrasonic waves to reconstruct an image of the interior of an opaque object for photography. Applications include nondestructive testing.

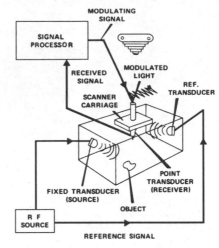

Ultrasonic-holography setup for an object immersed in liquid.

ultrasonic imaging *Acoustic imaging.*

ultrasonic inspection Nondestructive inspection of the interior of an object with ultrasonic techniques to obtain echoes from inner flaws.

ultrasonic level detector A level detector consisting of an ultrasonic transmitter and receiver that are set into one wall of a container or tank. When the level is below that of the ultrasonic beam, reflection occurs at the opposite wall. When the level of the liquid or other material reaches the beam, reflection occurs in the material, and the elapsed time of pulse travel is shorter.

ultrasonic light diffraction The formation of optical diffraction spectra when a beam of light is passed through a longitudinal ultrasonic wave field. The diffraction results from the periodic variation of the light refraction in the ultrasonic field.

ultrasonic light modulator A light modulator that exploits the action of ultrasonic waves on a light beam passing through a fluid.

ultrasonic material dispersion The production of suspensions or emulsions of one material in another because of the action of high-intensity ultrasonic waves.

ultrasonic modulation cell A cell with a laser beam that is frequency-modulated by the action of a longitudinal ultrasonic wave derived from a frequency-modulated video signal. The ultrasonic wave scatters the incident light into various orders that are frequency-shifted. It can transmit information to a photodetector which is feeding a television monitor.

ultrasonics The branch of acoustics concerned with vibrations and acoustic waves above about 20 kHz. It is also called ultrasound.

ultrasonic scanner A scanner that employs ultrasonic pulse-echo techniques for diagnostic imaging of tissue/organ interfaces.

ultrasonic sealing The sealing of thermoplastic packaging films by applying vibratory mechanical pressure at ultrasonic frequencies to develop localized heat that melts and fuses the mating plastic surfaces.

ultrasonic sounding See *depth sounder*.

ultrasonic space grating A periodic spatial variation of the index of refraction of a medium due to the presence of acoustic waves.

ultrasonic stroboscope A light interrupter whose action is based on the modulation of a light beam by an ultrasonic field.

ultrasonic therapy The use of ultrasonic vibrations for therapeutic purposes.

ultrasonic thickness gage A thickness gage that measures the time of travel of an ultrasonic beam through a sheet of material to determine the thickness of the material.

ultrasonic transducer A transducer that converts AC energy above 20 kHz to mechanical vibrations of the same frequency. It is generally either magnetostrictive or piezoelectric.

ultrasonic waves Elastic waves that have a frequency above about 20 kHz.

ultrasonic welding The use of ultrasonic energy to produce fusion of two mating pieces of metal without heat.

ultrasonic wire bonder *Ultrasonic bonding.*

ultrasound *Ultrasonics.*

ultraviolet lamp A lamp that provides ultraviolet radiation, such as various forms of mercury-vapor lamps.

ultraviolet radiation The band of the electromagnetic spectrum between visible violet light (about 380 nm, 0.38 μm, or 3800 Å) and the X-ray region at about 10 nm, 0.01 μm, or 100 Å. It is also called *black light*.

umbilical connector The quick-disconnect connector that terminates a missile umbilical cable and mates with the umbilical connector receptacle.

umbilical cord A cable that conveys power and test signals to a missile up to the instant of launching and might feed in last-minute target data. It is designed for quick release. Mating connectors are opened by a lanyard release mechanism.

umbra The region of total shadow behind an object in a beam of radiation. A straight line drawn from any point in this region to any point in the source passes through the object.

umbrella antenna An antenna with radiating wires that run downward at an angle in some or all directions from a central tower or from a wire running between two towers, somewhat like the ribs of an open umbrella. It is a low-frequency antenna.

unbalanced line A transmission line that conducts voltages on its two conductors which are not equal with respect to ground. A coaxial line is an example.

unbalanced output An output that has one of its two input terminals is substantially at ground potential.

unblanking pulse A pulse applied to the grid or cathode of a cathode-ray tube to turn on the beam for a particular length of time, normally the time of one sweep.

uncage To disconnect the erection circuit of a displacement gyroscope system.

uncertainty The estimated amount by which an observed or calculated value might depart from the true value.

uncertainty principle A principle of quantum mechanics, stating that it is impossible to obtain simultaneously and with high precision both the position and the momentum of a particle.

uncharged The property of a material having a normal number of electrons, and hence no electric charge.

undamped wave A continuous wave produced by oscillations that have constant amplitude.

underbunching A condition that represents less than optimum bunching in a velocity-modulation tube such as a klystron.

undercoupling A condition of coupling between tuned circuits when the degree of coupling is less than critical. The secondary current has a lower value, and the circuit has a narrower bandwidth. See also *critical coupling* and *overcoupling*.

undercurrent relay A relay that operates when its coil current falls below a predetermined value.

underdamping A circuit condition in which one or more output oscillations occur after a sudden change in input, before the output settles to its new value.

undermodulation Insufficient modulation at a transmitter, caused by electrical limitations or improper adjustment of the modulator.

undershoot The initial transient response to a unidirectional change in input. It precedes the main transition and is opposite in sense.

undervoltage alarm An alarm that gives an audible and/or visual indication when the voltage in equipment falls below a specified value.

undervoltage relay A relay that operates when its coil voltage falls below a predetermined value.

underwater ambient noise *Underwater background noise.*

underwater antenna An antenna located and used under water, as in surface-to-submarine communications.

underwater background noise Underwater sound other than the desired signal, including underwater ambient noise such as from sea animals including whales, porpoises or crabs. and that noise inherent in the hydrophone and its associated equipment.

underwater laser scanner A laser scanning system consisting of an argon laser, motor-driven rotating mirrors, and

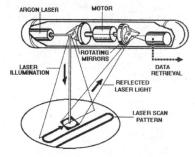

Underwater laser beam scans the ocean floor in a raster pattern, and reflected coherent light is returned to a receiver and signal processor for conversion to a video image.

light-to-digital converter circuitry that is mounted in a sealed, torpedo-shaped case that is towed behind a ship. The laser beam is scanned over the ocean floor in regions under surveillance, and reflected laser light is returned to the case for conversion into a raster scan signal that can be displayed on a cathode-ray tube monitor or converted into still pictures. It has applications in ocean engineering, salvage operations, archaeology and military surveillance. It can provide clear pictures of objects on the ocean floor than *side-looking sonar.*

underwater mine A mine located under water and exploded by propeller vibration, magnetic attraction, contact, or remote control.

underwater-mine coil A coil and associated equipment that detects changes in the magnetic field at an underwater mine caused by a passing ship.

underwater-mine depth compensator A hydrostatically actuated device that increases the sensitivity of the firing mechanism when an underwater mine exceeds a predetermined depth.

underwater signal An underwater disturbance that is to be detected, such as sonar transmissions from a projector, machinery noise, and ship propeller noise.

underwater sound communication set The components and apparatus required to provide underwater communication by sonic or ultrasonic waves in water.

underwater sound projector A transducer that produces sound waves in water. It is also called a projector.

Underwriters Laboratories, Inc. [UL] An independent organization that conducts safety testing of commercial and consumer products to established standards.

ungrounded An object or surface without a connection to ground.

uniconductor waveguide A waveguide that consists of a cylindrical or rectangular metallic surface which surrounds a homogeneous dielectric medium.

unidirectional 1. Flowing in only one direction, such as direct current. 2. Radiating in only one direction.

unidirectional antenna An antenna that has a single well-defined direction of maximum gain.

unidirectional coupler A directional coupler that samples only one direction of transmission.

unidirectional current A current that flows in the same direction at all times.

unidirectional hydrophone A unidirectional microphone that responds to waterborne sound waves.

unidirectional log-periodic antenna A broadband antenna that has the cut-out portions of a log-periodic antenna mounted at an angle to each other. It gives a unidirectional radiation pattern in which the major radiation is in the backward direction, off the apex of the antenna. Its impedance is essentially constant for all frequencies, as is its radiation pattern.

unidirectional microphone A microphone that is predominantly responsive to sound incident from one hemisphere, without picking up sounds from the sides or rear.

unidirectional pulse A pulse whose principal departures from the normally constant value occur in one direction only. It is also called a single-polarity pulse.

unidirectional transducer A transducer that senses stimuli in only one direction from a reference zero or rest position.

unidirectional voltage A voltage whose polarity, but not necessarily magnitude, is constant.

unifilar The property of having or using only one fiber, wire, or thread.

uniform line A line that has the same electrical properties along its entire length.

uniform plane wave A plane wave whose electric and magnetic field vectors have constant amplitude over the equiphase surfaces. This wave can be found only in free space at an infinite distance from the source.

uniform transmission line A transmission line whose physical and electrical characteristics do not change with distance along the axis of the guide.

uniform waveguide A waveguide whose physical and electrical characteristics do not change with distance along the axis of the guide.

unijunction transistor [UJT] A PN junction device that has the emitter connected to the PN junction on one side of a silicon die and connections for its two bases at opposite ends of the die. It can also be constructed by the planar process, with all three electrodes on one face of the silicon chip. The transistor has a stable negative resistance characteristic over a wide temperature range. It is primarily a switching device.

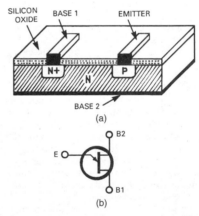

Unijunction transistor has two base electrodes and an emitter (*a*); its schematic symbol is shown in (*b*).

unilateral-area track A sound track that has only one edge of its opaque area modulated in accordance with the recorded signal. A second edge can be modulated by a noise-reduction device.

unilateral bearing A bearing obtained with a radio direction finder that has unilateral response, eliminating the chance of a 180° error.

unilateral conductivity Conductivity in only one direction, as in a perfect rectifier.

unilateral device A device that transmits energy in one direction only.

unilateralization The use of an external feedback circuit in a high-frequency transistor amplifier to prevent undesired oscillation by canceling both the resistive and reactive changes produced in the input circuit with internal voltage feedback. In neutralization, only the reactive changes are canceled.

unilateral transducer A transducer whose output waves cannot affect its input waves.

unimpeded harmonic operation The operation of a magnetic amplifier so that the impedance of the control circuit is substantially zero. This permits essentially unrestricted flow of all harmonic currents in the control circuit.

uninterruptible power supply (system) [UPS] A system that provides protection against primary AC power failure and variations in power-line frequency and voltage. The most common system for protecting a computer installation includes a storage battery, battery charger, solid-state inverter, and solid-state switching circuit. The system can be used on-line between the power line and load to provide voltage regulation and suppress transients. It can also be used off line, and switched on only when utility power fails.

unipolar The property of having only one pole, polarity, or direction.

unipolar machine *Homopolar generator.*

unipole *Isotropic antenna.*

unipotential cathode *Indirectly heated cathode.*

unit 1. An assembly or device capable of independent operation, such as a radio receiver, cathode-ray oscilloscope, or computer peripheral that performs some operation or function such as printing. 2. A quantity adopted as a standard of measurement.

unit charge The electric charge that will exert a repelling force of 1 dyn on an equal and like charge 1 cm away in a vacuum, assuming that each charge is concentrated at a point.

unitized construction The construction of equipment in subassemblies that can be manufactured and tested separately, and are readily replaceable as individual units within the equipment of which they are a part. See also *module.*

unit magnetic pole A magnetic pole that will repel an equal magnetic pole of the same sign with a force of 1 dyn if the two poles are placed 1 cm apart in a vacuum.

unit operator A symbolic operator that leaves every other operator unchanged.

unity coupling Perfect magnetic coupling between two coils so all the magnetic flux produced by the primary winding passes through the entire secondary winding.

unity-gain frequency The frequency at which the gain of an operational amplifier equals one.

unity power factor A power factor of 1.0, obtained when current and voltage are in phase, as in a circuit that contains only resistance.

universal asynchronous receiver-transmitter [UART] A communications circuit that converts parallel information to an asynchronous serial bit stream, and serial information to a parallel byte format. It is useful for connecting processors with parallel data buses to serial input/output (I/O) lines.

universal bridge A four-arm AC bridge for measuring capacitance. There are two versions: the *series resistance bridge* that measures equivalent series capacitance and the *parallel resistance bridge* that measures equivalent parallel resistance.

universal digital loop transceiver [UDLT] A high-speed voice-data transceiver circuit developed by Motorola capable of providing full-duplex data communication over No. 26 AWG and larger twisted-pair cable up to 2 km. Its primary use is in digital subscriber voice and data telephone systems.

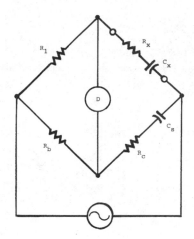

Universal bridge is a series resistance bridge for measuring capacitance.

universal joint A coupling that connects two shafts whose axes intersect at an angle.

universal motor A motor that can be operated at approximately the same speed and output on either direct current or single-phase alternating current.

universal output transformer An output transformer that has multiple taps on its winding. The taps permit its use between the audio output stage and the loudspeaker of a radio receiver by proper choice of connections.

universal product code [UPC] A 10-digit bar code that can be printed in ink on the outside of a package for laser or other electronic scanning at supermarket checkout counters. Each digit is represented by the ratio of the widths of adjacent stripes and white areas. The code is readable regardless of its orientation when it passes through the scanner system. The term also covers the corresponding combinations of binary digits into which the scanned bars are converted for computer processing that provides continuously updated inventory data and printout of the register tape at the checkout counter. The first five digits of the code identify the manufacturer; the second five digits identify the specific product. Current price and tax data for each product and descriptive terms for the register tape are stored in the computer.

universal product code scanner A scanner capable of reading the standard bar codes printed on packaged supermarket products. For annual theft-loss inventories of products on shelves, a scanner in the form of a manually moved electronic pencil can be used. It feeds a magnetic tape recorder which later serves as computer input.

universal receiver *AC/DC receiver.*

Universal Serial Bus [USB] A standard bus, connector, and associated software for connecting relatively slow computer input/output devices such as joysticks, keyboards, mice, modems, printers, scanners, speakers, and telephones to computers.

universal shunt *Ayrton shunt.*

universal synchronous-asynchronous receiver-transmitter [USART] A logic circuit that can interconnect a parallel input/output (I/O) bus to either an asynchronous or a synchronous serial I/O line.

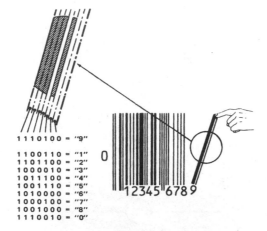

```
1110100 = "9"
1100110 = "1"
1101100 = "2"
1000010 = "3"
1011100 = "4"
1001110 = "5"
1010100 = "6"
1000100 = "7"
1001000 = "8"
1110010 = "0"
```

Universal product code. Each digit is allocated seven slots or bits, with black representing a 1 and white a 0, to give binary representation as shown for even-parity characters. Each digit has two runs of 1s separated by 0s.

universal time [UT]   Mean solar time at Greenwich, England, as reckoned from midnight. It was formerly called Greenwich mean time (GMT).

universal time coordinated [UTC]   A time scale coordinated by the International Bureau of Time to international atomic time (IAT) as derived from atomic clocks. It is also coordinated to universal time as corrected for polar motion.

univibrator   *Monostable multivibrator.*

unloaded antenna   An antenna that has no added inductance or capacitance.

unmodified scatter   Radiation that is scattered without a change in photon energy.

unmodulated   A signal that is not modulated as during moments of silence in a radio program or a disk recording.

unmodulated groove   A groove made in a mechanical recording medium when no signal is applied to the cutter.

unpack   To separate packed computer information into a sequence of separate words or elements.

unpolarized light   A beam of light in which the photon planes of vibration are oriented at random about the axis of the beam.

unquenched spark gap   A spark gap that has no special means for deionization.

unshielded twisted pair [UTP]   Paired copper wires without shielding.

unstabilized antenna   An antenna mounted directly on the structure of a vessel or aircraft, with no means for offsetting roll, pitch, and yaw.

unstable servo   A servo whose output drifts away from the input without limit.

untuned   A circuit or device that is not resonant at any of the frequencies of interest.

untuned antenna   *Aperiodic antenna.*

UPC   Abbreviation for *universal product code.*

upconverter   A converter that changes an incoming modulated or unmodulated carrier frequency to a higher frequency which is within the range of a receiver or radio test set.

up counter   A pulse counter that starts at zero and increases one count at a time to its design limit. Thus a modulus 16 binary up counter increases in 16 steps from 0000 to 1111.

update   1. To put into a computer master file the changes required by current information or transactions.   2. To modify an instruction so that the address numbers it contains are increased by a stated amount each time the instruction is performed.

up/down counter   A counter that has a control input which switches the direction of the counting mode without affecting the contents of the counter.

uplink   The radio or optical transmission path upward from the earth to a communication satellite, or from the earth to aircraft. The return path is the downlink.

upper sideband   The higher of two frequencies or groups of frequencies produced by a modulation process.

UPS   Abbreviation for *uninterruptible power supply.*

urea formaldehyde   A hard, strong plastic resin with excellent electrical properties that can be precision molded. It absorbs very little water and is used to make insulating electrical and electronic components and cases.

usable sensitivity   The signal strength required to produce a program level that is 30 dB greater in amplitude than the combined amplitudes of noise and distortion, in Institute of High Fidelity standards for FM tuners.

USAF   Abbreviation for United States Air Force.

USB   Abbreviation for *Universal Serial Bus.*

useful beam   That part of the primary radiation which passes through the aperture, cone, or other collimator in radiology.

UT   Abbreviation for *universal time.*

UTC   Abbreviation for *universal time coordinated.*

utility routine   A standard routine that assists in the operation of a computer, such as a conversion, printout, or tracing routine.

UTP   Abbreviation for *unshielded twisted pair,* a pair of unshielded twisted copper wires.

uvicon   A television camera tube that has a conventional vidicon scanning section preceded by an ultraviolet-sensitive photocathode, an electron-accelerating section, and a special target.

V

V 1. Abbreviation for *volt*. 2. Symbol used on diagrams to designate a voltmeter or an electron tube.

V.34 The international standard for communications is 28.8 kb/s. It will fall back or forward as data line quality changes.

V.35 The CCITT standard governing data transmission at 56 kb/s with 60- to 108-kHz group band circuits.

VA Abbreviation for *voltampere*.

VAC Abbreviation for *volts AC*.

vacancy A defect in the form of an unoccupied lattice position in a crystal.

vacuum An enclosed space from which air has been removed.

vacuum diffusion Diffusion of impurities into a semiconductor material in a continuously pumped hard vacuum.

vacuum evaporation Deposition of thin films of metal or other materials on a substrate, usually through openings in a mask, by evaporation from a boiling source in a hard vacuum.

vacuum fluorescent display [VFD] An alphanumeric or dot-matrix display panel whose individual dot or segmented elements are illuminated when a phosphor is excited by accelerated electrons in a hard vacuum envelope. The display is a triode vacuum tube, and it provides a blue-green color *cathodoluminescence*. It is used as a digital display in automobiles, point-of-sale terminals, test instruments, and desktop calculators because of the pleasing color of its characters. It consumes more power than a comparably sized liquid-crystal display (LCD).

vacuum forepump A vacuum pump capable of lowering the air pressure down to about 0.001 mmHg, which is low enough for operation of a diffusion pump.

vacuum gage A device that indicates the absolute gas pressure in a vacuum system, in millimeters of mercury or micrometers of mercury. One micrometer is the pressure that will support a column of mercury 0.001 mm high.

vacuum-impregnated An object impregnated with an insulating compound while in a vacuum, to ensure penetration of the compound into the layers of a capacitor or between the turns of a coil.

vacuum-leak detector An instrument that detects and locates leaks in a high-vacuum system. A mass spectrometer performs this function.

vacuum metallizing The deposition of a metal coating on a plastic or other object by evaporating the metal in a vacuum chamber that contains the object.

vacuum pencil A pencillike length of tubing connected to a small vacuum pump for picking up semiconductor dice or chips during the fabrication of solid-state devices.

vacuum phototube A phototube that has been evacuated sufficiently so that its electrical characteristics are essentially unaffected by gaseous ionization. In a gas phototube, some gas is intentionally introduced.

vacuum relay A relay that has its contacts mounted in an evacuated glass housing, to permit handling RF voltages as high as 20 kV without flashover between fixed and movable contacts.

vacuum seal An airtight junction.

vacuum spectrograph A spectrograph whose optical path is in a vacuum and a reflection grating usually replaces a dispersive prism. It can measure in the extreme infrared and ultraviolet ranges, where lenses and air would absorb the radiation.

vacuum switch A switch that has its contacts in an evacuated envelope to minimize sparking.

vacuum tank The airtight metal chamber that contains the electrodes of a mercury-arc rectifier or similar tube and in which the rectifying action takes place.

vacuum tube An electron tube evacuated to the extent that its electrical characteristics are essentially unaffected by the presence of residual gas or vapor.

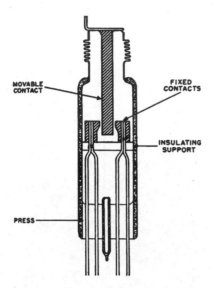

Vacuum switch, section view.

vacuum-tube amplifier An amplifier that includes one or more vacuum tubes to control the power obtained from a local source.

vacuum-tube electrometer An electrometer whose ionization current in an ionization chamber is amplified by a special vacuum triode that has an input resistance above 10 GΩ.

vacuum-tube modulator A modulator with a vacuum tube that is the modulating element for impressing an intelligence signal on a carrier.

vacuum-tube oscillator A circuit with a vacuum tube that converts DC power into AC power at a desired frequency.

vacuum-tube rectifier A rectifier that accomplishes rectification by the unidirectional passage of electrons from a heated electrode to one or more other electrodes within an evacuated space.

vacuum-tube transmitter A transmitter whose electron tubes convert the applied electric power into RF power.

vacuum-tube voltmeter [VTVM] An electronic voltmeter with a amplifier vacuum tube along with or without semiconductor devices. It is now an obsolete instrument.

valence A number that represents the proportion in which an atom is able to combine directly with other atoms. It depends on the number and arrangement of electrons in the outermost shell of each type of atom.

valence band The filled energy band that is the source of the excited electrons for the conduction band in a solid crystal such as a light-emitting diode (LED). The bands are separated by the *forbidden gap*.

valence bond The bond formed between the electrons of two or more atoms.

valence electron *Conduction electron.*

valence shell The electrons that form the outermost shell of an atom.

validity check A computer check of input data, based on known limits for variables in given fields. Thus a week might not have more than 168 working hours.

value The magnitude of a quantity.

valve British term for *electron tube*.

vanadium A metallic element with an atomic number of 23.

Van Allen belt One of the belts of ionizing radiation surrounding the earth, one centered about 2000 mi (3200 km) above the earth, one at about 10,000 mi (16,000 km), and one at about 20,000 mi (32,000 km). The radiation consists of protons and electrons that come chiefly from the sun as solar wind and are trapped by the earth's magnetic field. During massive solar storms, high-energy particles reach the belt directly from the sun and overload it, creating auroras as excess particles are dumped from the belt in the earth's polar regions. It is also called the *radiation belt*.

Van Atta array An antenna array that has pairs of corner reflectors or other elements equidistant from the center of the array. They are connected together by low-loss transmission line so that the received signal is reflected back to its source in a narrow beam, to give signal enhancement without amplification.

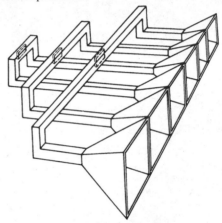

Van Atta array.

Van de Graaff generator An electrostatic generator that functions as an accelerator. An endless moving belt of insulating material collects electric charges by induction and discharges them inside a large hollow spherical electrode to produce voltages as high as 9 MV. It accelerates electrons, protons, and other nuclear particles.

vane attenuator *Flap attenuator.*

vane-type instrument A measuring instrument that depends on the force of repulsion between fixed and movable magnetized iron vanes, or the force existing between a coil and a pivoted vane-shaped piece of soft iron, to move the indicating pointer.

V antenna A horizontal bidirectional antenna system formed by positioning two long antenna wires in a V formation on three masts and feeding them 180° out of phase at the apex. If each long wire is terminated with a matching resistor (nonresonant antenna), the system is unidirectional. But if the ends of the long wire are open (resonant antenna), the system is bidirectional.

vapor deposition The process of depositing a thin film of metal on an object to be metallized by the condensation of molten metal vapor in a vacuum.

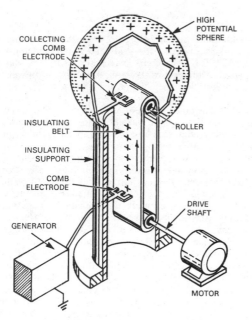

Van De Graaff generator accumulates very high static electric charges in its metal sphere from a moving belt.

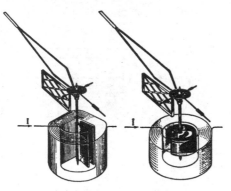

Vane-type instrument construction.

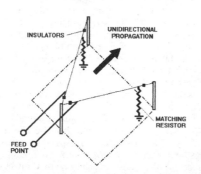

V antenna is unidirectional if it has grounded matching resistors at its two antenna ends, but it is bidirectional if the ends are left open.

vaporization-cooled A process of cooling by vaporization of a nonflammable liquid that has a low boiling point and high dielectric strength. The liquid is flowed or sprayed on hot electronic equipment in an enclosure, where it vaporizes, carrying the heat to the enclosure walls, radiators, or a heat exchanger. It is also called evaporative-cooled.

vapor-phase epitaxy [VPE] A material deposition process that employs vaporized chemicals to grow epitaxial films on crystal wafers in a furnace. It is also known as chemical-vapor deposition (CVD). Gases flow across heated wafers inside a furnace chamber and react (pyrolyze) at the crystal surface to deposit a thin film of material.

vapor phase reflow soldering *Vapor-phase soldering.*

vapor phase soldering A method for *reflow soldering* electronic components to circuit boards, primarily in *surface-mount technology* [SMT]. Components whose leads have been coated with lead-tin *eutectic* solder are positioned on the board with similarly prepared pads. The assembly is placed in a vapor-soldering chamber, and the 215°C (419°F) heat from the vapor of boiling fluorinated hydrocarbon liquid causes the solder to reflow and bond the components to the board. The melting temperature of the chamber is precisely controlled by the liquid's boiling point. Some systems use a secondary vapor "blanket" to reduce the evaporation loss of the primary vapor. It is also called *vapor-phase reflow soldering.*

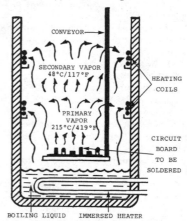

Vapor-phase soldering system. Heat from a boiling inert fluid melts lead-tin solder, causing it to flow and bond electronic components to a circuit board.

vapor pressure The pressure of the vapor of a liquid that is kept in confinement, as in a mercury-vapor rectifier tube.

var The name and symbol in the International System of Units (SI) for voltampere reactive, the unit of reactive power.

VAR Abbreviation for *visual-aural range.*

varactor A PN semiconductor diode whose capacitance varies with the applied voltage. It is a variable-reactance tuning element in oscillator and amplifier circuits, including parametric amplifiers. It is also called a varactor diode.

varactor diode *Varactor.*

varh Abbreviation for *voltampere-hour reactive.*

variable A quantity that can assume many distinct values.

variable-area track   A sound track divided laterally into opaque and transparent areas. A sharp line of demarcation between these areas corresponds to the waveform of the recorded signal.

variable attenuator   An attenuator that reduces the strength of an AC signal either continuously or in steps, without causing appreciable signal distortion, by maintaining an essentially constant impedance match.

variable-capacitance pickup   A phonograph pickup whose stylus produces a variation in capacitance that can frequency-modulate an oscillator. The output signal of the oscillator is converted to an audio voltage by a detector. It is also called an *FM pickup*.

variable-capacitance transducer   A transducer that measures a parameter or change in a parameter by a change in capacitance.

variable capacitor   A capacitor whose capacitance can be varied by moving one set of metal plates with respect to a fixed set. A *tuning capacitor* is an example.

variable-carrier modulation   *Controlled-carrier modulation.*

variable coupling   Inductive coupling that can be varied by moving one coil with respect to another.

variable-cycle operation   A computer operation in which the cycles of action can be of different lengths, as in an asynchronous computer.

variable-density sound track   A constant-width sound track in which the average light transmission varies along the longitudinal axis in proportion to some characteristic of the applied signal.

variable-depth sonar [VDS]   An antisubmarine shipboard sonar system whose projection and receiving transducer array is mounted inside a torpedo-shaped pod that is submerged and towed behind the ship. It is deployed by releasing the cable from a large winch on the deck of the antisubmarine ship, retrieved by reeling in the cable. The VDS pod is set to sink to a depth at which the effects of thermal layers in the sea will be minimized, and it is positioned far enough behind the towing ship to minimize any interference its internal machinery and propellor noise might have on sonar reception.

variable-erase recording   Recording on magnetic tape by selective erasure of a prerecorded signal.

variable-focal-length lens   A television camera lens system whose focal length can be changed continuously during use while maintaining sharp focusing and a constant aperture, to give the effect of gradually moving the camera toward or away from the subject. The Zoomar lens is an example.

variable-frequency oscillator [VFO]   An oscillator whose frequency can be varied over a given range.

variable-gain multiplier   A multiplier in which one variable is passed through an amplifier whose gain is controlled by the other variable. The output of such an amplifier is proportional to the required product.

variable inductance   A coil whose inductance value can be varied.

variable-inductance pickup   A phonograph pickup that depends for its operation on the variation of its inductance.

variable-inductance transducer   A transducer whose output voltage is a function of the change in a variable-inductance element.

variable-iris waveguide coupler   A microwave component that couples a waveguide to the external input or output cavity of a klystron. It permits simple matching adjustments over a wide tuning range without stub tuners or matching section.

variable-mu tube   An electron tube whose amplification factor varies with control-grid voltage in a predetermined manner. This characteristic is achieved by the variable spacing of the grid wires along the length of the grid, so that a very large negative grid bias is required to block anode current completely.

variable-reluctance microphone   A microphone whose operation depends on variations in the reluctance of a magnetic circuit. It is also called a magnetic microphone and a reluctance microphone.

variable-reluctance pickup   A phonograph pickup whose operation depends on variations in the reluctance of a magnetic circuit due to the movements of an iron stylus assembly, which is a part of the magnetic circuit. The reluctance variations alternately increase and decrease the flux through two series-connected coils, inducing in them the desired audio output voltage. It is also called a magnetic cartridge, magnetic pickup, or reluctance pickup.

variable-reluctance stepper motor   A stepper motor that has a soft iron rotor with teeth or poles so positioned that they cannot simultaneously align with all the stator poles. In one example, four of the rotor poles align with four stator poles, and the four other rotor poles fall between stator poles. Each actuation of a set of stator coils makes the rotor move half the angular distance between adjacent stator poles, to give stepping action. A wide range of speeds, torques, and stepping angles is possible by choosing the proper combination of stepper motor and gearbox designs.

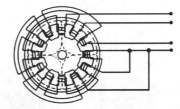

Variable-reluctance stepper motor with 12 stator poles and 8 rotor poles gives 15° stepping of rotor.

variable-reluctance transducer   A transducer that contains a slug of magnetic material which is moved between two coils by the displacement being monitored. This changes the reluctance of the coils, thereby changing their impedance.

variable-resistance pickup   A phonograph pickup whose operation depends on the variation of a resistance.

variable-resistance transducer   A transducer whose signal output depends on the change in a resistance element.

variable resistor   *Rheostat.*

variable speech control [VSC]   A method of removing small portions of speech from a tape recording at regular intervals and stretching the remaining sounds to fill the gaps. This permits recorded speech to be played back at twice or even 2½ times the original speed without changing pitch and without significant loss of intelligibility. It was developed for the blind or others who want to listen to speech at speeds approximating normal reading speed.

variable transformer   An iron-core transformer that has provisions for varying its output voltage over a limited range or continuously from zero to maximum output voltage. This is generally done with a contact arm that moves along exposed turns of the secondary winding. It can be an autotransformer.

Variac A trademark of General Radio Co. for its line of variable transformers.

varicap *Varactor.*

varindor A variable inductance in which a change in current varies the inductance value.

variocoupler An RF transformer that has provisions for varying the coupling between the two windings. Its construction is similar to that of a variometer, but its coils are not connected together.

variolosser A variable-loss circuit that improves the signal-to-noise ratio of a communication channel. At the transmitting end, the variolosser is connected so its loss increases as input signal strength increases. At the receiving terminal, the variolosser is in a circuit that restores the original dynamic range of the signal.

variometer A variable inductance that has two coils in series, one mounted inside the other, with provisions for rotating the inner coil to vary the total inductance of the unit over a wide range.

varistor A contraction of *variable resistor.* It is a nonlinear, two-electrode, voltage-dependent device that provides AC circuit transient suppression. Its resistance drops as the applied voltage increases. Its response also varies with temperature. Electrically equivalent to two back-to-back Zener diodes, it is formed from bulk zinc oxide. The device absorbs the potential destructive energy of incoming transients, thereby protecting vulnerable circuit components. See *metal-oxide varistor* [MOV] and *transient voltage suppressor* [TVS].

Varley loop test A method of using a Wheatstone bridge to determine the distance from the test point to a fault in a telephone or telegraph line or cable.

varmeter An instrument for measuring reactive power in vars. It is also called a reactive voltampere meter.

V band A band of frequencies in the millimeter region extending from 46 to 56 GHz, corresponding to wavelengths of 0.652 to 0.536 cm. Subdivisions of this obsolete U.S. military band designation (values in gigahertz) are

Va: 46–48
Vb: 48–50
Vc: 50–52
Vd: 52–54
Ve: 54–56

V-beam radar A volumetric radar system based on two fan beams that determine the distance, bearing, and height of a target. One beam is vertical and the other inclined. The beams intersect at ground level and rotate continuously about a vertical axis. The time difference between the arrivals of the echoes of the two beams is a measure of target elevation.

V.32 bis Additional requirements to the V.32 standard, it adds 14.4 and 12.0 kb/s to the V.32 9.6, 7.2, and 4.8 kb/s standard. All V.32 modems fall back and forward as data line quality deteriorates or improves.

V.34 bis The international standard for 33.6 kb/s, offering approximately 17% improvement over the V.34 standard. See *V.34.*

V.42 bis An extension of V.42 that includes 4× data compression that is compatible with MNP error control.

VCCS Abbreviation for *voltage-controlled current source.*

V-chip A dedicated microprocessor designed for installation in television sets or *set-top converters* that blocks the reception of objectionable cable TV programs. The programs must be encoded at the transmitting station with an electronic signal that designates a specified rating that can be identified by the V-chip for blocking. The chip is intended for installation in a manually controlled circuit that reads the signal code, compares it with the programming authorized by the TV set owner, and then blocks unwanted programming. V stands for violence, meaning that the device blocks TV programs with violent or salacious content.

VCO Abbreviation for *voltage-controlled oscillator.*

VCR Abbreviation for *videocassette recorder.*

VCVS Abbreviation for *voltage-controlled voltage source.*

VCXO Abbreviation for *voltage-controlled crystal oscillator.*

VDC Abbreviation for *volts DC.*

VDE Abbreviation for *Verband Deutscher Elektronotechniker.*

VDS Abbreviation for *variable-depth sonar.*

vector 1. In navigation, an arrow whose scaled length and angle (with respect to north) represents a ship or aircraft's speed and direction. 2. In mathematics, an arrow representing force and direction. 3. In electrical theory, an arrow that designates the magnitude of a variable (e.g., voltage, current, impedance) and its phase angle. 4. In computer science, an ordered sequence of numbers that simulates physical characteristics or quantities.

vector diagram An arrangement of vectors that shows the magnitude and phase relations between two or more alternating quantities which have the same frequency.

vectored interrupt A computer interrupt scheme in which each interrupting device causes the operating system to branch to a different interrupt routine. It is useful for very fast interrupt response.

vector plot format A system that stores graphic data for plotting navigational charts as a matrix of vectors, each with an *X-Y-Z* coordinate and vectors connecting it to adjacent points.

vector processing The ability of a computer to manipulate arrays of numbers simultaneously.

vector quantity *Vector.*

vectorscope A cathode-ray oscilloscope that displays both the phase and amplitude of an applied signal with respect to a reference signal.

vector-sum excited linear predictive [VSELP] **coding** A form of digital speech coding; also called *code-excited linear predictive* [CELP] *coding.*

Veitch diagram A diagram that simplifies the manipulation of logic associated with the design of counter circuits.

velocimeter A Doppler system that uses the Doppler shift of a continuous-wave carrier transmitted to and reflected from a moving target to measure radial velocity.

velocity A vector quantity having the same magnitude as its *speed,* but also including the direction of motion. Both speed and direction of motion must be included in stating a velocity. If either the speed or direction changes, the velocity of the body changes. When the direction is varied, the term angular velocity describes the motion.

velocity antiresonance The condition in which a small change in the frequency of a sinusoidal force applied to a body or system causes an increase in velocity at the driving point, or the frequency is such that the absolute value of the driving-point impedance is a maximum.

velocity control system A *closed-loop control system* that controls the velocity or speed of a motor or actuator. It is typically accomplished with a *tachometer* in a single loop.

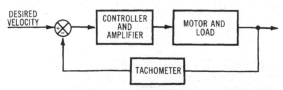

Velocity control system block diagram shows its single tachometer feedback loop.

velocity factor The ratio of the velocity of propagation in any medium to the velocity of propagation in free space. The velocity of an RF current in a conductor is slightly less than it would be in free space.

velocity-fluctuation noise The noise component developed in a traveling-wave tube or a weak-signal photodetector because of the wide thermal distribution of electron velocities in the beam.

velocity-focusing mass spectrograph *Velocity spectrograph.*

velocity hydrophone A velocity microphone that responds to waterborne sound waves.

velocity level A sound rating in decibels, equal to 20 times the logarithm to the base 10 of the ratio of the particle velocity of the sound to a specified reference particle velocity.

velocity-limiting servo A servomechanism whose maximum velocity is the chief limit on performance.

velocity microphone A microphone whose electric output depends on the velocity of the air particles that form a sound wave. Examples are hot-wire and ribbon microphones.

velocity-modulated oscillator An electron-tube structure that varies the velocity of an electron stream as the stream passes through a resonant cavity called a *buncher*. Energy is extracted from the bunched electron stream at a higher energy level in passing through a second cavity resonator called the catcher. Oscillations are sustained by coupling energy from the catcher cavity back to the buncher cavity.

velocity-modulated tube An electron-beam tube whose electron stream velocity is alternately increased and decreased within a period comparable to the local transit time.

velocity modulation Modulation in which a time variation in velocity is impressed on the electrons of a stream.

velocity of light A physical constant equal to 2.997925 × 10^{10} cm/s. This corresponds to 186,280 statute mi/s, 161,870 nautical mi/s, and 328 yd/μs. All electromagnetic radiation travels at this same speed in free space. The velocity of light is considered to be a vector quantity that represents the speed of light in a particular direction.

velocity of sound The approximate velocities of sound waves are 1100 ft/s (335 m/s) in air at 0°C, 4800 ft/s (1463 m/s) in water, and 16,400 ft/s (5000 m/s) in steel.

velocity sorting Any process of selecting electrons according to their velocities.

velocity spectrograph A mass spectrograph that selects positive ions which have enough velocity to pass through all three slits and enter a chamber where they are deflected by a magnetic field in proportion to their charge-to-mass ratio. It is also called a velocity-focusing mass spectrograph.

velocity transducer A transducer that generates an output which is proportional to its velocity.

Verband Deutscher Electrotechniker [VDE] A German agency that sets standards for product safety and noise emission, and tests and certifies products to those standards. Its *Geprufte Sicherheit* (G. S.) label is widely recognized in Europe, but products in compliance with VDE do not automatically assure compliance with the standards of other countries. See also *TUV.*

Verdet's constant A Faraday-effect constant that determines the angle of rotation of plane-polarized light when passing through certain materials in a magnetic field.

verification Automatic comparison of one data transcription with another transcription of the same data, to reveal errors.

vernier A control or scale used to obtain a fine adjustment or to increase the precision of a measurement.

vernier capacitor A small variable capacitor placed in parallel with a larger tuning capacitor to provide a finer adjustment after a larger unit has been set approximately to the desired position.

vernier dial A tuning dial that causes only a fraction of a revolution of the main shaft when its control knob is turned through a complete rotation. It permits fine and accurate measurement.

Veronica A computer utility that permits a search for key words in titles of files on the *Internet* that appear in *Gopher* menus.

vertical antenna A vertical metal tower, rod, or suspended wire that functions as an antenna.

vertical blanking Blanking of a television picture tube during the vertical retrace.

vertical blanking interval The brief time interval between television fields required for the scanning electron gun to retrace from the bottom of the image to the top, to begin scanning the next field.

vertical blanking pulse The rectangular pulse that is transmitted at the end of each field of a television signal to cut off the beam current of the picture tube while the beam is returning to the top of the screen for the start of the next field.

vertical centering control The centering control provided in a television receiver or cathode-ray oscilloscope to shift the position of the entire image vertically in either direction on the screen.

vertical compliance The ability of a stylus to move freely in a vertical direction while in the groove of a phonograph record.

vertical convergence control The control that adjusts the amplitude of the vertical dynamic convergence voltage in a color television receiver.

vertical definition *Vertical resolution.*

vertical deflection electrodes The pair of electrodes that moves the electron beam up and down on the fluorescent screen of an electrostatically deflected cathode-ray tube.

vertical deflection oscillator The oscillator that produces, under control of the vertical synchronizing signals, the sawtooth voltage waveform that is amplified to feed the vertical deflection coils on the picture tube of a television receiver. It is also called a vertical oscillator.

vertical field-effect transistor *Vertical metal-oxide semiconductor VMOS.*

vertical hold control The hold control that changes the free-running period of the vertical deflection oscillator in a television receiver. It keeps the picture steady in the vertical direction.

vertical-incidence sounder *Ionosonde.*

vertical interval reference [VIR] A reference signal inserted into a television program signal every ⅟₆₀ s, in line 19 of the vertical blanking period between television frames, to provide references for luminance amplitude, black-level amplitude, sync amplitude, chrominance amplitude, and color-burst amplitude and phase. In television transmitters and video recorders, it provides a reference for maintaining color, brightness, and contrast at exact specifications. It can also provide automatic adjustment of chroma and tint in receivers.

vertical linearity control A linearity control that permits narrowing or expanding the height of the image on the upper half of the screen of a television picture tube. It gives linearity in the vertical direction so circular objects appear as true circles. It is usually mounted at the rear of the receiver.

vertically polarized wave A linearly polarized wave in which the electric field vector is vertical.

vertical oscillator *Vertical deflection oscillator.*

vertical polarization Transmission of radio waves so that the electric lines of force are vertical, while the magnetic lines of force are horizontal. With this polarization, transmitting and receiving dipole antennas are placed in a vertical plane.

vertical quarter-wave stub An antenna with a vertical element that is electrically a quarter-wavelength long. It is generally used with a ground plane at the base of the stub.

vertical radiator A transmitting antenna that is perpendicular to the earth.

vertical resolution The number of distinct horizontal lines, alternately black and white, that can be seen in the reproduced image of a television or facsimile test pattern. Vertical resolution is primarily fixed by the number of horizontal lines used in scanning. It is also called vertical definition.

vertical retrace The return of the electron beam to the top of the screen at the end of each field in television.

vertical stylus force *Stylus force.*

vertical sweep The downward movement of the scanning beam from top to bottom of the picture being televised.

vertical synchronizing pulse One of the six pulses that are transmitted at the end of each field in a television system to keep the receiver in field-by-field synchronism with the transmitter. It is also called a *picture synchronizing pulse.*

very high-frequency [VHF] **band** A band of frequencies extending from 0.03 to 0.30 GHz (30 to 300 MHz), corresponding to wavelengths of 1000 and 100 cm, respectively, in accordance with IEEE Standard 521-1976. It corresponds with the A band (0.10 to 0.25 GHz) and part of the B band (0.25 to 0.50 GHz) of the U.S. Military Joint Chiefs of Staff (JCS) triservice frequency designations (1970).

very high-frequency omnidirectional range [VOR] A ground-based aircraft navigational system that operates in the VHF band. It is useful for high-flying aircraft out to distances of about 230 mi (370 km) and low-flying aircraft out to about 30 mi. (48 km) line-of-sight distance from a VOR station. There are more than 1000 VOR stations in the US and more than 1000 in the rest of the world. There is conventional VOR and Doppler VOR. It operates on 160 channels between 108 and 118 MHz.

Very High Speed Integrated Circuit [VHSIC] **Program** A U.S. Department of Defense program to extend integration levels and performance capabuilities for military integrated circuits to meet or exceed those available in commercial ICs.

very large integrated circuit [VLSI] An integrated circuit containing more than 1000 gates, elementary logic circuits that include a single transistor.

very long-range radar Radar whose maximum line-of-sight range is greater than 800 mi (1290 km) for a target that has an area of 1 m² perpendicular to the radar beam.

very low frequency [VLF] A Federal Communications Commission designation for a frequency in the band from 3 to 30 kHz, corresponding to a myriametric wave between 10 and 100 km.

very short-range radar A radar whose maximum line-of-sight range is less than 5 mi (8 km) for a target that has an area of 1 m² perpendicular to the radar beam.

very small aperture terminal [VSAT] A satellite TV receiver with a small-diameter antenna for commercial and consumer customers.

VESA Abbreviation for the *Video Electronics Standards Association.*

vessel traffic service [VTS] An integrated program for monitoring and surveillance of commercial shipping, to prevent vessel collisions in waterways with heavy traffic, or pollution or illegal dumping in the waterways near environmentally sensitive shorelines. Monitoring and surveillance is carried out with communications networks that include radar, direction finders, GPS receivers, local-area networks (LANs), and sonar in command and control centers.

vestigial sideband [VSB] The transmitted portion of an amplitude-modulated sideband that has been largely suppressed by a filter which has a gradual cutoff near the carrier frequency. The other sideband is transmitted without much suppression.

vestigial-sideband filter A filter that is inserted between a transmitter and its antenna to suppress part of one of the sidebands.

vestigial-sideband transmission A type of radio signal transmission for amplitude modulation in which the normal complete sideband on one side of the carrier is transmitted, but only a part of the other sideband is transmitted.

vestigial-sideband transmitter A transmitter that transmits one sideband and part of the other intentionally.

V/F Abbreviation for *voltage-to-frequency,* a reference to converters.

V.fast A proprietary 28 kb/s designation. V.fast modems that use a Rockwell International chip set are not completely compatible with V.34 modems.

VFC Abbreviation for *variable-frequency converter.*

V/F converter   Abbreviation for *voltage-to-frequency converter.*

VFET   Abbreviation for *vertical field-effect transistor.*

VFO   Abbreviation for *variable-frequency oscillator.*

VFR   Abbreviation for *visual flight rules.*

VFR conditions   Weather conditions equal to or better than the minimum prescribed for flights under visual flight rules.

VGA   Abbreviation for *video graphics array.*

VHF   Abbreviation for *very high frequency.*

VHF antenna   Any antenna designed for operation in the VHF band from 30 to 300 MHz. It is suitable for VHF television reception, FM radio reception and radio communication.

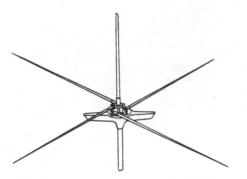

VHF antenna for shipboard radio communication is a vertical quarter-wave stub and a four-element ground plane.

VHF channel   One of the 6-MHz television channels designated by the numbers 2 through 13. Channels 2–4 cover 54–72 MHz; 5–6 cover 76–88 MHz; 7–13 cover 174–216 MHz.

VHF homing adapter   A homing adapter that operates in the VHF range.

VHF/UHF direction finder   A ground-based radio direction finder capable of stand-alone use or functioning in conjunction with airport surveillance radar.

VI   Abbreviation of *virtual instrument.*

via   1. A conductive path or feedthrough between metallization layers in an integrated circuit to interconnect two or more layers of metal separated by insulating layers. 2. In multilayer circuit boards, a plated-through hole providing a conductive path between two or more conductive layers. 3. In monolithic microwave integrated circuits (MMICs), a through-hole plated interconnect from the metallized ground plane on the lower surface to a metallized layer on top of the chip, typically the source metal. Vias in semiconductor materials are chemically drilled or etched through the substrate and then gold-plated to form the conductive path.

vibrating-reed magnetometer   An instrument that measures magnetic fields by noting their effect on the vibration of reeds excited by an alternating magnetic field.

vibrating-reed rectifier   An electromagnetic device that rectifies an alternating current by reversing the connections between the power line and load each time the alternating current reverses in direction. The reversing contacts are on a vibrating reed of magnetic material that is acted on by a coil carrying the alternating current, so the reed moves in synchronism with the current.

vibration   A periodic change in the position of an object, such as a pendulum or a diaphragm of a loudspeaker energized by a sinusoidal current. It is also called oscillation.

vibration meter   An instrument that measures the displacement, velocity, and acceleration associated with mechanical vibration. In one form it consists of a piezoelectric vibration pickup that has uniform response from 2 to 1000 Hz, feeding an amplifier which has an indicating meter at its output. It is also called a vibrometer.

vibration pickup   A pickup that responds to mechanical vibrations rather than to sound waves. In one type, twisting or bending of a Rochelle-salt crystal generates a voltage that varies in accordance with the vibration being analyzed.

vibrato   A musical embellishment that depends primarily on periodic variations of frequency, which are often accompanied by variations in amplitude and waveform.

video [Latin for "I see"]   1. A reference to picture signals or to the sections of a television system that carry these signals in either unmodulated or modulated form.   2. A reference to the demodulated radar receiver output that is applied to a radar indicator.

video amplifier   A wideband amplifier capable of amplifying video frequencies in radar and television.

video cable   Coaxial cable for signal transmission in cable television systems. Basic types include: (1) 75-Ω unbalanced indoor cable, (2) outdoor cable that has a single conductor centered in a shield, (3) 124-Ω balanced indoor cable, and (4) outdoor cable that has two parallel or twisted insulated conductors centered in a shield.

videocassette   A rectangular, flat plastic package containing two built-in tape reels (supply and takeup) and a single, ½-in-wide cobalt-alloy magnetic tape that is exchanged between them during recording, rewind, and playback. The audio and control signals are recorded linearly on the edges, but the video is recorded in diagonal tracks that contain information for one television frame. The cassette can record frequencies in excess of 4 MHz. The cassette is inserted in the videocassette recorder, and the VCR pulls

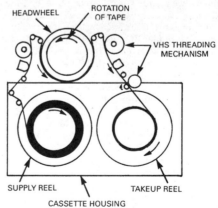

Videocassette reproduces recorded video and audio signals on tape. Host VCR's threading mechanism pulls out the tape and positions it against the rotating headwheel.

out the tape and positions it over rollers and around a rotating head drum at an angle for *helical recording*. The cassette for the most popular VHS (video home system) format measures approximately 7⅜ × 4 × 1 in and records at a rate of 580 m/s. See *videocassette recorder* [VCR].

videocassette recorder [VCR] A consumer electronics video recorder that can capture live television programs on videocassette tape for later replay. It is also capable of playing back prerecorded cassettes of personal events (from camcorders), commercial movies, or other televised entertainment or educational material through a standard TV receiver. VCRs permit the recording of on-air TV programs at times and for durations entered by the user, without supervision. Two recording formats were developed that dictated the architecture of the cassette and recorder: Betamax from Sony of Japan and VHS from JVC of Japan, but the VHS format is the most popular. Videocassette recording uses *helical scanning*. A typical VCR measures about 14 × 14 × 4 in and contains a shuttered window for inserting the tape cassette, a clock, visual indicators of the mode operation, and various manual controls. Most can be operated by remote, handheld, infrared controllers and are able to rewind the tapes. Some are able to stop action for close study. See also *videocassette*.

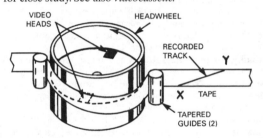

Videocassette recorder (VCR) includes a drumwheel that contains two or four read/write heads. Video is recorded diagonally in the center of the tape and audio is recorded linearly on the edges of the moving tape.

video circuit A broadband circuit that carries signals that can be converted to television pictures.

video compression Methods developed to overcome limitations in the transmission of video signals. Its effectiveness depends on both redundancies in the data and the nonlinearities of human vision.

video correlator A radar circuit that enhances automatic target-detection capability, provides data for digital target plotting, and gives improved immunity to noise, interference, and jamming.

video data digital processing Digital processing of video signals for pictures transmitted over a television link, to improve picture quality by reducing the effects of noise and distortion. The computer compares each scanned line with adjacent lines and eliminates extreme changes caused by electromagnetic interference.

video detector The detector that demodulates video IF signals in a television receiver.

video display terminal [VDS] A computer input/output (I/O) terminal with a CRT display and keyboard that interacts with a remote host computer. It replaced the teleprinter

as an I/O terminal. The first VDTs did not have microprocessors, but later versions included them for data and text formatting. Typical applications are as airline, bank, hotel, car rental, and travel agency computer terminals. The CRT monitor enclosure includes a power supply and electronics for managing the display, encoding and formatting data, and performing routine communications functions. Some VDS units permit the storage of data pages and editing.

video frequency One of the frequencies that exist in the output of a television camera when an image is scanned. It can be any value from almost zero to well over 4 MHz.

video-frequency amplifier An amplifier capable of amplifying the entire range of frequencies that comprise a periodic visual presentation in television, facsimile, or radar.

video gain control A control that adjusts the gain of a video amplifier, as for varying the intensity of the echoes on a radar PPI screen to get maximum contrast between desired echoes and undesired clutter.

video game 1. An electronic game that can be connected to or built into a television receiver, to use the television screen as a playing field or display showing player and ball movements, scores, or other actions called for by the game, race, or other type of activity which is controlled remotely by one or more players. Microprocessor-based models can also provide sound effects and interchangeable program cards for changing the rules of a game or changing the entire game. 2. Handheld units with liquid-crystal displays for displaying playing fields and moving stick characters or other events that can be influenced by the individual user. They are typically battery-powered and contain microcontrollers.

video graphics array [VGA] A circuit board for a personal computer color monitor that meets IBM's video standard. It supplies an analog red, green, blue (RGB) color signal that allows 64 color levels. At any one time 256 of 262,144 possible color combinations can be displayed. Characters are formed within a 9 × 16-character cell.

video integration A method of employing the redundancy of repetitive signals to improve the output signal-to-noise ratio, by summing the successive video signals.

video mixer A mixer that combines the output signals of two or more television cameras.

videophone *Video telephone*.

video player A general term that refers either to an analog video player/recorder called a *videocassette recorder* [VCR] or a *digital disk player*.

video recorder A general term that refers either to an analog video player/recorder called a *videocassette recorder* [VCR] or a digital disk recorder that is usually commercial audio studio equipment.

video recording The recording of information that has a bandwidth in excess of about 500 kHz, such as television or radar signals.

video signal A signal that contains periodic visual information together with blanking and synchronizing pulses, as in a radar or television system.

videotape Magnetic tape for recording the video signals of television programs. See also *videocassette*.

videotape recording [VTR] A method of recording television video signals on magnetic tape for later rebroadcasting of television programs. See also *videocassette*.

videotape replay A videotape recorder that uses a relatively short endless loop of magnetic tape to permit the repetition of a televised sports scene within seconds after the original action.

video telephone A combined telephone and video receiver that allows each party to see the other party while talking. It is also called videophone.

video terminal *Video display terminal* [VDS].

video waveform The part of the television signal waveform that corresponds to visual information. Synchronizing pulses are not included.

vidicon A camera tube whose charge-density pattern is formed by photoconduction and stored on a photoconductor surface that is scanned by an electron beam, usually of low-velocity electrons. Its chief applications are in industrial and other closed-circuit television cameras.

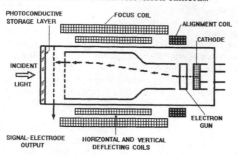

Vidicon camera tube for a closed-circuit television system.

viewfinder An auxiliary optical or optoelectronic device attached to a television camera so the operator can see the scene as the camera sees it.

viewing screen *Screen.*

Villari effect The change in magnetic induction that occurs when a magnetostrictive material is mechanically stressed.

vinyl resin A soft plastic material for making long-playing phonograph records.

VIR Abbreviation for *vertical interval reference.*

virtual address A binary address issued by a central processing unit (CPU) that refers to the location of information in a primary computer memory, such as main memory. When data is copied from disk to main memory, the physical address is changed to a virtual address.

virtual cathode The locus of a space-charge-potential minimum that permits only some of the electrons approaching to be transmitted. The remainder are reflected back to the electron-emitting cathode.

virtual height The apparent height of an ionized layer, as determined from the time interval between the transmitted signal and the ionospheric echo at vertical incidence.

virtual instrument [VI] An instrument formed by combining the hardware of a personal computer (microprocessor, memory, and display) and software (operating system) with instrumentation hardware, firmware, and applications software. The combination can emulate instruments such as the multimeter, oscilloscope, counter, and spectrum analyzer. Instrumentation hardware is typically on a plug-in circuit board and can include an analog-to-digital or digital-to analog converter, digital input/output timing, and signal-conditioning circuits. External leads and probes

might also be required. The applications software provides the data analysis, process communications, and graphical user interface (GUI).

virtual memory A combination of a computer's main memory and external computer memory that can be considered a single memory because the computer translates a program or virtual address to the actual hardware address. It permits the storage of programs and data outside a computer's main memory. In a multiuser machine, virtual memory also protects data and code when several programs are running simultaneously.

virtual reality A form of real-time three-dimensional computer simulation that permits the viewer to interact with all senses (sight, sound, touch, smell, and taste). Equipment includes head-mounted displays, sensing gloves, three-dimensional sound boards, and force-touch feedback tools. These are integrated with state-of-the-art graphics displayed on computers. Viewers gain a sense of being present in a generated scene, and they can look behind objects and obtain different perspectives as they "move" around in the scene. It has applications in electromechanical product design, military simulations, medical diagnosis and training, multimedia teleconferencing, architectural planning, and entertainment.

VIS Abbreviation for *viewable image size.*

viscous-damped arm A phonograph pickup arm is mechanically damped by a highly viscous liquid so that the arm floats gently down to a record when dropped.

visibility factor The ratio of the minimum signal input power detectable by an ideal instrument connected to the output of a receiver, to the minimum signal power detectable by a human operator through a display connected to the same receiver.

visible radiation See *standard observer curve.*

visual angle The pitch between picture elements (pixels) divided by the viewing distance and converted into minutes of arc.

visual-aural range [VAR] A VHF radio range that provides one course for display to the pilot on a zero-center left-right indicator and another course, at right angles to

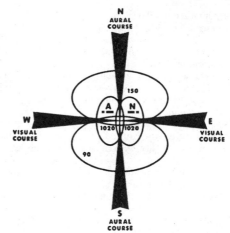

Visual-aural range radiation patterns, courses, and signals. Frequency values of tones are in hertz.

the first, in the form of aural A-N radio range signals. Either indication can be used to resolve the ambiguity of the other.

visual carrier frequency The frequency of the television carrier that is modulated by picture information.

visual flight rules [VFR] Regulations that govern aircraft flight when visibility is good up to a specified altitude.

visual radio range Any range facility whose course is flown by visual instrumentation not associated with audio reception.

visual signal The picture part of a television signal.

visual transmitter The radio equipment that transmits the video part of a television program. It is also called a picture transmitter. The visual and audio transmitters together are called a television transmitter.

visual transmitter power The peak power output when transmitting a standard television signal.

VLF Abbreviation for *very low frequency.*

VLSI Abbreviation for *very large-scale integration.*

VMOS [vertical metal-oxide semiconductor] An MOS technology in which four layers are diffused in silicon and a V-shaped groove is etched to a precisely controlled depth in the layers. This is followed by deposition of metal over silicon dioxide in the groove to form the gate electrode. It is now considered an obsolete technology.

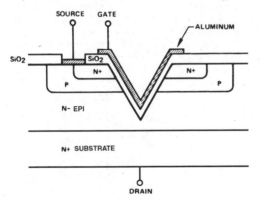

VMOS structure in an N-channel MOSFET.

vocoder A device that produces synthetic speech.

voice channel A telephone communication channel that has sufficient bandwidth to carry voice frequencies intelligibly. The minimum bandwidth for a voice channel is about 3 kHz.

voice coder A coder that converts speech input into a digital form which can be enciphered for secure transmission. The transmitted digital signals are deciphered and converted back into speech at the receiver.

voice coil The coil that is attached to the diaphragm of a moving-coil loudspeaker and moves through the air gap between the pole pieces because of interaction of the fixed magnetic field with that associated with the audio frequency current flowing through the voice coil.

voice digitization The conversion of analog voice signals to digital signals. Its advantages include improved communication because digital signals are relatively immune to noise, crosstalk, and distortion. Faded signals can be regen-

erated without loss of quality. Its techniques include various types of pulse-code modulation.

voice frequency An audio frequency in the range essential for transmission of speech of commercial quality, from about 300 to 3400 Hz. It is also called speech frequency.

voice-frequency carrier telegraphy Carrier telegraphy in which the modulated currents are transmitted over a voice-frequency telephone channel.

voice-grade line A public dial-up telephone line that has a bandwidth which extends from about 300 to 3400 Hz, as required for handling speech or digital data at speeds up to about 2400 b/s.

voice multiplexing The compression of two or more (typically four) simultaneous telephone conversations into one voice-grade channel so that it is undetectable to listeners. It is done with voice digitization.

voice-operated transmission [VOX] A method of radio communication in which the carrier is radiated only when a voice signal is present.

voiceprint A record of the distinctive patterns formed by a person's voice, as obtained with a sound spectrograph. It can be used for personal identification.

voice recognition unit A computer peripheral circuit that recognizes a limited number of spoken words and converts them into equivalent digital signals which can serve as computer input or initiate other desired actions.

voice response A computer-controlled recording system that stores basic sounds, numerals, words, and/or phrases individually for playback under computer control as the reply to a keyboarded query. Applications include stock market quotations and giving a checking account balance when an account number is entered into a remote computer terminal.

voice synthesizer A synthesizer that simulates speech in any language by assembling a language's elements or phonemes under digital control, each with the correct inflection, duration, pause, and other speech characteristics. Applications include computer-generated voice replies to queries by computers and Touch-Tone telephones. When combined with voice recognition, the system can provide oral replies to a limited variety of spoken queries. It is also called a *speech synthesizer.*

voice warning system An aircraft cockpit system that has a digitized or prerecorded vocabulary of warning words for such purposes as calling attention to possible trouble on the aircraft or warning of the possibility of an airborne collision.

volatile memory A semiconductor memory that does not retain its stored information (data) when the power is interrupted. Examples are the dynamic random-access memory (DRAM) and static random-access memory (SRAM). See also *nonvolatile memory.*

volt [V] The SI unit of voltage or potential difference. The difference in electrical potential between two points of a conducting wire carrying a constant current of 1 A when the power dissipated between these points is equal to 1 W. It was named after the Italian physicist Alessandro Volta, 1745–1829.

Volta effect *Contact potential.*

voltage [E] The term most often used to designate electric pressure that exists between two points and is capable of producing a flow of current when a closed circuit is connected between the two points. Voltage is measured in volts, millivolts, microvolts, and kilovolts. The terms electromotive force, potential, potential difference, and voltage drop are all often called voltage.

voltage amplification The ratio of the magnitude of the voltage across a specified load impedance to the magnitude of the input voltage of the amplifier or other transducer feeding that load. It is often expressed in decibels by multiplying the common logarithm of the ratio by 20.

voltage amplifier An amplifier intended primarily to increase the voltage of a signal, without supplying appreciable power.

voltage attenuation The ratio of the magnitude of the voltage across the input of a transducer to the magnitude of the voltage delivered to a specified load impedance connected to the transducer. It is expressed in decibels by multiplying the common logarithm of the ratio by 20.

voltage calibrator A voltage source that provides an adjustable, high-accuracy calibration voltage for calibrating measuring instruments.

voltage-controlled capacitor A capacitor whose capacitance value can be changed by varying an externally applied bias voltage, as in a silicon *varactor diode*.

voltage-controlled crystal oscillator [VCXO] A crystal oscillator circuit whose oscillator output frequency can be varied or swept over a range of frequencies by varying a DC modulating voltage.

voltage-controlled oscillator [VCO] An oscillator whose frequency of oscillation can be varied by changing an applied voltage.

voltage cutoff The electrode voltage that reduces the anode current, beam current, or some other electron-tube or transistor characteristic to a specified low value.

voltage-dependent resistor *Varistor.*

voltage divider A tapped resistor, adjustable resistor, potentiometer, or a series arrangement of two or more fixed resistors connected across a voltage source. A desired fraction of the total voltage is obtained from the intermediate tap, movable contact, or resistor junction.

voltage doubler A transformerless rectifier circuit that gives approximately double the output voltage of a conventional half-wave rectifier by charging a capacitor during the normally wasted half-cycle and discharging it in series with the output voltage during the next half-cycle. See *cascade voltage doubler, conventional voltage doubler* and *bridge voltage doubler.*

voltage drop The voltage developed across a component or conductor by the flow of current through the resistance or impedance of that component or conductor.

voltage feed Excitation of a transmitting antenna by applying voltage at a voltage loop or antinode.

voltage feedback Feedback in which the voltage drop across part of the load impedance acts in series with the input signal voltage.

voltage follower An operational amplifier that has no feedback components but has a direct feedback connection from the output to the inverting input. This gives unity gain so the output voltage follows the noninverting input voltage. A voltage follower has a very high input impedance and a very low output impedance.

voltage gain The difference between the output signal voltage level in decibels and the input signal voltage level

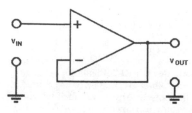

Voltage follower is an operational amplifier with an external feedback loop from the output terminal to the inverting terminal.

in decibels. This value is equal to 20 times the common logarithm of the ratio of the output voltage to the input voltage. The voltage gain is equal to the amplification factor of the tube or transistor only for a matched load.

voltage generator A two-terminal circuit component whose terminal voltage is independent of the current through the component.

voltage gradient The voltage per unit length along a resistor or other conductive path.

voltage jump An abrupt change or discontinuity in tube voltage drop during operation of a glow-discharge tube.

voltage loop An antinode at which voltage is a maximum.

voltage multiplier A source of DC voltages that are higher than the input AC voltage made from two or more *cascade voltage doubler* circuits. The number of multiplier stages is limited by the increasing size of capacitors required and the deterioration in regulation. See also *conventional voltage doubler* and *bridge voltage doubler.*

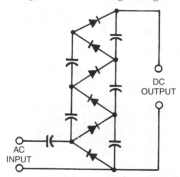

Voltage multiplier. Two or more cascade voltage-doubler stages are combined to permit higher DC voltages than can be obtained from a single stage. Its output is proportional to the number of stages.

voltage node A point that has zero voltage in a stationary wave system, as in an antenna or transmission line. A voltage node exists at the center of a half-wave antenna.

voltage-range multiplier *Instrument multiplier.*

voltage rating The maximum sustained voltage that can safely be applied to an electric device without risking the possibility of electric breakdown. It is also called working voltage.

voltage-reference diode A PN junction diode that has a sufficiently stable breakdown voltage to permit it to develop a reference voltage. See also *zener diode.*

voltage-reference tube A gas tube whose voltage drop is approximately constant over the operating range of current and is also relatively stable with time at fixed values of current and temperature. See also *neon diode.*

voltage reflection coefficient The ratio of the complex electric field strength or voltage of a reflected wave to that of the incident wave.

voltage-regulated AC power supply A power supply that operates from an AC line and delivers a regulated AC output voltage, usually adjustable, at the same frequency or at some other frequency.

voltage-regulating transformer A power transformer that delivers an essentially constant output voltage over a wide range of input voltage values.

voltage regulation The ratio of the difference between no-load and full-load output voltage of a device to the full-load output voltage, expressed as a percentage.

voltage regulator A circuit that includes a sensor capable of monitoring the load and restoring the output voltage to close tolerance limits despite changes in both the load and input voltage. This circuitry is now available in low-cost integrated circuits capable of holding DC output voltage levels of 3 to 30 V constant within ±2%. See also *three-terminal regulator.*

voltage-regulator diode A diode that maintains an essentially constant direct voltage in a circuit despite changes in line voltage or load. A Zener diode provides this regulation, but not all voltage-regulator diodes are based on the Zener effect.

voltage-regulator [VR] tube A glow-discharge tube whose voltage drop is approximately constant over the operating range of current. It maintains an essentially constant direct voltage in a circuit despite changes in line voltage or load. It is also called a *VR tube.*

voltage relay A relay that functions at a predetermined value of voltage.

voltage saturation *Anode saturation.*

voltage-sensitive resistor A resistor whose value varies markedly with applied voltage over at least a portion of its voltage range. It might consist of one or more mineral crystals or two or more metallic oxide disks, but it does not have rectifying properties.

voltage stabilizer A Zener diode or other device that suppresses variations in a DC voltage. It can replace a capacitor across a cathode or base biasing resistor.

voltage-stabilizing tube A gas-filled tube normally working with a glow discharge in that part of the characteristic where the voltage is practically independent of current.

voltage standard A voltage source whose value is known to a high degree of accuracy. A standard cell is an example.

voltage standing-wave ratio [VSWR] The ratio of the amplitude of the electric field or voltage at a voltage minimum to that at an adjacent maximum in a stationary-wave system, as in a waveguide, coaxial cable, or other transmission line.

voltage standing-wave ratio meter [VSWR meter] An electrically operated instrument that indicates voltage standing-wave ratios and is calibrated in voltage ratios.

voltage-to-frequency converter [V/F converter] A converter whose output frequency is a function of some reference or control signal. One version accepts an analog input from a sensor, transducer, or other analog device and generates a train of digital output pulses at a rate directly proportional to the instantaneous amplitude of the input signal. This digital output can be fed into a computer for

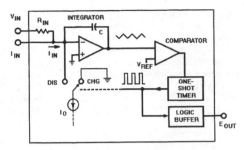

Voltage-to-frequency (V/F) converter. Its charge-balance circuit provides high linearity and stable conversion.

process control or other applications. Voltage-to-frequency converters are available as integrated circuits.

voltage transformer An instrument transformer whose primary winding is connected in parallel with a circuit in which the voltage is to be measured or controlled. It is also called a potential transformer.

voltage-tunable magnetron A magnetron with the cathode-anode geometry of the conventional magnetron, but with an anode that is designed so that it can be tuned by varying the plate voltage.

voltage-variable capacitor *Varactor.*

voltaic cell A primary cell that consists of two dissimilar metal electrodes in a solution which acts chemically on one or both of them to produce a voltage.

voltaic pile An early form of primary battery; it consists of a pile of alternate pairs of dissimilar metal disks, with moistened pads between pairs.

voltammeter An instrument that can function either as a voltmeter or ammeter.

voltampere [VA] The unit of apparent power in an AC circuit that contains reactance. Apparent power is equal to the voltage in volts multiplied by the current in amperes, without considering phase.

voltampere-hour reactive [varh] The unit of the integral of reactive power over time in the International System of Units (SI), equal to a reactive power of 1 var integrated over 1 h. It is also called a reactive voltampere-hour.

voltampere meter An instrument that measures the apparent power in an AC circuit.

voltampere reactive [var] The unit of reactive power in the International System of Units (SI), as adopted by the International Electrotechnical Commission in 1930. It is also called a reactive voltampere.

Volta's law The contact voltage developed between two dissimilar conductors is the same whether the contact is direct or through one or more intermediate conductors.

voltmeter An instrument that measures voltage. Its scale can be calibrated in volts or related smaller or larger units. A voltmeter that indicates millivolt values is often called a millivoltmeter. Similarly, a microvoltmeter indicates microvolt values, and a kilovoltmeter indicates kilovolt values.

voltmeter-ammeter A voltmeter and an ammeter combined in a single case, but with separate terminals.

voltmeter sensitivity The ratio of the total resistance of a voltmeter to its full-scale reading in volts, expressed in ohms per volt.

volt-ohm-milliammeter [VOM]   A test instrument that has different ranges for measuring voltage, current, and resistance. It is also called a multimeter.

volts AC [VAC]   The AC voltage in volts. Usage and common practice determine the exact meaning of the rating.

volts DC [VDC]   The DC voltage in volts. Usage and common practice determine the exact meaning of the rating. It is also called DC working volts and working volts DC.

volts root-mean-square [VRMS]   The root-mean-square voltage rating in volts.

volume   1. The magnitude of a complex audio current as measured in volume units (VU) on a standard volume indicator.   2. The intensity of a sound.

volume acoustic wave   *Bulk acoustic wave.*

volume compressor   An audio frequency control circuit that limits the volume range of a radio program at the transmitter, to permit using a higher average percent modulation without risk of overmodulation.

volume control   A potentiometer that varies the loudness of a reproduced sound by varying the audio-frequency signal voltage at the input of the audio amplifier.

volume expander   An audio frequency control circuit that increases the volume range of a radio program or recording by making weak sounds weaker and loud sounds louder. The expander counteracts volume compression at the transmitter or recording studio. It is also called an automatic volume expander.

volume indicator   An instrument for indicating the volume of a complex electric wave such as that for speech or music. The reading in volume units (VU) is equal to the number of decibels above a reference level. The sensitivity is adjusted so the reference level of 0 VU is indicated when the instrument is connected across a 600-Ω resistor that is dissipating 1 mW at 1 kHz.

volume ionization   The average ionization density in a given volume of ionizing particles.

volume lifetime   The average time interval between the generation and recombination of minority carriers in a homogeneous semiconductor.

volume-limiting amplifier   An amplifier that contains an automatic circuit which functions only when the input volume exceeds a predetermined level and then reduces the gain so that the output volume stays substantially constant despite further increases in input volume. The normal gain of the amplifier is restored when the input volume returns below the predetermined limiting level.

volume magnetostriction   The change in the volume of a magnetostrictive material when it is subjected to a magnetic field.

volume recombination   Recombination of positive and negative ions at low energies throughout the volume of an ionization chamber, or recombination of free electrons and holes in the volume of a semiconductor.

volume recombination rate   The time rate at which free electrons and holes recombine within the volume of a semiconductor.

volume unit [VU]   A unit for specifying the audio-frequency power level in decibels above a reference level of 1 mW (0.001 W), as measured with a standard volume indicator. A volume unit is equal to a decibel only when changes in power are involved or when the decibel value has this same reference level. It is unnecessary to specify the reference level when measuring in volume units because the level is a part of the definition.

volume-unit meter [VU meter]   A meter calibrated to read audio frequency power levels directly in volume units.

VOM   Abbreviation for *volt-ohm-milliammeter.*

Von Neumann machine   A conventional digital computer that carries out its instructions sequentially.

VOR   Abbreviation for *very high frequency omnidirectional range.*

VORDAC [VHF Omnidirectional Range and Distance-measuring equipment for Area Coverage]   A precision air navigation system for high-density air traffic routes; it consists of standard distance-measuring equipment and a high-accuracy VHF omnidirectional range.

VOR receiver   An aircraft radio receiver that receives signals from a VHF omnirange and interprets them for the pilot.

vortac [VHF Omnidirectional Range TA-Can]   An air navigation system based on civilian VHF omnirange facilities for directional guidance of aircraft, and military UHF tacan for measuring distance.

VOX   Abbreviation for *voice-operated transmission.*

Voyager   A NASA spacecraft designed for interplanetary exploration. Two were built and launched in 1977. After completing their assignments of moving in close to planets and their satellites to televise their surfaces and send the signals back to earth, they are proceeding on into outer space on different courses. In 1993 both Voyager 1 and Voyager 2 were sending useful data back to earth on solar winds and their resulting magnetic fields from positions well beyond the outer planets.

VRAM   Abbreviation for *video random-access memory,* computer memory dedicated to video applications.

VRMS   Abbreviation for *volts root-mean-square.*

VR tube   *Voltage-regulator tube.*

VSAT   Abbreviation for *very small-aperture terminal.*

VSB   Abbreviation for *vestigial sideband.*

VSC   Abbreviation for *variable speech control.*

VSELP   Abbreviation for *vector-sum excited linear predictive.*

VSWR   Abbreviation for *voltage standing-wave ratio.*

VSWR meter   Abbreviation for *voltage standing-wave ratio meter.*

VTR   Abbreviation for *videotape recording.*

VTS   Abbreviation for *vessel traffic service.*

VTVM   Abbreviation for *vacuum-tube voltmeter.*

VU   Abbreviation for *volume unit.*

VU meter   Abbreviation for *volume-unit meter.*

W

W Abbreviation for *watt*.

wafer A thin polished slice of semiconductor material on which discrete devices or integrated circuits can be fabricated. It is typically 10 to 30 mils thick, sawed from a cylindrical ingot *boule,* 4 to 8 in in diameter. The material is typically silicon or gallium arsenide.

wafer fab A slang term for *wafer fabrication facilities,* essentially the first part (front end) in the manufacture of diode and transistor dies or integrated-circuit chips prior to testing and packaging.

Wagner ground A ground connection for an AC bridge that minimizes stray capacitance errors when measuring high impedances. A potentiometer is connected across the bridge supply oscillator, with its movable tap grounded.

WAIS Abbreviation for *wide-area information server*.

wall outlet An telephone or power outlet mounted on a wall. Telephone service and electric power can be obtained by inserting the proper plug of a line cord.

warble-tone generator An audio frequency signal generator whose frequency is varied cyclically at a subaudio rate over a fixed range.

water calorimeter A calorimeter that measures RF power in terms of the rise in temperature of water which absorbs the RF energy.

water-cooled tube An electron tube that is cooled by circulating water through or around the anode structure.

water load A matched waveguide termination that absorbs electromagnetic energy with water. The resulting rise in the temperature of the water is a measure of the output power.

watt [W] The SI unit of electric power that in one second gives rise to energy of one joule. In a DC circuit, the power in watts is equal to volts multiplied by amperes. In an AC circuit, the true power is effective volts multiplied by effective amperes, then multiplied by the circuit power factor. There are 746 watts in one horsepower (hp). It is named after the Scottish inventor James Watt, 1736–1819.

wattage rating A rating that expresses the maximum power which a device can safely accept continuously.

watthour [Wh] The practical unit of electric energy, equal to a power of 1 W absorbed continuously for 1 h. One kilowatthour is equal to 1000 Wh.

watthour meter A meter that measures and registers power consumption of a home, office, or factory for billing purposes by the power utility. The power meter has a three-pole core with two sets of windings. One winding is across the AC line, and the other two are in series with the line. An aluminum disk is free to rotate between the poles. The interaction of flux produced by the three poles with currents that are induced in the aluminum disk causes the disk to rotate in proportion to the power consumed. The revolutions that are counted by the register represent watthours.

wattless power *Reactive power.*

wattmeter A meter that measures electric power in watts.

watt per steradian [W/sr] The SI unit of radiant intensity.

wattsecond [Ws] The amount of electric energy corresponding to 1 W acting for 1 second. One wattsecond is equal to 1 J.

wave A propagated disturbance whose intensity at any point in the medium is a function of time, and the intensity at a given instant is a function of the position of the point. A wave can be electric, electromagnetic, acoustic, or mechanical.

wave amplitude The magnitude of the maximum change from zero of a characteristic of a wave.

wave analyzer An instrument that measures the amplitude and frequency of the various components of a complex current or voltage wave.

wave angle The angle in bearing and elevation at which a radio wave arrives at a receiving antenna or leaves a transmitting antenna.

wave antenna   *Beverage antenna.*

wave equation   An equation that describes a particular wave motion through a medium.

wave filter   A transducer that separates waves on the basis of their frequency. It introduces relatively small insertion loss to waves in one or more frequency bands and relatively large insertion loss to waves of other frequencies.

waveform   The shape of a wave, as obtained by plotting a characteristic of the wave with respect to time.

waveform-amplitude distortion   *Frequency distortion.*

waveform analyzer   A frequency-selective voltmeter that measures the amplitude and frequency of each component of a complex waveform.

waveform monitor   A cathode-ray oscilloscope that has a time base suitable for viewing the waveform of the video signal in a television system.

waveform synthesizer   A signal generator whose output is variable in frequency, phase, harmonic content, and harmonic amplitude.

wavefront   The part of a wave envelope that is between the virtual zero point and the point where the wave reaches its crest value, as measured in either time or distance.

wave function   A set of solutions to Maxwell's equations for wave propagation in a homogeneous isotropic region.

wave group   The resultant of two or more wave trains that have different frequencies traversing the same path.

waveguide   A rectangular or circular metal pipe that has a predetermined cross section. It is specifically designed to guide or conduct high-frequency electromagnetic waves through its interior.

waveguide attenuator   An attenuator built into a waveguide to attenuate electromagnetic waves by absorption and reflection. See *fixed attenuator,* flap *attenuator* and *rotary-vane attenuator.*

waveguide bend   A section of waveguide that is bent along its longitudinal axis. An E-plane bend in a rectangular waveguide is bent along the narrow dimension, and an H-plane bend is bent along the wide dimension. It is also called a waveguide elbow.

waveguide component   A device connected at specified ports in a waveguide system.

waveguide connector   A mechanical fitting for electrically joining and locking together separable mating parts of a waveguide system. It is also called a waveguide coupling.

waveguide coupling   *Waveguide connector.*

waveguide cutoff wavelength   The wavelength that corresponds to the cutoff frequency of a waveguide. Below this frequency the attenuation rises rapidly.

waveguide directional coupler   A directional coupler made of two parallel waveguides that have slots cut in their common walls. These permit extraction of the RF energy traveling in one direction through the main waveguide while rejecting energy traveling in the opposite direction.

waveguide elbow   *Waveguide bend.*

waveguide filter   A filter made from waveguide components that can change the amplitude-frequency response characteristic of a waveguide system.

waveguide flange   *Flange.*

waveguide gasket   A gasket that maintains electric continuity between two mating sections of waveguide. It is typically made of metal wire.

waveguide iris   A narrowing of the cross-sectional area of a waveguide by the placement of thin metallic plates perpendicular to the guide walls and joined to them at the edges, with an opening between them. When the opening is parallel to the narrow walls of the guide, it is called an *inductive iris* and it presents an inductive susceptance. When it is parallel to the wide walls, it is called a *capacitive iris* and it presents a capacitive susceptance.

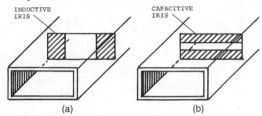

Waveguide irises in rectangular waveguides can match circuit impedances: (*a*) inductive iris and (*b*) capacitive iris.

waveguide junction   *Junction.*

waveguide lens   A microwave lens with waveguide elements which causes phase changes in the electromagnetic waves that pass through it.

waveguide mode suppressor   A waveguide filter that suppresses undesired modes of electromagnetic wave propagation in a waveguide.

waveguide phase shifter   A device that adjusts the phase of the output signal of a waveguide system with respect to the phase of the input signal.

waveguide plunger   *Piston.*

waveguide post   A post fastened across the narrow dimension of a rectangular waveguide that acts as an inductive shunt susceptance. Its value depends on the post diameter and its position in the transverse plane of the waveguide.

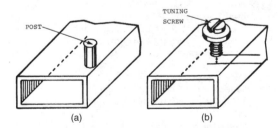

Waveguide post and screw perform impedance matching. Post (*a*) across the narrow dimension acts as an inductive shunt susceptance, and the tuning screw (*b*) (projecting part way across the narrow dimension) acts as a capacitive susceptance.

waveguide probe   *Probe.*

waveguide radiator   An open-ended waveguide, with or without a flaring horn, that radiates electromagnetic energy to a reflector or out into space.

waveguide resonator   *Cavity resonator.*

waveguide seal A seal over the end of a waveguide to prevent the entrance of moisture without appreciably attenuating radio frequencies.

waveguide shim A thin, resilient, metal sheet inserted between waveguide components to ensure electric continuity.

waveguide shutter A waveguide section that contains an adjustable mechanical barrier which can be set to block or divert RF energy.

waveguide slug tuner A quarter-wavelength dielectric slug that projects into a waveguide for tuning purposes. It is adjustable in position and depth of penetration.

waveguide stub An auxiliary section of waveguide that has an essentially nondissipative termination, joined at some angle with a main section of waveguide.

waveguide stub tuner A waveguide tuning or detuning device that consists of an adjustable piston mounted in a waveguide stub.

waveguide switch A switch for mechanically positioning a waveguide section, to couple it to one of several other sections in a waveguide system.

waveguide taper A section of tapered waveguide.

waveguide transformer A waveguide component that provides impedance transformation.

waveguide tuner An adjustable tuner that provides impedance transformation in a waveguide system.

waveguide-tuning screw A screw that projects part of the way across the narrow dimension of a rectangular waveguide and acts as a capacitive susceptance. The penetration of the screw can be adjusted externally for precise tuning.

waveguide twist A waveguide section whose cross section exhibits a progressive rotation about its longitudinal axis.

waveguide wavelength The distance along a uniform waveguide, at a given frequency and for a given mode, between similar points at which a signal component differs in phase by 2π radians.

waveguide window A thin conducting metal window placed transversely in a waveguide for impedance matching. For an inductive window, the edges of the slit in the window are parallel to the electric field in the lowest mode in the waveguide. For a capacitive window, the edges of the slit are perpendicular to the electric field.

wave heating Heating of a material by energy absorption from a traveling electromagnetic wave.

wave impedance The ratio of the transverse electric field to the transverse magnetic field in a waveguide.

wave interference The variation of wave amplitude with distance or time, caused by the superposition of two or more waves that have the same or nearly the same frequency.

wave-interference microphone A highly directional microphone that is coupled to the sound field over an area and responds to the sum of the pressures over this area. Its chief drawbacks are variation of polar response with frequency and its large size. Examples include line microphones and microphones with parabolic reflectors.

wavelength The distance between points that have corresponding phase in two consecutive cycles of a periodic wave. The wavelength in meters is approximately equal to 300 divided by the frequency in megahertz. Wavelengths of light are specified in micrometers or nanometers [1 μm (formerly 1 μ) = 10^{-6} m = 1000 nm = 10,000 Å]. Multiply angstroms by 0.1 to get nanometers, or by 0.0001 to get micrometers.

wavelength shifter A photofluorescent compound that combined with a scintillator material increases the wavelengths of the optical photons, thereby permitting more efficient use of the photons by the phototube or photocell.

wave mechanics A theory that assigns wave characteristics to the components of atomic structure and seeks to interpret physical phenomena in terms of hypothetical waveforms. It was introduced by the German scientist Schroedinger in 1926.

wavemeter An instrument that measures the wavelength of an RF wave. Because wavelength is related to frequency, a wavemeter also serves as a frequency meter.

wave normal A unit vector that is perpendicular to an equiphase surface, with its positive direction taken on the same side of the surface as the direction of propagation. In isotropic media, the wave normal is in the direction of propagation.

wave packet A wave function that describes a localized particle whose position is known within fairly narrow limits.

wave propagation *Propagation.*

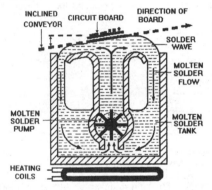

Wave soldering. The solder side of an assembled board passes through a molten solder wave on an inclined conveyor to bond components to the circuit board.

wave soldering An automated process for soldering electronic components to circuit boards by passing a board with its components attached solder-side down through the molten solder wave. Solder connections are formed between the pretinned leads and prepared presoldered board pads. The solder wave is formed by pumping molten solder through a nozzle to form the wave before it flows back to the bottom of a holding tank, where it is reheated and recirculated continuously. The boards are positioned on an inclined conveyor that moves at speeds which can exceed 20 ft/min. The process includes a preheat stage, an active wave stage, and an exit or postheat stage. The velocity profile of the wave can be controlled to optimize the soldering task. The process is widely used in surface-mount soldering of leadless components that are first bonded to the board with an adhesive. Both complete packaged components and their pretinned leads pass through the solder wave. The process is also called *flow soldering*.

wavetable sound   *Wavetable synthesis.*

wavetable synthesis   A process for creating more realistic sound than *MIDI* sound synthesis for games and applications. The sounds of actual musical instruments are recorded, digitized, compressed, and programmed into a semiconductor memory. The memory is then put on a computer plug-in sound card. When a program requires a specific instrument sound, software sends a request for that sound to the system microprocessor (MPU) or microcontroller (MCU). The memory device does not contain all the notes of a musical instrument, but if it contains an A-flat and an A-sharp, it can interpolate an A. The sound IC then generates the recorded music. It is also called *wavetable sound.*

wave tail   The part of a wave envelope that is between the crest and the end of the envelope.

wave tilt   The forward inclination of the waveform of radio waves arriving along the ground. Its value depends on the electric constants of the ground.

wave train   A series of wave cycles produced by the same disturbance.

wave trap   A resonant circuit connected to the antenna system of a receiver to suppress signals at a particular frequency, such as that of a powerful local station which is interfering with reception of other stations.

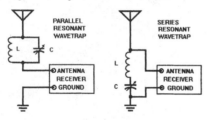

Wave traps are resonant circuits that suppress unwanted frequencies before they reach the receiver.

wave trough   The minimum value of the envelope of a progressive wave.

wave-type microphone   A microphone that depends on wave interference for its directivity.

wave velocity   The velocity of propagation of an electromagnetic wave, equal to 3×10^{10} cm/s in free space. In a waveguide, the rate of energy transfer is called the group velocity and is less than the wave velocity. By contrast, the velocity of the electric wave is called the phase velocity and it can be greater than the wave velocity.

way point   A selected point on a radio navigation course line that has some particular significance to the navigator.

Wb   Abbreviation for *weber.*

weather radar   A radar that is capable of detecting echoes from clouds or rain and is normally assigned to that duty.

weber [Wb]   The SI unit of magnetic flux. It is the amount of flux that, when linked with a single turn of wire, will induce 1 V in the turn as it decreases uniformly to zero in 1 second. One weber is equal to 10^8 maxwells.

wedge   1. A waveguide termination that consists of a tapered length of dissipative material such as carbon introduced into the guide.   2. A convergent pattern of equally spaced black and white lines in a television test pattern that indicates resolution.   3. An optical filter that causes the transmission to decrease continuously or in steps from one end to the other.

wedge-base lamp   A small incandescent indicator lamp that has wire leads folded back on opposite sides of its flat glass base. It is designed for insertion into a socket that has wedge-shaped spring contacts.

Wedge-base incandescent lamps.

wedge bond   See *wire bonding.*

wedge bonding   A method for *wire bonding* semiconductor chip pads to package leads with fine (0.7- to 1.0-mil) aluminum wire. It was formerly called *stitch bonding.* See also *ball bonding* and *wire bonding.*

wedge filter   A radiation filter constructed so that its thickness or transmission characteristics vary continuously or in steps from one edge to the other to increase the uniformity of radiation in certain applications.

wedge spectrograph   A spectrograph that permits the density of the radiation passing through the entrance slit to be varied by moving an optical wedge.

Wehnelt cathode   *Oxide-coated cathode.*

Wehnelt cylinder   The metal tube that encloses the cathode of a cathode-ray tube and concentrates the electrons emitted in all directions from the cathode.

weighted noise level   The noise level weighted in accordance with the 70-dB equal-loudness contour of the human ear, expressed in decibels above 1 mW.

weighting   The artificial adjustment of measurements to account for factors that, in the normal use of the product, would otherwise be different from conditions encountered during the measurements. As an example, background noise measurements can be weighted by applying factors or by introducing networks to reduce measured values in inverse ratio to their interfering effects.

weighting network   A network whose loss varies with frequency in a predetermined manner.

weightlessness   A condition in which no acceleration, whether of gravity or other force, can be detected by an observer within the system. An unaccelerated satellite orbiting the earth is "weightless," although gravity affects its orbit. Weightlessness can be produced within the atmosphere in aircraft that is flying a parabolic flight path.

Weissenberg method   An X-ray crystal analysis method in which the crystal is rotated in the beam of X-rays and the film is moved parallel to the axis of rotation. The crystal is surrounded by a sleeve that has a slot which passes a line-shaped beam of X-rays to give positive identification of each spot or line on the pattern.

welding   A process of joining metals by the application of heat, pressure, or both.

welding current   The current that is sent through a joint to produce the heat needed to make a weld.

welding cycle The complete series of events involved in making a weld.

weld interval The total of all heating and cooling times when making a single multiple-impulse weld by resistance welding.

weld-interval timer A timer that controls the heating and cooling intervals when multiple-impulse welds are being made.

weld time The time that welding current flows through the work when a weld is being made.

weld timer A timer that controls only the weld time.

Wertheim effect An effect that occurs when a ferromagnetic wire is twisted in a longitudinal magnetic field. A voltage is produced between the ends of the wire.

Western Union joint A joint or splice that has good mechanical strength as well as good conductivity. It is made by crossing the cleaned ends of two wires and then winding the end of each wire around the other wire and soldering the joint.

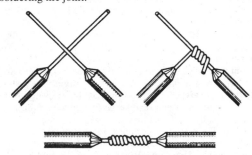

Western Union joint.

Weston standard cell A standard cell that is a highly accurate voltage source for calibration. The positive electrode is mercury, the negative electrode is cadmium, and the electrolyte is a saturated cadmium sulfate solution. The Weston standard cell has a voltage of 1.018636 V at 20°C.

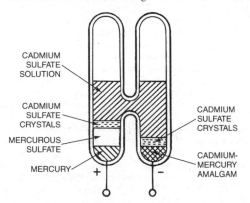

Weston standard cell produces a stable 1.018636 V for instrument calibration.

wet cell A cell whose electrolyte is in liquid form.

wet electrolytic capacitor An electrolytic capacitor that has a liquid electrolyte.

wetting 1. The coating of a contact surface with an adherent film of mercury. 2. The coating of a surface with molten solder.

wetting agent A substance that decreases the surface tension of a liquid, to make the liquid spread and adhere better.

Wh Abbreviation for *watthour*.

Wheatstone bridge A four-arm bridge with arms that are predominantly resistive. It can measure resistance. It is also called a *resistance bridge*.

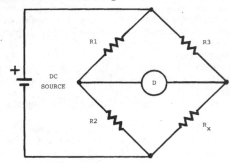

Wheatstone bridge can measure DC resistance with a DC voltage source. Many different versions have been developed for measuring electrical values.

wheel printer A line printer, typically a computer peripheral, that uses a printhead shaped as a disk with the raised type characters arranged around its rim. Electronic control circuitry switches the selected character into position for impact printing. This printer can make multiple copies with carbon or other duplicating paper inserted between individual sheets. An example is the "daisy wheel" printer.

whip antenna A flexible vertical-rod antenna, primarily for vehicles.

whisker 1. An extremely fine single-crystal filament of a metal or inorganic compound. It is produced by the reduction of metal halides in hydrogen at about 700°C to initiate crystal growth and reaction of molten metal vapor with an oxidizing, nitriding, or other atmosphere conducive to crystal growth. Undesired whisker growth in hot and humid environments causes shorting between conductors in electronic equipment. Whiskers have much greater tensile strength and higher elasticity than the corresponding bulk form of the metal. 2. *Catwhisker.*

whistler An effect caused when an electric disturbance produced by a lightning discharge travels out along lines of magnetic force of the earth's field and is reflected back to its origin from a magnetically conjugate point on the earth's surface. The characteristic drawn-out descending pitch of the whistler is a dispersion effect caused by the greater velocity of the higher-frequency components of the disturbance. Radio signals can be transmitted along whistler paths from the northern to the southern hemisphere.

white compression The reduction in picture-signal gain at levels corresponding to light areas with respect to the gain at the level for midrange light values. The overall effect of white compression is to reduce contrast in the highlights of the picture. It is also called white saturation.

white level The carrier signal level that corresponds to maximum picture brightness in television and facsimile.

white light Light that is comparable in wavelength content to average noon sunlight.

white noise Random acoustic or electric noise that has equal energy per cycle over a specified total frequency band. The electric disturbance caused by random movements of free electrons in a conductor or semiconductor is one example; another is the frequency spectrum of white light.

white-noise record A tape recording for testing the frequency response of audio reproduction systems. A recording of acoustic white noise extending over the entire band of audio frequencies is reduced in bandwidth progressively in steps as the top limit of the bandwidth at each step is announced.

white object An object that reflects all wavelengths of light with substantially equal high efficiencies and considerable diffusion.

white peak A peak excursion of the picture signal in the white direction.

white recording A form of amplitude-modulation recording in which the maximum received power corresponds to the minimum density of the record medium in a facsimile system. In a frequency-modulated white recording the lowest received frequency corresponds to the minimum density of the record medium.

white room *Clean room.*

white saturation *White compression.*

white signal The signal produced at any point in a facsimile system by scanning a minimum-density area of the subject copy.

white-to-black amplitude range The ratio of signal voltage for picture white to signal voltage for picture black at a given point in a facsimile system that uses positive amplitude modulation. It is expressed in decibels. For negative amplitude modulation the reverse ratio, of black to white, is used.

white-to-black frequency swing The difference between the signal frequencies corresponding to picture white and picture black at any point in a facsimile system that is frequency modulated.

white transmission Amplitude-modulated transmission whose maximum transmitted power corresponds to minimum density of the subject copy, or frequency-modulated transmission whose lowest transmitted frequency corresponds to the minimum density of the subject copy in a facsimile system.

whole step *Whole tone.*

whole tone The interval between two sounds whose basic frequency ratio is approximately equal to the sixth root of 2.

wicking The flow of molten solder up under the insulation of covered wire.

wide-angle lens An optical lens that has a large angular field, generally greater than 80°.

Wide-Area Augmentation System A system for improving the accuracy of location fixes obtained from the Global Positioning System (GPS-C). The degraded commercial signals are corrected in real time with a supplementary signal that is generated locally based on the knowledge of the transmitter's precise location in latitude and longitude coordinates. The system is expected to be mandatory for commercial aircraft flying over the United States, but commercial and private aircraft, ships, and vehicles of all kinds can take advantage of the system if they install the necessary supplementary receiver. This system will be able to pinpoint the location of a GPS-C receiver to within about 21 ft (7 m), anywhere in the continental United States. This compares to an accuracy of within about 300 ft (100 m) now obtainable with uncompensated consumer-grade GPS-C signal receivers. By contrast, the GPS military signals can fix the receiver's position to within 1 cm. See *Differential Global Positioning System* [DGPS].

wide-area information server [WAIS] Software for finding and retrieving documents based on user-defined key words.

wideband A reference to the ability of a circuit or transmission line to transmit and receive a wide range of frequencies, typically superior to voice-grade transmission lines.

wideband amplifier *Broadband amplifier.*

wideband axis The direction of the phasor that represents the fine chrominance primary (the I signal) in color television.

wideband dipole A dipole that has a low ratio of length to diameter, to give resonance over a relatively wide frequency band.

wideband ratio The ratio of the occupied frequency bandwidth to the intelligence bandwidth in a multiplex system.

wide-open receiver A receiver that has essentially no tuned circuits so it can receive all frequencies simultaneously in the band of coverage.

width 1. The horizontal dimension of a television or facsimile picture. 2. The time duration of a pulse.

width control The control that adjusts the width of the pattern on the screen of a cathode-ray tube in a television receiver or oscilloscope.

Wiedemann effect The twist produced in a current-carrying wire when the wire is placed in a longitudinal magnetic field.

Wiedemann-Franz law The ratio of the thermal conductivity to the electric conductivity is proportional to the absolute temperature for all metals.

Wiegand effect The ability of a mechanically stressed ferromagnetic wire to recognize rapid switching of magnetization when subjected to a DC magnetic field.

Wien bridge A four-arm AC bridge that can measure the equivalent capacitance and parallel loss resistance of an imperfect capacitor. It could also be a sample of insulation or a length of cable. Balance depends on frequency. It is used as a frequency-determining network in RC oscillators. See *Wien-bridge oscillator.*

Wien-bridge oscillator A phase-shift feedback oscillator that includes a Wien bridge as its frequency-determining element.

Wien's displacement law The wavelength of the peak radiation is inversely proportional to the absolute temperature of a blackbody. As temperature rises, the peak of the spectral energy distribution curve is shifted toward the short-wavelength end of the spectrum.

willemite A natural fluorescent mineral that consists chiefly of zinc orthosilicate which is applied to cathode-ray tube screens.

Wimshurst machine An electrostatic generator that consists of two glass disks rotating in opposite directions. The

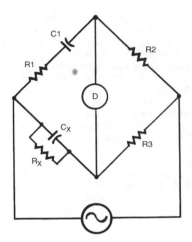

Wien bridge can determine the equivalent capacitance and parallel loss resistance of a capacitor.

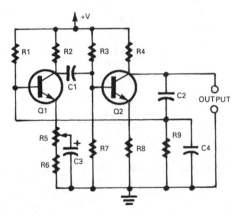

Wien bridge oscillator is a phase-shift oscillator that has a Wien bridge as its frequency-determining element.

winding 1. One or more turns of wire forming a continuous coil for a transformer, relay, rotating machine, or other electric device. 2. A conductive path, usually of wire, that is inductively coupled to a magnetic storage core or cell.

Windom antenna A multiband transmitting antenna that provides satisfactory operation on even harmonics of its fundamental frequency. An example is operation on several harmonically related amateur bands. One version is a horizontal wire one half-wavelength long at the fundamental frequency, with a 300-Ω twin-line feeder connected about 35% off center.

window 1. A radar countermeasure reflector that consists of metal foil or wires that are cut to lengths that will resonate at expected enemy radar frequencies. They are dropped in clusters from aircraft or expelled from shells or rockets. They were first called window because the first pieces were the size of small panes of glass. Later it was found that strips worked just as well and consumed less foil. 2. A hole in a partition between two cavities or waveguides, used for coupling. 3. A material that absorbs and reflects a minimum amount of radiant energy. It is sealed onto the vacuum envelope of a microwave or other electron tube to permit passage of the desired radiation through the envelope to the output system. Alumina, sapphire, beryllia, and quartz are examples of window materials that will pass microwave energy. Other materials are used for infrared energy output. 4. Atmospheric radio window.

window comparator A comparator that detects signal voltages at two different levels by comparing them to fixed references.

wiper A wiping contact that moves over stationary contacts in a selector switch or stepping relay, or over a resistive element in a potentiometer. See also *slider*.

wiping contact A switch or relay contact that moves laterally with a wiping motion after it touches a mating contact. It is also called a self-cleaning contact or sliding contact.

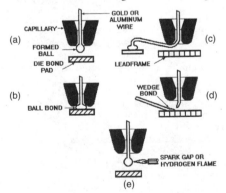

Wire bonding process. A molten ball (*a*) is formed on the end of the wire and the ball is bonded (*b*) to a semiconductor device die or chip. Wire (*c*) is fed out and bonded to an adjacent surface and then cut (*d*), leaving a wedge bond. The end of the wire is then heated with a spark gap or hydrogen flame to form a new ball and the process is repeated.

disks have sectors covered with metal foil and collecting combs so arranged that static electricity is produced for charging Leyden jars or discharging across a gap.

Winchester drive A reference to the Winchester hard-disk memory drive developed by IBM and characterized by one to four or more magnetic disks or platters that have been coated with a magnetic material. The disks can store data on both sides, and data is written and read by *flying heads* that remain a few thousandths of an inch from the surface of the disk to avoid touching it during operation. Disk diameters now available range in size from 1.3 to 8 in in diameter.

Winchester formatter A circuit that organizes data for storage on a hard disk.

wind The manner in which magnetic tape is wound onto a reel. In an *A* wind, the coated surface faces the hub. In a *B* wind, the coated surface faces away from the hub.

wind-driven generator A generator that derives its power from wind acting on a propeller.

wire bonding A method for wiring an electrical connection between the pads of a semiconductor chip or die and the terminals or pins of the package. Thin gold wire is joined in a loop between two contact points by bonds on each end with *ball* or *ball and wedge bonding* performed by a *thermocompression bonder.* An alternative method for thin aluminum wire is *wedge bonding* with a *thermosonic wire bonder.*

wire bonder   Equipment for *wire bonding.* There are two types: *thermocompression* and *ultrasonic* (also called *thermosonic*). Thermocompression bonders form a bond between the wire and bond pads with heat, and ultrasonic bonders form it with ultrasonic vibration. The choice depends on the desired die-attachment process. The *capillary* feeds the wire and also acts as a compression tool to bond the wire loop between the chip and the package pads. Modern equipment is guided by *pattern recognition* techniques and makes bonds at speeds up to 6 wire loop connections per second. See *wire bonding.*

wireframe   A technique for illustrating solid three-dimensional objects in computer-aided graphics in which solid lines that look like "wires" or "sticks" frame the outline of the object. It has no curved lines. In a wireframe rendering, lines on the far side of the object that ordinarily would not be visible can be seen because the spaces between the "wires" are not filled in by opaque medium. Wireframe is often a preliminary step to a complete colored and shaded rendering.

wire gage   A system of numerical designations of wire sizes. The American wire gage [AWG] starts with 0000 as the largest size, going to 000. 00. 0, 1, 2, and up to 40 and beyond for the smallest sizes.

wire-grid lens antenna   A high-frequency lens antenna that consists of two circular grids suspended one over the other, with the combination surrounded by a radial wire horn. It is used for radio communication in the range of 3 to 30 MHz.

wire guidance   Control of a guided missile or torpedo by sending control signals over wires pulled by the torpedo or missile.

wireless   British term for radio.

wireless LAN   A local area network (LAN) linked by radio-frequency or infrared signals rather than hard-wired coaxial or twisted-pair cable.

wireless stereo headphones   Stereo headphones that include a built-in radio receiver. They receive television and stereo signals at a frequency of about 900 MHz to permit listening at distances up to about 150 ft (45 m) from the transmitter, which is connected to the stereo system or TV receiver.

wireless stereo speakers   Stereo speakers that include built-in radio receivers and amplifiers. They receive television and stereo signals at a frequency of about 900 MHz to permit listening at distances up to about 150 ft (45 m) from the transmitter, which is connected to the stereo system or TV receiver.

wirephoto   1. A photograph transmitted over wires to a facsimile receiver.   2. *Facsimile.*

wire recorder   A magnetic recorder that records data on round stainless steel wire about 0.004 in (0.1 mm) in diameter instead of magnetic tape. It is now an obsolete technology of historic interest in the recording field.

wire stripper   A hand-operated tool or special machine designed to cut and remove the insulation for a predeter-

mined distance from the end of an insulated wire without damaging the solid or stranded wire inside.

wiretap   A secretly made and concealed connection to a telephone line, office intercommunication line, or other wiring system, for monitoring conversations and activities in a room from a remote location without the knowledge of the participants under surveillance. It is done both legally and illegally.

wirewound resistor [WW]   A resistor whose resistance element is a length of high-resistance wire or ribbon, usually Nichrome, wound on an insulating form or mandrel. See *resistor.*

wirewound rheostat   A rheostat that has a sliding or rolling contact which moves over a bare section of resistance wire that has been wound on an insulating core.

wire-wrap connection   A solderless connection made by wrapping several turns of insulated wire around a terminal that has a square cross section under tension, using either a power or hand tool. The tension opens the insulation to form a pin-to-wire connection. It is also called a solderless wrapped connection or a wrapped connection.

Wire-wrap connection.

wire-wrap pin   A rectangular cross-section, sharp-cornered terminal to which a connection can be made with a wire-wrap tool. The end of a wire is wound around the terminal several times, under tension, to make electrical contact.

wiring harness   An array of insulated conductors bound together by lacing cord, plastic straps or other bindings. It is typically custom-made for specific equipment or systems.

withstand voltage   The maximum voltage that can be applied between circuits or components without causing a dielectric breakdown.

wolfram   The term for *tungsten* outside of the United States.

woofer   A large loudspeaker that reproduces low audio frequencies at relatively high power levels. It is usually used in combination with a crossover network and a high-frequency loudspeaker called a tweeter.

word   A set of characters that is treated, stored, and transported by computers as a unit. The latest microprocessors can handle a word size of 32 bits or 4 bytes.

word generator   A special pulse generator that generates a programmable train of pulses (called a word) for testing data-processing and telemetry equipment.

word processing [WP]   the generation of letters, notes, manuscripts, and other written documentation by entering words and numbers on a keyboard for processing by a digital computer or a specialized word-processing machine that includes elements of a computer under the direction of word-processing software. The text can be viewed as it is written on a computer monitor, liquid-crystal, or other alphanumeric display for editing and revision before it is printed out, typically on a separate printer. The data for forming the document can be stored in computer memory for reference, reuse, or later revision or be transmitted over a modem to other computers or word-processing machines.

work The product of force times displacement, with the force in the direction of the displacement.

work function The minimum energy needed to remove an electron from the Fermi level of a metal to infinity. It is expressed in electronvolts.

working memory A portion of the internal memory of a computer that is reserved for the data upon which operations are being performed, including partial results.

working voltage (WVDC) *Voltage rating.* Working voltage is generally used in reference to DC voltage ratings for capacitors. The exact meaning of this rating in a particular usage should always be specified. The preferred abbreviation for the value of the voltage rating in volts is VDC.

workstation A desktop computer optimized to run computer-aided design (CAD) programs for engineering applications. It typically has a high-resolution, multicolor cathode-ray tube (1000 or more lines) to make high-definition drawings of complex integrated circuits and circuit boards. Three-dimensional, multicolor images can be manipulated on the screen to give different views. Other features include multitasking (through a multitasking operating system) that allows independent operation of different areas of the screen, and networking for distributed computing—the acceptance of work from other computers.

World Wide Web [WWW] A service provided to personal computer users with the appropriate software installed and a high-speed modem. It provides a wide selection of text files, graphics, and audio for commercial promotion, sales, and educational purposes. Specific *sites* can be found on the network of servers and accessed by *hypertext* links. See also *Internet.*

WORM Acronym for *write-once, read-many.* An optical data-recording medium consisting of an optically sensitive rotating disk on whose surface digital data is recorded with a high-resolution laser beam. The focused laser beam permanently modifies the surface of the medium by burning microscopic pits in the disk corresponding to a binary code. The data can then be read by a laser beam of lower, nondestructive intensity. Once recorded, data can never be erased or modified. See also *CD-ROM.*

wow A low-frequency flutter. When caused by an off-center hole in a disk record, it occurs once per revolution of the turntable.

WP Abbreviation for *word processing.*

WPM Abbreviation for *words per minute.*

wrapped connection *Wire-wrap connection.*

write To introduce information into some form of memory in a computer.

write pulse A pulse that causes information to be introduced into a memory cell or cells for storage in a computer.

write time The time interval between the instant at which information is ready for storage and the instant at which storage is completed in a computer. It is also called *access time.*

writing speed The rate of writing on successive storage elements in a charge-storage tube.

Ws Abbreviation for *wattsecond.*

W/sr Abbreviation for *watt per steradian.*

Wullenweber antenna A wide-aperture multiport antenna with many vertical elements arranged in a circle for direction finding. The rotation of the response pattern is obtained by a rotating, capacity-coupled goniomenter. It typically combines responses from adjacent antenna elements to form highly directive sum-and-difference beams. The system can only determine the azimuth angle of the emitter and not its range.

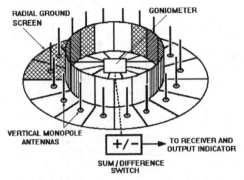

Wullenweber antenna has an array of monopoles positioned around vertical and horizontal metal screening to provide a wide aperture.

WVDC Abbreviation for *working volts DC.*

WWV The call letters of a radio station maintained by the National Institute of Standards and Technology (NIST) to provide standard radio and audio frequencies and other technical services such as precision time signals and radio-propagation disturbance warnings. The station broadcasts from Boulder, Colorado, on 2.5, 5, 10, 15, 20, and 25 MHz.

WWW Abbreviation for *World Wide Web.*

WWVH The National Institute of Standards and Technology (NIST) station on Kauai, Hawaii, broadcasting services similar to those of WWV on 5, 10, and 15 MHz.

X Symbol for *reactance*. Inductive reactance is designated as X_L and capacitive reactance as X_C.

X axis 1. A reference axis in a quartz crystal. 2. The horizontal axis on a cathode-ray oscilloscope screen or on a graph. The corresponding vertical axis is the Y axis.

X band A frequency band from 8.0 to 12.0 GHz with wavelengths of 3.75 to 2.50 cm, respectively, in accordance with IEEE Standard 521-1976. It was formerly known as the *three centimeter band*. It corresponds with the I band (8.0 to 10.0 GHz) and part of the J band (10.0 to 20.0 GHz) of the U.S. Military Joint Chiefs of Staff (JCS) triservice frequency designations (1970).

X cut A quartz-crystal cut made so that the X axis is perpendicular to the faces of the resulting slab.

xenon [Xe] A gaseous element. It is one of the rare gases used in some thyratrons and other gas-discharge tubes. It has an atomic number of 54.

xenon flash tube A flash tube that contains xenon gas. It produces an intense peak of radiant energy at a wavelength of 0.57 μm (white light) when a high DC pulsed voltage is applied between electrodes at opposite ends of the tube.

xerography The original form of electrophotography in which an electrostatic image was formed on a light-sensitive selenium-coated surface when exposed to an optical image. The charged image areas attract and hold a fine black or colored resinous powder called a toner. The powder image is transferred to a sheet of paper and fused by heat to make it permanent.

xeroprinting An electrostatic image-forming process. The printing plate is a metal substrate that has a permanent electrically insulating image. When the plate is moved under corona-charging wires, the image areas retain a static electric charge, and nonimage areas dissipate the charge. The charged areas attract and hold the black or colored powder toner. Another corona-charging process then transfers the powder image to a sheet of paper. The image is then fixed by applying heat to fuse the powder to the paper.

xeroradiography An electrostatic image-forming process in which X-rays or gamma rays form an electrostatic image on a photoconductive insulating medium. The charged image areas attract and hold a fine powder called a toner. The powder image is then transferred to paper and fused there by heat. Xeroradiography is one branch of electrophotography; the other is xerography, in which the images are formed by infrared, visible, or ultraviolet radiation.

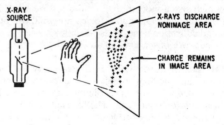

Xeroradiography gives a permanent black image of the hand on paper, in contrast to radiography, where image on photographic X-ray film requires chemical developing.

X guide A surface-wave transmission line that is a dielectric structure with an X-shaped cross section.

X-on/X-off An abbreviation for *transmitter on/transmitter off.*

XOR gate *EXCLUSIVE-OR gate.*

XPS Abbreviation for *X-ray photoelectron spectroscopy.*

X plate One of the two deflection electrodes that deflect the electron beam horizontally in an electrostatic cathode-ray tube.

X-radiation Radiation of X-rays.

X-ray 1. A penetrating electromagnetic radiation with properties similar to light, but with much shorter wavelengths (from about 10^{-7} to 10^{-10} cm or 0.1 to 100 Å), between ultraviolet and gamma rays. X-rays are usually generated by accelerating electrons to a high velocity and suddenly stopping them by collision with a metal target. The resulting bombardment of the atoms in the target causes the atoms to lose energy, and this energy is radiated as X-rays of definite wavelength. X-rays are also produced by transitions of atoms from higher to lower energy states. Properties of X-rays include ionization of a gas through which they pass, penetration of all solids in varying degrees, production of secondary X-rays when stopped by material bodies, and action on fluorescent screens and photographic film. Photons that originate in the nucleus of an atom are generally called gamma rays, and photons originating outside the nucleus are called X-rays. They are also called roentgen rays. 2. To photograph with X-rays.

X-ray analysis Determination of the internal structure of crystalline solids by the diffraction pattern produced when X-rays are passed through the material.

X-ray diffraction Diffraction of a beam of X-rays by the regular atomic lattice of a crystal. A characteristic diffraction pattern is obtained for each crystalline material.

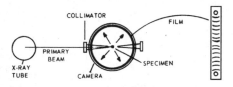

X-ray diffraction: the atoms of a specimen produce characteristic curved lines on film.

X-ray diffractometer An instrument for X-ray analysis that measures the intensities of the diffracted beams at different angles.

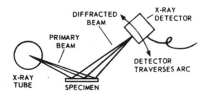

X-ray diffractometer. The output of a detector can be fed to recorder to give a permanent record.

X-ray film A film base coated, usually on both sides, with an emulsion sensitive to X-rays.

X-ray fluorescence absorptiometer An absorptiometer that measures the thickness of a coating, such as tin on steel plate. The primary X-ray beam is directed at the coating, causing the base metal to fluoresce. The resulting secondary radiation from the fluorescence is partially absorbed by the coating, according to its thickness. The measured intensity of the secondary radiation reaching the detector is therefore a measure of the thickness of the coating.

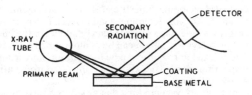

X-ray fluorescence absorptiometer.

X-ray goniometer An instrument that determines the positions of the electrical axes of a quartz crystal by reflecting X-rays from the atomic planes of the crystal.

X-ray hardness The penetrating ability of X-rays. It is an inverse function of the wavelength.

X-ray lithography A lithographic process for transferring mask patterns to a resist-covered semiconductor (silicon or gallium arsenide) wafer that employs X-ray radiation rather than ultraviolet light or an electron beam. The shorter X-ray wavelengths of 10 to 50 Å vs. 2000 to 3000 Å for ultraviolet light minimize diffraction and extend the useful range of lithography toward 0.1 monolithic IC feature sizes of 0.1 μm. Optical lithography is expected to reach its limit at feature sizes shorter than 0.25 μm. See *lithography* and *electron-beam lithography*.

X-ray machine The X-ray tube, power supply, and associated equipment required for producing X-ray photographs.

X-ray photograph A radiograph made with X-rays. It is also called an X-ray.

X-ray spectrogram A record of an X-ray diffraction pattern.

X-ray spectrograph An X-ray spectrometer equipped with photographic or other recording apparatus.

X-ray spectrometer An instrument that produces the X-ray spectrum of a material and measures the wavelengths of the various components.

X-ray spectrum The spectrum of X-rays arranged according to wavelength, produced by electron bombardment of a target, as in an X-ray tube. It consists of a continuous spectrum on which are superimposed certain groups of much sharper lines characteristic of the element in the target. These lines, such as the K, L, and M lines, correspond to transitions between the inner energy levels of the atom.

X-ray television A closed-circuit television system replaces photographic film during X-ray inspection of welded joints and other industrial X-ray applications. The technique gives instant images, without the time and

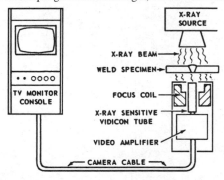

X-ray television for inspecting welding joints.

cost of developing film. It can show enlargements of as much as 50×, for detecting small defects. For permanent records, a videotape recorder can be added, or the images on the television screen can be photographed selectively. The television monitor can be located remotely from the inspection area, so personnel are protected from harmful X-ray radiation.

X-ray tube   A vacuum tube that produces X-rays by accelerating electrons to a high velocity with an electrostatic field and stopping them suddenly when they collide with a target.

X unit   A unit of wavelength equal to 0.001 Å or 10^{-11} cm. It can specify wavelengths of X-rays and other highly penetrating radiations.

X-wave   *Extraordinary-wave component.*

XY recorder   A recorder that traces the relationships between two variables on a moving chart. An example is

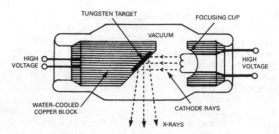

X-ray tube has a heavy water-cooled nonrotating anode to dissipate heat.

the plot of voltage vs. current for a diode. The traces might be made by ink pens on the ends of moving arms or heated styli on heat-sensitive paper.

Y Symbol for *admittance.*
YAG Abbreviation for *yttrium-aluminum garnet.*
Yagi antenna *Yagi-Uda antenna.*

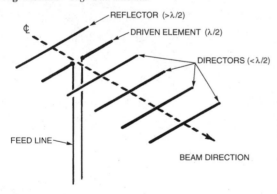

Yagi-Uda antenna is a parasitic end-fire array with one driven element, one reflector, and one or more directors.

Yagi-Uda antenna A general-purpose *parasitic end-fire array* that usually consists of one driven half-wavelength dipole with one parasitic reflector on one side of it and three or more parasitic directors on the other side of it. All of the elements are mounted in parallel in the same plane, about a quarter wavelength apart. Typically, the *reflector* is slightly longer than a half wavelength, and the *directors* are slightly shorter than a half wavelength. The antenna offers the advantages of a unidirectional beam of moderate directivity with a simple feed. It is classified as a *surface-wave antenna* that is useful up to 2.5 GHz. It can be made light in weight at low cost. It is also called a *Yagi antenna, Yagi array,* and a *Yagi.*

Yag-laser *Neodymium-doped yttrium-aluminum garnet laser.*
Y antenna *Delta-matched antenna.*
Y axis 1. A reference axis in a quartz crystal. 2. The vertical axis on a cathode-ray oscilloscope screen or on a graph.
Y circulator A circulator that has three identical rectangular waveguides which are joined to form a symmetrical Y-shaped configuration with a ferrite post or wedge at its center. Power entering any of the three waveguides will emerge from only one adjacent waveguide.
Y connection *Y network.*
Y cut A quartz-crystal cut so that the *Y* axis is perpendicular to the faces of the resulting slab.
Y factor A noise measurement factor for specifying the noise figure of a receiver. It is based on known cold and hot reference temperatures.
yield A term that expresses the percentage of products started that conform to both mechanical and electrical specifications. It applies to passive as well as active components. In the semiconductor industry yields are classified by type: wafer fab (percentage of wafers that complete processing), wafer probe (the fraction of dice or chips per wafer that meet specifications), assembly (percent of units that are assembled correctly), and final test yield (percentage of packaged units that meet performance and mechanical specifications).
YIG Abbreviation for *yttrium-iron garnet.*
YIG filter A filter that consists of an yttrium-iron garnet crystal positioned in a magnetic field provided by a permanent magnet and a solenoid. Tuning is achieved by varying the amount of direct current through the solenoid. The bias magnet tunes the filter to the center of the band, thus minimizing the solenoid power required to tune over wide bandwidths.
YIG-tuned oscillator A tunable microwave-frequency oscillator with a high-*Q* resonant circuit formed by a YIG sphere in a DC magnetic field, acting as a shunt-resonant

YIG-tuned parametric amplifier

tank. RF signals are coupled from a transistor into the YIG sphere by a wire loop. The resonant frequency is a function of the magnetic field strength. Tuning is linear over several octaves, and frequencies of 2 to 40 GHz have been achieved.

YIG-tuned parametric amplifier A parametric amplifier that is tuned by varying the amount of direct current flowing through the solenoid of a YIG filter.

Y junction A waveguide whose longitudinal axes form a Y.

Y network A star network that has three branches. It is also called a Y connection.

yoke *Deflection yoke.*

Y plate One of the two deflection electrodes that deflect the electron beam vertically in an electrostatic cathode-ray tube.

Y signal *Luminance signal.*

yttrium [Y] A rare-earth metallic element that has an atomic number of 39.

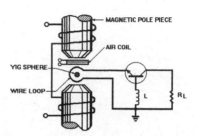

YIG-tuned oscillator includes a transistor amplifier.

yttrium-aluminum garnet [YAG] A crystalline material used in some solid-state lasers.

yttrium-iron garnet [YIG] A crystalline material used in microwave devices.

Z   1. Symbol for *atomic number*.   2. Symbol for *impedance*.

Z axis   The optical axis of a quartz crystal. It is perpendicular to both the *X* and *Y* axes.

Z-axis modulation   *Intensity modulation.*

Z-cut crystal   A quartz crystal cut so that the *Z* axis is perpendicular to the face of the resulting slab.

Zebra time   Mean time at the Greenwich meridian used in communication and for synchronized reckonings. The hour 2400 Zebra time is 1900 EST, 1800 CST, 1700 MST, 1600 PST, 1400 Hawaiian standard time, 1000 Sydney standard time, 0900 Tokyo standard time, 0800 Manila standard time, 0300 Moscow standard time, and 0100 Berlin standard time. It is also called Z time.

Zeeman effect   The increase in the number of spectrum lines produced by a light source when it is in a strong magnetic field.

Zener breakdown   Nondestructive breakdown in a semiconductor. It occurs when the electric field across the barrier region becomes high enough to produce a form of field emission that suddenly increases the number of carriers in this region.

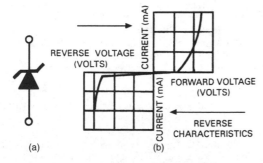

(a)   (b)

Zener diode symbol (*a*) and its characteristic curve (*b*).

Zener diode   A silicon PN junction or reference diode that provides a specified reverse reference voltage when it is operated into its reverse-bias avalanche breakdown region. It exhibits a sharp reverse breakdown at less than 6 V. It is widely used as a general-purpose regulator, and for clipping or bypassing transient voltage above a specified level. Specialized versions of the Zener diode are specified for circuit protection. See *avalanche diode* and *transient voltage suppressor* (TVS).

Zener effect   The effect that is responsible for Zener breakdown in a semiconductor.

Zener noise   Noise generated as a result of the breakdown phenomenon in a Zener diode. It is largely white noise.

Zener voltage   *Breakdown voltage.*

zepp antenna   A horizontal antenna that is some multiple of a half-wavelength long. It is fed at one end by one lead of a two-wire transmission line, which is also some multiple of a half-wavelength long.

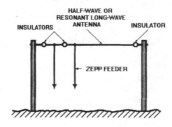

Zepp antenna is fed standing waves at one end by a two-wire feeder.

zero   Nothing. Most computers use both plus and minus zero. A positive binary zero is usually indicated by the absence of digits or pulses in a word, and negative binary zero (in a computer operating on one's complements) is indicated by a pulse in every pulse position in a word. In a coded decimal machine, decimal zero and binary zero might not have the same representation.

zero-access memory A computer memory whose access time is so short that it is negligible.

zero-address instruction A computer instruction that specifies an operation in which the locations of the operands are defined by the computer code, so no address is needed.

zero adjuster A device that adjusts the pointer position of an instrument or meter to read zero when the electrical quantity is zero.

zero beat The condition in which a circuit is oscillating at the exact frequency of an input signal, so no beat tone is produced or heard.

zero-beat reception *Homodyne reception.*

zero bias The condition in a bipolar junction transistor when the base and emitter are at the same DC voltage, or in an FET when the source and gate are at the same DC voltage.

zero compression Any of a number of techniques that eliminate the storage of nonsignificant leading zeros during data processing in a computer.

zero-current switching AC circuit switching under the control of a stage that senses when the current waveform is passing through the zero reference before permitting the circuit to be turned on or off.

zero-crossing detector A comparator that determines whether the input signal is greater than or less than zero.

zero-cut crystal A quartz crystal that has been cut so that its temperature coefficient with respect to frequency is essentially zero.

zero defects A program for improving product quality to the point of perfection, so there will be no failures due to defects in construction.

zero-frequency component The DC component of a complex waveform.

zero gravity The complete absence of gravitational effects; it exists when gravitational attraction is exactly nullified or is counterbalanced by centrifugal force, as in an orbiting satellite.

zero-gravity switch A switch that closes as weightlessness or zero gravity is approached. In one version, a conductive sphere of mercury encompasses two contacts at zero gravity but it flattens away from the upper contact under the influence of gravity. It is also called a *weightlessness switch.*

zero insertion force [ZIF] **socket** A socket or connector that exerts no stress on its mating part or pins during insertion and removal. It is generally achieved with a clamping mechanism that can be manually opened and closed with a lever.

zero level The reference level for comparing sound or signal intensities. At audio frequencies, a power of 6 mW is generally considered to be zero level. In sound, the threshold of hearing is generally assumed as the zero level.

zero-point energy The kinetic energy remaining in a substance at a temperature of absolute zero.

zero potential An expression usually applied to the potential of the earth, as a convenient reference for comparison.

zero shift The output of a balanced magnetic amplifier for zero control signal, due to drift.

zero stability The maximum zero shift that occurs over a given period of time during given changes in operating conditions of a balanced magnetic amplifier.

zero state A state of a magnetic cell in which the magnetic flux through a specified cross-sectional area has a negative value, when determined from an arbitrarily specified direction for negative flux. The opposite state in which the magnetic flux has a positive value, when similarly determined, is called a one state.

zero-subcarrier chromaticity The chromaticity that is displayed when the subcarrier amplitude is zero in a color television system.

zero suppression The elimination of nonsignificant zeros to the left of the integral part of a quantity before the results of a computer operation are printed.

zero-voltage activated circuit [ZVA] A circuit that is designed to turn on at zero AC voltage and turn off at zero AC current.

zero-voltage switch A circuit with a sensor that applies voltage to a load when its AC voltage waveform reaches zero and turns it off under the same conditions. This avoids the application of a reactive voltage across the load at startup or shutdown.

ZIF Abbreviation for *zero insertion force.*

zinc [Zn] A bluish white metallic element that has an atomic number of 30 and an atomic weight of 65.37.

zinc-air cell A *primary* low-voltage *air cell* that provides high energy density, and is typically made as a *button cell.* It is used to power hearing aids, pagers, and other low-voltage portable circuits. The *anode* is zinc and the *cathode* is oxygen (air). It has a nominal voltage of 1.5 V and an energy density of 340 Wh/kg or 1050 Wh/L.

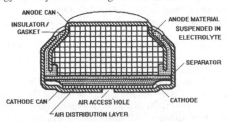

Zinc-air cell takes oxygen from the air as a reactant to generate electrical energy for powering small instruments and hearing aids.

zinc-chloride cell A primary cell that includes carbon and zinc electrodes with an electrolyte which contains only zinc chloride. Its nominal voltage is 1.3 V, just as for common Leclanche zinc-carbon cells, but current output is much higher even at low temperatures. Its life is about twice that of a zinc-carbon cell under comparable conditions.

zinc-mercuric oxide cell *Mercury cell.*

zinc orthosilicate The mineral, also called willemite, is used to make fluorescent screens for cathode-ray tubes. When bombarded by an electron beam, it glows with a green tint.

zinc-silver chloride primary cell A reserve primary cell that is activated by adding water. It can have a capacity, up to 40 Wh/lb, and long life after activation.

zinc-silver-oxide cell *Silver-oxide cell.*

zinc telluride [ZnTe] A semiconductor that has a forbidden-band gap of 2.2 eV and a maximum operating temperature of 780°C when used in a device.

Zip drive An auxiliary memory drive for a personal computer to supplement the capacity of its hard-disk drive or act as memory file backup. It uses removable disks with capacities of 100 Mbytes. *Zip* is a trademark of Iomega Corp.

zirconium [Zr] A metallic element with an atomic number of 40 and an atomic weight of 91.22.

zirconium lamp A lamp with a zirconium-oxide cathode that can provide a point source of high-intensity light concentrated in the long visible wavelengths.

ZnTe Abbreviation for *zinc telluride*.

zone leveling In the crystal-growth process, the passage of one or more molten zones along a semiconductor body for distributing impurities uniformly throughout the material.

zone marker A radio station that radiates signals vertically in a cone-shaped pattern to define a zone above a radio range station.

zone of silence *Skip zone.*

zone purification In crystal growth, the passage of one or more molten zones along a semiconductor for reducing the impurity concentration of part of the ingot. The semiconductor crystal is slowly moved through zones of intense heat. The crystal melts a section at a time. As the molten region moves from one end of the crystal to the other, the impurities move with it and congregate at one end of the crystal. Later that end can be sawed off. It is also called zone refining.

zoning The displacement of various parts of the lens or surface of a microwave reflector so the resulting phase front in the near field remains unchanged. It is also called *stepping.*

zoom To enlarge a section of a television picture or computer monitor screen electronically.

Z time *Zebra time.*

ZVA Abbreviation for *zero-voltage activated circuit.*